Some Fundamental Constants[a]

Quantity	Symbol	Value[b]
Atomic mass unit	u	$1.660\ 540\ 2(10) \times 10^{-27}$ kg
		$931.494\ 32(28)$ MeV/c^2
Avogadro's number	N_A	$6.022\ 136\ 7(36) \times 10^{23}$ (mol)$^{-1}$
Bohr magneton	$\mu_B = \dfrac{e\hbar}{2m_e}$	$9.274\ 015\ 4(31) \times 10^{-24}$ J/T
Bohr radius	$a_0 = \dfrac{\hbar^2}{m_e e^2 k_e}$	$0.529\ 177\ 249(24) \times 10^{-10}$ m
Boltzmann's constant	$k_B = R/N_A$	$1.380\ 658(12) \times 10^{-23}$ J/K
Compton wavelength	$\lambda_C = \dfrac{h}{m_e c}$	$2.426\ 310\ 58(22) \times 10^{-12}$ m
Deuteron mass	m_d	$3.343\ 586\ 0(20) \times 10^{-27}$ kg
		$2.013\ 553\ 214(24)$ u
Electron mass	m_e	$9.109\ 389\ 7(54) \times 10^{-31}$ kg
		$5.485\ 799\ 03(13) \times 10^{-4}$ u
		$0.510\ 999\ 06(15)$ MeV/c^2
Electron-volt	eV	$1.602\ 177\ 33(49) \times 10^{-19}$ J
Elementary charge	e	$1.602\ 177\ 33(49) \times 10^{-19}$ C
Gas constant	R	$8.314\ 510(70)$ J/K·mol
Gravitational constant	G	$6.672\ 59(85) \times 10^{-11}$ N·m^2/kg^2
Hydrogen ground state	$E_0 = \dfrac{m_e e^4 k_e{}^2}{2\hbar^2} = \dfrac{e^2 k_e}{2a_0}$	$13.605\ 698(40)$ eV
Josephson frequency-voltage ratio	$2e/h$	$4.835\ 976\ 7(14) \times 10^{14}$ Hz/V
Magnetic flux quantum	$\Phi_0 = \dfrac{h}{2e}$	$2.067\ 834\ 61(61) \times 10^{-15}$ Wb
Neutron mass	m_n	$1.674\ 928\ 6(10) \times 10^{-27}$ kg
		$1.008\ 664\ 904(14)$ u
		$939.565\ 63(28)$ MeV/c^2
Nuclear magneton	$\mu_n = \dfrac{e\hbar}{2m_p}$	$5.050\ 786\ 6(17) \times 10^{-27}$ J/T
Permeability of free space	μ_0	$4\pi \times 10^{-7}$ N/A^2 (exact)
Permittivity of free space	$\epsilon_0 = 1/\mu_0 c^2$	$8.854\ 187\ 817 \times 10^{-12}$ C^2/N·m^2 (exact)
Planck's constant	h	$6.626\ 075(40) \times 10^{-34}$ J·s
	$\hbar = h/2\pi$	$1.054\ 572\ 66(63) \times 10^{-34}$ J·s
Proton mass	m_p	$1.672\ 623(10) \times 10^{-27}$ kg
		$1.007\ 276\ 470(12)$ u
		$938.272\ 3(28)$ MeV/c^2
Quantized Hall resistance	h/e^2	$25812.805\ 6(12)$ Ω
Rydberg constant	R_H	$1.097\ 373\ 153\ 4(13) \times 10^7$ m^{-1}
Speed of light in vacuum	c	$2.997\ 924\ 58 \times 10^8$ m/s (exact)

[a] These constants are the values recommended in 1986 by CODATA, based on a least-squares adjustment of data from different measurements. For a more complete list, see Cohen, E. Richard, and Barry N. Taylor, *Rev. Mod. Phys.* **59**:1121, 1987.

[b] The numbers in parentheses for the values below represent the uncertainties in the last two digits.

Principles of Physics

SECOND EDITION

Principles
of Physics

SECOND EDITION

Raymond A. Serway

James Madison University

with contributions by

John W. Jewett, Jr.

California State Polytechnic University, Pomona

SAUNDERS COLLEGE PUBLISHING

Harcourt Brace College Publishers

Fort Worth   Philadelphia   San Diego   New York   Orlando   Austin
San Antonio   Toronto   Montreal   London   Sydney   Tokyo

Requests for permission to make copies of any part of the work should be mailed to:
Permissions Department, Harcourt Brace & Company, 6277 Sea Harbor Drive, Orlando, Florida 32887-6777

Publisher: Emily Barrosse
Publisher: John Vondeling
Product Manager: Angus McDonald
Developmental Editor: Susan Dust Pashos
Project Editor: Elizabeth Ahrens
Production Manager: Charlene Catlett Squibb
Art Director and Cover Designer: Carol Bleistine

Cover Credit: Wolf Howling, Aurora Borealis. (© 1997 Michael DeYoung/Alaska Stock)
Frontispiece Credit: David Malin, Anglo-Australian Observatory

Printed in the United States of America

PRINCIPLES OF PHYSICS, Second Edition
0-03-020457-7

Library of Congress Catalog Card Number: 97-65256

890123456   032   10 987654

Preface

Principles of Physics is designed for a one-year introductory calculus-based physics course for engineering and science students. This second edition contains many new pedagogical features—most notably, an increased emphasis on physical concepts through the use of conceptual examples and conceptual problems. Based on comments from users of the first edition and reviewers' suggestions, a major effort was made to improve organization, clarity of presentation, precision of language, and accuracy throughout.

This project was conceived because of the well-known problem we continue to wrestle with in teaching the introductory calculus-based physics course. The course content (and hence the size of textbooks) continues to grow, while the number of contact hours with students has either dropped or remained unchanged. Furthermore, traditional one-year courses cover little if any 20th-century physics.

In preparing this book, I was motivated by the spreading interest in reforming this course, primarily through efforts of the Introductory University Physics Project (IUPP) sponsored by the American Association of Physics Teachers and American Institute of Physics. The primary goals and guidelines of this project are to:

Bruce Byers, FPG

- reduce course content following the "less may be more" theme
- incorporate contemporary physics naturally into the course
- organize the course in the context of one or more "story lines"
- treat all student constituents equitably

Recognizing a need for a textbook that could meet these guidelines several years ago, I studied the various proposed IUPP models and the many reports from IUPP committees. Eventually, I became actively involved in the review and planning of one specific model, initially developed at the U.S. Air Force Academy, entitled "A Particles Approach to Introductory Physics." I spent part of the summer of 1990 at the Academy working with Colonel

James Head and Lt. Col. Rolf Enger, the primary authors of the IUPP model, and other members of that department. This most useful collaboration was the starting point of this project.

In my opinion, the IUPP model developed at the U.S. Air Force Academy, and modified by a team of people over the last few years, was an excellent choice for designing a reformed curriculum for several reasons:

- It is an evolutionary approach (rather than a revolutionary approach), which should meet the current demands of the physics community.
- It deletes many topics in classical physics (such as alternating current circuits and optical instruments), and places less emphasis on rigid body motion, optics, and thermodynamics.
- Some topics in 20th century physics are introduced early in the textbook such as special relativity, energy and momentum quantization, and the Bohr model of the hydrogen atom.
- A deliberate attempt is made at showing the unity of physics.

CHANGES TO THE SECOND EDITION

A number of changes and improvements have been made in the second edition of this text. Many changes are in response to comments and suggestions provided by reviewers of the manuscript and instructors using the first edition. The following represent the major changes in the second edition:

Superstock

Organization   The organization of the textbook is slightly different from that of the first edition. Chapter 6 in the first edition has been deleted, and the sections dealing with the fundamental forces in nature and the gravitational field are now incorporated in Chapter 5. The material on oscillatory motion and waves (Chapters 21, 22, and 23 in the first edition) is now covered earlier (Chapters 12, 13, and 14 in the second edition), following mechanics. The two chapters in the first edition covering interference and diffraction of light waves (Chapters 27 and 28) have been combined and condensed into one chapter entitled Wave Optics (Chapter 27). The section on polarization of light waves is now treated earlier in Chapter 24 entitled Electromagnetic Waves. Finally, by popular demand, Chapter 15 entitled Fluid Mechanics is new to this edition.

Content   In this second edition, a concerted effort was made to place more emphasis on critical thinking and teaching physical concepts. This was accomplished by the addition of approximately 150 conceptual examples, called **Thinking Physics,** and over 200 **Conceptual Problems.** Most of the **Thinking Physics** examples, which include reasoning statements, provide students with a means of reviewing the concepts presented in that section. Some examples demonstrate the connection between the content of that chapter and other scientific disciplines. These examples also serve as models for students to respond to the **Conceptual Problems** posed in the text and end-of-chapter conceptual questions. The answers to all **Conceptual Problems** are contained at the end of each chapter.

Problems and Conceptual Questions   A substantial revision of the end-of-chapter problems and conceptual questions was made in this second edition. Most of the new problems that have been added are intermediate in level, and many of them are problems that require students to make order-of-magnitude calculations or estimates. All problems have been carefully edited and reworded where necessary. Solutions to approximately 25 percent of the end-of-chapter problems and selected conceptual questions are included in the Study Guide and Student Solutions Manual. These problems and questions are identified by boxes around their numbers.

Significant Figures   Significant figures in both worked examples and end-of-chapter problems have been handled with care. Most numerical examples and problems are worked out to three significant figures.

TEXTBOOK FEATURES

The textbook includes many features which are intended to enhance its usefulness to both the student and instructor:

Style   I have attempted to write the book in a style that is clear, logical, and succinct. At the same time, my relaxed writing style is meant to make the reading more enjoyable and less intimidating.

Previews   Most chapters begin with a chapter preview, which includes a brief discussion of chapter objectives and content. This feature enables the student to better understand how the topics covered in that chapter fit into the overall structure and objectives of the course.

Mathematical Level   Calculus is introduced gradually, where it is needed, keeping in mind that many students often take a course in calculus concurrently. Most steps are shown when basic equations are developed, and reference is made to mathematical appendices provided at the end of the text. Vector products are covered where they are needed in physical applications.

Conceptual Examples and Problems   As mentioned earlier, one of the objectives of this edition is to place more emphasis on critical thinking and conceptual understanding. To meet this objective, I have included many **Thinking Physics** examples and **Conceptual Problems** throughout each chapter. The reasoning or responses to all examples immediately follow the example statement, while the answers to all conceptual problems are included at the end of each chapter. Ideally, the student will use these features to better understand physical concepts before being presented with quantitative examples and working homework problems.

Worked Examples   Every chapter includes several worked examples of varying difficulty which are intended to reinforce conceptual understanding and to serve as models for solving end-of-chapter problems.

Exercises   Many examples are followed immediately by exercises with answers. Those exercises contained in an example box represent extensions of the worked examples. Other related exercises with answers are included at the end of many sections. All exercises are intended to encourage students to practice problem solving for immediate reinforcement, and to test the student's understanding of concepts and problem-solving skills. Students who work through these exercises on a regular basis should find the end-of-chapter problems less intimidating.

Important Statements and Equations   Most important statements and definitions are set in **boldface** type for added emphasis and ease of review. Important equations are highlighted with a tan screen for review or reference.

Illustrations   The text material, worked examples, and end-of-chapter questions and problems are accompanied by numerous figures, photographs, and tables. Full color is used to add clarity to the figures and to make the visual presentations as realistic as possible. Three-dimensional effects are produced with the use of color airbrushed areas, where appropriate. Vectors are color-coded, and curves in *xy*-plots are drawn in color. Color photographs have been carefully selected, and their accompanying captions have been written to serve as an added instructional tool.

Russell Schlepman

Margin Notes   Margin notes are used to help students locate and review important statements, concepts, and equations.

Units   The international system of units (SI) is used throughout the text. The British engineering system of units is used only to a limited extent in chapters on mechanics, heat, and thermodynamics.

Problem-Solving Strategies and Hints   I have included general strategies and hints for solving the types of problems featured in both the worked examples and in the end-of-chapter problems. This feature will help students identify important steps and understand the logic for solving specific classes of problems.

Summaries   Each chapter contains a summary which reviews the important concepts and equations discussed in that chapter.

Conceptual Questions   A list of conceptual questions is provided at the end of each chapter. These questions, which require verbal or written responses, provide the student with a means of self-testing the concepts presented in the chapter. Others could serve as a basis for initiating classroom discussions. Answers to selected questions, indicated by boxes around their numbers, are included in the Study Guide and Student Solutions Manual.

End-of-Chapter Problems   An extensive set of problems is included at the end of each chapter. Answers to odd-numbered problems are given at the end of the book; these pages have colored edges for ease of location. For the convenience of both the student and the instructor, about two thirds of the problems are keyed to specific sections of the chapter. The remaining problems, labeled "Additional Problems," are not keyed to specific sections. There are three levels of problems according to their level of difficulty. Straightforward problems are numbered in black, intermediate level problems are numbered in blue, and the most challenging problems are numbered in magenta.

Spreadsheet Problems　　Most chapters include several spreadsheet problems following the end-of-chapter problem sets. Spreadsheet modeling of physical phenomena enables the student to obtain graphical representations of physical phenomena and perform numerical analyses without the burden of having to learn a high-level computer language. Spreadsheets are particularly useful in exploratory investigations; "what if" questions can be addressed easily and depicted graphically.

　　The level of difficulty in the spreadsheet problems, as with all end-of-chapter problems, is indicated by the color of the problem number. For the most straightforward problems (black), a disk with spreadsheet problems is provided. The student must enter the pertinent data, vary the parameters, and interpret the results. Intermediate level problems (blue) usually require students to modify an existing template to perform the required analysis. The more challenging problems (magenta) require students to develop their own spreadsheet templates. Brief instructions on using the templates are provided in Appendix E.

Appendices and Endpapers　　Several appendices are provided at the end of the textbook, including the new appendix with instructions for problem-solving with spreadsheets. Most of the appendix material represents a review of mathematical techniques used in the text, including scientific notation, algebra, geometry, trigonometry, differential calculus, and integral calculus. Reference to these appendices is made throughout the text. Most mathematical sections include worked examples and exercises with answers. In addition to the mathematical reviews, the appendices contain tables of physical data, conversion factors, atomic masses, the SI units of physical quantities, and the periodic chart. The endpapers contain other useful information including the color code used in the text, lists of fundamental constants and physical data, conversions, and the Greek alphabet.

SAUNDERS CORE CONCEPTS IN PHYSICS CD-ROM

Saunders Core Concepts in Physics CD-ROM applies the power of multimedia to the calculus-based course, offering unparalleled full-motion animation and video, engaging interactive graphics, clear and concise text, and guiding narration. *Saunders Core Concepts in Physics* focuses on those concepts students typically find most difficult in the course, drawing from mechanics, thermodynamics, electromagnetism, and optics. The CD-ROM also presents step-by-step explorations of essential mathematics, problem-solving strategies, and animations of problems to promote conceptual understanding and sharpen problem-solving skills. Problems on this CD-ROM are taken primarily from Serway's *Principles of Physics,* 2/e. Available for Macintosh and Windows.

STUDENT ANCILLARIES

Study Guide and Student Solutions Manual by John R. Gordon, Ralph McGrew, Steve Van Wyk, and Ray Serway. The manual features detailed solutions to 25 percent of the end-of-chapter problems and selected conceptual questions from the text. These are indicated in the text with boxed numbers. The manual also features a skills section, important notes from key sections of the text, and a list of important equations and concepts.

Courtesy of NASA

Spreadsheet Templates   The Spreadsheet Template Disk contains spreadsheet files designed to be used with the end-of-chapter problems entitled Spreadsheet Problems. The files have been developed in Microsoft Excel 5.0 for the Macintosh, and in several formats for Windows and DOS users (see Appendix E). These can be used with most spreadsheet programs including all the recent versions of Lotus 1-2-3, Excel for Windows, Quattro Pro, and f(g) Scholar. Over 30 templates are provided on the disk.

Spreadsheet Investigations in Physics by Lawrence B. Golden and James R. Klein. This workbook with the accompanying disk illustrates how spreadsheets can be used for solving many physics problems. The workbook is divided into two parts. The first part consists of spreadsheet tutorials, while the second part is a short introduction to numerical methods. The tutorials include basic spreadsheet techniques and emphasize navigating the spreadsheet, entering data, constructing formulas, and graphing. The numerical methods include differentiation, integration, interpolation, and the solution of differential equations. Many examples and exercises are provided. Step-by-step instructions are given for constructing numerical models of selected physics problems. The exercises and examples used to illustrate the numerical methods are chosen from introductory physics and mathematics.

Pocket Guide by V. Gordon Lind. This 5″ by 7″ notebook is a section-by-section capsule of the textbook that provides a handy guide for looking up important concepts, formulas, and problem-solving hints.

Mathematical Methods for Introductory Physics with Calculus by Ronald C. Davidson. This brief book is designed for students who find themselves unable to keep pace in their physics class because of a lack of familiarity with the necessary mathematical tools. *Mathematical Methods* provides an overview of those mathematical topics that are needed in an introductory-level physics course through the use of many worked examples and exercises.

So You Want to Take Physics: A Preparation for Scientists and Engineers by Rodney Cole. This text is useful to those students who need additional preparation before or during a course in physics. The book includes typical problems with worked solutions, and a review of techniques in mathematics and physics. The friendly, straightforward style makes it easier to understand how mathematics is used in the context of physics.

Life Science Applications for Physics compiled by Jerry Faughn. This supplement provides examples, readings, and problems from the biological sciences as they relate to physics. Topics include "Friction in Human Joints," "Physics of the Human Circulatory System," "Physics of the Nervous System," and "Ultrasound and Its Applications." This supplement is useful in those courses having a significant number of pre-med students.

Physics Laboratory Manual by David Loyd supplements the learning of basic physical principles while introducing laboratory procedures and equipment. Each chapter of the laboratory manual includes a pre-laboratory assignment, a list of objectives, an equipment list, the theory behind the experiment, the experimental procedure to be followed, calculations, graphs, and questions. In addition, laboratory report templates are provided for each experiment so the student can record data, calculations, and experimental results.

INSTRUCTOR ANCILLARIES

Instructor's Solutions Manual by Steve Van Wyk, Ralph McGrew, and Ray Serway. This manual contains solutions to all of the problems in the text and provides answers to even-numbered problems. All solutions have been carefully reviewed for accuracy.

Printed Test Bank by Myron Schneiderwent. This test bank offers over 1500 multiple choice and conceptually oriented critical thinking questions.

Computerized Test Bank   Available for the IBM PC (DOS and Windows) and Macintosh computers, this computerized version of the printed test bank also contains over 1500 questions. The test bank enables the instructor to create unique tests and permits the editing of questions as well as addition of new questions. The program gives answers to all problems and prints each answer on a separate grading key.

Guy Sauvage, Photo Researchers, Inc.

Overhead Transparency Acetates   A set of 175 transparencies with approximately 300 full-color figures from the text.

Instructor's Manual to Accompany Physics Laboratory Manual by David Loyd. Each chapter contains a discussion of the experiment, teaching hints, answers to selected questions, and a post-laboratory quiz with short answer and essay questions. Also included is a list of suppliers of scientific equipment and a summary of all the equipment needed for the laboratory experiments in the manual.

Physics Demonstrations Videotape by J. C. Sprott. This video features two hours of demonstrations divided into 12 primary topics. Each topic contains between four and nine demonstrations for a total of 70 physics demonstrations.

TEACHING OPTIONS

Although many topics found in traditional textbooks have been omitted from this textbook, instructors may find that the current text still contains more material than can be covered in a two-semester sequence. For this reason, I would like to offer the following suggestions. If you wish to place more emphasis on contemporary topics in physics, you should consider omitting parts or all of Chapters 15, 16, 17, 18, 24, 25, and 26. On the other hand, if you wish to follow a more traditional approach which places more emphasis on classical physics, you could omit Chapters 9, 11, 28, 29, 30, and 31. Either approach can be used without any loss in continuity. Other teaching options would fall somewhere in between these two extremes by omitting the sections labeled "Optional."

ACKNOWLEDGMENTS

As any author well knows, a project of this magnitude would be an impossible task without the assistance of many talented and dedicated individuals whose comments, criticisms, and suggestions are so important in developing a finished product. In preparing the second edition of this textbook, I have been guided by the expertise of many people who reviewed part or all of the manuscript. Special thanks go to John Jewett, Jr., Ron Bieniek, and Ralph McGrew who provided in-depth reviews of parts or all of the manuscript at various stages and offered many suggestions for its improvement. Laurent Hodges wrote several new end-of-chapter problems. I am most grateful to the following reviewers of the manuscript for their valuable feedback:

Ralph V. Chamberlin, *Arizona State University*

Gary G. DeLeo, *Lehigh University*

Alan J. DeWeerd, *Creighton University*

Philip Fraundorf, *University of Missouri—St. Louis*

Todd Hann, *United States Military Academy*

Richard W. Henry, *Bucknell University*

Laurent Hodges, *Iowa State University*

Herb Jaeger, *Miami University*

Thomas H. Keil, *Worcester Polytechnic Institute*

John W. McClory, *United States Military Academy*

L. C. McIntyre, Jr., *University of Arizona*

Alan S. Meltzer, *Rensselaer Polytechnic Institute*

Perry Rice, *Miami University*

Janet E. Seger, *Creighton University*

Antony Simpson, *Dalhousie University*

Harold Slusher, *University of Texas at El Paso*

Chris Vuille, *Embry-Riddle Aeronautical University*

James Whitmore, *Pennsylvania State University*

I also thank the following individuals for their suggestions during the development of the first edition of this textbook: Edward Adelson, Ohio State University; Subash Antani, Edgewood College; Harry Bingham, University of California, Berkeley; Anthony Buffa, California Polytechnic State University, San Luis Obispo; Gordon Emslie, University of Alabama at Huntsville; Donald Erbsloe, United States Air Force Academy; Gerald Hart, Moorhead State University; Joey Huston, Michigan State University; David Judd, Broward Community College; V. Gordon Lind, Utah State University; David Markowitz, University of Connecticut; Roy Middleton, University of Pennsylvania; Clement J. Moses, Utica College of Syracuse University; Anthony Novaco, Lafayette College; Desmond Penny, Southern Utah University; Prabha Ramakrishnan, North Carolina State University; Rogers Redding, University of North Texas; J. Clinton Sprott, University of Wisconsin at Madison; Cecil Thompson, University of Texas at Arlington.

Some end-of-chapter problems in Chapters 3, 6, and 8 of this second edition came from *University Physics*, 2nd edition by Alvin Hudson and Rex Nelson (Saunders College Publishing, 1990) or from *University Physics*, 2nd edition by George Arfken, David Griffing, Donald Kelly and Joseph Priest (Harcourt Brace Jovanovich, 1989). I thank the authors of these problems.

I am indebted to Colonel James Head and Lt. Col. Rolf Enger of the U.S. Air Force Academy for inspiring me to begin work on this project while I was at the Academy during the summer of 1990, and for their warm hospitality during my visit. They and their colleagues designed the original IUPP model upon which this textbook is based. Special thanks go to Ralph McGrew for assembling the problem sets, writing new problems, and checking solutions to all problems. I appreciate the tedious work of Ralph McGrew and Steve Van Wyk in writing the Instructor's Solutions Manual, and I thank Gloria Langer for her fine work in preparing this manual. I am grateful to John R. Gordon, Ralph McGrew, and Steve Van Wyk who co-authored the Study Guide and Student Solutions Manual, and to Linda Miller and Michael Rudmin for their assistance in its preparation. Dena Digilio-Betz was very helpful in locating many of the photographs used in this text. During the development of this text, I benefited from many useful discussions with my colleagues and other physics instructors including Robert Bauman, Don Chodrow, Jerry Faughn, John R. Gordon, Kevin Giovanetti, Dick Jacobs, Clem Moses, Dorn Peterson, Joseph Rudmin, and Gerald Taylor. I owe a debt of gratitude to Irene Nunes for clarifying much of the material in the first edition, and for eliminating many inconsistencies, redundancies, and other nightmares that authors must face. Special thanks and recognition go to the professional staff at Saunders College Publishing for their careful attention to detail and for their dedication to producing first quality textbooks, especially Susan Dust Pashos, Beth Ahrens, Alexandra Buczek, Sally Kusch, Carol Bleistine, and Angus McDonald. I sincerely appreciate the wisdom and enthusiasm of my publisher and good friend John Vondeling, who continues to publish high quality instructional products for science education.

Finally, I thank my entire family for their endless love, understanding, and continued encouragement, especially my beautiful and devoted wife Elizabeth, who is always at my side to share my joys and sorrows.

Michel Gouverneur, Photo News, Gamma Sport

Ray Serway
Harrisonburg, VA

JULY 1997

To the Student

I feel it is appropriate to offer some words of advice which should be of benefit to you, the student. Before doing so, I will assume that you have read the Preface, which describes the various features of the text that will help you through the course.

HOW TO STUDY

Very often instructors are asked "How should I study physics and prepare for examinations?" There is no simple answer to this question, but I would like to offer some suggestions based on my own experiences in learning and teaching over the years.

First and foremost, maintain a positive attitude towards the subject matter, keeping in mind that physics is the most fundamental of all natural sciences. Other science courses that follow will use the same physical principles, so it is important that you understand and be able to apply the various concepts and theories discussed in the text.

CONCEPTS AND PRINCIPLES

It is essential that you understand the basic concepts and principles *before* attempting to solve assigned problems. This is best accomplished through a careful reading of the textbook before attending your lecture on that material. In the process, it is useful to jot down certain points which are not clear to you. Take careful notes in class, and then ask questions pertaining to those ideas that require clarification. Keep in mind that few people are able to absorb the full meaning of scientific material after one reading. Several readings of the text and notes may be necessary. Your lectures and laboratory work should supplement the text and clarify some of the more difficult material. You should reduce memorization of material to a minimum. Memorizing passages from a text, equations, and derivations does not necessarily mean you understand the material. Your understanding of the material will be enhanced through a combination of efficient study habits, discussions with other students and instructors, and your ability to solve the problems in the text. Ask questions whenever you feel it is necessary.

STUDY SCHEDULE

It is important to set up a regular study schedule, preferably on a daily basis. Make sure to read the syllabus for the course and adhere to the schedule set by your instructor. The lectures will be much more meaningful if you read the corresponding textual material before attending the lecture. As a general rule, you should devote about two hours of study time for every hour in class. If you are having trouble with the course, seek the advice of the instructor or students who have taken the course. You may find it necessary to seek further instruction from experienced students. Very often, instructors will offer review sessions in addition to regular class periods. It is important that you avoid the practice of delaying study until a day or two before an exam. More often than not, this will lead to disastrous results. Rather than an all-night study session, it is better to briefly review the basic concepts and equations, followed by a good night's rest. If you feel in need of additional help in understanding the concepts, preparing for exams, or in problem-solving, we suggest that you acquire a copy of the Study Guide/Student Solutions Manual which accompanies the text and should be available at your college bookstore.

USE THE FEATURES

You should make full use of the various features of the text discussed in the Preface. For example, marginal notes are useful for locating and describing important equations and concepts, while important statements and definitions are highlighted in **boldface** type or with a blue box. Many useful tables are contained in appendices, but most are incorporated in the text where they are used most often. Appendix B is a convenient review of mathematical techniques. Answers to odd-numbered problems are provided at the end of the book. Exercises (with answers), which follow some worked examples, represent extensions of those examples, and in most cases you are expected to perform a simple calculation. Their purpose is to test your problem-solving skills as you read through the text. Problem-Solving Strategies and Hints are included in selected chapters throughout the text to give you additional information to help you solve problems. An overview of the entire text is given in the table of contents, while the index will enable you to locate specific material quickly. Footnotes are sometimes used to supplement the discussion or to cite other references on the subject. Many chapters include problems that require the use of programmable calculators, computers, or simulation software. Consult your instructor; it may be possible to obtain some of the relevant software from the Saunders physics Web site. Spreadsheet problems can be solved either analytically or with the use of the *Spreadsheet Investigations* supplement. These are intended for those courses that place some emphasis on numerical methods. You may want to develop appropriate programs for some of these problems even if they are not assigned by your instructor.

Courtesy of NASA

After reading a chapter, you should be able to define any new quantities introduced in that chapter, and discuss the principles and assumptions that were used to arrive at certain key relations. The chapter summaries and the review sections of the Study Guide/Student Solutions Manual should help you in this regard. In some cases, it will be necessary to refer to the index of the text to locate certain topics. You should be able to correctly associate with each physical quantity a symbol used to represent that quantity and the unit in which the quantity is specified. Furthermore, you should be able to express each important relation in a concise and accurate prose statement.

THE IMPORTANCE OF PROBLEM SOLVING

R.P. Feynman, Nobel laureate in physics, once said, "You do not know anything until you have practiced." In keeping with this statement, I strongly advise you to develop the skills necessary to solve a wide range of problems. Your ability to solve problems will be one of the main tests of your knowledge of physics; therefore, you should try to solve as many problems as possible. It is essential that you understand basic concepts and principles before attempting to solve problems. It is good practice to try to find alternate solutions to the same problem. For example, problems in mechanics can be solved using Newton's laws, but very often an alternative method using energy considerations is more direct. You should not deceive yourself into thinking you understand the problem after seeing its solution in class. You must be able to solve the problem and similar problems on your own.

The method of solving problems should be carefully planned. A systematic plan is especially important when a problem involves several concepts. First, read the problem several times until you are confident you understand what is being asked. Look for any key words that will help you interpret the problem, and perhaps allow you to make certain assumptions. Your ability to interpret the question properly is an integral part of problem solving. You should acquire the habit of writing down the information given in a problem, and decide what quantities need to be found. You might want to construct a table listing quantities given and quantities to be found. This procedure is sometimes used in the worked examples of the text. After you have decided on the method you feel is appropriate for the situation, proceed with your solution. General problem-solving strategies of this type are included in the text and are highlighted by a tan screen.

I often find that students fail to recognize the limitations of certain formulas or physical laws in a particular situation. It is very important that you understand and remember the assumptions which underlie a particular theory or formalism. For example, certain equations in kinematics apply only to a particle moving with constant acceleration. These equations are not valid for situations in which the acceleration is not constant, such as the motion of an object connected to a spring, or the motion of an object through a fluid.

GENERAL PROBLEM-SOLVING STRATEGY

Most courses in general physics require the student to learn the skills of problem solving, and examinations are largely composed of problems that test such skills. This brief section describes some useful ideas which will enable you to increase your accuracy in solving problems, enhance your understanding of physical concepts, eliminate initial panic or lack of direction in aproaching a problem, and organize your work. One way to help accomplish these goals is to adopt a problem-solving strategy. Many chapters in this text will include a section labeled "Problem-Solving Strategies and Hints" which should help you through the "rough spots."

In developing problem-solving strategies, five basic steps are commonly used, and are exemplified in the sample Example below.

1. Draw a suitable diagram with appropriate labels and coordinate axes if needed.
2. As you examine what is being asked in the problem, identify the basic physical principle (or principles) that are involved, listing the knowns and unknowns.
3. Select a basic relationship or derive an equation that can be used to find the unknown, and solve the equation for the unknown symbolically.
4. Substitute the given values along with the appropriate units into the equation.
5. Obtain a numerical value for the unknown. The problem is verified and receives a check mark if the following questions can be properly answered: Do the units match? Is the answer reasonable? Is the plus or minus sign proper or meaningful?

One of the purposes of this strategy is to promote accuracy. Properly drawn diagrams can eliminate many sign errors. Diagrams also help to isolate the physical principles of the problem. Symbolic solutions and carefully labeled knowns and unknowns will help eliminate other careless errors. The use of symbolic solutions should help you think in terms of the physics of the problem. A check of units at the end of the problem can indicate a possible algebraic error. The physical layout and organization of your problem will make the final product more understandable and easier to follow. Once you have developed an organized system for examining problems and extracting relevant information, you will become a more confident problem solver.

Example

A person driving in a car at a speed of 20 m/s applies the brakes and stops in a distance of 100 m. What was the acceleration of the car?

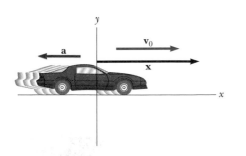

Given:

$x_0 = 0$ m

$x = 100$ m

$v_0 = 20$ m/s

$v = 0$ m/s

$a = ?$

$$v^2 = v_0^2 + 2a(x - x_0)$$

$$a = \frac{v^2 - v_0^2}{2(x - x_0)}$$

$$a = \frac{(0 \text{ m/s})^2 - (20 \text{ m/s})^2}{2(100 \text{ m})} = -2.0 \text{ m/s}^2$$

Check units for consistency:

$$\frac{\text{m}^2/\text{s}^2}{\text{m}} = \frac{\text{m}}{\text{s}^2}$$

EXPERIMENTS

Physics is a science based upon experimental observations. In view of this fact, I recommend that you try to supplement the text through various type of "hands-on" experiments, either at home or in the laboratory. These can be used to test ideas and models discussed in class or in the text. For example, the common "Slinky" toy is excellent for studying traveling waves; a ball swinging on the end of a long string can be used to investigate pendulum motion; various masses attached to the end of a vertical spring or rubber band can be used to determine their elastic nature; an old pair of Polaroid sunglasses and some discarded lenses and magnifying glass are the components of various experiments in optics; you can get an approximate measure of the acceleration of gravity simply by dropping a ball from a known height and measuring the time of its fall with a stopwatch. The list is endless. When physical models are not available, be imaginative and try to develop models of your own.

Tom Raymond / Tony Stone Images

THINKING PHYSICS AND CONCEPTUAL PROBLEMS

These new features are intended to develop your conceptual understanding of physics. You should be able to answer all **Conceptual Problems** based on previous content in the chapter. (Answers are provided at the chapter's end, but please try to answer them on your own first.) **Thinking Physics** examples provide the model for the form your answers should take. The answers ("Reasoning") for **Thinking Physics** are provided immediately for you because these examples often extend and further develop the chapter content. They may, for instance, illuminate the connection between physics and other scientific disciplines.

AN INVITATION TO PHYSICS

It is my sincere hope that you too will find physics an exciting and enjoyable experience, and that you will profit from this experience, regardless of your chosen profession. Welcome to the exciting world of physics.

> *The scientist does not study nature because it is useful; he studies it because he delights in it, and he delights in it because it is beautiful. If nature were not beautiful, it would not be worth knowing, and if nature were not worth knowing, life would not be worth living.*
>
> Henri Poincaré

Contents Overview

NASA

Contents

David Madison / Tony Stone Images

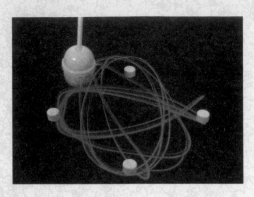

© *Yoav Levy/Phototake*

Martin Dohrn/SPL/Photo Researchers

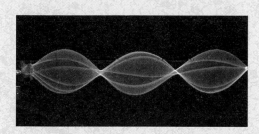

Richard Megna, Fundamental Photographs, NYC

© Ben Rose 1992/The IMAGE Bank

Kim Vandiver and Harold Edgerton,
Palm Press, Inc.

12   Oscillatory Motion   332

13   Wave Motion   360

14   Superposition and Standing Waves   389

Earl Young/FPG

Courtesy of Central Scientific Company

Tom Mareschel, The IMAGE Bank

19   Electric Forces and Electric Fields   521

© 1968 Fundamental Photographs

20   Electric Potential and Capacitance   558

Courtesy of IBM Research

Courtesy of Leon Lewandowski

Ron Chapple/FPG

Peter Aprahamian/Science Photo Library

26   Mirrors and Lenses   756

25   Reflection and Refraction of Light   730

© Richard Megna 1990, Fundamental Photographs

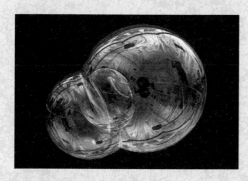

Dr. Jeremy Burgess/Science Photo Library

Courtesy of Central Scientific Company

David Parker/Science Photo Library/Photo Researchers

An Invitation to Physics

Physics, the most fundamental physical science, is concerned with the basic principles of the Universe. It is the foundation on which the other physical sciences—astronomy, chemistry, and geology—are based. The beauty of physics lies in the simplicity of its fundamental theories and in the manner in which a small number of basic concepts, equations, and assumptions can alter and expand our view of the world.

All physical phenomena in our world can be described in terms of one or more of the following theories:

- Classical mechanics, which is concerned with the motion of objects moving at speeds that are low compared to the speed of light;
- Relativity, which is a theory describing particles moving at any speed, even those whose speeds approach the speed of light;
- Thermodynamics, which deals with heat, temperature, and the behavior of large numbers of particles;
- Electromagnetism, which involves the theory of electricity, magnetism, and electromagnetic fields;
- Quantum mechanics, a theory dealing with the behavior of submicroscopic particles as well as the macroscopic world.

When a discrepancy between theory and experiment arises in any of the theories listed above, the theories must be modified and experiments must be performed to test the predictions of the modified theories. Many times a theory is satisfactory under limited conditions; a more general theory might be satisfactory without such limi-

◀ **White light directed through a triangular glass prism creates "Pheaenomena of Colours" on the manuscript of Isaac Newton's Opticks.** *(© Erich Lessing, Magnum Photos)*

tations. A classic example is Newtonian mechanics, which accurately describes the motions of objects at low speeds but does not apply to objects moving at speeds comparable to the speed of light. The special theory of relativity developed by Albert Einstein (1879–1955) successfully predicts the motions of objects at speeds approaching the speed of light and hence is a more general theory of motion.

Classical physics, developed prior to 1900, includes the theories, concepts, laws, and experiments in classical mechanics, thermodynamics, and electromagnetism. For example, Galileo Galilei (1564–1642) made significant contributions to classical mechanics through his work on the laws of motion with constant acceleration. In the same era, Johannes Kepler (1571–1630) used astronomical observations to develop empirical laws for the motions of planetary bodies.

The most important contributions to classical mechanics were provided by Isaac Newton (1642–1727), who developed classical mechanics as a systematic theory and was one of the originators of the calculus as a mathematical tool. Although major developments in classical physics continued in the 18th century, thermodynamics and electromagnetism were not developed until the latter part of the 19th century, principally because the apparatus for controlled experiments was either too crude or unavailable. Although many electric and magnetic phenomena had been studied earlier, it was the work of James Clerk Maxwell (1831–1879) that provided a unified theory of electromagnetism. In this book we shall discuss the various disciplines of classical physics in separate sections; however, we will see that the disciplines of mechanics and electromagnetism are basic to all the branches of classical and modern physics.

A major revolution in physics, usually referred to as *modern physics*, began near the end of the 19th century. Modern physics developed in response to the discovery that many physical phenomena could not be explained by classical physics. The two most important developments in this modern era were the theories of relativity and quantum mechanics. Einstein's theory of relativity completely revolutionized the traditional concepts of space, time, and energy. Einstein's theory correctly described the motion of objects moving at speeds comparable to the speed of light. Newton's laws fail to make correct predictions for objects moving at such high speeds. The theory of relativity also assumes that the speed of light is the upper limit of the speed of an object or signal and shows that mass and energy are related. Quantum mechanics was formulated by a number of distinguished scientists to provide descriptions of physical phenomena at the atomic level.

Scientists continually work at improving our understanding of fundamental laws, and new discoveries are made every day. In many research areas there is a great deal of overlap among physics, chemistry, and biology. The many technological advances in recent times are the result of the efforts of many scientists, engineers, and technicians. Some of the most notable recent developments are (1) unmanned space missions and manned moon landings, (2) microcircuitry and high-speed computers, and (3) sophisticated imaging techniques used in scientific research and medicine. The impacts of such developments and discoveries on society have indeed been great, and it is very likely that future discoveries and developments will be exciting, challenging, and of great benefit to humanity.

1

Introduction and Vectors

The goal of physics is to provide a quantitative understanding of certain basic phenomena that occur in our Universe. Physics is a science based on experimental observations and mathematical analyses. The main objective behind such experiments and analyses is to develop theories that explain the phenomenon being studied and to relate those theories to other established theories. Fortunately, it is possible to explain the behavior of various physical systems using relatively few fundamental laws. Analytical procedures require the expression of those laws in the language of mathematics, the tool that provides a bridge between theory and experiment. In this chapter we shall discuss a few mathematical concepts and techniques that will be used throughout the book.

Because later chapters will be concerned with the laws of physics, it is necessary to provide clear definitions of the basic quantities involved in these laws. For example, such physical quantities as force, velocity, volume, and acceleration can be described in terms of more fundamental quantities. In the next several chapters we will encounter three of the latter: **length** (L), **time** (T), and **mass** (M). Later in the book we shall add units for two other standard quantities to our list, one for temperature (the kelvin), the other for electric current (the ampere). In our study of mechanics, however, we shall be concerned only with the units of mass, length, and time.

Some physical quantities require the measurement of more than one attribute. For example, physical quantities that have both numerical and direc-

◄ **This photograph was taken in Kinsley, Kansas, located midway between New York City and San Francisco.** *(Mike & Carol Werner/Comstock)*

tional properties—such as force, velocity, and acceleration—are represented by vectors. A large part of this chapter is concerned with vector algebra and the general properties of vectors. The addition and subtraction of vectors will be discussed, together with some common applications to physical situations. Because vectors will be used throughout this book, it is imperative that you understand both their graphical and their algebraic properties.

1.1 • STANDARDS OF LENGTH, MASS, AND TIME

If we make a measurement of a certain quantity and wish to describe it to someone, a unit for the quantity must be defined. For example, it would be meaningless for a visitor from another planet to talk to us about a length of 8 "glitches" if we did not know the meaning of the unit "glitch." However, if someone familiar with our system of measurement and weights reports that a wall is 2.0 meters high and our unit of length is defined to be 1.0 meter, we then know that the height of the wall is twice our fundamental unit of length. Likewise, if we are told that a person has a mass of 75 kilograms and our unit of mass is defined as 1.0 kilogram, then that person has a mass 75 times larger than our fundamental unit of mass. In 1960 an international committee agreed on a system of definitions and standards to describe fundamental physical quantities. It is called the **SI system** (Système International) of units. Its units of length, mass, and time are the meter, kilogram, and second, respectively.

Length

In A.D. 1120 the king of England decreed that the standard of length in his country would be the yard and that the yard would be precisely equal to the distance from the tip of his nose to the end of his outstretched arm. In a similar way, the original standard for the foot adopted by the French was the length of the royal foot of King Louis XIV. This standard prevailed until 1799, when the legal standard of length in France became the meter, defined as one ten-millionth of the distance from the Equator to the North Pole.

Many other systems have been developed in addition to those just discussed, but the French system has prevailed in most countries and in scientific circles everywhere. As recently as 1960, the length of the meter was still defined as the distance between two lines on a specific bar of platinum-iridium alloy stored under controlled conditions. This standard was abandoned for several reasons, a principal one being that the limited accuracy with which the separation between the lines can be determined does not meet the current requirements of science and technology. Until recently, the meter was defined as 1 650 763.73 wavelengths of orange-red light emitted from a krypton-86 lamp. However, in October 1983 the **meter**

Definition of the meter •   **was redefined to be the distance traveled by light in a vacuum during a time of 1/299 792 458 s.** In effect, this latest definition establishes that the speed of light in a vacuum is 299 792 458 m/s.

Mass

Definition of the kilogram •   The SI unit of mass, **the kilogram, is defined as the mass of a specific platinum-iridium alloy cylinder kept at the International Bureau of Weights and Measures at Sèvres, France.** At this point, we should add a word of caution. Most

TABLE 1.1   Approximate Values of Some Measured Lengths

	Length (m)
Distance from Earth to most remote quasar known	1.4×10^{26}
Distance from Earth to most remote normal galaxies known	4×10^{25}
Distance from Earth to nearest large galaxy (M 31 in Andromeda)	2×10^{22}
Distance from Sun to nearest star (Proxima Centauri)	4×10^{16}
One lightyear	9.46×10^{15}
Mean orbit radius of Earth	1.5×10^{11}
Mean distance from Earth to Moon	3.8×10^{8}
Distance from Equator to North Pole	1×10^{7}
Mean radius of Earth	6.4×10^{6}
Typical altitude of orbiting Earth satellite	2×10^{5}
Length of a football field	9.1×10^{1}
Length of a housefly	5×10^{-3}
Size of smallest dust particles	1×10^{-4}
Size of cells of most living organisms	1×10^{-5}
Diameter of a hydrogen atom	1×10^{-10}
Diameter of a uranium nucleus	1.4×10^{-14}
Diameter of a proton	1×10^{-15}

beginning students of physics tend to confuse the physical quantities called *weight* and *mass*. For the present we shall not discuss the distinction between them; they will be clearly defined in later chapters. For now you should note that they are distinctly different quantities.

Time

Before 1960 the standard of time was defined in terms of the average length of a solar day in the year 1900. (A solar day is the time interval between successive appearances of the Sun at the highest point it reaches in the sky each day.) The basic unit of time, the second, was defined to be $(1/60)(1/60)(1/24) = 1/86\ 400$ of the average solar day. In 1967 the second was redefined to take advantage of the great precision obtainable with a device known as an atomic clock, which uses the characteristic frequency of the cesium-133 atom as the "reference clock." **The second is now defined as 9 192 631 770 times the period of oscillation of radiation from the cesium atom.**

Approximate Values for Length, Mass, and Time

Approximate values of various lengths, masses, and time intervals are presented in Tables 1.1, 1.2, and 1.3, respectively. Note the wide range of values for these quantities.[1] You should study the tables and get a feel for what is meant by a mass of 100 kilograms, for example, or by a time interval of 3.2×10^{7} seconds. Systems of units commonly used are the SI system in which the units of length, mass, and

TABLE 1.2   Masses of Various Bodies (approximate values)

	Mass (kg)
Universe	10^{52}
Milky Way galaxy	10^{42}
Sun	2×10^{30}
Earth	6×10^{24}
Moon	7×10^{22}
Shark	3×10^{2}
Human	7×10^{1}
Frog	1×10^{-1}
Mosquito	1×10^{-5}
Bacterium	1×10^{-15}
Hydrogen atom	1.67×10^{-27}
Electron	9.11×10^{-31}

[1]If you are unfamiliar with the use of powers of ten (scientific notation), you should review Appendix B.1.

TABLE **1.3**   **Approximate Values of Some Time Intervals**

	Interval (s)
Age of the Universe	5×10^{17}
Age of the Earth	1.3×10^{17}
Time since the fall of the Roman Empire	5×10^{12}
Average age of a college student	6.3×10^8
One year	3.2×10^7
One day (time for one revolution of Earth around its axis)	8.6×10^4
Time between normal heartbeats	8×10^{-1}
Period[a] of audible sound waves	1×10^{-3}
Period of typical radio waves	1×10^{-6}
Period of vibration of an atom in a solid	1×10^{-13}
Period of visible light waves	2×10^{-15}
Duration of a nuclear collision	1×10^{-22}
Time for light to cross a proton	3.3×10^{-24}

[a] A period is defined as the time interval required for one complete vibration.

TABLE **1.4**   **Some Prefixes for Powers of Ten**

Power	Prefix	Abbreviation
10^{-18}	atto	a
10^{-15}	femto	f
10^{-12}	pico	p
10^{-9}	nano	n
10^{-6}	micro	μ
10^{-3}	milli	m
10^{-2}	centi	c
10^{-1}	deci	d
10^3	kilo	k
10^6	mega	M
10^9	giga	G
10^{12}	tera	T
10^{15}	peta	P
10^{18}	exa	E

time are the meter (m), kilogram (kg), and second (s), respectively; the cgs or Gaussian system, in which the units of length, mass, and time are the centimeter (cm), gram (g), and second (s), respectively; and the British engineering system (sometimes called the *conventional system*), in which the units of length, mass, and time are the foot (ft), slug, and second, respectively. Throughout most of this book we shall use SI units, because they are almost universally accepted in science and industry. We will make limited use of conventional units in the study of classical mechanics.

Some of the most frequently used prefixes for the powers of ten and their abbreviations are listed in Table 1.4. For example, 10^{-3} m is equivalent to 1 millimeter (mm), and 10^3 m is 1 kilometer (km). Likewise, 1 kg is 10^3 g, and 1 megavolt (MV) is 10^6 volts.

1.2 • DENSITY AND ATOMIC MASS

The **density** ρ (Greek letter rho) of any substance is defined as its *mass per unit volume* (a table of the letters in the Greek alphabet is provided at the back of the book):

Density •

$$\rho \equiv \frac{m}{V} \qquad [1.1]$$

For example, aluminum has a density of 2.70×10^3 kg/m^3 and lead has a density of 11.3×10^3 kg/m^3. A list of densities for various substances is given in Table 1.5.

Atomic mass •

The difference in density between aluminum and lead is due in part to their different *atomic masses;* the atomic mass of lead is 207 and that of aluminum is 27. However, the ratio of atomic masses, $207/27 = 7.67$, does not correspond to the ratio of densities, $11.3/2.70 = 4.19$. The discrepancy is due to the difference in atomic spacings and atomic arrangements in the crystal structures of the two types of elements.

All ordinary matter consists of atoms, and each atom is made up of electrons and a nucleus. Practically all of the mass of an atom is contained in the nucleus, which consists of protons and neutrons. Thus, we can understand why the atomic masses of the elements differ. The mass of a nucleus is measured relative to the mass of an atom of carbon-12 (this isotope of carbon has six protons and six neutrons).

The mass of ^{12}C is defined to be exactly 12 atomic mass units (u), where $1\ u = 1.6605402 \times 10^{-27}$ kg. In these units, the proton and neutron have masses of about 1 u. To be more precise,

$$\text{Mass of proton} = 1.007\ 276\ u$$

$$\text{Mass of neutron} = 1.008\ 664\ u$$

The mass of the nucleus of ^{27}Al is approximately 27 u. In fact, a more precise measurement shows that the nuclear mass is always slightly *less* than the combined mass of the protons and neutrons making up the nucleus. The processes of nuclear fission and nuclear fusion are based on this mass difference.

One **mole** (mol) of any element (or compound) consists of Avogadro's number, N_A, of molecules of the substance. Avogadro's number, $N_A = 6.02 \times 10^{23}$, was defined so that 1 mole of carbon-12 atoms would have a mass of exactly 12 g. A mole of one element differs in mass from a mole of another. For example, 1 mole of aluminum has a mass of 27 g, and 1 mole of lead has a mass of 207 g. But 1 mole of aluminum contains the same number of atoms as 1 mole of lead, because there are 6.02×10^{23} atoms in 1 mole of *any* element. The mass of one molecule (or atom, if the substance is an element) is then given by

$$m = \frac{\text{molar mass}}{N_A} \qquad \text{[1.2]}$$

For example, the mass of an aluminum atom is

$$m = \frac{27\ \text{g/mol}}{6.02 \times 10^{23}\ \text{atoms/mol}} = 4.5 \times 10^{-23}\ \text{g/atom}$$

Note that 1 u is equal to N_A^{-1} g.

TABLE 1.5 Densities of Various Substances

Substance	Density $\rho(\text{kg/m}^3)$
Platinum	21.45×10^3
Gold	19.3×10^3
Uranium	18.7×10^3
Lead	11.3×10^3
Copper	8.93×10^3
Iron	7.86×10^3
Aluminum	2.70×10^3
Magnesium	1.75×10^3
Water	1.00×10^3
Air at atmospheric pressure	0.0012×10^3

Example 1.1   How Many Atoms in the Cube?

A solid cube of aluminum (density 2.7 g/cm^3) has a volume of 0.20 cm^3. How many aluminum atoms are contained in the cube?

Solution   Because density equals mass per unit volume, the mass of the cube is

$$M = \rho V = (2.7\ \text{g/cm}^3)(0.20\ \text{cm}^3) = 0.54\ \text{g}$$

To find the number of atoms, N, we can set up a proportion using the fact that 1 mole of aluminum (27 g) contains 6.02×10^{23} atoms:

$$\frac{6.02 \times 10^{23}\ \text{atoms}}{27\ \text{g}} = \frac{N}{0.54\ \text{g}}$$

$$N = \frac{(0.54\ \text{g})(6.02 \times 10^{23}\ \text{atoms})}{27\ \text{g}} = 1.2 \times 10^{22}\ \text{atoms}$$

1.3 • DIMENSIONAL ANALYSIS

The word *dimension* has a special meaning in physics. It usually denotes the physical nature of a quantity. Whether a distance is measured in units of feet or meters or furlongs, it is a distance. We say its dimension is *length.*

The symbols that will be used in this section to specify length, mass, and time are L, M, and T, respectively. We will often use square brackets [ ] to denote the dimensions of a physical quantity. For example, in this notation the dimensions of velocity, v, are written $[v] = L/T$, and the dimensions of area, A, are $[A] = L^2$. The dimensions of area, volume, velocity, and acceleration are listed in Table 1.6, along with their units in the three common systems of measurement. The dimensions of other quantities, such as force and energy, will be described as they are introduced in the book.

In many situations, you may be faced with having to derive or check a specific formula. Although you may have forgotten the details of the derivation, there is a useful and powerful procedure called *dimensional analysis* that can be used as a consistency check, to assist in the derivation, or to check your final expression. This procedure should always be used and should help minimize the rote memorization of equations. Dimensional analysis makes use of the fact that **dimensions can be treated as algebraic quantities.** That is, quantities can be added or subtracted only if they have the same dimensions. Furthermore, the terms on both sides of an equation must have the same dimensions. By following these simple rules, you can use dimensional analysis to help determine whether or not an expression has the correct form, because the relationship can be correct only if the dimensions on the two sides of the equation are the same.

To illustrate this procedure, suppose you wish to derive a formula for the distance x traveled by a car in a time t if the car starts from rest and moves with constant acceleration a. In Chapter 2 we shall find that the correct expression for this special case is $x = \frac{1}{2}at^2$. Let us check the validity of this expression from a dimensional analysis approach.

The quantity x on the left side has the dimension of length. In order for the equation to be dimensionally correct, the quantity on the right side must also have the dimension of length. We can perform a dimensional check by substituting the basic dimensions for acceleration, L/T^2, and time, T, into the equation. That is, the dimensional form of the equation $x = \frac{1}{2}at^2$ can be written as

$$L = \frac{L}{T^2} \cdot T^2 = L$$

The units of time cancel as shown, leaving the unit of length.

TABLE 1.6   Dimensions of Area, Volume, Velocity, and Acceleration

System	Area (L^2)	Volume (L^3)	Velocity (L/T)	Acceleration (L/T^2)
SI	m^2	m^3	m/s	m/s^2
cgs	cm^2	cm^3	cm/s	cm/s^2
British engineering	ft^2	ft^3	ft/s	ft/s^2

Example 1.2 **Analysis of an Equation**

Show that the expression $v = v_0 + at$ is dimensionally correct, where v and v_0 represent velocities, a is acceleration, and t is a time interval.

Solution Because

$$[v] = [v_0] = \frac{L}{T}$$

and the dimensions of acceleration are L/T^2, the dimensions of at are

$$[at] = \frac{L}{T^2} \cdot T = \frac{L}{T}$$

and the expression is dimensionally correct. However, if the expression were given as $v = v_0 + at^2$, it would be dimensionally *incorrect*. Try it and see!

1.4 • CONVERSION OF UNITS

Sometimes it is necessary to convert units from one system to another. Conversion factors between SI and conventional units of length are as follows:

1 mile = 1609 m = 1.609 km   1 ft = 0.3048 m = 30.48 cm

1 m = 39.37 in. = 3.281 ft   1 in. = 0.0254 m = 2.54 cm

A more complete list of conversion factors can be found in Appendix A.

Units can be treated as algebraic quantities that can cancel each other. For example, suppose we wish to convert 15.0 in. to centimeters. Because 1 in. = 2.54 cm (exactly), we find that

$$15.0 \text{ in.} = (15.0 \text{ in.}) \left(2.54 \frac{\text{cm}}{\text{in.}} \right) = 38.1 \text{ cm}$$

(*Left*) Conversion of miles to kilometers.   (*Right*) This modern car speedometer gives speed readings in both miles per hour and kilometers per hour. You should confirm the conversion between the two units for a few readings on the dial.   *(Paul Silverman, Fundamental Photographs)*

Example 1.3   The Density of a Cube

The mass of a solid cube is 856 g, and each edge has a length of 5.35 cm. Determine the density ρ of the cube in SI units.

Solution Because $1\ g = 10^{-3}\ kg$ and $1\ cm = 10^{-2}\ m$, the mass, m, and volume, V, in SI units are given by

$$m = 856\ g \times 10^{-3}\ kg/g = 0.856\ kg$$

$$V = L^3 = (5.35\ cm \times 10^{-2}\ m/cm)^3$$
$$= (5.35)^3 \times 10^{-6}\ m^3 = 1.53 \times 10^{-4}\ m^3$$

Therefore

$$\rho = \frac{m}{V} = \frac{0.856\ kg}{1.53 \times 10^{-4}\ m^3} = 5.59 \times 10^3\ kg/m^3$$

1.5 • ORDER-OF-MAGNITUDE CALCULATIONS

It is often useful to compute an approximate answer to a given physical problem even in instances in which little information is available. This answer can then be used to determine whether or not a more precise calculation is necessary. Such an approximation is usually based on certain assumptions, which must be modified if greater precision is needed. Thus, we will sometimes refer to an *order of magnitude* of a certain quantity as the power of ten of the number that describes that quantity. Usually, when an order-of-magnitude calculation is made, the results are reliable to within a factor of 10. If a quantity increases in value by three orders of magnitude, this means that its value increases by a factor of $10^3 = 1000$. We use the symbol $\sim$ for *is on the order of*. Thus,

$$0.0086 \sim 10^{-2} \qquad 0.0031 \sim 10^{-3} \qquad 700 \sim 10^3$$

The spirit of attempting order-of-magnitude calculations, sometimes referred to as '*guesstimates*' or '*ball-park figures*' is captured by the following quotation: "Make an estimate before every calculation, try a simple physical argument . . . before every derivation, guess the answer to every puzzle. Courage: No one else needs to know what the guess is."[2]

Example 1.4   The Number of Atoms in a Solid

Estimate the number of atoms in 1 cm³ of a solid.

Solution From Table 1.1 we note that the diameter of an atom is about 10^{-10} m. Thus, if in our model we assume that the atoms in the solid are solid spheres of this diameter, then the volume of each sphere is about 10^{-30} m³ (more precisely, volume $= 4\pi r^3/3 = \pi d^3/6$, where $r = d/2$). Therefore, be-cause 1 cm³ $= 10^{-6}$ m³, the number of atoms in the solid is on the order of $10^{-6}/10^{-30} = 10^{24}$ atoms.

A more precise calculation would require knowledge of the density of the solid and the mass of each atom. However, our estimate agrees with the more precise calculation to within a factor of ten.

[2]E. Taylor and J. A. Wheeler, *Spacetime Physics*, San Francisco, W. H. Freeman, 1966, p. 60.

Example 1.5   How Much Gas Do We Use?

Estimate the number of gallons of gasoline used by all cars in the United States each year.

Solution   Because there are about 240 million people in the United States, an estimate of the number of cars in the country is 60 million (assuming one car and four people per family). We also estimate that the average distance traveled per year is 10 000 miles. If we assume gasoline consumption of 0.05 gal/mi, each car uses about 500 gal/year. Multiplying this by the total number of cars in the United States gives an estimated total consumption of 3×10^{10} gal, which corresponds to a yearly consumer expenditure of more than $30 billion! This is probably a low estimate because we have not accounted for commercial consumption.

1.6 · SIGNIFICANT FIGURES

When certain quantities are measured, the measured values are known only to within the limits of the experimental uncertainty. The value of the uncertainty can depend on various factors, such as the quality of the apparatus, the skill of the experimenter, and the number of measurements performed.

Suppose that in a laboratory experiment we are asked to measure the area of a rectangular plate using a meter stick as a measuring instrument. Let us assume that the accuracy to which we can measure a particular dimension of the plate is ±0.1 cm. If the length of the plate is measured to be 16.3 cm, we can claim only that its length lies somewhere between 16.2 cm and 16.4 cm. In this case, we say that the measured value has three significant figures. Likewise, if its width is measured to be 4.5 cm, the actual value lies between 4.4 cm and 4.6 cm. This measured value has only two significant figures. Note that the significant figures include the first reliable digit. Thus, we could write the measured values as 16.3 ± 0.1 cm and 4.5 ± 0.1 cm.

Suppose now that we would like to find the area of the plate by multiplying the two measured values. If we were to claim that the area is (16.3 cm)(4.5 cm) = 73.35 cm², our answer would be unjustifiable because it contains four significant figures, which is greater than the number of significant figures in either of the measured lengths. A good rule of thumb in determining the number of significant figures that can be claimed is as follows:

> When multiplying several quantities, the number of significant figures in the final answer is the same as the number of significant figures in the *least* accurate of the quantities being multiplied, where *least accurate* means *having the lowest number of significant figures*. The same rule applies to division.

Applying this rule to the multiplication example given previously, we see that the answer for the area can have only two significant figures because the length of 4.5 cm has only two significant figures. Thus, all we can claim is that the area is 73 cm², realizing that the value can range between (16.2 cm)(4.4 cm) = 71 cm² and (16.4 cm)(4.6 cm) = 75 cm².

Zeros may or may not be significant figures. Those used to position the decimal point in such numbers as 0.03 and 0.0075 are not significant. Thus there are one and two significant figures, respectively, in these two values. When the positioning of zeros comes after other digits, however, there is the possibility of misinterpretation. For example, suppose the mass of an object is given as 1500 g. This value is ambiguous because we do not know whether the last two zeros are being used to locate the decimal point or whether they represent significant figures in the measurement. In order to remove this ambiguity, it is common to use scientific notation to indicate the number of significant figures. In this case, we would express the mass as 1.5×10^3 g if there are two significant figures in the measured value, 1.50×10^3 g if there are three significant figures, and 1.500×10^3 g if there are four. Likewise, 0.000 15 should be expressed in scientific notation as 1.5×10^{-4} if it has two significant figures or as 1.50×10^{-4} if it has three significant figures. The three zeros between the decimal point and the digit 1 in the number 0.000 15 are not counted as significant figures, because they are present only to locate the decimal point. In general, a **significant figure** is a reliably known digit (other than a zero used to locate the decimal point).

For addition and subtraction, the number of decimal places must be considered when you are determining how many significant figures to report.

> When numbers are **added** or **subtracted,** the number of decimal places in the result should equal the smallest number of decimal places of any term in the sum.

For example, if we wish to compute $123 + 5.35$, the answer is 128 and not 128.35. If we compute the sum $1.0001 + 0.0003 = 1.0004$, the result has the correct number of decimal places; consequently, it has five significant figures even though one of the terms in the sum, 0.0003, has only one significant figure. Likewise, if we perform the subtraction $1.002 - 0.998 = 0.004$, the result has only one significant figure even though one term has four significant figures and the other has three. In this book, **most of the numerical examples and end-of-chapter problems will yield answers having either two or three significant figures.**

Example 1.6   Installing a Carpet

A carpet is to be installed in a room, the length of which is measured to be 12.71 m (four significant figures) and the width of which is measured to be 3.46 m (three significant figures). Find the area of the room.

Solution   If you multiply 12.71 m by 3.46 m on your calculator, you will get an answer of 43.9766 m². How many of these numbers should you claim? Our rule of thumb for multiplication tells us that you can claim only the number of significant figures in the least accurate of the quantities being measured. In this example, we have only three significant figures in our least accurate measurement, so we should express our final answer as 44.0 m². Note that in the answer given, we used a general rule for rounding off numbers, which states that the last digit retained is to be increased by 1 if the first digit dropped was equal to 5 or greater. If the last digit is 5, the result should be rounded to the nearest even number. (This helps avoid accumulation of errors.)

1.7 • COORDINATE SYSTEMS AND FRAMES OF REFERENCE

Many aspects of physics deal in some way or another with locations in space. For example, the mathematical description of the motion of an object requires a method for describing the position of the object. Thus, it is fitting that we first discuss how to describe the position of a point in space. It is done by means of coordinates. A point on a line can be located with one coordinate; a point in a plane is located with two coordinates; and three coordinates are required to locate a point in space.

A coordinate system used to specify locations in space consists of:

- A fixed reference point O, called the *origin*,
- A set of specified axes or directions with an appropriate scale and labels on the axes,
- Instructions that tell us how to label a point in space relative to the origin and axes.

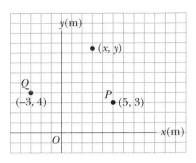

Figure 1.1 Designation of points in a cartesian coordinate system. Every point is labeled with coordinates (x, y).

One convenient coordinate system that we will use frequently is the *cartesian coordinate system,* sometimes called the *rectangular coordinate system*. Such a system in two dimensions is illustrated in Figure 1.1. An arbitrary point in this system is labeled with the coordinates (x, y). Positive x is taken to the right of the origin, and positive y is upward from the origin. Negative x is to the left of the origin, and negative y is downward from the origin. For example, the point P, which has coordinates $(5, 3)$, may be reached by going first 5 meters to the right of the origin and then 3 meters above the origin. In the same way, the point Q has coordinates $(-3, 4)$, which correspond to going 3 meters to the left of the origin and 4 meters above the origin.

Sometimes it is more convenient to represent a point in a plane by its *plane polar coordinates* (r, θ), as in Figure 1.2a. In this coordinate system, r is the length of the line from the origin to the point, and θ is the angle between that line and a fixed axis, usually the positive x axis, with θ measured counterclockwise. From the right triangle in Figure 1.2b, we find $\sin \theta = y/r$ and $\cos \theta = x/r$. (A review of trigonometric functions is given in Appendix B.4.) Therefore, starting with plane polar coordinates, one can obtain the cartesian coordinates through the equations

$$x = r \cos \theta \qquad [1.3]$$

$$y = r \sin \theta \qquad [1.4]$$

Furthermore, it follows that

$$\tan \theta = \frac{y}{x} \qquad [1.5]$$

and

$$r = \sqrt{x^2 + y^2} \qquad [1.6]$$

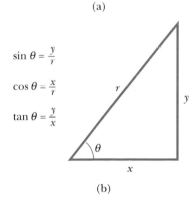

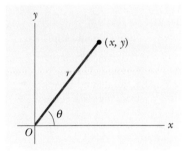

Figure 1.2 (a) The plane polar coordinates of a point are represented by the distance r and the angle θ. (b) The right triangle used to relate (x, y) to (r, θ).

You should note that these expressions relating the coordinates (x, y) to the coordinates (r, θ) apply only when θ is defined as in Figure 1.2a, in situations in which positive θ is an angle measured *counterclockwise* from the positive x axis. Other choices are made in navigation and astronomy. If the reference axis for the polar angle θ is chosen to be other than the positive x axis, or the sense of increasing θ is chosen differently, then the corresponding expressions relating the two sets of coordinates will change.

1.8 • PROBLEM-SOLVING STRATEGY

Most courses in general physics require the student to learn the skills of problem solving, and examinations usually include problems that test such skills. This brief section describes some useful suggestions that will enable you to increase your accuracy in solving problems, enhance your understanding of physical concepts, eliminate initial panic or lack of direction in approaching a problem, and organize your work. One way to help accomplish these goals is to adopt a problem-solving strategy. Many chapters in this book will include a section labeled "Problem-Solving Strategy," which should help you through the rough spots.

The following steps are commonly used to develop a problem-solving strategy:

A MENU FOR
PROBLEM SOLVING

Read Problem

Draw Diagram

Identify Data

Choose Equation(s)

Solve Equation(s)

Evaluate and Check
Answer

1. Read the problem carefully at least twice. Be sure you understand the nature of the problem before proceeding further.
2. Draw a suitable diagram with appropriate labels and coordinate axes if needed.
3. Imagine a movie, running in your mind, of what happens in the problem.
4. As you examine what is being asked in the problem, identify the basic physical principle or principles that are involved, listing the knowns and unknowns.
5. Select a basic relationship or derive an equation that can be used to find the unknown and symbolically solve the equation for the unknown.
6. Substitute the given values, along with the appropriate units, into the equation.
7. Obtain a numerical value with units for the unknown. You can have confidence in your result if the following questions can be properly answered: Do the units match? Is the answer reasonable? Is the plus or minus sign proper or meaningful?

One of the purposes of this strategy is to promote accuracy. Properly drawn diagrams can eliminate many sign errors. Diagrams also help to isolate the physical principles of the problem. Symbolic solutions and carefully labeled knowns and unknowns will help eliminate other careless errors. The use of symbolic solutions should help you think in terms of the physics of the problem. A check of units at the end of the problem can indicate a possible algebraic error. The physical layout and organization of your problem will make the final product more understandable and easier to follow. Once you have developed an organized system for examining problems and extracting relevant information, you will become a more confident problem solver.

1.9 · VECTORS AND SCALARS

Each of the physical quantities that we shall encounter in this book can be placed in one of two categories: It is either a scalar or a vector. A scalar is a quantity that is completely specified by a positive or negative number with appropriate units:

> A **scalar** quantity has a positive or negative value and no direction. However, a **vector** is a physical quantity that must be specified by both magnitude and direction.

The number of apples in a basket is an example of a scalar quantity. If you are told there are 38 apples in the basket, this completes the required information; no specification of direction is required. Other examples of scalars are temperature,

(*Left*) The number of apples in the basket is one example of a scalar quantity. Can you think of other examples? (*Superstock*)   (*Right*) Jennifer pointing in the right direction tells us to travel 5 blocks to the north to reach the courthouse. A vector is a physical quantity that must be specified by both magnitude and direction. (*Photo by Raymond A. Serway*)

volume, mass, and time intervals. The rules of ordinary arithmetic are used to manipulate scalar quantities.

Force is an example of a vector quantity. To completely describe the force on an object, we must specify both the direction of the applied force, the magnitude of the force, and the location of the force. When the motion (velocity) of an object is described, we must specify both the speed and the direction of its motion.

Another simple example of a vector quantity is the **displacement** of a particle, defined as its *change in position*. Suppose the particle moves from some point O to a point P along a straight path, as in Figure 1.3. We represent this displacement by drawing an arrow from O to P, where the tip of the arrow represents the direction of the displacement and the length of the arrow represents the magnitude of the displacement. If the particle travels along some other path from O to P, such as the broken line in Figure 1.3, its displacement is still OP. The vector displacement along any indirect path from O to P is defined as being equivalent to the displacement represented by the direct path from O to P. The magnitude of the displacement is the shortest distance between the end points. Thus, **the displacement of a particle is completely known if its initial and final coordinates are known.** The path need not be specified. In other words, **the displacement is independent of the path** if the end points of the path are fixed.

It is important to note that the distance traveled by a particle is distinctly different from its displacement. The **distance** traveled (a scalar quantity) is the length of the path, which in general can be much greater than the magnitude of the displacement (see Fig. 1.3).

If the particle moves along the x axis from position x_i to position x_f, as in Figure 1.4, its displacement is given by $x_f - x_i$. (The indices i and f refer to the initial and final values.) We use the Greek letter delta (Δ) to denote the *change* in a quantity.

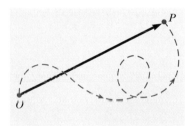

Figure 1.3 As a particle moves from O to P along the broken line, its displacement vector is the arrow drawn from O to P.

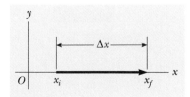

Figure 1.4 A particle moving along the x axis from x_i to x_f undergoes a displacement $\Delta x = x_f - x_i$.

Therefore, we define the change in the position of the particle (the displacement) as

$$\Delta x \equiv x_f - x_i \qquad \text{[1.7]}$$

From this definition we see that Δx is positive if x_f is greater than x_i and negative if x_f is less than x_i. For example, if a particle changes its position from $x_i = -3$ units to $x_f = 5$ units, its displacement is 8 units.

Many physical quantities in addition to displacement are vectors. They include velocity, acceleration, force, and momentum, all of which will be defined in later chapters. In this book we will use boldface letters, such as **A**, to represent arbitrary vectors. Another common method of notating vectors with which you should be familiar is to use an arrow over the letter: $\vec{A}$.

The magnitude of the vector **A** is written A or, alternatively, |**A**|. The magnitude of a vector is always positive and has physical units, such as meters for displacement or meters per second for velocity, as discussed earlier. Vectors combine according to special rules, which will be discussed in Sections 1.10 and 1.11.

Thinking Physics 1

Consider your commute to work or school in the morning. Which is larger, the distance you traveled or the magnitude of the displacement vector?

Reasoning The distance traveled, unless you have a very unusual commute, *must* be larger than the magnitude of the displacement vector. The distance includes all of the twists and turns that you made in following the roads from home to work or school. However, the magnitude of the displacement vector is the length of a straight line from your home to work or school. This is often described informally as "the distance as the crow flies." The only way that the distance could be the same as the magnitude of the displacement vector is if your commute is a perfect straight line, which is highly unlikely! The distance could *never* be less than the magnitude of the displacement vector, because the shortest distance between two points is a straight line.

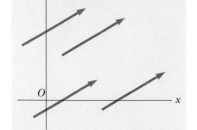

Figure 1.5 Four representations of the same vector.

1.10 • SOME PROPERTIES OF VECTORS

Equality of Two Vectors   Two vectors **A** and **B** are defined to be equal if they have the same magnitude and the same direction. That is, **A** = **B** only if $A = B$ *and* **A** and **B** act in parallel. For example, all the vectors in Figure 1.5 are equal even though they have different starting points. This property allows us to translate a vector parallel to itself in a diagram without affecting the vector.

Addition   When two or more vectors are added together, they must *all* have the same units. For example, it would be meaningless to add a velocity vector to a displacement vector, because they are different physical quantities. Scalars obey the same rule. For example, it would be meaningless to add time intervals and temperatures.

The rules for vector sums are conveniently described by geometric methods. To add vector **B** to vector **A**, first draw vector **A**, with its magnitude represented by a convenient scale, on graph paper and then draw vector **B** to the same scale with its tail starting from the tip of **A**, as in Figure 1.6. The *resultant vector* **R** = **A** + **B** is the vector drawn from the tail of **A** to the tip of **B**. This is known as the *triangle*

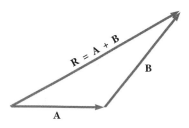

Figure 1.6 When vector **A** is added to vector **B**, the resultant **R** runs from the tail of **A** to the tip of **B**.

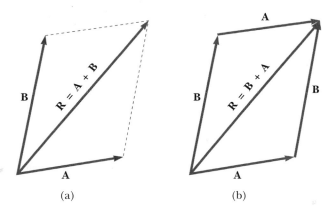

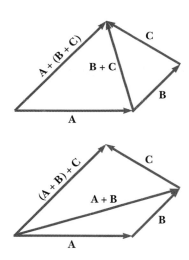

Figure 1.7 (a) In this construction, the resultant **R** is the diagonal of a parallelogram with sides **A** and **B**. (b) This construction shows that **A** + **B** = **B** + **A**.

Figure 1.8 Geometric constructions for verifying the associative law of addition.

method of addition. An alternative graphical procedure for adding two vectors, known as the **parallelogram rule of addition,** is shown in Figure 1.7a. In this construction, the tails of the two vectors **A** and **B** are together, and the resultant vector **R** is the diagonal of a parallelogram formed with **A** and **B** as its sides.

When two vectors are added, **the sum is independent of the order of the addition.** This can be seen from the geometric construction in Figure 1.7b and is known as the **commutative law of addition:**

$$\mathbf{A} + \mathbf{B} = \mathbf{B} + \mathbf{A} \qquad [1.8]$$

- *Commutative law*

If three or more vectors are added, **their sum is independent of the way in which they are grouped.** A geometric proof of this for three vectors is given in Figure 1.8. This is called the **associative law of addition:**

$$\mathbf{A} + (\mathbf{B} + \mathbf{C}) = (\mathbf{A} + \mathbf{B}) + \mathbf{C} \qquad [1.9]$$

- *Associative law*

Geometric constructions can also be used to add more than three vectors. This is shown in Figure 1.9 for the case of four vectors. The resultant vector sum **R** = **A** + **B** + **C** + **D** is **the vector that completes the polygon.** In other words, **R is the vector drawn from the tail of the first vector to the tip of the last vector.** Again, the order of the summation is unimportant.

Thus we conclude that **a vector is a quantity that has both magnitude and direction and also obeys the laws of vector addition,** as described in Figures 1.6 through 1.9.

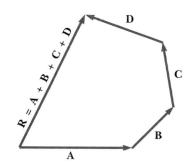

Negative of a Vector The negative of the vector **A** is defined as the vector that, when added to **A**, gives zero for the vector sum. That is, **A** + (−**A**) = 0. The vectors **A** and −**A** have the same magnitude but opposite directions.

Figure 1.9 Geometric construction for summing four vectors. The resultant vector **R** completes the polygon.

Subtraction of Vectors The operation of vector subtraction makes use of the definition of the negative of a vector. We define the operation **A** − **B** as vector −**B** added to vector **A**:

$$\mathbf{A} - \mathbf{B} = \mathbf{A} + (-\mathbf{B}) \qquad [1.10]$$

The geometric construction for subtracting two vectors is shown in Figure 1.10.

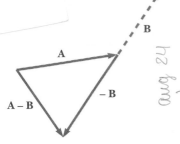

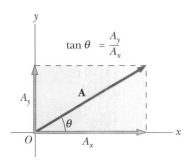

Figure 1.10 This construction shows how to subtract vector **B** from vector **A**. The vector $-\mathbf{B}$ is equal in magnitude and opposite to the vector **B**.

Figure 1.11 Any vector **A** lying in the xy plane can be represented by its components A_x and A_y.

Multiplication of a Vector by a Scalar   If a vector **A** is multiplied by a positive scalar quantity m, the product $m\mathbf{A}$ is a vector that has the same direction as **A** and magnitude mA. If m is a negative scalar quantity, the vector $m\mathbf{A}$ is directed opposite to **A**. For example, the vector $5\mathbf{A}$ is five times as great in magnitude as **A** and has the same direction as **A**. By contrast, the vector $-\frac{1}{3}\mathbf{A}$ has one third the magnitude of **A** and points in the direction opposite **A** (because of the negative sign).

Multiplication of Two Vectors   Two vectors **A** and **B** can be multiplied to produce either a scalar or a vector quantity. The **scalar product** (or dot product) $\mathbf{A} \cdot \mathbf{B}$ is a scalar quantity equal to $AB \cos \theta$, where θ is the angle between **A** and **B**. The **vector product** (or cross product) $\mathbf{A} \times \mathbf{B}$ is a vector quantity, the magnitude of which is equal to $AB \sin \theta$. We shall discuss these products more fully in Chapters 6 and 10, where they are first used.

Each chapter of this textbook contains a number of exercises. The purpose of such exercises is to test your understanding of the material just discussed by asking you to do a calculation or answer some question related to a worked example. Here is your first exercise.

EXERCISE 1   The magnitudes of two vectors **A** and **B** are $A = 8$ units and $B = 3$ units. Find the largest and smallest possible values for the magnitude of the resultant vector $\mathbf{R} = \mathbf{A} + \mathbf{B}$.   Answer   The largest value of R is 11 units, corresponding to the case when **A** and **B** point in the same direction. The smallest value of R is 5 units, corresponding to the case when **A** is directed opposite to **B**.

1.11 • COMPONENTS OF A VECTOR AND UNIT VECTORS

The geometric method of adding vectors is not the recommended procedure in situations in which great precision is required or in three-dimensional problems. In this section we describe a method of adding vectors that makes use of the *projections* of a vector along the axes of a rectangular coordinate system.

Consider a vector **A** lying in the xy plane and making an arbitrary angle θ with the positive x axis, as in Figure 1.11. The vector **A** can be represented by its **rectangular components,** A_x and A_y. The component A_x represents the projection of **A** along the x axis, whereas A_y represents the projection of **A** along the y axis. The components of a vector, which are scalar quantities, can be positive or negative. For example, in Figure 1.11, A_x and A_y are both positive.

From Figure 1.11 and the definition of the sine and cosine of an angle, we see that $\cos \theta = A_x/A$ and $\sin \theta = A_y/A$. Hence, the components of **A** are given by

*Components of the vector **A*** •

$$A_x = A \cos \theta \qquad \text{and} \qquad A_y = A \sin \theta \qquad \text{[1.11]}$$

It is important to note that when using these component equations, θ must be measured counterclockwise from the positive x axis. These components form two sides of a right triangle, the hypotenuse of which has a magnitude A. Thus, it follows that the magnitude of **A** and its direction are related to its components through the expressions

*Magnitude of **A*** •

$$A = \sqrt{A_x^{\,2} + A_y^{\,2}} \qquad \text{[1.12]}$$

*Direction of **A*** •

$$\tan \theta = \frac{A_y}{A_x} \qquad \text{[1.13]}$$

To solve for θ, we can write $\theta = \tan^{-1}(A_y/A_x)$, which is read "$\theta$ equals the angle whose tangent is the ratio A_y/A_x." **Note that the signs of the components A_x and A_y depend on the angle θ.** For example, if $\theta = 120°$, A_x is negative and A_y is positive. By contrast, if $\theta = 225°$, both A_x and A_y are negative. If both components are negative, the angle measured from the positive x axis will be the angle you determine on your calculator plus 180°. Figure 1.12 summarizes the signs of the components when **A** lies in the various quadrants.

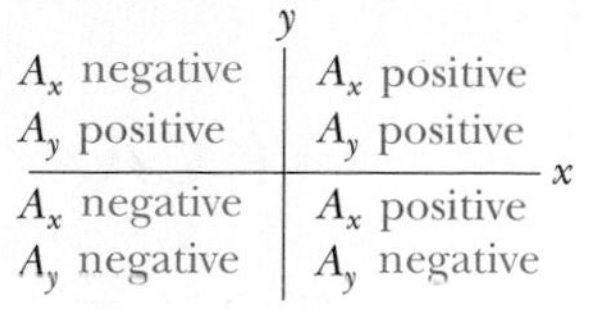

If you choose reference axes or an angle other than those shown in Figure 1.11, the components of the vector must be modified accordingly. In many applications it is more convenient to express the components of a vector in a coordinate system having axes that are not horizontal and vertical but that are still perpendicular to each other. Suppose a vector **B** makes an angle θ' with the x' axis defined in Figure 1.13. The components of **B** along these axes are given by $B_{x'} = B\cos\theta'$ and $B_{y'} = B\sin\theta'$, as in Equation 1.9. The magnitude and direction of **B** are obtained from expressions equivalent to Equations 1.10 and 1.11. Thus, we can express the components of a vector in *any* coordinate system that is convenient for a particular situation.

The components of a vector differ when viewed from different coordinate systems. Furthermore, the components of a vector can change with respect to a fixed coordinate system if the vector changes in magnitude, orientation, or both.

Vector quantities are often expressed in terms of unit vectors. **A unit vector is a dimensionless vector 1 unit in length used to specify a given direction.** Unit vectors have no other physical significance. They are used simply as a convenience in describing a direction in space. We will use the symbols **i**, **j**, and **k** to represent unit vectors pointing in the x, y, and z directions, respectively. Thus, the unit vectors **i**, **j**, and **k** form a set of mutually perpendicular vectors as shown in Figure 1.14a, where the magnitude of the unit vectors equals unity. That is, $|\mathbf{i}| = |\mathbf{j}| = |\mathbf{k}| = 1$.

Consider a vector **A** lying in the xy plane, as in Figure 1.14b. The product of the component A_x and the unit vector **i** is the vector $A_x\mathbf{i}$ parallel to the x axis with magnitude A_x. Likewise, $A_y\mathbf{j}$ is a vector of magnitude A_y parallel to the y axis. When

A_x negative	A_x positive
A_y positive	A_y positive
A_x negative	A_x positive
A_y negative	A_y negative

Figure 1.12   The signs of the components of a vector **A** depend on the quadrant in which the vector is located.

• *Unit vectors*

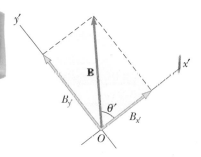

Figure 1.13   The components of vector **B** in a coordinate system that is tilted.

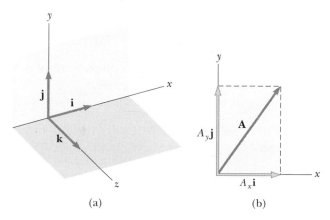

(a)   (b)

Figure 1.14   (a) The unit vectors **i**, **j**, and **k** are directed along the x, y, and z axes, respectively. (b) A vector **A** lying in the xy plane has component vectors $A_x\mathbf{i}$ and $A_y\mathbf{j}$ where A_x and A_y are the components of **A**.

using the unit form of a vector, we are simply multiplying a vector (the unit vector) by a scalar. Thus, the unit-vector notation for the vector **A** is written

$$\mathbf{A} = A_x\mathbf{i} + A_y\mathbf{j} \tag{1.14}$$

The vectors $A_x\mathbf{i}$ and $A_y\mathbf{j}$ are the component vectors of **A**. These should not be confused with A_x and A_y, which we shall always refer to as the components of **A**.

Now suppose we wish to add vector **B** to vector **A**, where **B** has components B_x and B_y. The procedure for performing this sum is simply to add the x and y components separately. The resultant vector $\mathbf{R} = \mathbf{A} + \mathbf{B}$ is therefore

$$\mathbf{R} = (A_x + B_x)\mathbf{i} + (A_y + B_y)\mathbf{j} \tag{1.15}$$

Thus, the components of the resultant vector are given by

$$R_x = A_x + B_x$$
$$R_y = A_y + B_y \tag{1.16}$$

The magnitude of **R** and the angle it makes with the x axis can then be obtained from its components using the relationships

$$R = \sqrt{R_x^2 + R_y^2} = \sqrt{(A_x + B_x)^2 + (A_y + B_y)^2} \tag{1.17}$$

$$\tan\theta = \frac{R_y}{R_x} = \frac{A_y + B_y}{A_x + B_x} \tag{1.18}$$

The procedure just described for adding two vectors **A** and **B** using the component method can be checked using a geometric construction, as in Figure 1.15. Again you must take note of the *signs* of the components when using either the algebraic or the geometric method.

The extension of these methods to three-dimensional vectors is straightforward. If **A** and **B** both have x, y, and z components, we express them in the form

$$\mathbf{A} = A_x\mathbf{i} + A_y\mathbf{j} + A_z\mathbf{k} \qquad \text{and} \qquad \mathbf{B} = B_x\mathbf{i} + B_y\mathbf{j} + B_z\mathbf{k}$$

The sum of **A** and **B** is

$$\mathbf{R} = \mathbf{A} + \mathbf{B} = (A_x + B_x)\mathbf{i} + (A_y + B_y)\mathbf{j} + (A_z + B_z)\mathbf{k} \tag{1.19}$$

Thus, the resultant vector also has a z component, given by $R_z = A_z + B_z$. The same procedure can be used to sum up three or more vectors.

If a vector **R** has x, y, and z components, the magnitude of the vector is

$$R = \sqrt{R_x^2 + R_y^2 + R_z^2}$$

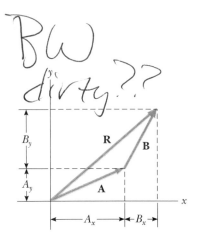

Figure 1.15 A geometric construction showing the relation between the components of the resultant **R** of two vectors and the individual component vectors.

Thinking Physics 2

You may have asked someone directions to a destination in a city and have been told something like, "Walk 3 blocks east and then 5 blocks south." If so, are you experienced with vector components?

Reasoning   Yes, you are! Although you may not have thought of vector component language when you heard these directions, this is exactly what the directions represent. The perpendicular streets of the city reflect an *x–y* coordinate system—we can assign the *x* axis to the east–west streets and the *y* axis to the north–south streets. Thus, the comment of the person giving you directions can be translated as, "Undergo a displacement vector that has an *x* component of $+3$ blocks and a *y* component of -5 blocks." You would arrive at the same destination by undergoing the *y* component first, followed by the *x* component, demonstrating the commutative law of addition.

Many sections of this book will include conceptual problems that are nonnumerical like the Thinking Physics example above. The answers to all conceptual problems are found at the end of each chapter. Here is your first set of conceptual problems.

CONCEPTUAL PROBLEM 1

If one component of a vector is not zero, can its magnitude be zero? Explain.

CONCEPTUAL PROBLEM 2

If $\mathbf{A} + \mathbf{B} = 0$, what can you say about the components of the two vectors?

CONCEPTUAL PROBLEM 3

Figure CP1.3 shows two vectors lying in the *xy*-plane. Determine the signs of (a) the *x* components of **A** and **B**, (b) the *y* components of **A** and **B**, and (c) the *x* and *y* components of **A + B**.

CONCEPTUAL PROBLEM 4

Can the component of a vector ever be equal to the magnitude of the vector?

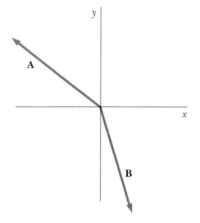

Figure CP1.9

PROBLEM-SOLVING STRATEGY • **Adding Vectors, Using Components**

When two or more vectors are to be added, the following steps are recommended:

1. Select a coordinate system.
2. Draw a sketch of the vectors to be added (or subtracted) and label each vector.
3. Find the *x* and *y* components of all vectors.
4. Find the resultant components (the algebraic sum of the components) in both the *x* and *y* directions.
5. Use the Pythagorean theorem to find the magnitude of the resultant vector.
6. Use a suitable trigonometric function to find the angle the resultant vector makes with the *x* axis.

Example 1.7   The Sum of Two Vectors

Find the sum of two vectors **A** and **B** lying in the *xy* plane and given by

$$\mathbf{A} = 2.0\mathbf{i} + 2.0\mathbf{j} \quad \text{and} \quad \mathbf{B} = 2.0\mathbf{i} - 4.0\mathbf{j}$$

Solution   Note that $A_x = 2.0$, $A_y = 2.0$, $B_x = 2.0$, and $B_y = -4.0$. Therefore, the resultant vector **R** is

$$\mathbf{R} = \mathbf{A} + \mathbf{B} = (2.0 + 2.0)\mathbf{i} + (2.0 - 4.0)\mathbf{j} = 4.0\mathbf{i} - 2.0\mathbf{j}$$

or

$$R_x = 4.0, \qquad R_y = -2.0$$

The magnitude of **R** is

$$R = \sqrt{R_x^2 + R_y^2} = \sqrt{(4.0)^2 + (-2.0)^2} = \sqrt{20} = 4.5$$

EXERCISE 2   Find the angle θ that the resultant vector **R** makes with the positive *x* axis.   Answer   330°

Example 1.8   The Resultant Displacement

A particle undergoes three consecutive displacements: $\mathbf{d}_1 = (1.5\mathbf{i} + 3.0\mathbf{j} - 1.2\mathbf{k})$ cm, $\mathbf{d}_2 = (2.3\mathbf{i} - 1.4\mathbf{j} - 3.6\mathbf{k})$ cm, and $\mathbf{d}_3 = (-1.3\mathbf{i} + 1.5\mathbf{j})$ cm. Find the components of the resultant displacement and its magnitude.

Solution

$$\mathbf{R} = \mathbf{d}_1 + \mathbf{d}_2 + \mathbf{d}_3$$
$$= (1.5 + 2.3 - 1.3)\mathbf{i} + (3.0 - 1.4 + 1.5)\mathbf{j}$$
$$+ (-1.2 - 3.6 + 0)\mathbf{k}$$
$$= (2.5\mathbf{i} + 3.1\mathbf{j} - 4.8\mathbf{k})\,\text{cm}$$

That is, the resultant displacement has components $R_x = 2.5$ cm, $R_y = 3.1$ cm, and $R_z = -4.8$ cm. Its magnitude is

$$R = \sqrt{R_x^2 + R_y^2 + R_z^2}$$
$$= \sqrt{(2.5 \text{ cm})^2 + (3.1 \text{ cm})^2 + (-4.8 \text{ cm})^2} = 6.2 \text{ cm}$$

Example 1.9   Taking a Hike

A hiker begins a trip by first walking 25.0 km due southeast from her base camp. On the second day she walks 40.0 km in a direction 60.0° north of east, at which point she discovers a forest ranger's tower.

(a) Determine the components of the hiker's displacements in the first and second days.

Solution   If we denote the displacement vectors on the first and second days by **A** and **B**, respectively, and use the camp as the origin of coordinates, we get the vectors shown in Figure 1.16. Displacement **A** has a magnitude of 25.0 km and is 45.0° southeast. Its components are

$$A_x = A \cos(-45.0°) = (25.0 \text{ km})(0.707) = 17.7 \text{ km}$$

$$A_y = A \sin(-45.0°) = -(25.0 \text{ km})(0.707) = -17.7 \text{ km}$$

The negative value of A_y indicates that the *y* coordinate decreased in this displacement. The signs of A_x and A_y are also evident from Figure 1.16.

The second displacement, **B**, has a magnitude of 40.0 km and is 60.0° north of east. Its components are

$$B_x = B \cos 60.0° = (40.0 \text{ km})(0.500) = 20.0 \text{ km}$$

$$B_y = B \sin 60.0° = (40.0 \text{ km})(0.866) = 34.6 \text{ km}$$

(b) Determine the components of the hiker's total displacement for the trip.

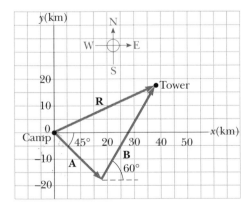

Figure 1.16   The total displacement of the hiker is the vector $\mathbf{R} = \mathbf{A} + \mathbf{B}$.

Solution The resultant displacement vector for the trip, **R** = **A** + **B**, has components

$$R_x = A_x + B_x = 17.7 \text{ km} + 20.0 \text{ km} = 37.7 \text{ km}$$

$$R_y = A_y + B_y = 17.7 \text{ km} + 34.6 \text{ km} = 16.9 \text{ km}$$

In unit-vector form, we can write the total displacement as **R** = (37.7**i** + 16.9**j**) km.

EXERCISE 3 Determine the magnitude and direction of the total displacement.   Answer   41.3 km, 24.1° north of east from the base camp

SUMMARY

Mechanical quantities can be expressed in terms of three fundamental quantities, *length, mass,* and *time,* which in the SI system have the units *meters* (m), *kilograms* (kg), and *seconds* (s), respectively. It is often useful to use the *method of dimensional analysis* to check equations and to assist in deriving expressions.

The density of a substance is defined as its mass per unit volume. Different substances have different densities mainly because of differences in their atomic masses and atomic arrangements.

The number of molecules in 1 mole of any element or compound, called **Avogadro's number** (N_A), is 6.02×10^{23}.

Vectors are quantities that have both magnitude and direction and obey the vector law of addition. **Scalars** are quantities that have only magnitude.

Two vectors **A** and **B** can be added using either the triangle method or the parallelogram rule. In the triangle method (Fig. 1.6), the vector **C** = **A** + **B** runs from the tail of **A** to the tip of **B**. In the parallelogram method (Fig. 1.7a), **C** is the diagonal of a parallelogram having **A** and **B** as its sides.

The x component of the vector **A**, A_x, is equal to its projection along the x axis of a coordinate system as in Figure 1.11, where $A_x = A \cos \theta$ and θ is the angle **A** makes with the x axis. Likewise, the y component of **A**, A_y, is its projection along the y axis, where $A_y = A \sin \theta$. The resultant of two or more vectors can be found by resolving all vectors into their x and y components, adding their resultant x and y components, and then using the Pythagorean theorem to find the magnitude of the resultant vector. The angle that the resultant vector makes with the x axis can be found by the use of a suitable trigonometric function.

If a vector **A** has an x component equal to A_x and a y component equal to A_y, the vector can be expressed in unit-vector form as **A** = A_x**i** + A_y**j**. In this notation, **i** is a unit vector in the positive x direction and **j** is a unit vector in the positive y direction. Because **i** and **j** are unit vectors, $|\mathbf{i}| = |\mathbf{j}| = 1$.

CONCEPTUAL QUESTIONS

1. Suppose that the three fundamental standards of the metric system were length, *density*, and time rather than length, *mass*, and time. The standard of density in this system is to be defined as that of water. What considerations about water would need to be addressed to make sure that the standard unit of density were to be as accurate as possible?

2. What types of natural phenomena could serve as alternative time standards?

3. The height of a horse is sometimes given in units of "hands." Why is this a poor standard of length?

4. Express the following quantities using the prefixes given in Table 1.4: (a) 3×10^{-4} m, (b) 5×10^{-5} s, (c) 72×10^2 g.

5. As one moves upward in atomic number in the periodic table, the atomic masses of the elements increases. Because the atoms are becoming more and more massive, why doesn't the *density* of elemental materials increase in the same way?

6. Suppose that two quantities A and B have different dimensions. Determine which of the following arithmetic opera-

Boxed numbers indicate questions that have answers available in the Student Solutions Manual and Study Guide.

tions *could* be physically meaningful: (a) *A* + *B*, (b) *A*/*B*, (c) *B* − *A*, (d) *AB*.

7. What accuracy is implied in an order-of-magnitude calculation?

8. Apply an order-of-magnitude calculation to an everyday situation you might encounter. For example, how far do you walk or drive each day?

9. Which of the following are vectors and which are not: force, temperature, the volume of water in a can, the ratings of a TV show, the height of a building, the velocity of a sports car, the age of the Universe?

10. A vector **A** lies in the *xy* plane. For what orientations of **A** will both of its rectangular components be negative? For what orientations will its components have opposite signs?

11. A book is moved once around the perimeter of a tabletop with the dimensions 1.0 m × 2.0 m. If the book ends up at its initial position, what is its displacement? What is the distance traveled?

12. While traveling along a straight interstate highway you notice that the mile marker reads 260. You travel until you reach the 150-mile marker and then retrace your path to the 175-mile marker. What is the magnitude of your resultant displacement from the 260-mile marker?

13. If the component of vector **A** along the direction of vector **B** is zero, what can you conclude about the two vectors?

14. If **A** = **B**, what can you conclude about the components of **A** and **B**?

15. Is it possible to add a vector quantity to a scalar quantity? Explain.

16. A roller coaster travels 135 ft at an angle of 40.0° above the horizontal. How far does it move horizontally and vertically?

17. The resolution of vectors into components is equivalent to replacing the original vector with the sum of two vectors, whose sum is the same as the original vector. There are an infinite number of pairs of vectors that will satisfy this condition; we choose that pair with one vector parallel to the *x* axis and the second parallel to the *y* axis. What difficulties would be introduced by defining components relative to axes that are not perpendicular—for example, the *x* axis and a *y* axis oriented at 45° to the *x* axis?

PROBLEMS

Section 1.2   Density and Atomic Mass

1. Calculate the mass of an atom of (a) helium, (b) iron, (c) lead. Give your answers in atomic mass units and in grams. The atomic masses of the atoms given are 4.00, 55.9, and 207, respectively.

2. The standard kilogram is a platinum-iridium cylinder 39.0 mm in height and 39.0 mm in diameter. What is the density of the material?

3. One cubic meter (1.00 m³) of aluminum has a mass of 2.70 × 10³ kg, and one cubic meter of iron has a mass of 7.86 × 10³ kg. Find the radius of a solid aluminum sphere that will balance a solid iron sphere of radius 2.00 cm on an equal-arm balance.

4. Let ρ_{Al} represent the density of aluminum and ρ_{Fe} that of iron. Find the radius of a solid aluminum sphere that balances a solid iron sphere of radius r_{Fe} on an equal-arm balance.

5. The mass of the planet Saturn is 5.64 × 10²⁶ kg, and its radius is 6.00 × 10⁷ m. (a) Calculate its density. (b) If Saturn were placed in a large enough ocean of water, would it float? Explain.

Section 1.3   Dimensional Analysis

6. The radius *r* of a circle inscribed in any triangle whose sides are *a*, *b*, and *c* is given by $r = [(s − a)(s − b)(s − c)/s]^{1/2}$, where *s* is an abbreviation for (*a* + *b* + *c*)/2. Check this formula for dimensional consistency.

7. The consumption of natural gas by a company satisfies the empirical equation $V = 1.50t + 0.00800t^2$, where *V* is the volume in millions of cubic feet and *t* the time in months. Express this equation in units of cubic feet and seconds. Put the proper units on the coefficients. Assume that a month is 30.0 days.

8. Newton's law of universal gravitation is given by

$$F = G\frac{Mm}{r^2}$$

Here *F* is the force of gravity, *M* and *m* are masses, and *r* is a length. Force has the units kg·m/s². What are the SI units of the proportionality constant *G*?

9. The period *T* of a simple pendulum (the time for one complete oscillation) is measured in time units and is given by

$$T = 2\pi\sqrt{\frac{\ell}{g}}$$

where ℓ is the length of the pendulum and *g* is the acceleration due to gravity in units of length divided by the square of time. Show that this equation is dimensionally consistent.

Boxed numbers indicate problems that have full solutions available in the Student Solutions Manual and Study Guide. Cyan numbers indicate intermediate level problems and magenta numbers indicate more challenging problems.

Section 1.4 Conversion of Units

10. An auditorium measures 40.0 m × 20.0 m × 12.0 m. The density of air is 1.20 kg/m^3. What are (a) the volume of the room in cubic feet and (b) the weight of air in the room in pounds?

11. An astronomical unit (AU) is defined as the average distance between the Earth and Sun. (a) How many astronomical units are there in one lightyear? (b) Determine the distance from Earth to the Andromeda galaxy in astronomical units.

12. Assume that it takes 7.00 minutes to fill a 30.0-gal gasoline tank. (a) Calculate the rate at which the tank is filled in gallons per second. (b) Calculate the rate at which the tank is filled in cubic meters per second. (c) Determine the time, in hours, required to fill a 1-cubic-meter volume at the same rate. (1 U.S. gal = 231 in.^{3})

13. One gallon of paint (volume = 3.78 × 10^{-3} m^3) covers an area of 25.0 m^2. What is the thickness of the paint on the wall?

14. You can obtain a rough estimate of the size of a molecule by the following simple experiment. Let a droplet of oil spread out on a smooth water surface. The resulting oil slick will be approximately 1 molecule thick. If an oil droplet of mass 9.00 × 10^{-7} kg and density 918 kg/m^3 spreads out into a circle of radius 41.8 cm on the water surface, calculate the diameter of an oil molecule.

15. (a) Find a conversion factor to convert from miles per hour to kilometers per hour. (b) Until recently, federal law mandated that highway speeds would be 55 mi/h. Use the conversion factor of part (a) to find the speed in kilometers per hour. (c) The maximum highway speed has been raised to 65 mi/h in some places. In kilometers per hour, how much increase is this over the 55-mi/h limit?

16. (a) How many seconds are there in a year? (b) If one micrometeorite (a sphere with a diameter of 10^{-6} m) struck each square meter of the Moon each second, how many years would it take to cover the Moon to a depth of 1.00 m? (*Hint:* Consider a cubic box on the Moon 1.00 m on a side and find how long it would take to fill the box.)

17. The mass of the Sun is about 1.99 × 10^{30} kg, and the mass of a hydrogen atom, of which the Sun is mostly composed, is 1.67 × 10^{-27} kg. How many atoms are there in the Sun?

Section 1.5 Order-of-Magnitude Calculations

18. Compute the order of magnitude of the mass of (a) a bathtub filled with water, and (b) a bathtub filled with pennies. In your solution list the quantities you estimate and the value you estimate for each.

19. Estimate the number of piano tuners living in New York City. This problem was posed by the physicist Enrico Fermi, who was well known for making order-of-magnitude calculations.

20. Soft drinks are commonly sold in aluminum containers. Estimate the number of such containers thrown away or re-

cycled each year by U.S. consumers. Approximately how many tons of aluminum does this represent?

Section 1.6 Significant Figures

21. Carry out the following arithmetic operations: (a) the sum of the numbers 756, 37.2, 0.83, and 2.5; (b) the product 3.2 × 3.563; (c) the product 5.6 × π.

22. How many significant figures are there in (a) 78.9 ± 0.2, (b) 3.788 × 10^9, (c) 2.46 × 10^{-6}, (d) 0.0053?

23. The radius of a circle is measured to be 10.5 ± 0.2 m. Calculate the (a) area and (b) circumference of the circle and give the uncertainty in each value.

24. The *radius* of a solid sphere is measured by a student to be (6.50 ± 0.20) cm, and its mass is measured to be (1.85 ± 0.02) kg. Determine the density of the sphere, in kilograms per cubic meter, and the uncertainty in the density.

Section 1.7 Coordinate Systems and Frames of Reference

25. The polar coordinates of a point are r = 5.50 m and θ = 240.0°. What are the cartesian coordinates of this point?

26. A point in the xy plane has cartesian coordinates (−3.00, 5.00) m. What are the polar coordinates of this point?

27. Two points in a plane have polar coordinates (2.50 m, 30.0°) and (3.80 m, 120.0°). Determine (a) the cartesian coordinates of these points and (b) the distance between them.

28. If the polar coordinates of the point (x, y) are (r, θ), determine the polar coordinates for the points: (a) (−x, y), (b) (−2x, −2y), and (c) (3x, −3y).

Section 1.9 Vectors and Scalars

Section 1.10 Some Properties of Vectors

29. A person walks along a circular path of radius 5.00 m, around one half of the circle. (a) Find the magnitude of the displacement vector. (b) How far did the person walk? (c) What is the magnitude of the displacement if the circle is completed?

30. A jogger runs 100 m due west, then changes direction for the second leg of the run. At the end of the run, she is 175 m away from the starting point at an angle of 15.0° north of west. What were the direction and magnitude of her second displacement? Use a graph to solve this problem.

31. Each of the displacement vectors **A** and **B** shown in Figure P1.31 on page 26 has a magnitude of 3.0 m. Graphically find (a) **A** + **B**, (b) **A** − **B**, (c) **B** − **A**, (d) **A** − 2**B**.

32. Indiana Jones is trapped in a maze. To find his way out, he walks 10.0 m, makes a 90.0° right turn, walks 5.00 m, makes

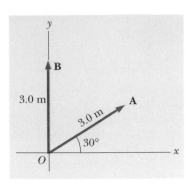

Figure P1.31

another 90.0° right turn, and walks 7.00 m. What is his displacement from his initial position?

Section 1.11   Components of a Vector and Unit Vectors

33. A vector has an x component of -25.0 units and a y component of 40.0 units. Find the magnitude and direction of this vector.

34. Find the horizontal and vertical components of the 100-m displacement of a superhero who flies from the top of a tall building along the path shown in Figure P1.34.

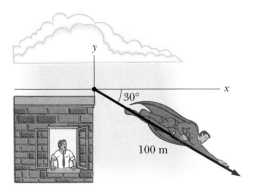

Figure P1.34

35. Instructions for finding a buried treasure include the following: Go 75 paces at 240°, turn to 135° and walk 125 paces, then travel 100 paces at 160°. Determine the resultant displacement from the starting point.

36. Given the vectors $\mathbf{A} = 2.00\mathbf{i} + 6.00\mathbf{j}$ and $\mathbf{B} = 3.00\mathbf{i} - 2.00\mathbf{j}$, (a) sketch the vector sum $\mathbf{C} = \mathbf{A} + \mathbf{B}$ and the vector

subtraction $\mathbf{D} = \mathbf{A} - \mathbf{B}$. (b) Find analytical solutions for $\mathbf{C}$ and $\mathbf{D}$ first in terms of unit vectors and then in terms of polar coordinates, with angles measured with respect to the positive x axis.

37. Three vectors are oriented as shown in Figure P1.37, where $|\mathbf{A}| = 20.0$, $|\mathbf{B}| = 40.0$, and $|\mathbf{C}| = 30.0$ units. Find (a) the x and y components of the resultant vector and (b) the magnitude and direction of the resultant vector.

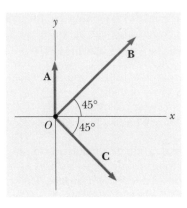

Figure P1.37

38. A novice golfer on the green takes three strokes to sink the ball. The successive displacements are 4.00 m due north, 2.00 m northeast, and 1.00 m 30.0° west of south. Starting at the same initial point, an expert golfer might make the hole in what single vector displacement?

39. A person going for a walk follows the path shown in Figure P1.39. The total trip consists of four straight-line paths. At

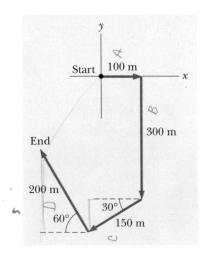

Figure P1.39

the end of the walk, what is the person's resultant displacement, measured from the starting point?

40. A vector is given by $\mathbf{R} = 2.00\mathbf{i} + 1.00\mathbf{j} + 3.00\mathbf{k}$. Find (a) the magnitudes of the x, y, and z components, (b) the magnitude of $\mathbf{R}$, and (c) the angles between $\mathbf{R}$ and the x, y, and z axes.

41. A particle undergoes two displacements. The first has a magnitude of 150 cm and makes an angle of 120.0° with the positive x axis. The *resultant* displacement has a magnitude of 140 cm and is directed at an angle of 35.0° to the positive x axis. Find the magnitude and direction of the second displacement.

42. A jet airliner moving initially at 300 mph due east enters a region where the wind is blowing at 100 mph in a direction 30.0° north of east. Determine the new speed and direction of the aircraft.

Additional Problems

43. A useful fact is that there are about $\pi \times 10^7$ s in one year. Use a calculator to find the percentage error in this approximation. *Note:*

$$\text{Percentage error} = \frac{|\text{assumed value} - \text{true value}|}{\text{true value}} \times 100\%$$

44. The eye of a hurricane passes over Grand Bahama Island. It is moving in a direction 60.0° north of west with a speed of 41.0 km/h. Three hours later the course of the hurricane suddenly shifts due north, and its speed slows to 25.0 km/h. How far is the hurricane from Grand Bahama 4.50 hours after it passes over the island?

45. Assume that there are 50 million passenger cars in the United States and that the average fuel consumption is 20 mi/gal of gasoline. If the average distance traveled by each car is 10 000 miles/year, how much gasoline would be saved per year if average fuel consumption could be increased to 25 mi/gal?

46. The basic function of the carburetor of an automobile is to atomize the gasoline and mix it with air to promote rapid combustion. As an example, assume that 30.0 cm³ of gasoline is atomized into N spherical droplets, each with a radius of 2.00×10^{-5} m. What is the total surface area of these N spherical droplets?

47. Assume that 70 percent of the Earth's surface is covered with water at an average depth of 1 mile, and **estimate** the mass of the water on Earth in kilograms.

48. The data in the following table represent measurements of the masses and dimensions of solid cylinders of aluminum, copper, brass, tin, and iron. Use these data to calculate the densities of the substances. Compare your results for aluminum, copper, and iron with those given in Table 1.5.

Substance	Mass (g)	Diameter (cm)	Length (cm)
Aluminum	51.5	2.52	3.75
Copper	56.3	1.23	5.06
Brass	94.4	1.54	5.69
Tin	69.1	1.75	3.74
Iron	216.1	1.89	9.77

49. An air-traffic controller notices two aircraft on his radar screen. The first is at altitude 800 m, horizontal distance 19.2 km, and 25.0° south of west. The second aircraft is at altitude 1100 m, horizontal distance 17.6 km, and 20.0° south of west. What is the distance between the two aircraft? (Place the x axis west, the y axis south, and the z axis vertical.)

50. Two people pull on a stubborn mule, as shown by the helicopter view in Figure P1.50. Find (a) the single force that is equivalent to the two forces shown, and (b) the force that a third person would have to exert on the mule to make the net force equal to zero.

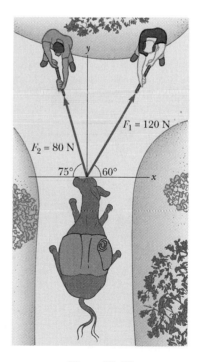

Figure P1.50

51. The distance from the Sun to the nearest star is 4×10^{16} m. The Milky Way galaxy is roughly a disk of radius $\sim 10^{21}$ m and thickness $\sim 10^{19}$ m. Find the **order of magnitude** of the number of stars in the Milky Way. Assume the

distance between the Sun and the nearest neighbor is typical.

52. A rectangular parallelepiped has dimensions a, b, and c, as in Figure P1.52. (a) Obtain a vector expression for the face diagonal vector $\mathbf{R}_1$. What is the magnitude of this vector? (b) Obtain a vector expression for the body diagonal vector $\mathbf{R}_2$. Note that $\mathbf{R}_1$, $c\mathbf{k}$, and $\mathbf{R}_2$ form a right triangle and prove that the magnitude of $\mathbf{R}_2$ is $\sqrt{a^2 + b^2 + c^2}$.

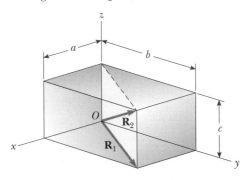

Figure P1.52

⚙ Spreadsheet Problems

S1. Spreadsheet programs are useful in examining data graphically. Most spreadsheet programs can fit the best straight line (a regression line) to a set of data. The following table gives experimental results for the measurement of the period T of a pendulum of length L. These data are consistent with an equation of the form $T = CL^n$, where C and n are constants and n is not necessarily an integer.

Length L(m)	Period T(s)
0.25	1.00
0.50	1.40
0.75	1.75
1.00	2.00
1.50	2.50
2.00	2.80

S1. (a) Use the least-squares or the regression-line procedure of your spreadsheet program to find the best fit to the data. Because $T = CL^n$, we see that

$$\log T = \log C + n \log L$$

First calculate a column of log T values and another of log L as the independent variable and log T as the dependent variable. The least-squares fit finds the slope n and the intercept log C. (b) Which data points deviate most from a straight line plot of T versus L^n? (c) Is the experimental value of n found in part (a) consistent with a dimensional analysis of $T = CL^n$?

S2. The period T and orbit radius R for the motions of four moons of Jupiter are

	Io	Europa	Ganymede	Callisto
T (days)	1.77	3.55	7.16	16.69
R (km)	422 000	671 000	1 070 000	1 880 000

(a) These data can be fitted by the formula $T = CR^n$. Follow the procedures in Problem S1a to find C and n. (b) A fifth satellite, Amalthea, has a period of 0.50 days. Use $T = CR^n$ to find the radius for its orbit.

S3. Use Spreadsheet 1.1 to find the total displacement corresponding to the sum of the three displacement vectors: $\mathbf{A} = 3.0\mathbf{i} + 4.0\mathbf{j}$, $\mathbf{B} = -2.3\mathbf{i} - 7.8\mathbf{j}$, and $\mathbf{C} = 5.0\mathbf{i} - 2.0\mathbf{j}$, where the units are in meters. (a) Give your answer in component form. (b) Give your answer in polar form. (c) Find a fourth displacement that returns you to the origin.

ANSWERS TO CONCEPTUAL PROBLEMS

1. No. The magnitude of a vector $\mathbf{A}$ is equal to $\sqrt{A_x^2 + A_y^2 + A_z^2}$. Therefore, if any component is nonzero, $\mathbf{A}$ cannot be zero.

2. $\mathbf{A} = -\mathbf{B}$, therefore the components of the two vectors must have opposite signs and equal magnitudes.

3. (a) A_x is negative, B_x is positive; (b) A_y is positive, B_y is negative; (c) both components are negative.

4. It is possible to have a situation in which one component of a vector is equal to the magnitude of the vector, as long as the other component is *zero*. For example, a velocity vector lying directly along the east direction has an eastward component that is equal to the magnitude of the velocity vector.

2

Motion in One Dimension

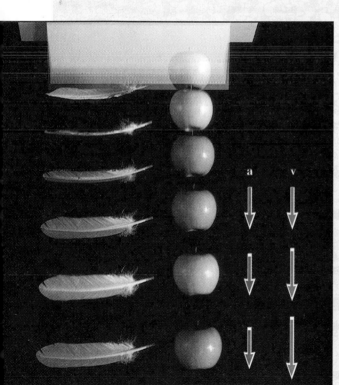

An apple and a feather, released from rest in a 4-ft vacuum chamber, fall at the same rate, regardless of their masses. Neglecting air resistance, all objects fall to Earth with the same acceleration, $a = g$, of magnitude 9.80 m/s^2 as indicated by the violet arrows in this multiflash photograph. The velocity, v, of the two objects increases linearly with time, as indicated by the series of red arrows.

Dynamics is the study of the motions of objects and the relationships of those motions to such physical concepts as force and mass. Before beginning our study of dynamics, however, it is convenient to describe motion using the concepts of space and time, without regard to the causes of the motion; this is a portion of mechanics called *kinematics*. In this chapter we shall consider motion along a straight line—that is, one-dimensional motion. Starting with the concept of displacement, discussed in Chapter 1, we shall define velocity and acceleration. Using these concepts, we shall proceed to study the motion of objects undergoing constant acceleration. In Chapter 3 we shall extend our discussion to two-dimensional motion.

From everyday experience we recognize that motion represents continuous change in the position of an object. The movement of an object through space may be accompanied by the rotation or vibration of the object. Such motions can be quite complex. However, it is sometimes possible to simplify matters by temporarily neglecting the internal motions of the moving object. In many situations, an object can be treated as a *particle* if the only motion being considered is translation through space.

Although an idealized particle is a mathematical point with no size, we can sometimes perform useful calculations by representing macroscopic objects as particles. For example, if we wish to describe the motion of the Earth around the Sun, we can approximate the Earth by treating it as a particle and thereby attain reasonable accuracy in predicting the Earth's orbit. This approximation is justified because the radius of the Earth's orbit is large compared with the dimensions of the Earth and Sun. By contrast, we could not use a particle description to explain the internal structure of the Earth or such phenomena as tides, earthquakes, and volcanic activity. On a much smaller scale, it is possible to explain the pressure exerted by a gas on the walls of a container by treating the gas molecules as particles. However, the particle description of the gas molecules is generally inadequate for understanding those properties of the gas that depend on the internal motions (vibrations) and rotations of the gas molecules.

2.1 · AVERAGE VELOCITY

The motion of a particle is completely known if the position of the particle in space is known at all times. Consider a particle moving along the x axis from point P to point Q. Let its position at point P be x_i at some time t_i, and let its position at point Q be x_f at time t_f. At times other than t_i and t_f, the position of the particle between these two points may vary, as in Figure 2.1. Such a plot is often called a *position–time graph*. In the time interval $\Delta t = t_f - t_i$, the displacement of the particle is $\Delta x = x_f - x_i$. (Recall that displacement is defined as the change in the position of the particle, which is equal to its final position value minus its initial position value.) Using unit-vector notation, the **displacement vector** can be expressed as $\Delta \mathbf{x} \equiv (x_f - x_i)\mathbf{i}$.

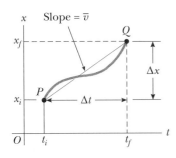

Figure 2.1   Position–time graph for a particle moving along the x axis. The average velocity $\bar{v}_x$ in the interval $\Delta t = t_f - t_i$ is obtained from the slope of the straight line connecting the points P and Q.

Definition of average velocity •

> The x component of the **average velocity** of the particle, $\bar{\mathbf{v}}_x$, is defined as the ratio of its displacement vector, $\Delta \mathbf{x}$, and the time interval, Δt:

$$\bar{\mathbf{v}}_x \equiv \frac{\Delta \mathbf{x}}{\Delta t} = \frac{(x_f - x_i)\mathbf{i}}{t_f - t_i} \qquad [2.1]$$

From this definition we see that the average velocity has the dimensions of length divided by time (L/T) — m/s in SI units and ft/s in conventional units. The average velocity is *independent* of the path taken between the points P and Q. This is true because the average velocity is proportional to the displacement, $\Delta \mathbf{x}$, which in turn depends only on the initial and final coordinates of the particle. It therefore follows that if a particle starts at some point and returns to the same point via any path, its average velocity for this trip is zero, because its displacement along such a path is zero. Displacement should not be confused with the distance traveled, because the distance traveled for any motion is clearly nonzero. Thus, average velocity gives us no details of the motion between points P and Q. (How we evaluate the velocity at some instant in time is discussed in the next section.) Finally, note that the average velocity in one dimension can be positive or negative, depending on the sign of the displacement vector. (The time interval, Δt, is always positive.) If the x coordinate of the particle increases in time (that is, if $x_f > x_i$), then $\Delta \mathbf{x}$ is positive and $\overline{\mathbf{v}}_x$ is positive. This corresponds to an average velocity in the positive x direction. By contrast, if the coordinate decreases in time ($x_f < x_i$), $\Delta \mathbf{x}$ is negative; hence $\overline{\mathbf{v}}_x$ is negative. This corresponds to an average velocity in the negative x direction.

The average velocity can also be interpreted geometrically. A straight line drawn between the points P and Q in Figure 2.1 forms the hypotenuse of a right triangle of height Δx and base Δt. The slope of this line is the ratio $\Delta x / \Delta t$. Therefore, we see that **the average velocity of the particle during the time interval t_i to t_f is equal to the slope of the straight line joining the initial and final points on the position–time graph**. (The word *slope* will often be used in reference to the graphs of physical data. Regardless of what data are plotted, the word *slope* will represent the ratio of the change in the quantity represented on the vertical axis to the change in the quantity represented on the horizontal axis.)

Example 2.1   Calculate the Average Velocity

A particle moving along the x axis is located at $x_i = 12$ m at $t_i = 1$ s and at $x_f = 4$ m at $t_f = 3$ s. Find its displacement and average velocity during this time interval.

Solution   The displacement is

$$\Delta \mathbf{x} = (x_f - x_i)\mathbf{i} = (4 \text{ m} - 12 \text{ m})\mathbf{i} = -8\mathbf{i} \text{ m}$$

The average velocity is

$$\overline{\mathbf{v}}_x = \frac{\Delta \mathbf{x}}{\Delta t} = \frac{(x_f - x_i)\mathbf{i}}{t_f - t_i} = \frac{(4 \text{ m} - 12 \text{ m})\mathbf{i}}{3 \text{ s} - 1 \text{ s}} = -4\mathbf{i} \text{ m/s}$$

Because the displacement is negative for this time interval, we conclude that the particle has moved to the left, toward decreasing values of x.

EXERCISE 1   A jogger runs in a straight line, with an average speed of 5.00 m/s for 4.00 min, and then with an average speed of 4.00 m/s for 3.00 min. (a) What is her total displacement? (b) What is her average speed during this time?
Answer   (a) 1920 m   (b) $\frac{32}{7}$ m/s

2.2 • INSTANTANEOUS VELOCITY

We would like to be able to define the velocity of a particle at a particular instant of time, rather than just during a finite interval of time. The velocity of a particle at any instant of time, or at some point on a position–time graph, is called the **instantaneous velocity.** This concept is especially important when the average velocity is *not constant* through different time intervals.

Consider the motion of a particle between the two points P and Q on the position–time graph shown in Figure 2.2. As the point Q is brought closer and closer to the point P, the time intervals ($\Delta t_1, \Delta t_2, \Delta t_3, \ldots$) get progressively smaller. The average velocity for each time interval is given by the slope of the appropriate dotted line in Figure 2.2. As the point Q approaches P, the time interval approaches zero, but at the same time the slope of the dotted line approaches that of the line tangent to the curve at the point P. **The slope of the line tangent to the curve at P gives the instantaneous velocity at the time t_i.** In other words,

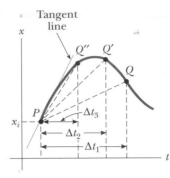

Figure 2.2 Position–time graph for a particle moving along the *x* axis. As the time intervals, starting at t_i, get smaller and smaller, the average velocity for one interval approaches the slope of the line tangent at *P*. The instantaneous velocity at *P* is obtained from the slope of the blue tangent line at the time t_i.

> the instantaneous velocity, $\mathbf{v}_x$, equals the limiting value of the ratio $\Delta \mathbf{x}/\Delta t$ as Δt approaches zero:[1]

$$\mathbf{v}_x \equiv \lim_{\Delta t \to 0} \frac{\Delta \mathbf{x}}{\Delta t} \qquad [2.2]$$

The instantaneous velocity vector is the derivative of the displacement vector with respect to time.

In the calculus notation, this limit is called the *derivative* of *x* with respect to *t*, written $d\mathbf{x}/dt$:

$$\mathbf{v}_x \equiv \lim_{\Delta t \to 0} \frac{\Delta \mathbf{x}}{\Delta t} = \frac{d\mathbf{x}}{dt} \qquad [2.3]$$

The instantaneous velocity can be positive, negative, or zero. When the slope of the position–time graph is positive, such as at the point *P* in Figure 2.3, *v* is positive. At point *R*, *v* is negative, because the slope is negative. Finally, the instantaneous velocity is zero at the peak *Q* (the turning point), where the slope is zero. **From here on, we shall usually use the word *velocity* to designate instantaneous velocity.**

The **instantaneous speed** of a particle is defined as the magnitude of the instantaneous velocity vector. Hence, by definition, *speed* can never be negative.

Those of you familiar with the calculus should recognize that there are specific rules for taking the derivatives of functions. These rules, which are listed in Appendix B.6, enable us to evaluate derivatives quickly.

Suppose *x* is proportional to some power of *t*, such as

$$x = At^n$$

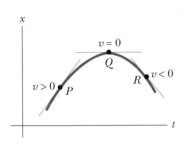

Figure 2.3 In the position–time graph shown, the velocity is positive at *P*, where the slope of the tangent line is positive; the velocity is zero at *Q*, where the slope of the tangent line is zero; and the velocity is negative at *R*, where the slope of the tangent line is negative.

where *A* and *n* are constants. (This is a very common functional form.) The derivative of *x* with respect to *t* is

$$\frac{dx}{dt} = nAt^{n-1}$$

For example, if $x = 5t^3$, we see that $dx/dt = 3(5)t^{3-1} = 15t^2$.

---

[1]*Note:* The magnitude of the displacement, Δx, also approaches zero as Δt approaches zero. However, as Δx and Δt become smaller and smaller, the ratio $\Delta x/\Delta t$ approaches a value equal to the *true* slope of the line tangent to the *x* versus *t* curve.

Thinking Physics 1

Consider the following motions of an object in one dimension: (a) A ball is thrown directly upward, rises to a highest point, and falls back into the thrower's hand. (b) A race car starts from rest and speeds up to 100 m/s. (c) A Voyager spacecraft drifts through space at constant velocity. Are there any points in the motion of these particles at which the average velocity (over the entire interval) and the instantaneous velocity (at an instant of time within the interval) are the same? If so, identify the point(s).

Reasoning (a) The average velocity for the thrown ball is zero—the ball returns to the starting point at the end of the time interval. There is one point at which the instantaneous velocity is zero—at the top of the motion. (b) The average velocity for the motion of the race car cannot be evaluated unambiguously with the information given, but it must be some value between 0 and 100 m/s. Because the car will have every instantaneous velocity between 0 and 100 m/s at some time during the interval, there must be some instant at which the instantaneous velocity is equal to the average velocity. (c) Because the instantaneous velocity of the spacecraft is constant, its instantaneous velocity over *any* time interval and its average velocity at *any* time are the same.

Thinking Physics 2

Does an automobile speedometer measure average or instantaneous speed?

Reasoning Based on the fact that changes in the speed of the automobile are reflected in changes in the speedometer reading, we might be tempted to say that a speedometer measures instantaneous speed. We would be wrong, however. The reading on a speedometer is related to the rotation of the wheels. Often, a rotating magnet, whose rotation is related to that of the wheels, is used in a speedometer. The reading is based on the *time interval* between rotations of the magnet. Thus, the reading is an average velocity over this time interval. This time interval is generally very short, so that the average velocity is close to the instantaneous velocity. As long as the velocity changes are gradual, this approximation is very good. If the automobile brakes to a quick stop, however, the speedometer may not be able to keep up with the rapid changes in speed, and the measured average speed over the time interval may be very different than the instantaneous speed at the end of the interval.

CONCEPTUAL PROBLEM 1

Under what conditions is the magnitude of the average velocity of a particle moving in one dimension smaller than the average speed?

Example 2.2 Average and Instantaneous Velocity

A particle moves along the x axis. Its x coordinate varies with time according to the expression $x = -4t + 2t^2$, where x is in meters and t is in seconds. The position–time graph for this motion is shown in Figure 2.4. Note that the particle moves in the negative x direction for the first second of motion, stops instantaneously at $t = 1$ s, and then heads back in the positive x direction for $t > 1$ s.

(a) Determine the displacement of the particle in the time intervals $t = 0$ to $t = 1$ s and $t = 1$ s to $t = 3$ s.

Solution In the first time interval we set $t_i = 0$ and $t_f = 1$ s. Because $x = -4t + 2t^2$, we get for the first displacement

$$\Delta x_{01} = (x_f - x_i)\mathbf{i} = [-4(1) + 2(1)^2]\mathbf{i} - [-4(0) + 2(0)^2]\mathbf{i}$$
$$= -2\mathbf{i} \text{ m}$$

Likewise, in the second time interval we can set $t_i = 1$ s and $t_f = 3$ s. Therefore, the displacement in this interval is

$$\Delta x_{13} = (x_f - x_i)\mathbf{i} = [-4(3) + 2(3)^2]\mathbf{i} - [-4(1) + 2(1)^2]\mathbf{i}$$
$$= 8\mathbf{i} \text{ m}$$

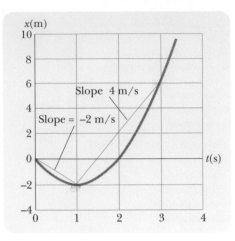

Figure 2.4 (Example 2.2) Position–time graph for a particle having an x coordinate that varies in time according to $x = -4t + 2t^2$.

Note that these displacements can also be read directly from the position–time graph (Fig. 2.4).

(b) Calculate the average velocity in the time intervals $t = 0$ to $t = 1$ s and $t = 1$ s to $t = 3$ s.

Solution In the first time interval, $\Delta t = t_f - t_i = 1$ s. Therefore, using Equation 2.1 and the results from (a) gives

$$\overline{\mathbf{v}}_x = \frac{\Delta \mathbf{x}_{01}}{\Delta t} = \frac{-2\mathbf{i} \text{ m}}{1 \text{ s}} = -2\mathbf{i} \text{ m/s}$$

Likewise, in the second time interval, $\Delta t = 2$ s; therefore,

$$\overline{\mathbf{v}}_x = \frac{\Delta \mathbf{x}_{13}}{\Delta t} = \frac{8\mathbf{i} \text{ m}}{2 \text{ s}} = 4\mathbf{i} \text{ m/s}$$

These values (the coefficients of $\mathbf{i}$) agree with the slopes of the lines joining these points in Figure 2.4.

(c) Find the instantaneous velocity of the particle at $t = 2.5$ s.

Solution We can find the instantaneous velocity at any time t by taking the first derivative of x with respect to t:

$$v = \frac{dx}{dt} = \frac{d}{dt}(-4t + 2t^2) = -4 + 4t$$

Thus, at $t = 2.5$ s, we find that $v = -4 + 4(2.5) = 6$ m/s.

We can also obtain this result by measuring the slope of the position–time graph at $t = 2.5$ s. (You should show that the velocity is $-4\mathbf{i}$ m/s at $t = 0$ and zero at $t = 1$ s.) Do you see any symmetry in the motion? For example, does the speed ever repeat itself?

Example 2.3   The Limiting Process

The position of a particle moving along the x axis varies in time according to the expression $x = (3 \text{ m/s}^2) \, t^2$, where x is in meters and t is in seconds. Find the velocity in terms of t at any time.

Solution The position–time graph for this motion is shown in Figure 2.5. We can compute the velocity at any time t by using the definition of the instantaneous velocity. If the initial coordinate of the particle at time t is $x_i = 3t^2$, then the coordinate at a later time, $t + \Delta t$, is

$$x_f = 3(t + \Delta t)^2 = 3[t^2 + 2t \Delta t + (\Delta t)^2]$$
$$= 3t^2 + 6t \Delta t + 3(\Delta t)^2$$

Therefore, the displacement in the time interval Δt is

$$\Delta \mathbf{x} = (x_f - x_i)\mathbf{i} = [3t^2 + 6t \Delta t + 3(\Delta t)^2 - 3t^2]\mathbf{i}$$
$$= (6t \Delta t + 3(\Delta t)^2)\mathbf{i}$$

The average velocity in this time interval is

$$\overline{\mathbf{v}}_x = \frac{\Delta \mathbf{x}}{\Delta t} = (6t + 3 \Delta t)\mathbf{i}$$

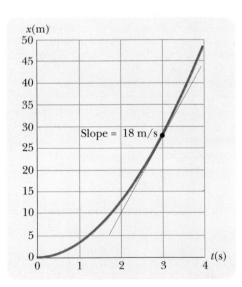

Figure 2.5 (Example 2.3) Position–time graph for a particle having an x coordinate that varies in time according to $x = 3t^2$. Note that the instantaneous velocity at $t = 3.0$ s is obtained from the slope of the blue line tangent to the curve at this point.

To find the instantaneous velocity, we take the limit of this expression as Δt approaches zero. In doing so, we see that the term $3\,\Delta t$ goes to zero; therefore,

$$\mathbf{v}_x = \lim_{\Delta t \to 0} \frac{\Delta \mathbf{x}}{\Delta t} = 6t\mathbf{i} \text{ m/s}$$

Notice that this expression gives us the velocity at *any* general time t. It tells us that $\mathbf{v}_x$ is increasing linearly in time. It is then

a straightforward matter to find the velocity at some specific time from the expression $\mathbf{v}_x = 6t\mathbf{i}$. For example, at $t = 3.0$ s, the velocity is $\mathbf{v}_x = 6(3)\mathbf{i} = 18\mathbf{i}$ m/s. Again, this can be checked from the slope of the graph at $t = 3.0$ s.

We can also find v by taking the first derivative of x with respect to time. In this example, $x = 3t^2$, and we see that $v_x = dx/dt = 6t$, in agreement with our result of taking the limit explicitly.

2.3 · ACCELERATION

When the velocity of a particle changes with time, the particle is said to be *accelerating*. For example, the speed of a car increases when you step on the gas. The car slows down when you apply the brakes, and it changes direction when you turn the wheel. However, we need a more precise definition of acceleration.

Suppose a particle moving along the x axis has a velocity $\mathbf{v}_{xi}$ at time t_i and a velocity $\mathbf{v}_{xf}$ at time t_f.

> The **average acceleration** of a particle in the time interval $\Delta t = t_f - t_i$ is defined as the ratio $\Delta \mathbf{v}_x / \Delta t$, where $\Delta \mathbf{v}_x = \mathbf{v}_{xf} - \mathbf{v}_{xi}$ is the *change* in velocity in this time interval:

$$\bar{\mathbf{a}}_x \equiv \frac{\mathbf{v}_{xf} - \mathbf{v}_{xi}}{t_f - t_i} = \frac{\Delta \mathbf{v}_x}{\Delta t} \qquad [2.4]$$

• *Definition of average acceleration*

Acceleration is a vector quantity having dimensions of length divided by (time)^2, or L/T^2. Some of the common units of acceleration are meters per second per second (m/s^2) and feet per second per second (ft/s^2).

In some situations, the value of the average acceleration may be different for different time intervals. It is therefore useful to define the **instantaneous acceleration** as the limit of the average acceleration as Δt approaches zero. This concept is analogous to the definition of instantaneous velocity discussed in the previous section. If we take the limit of the ratio $\Delta \mathbf{v}_x / \Delta t$ as Δt approaches zero, we get the **instantaneous acceleration:**

$$\mathbf{a}_x \equiv \lim_{\Delta t \to 0} \frac{\Delta \mathbf{v}_x}{\Delta t} = \frac{d\mathbf{v}_x}{dt} \qquad [2.5]$$

• *The instantaneous acceleration is the derivative of the velocity with respect to time.*

That is, **the instantaneous acceleration equals the derivative of the velocity with respect to time, which by definition is the slope of the velocity–time graph.** One can interpret the derivative of the velocity with respect to time as the *time rate of change of velocity*. Note that if $\mathbf{a}_x$ is positive, the acceleration is in the positive x direction, whereas negative $\mathbf{a}_x$ implies acceleration in the negative x direction. This does not mean that the particle is *moving* in the x direction. From now on we shall use the term *acceleration* to mean instantaneous acceleration.

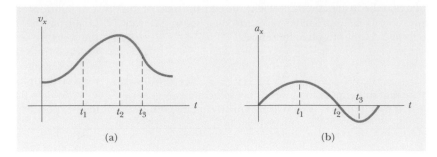

Figure 2.6 The instantaneous acceleration can be obtained from the velocity–time graph (a). At each instant, the acceleration in the a_x versus t graph (b) equals the slope of the line tangent to the v versus t curve.

Because $\mathbf{v}_x = d\mathbf{x}/dt$, the acceleration can also be written as

$$\mathbf{a}_x = \frac{d\mathbf{v}_x}{dt} = \frac{d}{dt}\left(\frac{d\mathbf{x}}{dt}\right) = \frac{d^2\mathbf{x}}{dt^2} \qquad [2.6]$$

That is, the acceleration equals the *second derivative* of the displacement with respect to time.

Figure 2.6 shows how the acceleration–time curve can be derived from the velocity–time curve. In these sketches, the acceleration at any time is simply the slope of the velocity–time graph at that time. Positive values of the acceleration correspond to those points at which the velocity is increasing in the positive x direction. The acceleration reaches a maximum at time t_1, when the slope of the velocity–time graph is a maximum. The acceleration then goes to zero at time t_2, when the velocity is a maximum (that is, when the velocity is momentarily not changing and the slope of the v versus t graph is zero). Finally, the acceleration is negative when the velocity in the positive x direction is decreasing in time.

As an example of the computation of acceleration, consider the car pictured in Figure 2.7. In this case, the velocity of the car has changed from an initial value of $30\mathbf{i}$ m/s to a final value of $15\mathbf{i}$ m/s in a time interval of 2.0 s. The average acceleration during this time interval is

$$\bar{\mathbf{a}}_x = \frac{15\mathbf{i}\ \text{m/s} - 30\mathbf{i}\ \text{m/s}}{2.0\ \text{s}} = -7.5\mathbf{i}\ \text{m/s}^2$$

Acceleration can mean speeding up, slowing down, or changing direction.

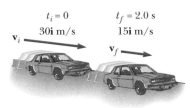

Figure 2.7 The velocity of the car decreases from $30\mathbf{i}$ m/s to $15\mathbf{i}$ m/s in a time interval of 2.0 s.

The minus sign in this example indicates that the acceleration vector is in the negative x direction (to the left). For the case of motion in a straight line, the direction of the velocity of an object and the direction of its acceleration are related as follows. **When the object's velocity and acceleration are in the same direction, the object is speeding up in that direction.** However, **when the object's velocity and acceleration are in opposite directions, the speed of the object decreases in time.**

CONCEPTUAL PROBLEM 2

If a car is traveling eastward, can its acceleration be westward? Explain.

CONCEPTUAL PROBLEM 3

Can a particle with constant acceleration ever stop and stay stopped?

Example 2.4 Average and Instantaneous Acceleration

The velocity of a particle moving along the x axis varies in time according to the expression $\mathbf{v}_x = (40 - 5t^2)\mathbf{i}$ m/s, where t is in seconds.

(a) Find the average acceleration in the time interval $t = 0$ to $t = 2.0$ s.

Solution The velocity–time graph for this function is given in Figure 2.8. The velocities at $t_i = 0$ and $t_f = 2.0$ s are found by substituting these values of t into the expression given for the velocity:

$$\mathbf{v}_{xi} = (40 - 5t_i^2)\mathbf{i} \text{ m/s} = [40 - 5(0)^2]\mathbf{i} \text{ m/s} = 40\mathbf{i} \text{ m/s}$$

$$\mathbf{v}_{xf} = (40 - 5t_f^2)\mathbf{i} \text{ m/s} = [40 - 5(2)^2]\mathbf{i} \text{ m/s} = 20\mathbf{i} \text{ m/s}$$

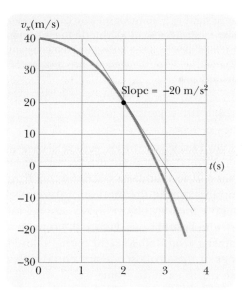

Therefore, the average acceleration in the specified time interval, $\Delta t = t_f - t_i = 2.0$ s, is

$$\overline{\mathbf{a}}_x = \frac{\mathbf{v}_{xf} - \mathbf{v}_{xi}}{t_f - t_i} = \frac{(20 - 40)\mathbf{i} \text{ m/s}}{(2 - 0)\text{s}} = -10\mathbf{i} \text{ m/s}^2$$

The negative sign is consistent with the fact that the slope of the line joining the initial and final points on the velocity–time graph is negative.

(b) Determine the acceleration at $t = 2.0$ s.

Solution The velocity at time t is $\mathbf{v}_{xi} = (40 - 5t^2)\mathbf{i}$ m/s, and the velocity at time $t + \Delta t$ is

$$\mathbf{v}_{xf} = 40\mathbf{i} - 5(t + \Delta t)^2\mathbf{i} = [40 - 5t^2 - 10t\,\Delta t - 5(\Delta t)^2]\mathbf{i}$$

Therefore, the change in velocity over the time interval Δt is

$$\Delta\mathbf{v}_x = \mathbf{v}_{xf} - \mathbf{v}_{xi} = [-10t\,\Delta t - 5(\Delta t)^2]\mathbf{i} \text{ m/s}$$

Dividing this expression by Δt and taking the limit of the result as Δt approaches zero, we get the acceleration at *any* time t:

$$\mathbf{a}_x = \lim_{\Delta t \to 0} \frac{\Delta\mathbf{v}_x}{\Delta t} = \lim_{\Delta t \to 0} (-10t - 5\,\Delta t)\mathbf{i} = -10t\mathbf{i} \text{ m/s}^2$$

Therefore, at $t = 2.0$ s we find that

$$\mathbf{a}_x = (-10)(2)\mathbf{i} \text{ m/s}^2 = -20\mathbf{i} \text{ m/s}^2$$

This result can also be obtained by measuring the slope of the velocity–time graph at $t = 2.0$ s. Note that the acceleration is not constant in this example. Situations involving constant acceleration will be treated in Section 2.5.

Figure 2.8 (Example 2.4) The velocity–time graph for a particle moving along the x axis according to the relation $v_x = (40 - 5t^2)$ m/s. Note that the acceleration at $t = 2.0$ s is obtained from the slope of the blue tangent line at that time.

EXERCISE 2 When struck by a club, a golf ball initially at rest acquires a speed of 31.0 m/s. If the ball is in contact with the club for 1.17 ms, what is the magnitude of the average acceleration of the ball?   Answer   26 500 m/s^2

2.4 • MOTION DIAGRAMS

The concepts of velocity and acceleration are often confused with each other, but in fact they are quite different quantities. It is instructive to make use of motion diagrams to describe the velocity and acceleration vectors as time progresses while

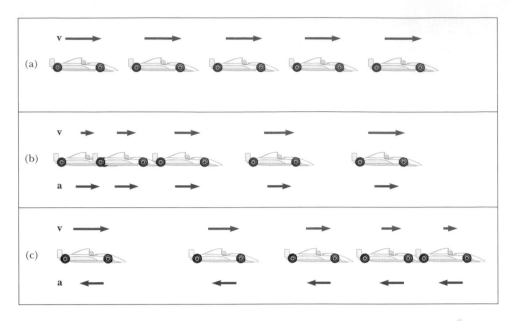

Figure 2.9   (a) Motion diagram for a car moving at constant velocity (zero acceleration). (b) Motion diagram for a car, the constant acceleration of which is in the direction of its velocity. The velocity vector at each instant is indicated by a red arrow, and the constant acceleration vector by a violet arrow. (c) Motion diagram for a car, the constant acceleration of which is in the direction *opposite* the velocity at each instant.

an object is in motion. In order not to confuse these two vector quantities, we use red for velocity vectors and violet for acceleration vectors as in Figure 2.9. The vectors are sketched at several instants during the motion of the object.

A stroboscopic photograph of a moving object shows several images of the object, each image taken as the strobe light flashes. Figure 2.9 represents three sets of strobe photographs of cars moving from left to right along a straight roadway. The time intervals between flashes of the stroboscope are equal in each diagram. Let us describe the motion of the car in each diagram.

In Figure 2.9a, the images of the car are equally spaced—the car moves the same distance in each time interval. Thus, the car moves with *constant positive velocity* and has zero acceleration.

In Figure 2.9b, the images of the car become farther apart as time progresses. In this case, the velocity vector increases in time because the car's displacement between adjacent positions increases as time progresses. Thus, the car is moving with a *positive velocity* and a *positive acceleration*.

In Figure 2.9c, the car slows as it moves to the right because its displacement between adjacent positions decreases as time progresses. In this case, the car moves initially to the right with a constant negative acceleration. The velocity vector decreases in time and eventually reaches zero. (This type of motion is exhibited by a car that skids to a stop after applying its brakes.) From this diagram we see that the acceleration and velocity vectors are *not* in the same direction.

2.5 • ONE-DIMENSIONAL MOTION WITH CONSTANT ACCELERATION

If the acceleration of a particle varies in time, the motion can be complex and difficult to analyze. A common and simple type of one-dimensional motion occurs when the acceleration is constant or uniform. In this case the average acceleration over any time interval equals the instantaneous acceleration at any instant of time within the interval. As a consequence, the velocity increases or decreases at the same rate throughout the motion.

If we replace $\bar{\mathbf{a}}_x$ with $\mathbf{a}_x$ in Equation 2.4, we find that

$$\mathbf{a}_x = \frac{\mathbf{v}_{xf} - \mathbf{v}_{xi}}{t_f - t_i}$$

Because we are dealing with motion in one dimension, we shall drop the boldface notation for vectors and use a minus sign when the vector is in the negative x direction. For convenience, let $t_i = 0$ and t_f be any arbitrary time t. Also, let $v_i = v_{x0}$ (the initial velocity at $t = 0$) and $v_f = v_x$ (the velocity at any arbitrary time t). With this notation, we can express the acceleration as

$$a_x = \frac{v_x - v_{x0}}{t}$$

or

$$v_x = v_{x0} + a_x t \qquad \text{(for constant } a_x) \qquad \qquad [2.7] \qquad \bullet \ \textit{Velocity as a function of time}$$

This expression enables us to predict the velocity at *any* time t if the initial velocity, acceleration, and elapsed time are known. A graph of position versus time for this motion is shown in Figure 2.10a. The velocity–time graph shown in Figure 2.10b is a straight line, the slope of which is the acceleration, a_x. This is consistent with the fact that $a_x = dv_x/dt$ is a constant. From this graph and from Equation 2.7, we see that the velocity at any time t is the sum of the initial velocity, v_{x0}, and the change in velocity, $a_x t$. The graph of acceleration versus time (Fig. 2.10c) is a straight line with a slope of zero, because the acceleration is constant. Note that if the acceleration were negative, the slope of Figure 2.10b would be negative.

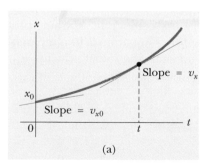

(a)

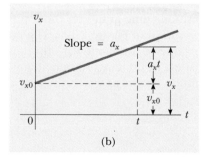

(b)

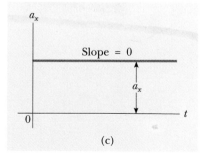

(c)

Figure 2.10 A particle moving along the x axis with constant acceleration a_x: (a) the position–time graph, (b) the velocity–time graph, and (c) the acceleration–time graph.

Because the velocity varies linearly with time according to Equation 2.7, we can express the average velocity in any time interval as the arithmetic mean of the initial velocity, v_{x0}, and the final velocity, v_x.

$$\bar{v}_x = \frac{v_{x0} + v_x}{2} \qquad \text{(for constant } a_x) \qquad \text{[2.8]}$$

Note that this expression is useful only when the acceleration is constant—that is, when the velocity varies linearly with time.

We can now use Equations 2.1 and 2.8 to obtain the displacement as a function of time. Again we choose $t_i = 0$, at which time the initial position is $x_i = x_0$. This gives

$$\Delta x = \bar{v}_x \Delta t = \left(\frac{v_{x0} + v_x}{2} \right) t$$

$$x - x_0 = \frac{1}{2}(v_{x0} + v_x)t \qquad \text{(for constant } a_x) \qquad \text{[2.9]}$$

We can obtain another useful expression for the displacement by substituting Equation 2.7 into Equation 2.9:

$$x - x_0 = \tfrac{1}{2}(v_{x0} + v_{x0} + a_x t)t$$

Displacement as a function of time •

$$x - x_0 = v_{x0}t + \tfrac{1}{2}a_x t^2 \qquad \text{(for constant } a_x) \qquad \text{[2.10]}$$

The validity of this expression can be checked by differentiating it with respect to time, to give

$$v_x = \frac{dx}{dt} = \frac{d}{dt}(x_0 + v_{x0}t + \tfrac{1}{2}a_x t^2) = v_{x0} + a_x t$$

Finally, we can obtain an expression that does not contain the time by substituting the value of t from Equation 2.7 into Equation 2.9. This gives

$$x - x_0 = \tfrac{1}{2}(v_{x0} + v_x)\left(\frac{v_x - v_{x0}}{a_x} \right) = \frac{v_x^2 - v_{x0}^2}{2a_x}$$

Velocity as a function of displacement •

$$v_x^2 = v_{x0}^2 + 2a_x(x - x_0) \qquad \text{(for constant } a_x) \qquad \text{[2.11]}$$

A position–time graph for motion under constant acceleration, assuming positive a_x, is shown in Figure 2.10a. The curve representing Equation 2.10 is a parabola. The slope of the tangent to this curve at $t = 0$ equals the initial velocity, v_{x0}, and the slope of the tangent line at any time t equals the velocity at that time.

If motion occurs in which the acceleration is *zero*, then we see that

$$\left. \begin{array}{l} v_x = v_{x0} \\ x - x_0 = v_x t \end{array} \right\} \qquad \text{when } a_x = 0$$

That is, when the acceleration is zero, the velocity remains constant and the displacement changes linearly with time.

Equations 2.7 through 2.11 **may be used to solve any problem in one-dimensional motion with constant acceleration.** Keep in mind that these rela-

**TABLE 2.1   Kinematic Equations for Motion in a Straight
Line Under Constant Acceleration**

Equation	Information Given by Equation
$v_x = v_{x0} + a_x t$	Velocity as a function of time
$x - x_0 = \frac{1}{2}(v_x + v_{x0})t$	Displacement as a function of velocity and time
$x - x_0 = v_{x0}t + \frac{1}{2}a_x t^2$	Displacement as a function of time
$v_x^2 = v_{x0}^2 + 2a_x(x - x_0)$	Velocity as a function of displacement

Note: Motion is along the x axis. At $t = 0$, the position of the particle is x_0 and its velocity is v_{x0}.

tionships were derived from the definitions of velocity and acceleration, together with some simple algebraic manipulations and the requirement that the acceleration be constant. It is often convenient to choose the initial position of the particle as the origin of the motion, so that $x_0 = 0$ at $t = 0$. In such a case, the displacement is simply x.

The four kinematic equations that are used most often are listed in Table 2.1. The choice of which kinematic equation or equations you should use in a given situation depends on what is known beforehand. Sometimes it is necessary to use two of these equations to solve for two unknowns, such as the displacement and velocity at some instant. For example, suppose the initial velocity, v_{x0}, and acceleration, a_x, are given. You can then find (1) the velocity after a time t has elapsed using $v_x = v_{x0} + a_x t$, and (2) the displacement after a time t has elapsed using $x - x_0 = v_{x0}t + \frac{1}{2}a_x t^2$. You should recognize that the quantities that vary during the motion are velocity, displacement, and time.

You will get a great deal of practice in the use of these equations by solving exercises and problems. You will discover again and again that there is more than one method for obtaining a solution.

PROBLEM-SOLVING STRATEGY • Accelerated Motion

The following procedure is recommended for solving problems that involve accelerated motion:

1. Make sure all the units in the problem are consistent. That is, if distances are measured in meters, be sure that velocities have units of meters per second and accelerations have units of meters per second per second.
2. Choose a coordinate system.
3. Choose an instant to call the *initial point* and another the *final point*. Make a list of all the quantities given in the problem and a separate list of those to be determined.
4. Think about what is going on physically in the problem and then select from the list of kinematic equations the one or ones that will enable you to determine the unknowns.
5. Construct an appropriate motion diagram and check to see if your answers are consistent with the diagram.

Example 2.5　Watch Out for the Speed Limit

A car traveling at a constant speed of 30.0 m/s ($\approx$67 mi/h) passes a trooper hidden behind a billboard. One second after the speeding car passes the billboard, the trooper sets off in chase with a constant acceleration of 3.00 m/s². How long does it take the trooper to overtake the speeding car?

Reasoning　To solve this problem algebraically, we write an expression for the position of each vehicle as a function of time. It is convenient to choose the origin at the position of the billboard and take $t = 0$ as the time the trooper begins moving. At that instant, the speeding car has already traveled a distance of 30.0 m, because it travels at a constant speed of 30.0 m/s. Thus, the initial position of the speeding car is $x_0 = 30.0$ m.

Solution　Because the car moves with constant speed, its acceleration is zero, and applying Equation 2.10 gives

$$x_C = 30.0 \text{ m} + (30.0 \text{ m/s})t$$

Note that at $t = 0$, this expression does give the car's correct initial position, $x_C = x_0 = 30.0$ m.

For the trooper, who starts from the origin at $t = 0$, we have $x_0 = 0$, $v_{x0} = 0$, and $a_x = 3.00$ m/s². Hence, the position of the trooper as a function of time is

$$x_T = \tfrac{1}{2}a_x t^2 = \tfrac{1}{2}(3.00 \text{ m/s}^2)t^2$$

The trooper overtakes the car at the instant that $x_T = x_C$, or

$$\tfrac{1}{2}(3.00 \text{ m/s}^2)t^2 = 30.0 \text{ m} + (30.0 \text{ m/s})t$$

This gives the quadratic equation

$$1.50t^2 - 30.0t - 30.0 = 0$$

the positive solution of which is $t = 21.0$ s. Note that in this time interval, the trooper has traveled a distance of about 660 m.

EXERCISE 3　This problem can also be easily solved graphically. On the *same* graph, plot the position versus time for each vehicle, and from the intersection of the two curves determine the time at which the trooper overtakes the speeding car.

EXERCISE 4　A hockey puck sliding on a frozen lake comes to rest after traveling 200 m. Its initial speed was 3.00 m/s. (a) What was its acceleration if it is assumed to have been constant? (b) How long was it in motion? (c) What was its speed after traveling 150 m?
Answer　(a) -2.25×10^{-2} m/s²　(b) 133 s　(c) 1.50 m/s

2.6 · FREELY FALLING OBJECTS

It is well known that all objects, when dropped, fall toward the Earth with nearly constant acceleration in the absence of air resistance. There is a legend that Galileo Galilei first discovered this fact by observing that two different weights dropped simultaneously from the Tower of Pisa hit the ground at approximately the same time. Although there is some doubt that this particular experiment was carried out, it is well established that Galileo did perform many systematic experiments on objects moving on inclined planes. Through careful measurements of distances and time intervals, he was able to show that the displacement of an object starting from rest is proportional to the square of the time the object is in motion. This observation is consistent with one of the kinematic equations we derived for motion under constant acceleration (Eq. 2.10). Galileo's achievements in the science of mechanics paved the way for Newton in his development of the laws of motion.

You might want to try the following experiment. Drop a coin and a crumpled-up piece of paper simultaneously from the same height. In the absence of air resistance, the two experience the same motion and hit the floor at the same time. In a real (nonideal) experiment, air resistance cannot be neglected. In the idealized case, however, where air resistance *is* neglected, such motion is referred to as *free*

Galileo Galilei (1564–1642).

fall. If this same experiment could be conducted in a good vacuum, where air friction is truly negligible, the paper and coin would fall with the same acceleration, regardless of the shape or weight of the paper. This point is illustrated convincingly on page 29, in the photograph of the apple and feather falling in a vacuum. On August 2, 1971, such an experiment was conducted on the Moon by astronaut David Scott. He simultaneously released a geologist's hammer and a falcon's feather, and in unison they fell to the lunar surface. This demonstration surely would have pleased Galileo!

We shall denote the free-fall acceleration with the symbol **g**. The magnitude of **g** decreases with increasing altitude. Furthermore, there are slight variations in the magnitude of **g** with latitude. However, at the surface of the Earth the magnitude of **g** is approximately 9.80 m/s², or 980 cm/s², or 32 ft/s². Unless stated otherwise, we shall use the value 9.80 m/s² when doing calculations. Furthermore, we shall assume that the vector **g** is directed downward toward the center of the Earth.

When we use the expression *freely falling object*, we do not necessarily mean an object dropped from rest.

A freely falling object is an object moving freely under the influence of gravity only, regardless of its initial motion. Objects thrown upward or downward and those released from rest are all falling freely once they are released.

It is important to emphasize that any freely falling object experiences an acceleration directed downward, as shown in the multiflash photograph of a falling billiard

Figure 2.11 A multiflash photograph of a falling billiard ball. As the ball falls, the spacing between successive images increases, indicating that the ball accelerates downward. The motion diagram shows that the ball's velocity (red arrows) increases with time, whereas its acceleration (violet arrows) remains constant. *(© Richard Megna 1990, Fundamental Photographs)*

ball (Fig. 2.11). This is true regardless of the initial motion of the object. An object thrown upward (or downward) experiences the same acceleration as an object released from rest. **Once they are in free fall, all objects have a downward acceleration equal to the free-fall acceleration.**

If we neglect air resistance and assume that the gravitational acceleration does not vary with altitude, then a freely falling object moves in one dimension with constant acceleration. Therefore, the equations developed in Section 2.5 for objects moving with constant acceleration can be applied. The only modification that we need to make in these equations for freely falling objects is to note that the motion is in the vertical direction (the y direction) rather than the horizontal direction (the x direction) and that the acceleration is downward and has a magnitude of 9.80 m/s^2. Thus, for a freely falling object we commonly take $a_y = -g = -9.80$ m/s^2, where the minus sign means that the acceleration of the object is downward.

Thinking Physics 3

A skydiver jumps out of a helicopter. A few seconds later, another skydiver jumps out, so that they both fall along the same vertical line. Ignore air resistance, so that both skydivers fall with the same acceleration. Does the vertical distance between them stay the same? Does the difference in their velocities stay the same? If they were connected by a long bungee cord, would the tension in the cord become greater, less, or stay the same?

Reasoning At any given instant of time, the velocities of the jumpers are definitely different, because one had a head start. In a time interval after this instant, however, each jumper increases his or her velocity by the same amount, because they have the same acceleration. Thus, the difference in velocities remains the same. The first jumper will always be moving with a higher velocity than the second. Thus, in a given time interval, the first jumper will cover more distance than the second. Thus, the separation distance between them increases. As a result, once the distance between the divers reaches the value at which the bungee cord becomes straight, the tension in the bungee cord will increase as the jumpers move apart. Of course, the bungee cord will pull downward on the second jumper and upward on the first, resulting in a force in addition to gravity, which will affect their motion. If the bungee cord is robust enough, this will result in the separation distance of the divers becoming smaller as the cord pulls them together.

CONCEPTUAL PROBLEM 4

A child throws a ball into the air with some initial velocity. Another child drops a toy at the same instant. Compare the accelerations of the two objects while they are in the air.

CONCEPTUAL PROBLEM 5

A ball is thrown upward. While the ball is in the air, (a) what happens to its velocity? (b) Does its acceleration increase, decrease, or remain constant?

Example 2.6   Try to Catch the Dollar

Emily challenges David to catch a dollar bill as follows. She holds the bill vertically, as in Figure 2.12, with the center of the bill between David's index finger and thumb. David must catch the bill after Emily releases it without moving his hand downward. Who would you bet on?

Reasoning   Place your bets on Emily. There is a time delay between the instant Emily releases the bill and the time David reacts and closes his fingers. The reaction time of most people is at best about 0.2 s. Because the bill is in free fall and undergoes a downward acceleration of 9.80 m/s², in 0.2 s it falls a distance of $\frac{1}{2}gt^2 \cong 0.2$ m $= 20$ cm. This distance is about twice the distance between the center of the bill and its top edge ($\cong 8$ cm). Thus, David will be unsuccessful. You might want to try this on one of your friends.

Figure 2.12   (Example 2.6)

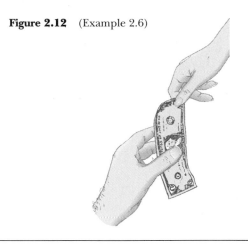

Example 2.7   Not a Bad Throw for a Rookie!

A stone is thrown from the top of a building with an initial velocity of 20.0 m/s straight upward. The building is 50.0 m high, and the stone just misses the edge of the roof on its way down, as in Figure 2.13. Determine (a) the time needed for the stone to reach its maximum height, (b) the maximum height, (c) the time needed for the stone to return to the level of the thrower, (d) the velocity of the stone at this instant, and (e) the velocity and position of the stone at $t = 5.00$ s.

Solution   (a) To find the time necessary to reach the maximum height, use Equation 2.7, $v_y = v_{y0} + a_y t$, noting that $v_y = 0$ at maximum height:

$$20.0 \text{ m/s} + (-9.80 \text{ m/s}^2)t_1 = 0$$

$$t_1 = \frac{20.0 \text{ m/s}}{9.80 \text{ m/s}^2} = 2.04 \text{ s}$$

(b) This value of time can be substituted into Equation 2.10, $y = v_{y0}t + \frac{1}{2}a_y t^2$, to give the maximum height measured from the position of the thrower:

$$y_{max} = (20.0 \text{ m/s})(2.04 \text{ s}) + \frac{1}{2}(-9.80 \text{ m/s}^2)(2.04 \text{ s})^2$$

$$= 20.4 \text{ m}$$

(c) When the stone is back at the height of the thrower, the y coordinate is zero. From the expression $y = v_{y0}t + \frac{1}{2}a_y t^2$ (Eq. 2.10), with $y = 0$, we obtain the expression

$$20.0t - 4.90t^2 = 0$$

This is a quadratic equation and has two solutions for t. The equation can be factored to give

$$t(20.0 - 4.90t) = 0$$

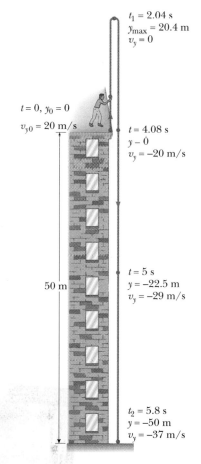

$t_1 = 2.04$ s
$y_{max} = 20.4$ m
$v_y = 0$

$t = 0, y_0 = 0$
$v_{y0} = 20$ m/s

$t = 4.08$ s
$y = 0$
$v_y = -20$ m/s

$t = 5$ s
$y = -22.5$ m
$v_y = -29$ m/s

50 m

$t_2 = 5.8$ s
$y = -50$ m
$v_y = -37$ m/s

Figure 2.13   (Example 2.7) Position and velocity versus time for a freely falling particle initially thrown upward with a velocity $v_{y0} = 20$ m/s. (Not drawn to scale.)

One solution is $t = 0$, corresponding to the time the stone starts its motion. The other solution—the one we are after—is $t = 4.08$ s.

(d) The value for t found in (c) can be inserted into $v_y = v_{y0} + a_y t$ (Eq. 2.7) to give

$$v_y = 20.0 \text{ m/s} + (-9.80 \text{ m/s}^2)(4.08 \text{ s}) = -20.0 \text{ m/s}$$

Note that the velocity of the stone when it arrives back at its original height is equal in magnitude to its initial velocity but opposite in direction. This indicates that the motion is symmetric.

(e) From $v_y = v_{y0} + a_y t$ (Eq. 2.7), the velocity after 5.00 s is

$$v_y = 20 \text{ m/s} + (-9.80 \text{ m/s}^2)(5.00 \text{ s}) = -29.0 \text{ m/s}$$

We can use $y = v_{y0}t + \frac{1}{2}a_y t^2$ (Eq. 2.10) to find the position of the particle at $t = 5.00$ s:

$$y = (20.0 \text{ m/s})(5.00 \text{ s}) + \frac{1}{2}(-9.80 \text{ m/s}^2)(5.00 \text{ s})^2$$

$$= -22.5 \text{ m}$$

EXERCISE 5   Find (a) the velocity of the stone just before it hits the ground and (b) the total time the stone is in the air.
Answer   (a) -37.1 m/s   (b) 5.83 s

SUMMARY

The **average velocity** of a particle during some time interval is equal to the ratio of the displacement vector, $\Delta \mathbf{x}$, and the time interval, Δt:

$$\overline{\mathbf{v}}_x \equiv \frac{\Delta \mathbf{x}}{\Delta t} \qquad [2.1]$$

The **instantaneous velocity** of a particle is defined as the limit of the ratio $\Delta \mathbf{x}/\Delta t$ as Δt approaches zero:

$$\mathbf{v}_x \equiv \lim_{\Delta t \to 0} \frac{\Delta \mathbf{x}}{\Delta t} = \frac{d\mathbf{x}}{dt} \qquad [2.3]$$

The **speed** of a particle is defined as the magnitude of the instantaneous velocity vector.

The **average acceleration** of a particle during some time interval is defined as the ratio of the change in its velocity, $\Delta \mathbf{v}_x$, and the time interval, Δt:

$$\overline{\mathbf{a}}_x \equiv \frac{\Delta \mathbf{v}_x}{\Delta t} \qquad [2.4]$$

The **instantaneous acceleration** is equal to the limit of the ratio $\Delta \mathbf{v}_x/\Delta t$ as $\Delta t \to 0$. By definition, this equals the derivative of $\mathbf{v}_x$ with respect to t or the time rate of change of the velocity:

$$\mathbf{a}_x \equiv \lim_{\Delta t \to 0} \frac{\Delta \mathbf{v}_x}{\Delta t} = \frac{d\mathbf{v}_x}{dt} \qquad [2.5]$$

The slope of the tangent to the x versus t curve at any instant gives the instantaneous velocity of the particle. The slope of the tangent to the v versus t curve gives the instantaneous acceleration of the particle.

The **equations of kinematics** for a particle moving along the x axis with uniform acceleration a (constant in magnitude and direction) are

$$\left. \begin{array}{l} v_x = v_{x0} + a_x t \\[4pt] x - x_0 = \frac{1}{2}(v_{x0} + v_x)t \\[4pt] x - x_0 = v_{x0}t + \frac{1}{2}a_x t^2 \\[4pt] v_x^2 = v_{x0}^2 + 2a_x(x - x_0) \end{array} \right\} \quad \text{(constant } a_x \text{ only)} \qquad \begin{array}{l} [2.7] \\[4pt] [2.9] \\[4pt] [2.10] \\[4pt] [2.11] \end{array}$$

A body falling freely experiences an acceleration directed toward the center of the Earth. If air friction is neglected, and if the altitude of the motion is small compared with the Earth's radius, then one can assume that the free-fall acceleration, **g**, is constant over the range of motion, where g is equal to 9.80 m/s^2, or 32 ft/s^2. Assuming y to be positive upward, the acceleration is given by $-g$, and the equations of kinematics for a body in free fall are the same as those already given, with the substitutions $x \rightarrow y$ and $a_y \rightarrow -g$.

CONCEPTUAL QUESTIONS

1. Are there any conditions under which the average velocity of a particle moving in one dimension is larger than the average speed?

2. The average velocity for a particle moving in one dimension has a positive value. Is it possible for the instantaneous velocity of the particle to have been negative at any time in the interval? Suppose the particle started at the origin, $x = 0$. If the average velocity is positive, could the particle ever have been in the $-x$ region?

3. If the average velocity of an object is zero in some time interval, what can you say about the displacement of the object for that interval?

4. Can the instantaneous velocity of an object ever be greater in magnitude than the average velocity? Can it ever be less?

5. Consider the following combinations of signs and values for velocity and acceleration of a particle with respect to a one-dimensional x axis:

Velocity	Acceleration
a. Positive	Positive
b. Positive	Negative
c. Positive	Zero
d. Negative	Positive
e. Negative	Negative
f. Negative	Zero
g. Zero	Positive
h. Zero	Negative

Describe what a particle is doing in each case, and give a real-life example for an automobile on an east–west one-dimensional axis, with east considered the positive direction.

6. Can the equations of kinematics (Eqs. 2.7–2.11) be used in a situation in which the acceleration varies in time? Can they be used when the acceleration is zero?

7. A student at the top of a building of height h throws one ball upward with an initial speed v_0 and then throws a second ball downward with the same initial speed. How do the final velocities of the balls compare when they reach the ground?

8. Two cars are moving in the same direction in parallel lanes along a highway. At some instant, the velocity of car A exceeds the velocity of car B. Does this mean that the acceleration of A is greater than that of B? Explain.

9. Car A traveling south from New York to Miami has a speed of 25 m/s. Car B traveling west from New York to Chicago also has a speed of 25 m/s. Are their velocities equal? Explain.

10. You drop a ball from a window on an upper floor of a building. It strikes the ground with velocity v. You now repeat the drop, but you have a friend down on the street who throws another ball upward at velocity v. Your friend throws the ball upward at exactly the same time that you drop yours from the window. At some location, the balls pass each other. Is this location *at* the halfway point between window and ground, *above* this point, or *below* this point?

11. Galileo experimented with balls rolling down inclined planes in order to reduce the acceleration along the plane and thus reduce the rate of descent of the balls. Suppose the angle that the incline makes with the horizontal is θ. How would you expect the acceleration along the plane to decrease as θ decreases? What specific trigonometric dependence on θ would you expect for the acceleration?

12. A ball rolls in a straight line along the horizontal direction. Using motion diagrams (or multiflash photographs) as in Figure 2.9, describe the velocity and acceleration of the ball for each of the following situations: (a) The ball moves to the right at a constant speed. (b) The ball moves from right to left and continually slows down. (c) The ball moves from right to left and continually speeds up. (d) The ball moves to the right, first speeding up at a constant rate and then slowing down at a constant rate.

Boxed numbers indicate questions that have answers available in the Student Solutions Manual and Study Guide.

Figure Q2.14 See Question 14.

13. A rapidly growing plant doubles in height each week. At the end of the 25th day, the plant reaches the height of the building. At what time was the plant one fourth the height of the building?

14. A pebble is dropped into a water well, and the splash is heard 16 s later, as illustrated in the cartoon strip. What is the *approximate* distance from the rim of the well to the water's surface?

15. The motion of the Earth's crustal plates is described by a model referred to as *plate tectonic motion*. Measurements indicate that coastal portions of southern California have northward plate tectonic motions of 2.5 cm per year. Estimate the time it would take for this motion to carry southern California to Alaska.

16. A heavy object falls from a height h under the influence of gravity. It is released at $t = 0$ and strikes the ground at time t. Ignore air resistance. (a) When the object is at height $0.5\ h$, is the time earlier than $0.5\ t$, later than $0.5\ t$, or equal to $0.5\ t$? (b) When the time is $0.5\ t$, is the height of the object greater than $0.5\ h$, less than $0.5\ h$, or equal to $0.5\ h$?

PROBLEMS

Section 2.1   Average Velocity

1. The position of a pinewood derby car was observed at various times; the results are summarized in the table. Find the average velocity of the car for (a) the first second, (b) the last 3 seconds, and (c) the entire period of observation.

x (m)	0	2.3	9.2	20.7	36.8	57.5
t (s)	0	1.0	2.0	3.0	4.0	5.0

2. A motorist drives north for 35.0 minutes at 85.0 km/h and then stops for 15.0 minutes. He then continues north, traveling 130 km in 2.00 h. (a) What is his total displacement? (b) What is his average velocity?

3. The displacement versus time for a certain particle moving along the x axis is shown in Figure P2.3. Find the average velocity in the time intervals (a) 0 to 2 s, (b) 0 to 4 s, (c) 2 s to 4 s, (d) 4 s to 7 s, (e) 0 to 8 s.

4. A person walks first at a constant speed of 5.00 m/s along a straight line from point A to point B and then back along the line from B to A at a constant speed of 3.00 m/s. (a) What is her average speed over the entire trip? (b) Her average velocity over the entire trip?

5. A person first walks at a constant speed v_1 along a straight line from A to B and then back along the line from B to A at a constant speed v_2. (a) What is her average speed over the entire trip? (b) Her average velocity over the entire trip?

Boxed numbers indicate problems that have full solutions available in the Student Solutions Manual and Study Guide. Cyan numbers indicate intermediate level problems and magenta numbers indicate more challenging problems.

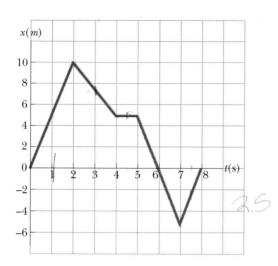

Figure P2.3

Section 2.2 Instantaneous Velocity

6. At $t = 1.00$ s, a particle moving with constant velocity is located at $x = -3.00$ m, and at $t = 6.00$ s the particle is located at $x = 5.00$ m. (a) From this information, plot the position as a function of time. (b) Determine the velocity of the particle from the slope of this graph.

7. A position–time graph for a particle moving along the x axis is shown in Figure P2.7. (a) Find the average velocity in the time interval $t = 1.5$ s to $t = 4.0$ s. (b) Determine the instantaneous velocity at $t = 2.0$ s by measuring the slope of the tangent line shown in the graph. (c) At what value of t is the velocity zero?

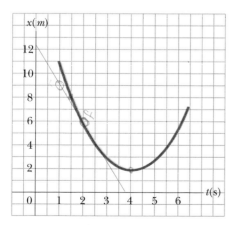

Figure P2.7

8. (a) Use the data in Problem 1 to construct a smooth graph of position versus time. (b) By constructing tangents to the $x(t)$ curve, find the instantaneous velocity of the car at several instants. (c) Plot the instantaneous velocity versus time and, from this, determine the average acceleration of the car. (d) What was the initial velocity of the car?

9. Find the instantaneous velocity of the particle described in Figure P2.3 at the following times: (a) $t = 1.0$ s, (b) $t = 3.0$ s, (c) $t = 4.5$ s, and (d) $t = 7.5$ s.

Section 2.3 Acceleration

10. A particle is moving with a velocity $\mathbf{v}_0 = 60.0\mathbf{i}$ m/s at $t = 0$. Between $t = 0$ and $t = 15.0$ s the velocity decreases uniformly to zero. What was the acceleration during this 15.0-s interval? What is the significance of the sign of your answer?

11. A 50-g superball traveling at 25.0 m/s bounces off a brick wall and rebounds at 22.0 m/s. A high-speed camera records this event. If the ball is in contact with the wall for 3.50 ms, what is the magnitude of the average acceleration of the ball during this time interval? (*Note:* 1 ms = 10^{-3} s.)

12. A velocity–time graph for an object moving along the x axis is shown in Figure P2.12. (a) Plot a graph of the acceleration versus time. (b) Determine the average acceleration of the object in the time intervals $t = 5$ s to $t = 15$ s and $t = 0$ to $t = 20$ s.

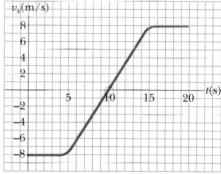

Figure P2.12

13. A particle moves along the x axis according to the equation $x = 2.0 + 3.0t - t^2$, where x is in meters and t is in seconds. At $t = 3.00$ s, find (a) the position of the particle, (b) its velocity, and (c) its acceleration.

14. When struck by a club, a golf ball initially at rest acquires a speed of 31.0 m/s. If the ball is in contact with the club for 1.17 ms, what is the magnitude of the average acceleration of the ball?

15. Figure P2.15 shows a graph of v versus t for the motion of a motorcyclist as he starts from rest and moves along the road in a straight line. (a) Find the average acceleration for the time interval $t_0 = 0$ to $t_1 = 6.0$ s. (b) Estimate the time at which the acceleration has its greatest positive value and the value of the acceleration at that instant. (c) When is the acceleration zero? (d) Estimate the maximum negative value of the acceleration and the time at which it occurs.

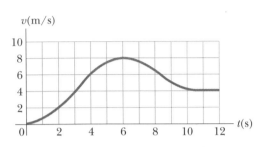

Figure P2.15

Section 2.4   Motion Diagrams

Section 2.5   One-Dimensional Motion with Constant Acceleration

Section 2.6   Freely Falling Objects

16. A truck travels 40.0 m in 8.50 s while uniformly slowing down to a final speed of 2.80 m/s. Find (a) its original speed and (b) its acceleration.

17. A body moving with uniform acceleration has a velocity of 12.0 cm/s in the positive x direction when its x coordinate is 3.00 cm. If its x coordinate 2.00 s later is -5.00 cm, what is the magnitude of its acceleration?

18. The new BMW M3 can accelerate from 0 to 60.0 mi/h in 5.00 s. (a) What is this acceleration in m/s²? (b) How long does it take this car to go from 60.0 mi/h to 130 mi/h?

19. The minimum distance required to stop a car moving at 35.0 mi/h is 40.0 ft. What is the minimum stopping distance for the same car moving at 70.0 mi/h, assuming the same rate of acceleration?

20. A drag racer starts her car from rest and accelerates at 10.0 m/s² for the entire distance of 400 m ($\frac{1}{4}$ mile). (a) How long did it take the race car to travel this distance? (b) What is the speed of the race car at the end of the run?

21. A jet plane lands with a speed of 100 m/s and can accelerate at a maximum rate of -5.00 m/s² as it comes to rest. (a) From the instant the plane touches the runway, what is the minimum time needed before it can come to rest? (b) Can this plane land on a small tropical island airport where the runway is 0.800 km long?

22. A locomotive slows from a speed of 26.0 m/s to 0 in 18.0 s. What distance does it travel?

23. The driver of a car slams on the brakes when he sees a tree blocking the road. The car slows uniformly with acceleration -5.60 m/s² for 4.20 s, making straight skid marks 62.4 m long. With what speed does the car then strike the tree?

24. *Help! One of our equations is missing!* We describe motion under constant acceleration with the parameters v_x, v_{x0}, a_x, t, and $x - x_0$. Of the equations listed in Table 2.1, the first does not involve $x - x_0$; the second does not contain a_x; the third omits v_x; and the last does not contain t. To complete the set, there should be an equation not involving v_{x0}. Derive this equation from the others. Use it to solve Problem 23 in one step.

25. Until recently, the world's land-speed record was held by Col. John P. Stapp, USAF. On March 19, 1954, he rode a rocket-propelled sled that moved down a track at 632 mi/h. He and the sled were safely brought to rest in 1.40 s. Determine (a) the negative acceleration he experienced and (b) the distance he traveled during this negative acceleration.

26. An electron in a cathode-ray-tube (CRT) accelerates from 2.00×10^4 m/s to 6.00×10^6 m/s over 1.50 cm. (a) How long does the electron take to travel this 1.50 cm? (b) What is its acceleration?

27. A student throws a set of keys vertically upward to her sorority sister, who is in a window 4.00 m above. The keys are caught 1.50 s later by the sister's outstretched hand. (a) With what initial velocity were the keys thrown? (b) What was the velocity of the keys just before they were caught?

28. A ball is thrown directly downward, with an initial speed of 8.00 m/s, from a height of 30.0 m. After what interval does the ball strike the ground?

29. A baseball is hit so that it travels straight upward after being struck by the bat. A fan observes that it takes 3.00 s for the ball to reach its maximum height. Find (a) its initial velocity and (b) the height reached by the ball. Ignore the effects of air resistance.

30. A hot-air balloon is traveling vertically upward at a constant speed of 5.00 m/s. When it is 21.0 m above the ground, a package is released from the balloon. (a) For how long after being released is the package in the air? (b) What is the velocity of the package just before impact with the ground? (c) Repeat (a) and (b) for the case of the balloon descending at 5.00 m/s.

31. A daring stunt woman sitting on a tree limb wishes to drop vertically onto a horse galloping under the tree. The speed of the horse is 10.0 m/s, and the distance from the limb to the saddle is 3.00 m. (a) What must be the horizontal distance between the saddle and limb when the woman makes her move? (b) How long is she in the air?

sters can have different accelerations A1 and A2 as well as different maximum speeds V1 and V2. Both cars start from rest at the same starting position. However, a time delay t' in starting times may be introduced if the cars are quite different. Enter the following data:

	Acceleration (ft/s²)	Maximum Speed (ft/s)	Delay Time t' (s)
Car 1	5	300	—
Car 2	6	250	0.5

(a) Which car wins a 1/4-mile race? (b) If the acceleration of Car 1 is increased to 5.2 ft/s², which car now wins?

S3. Modify Spreadsheet 2.1 to solve this problem. The police have set up a "speed trap" on the highway. From a police car hidden behind a billboard, an officer with a radar gun measures a motorist's speed to be 35.0 m/s. Three seconds later, she alerts her partner, who is in another police car 100 m down the road. The second police car starts from rest accelerating at 2.00 m/s² in pursuit of the speeder 2.00 s after receiving the alert. (a) How much time elapses before the speeder is overtaken? (b) What is the police car's speed when it overtakes the speeder? (c) How far does the second police car travel before overtaking the speeder?

ANSWERS TO CONCEPTUAL PROBLEMS

1. If the particle moves along a line without changing direction, the displacement and distance over any time interval will be the same. As a result, the magnitude of the average velocity and the average speed will be the same. If the particle reverses direction, however, then the displacement will be less than the distance. In turn, the magnitude of the average velocity will be smaller than the average speed.

2. Yes. This occurs when a car is slowing down, so that the direction of its acceleration is opposite to its direction of motion.

3. For an accelerating particle to stop at all, the velocity and acceleration must have opposite signs, so that the speed is decreasing. Given that this is the case, the particle will eventually come to rest. If the acceleration remains constant, however, the particle must begin to move again, opposite to the direction of its original motion. If the particle comes to rest and then stays at rest, the acceleration has disappeared at the moment the motion stops. This is the case for a braking car—the acceleration is negative and goes to zero as the car comes to rest.

4. Once the objects leave the hand, both are freely falling objects, and both experience the same downward acceleration equal to the free-fall acceleration of magnitude 9.80 m/s².

5. (a) As it travels upward, its speed decreases by 9.80 m/s during each second of its motion. When it reaches the peak of its motion, its speed becomes zero. As the ball moves downward, its speed increases by 9.80 m/s each second. (b) The acceleration of the ball remains constant while it is in the air. Its magnitude of its acceleration is the free-fall acceleration, $g - 9.80$ m/s².

3

Motion in Two Dimensions

In this chapter we deal with the kinematics of a particle moving in a plane, which is two-dimensional motion. Some common examples of motion in a plane are the motion of projectiles and satellites and the motion of charged particles in uniform electric fields. We begin by showing that displacement, velocity, and acceleration are vector quantities. As in the case of one-dimensional motion, we derive the kinematic equations for two-dimensional motion from the fundamental definitions of displacement,

A multiflash photograph of a popular lecture demonstration in which a projectile is fired at a target that is being held by a magnet in the device at the top right of the photograph. The conditions of the experiment are that the gun is aimed at the target and the projectile leaves the gun at the instant the target is released from rest. Under these conditions the projectile will hit the target, independent of the initial speed of the projectile. The reason is that they both experience the same downward acceleration, and hence the velocities of the projectile and target change by the same amount in the same time interval. Note that the velocity of the projectile (red arrows) changes in direction and magnitude, whereas its downward acceleration (violet arrows) remains constant. (Central Scientific Company)

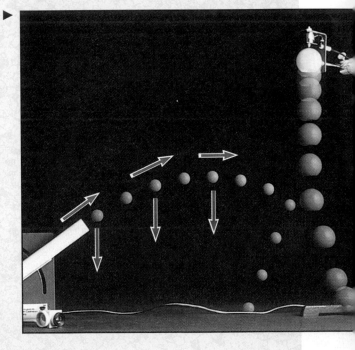

velocity, and acceleration. As special cases of motion in two dimensions, we then treat constant-acceleration motion in a plane and uniform circular motion.

3.1 • THE DISPLACEMENT, VELOCITY, AND ACCELERATION VECTORS

In Chapter 2 we found that the motion of a particle moving along a straight line is completely known if its coordinate is known as a function of time. Now let us extend this idea to motion in the *xy* plane. We begin by describing the position of a particle with a *position vector* **r**, drawn from the origin of some reference frame to the particle located in the *xy* plane, as in Figure 3.1. At time t_i, the particle is at the point *P*, and at some later time, t_f, the particle is at *Q*, where the indices *i* and *f* refer to initial and final values. As the particle moves from *P* to *Q* in the time interval $\Delta t = t_f - t_i$, the position vector changes from $\mathbf{r}_i$ to $\mathbf{r}_f$. As we learned in Chapter 2, the displacement of a particle is the difference between its final position and initial position. Therefore, the **displacement vector** for the particle of Figure 3.1 equals the difference between its final position vector and its initial position vector:

$$\Delta\mathbf{r} \equiv \mathbf{r}_f - \mathbf{r}_i \qquad [3.1]$$

The direction of $\Delta\mathbf{r}$ is indicated in Figure 3.1. As we see from the figure, the magnitude of the displacement vector is *less* than the distance traveled along the curved path.

> We define the **average velocity** of the particle during the time interval Δt as the ratio of the displacement to that time interval:

$$\bar{\mathbf{v}} \equiv \frac{\Delta\mathbf{r}}{\Delta t} \qquad [3.2]$$

Because displacement is a vector quantity and the time interval is a scalar quantity, we conclude that the average velocity is a *vector* quantity directed along $\Delta\mathbf{r}$. Note that the average velocity between points *P* and *Q* is *independent of the path* between the two points. This is because the average velocity is proportional to the displacement, which in turn depends only on the initial and final position vectors and not on the path taken between those two points. As we did with one-dimensional motion, we conclude that if a particle starts its motion at some point and returns to this point via any path, its average velocity is zero for this trip, because its displacement is zero. Consider again the motion of a particle between two points in the *xy* plane, as in Figure 3.2. As the time intervals over which we observe the motion become smaller and smaller, the direction of the displacement approaches that of the line tangent to the path at the point *P*.

> The **instantaneous velocity, v**, is defined as the limit of the average velocity, $\Delta\mathbf{r}/\Delta t$, as Δt approaches zero:

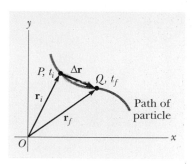

Figure 3.1 A particle moving in the *xy* plane is located with the position vector **r** drawn from the origin to the particle. The displacement of the particle as it moves from *P* to *Q* in the time interval $\Delta t = t_f - t_i$ is equal to the vector $\Delta\mathbf{r} = \mathbf{r}_f - \mathbf{r}_i$.

• *Definition of the displacement vector*

• *Definition of average velocity*

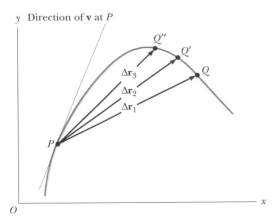

Figure 3.2   As a particle moves between two points, its average velocity is in the direction of the displacement vector $\Delta \mathbf{r}$. As the end point of the path is moved from Q to Q' to Q'', the respective displacements and corresponding time intervals become smaller and smaller. In the limit that the end point approaches P, Δt approaches zero and the direction of $\Delta \mathbf{r}$ approaches that of the line tangent to the curve at P. By definition, the instantaneous velocity at P is in the direction of this tangent line.

Definition of instantaneous velocity •

$$\mathbf{v} \equiv \lim_{\Delta t \to 0} \frac{\Delta \mathbf{r}}{\Delta t} = \frac{d\mathbf{r}}{dt} \qquad \text{[3.3]}$$

That is, the instantaneous velocity equals the derivative of the position vector with respect to time. The direction of the instantaneous velocity vector at any point in a particle's path is along a line that is tangent to the path at that point and in the direction of motion; this is illustrated in Figure 3.3. The magnitude of the instantaneous velocity vector is called the *speed*.

As a particle moves from one point to another along some path, as in Figure 3.3, its instantaneous velocity vector changes from $\mathbf{v}_i$ at time t_i to $\mathbf{v}_f$ at time t_f.

> The **average acceleration** of a particle as it moves from P to Q is defined as the ratio of the change in the instantaneous velocity vector, $\Delta \mathbf{v}$, to the elapsed time, Δt:

Definition of average acceleration •

$$\bar{\mathbf{a}} \equiv \frac{\mathbf{v}_f - \mathbf{v}_i}{t_f - t_i} = \frac{\Delta \mathbf{v}}{\Delta t} \qquad \text{[3.4]}$$

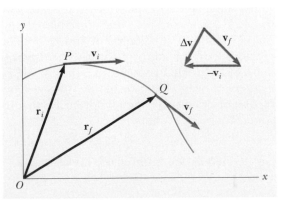

Figure 3.3   The average acceleration vector $\bar{\mathbf{a}}$ for a particle moving from P to Q is in the direction of the change in velocity, $\Delta \mathbf{v} = \mathbf{v}_f - \mathbf{v}_i$.

Because the average acceleration is the ratio of a vector quantity, $\Delta\mathbf{v}$, and a scalar quantity, Δt, we conclude that $\overline{\mathbf{a}}$ is a vector quantity directed along $\Delta\mathbf{v}$. As is indicated in Figure 3.3, the direction of $\Delta\mathbf{v}$ is found by adding the vector $-\mathbf{v}_i$ (the negative of $\mathbf{v}_i$) to the vector $\mathbf{v}_f$, because by definition $\Delta\mathbf{v} = \mathbf{v}_f - \mathbf{v}_i$.

> The **instantaneous acceleration, a,** is defined as the limiting value of the ratio $\Delta\mathbf{v}/\Delta t$ as Δt approaches zero:

$$\mathbf{a} \equiv \lim_{\Delta t \to 0} \frac{\Delta\mathbf{v}}{\Delta t} = \frac{d\mathbf{v}}{dt} \qquad\qquad [3.5]$$

• *Definition of instantaneous acceleration*

In other words, the instantaneous acceleration equals the derivative of the velocity vector with respect to time.

It is important to recognize that various changes can cause a particle to accelerate. First, the magnitude of the velocity vector (the speed) may change with time as in straight-line (one-dimensional) motion. Second, only the direction of the velocity vector may change with time as its magnitude (speed) remains constant, as in curved-path (two-dimensional) motion. Third, both the magnitude and the direction of the velocity vector may change.

Thinking Physics 1

The gas pedal in an automobile is called the *accelerator*. Are there any other automobile controls that could also be considered "accelerators"?

Reasoning The gas pedal is called the accelerator because the cultural usage of the word *acceleration* refers to an *increase in speed*. The scientific definition, however, is that acceleration occurs *whenever the velocity changes* in any way. Thus, the *brake pedal* can also be considered an accelerator, because it causes the car to slow down. The *steering wheel* is also an accelerator, because it changes the direction of the velocity vector.

CONCEPTUAL PROBLEM 1

(a) Can an object accelerate if its speed is constant? (b) Can an object accelerate if its velocity is constant?

CONCEPTUAL PROBLEM 2

Can a particle have a constant velocity and varying speed? Can a particle have a constant speed and a varying velocity? Give examples of a positive answer and explain a negative answer.

3.2 • TWO-DIMENSIONAL MOTION WITH CONSTANT ACCELERATION

Let us consider two-dimensional motion during which the acceleration remains constant. That is, we assume that the magnitude and direction of the acceleration remain unchanged during the motion.

The motion of a particle can be determined by its position vector **r**. The position vector for a particle moving in the xy plane can be written

$$\mathbf{r} = x\mathbf{i} + y\mathbf{j} \qquad\qquad [3.6]$$

where x, y, and **r** change with time as the particle moves. If the position vector is known, the velocity of the particle can be obtained from Equations 3.3 and 3.6, which give

$$\mathbf{v} = v_x\mathbf{i} + v_y\mathbf{j} \qquad\qquad [3.7]$$

Because **a** is assumed constant, its components a_x and a_y are also constants. Therefore, we can apply the equations of kinematics to the x and y components of the velocity vector. Substituting $v_x = v_{x0} + a_x t$ and $v_y = v_{y0} + a_y t$ into Equation 3.7 gives

$$\mathbf{v} = (v_{x0} + a_x t)\mathbf{i} + (v_{y0} + a_y t)\mathbf{j}$$
$$= (v_{x0}\mathbf{i} + v_{y0}\mathbf{j}) + (a_x\mathbf{i} + a_y\mathbf{j})t$$

$$\mathbf{v} = \mathbf{v}_0 + \mathbf{a}t \qquad\qquad [3.8]$$

Velocity vector as a function of time •

This result states that the velocity of a particle at some time t equals the vector sum of its initial velocity, $\mathbf{v}_0$, and the additional velocity $\mathbf{a}t$ acquired in the time t as a result of its constant acceleration.

In a similar way, from Equation 2.9 we know that the x and y coordinates of a particle moving with constant acceleration are

$$x = x_0 + v_{x0}t + \tfrac{1}{2}a_x t^2 \qquad \text{and} \qquad y = y_0 + v_{y0}t + \tfrac{1}{2}a_y t^2$$

Substituting these expressions into Equation 3.6 gives

$$\mathbf{r} = (x_0 + v_{x0}t + \tfrac{1}{2}a_x t^2)\mathbf{i} + (y_0 + v_{y0}t + \tfrac{1}{2}a_y t^2)\mathbf{j}$$
$$= (x_0\mathbf{i} + y_0\mathbf{j}) + (v_{x0}\mathbf{i} + v_{y0}\mathbf{j})t + \tfrac{1}{2}(a_x\mathbf{i} + a_y\mathbf{j})t^2$$

Position vector as a function of time •

$$\mathbf{r} = \mathbf{r}_0 + \mathbf{v}_0 t + \tfrac{1}{2}\mathbf{a}t^2 \qquad\qquad [3.9]$$

This equation implies that the displacement vector $\mathbf{r} - \mathbf{r}_0$ is the vector sum of a displacement $\mathbf{v}_0 t$, arising from the initial velocity of the particle, and a displacement $\tfrac{1}{2}\mathbf{a}t^2$, resulting from the uniform acceleration of the particle. Graphical representations of Equations 3.8 and 3.9 are shown in Figure 3.4a and 3.4b. For simplicity in drawing the figure, we have taken $\mathbf{r}_0 = 0$ in Figure 3.4b. That is, we assume that the particle is at the origin at $t = 0$. Note from Figure 3.4b that **r** is generally not along the direction of $\mathbf{v}_0$ or **a**, because the relationship between these quantities is a vector expression. For the same reason, from Figure 3.4a we see that **v** is generally not along the direction of $\mathbf{v}_0$ or **a**. Finally, if we compare the two figures, we see that **v** and **r** are not in the same direction.

Because Equations 3.8 and 3.9 are *vector* expressions, we may write their x and y component forms with $\mathbf{r}_0 = 0$:

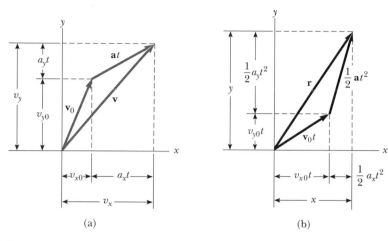

(a)   (b)

Figure 3.4   Vector representations and rectangular components of (a) the velocity and (b) the displacement of a particle moving with a uniform acceleration **a**.

$$\mathbf{v} = \mathbf{v}_0 + \mathbf{a}t \qquad \begin{cases} v_x = v_{x0} + a_x t \\ v_y = v_{y0} + a_y t \end{cases}$$

$$\mathbf{r} = \mathbf{v}_0 t + \tfrac{1}{2}\mathbf{a}t^2 \qquad \begin{cases} x = v_{x0}t + \tfrac{1}{2}a_x t^2 \\ y = v_{y0}t + \tfrac{1}{2}a_y t^2 \end{cases}$$

These components are illustrated in Figure 3.4. In other words, two-dimensional motion having constant acceleration is equivalent to two independent motions in the x and y directions having constant accelerations a_x and a_y.

Example 3.1   Motion in a Plane

A particle starts from the origin at $t = 0$ with an initial velocity having an x component of 20 m/s and a y component of -15 m/s. The particle moves in the xy plane with an x component of acceleration only, given by $a_x = 4.0$ m/s². (a) Determine the components of velocity as a function of time and the total velocity vector at any time.

Solution   With $v_{x0} = 20$ m/s and $a_x = 4.0$ m/s², the equations of kinematics give

$$v_x = v_{x0} + a_x t = (20 + 4.0t) \text{ m/s}$$

Also, with $v_{y0} = -15$ m/s and $a_y = 0$,

$$v_y = v_{y0} = -15 \text{ m/s}$$

Therefore, using these results and noting that the velocity vector **v** has two components, we get

$$\mathbf{v} = v_x\mathbf{i} + v_y\mathbf{j} = [(20 + 4.0t)\mathbf{i} - 15\mathbf{j}] \text{ m/s}$$

We could also obtain this result using Equation 3.8 directly, noting that $\mathbf{a} = 4.0\mathbf{i}$ m/s² and $\mathbf{v}_0 = (20\mathbf{i} - 15\mathbf{j})$ m/s. Try it!

(b) Calculate the velocity and speed of the particle at $t = 5.0$ s.

Solution   With $t = 5.0$ s, the result from (a) gives

$$\mathbf{v} = \{[20 + 4(5.0)]\mathbf{i} - 15\mathbf{j}\} \text{ m/s} = (40\mathbf{i} - 15\mathbf{j}) \text{ m/s}$$

That is, at $t = 5.0$ s, $v_x = 40$ m/s, and $v_y = -15$ m/s. Knowing these two components for this two-dimensional motion, we know the numerical value of the velocity vector. To determine the angle θ that **v** makes with the x axis, we use the fact that $\tan\theta = v_y/v_x$, or

$$\theta = \tan^{-1}\left(\frac{v_y}{v_x}\right) = \tan^{-1}\left(\frac{-15 \text{ m/s}}{40 \text{ m/s}}\right) = -21°$$

The speed is the magnitude of **v**:

$$v = |\mathbf{v}| = \sqrt{v_x^2 + v_y^2} = \sqrt{(40)^2 + (-15)^2} \text{ m/s} = 43 \text{ m/s}$$

(*Note:* If you calculate v_0 from the x and y components of $\mathbf{v}_0$, you will find that $v > v_0$. Why?)

EXERCISE 1 Determine the x and y coordinates of the particle at any time t and the displacement vector at this time.   Answer $x = (20t + 2.0t^2)$ m; $y = (-15t)$ m; $\mathbf{r} = [(20t + 2.0t^2)\mathbf{i} - 15t\mathbf{j}]$ m

EXERCISE 2 A particle starts from rest at $t = 0$ at the origin and moves in the xy plane with a constant acceleration of $\mathbf{a} = (2.0\mathbf{i} + 4.0\mathbf{j})$ m/s^2. After a time t has elapsed, determine (a) the x and y components of velocity, (b) the coordinates of the particle, and (c) the speed of the particle.

Answer   (a) $v_x = 2t$ m/s, $v_y = 4t$ m/s (b) $x = t^2$ m, $y = 2t^2$ m   (c) $4.47t$ m/s

3.3 • PROJECTILE MOTION

Anyone who has observed a baseball in motion (or any object thrown into the air) has observed projectile motion. The ball moves in a curved path when thrown at some angle with respect to the Earth's surface. This very common form of motion is surprisingly simple to analyze if the following two assumptions are made: (1) the free-fall acceleration, g, is constant over the range of motion and is directed downward,[1] and (2) the effect of air resistance is negligible.[2] With these assumptions, we find that the path of a projectile, which we call its *trajectory*, is *always* a parabola. **We use these assumptions throughout this chapter.**

Assumptions of projectile motion •

If we choose our reference frame such that the y direction is vertical and positive upward, then $a_y = -g$ (as in one-dimensional free fall) and $a_x = 0$ (because air resistance is neglected). Furthermore, let us assume that at $t = 0$, the projectile leaves the origin ($x_0 = y_0 = 0$) with speed v_0, as in Figure 3.5. If the vector $\mathbf{v}_0$ makes an angle θ_0 with the horizontal, where θ_0 is the angle at which the projectile leaves the origin as in Figure 3.5, then from the definitions of the cosine and sine functions we have

$$\cos \theta_0 = \frac{v_{x0}}{v_0} \quad \text{and} \quad \sin \theta_0 = \frac{v_{y0}}{v_0}$$

Therefore, the initial x and y components of velocity are

$$v_{x0} = v_0 \cos \theta_0 \quad \text{and} \quad v_{y0} = v_0 \sin \theta_0$$

Substituting these expressions into Equations 3.8 and 3.9 with $a_x = 0$ and $a_y = -g$ gives the velocity components and coordinates for the projectile at any time t:

Horizontal velocity component • $v_x = v_{x0} = v_0 \cos \theta_0 = \text{constant}$ **[3.10]**

Vertical velocity component • $v_y = v_{y0} - gt = v_0 \sin \theta_0 - gt$ **[3.11]**

Horizontal position component • $x = v_{x0}t = (v_0 \cos \theta_0)\, t$ **[3.12]**

Vertical position component • $y = v_{y0}t - \frac{1}{2}gt^2 = (v_0 \sin \theta_0)\, t - \frac{1}{2}gt^2$ **[3.13]**

[1]This approximation is reasonable as long as the range of motion is small compared with the radius of the Earth (6.4×10^6 m). In effect, this approximation is equivalent to assuming that the earth is flat within the range of motion considered.

[2]This approximation is generally *not* justified, especially at high velocities. In addition, the spin of a projectile, such as a baseball, can give rise to some very interesting effects associated with aerodynamic forces (for example, a curveball thrown by a pitcher).

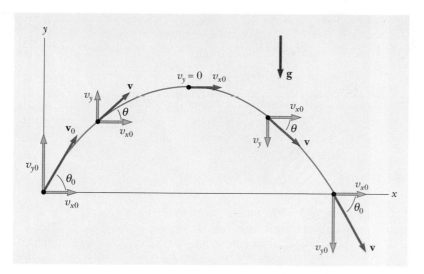

Figure 3.5   The parabolic path of a projectile that leaves the origin with a velocity $\mathbf{v}_0$. The velocity vector $\mathbf{v}$ changes with time in both magnitude and direction. The change in the velocity vector is the result of acceleration in the negative y direction. The x component of velocity remains constant in time because there is no acceleration along the horizontal direction. Also, the y component of velocity is zero at the peak of the path.

From Equation 3.10 we see that v_x remains constant in time and is equal to v_{x0}; there is no horizontal component of acceleration. For the y motion note that v_y and y are similar to Equations 2.12 and 2.14 for freely falling bodies. In fact, *all* of the equations of kinematics developed in Chapter 2 are applicable to projectile motion.

If we solve for t in Equation 3.12 and substitute this expression for t into Equation 3.13, we find that

$$y = (\tan \theta_0)x - \left(\frac{g}{2v_0^2 \cos^2 \theta_0}\right)x^2 \qquad \textbf{[3.14]}$$

which is valid for the angles in the range $0 < \theta_0 < \pi/2$. This expression is of the form $y = ax - bx^2$, which is the equation of a parabola that passes through the origin. Thus, we have proved that the trajectory of a projectile is a parabola. Note that the trajectory is *completely* specified if v_0 and θ_0 are known.

One can obtain the speed, v, of the projectile as a function of time by noting that Equations 3.10 and 3.11 give the x and y components of velocity at any instant. Therefore, by definition, because v is equal to the magnitude of $\mathbf{v}$,

$$v = \sqrt{v_x^2 + v_y^2} \qquad \textbf{[3.15]} \qquad \bullet \quad \textit{Speed}$$

Also, because the velocity vector is tangent to the path at any instant, as shown in Figure 3.5, the angle θ that v makes with the horizontal can be obtained from v_x and v_y through the expression

$$\tan \theta = \frac{v_y}{v_x} \qquad \textbf{[3.16]}$$

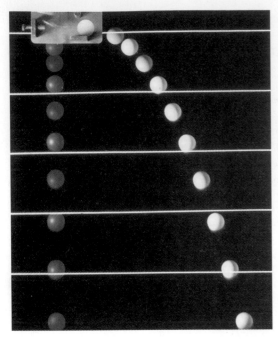

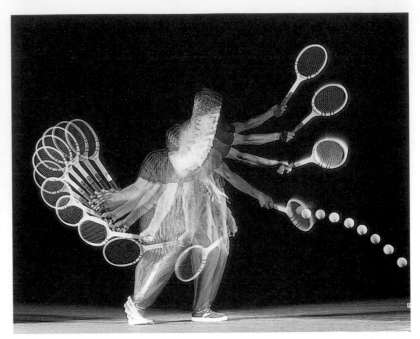

(Left) This multiflash photograph of two balls released simultaneously illustrates both free fall (red ball) and projectile motion (yellow ball). The yellow ball was projected horizontally, whereas the red ball was released from rest. Can you explain why both balls reach the floor simultaneously? *(© Richard Megna 1990, Fundamental Photographs)*   *(Right)* Multiflash exposure of a tennis player executing a backhand swing. Note that the ball follows a parabolic path characteristic of a projectile. Such photographs can be used to study the quality of sports equipment and the performance of an athlete. *(© Zimmerman, FPG International)*

The vector expression for the position vector of the projectile as a function of time follows directly from Equation 3.9, with $\mathbf{a} = \mathbf{g}$:

$$\mathbf{r} = \mathbf{v}_0 t + \tfrac{1}{2}\mathbf{g}t^2$$

This expression gives the same information as the combination of Equations 3.12 and 3.13 and is plotted in Figure 3.6. Note that this expression for $\mathbf{r}$ is consistent

Figure 3.6 The displacement vector $\mathbf{r}$ of a projectile, the initial velocity of which is $\mathbf{v}_0$ at the origin. The vector $\mathbf{v}_0 t$ would be the displacement of the projectile if gravity were absent, and the vector $\tfrac{1}{2}\mathbf{g}t^2$ is its vertical displacement due to its downward gravitational acceleration.

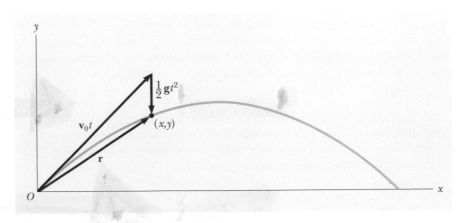

with Equation 3.13, because the expression for **r** is a vector equation and $\mathbf{a} = \mathbf{g} = -g\mathbf{j}$ when the upward direction is taken to be positive.

It is interesting to note that the motion of a particle can be considered the superposition of the term $\mathbf{v}_0 t$, which would be the displacement if no acceleration were present, and the term $\frac{1}{2}\mathbf{g}t^2$, which arises from the acceleration due to gravity. In other words, if there were no gravitational acceleration, the particle would continue to move along a straight path in the direction of $\mathbf{v}_0$. Therefore, the vertical distance $\frac{1}{2}gt^2$ through which the particle "falls" off the straight-path line is the same distance a freely falling body would fall during the same time interval. **We conclude that projectile motion is the superposition of two motions: (1) constant velocity motion in the initial direction and (2) the motion of a particle freely falling in the vertical direction under constant acceleration.**

Horizontal Range and Maximum Height of a Projectile   Let us assume that a projectile is fired from the origin at $t = 0$ with a positive v_y component, as in Figure 3.7. There are two special points that are interesting to analyze: the peak that has Cartesian coordinates $(R/2, h)$ and the point having coordinates $(R, 0)$. The distance R is called the *horizontal range* of the projectile and h is its *maximum height*. Let us find h and R in terms of v_0, θ_0, and g.

We can determine h by noting that at the peak $v_y = 0$. Therefore, Equation 3.11 can be used to determine the time t_1 it takes to reach the peak:

$$t_1 = \frac{v_0 \sin \theta_0}{g}$$

Substituting this expression for t_1 into Equation 3.13 and replacing y with h gives h in terms of v_0 and θ_0:

$$h = (v_0 \sin \theta_0) \frac{v_0 \sin \theta_0}{g} - \frac{1}{2}g \left(\frac{v_0 \sin \theta_0}{g} \right)^2$$

$$h = \frac{v_0^2 \sin^2 \theta_0}{2g} \qquad \text{[3.17]}$$

The range, R, is the horizontal distance traveled in twice the time it takes to reach the peak—that is, in a time $2t_1$. (This can be seen by setting $y = 0$ in Equation 3.13 and solving the quadratic for t. One solution of this quadratic is $t = 0$, and the other is $t = 2t_1$.) Using Equation 3.12 and noting that $x = R$ at $t = 2t_1$, we find that

$$R = (v_0 \cos \theta_0) 2t_1 = (v_0 \cos \theta_0) \frac{2v_0 \sin \theta_0}{g} = \frac{2v_0^2 \sin \theta_0 \cos \theta_0}{g}$$

Because $\sin 2\theta = 2 \sin \theta \cos \theta$, R can be written in the most compact form

$$R = \frac{v_0^2 \sin 2\theta_0}{g} \qquad \text{[3.18]}$$

Keep in mind that Equations 3.17 and 3.18 are useful for calculating h and R only if v_0 and θ_0 are known and only for a symmetric path, as shown in Figure 3.7 (which means that only $\mathbf{v}_0$ has to be specified). The general expressions given by Equations 3.10 through 3.13 are the *more important* results, because they give the coordinates and velocity components of the projectile at *any* time t.

You should note that the maximum value of R from Equation 3.18 is given by $R_{\max} = v_0^2/g$. This result follows from the fact that the maximum value of $\sin 2\theta_0$

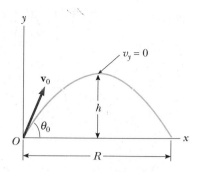

Figure 3.7   A projectile fired from the origin at $t = 0$ with an initial velocity $\mathbf{v}_0$. The maximum height of the projectile is h, and its horizontal range is R. At the peak of the trajectory, the y component of its velocity is zero and the projectile has coordinates $(\frac{R}{2}, h)$.

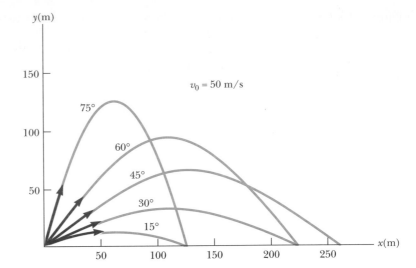

Figure 3.8 A projectile fired from the origin with an initial speed of 50 m/s at various angles of projection. Note that complementary values of θ_0 will result in the same value of x (range of the projectile).

is unity, which occurs when $2\theta_0 = 90°$. Therefore, we see that R is a maximum when $\theta_0 = 45°$, as you would expect if air resistance is neglected.

Figure 3.8 illustrates various trajectories for a projectile of a given initial speed. As you can see, the range is a maximum for $\theta_0 = 45°$. In addition, for any θ_0 other than 45°, a point with coordinates $(R, 0)$ can be reached by using either one of two complementary values of θ_0, such as 75° and 15°. Of course, the maximum height and the time of flight will be different for these two values of θ_0.

Before we look at some numerical examples dealing with projectile motion, let us pause to summarize what we have learned so far about this kind of motion:

- Provided air resistance is negligible, the horizontal component of velocity, v_x, remains constant because there is no horizontal component of acceleration.
- The vertical component of acceleration is equal to the free-fall acceleration, **g**.
- The vertical component of velocity, v_y, and the displacement in the y direction are identical to those of a freely falling body.
- Projectile motion can be described as a superposition of the two motions in the x and y directions.

PROBLEM-SOLVING STRATEGY • Projectile Motion

We suggest that you use the following approach to solving projectile motion problems:

1. Select a coordinate system.
2. Resolve the initial velocity vector into x and y components.
3. Treat the horizontal motion and the vertical motion independently.

4. Follow the techniques for solving problems with constant velocity to analyze the horizontal motion of the projectile.
5. Follow the techniques for solving problems with constant acceleration to analyze the vertical motion of the projectile.

Thinking Physics 2

A home run is hit in a baseball game. The ball is hit from home plate into the center-field stands along a long, parabolic path. What is the acceleration of the ball (a) while it is rising, (b) at the highest point of the trajectory, and (c) while it is descending after reaching the highest point? Ignore air resistance.

Reasoning The answers to all three parts are the same—the acceleration is that due to gravity, 9.8 m/s^2, because the force of gravity is pulling downward on the ball during the entire motion. During the rising part of the trajectory, the downward acceleration results in the decreasing positive values of the vertical component of the velocity of the ball. During the falling part of the trajectory, the downward acceleration results in the increasing negative values of the vertical component of the velocity. Many people have trouble with the topmost point, mistakenly thinking that the acceleration of the ball at the highest point is zero. This interpretation arises from a confusion between zero vertical velocity and zero acceleration. Even though the ball has momentarily come to rest at the highest point (vertically; it is still moving horizontally), it is still accelerating, because the force of gravity is still acting. If the ball were to come to rest and experience zero acceleration, then the velocity would not change—the ball would remain suspended at the highest point! We don't see that happening, because the acceleration is *not* equal to zero.

CONCEPTUAL PROBLEM 3

Suppose that as you are running at constant speed, you wish to throw a ball such that you will catch it as it comes back down. How should you throw the ball?

CONCEPTUAL PROBLEM 4

As a projectile moves in its parabolic path, is there any point along its path at which the velocity and acceleration vectors are (a) perpendicular to each other? (b) parallel to each other?

Example 3.2 That's Quite an Arm

A stone is thrown from the top of a building upward at an angle of 30.0° to the horizontal and with an initial speed of 20.0 m/s, as in Figure 3.9. If the height of the building is 45.0 m, (a) how long is the stone "in flight"?

Solution The initial x and y components of the velocity are

$$v_{x0} = v_0 \cos \theta_0 = (20.0 \text{ m/s})(\cos 30.0°) = 17.3 \text{ m/s}$$

$$v_{y0} = v_0 \sin \theta_0 = (20.0 \text{ m/s})(\sin 30.0°) = 10.0 \text{ m/s}$$

To find t, we can use $y = v_{y0}t - \frac{1}{2}gt^2$ (Eq. 3.13) with $y = -45.0$ m and $v_{y0} = 10.0$ m/s (we have chosen the top of the building as the origin, as in Fig. 3.9):

$$-45.0 \text{ m} = (10.0 \text{ m/s})t - \tfrac{1}{2}(9.80 \text{ m/s}^2)t^2$$

Solving the quadratic equation for t gives, for the positive root, $t = 4.22$ s. Does the negative root have any physical

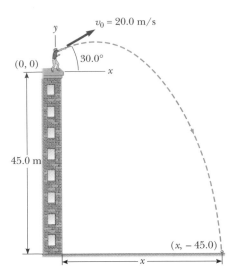

$v_0 = 20.0$ m/s

$30.0°$

$(0, 0)$

45.0 m

$(x, -45.0)$

x

Figure 3.9   (Example 3.2)

meaning? (Can you think of another way of finding t from the information given?)

(b) What is the speed of the stone just before it strikes the ground?

Solution   The y component of the velocity just before the stone strikes the ground can be obtained using the equation $v_y = v_{y0} - gt$ (Eq. 3.11) with $t = 4.22$ s:

$$v_y = 10.0 \text{ m/s} - (9.80 \text{ m/s}^2)(4.22 \text{ s}) = -31.4 \text{ m/s}$$

Because $v_x = v_{x0} = 17.3$ m/s, the required speed is

$$v = \sqrt{v_x^2 + v_y^2} = \sqrt{(17.3)^2 + (-31.4)^2} \text{ m/s} = 35.9 \text{ m/s}$$

EXERCISE 3   Where does the stone strike the ground?
Answer   73.0 m from the base of the building.

Example 3.3   The Stranded Explorers

An Alaskan rescue plane drops a package of emergency rations to a stranded party of explorers, as shown in Figure 3.10. If the plane is traveling horizontally at 40.0 m/s at a height of 100 m above the ground, where does the package strike the ground relative to the point at which it is released?

Reasoning and Solution   The coordinate system for this problem is selected as shown in Figure 3.10, with the positive x direction to the right and the positive y direction upward.

Consider first the horizontal motion of the package. The only equation available to us is $x = v_{x0}t$ (Eq. 3.12).

The initial x component of the package velocity is the same as that of the plane when the package is released, 40.0 m/s. Thus, we have

$$x = (40.0 \text{ m/s})t$$

If we know t, the length of time the package is in the air, we can determine x, the distance traveled by the package in the horizontal direction. To find t, we move to the equations for the vertical motion of the package. We know that at the instant the package hits the ground, its y coordinate is -100 m. We also know that the initial component of velocity of the package in the vertical direction, v_{y0}, is zero because the package was released with only a horizontal component of velocity. From Equation 3.13, we have

$$y = -\tfrac{1}{2}gt^2$$
$$-100 \text{ m} = -\tfrac{1}{2}(9.80 \text{ m/s}^2)t^2$$
$$t^2 = 20.4 \text{ s}^2$$
$$t = 4.51 \text{ s}$$

The value for the time of flight substituted into the equation for the x coordinate gives

$$x = (40.0 \text{ m/s})(4.51 \text{ s}) = 180 \text{ m}$$

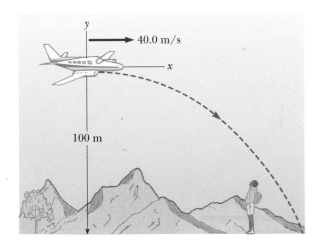

y

40.0 m/s

x

100 m

Figure 3.10   (Example 3.3) According to a ground observer, a package released from the rescue plane travels along the path shown. How does the path followed by the package appear to an observer on the plane (assumed to be moving at constant speed)?

The package hits the ground not directly under the drop point but 180 m to the right of that point.

EXERCISE 4 What are the horizontal and vertical components of the velocity of the package just before it hits the ground? Answer $v_x = 40.0$ m/s; $v_y = -44.1$ m/s

EXERCISE 5 It has been said that in his youth George Washington threw a silver dollar across a river. Assuming that the river was 75 m wide, (a) what *minimum initial* speed was necessary to get the coin across the river, and (b) how long was the coin in flight? Answer (a) 27.1 m/s (b) 3.91 s

3.4 • UNIFORM CIRCULAR MOTION

Figure 3.11a shows an object moving in a circular path with *constant linear speed v.* Such motion is called **uniform circular motion.** It is often surprising to students to find that **even though the object moves at a constant speed, it still has an acceleration.** To see why, consider the defining equation for average acceleration, $\bar{\mathbf{a}} = \Delta\mathbf{v}/\Delta t$ (Eq. 3.4).

change in velocity vector

Note that the acceleration depends on *the change in the velocity vector*. Because velocity is a vector quantity, there are two ways in which an acceleration can be produced, as mentioned in Section 3.1: by a change in the *magnitude* of the velocity and by a change in the *direction* of the velocity. It is the latter situation that is occurring for an object moving with constant speed in a circular path. The velocity vector is always tangent to the path of the object and perpendicular to the radius r of the circular path. We now show that the acceleration vector in uniform circular motion is always perpendicular to the path and always points toward the center of the circle. An acceleration of this nature is called a *centripetal acceleration* (center seeking) and its magnitude is

$$a_r = \frac{v^2}{r} \qquad\qquad [3.19]$$

• *Magnitude of centripetal acceleration*

where r is the radius of the circle.

To derive Equation 3.19, consider Figure 3.11b. Here an object is seen first at point P and then at point Q. The particle is at P at time t_i, and its velocity at that time is $\mathbf{v}_i$; it is at Q at some later time t_f, and its velocity at that time is $\mathbf{v}_f$. Let us

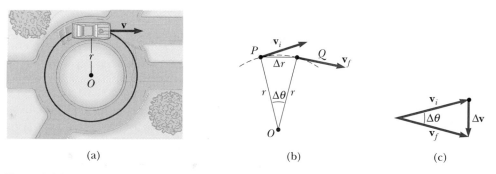

(a)   (b)   (c)

Figure 3.11 (a) An object moving along a circular path at constant speed experiences uniform circular motion. (b) As the particle moves from P to Q, the direction of its velocity vector changes from $\mathbf{v}_i$ to $\mathbf{v}_f$. (c) The construction for determining the direction of the change in velocity $\Delta\mathbf{v}$ that is toward the center of the circle.

also assume here that $\mathbf{v}_i$ and $\mathbf{v}_f$ differ only in direction; their magnitudes are the same (that is, $v_i = v_f = v$). In order to calculate the acceleration of the particle, let us begin with the defining equation for average acceleration (Eq. 3.4):

$$\bar{\mathbf{a}} = \frac{\mathbf{v}_f - \mathbf{v}_i}{t_f - t_i} = \frac{\Delta\mathbf{v}}{\Delta t}$$

This equation indicates that we must vectorially subtract $\mathbf{v}_i$ from $\mathbf{v}_f$, where $\Delta\mathbf{v} = \mathbf{v}_f - \mathbf{v}_i$ is the change in the velocity. Because $\mathbf{v}_i + \Delta\mathbf{v} = \mathbf{v}_f$, the vector $\Delta\mathbf{v}$ can be found using the vector triangle in Figure 3.11c. When Δt is very small, Δr and $\Delta\theta$ are also very small. In this case, $\mathbf{v}_f$ is almost parallel to $\mathbf{v}_i$ and the vector $\Delta\mathbf{v}$ is approximately perpendicular to them, pointing toward the center of the circle.

Now consider the triangle in Figure 3.11b, which has sides Δr and r. This triangle and the one in Figure 3.11c that has sides Δv and v are similar. (Two triangles are *similar* if the angle between any two sides is the same for both triangles and if the ratio of lengths of these sides is the same.) This enables us to write a relationship between the lengths of the sides:

$$\frac{\Delta v}{v} = \frac{\Delta r}{r}$$

This equation can be solved for Δv and the expression obtained from it can be substituted into $\bar{a} = \Delta v/\Delta t$ (Eq. 3.4) to give

$$\bar{a} = \frac{v\,\Delta r}{r\Delta t}$$

Now imagine that points P and Q in Figure 3.11b come extremely close together. In this case $\Delta\mathbf{v}$ would point toward the center of the circular path, and because the acceleration is in the direction of $\Delta\mathbf{v}$, it too is toward the center. Furthermore, as P and Q approach each other, Δt approaches zero, and the ratio $\Delta r/\Delta t$ approaches the speed v. Hence, in the limit $\Delta t \to 0$, the magnitude of the acceleration is

$$a_r = \frac{v^2}{r}$$

Thus, we conclude that in uniform circular motion the acceleration is directed inward toward the center of the circle and has magnitude v^2/r. You should show that the dimensions of a_r are L/T^2—as required because this is a true acceleration.

In many situations it is convenient to describe the motion of a particle moving with constant speed in a circle of radius r in terms of the **period,** T, which is defined as the time required for one complete revolution. In the time T the particle moves a distance of $2\pi r$, which is equal to the circumference of the particle's circular path. Therefore, because its speed is equal to the circumference of the circular path divided by the period, or $v = 2\pi r/T$, it follows that

$$T \equiv \frac{2\pi r}{v} \qquad\qquad [3.20]$$

Thinking Physics 3

An airplane travels from Los Angeles to Sydney, Australia. For 15 hours after cruising altitude is reached, the instruments on the plane indicate that the ground speed holds rock-steady at 700 km/hr and that the heading of the airplane does not change. Is the velocity of the airplane constant?

Reasoning   The velocity is not constant, due to the curvature of the Earth. Even though the speed does not change and the heading is always toward Sydney (is this actually true?), the airplane is traveling around a significant portion of the circumference of the Earth. Thus, the direction of the velocity vector does change. We could extend this by imagining that the airplane passes over Sydney and continues (assuming it has enough fuel!) around the Earth until it arrives at Los Angeles again. It is impossible for an airplane to have a constant velocity (relative to the Universe, not to the Earth's surface) and return to its starting point.

EXERCISE 6   Find the acceleration of a particle moving with a constant speed of 8.0 m/s in a circle 2.0 m in radius.   Answer   32 m/s² toward the center of the circle

3.5 • TANGENTIAL AND RADIAL ACCELERATION

Let us consider the motion of a particle along a curved path where the velocity changes both in direction and in magnitude, as described in Figure 3.12. In this situation, the velocity vector is always tangent to the path; however, the acceleration vector **a** is at some angle to the path.

As the particle moves along the curved path in Figure 3.12, we see that the direction of the total acceleration vector, **a**, changes from point to point. This vector can be resolved into two components: a radial component, a_r, and a tangential component, a_t. The *total* acceleration vector, **a**, can be written as the vector sum of the component vectors:

$$\mathbf{a} = \mathbf{a}_r + \mathbf{a}_t \qquad\qquad [3.21]$$

• *Total acceleration*

The tangential acceleration arises from the change in the speed of the particle, and the projection of the tangential acceleration along the direction of the velocity is

$$a_t = \frac{d\,|\mathbf{v}|}{dt} \qquad\qquad [3.22]$$

• *Tangential acceleration*

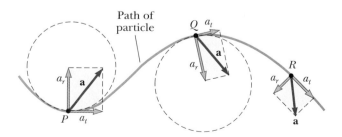

Path of particle

Figure 3.12   The motion of a particle along an arbitrary curved path lying in the *xy* plane. If the velocity vector **v** (always tangent to the path) changes in direction and magnitude, the acceleration **a** has a tangential component a_t and a radial component a_r.

The radial acceleration is due to the change in direction of the velocity vector and has an absolute magnitude given by

Radial or centripetal •
acceleration

$$a_r = \frac{v^2}{r}$$ [3.23]

where r is the radius of curvature of the path at the point in question. Because $\mathbf{a}_r$ and $\mathbf{a}_t$ are perpendicular component vectors of $\mathbf{a}$, it follows that $a - \sqrt{a_r^2 + a_t^2}$. As in the case of uniform circular motion, $\mathbf{a}_r$ always points toward the center of curvature, as shown in Figure 3.12. Also, at a given speed, a_r is large when the radius of curvature is small (as at points P and Q in Fig. 3.12) and small when r is large (such as at point R). The direction of $\mathbf{a}_t$ is either in the same direction as $\mathbf{v}$ (if v is increasing) or opposite $\mathbf{v}$ (if v is decreasing).

Note that in the case of uniform circular motion, where v is constant, $a_t = 0$ and the acceleration is always radial, as we described in Section 3.4. (*Note:* Eq. 3.23 is identical to Eq. 3.19.) In other words, uniform circular motion is a special case of motion along a curved path. Furthermore, if the direction of $\mathbf{v}$ does not change, then there is no radial acceleration and the motion is one-dimensional ($a_r = 0$, but a_t may not be zero).

Thinking Physics 4

In uniform circular motion, the acceleration vector is always perpendicular to the velocity vector.   (a) Suppose a particle is moving at uniform speed around an *elliptical* path. What is the relationship between the acceleration and velocity vectors in this case?   (b) Now, let the elliptical path be that of Halley's Comet in orbit around the Sun. This is *not* a case of uniform speed—the comet moves slowly when it is far from the Sun and rapidly when it is close. Are the acceleration and velocity vectors perpendicular in this case?

Reasoning   (a) We consider first the case of the particle moving at constant speed around the elliptical path (Fig. 3.13a). Because the speed is constant, there must be no component of the acceleration tangent to the path. Thus, the acceleration must be

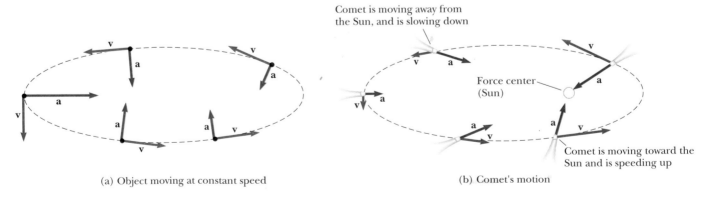

(a) Object moving at constant speed

(b) Comet's motion

Figure 3.13   (Thinking Physics 4)

perpendicular to the path and, therefore, always perpendicular to the velocity vector. The magnitude of the acceleration vector is not constant—it will be largest at the ends of the ellipse, because the velocity is changing direction the most rapidly there. Figure 3.13a shows the velocity and acceleration vectors at several locations around the path.

(b) In the case of Halley's Comet, or any other object in an elliptical orbit around a gravitational-force center, the direction of the acceleration is the same as the direction of the gravitational force on the object, which is toward the *focus* of the ellipse, where the gravitational force center lies (Fig. 3.13b). Thus, the acceleration and velocity vectors are not perpendicular, except for the points at the ends of the ellipse. As the comet moves away from the Sun, the acceleration has a component opposite to the velocity. As it moves toward the Sun, the acceleration has a component parallel to the velocity. This results in the slowing down of the comet as it moves away from the Sun and its speeding up as it moves toward the Sun. Notice in Figure 3.13b that the velocity vectors have large magnitudes when the comet is near the Sun and small magnitudes when it is far away.

Example 3.4   The Swinging Ball

A ball tied to the end of a string 0.50 m in length swings in a vertical circle under the influence of gravity, as in Figure 3.14. When the string makes an angle of $\theta = 20°$ with the vertical, the ball has a speed of 1.5 m/s. (a) Find the magnitude of the radial component of acceleration at this instant.

Solution   Because $v = 1.5$ m/s and $r = 0.50$ m, we find that

$$a_r = \frac{v^2}{r} = \frac{(1.5 \text{ m/s})^2}{0.50 \text{ m}} = 4.5 \text{ m/s}^2$$

(b) When the ball is at an angle θ to the vertical, it has a tangential acceleration of magnitude $g \sin \theta$ (the component of $\mathbf{g}$ tangent to the circle). Therefore, at $\theta = 20°$, $a_t = g \sin 20° = 3.4$ m/s^2. Find the magnitude and direction of the *total* acceleration at $\theta = 20°$.

Solution   Because $\mathbf{a} = \mathbf{a}_r + \mathbf{a}_t$, the magnitude of $\mathbf{a}$ at $\theta = 20°$ is

$$a = \sqrt{a_r^2 + a_t^2} = \sqrt{(4.5)^2 + (3.4)^2} \text{ m/s}^2 = 5.6 \text{ m/s}^2$$

If ϕ is the angle between $\mathbf{a}$ and the string, then

$$\phi = \tan^{-1} \frac{a_t}{a_r} = \tan^{-1} \left(\frac{3.4 \text{ m/s}^2}{4.5 \text{ m/s}^2} \right) = 36°$$

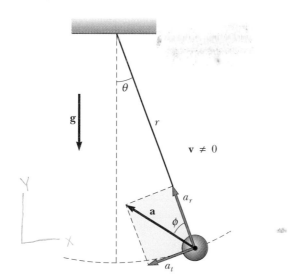

Figure 3.14   (Example 3.4) Motion of a ball suspended by a string of length r. The ball swings with nonuniform circular motion in a vertical plane, and its acceleration $\mathbf{a}$ has a radial component a_r and a tangential component a_t.

Note that all of the vectors—$\mathbf{a}$, $\mathbf{a}_t$, and $\mathbf{a}_r$—change in direction *and* magnitude as the ball swings through the circle. When the ball is at its lowest elevation ($\theta = 0$), $a_t = 0$, because there is no tangential component of $\mathbf{g}$ at this angle, and a_r is a *maximum* because v is a maximum. If the ball has enough speed to reach its highest position ($\theta = 180°$), a_t is again zero, but a_r is a minimum because v is now a minimum. Finally, in the two horizontal positions ($\theta = 90°$ and 270°), $|\mathbf{a}_t| = g$ and a_r has a value between its minimum and maximum values.

SUMMARY

If a particle moves with *constant* acceleration $\mathbf{a}$ and has velocity $\mathbf{v}_0$ and position $\mathbf{r}_0$ at $t = 0$, its velocity and position vectors at some later time t are

$$\mathbf{v} = \mathbf{v}_0 + \mathbf{a}t \qquad [3.8]$$

$$\mathbf{r} = \mathbf{r}_0 + \mathbf{v}_0 t + \tfrac{1}{2}\mathbf{a}t^2 \qquad [3.9]$$

For two-dimensional motion in the *xy* plane under constant acceleration, these vector expressions are equivalent to two component expressions, one for the motion along *x* and one for the motion along *y*.

Projectile motion is two-dimensional motion under constant acceleration, where $a_x = 0$ and $a_y = -g$. In this case, if $x_0 = y_0 = 0$, the components of Equations 3.8 and 3.9 reduce to

$$v_x = v_{x0} = \text{constant} \qquad [3.10]$$

$$v_y = v_{y0} - gt \qquad [3.11]$$

$$x = v_{x0}t \qquad [3.12]$$

$$y = v_{y0}t - \tfrac{1}{2}gt^2 \qquad [3.13]$$

where $v_{x0} = v_0 \cos \theta_0$, $v_{y0} = v_0 \sin \theta_0$, v_0 is the initial speed of the projectile, and θ_0 is the angle $\mathbf{v}_0$ makes with the positive *x* axis. Note that these expressions give the velocity components (and hence the velocity vector) and the coordinates (and hence the position vector) at *any* time t that the projectile is in motion.

It is useful to think of projectile motion as the superposition of two motions: (1) uniform motion in the *x* direction and (2) motion in the vertical direction subject to a constant downward acceleration of magnitude $g = 9.80 \text{ m/s}^2$.

A particle moving in a circle of radius *r* with constant speed *v* undergoes a centripetal (or radial) acceleration, $\mathbf{a}_r$, because the direction of $\mathbf{v}$ changes in time. The magnitude of $\mathbf{a}_r$ is

$$a_r = \frac{v^2}{r} \qquad [3.19]$$

and its direction is always toward the center of the circle.

If a particle moves along a curved path in such a way that the magnitude and direction of $\mathbf{v}$ change in time, the particle has an acceleration vector that can be described by two components: (1) a radial component, a_r, arising from the change in direction of $\mathbf{v}$, and (2) a tangential component, a_t, arising from the change in magnitude of $\mathbf{v}$.

CONCEPTUAL QUESTIONS

1. If you know the position vectors of a particle at two points along its path and also know the time it took to get from one point to the other, can you determine the particle's instantaneous velocity? Its average velocity? Explain.

2. Explain whether or not the following particles have an acceleration: (a) a particle moving in a straight line with constant speed and (b) a particle moving around a curve with constant speed.

3. Correct the following statement: "The racing car rounds the turn at a constant velocity of 90 miles per hour."

4. A spacecraft drifts through space at a constant velocity. Suddenly a gas leak in the side of the spacecraft causes a constant acceleration of the spacecraft in a direction perpendicular to the initial velocity. The orientation of the spacecraft does not change, so that the acceleration remains perpendicular to the original direction of the velocity. What

Boxed numbers indicate questions that have answers available in the Student Solutions Manual and Study Guide.

is the shape of the path followed by the spacecraft in this situation?

5. A ball is projected horizontally from the top of a building. One second later another ball is projected horizontally from the same point with the same velocity. At what point in the motion will the balls be closest to each other? Will the first ball always be traveling faster than the second ball? What will be the time difference between when the balls hit the ground? Can the horizontal projection velocity of the second ball be changed so that the balls arrive at the ground at the same time?

6. At the end of a pendulum's arc, its velocity is zero. Is its acceleration also zero at that point?

7. If a rock is dropped from the top of a sailboat's mast, will it hit the deck at the same point whether the boat is at rest or in motion at constant velocity?

8. Two projectiles are thrown with the same magnitude of initial velocity, one at an angle θ with respect to the level ground and the other at angle $90° - \theta$. Both projectiles will strike the ground at the same distance from the projection point. Will both projectiles be in the air for the same time interval?

9. A projectile is fired at some angle to the horizontal with some initial speed v_0, and air resistance is neglected. Is the projectile a freely falling body? What is its acceleration in the vertical direction? What is its acceleration in the horizontal direction?

10. State which of the following quantities, if any, remain constant as a projectile moves through its parabolic trajectory: (a) speed, (b) acceleration, (c) horizontal component of velocity, (d) vertical component of velocity.

11. The maximum range of a projectile occurs when it is launched at an angle of 45.0° with the horizontal, if air resistance is neglected. If air resistance is not neglected, will the optimum angle be greater or less than 45.0°? Explain.

12. An object moves in a circular path with constant speed v. (a) Is the velocity of the object constant? (b) Is its acceleration constant? Explain.

13. A projectile is fired on the Earth with some initial velocity. Another projectile is fired on the Moon with the *same* initial velocity. Neglecting air resistance, which projectile has the greater range? Which reaches the greater altitude? (Note that the free-fall acceleration on the Moon is about 1.6 m/s².)

14. A coin on a table is given an initial horizontal velocity such that it ultimately leaves the end of the table and hits the floor. At the instant the coin leaves the end of the table, a ball is released from the same height and falls to the floor. Explain why the two objects hit the floor simultaneously, even though the coin has an initial velocity.

15. Describe how a driver can steer a car traveling at constant speed so that (a) the acceleration is zero or (b) the magnitude of the acceleration remains constant.

16. An ice skater is executing a figure eight, consisting of two equal, tangent circular paths. Throughout the first loop she increases her speed uniformly, and during the second loop she moves at a constant speed. Make a sketch of her acceleration vector at several points along the path of motion.

17. Construct motion diagrams showing the velocity and acceleration of a projectile at several points along its path if (a) the projectile is fired horizontally and (a) the projectile is fired at an angle θ with the horizontal.

18. A baseball is thrown such that its initial x and y components of velocity are known. Neglecting air resistance, describe how you would calculate, at the instant the ball reaches the top of its trajectory, (a) its coordinates, (b) its velocity, and (c) its acceleration. How would these results change if air resistance were taken into account?

PROBLEMS

Section 3.1 The Displacement, Velocity, and Acceleration Vectors

1. A motorist drives south at 20.0 m/s for 3.00 min, then turns west and travels at 25.0 m/s for 2.00 min, and finally travels northwest at 30.0 m/s for 1.00 min. For this 6.00-min trip, find (a) the net vector displacement of the motorist, (b) the motorist's average speed, and (c) the average velocity of the motorist.

2. Suppose that the position-vector function for a particle is given as $\mathbf{r}(t) = x(t)\mathbf{i} + y(t)\mathbf{j}$, with $x(t) = at + b$ and $y(t) = ct^2 + d$, where $a = 1.00$ m/s, $b = 1.00$ m, $c =$

0.125 m/s², and $d = 1.00$ m. (a) Calculate the average velocity during the time interval $t = 2.00$ s to $t = 4.00$ s. (b) Determine the velocity and speed at $t = 2.00$ s.

3. A golf ball is hit off a tee at the edge of a cliff. Its x and y coordinates versus time are given by the following expressions:

$$x = (18.0 \text{ m/s})t \qquad y = (4.00 \text{ m/s})t - (4.90 \text{ m/s}^2)t^2$$

(a) Write a vector expression for the position $\mathbf{r}$ versus time t, using the unit vectors $\mathbf{i}$ and $\mathbf{j}$. By using derivatives, repeat

Boxed numbers indicate problems that have full solutions available in the Student Solutions Manual and Study Guide. Cyan numbers indicate intermediate level problems and magenta numbers indicate more challenging problems.

for (b) the velocity vector **v** versus time and (c) the acceleration vector **a** versus time. (d) Find the x and y coordinates of the golf ball at $t = 3.00$ s. Using the unit vectors **i** and **j**, write expressions for (e) the velocity **v** and (f) the acceleration **a** at the instant $t = 3.00$ s.

4. The coordinates of an object moving in the xy plane vary with time according to the equations $x = -(5.00$ m$) \sin(t)$ and $y = (4.00$ m$) - (5.00$ m$) \cos(t)$, where t is in seconds. (a) Determine the components of velocity and components of acceleration at $t = 0$ s. (b) Write expressions for the position vector, the velocity vector, and the acceleration vector at any time $t > 0$. (c) Describe the path of the object in an xy plot.

Section 3.2   Two-Dimensional Motion with Constant Acceleration

5. At $t = 0$, a particle moving in the xy plane with constant acceleration has a velocity of $\mathbf{v}_0 = (3.00\mathbf{i} - 2.00\mathbf{j})$ m/s at the origin. At $t = 3.00$ s, the particle's velocity is $\mathbf{v} = (9.00\mathbf{i} + 7.00\mathbf{j})$ m/s. Find (a) the acceleration of the particle and (b) its coordinates at any time t.

6. The vector position of a particle varies in time according to the expression $\mathbf{r} = (3.00\mathbf{i} - 6.00t^2\mathbf{j})$ m. (a) Find expressions for the velocity and acceleration as functions of time. (b) Determine the particle's position and velocity at $t = 1.00$ s.

7. A fish swimming in a horizontal plane has velocity $\mathbf{v}_0 = (4.00\mathbf{i} + 1.00\mathbf{j})$ m/s when its displacement from a certain rock is $\mathbf{r}_0 = (10.0\mathbf{i} - 4.00\mathbf{j})$ m. After the fish swims with constant acceleration for 20.0 s, its velocity is $\mathbf{v} = (20.0\mathbf{i} - 5.00\mathbf{j})$ m/s. (a) What are the components of the acceleration? (b) What is the direction of the acceleration with respect to unit vector **i**? (c) Where is the fish at $t = 25.0$ s and in what direction is it moving?

8. A particle initially located at the origin has an acceleration of $\mathbf{a} = 3.00\mathbf{j}$ m/s^2 and an initial velocity of $\mathbf{v}_0 = 5.00\mathbf{i}$ m/s. Find (a) the vector position and velocity at any time t and (b) the coordinates and speed of the particle at $t = 2.00$ s.

Section 3.3   Projectile Motion

(Neglect air resistance in all problems. Use $g = 9.80$ m/s^2.)

9. In a local bar, a customer slides an empty beer mug on the counter for a refill. The bartender is momentarily distracted and does not see the mug, which slides off the counter and strikes the floor 1.40 m from the base of the counter. If the height of the counter is 0.860 m, (a) with what velocity did the mug leave the counter, and (b) what was the direction of the mug's velocity just before it hit the floor?

10. A student decides to measure the muzzle velocity of a pellet from his gun. He points the gun horizontally. He places a target on a vertical wall a distance x away from the gun. The pellet hits the target a vertical distance y below the gun. (a) Show that the position of the pellet when traveling through the air is given by $y = Ax^2$, where A is a constant. (b) Express the constant A in terms of the initial velocity and the free-fall acceleration. (c) If $x = 3.00$ m and $y = 0.210$ m, what is the initial speed of the pellet?

11. A golfer wants to drive a golf ball a distance of 310 yards (283 m). If the four-wood launches the ball at 15.0° above the horizontal, what must be the initial speed of the ball to achieve the required distance?

12. One strategy in a snowball fight is to throw a snowball at a high angle over level ground. While your opponent is watching the first one, you throw a second snowball at a low angle timed to arrive before or at the same time as the first one. Assume both snowballs are thrown with a speed of 25.0 m/s. The first one is thrown at an angle of 70.0° with respect to the horizontal. (a) At what angle should the second snowball be thrown to arrive at the same point as the first? (b) How many seconds later should the second snowball be thrown after the first to arrive at the same time?

13. A tennis player standing 12.6 m from the net hits the ball at 3.00° above the horizontal. To clear the net, the ball must rise at least 0.330 m. If the ball just clears the net at the apex of its trajectory, how fast was the ball moving when it left the racket?

14. An artillery shell is fired with an initial velocity of 300 m/s at 55.0° above the horizontal. It explodes on a mountainside 42.0 s after firing. What are the x and y coordinates of the shell where it explodes relative to its firing point?

15. An astronaut on a strange planet finds that she can jump a *maximum* horizontal distance of 15.0 m if her initial speed is 3.00 m/s. What is the free-fall acceleration on the planet?

16. A ball is tossed from an upper-story window. The ball is given an initial velocity of 8.00 m/s at an angle of 20.0° below the horizontal. It strikes the ground 3.00 s later. (a) How far horizontally from the base of the building does the ball strike the ground? (b) Find the height from which the ball was thrown. (c) How long does it take the ball to reach a point 10.0 m below the level of launching?

17. A ball is thrown horizontally from the top of a building 35.0 m high. The ball strikes the ground at a point 80.0 m from the base of the building. Find (a) the time the ball is in flight, (b) its initial velocity, and (c) the x and y components of velocity just before the ball strikes the ground.

18. A cannon with a muzzle speed of 1000 m/s is used to destroy a target on a mountain top. The target is 2000 m from the cannon horizontally and 800 m above the ground. At what angle, relative to the ground, should the cannon be fired?

19. A placekicker must kick a football from a point 36.0 m (about 40 yards) from the goal and the ball must clear the crossbar, which is 3.05 m high. The ball leaves the ground when kicked with a speed of 20.0 m/s at an angle of 53.0°

to the horizontal. (a) By how much does the ball clear or fall short of clearing the crossbar? (b) Does the ball approach the crossbar while still rising or while falling?

20. A fireman, 50.0 m away from a burning building, directs a stream of water from a fire hose at an angle of 30.0° above the horizontal. If the speed of the stream is 40.0 m/s, at what height will the stream of water strike the building?

21. The speed of a projectile when it reaches its maximum height is one half of its speed when it is at half its maximum height. What is the initial projection angle of the projectile?

22. A soccer player kicks a rock horizontally off a 40.0-m–high cliff into a pool of water. If the player hears the sound of the splash 3.00 s later, what was the initial speed given to the rock? Assume the speed of sound in air to be 343 m/s.

23. During the 1968 Olympics in Mexico City, Bob Beamon executed a record long jump. The horizontal distance he achieved was 8.90 m. His center of gravity started at an elevation of 1.00 m, reached a maximum height of 1.90 m, and finished at 0.150 m. From these data determine (a) his time of flight, (b) his horizontal and vertical velocity components at the takeoff time, and (c) his takeoff angle.

Section 3.4   Uniform Circular Motion

24. The orbit of the Moon around the Earth is approximately circular, with a mean radius of 3.84×10^8 m. It takes 27.3 days for the Moon to complete one revolution around the Earth. Find (a) the mean orbital speed of the Moon and (b) its centripetal acceleration.

25. An athlete rotates a 1.00-kg discus along a circular path of radius 1.06 m (Figure P3.25). The maximum speed of the discus is 20.0 m/s. Determine the magnitude of the maximum radial acceleration of the discus.

26. A tire 0.500 m in radius rotates at a constant rate of 200 rev/min. Find the speed and magnitude of acceleration of a small stone lodged in the outer edge of the tread of the tire.

27. Young David who slew Goliath experimented with slings before tackling the giant. He found that he could revolve a sling of length 0.600 m at the rate of 8.00 rev/s. If he increased the length to 0.900 m, he could revolve the sling only 6.00 times per second. (a) Which rate of rotation gives the greater linear speed? (b) What is the centripetal acceleration at 8.00 rev/s? (c) What is the centripetal acceleration at 6.00 rev/s?

28. An astronaut standing on the Moon fires a gun so that the bullet leaves the barrel initially moving in a horizontal direction. (a) What must be the muzzle speed of the bullet so that it travels completely around the Moon and returns to its original location? (b) How long is the bullet in flight? Assume that the free-fall acceleration on the Moon is one sixth that on the Earth.

29. If the rotation of the Earth increased so that the centripetal acceleration was equal to the gravitational acceleration at

Figure P3.25

the Equator, (a) what would be the tangential speed of a person standing at the Equator, and (b) how long would a day be?

Section 3.5   Tangential and Radial Acceleration

30. An automobile, the speed of which is increasing at a rate of 0.600 m/s², travels along a circular road of radius 20.0 m. When the instantaneous speed of the automobile is 4.00 m/s, find (a) the tangential acceleration component, (b) the centripetal acceleration component, and (c) the magnitude and direction of the total acceleration.

31. A train slows down as it rounds a sharp horizontal turn, slowing from 90.0 km/h to 50.0 km/h in the 15.0 s that it takes to round the bend. The radius of the curve is 150 m. Compute the acceleration at the moment the train speed reaches 50.0 km/h, assuming it continues to slow down at this time.

32. A point on a rotating turntable 20.0 cm from the center accelerates from rest to a final speed of 0.700 m/s in 1.75 s. At $t = 1.25$ s, find the magnitude and direction of

(a) the centripetal acceleration, (b) the tangential acceleration, and (c) the total acceleration of the point.

33. Figure P3.33 represents the total acceleration of a particle moving clockwise in a circle of radius 2.50 m at a given instant of time. At this instant, find (a) the centripetal acceleration, (b) the speed of the particle, and (c) its tangential acceleration.

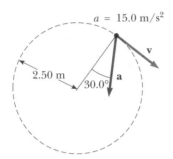

$a = 15.0 \text{ m/s}^2$

2.50 m

30.0°

Figure P3.33

34. A ball swings in a vertical circle at the end of a 1.50-m long rope. When it is 36.9° past the lowest point on its way up, the ball's total acceleration is $(-22.5\mathbf{i} + 20.2\mathbf{j})$ m/s². At that instant, (a) sketch a vector diagram showing the components of its acceleration, (b) determine the magnitude of its centripetal acceleration, and (c) determine the speed and velocity of the ball.

Additional Problems

35. At $t = 0$ a particle leaves the origin with a velocity of 6.00 m/s in the positive y direction. Its acceleration is given by $\mathbf{a} = (2.00\mathbf{i} - 3.00\mathbf{j})$ m/s². When the particle reaches its *maximum* y coordinate, its y component of velocity is zero. At this instant, find (a) the velocity of the particle and (b) its x and y coordinates.

36. A ball on the end of a string is whirled around in a horizontal circle of radius 0.300 m. The plane of the circle is 1.20 m above the ground. The string breaks and the ball lands 2.00 m (horizontally) away from the point on the ground directly beneath the ball's location when the string breaks. Find the centripetal acceleration of the ball during its circular motion.

37. A home run is hit in such a way that the baseball just clears a wall 21.0 m high, located 130 m from home plate. The ball is hit at an angle of 35.0° to the horizontal, and air resistance is negligible. Find (a) the initial speed of the ball, (b) the time it takes the ball to reach the wall, and (c) the velocity components and the speed of the ball when it reaches the wall. (Assume the ball is hit at a height of 1.00 m above the ground.)

38. The astronaut orbiting the Earth in the photograph is preparing to dock with a spinning Westar VII satellite. The satellite is in a circular orbit 600 km above the Earth's surface, where the free-fall acceleration is 8.21 m/s². The radius of the Earth is 6400 km. Determine the speed of the satellite and the time required to complete one orbit around the Earth.

Figure P3.38   *(Courtesy of NASA)*

39. A particle has velocity components

$$v_x = +4.00 \text{ m/s} \qquad v_y = -(6.00 \text{ m/s}^2)t + 4.00 \text{ m/s}$$

Calculate the speed of the particle and the direction $\theta = \tan^{-1}(v_y/v_x)$ of the velocity vector at $t = 2.00$ s.

40. When baseball outfielders throw the ball in, they usually allow it to take one bounce on the theory that the ball arrives sooner this way. Suppose that after the bounce the ball rebounds at the same angle θ as before, as in Figure P3.40 but loses half its speed. (a) Assuming the ball is always

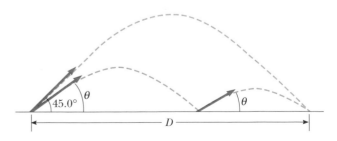

45.0° θ θ D

Figure P3.40

thrown with the same initial speed, at what angle θ should the ball be thrown in order to go the same distance D with one bounce (blue path) as one thrown upward at 45.0° with no bounce (green path)? (b) Determine the ratio of the times for the one-bounce and no-bounce throws.

41. A girl can throw a ball a maximum horizontal distance of 40.0 m on a level field. How far can she throw the same ball vertically upward? Assume that her muscles give the ball the same speed in each case.

42. A girl can throw a ball a maximum horizontal distance of R on a level field. How far can she throw the same ball vertically upward? Assume that her muscles give the ball the same speed in each case.

43. A bomber is flown horizontally, with a ground speed of 275 m/s, at an altitude of 3000 m over level terrain. Neglect the effects of air resistance. (a) How far from the point that is vertically under the point of release will a bomb hit the ground? (b) If the plane maintains its original course and speed, where will it be when the bomb hits the ground? (c) For the preceding conditions, at what angle from the vertical at the point of release must the telescopic bomb sight be set so that the bomb will hit the target seen in the sight at the time of release?

44. A football is thrown toward a receiver with an initial speed of 20.0 m/s, at an angle of 30.0° above the horizontal. At that instant, the receiver is 20.0 m from the quarterback. In what direction and with what constant speed should the receiver run in order to catch the football at the level at which it was thrown?

45. A hawk is flying horizontally at 10.0 m/s in a straight line, 200 m above the ground. A mouse it has been carrying is released from its grasp. The hawk continues on its path at the same speed for 2.00 s before attempting to retrieve its prey. To accomplish the retrieval, it dives in a straight line at constant speed and recaptures the mouse 3.00 m above the ground. (a) Assuming no air resistance, find the diving speed of the hawk. (b) What angle did the hawk make with the horizontal during its descent? (c) For how long did the mouse "enjoy" free fall?

46. A rocket is launched at an angle of 53.0° to the horizontal with an initial speed of 100 m/s. For 3.00 s it moves along its initial line of motion with an acceleration of 30.0 m/s². Then its engines fail and the rocket proceeds to move in free fall. Find (a) the maximum altitude reached by the rocket, (b) its total time of flight, and (c) its horizontal range.

47. A car is parked on a steep incline overlooking the ocean, where the incline makes an angle of 37.0° with the horizontal. The negligent driver leaves the car in neutral, and the parking brakes are defective. The car rolls from rest down the incline with a constant acceleration of 4.00 m/s², traveling 50.0 m to the edge of a cliff. The cliff is 30.0 m above the ocean. Find (a) the speed of the car when it reaches the edge of the cliff and the time it takes to get there,

(b) the velocity of the car when it lands in the ocean, (c) the total time the car is in motion, and (d) the position of the car when it lands in the ocean, relative to the base of the cliff.

48. The determined coyote is out once more to try to capture the elusive roadrunner. The coyote wears a pair of Acme jet-powered roller skates, which provide a constant horizontal acceleration of 15.0 m/s², as pictured in Figure P3.48. The coyote starts off at rest 70.0 m from the edge of a cliff at the instant the roadrunner zips past him in the direction of the cliff. (a) If the roadrunner moves with constant speed, determine the minimum speed he must have in order to reach the cliff before the coyote. (b) If the cliff is 100 m above the base of a canyon, determine where the coyote lands in the canyon (assume his skates are still in operation when he is in flight). (c) Determine the coyote's velocity components just before he lands in the canyon. (As usual, the roadrunner saves himself by making a sudden turn at the cliff.)

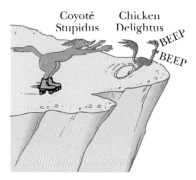

Figure P3.48

49. A skier leaves the ramp of a ski jump with a velocity of 10.0 m/s, 15.0° above the horizontal, as in Figure P3.49. The slope is inclined at 50.0°, and air resistance is negligi-

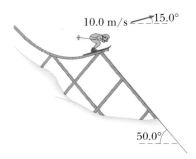

Figure P3.49

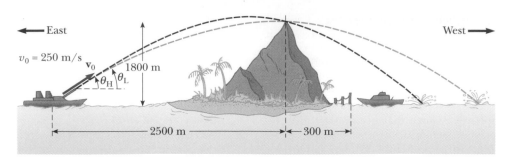

Figure P3.51

ble. Find (a) the distance from the ramp to where the jumper lands and (b) the velocity components just before the landing. (How do you think the results might be affected if air resistance were included? Note that jumpers lean forward in the shape of an airfoil, with their hands at their sides, to increase their distance. Why does this work?)

50. Describe what you might do to give your hand a large acceleration without hurting yourself or striking your hand against anything. Compute an order-of-magnitude estimate of this acceleration, stating the quantities you measure or estimate and their values.

51. An enemy ship is on the east side of a mountain island, as shown in Figure P3.51. The enemy ship can maneuver to within 2500 m of the 1800-m–high mountain peak and can shoot projectiles with an initial speed of 250 m/s. If the western shoreline is horizontally 300 m from the peak, what are the distances from the western shore at which a ship can be safe from the bombardment of the enemy ship?

▣ Spreadsheet Problems

S1. The punter for the St. Louis Rams needs to kick the football 40 yd to 45 yd (36.6 m to 41.1 m) to keep the opposing team inside the 5-yd line. Use Spreadsheet 3.1 to estimate the angle at which he should kick the ball. *Note:* He must keep the ball in the air as long as possible in order to help prevent a runback (in other words, he wants a long "hang time"). The initial speed of the ball should be in the range of 20 m/s to 30 m/s.

S2. Modify Spreadsheet 3.1 by adding two columns and then calculate the maximum height reached by the football in Problem S1 and the total time it is in the air.

S3. The x and y coordinates of a projectile are given by Equations 3.12 and 3.13. Using these equations, set up a spreadsheet to calculate the x and y coordinates of a projectile and the components of its velocity v_x and v_y as functions of time. The initial speed and initial angle of the projectile should be input parameters. Use the graph function of the spreadsheet to plot the x and y coordinates as functions of time. Also graph v_x and v_y as functions of time. Use this spreadsheet to solve the following problem.

The New York Giants are tied with the Chicago Bears with only a few seconds left in the game. The Giants have the football and call the placekicker into the game. He must kick the ball 52 yd (47.5 m) for a field goal. If the crossbar of the goal is 10 ft (3.05 m) high, at what angle and speed should he kick the ball to score the winning points? Is there only one solution to this problem? Solutions can be found by varying the parameters and studying the graph of y versus x.

ANSWERS TO CONCEPTUAL PROBLEMS

1. (a) Yes. Although its speed may be constant, the direction of its motion (that is, the direction of **v**) may change, causing an acceleration. For example, an object moving in a circle with constant speed has an acceleration directed toward the center of the circle. (b) No. An object that moves with constant velocity has zero acceleration. Constant velocity means that both the direction and magnitude of **v** remain constant.

2. A constant velocity is interpreted as constant in both magnitude and direction. Thus, a constant velocity clearly implies a nonvarying speed. On the other hand, a constant speed indicates that only the magnitude of the velocity vector is constant—the direction can change. Thus, as long as a particle covers the same distance in equal time intervals, the speed is constant, and its direction can change continuously, giving a varying velocity. A familiar example is a particle moving in a circular path at constant speed, such that the velocity is constantly changing direction.

3. You should simply throw it straight up in the air. Because the ball was moving along with you, it would follow a parabolic trajectory with a horizontal component of velocity that is the same as yours.

4. (a) At the top of its flight, **v** is horizontal and **a** is vertical. This is the only point at which the velocity and acceleration vectors are perpendicular. (b) If the object is thrown straight up or down, then **v** and **a** will be parallel throughout the downward motion. Otherwise, the velocity and acceleration vectors are never parallel.

The Wizard of Id   **by Parker and Hart**

By permission of John Hart and Field Enterprises, Inc.

4

The Laws of Motion

In the preceding two chapters on kinematics, we described the motions of particles based on the definitions of displacement, velocity, and acceleration. However, we would like to be able to answer specific questions related to the causes of motion, such as, "What mechanism causes motion?" and, "Why do some objects accelerate at higher rates than others?" In this chapter we shall describe the change in motion of particles using the concepts of force and mass. We then discuss the three fundamental laws of motion, which are based on experimental observations and were formulated about three centuries ago by Sir Isaac Newton.

Classical mechanics describes the relationship between the motion of a body and the forces acting on it. Classical mechanics deals only with objects that are large compared with the dimensions of atoms ($\sim 10^{-10}$ m) and that move at speeds much slower than the speed of light (3.00×10^8 m/s).

We learn in this chapter how to describe the acceleration of an object in terms of the resultant force acting on the object and its mass. This force represents the interaction of the object with its environment. Mass is a measure of the object's tendency to resist an acceleration when a force acts on it.

We also discuss *force laws*, which describe how to

Calf-roping, as shown in this photograph taken in Steamboat, Colorado, is a standard rodeo event. The external forces exerted on the horse are the force of friction between the horse and ground, the downward force of gravity, the force exerted by the rope attached to the calf, the downward force exerted by the cowboy on the horse, and the upward force of the ground. Can you identify the forces acting on the calf? *(Fourbyfive, Inc.)*

calculate the force on an object if its environment is known. We shall see that, although the force laws are simple in form, they successfully explain a wide variety of phenomena and experimental observations. They, together with the laws of motion, are the foundations of classical mechanics.

4.1 • THE CONCEPT OF FORCE

Everyone has a basic understanding of the concept of force as a result of everyday experiences. When you push or pull an object, you exert a force on it. You exert a force when you throw or kick a ball. In these examples, the word *force* is associated with the result of muscular activity and with some change in the state of motion of an object. Forces do not always cause an object to move, however. For example, as you sit reading this book, the force of gravity acts on your body, and yet you remain stationary. Likewise, you can push on a block of stone and yet fail to move it.

What force (if any) causes a distant star to drift freely through space? Newton answered such questions by stating that the change in velocity of an object is caused by unbalanced forces. Therefore, if an object moves with uniform motion (constant velocity), no force is required to maintain the motion. Because only a force can cause a change in velocity, we can think of force as that which causes a body to accelerate.

• *A body accelerates due to an external force.*

Now consider a situation in which several forces act simultaneously on an object. In this case, the object accelerates only if the **net force** acting on it is not equal to zero. We shall often refer to the net force as the *resultant force* or as the *unbalanced force*. **If the net force is zero, the acceleration is zero and the velocity of the object remains constant.** That is, if the net force acting on the object is zero, either the object will be at rest or it will move with constant velocity. **When a body has constant velocity or is at rest, it is said to be in equilibrium.**

• *Definition of equilibrium*

This chapter is concerned with the relation between the force on an object and the acceleration of that object. If you pull on a spring, as in Figure 4.1a, the spring stretches. If the spring is calibrated, the distance it stretches can be used to measure the strength of the force. If a child pulls hard enough on a cart to overcome friction, as in Figure 4.1b, the cart moves. When a football is kicked, as in Figure 4.1c, it is both deformed and set in motion. These are all examples of a class of forces called *contact forces*. That is, they represent the result of physical contact between two objects.

Another class of forces, which do not involve physical contact between two objects but act through empty space, are known as *field forces*. The force of gravitational attraction between two objects is an example of this class of force, illustrated in Figure 4.1d. This gravitational force keeps objects bound to the Earth and gives rise to what we commonly call the *weight* of an object. The planets of our solar system are bound under the action of gravitational forces. Another common example of a field force is the electric force that one electric charge exerts on another electric charge, as in Figure 4.1e. These charges might be an electron and proton forming a hydrogen atom. A third example of a field force is the force that a bar magnet exerts on a piece of iron, as shown in Figure 4.1f. The forces holding an atomic nucleus together are also field forces but are usually very short-range. They are the dominating interaction for particle separations of the order of 10^{-15} m.

Early scientists, including Newton, were uneasy with the concept of a force acting between two disconnected objects. To overcome this conceptual problem,

A football is set in motion as a result of the contact force **F** on it due to the kicker's foot. The ball is distorted in the short time it is in contact with the foot.   *(Ralph Cowan, Tony Stone Worldwide)*

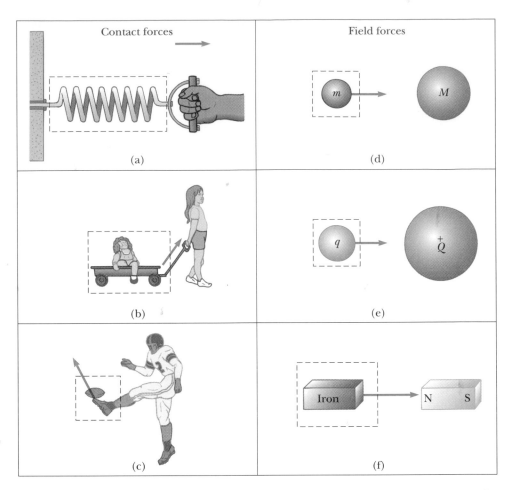

Figure 4.1   Some examples of forces applied to various objects. In each case a force is exerted on the particle or object within the boxed area. The environment external to the boxed area provides this force.

Michael Faraday (1791–1867) introduced the concept of a *field*. According to this approach, when a mass m_1 is placed at some point P near a mass m_2, one can say that m_1 interacts with m_2 by virtue of the gravitational field that exists at P. The field at P is created by mass m_2. Likewise, a field exists at the position of m_2 created by m_1. In fact, all objects create a gravitational field in the space around them.

The distinction between contact forces and field forces is not as sharp, however, as the previous section might lead you to believe. At the atomic level, all the forces we classify as contact forces turn out to be due to repulsive electrical (field) forces of the type illustrated in Figure 4.1e. Nevertheless, in developing models for macroscopic phenomena, it is convenient to use both classifications of forces. However, *Fundamental forces in nature* • the only known *fundamental* forces in nature are all field forces: (1) gravitational attractions between objects, (2) electromagnetic forces between charges, (3) strong nuclear forces between subatomic particles, and (4) weak nuclear forces that arise in certain radioactive decay processes. In classical physics we are concerned only with gravitational and electromagnetic forces.

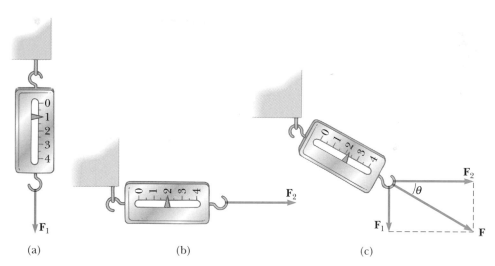

Figure 4.2   The vector nature of a force is tested with a spring scale. (a) The downward vertical force $\mathbf{F}_1$ elongates the spring 1 unit. (b) The horizontal force $\mathbf{F}_2$ elongates the spring 2 units. (c) The combination of $\mathbf{F}_1$ and $\mathbf{F}_2$ elongates the spring $\sqrt{1^2 + 2^2} = \sqrt{5}$ units.

It is sometimes convenient to use the deformation of a spring to measure force. Suppose a force is applied vertically to a spring that has a fixed upper end, as in Figure 4.2a. We can calibrate the spring by defining the unit force, $\mathbf{F}_1$, as the force that produces an elongation of 1.00 cm. If a force $\mathbf{F}_2$, applied horizontally as in Figure 4.2b, produces an elongation of 2.00 cm, the magnitude of $\mathbf{F}_2$ is 2 units. If the two forces $\mathbf{F}_1$ and $\mathbf{F}_2$ are applied simultaneously, as in Figure 4.2c, the elongation of the spring is $\sqrt{5} = 2.24$ cm. The single force $\mathbf{F}$ that would produce this same elongation is the vector sum of $\mathbf{F}_1$ and $\mathbf{F}_2$, as described in Figure 4.2c. That is, $|\mathbf{F}| = \sqrt{F_1{}^2 + F_2{}^2} = \sqrt{5}$ units, and its direction is $\theta = \arctan(-0.500) = -26.6°$. **Because forces are vectors, you must use the rules of vector addition to get the resultant force on a body.**

Springs that elongate in proportion to an applied force are said to obey **Hooke's law.** Such springs can be constructed and calibrated to measure unknown forces.

4.2 • NEWTON'S FIRST LAW AND INERTIAL FRAMES

Before about 1600, scientists believed that the natural state of matter was the state of rest. Galileo was the first to take a different approach to motion and the natural state of matter. He devised thought experiments, such as an object moving on a frictionless surface, and concluded that it is not the nature of an object to stop once it is set in motion; rather, it is an object's nature to resist changes in its motion. In his words, "Any velocity, once imparted to a moving body, will be rigidly maintained as long as the external causes of retardation are removed."

This new approach to motion was later formalized by Newton in a statement that has come to be known as **Newton's first law of motion:**

An object at rest remains at rest, and an object in motion continues in motion with a constant velocity (that is, constant speed in a straight line), unless it experiences a net external force.

A statement of Newton's first •
law

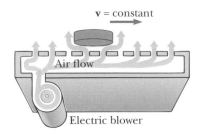

v = constant

Air flow

Electric blower

Figure 4.3   A disk moving on a layer of air is an example of uniform motion—that is, motion in which the acceleration is zero.

In simpler terms, we can say that **when the net force on a body is zero, its acceleration is zero.** That is, when $\Sigma \mathbf{F} = 0$, then $\mathbf{a} = 0$. From the first law, we conclude that an isolated body (a body that does not interact with its environment) is either at rest or moving with constant velocity.

One example of uniform motion on a nearly frictionless plane is the motion of a light disk on a layer of air, as in Figure 4.3. If the disk is given an initial velocity, it coasts a great distance before coming to rest. This idea is used in the game of air hockey, where the disk makes many collisions with the walls before coming to rest.

As a second example, consider a spaceship traveling in space, far removed from any planets or other matter. The spaceship requires some propulsion system to change its velocity. However, if the propulsion system is turned off when the spaceship reaches a velocity $\mathbf{v}$, the spaceship "coasts" in space with that velocity, and the astronauts get a free ride (that is, no propulsion system is required to keep them moving at the velocity $\mathbf{v}$).

Inertial Frames

Inertial frame •

Newton's first law, sometimes called the *law of inertia,* defines a special set of reference frames called inertial frames. **An inertial frame of reference is one in which Newton's first law is valid.** Any reference frame that moves with constant velocity with respect to an inertial frame is itself an inertial frame. A reference frame that moves with constant velocity relative to the distant stars is the best approximation of an inertial frame. The Earth is not an inertial frame because of its orbital motion around the Sun and rotational motion around its own axis. As the Earth travels in its nearly circular orbit around the Sun, it experiences a centripetal acceleration of about $4.4 \times 10^{-3} \text{ m/s}^2$ toward the Sun. In addition, because the Earth rotates around its own axis once every 24 h, a point on the Equator experiences an additional centripetal acceleration of $3.37 \times 10^{-2} \text{ m/s}^2$ toward the center of the Earth. However, these accelerations are small compared with g and can often be neglected. In most situations **we shall assume that a frame on or near the Earth's surface is an inertial frame.**

Thus, if an object is moving with constant velocity, an observer in one inertial frame (say, one at rest with respect to the object) will claim that the acceleration and the resultant force on the object are zero. An observer in *any other* inertial frame will also find that $\mathbf{a} = 0$ and $\mathbf{F} = 0$ for the object. According to the first law, a body at rest and one moving with constant velocity are equivalent. Unless stated otherwise, we shall usually write the laws of motion with respect to an observer "at rest" in an inertial frame.

Thinking Physics 1

When the Copernican Theory of the Solar System was proposed, a natural question arose: "What keeps the Earth and other planets moving in their paths around the Sun?" An interesting response to this question comes from Richard Feynman[1]: "In those days, one of the theories proposed was that the planets went around because behind them there were invisible angels, beating their wings and driving the planets forward. . . . It turns out that in order to keep the planets going around, the invisible angels must fly in a different direction. . . ." What did Feynman mean by this?

Reasoning First, the question asked by those at the time of Copernicus indicates that they did not have a proper understanding of inertia, as described by Newton's first law. At that time in history, before Galileo and Newton, the interpretation was that *motion* was caused by force. This is different from our current understanding that *changes in motion* are caused by force. Thus, it was natural for Copernicus' contemporaries to ask what force propelled a planet in its orbit. According to our current understanding, it is equally natural for us to realize that no force is necessary—the motion simply continues due to inertia. Thus, in Feynman's imagery, we do not need the angels pushing the planet *from behind*. We do need the angels to push *inward*, however, to provide the centripetal acceleration needed to change its direction into a circle. Of course, the angels are not real, from a scientific point of view—they are a metaphor for the *gravitational force*.

CONCEPTUAL PROBLEM 1

Is it possible to have motion in the absence of a force?

CONCEPTUAL PROBLEM 2

If a single force acts on an object, is there an acceleration? If an object experiences an acceleration, is a force acting on it? If an object experiences no acceleration, is there no force acting?

4.3 • INERTIAL MASS

If you attempt to change the velocity of an object, the object resists this change. **Inertia** is solely a property of an individual object; it is a measure of the response of an object to an external force. For instance, consider two large, solid cylinders of equal size, one balsa wood and the other steel. If you were to push the cylinders along a horizontal, rough surface, the force required to give the steel cylinder some acceleration would be larger than the force needed to give the balsa wood cylinder the same acceleration. Therefore, we say that the steel cylinder has more inertia than the balsa wood cylinder.

• *Inertia*

Mass is used to measure inertia, and the SI unit of mass is the kilogram. The greater the mass of a body, the less it accelerates (changes its state of motion) under the action of an applied force.

A quantitative measurement of mass can be made by comparing the accelera-

[1]R. P. Feynman, R. B. Leighton, and M. Sands, *The Feynman Lectures on Physics*, Vol. 1, Reading, Massachusetts, 1963, Addison-Wesley Publishing Co., p. 7-2.

tions that a given force produces on different bodies. Suppose a force acting on a body of mass m_1 produces an acceleration $\mathbf{a}_1$, and the *same force* acting on a body of mass m_2 produces an acceleration $\mathbf{a}_2$. The ratio of the two masses is defined as the *inverse* ratio of the magnitudes of the accelerations produced by the same force:

$$\frac{m_1}{m_2} \equiv \frac{a_2}{a_1}$$

[4.1]

If one mass is standard and known—say, 1 kg—the mass of an unknown object can be obtained from acceleration measurements. For example, if the standard 1-kg mass undergoes an acceleration of 3 m/s^2 under the influence of some force, a 2-kg mass will undergo an acceleration of 1.5 m/s^2 under the action of the same force.

Mass is an inherent property of a body, independent of the body's surroundings and of the method used to measure it. It is an experimental fact that **mass is a scalar quantity.** Finally, **mass is a quantity that obeys the rules of ordinary arithmetic.** That is, several masses can be combined in simple numerical fashion. For example, if you combine a 3-kg mass with a 5-kg mass, their total mass is 8 kg. This can be verified experimentally by comparing the acceleration of each object produced by a known force with the acceleration of the combined system using the same force.

Mass should not be confused with weight. **Mass and weight are two different quantities.** The weight of a body is equal to the magnitude of the force exerted by the Earth on the body and varies with location. For example, a person who weighs 180 lb on Earth weighs only about 30 lb on the Moon. However, the mass of a body is the same everywhere. An object having a mass of 2 kg on Earth also has a mass of 2 kg on the Moon.

Isaac Newton (1642–1727)

An English physicist and mathematician, Newton was one of the most brilliant scientists in history. Before the age of 30, he formulated the basic concepts and laws of mechanics, discovered the law of universal gravitation, and invented the mathematical methods of calculus. As a consequence of his theories, Newton was able to explain the motions of the planets, the ebb and flow of the tides, and many special features of the motions of the Moon and the Earth. He also interpreted many fundamental observations concerning the nature of light. His contributions to physical theories dominated scientific thought for two centuries and remain important today.   *(Giraudon/Art Resource)*

4.4 • NEWTON'S SECOND LAW

Newton's first law explains what happens to an object when the net external force on it is zero: It either remains at rest or moves in a straight line with constant speed. Newton's second law answers the question of what happens to an object that has a nonzero net force acting on it.

Imagine you are pushing a block of ice across a frictionless horizontal surface. When you exert some horizontal force $\mathbf{F}$, the block moves with some acceleration $\mathbf{a}$. If you apply a force twice as large, the acceleration doubles. If you increase the applied force to $3\mathbf{F}$, the original acceleration is tripled, and so on. From such observations, we conclude that **the acceleration of an object is directly proportional to the net force acting on it.**

As stated in the preceding section, the acceleration of an object also depends on its mass. This can be understood by considering the following set of experiments. If you apply a force $\mathbf{F}$ to a block of ice on a frictionless surface, the block will undergo some acceleration $\mathbf{a}$. If the mass of the block is doubled, the same applied force will produce an acceleration $\mathbf{a}/2$. If the mass is tripled, the same applied force will produce an acceleration $\mathbf{a}/3$, and so on. We conclude that **the acceleration of an object is inversely proportional to its mass.**

These observations are summarized in **Newton's second law:**

> The acceleration of an object is directly proportional to the net force acting on it and inversely proportional to its mass.

Thus, we can relate mass and force through the following mathematical statement of Newton's second law:[2]

$$\Sigma \mathbf{F} = m\mathbf{a} \qquad [4.2]$$

• *Newton's second law*

You should note that Equation 4.2 is a *vector* expression and hence is equivalent to the following three component equations:

$$\Sigma F_x = ma_x \qquad \Sigma F_y = ma_y \qquad \Sigma F_z = ma_z \qquad [4.3]$$

• *Newton's second law — component form*

Units of Force and Mass

The SI unit of force is the **newton,** which is defined as the force that, when acting on a 1-kg mass, produces an acceleration of 1 m/s². From this definition and Newton's second law, we see that the newton can be expressed in terms of the fundamental units of mass, length, and time:

$$1 \text{ N} \equiv 1 \text{ kg} \cdot \text{m/s}^2 \qquad [4.4]$$

• *Definition of a newton*

The units of force and mass are summarized in Table 4.1. Most of the calculations we shall make in our study of mechanics will be in SI units. Conversion factors between the three systems are given in Appendix A.

Thinking Physics 2

You have most likely had the experience of standing in an elevator that accelerates upward as it leaves to move toward a higher floor. In this case, you *feel* heavier. If you are standing on a bathroom scale at the time, it will *measure* a force larger than your weight. Thus, you have tactile and measured evidence that lead you to believe that you are heavier in this situation. *Are* you heavier?

Reasoning No, you are not—your weight is unchanged. In order to provide the acceleration upward, the floor or the bathroom scale must apply an upward force larger than your weight. It is this larger force that you feel, which you interpret as feeling heavier. A bathroom scale reads this upward force, not your weight, so its reading also increases.

TABLE 4.1 Units of Force, Mass, and Acceleration[a]

Systems of Units	Mass	Acceleration	Force
SI	kg	m/s²	N = kg·m/s²
cgs	g	cm/s²	dyne = g·cm/s²
British engineering	slug	ft/s²	lb = slug·ft/s²

[a] 1 N = 10⁵ dyne = 0.225 lb.

---

[2]Equation 4.2 is valid only when the speed of the object is much less than the speed of light. We will treat the relativistic situation in Chapter 9.

Thinking Physics 3

In a train, the cars are connected by *couplers*. The couplers between the cars are under tension as the train is pulled by the locomotive in the front. As you move from the locomotive to the caboose, does the tension force in the couplers *increase, decrease,* or *stay the same* as the train speeds up? What if the engineer applies the brakes, so that the couplers are under compression? How does the compression force vary from locomotive to caboose in this case?

Reasoning   The tension force *decreases* from the front of the train to the back. The coupler between the locomotive and the first car must apply enough force to accelerate all of the remaining cars. As we move back along the train, each coupler is accelerating less mass behind it. The last coupler only has to accelerate the caboose, so it is under the least tension. If the brakes are applied, the couplers are under compression. The force of compression decreases from the front to the back of the train in this case also. The first coupler, at the back of the locomotive, must apply a large force to slow down all of the remaining cars. The final coupler must only apply a force large enough to slow down the mass of the caboose.

Example 4.1   An Accelerating Hockey Puck

A hockey puck with a mass of 0.30 kg slides on the horizontal frictionless surface of an ice rink. Two forces act on the puck as shown in Figure 4.4. The force $\mathbf{F}_1$ has a magnitude of 5.0 N, and $\mathbf{F}_2$ has a magnitude of 8.0 N. Determine the acceleration of the puck.

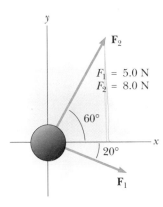

Figure 4.4   (Example 4.1) An object moving on a frictionless surface accelerates in the direction of the *resultant* force, $\mathbf{F}_1 + \mathbf{F}_2$.

Solution   The resultant force in the *x* direction is

$$\sum F_x = F_{1x} + F_{2x} = F_1 \cos 20° + F_2 \cos 60°$$
$$= (5.0\ N)(0.940) + (8.0\ N)(0.500) = 8.7\ N$$

The resultant force in the *y* direction is

$$\sum F_y = F_{1y} + F_{2y} = -F_1 \sin 20° + F_2 \sin 60°$$
$$= -(5.0\ N)(0.342) + (8.0\ N)(0.866) = 5.2\ N$$

Now we can use Newton's second law in component form to find the *x* and *y* components of acceleration:

$$a_x = \frac{\sum F_x}{m} = \frac{8.7\ N}{0.30\ kg} = 29\ m/s^2$$

$$a_y = \frac{\sum F_y}{m} - \frac{5.2\ N}{0.30\ kg} = 17\ m/s^2$$

The acceleration has a magnitude of

$$a = \sqrt{(29)^2 + (17)^2}\ m/s^2 = 34\ m/s^2$$

and its direction is

$$\theta = \tan^{-1}\left(\frac{a_y}{a_x}\right) = \tan^{-1}\left(\frac{17}{29}\right) = 31°$$

relative to the positive *x* axis.

EXERCISE 1   Determine the components of a third force that, when applied to the puck, will cause it to be in equilibrium.   Answer $F_x = -8.7\ N, F_y = -5.2\ N$

EXERCISE 2   A 6.0-kg object undergoes an acceleration of 2.0 m/s². (a) What is the magnitude of the resultant force acting on the object? (b) If this same force is applied to a 4.0-kg object, what acceleration does it produce?   Answer   (a) 12 N   (b) 3.0 m/s²

EXERCISE 3   An 1800-kg car is traveling in a straight line with a speed of 25.0 m/s. What is the magnitude of the constant horizontal force that is needed to bring the car to rest in a distance of 80.0 m?   Answer   7030 N

4.5 • THE GRAVITATIONAL FORCE AND WEIGHT

We are well aware of the fact that all objects are attracted to the Earth. The force exerted by the Earth on an object is the gravitational force $\mathbf{F}_g$ (Fig. 4.5). This force is directed toward the center of the Earth.[3] The magnitude of the gravitational force is called the **weight** of the object, w.

We have seen that a freely falling object experiences an acceleration $\mathbf{g}$ directed toward the center of the Earth. Applying Newton's second law to a freely falling object of mass m, we have $\mathbf{F} = m\mathbf{a}$. Because $\mathbf{F}_g = m\mathbf{g}$, it follows that $\mathbf{a} = \mathbf{g}$ and

$$w = mg \qquad [4.5]$$

Because it depends on g, weight varies with geographic location. (You should not confuse the italicized symbol g that we use for gravitational acceleration with the symbol g that is used for grams.) Bodies weigh less at higher altitudes than at sea level because g decreases with increasing distance from the center of the Earth. Hence, weight, unlike mass, is not an inherent property of a body. For example, if a body has a mass of 70 kg, then its weight in a location where $g = 9.80$ m/s^2 is $mg = 686$ N (about 154 lb). At the top of a mountain where $g = 9.76$ m/s^2, the

Astronaut Edwin E. Aldrin, Jr., walking on the Moon after the Apollo 11 lunar landing. The weight of the astronaut on the Moon is less than it is on Earth but his mass remains the same.   *(Courtesy of NASA)*

Figure 4.5   The only force acting on an object in free fall is the gravitational force, $\mathbf{F}_g$. The magnitude of this force is the weight of the object, mg

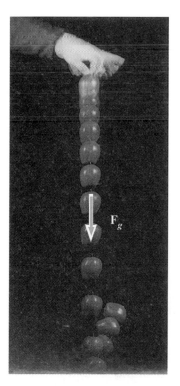

[3]This statement is a simplification in that it ignores the fact that the mass distribution of the Earth is not perfectly spherical.

body's weight would be 683 N. Thus, if you want to lose weight without going on a diet, climb a mountain or weigh yourself at 30 000 ft during an airplane flight.

Because $w = mg$, we can compare the masses of two bodies by measuring their weights with a spring scale or balance. At a given location the ratio of the weights of two bodies equals the ratio of their masses.

Thinking Physics 4

In the absence of air friction, it is claimed that all objects fall with the same acceleration. A heavier object is pulled to the Earth with more force than a light object. Why does the heavier object not fall faster?

Reasoning It is indeed true that the heavier object is pulled with a larger force. The *strength* of the force is determined by the gravitational mass of the object. But the *resistance* to the force and, therefore, to the change in motion of the object, is represented by the object's inertial mass. Inertial mass and gravitational mass are chosen to be equal in Newtonian mechanics. Thus, if an object has twice as much mass as another, it is pulled to the Earth with twice the force, but it also exhibits twice the resistance to having its motion changed. These factors cancel, so that the change in motion, the acceleration, is the same for all objects, regardless of mass.

CONCEPTUAL PROBLEM 3

Suppose you are talking by interplanetary telephone to your friend, who lives on the Moon. He tells you that he has just won a newton of gold in a contest. You tell him that you entered the Earth-version of the same contest and also won a newton of gold! Who is richer?

CONCEPTUAL PROBLEM 4

A baseball of mass m is thrown upward with some initial speed. If air resistance is neglected, what is the force on the ball (a) when it reaches half its maximum height and (b) when it reaches its peak?

4.6 · NEWTON'S THIRD LAW

A statement of Newton's third law •

> Newton's third law states that if two bodies interact, the force exerted on body 1 by body 2 is equal in magnitude but opposite in direction to the force exerted on body 2 by body 1.

$$\mathbf{F}_{12} = -\mathbf{F}_{21} \qquad [4.6]$$

This law, which is illustrated in Figure 4.6a, is equivalent to stating that **forces always occur in pairs,** or that **a single isolated force cannot exist.** The force that body 1 exerts on body 2 is sometimes called the *action force,* and the force of body 2 on body 1 is called the *reaction force.* In reality, either force can be labeled the action or reaction force. **The action force is equal in magnitude to the reaction force and opposite in direction. In all cases, the action and reaction forces act on different objects and must be of the same type.** For example, the force acting on a freely falling projectile is the force of the Earth on the projectile, $\mathbf{F}_g$, and the

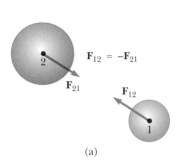

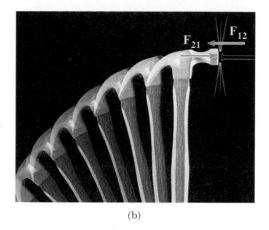

$$\mathbf{F}_{12} = -\mathbf{F}_{21}$$

(a)

(b)

Figure 4.6 Newton's third law. (a) The force exerted by object 1 on object 2 is equal in magnitude and opposite the force exerted by object 2 on object 1. (b) The force exerted by the hammer on the nail is equal in magnitude and opposite the force exerted by the nail on the hammer.   (*John Gillmoure, The Stock Market*)

magnitude of this force is *mg*. The reaction to this force is the force of the projectile on the Earth, $\mathbf{F}'_g = -\mathbf{F}_g$. The reaction force, $\mathbf{F}'_g$, must accelerate the Earth toward the projectile just as the action force, $\mathbf{F}_g$, accelerates the projectile toward the Earth. However, because the Earth has such a large mass, its acceleration due to this reaction force is negligibly small.

Another example of Newton's third law in action is shown in Figure 4.6b. The force exerted by the hammer on the nail (the action) is equal to and opposite the force exerted by the nail on the hammer (the reaction). You directly experience the law if you slam your fist against a wall or kick a football with your bare foot. You should be able to identify the action and reaction forces in these cases.

As we mentioned earlier, the Earth exerts a force $\mathbf{F}_g$ on any object. If the object is a TV at rest on a table, as in Figure 4.7a, the reaction force to $\mathbf{F}_g$ is the force the TV exerts on the Earth, $\mathbf{F}'_g$. The TV does not accelerate, because it is held up by the table. The table, therefore, exerts on the TV an upward action force, **n**, called

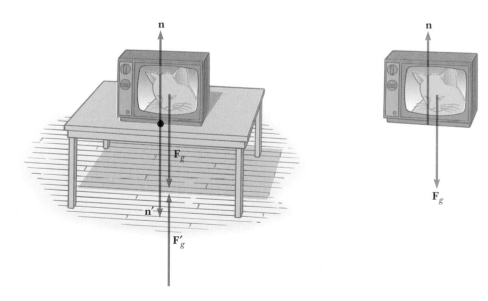

(a)

(b)

Figure 4.7   (a) When a TV set is sitting on a table, the forces acting on the set are the normal force, **n**, and the force of gravity, $\mathbf{F}_g$. The reaction to **n** is the force of the TV set on the table, **n**′. The reaction to $\mathbf{F}_g$ is the force of the TV set on the Earth, $\mathbf{F}'_g$. (b) The free-body diagram for the TV set.

Normal force • the **normal force.**[4] This is the force that prevents the TV from falling through the table; it can have any value needed, up to the point of breaking the table. The normal force balances the weight and provides equilibrium. The reaction to **n** is the force of the TV on the table, **n'**. Therefore, we conclude that

$$\mathbf{F}_g = -\mathbf{F}'_g \quad \text{and} \quad \mathbf{n} = -\mathbf{n}'$$

The forces **n** and **n'** have the same magnitude, which is the same as $\mathbf{F}_g$ unless the table has broken. Note that the forces acting on the TV are $\mathbf{F}_g$ and **n**, as shown in Figure 4.7b. The two reaction forces, $\mathbf{F}'_g$ and **n'**, are exerted on objects other than the TV. Remember, the two forces in an action–reaction pair always act on two different objects.

From Newton's second law, we see that, because the TV is in equilibrium (**a** = 0), it follows that $F_g = n = mg$.

Thinking Physics 5

A horse pulls a sled with a horizontal force, causing it to accelerate as in Figure 4.8a. Newton's third law says that the sled exerts an equal and opposite force on the horse. In view of this, how can the sled accelerate? Under what condition does the system (horse plus sled) move with constant velocity?

Reasoning When applying Newton's third law, it is important to remember that the forces involved act on different objects. When you are determining the motion of an object, you must add the forces on that object alone. The force that accelerates the system (horse and sled) is the force exerted by the Earth on the horse's feet. Newton's third law tells us that the horse exerts an equal and opposite force on the Earth. The horizontal forces exerted on the sled are the forward force exerted by the horse and the backward force of friction between sled and surface (Fig. 4.8b). When the forward force exerted by the horse on the sled exceeds the backward force, the sled accelerates to the right. The horizontal forces exerted on the horse are the forward force of the Earth and the backward force of the sled (Fig. 4.8c). The resultant of these two forces causes the horse to accelerate. When the forward force of the Earth on the horse balances the force of friction between sled and surface, the system moves with constant velocity.

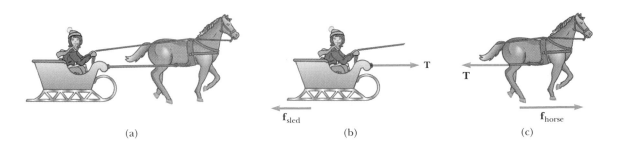

(a)   (b)   (c)

Figure 4.8   (Thinking Physics 5)

[4]The word *normal* is used because the direction of **n** is always *perpendicular* to the surface.

CONCEPTUAL PROBLEM 5

If a small sports car collides head-on with a massive truck, which vehicle experiences the greater impact force? Which vehicle experiences the greater acceleration?

CONCEPTUAL PROBLEM 6

Two forces, $\mathbf{F}_1$ and $\mathbf{F}_2$, are exerted on an object of mass m. $\mathbf{F}_1$ is directed toward the right, whereas $\mathbf{F}_2$ is directed toward the left. Under what condition does the object accelerate toward the right? Toward the left? Can its acceleration ever be zero?

CONCEPTUAL PROBLEM 7

What force causes an automobile to move? A propeller-driven airplane? A rocket? A rowboat?

CONCEPTUAL PROBLEM 8

A large man and a small boy stand facing each other on frictionless ice. They put their hands together and push toward each other, so that they move away from each other. Who exerts the larger force? Who experiences the larger acceleration? Who moves away with the higher velocity? Who moves over a longer distance while their hands are in contact?

4.7 • SOME APPLICATIONS OF NEWTON'S LAWS

In this section we present some simple applications of Newton's laws to bodies that are either in equilibrium ($\mathbf{a} = 0$) or moving linearly under the action of constant external forces. For our model, we shall assume that the bodies behave as particles so that we need not worry about rotational motions. In this section we also neglect the effects of friction for those problems involving motion. This is equivalent to stating that the surfaces are *frictionless*. Finally, we usually neglect the masses of any ropes involved. In this approximation, the magnitude of the force exerted at any point along the rope is the same at all points along the rope. In problem statements, the terms *light* and *of negligible mass* are used to indicate that a mass is to be ignored when you work the problem. These two terms are synonymous in this context.

When we apply Newton's laws to an object, we shall be interested only in those external forces that act *on the object*. For example, in Figure 4.7 the only external forces acting on the TV are $\mathbf{n}$ and $\mathbf{F}_g$. The reactions to these forces, $\mathbf{n}'$ and $\mathbf{F}_g'$, act on the table and on the Earth, respectively, and do not appear in Newton's second law as applied to the TV.

When an object such as a block is being pulled by a rope attached to it, the rope exerts a force on the object. In general, **tension** is a scalar and is defined as the magnitude of the force that the rope exerts on whatever is attached to it. • *Tension*

Consider a crate being pulled to the right on the frictionless, horizontal surface, as in Figure 4.9a. Suppose you are asked to find the acceleration of the crate and the force the floor exerts on it. First, note that the horizontal force being applied to the crate acts through the rope. The force the rope exerts on the crate is denoted by the symbol $\mathbf{T}$. The magnitude of $\mathbf{T}$ is equal to the tension in the rope. In Figure 4.9a, a dotted circle is drawn around the block to remind you to isolate it from its surroundings.

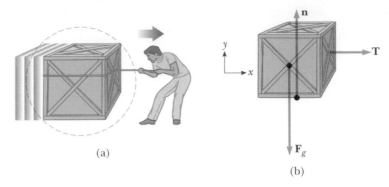

Figure 4.9   (a) A crate being pulled to the right on a frictionless surface. (b) The free-body diagram that represents the external forces on the crate.

Because we are interested only in the motion of the block, we must be able to identify all external forces acting on it. These are illustrated in Figure 4.9b. In addition to the force **T**, the force diagram for the block includes the force of gravity, $\mathbf{F}_g$, and the normal force, **n**, exerted by the floor on the crate. Such a force diagram is referred to as a **free-body diagram.** The construction of a correct free-body diagram is an essential step in applying Newton's laws. The *reactions* to the forces we have listed—namely, the force exerted by the rope on the hand, the force exerted by the crate on the Earth, and the force exerted by the crate on the floor—are not included in the free-body diagram, because they act on *other* bodies and not on the crate.

Now let us apply Newton's second law to the block. First we must choose an appropriate coordinate system. In this case it is convenient to use the coordinate system shown in Figure 4.9b, with the *x* axis horizontal and the *y* axis vertical. We can apply Newton's second law in the *x* direction, *y* direction, or both, depending on what we are asked to find in the problem. In addition, we may be able to use the equations of motion for constant acceleration that are found in Chapter 2. However, you should use these equations only when the acceleration is constant. For example, if the force **T** in Figure 4.9 is constant, then it follows that the acceleration in the *x* direction is also constant, because $\mathbf{a}_x = \mathbf{T}/m$. Hence, if we need to find the displacement or the velocity of the block at some instant of time, we can use the equations of motion with constant acceleration.

Objects in Equilibrium and Newton's First Law

Objects that are either at rest or moving with constant velocity are said to be in equilibrium, and Newton's first law is a statement of one condition that must be true for equilibrium conditions to prevail. In equation form, this condition of equilibrium can be expressed as

$$\sum \mathbf{F} = 0 \qquad \qquad \textbf{[4.7]}$$

This statement signifies that the *vector* sum of all the forces (the net force) acting on an object in equilibrium is zero.[5]

[5]This is only one condition of equilibrium. A second condition of equilibrium is a statement of rotational equilibrium. This condition will be discussed in Chapter 10.

Usually, the problems we encounter in our study of equilibrium are easier to solve if we work with Equation 4.7 in terms of the components of the external forces acting on an object. By this we mean that, in a two-dimensional problem, the sum of all the external forces in the x and y directions must separately equal zero; that is,

$$\sum F_x = 0 \quad \text{and} \quad \sum F_y = 0 \qquad [4.8]$$

This set of equations is often referred to as the **first condition for equilibrium.** We shall not consider three-dimensional problems in this book, but the extension of Equations 4.8 to a three-dimensional situation can be made by adding a third equation, $\Sigma F_z = 0$.

PROBLEM-SOLVING STRATEGY • Objects in Equilibrium

The following procedure is recommended for problems involving objects in equilibrium:

1. Make a sketch of the object under consideration.
2. Draw a free-body diagram for the *isolated* object, and label all external forces acting on the object. Assume a direction for each force. If you select a direction that leads to a negative sign in your solution for a force, do not be alarmed; this merely means that the direction of the force is the opposite of what you assumed.
3. Resolve all forces into x and y components, choosing a convenient coordinate system.
4. Use the equations $\Sigma F_x = 0$ and $\Sigma F_y = 0$. Remember to keep track of the signs of the various force components.
5. Application of the last step leads to a set of equations with several unknowns. Solve the simultaneous equations for the unknowns in terms of the known quantities.

$x = r\cos\theta$
$y = r\sin\theta$

Example 4.2 A Traffic Light at Rest

A traffic light weighing 125 N hangs from a cable tied to two other cables fastened to a support, as in Figure 4.10a. The upper cables make angles of 37.0° and 53.0° with the horizontal. Find the tension in the three cables.

Reasoning We must construct two free-body diagrams in order to work this problem. The first of these is for the traffic light, shown in Figure 4.10b; the second is for the knot that holds the three cables together, as in Figure 4.10c. This knot is a convenient point to choose because all the forces we are interested in act through this point. Because the acceleration of the system is zero, we can use the condition that the net force on the light is zero, and the net force on the knot is zero.

Solution First we construct a free-body diagram for the traffic light, as in Figure 4.10b. The force exerted by the vertical cable, $\mathbf{T}_3$, supports the light, and so $T_3 = w = 125$ N. Next, we choose the coordinate axes as shown in Figure 4.10c and resolve the forces into their x and y components:

Force	x component	y component
$\mathbf{T}_1$	$-T_1 \cos 37.0°$	$T_1 \sin 37.0°$
$\mathbf{T}_2$	$T_2 \cos 53.0°$	$T_2 \sin 53.0°$
$\mathbf{T}_3$	0	-125 N

The first condition for equilibrium gives us the equations

(1) $\sum F_x = T_2 \cos 53.0° - T_1 \cos 37.0° = 0$

(2) $\sum F_y = T_1 \sin 37.0° + T_2 \sin 53.0° - 125 \text{ N} = 0$

From (1) we see that the horizontal components of $\mathbf{T}_1$ and $\mathbf{T}_2$ must be equal in magnitude, and from (2) we see that the sum of the vertical components of $\mathbf{T}_1$ and $\mathbf{T}_2$ must balance the weight of the light. We can solve (1) for T_2 in terms of T_1 to give

$$T_2 = T_1 \left(\frac{\cos 37.0°}{\cos 53.0°} \right) = 1.33 T_1$$

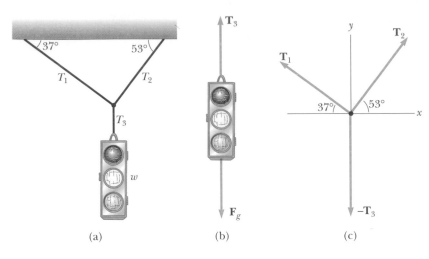

(a)   (b)   (c)

Figure 4.10   (Example 4.2) (a) A traffic light suspended by cables. (b) Free-body diagram for the traffic light. (c) Free-body diagram for the knot in the cable.

This value for T_2 can be substituted into (2) to give

$$T_1 \sin 37.0° + (1.33T_1)(\sin 53.0°) - 125 \text{ N} = 0$$

$$T_1 = 75.1 \text{ N}$$

$$T_2 = 1.33T_1 = 99.9 \text{ N}$$

EXERCISE 4   In what situation will $T_1 = T_2$?
Answer   The angles must be equal.

Figure 4.11   (Exercise 5) A child holding a sled on a frictionless hill.

EXERCISE 5   A child holds a sled weighing 77.0 N at rest on a frictionless snow-covered incline as in Figure 4.11. Find (a) the magnitude of the force the child must exert on the rope, and (b) the magnitude of the force the incline exerts on the sled. (c) What happens to the normal force as the angle of incline increases? (d) Under what conditions would the normal force equal the weight of the sled?
Answer   (a) 38.5 N   (b) 66.7 N   (c) It decreases.   (d) If the sled were on a horizontal surface and the applied force were either zero or along the horizontal

Accelerating Objects and Newton's Second Law

In a situation in which a net force is acting on an object, the object is accelerated, and we use Newton's second law in order to determine the features of the motion. The representative problems and suggestions that follow should help you to solve problems of this kind.

PROBLEM-SOLVING STRATEGY • Newton's Second Law

The following procedure is recommended when dealing with problems involving the application of Newton's second law:

1. Draw a diagram of the system.
2. Isolate the object the motion of which is being analyzed. Draw a free-body diagram for this object, showing *all external forces acting on it*. For

systems containing more than one object, draw a *separate* diagram for each object.

3. Establish convenient coordinate axes for each object and find the components of the forces along those axes. Apply Newton's second law, $\Sigma \mathbf{F} = m\mathbf{a}$, in the x and y directions for each object.

4. Solve the component equations for the unknowns. Remember that in order to obtain a complete solution, you must have as many independent equations as you have unknowns.

5. If necessary, use the equations of kinematics (motion with constant acceleration) from Chapter 2 to find all the unknowns.

Example 4.3   A Crate on a Frictionless Incline

A crate of mass m is placed on a frictionless, inclined plane of angle θ, as in Figure 4.12a. (a) Determine the acceleration of the crate after it is released.

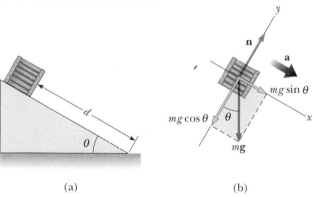

(a)   (b)

Figure 4.12   (Example 4.3) (a) A crate sliding down a frictionless incline. (b) Free-body diagram for the crate. Note that the magnitude of its acceleration along the incline is $g \sin \theta$.

Reasoning   Because the forces acting on the crate are known, Newton's second law can be used to determine its acceleration. First, we construct the free-body diagram for the crate as in Figure 4.12b. The only forces on the crate are the normal force $\mathbf{n}$ acting perpendicular to the plane and the force of gravity $\mathbf{F}_g$ acting vertically downward. For problems of this type involving inclined planes, **it is convenient to choose the coordinate axes with x along the incline and y perpendicular to it.** Then, we replace $\mathbf{F}_g$ by a component of magnitude $mg \sin \theta$ along the *positive* x axis and one of magnitude $mg \cos \theta$ in the *negative* y direction.

Solution   Applying Newton's second law to the crate in component form while noting that $a_y = 0$ gives

$$(1) \quad \sum F_x = mg \sin \theta = ma_x$$

$$(2) \quad \sum F_y = n - mg \cos \theta = 0$$

From (1) we see that the acceleration along the incline is provided by the component of the force of gravity down the incline:

$$(3) \quad a_x = \boxed{g \sin \theta}$$

From (2) we conclude that the component of the force of gravity perpendicular to the incline is *balanced* by the normal force; that is, $n = mg \cos \theta$. Note that the acceleration given by (3) is *independent* of the mass of the crate—it depends only on the angle of inclination and on g.

Special Cases   When $\theta = 90°$, $a = g$ and $n = 0$. This case corresponds to the crate in free fall. When $\theta = 0$, $a_x = 0$ and $n = mg$ (its maximum value).

(b) Suppose the crate is released from rest at the top, and the distance from the front edge of the crate to the bottom is d. How long does it take the front edge of the crate to reach the bottom, and what is its speed just as it gets there?

Solution   Because a_x = constant, we can apply the Equation 2.10, $x - x_0 = v_{x0}t + \frac{1}{2}a_x t^2$ to the crate. Because the displacement $x - x_0 = d$ and $v_{x0} = 0$, we get $d = \frac{1}{2}a_x t^2$, or

$$(4) \quad t = \sqrt{\frac{2d}{a_x}} = \sqrt{\frac{2d}{g \sin \theta}}$$

Using Equation 2.11, $v_x^2 = v_{x0}^2 + 2a_x(x - x_0)$, with $v_{x0} = 0$, we find that $v_x^2 = 2a_x d$, or

$$(5) \quad v_x = \sqrt{2a_x d} = \sqrt{2gd \sin \theta}$$

Again, t and v_x are *independent* of the mass of the crate. This fact suggests a simple method of measuring g using an inclined air track or some other frictionless incline. Simply measure the angle of inclination, the distance traveled by the crate, and the time it takes to reach the bottom. The value of g can then be calculated from (4) and (5).

Example 4.4   Atwood's Machine

When two unequal masses are hung vertically over a light, frictionless pulley as in Figure 4.13a, the arrangement is called *Atwood's machine*. The device is sometimes used in the laboratory to measure the free-fall acceleration. Calculate the magnitude of the acceleration of the two masses and the tension in the string.

Reasoning   The free-body diagrams for the two masses are shown in Figure 4.13b. Two forces act on each block: the upward force exerted by the string, **T**, and the downward force of gravity. Thus, the magnitude of the net force exerted on m_1 is $T - m_1g$, whereas the magnitude of the net force exerted on m_2 is $T - m_2g$. Because the blocks are connected by a string, their accelerations must be equal in magnitude. If we assume that $m_2 > m_1$, then m_1 must accelerate upward, whereas m_2 must accelerate downward.

Solution   When Newton's second law is applied to m_1, with **a** upward for this mass (because $m_2 > m_1$), we find (taking upward to be the positive y direction)

$$(1) \quad \sum F_y = T - m_1g = m_1a$$

In a similar way, for m_2 we find

$$(2) \quad \sum F_y = T - m_2g = -m_2a$$

The negative sign on the right-hand side of (2) indicates that m_2 accelerates downward in the negative y direction.

When (2) is subtracted from (1), T drops out and we get

$$-m_1g + m_2g = m_1a + m_2a$$

$$(3) \quad a = \left(\frac{m_2 - m_1}{m_1 + m_2}\right)g$$

If (3) is substituted into (1), we get

$$(4) \quad T = \left(\frac{2m_1m_2}{m_1 + m_2}\right)g$$

Special Cases   When $m_1 = m_2$, $a = 0$ and $T = m_1g = m_2g$, as we would expect for the balanced case. Also, if $m_2 \gg m_1$, $a \approx g$ (a freely falling body) and $T \approx 2m_1g$.

EXERCISE 6   Find the acceleration and tension of an Atwood's machine in which $m_1 = 2.00$ kg and $m_2 = 4.00$ kg.
Answer $a = 3.27$ m/s^2; $T = 26.1$ N

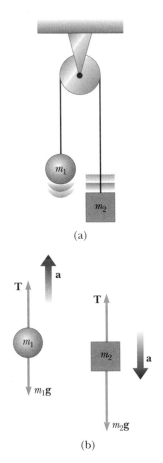

(a)

(b)

Figure 4.13   (Example 4.4) Atwood's machine. (a) Two masses connected by a light string over a frictionless pulley. (b) Free-body diagrams for m_1 and m_2.

Example 4.5   One Block Pushes Another

Two blocks of masses m_1 and m_2 are placed in contact with each other on a frictionless, horizontal surface, as in Figure 4.14a. A constant horizontal force **F** is applied to m_1 as shown. (a) Find the magnitude of the acceleration of the system.

Reasoning and Solution   Both blocks must experience the *same* acceleration because they are in contact with each other. Because **F** is the only horizontal force exerted on the system (the two blocks), we have

$$\sum F_x(\text{system}) = F = (m_1 + m_2)a$$

$$(1) \quad a = \frac{F}{m_1 + m_2}$$

(b) Determine the magnitude of the contact force between the two blocks.

Reasoning and Solution   To solve this part of the problem, first construct a free-body diagram for each block, as shown

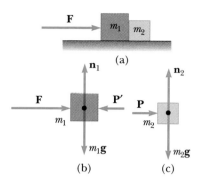

Figure 4.14   (Example 4.5)

in Figures 4.14b and 4.14c, where the contact force is denoted by **P**. From Figure 4.14c we see that the only horizontal force acting on m_2 is the contact force **P** (the force exerted by m_1 on m_2), which is directed to the right. Applying Newton's second law to m_2 gives

$$(2) \quad \sum F_x = P = m_2 a$$

Substituting the value of the acceleration a given by (1) into (2) gives

$$(3) \quad P = m_2 a = \left(\frac{m_2}{m_1 + m_2}\right) F$$

From this result, we see that the contact force **P** is *less* than the applied force **F**. This is consistent with the fact that the force required to accelerate m_2 alone must be less than the force required to produce the same acceleration for the system of two blocks.

It is instructive to check this expression for P by considering the forces acting on m_1, shown in Figure 4.14b. The horizontal forces acting on m_1 are the applied force **F** to the right and the contact force **P'** to the left (the force exerted by m_2 on m_1). From Newton's third law, **P'** is the reaction to **P**, so $|P'| = |P|$. Applying Newton's second law to m_2 gives

$$(4) \quad \sum F_x = F - P' = F - P = m_2 a$$

Substituting the value of a from (2) into (4) gives

$$P = F - m_2 a = F - \frac{m_1 F}{m_1 + m_2} = \left(\frac{m_2}{m_1 + m_2}\right) F$$

This agrees with (3), as it must.

EXERCISE 7   If $m_1 = 4.00$ kg, $m_2 - 3.00$ kg, and $F = 9.00$ N, find the magnitude of the acceleration of the system and that of the contact force.   Answer $a = 1.29$ m/s²; $P = 3.86$ N

Example 4.6   Weighing a Fish in an Elevator

A person weighs a fish on a spring scale attached to the ceiling of an elevator, as shown in Figure 4.15. Show that if the elevator accelerates or decelerates, the spring scale reads a value different from the weight of the fish.

Reasoning   The external forces acting on the fish are the downward force of gravity mg and the upward force **T** exerted on it by the scale. By Newton's third law, the tension T is also the reading of the spring scale. If the elevator is either at rest or moving at constant velocity, then the fish is not accelerating and $T = mg$. However, if the elevator accelerates in either direction, the tension is no longer equal to the weight of the fish.

Solution   If the elevator accelerates with an acceleration **a** relative to an observer outside the elevator in an inertial frame, then the second law applied to the fish gives the total force on the fish:

$$\sum F = T - mg = ma_y$$

which leads to

$$(1) \quad T = mg + ma_y$$

Thus, we conclude from (1) that the scale reading, T, is greater than the weight, mg, if **a** is upward, as in Figure 4.15a. Furthermore, we see that T is less than mg if **a** is downward, as in Figure 4.15b.

For example, if the weight of the fish is 40.0 N, and a_y is 2.00 m/s² upward, then the scale reading is

$$T = mg + ma = mg\left(\frac{a}{g} + 1\right)$$

$$= w\left(\frac{a}{g} + 1\right) = (40.0 \text{ N})\left(\frac{2.00 \text{ m/s}^2}{9.80 \text{ m/s}^2} + 1\right)$$

$$= 48.2 \text{ N}$$

If a_y is 2.00 m/s² downward, then

$$T = mg - ma = mg\left(1 - \frac{a}{g}\right) = 31.8 \text{ N}$$

Hence, if you buy a fish in an elevator, make sure the fish is weighed while the elevator is at rest or accelerating down-

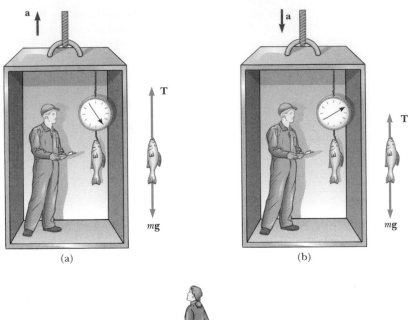

(a)   (b)

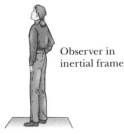

Observer in
inertial frame

Figure 4.15   (Example 4.6). (a) When the elevator accelerates *upward,* the spring scale reads a value *greater* than the weight.
(b) When the elevator accelerates *downward,* the spring scale reads a value *less* than the weight.

ward! Furthermore, note that one cannot determine the *direction* of motion of the elevator from the information given here.

Special Cases   If the cable breaks, then the elevator falls freely and $a_y = -g$, and from (1) we see that the tension, T, is zero;

that is, the fish appears to be weightless. If the elevator accelerates *downward* with an acceleration *greater* than g, the fish (along with the person in the elevator) will eventually hit the ceiling, because the acceleration of the fish will still be that of a freely falling body relative to an outside observer.

SUMMARY

Newton's first law states that a body at rest will remain at rest, and a body in uniform motion in a straight line will maintain that motion, unless an external resultant force acts on the body.

Newton's second law states that the acceleration of an object is directly proportional to the resultant force acting on the object and inversely proportional to the object's mass. If the mass of the body is constant, the net force equals the product of the mass and its acceleration, or $\Sigma \mathbf{F} = m\mathbf{a}$.

Newton's first and second laws are valid in an inertial frame of reference. An **inertial frame** is one in which Newton's first law is valid.

The **weight** of a body is equal to the product of its mass (a scalar quantity) and the magnitude of the free-fall acceleration, or $w = mg$.

Newton's third law states that if two bodies interact, the force exerted on body 1 by body 2 is equal in magnitude but opposite in direction to the force exerted on body 2 by body 1. Thus, an isolated force cannot exist in nature.

CONCEPTUAL QUESTIONS

1. Draw a free-body diagram for each of the following objects: (a) a projectile in motion in the presence of air resistance, (b) a rocket leaving the launch pad with its engines operating, (c) an athlete running along a horizontal track.

2. In the motion picture *It Happened One Night* (Columbia Pictures, 1934), Clark Gable is standing inside a stationary bus in front of Claudette Colbert, who is seated. The bus suddenly starts moving forward and Clark falls into Claudette's lap. Why did this happen?

3. As you sit in a chair, the chair pushes up on you with a normal force. This force is equal to your weight and in the opposite direction. Is it the Newton's third law reaction force to your weight?

4. The observer in the elevator of Example 4.6 would claim that the "weight" of the fish is *T*, the scale reading. This is obviously wrong. Why does this observation differ from that of a person outside the elevator, at rest with respect to the Earth?

5. Identify the action–reaction pairs in the following situations: a man takes a step; a snowball hits a girl in the back; a baseball player catches a ball; a gust of wind strikes a window.

6. While a football is in flight, what forces act on it? What are the action–reaction pairs while the football is being kicked and while it is in flight?

7. A rubber ball is dropped onto the floor. What force causes the ball to bounce back into the air?

8. What is wrong with the statement "Because the car is at rest, there are no forces acting on it"? How would you correct this sentence?

9. A weightlifter stands on a bathroom scale. He pumps a barbell up and down. What happens to the reading on the bathroom scale as this is done? Suppose he is strong enough to actually *throw* the barbell upward. How does the reading on the scale vary now?

10. The mayor of a city decides to fire some city employees because they will not remove the sags from the cables that support the city traffic lights. If you were a lawyer, what defense would you give on behalf of the employees? Who do you think would win the case in court?

11. In a tug-of-war between two athletes, each athlete pulls on the rope with a force of 200 N. What is the tension in the rope? If the rope does not move what force does each athlete exert against the ground?

12. Suppose a truck loaded with sand accelerates at 0.5 m/s² on a highway. If the driving force on the truck remains constant, what happens to the truck's acceleration if its trailer leaks sand at a constant rate through a hole in its bottom?

PROBLEMS

Sections 4.1 Through 4.6

1. A force, **F**, applied to an object of mass m_1 produces an acceleration of 3.00 m/s². The same force applied to a second object of mass m_2 produces an acceleration of 1.00 m/s². (a) What is the value of the ratio m_1/m_2? (b) If m_1 and m_2 are combined, find their acceleration under the action of the force **F**.

2. A force of 10.0 N acts on a body of mass 2.00 kg. What are (a) the body's acceleration, (b) its weight in newtons, and (c) its acceleration if the force is doubled?

3. A 3.00-kg mass undergoes an acceleration given by **a** = $(2.00\mathbf{i} + 5.00\mathbf{j})$ m/s². Find the resultant force, **F**, and its magnitude.

4. A heavy freight train has a mass of 15 000 metric tons. If the locomotives can exert a pulling force of 750 000 N, how long does it take to increase the speed from 0 to 80.0 km/h?

5. A pitcher throws a baseball of weight 1.40 N with velocity $(32.0\mathbf{i})$ m/s by uniformly accelerating her arm for 0.0900 s. If the ball starts from rest, (a) through what distance does the ball accelerate before its release? (b) What vector force does she exert on it?

6. A pitcher throws a baseball of weight $-w\mathbf{j}$ with velocity $v\mathbf{i}$ by uniformly accelerating her arm for time *t*. If the ball starts from rest, (a) through what distance does the ball accelerate before its release? (b) What vector force does she exert on it?

7. A 4.00-kg object has a velocity of $3.00\mathbf{i}$ m/s at one instant. Eight seconds later, its velocity is $(8.00\mathbf{i} + 10.0\mathbf{j})$ m/s. Assuming the object was subject to a constant net force, find (a) the components of the force and (b) its magnitude.

8. The average speed of a nitrogen molecule in air is about 6.70×10^2 m/s, and its mass is about 4.68×10^{-26} kg. (a) If it takes 3.00×10^{-13} s for a nitrogen molecule to hit a wall and rebound with the same speed but in an opposite direction, what is the average acceleration of the molecule during this time interval? (b) What average force does the molecule exert on the wall?

9. An electron of mass 9.11×10^{-31} kg has an initial speed of 3.00×10^5 m/s. It travels in a straight line, and its speed

increases to 7.00×10^5 m/s in a distance of 5.00 cm. Assuming its acceleration is constant, (a) determine the force on the electron and (b) compare this force with the weight of the electron, which we neglected.

10. (a) A car with a mass of 850 kg is moving to the right with a constant speed of 1.44 m/s. What is the total force on the car? (b) What is the total force on the car if it is moving to the left?

11. A woman weighs 120 lb. Determine (a) her weight in newtons and (b) her mass in kilograms.

12. If a man weighs 900 N on Earth, what would he weigh on Jupiter, where the acceleration due to gravity is 25.9 m/s²?

13. Two forces, $\mathbf{F}_1$ and $\mathbf{F}_2$, act on a 5.00-kg mass. If $F_1 = 20.0$ N and $F_2 = 15.0$ N, find the accelerations in (a) and (b) of Figure P4.13.

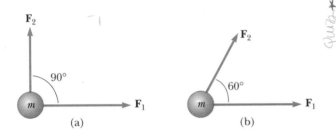

(a)

(b)

Figure P4.13

14. Besides its weight, a 2.80-kg object is subjected to one other constant force. The object starts from rest and in 1.20 s experiences a displacement of $(4.20 \text{ m})\mathbf{i} - (3.30 \text{ m})\mathbf{j}$, where the direction of $\mathbf{j}$ is the upward vertical direction. Determine the other force.

15. You stand on the seat of a chair and then hop off. (a) During the time you are in flight down to the floor, estimate the acceleration of the Earth as it lurches up toward you. (b) Estimate the upward distance through which the Earth moves as you fall to the floor. Visualize the Earth as a perfectly solid object.

16. Forces of 10.0 N north, 20.0 N east, and 15.0 N south are simultaneously applied to a 4.00-kg mass. Obtain the object's acceleration.

17. A 1000-kg boat moves through the water with two forces acting on it. One is a 2000-N forward push by the propeller; the other is an 1800-N resistive force due to the water. (a) What is the acceleration of the boat? (b) If it starts from rest, how far will it move in 10.0 s? (c) What will be its speed at the end of this time?

18. Three forces, $\mathbf{F}_1 = (-2.00\mathbf{i} + 2.00\mathbf{j})$ N, $\mathbf{F}_2 = (5.00\mathbf{i} - 3.00\mathbf{j})$ N, and $\mathbf{F}_3 = (-45.0\mathbf{i})$ N, act on an object to give it

an acceleration of magnitude 3.75 m/s². (a) What is the direction of the acceleration? (b) What is the mass of the object? (c) If the object is initially at rest, what is its speed after 10.0 s? (d) What are the velocity components of the object after 10.0 s?

19. A 15.0-lb block rests on the floor. (a) What force does the floor exert on the block? (b) If a rope is tied to the block and run vertically over a pulley and the other end is attached to a free-hanging 10.0-lb weight, what is the force of the floor on the 15.0-lb block? (c) If we replace the 10.0-lb weight in part (b) with a 20.0-lb weight, what is the force of the floor on the 15.0-lb block?

Section 4.7   Some Applications of Newton's Laws

20. A 3.00-kg mass is moving in a plane, with its x and y coordinates given by $x = 5t^2 - 1$ and $y = 3t^3 + 2$ (x and y are in meters and t is in seconds). Find the magnitude of the net force acting on this mass at $t = 2.00$ s.

21. Find the tension in each cord of the systems described in Figure P4.21. (Neglect the masses of the cords.)

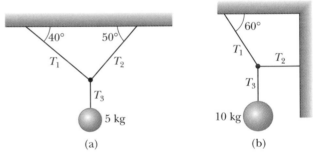

(a)

(b)

Figure P4.21

22. The distance between two telephone poles is 50.0 m. When a 1.00-kg bird lands on the telephone wire midway between the poles, the wire sags 0.200 m. Draw a free-body diagram of the bird. How much tension does the bird produce in the wire? Ignore the weight of the wire.

23. A bag of cement hangs from three wires as shown in Figure P4.23. Two of the wires make angles θ_1 and θ_2 with the horizontal. If the system is in equilibrium, (a) show that

$$T_1 = \frac{w \cos \theta_2}{\sin (\theta_1 + \theta_2)}$$

(b) Given that $w = 325$ N, $\theta_1 = 10.0°$ and $\theta_2 = 25.0°$, find the tensions T_1, T_2, and T_3 in the wires.

24. The systems shown in Figure P4.24 are in equilibrium. If

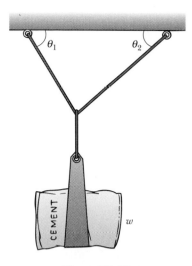

Figure P4.23

the spring scales are calibrated in newtons, what do they read? (Neglect the masses of the pulleys and strings, and assume the incline is frictionless.)

25. A fire helicopter carries a 620-kg bucket of water at the end of a 20.0-m long cable. Flying back from a fire at a constant speed of 40.0 m/s, the cable makes an angle of 40.0° with respect to the vertical. (a) Determine the force of air resistance on the bucket. (b) After filling the bucket with sea water, the helicopter returns to the fire at the same speed with the bucket now making an angle of 7.00° with the vertical. What is the mass of the water in the bucket?

26. A simple accelerometer is constructed by suspending a mass m from a string of length L that is tied to the top of a cart. As the cart is accelerated the string-mass system makes an angle of θ with the vertical. (a) Assuming that the string mass is negligible compared to m, derive an expression

for the cart's acceleration in terms of θ, and show that it is independent of the mass m and the length L. (b) Determine the acceleration of the cart when $\theta = 23.0°$.

27. A 1.00-kg mass is observed to accelerate at 10.0 m/s^2 in a direction 30.0° north of east (Fig. P4.27). The force $\mathbf{F}_2$ acting on the mass has a magnitude of 5.00 N and is directed north. Determine the magnitude and direction of the force $\mathbf{F}_1$ acting on the mass.

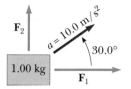

Figure P4.27

28. Draw a free-body diagram of a block that slides down a frictionless plane having an inclination of $\theta = 15.0°$ (Fig. P4.28). If the block starts from rest at the top and the length of the incline is 2.00 m, find (a) the acceleration of the block and (b) its speed when it reaches the bottom of the incline.

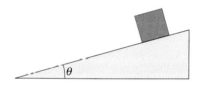

Figure P4.28

29. A block is given an initial velocity of 5.00 m/s up a frictionless 20.0° incline (Fig. P4.28). How far up the incline does the block slide before coming to rest?

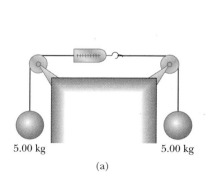

(a)

(b)

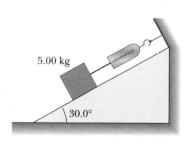

(c)

Figure P4.24

30. Two masses are connected by a light string that passes over a frictionless pulley, as in Figure P4.30. The incline is frictionless, $m_1 = 2.00$ kg, $m_2 = 6.00$ kg, and $\theta = 55.0°$. (a) Draw free-body diagrams of both masses. Find (b) the accelerations of the masses, (c) the tension in the string, and (d) the speed of each mass 2.00 s after being released from rest.

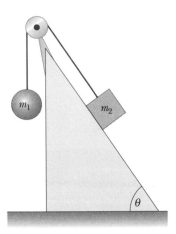

Figure P4.30

31. Two masses, m_1 and m_2, situated on a frictionless, horizontal surface, are connected by a light string. A force, **F**, is exerted on one of the masses to the right (Fig. P4.31). Determine the acceleration of the system and the tension, T, in the string.

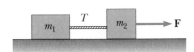

Figure P4.31

32. Two masses of 3.00 kg and 5.00 kg are connected by a light string that passes over a frictionless pulley, as in Figure 4.13a (page 98). Determine (a) the tension in the string, (b) the acceleration of each mass, and (c) the distance each mass will move in the first second of motion if they start from rest.

33. Mass m_1 on a frictionless horizontal table is connected to mass m_2 through a very light pulley, P_1, and a light fixed pulley, P_2, as shown in Figure P4.33. (a) If a_1 and a_2 are the accelerations of m_1 and m_2, respectively, what is the relation between these accelerations? Express (b) the tensions in the strings and (c) the accelerations a_1 and a_2 in terms of the masses m_1, and m_2, and g.

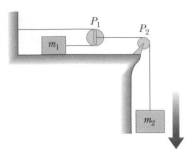

Figure P4.33

34. A 5.00-kg mass placed on a frictionless, horizontal table is connected to a cable that passes over a pulley and then is fastened to a hanging 10.0-kg mass, as in Figure P4.34. Draw free-body diagrams of both masses. Find the acceleration of the two objects and the tension in the string.

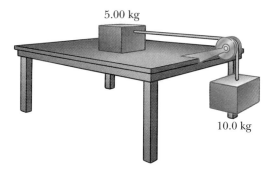

Figure P4.34

35. A 72.0-kg man stands on a spring scale in an elevator. Starting from rest, the elevator ascends, attaining its maximum speed of 1.20 m/s in 0.800 s. It travels with this constant speed for the next 5.00 s. The elevator then undergoes a uniform acceleration in the negative y direction for 1.50 s and comes to rest. What does the spring scale register (a) before the elevator starts to move? (b) during the first 0.800 s? (c) while the elevator is traveling at constant speed? (d) during the time it is slowing down?

36. Two people pull as hard as they can on ropes attached to a boat that has a mass of 200 kg. If they pull in the same direction, the boat has an acceleration of 1.52 m/s² to the right. If they pull in opposite directions, the boat has an acceleration of 0.518 m/s² to the left. What is the force exerted by each person on the boat? (Disregard any other forces on the boat.)

Additional Problems

37. A time-dependent force, $\mathbf{F} = (8.00\mathbf{i} - 4.00t\mathbf{j})$ N (where t is in seconds) is applied to a 2.00-kg object initially at rest. (a) At what time will the object be moving with a speed of 15.0 m/s? (b) Through what total displacement has the object traveled at this time? (c) How far is the object from its initial position when its speed is 15.0 m/s?

38. An elevator accelerates upward at 1.50 m/s^2. If the elevator has a mass of 200 kg, find the tension in the supporting cable.

39. The largest-caliber anti-aircraft gun operated by the German air force during World War II was the 12.8-cm Flak 40. This weapon fired a 25.8-kg shell with a muzzle speed of 880 m/s. What propulsive force was necessary to attain the muzzle speed within the 6.00-m barrel? (Assume constant acceleration and neglect the Earth's gravitational effect.)

40. One of the great dangers to mountain climbers is the *avalanche*, in which a mass of snow and ice breaks loose and goes on an essentially frictionless "ride" down the mountain on a cushion of compressed air. If you were on a 30.0° mountain slope and an avalanche started 400 m up the slope, how much time would you have to get out of the way?

41. An inventive child named Brian wants to reach an apple in a tree without climbing the tree. Sitting in a chair connected to a rope that passes over a frictionless pulley (Fig. P4.41), Brian pulls on the loose end of the rope with such a force that the spring scale reads 250 N. Brian's weight is 320 N, and the chair weighs 160 N. (a) Draw free-body diagrams for Brian and the chair considered as separate systems and another diagram for Brian and the chair considered as one system. (b) Show that the acceleration of the system is *upward* and find its magnitude. (c) Find the force Brian exerts on the chair.

42. Three blocks are in contact with each other on a frictionless, horizontal surface, as in Figure P4.42. A horizontal force $\mathbf{F}$ is applied to m_1. If $m_1 = 2.00$ kg, $m_2 = 3.00$ kg, $m_3 = 4.00$ kg, and $F = 18.0$ N, draw free-body diagrams of each block and find (a) the acceleration of the blocks, (b) the *resultant* force on each block, and (c) the magnitudes of the contact forces between the blocks.

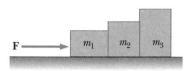

Figure P4.42

43. A high diver of mass 70.0 kg jumps off a board 10.0 m above the water. If her downward motion is stopped 2.00 s after she enters the water, what average upward force did the water exert on the diver?

44. Two forces, $\mathbf{F}_1 = (-6\mathbf{i} - 4\mathbf{j})$ N and $\mathbf{F}_2 = (-3\mathbf{i} + 7\mathbf{j})$ N, act on a particle of mass 2.00 kg that is initially at rest at coordinates $(-2.00$ m, $+4.00$ m$)$. (a) What are the components of the particle's velocity at $t = 10.0$ s? (b) In what direction is the particle moving at $t = 10.0$ s? (c) What displacement does the particle undergo during the first 10.0 s? (d) What are the coordinates of the particle at $t = 10.0$ s?

45. A mass M is held in place by an applied force $\mathbf{F}$ and a pulley system as shown in Figure P4.45. The pulleys are massless and frictionless. Find (a) the tension in each section of rope, T_1, T_2, T_3, T_4, and T_5 and (b) the magnitude of $\mathbf{F}$.

46. A student is asked to measure the acceleration of a cart on a "frictionless" inclined plane, as in Figure 4.12, using an air track, a stopwatch, and a meter stick. The height of the incline is measured to be 1.774 cm, and the total length of the incline is measured to be $d = 127.1$ cm. Hence, the angle of inclination θ is determined from the relation $\sin \theta = 1.774/127.1$. The cart is released from rest at the top of the incline, and its displacement along the incline, x, is measured versus time, where $x = 0$ refers to the initial position of the cart. For x values of 10.0 cm, 20.0 cm, 35.0 cm, 50.0 cm, 75.0 cm, and 100 cm, the measured times to undergo these displacements (averaged over five runs) are 1.02 s, 1.53 s, 2.01 s, 2.64 s, 3.30 s, and 3.75 s, respec-

Figure P4.41

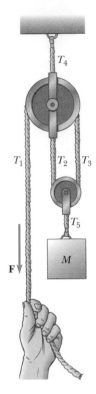

Figure P4.45

road exert on the car during this time, and how far did the car travel during the deceleration?

49. A van accelerates down a hill (Fig. P4.49), going from rest to 30.0 m/s in 6.00 s. During the acceleration, a toy ($m = 0.100$ kg) hangs by a string from the van's ceiling. The acceleration is such that the string remains perpendicular to the ceiling. Determine (a) the angle θ and (b) the tension in the string.

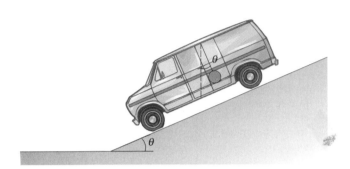

Figure P4.49

tively. Construct a graph of x versus t^2, and perform a linear least-squares fit to the data. Determine the acceleration of the cart from the slope of this graph and compare it with the value you would get using $a' = g \sin \theta$, where $g = 9.80$ m/s².

47. What horizontal force must be applied to the cart shown in Figure P4.47 in order that the blocks remain stationary relative to the cart? Assume all surfaces, wheels, and pulley are frictionless. (*Hint:* Note that the force exerted by the string accelerates m_1 and that m_2 is in contact with the cart.)

50. A car is at rest at the top of a driveway that has a slope of 20.0°. If the brake of the car is released with the car in neutral gear, find (a) the acceleration of the car down the drive and (b) the time it takes for the car to reach the street 10.0 m away.

Spreadsheet Problems

S1. An 8.4-kg mass slides down a fixed, frictionless inclined plane. Design and write a spreadsheet to determine the normal force exerted on the mass and its acceleration for a series of incline angles (measured from the horizontal) ranging from 0° to 90° in 5° increments. Use the graphing capability of your spreadsheet data to plot the normal force and the acceleration as functions of the incline angle. In the limiting cases of 0° and 90°, are your results consistent with the known behavior?

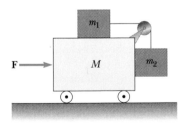

Figure P4.47

48. A 2000-kg car is slowed down uniformly from a speed of 20.0 m/s to 5.00 m/s in a time of 4.00 s. What force did the

ANSWERS TO CONCEPTUAL PROBLEMS

1. Motion requires no force. Newton's first law says an object in motion continues to move by itself in the absence of external forces.

2. If a single force acts on an object, it must accelerate. (From Newton's second law, $a = F/m$.) If an object accelerates, at least one force must act on it. However, if an object has no acceleration, you cannot conclude that no forces act on it. In this case, you can only say that the *net* force on the object is zero.

3. Because the gravitational field g_M is smaller on the Moon than on the Earth, more mass of gold would be required to represent 1 newton of weight. Thus, your friend on the Moon is richer, by about a factor of 6!

4. The only force acting on the ball during its motion is the downward force of gravity. The magnitude of this force is *mg*.

5. The car and truck experience forces that are equal in magnitude but in opposite directions. A calibrated spring scale placed between the colliding vehicles reads the same whichever way it faces. Because the car has the smaller mass, it experiences greater acceleration.

6. The *net* force on the object is $\mathbf{F}_1 + \mathbf{F}_2$, thus from Newton's second law, its acceleration is $(\mathbf{F}_1 + \mathbf{F}_2)/m$. The object accelerates toward the right when $|\mathbf{F}_1| > |\mathbf{F}_2|$, and toward the left when $|\mathbf{F}_2| > |\mathbf{F}_1|$. The acceleration of the object is zero when the two forces have equal magnitudes, that is, when $|\mathbf{F}_1| = |\mathbf{F}_2|$.

7. The force causing an automobile to move is the friction between the tires and the roadway as the automobile attempts to push the roadway backward. Notice that forces from the engine are not directly what accelerate the automobile. The force driving a propeller-driven airplane forward is the force of the air on the propeller as the rotating propeller pushes the air backward. The force propelling a rocket is the force of the exhausted gases from the back of the rocket on the rocket, as the rocket pushes the gases backward. In a rowboat, the rower pushes the water backward with the oars. As a result, the water pushes forward on the boat, and the force is applied at the oarlocks. Notice that all of these situations involve the propulsion of an object by Newton's third law.

8. According to Newton's third law, the force of the man on the boy and the force of the boy on the man are third law reaction forces, so they must be equal in magnitude. Both individuals experience the same force, so the boy, with the smaller mass, experiences the larger acceleration. Both individuals experience an acceleration over the same time interval. The larger acceleration of the boy will result in his moving away from the interaction with a larger velocity. Again, because of the larger acceleration, the boy moves through a larger distance during the interval while the hands are in contact.

B.C.   By John Hart

By permission of John Hart and Field Enterprises, Inc.

5

More Applications of Newton's Laws

In Chapter 4 we introduced Newton's laws of motion and applied them to some linear motion situations in which we were able to neglect frictional forces. In this chapter we shall expand our investigation to systems moving in the presence of friction forces. These systems include objects moving on rough surfaces and objects moving through viscous media such as liquids and air. In one section we discuss how numerical methods can be used to solve such real-world problems as motion in which the resistive force is velocity-dependent. We also apply Newton's laws to the dynamics of circular motion. Finally, we conclude this chapter with a discussion of the fundamental forces in nature.

5.1 • FORCES OF FRICTION

When a body is in motion either on a surface or through a viscous medium such as air or water, there is resistance to the motion because the body interacts with its surroundings. We call such resistance a **force of friction.** Forces of fric-

The passengers on this corkscrew rollercoaster experience the thrill of various forces as they travel along the curved track. The forces on one of the passenger cars include the force exerted by the track, the force of gravity, and the force of air resistance. *(Robin Smith/ Tony Stone Images)*

tion are very important in our everyday lives. They allow us to walk or run and are necessary for the motion of wheeled vehicles.

Consider a block on a horizontal table, as in Figure 5.1a. If we apply an external horizontal force **F** to the block, acting to the right, the block remains stationary if **F** is not too large. The force that counteracts **F** and keeps the block from moving acts to the left and is called the force of static friction, $\mathbf{f}_s$. As long as the block is not moving, $f_s = F$. Thus, if **F** is increased, $\mathbf{f}_s$ also increases. Likewise, if **F** decreases, $\mathbf{f}_s$ also decreases. Experiments show that the frictional force arises from the nature of the two surfaces: Because of their roughness, contact is made only at a few points, as shown in the magnified view of the surface in Figure 5.1a. Actually, the frictional force is much more complicated than presented here, because it ultimately involves the electrostatic force between atoms or molecules.

If we increase the magnitude of **F**, as in Figure 5.1b, the block eventually slips. When the block is on the verge of slipping, f_s is a maximum as shown in Figure 5.1c. When F exceeds $f_{s,\,max}$, the block moves and accelerates to the right. When

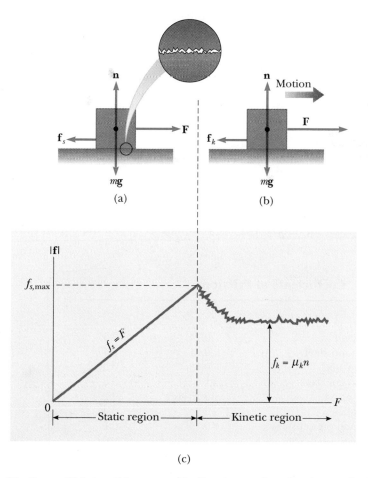

(a)

(b)

(c)

Figure 5.1   The force of friction, **f**, between a block and a rough surface is opposite the applied force, **F**. (a) The force of static friction equals the applied force. (b) When the applied force exceeds the force of kinetic friction, the block accelerates to the right. (c) A graph of the magnitude of the frictional force versus the applied force. Note that $f_{s,\,max} > f_k$.

the block is in motion, the frictional force becomes less than $f_{s, \text{max}}$ (Fig. 5.1c). We call the frictional force for an object in motion the **force of kinetic friction, f_k.** The unbalanced force in the x direction, $F - f_k$, produces an acceleration to the right. If $F = f_k$, the block moves to the right with constant speed. If the applied force is removed, then the frictional force acting to the left accelerates the block in the negative x direction and eventually brings it to rest.

Experimentally, one finds that, to a good approximation, both $f_{s, \text{max}}$ and f_k are proportional to the normal force acting on the block. The experimental observations can be summarized as follows:

- The magnitude of the force of static friction between any two surfaces in contact can have the values

Force of static friction •

$$f_s \leq \mu_s n \tag{5.1}$$

where the dimensionless constant μ_s is called the **coefficient of static friction** and n is the magnitude of the normal force. The equality in Equation 5.1 holds when the block is on the verge of slipping—that is, when $f_s = f_{s, \text{max}} \equiv \mu_s n$. The inequality holds when the applied force is less than this value.

- The magnitude of the force of kinetic friction acting on an object is

Force of kinetic friction •

$$f_k = \mu_k n \tag{5.2}$$

where μ_k is the **coefficient of kinetic friction.**

- The values of μ_k and μ_s depend on the nature of the surfaces, but μ_k is generally less than μ_s. Typical values of μ range from around 0.05 to 1.5. Table 5.1 lists some reported values.
- The coefficients of friction are nearly independent of the area of contact between the surfaces.

Finally, although the coefficient of kinetic friction varies with speed, we shall neglect any such variations. The approximate nature of the equations is easily dem-

TABLE 5.1   Coefficients of Friction[a]

	μ_s	μ_k
Steel on steel	0.74	0.57
Aluminum on steel	0.61	0.47
Copper on steel	0.53	0.36
Rubber on concrete	1.0	0.8
Wood on wood	0.25–0.5	0.2
Glass on glass	0.94	0.4
Waxed wood on wet snow	0.14	0.1
Waxed wood on dry snow	—	0.04
Metal on metal (lubricated)	0.15	0.06
Ice on ice	0.1	0.03
Teflon on Teflon	0.04	0.04
Synovial joints in humans	0.01	0.003

[a] All values are approximate.

onstrated by trying to get a block to slip down an incline at constant speed. Especially at low speeds, the motion is likely to be characterized by alternate *stick* and *slip* episodes.

Thinking Physics 1

In the motion picture *The Abyss* (Twentieth Century Fox, 1990), an underwater oil exploration rig sits at the bottom of very deep water. It is connected to a ship on the ocean surface by an "umbilical cord," as suggested in Figure 5.2a. On the ship, the umbilical cord is attached to a gantry. During a hurricane, the gantry structure breaks loose from the ship, falls into the water, and sinks to the bottom, passing over the edge of an extremely deep abyss. As a result, the rig is dragged by the umbilical cord along the ocean bottom, as described in Figure 5.2b. As the rig approaches the edge of the abyss, however, it is not pulled over the edge but stops just short of the edge, as shown in Figure 5.2c. Is this purely a cinematic edge-of-the-seat scenario, or is there a reason within the principles of physics for why the rig does not go over the edge?

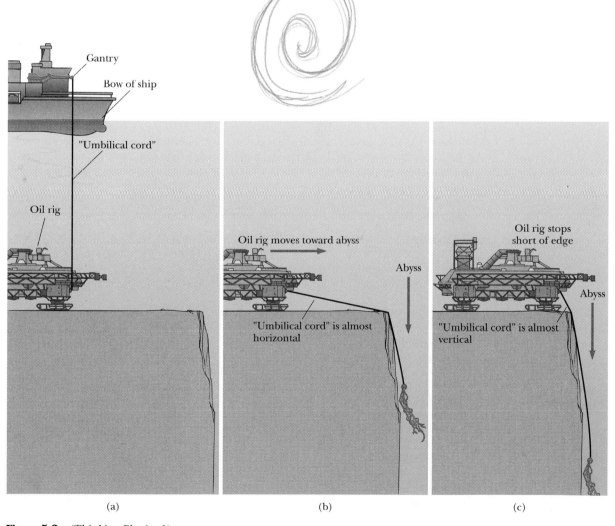

Figure 5.2   (Thinking Physics 1)

Reasoning There is a physics reason for this phenomenon. While the rig is being pulled across the ocean floor, it is pulled by the section of the umbilical cord that is almost horizontal—almost parallel to the ocean floor. Thus, the rig is subject to two horizontal forces—the tension in the umbilical cord pulling it forward and friction with the ocean floor pulling back. Let us assume that these forces are equal, so that the rig moves with constant velocity. As the rig nears the edge of the abyss, the angle that the umbilical cord makes with the horizontal increases. As a result, the component of force parallel to the ocean floor decreases and the downward vertical component increases. As a result of the increased vertical force, the rig is pulled downward more strongly into the ocean floor, increasing the normal force on it and, in turn, increasing the friction force between the rig and the ocean floor. Thus, with less force pulling it forward (from the umbilical cord) and more force pulling it back (due to friction), the rig slows down. By the time that the rig reaches the edge of the abyss, the force from the umbilical cord is almost *straight down*, resulting in little forward force. This downward force pulls the rig hard into the ocean floor, so that the friction force is large, and the rig stops.

This is not to say that this would happen in any situation. The characters on the rig had some luck in their favor. For example, if the umbilical cord had been attached at a lower point on the rig or if the coefficient of friction between the rig feet and the ocean floor were smaller, then the forward velocity of the rig could have been large enough that the friction force would not be sufficient to stop it, and it would have been propelled off the edge.

Thinking Physics 2

Sometimes, an incorrect statement about the frictional force between an object and a surface is made—"the friction force on an object is opposite to its motion or impending motion (motion that is about to occur)"—rather than the correct phrasing, "the friction force on an object is opposite to its motion or impending motion *relative to the surface*." To emphasize the incorrect nature of the first statement, give some examples in which the friction force is in the *same direction* as the motion.

Reasoning Walking is a familiar example. To walk, you push backward on the ground, and the friction force pushes you forward, in the direction of your walking motion. In this case, the impending motion (because this is static friction) of the object, your foot, is backward relative to the ground, and the friction force exerted on your foot is forward.

Another example is a package sitting in the back of a pickup truck. As the truck accelerates forward, the natural inertia of the package tends to "leave it behind." Thus, the impending motion relative to the floor of the truck is toward the rear, and the friction force exerted on the package is forward, in the same direction as it moves. If the truck accelerates rapidly enough, the package will slide backward in the bed of the truck. In this case, we have real motion of the object rather than impending motion (relative to the surface), and the motion is backward. The friction force, then, is forward. Even though the package is sliding backward in the truck, an observer on the ground will see the package moving forward relative to the ground, so the friction force exerted on the package is again in the same direction as its motion.

As a final example, consider the popular dish-and-tablecloth trick, in which the demonstrator quickly pulls a tablecloth out from under some dishes. It may appear that the dishes do not move. Careful measurement, however, will show that the dishes are slightly closer to the edge of the table after the trick than before. As the tablecloth is pulled, the motion of the dishes relative to the tablecloth is away from the puller. Thus, the friction force on the dishes is toward the puller. This force causes the dishes to move toward the puller. Once the tablecloth is completely out from under the dishes, the tabletop becomes the new surface. The motion of the dishes relative to the tabletop is toward the puller. Thus, the friction force on the dishes is now away from the puller and opposite to the motion of the dishes, causing them to come to rest. The success of the trick depends on a relatively small friction force between the dishes and the tablecloth, so that the dishes do not achieve a very large velocity toward the puller, and a relatively large friction force between the dishes and the tabletop, so that they stop quickly.

CONCEPTUAL PROBLEM 1

You are playing with your daughter in the snow. She is sitting on a sled and asking you to slide her across a flat, horizontal field. You have a choice of pushing her from behind, by applying a force at 30° below the horizontal, or attaching a rope to the front of the sled and pulling with a force at 30° above the horizontal. Which would be easier for you and why?

done in class — see notes

CONCEPTUAL PROBLEM 2

A book is given a push at the bottom of a nonfrictionless ramp so that it slides up to a point and then slides back down to the starting point. Does it take the same time to go up as it does to come down?

longer to come down

CONCEPTUAL PROBLEM 3

The driver of an empty speeding truck slams on the brakes and skids to a stop through a distance *d*. (a) If the truck carried a heavy load such that its mass were doubled, what would be its skidding distance? (b) If the initial speed of the truck is halved, what would be its skidding distance?

Example 5.1   The Sliding Hockey Puck

A hockey puck on a frozen pond is hit and given an initial speed of 20.0 m/s, as in Figure 5.3. If the puck always remains on the ice and slides 115 m before coming to rest, determine the coefficient of kinetic friction between the puck and the ice.

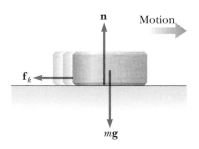

Figure 5.3   (Example 5.1) *After* the puck is given an initial velocity, the external forces acting on it are the weight, *m***g**, the normal force, **n**, and the force of kinetic friction, **f**$_k$.

Reasoning   The forces acting on the puck after it is in motion are shown in Figure 5.3. If we assume that the force of friction, f_k, remains constant, then this force produces a uniform negative acceleration of the puck. First, we find the acceleration using Newton's second law. Knowing the acceleration of the puck and the distance it travels, we can then use kinematics to find the coefficient of kinetic friction.

Solution   Applying Newton's second law in component form to the puck gives

$$(1) \quad \sum F_x = -f_k = ma$$

$$(2) \quad \sum F_y = n - mg = 0 \qquad (a_y = 0)$$

But $f_k = \mu_k n$, and from (2) we see that $n = mg$. Therefore, (1) becomes

$$-\mu_k n = -\mu_k mg = ma$$

$$a = -\mu_k g$$

The negative sign means that the acceleration is to the left, corresponding to a negative acceleration of the puck. The acceleration is independent of the mass of the puck and is constant, because we assume that μ_k remains constant.

Because the acceleration is constant, we can use Equation 2.12, $v^2 = v_0^2 + 2ax$, with the final speed $v = 0$:

$$v_0^2 + 2ax = v_0^2 - 2\mu_k gx = 0$$

$$\mu_k = \frac{v_0^2}{2gx}$$

$$\mu_k = \frac{(20.0\ \text{m/s})^2}{2(9.80\ \text{m/s}^2)(115\ \text{m})} = 0.177$$

Note that μ_k has no dimensions.

Example 5.2   Connected Objects

A ball and a cube are connected by a light string that passes over a frictionless pulley, as in Figure 5.4a. The coefficient of kinetic friction between the cube and the surface is 0.30. Find the acceleration of the two objects and the tension in the string.

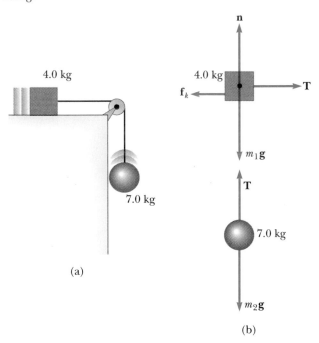

(a)

(b)

Figure 5.4   (Example 5.2) (a) Two objects connected by a light string that passes over a frictionless pulley. (b) Free-body diagrams for the objects.

Reasoning   First, we draw the free-body diagrams for the two objects as in Figure 5.4b. Next, we apply Newton's second law

in component form to each object and make use of the fact that the magnitude of the force of kinetic friction acting on the block is proportional to the normal force according to $f_k = \mu_k n$. Finally, we solve for the acceleration in terms of the parameters given.

Solution   Newton's second law applied to the cube in component form, with the positive x direction to the right, gives

$$\sum F_x = T - f_k = (4.0\ \text{kg})a$$

$$\sum F_y = n - (4.0\ \text{kg})g = 0$$

Because $n = mg = (4.0\ \text{kg})(9.80\ \text{m/s}^2) = 39.2\ \text{N}$ and $f_k = \mu_k n$, we have $f_k = \mu_k n = (0.30)(39.2\ \text{N}) = 11.8\ \text{N}$. Therefore,

$$(1)\quad T = f_k + (4.0\ \text{kg})a = 11.8\ \text{N} + (4.0\ \text{kg})a$$

Now we apply Newton's second law to the ball moving in the vertical direction, where the downward direction is selected as positive:

$$\sum F_y = (7.0\ \text{kg})g - T = (7.0\ \text{kg})a$$

or

$$(2)\quad T = 68.6\ \text{N} - (7.0\ \text{kg})a$$

Subtracting (1) from (2) eliminates T:

$$56.8\ \text{N} - (11\ \text{kg})a = 0$$

$$a = 5.2\ \text{m/s}^2$$

When this value for the acceleration is substituted into (1), we get

$$T = 33\ \text{N}$$

EXERCISE 1   A block moves up a 45° incline with constant speed under the action of a force of 15 N applied *parallel* to the incline. If the coefficient of kinetic friction is 0.30, determine (a) the weight of the block and (b) the minimum force required to allow it to move *down* the incline at constant speed.   Answer   (a) 16.3 N   (b) 8.07 N up the incline

EXERCISE 2   A block is placed on a plane that is inclined at 60° with respect to the horizontal. If the block slides down the plane with an acceleration of $g/2$, determine the coefficient of kinetic friction between the block and the plane.   Answer   0.732

5.2 • NEWTON'S SECOND LAW APPLIED TO UNIFORM CIRCULAR MOTION

Solving problems involving friction is just one of many applications of Newton's second law. Let us now apply Newton's second law to another common situation: uniform circular motion. In Chapter 3, we found that a particle moving in a circular path of radius r with uniform speed v experiences an acceleration of magnitude

$$a_r = \frac{v^2}{r}$$

• *Centripetal acceleration*

Because the velocity vector **v** changes its direction continuously during the motion, the acceleration vector $\mathbf{a}_r$ is directed toward the center of the circle and hence is called **centripetal acceleration.** Furthermore, $\mathbf{a}_r$ is *always* perpendicular to **v**.

Consider a ball of mass m tied to a string of length r and being whirled in a horizontal circular path on a table top as in Figure 5.5. Let us assume that the ball moves with constant speed. The ball tends to maintain the motion in a straight-line path; however, the string prevents this motion along a straight line by exerting a force on the ball to make it follow its circular path. This force is directed along the length of the string toward the center of the circle, as shown in Figure 5.5. This force can be any one of our familiar forces playing the role of causing an object to follow a circular path.

If we apply Newton's second law along the radial direction, we find that the value of the force causing the centripetal acceleration can be evaluated:

$$F_r = ma_r = m\frac{v^2}{r} \qquad\qquad [5.3]$$

Figure 5.5   An overhead view of a ball moving in a circular path in a horizontal plane. A force $\mathbf{F}_r$ directed toward the center of the circle keeps the ball moving in the circle with constant speed.

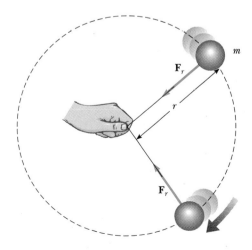

An athlete in the process of throwing the hammer. The force exerted by the chain on the hammer provides the centripetal acceleration. Only when the athlete releases the hammer will it move along a straight-line path tangent to the circle.   *(Focus on Sports)*

A force causing a centripetal acceleration acts toward the center of the circular path and causes a change in the direction of the velocity vector. In the case of a ball rotating at the end of a string, the force exerted by the string on the ball causes the centripetal acceleration of the ball. For a satellite in a circular orbit around the Earth, the force of gravity causes the centripetal acceleration of the satellite. The force acting on a car rounding a curve on a flat road to cause the centripetal acceleration of the car is the force of static friction between the tires and pavement, and so forth. In general, a body can move in a circular path under the influence of various types of forces or a combination of forces.

Regardless of the type of force causing the centripetal acceleration, if the force acting on an object should vanish, the object would no longer move in its circular path; instead it would move along a straight-line path tangent to the circle. This idea is illustrated in Figure 5.6 for the case of the ball whirling in a circle at the end of a string. If the string breaks at some instant, the ball will move along the straight-line path tangent to the circle at the point at which the string broke.

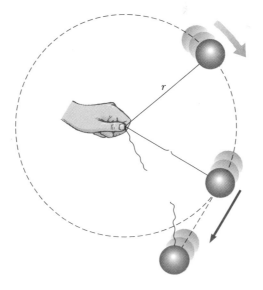

Figure 5.6 When the string breaks, the ball moves in the direction tangent to the circular path.

Thinking Physics 3

The force causing centripetal acceleration is often called a *centripetal force*. We are familiar with a variety of forces in nature—friction, gravity, normal forces, electrical forces, tension, and so on. Should we add *centripetal* force to this list?

Reasoning   Centripetal force should *not* be added to this list. This is a pitfall for many students. Giving the force causing circular motion a name—centripetal force—leads many students to consider this as a new *kind* of force rather than a new *role* for force. A common mistake in force diagrams is to draw in all of the usual forces and then add another vector for the centripetal force. But it is not a separate force—it is simply one of our familiar forces *acting in the role of causing a circular motion*. Consider some examples. For the motion of the Earth around the Sun, the centripetal force is *gravity*. For an object sitting on a rotating turntable, the centripetal force is *friction*. For a rock whirled on the end of a string, the centripetal force is the *tension* in the string. For an amusement park patron pressed against the inner wall of a rapidly rotating circular room, the centripetal force is the *normal force* from the wall. In addition, the centripetal force could be a combination of two or more forces. For example, as a Ferris-wheel rider passes through the lowest point of the ride, the centripetal force on the rider is the difference between the normal force from the seat and the force of gravity.

Thinking Physics 4

High-speed, curved roadways are *banked*—that is, the roadway is tilted toward the inside of tight curves. Why is this?

Reasoning   If an automobile is rounding a curve, there must be a force providing the centripetal acceleration. Normally, this is the static friction force between the tires and the roadway, parallel to the roadway surface and perpendicular to the centerline of the car. This is the only horizontal force—the weight and normal force are both vertical. If the curve is tight or the speed is high, the required friction force may be larger than the maximum possible friction force. In this case, Newton's first law takes over and the car skids along a straight line, which is off the side of the curved road. This situation is even more likely if the road is icy, which reduces the friction coefficient. Banking the roadway tilts the normal force away from the vertical, so that it has a horizontal component toward the center of the circular roadway. Thus, the normal force and friction work together to provide the needed centripetal acceleration. The larger the acceleration, the stronger the banking, in order to provide a larger component of the normal force in the horizontal direction.

As these cyclists in the Tour de France negotiate a curve on a flat racing track, the centripetal acceleration is provided by the force of static friction between the tires and the track surface. *(Michel Gouverneur, Photo News, Gamma Sport)*

CONCEPTUAL PROBLEM 4

An object executes circular motion with a constant speed whenever a net force of constant magnitude acts perpendicular to the velocity. What happens to the speed if the force is not perpendicular to the velocity?

Example 5.3   How Fast Can It Spin?

A ball of mass 0.500 kg is attached to the end of a cord, the length of which is 1.50 m. The ball is whirled in a horizontal circle as in Figure 5.6. If the cord can withstand a maximum tension of 50.0 N, what is the maximum speed the ball can have before the cord breaks?

Solution   Because the force that provides the centripetal acceleration in this case is the force **T** exerted by the cord on the ball, Equation 5.3 gives

$$T = m\frac{v^2}{r}$$

Solving for v, we have

$$v = \sqrt{\frac{Tr}{m}}$$

The maximum speed that the ball can have corresponds to the maximum value of the tension. Hence, we find

$$v_{max} = \sqrt{\frac{T_{max}r}{m}} = \sqrt{\frac{(50.0 \text{ N})(1.50 \text{ m})}{0.500 \text{ kg}}} = 12.2 \text{ m/s}$$

EXERCISE 3   Calculate the tension in the cord if the speed of the ball is 5.00 m/s.   Answer   8.33 N

Example 5.4   The Conical Pendulum

A small body of mass m is suspended from a string of length L. The body revolves in a horizontal circle of radius r with constant speed v, as in Figure 5.7. (Because the string sweeps out the surface of a cone, the system is known as a *conical pendulum*.) Find the speed of the body and the period of revolution, T_P, defined as the time needed to complete one revolution.

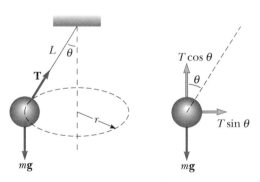

Figure 5.7   (Example 5.4) The conical pendulum and its free-body diagram.

Solution   The free-body diagram for the mass m is shown in Figure 5.7, where the force exerted by the string, **T**, has been resolved into a vertical component, $T\cos\theta$, and a component $T\sin\theta$ acting toward the center of rotation. Because the body does not accelerate in the vertical direction, the vertical component of **T** must balance the weight. Therefore,

$$(1) \quad T\cos\theta = mg$$

Because the force that provides the centripetal acceleration in this example is the component $T\sin\theta$, from Newton's second law we get

$$(2) \quad T\sin\theta = ma_r = \frac{mv^2}{r}$$

By dividing (2) by (1), we eliminate T and find that

$$\tan\theta = \frac{v^2}{rg}$$

But from the geometry we note that $r = L\sin\theta$; therefore,

$$v = \sqrt{rg\tan\theta} = \sqrt{Lg\sin\theta\tan\theta}$$

Because the ball travels a distance of $2\pi r$ (the circumference of the circular path) in a time equal to the period of revolution, T_P (not to be confused with the force **T**), we find

$$(3) \quad T_P = \frac{2\pi r}{v} = \frac{2\pi r}{\sqrt{rg\tan\theta}} = 2\pi\sqrt{\frac{L\cos\theta}{g}}$$

The intermediate algebraic steps used in obtaining (3) are left to the reader. Note that T_P is independent of m. If we take $L = 1.00$ m and $\theta = 20.0°$, using (3) we find that

$$T_P = 2\pi\sqrt{\frac{(1.00 \text{ m})(\cos 20.0°)}{9.80 \text{ m/s}^2}} = 1.95 \text{ s}$$

Is it physically possible to have a conical pendulum with $\theta = 90°$?

Example 5.5 What Is the Maximum Speed of the Car?

A 1500-kg car moving on a flat, horizontal road negotiates a curve, the radius of which is 35.0 m, as in Figure 5.8. If the coefficient of static friction between the tires and the dry pavement is 0.500, find the maximum speed the car can have in order to make the turn successfully.

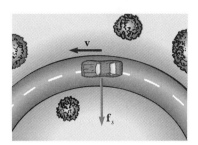

Figure 5.8 (Example 5.5) The force of static friction directed toward the center of the arc keeps the car moving in a circle.

Solution In this case, the force that provides the centripetal acceleration and enables the car to remain in its circular path is the force of static friction. Hence, from Equation 5.3 we have

$$(1) \quad f_s = m\frac{v^2}{r}$$

The maximum speed that the car can have around the curve corresponds to the speed at which it is on the verge of skidding outward. At this point, the friction force has its maximum value

$$f_{s,\,max} = \mu_s n$$

Because the magnitude of the normal force equals the weight in this case, we find

$$f_{s,\,max} = \mu_s mg = (0.500)(1500\text{ kg})(9.80\text{ m/s}^2) = 7350\text{ N}$$

Substituting this value into (1), we find that the maximum speed is

$$v_{max} = \sqrt{\frac{f_{s,\,max}\,r}{m}} = \sqrt{\frac{(7350\text{ N})(35.0\text{ m})}{1500\text{ kg}}} = \boxed{13.1\text{ m/s}}$$

EXERCISE 4 On a wet day, the car described in this example begins to skid on the curve when its speed reaches 8.00 m/s. What is the coefficient of static friction in this case?
Answer 0.187

Example 5.6 Let's Go Loop-the-loop

A pilot of mass m in a jet aircraft executes a loop-the-loop maneuver, as illustrated in Figure 5.9a. In this flying pattern, the aircraft moves in a vertical circle of radius 2.70 km at a *constant speed* of 225 m/s. Determine the force exerted by the seat on the pilot at (a) the bottom of the loop and (b) the top of the loop. Express the answers in terms of the weight of the pilot, *mg*.

Solution (a) The free-body diagram for the pilot at the bottom of the loop is shown in Figure 5.9b. The only forces acting on the pilot are the downward force of gravity, *m***g**, and the upward force $\mathbf{n}_{bot}$ exerted by the seat on the pilot. Because the net upward force that provides the centripetal acceleration has a magnitude $n_{bot} - mg$, Newton's second law for the radial direction gives

$$n_{bot} - mg = m\frac{v^2}{r}$$

$$n_{bot} = mg + m\frac{v^2}{r} = mg\left[1 + \frac{v^2}{rg}\right]$$

Substituting the values given for the speed and radius gives

$$n_{bot} = mg\left[1 + \frac{(225\text{ m/s})^2}{(2.70 \times 10^3\text{ m})(9.80\text{ m/s}^2)}\right] = \boxed{2.91\,mg}$$

Hence, the force exerted by the seat on the pilot is *greater* than his weight by a factor of 2.91. The pilot experiences an apparent weight that is greater than his weight by the factor 2.91. This is discussed further in Section 5.3.

(b) The free-body diagram for the pilot at the top of the loop is shown in Figure 5.9c. At this point, both the force of gravity and the force exerted by the seat on the pilot, $\mathbf{n}_{top}$, act *downward*, so the net force downward that provides the centripetal acceleration has a magnitude $n_{top} + mg$. Applying Newton's second law gives

$$n_{top} + mg = m\frac{v^2}{r}$$

$$n_{top} = m\frac{v^2}{r} - mg = mg\left[\frac{v^2}{rg} - 1\right]$$

$$n_{top} = mg\left[\frac{(225\text{ m/s})^2}{(2.70 \times 10^3\text{ m})(9.80\text{ m/s}^2)} - 1\right]$$

$$= \boxed{0.911\,mg}$$

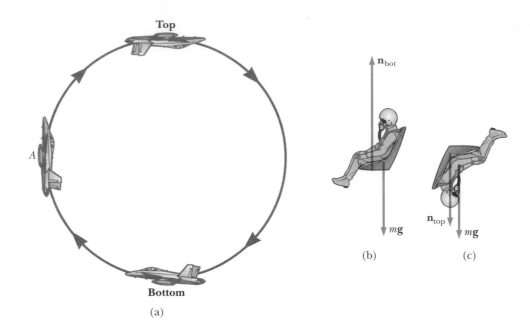

Figure 5.9   (Example 5.6)

In this case, the force exerted by the seat on the pilot is *less* than the weight by a factor of 0.911. Hence, the pilot will feel lighter at the top of the loop.

EXERCISE 5   Calculate the force on the pilot causing the centripetal acceleration if the aircraft is at point *A* in Figure 5.9a, midway up the loop.   Answer $1.911mg$ directed to the right

EXERCISE 6   An automobile moves at constant speed over the crest of a hill. The driver moves in a vertical circle of radius 18.0 m. At the top of the hill, she notices that she barely remains in contact with the seat. Find the speed of the vehicle.   Answer   13.3 m/s

EXERCISE 7   A 1500-kg car rounds an unbanked curve with a radius of 52 m at a speed of 12 m/s. What minimum coefficient of friction must exist between the road and tires to prevent the car from slipping?   Answer   0.283

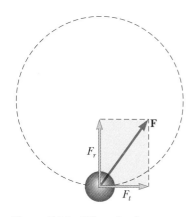

Figure 5.10   When the force acting on a particle moving in a circular path has a tangential component F_t, its speed changes. The total force on the particle also has a component F_r directed toward the center of the circular path. Thus, total force is $\mathbf{F} = \mathbf{F}_t + \mathbf{F}_r$.

5.3 • NONUNIFORM CIRCULAR MOTION

In Chapter 3 we found that if a particle moves with varying speed in a circular path, there is, in addition to the centripetal component of acceleration, a tangential component of magnitude dv/dt. Therefore, the force acting on the particle must also have a tangential and a radial component, as shown in Figure 5.10. That is, because the total acceleration is $\mathbf{a} = \mathbf{a}_r + \mathbf{a}_t$, the total force exerted on the particle is $\mathbf{F} = \mathbf{F}_r + \mathbf{F}_t$. The component F_r is directed toward the center of the circle and is responsible for the centripetal acceleration. The component F_t tangent to the circle is responsible for the tangential acceleration, which causes the speed of the particle to change with time. The following examples demonstrate this type of motion.

Thinking Physics 5

Suppose you are the only rider on a Ferris wheel. The drive mechanism breaks after the ride starts, so that you continue to rotate freely, without friction. As you pass through the lowest point (let us call it the 6-o'clock point), you will feel the heaviest—that is, the seat will press upward on you with the largest force. As you pass over the highest point (the 12-o'clock point), you will feel the lightest. At the 3-o'clock and 9-o'clock points, will you feel as if you have the "correct" weight? If not, where *do* you feel as if your weight is correct?

Reasoning If the wheel were rotating at a constant speed, the answer to the question would be *yes*. If you are the only rider, however, the wheel is *unbalanced* and would not rotate with a constant speed. As you come down from the top point, your velocity increases. After you pass through the lowest point, your velocity decreases as you move back to the top. Thus, you will experience tangential acceleration as well as centripetal acceleration. At the 3-o'clock position, at which you are moving downward, you have a downward tangential acceleration. Thus, just as in a downward accelerating elevator, you will feel lighter. At the 9-o'clock position, when you are moving upward, you are accelerating downward tangentially. Again, you will feel lighter. In order to counteract the *downward* component of tangential acceleration near both positions, you need some *upward* component from the centripetal acceleration. This will be present if you are at positions *below* 3 o'clock or 9 o'clock, so that the centripetal acceleration has an upward component. The exact positions will depend on your mass and the details of the Ferris wheel.

Example 5.7 Follow the Rotating Ball

A small sphere of mass m is attached to the end of a cord of length R, which rotates in a *vertical* circle about a fixed point O, as in Figure 5.11a. Let us determine the tension in the cord at any instant when the speed of the sphere is v and the cord makes an angle θ with the vertical.

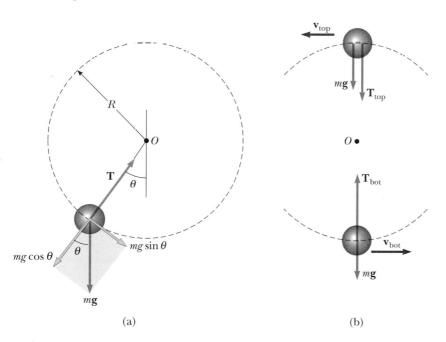

Figure 5.11 (Example 5.7) (a) Forces acting on a mass m connected to a string of length R and rotating in a vertical circle centered at O. (b) Forces acting on m when it is at the top and bottom of the circle. Note that the tension at the bottom is a maximum and the tension at the top is a minimum.

(a)

(b)

Solution   First we note that the speed is *not* uniform, because there is a tangential component of acceleration arising from the weight of the sphere. From the free-body diagram in Figure 5.11a, we see that the only forces acting on the sphere are the force of gravity, $m\mathbf{g}$, and the force exerted by the cord, $\mathbf{T}$. Now we resolve $m\mathbf{g}$ into a tangential component, $mg\sin\theta$, and a radial component, $mg\cos\theta$. Applying Newton's second law to the forces in the tangential direction gives

$$\sum F_t = mg\sin\theta = ma_t$$

$$a_t = g\sin\theta$$

This component causes v to change in time, because $a_t = dv/dt$.

Applying Newton's second law to the forces in the radial direction and noting that both $\mathbf{T}$ and $\mathbf{a}_r$ are directed toward O, we get

$$\sum F_r = T - mg\cos\theta = \frac{mv^2}{R}$$

$$T = m\left(\frac{v^2}{R} + g\cos\theta\right)$$

Limiting Cases   At the top of the path, where $\theta = 180°$, we have $\cos 180° = -1$, and the tension equation becomes

$$T_{\text{top}} = m\left(\frac{v_{\text{top}}{}^2}{R} - g\right)$$

This is the *minimum* value of T. Note that at this point $a_t = 0$ and therefore the acceleration is radial and directed downward, as in Figure 5.11b.

At the bottom of the path, where $\theta = 0$, we see that because $\cos 0 = 1$,

$$T_{\text{bot}} = m\left(\frac{v_{\text{bot}}{}^2}{R} + g\right)$$

This is the *maximum* value of T. Again, at this point $a_t = 0$ and the acceleration is radial and directed upward.

EXERCISE 8   At what orientation of the system would the cord most likely break if the average speed increased?   Answer   At the bottom of the path, where T has its maximum value.

EXERCISE 9   A 0.40-kg object is swung in a vertical circular path on a string 0.50 m long. If a constant speed of 4.0 m/s is maintained, what is the tension in the string when the object is at the top of the circle?   Answer   8.88 N

EXERCISE 10   A car traveling on a straight road at 9.00 m/s goes over a hump in the road. The hump may be regarded as an arc of a circle of radius 11.0 m. What must be the speed of the car over the hump if a 600-N passenger is to experience weightlessness?   Answer   10.4 m/s

5.4 • MOTION IN THE PRESENCE OF VELOCITY-DEPENDENT RESISTIVE FORCES

Earlier we described the interaction between a moving object and the surface along which it moves. We completely ignored any interaction between the object and the medium through which it moves. Now let us consider the effect of a medium such as a liquid or gas. The medium exerts a **resistive force, R,** on the object moving through it. The magnitude of this force depends on the speed of the object, and the direction of **R** is always opposite the direction of motion of the object relative to the medium. In general, the magnitude of the resistive force increases with increasing speed. Some examples are the air resistance associated with moving vehicles (sometimes called air drag) and the viscous forces that act on objects moving through a liquid.

The resistive force can have a complicated speed dependence. In the following discussions, we consider two situations. First, we assume that the resistive force is proportional to the speed; this is the case for objects that fall through a liquid with low speed and for very small objects, such as dust particles, that move through air. Second, we treat situations for which the resistive force is proportional to the square of the speed of the object; large objects, such as a skydiver moving through air in free fall, experience such a force.

Resistive Force Proportional to Object Speed

If we assume that the resistive force acting on an object that is moving through a viscous medium is proportional to the object's velocity, then the resistive force can be expressed as

$$\mathbf{R} = -b\mathbf{v} \qquad [5.4]$$

where $\mathbf{v}$ is the velocity of the object and b is a constant that depends on the properties of the medium and on the shape and dimensions of the object. If the object is a sphere of radius r, then b is proportional to r.

Consider a sphere of mass m released from rest in a liquid, as in Figure 5.12a. Assuming the only forces acting on the sphere are the resistive force, $-b\mathbf{v}$, and the gravitational force, $m\mathbf{g}$, let us describe its motion.[1]

Applying Newton's second law to the vertical motion, choosing the downward direction to be positive, and noting that $\Sigma F_y = mg - bv$, we get

$$mg - bv = m\frac{dv}{dt}$$

where the acceleration is downward. Simplifying this expression gives

$$\frac{dv}{dt} = g - \frac{b}{m}v \qquad [5.5]$$

Equation 5.5 is called a *differential equation*, and the methods of solving such an equation may not be familiar to you as yet. However, note that initially, when $v = 0$, the resistive force is zero and the acceleration, dv/dt, is simply g. As t increases, the resistive force increases and the acceleration decreases. Eventually, the acceleration becomes zero when the resistive force balances the weight. At this point, the object reaches its **terminal speed**, v_t, and from then on it continues to move with zero acceleration. The terminal speed can be obtained from Equation 5.5 by setting $a = dv/dt = 0$. This gives

$$mg - bv_t = 0 \qquad \text{or} \qquad v_t = \frac{mg}{b}$$

The expression for v that satisfies Equation 5.5 with $v = 0$ at $t = 0$ is

$$v = \frac{mg}{b}(1 - e^{-bt/m}) = v_t(1 - e^{-t/\tau}) \qquad [5.6]$$

This function is plotted in Figure 5.12b. The time constant $\tau = m/b$ is the time it takes the object to reach 63.2% of its terminal speed. This can be seen by noting that when $t = \tau$, Equation 5.6 gives $v = 0.632v_t$. We can check that Equation 5.6 is a solution to Equation 5.5 by direct differentiation:

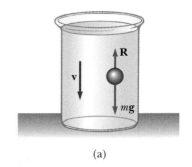

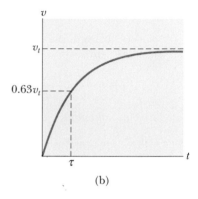

Figure 5.12   (a) A small sphere falling through a viscous fluid. (b) The speed–time graph for an object falling through a viscous medium. The object reaches a maximum, or terminal speed v_t, and τ is the time it takes to reach $0.63v_t$.

• *Terminal speed*

---

[1]There is also a *buoyant* force acting on the submerged object, and this force is constant and equal to the weight of the displaced fluid, as will be discussed in Chapter 12. This force will only change the apparent weight of the sphere by a constant factor, so we can ignore it here.

$$\frac{dv}{dt} = \frac{d}{dt}\left(\frac{mg}{b} - \frac{mg}{b}\,e^{-bt/m}\right) = -\frac{mg}{b}\frac{d}{dt}\,e^{-bt/m} = ge^{-bt/m}$$

Substituting this expression and Equation 5.6 into Equation 5.5 shows that our solution satisfies the differential equation.

Example 5.8   A Sphere Falling in Oil

A small sphere of mass 2.00 g is released from rest in a large vessel filled with oil. The sphere reaches a terminal speed of 5.00 cm/s. Determine the time constant τ and the time it takes the sphere to reach 90% of its terminal speed.

Solution   Because the terminal speed is given by $v_t = mg/b$, the coefficient b is

$$b = \frac{mg}{v_t} = \frac{(2.00 \text{ g})(980 \text{ cm/s}^2)}{5.00 \text{ cm/s}} = 392 \text{ g/s}$$

Therefore, the time τ is

$$\tau = \frac{m}{b} = \frac{2.00 \text{ g}}{392 \text{ g/s}} = 5.10 \times 10^{-3} \text{ s}$$

The speed of the sphere as a function of time is given by Equation 5.6. To find the time t it takes the sphere to reach a speed of $0.900v_t$, we set $v = 0.900v_t$ into the expression and solve for t:

$$0.900v_t = v_t(1 - e^{-t/\tau})$$

$$1 - e^{-t/\tau} = 0.900$$

$$e^{-t/\tau} = 0.100$$

$$-\frac{t}{\tau} = -2.30$$

$$t = 2.30\tau = 2.30(5.10 \times 10^{-3} \text{ s})$$

$$= 11.7 \times 10^{-3} \text{ s} = 11.7 \text{ ms}$$

EXERCISE 11   What is the sphere's speed through the oil at $t = 11.7$ ms? Compare this value with the speed the sphere would have if it were falling in a vacuum and so influenced only by gravity.   Answer   4.50 cm/s in oil versus 11.5 cm/s in free fall

Air Drag at High Speeds

For large objects moving at high speeds through air, such as airplanes, skydivers, and baseballs, the resistive force is approximately proportional to the square of the speed. In these situations, the magnitude of the resistive force can be expressed as

$$R = \tfrac{1}{2}D\rho Av^2 \qquad\qquad \textbf{[5.7]}$$

where ρ is the density of air, A is the cross-sectional area of the falling object measured in a plane perpendicular to its motion, and D is a dimensionless empirical quantity called the *drag coefficient*. The drag coefficient has a value of about 0.5 for spherical objects but can be as high as 2.0 for irregularly shaped objects.

Aerodynamic car. Streamlined bodies are used for sports cars and other vehicles to reduce air drag and increase fuel efficiency.   *(© 1992 Dick Kelley)*

Consider an airplane in flight that experiences such a resistive force. Equation 5.7 shows that the force is proportional to the density of air and hence decreases with decreasing air density. Because air density decreases with increasing altitude, the resistive force on a jet airplane flying at a given speed must also decrease with increasing altitude. However, if at given altitude the plane's speed is doubled, the resistive force increases by a factor of four. In order to maintain this increased speed, the propulsive force must also increase by a factor of four.

Now let us analyze the motion of a mass in free fall subject to an upward air resistive force, the magnitude of which is $R = \frac{1}{2}D\rho Av^2$. Suppose a mass m is released from rest from the position $y = 0$, as in Figure 5.13. The mass experiences two external forces: the downward force of gravity, $m\mathbf{g}$, and the resistive force, $\mathbf{R}$, upward. (There is also an upward buoyant force that we neglect.) Hence, the magnitude of the net force is

$$F_{\text{net}} = mg - \tfrac{1}{2}D\rho Av^2 \qquad [5.8]$$

Substituting $F_{\text{net}} = ma$ into Equation 5.8, we find that the mass has a downward acceleration of magnitude

$$a = g - \left(\frac{D\rho A}{2m}\right)v^2 \qquad [5.9]$$

Again, we can calculate the terminal speed, v_t, using the fact that when the force of gravity is balanced by the resistive force, the net force is zero and therefore the acceleration is zero. Setting $a = 0$ in Equation 5.9 gives

$$g - \left(\frac{D\rho A}{2m}\right)v_t^2 = 0$$

$$v_t = \sqrt{\frac{2mg}{D\rho A}} \qquad [5.10]$$

Using this expression, we can determine how the terminal speed depends on the dimensions of the object. Suppose the object is a sphere of radius r. In this case, $A \propto r^2$ and $m \propto r^3$ (because the mass is proportional to the volume). Therefore, $v_t \propto \sqrt{r}$.

Table 5.2 lists the terminal speeds for several objects falling through air, all computed on the assumption that the drag coefficient is 0.5

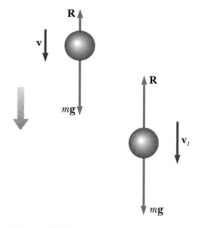

Figure 5.13 An object falling through air experiences a resistive drag force $\mathbf{R}$ and a gravitational force $m\mathbf{g}$. The object reaches terminal speed (on the right) when the net force is zero, that is, when $\mathbf{R} = -m\mathbf{g}$, or $R = mg$. Before this occurs, the acceleration varies with speed according to Equation 5.9.

TABLE 5.2   Terminal Speeds for Various Objects Falling Through Air

Object	Mass (kg)	Cross-sectional Area (m²)	v_t(m/s)[a]
Skydiver	75.0	0.70	60
Baseball (radius 3.7 cm)	0.145	4.2×10^{-3}	33
Golf ball (radius 2.1 cm)	0.046	1.4×10^{-3}	32
Hailstone (radius 0.50 cm)	4.8×10^{-4}	7.9×10^{-5}	14
Raindrop (radius 0.20 cm)	3.4×10^{-5}	1.3×10^{-5}	9

[a] The drag coefficient, D, is assumed to be 0.5 in each case.

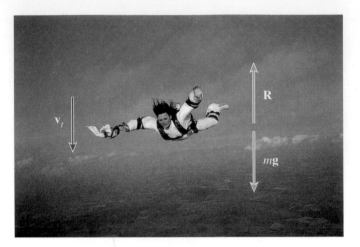

Figure 5.14   (Conceptual Problem 5) By spreading their arms and legs and by keeping their bodies parallel to the ground, skydivers experience maximum air resistance resulting in a minimum terminal speed.   *(Guy Sauvage, Photo Researchers, Inc.)*

CONCEPTUAL PROBLEM 5

Consider a skydiver falling through air before reaching her terminal speed, as in Figure 5.14. As the speed of the skydiver increases, what happens to her acceleration? What is her acceleration once she reaches terminal speed?

5.5 • NUMERICAL MODELING IN PARTICLE DYNAMICS[2]

As we have seen in this and the preceding chapter, the study of dynamics of a particle focuses on describing the position, velocity, and acceleration as functions of time. Cause-and-effect relationships exist between these quantities: A velocity causes position to change, an acceleration causes velocity to change, and an acceleration is the direct result of applied forces. Therefore, the study of motion usually begins with an evaluation of the net force on the particle.

In this section, we confine our discussion to one-dimensional motion, so boldface notation will not be used for vector quantities. If a particle of mass m moves under the influence of a net force F, Newton's second law tells us that the acceleration of the particle is given by $a = F/m$. In general, we can then obtain the solution to a dynamics problem by using the following procedure:

1. Sum all the forces on the particle to get the net force F.
2. Use this force to determine the acceleration, using $a = F/m$.
3. Use this acceleration to determine the velocity from $dv/dt = a$.
4. Use this velocity to determine the position from $dx/dt = v$.

---

[2]The author is most grateful to Colonel James Head of the U.S. Air Force Academy for preparing this section.

Example 5.9   An Object Falling in a Vacuum

We can illustrate the procedure just described by considering a particle falling in a vacuum under the influence of the force of gravity, as in Figure 5.15.

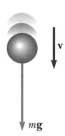

Figure 5.15   (Example 5.9) An object falling in vacuum under the influence of gravity.

Reasoning and Solution   Applying Newton's second law, we set the sum of the external forces equal to the mass of the particle times its acceleration (taking upward to be the positive direction):

$$F = ma = -mg$$

Thus, $a = -g$, a constant. Because $dv/dt = a$, the resulting differential equation for velocity is $dv/dt = -g$, which may be integrated to give

$$v(t) = v_0 - gt$$

Then, because $dx/dt = v$, the position of the particle is obtained from another integration, which yields the well-known result

$$x(t) = x_0 + v_0 t - \tfrac{1}{2}gt^2$$

In this expression, x_0 and v_0 represent the position and speed of the particle at $t = 0$.

The procedure just described is straightforward for some physical situations, such as the one described in the previous example. In the real world, however, complications often arise that make analytical solutions for many practical situations difficult and perhaps beyond the mathematical abilities of most students taking introductory physics. For example, the force may depend on the position of the particle, as in cases in which the variation in gravitational acceleration with height must be taken into account. The force may also vary with velocity, as we have seen in cases of resistive forces caused by motion through a liquid or gas. The force may depend on both position and velocity, as in the case of an object falling through air where the resistive force depends on velocity and on height (air density). In rocket motion, the mass changes with time, so even if the force is constant, the acceleration is not.

Another complication arises because the equations relating acceleration, velocity, position, and time are differential equations, not algebraic equations. Differential equations are usually solved using integral calculus and other special techniques that introductory students may not have mastered.

So how does one proceed to solve real-world problems without advanced mathematics? One answer is to solve such problems on personal computers, using elementary numerical methods. The simplest of these is the Euler method, named after the Swiss mathematician Leonhard Euler (1707–1783).

The Euler Method

In the **Euler method** of solving differential equations, derivatives are approximated as finite differences. Considering a small increment of time, Δt, the relationship between speed and acceleration may be approximated as

$$a(t) = \frac{\Delta v}{\Delta t} = \frac{v(t + \Delta t) - v(t)}{\Delta t}$$

Then the speed of the particle at the end of the period Δt is approximately equal to the speed at the beginning of the period, plus the acceleration during the interval multiplied by Δt:

$$v(t + \Delta t) = v(t) + a(t)\,\Delta t \qquad \text{[5.11]}$$

Because the acceleration is a function of time, this estimate of $v(t + \Delta t)$ will be accurate only if the time interval Δt is short enough that the change in acceleration during it is very small (as will be discussed later).

The position can be found in the same manner:

$$v(t) = \frac{x(t + \Delta t) - x(t)}{\Delta t}$$

$$x(t + \Delta t) = x(t) + v(t)\,\Delta t \qquad \text{[5.12]}$$

It may be tempting to add the term $\frac{1}{2}a(\Delta t)^2$ to this result to make it look like the familiar kinematics equation, but this term is not included in the Euler method of integration because Δt is assumed to be so small that $(\Delta t)^2$ is nearly zero.

If the acceleration at any instant t is known, the particle's velocity and position at $(t + \Delta t)$ can be calculated from Equations 5.11 and 5.12. The calculation can then proceed in a series of finite steps to determine the velocity and position at any later time. The acceleration is determined by the net force acting on the object,

$$a(x, v, t) = \frac{F(x, v, t)}{m} \qquad \text{[5.13]}$$

which may depend explicitly on the position, velocity, or time.

It is convenient to set up the numerical solution to this kind of problem by numbering the steps and entering the calculations in a table. Table 5.3 illustrates how to do this in an orderly way. The equations provided in the table can be entered into a spreadsheet and the calculations performed row by row to determine the velocity, position, and acceleration as functions of time. The calculations can also be carried out using a program written in BASIC, PASCAL, or FORTRAN, or with commercially available mathematics packages for personal computers. Many small increments can be taken, and accurate results can usually be obtained with the help of a computer. Graphs of velocity versus time or position versus time can be displayed to help you visualize the motion.

TABLE 5.3   The Euler Method for Solving Dynamics Problems

Step	Time	Position	Velocity	Acceleration
0	t_0	x_0	v_0	$a_0 = F(x_0, v_0, t_0)/m$
1	$t_1 = t_0 + \Delta t$	$x_1 = x_0 + v_0\,\Delta t$	$v_1 = v_0 + a_0\,\Delta t$	$a_1 = F(x_1, v_1, t_1)/m$
2	$t_2 = t_1 + \Delta t$	$x_2 = x_1 + v_1\,\Delta t$	$v_2 = v_1 + a_1\,\Delta t$	$a_2 = F(x_2, v_2, t_2)/m$
3	$t_3 = t_2 + \Delta t$	$x_3 = x_2 + v_2\,\Delta t$	$v_3 = v_2 + a_2\,\Delta t$	$a_3 = F(x_3, v_3, t_3)/m$
.	.	.	.	.
.	.	.	.	.
.	.	.	.	.
n	t_n	x_n	v_n	a_n

The Euler method has the advantage that the dynamics are not obscured—the fundamental relationships of acceleration to force, velocity to acceleration, and position to velocity are clearly evident. Indeed, these relationships form the heart of the calculations. There is no need to use advanced mathematics, and the basic physics governs the dynamics.

The Euler method is completely reliable for infinitesimally small time increments, but for practical reasons a finite increment size must be chosen. In order for the finite difference approximation of Equation 5.11 to be valid, the time increment must be small enough that the acceleration can be approximated as being constant during the increment. We can determine an appropriate size for the time increment by examining the particular problem that is being investigated. The criterion for the size of the time increment may need to be changed during the course of the motion. In practice, however, we usually choose a time increment appropriate to the initial conditions and use the same value throughout the calculations.

The size of the time increment influences the accuracy of the result, but unfortunately it is not easy to determine the accuracy of a solution by the Euler method without a knowledge of the correct analytical solution. One method of determining the accuracy of the numerical solution is to repeat the calculations with a smaller time increment and compare results. If the two calculations agree to a certain number of significant figures, you can assume that the results are correct to that precision.

5.6 • THE FUNDAMENTAL FORCES OF NATURE

We have described a variety of forces that are experienced in our everyday activities, such as the force of gravity that acts on all objects at or near the Earth's surface and the force of friction as one surface slides over another. Forces also act in the atomic and subatomic world. For example, atomic forces within the atom are responsible for holding its constituents together, and nuclear forces act on different parts of the nucleus to keep its parts from separating.

Until recently, physicists believed that there were four fundamental forces in nature: the gravitational force, the electromagnetic force, the strong nuclear force, and the weak nuclear force.

The Gravitational Force

The gravitational force is the mutual force of attraction between any two objects in the Universe. It is interesting and rather curious that, although the gravitational force can be very strong between macroscopic objects, it is the weakest of all the fundamental forces. For example, the gravitational force between the electron and proton in the hydrogen atom is only about 10^{-47} N, whereas the electromagnetic force between these same two particles is about 10^{-7} N.

Newton's law of gravitation states that every particle in the Universe attracts every other particle with a force that is directly proportional to the product of the two masses and inversely proportional to the square of the distance between them.

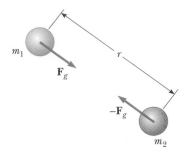

Figure 5.16 Two particles of masses m_1 and m_2 attract each other with a force of magnitude Gm_1m_2/r^2.

If the particles have masses m_1 and m_2 and are separated by a distance r, as in Figure 5.16, the magnitude of the gravitational force is

$$F_g = G\frac{m_1 m_2}{r^2} \qquad [5.14]$$

where $G = 6.67 \times 10^{-11}$ N·m²/kg² is the universal gravitational constant.

The Electromagnetic Force

The electromagnetic force is the force that binds atoms and molecules in compounds to form ordinary matter. It is much stronger than the gravitational force. The force that causes a rubbed comb to attract bits of paper and the force that a magnet exerts on an iron nail are electromagnetic forces. Essentially all forces at work in our macroscopic world (apart from the gravitational force) are manifestations of the electromagnetic force. For example, friction forces, contact forces, tension forces, and forces in elongated springs are consequences of electromagnetic forces between charged particles in proximity.

The electromagnetic force involves two types of particles: those with positive charge and those with negative charge. (The properties of these two types of charges will be discussed shortly.) Unlike the gravitational force, which is always an attractive interaction, the electromagnetic force can be either attractive or repulsive, depending on the charges on the particles. In general, the electromagnetic force acts between two charged particles that are in relative motion.

When the charged particles are at rest relative to each other, we refer to the electromagnetic force between them as the electrostatic force. Coulomb's law expresses the magnitude of the electrostatic force F_e between two charged particles separated by a distance r:

$$F_e = k_e\frac{q_1 q_2}{r^2} \qquad [5.15]$$

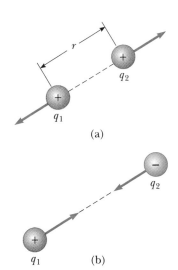

Figure 5.17 Two point charges separated by a distance r exert an electrostatic force on each other given by Coulomb's law. (a) When the charges are of the same sign, the charges repel each other. (b) When the charges are of opposite sign, the charges attract each other.

where q_1 and q_2 are the charges on the two particles, measured in coulombs (C), and k_e is the Coulomb constant, with the value $k_e = 8.99 \times 10^9$ N·m²/C². Note that the electrostatic force has the same mathematical form as Newton's law of gravity (Eq. 5.14); however, charge replaces mass, and the constants are different. The electrostatic force is attractive if the two charges have opposite signs and repulsive if the two charges have the same sign, as indicated in Figure 5.17. In other words, opposite charges attract each other, and like charges repel each other.

The smallest amount of isolated charge found (so far) in nature is the charge on an electron or proton. This fundamental unit of charge has the magnitude $e = 1.60 \times 10^{-19}$ C. Theories developed in the 1970s and 1980s propose that protons and neutrons are made up of smaller particles called quarks, which have charges of either $2e/3$ or $-e/3$ (discussed further in Chapter 31). Although experimental evidence has been found for such particles inside nuclear matter, free quarks have never been detected.

The most important properties of electric charges can be summarized as follows:

- Charges are scalar quantities; hence, their values are additive.
- The two kinds of charges that exist in nature are labeled positive and negative. Charges of like sign repel each other, and charges of opposite sign attract each other.
- The net charge on any object has discrete values; in other words, charge is said to be quantized. Any isolated elementary charged particle has a charge of e. For example, electrons have a charge $-e$ and protons have a charge $+e$. (Neutrons have no charge.) Because e is the fundamental unit of charge, the net charge of an object is Ne, where N is an integer. For macroscopic objects, N is very large and the quantization of charge is not noticed.
- Electric charge is always conserved. This means that in any kind of process—such as a collision event, a chemical reaction, or nuclear decay—the total charge of an isolated system remains constant.

• *Properties of electric charge*

The Strong Force

An atom, as we currently understand it, consists of an extremely dense, positively charged nucleus surrounded by a cloud of negatively charged electrons, with the electrons attracted to the nucleus by the Coulomb force. Because all nuclei except those of hydrogen are combinations of positively charged protons and neutral neutrons (collectively called nucleons), why does the repulsive electrostatic force between the protons not cause nuclei to break apart? Clearly, there must be an attractive force that overcomes the strong electrostatic repulsive force and is responsible for the stability of nuclei. This great force that binds the nucleons to form a nucleus is called the *strong force*. It should be noted that electrons and certain other particles are immune to the strong force. Unlike the gravitational and electromagnetic forces, which depend on distance in an inverse-square fashion, the strong force is extremely short-range; its strength decreases very rapidly outside the nucleus and is negligible for separations greater than approximately 10^{-14} m. For separations of approximately 10^{-15} m (a typical nuclear dimension), the strong force is about two orders of magnitude stronger than the electromagnetic force.

The Weak Force

The *weak force* is a short-range force that tends to produce instability in certain nuclei. It was first observed in naturally occurring radioactive substances and was later found to play a key role in most radioactive decay reactions. The weak force is about 10^{25} times stronger than the gravitational force and about 10^{12} times weaker than the electromagnetic force.

The Current View of Fundamental Forces

For years physicists have searched for a simplification scheme that would reduce the number of fundamental forces in nature. In 1967 physicists predicted that the electromagnetic force and the weak force, originally thought to be independent of each other and both fundamental, are in fact manifestations of one force, now called the *electroweak force*. The prediction was confirmed experimentally in 1984. We also now know that protons and neutrons are not fundamental particles but are composed of simpler particles called quarks. For example, the proton consists of

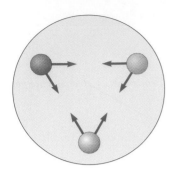

Figure 5.18 A proton is composed of three quarks that are bound together by a very strong force called the color force.

three quarks as shown in Figure 5.18. Quarks are bound together by a strong force called the **color force.**

Scientists believe that the fundamental forces of nature are closely related to the origin of the Universe. The Big Bang theory holds that the Universe began with a cataclysmic explosion 15 billion to 20 billion years ago. According to this theory, the first moments after the Big Bang saw such extremes of energy that all the fundamental forces were unified into one force. Physicists are continuing their search for connections among the known fundamental forces—connections that could eventually prove that the forces are all merely different forms of a single superforce. The recent success linking the electromagnetic and weak nuclear forces has spurred greater efforts at a unification scheme (yet to be proved) called the Grand Unified Theory. This fascinating subject continues to be at the forefront of physics.

5.7 • THE GRAVITATIONAL FIELD

When Newton first published his theory of gravitation, his contemporaries found it difficult to accept the concept of a force that one object could exert on another without anything happening in the space between them. They asked how it was possible for two masses to interact even though they were not in contact with each other. Although Newton himself could not answer this question, his theory was considered a success because it satisfactorily explained the motions of the planets.

An alternative approach is to think of the gravitational interaction as a two-step process. First, one object creates a gravitational field **g** throughout the space around it. Then a second object of mass m experiences a force $\mathbf{F}_g = m\mathbf{g}$. In other words, the field **g** exerts a force on the particle. Hence, the gravitational field is defined by

Definition of gravitational field •

$$\mathbf{g} \equiv \frac{\mathbf{F}_g}{m} \qquad [5.16]$$

That is, the gravitational field at a point in space equals the gravitational force that a test mass experiences at that point divided by the mass. As a consequence, if **g** is known at some point in space, a test particle of mass m experiences a gravitational force $m\mathbf{g}$ when placed at that point.

As an example, consider an object of mass m near the Earth's surface. The gravitational force on the object is directed toward the center of the Earth and has a magnitude mg. Thus we see that the gravitational field experienced by the object at some point has a magnitude equal to the acceleration of gravity at that point. Because the gravitational force on the object has a magnitude $GM_E m/r^2$ (where M_E is the mass of the Earth), the field **g** at a distance r from the center of the Earth is given by

Gravitational field of the •
Earth

$$\mathbf{g} = \frac{\mathbf{F}_g}{m} = -\frac{GM_E}{r^2}\,\hat{\mathbf{r}} \qquad [5.17]$$

where $\hat{\mathbf{r}}$ is a unit vector pointing radially outward from the Earth, and the minus sign indicates that the field points toward the center of the Earth, as shown in Figure

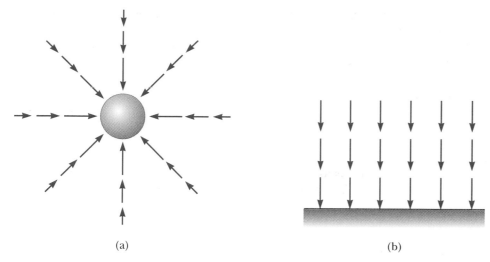

(a)   (b)

Figure 5.19   (a) The gravitational field vectors in the vicinity of a uniform spherical mass vary in both direction and magnitude. (b) The gravitational field vectors in a small region near the Earth's surface are uniform—that is, they match in both direction and magnitude.

5.19a. Note that the field vectors at different points surrounding the spherical mass vary in both direction and magnitude. In a small region near the Earth's surface, **g** is approximately constant and the downward field is uniform, as indicated in Figure 5.19b. This expression is valid at all points outside the Earth's surface, assuming that the Earth is spherical and that rotation can be neglected. At the Earth's surface, where $r = R_E$, **g** has a magnitude of 9.80 m/s².

The field concept is used in many other areas of physics. In fact, it was introduced by Michael Faraday (1791–1867) in his study of electromagnetism. In spite of its abstract nature, the field concept is particularly useful for describing electric and magnetic interactions.

Gravitational, electric, and magnetic fields are vector fields, because a vector is associated with each point in space. In the same way, a scalar field is one in which a scalar quantity is used to describe each point in space. For example, the variation in temperature over a given region can be described by a scalar temperature field.

SUMMARY

The maximum force of static friction, $f_{s,\ max}$, between a body and a surface is proportional to the normal force acting on the body. This maximum force occurs when the body is on the verge of slipping. In general, $f_s \leq \mu_s n$, where μ_s is the *coefficient of static friction* and **n** is the normal force. When a body slides over a rough surface, the *force of kinetic friction*, $\mathbf{f}_k$, is opposite the motion and is also proportional to the normal force. The magnitude of this force is given by $f_k = \mu_k n$, where μ_k is the *coefficient of kinetic friction*. Usually, $\mu_k < \mu_s$.

Newton's second law, applied to a particle moving in **uniform circular motion,** states that the net force in the radial direction must equal the product of the mass and the centripetal acceleration:

$$F_r = ma_r = \frac{mv^2}{r} \qquad [5.3]$$

A particle moving in nonuniform circular motion has both a centripetal acceleration and a tangential component of acceleration. In the case of a particle rotating in a vertical circle, the force of gravity provides the tangential acceleration and part or all of the centripetal acceleration.

A body moving through a liquid or gas experiences a **resistive force** that is velocity-dependent. This resistive force, which opposes the motion, generally increases with velocity. The force depends on the shape of the body and on the properties of the medium through which the body is moving. In the limiting case for a falling body, when the resistive force balances the weight ($a = 0$), the body reaches its **terminal speed**.

There are four fundamental forces in nature: the strong nuclear force, the electromagnetic force, the gravitational force, and the weak nuclear force.

CONCEPTUAL QUESTIONS

1. If you push on a heavy box that is at rest, it requires some force **F** to start its motion. However, once it is sliding, it requires a smaller force to maintain that motion. Why?

2. Suppose you are driving a car along a highway at a high speed. Why should you avoid slamming on your brakes if you want to stop in the shortest distance? That is, why should you keep the wheels turning as you brake?

3. Twenty people are playing tug-of-war, with ten people on a team. The teams are so evenly matched that neither team wins. After they give up the game, they notice that one of the team member's cars is mired in mud. They attach the tug-of-war rope to the bumper of the car and the 20 people try to pull the car from the mud. The friction force between the car and the mud is huge, the car does not move, and the rope breaks. Why did the rope break in this situation when it did not break when 20 people pulled on it in the tug-of-war?

4. What causes a rotary lawn sprinkler to turn?

5. It has been suggested that rotating cylinders about 10 mi in length and 5 mi in diameter be placed in space and used as colonies. The purpose of the rotation is to simulate gravity for the inhabitants. Explain this concept for producing an effective gravity.

6. Consider a rotating space station, spinning with just the right speed such that the centripetal acceleration on the inner surface is *g*. Thus, astronauts standing on this inner surface would feel pressed to the surface as if they were pressed into the floor because of the Earth's gravitational force. Suppose an astronaut in this station holds a ball above her head and "drops" it to the floor. Will the ball fall just like it would on the Earth?

7. A pail of water can be whirled in a vertical path such that none is spilled. Why does the water stay in, even when the pail is above your head?

8. How would you explain the force that pushes a rider toward the side of a car as the car rounds a corner?

9. Why does a pilot tend to black out when pulling out of a steep dive?

10. Suppose a race track is built on the Moon. For a given speed of a car around the track, would the roadway have to be banked at a steeper angle or a shallower angle compared to the same roadway on the Earth?

11. A skydiver in free fall reaches terminal speed. After the parachute is opened, what parameters change to decrease this terminal speed?

12. Consider a small raindrop and a large raindrop falling through the atmosphere. Compare their terminal speeds. What are their accelerations when they reach terminal speed?

13. On long journeys, jet aircraft usually fly at high altitudes of about 30 000 ft. What is the main advantage of flying at these altitudes from an economic viewpoint?

14. If someone told you that astronauts are weightless in orbit because they are beyond the pull of gravity, would you accept the statement? Explain.

PROBLEMS

Section 5.1   Forces of Friction

1. A 25.0-kg block is initially at rest on a horizontal surface. A horizontal force of 75.0 N is required to set the block in motion. After it is in motion, a horizontal force of 60.0 N is required to keep the block moving with constant speed. Find the coefficients of static and kinetic friction from this information.

2. A racing car accelerates uniformly from 0 to 80.0 mi/h in 8.00 s. The external force that accelerates the car is the frictional force between the tires and the road. If the tires do

not spin, determine the *minimum* coefficient of friction between the tires and the road.

3. A car is traveling at 50.0 mi/h on a horizontal highway. (a) If the coefficient of friction between road and tires on a rainy day is 0.100, what is the minimum distance in which the car will stop? (b) What is the stopping distance when the surface is dry and $\mu_s = 0.600$?

4. A woman at an airport is towing her 20.0-kg suitcase at constant speed by pulling on a strap at an angle of θ above the horizontal (Fig. P5.4). She pulls on the strap with a 35.0-N force, and the friction force on the suitcase is 20.0 N. (a) Draw a free-body diagram of the suitcase. (b) What angle does the strap make with the horizontal? (c) What normal force does the ground exert on the suitcase?

Figure P5.4

5. A 3.00-kg block is released from rest at the top of a 30.0° incline and slides a distance of 2.00 m down the incline in 1.50 s. Find (a) the magnitude of the acceleration of the block, (b) the coefficient of kinetic friction between block and plane, (c) the frictional force acting on the block, and (d) the speed of the block after it has slid 2.00 m.

6. In order to determine the coefficients of friction between rubber and various surfaces, a student uses a rubber eraser and an incline. In one experiment the eraser slips down the incline when the angle of inclination is 36.0° and then moves down the incline with constant speed when the angle is reduced to 30.0°. From these data, determine the coefficients of static and kinetic friction for this experiment.

7. A boy drags his 60.0-N sled at constant speed up a 15.0° hill. He does so by pulling with a 25.0-N force on a rope attached

to the sled. If the rope is inclined at 35.0° to the horizontal, (a) what is the coefficient of kinetic friction between sled and snow? (b) At the top of the hill, he jumps on the sled and slides down the hill. What is the magnitude of his acceleration down the slope?

8. Determine the stopping distance for a skier with a speed of 20.0 m/s (Fig. P5.8). Assume $\mu_k = 0.180$ and $\theta = 5.00°$.

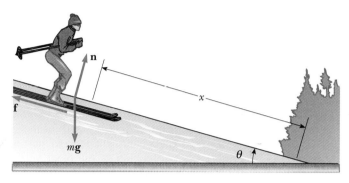

Figure P5.8

9. A 9.00-kg hanging weight is connected by a string over a pulley to a 5.00-kg block that is sliding on a flat table (Fig. P5.9). If the coefficient of kinetic friction is 0.200, find the tension in the string.

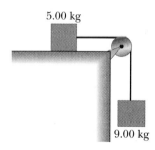

Figure P5.9

10. Three masses are connected on the table as shown in Figure P5.10. The table is rough and has a coefficient of sliding friction of 0.350. The three masses are $m_1 = 4.00$ kg, $m_2 = 1.00$ kg, and $m_3 = 2.00$ kg, respectively, and the pulleys are frictionless. (a) Draw free-body diagrams for each of the masses. (b) Determine the acceleration of each block and their directions. (c) Determine the tensions in the two cords.

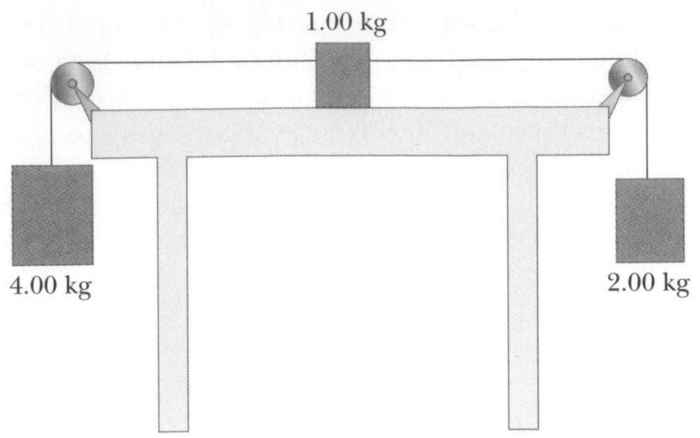

1.00 kg

4.00 kg   2.00 kg

Figure P5.10

11. A mass $M = 2.20$ kg is accelerated across a rough surface by a rope passing over a pulley, as shown in Figure P5.11. The tension in the rope is 10.0 N and the pulley is 10.0 cm above the top of the block. The coefficient of kinetic friction is 0.400. (a) Determine the acceleration of the block when $x = 40.0$ cm. (b) Find the value of x at which the acceleration becomes zero.

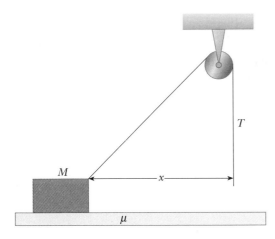

T

M

x

μ

Figure P5.11

Section 5.2 Newton's Second Law Applied to Uniform Circular Motion

12. A toy car moving at constant speed completes one lap around a circular track (a distance of 200 m) in 25.0 s. (a) What is the average speed? (b) If the mass of the car is 1.50 kg, what is the magnitude of the force that keeps it moving in a circle?

13. A 3.00-kg mass attached to a light string rotates on a horizontal, frictionless table. The radius of the circle is 0.800 m, and the string can support a mass of 25.0 kg before

breaking. What range of speeds can the mass have before the string breaks?

14. A 55.0-kg ice skater is moving at 4.00 m/s when she grabs the loose end of a rope the opposite end of which is tied to a pole. She then moves in a circle of radius 0.800 m around the pole. (a) Determine the force exerted by the rope on her arms. (b) Compare this force with her weight.

15. In the Bohr model of the hydrogen atom, the speed of the electron is approximately 2.20×10^6 m/s. Find (a) the centripetal acceleration of the electron and (b) the force acting on the electron as it revolves in a circular orbit of radius 0.530×10^{-10} m.

16. A crate of eggs is located in the middle of the flat bed of a pickup truck as the truck negotiates an unbanked curve in the road. The curve may be regarded as an arc of a circle of radius 35.0 m. If the coefficient of static friction between crate and truck is 0.600, what must be the maximum speed of the truck if the crate is not to slide?

17. A coin placed 30.0 cm from the center of a rotating, horizontal turntable slips when its speed is 50.0 cm/s. (a) What force provides the centripetal acceleration when the coin is stationary relative to the turntable? (b) What is the coefficient of static friction between coin and turntable?

18. A small turtle, appropriately named Dizzy, is placed on a horizontal, rotating turntable 20.0 cm from its center. Dizzy's mass is 50.0 g, and the coefficient of static friction between his feet and the turntable is 0.300. Find (a) the maximum number of revolutions per second the turntable can have if Dizzy is to remain stationary relative to the turntable and (b) Dizzy's speed and centripetal acceleration when he is on the verge of slipping.

19. Consider a conical pendulum with an 80.0-kg bob on a 10.0-m wire making an angle of 5.00° with the vertical (Fig. P5.19). Determine (a) the horizontal and vertical components of the force exerted by the wire on the pendulum, and (b) the centripetal acceleration of the bob.

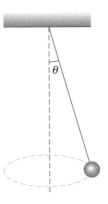

θ

Figure P5.19

Section 5.3   Nonuniform Circular Motion

20. A 40.0-kg child sits in a swing supported by two chains, each 3.00 m long. If the tension in each chain at the lowest point is 350 N, find (a) the child's speed at the lowest point and (b) the force of the seat on the child at the lowest point. (Neglect the mass of the seat.)

21. Tarzan ($m = 85.0$ kg) tries to cross a river by swinging from a vine. The vine is 10.0 m long, and his speed at the bottom of the swing (as he just clears the water) is 8.00 m/s. Tarzan doesn't know that the vine has a breaking strength of 1000 N. Does he make it safely across the river?

22. At the Six Flags Great America amusement park in Gurnee, Illinois, there is a roller coaster that incorporates some of the latest design technology and some basic physics. Each vertical loop, instead of being circular, is shaped like a teardrop (Fig. P5.22). The cars ride on the inside of the loop at the top, and the speeds are high enough to ensure that the cars remain on the track. The biggest loop is 40.0 m high, with a maximum speed of 31.0 m/s (nearly 70 mph) at the bottom. Suppose the speed at the top is 13.0 m/s and the corresponding centripetal acceleration is 2g. (a) What is the radius of the arc of the teardrop at the top? (b) If the total mass of the cars plus people is M, what force does the rail exert on it at the top? (c) Suppose the roller coaster had a loop of radius 20.0 m. If the cars have the same speed, 13.0 m/s at the top, what is the centripetal acceleration at the top? Comment on the normal force at the top in this situation.

Figure P5.22   *(Frank Cezus, FPG International)*

23. An 1800-kg car passes over a hump in a road that follows the arc of a circle of radius 42.0 m, as in Figure P5.23. (a) What force does the road exert on the car as the car passes the highest point of the hump if the car travels at 16.0 m/s? (b) What is the maximum speed the car can have as it passes this highest point before losing contact with the road?

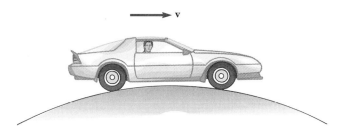

Figure P5.23

24. A car of mass m passes over a hump in a road that follows the arc of a circle of radius R, as in Figure P5.23. (a) What force does the road exert on the car as the car passes the highest point of the hump if the car travels at a speed v? (b) What is the maximum speed the car can have as it passes this highest point before losing contact with the road?

25. A pail of water is rotated in a vertical circle of radius 1.00 m. What is the minimum speed of the pail at the top of the circle if no water is to spill out?

Section 5.4   Motion in the Presence of Velocity-Dependent Resistive Forces

26. A sky diver of mass 80.0 kg jumps from a slow-moving aircraft and reaches a terminal speed of 50.0 m/s. (a) What is the acceleration of the sky diver when her speed is 30.0 m/s? What is the drag force on the diver when her speed is (b) 50.0 m/s? (c) 30.0 m/s?

27. A small piece of Styrofoam packing material is dropped from a height of 2.00 m above the ground. Until the terminal speed is reached, the magnitude of the acceleration is given by $a = g - cv$. After falling 0.500 m, it reaches terminal speed, and then the Styrofoam takes an extra 5.00 s to reach the ground. (a) What is the value of the constant c? (b) What is the acceleration at $t = 0$? (c) What is the acceleration when the speed is 0.150 m/s?

28. (a) Estimate the terminal speed of a wooden sphere (density 0.830 g/cm³) moving in air if its radius is 8.00 cm. (b) From what height would a freely falling object reach this speed in the absence of air resistance? Note that the density of air is 1.20 kg/m³.

29. A motorboat cuts its engine when its speed is 10.0 m/s and coasts to rest. The equation governing the motion of the motorboat during this period is $v = v_0 e^{-ct}$, where v is the

speed at time t, v_0 is the initial speed, and c is a constant. At $t = 20.0$ s, the speed is 5.00 m/s. (a) Find the constant c. (b) What is the speed at $t = 40.0$ s? (c) Differentiate the expression for $v(t)$ and thus show that the acceleration of the boat is proportional to the speed at any time.

30. Assume that the resistive force exerted on a speed skater is $f = -kmv^2$, where k is a constant and m is the skater's mass. The skater crosses the finish line of a straight-line race with speed v_f and then slows down by coasting without skating. Show that the skater's speed at any time t after crossing the finish line is $v(t) = v_f/(1 + ktv_f)$.

Section 5.5   Numerical Modeling in Particle Dynamics

31. A hailstone of mass 4.80×10^{-4} kg falls through the air and experiences a net force given by

$$F = -mg + Dv^2$$

where $D = 2.50 \times 10^{-5}$ kg/m. (a) Calculate the terminal speed of the hailstone. (b) Use Euler's method of numerical analysis to find the speed and position of the hailstone at 0.2-s intervals, taking the initial speed to be zero. Continue the calculation until the hailstone reaches 99% of terminal speed.

32. A 3.00-g leaf is dropped from a height of 2.00 m above the ground. Assume the net downward force on the leaf is $F = mg - bv$, where the drag factor is $b = 0.0300$ kg/s. (a) Calculate the terminal speed of the leaf. (b) Use Euler's method of numerical analysis to find the speed and position of the leaf, as functions of time, from the instant it is released until 99% of terminal speed is reached. (*Hint:* Try $\Delta t = 0.005$ s.)

33. A 50.0-kg parachutist jumps from an airplane and falls to Earth with a drag force proportional to the square of the speed, $R = Dv^2$. Take $D = 0.200$ kg/m (with the parachute closed) and $D = 20.0$ kg/m (with the chute open). (a) Determine the terminal speed of the parachutist in both configurations, before and after the chute is opened. (b) Set up a numerical analysis of the motion, and compute the speed and position as functions of time, assuming the jumper begins the descent at 1000 m above the ground and is in free fall for 10.0 s before opening the parachute. (*Hint:* When the parachute opens, a sudden large acceleration takes place; a smaller time step may be necessary in this region.)

34. Consider a 10.0-kg projectile launched with an initial speed of 100 m/s, at an angle of 35.0° elevation. The resistive force is $\mathbf{R} = -b\mathbf{v}$, where $b = 10.0$ kg/s. (a) Use a numerical method to determine the horizontal and vertical positions of the projectile as functions of time. (b) What is the range of this projectile? (c) Determine the elevation angle that gives the maximum range for the projectile. (*Hint:* Adjust the elevation angle by trial and error to find the greatest range.)

Section 5.6   The Fundamental Forces of Nature

35. Two identical isolated particles, each of mass 2.00 kg, are separated by a distance of 30.0 cm. What is the magnitude of the gravitational force of one particle on the other?

36. In a thundercloud there may be electric charges of $+40.0$ C near the top of the cloud and -40.0 C near the bottom of the cloud. These charges are separated by 2.00 km. What is the electric force on the top charge?

Section 5.7   The Gravitational Field

37. When a falling meteor is at a distance above the Earth's surface of 3.00 times the Earth's radius, what is its free-fall acceleration due to the gravitational force exerted on it?

Additional Problems

38. A crate of weight w is pushed by a force $\mathbf{F}$ on a horizontal floor. (a) If the coefficient of static friction is μ_s and $\mathbf{F}$ is directed at angle ϕ below the horizontal, show that the minimum value of F that will move the crate is given by

$$F = \frac{\mu_s w \sec \phi}{1 - \mu_s \tan \phi}$$

(b) Find the minimum value of F that can produce motion when $\mu_s = 0.400$, $w = 100$ N, and $\phi = 0°, 15.0°, 30.0°, 45.0°$, and $60.0°$.

39. A 1.30-kg toaster is not plugged in. The coefficient of static friction between the toaster and a horizontal countertop is 0.350. To make the toaster start moving, you carelessly pull on its electric cord. (a) In order for the cord tension to be as small as possible, you should pull at what angle above the horizontal? (b) Then how large must the tension be?

40. A student builds and calibrates an accelerometer, which she uses to determine the speed of her car around a certain unbanked highway curve. The accelerometer is a plumb bob with a protractor that she attaches to the roof of her car. A friend riding in the car with her observes that the plumb bob hangs at an angle of 15.0° from the vertical when the car has a speed of 23.0 m/s. (a) What is the centripetal acceleration of the car rounding the curve? (b) What is the radius of the curve? (c) What is speed of the car if the plumb bob deflection is 9.00° while rounding the same curve?

41. A space station, in the form of a large wheel 120 m in diameter, rotates to provide an artificial gravity of 3.00 m/s² for persons situated at the outer rim. Find the rotational frequency of the wheel (in revolutions per minute) that will produce this effect.

42. An amusement park ride consists of a rotating circular platform 8.00 m in diameter from which 10.0-kg seats are suspended at the end of 2.50-m massless chains (Fig. P5.42). When the system rotates, the chains make an angle $\theta = 28.0°$ with the vertical. (a) What is the speed of each seat?

(b) Draw a free-body diagram of a 40.0-kg child riding in a seat and find the tension in the chain.

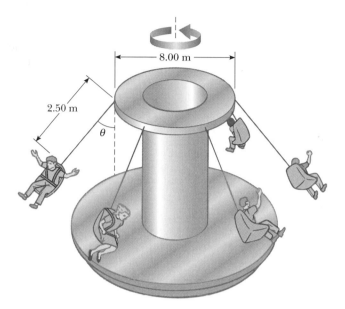

Figure P5.42

43. Because the Earth rotates about its axis, a point on the Equator experiences a centripetal acceleration of 0.0337 m/s², while a point at the poles experiences no centripetal acceleration. (a) Show that at the Equator the magnitude of the gravitational force on an object (its weight) must exceed the object's apparent weight. (b) What is the apparent weight at the Equator and at the poles of a person having a mass of 75.0 kg? (Assume the Earth is a uniform sphere and take $g = 9.800$ N/kg.)

44. The Earth rotates around its axis with a period of 24.0 h. Imagine that the rotational speed can be increased. If an object at the Equator is to have zero apparent weight, (a) what must the new period be? (b) By what factor would the speed of the object be increased when the planet is rotating at the higher speed? (*Hint:* See Problem 43 and note that the apparent weight of the object becomes zero when the normal force exerted on it is zero. Also, the distance traveled during one period is $2\pi R$, where R is the Earth's radius.)

45. An engineer wishes to design a curved exit ramp for a toll road in such a way that a car will not have to rely on friction to round the curve without skidding. Suppose that a typical car rounds the curve with a speed of 30 mi/h (13.4 m/s) and that the radius of the curve is 50.0 m. At what angle should the curve be banked? (See Fig. P5.45.)

46. A car rounds a banked curve as in Figure P5.45. The radius of curvature of the road is R, the banking angle is θ, and the coefficient of static friction is μ_s. (a) Determine the

range of speeds the car can have without slipping up or down the road. (b) Find the minimum value for μ_s such that the minimum speed is zero. (c) What is the range of speeds possible if $R = 100$ m, $\theta = 10.0°$, and $\mu_s = 0.100$ (slippery conditions)?

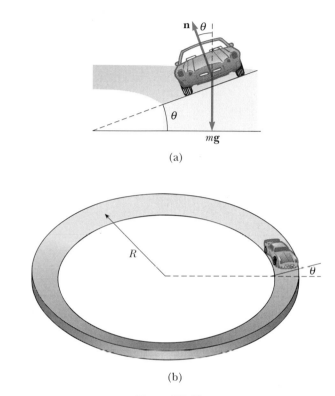

Figure P5.45

47. An amusement park ride consists of a large vertical cylinder that spins around its axis fast enough that any person inside is held up against the wall when the floor drops away (Fig. P5.47). The coefficient of static friction between person and wall is μ_s, and the radius of the cylinder is R. (a) Show that the maximum period of revolution necessary to keep the person from falling is $T = (4\pi^2 R\mu_s/g)^{1/2}$. (b) Obtain a numerical value for T if $R = 4.00$ m and $\mu_s = 0.400$. How many revolutions per minute does the cylinder make?

48. The expression $F = arv + br^2v^2$ gives the magnitude of the resistive force (in newtons) on a sphere of radius r (in meters) exerted by a stream of air with speed v (in meters per second), where a and b are constants with appropriate SI units. Their numerical values are $a = 3.10 \times 10^{-4}$ and $b = 0.870$. Using this formula, find the terminal speed for water droplets falling under their own weight in air, taking the following values for the drop radii: (a) 10.0 μm, (b) 100 μm, (c) 1.00 mm. Note that for (a) and (c) you can obtain accurate answers without solving a quadratic equa-

Figure P5.47

tion, by considering which of the two contributions to the air resistance is dominant and ignoring the lesser contribution.

49. A model airplane of mass 0.750 kg flies in a horizontal circle at the end of a 60.0-m control wire, with a speed of 35.0 m/s. Compute the tension in the wire if it makes a constant angle of 20.0° with the horizontal. The forces exerted on the airplane are the pull of the control wire, the force of gravity, and aerodynamic lift, which acts at 20.0° inward from the vertical as shown in Figure P5.49.

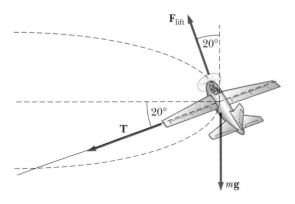

Figure P5.49

50. A 9.00-kg object starting from rest falls through a viscous medium and experiences a resistive force $\mathbf{R} = -b\mathbf{v}$, where $\mathbf{v}$ is the velocity of the object. If the object's speed reaches one half its terminal speed in 5.54 s, (a) determine the terminal speed. (b) At what time is the speed of the object three fourths the terminal speed? (c) How far has the object traveled in the first 5.54 s of motion?

51. Plaskett's binary system consists of two stars that revolve in a circular orbit around a center of mass midway between them. This means that the masses of the two stars are equal (Fig. P5.51). If the orbital velocity of each star is 220 km/s and the orbital period of each is 14.4 days, find the mass M of each star. (For comparison, the mass of our Sun is 1.99×10^{30} kg.)

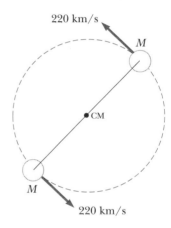

Figure P5.51

52. One block of mass 5.00 kg sits on top of a second rectangular block of mass 15.0 kg, which in turn is on a horizontal table. The coefficients of friction between the two blocks are $\mu_s = 0.300$ and $\mu_k = 0.100$. The coefficients of friction between the lower block and the rough table are $\mu_s = 0.500$ and $\mu_k = 0.400$. You apply a constant horizontal force to the lower block, just large enough to make this block start sliding out from between the upper block and the table. (a) Draw a free-body diagram of each block, naming the forces on each. (b) Determine the magnitude of each force on each block at the instant at which you have started pushing but motion has not yet started. In particular, what force must you apply? (c) Determine the acceleration you measure for each block.

53. Determine the order of magnitude of the gravitational force that you exert on another person 2 m away. In your solution state the quantities you estimate and the values you estimate for them.

Spreadsheet Problems

S1. A 0.142-kg baseball has a terminal speed of 42.5 m/s (95 mph). (a) If a baseball experiences a drag force of magnitude $R = Dv^2$, what is the value of D? (b) What is the magnitude of the drag force when the speed of the baseball is 36 m/s? (c) Set up a spreadsheet to determine the motion of a baseball thrown vertically upward at an initial speed of 36 m/s. What maximum height does the ball reach? How long is it in the air? What is its speed just before it hits the ground?

S2. A 10.0-kg projectile is launched with an initial speed of 150 m/s at an elevation angle of 35.0° with the horizontal. The resistive force acting on the projectile is $\mathbf{R} = -b\mathbf{v}$, where $b = 15$ kg/s. (a) Set up a spreadsheet to determine the horizontal and vertical positions of the projectile as functions of time. (b) Find the range of this projectile. (c) Determine the elevation angle that gives the maximum range for this projectile. (*Hint:* Adjust the elevation angle by trial and error to find the greatest range.)

S3. A professional golfer hits her 5-iron 155 m (170 yd). A 46-g golf ball experiences a drag force of magnitude $R = Dv^2$ and has a terminal speed of 44 m/s. (a) Calculate the drag coefficient D for the golf ball. (b) Set up a spreadsheet to calculate the trajectory of this shot. If the initial velocity of the ball makes an angle of 31° (the loft angle) with the horizontal, what initial speed must the ball have to reach the 155-m distance? (c) If this same golfer hits her 9-iron (47° loft) a distance of 119 m, what is the initial speed of the ball in this case? Discuss the differences in trajectories between the two shots.

ANSWERS TO CONCEPTUAL PROBLEMS

1. It is easier to attach the rope and pull. In this case, there is a component of your applied force which is upward. This reduces the normal force between the sled and the snow. In turn, this reduces the friction force between the sled and the snow, making it easier to move. If you push from behind, with a force with a downward component, the normal force is larger, the friction force is larger, and the sled is harder to move.

2. On the way up the ramp, the friction force and the component of the gravitational force on the book are both directed downhill. When the book returns back down the ramp, the component of the gravitational force is still downhill, but the friction force is uphill. Thus, there is a smaller net force when the book is sliding downhill, and according to Newton's second law, a smaller magnitude of acceleration. With a smaller acceleration, the velocity of the book at any point on the ramp is smaller when coming downhill than it was when going uphill. As a result, it takes longer to come down than to go up. The extreme case would be if the angle of the hill is such that once the book stops at the top of its motion, the maximum static friction force is larger than the component of the gravitational force, and the book "sticks" to the ramp. Then it takes *forever* to come down!

3. (a) Doubling the truck's mass would double the normal force the road exerts and double the backward force of kinetic friction. With twice the friction force acting on the mass, its acceleration would be the same and so would all other parameters of its motion, including distance. (b) With the same acceleration but half its original speed, the time required to stop would be halved, and moving with one-half the average speed for one-half the time, the truck would travel one-quarter the distance.

4. An object can move in a circle even if the total force on it is not perpendicular to its velocity, but then its speed will change. Resolve the total force into an inward radial component and a perpendicular tangential component. If the tangential force is forward, the object will speed up, and if the tangential force acts backward, it will slow down.

5. The forces exerted on the sky diver are the downward force of gravity, $m\mathbf{g}$, and an upward force of air resistance, $\mathbf{R}$, the magnitude of which is less than her weight before she reaches terminal speed. As her downward speed increases, the force of air resistance increases. The vector sum of the force of gravity and the force of air resistance gives a total force that decreases with time, so her acceleration decreases. Once she reaches terminal speed, the two forces balance each other, the net force is zero, and her acceleration is zero.

6

Work and Energy

Energy is present in the Universe in various forms, including mechanical energy, electromagnetic energy, chemical energy, thermal energy, and nuclear energy. When energy is changed from one form to another, its total amount remains the same. Conservation of energy says that if an isolated system loses energy in some form, it will gain an equal amount of energy in other forms.

In this chapter we shall concern ourselves only with the mechanical form of energy. We shall see that the concepts of work and energy can be applied to the dynamics of a mechanical system without resorting to Newton's laws. (However, it is important to note that the work–energy concepts are based on Newton's laws and therefore do not involve any new physical principles.) This energy approach to describing motion is especially useful when the force acting on a particle is not constant. Because the acceleration is not constant, we cannot apply the kinematic equations we developed in Chapter 2. Particles in nature are often subject to forces that vary with the particles' positions. These forces include gravitational forces and the force exerted on a body attached to a spring. We shall describe techniques for treating such systems with the help of an extremely important concept called the *work-kinetic energy theorem*, which is the central topic of this chapter.

We begin by defining work, a concept that provides a link between the concepts of force and energy.

These cyclists are working hard and expending energy as they pedal uphill in Marin County, California. *(David Madison, Tony Stone Images)*

142

6.1 · WORK DONE BY A CONSTANT FORCE

Almost all of the terms we have used thus far—*velocity, acceleration, force,* and so on—have conveyed the same meaning in physics as they do in everyday life. Now, however, we encounter a term the meaning of which in physics is distinctly different from its everyday meaning. This new term is **work**. Consider a particle that under goes a displacement **d** along a straight line while acted on by a constant force **F**, which makes an angle θ with **d**, as in Figure 6.1.

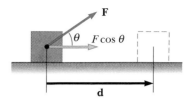

Figure 6.1 If an object undergoes a displacement **d**, the work done by the force **F** is $(F \cos \theta) d$.

> The work W done by an agent exerting a constant force is the product of the component of the force along the direction of the displacement of the point of application of the force and the magnitude of the displacement:

$$W \equiv Fd \cos \theta \qquad \text{[6.1]}$$

From this definition, we see that a force does no work on a particle if the particle does not move: If $d = 0$, Equation 6.1 gives $W = 0$. Also note from Equation 6.1 that the work done by a force is zero when the force is perpendicular to the displacement. That is, if $\theta = 90°$, then $\cos 90° = 0$, and $W = 0$. For example, in Figure 6.2, the work done by the normal force and the work done by the force of gravity during the horizontal displacement are zero for the same reason. In general, the particle may be moving with a constant or varying velocity under the influence of several forces. In that case, because work is a scalar quantity, the total work done as the particle undergoes some displacement is the algebraic sum of the work done by each of the forces.

The sign of the work also depends on the direction of **F** relative to **d**. The work done by the applied force is positive when the vector associated with the component $F \cos \theta$ is in the *same direction* as the displacement. For example, when an object is lifted, the work done by the applied force is positive because the lifting force is upward—that is, in the same direction as the displacement.

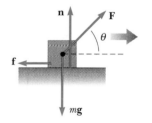

Figure 6.2 When an object is displaced horizontally, the normal force **n** and the force of gravity $m\mathbf{g}$ do no work.

When the vector associated with the component $F \cos \theta$ is in the direction *opposite* the displacement, W is *negative*. In the case of the object being lifted, for instance, the work done by the gravitational force is negative. It is important to note that work is an energy transfer; if energy is transferred *to* the system, W is positive; if energy is transferred *from* the system, W is negative.

If an applied force **F** acts along the direction of the displacement, then $\theta = 0$ and $\cos 0 = 1$. In this case, Equation 6.1 gives

$$W = Fd \qquad \text{[6.2]}$$

Work is a scalar quantity, and its units are force multiplied by length. Therefore, the SI unit of work is the **newton·meter** (N·m). The newton·meter, when it refers to work or energy, is called the **joule** (J). The unit of work in the cgs system is the **dyne·cm**, also called the **erg**; the unit in the British engineering system is the **ft·lb**. These are summarized in Table 6.1. Note that $1 \text{ J} = 10^7$ ergs.

TABLE 6.1   Units of Work in the Three Common Systems of Measurement

System	Unit of Work	Name of Combined Unit
SI	newton·meter (N·m)	joule (J)
cgs	dyne·centimeter (dyne·cm)	erg
British engineering (conventional)	foot·pound (ft·lb)	foot·pound (ft·lb)

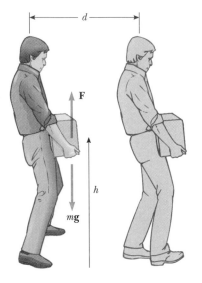

Figure 6.3   (Thinking Physics 1) A person lifts a cement block of mass m a vertical distance h and then walks horizontally a distance d.

Thinking Physics 1

A person slowly lifts a cement block of mass m a vertical height h and then walks horizontally a distance d while holding the block, as in Figure 6.3. Determine the work done by the person and by the force of gravity in this process.

Reasoning   Assuming that the person lifts the block with a force of magnitude equal to the weight of the block, mg, the work done by the person during the vertical displacement is mgh, because the force in this case is in the direction of the displacement. The work done by the person during the horizontal displacement of the block is zero, because the applied force in this process is perpendicular to the displacement. Thus, the net work done by the person is mgh. The work done by the force of gravity during the vertical displacement of the block is $-mgh$, because this force is opposite the displacement. The work done by the force of gravity is zero during the horizontal displacement because this force is also perpendicular to the displacement. Hence, the net work done by the force of gravity is $-mgh$. The net work on the block is zero $(+mgh - mgh = 0)$.

Thinking Physics 2

A planet moves in a perfectly circular orbit around the Sun. What can be said about the work done by the gravitational force on the planet? Suppose the orbit is elliptical; what can be said now about the work done by the gravitational force?

Reasoning   In the case of the circular orbit, the gravitational force is always perpendicular to the displacement of the planet as it moves around the orbit. Thus, there is no work done over any portion of the orbit, and no net work done in one complete orbit. In the case of the elliptical orbit, the Sun is at one focus of the ellipse. Thus, the force vector is not, in general, perpendicular to the displacement. While the planet is moving away from the Sun toward its farthest distance, called the aphelion, the force vector has a component opposite to the displacement. Thus, negative work is done on the planet and it slows down. As the planet passes aphelion and moves back toward its closest distance to the Sun, called the perihelion, there is a component of the gravitational force that is parallel to the displacement. The planet thus has positive work done on it and speeds up. The total negative work done on the outbound trip is canceled by the positive work done on the return trip. Thus, in the elliptical orbit, there is work done over any given portion of the orbit (except for portions symmetric about the aphelion or perihelion), but no net work is done over one complete orbit.

CONCEPTUAL PROBLEM 1

Consider a tug-of-war, in which the two teams pulling on the rope are evenly matched, so that no motion takes place. Is work done on the rope? On the pullers? On the ground? Is work done on anything?

No work on Anything

CONCEPTUAL PROBLEM 2

A team of furniture movers wishes to load a truck using a ramp from the ground to the rear of the truck. One of the movers claims that less work would be required to load the truck if the length of the ramp were increased, reducing the angle of the ramp with respect to the horizontal. Is his claim valid? Explain.

no, same amount of work just over different distances

CONCEPTUAL PROBLEM 3

Roads going up mountains are formed into *switchbacks*, with the road weaving back and forth along the face of the slope, such that there is only a gentle rise on any portion of the roadway. Does this require any less work to be done by an automobile climbing the mountain, compared to driving on a roadway that runs straight up the slope? Why are the switchbacks used?

same amount of work but used because the force on engine is smaller.

Example 6.1   Mr. Clean

A man cleaning his apartment pulls a vacuum cleaner with a force of magnitude $F = 50$ N. The force makes an angle of 30° with the horizontal as shown in Figure 6.4. The vacuum cleaner is displaced 3.0 m to the right. Calculate the work done by the 50-N force.

Solution   Using the definition of work (Equation 6.1), we have

$$W_F = (F \cos \theta)d = (50 \text{ N})(\cos 30°)(3.0 \text{ m})$$
$$= 130 \text{ N} \cdot \text{m} = 130 \text{ J}$$

Note that the normal force, **n**, the force of gravity, $m\mathbf{g}$, and the upward component of the applied force, $(50 \text{ N}) \sin 30°$, do *no* work because they are perpendicular to the displacement.

EXERCISE 1   Find the work done by the man on the vacuum cleaner if he pulls it 3.0 m with a horizontal force of 32 N.
Answer   96 J

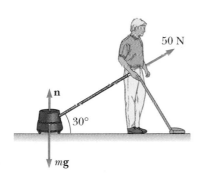

50 N

n

30°

$m\mathbf{g}$

Figure 6.4   (Example 6.1) A vacuum cleaner being pulled at an angle of 30° with the horizontal.

EXERCISE 2   If a person lifts a 20.0-kg bucket from a well and does 6.00 kJ of work, how deep is the well? Assume the speed of the bucket remains constant as it is lifted.
Answer   30.6 m

EXERCISE 3   A 65-kg woman climbs a flight of 20 stairs, each 23 cm high. How much work is done against the force of gravity in the process?   Answer   2.93 kJ

6.2 · THE SCALAR PRODUCT OF TWO VECTORS

We have defined work as a *scalar* quantity given by the product of the magnitude of the displacement and the component of a force in the direction of the displacement. It is convenient to express Equation 6.1 in terms of a **scalar product** of the two vectors **F** and **d**. We write this scalar product **F·d**. Because of the dot symbol, the scalar product is often called the *dot product*. Thus, we can express Equation 6.1 as a scalar product:

• *Work expressed as a scalar product*

$$W = \mathbf{F} \cdot \mathbf{d} = Fd \cos \theta \qquad \text{[6.3]}$$

In other words, **F·d** (read "F dot d") is a shorthand notation for $Fd \cos \theta$.

> In general, the scalar product of any two vectors **A** and **B** is a scalar quantity equal to the product of the magnitudes of the two vectors and the cosine of the angle θ between them:
>
> $$\mathbf{A} \cdot \mathbf{B} \equiv AB \cos \theta \qquad \text{[6.4]}$$

• *Scalar product of any two vectors* **A** *and* **B**

where θ is the smaller angle between **A** and **B**, as in Figure 6.5. Note that **A** and **B** need not have the same units.

In Figure 6.5, $B \cos \theta$ is the projection of **B** onto **A**. Therefore, Equation 6.4 says that **A·B** is the product of the magnitude of **A** and the projection of **B** onto **A**.[1]

From Equation 6.4 we also see that the scalar product is *commutative*. That is,

• *The order of the scalar product can be reversed.*

$$\mathbf{A} \cdot \mathbf{B} = \mathbf{B} \cdot \mathbf{A} \qquad \text{[6.5]}$$

Finally, the scalar product obeys the *distributive law of multiplication*, so that

$$\mathbf{A} \cdot (\mathbf{B} + \mathbf{C}) = \mathbf{A} \cdot \mathbf{B} + \mathbf{A} \cdot \mathbf{C} \qquad \text{[6.6]}$$

The dot product is simple to evaluate from Equation 6.4 when **A** is either perpendicular or parallel to **B**. If **A** is perpendicular to **B** ($\theta = 90°$), then **A·B** = 0. (The equality **A·B** = 0 also holds in the more trivial case when either **A** or **B** is zero.) If **A** and **B** point in the same direction ($\theta = 0°$), then **A·B** = AB. If **A** and **B** point in opposite directions ($\theta = 180°$), then **A·B** = $-AB$. The scalar product is negative when $90° < \theta < 180°$.

The unit vectors **i**, **j**, and **k**, which were defined in Chapter 1, lie in the positive x, y, and z directions, respectively, of a right-handed coordinate system. Therefore, it follows from the definition of **A·B** that the scalar products of these unit vectors are given by

$$\mathbf{i} \cdot \mathbf{i} = \mathbf{j} \cdot \mathbf{j} = \mathbf{k} \cdot \mathbf{k} = 1 \qquad \text{[6.7]}$$

$$\mathbf{i} \cdot \mathbf{j} = \mathbf{i} \cdot \mathbf{k} = \mathbf{j} \cdot \mathbf{k} = 0 \qquad \text{[6.8]}$$

• *Scalar products of unit vectors*

Two vectors **A** and **B** can be expressed in component form as

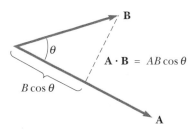

Figure 6.5 The scalar product **A·B** equals the magnitude of **A** multiplied by the projection of **B** onto **A**.

$\mathbf{A} \cdot \mathbf{B} = AB \cos \theta$

$B \cos \theta$

---

[1]This is equivalent to stating that **A·B** equals the product of the magnitude of **B** and the projection of **A** onto **B** or, vice versa, the product of the magnitude of **A** and the projection of **B** onto **A.**

$$\mathbf{A} = A_x\mathbf{i} + A_y\mathbf{j} + A_z\mathbf{k}$$

$$\mathbf{B} = B_x\mathbf{i} + B_y\mathbf{j} + B_z\mathbf{k}$$

Therefore, Equations 6.7 and 6.8 reduce the scalar product of **A** and **B** to

$$\mathbf{A}\cdot\mathbf{B} = A_xB_x + A_yB_y + A_zB_z \qquad [6.9]$$

In the special case where **A** = **B**, we see that

$$\mathbf{A}\cdot\mathbf{A} = A_x^2 + A_y^2 + A_z^2 = A^2$$

Example 6.2   The Scalar Product

The vectors **A** and **B** are given by **A** = 2**i** + 3**j** and **B** = −**i** + 2**j**.
(a) Determine the scalar product **A**·**B**.

Solution
$$\mathbf{A}\cdot\mathbf{B} = (2\mathbf{i} + 3\mathbf{j})\cdot(-\mathbf{i} + 2\mathbf{j})$$
$$= -2\mathbf{i}\cdot\mathbf{i} + 2\mathbf{i}\cdot2\mathbf{j} - 3\mathbf{j}\cdot\mathbf{i} + 3\mathbf{j}\cdot2\mathbf{j}$$
$$= -2 + 6 = 4$$

where we have used the facts that $\mathbf{i}\cdot\mathbf{i} = \mathbf{j}\cdot\mathbf{j} = 1$ and $\mathbf{i}\cdot\mathbf{j} = \mathbf{j}\cdot\mathbf{i} = 0$. The same result is obtained using Equation 6.9 directly, where $A_x = 2$, $A_y = 3$, $B_x = -1$, and $B_y = 2$.
(b) Find the angle θ between **A** and **B**.

Solution   The magnitudes of **A** and **B** are given by
$$A = \sqrt{A_x^2 + A_y^2} = \sqrt{(2)^2 + (3)^2} = \sqrt{13}$$
$$B = \sqrt{B_x^2 + B_y^2} = \sqrt{(-1)^2 + (2)^2} = \sqrt{5}$$

Using Equation 6.4 and the result from (a) gives
$$\cos\theta = \frac{\mathbf{A}\cdot\mathbf{B}}{AB} = \frac{4}{\sqrt{13}\sqrt{5}} = \frac{4}{\sqrt{65}}$$
$$\theta = \cos^{-1}\frac{4}{8.06} = 60.2°$$

EXERCISE 4   For **A** = 4**i** + 3**j** and **B** = −**i** + 3**j**, find (a) **A**·**B** and (b) the angle between **A** and **B**.   Answer   (a) 5.00   (b) 71.6°

EXERCISE 5   As a particle moves from the origin to (3**i** − 4**j**) m, it is acted on by a force given by (4**i** − 5**j**) N. Calculate the work done by this force as the particle moves through the given displacement.   Answer   32 J

6.3 • WORK DONE BY A VARYING FORCE

Consider a particle being displaced along the *x* axis under the action of a varying force, as in Figure 6.6. The particle is displaced in the direction of increasing *x* from $x = x_i$ to $x = x_f$. In such a situation, we cannot use $W = (F\cos\theta)d$ to calculate the work done by the force, because this relationship applies only when **F** is constant in magnitude and direction. However, if we imagine that the particle undergoes a small displacement, Δx, shown in Figure 6.6a, then the *x* component of the force, F_x, is approximately constant over this interval and we can express the work done by the force for this small displacement as

$$W_1 = F_x\,\Delta x \qquad [6.10]$$

This quantity is just the area of the shaded rectangle in Figure 6.6a. If we imagine that the F_x versus *x* curve is divided into a large number of such intervals, then the

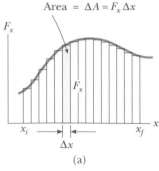

Area = $\Delta A = F_x \Delta x$

(a)

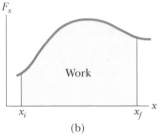

Work

(b)

Figure 6.6 (a) The work done by the force F_x for the small displacement Δx is $F_x \Delta x$, which equals the area of the shaded rectangle. The total work done for the displacement from x_i to x_f is approximately equal to the sum of the areas of all the rectangles. (b) The work done by the variable force F_x as the particle moves from x_i to x_f is *exactly* equal to the area under this curve.

Does the weight lifter do any work as he holds the weight on his shoulders? Does he do any work as he raises the weight?

total work done for the displacement from x_i to x_f is approximately equal to the sum of a large number of such terms:

$$W \cong \sum_{x_i}^{x_f} F_x \Delta x$$

If the displacements Δx are allowed to approach zero, then the number of terms in the sum increases without limit, but the value of the sum approaches a definite value equal to the area under the curve bounded by F_x and the x axis as in Figure 6.6b. As you probably have learned in calculus, this limit of the sum is called an *integral* and is represented by

$$\lim_{\Delta x \to 0} \sum_{x_i}^{x_f} F_x \Delta x = \int_{x_i}^{x_f} F_x \, dx$$

The limits on the integral, $x = x_i$ to $x = x_f$, define what is called a **definite integral.** (An *indefinite integral* is the limit of a sum over an unspecified interval. Appendix B.7 gives a brief description of integration.) This definite integral is numerically equal to the area under the F_x versus x curve between x_i and x_f. Therefore, we can express the work done by F_x for the displacement of the particle from x_i to x_f as

$$W = \int_{x_i}^{x_f} F_x \, dx \qquad \text{[6.11]}$$

This equation reduces to Equation 6.1 when $F_x = F \cos \theta$ is constant.

If more than one force acts on a particle, the total work done is just the work done by the resultant force. If we express the resultant force in the x direction as ΣF_x, then the *net work* done as the particle moves from x_i to x_f is

$$W_{\text{net}} = \int_{x_i}^{x_f} \left(\sum F_x \right) dx \qquad \text{[6.12]}$$

Thinking Physics 3

Consider a single force that causes an object to follow a path. If the force is constant, under what conditions can the work done over the entire path be zero? If the force is varying, under what conditions can the work done over the entire path be zero?

Reasoning In the case of the constant force, there are several possibilities that will give a result of zero work. One possibility is that the constant value of the force is zero, in which case we have a trivial result of nothing pushing. The second possibility for zero work is that a nonzero force does act, but it does not succeed in moving the object, although this does not fit with the question, because it is stated that the object *does* move over a path. The third possibility is that the force is always perpendicular to the path, as in the case of a planet moving in a circular orbit around a gravitational-force center.

For the varying force, only one of the above possibilities is consistent with the question. The zero-force possibility is not available, because a force whose value is *always* zero is not a *varying* force. The second possibility, that of zero displacement, as is the case above, does not fit with the statement that the object does move through a path. The third possibility, in which the force is perpendicular to the displacement, will also give zero work in the case of a varying force.

We have additional possibilities in the case of the varying force that were not available in the constant force case. One possibility is that the varying force (or a constant force applied opposite the velocity) carries the object around a path such that the object ends up back where it started. In this case the net work is zero, because there was no net displacement. This is the case for a comet in an elliptical orbit around the Sun—no net work is done by the gravitational force during one complete orbit. Another possibility is that the relative direction of the displacement and the force reverse during the path in such a way that the area above the axis in the force-displacement graph exactly equals the area below the axis. This is the case for an automobile accelerating to some speed and then braking to a stop. The positive work done while the automobile is speeding up exactly equals the negative work done as the automobile slows back down (the equality of the positive and negative work is probably not obvious at this point, but will be after Section 6.4).

Example 6.3   Calculating Total Work Done from a Graph

A force acting on a particle varies with x, as shown in Figure 6.7. Calculate the work done by the force as the particle moves from $x = 0$ to $x = 6.0$ m.

Solution   The work done by the force is equal to the area under the curve from $x = 0$ to $x = 6.0$ m. This area is equal to the area of the rectangular section from $x = 0$ to $x = 4.0$ m plus the area of the triangular section that extends from $x = 4.0$ m to $x = 6.0$ m. The area of the rectangle is $(4.0)(5.0)$ N·m $= 20$ J, and the area of the triangle is $\frac{1}{2}(2.0)(5.0)$ N·m $= 5.0$ J. Therefore, the total work done is 25 J.

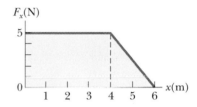

Figure 6.7   (Example 6.3) The force acting on a particle is constant for the first 4.0 m of motion and then decreases linearly with x from $x = 4.0$ m to $x = 6.0$ m. The net work done by this force is the area under this curve.

Work Done by a Spring

A common physical system for which the force varies with position is shown in Figure 6.8. A block on a horizontal, frictionless surface is connected to a spring. If the spring is stretched or compressed a small distance from its unstretched, or equilibrium, configuration, the spring will exert a force on the block given by

$$F_s = -kx \qquad\qquad\qquad [6.13]$$

• *Spring force*

where x is the displacement of the block from its unstretched ($x = 0$) position and k is a positive constant called the *force constant* of the spring. As mentioned in Chapter 4, Section 4.2, this force law for springs is known as **Hooke's law.** For many materials, Hooke's law can be very accurate, provided the displacement is not too large. The value of k is a measure of the stiffness of the spring. Stiff springs have large k values, and soft springs have small k values.

The negative sign in Equation 6.13 signifies that the force exerted by the spring is always directed *opposite* the displacement. For example, when $x > 0$, as in Figure 6.8a, the spring force is to the left, or negative. When $x < 0$, as in Figure 6.8c, the

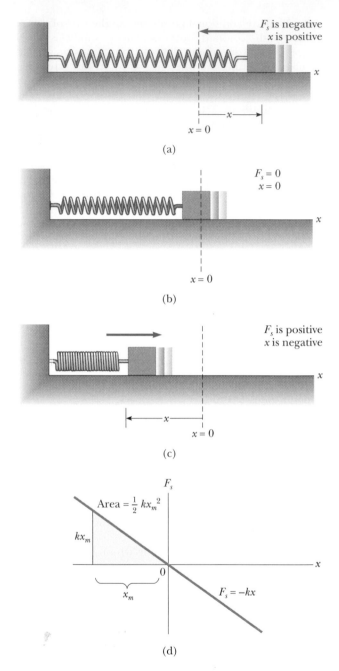

Figure 6.8   The force exerted by a spring on a block varies with the block's displacement from the equilibrium position $x = 0$. (a) When x is positive (stretched spring), the spring force is to the left. (b) When x is zero, the spring force is zero (natural length of the spring). (c) When x is negative (compressed spring), the spring force is to the right. (d) Graph of F_s versus x for the mass–spring system. The work done by the spring force as the block moves from $-x_m$ to 0 is the area of the shaded triangle, $\frac{1}{2}kx_m^2$.

spring force is to the right, or positive. Of course, when $x = 0$, as in Figure 6.8b, the spring is unstretched and $F_s = 0$. Because the spring force always acts toward the equilibrium position, it is sometimes called a *restoring force*. Once the block is displaced some distance x_m from equilibrium and then released, it moves from $-x_m$ through zero to $+x_m$. The details of the ensuing oscillating motion will be given in Chapter 12.

Suppose that the block is pushed to the left a distance x_m from equilibrium, as in Figure 6.8c, and then released. Let us calculate the *work done by the spring force* as the block moves from $x_i = -x_m$ to $x_f = 0$. Applying Equation 6.11 assuming the block may be treated as a particle, we get

$$W_s = \int_{x_i}^{x_f} F_s \, dx = \int_{-x_m}^{0} (-kx) \, dx = \tfrac{1}{2}kx_m^2 \qquad \text{[6.14]}$$

• *Work done by a spring*

where we have used the indefinite integral $\int x \, dx = x^2/2$. That is, the work done by the spring force is positive because the spring force is in the same direction as the displacement caused by the spring force (both are to the right). However, if we consider the work done by the spring force as the block moves from $x_i = 0$ to $x_f = x_m$, we find that $W_s = -\tfrac{1}{2}kx_m^2$, because for this part of the motion the displacement caused by the spring force is to the right and the spring force is to the left. Therefore, the *net* work done by the spring force as the block moves from $x_i = -x_m$ to $x_f = x_m$ is *zero*.

If we plot F_s versus x, as in Figure 6.8d, we arrive at the same results. Note that the work calculated in Equation 6.14 is equal to the area of the shaded triangle in Figure 6.8d, with base x_m and height kx_m. This area is $\tfrac{1}{2}kx_m^2$.

If the block undergoes an *arbitrary* displacement from $x = x_i$ to $x = x_f$, the work done by the spring force is

$$W_s = \int_{x_i}^{x_f} (-kx) \, dx = \tfrac{1}{2}kx_i^2 - \tfrac{1}{2}kx_f^2 \qquad \text{[6.15]}$$

From this equation we see that the work done by the spring force is zero for any motion that ends where it began $(x_i = x_f)$. We shall make use of this important result in Chapter 7, where we describe the motion of this system in more detail.

Equations 6.14 and 6.15 describe the work done by the spring force exerted on the block. Now let us consider the work done on the spring by an *external agent* as the spring is stretched *very slowly* from $x_i = 0$ to $x_f = x_m$, as in Figure 6.9. This work can be easily calculated by noting that the *applied force*, $\mathbf{F}_{app}$, is equal to and opposite the spring force, $\mathbf{F}_s$, at any value of the displacement, so that $F_{app} = -(-kx) = kx$. Therefore, the work done on the spring by this applied force (the external agent) is

$$W_{F_{app}} = \int_{0}^{x_m} F_{app} \, dx = \int_{0}^{x_m} kx \, dx = \tfrac{1}{2}kx_m^2$$

You should note that this work is equal to the negative of the work done by the spring force for this displacement.

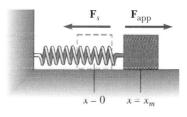

Figure 6.9 A block being pulled from $x = 0$ to $x = x_m$ on a frictionless surface by a force $\mathbf{F}_{app}$. If the process is carried out very slowly, the applied force is equal to and opposite the spring force at all times.

CONCEPTUAL PROBLEM 4

A spring is stretched by a distance x_m, resulting in work W being done by the external agent. The spring is then relaxed, cut in half, and one of the halves stretched by the same distance. How much work is done in this case?

Example 6.4 Work Required to Stretch a Spring

One end of a horizontal Hooke's-law spring ($k = 80$ N/m) is held fixed while an external force is applied to the free end, stretching it from $x_0 = 0$ to $x_1 = 4.0$ cm. (a) Find the work done by the external force.

$W_{Fapp} = \int_{x_i}^{x_f} kx\,dx = \frac{1}{2}kx_f^2 - \frac{1}{2}kx_i^2$
$\frac{1}{2}(80)(7)^2 - \frac{1}{2}(80)(4)^2$

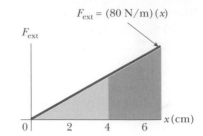

$F_{ext} = (80\ N/m)(x)$

Figure 6.10   (Example 6.4) A graph of the external force required to stretch a spring that obeys Hooke's law versus the elongation of the spring.

Solution   We place the zero reference of the coordinate axis at the free end of the unstretched spring (Fig. 6.10). The external force is $F_{ext} = (80\ N/m)(x)$. The work done by F_{ext} is the area of the triangle from 0 to 4.0 cm:

$$W = \tfrac{1}{2}kx_1^2 = \tfrac{1}{2}(80\ N/m)(0.040\ m)^2 = \boxed{64\ mJ}$$

(b) Find the additional work done in stretching the spring from $x_1 = 4.0$ cm to $x_2 = 7.0$ cm.

Solution   The work done in stretching the spring the additional amount, from $x_1 = 0.040$ m to $x_2 = 0.070$ m, is the darker-shaded area between these limits. Geometrically, it is

$$W = \tfrac{1}{2}kx_2^2 - \tfrac{1}{2}kx_1^2$$
$$= \tfrac{1}{2}(80\ N/m)\,[(0.070\ m)^2 - (0.040\ m)^2] = 130\ mJ$$

Using calculus, we find that

$$W = \int_{x_1}^{x_2} F_{ext}\,dx = \int_{0.04\,m}^{0.07\,m} (80\ N/m)\,dx$$

$$= \tfrac{1}{2}(80\ N/m)(x^2)\Big|_{0.04\,m}^{0.07\,m}$$

$$W = \tfrac{1}{2}(80\ N/m)\,[(0.070\ m)^2 - (0.040\ m)^2] = \boxed{130\ mJ}$$

EXERCISE 6   If an applied force varies with position according to $F_x = 3x^2 - 5$, where x is in meters, how much work is done by this force on an object that moves from $x = 4$ m to $x = 7$ m?   Answer   264 J

6.4 • KINETIC ENERGY AND THE WORK-KINETIC ENERGY THEOREM

Solutions using Newton's second law can be difficult if the forces in the problem are complex. An alternative approach that enables us to understand and solve such motion problems is to relate the speed of the particle to its displacement under the influence of some net force. As we shall see in this section, if the work done by the net force on a particle can be calculated for a given displacement, the change in the particle's speed will be easy to evaluate.

Figure 6.11 shows a particle of mass m moving to the right under the action of a constant net force **F**. Because the force is constant, we know from Newton's second law that the particle will move with a constant acceleration **a**. If the particle is displaced a distance d, the work done by the force **F** is

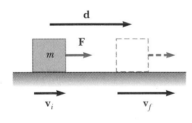

Figure 6.11   A particle undergoing a displacement and change in velocity under the action of a constant net force **F**.

$$W_{net} = Fd = (ma)d \qquad\qquad [6.16]$$

In Chapter 2 (Eqs. 2.4 and 2.9) we found that the following relationships are valid when a particle undergoes constant acceleration:

$$d = \tfrac{1}{2}(v_i + v_f)t \qquad a = \frac{v_f - v_i}{t}$$

where v_i is the speed at $t = 0$ and v_f is the speed at time t. Substituting these expressions into Equation 6.16 gives

$$W_{net} = m \left(\frac{v_f - v_i}{t} \right) \frac{1}{2} (v_i + v_f) t$$

$$W_{net} = \frac{1}{2} m v_f^2 - \frac{1}{2} m v_i^2 \qquad [6.17]$$

The quantity $\frac{1}{2} m v^2$ represents the energy associated with the motion of a particle. It is so important that it has been given a special name—**kinetic energy.** The kinetic energy, K, of a particle of mass, m, moving with a speed, v, is defined as

$$K \equiv \frac{1}{2} m v^2 \qquad [6.18]$$

Kinetic energy is a scalar quantity and has the same units as work. For example, a 2.0-kg mass moving with a speed of 4.0 m/s has a kinetic energy of 32 J. It is often convenient to write Equation 6.17 in the form

$$W_{net} = K_f - K_i = \Delta K \qquad [6.19]$$

That is,

> the work done by the constant net force $\mathbf{F}_{net}$ in displacing a particle equals the change in kinetic energy of the particle.

Equation 6.19 is an important result known as the **work-kinetic energy theorem.** For convenience, it was derived under the assumption that the net force acting on the particle was constant. A more general derivation would show that this equation is also valid for a variable force.

The work-kinetic energy theorem also says that the speed of the particle will increase if the net work done on it is positive, because the final kinetic energy will be greater than the initial kinetic energy. The speed will decrease if the net work is negative, because the final kinetic energy will be less than the initial kinetic energy. The speed and kinetic energy of a particle change only if work is done on the particle by some external force.

Consider the relationship between the work done on a particle and the change in its kinetic energy as expressed by Equation 6.19. Because of this connection, we can also think of kinetic energy as the work the particle can do in coming to rest. For example, suppose a hammer is on the verge of striking a nail, as in Figure 6.12. The moving hammer has kinetic energy and can do work on the nail. The work done on the nail appears as the product Fd, where F is the average force exerted on the nail by the hammer and d is the distance the nail is driven into the wall. However, the hammer and the nail are not particles, so part of the kinetic energy of the hammer goes into warming the hammer and nail, and all of the work done on the nail goes into warming the nail and wall and locally deforming the wall.

• *Kinetic energy is energy associated with the motion of a particle.*

• *The work-kinetic energy theorem states that the work done on a particle equals the change in its kinetic energy.*

Figure 6.12   The hammer has kinetic energy associated with its motion and can do work on the nail, driving it into the wall.

CONCEPTUAL PROBLEM 5

You are working in a library, reshelving books. You lift a book from the floor to the top shelf. The kinetic energy of the book on the floor was zero, and the kinetic energy of the book on the top shelf is zero, so there is no change in kinetic energy. Yet you did some work in lifting the book. Is the work-kinetic energy theorem violated?

Situations Involving Kinetic Friction

When dealing with a force acting on an extended object, one must be careful in calculating the work done by that force because the displacement of the object is generally not equal to the displacement of the point of application of that force. Suppose that an object of mass m sliding on a horizontal surface is pulled with a constant horizontal external force $\mathbf{F}$ to the right and a kinetic frictional force $\mathbf{f}$ acts to the left, where $\mathbf{F} > \mathbf{f}$. In this case, the net force is to the right as in Figure 6.11, and the net work done on the object as it undergoes a displacement $\mathbf{d}$ to the right is

$$W_{\text{net}} = (\mathbf{F} - \mathbf{f}) \cdot \mathbf{d} = Fd - fd \qquad [6.20]$$

The quantity Fd is the work done on the object by the constant force $\mathbf{F}$. The quantity $-fd$ is negative because the force of kinetic friction is opposite the displacement. The work done by kinetic friction depends on both the displacement of the object and on the details of the motion between the initial and final positions. In fact, the work done by kinetic friction on an extended object cannot be explicitly evaluated because friction forces and their individual displacements are complex.

Now suppose that a block moving on a horizontal surface and given an initial horizontal velocity $\mathbf{v}_i$ slides a distance d before reaching a final velocity $\mathbf{v}_f$ as in Figure 6.13. The external force that causes the block to undergo an acceleration in the negative x direction is the force of kinetic friction $\mathbf{f}$ acting to the left, opposite the motion. The initial kinetic energy of the block is $\frac{1}{2}mv_i^2$ and its final kinetic energy is $\frac{1}{2}mv_f^2$. The change in kinetic energy of the block is equal to $-fd$. This can be shown by applying Newton's second law to the block. (Newton's second law gives the acceleration of the center of mass of any object regardless of how or where the forces act.) Because the net force on the block in the x direction is the friction force, Newton's second law gives $-f = ma$. Multiplying both sides of this expression by d and using the expression $v_f^2 - v_i^2 = 2ad$ for motion under constant acceleration gives $-fd = (ma)d = \frac{1}{2}mv_f^2 - \frac{1}{2}mv_i^2$ or

$$\Delta K = -fd \qquad [6.21]$$

This result says that the change in kinetic energy of the block is equal to $-fd$, which corresponds to the energy dissipated by the force of kinetic friction. Part of this energy is transferred to internal thermal energy of the block, and part is transferred from the block to the surface.[2] In effect, the loss in kinetic energy of the block results in an increase in thermal energy of both the block and surface. For example, if the loss in kinetic energy of the block is 300 J, and 100 J appears as an increase in thermal energy of the block, then the remaining 200 J must have been transferred from the block to the surface.

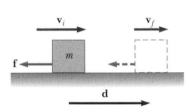

Figure 6.13 A block sliding to the right on a horizontal surface slows down in the presence of a force of kinetic friction acting to the left. The initial velocity of the block is $\mathbf{v}_i$, and its final velocity is $\mathbf{v}_f$. The normal force and force of gravity are not included in the diagram because they are perpendicular to the direction of motion and therefore do not influence the change in velocity of the block.

Thinking Physics 4

A car traveling at a speed v skids a distance d after its brakes lock. Estimate how far it will skid if its brakes lock when its initial speed is $2v$. What happens to the car's kinetic energy as it stops?

[2]For more details on energy transfer situations involving forces of kinetic friction, see B. A. Sherwood and W. H. Bernard, *American Journal of Physics*, 52:1001, 1984, and R. P. Bauman, *The Physics Teacher*, 30:264, 1992.

Reasoning Let us assume that the force of kinetic friction between car and road surface is constant and the same in both cases. The net force times the displacement of the car is equal to its initial kinetic energy. If the speed is doubled as in this example, the kinetic energy of the car is quadrupled. For a given applied force (in this case, the frictional force), the distance traveled is four times as great when the initial speed is doubled, so the estimated distance it skids is 4*d*. The kinetic energy of the car is changed into internal energy associated with the tires, brake pads, and road as they increase in temperature.

Thinking Physics 5

In most situations we have encountered in this chapter, frictional forces tend to reduce the kinetic energy of an object. However, frictional forces can sometimes increase an object's kinetic energy. Describe a few situations in which friction causes an increase in kinetic energy.

Reasoning If a crate is located on the bed of a truck and the truck accelerates to the east, the static friction force exerted on the crate by the truck acts to the east to give the crate the same acceleration as the truck (assuming the crate doesn't slip). Another example is a car that accelerates because of the frictional forces exerted on the car's tires by the road. These forces act in the direction of the car's motion, and the sum of these forces causes an increase in the car's kinetic energy.

CONCEPTUAL PROBLEM 6

It is a known fact that more energy is expended by walking downstairs than by walking horizontally at the same speed. Why do you think this is so?

Example 6.5   A Block Pulled on a Frictionless Surface

A 6.0-kg block initially at rest is pulled to the right along a horizontal, frictionless surface by a constant, horizontal force of 12 N, as in Figure 6.14. Find the speed of the block after it has moved 3.0 m.

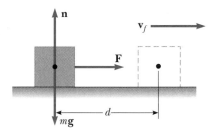

Figure 6.14   (Example 6.5)

Solution The gravitational force acting on the block is balanced by the normal force, and neither of these forces does work because the displacement is horizontal. Because there is no friction, the resultant external force is the 12-N force. The work done by this force is

$$W = Fd = (12 \text{ N})(3.0 \text{ m}) = 36 \text{ N} \cdot \text{m} = 36 \text{ J}$$

Using the work-kinetic energy theorem and noting that the initial kinetic energy is zero, we get

$$W = K_f - K_i = \tfrac{1}{2}mv_f^2 - 0$$

$$v_f^2 = \frac{2W}{m} = \frac{2(36 \text{ J})}{6.0 \text{ kg}} = 12 \text{ m}^2/\text{s}^2$$

$$v_f = \boxed{3.5 \text{ m/s}}$$

EXERCISE 7 Find the acceleration of the block using the kinematic equation $v_f^2 = v_i^2 + 2ax$.

Answer $a = 2.0 \text{ m/s}^2$

Example 6.6   A Block Pulled on a Rough Surface

Find the final speed of the block described in Example 6.5 if the surface is rough and the coefficient of kinetic friction is 0.15.

Reasoning In this case, we must use Equation 6.21 to calculate the change in kinetic energy, ΔK. The net force exerted on the block is the sum of the applied 12-N force and the frictional force, as in Figure 6.15. Because the frictional force is in the direction opposite the displacement, it must be subtracted.

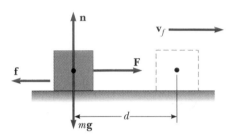

Figure 6.15   (Example 6.6)

Solution The magnitude of the frictional force is $f = \mu_k n = \mu_k mg$. Therefore the net force acting on the block is

$$F_{net} = F - \mu_k mg = 12 \text{ N} - (0.15)(6.0 \text{ kg})(9.80 \text{ m/s}^2)$$
$$= 12 \text{ N} - 8.82 \text{ N} = 3.18 \text{ N}$$

Multiplying this constant force by the displacement and using Equation 6.21 gives

$$\Delta K = F_{net}d = (3.18 \text{ N})(3.0 \text{ m}) = 9.54 \text{ J} = \tfrac{1}{2}mv_f^2$$

using the information that $v_i = 0$. Therefore,

$$v_f^2 = \frac{2(9.54 \text{ J})}{6.0 \text{ kg}} = 3.18 \text{ m}^2/\text{s}^2$$

$$v_f = \boxed{1.8 \text{ m/s}}$$

EXERCISE 8 Find the acceleration of the block from Newton's second law.   Answer $a = 0.53 \text{ m/s}^2$

Example 6.7   A Mass-Spring System

A block of mass 1.6 kg is attached to a spring that has a force constant of 1.0×10^3 N/m as in Figure 6.16. The spring is compressed a distance of 2.0 cm, and the block is released from rest. (a) Calculate the speed of the block as it passes through the equilibrium position $x = 0$ if the surface is frictionless.

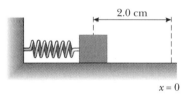

Figure 6.16   (Example 6.7)

Solution We use Equation 6.14 to find the work done by the spring with $x_m = -2.0$ cm $= -2.0 \times 10^{-2}$ m:

$$W_s = \tfrac{1}{2}kx_m^2 = \tfrac{1}{2}(1.0 \times 10^3 \text{ N/m})(-2.0 \times 10^{-2} \text{ m})^2 = 0.20 \text{ J}$$

Using the work-kinetic energy theorem with $v_i = 0$ gives

$$W_s = \tfrac{1}{2}mv_f^2 - \tfrac{1}{2}mv_i^2$$

$$0.20 \text{ J} = \tfrac{1}{2}(1.6 \text{ kg})v_f^2 - 0$$

$$v_f^2 = \frac{0.40 \text{ J}}{1.6 \text{ kg}} = 0.25 \text{ m}^2/\text{s}^2$$

$$v_f = \boxed{0.50 \text{ m/s}}$$

(b) Calculate the speed of the block as it passes through the equilibrium position if a constant frictional force of 4.0 N retards its motion.

Solution We use Equation 6.21 to calculate the kinetic energy lost due to friction and add this to the kinetic energy found in the absence of friction. Considering only the frictional force, the kinetic energy lost due to friction is

$$-fd = -(4.0 \text{ N})(2.0 \times 10^{-2} \text{ m}) = -0.08 \text{ J}$$

The final kinetic energy, without this loss, was found in part (a) to be 0.20 J. Therefore, the final kinetic energy in the presence of friction is

$$K_f = 0.20 \text{ J} - 0.08 \text{ J} = 0.12 \text{ J} = \tfrac{1}{2}mv_f^2$$

$$\tfrac{1}{2}(1.6 \text{ kg})v_f^2 = 0.12 \text{ J}$$

$$v_f^2 = \frac{0.24 \text{ J}}{1.6 \text{ kg}} = 0.15 \text{ m}^2/\text{s}^2$$

$$v_f = \boxed{0.39 \text{ m/s}}$$

Note that this value for v_f is less than that obtained in the frictionless case. Does this result surprise you?

EXERCISE 9   A 100-kg sled is dragged by a team of dogs a distance of 2.00 km over a horizontal surface at a constant velocity. If the coefficient of friction between the sled and snow is 0.150, find (a) the work done by the team of dogs and (b) the energy lost due to friction.
Answer   (a) 2.94×10^5 J   (b) All of this energy is dissipated due to friction.

EXERCISE 10   A 15-kg block is dragged over a horizontal surface by a constant force of 70 N acting at an angle of 20° above the horizontal. The block is displaced 5.0 m, and the coefficient of kinetic friction between the block and surface is 0.30. Find the work done by (a) the 70-N force, (b) the normal force, and (c) the force of gravity. (d) Calculate the energy lost due to friction.   Answer   (a) 330 J   (b) 0   (c) 0   (d) -185 J

6.5 • POWER

From a practical viewpoint, it is interesting to know not only the work done on an object but also the rate at which the work is being done. The time rate at which work is done, or the time rate of energy transfer, is called **power**.

If an external force is applied to an object (which we assume acts as a particle), and if the work done by this force is W in the time interval Δt, then the **average power** during this interval is defined as

$$\overline{P} \equiv \frac{W}{\Delta t}$$

[6.22]   • *Average power*

The work done on the object contributes to increasing the energy of the object. A more general definition of power is the *time rate of energy transfer*. The **instantaneous power,** P, is the limiting value of the average power as Δt approaches zero:

$$P \equiv \lim_{\Delta t \to 0} \frac{W}{\Delta t} - \frac{dW}{dt}$$

[6.23]   • *Instantaneous power*

where we have represented the infinitesimal value of the work done by dW (even though it is not a change and therefore not a differential). We know from Equation 6.3 that $dW = \mathbf{F} \cdot d\mathbf{x}$. Therefore, the instantaneous power can be written

$$P = \frac{dW}{dt} = \mathbf{F} \cdot \frac{d\mathbf{x}}{dt} = \mathbf{F} \cdot \mathbf{v}$$

[6.24]   • *The watt*

where we have used the fact that $\mathbf{v} = d\mathbf{x}/dt$.

The SI unit of power is joules per second (J/s), also called a *watt* (W) (after James Watt):

$$1 \text{ W} = 1 \text{ J/s} = 1 \text{ kg} \cdot \text{m}^2/\text{s}^3$$

The upright symbol W for watt should not be confused with the italic symbol W for work.

The unit of power in the British engineering system is the horsepower (hp):

$$1 \text{ hp} \equiv 550 \text{ ft} \cdot \text{lb/s} = 746 \text{ W}$$

A new unit of energy (or work) can now be defined in terms of the unit of power. One kilowatt hour (kWh) is the energy converted or consumed in 1 h at the constant rate of 1 kW. The numerical value of 1 kWh is

$$1 \text{ kWh} = (10^3 \text{ W})(3600 \text{ s}) = 3.60 \times 10^6 \text{ J}$$

It is important to realize that a kilowatt hour is a unit of energy, not power. When you pay your electric bill, you are buying energy, and the amount of electricity used by an appliance is usually expressed in kilowatt hours. For example, an electric bulb rated at 100 W would "consume" 3.60×10^5 J of energy in 1 h.

Thinking Physics 6

A light bulb is described by some individuals as "having 60 watts." What's wrong with this phrase?

Reasoning The number given represents the power of the light bulb, which is the rate at which energy passes through it, entering as electrical energy from the power source. It is not a value that the light bulb "possesses," however. It is only the approximate power rating of the bulb when it is connected to a normal 120-volt household supply of electricity. If the bulb were connected to a 12-volt supply, energy would enter the bulb at a rate other than 60 watts. Thus, the 60 watts is not an intrinsic property of the bulb, such as density is for a given solid material.

CONCEPTUAL PROBLEM 7

An older model car accelerates from 0 to speed v in 10 seconds. A newer, more powerful sports car accelerates from 0 to $2v$ in the same time period. What is the ratio of powers expended by the two cars? Consider the energy coming from the engine to appear only as kinetic energy of the car.

Example 6.8   Power Delivered by an Elevator Motor

An elevator has a mass of 1000 kg and carries a maximum load of 800 kg. A constant frictional force of 4000 N retards its motion upward, as in Figure 6.17. (a) What must be the minimum power delivered by the motor to lift the elevator at a constant speed of 3.00 m/s?

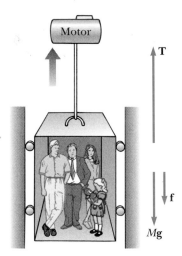

Figure 6.17   (Example 6.8) A motor exerts an upward force, equal in magnitude to the tension, T, on the elevator. A frictional force **f** and the force of gravity $M\mathbf{g}$ act downward.

Solution The motor must supply the force **T** that pulls the elevator upward. From Newton's second law and from the fact that $a = 0$ because v is constant, we get

$$T - f - Mg = 0$$

where M is the *total* mass (elevator plus load), equal to 1800 kg. Therefore,

$$T = f + Mg$$
$$= 4.00 \times 10^3 \text{ N} + (1.80 \times 10^3 \text{ kg})(9.80 \text{ m/s}^2)$$
$$= 2.16 \times 10^4 \text{ N}$$

Using Equation 6.24 and the fact that **T** is in the same direction as **v** gives

$$P = \mathbf{T} \cdot \mathbf{v} = Tv$$
$$= (2.16 \times 10^4 \text{ N})(3.00 \text{ m/s}) = 6.48 \times 10^4 \text{ W}$$
$$= 64.8 \text{ kW}$$

(b) What power must the motor deliver at any instant if it is designed to provide an upward acceleration of 1.00 m/s²?

Solution Applying Newton's second law to the elevator gives

$$T - f - Mg = Ma$$
$$T = M(a + g) + f$$
$$= (1.80 \times 10^3 \text{ kg})(1.00 + 9.80) \text{ m/s}^2$$
$$+ 4.00 \times 10^3 \text{ N}$$
$$= 2.34 \times 10^4 \text{ N}$$

Therefore, using Equation 6.24, for the required power we get

$$P = Tv = (2.34 \times 10^4 \ v) \text{ W}$$

where v is the instantaneous speed of the elevator in meters per second. Hence, the power required increases with increasing speed.

EXERCISE 11 A 65-kg athlete runs a distance of 600 m up a mountainside that is inclined at 20° to the horizontal. He performs this feat in 80 s. Assuming that air resistance is negligible, (a) how much work is done against gravity, and (b) what is his average power output during the run?   Answer   (a) 1.3×10^5 J   (b) 1.6 kW

SUMMARY

The **work** done by a *constant* force **F** acting on a particle is defined as the product of the component of the force in the direction of the particle's displacement and the magnitude of the displacement. If **F** makes an angle θ with the displacement **d**, the work done by **F** is

$$W \equiv Fd \cos \theta \qquad \text{[6.1]}$$

The **scalar**, or dot **product** of any two vectors **A** and **B** is defined by the relationship

$$\mathbf{A} \cdot \mathbf{B} \equiv AB \cos \theta \qquad \text{[6.5]}$$

where the result is a scalar quantity and θ is the angle between the directions of the two vectors. The scalar product obeys the commutative and distributive laws.

The **work** done by a *varying* force acting on a particle moving along the x axis from x_i to x_f is

$$W \equiv \int_{x_i}^{x_f} F_x \, dx \qquad \text{[6.11]}$$

where F_x is the component of force in the x direction. If there are several forces acting on the particle, the net work done by all forces is the sum of the individual amounts of work done by each force.

The **kinetic energy** of a particle of mass m moving with a speed v (where v is small compared with the speed of light) is

$$K \equiv \tfrac{1}{2}mv^2 \qquad \text{[6.18]}$$

The **work–kinetic energy theorem** states that the net work done on a particle by external forces equals the change in kinetic energy of the particle:

$$W_{\text{net}} = K_f - K_i = \tfrac{1}{2}mv_f^2 - \tfrac{1}{2}mv_i^2 \qquad \text{[6.19]}$$

Average power is the time rate of doing work:

$$\overline{P} \equiv \frac{W}{\Delta t} \qquad \text{[6.22]}$$

If an agent applies a force **F** to an object moving with a velocity **v**, the **instantaneous power** delivered by that agent is

$$P \equiv \frac{dW}{dt} = \mathbf{F} \cdot \mathbf{v} \qquad \text{[6.24]}$$

CONCEPTUAL QUESTIONS

1. Explain why the work done by the force of sliding friction is negative when an object undergoes a displacement on a rough surface.

2. When a particle rotates in a circle, a force acts on it directed toward the center of rotation. Why is it that this force does no work on the particle?

3. When a punter kicks a football, is he doing any work on the ball while his toe is in contact with it? Is he doing any work on the ball after it loses contact with his toe? Are any forces doing work on the ball while it is in flight?

4. Cite two examples in which a force is exerted on an object without doing any work on the object.

5. As a simple pendulum swings back and forth, the forces acting on the suspended mass are the force of gravity, the tension in the supporting cord, and air resistance. (a) Which of these forces, if any, does no work on the pendulum? (b) Which of these forces does negative work at all times during its motion? (c) Describe the work done by the force of gravity while the pendulum is swinging.

6. If the dot product of two vectors is positive, does this imply that the vectors must have positive rectangular components?

7. A hockey player pushes a puck with his hockey stick over frictionless ice. If the puck starts from rest and ends up moving at speed v, has the hockey player performed any work in his reference frame? An ant is on the puck and hangs on tightly while the puck is accelerated. In the reference frame of the ant, has any work been done? Is the work-kinetic energy theorem satisfied for each observer?

8. Can kinetic energy be negative? Explain.

9. One bullet has twice the mass of a second bullet. If both are fired so that they have the same speed, which has more kinetic energy? What is the ratio of the kinetic energies of the two bullets?

10. If the speed of a particle is doubled, what happens to its kinetic energy?

11. What can be said about the speed of a particle if the net work done on it is zero?

12. Can the average power ever equal the instantaneous power? Explain.

13. In Example 6.8, does the required power increase or decrease as the force of friction is reduced?

14. Sometimes physics words are used in popular literature in interesting ways. For example, consider a description of a rock falling from the top of a cliff as "gathering force as it falls to the beach below." What does the phrase "gathering force" mean and can you repair this phrase?

PROBLEMS

Section 6.1   Work Done by a Constant Force

1. A tugboat exerts a constant force of 5000 N on a ship moving at constant speed through a harbor. How much work does the tugboat do on the ship in a distance of 3.00 km?

2. A shopper in a supermarket pushes a cart with a force of 35.0 N directed at an angle of 25.0° downward from the horizontal. Find the work done by the shopper as he moves down a 50.0-m length of aisle.

3. A raindrop ($m = 3.35 \times 10^{-5}$ kg) falls vertically at constant speed under the influence of gravity and air resistance. After the drop has fallen 100 m, what is the work done (a) by gravity and (b) by air resistance?

4. A block of mass 2.50 kg is pushed 2.20 m along a frictionless horizontal table by a constant 16.0-N force directed 25.0° below the horizontal. Determine the work done by (a) the applied force, (b) the normal force exerted by the table, (c) the force of gravity, and (d) the net force on the block.

5. Batman, whose mass is 80.0 kg, is holding on to the free end of a 12.0-m rope, the other end of which is fixed to a tree limb directly above. He is able to get the rope in motion as only Batman knows how, eventually getting it to swing enough that he can reach a ledge when the rope makes a 60.0° angle with the vertical. How much work was done by him against the force of gravity in this maneuver?

Section 6.2   The Scalar Product of Two Vectors

6. Vector **A** has a magnitude of 5.00 units, and **B** has a magnitude of 9.00 units. The two vectors make an angle of 50.0° with each other. Find $\mathbf{A} \cdot \mathbf{B}$.

7. Vector **A** extends from the origin to a point having polar coordinates (7, 70°), and vector **B** extends from the origin to a point having polar coordinates (4, 130°). Find $\mathbf{A} \cdot \mathbf{B}$.

8. Given two arbitrary vectors **A** and **B**, show that $\mathbf{A} \cdot \mathbf{B} = A_x B_x + A_y B_y + A_z B_z$. (*Hint:* Write **A** and **B** in unit vector form and use Equations 6.7 and 6.8.)

9. A force $\mathbf{F} = (6\mathbf{i} - 2\mathbf{j})$ N acts on a particle that undergoes a displacement $\mathbf{d} = (3\mathbf{i} + \mathbf{j})$ m. Find (a) the work done by the force on the particle and (b) the angle between **F** and **d**.

10. For $\mathbf{A} = 3\mathbf{i} + \mathbf{j} - \mathbf{k}$, $\mathbf{B} = -\mathbf{i} + 2\mathbf{j} + 5\mathbf{k}$, and $\mathbf{C} = 2\mathbf{j} - 3\mathbf{k}$, find $\mathbf{C} \cdot (\mathbf{A} - \mathbf{B})$.

11. Using the definition of the scalar product, find the angles between (a) $\mathbf{A} = 3\mathbf{i} - 2\mathbf{j}$ and $\mathbf{B} = 4\mathbf{i} - 4\mathbf{j}$; (b) $\mathbf{A} = -2\mathbf{i} +$

4**j** and **B** = 3**i** − 4**j** + 2**k**; (c) **A** = **i** − 2**j** + 2**k** and **B** = 3**j** + 4**k**.

Section 6.3   Work Done by a Varying Force

12. The force acting on a particle varies as in Figure P6.12. Find the work done by the force as the particle moves (a) from $x = 0$ to $x = 8.00$ m, (b) from $x = 8.00$ m to $x = 10.0$ m, and (c) from $x = 0$ to $x = 10.0$ m.

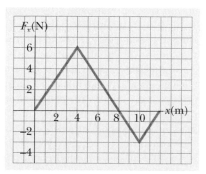

Figure P6.12

13. A particle is subject to a force F_x that varies with position as in Figure P6.13. Find the work done by the force on the body as it moves (a) from $x = 0$ to $x = 5.00$ m, (b) from $x = 5.00$ m to $x = 10.0$ m, and (c) from $x = 10.0$ m to $x = 15.0$ m. (d) What is the total work done by the force over the distance $x = 0$ to $x = 15.0$ m?

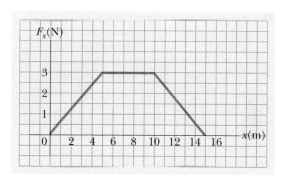

Figure P6.13

14. The force acting on a particle is $F_x = (8x − 16)$ N, where x is in meters. (a) Make a plot of this force versus x from $x = 0$ to $x = 3.00$ m. (b) From your graph, find the net work done by this force as the particle moves from $x = 0$ to $x = 3.00$ m.

15. A force **F** = $(4.0x\mathbf{i} + 3.0y\mathbf{j})$ N acts on an object as the object

moves in the x direction from the origin to $x = 5.00$ m. Find the work done on the object by the force.

16. When a 4.00-kg mass is hung vertically on a certain light spring that obeys Hooke's law, the spring stretches 2.50 cm. If the 4.00-kg mass is removed, (a) how far will the spring stretch if a 1.50-kg mass is hung on it, and (b) how much work must an external agent do to stretch the same spring 4.00 cm from its unstretched position?

17. An archer pulls her bow string back 0.400 m by exerting a force that increases uniformly from 0 to 230 N. (a) What is the equivalent spring constant of the bow? (b) How much work is done in pulling the bow?

18. A 100-g bullet is fired from a rifle having a barrel 0.600 m long. Assuming the origin is placed where the bullet begins to move, the force (in newtons) exerted on the bullet by the expanding gas is $15\,000 + 10\,000x − 25\,000x^2$, where x is in meters. (a) Determine the work done by the gas on the bullet as the bullet travels the length of the barrel. (b) If the barrel is 1.00 m long, how much work is done, and how does this value compare to the work calculated in (a)?

19. If it takes 4.00 J of work to stretch a Hooke's-law spring 10.0 cm from its unstressed length, determine the extra work required to stretch it an additional 10.0 cm.

20. If it takes work W to stretch a Hooke's-law spring a distance d from its unstressed length, determine the extra work required to stretch it an additional distance d.

Section 6.4   Kinetic Energy and the Work-Kinetic Energy Theorem

21. A 0.600-kg particle has a speed of 2.00 m/s at point A and kinetic energy of 7.50 J at point B. What is (a) its kinetic energy at A? (b) its speed at B? (c) the total work done on the particle as it moves from A to B?

22. A 0.300-kg ball has a speed of 15.0 m/s. (a) What is its kinetic energy? (b) If its speed is doubled, what is its kinetic energy?

23. A 3.00-kg mass has an initial velocity $\mathbf{v}_0 = (6.00\mathbf{i} − 2.00\mathbf{j})$ m/s. (a) What is its kinetic energy at this time? (b) Find the total work done on the object if its velocity changes to $(8.00\mathbf{i} + 4.00\mathbf{j})$ m/s. (*Hint*: Remember that $v^2 = \mathbf{v} \cdot \mathbf{v}$.)

24. You can think of the work-kinetic energy theorem as a second theory of motion, parallel to Newton's laws in describing how outside influences affect the motion of an object. In this problem work out parts (a) and (b) separately from parts (c) and (d) to compare the predictions of the two theories. In a rifle barrel, a 15.0-g bullet is accelerated from rest to a speed of 780 m/s. (a) Find the work done on it. (b) If the rifle barrel is 72.0 cm long, find the magnitude of the average total force that acted on it, as $F = W/(d \cos \theta)$. (c) Find the constant acceleration of a bullet that starts from rest and gains speed 780 m/s over a distance

of 72.0 cm. (d) If the bullet has mass 15.0 g, find the total force that acted on it as $\Sigma F = ma$.

25. A cart loaded with bricks has a total mass of 18.0 kg and is pulled at constant speed by a rope. The rope is inclined at 20.0° above the horizontal, and the cart moves a distance of 20.0 m on a horizontal surface. The coefficient of kinetic friction between the cart and surface is 0.500. (a) What is the tension of the rope? (b) How much work is done on the cart by the rope? (c) What is the energy lost due to friction?

26. A 40.0-kg box initially at rest is pushed 5.00 m along a rough, horizontal floor with a constant applied horizontal force of 130 N. If the coefficient of friction between box and floor is 0.300, find (a) the work done by the applied force, (b) the energy lost due to friction, (c) the change in kinetic energy of the box, and (d) the final speed of the box.

27. A sled of mass m is given a kick on a frozen pond, imparting to it an initial speed $v_i = 2.00$ m/s. The coefficient of kinetic friction between sled and ice is $\mu_k = 0.100$. Use energy considerations to find the distance the sled moves before stopping.

28. A block of mass 12.0 kg slides from rest down a frictionless 35.0° incline and is stopped by a strong spring with $k = 3.00 \times 10^4$ N/m. The block slides 3.00 m from the point of release to the point where it comes to rest against the spring. When the block comes to rest, how far has the spring been compressed?

29. A crate of mass 10.0 kg is pulled up a rough incline with an initial speed of 1.50 m/s. The pulling force is 100 N parallel to the incline, which makes an angle of 20.0° with the horizontal. The coefficient of kinetic friction is 0.400, and the crate is pulled 5.00 m. (a) How much work is done by gravity? (b) How much energy is lost due to friction? (c) How much work is done by the 100-N force? (d) What is the change in kinetic energy of the crate? (e) What is the speed of the crate after being pulled 5.00 m?

30. A picture tube in a certain television set is 36.0 cm long. The electrical force accelerates an electron in the tube from rest to 1.00% of the speed of light over this distance. Determine (a) the kinetic energy of the electron as it strikes the screen at the end of the tube, (b) the magnitude of the average electrical force acting on the electron over this distance, (c) the magnitude of the average acceleration of the electron over this distance, and (d) the time of flight.

31. A time-varying net force acting on a 4.00-kg object causes the object to have a displacement given by $x = 2.0t - 3.0t^2 + 1.0t^3$, where x is in meters and t is in seconds. Find the work done on the object in the first 3.00 s of motion.

32. A 5.00-kg steel ball is dropped onto a copper plate from a height of 10.0 m. If the ball leaves a dent 0.320 cm deep, what is the average force exerted on the ball by the plate during the impact?

Section 6.5   Power

33. A 700-N Marine in basic training climbs a 10.0-m vertical rope at a constant speed in 8.00 s. What is her power output?

34. Make an order-of-magnitude estimate of the power your car engine puts into speeding the car up to highway speed. In your solution state the physical quantities you take as data and the values you measure or estimate for them. The mass of the vehicle is given in the owner's manual. If not a car, consider a bus or truck that you specify.

35. If a certain horse can maintain 1.00 hp of output for 2.00 h, how many 70.0-kg bundles of shingles can that horse hoist (via some pulley arrangement) to the roof of a house 8.00 m tall, assuming 70.0% efficiency?

36. A certain automobile engine delivers 30.0 hp (2.24 × 10⁴ W) to its wheels when moving at a constant speed of 27.0 m/s (≈60 mi/h). What is the resistive force acting on the automobile at that speed?

37. A skier of mass 70.0 kg is pulled up a slope by a motor-driven cable. (a) How much work is required to pull him a distance of 60.0 m up a 30.0° slope (assumed frictionless) at a constant speed of 2.00 m/s? (b) A motor of what power is required to perform this task?

38. A 650-kg elevator starts from rest. It moves upward for 3.00 s with constant acceleration until it reaches its cruising speed of 1.75 m/s. (a) What is the average power of the elevator motor during this period? (b) How does this power compare with its power when it moves at its cruising speed?

Additional Problems

39. A baseball outfielder throws a 0.150-kg baseball at a speed of 40.0 m/s and an initial angle of 30.0°. What is the kinetic energy of the baseball at the highest point of the trajectory?

40. While running, a person dissipates about 0.600 J of mechanical energy per step per kilogram of body mass. If a 60.0-kg runner dissipates a power of 70.0 W during a race, how fast is the person running? Assume a running step is 1.50 m long.

41. A particle of mass m moves with a constant acceleration **a**. If the initial position vector and velocity of the particle are $\mathbf{r}_0$ and $\mathbf{v}_0$, respectively, show that its speed v at any time satisfies the equation

$$v^2 = v_0{}^2 + 2\mathbf{a} \cdot (\mathbf{r} - \mathbf{r}_0)$$

where **r** is the position vector of the particle at that same time.

42. The direction of an arbitrary vector **A** can be completely specified with the angles α, β, and γ that the vector makes with the x, y, and z axes, respectively. If $\mathbf{A} = A_x\mathbf{i} + A_y\mathbf{j} + A_z\mathbf{k}$, (a) find expressions for $\cos \alpha$, $\cos \beta$, and $\cos \gamma$ (these

are known as *direction cosines*), and (b) show that these angles satisfy the relation $\cos^2 \alpha + \cos^2 \beta + \cos^2 \gamma = 1$. (*Hint:* Take the scalar product of **A** with **i, j**, and **k** separately.)

43. A 4.00-kg particle moves along the x axis. Its position varies with time according to $x = t + 2.0t^3$, where x is in meters and t is in seconds. Find (a) the kinetic energy at any time t, (b) the acceleration of the particle and the force acting on it at time t, (c) the power being delivered to the particle at time t, and (d) the work done on the particle in the interval $t = 0$ to $t = 2.00$ s.

44. When a spring is stretched beyond its elastic limit, the restoring force satisfies the equation $F = -kx + \beta x^3$. If $k = 10.0$ N/m and $\beta = 100$ N/m³, calculate the work done by this force when the spring is stretched 0.100 m.

45. A 2100-kg pile driver is used to drive a steel I-beam into the ground. The pile driver falls 5.00 m before contacting the beam, and it drives the beam 12.0 cm into the ground before coming to rest. Using energy considerations, calculate the average force the beam exerts on the pile driver while the pile driver is brought to rest after each hit.

46. A cyclist and her bicycle have a combined mass of 75.0 kg. She coasts down a road inclined at 2.00° with the horizontal at 4.00 m/s and down a road inclined at 4.00° at 8.00 m/s. She then holds on to a moving vehicle and coasts on a level road. What power must the vehicle expend to maintain her speed at 3.00 m/s? Assume that the force of air resistance is proportional to her speed and assume that other frictional forces remain constant.

47. A 200-g block is pressed against a spring of force constant 1.40 kN/m until the block compresses the spring 10.0 cm. The spring rests at the bottom of a ramp inclined at 60.0° to the horizontal. Use energy considerations to determine how far up the incline the block moves before it stops if (a) there is no friction between block and ramp and if (b) the coefficient of kinetic friction is 0.400.

48. A 0.400-kg particle slides on a horizontal circular track 1.50 m in radius. It is given an initial speed of 8.00 m/s. After one revolution, its speed drops to 6.00 m/s because of friction. (a) Find the energy lost due to friction in one revolution. (b) Calculate the coefficient of kinetic friction. (c) What is the total number of revolutions the particle makes before stopping?

49. The ball launcher in a pinball machine has a spring that has a force constant of 1.20 N/cm (Fig. P6.49). The surface on which the ball moves is inclined 10.0° with respect to the horizontal. If the spring is initially compressed 5.00 cm, find the launching speed of a 100-g ball when the plunger

Figure P6.49

is released. Friction and the mass of the plunger are negligible.

50. In diatomic molecules, the constituent atoms exert attractive forces on each other at large distances and repulsive forces at short distances. For many molecules, the Lennard-Jones law is a good approximation to the magnitude of these forces:

$$F = F_0 \left[2 \left(\frac{\sigma}{r} \right)^{13} - \left(\frac{\sigma}{r} \right)^{7} \right]$$

where r is the center-to-center distance between the atoms in the molecule, σ is a length parameter, and F_0 is the force when $r = \sigma$. For an oxygen molecule, the values of F_0 and σ are $F_0 = 9.60 \times 10^{-11}$ N and $\sigma = 3.50 \times 10^{-10}$ m. Determine the work done by this force from $r = 4.00 \times 10^{-10}$ m to $r = 9.00 \times 10^{-10}$ m.

51. Suppose a car is modeled as a cylinder with cross-sectional area A moving with a speed v, as in Figure P6.51. In a time Δt, a column of air of mass Δm must be moved a distance $v \Delta t$ and given a kinetic energy $\frac{1}{2}(\Delta m) v^2$. Using this model, show that the power loss due to air resistance is $\frac{1}{2}\rho A v^3$ and the resistive force is $\frac{1}{2}\rho A v^2$, where ρ is the density of air.

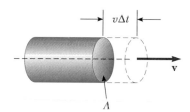

Figure P6.51

💠 Spreadsheet Problems

S1. When different weights are hung on a spring, the spring stretches to different lengths as shown in the table that follows. (a) Use a spreadsheet to plot the length of the spring versus applied force. Use the spreadsheet's least-squares fitting features to determine the straight line that best fits the data. (You may not want to use all the data points.) (b) From the slope of the least-squares fit line, find the spring constant k. (c) If the spring is extended to 105 mm, what force does it exert on the suspended weight?

F(N)	L(mm)	F(N)	L(mm)
2.0	15	14	112
4.0	32	16	126
6.0	49	18	149
8.0	64	20	175
10	79	22	190
12	98		

S2. A 0.178-kg particle moves along the x axis from $x = 12.8$ m to $x = 23.7$ m under the influence of a force

$$F = \frac{375}{x^3 + 3.75x}$$

where F is in newtons and x is in meters. Use numerical integration and set up a spreadsheet to determine the total work done by this force during this displacement. Your calculations should have an accuracy of at least 2%.

ANSWERS TO CONCEPTUAL PROBLEMS

1. Since there is no motion taking place, the rope experiences no displacement. Thus, no work is done on it. For the same reason, no work is being done on the pullers or the ground. Work is only being done within the bodies of the pullers. For example, the heart of each puller is applying forces on the blood to move it through the body.

2. Less force will be necessary with a longer ramp, but the force must act over a longer distance to do the same amount of work. Suppose the refrigerator is rolled up the ramp at constant speed. The normal force does no work because it acts at 90° to the motion. The work by gravity is just the weight of the refrigerator times the vertical height through which it is displaced times (cos 180°), or $Wg = -mgh$. Therefore, the movers must do work mgh on the refrigerator, however long the ramp.

3. If we ignore the effects of rolling friction on the tires of the car, the same amount of work would be done in driving up the switchback and driving straight up the mountain, since the weight of the car is moved upward against gravity by the same vertical distance in each case. If we include friction, there is more work done in driving the switchback, since the distance over which the friction force acts is much longer. So why do we use the switchback? The answer lies in the force required, not the work. The force from the engine required to follow a gentle rise is much smaller than that required to drive straight up the hill. Roadways running straight uphill would require redesigning engines so as to be able to apply much larger forces. This is similar to the ease with which heavy objects can be rolled up ramps into moving trucks, compared to lifting the object straight up from the ground.

4. If the spring is cut in half and stretched by the same distance, each adjacent pair of coils must now be twice as far apart as in the original stretched spring. Thus, the force applied by the spring is twice as great for any given extension. This leads to an increase in the spring constant by a factor of 2. Thus,

if the cut spring is stretched by the same amount, it must require twice as much work.

5. We focus on the book as the system. You indeed did some work in applying an upward force to lift the book. But there is another force on the book—gravity. The work done by gravity is negative, since the force of gravity is opposite to the displacement. Assuming that you lifted the book slowly, the upward force that you applied is approximately the same in magnitude as the weight of the book. Thus, the positive work done by you and the negative work done by gravity cancel. Thus, there is no net work performed, and no net change in the kinetic energy—the work-kinetic energy theorem is satisfied. If we consider the book and the Earth to be the system, then there is net work done by you on this system. This work appears in the system as potential energy, which will be investigated in Chapter 7.

6. When walking downstairs, a person has to stop the motion of the entire body at each step. This requires work to be done by the muscles. The work done by the muscles is negative, since the forces applied are opposite to the displacements representing the movement of the body. Negative work represents a transfer of energy out of the muscles (from the store of energy from the food eaten by the walker). When walking horizontally, the body can be kept in motion with an almost constant horizontal speed—only the feet and legs start and stop. Thus, more energy is transferred from the body in bringing the entire body to a stop in the trip downstairs than in walking.

7. Since the time periods are the same for both cars, we need only to compare the work done. Since the sports car is moving twice as fast as the older car at the end of the time interval, it has four times the kinetic energy. Thus, according to the work-kinetic energy theorem, four times as much work was done, and the engine must have expended four times the power.

7

Potential Energy and Conservation of Energy

In Chapter 6 we introduced the concept of kinetic energy, which is associated with the motion of an object. In this chapter we introduce another form of mechanical energy, called *potential energy*, that is associated with the position or configuration of a system. The potential energy of a system can be thought of as stored energy that can be converted to kinetic energy or other forms of energy.

The potential energy concept can be used only with a special class of forces called *conservative forces*. When only internal conservative forces, such as gravitational or spring forces, act within a system, the kinetic energy gained (or lost) by the system as its members change their relative positions is compensated by an equal loss (or gain) in potential energy.

◄ Twin Falls on the Island of Kauai, Hawaii. The gravitational potential energy of the water-Earth system when the water is at the top of the falls is converted to kinetic energy of the water at the bottom. In many locations, this mechanical energy is used to produce electrical energy. *(Bruce Byers, FPG)*

165

7.1 • POTENTIAL ENERGY

In Chapter 6 we saw that an object with kinetic energy can do work on another object, as illustrated by the moving hammer driving a nail into a wall. Now we shall see that an object can also do work because of the energy resulting from its *position* in space.

As an object falls in a gravitational field, the field exerts a force on the object, does work on it, and thereby changes its kinetic energy. Consider a brick dropped from rest directly above a nail in a board that is lying horizontally on the ground. When the brick is released, it falls toward the ground, gaining speed and therefore gaining kinetic energy. The potential energy of the brick–Earth system is converted into kinetic energy as the brick falls. When the brick reaches the ground, it does work on the nail, driving it into the board. The potential energy of a system consisting of any object and the Earth is called **gravitational potential energy.**

Gravitational potential •
energy

Let us now derive an expression for the gravitational potential energy associated with an object at a given location in space. To do this, consider a block of mass m at an initial height y_i above the ground, as in Figure 7.1. With air resistance neglected as the block falls, the only force that does work on it is the gravitational force, $m\mathbf{g}$. The work done by the gravitational force as the block undergoes a downward displacement $\mathbf{d}$ is given by the product of the downward force $m\mathbf{g}$ and the displacement, or

$$W_g = (m\mathbf{g}) \cdot \mathbf{d} = (-mg\mathbf{j}) \cdot [(y_f - y_i)\mathbf{j}] = mgy_i - mgy_f$$

We can represent the quantity mgy to be the gravitational potential energy, U_g:

$$U_g = mgy \qquad\qquad \text{[7.1]}$$

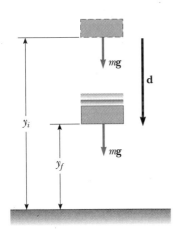

Figure 7.1 The work done by the gravitational force as the block falls from y_i to y_f is equal to $mgy_i - mgy_f$.

In this representation, the gravitational potential energy associated with an object at any point in space is the product of the object's weight and its vertical coordinate. The origin of the coordinate system could be located at the surface of the Earth or at any other convenient point.

If we substitute U for the mgy terms in the expression for W_g, we have

$$W_g = U_i - U_f \qquad\qquad \text{[7.2]}$$

From this result, we see that the work done on any object by the force of gravity —that is, the energy transferred to the object from the gravitational field—is equal to the initial value of the potential energy minus the final value of potential energy.

The units of gravitational potential energy are the same as those of work. That is, potential energy may be expressed in joules, ergs, or foot-pounds. Potential energy, like work and kinetic energy, is a scalar quantity.

Note that the gravitational potential energy depends only on the vertical height of the object above the surface of the Earth. From this result, we see that the same amount of work is done on an object if it falls vertically to the Earth as if it starts at the same point and slides down a frictionless incline to the Earth. Also note that Equation 7.1 is valid only for objects near the surface of the Earth, where $\mathbf{g}$ is approximately constant.

In working problems involving gravitational potential energy, it is always necessary to choose a location at which to set the gravitational potential energy equal to zero. The choice of zero level is completely arbitrary, because the important quantity is the *difference* in potential energy, and this difference is independent of the choice of zero level.

It is often convenient to choose the surface of the Earth as the reference position for zero potential energy, but this is not essential. Often, the statement of the problem suggests a convenient level to use.

Thinking Physics 1

Elevators that are lifted by a cable have a counterweight, as shown in Figure 7.2. The elevator and the counterweight are connected by a cable that passes over a drive wheel. The counterweight moves in the direction opposite to that of the elevator. What is the purpose of the counterweight?

Reasoning Without the counterweight, the elevator motor would need to lift the weight of the elevator and the people riding it. This would represent a large increase in gravitational potential energy, which means a large expenditure of energy from the motor. With the counterweight, as the elevator moves upward, the counterweight moves down. Thus, the center of mass of the combined system of the elevator and counterweight moves only over a short distance near the middle region of the vertical shaft. This results in a much smaller variation of potential energy of the system than without the counterweight. In the particular case of the elevator and occupants having exactly the same mass as the counterweight, the center of mass is midway between them and never moves as the elevator changes height. In this case, once the elevator is in motion, the motor is no longer required—the system will continue to move as described by Newton's first law (as long as we ignore friction). The motor is needed again to stop the elevator at the desired floor. Thus, the presence of the counterweight reduces the energy requirements on the motor—instead of having to pull the elevator through its entire motion, it is needed only to start and stop the elevator, to overcome friction, and to account for any difference in mass between the counterweight and the elevator, associated with the number of passengers on the elevator.

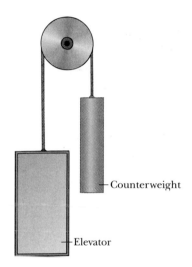

Figure 7.2   (Thinking Physics 1)

EXERCISE 1 What is the gravitational potential energy, relative to the ground, of a 0.15-kg baseball at the top of a 100–m-tall building?   Answer   147 J

7.2 • CONSERVATIVE AND NONCONSERVATIVE FORCES

Forces found in nature can be divided into two categories: conservative and nonconservative. We shall describe the properties of conservative and nonconservative forces separately in this section.

Conservative Forces

A force is conservative if the work it does on an object moving between any two points is independent of the path taken by the object. The work done by a conservative force depends only on the initial and final coordinates of the object. A conservative force can also be defined in a second way. **A force is conservative if the work it does on an object moving through any closed path is zero.**

• *Definition of a conservative force*

The force of gravity is conservative. As we learned in the preceding section, the work done by the gravitational force on an object moving between any two points near the Earth's surface is

$$W_g = mgy_i - mgy_f$$

From this, we see that W_g depends only on the initial and final coordinates of the object and hence is independent of path. Furthermore, W_g is zero when the object moves over any closed path (where $y_i = y_f$).

We can associate a potential energy function with any conservative force. In the preceding section, the potential energy function associated with the gravitational force was found to be

$$U_g = mgy$$

Potential energy functions can be defined only for conservative forces. In general, the work, W_c, done on an object by a conservative force is given by the initial value of the potential energy associated with the object minus the final value:

$$W_c = U_i - U_f \qquad\qquad [7.3]$$

Another example of a conservative force is the force of a spring on an object attached to the spring, where the spring force is given by $F_s = -kx$. As we learned in Chapter 6 (Eq. 6.15), the work done by the spring force is

$$W_s = \tfrac{1}{2}kx_i^2 - \tfrac{1}{2}kx_f^2$$

where the initial and final coordinates of the object are measured from its equilibrium position, $x = 0$. Again we see that W_s depends only on the initial and final coordinates of the object and is zero for any closed path. Hence, the spring force is conservative. The **elastic potential energy** function associated with the spring force is defined by

Potential energy stored in a •
spring

$$U_s \equiv \tfrac{1}{2}kx^2 \qquad\qquad [7.4]$$

where x is measured from the uncompressed position of the spring. The elastic potential energy can be thought of as the energy stored in the deformed spring (one that is either compressed or stretched from its equilibrium position). To visualize this, consider Figure 7.3a, which shows an undeformed spring on a frictionless, horizontal surface. When the block is pushed against the spring (Fig. 7.3b), compressing the spring a distance x, the elastic potential energy stored in the spring is $kx^2/2$. When the block is released, the spring snaps back to its original length and the stored elastic potential energy is transformed into kinetic energy of the block (Fig. 7.3c). The elastic potential energy stored in the spring is zero whenever the spring is undeformed ($x = 0$). Energy is stored in the spring only when the spring is either stretched or compressed. Furthermore, the elastic potential energy is a maximum when the spring has reached its maximum compression or extension (that is, when $|x|$ is a maximum). Finally, because the elastic potential energy is proportional to x^2, we see that U_s is always positive in a deformed spring. If the spring and object are taken together as the system, then no work is done as the spring changes length because the forces are internal.

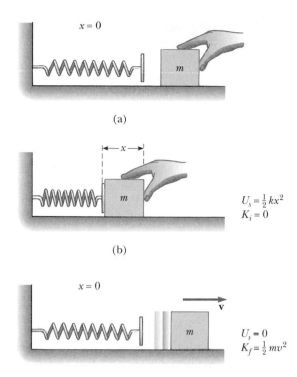

(a)

$U_s = \frac{1}{2} kx^2$
$K_i = 0$

(b)

Figure 7.3   A block of mass m on a frictionless horizontal surface is pushed against a spring and then released. If x is the compression in the spring as in (b), the elastic potential energy stored in the spring is $\frac{1}{2}kx^2$. This energy is transferred to the block in the form of kinetic energy as in (c).

$U_s = 0$
$K_f = \frac{1}{2} mv^2$

(c)

Nonconservative Forces

A force is called **nonconservative** if it leads to dissipation of mechanical energy. For example, if you move an object on a horizontal surface, returning to the same location and same state of motion, but you found that it was necessary to do a net amount of work on the object, then something must have dissipated that energy transferred to the object. That dissipative force is recognized as friction between the surface and the object. Friction is a dissipative, or nonconservative, force. By contrast, if the object is lifted, work is required, but that energy is recovered when the object is lowered. The gravitational force is a nondissipative or conservative force.

• *Definition of a nonconservative force*

Suppose you displace a book between two points on a table. If the book is displaced in a straight line along the blue path between points A and B in Figure 7.4, the loss in mechanical energy due to the friction force f is simply $-fd$, where d is the distance between the two points. However, if the book is moved along any other path between the two points, the loss in mechanical energy due to friction is greater (in absolute magnitude) than $-fd$. For example, the loss in mechanical energy due to friction along the red semicircular path in Figure 7.4 is equal to $-f(\pi d/2)$, where d is the diameter of the circle.

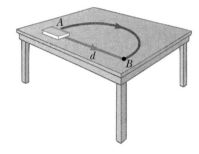

Figure 7.4   The decrease in mechanical energy due to the force of friction depends on the path taken as the book is moved from A to B, and hence friction is a nonconservative force. The decrease in mechanical energy is greater along the red path compared to the blue path.

7.3 • CONSERVATIVE FORCES AND POTENTIAL ENERGY

In the preceding section we found that the work done on a particle by a conservative force does not depend on the path taken by the particle and is independent of the particle's velocity. The work done is a function only of the particle's initial and final coordinates. As a consequence, we are able to define a **potential energy function** *U* such that the work done on a particle equals the decrease in the potential energy of the particle. The work done by a conservative force **F** as the particle moves along the *x* axis can be expressed as[1]

Work done by a conservative •
force

$$W_c = \int_{x_i}^{x_f} F_x \, dx = -\Delta U = U_i - U_f \qquad \text{[7.5]}$$

That is, **the work done by a conservative force equals the negative of the change in the potential energy associated with that force,** where the change in the potential energy is defined as $\Delta U = U_f - U_i$. For example, the work, W_c, done by the gravitational field on an object, as the object is lowered in the field, is $U_i - U_f$, which is positive, showing that energy has been transferred from the gravitational field to the object. This energy may appear as kinetic energy of the falling object or may be transferred to something else. We can also express Equation 7.5 as

Change in potential energy •

$$\Delta U = U_f - U_i = -\int_{x_i}^{x_f} F_x \, dx \qquad \text{[7.6]}$$

where F_x is the component of **F** in the direction of the displacement: **F** is the force exerted by the field on the object. Therefore ΔU is negative when F_x and dx are in the same direction, as when an object is lowered in a gravitational field or a spring pushes an object toward equilibrium.

The term *potential energy* implies that an object has the potential, or capability, of either gaining kinetic energy or doing work when released from some point under the influence of gravity. It is often convenient to establish some particular location, x_i, to be a reference point and measure all potential energy differences with respect to that point. We can then define the potential energy function as

$$U_f = -\int_{x_i}^{x_f} F_x \, dx + U_i \qquad \text{[7.7]}$$

Furthermore, as we discussed earlier, the value of U_i is often taken to be zero at some arbitrary reference point. It really doesn't matter what value we assign to U_i, because any value only shifts U_f by a constant, and it is only the *change* in potential energy that is physically meaningful. If the conservative force is known as a function of position, we can use Equation 7.7 to calculate the change in potential energy of a body as it moves from x_i to x_f.

The amount of mechanical energy dissipated by a nonconservative force depends on the path as an object moves from one position to another and can also depend on the object's speed or on other quantities. Because the work done by a nonconservative force is not simply a function of the initial and final coordinates, we conclude that there is no potential energy function associated with a nonconservative force.

---

[1] For a general displacement, the work done in two or three dimensions also equals $U_i - U_f$, where $U = U(x, y, z)$. We write this formally as $W = \int_i^f \mathbf{F} \cdot d\mathbf{s} = U_i - U_f$.

7.4 · CONSERVATION OF MECHANICAL ENERGY

An object held at some height h above the floor has no kinetic energy, but, as we learned earlier, there is an associated gravitational potential energy equal to mgh, relative to the floor if the gravitational field is included as part of the system. If the object is dropped, it falls to the floor; as it falls, its speed and thus its kinetic energy increase while the potential energy decreases. If factors such as air resistance are ignored, whatever potential energy the object loses as it moves downward appears as kinetic energy. In other words, the sum of the kinetic and potential energies, called the *mechanical energy E,* remains constant in time. This is an example of principle of **conservation of mechanical energy.** For the case of an object in free-fall, this principle tells us that any increase (or decrease) in potential energy is accompanied by an equal decrease (or increase) in kinetic energy.

Because the total mechanical energy E is defined as the sum of the kinetic and potential energies, we can write

$$E \equiv K + U \qquad [7.8]$$

• *Total mechanical energy*

Therefore, we can apply conservation of mechanical energy in the form $E_i = E_f$, or

$$K_i + U_i = K_f + U_f \qquad [7.9]$$

• *Conservation of mechanical energy*

We can state this in a more formal way: **conservation of mechanical energy requires that the total mechanical energy of a system remains constant in any isolated system of objects which interact only through conservative forces.** It is important to note that Equation 7.9 is valid *provided* no energy is added to or removed from the system. Furthermore, there must be no nonconservative forces within the system.

• *A formal statement of conservation of energy*

Because mechanical energy E remains constant with time, $dE/dt = 0$. Taking the derivative of Equation 7.8 with respect to time,

$$\frac{dE}{dt} = 0 = \frac{dK}{dt} + \frac{dU}{dt} \qquad [7.10]$$

Because $K = \frac{1}{2}mv^2$, then

$$\frac{dK}{dt} = \frac{d}{dt}(\tfrac{1}{2}mv^2) = mv\frac{dv}{dt} = mva = F_x v$$

Applying the chain rule (see Appendix B.6) to dU/dt, we have

$$\frac{dU}{dt} = \frac{dU}{dx}\frac{dx}{dt} = \left(\frac{dU}{dx}\right)v$$

Substituting these expressions for dK/dt and dU/dt into Equation 7.10 gives

$$F_x v + \left(\frac{dU}{dx}\right)v = 0$$

$$F_x = -\frac{dU}{dx} \qquad [7.11]$$

• *Relation between a conservative force and potential energy*

That is, **the conservative internal force acting between parts of a system equals the negative derivative of the potential energy associated with that system.**

We can easily check this relationship for the two instances already discussed. In the case of the deformed spring, $U_s = \frac{1}{2}kx^2$, and therefore

$$F_s = -\frac{dU_s}{dx} = -\frac{d}{dx}(\tfrac{1}{2}kx^2) = -kx$$

which corresponds to the restoring force exerted by the spring. In the case of an object located a distance y above some reference point, the gravitational potential energy function is given by $U_g = mgy$, and it follows from Equation 7.11 that $F_g = -mg$.

We now see that U is an important function; from it can be derived the conservative force acting in any system. Furthermore, Equation 7.11 should clarify the fact that adding a constant to the potential energy is unimportant, because the location of the reference point is arbitrary.

Equation 7.11 can also be written in the form $dU = -F\,dx$, which, when integrated between the initial and final position values, gives

$$U_f - U_i = -\int_{x_i}^{x_f} F\,dx \qquad \textbf{[7.12]}$$

This result, which is identical to Equation 7.6, tells us that if the conservative force F acting on an object within a system is known as a function of x, we can calculate the *difference* in the potential energy associated with the object between the initial and final positions.

If more than one conservative force acts on the object, then a potential energy function is associated with *each* force. In such a case, we can apply the law of conservation of energy for the system as

Conservation of mechanical • energy

$$K_i + \sum U_i = K_f + \sum U_f \qquad \textbf{[7.13]}$$

where the number of terms in the sums equals the number of conservative forces present. For example, if a mass connected to a spring oscillates vertically, two conservative forces act on it: the spring force and the force of gravity. (We will discuss this situation later in a worked example.)

If the force of gravity is the *only* force acting on a body, then the total mechanical energy of the body is constant. Therefore, the law of conservation of energy for a freely falling body can be written

Conservation of mechanical • energy for a freely falling body

$$\tfrac{1}{2}mv_i^2 + mgy_i = \tfrac{1}{2}mv_f^2 + mgy_f \qquad \textbf{[7.14]}$$

Thinking Physics 2

You have graduated from college and are designing roller coasters for a living. You design a roller coaster in which a car is pulled to the top of a hill of height h and then, starting from a momentary rest, rolls freely down the hill and upward toward the peak of the next hill, which is at height $1.1h$. Will you have a long career in this business?

Reasoning Your career will probably not be long, because this roller coaster will not work! At the top of the first hill, the roller coaster train has no kinetic energy, and gravitational potential energy associated with a height h. If it were to reach the top of the next hill, it would have higher potential energy, that associated with height $1.1h$.

This would violate the principle of conservation of mechanical energy. If this coaster were actually built, the car would move upward on the second hill to a height h (ignoring the effects of friction), stop short of the peak, and then start rolling backward, becoming trapped between the two peaks.

CONCEPTUAL PROBLEM 1

Discuss the energy transformations that occur during the pole vault event pictured in the multiflash photograph shown in Figure 7.5. Ignore rotational motion.

Figure 7.5 (Conceptual Problem 1) Multiflash photograph of a pole vault event. How many forms of energy can you identify in this picture? *(© Harold E. Edgerton, Courtesy of Palm Press, Inc.)*

CONCEPTUAL PROBLEM 2

Three identical balls are thrown from the top of a building, all with the same initial speed. The first ball is thrown horizontally, the second at some angle above the horizontal, and the third at some angle below the horizontal as in Figure 7.6. Neglecting air resistance, describe their motions and compare the speeds of the balls as they reach the ground.

Figure 7.6 (Conceptual Problem 2) Three identical balls are thrown with the same initial speed from the top of a building.

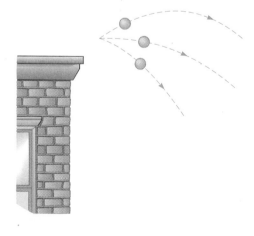

Example 7.1   Ball in Free-Fall

A ball of mass m is dropped from a height h above the ground, as in Figure 7.7. (a) Neglecting air resistance, determine the speed of the ball when it is at a height y above the ground.

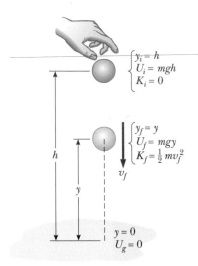

Figure 7.7   (Example 7.1) A ball is dropped from a height h above the ground. Initially, the total energy is gravitational potential energy, equal to mgh relative to the ground. At the elevation y, the total energy is the sum of kinetic and potential energies.

Reasoning   Because the ball is in free-fall, the only force acting on it is the gravitational force. Therefore, we can use the principle of conservation of mechanical energy. Initially, the ball has potential energy and no kinetic energy. As it falls, its total energy (the sum of kinetic and potential energies) remains constant and equal to its initial potential energy.

Solution   When the ball is released from rest at a height h above the ground, its kinetic energy is $K_i = 0$ and its potential energy is $U_i = mgh$, where the y coordinate is measured from ground level. When the ball is at a distance y above the ground, its kinetic energy is $K_f = \frac{1}{2}mv_f^2$ and its potential energy relative to the ground is $U_f = mgy$. Applying Equation 7.9, we get

$$K_i + U_i = K_f + U_f$$
$$0 + mgh = \tfrac{1}{2}mv_f^2 + mgy$$
$$v_f^2 = 2g(h - y)$$
$$v_f = \boxed{\sqrt{2g(h - y)}}$$

(b) Determine the speed of the ball at y if it is given an initial speed v_i at the initial altitude h.

Solution   In this case, the initial energy includes kinetic energy equal to $\frac{1}{2}mv_i^2$, and Equation 7.14 gives

$$\tfrac{1}{2}mv_i^2 + mgh = \tfrac{1}{2}mv_f^2 + mgy$$
$$v_f^2 = v_i^2 + 2g(h - y)$$
$$v_f = \sqrt{v_i^2 + 2g(h - y)}$$

This result is consistent with the expression from kinematics, $v_y^2 = v_{y0}^2 - 2g(y - y_0)$, where $y_0 = h$. Furthermore, this result is valid even if the initial velocity is at an angle to the horizontal (the projectile situation).

Example 7.2   One Way to Lift an Object

Two blocks are connected by a massless cord that passes over a frictionless pulley and a frictionless peg, as in Figure 7.8. One end of the cord is attached to a mass $m_1 = 3.00$ kg that is a distance $R = 1.20$ m from the peg. The other end of the cord is connected to a block of mass $m_2 = 6.00$ kg resting on a table. From what angle θ (measured from the vertical) must the 3.00-kg mass be released in order to just begin to lift the 6.00-kg block off the table?

Reasoning   It is necessary to use several concepts to solve this problem. First, we use conservation of energy to find the speed of the 3.00-kg mass at the bottom of the circular path as a function of θ and the radius of the path. Next, we apply Newton's second law to the 3.00-kg mass at the bottom of its path to find the tension as a function of the given parameters. Finally, we note that the 6.00-kg block lifts off the table when

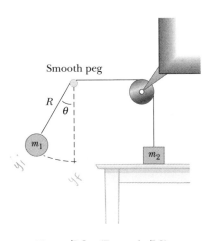

Figure 7.8   (Example 7.2)

the upward force exerted on it by the cord exceeds the force of gravity acting on the block. This procedure enables us to find the required angle.

Solution   Applying conservation of energy to the 3.00-kg mass gives

$$K_i + U_i = K_f + U_f$$

$$(1) \quad 0 + m_1 g y_i = \tfrac{1}{2} m_1 v^2 + 0$$

where v is the speed of the 3.00-kg mass at the bottom of its path. (Note that $K_i = 0$, because the 3.00-kg mass starts from rest and $U_f = 0$, because the bottom of the circle is the zero level of potential energy.) From the geometry in Figure 7.8, we see that $y_i = R - R \cos \theta = R(1 - \cos \theta)$. Using this relation in (1) gives

$$(2) \quad v^2 = 2gR(1 - \cos \theta)$$

Now we apply Newton's second law to the 3.00-kg mass when it is at the bottom of the circular path:

$$T - m_1 g = m_1 \frac{v^2}{R}$$

$$(3) \quad T = m_1 g + m_1 \frac{v^2}{R}$$

This same force is transmitted to the 6.00-kg block, and if it is to be just lifted off the table, the normal force on it becomes zero, and we require that $T = m_2 g$. Using this condition, together with (2) and (3), gives

$$m_2 g = m_1 g + m_1 \frac{2gR(1 - \cos \theta)}{R}$$

Solving for θ, and substituting in the given parameters, we get

$$\cos \theta = \frac{3m_1 - m_2}{2m_1} = \frac{3(3.00 \text{ kg}) - 6.00 \text{ kg}}{2(3.00 \text{ kg})} = \frac{1}{2}$$

$$\theta = 60.0°$$

EXERCISE 2   If the initial angle is $\theta = 40.0°$, find the speed of the 3.00-kg mass and the tension in the cord when the 3.00-kg mass is at the bottom of its circular path.
Answer   2.35 m/s; 43.2 N

EXERCISE 3   A 4.0-kg particle moves along the x axis under the influence of a single conservative force. If the work done on the particle is 80.0 J as it moves from the point $x = 2.0$ m to $x = 5.0$ m, find (a) the change in its kinetic energy, (b) the change in its potential energy, and (c) its speed at $x = 5.0$ m if it starts at rest at $x = 2.0$ m.   Answer   (a) 80.0 J (b) -80.0 J   (c) 6.32 m/s

EXERCISE 4   A rocket is launched at an angle of 53° to the horizontal from an altitude h with a speed v_0. Use energy methods to find its speed when its altitude is $h/2$.
Answer $v = \sqrt{v_0^2 + gh}$

7.5 • WORK DONE BY NONCONSERVATIVE FORCES

As we have seen, if the forces acting on a system are conservative, the mechanical energy of the system remains constant. However, if some of the forces acting on the system are not conservative, the mechanical energy of the system does not remain constant. Let us examine two types of nonconservative forces: an applied force and the force of kinetic friction.

Work Done by an Applied Force

When you lift a book through some distance by applying a force to it, the force you apply does work W_{app} on the book, and the force of gravity does work W_g on the book. If we treat the book as a particle, then the net work done on the book is related to the change in its kinetic energy through the work-kinetic energy theorem given by Equation 6.19:

$$W_{app} + W_g = \Delta K \qquad\qquad [7.15]$$

Because the force of gravity is conservative, we can use Equation 7.2 to express the work done by the force of gravity in terms of the change in gravitational potential energy or $W_g = -\Delta U$. Substituting this into Equation 7.15 gives

$$W_{app} = \Delta K + \Delta U \qquad [7.16]$$

Note that the right side of this equation is the change in the mechanical energy of the book–Earth system. This result says that your applied force transfers energy to the system in the form of kinetic energy of the book and to the gravitational potential energy of the book–Earth system. Thus, we conclude that if an object is part of a system, then **an applied force can transfer energy into or out of the system.**

Situations Involving Kinetic Friction

Kinetic friction is an example of a nonconservative force. If a book is given some initial velocity on a horizontal surface that is not frictionless, as in Figure 7.9, the force of kinetic friction acting on the book opposes its motion and negative work is done by the friction force. The book slows down and eventually stops after undergoing a displacement **d**. The force of friction reduces the kinetic energy of the book by transferring energy to thermal (internal) energy of the book and part of the horizontal surface. Note that only part of the book's kinetic energy is transferred to thermal energy in the book. The rest is transferred as thermal energy from the book to the surface.

As the book in Figure 7.9 moves through a distance d, the only force that does work is the force of kinetic friction. In this situation, we can apply Equation 7.16, where $\Delta U = 0$ and note that the force of kinetic friction is opposite the displacement. Thus, we see that the amount by which the force of friction decreases the kinetic energy is $-f_k d$, or

$$\Delta K = -f_k d \qquad [7.17]$$

If the book moves on an *incline* that is not frictionless, then a change in gravitational potential energy also occurs, and $-f_k d$ is the amount by which the force of kinetic friction reduces the mechanical energy $E = K + U$ of the system. In such cases,

$$\Delta K + \Delta U = -f_k d \qquad [7.18]$$

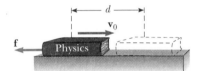

Figure 7.9   A book that is given an initial velocity $\mathbf{v}_0$ on a horizontal surface comes to rest due to the force of kinetic friction after traveling a distance d.

CONCEPTUAL PROBLEM 3

A rock is thrown vertically in the air. Ignoring air friction, it takes the same time to go up as it does to come down. Is this still true in the presence of air friction?

CONCEPTUAL PROBLEM 4

A driver brings an automobile to a stop. If the brakes lock, so that the car skids, where is the energy, and in what form is it, after the car stops? Answer the same question for the case in which the brakes do not lock, but the wheels continue to turn.

PROBLEM-SOLVING STRATEGIES • Conservation of Energy

Many problems in physics can be solved using the principle of conservation of energy. The following procedure should be used when you apply this principle:

1. Define your system, which may consist of more than one object and may or may not include fields, springs, or other sources of potential energy. Choose instants to call the initial and final points.

2. Select a reference position for the zero point of potential energy (both gravitational and spring), and use this throughout your analysis. If there is more than one conservative force, write an expression for the potential energy associated with each force.

3. Determine whether any nonconservative forces are present. Remember that if friction or air resistance is present, mechanical energy *is not constant*.

4. If mechanical energy is *constant*, you can write the total initial energy, E_i, at some point as the sum of the kinetic and potential energy at that point. Then write an expression for the total final energy, $E_f = K_f + U_f$, at the final point that is of interest. Because mechanical energy is *constant*, you can equate the two total energies and solve for the quantity that is unknown.

5. If nonconservative forces are present (and thus mechanical energy is not *constant*), first write expressions for the total initial and total final energies. In this case, the difference between the total final mechanical energy and the total initial mechanical energy equals the energy transferred to or from the system by the nonconservative forces. It is convenient to think of a general work-energy theorem as a combination of Equations 7.16 and 7.18, written as

$$K_i + U_i + W_{app} - f_k d = K_f + U_f$$

Example 7.3   Crate Sliding Down a Ramp

A 3.00-kg crate slides down a ramp at a loading dock. The ramp is 1.00 m in length, and inclined at an angle of 30.0°, as shown in Figure 7.10. The crate starts from rest at the top, experiences a constant frictional force of magnitude 5.00 N, and continues to move a short distance on the flat floor. Use energy methods to determine the speed of the crate when it reaches the bottom of the ramp.

Solution   Because $v_i = 0$, the initial kinetic energy is zero. If the *y* coordinate is measured from the bottom of the ramp, then $y_i = 0.500$ m. Therefore, the total mechanical energy of the crate–Earth system at the top is all gravitational potential energy:

$$U_i = mgy_i = (3.00 \text{ kg}) \left(9.80 \, \frac{\text{m}}{\text{s}^2} \right) (0.5000 \text{ m}) = 14.7 \text{ J}$$

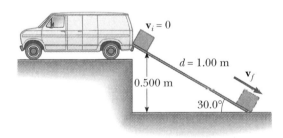

Figure 7.10   (Example 7.3) A crate slides down a ramp under the influence of gravity. The potential energy of the system decreases, and the kinetic energy of the crate increases.

When the crate reaches the bottom, the gravitational potential energy of the system is *zero*, because the elevation of the crate is $y_f = 0$. Therefore, the total mechanical energy at the bottom is all kinetic energy,

$$K_f = \tfrac{1}{2}mv_f^2$$

However, we cannot say that $U_i = K_f$ in this case, because there is an external nonconservative force that reduces the mechanical energy of the system: the force of kinetic friction. In this case, $\Delta E = -f_k d$ where d is the displacement along the ramp. (Remember that the forces normal to the ramp do no work on the crate because they are perpendicular to the displacement.) With $f_k = 5.00$ N and $d = 1.00$ m, we have

$$\Delta E = -f_k d = (-5.00 \text{ N})(1.00 \text{ m}) = -5.00 \text{ J}$$

This says that the decrease in mechanical energy is due to the presence of the force of kinetic friction that opposes the motion. Because $\Delta E = \tfrac{1}{2}mv_f^2 - mgy_i$ in this situation, Equation 7.18 gives

$$-f_k d = \tfrac{1}{2}mv_f^2 - mgy_i$$

$$\tfrac{1}{2}mv_f^2 = 14.7 \text{ J} - 5.00 \text{ J} = 9.70 \text{ J}$$

$$v_f^2 = \frac{19.4 \text{ J}}{3.00 \text{ kg}} = 6.47 \text{ m}^2/\text{s}^2$$

$$v_f = \boxed{2.54 \text{ m/s}}$$

EXERCISE 5 Use Newton's second law to find the acceleration of the crate along the ramp and the equations of kinematics to determine the final speed of the crate.
Answer   3.23 m/s²; 2.54 m/s

EXERCISE 6 If the ramp is assumed to be frictionless, find the final speed of the crate and its acceleration along the ramp.   Answer   3.13 m/s; 4.90 m/s²

Example 7.4   Motion on a Curved Track

A child of mass m takes a ride on an irregularly curved slide of height $h = 6.00$ m, as in Figure 7.11. The child starts from rest at the top. (a) Determine the speed of the child at the bottom, assuming no friction is present.

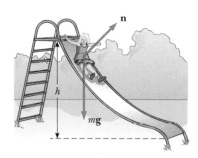

Figure 7.11 (Example 7.4) If the slide is frictionless, the speed of the child at the bottom depends only on the height of the slide.

Reasoning The normal force, **n**, does no work on the child, because this force is always perpendicular to each element of the displacement. Furthermore, because there is no friction, mechanical energy is constant—that is, $K + U = $ constant.

Solution If we measure the y coordinate from the bottom of the slide, then $y_i = h$, $y_f = 0$, and we get

$$K_i + U_i = K_f + U_f$$

$$0 + mgh = \tfrac{1}{2}mv_f^2 + 0$$

$$v_f = \sqrt{2gh}$$

Note that the result is the same as it would be if the child fell vertically through a distance h! In this example, $h = 6.00$ m, giving

$$v_f = \sqrt{2gh} = \sqrt{2\left(9.80 \frac{\text{m}}{\text{s}^2}\right)(6.00 \text{ m})} = \boxed{10.8 \text{ m/s}}$$

(b) If a frictional force acts on the child, how much mechanical energy is dissipated by this force? Assume that $v_f = 8.00$ m/s and $m = 20.0$ kg.

Solution In this case, $\Delta E \neq 0$ and mechanical energy is *not* constant. We can use Equation 7.18 to find the loss of mechanical energy due to friction, assuming the final speed at the bottom is known:

$$\Delta E = E_f - E_i = \tfrac{1}{2}mv_f^2 - mgh$$

$$\Delta E = \tfrac{1}{2}(20.0 \text{ kg})(8.00 \text{ m/s})^2 - (20.0 \text{ kg})\left(9.80 \frac{\text{m}}{\text{s}^2}\right)(6.00 \text{ m})$$

$$= -536 \text{ J}$$

Again, ΔE is negative because friction reduces the mechanical energy of the system. Note, however, that because the slide is curved, the normal force changes in magnitude and direction during the motion. Therefore, the frictional force, which is proportional to n, also changes during the motion. Do you think it would be possible to determine μ from these data?

Example 7.5   Mass–Spring Collision

A mass of 0.80 kg is given an initial velocity $v_i = 1.2$ m/s to the right and collides with a light spring of force constant $k = 50$ N/m, as in Figure 7.12. (a) If the surface is frictionless, calculate the initial maximum compression of the spring after the collision.

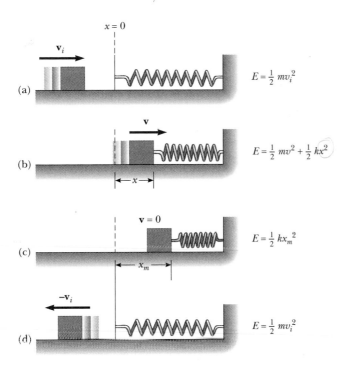

$E = \frac{1}{2} mv_i^2$

(a)

$E = \frac{1}{2} mv^2 + \frac{1}{2} kx^2$

(b)

$E = \frac{1}{2} kx_m^2$

(c)

$E = \frac{1}{2} mv_i^2$

(d)

Figure 7.12   (Example 7.12) A block sliding on a smooth, horizontal surface collides with a light spring. (a) Initially the mechanical energy is all kinetic energy. (b) The mechanical energy is the sum of the kinetic energy of the block and the elastic potential energy in the spring. (c) The energy is entirely potential energy. (d) The energy is converted back to the kinetic energy of the block. The total energy remains constant throughout the motion.

Reasoning   Before the collision, the mass has kinetic energy and the spring is uncompressed, so the energy stored in the spring is zero. Thus, the total energy of the system (mass plus spring) before the collision is $\frac{1}{2} mv_i^2$. After the collision, and

when the spring is fully compressed, the mass is momentarily at rest and has zero kinetic energy, and the energy stored in the spring has its maximum value, $\frac{1}{2} kx_f^2$. The total mechanical energy of the system is constant because no nonconservative forces act on the system.

Solution   Because mechanical energy is conserved, the kinetic energy of the mass before the collision must equal the maximum energy stored in the spring when it is fully compressed, or

$$\tfrac{1}{2} mv_i^2 = \tfrac{1}{2} kx_f^2$$

$$x_f = \sqrt{\frac{m}{k}}\, v_i = \sqrt{\frac{0.80 \text{ kg}}{50 \text{ N/m}}}\,(1.2 \text{ m/s}) = 0.15 \text{ m}$$

(b) If a constant force of kinetic friction acts between block and surface with $\mu_k = 0.50$ and if the speed of the block just as it collides with the spring is $v_i = 1.2$ m/s, what is the maximum compression in the spring?

Solution   In this case, mechanical energy is *not* conserved because of friction. The magnitude of the frictional force is

$$f_k = \mu_k n = \mu_k mg = 0.50(0.80 \text{ kg})\left(9.80\,\frac{\text{m}}{\text{s}^2}\right) = 3.9 \text{ N}$$

Therefore, the decrease in kinetic energy due to friction as the block is displaced from $x_i = 0$ to $x_f = x$ is

$$\Delta K = -f_k x = (-3.92x) \text{ J}$$

Substituting this into Equation 7.16 gives

$$\Delta K = (0 + \tfrac{1}{2} kx^2) - (\tfrac{1}{2} mv_i^2 + 0)$$

$$-3.92x = \tfrac{50}{2} x^2 - \tfrac{1}{2}(0.80)(1.2)^2$$

$$25x^2 + 3.92x - 0.576 = 0$$

Solving the quadratic equation for x gives $x = 0.092$ m and $x = -0.25$ m. The physically meaningful root is $x = 0.092$ m $= 9.2$ cm. The negative root is meaningless because the block must be to the right of the origin when it comes to rest. Note that 9.2 cm is less than the distance obtained in the frictionless case (a). This result is what we expect because friction retards the motion of the system.

EXERCISE 7   A 3.0-kg block starts at a height $h = 60$ cm on a plane that has an inclination angle of 30°. On reaching the bottom, the block slides along a horizontal surface. If the coefficient of friction on both surfaces is $\mu_k = 0.20$, how far does the block slide on the horizontal surface before coming to rest?   Answer   1.96 m

EXERCISE 8   A child starts from rest at the top of a slide of height 4.0 m. (a) What is her speed at the bottom if the incline is frictionless? (b) If she reaches the bottom with a speed of 6.0 m/s, what percentage of her total energy at the top of the slide was lost as a result of friction?   Answer   (a) 8.85 m/s   (b) 54.1%

7.6 • CONSERVATION OF ENERGY IN GENERAL

We have seen that the total mechanical energy of a system is constant when only conservative internal forces act within the system. Furthermore, we were able to associate a potential energy function with each conservative force. However, mechanical energy is not constant when nonconservative forces, such as friction, are present.

Whenever the mechanical energy of a system decreases, the energy does not disappear. Instead, the mechanical energy is transformed into other forms of energy. In the study of thermodynamics we shall find that mechanical energy can be transformed into internal energy of the system. For example, when a block slides over a surface, part of the mechanical energy is transformed into internal energy stored in the block and the surface, as evidenced by a measurable increase in the block's temperature. We shall see that, on a submicroscopic scale, this internal energy is associated with the vibration of atoms about their equilibrium positions. The energy acquired by the atoms is unevenly and randomly distributed among the atoms. This disorderly internal form of energy is often called **thermal energy.** Such internal atomic motion has kinetic and potential energy, and so one can say that frictional forces arise fundamentally from conservative atomic forces. Therefore, if we include this increase in the internal energy of the system in our energy expression, the total energy is conserved.

This is just one example of how you can analyze an isolated system and always find that its total energy does not change, as long as you account for all forms of energy. That is, **energy can never be created or destroyed. Energy may be transformed from one form to another, but the total energy of an isolated system is always constant.** From a universal point of view, we can say that **the total energy of the Universe is constant:** If one part of the Universe gains energy in some form, another part must lose an equal amount of energy. No violation of this principle has been found.

Total energy is always • conserved.

Other examples of energy transformations include the energy carried by sound waves resulting from the collision of two objects, the energy radiated by an accelerating charge in the form of electromagnetic waves (a radio antenna), and the elaborate sequence of energy conversions in a thermonuclear reaction.

In subsequent chapters we shall see that the energy concept, and especially transformations of energy, unite the branches of physics. The subjects of mechanics, thermodynamics, and electromagnetism cannot really be separated. In practical terms, all mechanical and electronic devices rely on energy transformations.

Thinking Physics 3

An automobile with kinetic energy and potential energy carried within the gasoline strikes a tree and comes to rest. The total mechanical energy in the system (the car) is now less than before. How did the energy leave the system, and in what form is energy left in the system?

Reasoning There are a number of ways that energy transferred out of the system. We shall describe the main mechanisms. The automobile did some *work* on the tree during the collision, causing the tree to become warmer and deform. There was a large crash during the collision, representing transfer of energy by *sound*. If the gasoline tank leaks after the collision, the automobile is losing energy by means of *mass transfer*. After the car has come to rest, there is no more kinetic energy. There is some *potential energy* left in any gas remaining within the tank. There is also more *internal energy* in the automobile, because its temperature is likely to be higher after the collision than before.

CONCEPTUAL PROBLEM 5

A toaster is turned on. Discuss the forms of energy and energy transfer occurring in the coils of the toaster.

CONCEPTUAL PROBLEM 6

Soft steel can be made red hot by continued hammering on it. This doesn't work as well for hard steel. Why is there this difference?

7.7 · GRAVITATIONAL POTENTIAL ENERGY REVISITED

Earlier in this chapter we introduced the concept of gravitational potential energy, that is, the energy associated with the position of a particle. We emphasized the fact that the gravitational potential energy function, $U = mgy$, is valid only when a particle is near the Earth's surface. Because the gravitational force between two particles varies as $1/r^2$, it follows that the correct potential energy function of the system depends on the amount of separation between the particles.

Consider a particle of mass m moving between two points P and Q above the Earth's surface, as in Figure 7.13. The gravitational force acting on m is

$$\mathbf{F}_g = -\frac{GM_em}{r^2}\,\hat{\mathbf{r}} \qquad [7.19]$$

where $\hat{\mathbf{r}}$ is a unit vector directed from the Earth to the particle and the negative sign indicates that the force is attractive. This expression shows that the gravitational force depends only on the polar coordinate r. Furthermore, the gravitational force is conservative. Because the change in potential energy associated with a given displacement of the particle is defined as the negative of the work done by the conservative gravitational force during that displacement, Equation 7.12 gives

$$U_f - U_i = -\int_{r_i}^{r_f} F(r)\,dr = GM_em\int_{r_i}^{r_f}\frac{dr}{r^2} = GM_em\left[-\frac{1}{r}\right]_{r_i}^{r_f}$$

or

$$U_f - U_i = -GM_em\left(\frac{1}{r_f} - \frac{1}{r_i}\right) \qquad [7.20]$$

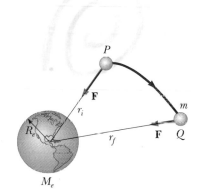

Figure 7.13 As a particle of mass m moves from P to Q above the Earth's surface, the potential energy changes according to Equation 7.20.

As always, the choice of a reference point for the potential energy is completely arbitrary. It is customary to locate the reference point where the force is zero. Taking $U_i = 0$ at $r_i = \infty$, we obtain the important result

Gravitational potential •
energy r > R_e

$$U(r) = -\frac{GM_e m}{r} \qquad \text{[7.21]}$$

This equation applies to the Earth–particle system separated by a distance r, provided that $r > R_e$. The result is not valid for particles moving inside the Earth, where $r < R_e$. Because of our choice of U_i, the function $U(r)$ is always negative (Fig. 7.14).

Although Equation 7.21 was derived for the particle–Earth system, it can be applied to *any* two particles. That is, the gravitational potential energy associated with *any pair* of particles of masses m_1 and m_2 separated by a distance r is

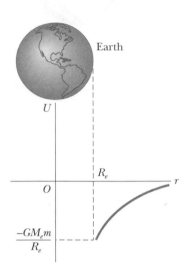

$$U_g = -\frac{Gm_1 m_2}{r} \qquad \text{[7.22]}$$

This expression also applies to larger objects *if they are spherically symmetric*, as first shown by Newton using integral calculus. Equation 7.22 shows that the gravitational potential energy for any pair of particles varies as $1/r$ (whereas the force between them varies as $1/r^2$). Furthermore, the potential energy is *negative*, because the force is attractive, and we have taken the potential energy as zero when the particle separation is infinity. Because the force between the particles is attractive, we know that an external agent must do positive work to increase the separation between the two particles. The work done by the external agent produces an increase in the potential energy as the two particles are separated. That is, U_g becomes less negative as r increases. (Note that part of the work done can also produce a change in kinetic energy of the system. That is, if the work done in separating the particles exceeds the increase in potential energy, the excess energy is accounted for by the increase in kinetic energy of the system.) When the two particles are separated by a distance r, an external agent would have to supply an energy *at least* equal to $+Gm_1 m_2 / r$ in order to separate the particles by an infinite distance.

Figure 7.14 Graph of the gravitational potential energy, U_g, versus r for a particle above the Earth's surface. The potential energy of the system goes to zero as r approaches ∞.

It is convenient to think of the absolute value of the gravitational potential energy defined this way as the **binding energy** of the system. If the external agent supplies an energy *greater than* the binding energy, $Gm_1 m_2 / r$, the additional energy of the system is in the form of kinetic energy when the particles are at an infinite separation.

We can extend this concept to three or more particles. In this case, the total potential energy of the system is the sum over all *pairs* of particles.[2] Each pair contributes a term of the form given by Equation 7.22. For example, if the system contains three particles, as in Figure 7.15, we find that

$$U_{\text{total}} = U_{12} + U_{13} + U_{23} = -G\left(\frac{m_1 m_2}{r_{12}} + \frac{m_1 m_3}{r_{13}} + \frac{m_2 m_3}{r_{23}}\right) \qquad \text{[7.23]}$$

The absolute value of U_{total} represents the work needed to separate the particles by an infinite distance. If the system consists of four particles, there are six terms in the sum, corresponding to the six distinct pairs of interaction forces. The total mechanical energy of this system of four particles includes ten terms: four kinetic energy terms (one for each particle) and six potential energy terms.

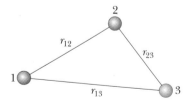

Figure 7.15 Diagram of three interacting particles.

[2]The fact that one can add potential energy terms for all pairs of particles stems from the experimental fact that gravitational forces obey the superposition principle. That is, if $\Sigma \mathbf{F} = \mathbf{F}_{12} + \mathbf{F}_{13} + \mathbf{F}_{23} + \dots$, then there exists a potential energy term for each interaction $\mathbf{F}_{ij}$.

Thinking Physics 4

Why is the Sun hot?

Reasoning   The Sun was formed when a cloud of gas and dust coalesced, due to gravitational attraction, into a massive astronomical object. Before this occurred, the particles were widely scattered, representing a large amount of gravitational potential energy. As the particles came together to form the Sun, the gravitational potential energy decreased. According to the principle of conservation of energy, this potential energy must be transformed to another form. The form to which it transformed was internal energy, representing an increase in temperature. If enough particles come together, the temperature can rise to a point at which nuclear fusion occurs, and the object becomes a star. If there are not enough particles, the temperature rises, but not to a point at which fusion occurs. The object is a planet if it is in orbit around a star. Jupiter is an example of a large planet that might have been a star if more particles were available to be collected.

Example 7.6   The Change in Potential Energy

A particle of mass m is displaced through a small vertical distance Δy near the Earth's surface. Let us show that the general expression for the change in gravitational potential energy given by Equation 7.20 reduces to the familiar relationship $\Delta U_g = mg\,\Delta y$.

Solution   We can express Equation 7.20 in the form

$$\Delta U_g = -GM_e m \left(\frac{1}{r_f} - \frac{1}{r_i}\right) = GM_e m \left(\frac{r_f - r_i}{r_i r_f}\right)$$

If both the initial and final positions of the particle are close to the Earth's surface, then $r_f - r_i = \Delta y$ and $r_i r_f \approx R_e^2$. (Recall that r is measured from the center of the Earth.) Therefore, the *change* in potential energy becomes

$$\Delta U_g \approx \frac{GM_e m}{R_e^2} \Delta y = mg\,\Delta y$$

where we have used the fact that $g = GM_e/R_e^2$. Keep in mind that the reference point is arbitrary, because it is the *change* in potential energy that is meaningful.

EXERCISE 9   A satellite of the Earth has a mass of 100 kg and its altitude is 2.0×10^6 m. (a) What is the gravitational potential energy of the satellite–Earth system? (b) What is the magnitude of the force on the satellite?   Answer   (a) -4.8×10^9 J   (b) 570 N

7.8 • ENERGY DIAGRAMS AND STABILITY OF EQUILIBRIUM   OPTIONAL

The motion of a system can often be understood qualitatively through an analysis of the system's potential energy curve. Consider the potential energy function for the mass–spring system, given by $U_s = \frac{1}{2}kx^2$. This function is plotted versus x in Figure 7.16a. The spring force is related to U through Equation 7.11:

$$F_s = -\frac{dU_s}{dx} = -kx$$

That is, the force is equal to the negative of the *slope* of the U-versus-x curve. When the mass is placed at rest at the equilibrium position ($x = 0$), where $F = 0$, it will remain there unless some external force acts on it. If the spring is stretched from equilibrium, x is positive and the slope dU/dx is positive; therefore, F_s is negative and the mass accelerates back toward $x = 0$. If the spring is compressed, x is negative

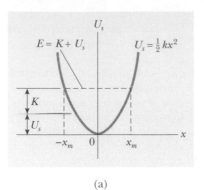

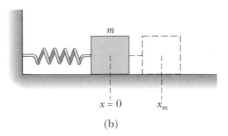

Figure 7.16   (a) The potential energy as a function of x for the mass–spring system shown in (b). The mass oscillates between the turning points, which have the coordinates $x = \pm x_m$. Note that the restoring force of the spring always acts toward $x = 0$, the position of stable equilibrium.

and the slope is negative; therefore, F_s is positive and again the mass accelerates toward $x = 0$.

Stable equilibrium •   From this analysis we conclude that the $x = 0$ position is one of **stable equilibrium.** That is, any movement away from this position results in a force that is directed back toward $x = 0$. In general, **positions of stable equilibrium correspond to those points for which $U(x)$ has a relative minimum value.**

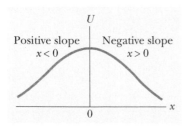

Figure 7.17   A plot of U versus x for a particle that has a position of unstable equilibrium, located at $x = 0$. For any finite displacement of the particle, the force on the particle is directed away from $x = 0$.

Unstable equilibrium •

From Figure 7.16 we see that if the mass is given an initial displacement x_m and released from rest, its total energy initially is the potential energy stored in the spring, given by $\frac{1}{2}kx_m{}^2$. As motion commences, the system acquires kinetic energy and loses an equal amount of potential energy. Because the total energy must remain constant, the mass oscillates between the two points $x = \pm x_m$, called the *turning points*. In fact, because no energy loss (no friction) takes place, the mass oscillates between $-x_m$ and $+x_m$ forever. (We shall discuss these oscillations further in Chapter 12). From an energy viewpoint, the energy of the system cannot exceed $\frac{1}{2}kx_m{}^2$; therefore, the mass must stop at these points and, because of the spring force, accelerate toward $x = 0$.

Another simple mechanical system that has a position of stable equilibrium is a ball rolling around in the bottom of a spherical bowl. If the ball is displaced from its lowest position, it always tends to return to that position when released.

Now consider an example in which the U-versus-x curve is as shown in Figure 7.17. In this case, $F_x = 0$ at $x = 0$, and so the particle is in equilibrium at this point. However, this is a position of **unstable equilibrium** for the following reason. Suppose the particle is displaced to the *right* ($x > 0$). Because the slope is negative for $x > 0$, $F_x = -dU/dx$ is positive and the particle accelerates away from $x = 0$. Now suppose the particle is displaced to the left ($x < 0$). In this case the force is *negative*, because the slope is positive for $x < 0$, and the particle again accelerates away from the equilibrium position. The $x = 0$ position in this situation is called a position of *unstable equilibrium* because, for any displacement from this point, the force pushes the particle farther away from equilibrium. In fact, the force pushes the particle toward a position of lower potential energy. A ball placed on the top of an inverted spherical bowl is obviously in a position of unstable equilibrium. If the ball is displaced slightly from the top and released, it will surely roll off the bowl. In general, **positions of unstable equilibrium correspond to those points for which $U(x)$ has a relative maximum value.**[3]

Finally, a situation may arise where U is constant over some region, and hence $F = 0$. This is called a position of **neutral equilibrium.** Small displacements from this position produce neither restoring nor disrupting forces. A ball lying on a flat horizontal surface is an example of an object in neutral equilibrium.

• *Neutral equilibrium*

SUMMARY

If a particle of mass m is elevated a distance y near the Earth's surface, the **gravitational potential energy** of the particle–Earth system can be represented as

$$U_g = mgy \qquad [7.1]$$

The **elastic potential energy** stored in a spring of force constant k is

$$U_s \equiv \tfrac{1}{2}kx^2 \qquad [7.4]$$

A force is **conservative** if the work it does on a particle is independent of the path the particle takes between two given points. Alternatively, a force is conservative if the work it does is zero when the particle moves through an arbitrary closed path and returns to its initial position. A force that does not meet these criteria is said to be **nonconservative.**

A **potential energy** function U can be associated only with a conservative force. If a conservative force $\mathbf{F}$ acts on a particle that moves along the x axis from x_i to x_f, *the change in the potential energy equals the negative of the work done by that force:*

$$U_f - U_i = -\int_{x_i}^{x_f} F_x \, dx \qquad [7.6]$$

The **total mechanical energy of a system** is defined as the sum of the kinetic energy and potential energy:

$$E \equiv K + U \qquad [7.8]$$

The principle of **conservation of mechanical energy** states that if no external forces do work on the system, and there are no nonconservative forces, the total mechanical energy is constant:

$$K_i + U_i = K_f + U_f \qquad [7.9]$$

If some of the forces acting on a system are not conservative, the mechanical energy of the system does not remain constant. If an object is part of a system, the work done by an applied force is equal to the change in mechanical energy of the system:

$$W_{\text{app}} = \Delta K + \Delta U \qquad [7.16]$$

An applied force transfers energy to or from the system in the form of changes in kinetic energy and gravitational potential energy.

The gravitational force is conservative, and therefore a potential energy function can be

---

[3]You can test mathematically whether an extreme of U is stable or unstable by examining the sign of d^2U/dx^2. A positive sign gives stable equilibrium and a negative sign gives unstable equilibrium.

defined. The **gravitational potential energy** associated with two particles separated by a distance r is

$$U_g = -\frac{Gm_1 m_2}{r} \qquad [7.22]$$

where U_g is taken to be zero at $r = \infty$. The total gravitational potential energy for a system of particles is the sum of energies for all pairs of particles, with each pair represented by a term of the form given by Equation 7.22.

CONCEPTUAL QUESTIONS

1. One person drops a ball from the top of a building, while another person at the bottom observes its motion. Will these two people agree on the value of the gravitational potential energy? On the change in potential energy? On its kinetic energy?

2. Discuss the production and dissipation of mechanical energy in (a) lifting a weight, (b) holding the weight up, and (c) lowering the weight slowly. Include the muscles in your discussion.

3. A skier prepares to take off down a snow-covered hill. Is it correct to say that gravity provides the energy for the trip down the hill?

4. You walk leisurely up a flight of stairs. You then return to the bottom and run up the same set of stairs as fast as you can. Have you performed a different amount of work in these two cases?

5. Many mountain roads are built so that they spiral around the mountain rather than go straight up the slope. Discuss this design from the viewpoint of energy and power.

6. In an earthquake, a large amount of energy is "released," and spreads outward, potentially causing severe damage. In what form does this energy exist before the earthquake, and by what energy transfer mechanism does it travel from the focus?

7. You ride a bicycle. In what sense is your bicycle solar-powered?

8. Can the gravitational potential energy of a system ever have a negative value? Explain.

9. A bowling ball is suspended from the ceiling of a lecture hall by a strong cord. The bowling ball is drawn away from its equilibrium position and released from rest at the tip of the demonstrator's nose. If the demonstrator remains stationary, explain why she will not be struck by the ball on its return swing. Would the demonstrator be safe if she pushed the ball as she released it?

10. A pile driver is a device used to drive objects into the Earth by repeatedly dropping a heavy weight on them. By how much does the energy of a pile driver increase when the weight it drops is doubled? (Assume the weight is dropped from the same height each time.)

11. Our body muscles exert forces when we lift, push, run, jump, and so forth. Are these forces conservative?

12. When nonconservative forces act on a system, does the total mechanical energy remain constant?

13. A block is connected to a spring that is suspended from the ceiling. If the block is set in motion and air resistance is neglected, describe the energy transformations that occur within the system consisting of the block and spring.

14. What would the curve of U versus x look like if a particle were in a region of neutral equilibrium?

15. A ball rolls on a horizontal surface. Is the ball in stable, unstable, or neutral equilibrium?

16. Discuss the energy transformations that occur during the operation of an automobile.

17. A ball is thrown straight up into the air. At what position is its kinetic energy a maximum? At what position is the gravitational potential energy a maximum?

PROBLEMS

Section 7.1   Potential Energy

Section 7.2   Conservative and Nonconservative Forces

1. A 1000-kg roller coaster is initially at the top of a rise, at point A. It then moves 135 ft, at an angle of 40.0° below the horizontal, to a lower point, B. (a) Choose point B to be the zero level for gravitational potential energy, and find the potential energy at points A and B and the difference in potential energy, $U_A - U_B$, between these points. (b) Repeat part (a), setting the zero reference level at point A.

2. A 2.00-kg ball is attached to the bottom end of a 1.00-m–long string hanging from the ceiling of a room. The height of the room is 3.00 m. What is the gravitational potential energy relative to (a) the ceiling, (b) the floor, and (c) a point at the same elevation as the ball?

3. A 40.0-N child is in a swing that is attached to ropes

2.00 m long. Find the gravitational potential energy of the system relative to the child's lowest position when (a) the ropes are horizontal, (b) the ropes make a 30.0° angle with the vertical, and (c) the child is at the bottom of the circular arc.

4. A 4.00-kg particle moves from the origin to the position having coordinates $x = 5.00$ m and $y = 5.00$ m under the influence of gravity acting in the negative y direction (Fig. P7.4). Using Equation 6.3, calculate the work done by gravity in going from O to C along (a) OAC, (b) OBC, (c) OC. Your results should all be identical. Why?

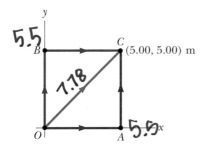

Figure P7.4

Section 7.3 Conservative Forces and Potential Energy

Section 7.4 Conservation of Mechanical Energy

5. A single conservative force $F_x = (2.0x + 4.0)$ N acts on a 5.00-kg particle, where x is in meters. As the particle moves along the x axis from $x = 1.00$ m to $x = 5.00$ m, calculate (a) the work done by this force, (b) the change in the potential energy of the system, and (c) the kinetic energy of the particle at $x = 5.00$ m if its speed at $x = 1.00$ m is 3.00 m/s.

6. A single constant force $\mathbf{F} = (3.0\mathbf{i} + 5.0\mathbf{j})$ N acts on a 4.00-kg particle. (a) Calculate the work done by this force if the particle moves from the origin to the point having the vector position $\mathbf{r} = (2.0\mathbf{i} - 3.0\mathbf{j})$ m. Does this result depend on the path? Explain. (b) What is the speed of the particle at $\mathbf{r}$ if its speed at the origin is 4.00 m/s? (c) What is the change in potential energy?

7. At time t_i, the kinetic energy of a particle is 30.0 J and the potential energy is 10.0 J. At some later time t_f, its kinetic energy is 18.0 J. (a) If only conservative forces act on the particle, what are the potential energy and the total energy at time t_f? (b) If the potential energy at time t_f is 5.00 J, are there any nonconservative forces acting on the particle? Explain.

8. A particle of mass 0.500 kg is shot from P as shown in Figure P7.8 with an initial velocity $\mathbf{v}_0$ having a horizontal component of 30.0 m/s. The particle rises to a maximum height of 20.0 m above P. Using conservation of energy, determine (a) the vertical component of $\mathbf{v}_0$, (b) the work done by the gravitational force on the particle during its motion from P

to B, and (c) the horizontal and the vertical components of the velocity vector when the particle reaches B.

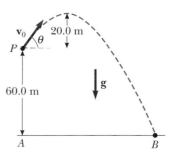

Figure P7.8

9. A bead slides without friction around a loop-the-loop (Fig. P7.9). If the bead is released from a height $h = 3.50R$, what is its speed at point A? How large is the normal force on it if its mass is 5.00 g?

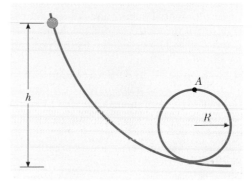

Figure P7.9

10. A simple 2.00-m–long pendulum is released from rest when the support string is at an angle of 25.0° from the vertical. What is the speed of the suspended mass at the bottom of the swing?

11. Two masses are connected by a light string passing over a light frictionless pulley as shown in Figure P7.11. The 5.00-kg mass is released from rest at a height of 4.00 m above the ground. Using the law of conservation of energy, (a) determine the speed of the 3.00-kg mass just as the 5.00-kg mass hits the ground. (b) Find the maximum height to which the 3.00-kg mass rises above the ground.

12. Two masses are connected by a light string passing over a light frictionless pulley as in Figure P7.11. The mass m_1 is released from rest at height h. Using the law of conservation of energy, (a) determine the speed of m_2 just as m_1 hits the ground. (b) Find the maximum height to which m_2 rises.

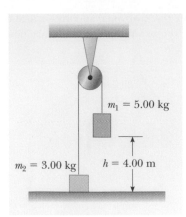

Figure P7.11

13. Dave Johnson, the bronze medalist at the 1992 Olympic decathlon in Barcelona, leaves the ground at the high jump with a vertical velocity of 6.00 m/s. How far does his center of gravity move up as he makes the jump?

14. A 0.400-kg ball is thrown into the air and reaches a maximum altitude of 20.0 m. Taking its initial position as the point of zero potential energy and using energy methods, find (a) its initial speed, (b) its total mechanical energy, and (c) the ratio of its kinetic energy to the potential energy when its altitude is 10.0 m.

15. A 20.0-kg cannon ball is fired from a cannon with the muzzle speed of 1000 m/s at an angle of 37.0° with the horizontal. A second ball is fired at an angle of 90.0°. Use the conservation of mechanical energy to find (a) the maximum height reached by each ball and (b) the total mechanical energy at the maximum height for each ball.

16. A 2.00-kg ball is attached to a 10-lb (44.5 N) fishing line. The ball is released from rest at the horizontal position ($\theta = 90.0°$). At what angle, θ, (measured from the vertical) will the fishing line break?

17. A child starts from rest and slides down the frictionless slide

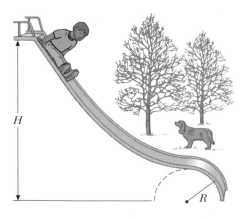

Figure P7.17

shown in Figure P7.17. In terms of R and H, at what height h will he lose contact with the section of radius R?

Section 7.5   Work Done by Nonconservative Forces

18. A 70.0-kg diver steps off a 10.0-m tower and drops straight down into the water. If he comes to rest 5.00 m beneath the surface of the water, determine the average resistance force exerted on the diver by the water.

19. A force F_x, shown as a function of distance in Figure P7.19, acts on a 5.00-kg mass. If the particle starts from rest at $x = 0$ m, determine the speed of the particle at $x = 2.00$ m, 4.00 m, and 6.00 m.

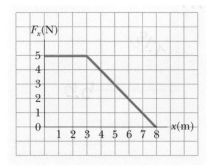

Figure P7.19

20. A softball pitcher swings in her hand a ball of mass 0.250 kg around a vertical circular path of radius 60.0 cm before releasing it. The pitcher maintains a component of force on the ball of constant magnitude 30.0 N in the direction of motion around the complete path. The speed of the ball at the top of the circle is 15.0 m/s. If the ball is released at the bottom of the circle, what is its speed on release?

21. The coefficient of friction between the 3.00-kg mass and surface in Figure P7.21 is 0.400. The system starts from rest. What is the speed of the 5.00-kg mass when it has fallen 1.50 m?

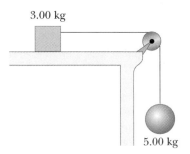

Figure P7.21

22. A 2000-kg car starts from rest at the top of a 5.00–m-long driveway that is sloped at an angle of 20.0° with the horizontal. If an average friction force of 4000 N impedes the motion of the car, find the speed of the car at the bottom of the driveway.

23. A 5.00-kg block is set into motion up an inclined plane with an initial speed of 8.00 m/s (Fig. P7.23). The block comes to rest after traveling 3.00 m along the plane, which is inclined at an angle of 30.0° to the horizontal. Determine (a) the change in the block's kinetic energy, (b) the change in potential energy, and (c) the frictional force exerted on it (assumed to be constant). (d) What is the coefficient of kinetic friction?

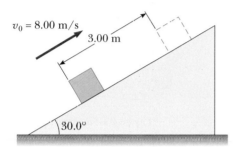

Figure P7.23

24. A parachutist of mass 50.0 kg jumps out of an airplane at a height of 1000 m and lands on the ground with a speed of 5.00 m/s. How much mechanical energy was lost to air friction during this jump?

25. A 80.0-kg skydiver jumps out of an airplane at an altitude of 1000 m and opens the parachute at an altitude of 200.0 m. (a) Assuming that the total retarding force on the diver is constant at 50.0 N with the parachute closed and constant at 3600 N with the parachute open, what is the speed of the diver when he lands on the ground? (b) Do you think the skydiver will get hurt? Explain. (c) At what height should the parachute be opened so that the final speed of the skydiver when he hits the ground is 5.00 m/s? (d) How realistic is the assumption that the total retarding force is constant? Explain.

26. A toy gun uses a spring to project a 5.30-g soft rubber ball. The spring is originally compressed by 5.00 cm and has stiffness constant 8.00 N/m. When it is fired, the ball moves 15.0 cm through the barrel of the gun, and there is a constant frictional force of 0.0320 N between barrel and ball. (a) With what speed does the projectile leave the barrel of the gun? (b) Just where does the ball have maximum speed? (c) What is this maximum speed?

27. A block slides down a curved frictionless track and then up an inclined plane, as in Figure P7.27. The coefficient of kinetic friction between block and incline is μ_k. Use energy

methods to show that the maximum height reached by the block is

$$y_{max} = \frac{h}{1 + \mu_k \cot \theta}$$

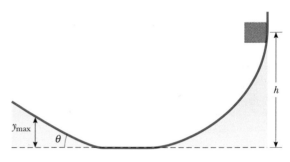

Figure P7.27

28. A 1.50-kg mass is held 1.20 m above a relaxed massless spring with a spring constant of 320 N/m. The mass is dropped onto the spring. (a) How far does it compress the spring? (b) How far does it compress the spring if the same experiment is performed on the moon where $g = 1.63$ m/s²? (c) Repeat part (a), but this time assume a constant air-resistance force of 0.700 N acts on the mass during its motion.

29. A skier starts from rest at the top of a hill that is inclined at an angle of 10.5° to the horizontal. The hill is 200 m long, and the coefficient of friction between the snow and the skis is 0.0750. At the bottom of the hill the snow is level and the coefficient of friction is unchanged. How far does the skier move along the horizontal portion of the snow before coming to rest?

Section 7.7 Gravitational Potential Energy Revisited

(Assume $U = 0$ at $r = \infty$.)

30. How much energy is required to move a 1000-kg mass from the Earth's surface to an altitude twice the Earth's radius?

31. After our Sun exhausts its nuclear fuel, its ultimate fate is possibly to collapse to a *white-dwarf* state, in which it has approximately the mass of the Sun but the radius of the Earth. Calculate (a) the average density of the white dwarf, (b) the free-fall acceleration at its surface, and (c) the gravitational potential energy of a 1.00-kg object at its surface. (Take $U_g = 0$ at infinity.)

32. At the Earth's surface a projectile is launched straight up at a speed of 10.0 km/s. To what height will it rise? Ignore air resistance.

33. A system consists of three particles, each of mass 5.00 g, located at the corners of an equilateral triangle with sides of 30.0 cm. (a) Calculate the potential energy of the system.

(b) If the particles are released simultaneously, where will they collide?

Section 7.8   Energy Diagrams and Stability of Equilibrium (Optional)

34. A right circular cone can be balanced on a horizontal surface in three different ways. Sketch these three equilibrium configurations and identify them as positions of stable, unstable, or neutral equilibrium.

35. For the potential energy curve shown in Figure P7.35, (a) determine whether the force F_x is positive, negative, or zero at the five points indicated. (b) Indicate points of stable, unstable, and neutral equilibrium. (c) Sketch the curve F_x versus x from $x = 0$ to $x = 8$ m.

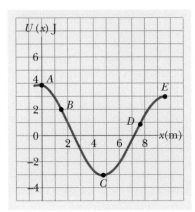

Figure P7.35

36. A hollow pipe has one or two weights attached to its inner surface as shown in Figure P7.36. Characterize each configuration as being stable, unstable, or neutral equilibrium and explain each of your choices.

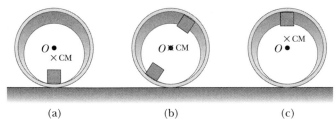

(a)   (b)   (c)

Figure P7.36

37. The potential energy of a two-particle system separated by a distance r is given by $U(r) = A/r$, where A is a constant. Find the radial force $\mathbf{F}_r$.

Additional Problems

38. A 200-g particle is released from rest at point A along the horizontal diameter on the inside of a frictionless, hemispherical bowl of radius $R = 30.0$ cm (Fig. P7.38). Calculate (a) the gravitational potential energy at point A relative to point B, (b) the kinetic energy of the particle at point B, (c) its speed at point B, and (d) its kinetic energy and the potential energy at point C.

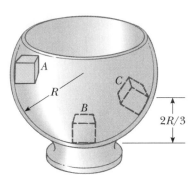

Figure P7.38

39. The particle described in Problem 38 (Fig. P7.38) is released from rest at A, and the surface of the bowl is rough. The speed of the particle at B is 1.50 m/s. (a) What is its kinetic energy at B? (b) How much mechanical energy is lost due to friction as the particle moves from A to B? (c) Is it possible to determine μ from these results in any simple manner? Explain.

40. Make an order-of-magnitude estimate of your power output. Let the work you do be climbing stairs. In your solution state the physical quantities you take as data and the values you measure or estimate for them. Do you consider your peak power or your sustainable power?

41. A 10.0-kg block is released from point A in Figure P7.41. The track is frictionless except for the portion BC, of length 6.00 m. The block travels down the track, hits a spring of force constant $k = 2250$ N/m, and compresses it 0.300 m from its equilibrium position before coming to rest momentarily. Determine the coefficient of kinetic friction between surface BC and block.

42. A child's pogo stick (Fig. P7.42) stores energy in a spring ($k = 2.50 \times 10^4$ N/m). At position A ($x_1 = -0.100$ m), the spring compression is a maximum and the child is momentarily at rest. At position B ($x = 0$), the spring is relaxed and the child is moving upward. At position C, the child is again momentarily at rest at the top of the jump. Assuming that the combined mass of child and pogo stick is 25.0 kg, (a) calculate the total energy of the system if both potential energies are zero at $x = 0$, (b) determine x_2, (c) calculate the speed of the child at $x = 0$, (d) determine the

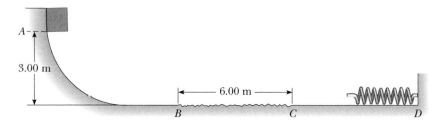

Figure P7.41

value of *x* for which the kinetic energy of the system is a maximum, and (e) obtain the child's maximum upward speed.

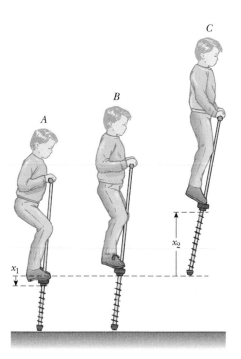

Figure P7.42

43. A 2.00-kg block situated on a rough incline is connected to a spring of negligible mass having a spring constant of 100 N/m (Fig. P7.43). The block is released from rest when the spring is unstretched, and the pulley is frictionless. The block moves 20.0 cm down the incline before coming to rest. Find the coefficient of kinetic friction between block and incline.

44. Suppose the incline is frictionless for the system described in Problem 43 (Fig. P7.43). The block is released from rest with the spring initially unstretched. (a) How far does it move down the incline before coming to rest? (b) What is its acceleration at its lowest point? Is the acceleration constant? (c) Describe the energy transformations that occur during the descent.

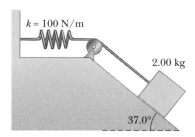

Figure P7.43

45. A 20.0-kg block is connected to a 30.0-kg block by a string that passes over a frictionless pulley. The 30.0-kg block is connected to a spring that has negligible mass and a force constant of 250 N/m, as in Figure P7.45. The spring is unstretched when the system is as shown in the figure, and the incline is frictionless. The 20.0-kg block is pulled 20.0 cm down the incline (so that the 30.0-kg block is 40.0 cm above the floor) and released from rest. Find the speed of each block when the 30.0-kg block is again 20.0 cm above the floor (that is, when the spring is unstretched).

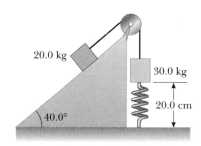

Figure P7.45

46. A potential energy function for a system is given by $U(x) = -x^3 + 2x^2 + 3x$. (a) Determine the force F_x as a function of *x*. (b) For what values of *x* is the force equal to zero? (c) Plot $U(x)$ versus *x* and F_x versus *x*, and indicate points of stable and unstable equilibrium.

47. A block of mass 0.500 kg is pushed against a horizontal spring of negligible mass, compressing the spring a distance of Δx (Fig. P7.47). The spring constant is 450 N/m. When

released, the block travels along a frictionless, horizontal surface to point B, the bottom of a vertical circular track of radius $R = 1.00$ m, and continues to move up the track. The speed of the block at the bottom of the track is $v_B = 12.0$ m/s, and the block experiences an average frictional force of magnitude 7.00 N while sliding up the track. (a) What is Δx? (b) What is the speed of the block at the top of the track? (c) Does the block reach the top of the track, or does it fall off before reaching the top?

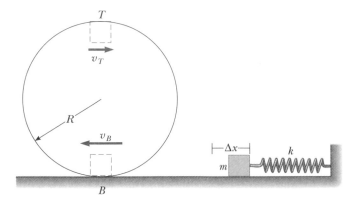

Figure P7.47

48. A 1.00-kg mass slides to the right on a surface having a coefficient of kinetic friction $\mu_k = 0.250$ (Fig. P7.48). The mass has a speed of $v_i = 3.00$ m/s when contact is made with a spring that has a spring constant $k = 50.0$ N/m. The

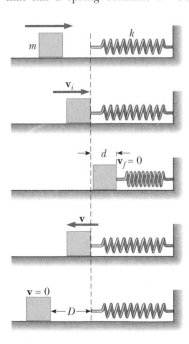

Figure P7.48

mass comes to rest after the spring has been compressed a distance d. The mass is then forced toward the left by the spring and continues to move in that direction beyond the unstretched position. Finally the mass comes to rest a distance D to the left of the unstretched spring. Find (a) the compressed distance d, (b) the speed v at the unstretched position when the system is moving to the left, and (c) the distance D where the mass comes to rest.

49. In the dangerous sport of bungee jumping, a student jumps from a balloon with a specially designed elastic cord attached to his ankles, as shown in the photograph. The unstretched length of the cord is 25.0 m, the student weighs 700 N, and the balloon is 36.0 m above the surface of a river below. Calculate the required force constant of the cord if the student is to stop safely 4.00 m above the river.

Bungee jumping (Problem 49).   *(Gamma)*

50. A uniform chain of length 8.00 m initially lies stretched out on a horizontal table. (a) If the coefficient of static friction between chain and table is 0.600, show that the chain will begin to slide off the table if at least 3.00 m of it hangs over the edge of the table. (b) Determine the speed of the chain as all of it leaves the table, given that the coefficient of kinetic friction between chain and table is 0.400.

51. An object of mass m is suspended from the top of a cart by a string of length L as in Figure P7.51a. The cart and object are initially moving to the right at constant speed v_0. The cart comes to rest after colliding and sticking to a bumper as in Figure P7.51b, and the suspended object swings through an angle θ. (a) Show that $v_0 = \sqrt{2gL(1 - \cos\theta)}$. (b) If $L = 1.20$ m and $\theta = 35.0°$, find the initial speed of the cart. (*Hint:* The force exerted by the string on the object does no work on the object.)

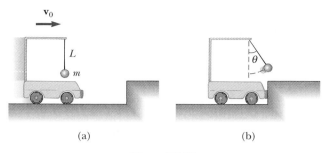

(a)　　　　　　　(b)

Figure P7.51

▣ Spreadsheet Problems

S1. The potential energy function of a particle is

$$U(x) = \tfrac{1}{2}kx^2 + bx^3 + c$$

where $k = 300$ N/m, $b = -12.0$ N/m^2, and $c = -1000$ J. Use Spreadsheet 7.1 to plot the function from $x =$ -5.00 m to $x = +15.0$ m. Describe the general motion of a particle under the influence of this potential energy function. How does the motion change when the initial energy of the object is increased?

S2. The potential energy function associated with the force between two atoms is modeled by the Lennard-Jones potential

$$U(x) = 4\epsilon \left[\left(\frac{\sigma}{x} \right)^{12} - \left(\frac{\sigma}{x} \right)^{6} \right]$$

In this model, there are two adjustable parameters, σ and ϵ, that are determined from experiments (σ is a range parameter and ϵ is the well depth). Modify Spreadsheet 7.1 to plot $U(x)$ versus x for $\sigma = 0.263$ nm and $\epsilon = 1.51 \times 10^{-22}$ J. Numerically integrate $U(x)$ to find the force F_x.

S3. Using Spreadsheet 7.1, numerically differentiate the potential energy function given in Problem S1. Find the equilibrium points. Are they stable or unstable?

ANSWERS TO CONCEPTUAL PROBLEMS

1. As an athlete runs, chemical energy in the body is converted mostly into thermal energy but also into kinetic energy. The kinetic energy of the athlete just before ascending becomes elastic potential energy stored in the bent pole at the bottom, then gravitational potential energy in the elevated body of the vaulter, and then kinetic energy as he falls. When the vaulter lands on the pad, this energy turns into additional thermal energy, mostly in the pad.

2. The first and third balls speed up after they are thrown, while the second ball first slows down and then speeds up after reaching its peak. The paths of all three are portions of parabolas. The three take different times to reach the ground. However, all have the same impact speed because all start with the same kinetic energy and undergo the same change in gravitational potential energy. In other words, $F_{total} = \tfrac{1}{2}mv^2 + mgy$ is the same for all three balls.

3. Air friction is a nonconservative force, so it will result in a transformation of mechanical energy to internal energy. On the way up, the rock slows down from its projection speed of v to zero. On the way down, it speeds up to v_f. Because of the air friction, the mechanical energy of the system as the rock lands must be less than that when it was projected. The potential energy of the rock-Earth system is the same at the initial and final points, since the two points are at the same height. Thus, the final kinetic energy must be less than the initial kinetic energy. In turn, the final speed, v_f, must be less than the initial speed, v. Finally, since the trip on the way down went from zero velocity to a speed lower than v, the average velocity on the way down must be less than that on the way up, and it must have taken longer for the rock to come down than to go up.

4. If the brakes lock, then the car skids over the roadway. The friction force between the roadway and the tires brings the car to a stop. In this case, the energy is transformed to internal energy in the tires and is also laid out as internal energy along the skid path of the tires. If the brakes do not lock, then the major friction force is between the brake pads and the discs (or the brake shoes and the drums), so that the energy appears as internal energy within the brake structure of the car.

5. The coils change their *internal energy*, since the temperature of the coils rises. The energy transfer mechanism for energy coming into the coils is *electrical transmission*, through the wire plugged into the wall. Energy is transferring out of the coils by *electromagnetic radiation*, since the coils are hot and glowing. There is also some transfer of thermal energy from the hot surfaces of the coils into the air. After a short warm-up period, the coils will stabilize in temperature, and the internal energy will no longer change. In this situation, the energy input and output will be balanced.

6. The rise in temperature of the soft steel is an example of transferring energy into a system by work and having it appear as an increase in the internal energy in the system. This works well for the soft steel because it is *soft*. This softness results in a deformation of the steel under the blow of the hammer. Thus, there is a displacement of the point of application of the force from the hammer, and work is done. With the hard steel, there is much less deformation. Thus, there is less displacement of the point of application of the force, and less work done. The soft steel is thus more efficient at absorbing energy from the hammer by means of work, and its temperature rises more rapidly.

8

Momentum and Collisions

Consider what happens when a golf ball is struck by a club. The ball is given a very large initial velocity as a result of the collision; as a consequence, it is able to travel more than 100 m through the air. The ball experiences a large change in velocity and a correspondingly large acceleration. Furthermore, because the ball experiences this acceleration over a short time interval, the average force on it during the collision is very large. As Newton's third law predicts, the club experiences a reaction force that is equal to and opposite the force on the ball. This reaction force produces a change in the velocity of the club. Because the club is much more massive than the ball, however, the change in velocity of the club is much less than the change in velocity of the ball.

One of the main objectives of this chapter is to enable you to understand and analyze such events. As a first step, we shall introduce the concept of *momentum,* a term that is used in describing objects in motion. For example, a massive football player is often said to have a great deal of momentum as he runs down the field. A much less massive player, such as a halfback, can have equal or greater momentum if his speed is greater than that of the more massive player. This follows from the fact that momentum is defined as the product of mass and velocity.

The concept of momentum leads us to a second conservation law, that

As a result of the collision between the bowling ball and pin, part of the ball's momentum is transferred to the pin. As a result, the pin acquires momentum and kinetic energy, and the ball loses momentum and kinetic energy. However, the total momentum of the system (ball and pin) remains constant. *(Ben Rose/The Image Bank)*

of conservation of momentum. This law is especially useful for treating problems that involve collisions between objects.

8.1 • LINEAR MOMENTUM AND ITS CONSERVATION

The **linear momentum** of a particle of mass m moving with a velocity $\mathbf{v}$ is defined to be the product of the mass and velocity:[1]

$$\mathbf{p} \equiv m\mathbf{v} \qquad [8.1]$$

• *Definition of linear momentum of a particle*

Because momentum equals the product of a scalar, m, and a vector, $\mathbf{v}$, it is a vector quantity. Its direction is along $\mathbf{v}$, and it has dimensions of ML/T. In the SI system, momentum has the units kg·m/s.

If a particle is moving in an arbitrary direction in three-dimensional space, $\mathbf{p}$ will have three components, and Equation 8.1 is equivalent to the component equations

$$p_x = mv_x \qquad p_y = mv_y \qquad p_z = mv_z \qquad [8.2]$$

As you can see from its definition, the concept of momentum provides a quantitative distinction between heavy and light particles moving at the same velocity. For example, the momentum of a bowling ball moving at 10 m/s is much greater than that of a tennis ball moving at the same speed. Newton called the product $m\mathbf{v}$ *quantity of motion*, perhaps a more graphic description than *momentum*, which comes from the Latin word for movement.

By using Newton's second law of motion, we can relate the linear momentum of a particle to the resultant force acting on the particle. In Chapter 4 we learned that Newton's second law can be written as $\Sigma\mathbf{F} = m\mathbf{a}$. However, this form applies only when the mass of the system remains constant. In situations in which the mass is changing with time, one must use an alternative statement of Newton's second law: **The time rate of change of momentum of a particle is equal to the resultant force acting on the particle.**

$$\Sigma\mathbf{F} = \frac{d\mathbf{p}}{dt} \qquad [8.3]$$

• *Newton's second law for a particle*

From Equation 8.3 we see that if the resultant force is zero, the time derivative of the momentum is zero, and therefore the momentum of any object must be constant. In other words, the linear momentum of an object is *constant* when $\Sigma\mathbf{F} = 0$. Of course, if the particle is *isolated* (that is, if it does not interact with its environment), then by necessity $\Sigma\mathbf{F} = 0$ and $\mathbf{p}$ remains unchanged.

In a certain sense, conservation of linear momentum is just another way of stating Newton's first law. If an object is in motion, its linear momentum and, as a consequence, its velocity do not change unless an external force acts on the system. A good example is a rocket with its engines disengaged, which coasts through space far from any gravitational sources.

---

[1]This expression is nonrelativistic and is valid only when $v \ll c$, where c is the speed of light.

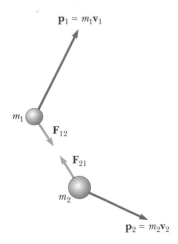

Figure 8.1 At some instant, the momentum of m_1 is $\mathbf{p}_1 = m_1\mathbf{v}_1$ and the momentum of m_2 is $\mathbf{p}_2 = m_2\mathbf{v}_2$. Note that $\mathbf{F}_{12} = -\mathbf{F}_{21}$.

Conservation of Linear Momentum for a Two-Particle System

To see how to apply the principle of *conservation of linear momentum*, consider a system of two particles that can interact with each other but are isolated from their surroundings (Fig. 8.1). That is, the particles may exert a force on each other, but no *external* forces are present. It is important to note the impact of Newton's third law on this analysis. Recall from Chapter 4 that Newton's third law states that the forces on these two particles are equal in magnitude and opposite in direction; that is, *forces always occur in pairs*. Thus, if an *internal* force (say a gravitational force) acts on particle 1, then there must be a second *internal* force, equal in magnitude but opposite in direction, that acts on particle 2.

Suppose that at some instant, the momentum of particle 1 is $\mathbf{p}_1$ and the momentum of particle 2 is $\mathbf{p}_2$. Applying Newton's second law to each particle, we can write

$$\mathbf{F}_{12} = \frac{d\mathbf{p}_1}{dt} \qquad \text{and} \qquad \mathbf{F}_{21} = \frac{d\mathbf{p}_2}{dt}$$

where $\mathbf{F}_{12}$ is the force exerted on particle 1 by particle 2 and $\mathbf{F}_{21}$ is the force exerted on particle 2 by particle 1. (These forces could be gravitational forces, or they could have some other origin. The source of the forces isn't important for the present discussion.) Newton's third law tells us that $\mathbf{F}_{12}$ and $\mathbf{F}_{21}$ are equal in magnitude and opposite in direction. That is, they form an action–reaction pair, and $\mathbf{F}_{12} = -\mathbf{F}_{21}$. We can also express this condition as

$$\mathbf{F}_{12} + \mathbf{F}_{21} = 0 \qquad \text{or as} \qquad \frac{d\mathbf{p}_1}{dt} + \frac{d\mathbf{p}_2}{dt} = \frac{d}{dt}(\mathbf{p}_1 + \mathbf{p}_2) = 0$$

Because the time derivative of the total momentum, $\mathbf{p}_{tot} = \mathbf{p}_1 + \mathbf{p}_2$, is *zero*, we conclude that the *total* momentum, $\mathbf{p}_{tot}$, must remain constant:

$$\mathbf{p}_{tot} = \mathbf{p}_1 + \mathbf{p}_2 = \text{constant} \qquad \qquad \textbf{[8.4]}$$

or, equivalently,

$$\mathbf{p}_{1i} + \mathbf{p}_{2i} = \mathbf{p}_{1f} + \mathbf{p}_{2f} \qquad \qquad \textbf{[8.5]}$$

where $\mathbf{p}_{1i}$ and $\mathbf{p}_{2i}$ are initial values and $\mathbf{p}_{1f}$ and $\mathbf{p}_{2f}$ are final values of the momentum during a time period, dt, over which the reaction pair interacts. Equation 8.5 in component form says that the total momenta in the x, y, and z directions are all *independently conserved;* that is,

$$p_{ix} = p_{fx} \qquad p_{iy} = p_{fy} \qquad p_{iz} = p_{fz} \qquad \qquad \textbf{[8.6]}$$

This result is known as the law of **conservation of linear momentum.** It is considered to be one of the most important laws of mechanics. We can state it as follows:

Conservation of momentum •

> Whenever two isolated particles interact with each other, their total momentum remains constant.

That is, **the total momentum of an isolated system at all times equals its initial total momentum.**

We can also describe the law of conservation of momentum in another way. Because we require that the system be isolated, no external forces are present, and the total momentum of the system remains constant. Therefore, momentum conservation is an alternative and more general statement of Newton's third law.

Notice that we have made no statement concerning the nature of the forces acting on the system. The only requirement was that the forces must be *internal* to the system. Thus, momentum is constant for a two-particle system *regardless* of the nature of the internal forces. One can use a similar and equivalent argument to show that the law of conservation of momentum also applies to a system of many particles.

Thinking Physics 1

A baseball is projected into the air at an upward angle to the ground. As it moves through its trajectory, its velocity and, therefore, its momentum constantly changes. Is this a violation of conservation of momentum?

Reasoning   The principle of conservation of momentum states that the momentum of a particle or a system of particles is conserved *in the absence of external forces*. A ball projected through the air is subject to the external force of gravity, so we would not expect its momentum to be conserved. If we consider this in a little more detail, the gravitational force is in the vertical direction, so it is actually only the vertical component of the momentum that changes due to this force. In the horizontal direction, there is no force (ignoring air friction), so the horizontal component of the momentum is conserved. In Chapter 3, we used the same idea in stating that the horizontal component of the *velocity* of a projectile remains constant.

If we consider the baseball and the Earth as a system of particles, then the gravitational force is an internal force to this system. The momentum of the ball–Earth system remains unchanged. The downward impulse due to the thrower's feet gives the Earth an initial motion. As the ball rises and falls, the Earth initially sinks but then rises (although imperceptibly) due to the upward gravitational force from the ball, so that the total momentum of the system remains unchanged.

Example 8.1   The Recoiling Pitching Machine

A baseball player uses a pitching machine to help him improve his batting average. He places the 50-kg machine on a frozen pond as in Figure 8.2. The machine fires a 0.15-kg baseball horizontally with a velocity of $36\mathbf{i}$ m/s. What is the recoil velocity of the machine?

Reasoning   We take the system to consist of the baseball and the pitching machine. Because of the force of gravity and the normal force, the system is not really isolated. However, both of these forces are directed perpendicularly to the motion of the system. Therefore, momentum is constant in the x direction because there are no external forces in this direction (assuming the surface is frictionless).

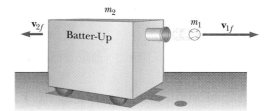

Figure 8.2   (Example 8.1) When the baseball is fired horizontally to the right, the pitching machine recoils to the left. The total momentum of the system before and after firing is zero.

Solution   The total momentum of the system before firing is zero $(m_1\mathbf{v}_{1i} + m_2\mathbf{v}_{2i} = 0)$. Therefore, the total momentum after firing must be zero; that is,

$$m_1\mathbf{v}_{1f} + m_2\mathbf{v}_{2f} = 0$$

With $m_1 = 0.15$ kg, $\mathbf{v}_{1i} = 36\mathbf{i}$ m/s, and $m_2 = 50$ kg, solving for $\mathbf{v}_{2f}$, we find the recoil velocity of the pitching machine to be

$$\mathbf{v}_{2f} = -\frac{m_1}{m_2}\mathbf{v}_{1f} = -\left(\frac{0.15 \text{ kg}}{50 \text{ kg}}\right)(36\mathbf{i} \text{ m/s}) = -0.11\mathbf{i} \text{ m/s}$$

The negative sign for $\mathbf{v}_{2f}$ indicates that the pitching machine is moving to the left after firing, in the direction opposite the direction of motion of the baseball. In the words of Newton's third law, for every force (to the left) on the pitching machine, there is an equal but opposite force (to the right) on the ball. Because the pitching machine is much more massive than the ball, the acceleration and consequent speed of the pitching machine are much smaller than the acceleration and speed of the ball.

Example 8.2   Decay of the Kaon at Rest

A meson is a nuclear particle that is more massive than an electron but less massive than a proton or neutron. One type of meson, called the neutral kaon (K^0), decays into a pair of charged pions (π^+ and π^-) that are oppositely charged but equal in mass, as in Figure 8.3. A pion is a particle associated with the strong nuclear force that binds the protons and neutrons together in the nucleus. Assuming the kaon is initially at rest, prove that after the decay, the two pions must have momenta that are equal in magnitude and opposite in direction.

Solution   The decay of the kaon, represented in Figure 8.3, can be written

$$K^0 \longrightarrow \pi^+ + \pi^-$$

If we let $\mathbf{p}_+$ be the momentum of the positive pion and $\mathbf{p}_-$ be the momentum of the negative pion after the decay, then the final momentum of the system can be written

$$\mathbf{p}_f = \mathbf{p}_+ + \mathbf{p}_-$$

Because the kaon is at rest before the decay, we know that $\mathbf{p}_i = 0$. Furthermore, because momentum is conserved, $\mathbf{p}_i = \mathbf{p}_f = 0$, so that $\mathbf{p}_+ + \mathbf{p}_- = 0$, or

$$\mathbf{p}_+ = -\mathbf{p}_-$$

Thus, we see that the two momentum vectors of the pions are equal in magnitude and opposite in direction. Furthermore, because the pions have equal mass, they will have equal and opposite velocities.

After decay

Figure 8.3   (Example 8.2) A kaon at rest decays spontaneously into a pair of oppositely charged pions. The pions move apart with momenta of equal magnitudes but opposite directions.

EXERCISE 1   The momentum of a 1250-kg car is equal to the momentum of a 5000-kg truck traveling at a speed of 10 m/s. What is the speed of the car?   Answer   40.0 m/s

EXERCISE 2   A 1500-kg car moving with a speed of 15 m/s collides with a utility pole and is brought to rest in 0.3 s. Find the average force exerted on the car during the collision.   Answer 7.5×10^4 N

EXERCISE 3   A 60-kg boy and a 40-kg girl, both wearing skates, face each other at rest. The girl pushes the boy, sending him eastward with a speed of 4 m/s. Describe the subsequent motion of the girl. (Neglect friction.)   Answer   She moves westward at 6 m/s.

8.2 • IMPULSE AND MOMENTUM

As we have seen, the momentum of a particle changes if a net force acts on the particle. Let us assume that a force **F** acts on a particle and that this force may vary with time. According to Newton's second law, $\mathbf{F} = d\mathbf{p}/dt$, or

$$d\mathbf{p} = \mathbf{F}\, dt \qquad [8.7]$$

We can integrate this expression to find the change in the momentum of a particle as it changes from $\mathbf{p}_i$ at time t_i to $\mathbf{p}_f$ at time t_f. Integrating Equation 8.7 gives

$$\Delta \mathbf{p} = \mathbf{p}_f - \mathbf{p}_i = \int_{t_i}^{t_f} \mathbf{F}\, dt \qquad [8.8]$$

The integral of a force over the time for which it acts is called the **impulse** of the force. The impulse of the force **F** for the time interval $\Delta t = t_f - t_i$ is a vector defined by

$$\mathbf{I} \equiv \int_{t_i}^{t_f} \mathbf{F}\, dt = \Delta \mathbf{p} \qquad [8.9]$$

• *Impulse of a force*

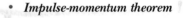

• *Impulse-momentum theorem*

That is, the **impulse** of the force **F** equals the change in the momentum of any object.[2] This statement, known as the **impulse-momentum theorem**, is equivalent to Newton's second law. From this definition we see that impulse is a vector quantity having a magnitude equal to the area under the force–time curve, as described in Figure 8.4. In this figure it is assumed that the force varies in time in the general manner shown and is nonzero in the time interval $\Delta t = t_f - t_i$. The direction of the impulse vector is the same as the direction of the change in momentum. Impulse has the dimensions of momentum, ML/T. Note that impulse is *not* a property of the particle itself; rather, it is a measure of the degree to which an external force changes the momentum of the particle. Therefore, when we say that an impulse is given to a particle, it is implied that momentum is transferred from an external agent to that particle.

Because the force can generally vary in time as in Figure 8.4a, it is convenient to define a time-averaged force $\overline{\mathbf{F}}$, given by

$$\overline{\mathbf{F}} \equiv \frac{1}{\Delta t} \int_{t_i}^{t_f} \mathbf{F}\, dt \qquad [8.10]$$

where $\Delta t = t_f - t_i$. Therefore, we can express Equation 8.9 as

$$\mathbf{I} = \Delta \mathbf{p} = \overline{\mathbf{F}}\, \Delta t \qquad [8.11]$$

This average force, described in Figure 8.4b, can be thought of as the constant force that would give the same impulse to the particle in the time interval Δt as the actual time-varying force gives over this same interval.

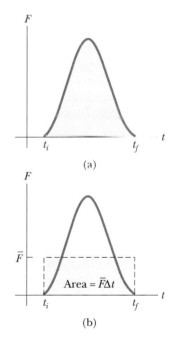

Figure 8.4 (a) A force acting on a particle may vary in time. The impulse is the area under the force-versus-time curve. (b) The average force (horizontal dashed line) gives the same impulse to the particle in the time Δt as the time-varying force described in (a).

---

[2]In general, if several forces act on a particle, the total impulse injected into the particle by the forces acting on it constitutes the change in momentum of the particle.

In principle, if **F** is known as a function of time, the impulse can be calculated from Equation 8.9. The calculation becomes especially simple if the force acting on the particle is constant. In this case, $\overline{\mathbf{F}} = \mathbf{F}$ and Equation 8.11 becomes

$$\mathbf{I} = \Delta\mathbf{p} = \mathbf{F}\,\Delta t \qquad\qquad [8.12]$$

In many physical situations, we shall use what is called the **impulse approximation: We assume that one of the forces exerted on a particle acts for a short time but is much greater than any other force present.** This approximation is especially useful in treating collisions, where the duration of the collision is very short. When this approximation is made, we refer to the force as an *impulsive force.* For example, when a baseball is struck with a bat, the duration of the collision is about 0.01 s, and the average force the bat exerts on the ball in this time is typically several thousand newtons. This is much greater than the force of gravity, so the impulse approximation is justified. It is important to remember that $\mathbf{p}_i$ and $\mathbf{p}_f$ represent the momenta *immediately* before and after the collision, respectively. Therefore, in the impulse approximation very little motion of the particle takes place during the collision.

Thinking Physics 2

A race car travels rapidly around a circular race track at constant speed. For a given portion of the track, what is the direction of the impulse vector? Once the car returns to the starting point, there is no net change in momentum. Thus, according to the impulse-momentum theorem, there must be no impulse. Yet a nonzero force acted over the time interval for one rotation around the track. How can the impulse be zero?

Reasoning For the movement around a short portion of the track, the direction of the impulse vector is the same as that of the change in momentum. The direction of the vector representing the change in momentum is the same as the direction of the vector representing the change in velocity. From our discussion of circular motion, we know that the direction of the vector representing the change in velocity is that of the acceleration vector, which is toward the center of the circle, because this is uniform circular motion. Thus, the impulse vector is directed toward the center of the circle.

If the car makes one rotation around the track, there is indeed no net change in momentum, so there must be zero net impulse. Keeping in mind that the force on the car causing the circular motion is the friction between the tires and the roadway, and that this force is always directed toward the center of the circle, we can argue the zero value of the total impulse. For every location of the car on the track, there is another point of its motion diametrically opposed across the circle, at which its force vector will be directed in the opposite direction. Thus, as we add up the vector impulses **F** *dt* around the circle, they will cancel in pairs for a net impulse of zero.

CONCEPTUAL PROBLEM 1

In boxing matches of the nineteenth century, bare fists were used. In modern boxing, fighters wear padded gloves. How does this better protect the brain of the boxer from injury? Boxers often "roll with the punch." How does this maneuver protect their health?

Example 8.3   How Good Are the Bumpers?

In a particular crash test, an automobile of mass 1500 kg collides with a wall, as in Figure 8.5. The initial and final velocities of the automobile are $\mathbf{v}_i = -15.0\mathbf{i}$ m/s and $\mathbf{v}_f = 2.6\mathbf{i}$ m/s. If the collision lasts for 0.150 s, find the impulse due to the collision and the average force exerted on the automobile.

Solution   The initial and final momenta of the automobile are

$$\mathbf{p}_i = m\mathbf{v}_i = (1500 \text{ kg})(-15.0\mathbf{i} \text{ m/s}) = -2.25 \times 10^4\mathbf{i} \text{ kg·m/s}$$

$$\mathbf{p}_f = m\mathbf{v}_f = (1500 \text{ kg})(2.6\mathbf{i} \text{ m/s}) = 0.39 \times 10^4\mathbf{i} \text{ kg·m/s}$$

Hence, the impulse is

$$\mathbf{I} = \Delta\mathbf{p} = \mathbf{p}_f - \mathbf{p}_i$$
$$= 0.39 \times 10^4\mathbf{i} \text{ kg·m/s} - (-2.25 \times 10^4\mathbf{i} \text{ kg·m/s})$$
$$\mathbf{I} = 2.64 \times 10^4\mathbf{i} \text{ kg·m/s}$$

The average force exerted on the automobile is

$$\overline{\mathbf{F}} = \frac{\Delta\mathbf{p}}{\Delta t} = \frac{2.64 \times 10^4\mathbf{i} \text{ kg·m/s}}{0.150 \text{ s}} = 1.76 \times 10^5\mathbf{i} \text{ N}$$

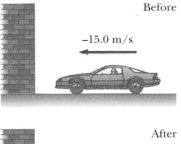

Before

–15.0 m/s

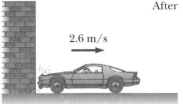

After

2.6 m/s

Figure 8.5   (Example 8.3)

EXERCISE 4   A child bounces a superball on the sidewalk. The linear impulse delivered by the sidewalk to the ball is 2.00 N·s during the 1/800 s of contact. What is the magnitude of the average force exerted on the ball by the sidewalk?   Answer   1.60 kN upward

EXERCISE 5   A 0.15-kg baseball is thrown with a speed of 40 m/s. It is hit straight back at the pitcher with a speed of 50 m/s. (a) What is the impulse delivered to the baseball? (b) Find the average force exerted by the bat on the ball if the two are in contact for 2.0×10^{-3} s.   Answer   (a) 13.5 kg·m/s   (b) 6.75 kN

8.3 • COLLISIONS

In this section we use the law of conservation of momentum to describe what happens when two objects collide. **The force due to the collision is assumed to be much larger than any external forces present.**

A collision may be the result of physical contact between two objects, as described in Figure 8.6a. This is a common observation when two macroscopic objects, such as two billiard balls or a baseball and a bat, collide. The notion of what we mean by *collision* must be generalized because "contact" on a submicroscopic scale is ill-defined and hence meaningless. More accurately, contact forces between two bodies arise from the electrostatic interaction of the electrons in the surface atoms of the bodies.

To understand this distinction between macroscopic and microscopic collisions, consider the collision of a proton with an alpha particle (the nucleus of the helium atom), such as occurs in Figure 8.6b. Because the two particles are positively charged, they repel each other.

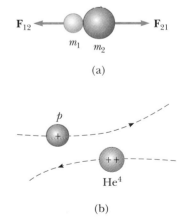

Figure 8.6   (a) The collision between two objects as the result of direct contact. (b) The "collision" between two charged particles.

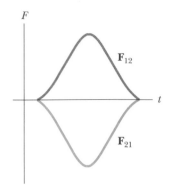

Figure 8.7   The impulse force as a function of time for the two colliding particles described in Figure 8.6a. Note that $\mathbf{F}_{12} = -\mathbf{F}_{21}$.

When two particles of masses m_1 and m_2 collide, the collision forces may vary in time in a complicated way (Fig. 8.7). If $\mathbf{F}_{12}$ is the force on m_1 due to m_2, then the change in momentum of m_1 due to the collision is given by Equation 8.8:

$$\Delta\mathbf{p}_1 = \int_{t_i}^{t_f} \mathbf{F}_{12}\, dt$$

Likewise, if $\mathbf{F}_{21}$ is the force exerted on m_2 by m_1, the change in momentum of m_2 is

$$\Delta\mathbf{p}_2 = \int_{t_i}^{t_f} \mathbf{F}_{21}\, dt$$

Newton's third law states that the force on m_1 by m_2 is equal to and opposite the force exerted on m_2 by m_1; that is, $\mathbf{F}_{12} = -\mathbf{F}_{21}$. (This is described graphically in Fig. 8.7.) Hence, we conclude that

$$\Delta\mathbf{p}_1 = \Delta\mathbf{p}_2$$
$$\Delta\mathbf{p}_1 + \Delta\mathbf{p}_2 = 0$$

Because the total momentum of the system is $\mathbf{p} = \mathbf{p}_1 + \mathbf{p}_2$, we conclude that the *change* in the momentum of the system due to the collision is zero:

$$\mathbf{p} = \mathbf{p}_1 + \mathbf{p}_2 = \text{constant}$$

This is precisely what we expect if no external forces are acting on the system (Section 8.2), or if the external forces are assumed to act for too short a time to make a significant change in momentum. However, the result is also valid if we consider the motion just before and just after the collision. Because the forces due to the collision are internal to the system, they do not change the total momentum of the system. Therefore, for any type of collision, the total momentum of a system just before the collision equals the total momentum of a system just after the collision.

CONCEPTUAL PROBLEM 2

You are watching a movie about a superhero and notice that the superhero hovers in the air and throws a piano at some bad guys while remaining stationary in the air. What's wrong with this scenario?

Example 8.4   Carry Collision Insurance

An 1800-kg car stopped at a traffic light is struck from the rear by a 900-kg car and the two become entangled. If the smaller car was moving at 20 m/s before the collision, what is the speed of the entangled mass after the collision?

Reasoning   The total momentum of the system (the two cars) before the collision equals the total momentum of the system after the collision because momentum is conserved.

Solution   The magnitude of the total momentum of the system before the collision is equal to that of the smaller car because the larger car is initially at rest:

$$p_i = m_i v_i = (900 \text{ kg})(20 \text{ m/s}) = 1.80 \times 10^4 \text{ kg} \cdot \text{m/s}$$

After the collision, the mass that moves is the sum of the masses of the cars. The magnitude of the momentum of the combination is

$$p_f = (m_1 + m_2)v_f = (2700 \text{ kg})v_f$$

Equating the momentum before to the momentum after and solving for v_f, the speed of the combined mass, we have

$$v_f = \frac{p_i}{m_1 + m_2} = \frac{1.80 \times 10^4 \text{ kg} \cdot \text{m/s}}{2700 \text{ kg}} = 6.67 \text{ m/s}$$

8.4 · ELASTIC AND INELASTIC COLLISIONS IN ONE DIMENSION

As we have seen, momentum is conserved in any type of collision. Kinetic energy, however, is generally *not* constant in a collision because some of it is converted to thermal energy, to internal elastic potential energy when the objects are deformed, and to rotational kinetic energy.

We define an inelastic collision as one in which total kinetic energy is not constant (even though momentum is constant). The collision of a rubber ball with a hard surface is inelastic, because some of the kinetic energy of the ball is lost when it is deformed while in contact with the surface. When two objects collide and stick together after a collision, some kinetic energy is lost, and the collision is called **perfectly inelastic.** For example, if two vehicles collide and become entangled, as in Example 8.4, they move with some common velocity after the perfectly inelastic collision. If a meteorite collides with the Earth, it becomes buried and the collision is perfectly inelastic.

• *Inelastic collision*

An elastic collision is defined as one in which total kinetic energy is constant (as well as momentum). Real collisions in the macroscopic world, such as those between billiard balls, are only approximately elastic because some deformation and loss of kinetic energy takes place. Billiard-ball collisions and the collisions of air molecules with the walls of a container at ordinary temperatures are highly elastic. Truly elastic collisions do occur, however, between atomic and subatomic particles. Elastic and perfectly inelastic collisions are *limiting* cases; most collisions fall in a category between them.

• *Elastic collision*

In the remainder of this section we treat collisions in one dimension and consider the two extreme cases—perfectly inelastic and elastic collisions. The important distinction between these two types of collisions is that **momentum is conserved in all cases, but kinetic energy is constant only in elastic collisions.**

Perfectly Inelastic Collisions

Consider two objects of masses m_1 and m_2 moving with initial velocities $\mathbf{v}_{1i}$ and $\mathbf{v}_{2i}$ along a straight line, as in Figure 8.8. If the two objects collide head-on, stick together, and move with some common velocity $\mathbf{v}_f$ after the collision, the collision is perfectly inelastic. Because the total momentum before the collision equals the total momentum of the composite system after the collision, we have

$$m_1\mathbf{v}_{1i} + m_2\mathbf{v}_{2i} = (m_1 + m_2)\mathbf{v}_f \qquad [8.13]$$

$$\mathbf{v}_f = \frac{m_1\mathbf{v}_{1i} + m_2\mathbf{v}_{2i}}{m_1 + m_2} \qquad [8.14]$$

Before collision

(a)

After collision

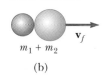

(b)

Figure 8.8   A perfectly inelastic head-on collision between two particles.

Elastic Collisions

Now consider two particles that undergo an elastic head-on collision (Fig. 8.9). In this case, both momentum and kinetic energy are constant; therefore, we can write

$$m_1\mathbf{v}_{1i} + m_2\mathbf{v}_{2i} = m_1\mathbf{v}_{1f} + m_2\mathbf{v}_{2f} \qquad [8.15]$$

$$\tfrac{1}{2}m_1v_{1i}^2 + \tfrac{1}{2}m_2v_{2i}^2 = \tfrac{1}{2}m_1v_{1f}^2 + \tfrac{1}{2}m_2v_{2f}^2 \qquad [8.16]$$

Before collision

(a)

After collision

(b)

Figure 8.9 An elastic head-on collision between two particles.

where v is positive if a particle moves to the right and negative if it moves to the left.

In a typical problem involving elastic collisions, there are two unknown quantities, and Equations 8.15 and 8.16 can be solved simultaneously to find them. An alternative approach, employing a little mathematical manipulation of Equation 8.16, often simplifies this process. To see this, let us cancel the factor of $\frac{1}{2}$ in Equation 8.16 and rewrite it as

$$m_1(v_{1i}^2 - v_{1f}^2) = m_2(v_{2f}^2 - v_{2i}^2)$$

Here we have moved the terms containing m_1 to one side of the equation and those containing m_2 to the other. Next, let us factor both sides:

$$m_1(v_{1i} - v_{1f})(v_{1i} + v_{1f}) = m_2(v_{2f} - v_{2i})(v_{2f} + v_{2i}) \qquad [8.17]$$

We now separate the terms containing m_1 and m_2 in the equation for the conservation of momentum (Eq. 8.15) to get

$$m_1(v_{1i} - v_{1f}) = m_2(v_{2f} - v_{2i}) \qquad [8.18]$$

To obtain our final result, we divide Equation 8.17 by Equation 8.18 and get

$$v_{1i} + v_{1f} = v_{2f} + v_{2i}$$

$$v_{1i} - v_{2i} = -(v_{1f} - v_{2f}) \qquad [8.19]$$

This equation, in combination with the condition for conservation of momentum, can be used to solve problems dealing with elastic collisions. According to Equation 8.19, the relative speed of the two objects before the collision, $v_{1i} - v_{2i}$, equals the negative of their relative speed after the collision, $-(v_{1f} - v_{2f})$.

Suppose that the masses and the initial velocities of both particles are known. Equations 8.15 and 8.16 can be solved for the final speeds in terms of the initial speeds, because there are two equations and two unknowns.

Elastic collision: relations •
between final and initial
velocities

$$v_{1f} = \left(\frac{m_1 - m_2}{m_1 + m_2}\right)v_{1i} + \left(\frac{2m_2}{m_1 + m_2}\right)v_{2i} \qquad [8.20]$$

$$v_{2f} = \left(\frac{2m_1}{m_1 + m_2}\right)v_{1i} + \left(\frac{m_2 - m_1}{m_1 + m_2}\right)v_{2i} \qquad [8.21]$$

It is important to remember that the appropriate signs for v_{1i} and v_{2i} must be included in Equations 8.20 and 8.21. For example, if m_2 is moving to the left initially, as in Figure 8.9, then v_{2i} is negative.

Let us consider some special cases. If $m_1 = m_2$, then $v_{1f} = v_{2i}$ and $v_{2f} = v_{1i}$. That is, the particles exchange speeds if they have equal masses. This is what one observes in head-on billiard ball collisions.

If m_2 is initially at rest, $v_{2i} = 0$, and Equations 8.20 and 8.21 become

$$v_{1f} = \left(\frac{m_1 - m_2}{m_1 + m_2}\right)v_{1i} \qquad [8.22]$$

$$v_{2f} = \left(\frac{2m_1}{m_1 + m_2}\right)v_{1i} \qquad [8.23]$$

If m_1 is very large compared with m_2, we see from Equations 8.22 and 8.23 that $v_{1f} \approx v_{1i}$ and $v_{2f} \approx 2v_{1i}$. That is, when a very heavy particle collides head-on with a very light one initially at rest, the heavy particle continues its motion unaltered after the collision, and the light particle rebounds with a speed equal to about twice the initial speed of the heavy particle. An example of such a collision would be that of a moving heavy atom, such as uranium, with a light atom, such as hydrogen.

If m_2 is much larger than m_1, and m_2 is initially at rest, then we find from Equations 8.22 and 8.23 that $v_{1f} \approx -v_{1i}$ and $v_{2f} \approx 0$. That is, when a very light particle collides head-on with a very heavy particle initially at rest, the velocity of the light particle is reversed, and the heavy particle remains approximately at rest. For example, imagine what happens when a marble hits a stationary bowling ball.

CONCEPTUAL PROBLEM 3

A great deal of expense is involved in repairing automobiles after collisions. It seems that much of this money could be saved if cars were built from hard rubber, so that they made approximately elastic collisions and simply bounced off each other without denting. Why are cars not made of hard rubber?

CONCEPTUAL PROBLEM 4

In perfectly inelastic collisions between two objects, under what conditions is all of the original kinetic energy transformed to forms other than kinetic?

The photograph of a car crash test (an inelastic collision) illustrates that much of the car's initial kinetic energy is transformed into the energy it took to damage the vehicle. Why do safety belts and air bags help prevent serious injury in such collisions? *(Courtesy of General Motors)*

Example 8.5   **Slowing Down Neutrons by Collisions**

In a nuclear reactor, neutrons are produced when a $^{235}_{92}$U atom splits in a process called fission. These neutrons are moving at about 10^7 m/s and must be slowed down to about 10^3 m/s before they take part in another fission event. They are slowed down by being passed through a solid or liquid material called a *moderator*. The slowing-down process involves elastic collisions. Let us show that a neutron can lose most of its kinetic energy if it collides elastically with a moderator containing light nuclei, such as deuterium (in "heavy water," D_2O) or carbon (in graphite).

Reasoning   Because momentum and energy are constant, Equations 8.22 and 8.23 can be applied to the head-on collision of a neutron with the moderator nucleus.

Solution   Let us assume that the moderator nucleus of mass m_m is at rest initially and that the neutron of mass m_n and initial speed v_{ni} collides head-on with it. The initial kinetic energy of the neutron is

$$K_{ni} = \tfrac{1}{2} m_n v_{ni}^2$$

After the collision, the neutron has a kinetic energy $\tfrac{1}{2} m_n v_{nf}^2$, where v_{nf} is given by Equation 8.22.

$$K_{nf} = \tfrac{1}{2} m_n v_{nf}^2 = \frac{m_n}{2} \left(\frac{m_n - m_m}{m_n + m_m} \right)^2 v_{ni}^2$$

Therefore, the fraction of the total kinetic energy possessed by the neutron after the collision is

$$(1) \quad f_n = \frac{K_{nf}}{K_{ni}} = \left(\frac{m_n - m_m}{m_n + m_m} \right)^2$$

From this result, we see that the final kinetic energy of the neutron is small when m_m is close to m_n and is zero when $m_n = m_m$.

We can calculate the kinetic energy of the moderator nucleus after the collision using Equation 8.23:

$$K_{mf} = \tfrac{1}{2} m_m v_{mf}^2 = \frac{2 m_n^2 m_m}{(m_n + m_m)^2} v_{ni}^2$$

Hence, the fraction of the total kinetic energy transferred to the moderator nucleus is

$$(2) \quad f_m = \frac{K_{mf}}{K_{ni}} = \frac{4 m_n m_m}{(m_n + m_m)^2}$$

Because the total energy is constant, (2) can also be obtained from (1) with the condition that $f_n + f_m = 1$, so that $f_m = 1 - f_n$.

Suppose that heavy water is used for the moderator. For collisions of the neutrons with deuterium nuclei in D_2O ($m_m = 2 m_n$), $f_n = 1/9$ and $f_m = 8/9$. That is, 89% of the neutron's kinetic energy is transferred to the deuterium nucleus. In practice, the moderator efficiency is reduced because head-on collisions are very unlikely to occur.

How do the results differ with graphite as the moderator?

Example 8.6   A Two-Body Collision with Spring

A block of mass $m_1 = 1.60$ kg, initially moving to the right with a speed of 4.00 m/s on a frictionless horizontal track, collides with a spring attached to a second block of mass $m_2 = 2.10$ kg, moving to the left with a speed of 2.50 m/s, as in Figure 8.10a. The spring has a spring constant of 600 N/m. (a) At the instant when m_1 is moving to the right with a speed of 3.00 m/s, as in Figure 8.10b, determine the speed of m_2.

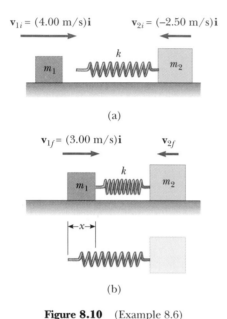

$\mathbf{v}_{1i} = (4.00 \text{ m/s})\mathbf{i}$ $\mathbf{v}_{2i} = (-2.50 \text{ m/s})\mathbf{i}$

k

m_1 m_2

(a)

$\mathbf{v}_{1f} = (3.00 \text{ m/s})\mathbf{i}$ $\mathbf{v}_{2f}$

k

m_1 m_2

$\leftarrow x \rightarrow$

(b)

Figure 8.10   (Example 8.6)

Solution   First, note that the initial velocity of m_2 is -2.50 m/s because its direction is to the left. Because the total momentum is conserved, we have

$$m_1 v_{1i} + m_2 v_{2i} = m_1 v_{1f} + m_2 v_{2f}$$

$$(1.60 \text{ kg})(4.00 \text{ m/s}) + (2.10 \text{ kg})(-2.50 \text{ m/s})$$

$$= (1.60 \text{ kg})(3.00 \text{ m/s}) + (2.10 \text{ kg})v_{2f}$$

$$v_{2f} = -1.74 \text{ m/s}$$

The negative value for v_{2f} means that m_2 is still moving to the left at the instant we are considering.

(b) Determine the distance the spring is compressed at that instant.

Solution   To determine the compression in the spring, x, shown in Figure 8.10b, we use conservation of energy, because there are no friction or other nonconservative forces acting on the system. Thus, we have

$$\tfrac{1}{2}m_1 v_{1i}^2 + \tfrac{1}{2}m_2 v_{2i}^2 = \tfrac{1}{2}m_1 v_{1f}^2 + \tfrac{1}{2}m_2 v_{2f}^2 + \tfrac{1}{2}kx^2$$

Substituting the given values and the result to part (a) into this expression gives

$$x = 0.173 \text{ m}$$

EXERCISE 6   Find the velocity of m_1 and the compression in the spring at the instant that m_2 is at rest.
Answer   0.719 m/s to the right; 0.251 m

EXERCISE 7   A 2.5-kg mass, moving initially with a speed of 10 m/s, makes a perfectly inelastic head-on collision with a 5.0-kg mass that is initially at rest. (a) Find the final velocity of the composite particle.   (b) How much kinetic energy is lost in the collision?
Answer   (a) 3.33 m/s   (b) 83.4 J

8.5 • TWO-DIMENSIONAL COLLISIONS

In Section 8.1 we showed that the total momentum of a system of particles is constant when the system is isolated (that is, when no external forces act on the system). For a general collision of two particles in three-dimensional space, the conservation-of-momentum law implies that the total momentum in each direction is constant. However, an important subset of collisions takes place in a plane. The game of billiards is a familiar example involving multiple collisions of objects moving on a two-dimensional surface. For such two-dimensional collisions, we obtain two component equations for the conservation of momentum:

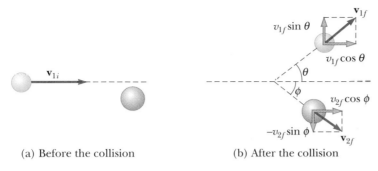

(a) Before the collision (b) After the collision

Figure 8.11 An elastic glancing collision between two particles.

$$m_1 v_{1ix} + m_2 v_{2ix} = m_1 v_{1fx} + m_2 v_{2fx}$$

$$m_1 v_{1iy} + m_2 v_{2iy} = m_1 v_{1fy} + m_2 v_{2fy}$$

Consider a two-dimensional problem in which a particle of mass m_1 collides elastically with a particle of mass m_2 that is initially at rest, as in Figure 8.11. After the collision, m_1 moves at an angle θ with respect to the horizontal, and m_2 moves at an angle ϕ with respect to the horizontal. This is called a *glancing* collision. Applying the law of conservation of momentum in component form, and noting that the total y component of momentum is zero, we get

x component: $m_1 v_{1i} + 0 = m_1 v_{1f} \cos\theta + m_2 v_{2f} \cos\phi$ **[8.24]**   • *Conservation of momentum*

y component: $0 + 0 = m_1 v_{1f} \sin\theta - m_2 v_{2f} \sin\phi$ **[8.25]**

We now have two independent equations. So long as only two of the preceding quantities are unknown, we can solve the problem completely.

Because the collision is elastic, we can write a third equation for kinetic energy, in the form

$$\tfrac{1}{2} m_1 v_{1i}^2 = \tfrac{1}{2} m_1 v_{1f}^2 + \tfrac{1}{2} m_2 v_{2f}^2 \qquad \textbf{[8.26]}$$

• *Conservation of energy*

If we know the initial velocity, v_{1i}, and the masses, we are left with four unknowns. Because we have only three equations, one of the four remaining quantities (v_{1f}, v_{2f}, θ, or ϕ) must be given to determine the motion after the collision from conservation principles alone.

If the collision is inelastic, kinetic energy is *not* constant, and Equation 8.26 does *not* apply.

PROBLEM-SOLVING STRATEGY • **Collisions**

The following procedure is recommended when dealing with problems involving collisions between two objects.

1. Set up a coordinate system and define your velocities with respect to that system. It is convenient to have the x axis coincide with one of the initial velocities.
2. In your sketch of the coordinate system, draw and label all velocity vectors and include all the given information. Write expressions for the x and y

components of the momentum of each object before and after the collision. Remember to include the appropriate signs for the components of the velocity vectors. For example, if an object is moving in the negative x direction, its x component of velocity must be taken to be negative. It is essential that you pay careful attention to signs.

3. Write expressions for the *total* momentum in the x direction *before* and *after* the collision, and equate the two. Repeat this procedure for the total momentum in the y direction. These steps follow from the fact that, because the momentum is conserved in any collision, the total momentum in any direction must be constant. It is important to emphasize that it is the momentum of the *system* (the two colliding objects) that is constant, not the momenta of the individual objects.

4. If the collision is inelastic, kinetic energy is *not* constant, and additional information is probably required. If the collision is perfectly inelastic, the final velocities of the two objects are equal. Proceed to solve the momentum equations for the unknown quantities.

5. If the collision is elastic, kinetic energy is constant, and you can equate the total kinetic energy before the collision to the total kinetic energy after the collision. This provides an additional relationship between the velocities.

Example 8.7   Proton–Proton Collision

A proton collides in an elastic fashion with another proton that is initially at rest. The incoming proton has an initial speed of 3.5×10^5 m/s and makes a glancing collision with the second proton, as in Figure 8.11. (At close separations, the protons exert a repulsive electrostatic force on each other.) After the collision, one proton moves off at an angle of $37°$ to the original direction of motion, and the second deflects at an angle of ϕ to the same axis. Find the final speeds of the two protons and the angle ϕ.

Solution   Both momentum and kinetic energy are constant in this glancing elastic collision. Because $m_1 = m_2$, $\theta = 37°$, and we are given $v_{1i} = 3.5 \times 10^5$ m/s, Equations 8.24, 8.25, and 8.26 become

$$v_{1f} \cos 37° + v_{2f} \cos \phi = 3.5 \times 10^5$$

$$v_{1f} \sin 37° - v_{2f} \sin \phi = 0$$

$$v_{1f}^2 + v_{2f}^2 = (3.5 \times 10^5)^2$$

Simultaneous solution of these three equations with three unknowns gives

$$v_{1f} = 2.80 \times 10^5 \text{ m/s} \qquad v_{2f} = 2.11 \times 10^5 \text{ m/s}$$

$$\phi = 53.0°$$

It is interesting to note that $\theta + \phi = 90°$. This result is *not* accidental. **Whenever two equal masses collide elastically in a glancing collision and one of them is initially at rest, their final velocities are *always* at right angles to each other.**

Example 8.8   Collision at an Intersection

A 1500-kg car traveling east with a speed of 25.0 m/s collides at an intersection with a 2500-kg van traveling north at a speed of 20.0 m/s, as shown in Figure 8.12. Find the direction and magnitude of the velocity of the wreckage after the collision, assuming that the vehicles undergo a perfectly inelastic collision (that is, they stick together).

Solution   Let us choose east to be along the positive x direction and north to be along the positive y direction, as in Figure 8.12. Before the collision, the only object having momen-

tum in the x direction is the car. Thus, the magnitude of the total initial momentum of the system (car plus van) in the x direction is

$$\sum p_{xi} = (1500 \text{ kg})(25.0 \text{ m/s}) = 3.75 \times 10^4 \text{ kg} \cdot \text{m/s}$$

Now let us assume that the wreckage moves at an angle θ and speed v after the collision. The magnitude of the total momentum in the x direction after the collision is

$$\sum p_{xf} = (4000 \text{ kg}) v \cos \theta$$

Because the total momentum in the x direction is constant, we can equate these two equations to get

$$(1) \quad (3.75 \times 10^4) \text{ kg} \cdot \text{m/s} = (4000 \text{ kg})v \cos \theta$$

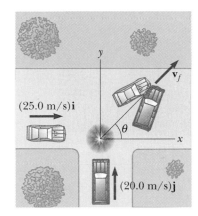

Figure 8.12   (Example 8.8) Top view of an eastbound car colliding with a northbound van.

In a similar way, the total initial momentum of the system in the y direction is that of the van, the magnitude of which is equal to $(2500 \text{ kg})(20.0 \text{ m/s})$. Applying conservation of momentum to the y direction, we have

$$\sum p_{yi} = \sum p_{yf}$$

$$(2500 \text{ kg})(20.0 \text{ m/s}) = (4000 \text{ kg})v \sin \theta$$

$$(2) \quad 5.00 \times 10^4 \text{ kg} \cdot \text{m/s} = (4000 \text{ kg})v \sin \theta$$

If we divide (2) by (1), we get

$$\tan \theta = \frac{5.00 \times 10^4}{3.75 \times 10^4} = 1.33$$

$$\theta = 53.1°$$

When this angle is substituted into (2), the value of v is

$$v = \frac{5.00 \times 10^4 \text{ kg} \cdot \text{m/s}}{(4000 \text{ kg}) \sin 53°} = 15.6 \text{ m/s}$$

EXERCISE 8   A 3.00-kg mass with an initial velocity of $5.00\mathbf{i}$ m/s collides with and sticks to a 2.00-kg mass with an initial velocity of $-3.00\mathbf{j}$ m/s. Find the final velocity of the composite mass.   Answer $(3.00\mathbf{i} - 1.20\mathbf{j})$ m/s

EXERCISE 9   A bird and a bee are approaching each other at right angles. The bird has a mass of 0.125 kg and a speed of 0.600 m/s, and the bee has a mass of 5.00×10^{-3} kg and a speed of 15.0 m/s. If the bird catches the bee, what is the new speed of the bird?   Answer   0.816 m/s

8.6 · THE CENTER OF MASS

In this section we describe the overall motion of a mechanical system in terms of a special point called the **center of mass** of the system. The mechanical system can be either a system of particles or any object. We shall see that the center of mass accelerates as if all of the system's mass were concentrated at that point and all external forces acted there. Furthermore, if the resultant external force on the system is **F** and the total mass of the system is M, the center of mass moves with an acceleration given by $\mathbf{a} = \mathbf{F}/M$. That is, the system moves as if the resultant external force were applied to a single particle of mass M located at the center of mass. This result was implicitly assumed in earlier chapters, because nearly all examples referred to the motion of extended objects.

Consider a system consisting of a pair of particles connected by a light, rigid rod (Fig. 8.13). The center of mass is located somewhere on the rod and is closer to the larger mass. If a single force is applied at some point on the rod that is closer to the smaller mass than to the larger one, the system rotates clockwise (Fig. 8.13a). If the force is applied at a point on the rod closer to the larger mass, the system rotates counterclockwise (Fig. 8.13b). If the force is applied at the center of mass,

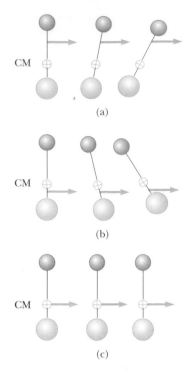

(a)

(b)

(c)

Figure 8.13 Two unequal masses are connected by a light, rigid rod. (a) The system rotates clockwise when a force is applied between the smaller mass and the center of mass. (b) The system rotates counterclockwise when a force is applied between the larger mass and the center of mass. (c) The system moves in the direction of the force without rotating when a force is applied at the center of mass.

This multiflash photograph shows that as the diver executes a back flip, his center of mass follows a parabolic path, the same path that a particle would follow.   (© *the Harold E. Edgerton 1992 Trust. Courtesy of Palm Press, Inc.*)

the system moves in the direction of **F** without rotating (Fig. 8.13c). Thus, the center of mass can be easily located.

The position of the center of mass of a system can be described as being the *average position* of the system's mass. For example, the center of mass of the pair of particles described in Figure 8.14 is located on the *x* axis, somewhere between the particles. The *x* coordinate of the center of mass in this case is

$$x_{CM} \equiv \frac{m_1 x_1 + m_2 x_2}{m_1 + m_2}$$ [8.27]

For example, if $x_1 = 0$, $x_2 = d$, and $m_2 = 2m_1$, we find that $x_{CM} = \frac{2}{3}d$. That is, the center of mass lies closer to the more massive particle. If the two masses are equal, the center of mass lies midway between the particles.

We can extend the center of mass concept to a system of many particles in three dimensions. The *x* coordinate of the center of mass of *n* particles is defined to be

$$x_{CM} \equiv \frac{m_1 x_1 + m_2 x_2 + m_3 x_3 + \ldots + m_n x_n}{m_1 + m_2 + m_3 + \ldots + m_n} = \frac{\Sigma m_i x_i}{\Sigma m_i}$$ [8.28]

where x_i is the *x* coordinate of the *i*th particle and Σm_i is the *total mass* of the system. For convenience, we shall express the total mass as $M = \Sigma m_i$, where the sum runs over all *n* particles. The *y* and *z* coordinates of the center of mass are similarly defined by the equations

$$y_{CM} \equiv \frac{\Sigma m_i y_i}{M} \quad \text{and} \quad z_{CM} \equiv \frac{\Sigma m_i z_i}{M}$$ [8.29]

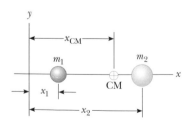

Figure 8.14 The center of mass of two particles on the *x* axis is located at x_{CM}, a point between the particles, closer to the larger mass.

The center of mass can also be located by its position vector, $\mathbf{r}_{CM}$. The rectangular coordinates of this vector are x_{CM}, y_{CM}, and z_{CM}, defined in Equations 8.28 and 8.29. Therefore,

$$\mathbf{r}_{CM} = x_{CM}\mathbf{i} + y_{CM}\mathbf{j} + z_{CM}\mathbf{k} = \frac{\Sigma m_i x_i \mathbf{i} + \Sigma m_i y_i \mathbf{j} + \Sigma m_i z_i \mathbf{k}}{M}$$

$$\mathbf{r}_{CM} \equiv \frac{\Sigma m_i \mathbf{r}_i}{M} \qquad [8.30]$$

where $\mathbf{r}_i$ is the position vector of the *i*th particle, defined by

$$\mathbf{r}_i \equiv x_i\mathbf{i} + y_i\mathbf{j} + z_i\mathbf{k}$$

The center of mass of a homogeneous, symmetric body must lie on an axis of symmetry. For example, the center of mass of a homogeneous rod must lie midway between the ends of the rod. The center of mass of a homogeneous sphere or a homogeneous cube must lie at the geometric center of the object. One can experimentally determine the center of mass of an irregularly shaped object, such as a wrench, by suspending the wrench from two different points (Fig. 8.15). The wrench is first hung from point *A*, and a vertical line, *AB*, is drawn (which can be established with a plumb bob) when the wrench is in equilibrium. The wrench is then hung from point *C*, and a second vertical line, *CD*, is drawn. The center of mass coincides with the intersection of these two lines. In fact, if the wrench is hung freely from any point, the vertical line through that point will pass through the center of mass.

Because a rigid body is a continuous distribution of mass, each portion is acted on by the force of gravity. The net effect of all of these forces is equivalent to the effect of a single force, $M\mathbf{g}$, acting through a special point called the **center of gravity.** If $\mathbf{g}$ is constant over the mass distribution, then the center of gravity coincides with the center of mass. If a rigid body is pivoted at its center of gravity, it will be balanced in any orientation.

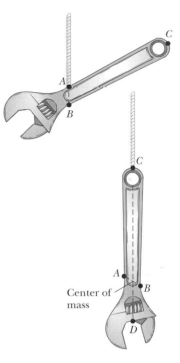

Figure 8.15 An experimental technique for determining the center of mass of a wrench. The wrench is hung freely from two different pivots, *A* and *C*. The intersection of the two vertical lines *AB* and *CD* locates the center of mass.

Example 8.9 The Center of Mass of Three Particles

A system consists of three particles located at the corners of a right triangle as in Figure 8.16. Find the center of mass of the system.

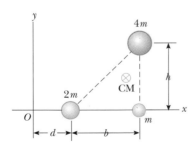

Figure 8.16 (Example 8.9) Locating the center of mass for a system of three particles.

Solution Using the basic defining equations for the coordinates of the center of mass, and noting that $z_{CM} = 0$, we get

$$x_{CM} = \frac{\Sigma m_i x_i}{M} = \frac{2md + m(d + b) + 4m(d + b)}{7m} = d + \tfrac{5}{7}b$$

$$y_{CM} = \frac{\Sigma m_i y_i}{M} = \frac{2m(0) + m(0) + 4mh}{7m} = \tfrac{4}{7}h$$

Therefore, we can express the position vector to the center of mass measured from the origin as

$$\mathbf{r}_{CM} = x_{CM}\mathbf{i} + y_{CM}\mathbf{j} + z_{CM}\mathbf{k} = \left(d + \tfrac{5}{7}b\right)\mathbf{i} + \tfrac{4}{7}h\mathbf{j}$$

8.7 · MOTION OF A SYSTEM OF PARTICLES

We can begin to understand the physical significance and utility of the center of mass concept by taking the time derivative of the position vector of the center of mass $\mathbf{r}_{CM}$, given by Equation 8.30. Assuming that M remains constant—that is, no particles enter or leave the system—we get the following expression for the **velocity of the center of mass:**

Velocity of the center of mass •

$$\mathbf{v}_{CM} = \frac{d\mathbf{r}_{CM}}{dt} = \frac{1}{M} \sum m_i \frac{d\mathbf{r}_i}{dt} = \frac{\sum m_i \mathbf{v}_i}{M} \qquad [8.31]$$

where $\mathbf{v}_i$ is the velocity of the ith particle. Rearranging Equation 8.31 gives

Total momentum of a system • *of particles*

$$M\mathbf{v}_{CM} = \sum m_i \mathbf{v}_i = \sum \mathbf{p}_i = \mathbf{p}_{tot} \qquad [8.32]$$

This result tells us that **the total momentum of the system equals the total mass multiplied by the velocity of the center of mass**—in other words, the total momentum of a single particle of mass M moving with a velocity $\mathbf{v}_{CM}$.

If we now differentiate Equation 8.32 with respect to time, we get the **acceleration of the center of mass:**

Acceleration of the center of • *mass for a system of particles*

$$\mathbf{a}_{CM} = \frac{d\mathbf{v}_{CM}}{dt} = \frac{1}{M} \sum m_i \frac{d\mathbf{v}_i}{dt} = \frac{1}{M} \sum m_i \mathbf{a}_i \qquad [8.33]$$

Rearranging this expression and using Newton's second law, we get

$$M\mathbf{a}_{CM} = \sum m_i \mathbf{a}_i = \sum \mathbf{F}_i \qquad [8.34]$$

where $\mathbf{F}_i$ is the force on particle i.

The forces on any particle in the system may include both external and internal forces. However, by Newton's third law, the force exerted by particle 1 on particle 2, for example, is equal to and opposite the force exerted by particle 2 on particle 1. When we sum over all internal forces in Equation 8.34, they cancel in pairs and the net force on the system is due *only* to external forces. Thus, we can write Equation 8.34 in the form

Newton's second law for a • *system of particles*

$$\sum \mathbf{F}_{ext} = M\mathbf{a}_{CM} = \frac{d\mathbf{p}_{tot}}{dt} \qquad [8.35]$$

That is, the resultant external force on the system of particles equals the total mass of the system multiplied by the acceleration of the center of mass. If we compare this to Newton's second law for a single particle, we see that the center of mass moves like an imaginary particle of mass M under the influence of the resultant external force on the system. In the absence of external forces, the center of mass moves with uniform velocity, as in the case of the rotating wrench in Figure 8.17. If the resultant force acts along a line through the center of mass of an extended body such as the wrench, the body is accelerated without rotation, and its kinetic energy is associated entirely with its linear motion. If the resultant force does not act through the center of mass, the body will be rotationally accelerated, and the body will acquire a rotational kinetic energy in addition to the kinetic energy of its linear motion. The linear acceleration of the center of mass is the same in either case, as given by Equation 8.35.

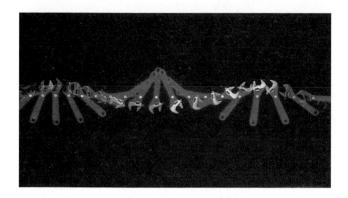

Figure 8.17 Multiflash photograph of a wrench moving on a horizontal surface. The center of mass of the wrench moves in a straight line as the wrench rotates about this point as marked with a white dot. *(Richard Megna, Fundamental Photographs)*

Finally, we see that if the resultant external force is zero, then from Equation 8.35 it follows that

$$\frac{d\mathbf{p}_{tot}}{dt} = M\mathbf{a}_{CM} = 0$$

so that

$$\mathbf{p}_{tot} = M\mathbf{v}_{CM} = \text{constant} \qquad \left(\text{when } \sum \mathbf{F}_{ext} = 0\right) \qquad \text{[8.36]}$$

That is, the total linear momentum of a system of particles is constant if there are no external forces acting on the system. It follows that, for an *isolated* system of particles, both the total momentum and the velocity of the center of mass are constant in time. The law of conservation of momentum that was derived in Section 8.1 for a two-particle system is thus generalized to a many-particle system.

Suppose an isolated system consisting of two or more members is at rest. The center of mass of such a system remains at rest unless acted on by an external force. For example, consider a system, initially at rest, made up of a swimmer and a raft. When the swimmer dives off the raft, the center of mass of the system remains at rest (if we neglect friction between raft and water). Furthermore, the horizontal momentum of the diver is equal in magnitude to the momentum of the raft, but opposite in direction.

As another example, suppose an unstable atom initially at rest suddenly decays into two fragments of masses M_1 and M_2, with velocities $\mathbf{v}_1$ and $\mathbf{v}_2$, respectively. Because the total momentum of the system before the decay is zero, the total momentum of the system after the decay must also be zero. Therefore, we see that $M_1\mathbf{v}_1 + M_2\mathbf{v}_2 = 0$. If the velocity of one of the fragments after the decay is known, the recoil velocity of the other fragment can be calculated.

Thinking Physics 3

A boy stands at one end of a floating canoe that is stationary relative to the shore (Fig. 8.18). He then walks to the opposite end of the canoe, away from the shore. Does the canoe move?

Reasoning Yes, the canoe moves toward the shore. Neglecting friction between the canoe and water, there are no horizontal forces acting on the system consisting of the boy and canoe. Therefore, the center of mass of the system remains fixed relative to the shore (or any stationary point). As the boy moves away from the shore, the canoe

Figure 8.18   (Thinking Physics 3)

must move toward the shore such that the center of mass of the system remains constant. An alternative explanation is that the momentum of the system remains constant if friction is neglected. As the boy acquires a momentum away from the shore, the canoe must acquire an equal momentum toward the shore such that the total momentum of the system is always zero.

CONCEPTUAL PROBLEM 5

Suppose you tranquilize a polar bear on a glacier as part of a research effort. How might you be able to estimate the weight of the polar bear using a measuring tape, a rope, and a knowledge of your own weight?

CONCEPTUAL PROBLEM 6

A projectile is fired into the air and suddenly explodes into several fragments (Fig. 8.19). What can be said about the motion of the center of mass of the fragments after the explosion?

EXERCISE 10   A 5.0-kg particle moves along the x axis with a velocity of 3.0 m/s. A 2.0-kg particle moves along the x axis with a velocity of -2.5 m/s. Find (a) the velocity of the center of mass and (b) the total momentum of the system.   Answer   (a) 1.4 m/s to the right (b) 10 kg·m/s to the right

EXERCISE 11   A 2.0-kg particle has a velocity $(2.0\mathbf{i} - 3.0\mathbf{j})$ m/s, and a 3.0-kg particle has a velocity $(1.0\mathbf{i} + 6.0\mathbf{j})$ m/s. Find (a) the velocity of the center of mass and (b) the total momentum of the system.   Answer   (a) $(1.4\mathbf{i} + 2.4\mathbf{j})$ m/s   (b) $(7.0\mathbf{i} + 12\mathbf{j})$ kg·m/s

Motion of
center of mass

Figure 8.19   (Conceptual Problem 6)

O P T I O N A L

8.8 • ROCKET PROPULSION

When ordinary vehicles, such as automobiles and locomotives, are propelled, the driving force for the motion is one of friction. In the case of the automobile, the driving force is the force exerted by the road on the car. A locomotive "pushes" against the tracks; hence, the driving force is the reaction force exerted by the tracks on the locomotive. However, a rocket moving in space has no road or tracks to "push" against. Therefore, the source of the propulsion of a rocket must be different. **The operation of a rocket depends on the law of conservation of**

Figure 8.20 Lift-off of the space shuttle *Columbia*. Enormous thrust is generated by the shuttle's liquid-fueled engines, aided by the two solid-fuel boosters. Many physical principles from mechanics, thermodynamics, and electricity and magnetism are involved in such a launch. *(Courtesy of NASA)*

momentum as applied to a system of particles, where the system is the rocket plus its ejected exhaust gases.

The propulsion of a rocket (Fig. 8.20) can be understood by first considering the mechanical system consisting of a machine gun mounted on a cart on wheels. As the machine gun is fired, each bullet receives a momentum $m\mathbf{v}$ in some direction, where $\mathbf{v}$ is the velocity of the bullet relative to the gun. For each bullet that is fired, the gun and cart must receive a compensating momentum in the opposite direction (neglecting the momentum imparted to the escaping gas in the barrel of the gun). That is, the reaction force exerted by the bullet on the gun accelerates the cart and gun. If there are n bullets fired each second, then the average force exerted on the gun is $\mathbf{F}_{av} - nm\mathbf{v}$.

In a similar manner, as a rocket moves in free space (a vacuum), **its momentum changes when some of its mass is released in the form of ejected gases. Because the ejected gases acquire some momentum, the rocket receives a compensating momentum in the opposite direction.** Therefore, **the rocket is accelerated as a result of the "push," or thrust, from the exhaust gases.** In free space, the center of mass of the system moves uniformly, independent of the propulsion process. It is interesting to note that the rocket and machine gun represent cases of the *inverse* of an inelastic collision—that is, momentum is conserved, but the kinetic energy of the system is *increased* (at the expense of potential energy carried in the fuel).

Suppose that at some time t, the momentum of the rocket plus the fuel is $(M + \Delta m)v$ (Fig. 8.21a). At some short time later, Δt, the rocket ejects fuel of mass Δm and the rocket's speed therefore increases to $v + \Delta v$ (Fig. 8.21b). If the fuel is ejected with a velocity v_e *relative to the rocket*, then the speed of the fuel relative to a stationary frame of reference is $v - v_e$. Thus, if we equate the total initial momentum of the system with the total final momentum, we get

$$(M + \Delta m)v = M(v + \Delta v) + \Delta m(v - v_e)$$

Simplifying this expression gives

$$M\,\Delta v = \Delta m(v_e)$$

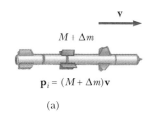

$$\mathbf{p}_i = (M + \Delta m)\mathbf{v}$$

(a)

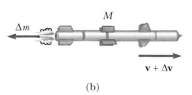

(b)

Figure 8.21 Rocket propulsion. (a) The initial mass of the rocket is $M + \Delta m$ at a time t, and its speed is v. (b) At a time $t + \Delta t$, the rocket's mass has reduced to M, and an amount of fuel Δm has been ejected. The rocket's speed increases by an amount Δv.

We can also arrive at this result by considering the system in the center-of-mass frame of reference—that is, a frame the velocity of which equals the center-of-mass velocity. In this frame, the total momentum is zero; therefore, if the rocket gains a momentum $M \Delta v$ by ejecting some fuel, the exhaust gases obtain a momentum $\Delta m(v_e)$ in the *opposite* direction, and so $M \Delta v - \Delta m(v_e) = 0$. If we now take the limit as Δt goes to zero, the $\Delta v \rightarrow dv$ and $\Delta m \rightarrow dm$. Furthermore, the increase in the exhaust mass, dm, corresponds to an equal decrease in the rocket mass, so that $dm = -dM$. Note that dM is given a negative sign because it represents a decrease in mass. Using this fact, we get

$$M\, dv = -v_e\, dM \qquad \qquad \text{[8.37]}$$

Integrating this equation and taking the initial mass of the rocket plus fuel to be M_i and the final mass of the rocket plus its remaining fuel to be M_f we get

$$\int_{v_i}^{v_f} dv = -v_e \int_{M_i}^{M_f} \frac{dM}{M}$$

Basic expression for rocket •
propulsion

$$v_f - v_i = v_e \ln\left(\frac{M_i}{M_f}\right) \qquad \qquad \text{[8.38]}$$

This is the basic expression of rocket propulsion. First, it tells us that the increase in speed is proportional to the exhaust speed, v_e. Therefore, the exhaust velocity should be very high. Second, the increase in speed is proportional to the logarithm of the ratio M_i/M_f. Therefore, this ratio should be as large as possible, which means that the rocket should carry as much fuel as possible.

The **thrust** on the rocket is the force exerted on the rocket by the ejected exhaust gases. We can obtain an expression for the thrust from Equation 8.37:

$$\text{Thrust} = M\frac{dv}{dt} = \left| v_e \frac{dM}{dt} \right| \qquad \qquad \text{[8.39]}$$

Here we see that the thrust increases as the exhaust speed increases and as the rate of change of mass (burn rate) increases.

Thinking Physics 4

When Robert Goddard proposed the possibility of rocket-propelled vehicles, the *New York Times* agreed that such vehicles would be useful and successful within the Earth's atmosphere ("Topics of the Times," *New York Times,* Jan. 13, 1920, p. 12). But the *Times* balked at the idea of using such a rocket in the vacuum of space, noting that, ". . . its flight would be neither accelerated nor maintained by the explosion of the charges it then might have left. To claim that it would be is to deny a fundamental law of dynamics . . . That Professor Goddard, with his 'chair' in Clark College and the countenancing of the Smithsonian Institution, does not know the relation of action to reaction, and of the need to have something better than a vacuum against which to react—to say that would be absurd."
What did the writer of these passages overlook?

Reasoning   The writer was making a common mistake in believing that a rocket works by expelling gases that *push* on something, propelling the rocket forward. With this belief, it is impossible to see how a rocket fired in empty space *would* work.
Gases do not need to push on anything—it is the act of expelling the gases itself that pushes the rocket forward. This can be argued from Newton's third law—the

rocket pushes the gases backward, resulting in the gases pushing the rocket forward. It can also be argued from conservation of momentum—as the gases gain momentum in one direction, the rocket must gain momentum in the opposite direction to conserve the original momentum of the rocket–gas system.

The *New York Times* did publish a retraction 49 years later ("A Correction," *New York Times,* July 17, 1969, p. 43), at the same time that the Apollo astronauts were on their way to the Moon. It appeared on a page with two other articles, one titled, "Fundamentals of Space Travel," and the other, "Spacecraft, Like Squid, Maneuver by 'Squirts.'" The following passages were part of these articles: ". . . an editorial feature of the *New York Times* dismissed the notion that a rocket could function in a vacuum and commented on the ideas of Robert H. Goddard. . . . Further investigation and experimentation have confirmed the findings of Isaac Newton in the 17th century, and it is now definitely established that a rocket can function in a vacuum as well as in an atmosphere. The *Times* regrets the error."

Example 8.10 A Rocket in Space

A rocket moving in free space has a speed of 3.0×10^3 m/s relative to Earth. Its engines are turned on, and exhaust is ejected in a direction opposite the rocket's motion at a speed of 5.0×10^3 m/s relative to the rocket. (a) What is the speed of the rocket relative to Earth once its mass is reduced to one half its mass before ignition?

Solution Applying Equation 8.38, we get

$$v_f = v_i + v_e \ln \left(\frac{M_i}{M_f} \right)$$

$$= 3.0 \times 10^3 + 5.0 \times 10^3 \ln \left(\frac{M_i}{0.5 M_i} \right)$$

$$= 6.5 \times 10^3 \text{ m/s}$$

(b) What is the thrust on the rocket if it burns fuel at the rate of 50 kg/s?

Solution

$$\text{Thrust} = \left| v_e \frac{dM}{dt} \right| = \left(5.0 \times 10^3 \frac{\text{m}}{\text{s}} \right) \left(50 \frac{\text{kg}}{\text{s}} \right)$$

$$= 2.5 \times 10^5 \text{ N}$$

EXERCISE 12 A rocket engine consumes 80 kg of fuel per second. If the exhaust speed is 2.5×10^3 m/s, calculate the thrust on the rocket.   Answer   200 kN

SUMMARY

The **linear momentum** of any object of mass m moving with a velocity $\mathbf{v}$ is

$$\mathbf{p} \equiv m\mathbf{v} \qquad \text{[8.1]}$$

Conservation of momentum applied to two interacting objects states that if the two objects form an isolated system, their total momentum is constant regardless of the nature of the force between them. Therefore, the total momentum of the system at all times equals its initial total momentum:

$$\mathbf{p}_{1i} + \mathbf{p}_{2i} = \mathbf{p}_{1f} + \mathbf{p}_{2f} \qquad \text{[8.5]}$$

The **impulse** of a force $\mathbf{F}$ on any object is equal to the change in the momentum of the object and is given by

$$\mathbf{I} \equiv \int_{t_i}^{t_f} \mathbf{F} \, dt = \Delta \mathbf{p} = \mathbf{p}_f - \mathbf{p}_i \qquad \text{[8.9]}$$

This is known as the **impulse-momentum theorem.**

Impulsive forces are forces that are very strong compared with other forces on the system. They usually act for a short time, as in the case of collisions.

When two bodies collide, the total momentum of the system before the collision always equals the total momentum after the collision, regardless of the nature of the collision. An **inelastic collision** is one for which kinetic energy is not constant, but momentum is. A perfectly inelastic collision is one in which the colliding bodies stick together after the collision. An **elastic collision** is one in which both momentum and kinetic energy are constant.

In a two- or three-dimensional collision, the components of momentum in each of the directions are conserved independently.

The **vector position of the center of mass of a system of particles** is defined as

$$\mathbf{r}_{CM} \equiv \frac{\sum m_i \mathbf{r}_i}{M} \qquad [8.30]$$

where $M = \sum m_i$ is the total mass of the system and $\mathbf{r}_i$ is the vector position of the ith particle. The **velocity of the center of mass for a system of particles** is

$$\mathbf{v}_{CM} = \frac{\sum m_i \mathbf{v}_i}{M} \qquad [8.31]$$

The total momentum of a system of particles equals the total mass multiplied by the velocity of the center of mass—that is, $\mathbf{p}_{tot} = M\mathbf{v}_{CM}$.

Newton's second law applied to a system of particles is

$$\sum \mathbf{F}_{ext} = M\mathbf{a}_{CM} = \frac{d\mathbf{p}_{tot}}{dt} \qquad [8.35]$$

where $\mathbf{a}_{CM}$ is the acceleration of the center of mass and the sum is over all external forces. Therefore, the center of mass moves like an imaginary particle of mass M under the influence of the resultant external force on the system. It follows from Equation 8.35 that the total momentum of the system is constant if there are no external forces acting on it.

CONCEPTUAL QUESTIONS

1. Does a large force always produce a larger impulse on a body than a smaller force does? Explain.

2. If two objects collide and one is initially at rest, is it possible for both to be at rest after the collision? Is it possible for one to be at rest after the collision? Explain.

3. Explain how linear momentum is conserved when a ball bounces from a floor.

4. Consider a perfectly inelastic collision between a car and a large truck. Which vehicle loses more kinetic energy as a result of the collision?

5. Your physical education teacher, who knows something about physics, throws you a tennis ball at a certain velocity, and you catch it. You are now given the following choice: The teacher can throw you a medicine ball, which is much more massive than the tennis ball, with the same velocity as the tennis ball, the same momentum, or the same kinetic energy. Which choice would you make in order to make the easiest catch, and why?

6. Can the center of mass of a body lie outside the body? If so, give examples.

7. Is the center of mass of a mountain higher or lower than the center of gravity?

8. In golf, novice players are often advised to be sure to "follow through" with their swing. Why does this make the ball travel a longer distance? What about a chip shot? There is very little follow-through here—why?

9. A sharpshooter fires a rifle while standing with the butt of the gun against his shoulder. If the forward momentum of a bullet is the same as the backward momentum of the gun, why isn't it as dangerous to be hit by the gun as by the bullet?

10. A pole-vaulter falls from a height of 6.0 m onto a foam rubber pad. Can you calculate his speed just before he reaches the pad? Could you calculate the force exerted on him due to the collision? Explain.

11. Does the center of mass of a rocket in free space accelerate? Explain. Can the speed of a rocket exceed the exhaust speed of the fuel? Explain.

12. An airbag is inflated when a collision occurs, which protects the passenger (the dummy, in this case) from serious injury (Fig. Q8.12). Why does the airbag soften the blow? Discuss the physics involved in this dramatic photograph.

13. A large bedsheet is held vertically by two students. A third student, who happens to be the star pitcher on the baseball team, throws a raw egg at the sheet. Explain why the egg does not break when it hits the sheet, regardless of its initial speed. (If you try this, make sure the pitcher hits the sheet near its center, and do not allow the egg to fall on the floor after being caught in the sheet.)

Figure Q8.12 *(Courtesy of Saab)*

PROBLEMS

Section 8.1 Linear Momentum and Its Conservation

1. A 3.00-kg particle has a velocity of $(3.00\mathbf{i} - 4.00\mathbf{j})$ m/s. Find its x and y components of momentum and the magnitude of its total momentum.

2. A 0.100-kg ball is thrown straight up into the air with an initial speed of 15.0 m/s. Find the momentum of the ball (a) at its maximum height and (b) halfway up to its maximum height.

3. A 40.0-kg child standing on a frozen pond throws a 0.500-kg stone to the east with a speed of 5.00 m/s. Neglecting friction between child and ice, find the recoil velocity of the child.

4. A pitcher claims she can throw a baseball with as much momentum as a 3.00-g bullet moving with a speed of 1500 m/s. A baseball has a mass of 0.145 kg. What must be its speed if the pitcher's claim is valid?

5. Two blocks of masses M and $3M$ are placed on a horizontal, frictionless surface. A light spring is attached to one of them, and the blocks are pushed together with the spring between them (Fig. P8.5). A cord holding them together is burned, after which the block of mass $3M$ moves to the right with a speed of 2.00 m/s. (a) What is the speed of the block of mass M? (b) Find the original elastic potential energy in the spring if $M = 0.350$ kg.

Section 8.2 Impulse and Momentum

6. A car is stopped for a traffic signal. When the light turns green, the car accelerates, increasing its speed from 0 to 5.20 m/s in 0.832 s. What linear impulse and average force does a 70.0-kg passenger in the car experience?

7. An estimated force–time curve for a baseball struck by a bat

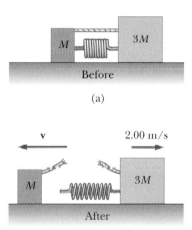

Figure P8.5

is shown in Figure P8.7. From this curve, determine (a) the impulse delivered to the ball, (b) the average force exerted on the ball, and (c) the peak force exerted on the ball.

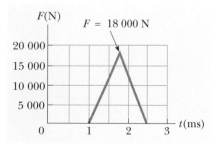

Figure P8.7

8. A tennis player receives a shot with the ball (0.0600 kg) traveling horizontally at 50.0 m/s and returns the shot with the ball traveling horizontally at 40.0 m/s in the opposite direction. (a) What is the impulse delivered to the ball by the racket? (b) What work does the racket do on the ball?

9. A 3.00-kg steel ball strikes a wall with a speed of 10.0 m/s at an angle of 60.0° with the surface. It bounces off with the same speed and angle (Fig. P8.9). If the ball is in contact with the wall for 0.200 s, what is the average force exerted on the ball by the wall?

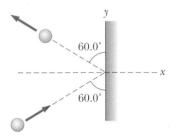

Figure P8.9

10. In a slow-pitch softball game, a 0.200-kg softball crossed the plate at 15.0 m/s at an angle of 45.0° below the horizontal. The ball was hit at 40.0 m/s, 30.0° above the horizontal. (a) Determine the impulse applied to the ball. (b) If the force on the ball increased linearly for 4.00 ms, held constant for 20.0 ms, then decreased to 0 linearly in another 4.00 ms, find the maximum force on the ball.

11. A machine gun fires 35.0 g bullets at a speed of 750.0 m/s. If the gun can fire 200 bullets/min, what is the average force the shooter must exert to keep the gun from moving?

Section 8.3   Collisions

Section 8.4   Elastic and Inelastic Collisions in One Dimension

12. High-speed stroboscopic photographs show that the head of a golf club of mass 200 g is traveling at 55.0 m/s just before it strikes a 46.0-g golf ball at rest on a tee. After the collision, the club head travels (in the same direction) at 40.0 m/s. Find the speed of the golf ball just after impact.

13. A 10.0-g bullet is stopped in a block of wood ($m = 5.00$ kg). The speed of the bullet-plus-wood combination immediately after the collision is 0.600 m/s. What was the original speed of the bullet?

14. A 75.0-kg ice skater, moving at 10.0 m/s, crashes into a stationary skater of equal mass. After the collision, the two skaters move as a unit at 5.00 m/s. The average force a skater can experience without breaking a bone is 4500 N. If the impact time is 0.100 s, does a bone break?

15. As shown in Figure P8.15, a bullet of mass m and speed v passes completely through a pendulum bob of mass M. The bullet emerges with a speed of $v/2$. The pendulum bob is suspended by a stiff rod of length ℓ and negligible mass. What is the minimum value of v such that the pendulum bob will barely swing through a complete vertical circle?

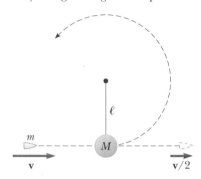

Figure P8.15

16. Gayle runs at a speed of 4.00 m/s and dives on a sled, initially at rest on the top of a frictionless snow-covered hill. After she has descended a vertical distance of 5.00 m, her brother, who is initially at rest, hops on her back and together they continue down the hill. What is their speed at the bottom of the hill if the total vertical drop is 15.0 m? Gayle's mass is 50.0 kg, the sled has a mass of 5.00 kg, and her brother has a mass of 30.0 kg.

17. A 1200-kg car traveling initially with a speed of 25.0 m/s in an easterly direction crashes into the rear end of a 9000-kg truck moving in the same direction at 20.0 m/s (Fig. P8.17). The velocity of the car right after the collision is 18.0 m/s to the east. (a) What is the velocity of the truck right after the collision? (b) How much mechanical energy is lost in the collision? Account for this loss in energy.

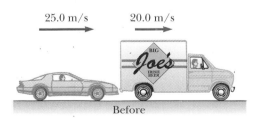

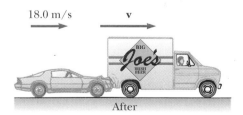

Figure P8.17

18. A railroad car of mass 2.50×10^4 kg, moving with a speed of 4.00 m/s, collides and couples with three other coupled railroad cars, each of the same mass as the single car and moving in the same direction with an initial speed of 2.00 m/s. (a) What is the speed of the four cars after the collision? (b) How much energy is lost in the collision?

19. A neutron in a reactor makes an elastic head-on collision with the nucleus of a carbon atom initially at rest. (a) What fraction of the neutron's kinetic energy is transferred to the carbon nucleus? (b) If the initial kinetic energy of the neutron is 1.60×10^{-13} J, find its final kinetic energy and the kinetic energy of the carbon nucleus after the collision. (The mass of the carbon nucleus is about 12.0 times the mass of the neutron.)

20. A 7.00-kg bowling ball collides head-on with a 2.00-kg bowling pin. The pin flies forward with a speed of 3.00 m/s. If the ball continues forward with a speed of 1.80 m/s, what was the initial speed of the ball? Ignore rotation of the ball.

21. A 12.0-g bullet is fired into a 100-g wooden block initially at rest on a horizontal surface. After impact, the block slides 7.50 m before coming to rest. If the coefficient of friction between block and surface is 0.650, what was the speed of the bullet immediately before impact?

22. A bullet of mass m_1 is fired into a wooden block of mass m_2 initially at rest on a horizontal surface. After impact, the block slides a distance d before coming to rest. If the coefficient of friction between block and surface is μ, what was the speed of the bullet immediately before impact?

23. Consider a frictionless track ABC as shown in Figure P8.23. A block of mass $m_1 = 5.00$ kg is released from A. It makes a head-on elastic collision with a block of mass $m_2 = 10.0$ kg at B, initially at rest. Calculate the maximum height to which m_1 rises after the collision.

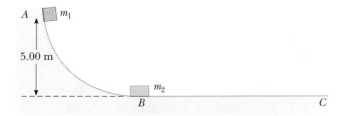

Figure P8.23

24. A 7.00-g bullet, when fired from a gun into a 1.00-kg block of wood held in a vise, penetrates the block to a depth of 8.00 cm. This block of wood is placed on a frictionless horizontal surface, and a 7.00-g bullet is fired from the gun into the block. To what depth will the bullet penetrate the block in this case?

Section 8.5   Two-Dimensional Collisions

25. A 90.0-kg fullback running east with a speed of 5.00 m/s is tackled by a 95.0-kg opponent running north with a speed of 3.00 m/s. If the collision is perfectly inelastic, (a) calculate the speed and direction of the players just after the tackle and (b) determine the energy lost as a result of the collision. Account for the missing energy.

26. The mass of the blue puck in Figure P8.26 is 20.0% greater than the mass of the green one. Before colliding, the pucks approach each other with equal and opposite momenta, and the green puck has an initial speed of 10.0 m/s. Find the speeds of the pucks after the collision if half the kinetic energy is lost during the collision.

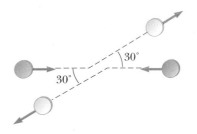

Figure P8.26

27. A billiard ball moving at 5.00 m/s strikes a stationary ball of the same mass. After the collision, the first ball moves at 4.33 m/s, at an angle of 30.0° with respect to the original line of motion. Assuming an elastic collision, find the struck ball's velocity after the collision.

28. A proton, moving with a velocity of $v_0\mathbf{i}$, collides elastically with another proton that is initially at rest. If the two protons have the same speed after the collision, find (a) the speed of each proton after the collision in terms of v_0 and (b) the direction of the velocity vectors after the collision.

29. A 0.300-kg puck, initially at rest on a horizontal, frictionless surface, is struck by a 0.200-kg puck moving initially along the x axis with a speed of 2.00 m/s. After the collision, the 0.200-kg puck has a speed of 1.00 m/s at an angle of $\theta = 53.0°$ to the positive x axis (Fig. 8.11). (a) Determine the velocity of the 0.300-kg puck after the collision. (b) Find the fraction of kinetic energy lost in the collision.

30. During the battle of Gettysburg, the gunfire was so intense that several bullets collided in mid-air and fused together. Assume a 5.00-g Union musket ball was moving to the right at a speed of 250 m/s, 20.0° above the horizontal, and that a 3.00-g Confederate ball was moving to the left at a speed of 280 m/s, 15.0° above the horizontal. Immediately after they fuse together, what is their velocity?

31. An unstable nucleus of mass 17.0×10^{-27} kg initially at rest disintegrates into three particles. One of the particles, of mass 5.00×10^{-27} kg, moves along the y axis with a velocity of 6.00×10^6 m/s. Another particle, of mass 8.40×10^{-27} kg, moves along the x axis with a speed of $4.00 \times$

10^6 m/s. Find (a) the velocity of the third particle and (b) the total energy given off in the process.

32. At an intersection, a 1500-kg car, traveling east with a speed of 20.0 m/s, collides with a 2500-kg van traveling south at a speed of 15.0 m/s. The vehicles undergo a perfectly inelastic collision, and the wreckage slides 6.00 m before coming to rest. Find the magnitude and direction of the constant force that decelerated them.

Section 8.6 The Center of Mass

33. Four objects are situated along the *y* axis as follows: a 2.00-kg object is at +3.00 m, a 3.00-kg object is at +2.50 m, a 2.50-kg object is at the origin, and a 4.00-kg object is at −0.500 m. Where is the center of mass of these objects?

34. A uniform carpenter's square has the shape of an L, as in Figure P8.34. Locate the center of mass relative to an origin at the lower left corner. (*Hint:* Note that the mass of each rectangular part is proportional to its area.)

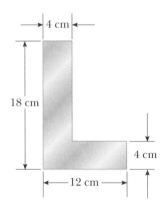

Figure P8.34

35. A uniform piece of sheet steel is shaped as in Figure P8.35. Compute the *x* and *y* coordinates of the center of mass of the piece.

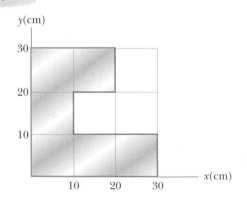

Figure P8.35

36. The mass of the Earth is 5.98×10^{24} kg, and the mass of the Moon is 7.36×10^{22} kg. The distance of separation, measured between their centers, is 3.84×10^8 m. Locate the center of mass of the Earth–Moon system as measured from the center of the Earth.

37. A water molecule consists of an oxygen atom with two hydrogen atoms bound to it (Fig. P8.37). The angle between the two bonds is 106°. If the bonds are 0.100 nm long, where is the center of mass of the molecule?

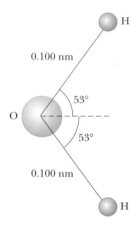

Figure P8.37

Section 8.7 Motion of a System of Particles

38. Consider a system of two particles in the *xy* plane: $m_1 = 2.00$ kg is at $\mathbf{r}_1 = (1.00\mathbf{i} + 2.00\mathbf{j})$ m and has velocity $(3.00\mathbf{i} + 0.500\mathbf{j})$ m/s; $m_2 = 3.00$ kg is at $\mathbf{r}_2 = (-4.00\mathbf{i} - 3.00\mathbf{j})$ m and has velocity $(3.00\,\mathbf{i} - 2.00\mathbf{j})$ m/s. (a) Plot these particles on a grid or graph paper. Draw their position vectors and show their velocities. (b) Find the position of the center of mass of the system and mark it on the grid. (c) Determine the velocity of the center of mass and also show it on the diagram. (d) What is the total linear momentum of the system?

39. Romeo (77.0 kg) entertains Juliet (55.0 kg) by playing his guitar from the rear of their boat in still water, 2.70 m away from Juliet in the front of the boat. After the serenade, Juliet carefully moves to the rear of the boat (away from shore) to plant a kiss on Romeo's cheek. How far does the 80.0-kg boat move toward the shore it is facing?

40. A ball of mass 0.200 kg has velocity $1.50\mathbf{i}$ m/s; a ball of mass 0.300 kg has velocity $-0.400\mathbf{i}$ m/s. Then they meet in a head-on elastic collision. (a) Find their velocities thereafter. (b) Find the velocity of their center of mass before the collision and afterward.

Section 8.8 Rocket Propulsion (Optional)

41. The first stage of a Saturn V space vehicle consumes fuel at the rate of 1.50×10^4 kg/s, with an exhaust speed of

2.60×10^3 m/s. (a) Calculate the thrust produced by these engines. (b) Find the initial acceleration of the vehicle on the launch pad if its initial mass is 3.00×10^6 kg. [*Hint:* You must include the force of gravity to solve part (b).]

42. A large rocket with an exhaust speed of $v_e = 3000$ m/s develops a thrust of 24.0 million newtons. (a) How much mass is being blasted out of the rocket exhaust per second? (b) What is the maximum speed the rocket can attain if it starts from rest in a force-free environment with $v_e = 3.00$ km/s and if 90.0% of its initial mass is fuel?

43. A rocket for use in deep space is to have the capability of accelerating a payload (including the rocket frame and engine) of 3.00 metric tons from rest to a final speed of 10 000 m/s with an engine and fuel designed to produce an exhaust speed of 2000 m/s. (a) How much fuel plus oxidizer is required? (b) If a different fuel and engine design could give an exhaust speed of 5000 m/s, what amount of fuel and oxidizer would be required for the same task?

Additional Problems

44. A golf ball ($m = 46.0$ g) is struck a blow that makes an angle of 45.0° with the horizontal. The drive lands 200 m away on a flat fairway. If the golf club and ball are in contact for 7.00 ms, what is the average force of impact? (Neglect air resistance.)

45. An 8.00-g bullet is fired into a 2.50-kg block initially at rest at the edge of a frictionless table of height 1.00 m (Fig. P8.45). The bullet remains in the block, and after impact the block lands 2.00 m from the bottom of the table. Determine the initial speed of the bullet.

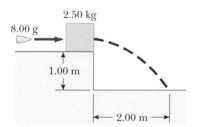

Figure P8.45

46. A spacecraft is stationary in deep space when its rocket engine is ignited for a 100-s "burn." Hot gases are ejected at a constant rate of 150 kg/s, with a speed of 3000 m/s relative to the spacecraft. The initial mass of the spacecraft (plus fuel and oxidizer) is 25 000 kg. Determine the thrust of the rocket engine and the initial acceleration in units of *g*. What is the final speed of the spacecraft?

47. A small block of mass $m_1 = 0.500$ kg is released from rest at the top of a curved-shaped frictionless wedge of mass $m_2 = 3.00$ kg, which sits on a frictionless horizontal surface as in Figure P8.47a. When the block leaves the wedge, its velocity is measured to be 4.00 m/s to the right, as in Figure

P8.47b. (a) What is the velocity of the wedge after the block reaches the horizontal surface? (b) What is the height, *h*, of the wedge?

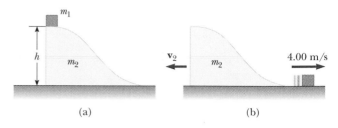

(a)   (b)

Figure P8.47

48. Tarzan, whose mass is 80.0 kg, swings from a 3.00-m vine that is horizontal when he starts. At the bottom of his arc, he picks up 60.0-kg Jane in a perfectly inelastic collision. What is the height of the highest tree limb they can reach on their upward swing?

49. A 80.0-kg astronaut is working on the engines of her ship, which is drifting through space with a constant velocity. The astronaut, wishing to get a better view of the Universe, pushes against the ship and much later finds herself 30.0 m behind the ship. Without a thruster, the only way to return to the ship is to throw her 0.500-kg wrench directly away from the ship. If she throws the wrench with a speed of 20.0 m/s relative to the ship, how long does it take the astronaut to reach the ship?

50. A 40.0-kg child stands at one end of a 70.0-kg boat that is 4.00 m in length (Fig. P8.50). The boat is initially 3.00 m from the pier. The child notices a turtle on a rock at the far end of the boat and proceeds to walk to that end to catch the turtle. Neglecting friction between boat and water, (a) describe the subsequent motion of the system (child plus boat). (b) Where is the child *relative to the pier* when he reaches the far end of the boat? (c) Will he catch the turtle? (Assume he can reach out 1.00 m from the end of the boat.)

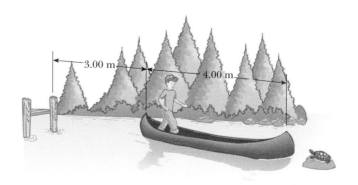

Figure P8.50

51. A 5.00-g bullet moving with an initial speed of 400 m/s is fired into and passes through a 1.00-kg block, as in Figure P8.51. The block, initially at rest on a frictionless, horizontal surface, is connected to a spring of force constant 900 N/m. If the block moves 5.00 cm to the right after impact, find (a) the speed at which the bullet emerges from the block and (b) the energy lost in the collision.

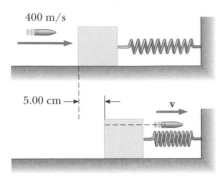

400 m/s

5.00 cm →

v

Figure P8.51

52. A student performs a ballistic pendulum experiment using the apparatus shown in Figure P8.52a. A ball of mass

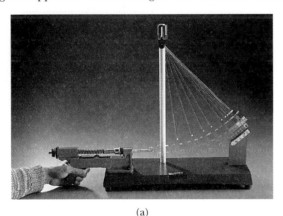

(a)

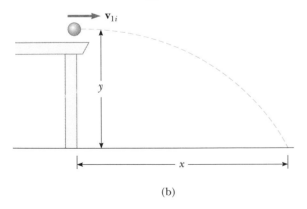

$\mathbf{v}_{1i}$

y

x

(b)

Figure P8.52 (a) and (b) *(Photo courtesy of CENCO)*

68.8 g is fired into a pendulum of mass 263 g hanging from a low friction pivot. The ball is caught inside the pendulum, which then swings up through a height 8.68 cm. (a) Determine the initial speed v_{1i} of the ball. (b) The second part of her experiment is to obtain v_{1i} by firing the same ball horizontally (with the pendulum removed from the path), by measuring its horizontal displacement, x, and vertical displacement, y (Fig. P8.52b). Show that the initial speed of the ball is related to x and y through the relation

$$v_{1i} = \frac{x}{\sqrt{2y/g}}$$

What numerical value does she obtain for v_{1i} based on her measured values of x = 257 cm and y = 85.3 cm? What factors might account for the difference in this value compared to that obtained in part (a)?

53. Two particles of masses m and 3m are moving toward each other along the x axis with the same initial speeds v_0. Mass m is traveling to the left, and mass 3m is traveling to the right. They undergo a head-on elastic collision. Find the final speeds of the particles.

54. Two particles of masses m and 3m are moving toward each other along the x axis with the same initial speeds v_0. Mass m is traveling to the left, and mass 3m is traveling to the right. They undergo an elastic glancing collision such that mass m is moving downward after the collision at right angles from its initial direction. (a) Find the final speeds of the two masses. (b) What is the angle θ at which the mass 3m is scattered?

55. How fast can you set the Earth moving? In particular, when you jump straight up as high as you can, you give the Earth a maximum recoil speed of what order of magnitude? Visualize the Earth as a perfectly rigid object. In your solution state the physical quantities you take as data and the values you measure or estimate for them.

56. There are (one can say) three co-equal theories of motion: Newton's second law, holding that the total force on an object causes its acceleration; the work-kinetic energy theorem, holding that the total work on an object causes its change in kinetic energy; and the impulse-momentum theorem, holding that the total impulse on an object causes its change in momentum. In this problem you compare the predictions of the three theories in one particular case. A 3.00-kg object has velocity $7.00\mathbf{j}$ m/s. Then a total force $12.0\mathbf{i}$ N acts on the object for 5.00 s. (a) Calculate its final velocity from the impulse-momentum theorem. (b) Calculate its acceleration from $\mathbf{a} = (\mathbf{v}_f - \mathbf{v}_0)/t$. (c) Calculate its acceleration from $\mathbf{a} = \Sigma\mathbf{F}/m$. (d) Find the object's vector displacement from $\mathbf{s} = \mathbf{v}_0 t + \frac{1}{2}\mathbf{a}t^2$. (e) Find the work done on the object from $W = \mathbf{F}\cdot\mathbf{s}$. (f) Find the final kinetic energy from $\frac{1}{2}mv_f^2 = \frac{1}{2}m\mathbf{v}_f\cdot\mathbf{v}_f$. (g) Find the final kinetic energy from $\frac{1}{2}mv_0^2 + W$.

Spreadsheet Problems

S1. A rocket has an initial mass of 20 000 kg, of which 20% is the payload. The rocket burns fuel at a rate of 200 kg/s and exhausts gas at a relative speed of 2.00 km/s. Its acceleration dv/dt is determined from the equation of motion

$$M\frac{dv}{dt} = v_e\left|\frac{dM}{dt}\right| + F_{ext}$$

Assume that there are no external forces and that the rocket's initial velocity is zero. When $F_{ext} = 0$, the rocket's velocity is $v(t) = v_e \ln (M_i/M)$, where M_i is the rocket's initial mass. Spreadsheet 8.1 calculates the acceleration and velocity as a function of time. (a) Using the given parameters, find the maximum acceleration and velocity. (b) At what time is the velocity equal to half its maximum value? Why isn't this time half of the burn time?

S2. Modify Spreadsheet 8.1 to calculate the distance traveled by the rocket in Problem S1. Add a new column to the spreadsheet to find the new position x_{i+1}. Estimate the new posi-

tion of the rocket by

$$x_{i+1} = x_i + \tfrac{1}{2}(v_{i+1} + v_i)\Delta t$$

where x_i is the old position, v_i is the old velocity, and v_{i+1} is the new velocity.

S3. If the rocket in the previous problem burns its fuel twice as fast (400 kg/s) while maintaining the same exhaust speed, how far does it travel before it runs out of fuel? The velocity reached by the rocket is the same in both cases; that is

$$v_f - v_i = v_e \ln\left(\frac{M_i}{M_f}\right)$$

Is there an advantage in burning the fuel faster? Is there a disadvantage? (*Hint:* It may be useful to plot the acceleration, velocities, and position for the same time intervals in both cases. In which case does the rocket travel farther? In which case does the rocket have to withstand the largest acceleration? By how much?)

ANSWERS TO CONCEPTUAL PROBLEMS

1. The brain is immersed in a cushioning fluid inside the skull. If the head is struck suddenly by a bare fist, there is a rapid acceleration of the skull. The brain matches this acceleration only because of the large impulsive force of the skull on the brain. This large and sudden force can severely injure the brain. Padded gloves extend the time over which the force is applied to the head. Thus, for a given impulse, the gloves provide a longer time interval than the bare fist, decreasing the average force. Because the average force is decreased, the acceleration of the skull is decreased, reducing (but not eliminating) the chance for brain injury. The same argument can be made for "rolling with the punch." If the head is held steady while being struck, the time interval over which the force is applied is relatively short and the average force is large. If the head is allowed to move in the same direction of the punch, the time interval is lengthened, and the average force reduced.

2. The scenario described violates the principle of conservation of momentum. If the superhero, initially at rest, throws the piano in one direction, he must recoil in the other direction, in order to maintain the zero momentum of the system.

3. Making cars out of hard rubber would save on automobile repair bills, but the results for the drivers and passengers would be disastrous. Imagine two identical cars traveling toward each other at identical speeds, and making a head-on collision. In the approximation that the collision is elastic, the cars bounce off each other and leave the collision with

their velocities reversed in direction. Thus, the inhabitants of the cars would suffer a large change in momentum in a short period of time. As a result, the force on the inhabitants would become very large, causing significant bodily injury. For real cars in the above situation, the cars would come to rest during the collision, as the front sections of the cars crumpled. Thus, the change in momentum is only half what it was in the elastic case, and the change occurs over the longer time period during which the cars crumpled. The resulting smaller force on the inhabitants will result in fewer injuries.

4. If all of the kinetic energy is transformed into other forms, then there is no motion of either of the objects after the collision. If nothing is moving, there is no momentum. Thus, if the final momentum of the system is zero, then the initial momentum of the system must have been zero. The required condition, then, is that the two objects are approaching each other so as to make a head-on collision, with equal, but oppositely directed, momenta.

5. Tie one end of the rope around the bear. Lay out the tape measure on the ice between the bear's original position and yours, as you hold the opposite end of the rope. Take off your spiked shoes and pull on the rope. Both you and the bear will slide over the ice until you meet. From the tape observe how far you have slid, x_y, and how far the bear has slid, x_b. The point where you meet the bear is the constant location of the center of mass of the system (bear plus you),

so you can determine the mass of the bear from $m_b x_b = m_y x_y$. (Unfortunately, if the bear now wakes up you cannot get back to your spiked shoes.)

6. Neglecting air resistance, the only external force on the projectile is the force of gravity. Thus, the projectile follows a parabolic path. If the projectile did not explode, it would continue to move along the parabolic path indicated by the broken line in Figure 8.19. Because the forces due to the explosion are internal, they do not affect the motion of the center of mass. Thus, after the explosion, and until the first fragment hits the ground, the center of mass of the fragments follows the same parabolic path the projectile would have followed if there had been no explosion.

9

Relativity

Most of our everyday experiences and observations have to do with objects that move at speeds much less than that of light. Newtonian mechanics and early ideas on space and time were formulated to describe the motion of such objects. This formalism is successful in describing a wide range of phenomena that occur at low speeds. It fails, however, when applied to particles the speeds of which approach that of light. The predictions of Newtonian theory can be tested experimentally at high speeds by accelerating electrons or other charged particles. For example, it is possible to accelerate an electron to a speed of $0.99c$ (where c is the speed of light) by using particle accelerators. According to Newtonian mechanics, if the energy transferred to an electron is increased by a factor of 4, the electron speed should jump to $1.98c$. However, experiments show that the speed of the electron—as well as the speeds of all other particles in the Universe—always remains less than the speed of light, regardless of the particle energy. Because it places no upper limit on speed, New-

◀ **Albert Einstein (1879–1955), one of the greatest physicists of all times, is best known for developing the theory of relativity. He is shown here in a playful mood riding a bicycle. The photograph was taken in 1933 in Santa Barbara, California.** *(From the California Institute of Technology archives)*

"Imagination is more important than knowledge."
Albert Einstein

227

tonian mechanics is contrary to modern experimental results and is clearly a limited theory.

In 1905, at the age of 26, Einstein published his special theory of relativity. Einstein wrote,

> The relativity theory arose from necessity, from serious and deep contradictions in the old theory from which there seemed no escape. The strength of the new theory lies in the consistency and simplicity with which it solves all these difficulties, using only a few very convincing assumptions. . . . [1]

Although Einstein made many important contributions to science, special relativity alone represents one of the greatest intellectual achievements of the twentieth century. With special relativity, experimental observations can be correctly predicted over the range of speeds from $v = 0$ to speeds approaching the speed of light. Newtonian mechanics, which was accepted for more than 200 years, is in fact a special case of special relativity. This chapter gives an introduction to special relativity, with emphasis on some of its consequences.

Special relativity covers phenomena such as the slowing down of clocks and the contraction of lengths in moving reference frames as measured by a laboratory observer. We also discuss the relativistic forms of momentum and energy, and some consequences of the famous mass–energy formula, $E = mc^2$.

In addition to its well known and essential role in theoretical physics, special relativity has practical applications, including the design of accelerators and other devices that use high-speed particles. These devices will not work if designed according to nonrelativistic principles. We shall have occasion to use special relativity in some subsequent chapters of this textbook, most often presenting only the outcome of relativistic effects.

9.1 · THE PRINCIPLE OF NEWTONIAN RELATIVITY

To describe a physical event, it is necessary to establish a frame of reference. You should recall from Chapter 5 that Newton's laws are valid in all inertial frames of reference. Because an inertial frame is defined as one in which Newton's first law is valid, it can be said that **an inertial system is one in which a free body experiences no acceleration.** Furthermore, any system moving with constant velocity with respect to an inertial system must also be an inertial system.

Inertial frame of reference •

There is no preferred frame. This means that the results of an experiment performed in a vehicle moving with uniform velocity will be identical to the results of the same experiment performed in a stationary vehicle. The formal statement of this result is called the principle of **Newtonian relativity:**

> The laws of mechanics must be the same in all inertial frames of reference.

Let us consider an observation that illustrates the equivalence of the laws of mechanics in different inertial frames. Consider a pickup truck moving with a constant velocity, as in Figure 9.1a. If a passenger in the truck throws a ball straight up in the air, the passenger observes that the ball moves in a vertical path. The motion

[1]A. Einstein and L. Infeld, *The Evolution of Physics*, New York, Simon and Schuster, 1961.

Figure 9.1   (a) The observer in the truck sees the ball move in a vertical path when thrown upward. (b) The Earth observer views the path of the ball to be a parabola.

of the ball appears to be precisely the same as if the ball were thrown by a person at rest on Earth. The law of gravity and the equations of kinematics are obeyed whether the truck is at rest or in uniform motion. Now consider the same experiment viewed by an observer at rest on Earth. This stationary observer sees the path of the ball as a parabola, as in Figure 9.1b. Furthermore, according to this observer, the ball has a horizontal component of velocity equal to the velocity of the truck. Although the two observers see different velocities and rates of energy change, they see the same forces and agree on the validity of Newton's laws as well as on principles like conservation of energy and conservation of momentum. All differences between the two views stem from the relative motion of one frame with respect to the other. That is, the notion of absolute motion through space plays no role in what is seen.

Suppose that some physical phenomenon that we call an *event* occurs in an inertial system. The event's location and time of occurrence can be specified by the coordinates (x, y, z, t). We would like to be able to transform these coordinates from one inertial system to another moving with uniform relative velocity. This is accomplished by using what is called a *Galilean transformation,* which owes its origin to Galileo.

Consider two inertial systems S and S' (Fig. 9.2). The system S' moves with a constant velocity $\mathbf{v}$ along the xx' axes, where $\mathbf{v}$ is measured relative to S. We assume that an event occurs at the point P and that the origins of S and S' coincide at $t = 0$. An observer in S describes the event with space–time coordinates (x, y, z, t), while an observer in S' uses (x', y', z', t') to describe the same event. As we can see from Figure 9.2, these coordinates are related by the equations

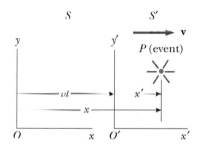

Figure 9.2   An event occurs at a point P. The event is observed by two observers in inertial frames S and S', where S' moves with a velocity $\mathbf{v}$ relative to S.

$$x' = x - vt$$

$$y' = y$$

$$z' = z$$

$$t' = t$$

[9.1]   • *Galilean transformation of coordinates*

These equations constitute what is known as a **Galilean transformation of coordinates.** Note that time is assumed to be the same in both inertial systems. That is, within the framework of classical mechanics, all clocks run at the same rate, regardless of their velocity, so that the time at which an event occurs for an observer in S is the same as the time for the same event in S'. As a consequence, the time interval between two successive events should be the same for both observers. Although this assumption may seem obvious, it turns out to be incorrect when treating situations in which v is comparable to the speed of light. This point of equal time intervals represents one of the most profound differences between Newtonian concepts and Einstein's special relativity.

Now suppose two events are separated by a distance dx and a time interval dt as measured by an observer in S. It follows from Equation 9.1 that the corresponding distance dx' measured by an observer in S' is $dx' = dx - vdt$, where dx is the distance between the two events measured by an observer in S. Because $dt = dt'$, we find that

$$\frac{dx'}{dt'} = \frac{dx}{dt} - v$$

or

Galilean addition law for • velocities

$$u'_x = u_x - v \qquad [9.2]$$

where u_x and u'_x are the instantaneous velocities of the object relative to S and S', respectively. This result, which is called the **Galilean addition law for velocities** (or Galilean velocity transformation), is used in everyday observations and is consistent with our intuitive notion of time and space. As we will soon see, however, it leads to serious contradictions when applied to electromagnetic waves.

9.2 • THE MICHELSON–MORLEY EXPERIMENT

Many experiments similar to the one described in the preceding section show us that the laws of mechanics are the same in all inertial frames of reference. When similar inquiries are made into the laws of other branches of physics, the results are contradictory. In particular, the laws of electricity and magnetism are found to depend on the frame of reference used. It might be argued that it is these laws that are wrong, but that is difficult to accept because the laws are in total agreement with all known experimental results. The Michelson–Morley experiment was an attempt to explain this dilemma.

The experiment stemmed from a misconception early physicists had concerning the manner in which light propagates. The properties of mechanical waves, such as water and sound waves, were well known, and all of these waves require a medium to support the disturbances. In the nineteenth century, physicists thought that electromagnetic waves also required a medium through which to propagate. They proposed that such a medium existed, and they named it the **luminiferous ether.** This ether was assumed to be present everywhere, even in a vacuum, and to have the unusual property of being a massless but rigid medium (a strange concept indeed!). In addition, it was found that the troublesome laws of electricity and magnetism took on their simplest forms in a frame of reference *at rest* with respect to the luminiferous ether; this frame was called the **absolute frame.** In any refer-

Albert A. Michelson (1852–1931)

A German American physicist, Michelson spent much of his life making accurate measurements of the speed of light. In 1907 he was the first American to be awarded the Nobel prize, which he received for his work in optics. His most famous experiment, conducted with Edward Morley in 1887, implied that it was impossible to measure the absolute velocity of the Earth with respect to the ether. *(AIP Emilio Segré Visual Archives, Michelson Collection)*

ence frame moving with respect to it, the laws of electricity and magnetism would remain valid but would have to be modified.

As a result of the importance attached to this absolute frame, experimental proof of its existence became of considerable interest in physics. However, all attempts to detect the presence of the ether (and hence the absolute frame) proved futile. The most famous experiment designed to show the presence of the ether was performed in 1887 by A. A. Michelson (1852–1931) and E. W. Morley (1838–1923). The objective was to determine the velocity of the Earth with respect to the ether, and the experimental tool used was a device called the interferometer, shown in Figure 9.3.

Suppose one arm of an interferometer ($M_0 M_2$ in Fig. 9.3) were aligned along the direction of the motion of the Earth through space. The Earth moving through the ether would be equivalent to the ether flowing past the Earth in the opposite direction. This "ether wind" blowing in the direction opposite the Earth's motion should cause the speed of light, as measured in the Earth's frame of reference, to be $c - v$ as the light approaches mirror M_2 in Figure 9.3, and $c + v$ after reflection, where c is the speed of light in the ether frame and v is the speed of the Earth through space and hence the speed of the ether wind. The incident and reflected beams of light would recombine, and a pattern would form as the two beams interfere with each other.

During the experiment, the interference pattern was observed while the interferometer was rotated through an angle of 90°. The idea was that this rotation would change the speed of the ether wind along the direction of the arms of the interferometer, and as a consequence the interference pattern would shift slightly but measurably. Measurements failed to show any change in the pattern. The Michelson–Morley experiment was repeated by other researchers under varying conditions and at different locations, but the results were always the same: *no interference-pattern shift of the magnitude required was ever observed.*[2]

The negative result of the Michelson–Morley experiment not only contradicted the ether hypothesis; it also meant that it was impossible to measure the absolute (orbital) velocity of the Earth with respect to the ether frame. From a theoretical viewpoint, this meant that it was impossible to find the absolute frame. However, as we shall see in the next section, Einstein offered a postulate that places a different interpretation on the negative result. In later years, when more was known about the nature of light, the idea of an ether that permeates all space was relegated to the ash heap of worn-out concepts. **Light is now understood to be an electromagnetic wave that requires no medium for its propagation.** As a result, an ether in which light could travel became an unnecessary construct.

Modern versions of the Michelson–Morley experiment have compared the frequencies of resonant laser cavities of identical length, oriented at right angles to each other. Most recently, Doppler shift experiments using gamma rays emitted by a radioactive sample of ^{57}Fe have placed an upper limit of about 5 cm/s on ether wind velocity. These results have shown quite conclusively that the motion of the Earth has no effect on the speed of light!

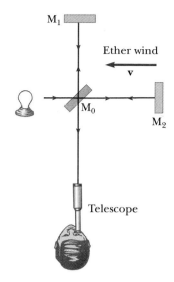

Figure 9.3 According to the ether wind theory, the speed of light should be $c - v$ as the beam approaches mirror M_2 and $c + v$ after reflection.

---

[2]From an Earth-observer's point of view, changes in the Earth's speed and direction of motion in the course of a year are viewed as ether wind shifts. Even if the speed of the Earth with respect to the ether were zero at some time, 6 months later the speed of the Earth would be 60 km/s with respect to the ether and a clear fringe shift should be found. None has ever been observed, however.

9.3 • EINSTEIN'S PRINCIPLE OF RELATIVITY

In the previous section we noted the impossibility of measuring the speed of the ether with respect to the Earth and the failure of the Galilean addition law for velocities in the case of light. Albert Einstein proposed a theory that boldly removed these difficulties and at the same time completely altered our notion of space and time.[3] Einstein based special relativity on two postulates:

The postulates of special relativity •

> 1. **The Principle of Relativity:** All the laws of physics are the same in all inertial reference frames.
> 2. **The Constancy of the Speed of Light:** The speed of light in vacuum has the same value, $c = 3.00 \times 10^8$ m/s, in all inertial frames, regardless of the velocity of the observer or the velocity of the source emitting the light.

The first postulate asserts that *all* the laws of physics, those dealing with mechanics, electricity and magnetism, optics, thermodynamics, and so on, are the same in all reference frames moving with constant velocity relative to each other. This postulate is a sweeping generalization of the principle of Newtonian relativity that only refers to the laws of mechanics. From an experimental point of view, Einstein's principle of relativity means that any kind of experiment (measuring the speed of light, for example), performed in a laboratory at rest must give the same result when performed in a laboratory moving at constant velocity past the first one. Hence no preferred inertial reference frame exists and it is impossible to detect absolute motion.

Note that postulate 2, the principle of the constancy of the speed of light, is required by postulate 1: If the speed of light were not the same in all inertial frames, it would be possible to distinguish between inertial frames and a preferred, absolute frame could be identified in contradiction to postulate 1. Postulate 2 also eliminates the problem of measuring the speed of the ether by denying the existence of the ether and boldly asserting that light always moves with speed c relative to all inertial observers.

Although the Michelson–Morley experiment was performed before Einstein published his work on relativity, it is not clear whether or not Einstein was aware of the details of the experiment. Nonetheless, the null result of the experiment can be readily understood within the framework of Einstein's theory. According to his principle of relativity, the premises of the Michelson–Morley experiment were incorrect. In the process of trying to explain the expected results, we stated that when light traveled against the ether wind its speed was $c - v$, in accordance with the Galilean addition law for velocities. However, if the state of motion of the observer or of the source has no influence on the value found for the speed of light, one will always measure the value to be c. Likewise, the light makes the return trip after reflection from the mirror at speed c, not at speed $c + v$. Thus, the motion of the Earth does not influence the interference pattern observed in the Michelson–Morley experiment, and a null result should be expected.

[3]A. Einstein, "On the Electrodynamics of Moving Bodies," *Ann. Physik* 17:891, 1905. For an English translation of this article and other publications by Einstein, see the book by H. Lorentz, A. Einstein, H. Minkowski, and H. Weyl, *The Principle of Relativity*, New York, Dover, 1958.

If we accept special relativity, we must conclude that relative motion is unimportant when measuring the speed of light. At the same time, we must alter our common-sense notion of space and time and be prepared for some bizarre consequences.

Thinking Physics 1

Imagine a very powerful lighthouse, with a rotating beacon. Imagine also drawing a horizontal circle around the lighthouse, with the lighthouse at the center. Along the circumference of the circle, the light beam lights up a portion of the circle and the lit portion of the circle moves around the circle at a certain tangential speed. If we now imagine a circle twice as big in radius, the tangential speed of the lit portion is faster, because it must travel a larger circumference in the time of one rotation of the light source. Imagine that we continue to make the circle larger and larger, eventually moving it out into space. The tangential speed of the lit portion will keep increasing. Is it possible that the tangential speed could become larger than the speed of light? Would this violate a principle of special relativity?

Reasoning   For a large enough circle, it is possible that the tangential speed of the lit portion of the circle could be larger than the speed of light. This does not violate a principle of special relativity, however, because no matter or information is traveling faster than the speed of light.

CONCEPTUAL PROBLEM 1

You are in a speedboat on a lake. You see ahead of you a wavefront, caused by the previous passage of another boat, moving away from you. You accelerate, catch up with, and pass the wavefront. Is this scenario possible if you are in a rocket and you detect a wavefront of *light* ahead of you?

9.4 • CONSEQUENCES OF SPECIAL RELATIVITY

Before we discuss the consequences of special relativity, we must first understand how an observer located in an inertial reference frame describes an event. As mentioned earlier, an event is an occurrence describable by three space coordinates and one time coordinate. Different observers in different inertial frames usually describe the same event with different space–time coordinates.

The reference frame used to describe an event consists of a coordinate grid and a set of synchronized clocks located at the grid intersections as shown in Figure 9.4 in two dimensions. The clocks can be synchronized in many ways with the help of light signals. For example, suppose the observer is located at the origin with his master clock and sends out a pulse of light at $t = 0$. The light pulse takes a time r/c to reach a second clock located a distance r from the origin. Hence, the second clock is synchronized with the clock at the origin if the second clock reads a time r/c at the instant the pulse reaches it. This procedure of synchronization assumes that the speed of light has the same value in all directions and in all inertial frames. Furthermore, the procedure concerns an event recorded by an observer in a specific inertial reference frame. An observer in some other inertial frame would assign different space–time coordinates to events being observed by using another coordinate grid and another array of clocks.

Albert Einstein (1879–1955)

One of the greatest physicists of all times, Einstein was born in Ulm, Germany. In 1905, at the age of 26, he published four scientific papers that revolutionized physics. Two of these papers were concerned with what is now considered his most important contribution of all, the special theory of relativity. In 1916, Einstein published his work on the general theory of relativity. The most dramatic prediction of this theory is the degree to which light is deflected by a gravitational field. Measurements made by astronomers on bright stars in the vicinity of the eclipsed Sun in 1919 confirmed Einstein's prediction, and Einstein suddenly became a world celebrity. Einstein was deeply disturbed by the development of quantum mechanics in the 1920s despite his own role as a scientific revolutionary. In particular, he could never accept the probabilistic view of events in nature that is a central feature of quantum theory. The last few decades of his life were devoted to an unsuccessful search for a unified theory that would combine gravitation and electromagnetism into one picture.

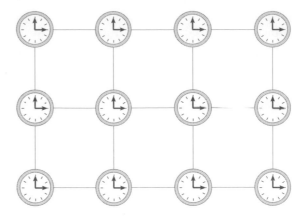

Figure 9.4   In relativity, we use a reference frame consisting of a coordinate grid and a set of synchronized clocks.

Almost everyone who has dabbled even superficially with science is aware of some of the startling predictions that arise because of Einstein's approach to relative motion. As we examine some of the consequences of relativity in the following three sections, we see that they conflict with our basic notions of space and time. We restrict our discussion to the concepts of length, time, and simultaneity, which are quite different in relativistic mechanics than in Newtonian mechanics. For example, **the distance between two points and the time interval between two events depend on the frame of reference in which they are measured.** That is, **there is no such thing as absolute length or absolute time in relativity.** Furthermore, **events at different locations that occur simultaneously in one frame are not simultaneous in another frame moving uniformly past the first.**

Simultaneity and the Relativity of Time

A basic premise of Newtonian mechanics is that a universal time scale exists that is the same for all observers. In fact, Newton wrote, "Absolute, true, and mathematical time, of itself, and from its own nature, flows equably without relation to anything external." Thus, Newton and his followers simply took simultaneity for granted. In his development of special relativity, Einstein abandoned this assumption. According to Einstein, **a time interval measurement depends on the reference frame in which the measurement is made.**

Einstein devised the following thought experiment to illustrate this point. A boxcar moves with uniform velocity, and two lightning bolts strike its ends, as in Figure 9.5a, leaving marks on the boxcar and on the ground. The marks on the boxcar are labeled A' and B', and those on the ground are labeled A and B. An observer at O' moving with the boxcar is midway between A' and B', and a ground observer at O is midway between A and B. The events recorded by the observers are the light signals from the lightning bolts.

The two light signals reach observer O at the same time, as indicated in Figure 9.5b. This observer realizes that the light signals have traveled at the same speed over equal distances, and so rightly concludes that the events at A and B occurred simultaneously. Now consider the same events as viewed by the observer on the boxcar at O'. The lightning strikes as A' passes A, O' passes O, and B' passes B. By

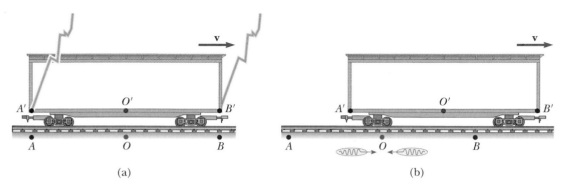

Figure 9.5 Two lightning bolts strike the end of a moving boxcar. (a) The events appear to be simultaneous to the stationary observer at O, who is midway between A and B. (b) The events do not appear to be simultaneous to the observer at O', who claims that the front of the train is struck before the rear.

the time the light has reached observer O, observer O' has moved as indicated in Figure 9.5b. Thus, the light signal from B' has already swept past O', and the light from A' has not yet reached O'. According to Einstein, **observer O' must find that light travels at the same speed as that measured by observer O.** Therefore, the observer O' concludes that the lightning struck the front of the boxcar before it struck the back. This thought experiment clearly demonstrates that the two events, which appear to be simultaneous to observer O, do not appear to be simultaneous to observer O'. In other words,

> two events that are simultaneous in one reference frame are in general not simultaneous in a second frame moving relative to the first. That is, simultaneity is not an absolute concept but one that depends on the state of motion of the observer.

At this point, you might wonder which observer is right concerning the two events. The answer is that *both are correct,* because the principle of relativity states that **there is no preferred inertial frame of reference.** Although the two observers reach different conclusions, both are correct in their own reference frame because the concept of simultaneity is not absolute. This, in fact, is the central point of relativity—any uniformly moving frame of reference can be used to describe events and do physics. However, observers in different inertial frames of reference always measure different time intervals with their clocks and different distances with their meter sticks. Nevertheless, all observers agree on the forms of the laws of physics in their respective frames, because these laws must be the same for all observers in uniform motion. It is the alteration of time and space that allows the laws of physics (including Maxwell's equations) to be the same for all observers in uniform motion.

Time Dilation

The fact that observers in different inertial frames always measure different time intervals between a pair of events can be illustrated by considering a vehicle moving to the right with a speed v, as in Figure 9.6a. A mirror is fixed to the ceiling of the

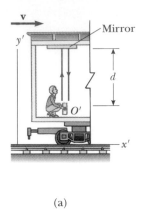

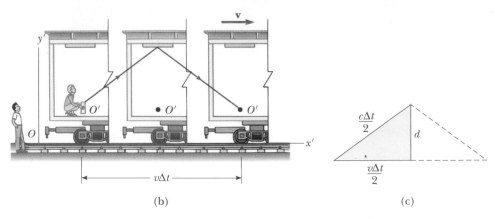

(a)   (b)   (c)

Figure 9.6   (a) A mirror is fixed to a moving vehicle, and a light pulse leaves O' at rest in the vehicle. (b) Relative to a stationary observer on Earth, the mirror and O' move with a speed v. Note that the distance the pulse travels is greater than $2d$ as measured by the stationary observer. (c) The right triangle for calculating the relationship between Δt and $\Delta t'$.

vehicle and observer O' at rest in this system holds a laser a distance d below the mirror. At some instant, the laser emits a pulse of light directed towards the mirror (event 1), and at some later time after reflecting from the mirror, the pulse arrives back at the laser (event 2). Observer O' carries a clock C' that she uses to measure the time interval Δt_p between these two events. Because the light pulse has a speed c, the time it takes the pulse to travel from O' to the mirror and back to O' can be found from the definition of speed:

$$\Delta t_p = \frac{\text{distance traveled}}{\text{speed}} = \frac{2d}{c} \qquad \textbf{[9.3]}$$

This time interval Δt_p measured by O', who is at rest in the moving vehicle, requires only a single clock C' located at the same place in this frame.

Now consider the same pair of events as viewed by observer O in a second frame, as in Figure 9.6b. According to this observer, the mirror and laser are moving to the right with a speed v. The sequence of events appears entirely different as viewed by this observer. By the time the light from the laser reaches the mirror, the mirror has moved a distance $v \, \Delta t/2$, where Δt is the time it takes the light to travel from O' to the mirror and back to O' as measured by observer O. In other words, the second observer concludes that, because of the motion of the vehicle, if the light is to hit the mirror, it must leave the laser at an angle with respect to the vertical direction. Comparing Figures 9.6a and 9.6b, we see that the light must travel farther in the second frame than in the first frame.

According to the second postulate of special relativity, both observers must measure c for the speed of light. Because the light travels farther in the second frame, it follows that the time interval Δt measured by the observer in the second frame is longer than the time interval Δt_p measured by the observer in the first frame. To obtain a relationship between these two time intervals, it is convenient to use the right triangle shown in Figure 9.6c. The Pythagorean theorem gives

$$\left(\frac{c \, \Delta t}{2}\right)^2 = \left(\frac{v \, \Delta t}{2}\right)^2 + d^2$$

Solving for Δt gives

$$\Delta t = \frac{2d}{\sqrt{c^2 - v^2}} = \frac{2d}{c\sqrt{1 - \dfrac{v^2}{c^2}}}$$ **[9.4]**

Because $\Delta t_p = 2d/c$, we can express Equation 9.4 as

$$\Delta t = \frac{\Delta t_p}{\sqrt{1 - \dfrac{v^2}{c^2}}} = \gamma\,\Delta t_p$$ **[9.5]**   • *Time dilation*

where $\gamma = (1 - v^2/c^2)^{-1/2}$. This result says that **the time interval Δt measured by an observer moving with respect to the clock is longer than the time interval Δt_p measured by an observer at rest with respect to the clock** because γ is always greater than unity. That is, $\Delta t > \Delta t_p$. This effect is known as **time dilation.**

The time interval Δt_p in Equation 9.5 is called the **proper time.** In general, proper time is defined as **the time interval between two events measured by an observer who sees the events occur at the same point in space.** In our case, observer O' measures the proper time. That is, **proper time is always the time interval measured with a single clock at rest in the frame in which the events take place at the same position.**

• *Proper time*

Because the time between ticks of a moving clock, $\gamma(2d/c)$, is observed to be longer than the time between ticks of an identical clock at rest, $2d/c$, it is often said, "*A moving clock runs slower than a clock at rest by a factor γ.*" This is true for ordinary mechanical clocks as well as for the light clock just described. In fact, we can generalize these results by stating that **all physical processes, including chemical and biological ones, slow down relative to a stationary clock when those processes occur in a moving frame.** For example, the heartbeat of an astronaut moving through space would keep time with a clock inside the spaceship. Both the astronaut's clock and heartbeat are slowed down relative to a stationary clock. The astronaut would not have any sensation of life slowing down in the spaceship. For the astronaut, it is the clock on Earth and the companions at Mission Control that are moving and therefore keep a slow time.

Time dilation is a verifiable phenomenon. For example, muons are unstable elementary particles that have a charge equal to that of the electron and a mass 207 times that of the electron. Muons can be produced by the collision of cosmic radiation with atoms high in the atmosphere. Slow-moving muons have a lifetime of only 2.2 μs in the laboratory. If we take 2.2 μs as the average lifetime of a muon and assume that its speed is close to the speed of light, we find that these particles can travel a distance of only approximately 600 m before they decay (Fig. 9.7a). Hence, they cannot reach the Earth from the upper atmosphere where they are produced. However, experiments show that a large number of muons *do* reach the Earth. The phenomenon of time dilation explains this effect. Relative to an observer on Earth, the muons have a lifetime equal to $\gamma\tau$, where $\tau = 2.2$ μs is the lifetime in a frame of reference traveling with the muons. For example, for $v = 0.99c$, $\gamma \approx 7.1$ and $\gamma\tau \approx 16$ μs. Hence, the average distance traveled as measured by an observer on Earth is $\gamma v\tau \approx 4800$ m, as indicated in Figure 9.7b.

In 1976, at the laboratory of the European Council for Nuclear Research in

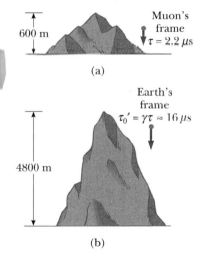

Figure 9.7   (a) Muons traveling with a speed of 0.99c travel only approximately 600 m as measured in the muons' reference frame, where their proper lifetime is about 2.2 μs. (b) The muons travel approximately 4800 m as measured by an observer on Earth. Because of time dilation, the muons' lifetime is longer as measured by the Earth observer.

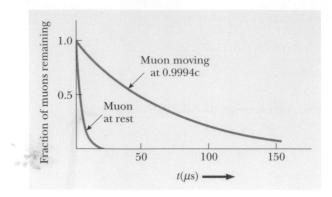

Figure 9.8   Decay curves for muons traveling at a speed of $0.9994c$ and for muons at rest.

Geneva, muons injected into a large storage ring reached speeds of approximately $0.9994c$. Electrons produced by the decaying muons were detected by counters around the ring, enabling scientists to measure the decay rate and hence the muon lifetime. The lifetime of the moving muons was measured to be approximately 30 times as long as that of the stationary muon (Fig. 9.8), in agreement with the prediction of relativity to within 2 parts in 1000.

The results of an experiment reported by Hafele and Keating provided direct evidence of time dilation.[4] The experiment involved the use of very stable cesium-beam atomic clocks. Time intervals measured with four such clocks in jet flight were compared with time intervals measured by reference clocks located at the U.S. Naval Observatory. To compare these results with the theory, many factors had to be considered, including periods of acceleration and deceleration relative to the Earth, variations in direction of travel, and the weaker gravitational field experienced by the flying clocks compared to the Earth-based clock. Their results were in good agreement with the predictions of special relativity and can be explained in terms of the relative motion between the Earth's rotation and the jet aircraft. In their paper, Hafele and Keating report the following: "Relative to the atomic time scale of the U.S. Naval Observatory, the flying clocks lost 59 ± 10 ns during the eastward trip and gained 273 ± 7 ns during the westward trip. . . . These results provide an unambiguous empirical resolution of the famous clock paradox with macroscopic clocks."

Thinking Physics 2

Suppose a student explains time dilation with the following argument: If you start running at $0.99c$ away from a clock at 12:00, you would not see the time change, because the light from the clock representing 12:01 would never reach you. What is the flaw in this argument?

Reasoning   The inference in this argument is that the velocity of light relative to the runner is approximately *zero*—"the light . . . would never reach you." This is a Galilean point of view, in which the relative velocity is a simple subtraction of running

[4]J. C. Hafele and R. E. Keating, "Around the World Atomic Clocks: Relativistic Time Gains Observed," *Science,* July 14, 1972, p. 168.

velocity from the light velocity. From the point of view of special relativity, one of the fundamental postulates is that the speed of light is the same for all observers, *including one running away from the light source at the speed of light.* Thus, the light from 12:01 will move toward the runner at the speed of light.

Example 9.1   What Is the Period of the Pendulum?

The period of a pendulum is measured to be 3.0 s in the rest frame of the pendulum's pivot. What is the period when measured by an observer moving at a speed of 0.95c relative to the pendulum?

Reasoning   Instead of the observer moving at 0.95c, we can take the equivalent point of view that the observer is at rest and that the pendulum is moving at 0.95c past the stationary observer. Hence the pendulum is an example of a moving clock.

Solution   The proper time is 3.0 s. Because a moving clock runs slower than a stationary clock by γ, Equation 9.5 gives

$$T = \gamma T_p = \frac{1}{\sqrt{1 - \dfrac{(0.95c)^2}{c^2}}} T_p = (3.2)(3.0\ \text{s}) = 9.6\ \text{s}$$

That is, a moving pendulum slows down or takes longer to complete a period compared to one at rest.

EXERCISE 1   Muons move in circular orbits at a speed of 0.9994c in a storage ring of radius 500 m. If a muon at rest decays into other particles after 2.2 μs (proper time), how many trips around the storage ring do we expect the muons to make before they decay?

Answer   6.1 revolutions

Length Contraction

The measured distance between two points also depends on the frame of reference. The **proper length** of an object is defined as **the length of the object measured by someone who is at rest relative to the object.** The length of an object measured by someone in a reference frame that is moving with respect to the object is always less than the proper length. This effect is known as **length contraction.**

• *Proper length*

Consider a spaceship traveling with a speed v from one star to another. There are two observers, one on Earth and the other in the spaceship. The observer at rest on Earth (and also assumed to be at rest with respect to the two stars) measures the distance between the stars to be L_p, the proper length. According to this observer, the time it takes the spaceship to complete the voyage is $\Delta t = L_p/v$. What does an observer in the moving spaceship measure for the distance between the stars? Because of time dilation, the space traveler measures a smaller time of travel: $\Delta t_p = \Delta t/\gamma$. The space traveler claims to be at rest and sees the destination star moving toward the spaceship with speed v. Because the space traveler reaches the star in the time Δt_p, she or he concludes that the distance, L, between the stars is shorter than L_p. This distance measured by the space traveler is

$$L = v\,\Delta t_p = v\,\frac{\Delta t}{\gamma}$$

Because $L_p = v\,\Delta t$, we see that $L = L_p/\gamma$ or

$$L = L_p\left(1 - \frac{v^2}{c^2}\right)^{1/2}$$ [9.6]   • *Length contraction*

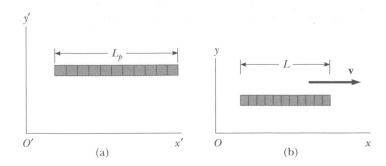

Figure 9.9 (a) A stick as viewed by an observer in a frame attached to the stick (i.e., both have the same velocity). (b) The stick as seen by an observer in a frame in which the stick has a velocity **v** relative to the frame. The length is *shorter* than the proper length, L_p, by a factor $(1 - v^2/c^2)^{1/2}$.

where $(1 - v^2/c^2)^{1/2}$ is a factor less than one. This result may be interpreted as follows:

> A moving observer measures the length L of an object (along the direction of motion) to be shorter than the length L_p measured by an observer at rest with respect to the object (the proper length).

Note that **length contraction takes place only along the direction of motion.** For example, suppose a stick moves past an Earth observer with speed v as in Figure 9.9. The length of the stick as measured by an observer in a frame attached to the stick is the proper length, L_p, as in Figure 9.9a. The length of the stick, L, measured by the Earth observer is shorter than L_p by the factor $(1 - v^2/c^2)^{1/2}$. Furthermore, length contraction is a symmetrical effect: If the stick is at rest on Earth, an observer in the moving frame would also measure its length to be shorter by the same factor $(1 - v^2/c^2)^{1/2}$.

It is important to emphasize that proper length and proper time are measured in different reference frames. As an example of this point, let us return to the decaying muons moving at speeds close to the speed of light. An observer in the muon's reference frame would measure the proper lifetime, whereas an Earth-based observer would measure the proper height of the mountain in Figure 9.7. In the muon's reference frame, there is no time dilation, but the distance of travel is observed to be shorter when measured in this frame. Likewise, in the Earth-based observer's reference frame, there is time dilation, but the distance of travel is measured to be the proper height of the mountain. Thus, when calculations on the muon are performed in both frames, the outcome of the experiment in one frame is the same as the outcome in the other frame!

Example 9.2   A Voyage to Sirius

An astronaut takes a trip to Sirius, located 8 lightyears from Earth. The astronaut measures the time of the one-way journey to be 6 y. If the spaceship moved at a constant speed of $0.8c$, how can the 8-lightyear distance be reconciled with the 6-y duration measured by the astronaut?

Reasoning The 8 lightyears (ly) represents the proper length (distance) from Earth to Sirius measured by an observer seeing both nearly at rest. The astronaut sees Sirius approaching her at $0.8c$ but also sees the distance contracted to

$$\frac{8 \text{ ly}}{\gamma} = (8 \text{ ly})\sqrt{1 - \frac{v^2}{c^2}} = (8 \text{ ly})\sqrt{1 - \frac{(0.8c)^2}{c^2}} = 5 \text{ ly}$$

So the travel time measured on the astronaut's clock is

$$\Delta t = \frac{d}{v} = \frac{5 \text{ ly}}{0.8c} = 6 \text{ y}$$

Example 9.3   The Triangular Spaceship

A spaceship in the form of a triangle flies by an observer with a speed of $0.95c$ along the x-direction. When the ship is at rest relative to the observer (Fig. 9.10a), the distances x and y are measured to be 52 m and 25 m, respectively. What is the shape of the ship as seen by an observer at rest when the ship is in motion along the direction shown in Figure 9.10b?

Solution   The observer sees the horizontal length of the ship to be contracted to a length

$$L = L_p \sqrt{1 - \frac{v^2}{c^2}} = (52 \text{ m}) \sqrt{1 - \frac{(0.95c)^2}{c^2}} = 16 \text{ m}$$

The 25-m vertical height is unchanged because it is perpendicular to the direction of relative motion between observer and spaceship. Figure 9.10b represents the shape of the spaceship as seen by the observer at rest.

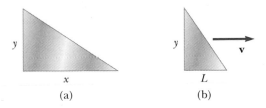

(a)   (b)

Figure 9.10   (Example 9.3) (a) When the spaceship is at rest, its shape is as shown. (b) The spaceship appears to look like this when it moves to the right with a speed v. Note that only its x dimension is contracted in this case.

EXERCISE 2   A spacecraft moves at $0.9c$. If its length is L_0 when measured from inside the spacecraft, what is its length measured by a ground observer?   Answer $0.436L_0$

EXERCISE 3   A spaceship is measured to be 120 m long while at rest relative to an observer. If this spaceship now flies by the observer with a speed $0.99c$, what length does the observer measure?   Answer   17 m

The Twins Paradox

O P T I O N A L

An intriguing consequence of time dilation is the so-called twins paradox. Consider an experiment involving a set of twins named Speedo and Goslo who are, say, 20 years old. The twins carry with them identical clocks that have been synchronized (Fig. 9.11). Speedo, the more adventuresome of the two, sets out on an epic journey

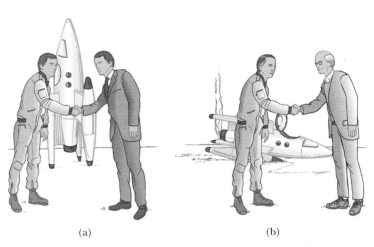

(a)   (b)

Figure 9.11   (a) As the twins depart, they are the same age. (b) When Speedo returns from his journey to Planet X, he is younger than his twin Goslo who remained on Earth.

to Planet X, located 10 lightyears from Earth. Furthermore, his spaceship is capable of reaching a speed of $0.500c$ relative to the inertial frame of his twin brother on Earth. After reaching Planet X, Speedo becomes homesick and immediately returns to Earth at the same high speed he had attained on the outbound journey. On his return, Speedo is shocked to discover that many things have changed in his absence. To Speedo, the most significant change to have occurred is that his twin brother Goslo has aged 40 years and is now 60 years old. Speedo, on the other hand, has aged by only 34.6 years.

At this point, it is fair to raise the following question—which twin is the traveler and which twin is really younger as a result of this experiment? From Goslo's frame of reference, he was at rest while his brother traveled at a high speed. From Speedo's perspective, it is he who is at rest while Goslo is on the high-speed space journey. According to Speedo, it is Goslo and the Earth that have raced away on a 17.3-year journey and then headed back for another 17.3 years. This leads to an apparent contradiction. Which twin has developed signs of excess aging?

To resolve this apparent paradox, recall that relative to any chosen inertial frame (like that of the stay-at-home twin), clocks in a moving object's frame move more slowly. However, the trip situation is not symmetrical. Speedo, the space traveler, must experience a series of accelerations during his journey. As a result, his speed is not always uniform and as a consequence Speedo is not in a single inertial frame. He cannot be regarded as always being at rest and Goslo to be in uniform motion because he changes his state of motion at least once. Therefore there is no paradox.

The conclusion that Speedo is in a noninertial frame is inescapable. The time required to accelerate and decelerate Speedo's spaceship may be made very small using large rockets, so that Speedo can claim that he spends most of his time traveling to Planet X at $0.500c$ in an inertial frame. However, Speedo must slow down, reverse his motion, and return to Earth in an altogether different inertial frame. At the very best, Speedo is in two different inertial frames during his journey. Only Goslo, who is in a single inertial frame, can apply the simple time dilation formula to Speedo's trip. Thus, Goslo finds that instead of aging 40 years, Speedo ages only $(1 - v^2/c^2)^{1/2}(40 \text{ years}) = 34.6$ years. However, Speedo spends 17.3 years traveling to Planet X and 17.3 years returning, for a total travel time of 34.6 years, in agreement with our earlier statement.

CONCEPTUAL PROBLEM 2

You are packing for a trip to another star, to which you will be traveling at $0.99c$. Should you buy smaller sizes of your clothing, since you will be skinnier on your trip? Can you sleep in a smaller cabin than usual, since you will be shorter when you lie down?

CONCEPTUAL PROBLEM 3

Suppose astronauts were paid according to the time spent traveling in space. After a long voyage traveling at a speed near that of light, a crew of astronauts return to Earth and open their pay envelopes. What will their reaction be?

CONCEPTUAL PROBLEM 4

You are observing a rocket moving away from you. You notice that it is measured to be shorter than when it was at rest on the ground next to you, and, through the rocket window, you can see a clock. You observe that the passage of time on the clock is measured to be slower than that of the watch on your wrist. What if the rocket turns around and comes toward you? Will it appear to be *longer* and will the rocketbound clock move *faster*?

9.5 · THE LORENTZ TRANSFORMATION EQUATIONS

We have seen that the Galilean transformation is not valid when v approaches the speed of light. In this section, we state the correct transformation equations that apply for all speeds in the range $0 \leq v < c$.

Suppose an event that occurs at some point P is reported by two observers, one at rest in a frame S and another in a frame S' that is moving to the right with speed v as in Figure 9.12. The observer in S reports the event with space–time coordinates (x, y, z, t), and the observer in S' reports the same event using the coordinates (x', y', z', t'). We would like to find a relationship between these coordinates that is valid for all speeds. In Section 9.1, we found that the Galilean transformation of coordinates, given by Equation 9.1, does not agree with experiment at speeds comparable to the speed of light.

The equations that are valid from $v = 0$ to $v = c$ and enable us to transform coordinates from S to S' are given by the **Lorentz transformation equations:**

$$x' = \gamma(x - vt)$$

$$y' = y$$

$$z' = z$$ *Lorentz transformation for* $S \rightarrow S'$ [9.7]

$$t' = \gamma\left(t - \frac{v}{c^2}x\right)$$

These transformation equations, known as the Lorentz transformation, were developed by Hendrik A. Lorentz (1853–1928) in 1890 in connection with electromagnetism. However, it was Einstein who recognized their physical significance and took the bold step of interpreting them within the framework of special relativity.

We see that the value for t' assigned to an event by an observer standing at O' depends both on the time t and on the coordinate x as measured by an observer at O. This is consistent with the notion that an event is characterized by four space–time coordinates (x, y, z, t). In other words, in relativity, space and time are not separate concepts but rather are closely interwoven with each other. This is unlike the case of the Galilean transformation in which $t = t'$.

If we wish to transform coordinates in the S' frame to coordinates in the S frame, we simply replace v by $-v$ and interchange the primed and unprimed coordinates in Equation 9.7:

$$x = \gamma(x' + vt')$$

$$y = y'$$

$$z = z'$$ *Inverse Lorentz transformation for* $S' \rightarrow S$ [9.8]

$$t = \gamma\left(t' + \frac{v}{c^2}x'\right)$$

When $v \ll c$, the Lorentz transformation should reduce to the Galilean transformation. To check this, note that as $v \rightarrow 0$, $v/c^2 \ll 1$ and $v^2/c^2 \ll 1$, so that

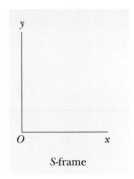

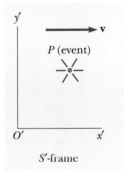

Figure 9.12 Representation of an event that occurs at some point P as observed by an observer at rest in the S frame and another in the S' frame, which is moving to the right with a speed v.

$\gamma = 1$ and Equation 9.7 reduces in this limit to the Galilean coordinate transformation equations

$$x' = x - vt \qquad y' = y \qquad z' = z \qquad t' = t$$

In many situations, we would like to know the difference in coordinates between two events or the time interval between two events as seen by observers at O and O'. This can be accomplished by writing the Lorentz equations in a form suitable for describing pairs of events. From Equations 9.7 and 9.8, we can express the differences between the four variables x, x', t, and t' in the form

$$\Delta x' = \gamma(\Delta x - v\,\Delta t)$$

$$\Delta t' = \gamma\left(\Delta t - \frac{v}{c^2}\,\Delta x\right) \qquad S \rightarrow S' \qquad \text{[9.9]}$$

$$\Delta x = \gamma(\Delta x' + v\,\Delta t')$$

$$\Delta t = \gamma\left(\Delta t' + \frac{v}{c^2}\,\Delta x'\right) \qquad S' \rightarrow S \qquad \text{[9.10]}$$

where $\Delta x' = x_2' - x_1'$ and $\Delta t' = t_2' - t_1'$ are the differences measured by the observer at O', and $\Delta x = x_2 - x_1$ and $\Delta t = t_2 - t_1$ are the differences measured by the observer at O. We have not included the expressions for relating the y and z coordinates because they are unaffected by motion along the x direction.[5]

Example 9.4   Simultaneity and Time Dilation Revisited

Use the Lorentz transformation equations in difference form to show that (a) simultaneity is not an absolute concept and (b) moving clocks run slower than stationary clocks.

Solution   (a) Suppose that two events are simultaneous according to a moving observer at O', so that $\Delta t' = 0$. From the expression for Δt given in Equation 9.10, we see that in this case, $\Delta t = \gamma v\,\Delta x'/c^2$. That is, the time interval for the same two events as measured by an observer at O is nonzero, and so they do not appear to be simultaneous in O.

(b) Suppose that an observer at O' finds that two events occur at the same place ($\Delta x' = 0$), but at different times ($\Delta t' \neq 0$). In this situation, the expression for Δt given in Equation 9.10 becomes $\Delta t = \gamma\,\Delta t'$. This is the equation for time dilation found earlier, Equation 9.5, where $\Delta t' = \Delta t$ is the proper time measured by the single clock located at O'.

EXERCISE 4   Use the Lorentz transformation equations in difference form to confirm that $L = L_p/\gamma$.

Lorentz Velocity Transformation

Let us now derive the Lorentz velocity transformation, which is the relativistic counterpart of the Galilean velocity transformation. Once again S is our stationary frame of reference and S' is our frame of reference that moves at a speed v relative to S. Suppose that an object is observed in the S' frame with an instantaneous speed u_x' measured in S' given by

$$u_x' = \frac{dx'}{dt'} \qquad \text{[9.11]}$$

---

[5]Although motion along x does not change y and z coordinates, it does change velocity components along y and z.

Using Equations 9.7, we have

$$dx' = \gamma(dx - v\,dt) \quad \text{and} \quad dt' = \gamma\left(dt - \frac{v}{c^2}\,dx\right)$$

Substituting these values into Equation 9.11 gives

$$u'_x = \frac{dx'}{dt'} = \frac{dx - v\,dt}{dt - \frac{v}{c^2}\,dx} = \frac{\dfrac{dx}{dt} - v}{1 - \dfrac{v}{c^2}\dfrac{dx}{dt}}$$

But dx/dt is just the velocity component u_x of the object measured in S, and so this expression becomes

$$u'_x = \frac{u_x - v}{1 - \dfrac{u_x v}{c^2}} \qquad \textbf{[9.12]}$$

- *Lorentz velocity transformation for $S \rightarrow S'$*

In a similar way, if the object has velocity components along y and z, the components in S' are

$$u'_y = \frac{u_y}{\gamma\left(1 - \dfrac{u_x v}{c^2}\right)} \quad \text{and} \quad u'_z = \frac{u_z}{\gamma\left(1 - \dfrac{u_x v}{c^2}\right)} \qquad \textbf{[9.13]}$$

When u_x and v are both much smaller than c (the nonrelativistic case), the denominator of Equation 9.12 approaches unity and so $u'_x \approx u_x - v$. This corresponds to the Galilean velocity transformations. In the other extreme, when $u_x = c$, Equation 9.12 becomes

$$u'_x = \frac{c - v}{1 - \dfrac{cv}{c^2}} = \frac{c\left(1 - \dfrac{v}{c}\right)}{1 - \dfrac{v}{c}} = c$$

The speed of light is the speed limit of the Universe.

From this result, we see that an object whose speed approaches c relative to an observer in S also has a speed approaching c relative to an observer in S'—independent of the relative motion of S and S'. Note that this conclusion is consistent with Einstein's second postulate—namely, that the speed of light must be c relative to all inertial frames of reference. Furthermore, the speed of an object can never exceed c. That is, the speed of light is the ultimate speed. We return to this point later when we consider the energy of a particle.

To obtain u_x in terms of u'_x, we replace v by $-v$ in Equation 9.12 and interchange the roles of u_x and u'_x:

$$u_x = \frac{u'_x + v}{1 + \dfrac{u'_x v}{c^2}} \qquad \textbf{[9.14]}$$

- *Inverse Lorentz velocity transformation for $S' \rightarrow S$*

Example 9.5   Relative Velocity of Spaceships

Two spaceships A and B are moving directly toward each other, as in Figure 9.13. An observer on Earth measures the speed of A to be $0.750c$ and the speed of B to be $0.850c$. Find the velocity of B with respect to A.

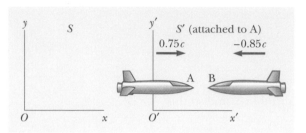

Figure 9.13 (Example 9.5) Two spaceships A and B move in opposite directions. The speed of B relative to A is *less* than *c* and is obtained by using the relativistic velocity transformation.

Solution This problem can be solved by taking the S' frame as being attached to A, so that $v = 0.750c$ relative to the Earth observer (the S frame). Spaceship B can be considered as an object moving with a velocity $u_x = -0.850c$ relative to the Earth observer. Hence, the velocity of B with respect to A can be obtained using Equation 9.12:

$$u'_x = \frac{u_x - v}{1 - \frac{u_x v}{c^2}} = \frac{-0.850c - 0.750c}{1 - \frac{(-0.850c)(0.750c)}{c^2}} = -0.980c$$

The negative sign indicates that spaceship B is moving in the negative x direction as observed by A. Note that the result is less than c. That is, a body whose speed is less than c in one frame of reference must have a speed less than c in any other frame. (If the Galilean velocity transformation were used in this example, we would find that $u'_x = u_x - v = -0.850c - 0.750c = -1.60c$, which is greater than c. The Galilean transformation does not work in relativistic situations.)

Example 9.6   Relativistic Leaders of the Pack

Two motorcycle pack leaders named David and Emily are racing at relativistic speeds along perpendicular paths, as in Figure 9.14. How fast does Emily recede over David's right shoulder as seen by David?

Solution Figure 9.14 represents the situation as seen by a police officer at rest in frame S, who observes the following:

$$\text{David: } u_x = 0.75c \qquad u_y = 0$$

$$\text{Emily: } u_x = 0 \qquad u_y = -0.90c$$

To get Emily's speed of recession as seen by David, we take

S' to move along with David and we calculate u'_x and u'_y for Emily using Equations 9.12 and 9.13:

$$u'_x = \frac{u_x - v}{1 - \frac{u_x v}{c^2}} = \frac{0 - 0.75c}{1 - \frac{(0)(0.75c)}{c^2}} = -0.75c$$

$$u'_y = \frac{u_y}{\gamma\left(1 - \frac{u_x v}{c^2}\right)} = \frac{\sqrt{1 - \frac{(0.75c)^2}{c^2}}\,(-0.90c)}{\left(1 - \frac{(0)(0.75c)}{c^2}\right)} = -0.60c$$

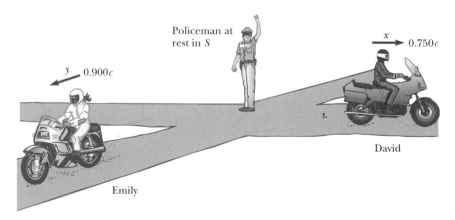

Figure 9.14 (Example 9.6) David moves to the east with a speed $0.750c$ relative to the policeman, while Emily travels south at a speed $0.900c$.

Thus, the speed of Emily as observed by David is

$$u' = \sqrt{(u'_x)^2 + (u'_y)^2} = \sqrt{(-0.75c)^2 + (-0.60c)^2} = 0.96c$$

Note that this speed is less than c as required by special relativity.

EXERCISE 5   Calculate the classical speed of recession for Emily as observed by David using a Galilean transformation. Answer $1.2c$

EXERCISE 6   A Klingon spaceship moves away from the Earth at a speed of $0.80c$. The starship *Enterprise* pursues at a speed of $0.90c$ relative to the Earth. Observers on Earth see the *Enterprise* overtaking the Klingon ship at a relative speed of $0.10c$. With what speed is the *Enterprise* overtaking the Klingon ship as seen by the crew of the *Enterprise*? Answer $0.36c$

9.6 • RELATIVISTIC MOMENTUM AND THE RELATIVISTIC FORM OF NEWTON'S LAWS

We have seen that in order to properly describe the motion of particles within the framework of special relativity, the Galilean transformation must be replaced by the Lorentz transformation. Because the laws of physics must remain unchanged under the Lorentz transformation, we must generalize Newton's laws and the definitions of momentum and energy to conform to the Lorentz transform and the principle of relativity. These generalized definitions should reduce to the classical (nonrelativistic) definitions for $v \ll c$.

First, recall that the law of conservation of momentum states that when two bodies collide, the total momentum remains constant, assuming the bodies are isolated. Suppose the collision is described in a reference frame S in which momentum is conserved. If the velocities in a second reference frame S' are calculated using the Lorentz transformation and the classical definition of momentum, $\mathbf{p} = m\mathbf{u}$, it is found that momentum is *not* conserved in the second reference frame. (We use the symbol $\mathbf{u}$ for particle velocity rather than $\mathbf{v}$, which is used for the relative velocity of two reference frames.) However, because the laws of physics are the same in all inertial frames, the momentum must be conserved in all systems. In view of this condition and assuming the Lorentz transformation is correct, we must modify the definition of momentum to satisfy the following conditions:

- $\mathbf{p}$ must be conserved in all collisions
- $\mathbf{p}$ must approach the classical value $m\mathbf{u}$ as $\mathbf{u} \to 0$

The correct relativistic equation for momentum that satisfies these conditions is

$$\mathbf{p} \equiv \frac{m\mathbf{u}}{\sqrt{1 - \dfrac{u^2}{c^2}}} \qquad [9.15]$$

• *Definition of relativistic momentum*

where $\mathbf{u}$ is the velocity of the particle. When u is much less than c, the denominator of Equation 9.15 approaches unity, so that $\mathbf{p}$ approaches $m\mathbf{u}$. Therefore, the relativistic equation for **p** reduces to the classical expression when u is small compared with c. Because it is simpler, Equation 9.15 is often written as

$$\mathbf{p} = \gamma m \mathbf{u} \qquad [9.16]$$

where $\gamma = (1 - u^2/c^2)^{-1/2}$. Note that γ has the same functional form as the γ in the Lorentz transformation. The transformation is from that of the particle to the frame of the observer moving at speed u relative to the particle.

The relativistic force **F** on a particle whose momentum is **p** is defined as

$$\mathbf{F} \equiv \frac{d\mathbf{p}}{dt} \qquad [9.17]$$

where **p** is given by Equation 9.15. This expression is reasonable because it preserves classical mechanics in the limit of low velocities and requires conservation of momentum for an isolated system (**F** = 0) both relativistically and classically.

It is left as an end-of-chapter problem (Problem 47) to show that the acceleration **a** of a particle decreases under the action of a constant force, in which case $a \propto (1 - u^2/c^2)^{3/2}$. From this formula note that as the particle's speed approaches c, the acceleration caused by any finite force approaches zero. Hence, it is impossible to accelerate a particle from rest to a speed $v \geq c$.

Example 9.7   Momentum of an Electron

An electron, which has a mass of 9.11×10^{-31} kg, moves with a speed of $0.750c$. Find its relativistic momentum and compare this with the momentum calculated from the classical expression.

Solution   Using Equation 9.15 with $u = 0.75c$, we have

$$p = \frac{mu}{\sqrt{1 - \dfrac{u^2}{c^2}}}$$

$$p = \frac{(9.11 \times 10^{-31}\ \text{kg})(0.750 \times 3.00 \times 10^8\ \text{m/s})}{\sqrt{1 - \dfrac{(0.750c)^2}{c^2}}}$$

$$= 3.10 \times 10^{-22}\ \text{kg} \cdot \text{m/s}$$

The incorrect classical expression gives

$$\text{Momentum} = m_e u = 2.05 \times 10^{-22}\ \text{kg} \cdot \text{m/s}$$

Hence, the correct relativistic result is 50% greater than the classical result!

9.7 · RELATIVISTIC ENERGY

We have seen that the definition of momentum and the laws of motion require generalization to make them compatible with the principle of relativity. This implies that the definition of kinetic energy must also be modified.

To derive the relativistic form of the work-kinetic energy theorem, let us start with the definition of the work done on a particle by a force F and use the definition of relativistic force, Equation 9.17:

$$W = \int_{x_1}^{x_2} F\, dx = \int_{x_1}^{x_2} \frac{dp}{dt}\, dx \qquad [9.18]$$

for force and motion both along the x axis. In order to perform this integration, and find the work done on a particle, or the relativistic kinetic energy as a function of u, we first evaluate dp/dt:

$$\frac{dp}{dt} = \frac{d}{dt} \frac{mu}{\sqrt{1 - \dfrac{u^2}{c^2}}} = \frac{m(du/dt)}{\left(1 - \dfrac{u^2}{c^2}\right)^{3/2}}$$

Substituting this expression for dp/dt and $dx = udt$ into Equation 9.18 gives

$$W = \int_0^t \frac{m(du/dt)\,udt}{\left(1 - \dfrac{u^2}{c^2}\right)^{3/2}} = m \int_0^u \frac{u}{\left(1 - \dfrac{u^2}{c^2}\right)^{3/2}}\,du$$

where we have assumed that the particle is accelerated from rest to some final speed u. Evaluating the integral, we find that

$$W = \frac{mc^2}{\sqrt{1 - \dfrac{u^2}{c^2}}} - mc^2 \qquad\qquad \textbf{[9.19]}$$

Recall from Chapter 6 that the work done by a force acting on a particle equals the change in kinetic energy of the particle. Because the initial kinetic energy is zero, we conclude that the work W is equivalent to the relativistic kinetic energy K:

$$K = \frac{mc^2}{\sqrt{1 - \dfrac{u^2}{c^2}}} - mc^2 = \gamma mc^2 - mc^2 \qquad\qquad \textbf{[9.20]} \qquad \bullet \text{ \textit{Relativistic kinetic energy}}$$

This equation is routinely confirmed by experiments using high-energy particle accelerators.

At low speeds, where $u/c \ll 1$, Equation 9.20 should reduce to the classical expression $K = \frac{1}{2}mu^2$. We can check this by using the binomial expansion $(1 - x^2)^{-1/2} \approx 1 + \frac{1}{2}x^2 + \ldots$ for $x \ll 1$, where the higher-order powers of x are neglected in the expansion. In our case, $x = u/c$, so that

$$\frac{1}{\sqrt{1 - \dfrac{u^2}{c^2}}} = \left(1 - \frac{u^2}{c^2}\right)^{-1/2} \approx 1 + \frac{1}{2}\frac{u^2}{c^2} + \cdots$$

Substituting this into Equation 9.20 gives

$$K \approx mc^2\left(1 + \frac{1}{2}\frac{u^2}{c^2} + \cdots\right) - mc^2 = \frac{1}{2}mu^2$$

which agrees with the classical result. Figure 9.15 shows a comparison of the speed–kinetic energy relationships for a particle using the nonrelativistic expression for K (the blue curve) and the relativistic expression for K (the brown curve). The curves are in good agreement at low speeds but deviate at higher speeds. The nonrelativistic expression indicates a violation of physical law because it suggests that sufficient energy can be added to the particle to accelerate it to a speed larger than c. In the relativistic case, the particle speed never exceeds c, regardless of the kinetic energy. When an object's speed is less than one tenth the speed of light, the classical kinetic energy equation differs by less than 1% from the relativistic

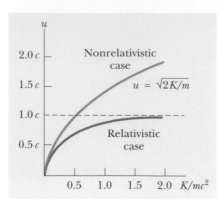

Figure 9.15   A graph comparing relativistic and nonrelativistic kinetic energy. The speeds are plotted versus energy. In the relativistic case, u is always less than c.

equation (which is experimentally verified at all speeds). Thus for practical calculations it is valid to use the classical equation when the object's speed is less than $0.1c$.

The constant term mc^2 in Equation 9.20, which is independent of the speed, is called the **rest energy** of the free particle E_R.

Rest energy •
$$E_R = mc^2 \qquad\qquad \textbf{[9.21]}$$

The term γmc^2 in Equation 9.20 depends on the particle speed and is the sum of the kinetic and rest energies. We define γmc^2 to be the **total energy** E—that is, total energy = kinetic energy + rest energy, or

$$E = \gamma mc^2 = K + mc^2 \qquad\qquad \textbf{[9.22]}$$

or, when γ is replaced by its equivalent,

Definition of total energy •
$$E = \frac{mc^2}{\sqrt{1 - \dfrac{u^2}{c^2}}} \qquad\qquad \textbf{[9.23]}$$

This, of course, is Einstein's famous mass–energy equivalence equation. The relation $E = \gamma mc^2 = \gamma E_R$ shows that **mass is one possible manifestation of energy.** Furthermore, this result shows that a small mass corresponds to an enormous amount of energy. This concept is fundamental to much of the field of nuclear physics.

In many situations, the momentum or energy of a particle is measured instead of its speed. It is therefore useful to have an expression relating the total energy E to the relativistic momentum p. This is accomplished by using the expressions $E = \gamma mc^2$ and $p = \gamma mu$. By squaring these equations and subtracting, we can eliminate u (Problem 35). The result, after some algebra, is

Energy–momentum •
relationship
$$E^2 = p^2c^2 + (mc^2)^2 \qquad\qquad \textbf{[9.24]}$$

When the particle is at rest, $p = 0$, and so $E = E_R = mc^2$. That is, the total energy equals the rest energy. For the case of particles that have zero mass, such as photons (massless, chargeless particles of light), we set $m = 0$ in Equation 9.24, and we see that

$$E = pc \qquad \text{[9.25]}$$

• *Energy of a photon*

This equation is an exact expression relating energy and momentum for photons, which always travel at the speed of light.

Finally, note that because the mass m of a particle is independent of its motion, m must have the same value in all reference frames. For this reason, m is often called the *invariant mass*. However, the total energy and momentum of a particle depend on the reference frame in which they are measured, because they both depend on velocity. Because m is a constant, then according to Equation 9.24 the quantity $E^2 - p^2c^2$ must have the same value in all reference frames. That is, $E^2 - p^2c^2$ is invariant (its value remains the same) under a Lorentz transformation. These equations do not yet make provision for potential energy.

When dealing with subatomic particles, it is convenient to express their energy in electron volts (eV), because the particles are usually given this energy in particle accelerators. The conversion factor is

$$1 \text{ eV} = 1.60 \times 10^{-19} \text{ J}$$

For example, the mass of an electron is 9.11×10^{-31} kg. Hence, the rest energy of the electron is

$$m_e c^2 = (9.11 \times 10^{-31} \text{ kg})(3.00 \times 10^8 \text{ m/s})^2 - 8.20 \times 10^{-14} \text{ J}$$

Converting this to eV, we have

$$m_e c^2 = (8.20 \times 10^{-14} \text{ J})(1 \text{ eV}/1.60 \times 10^{-19} \text{ J}) = 0.511 \text{ MeV}$$

Thinking Physics 3

A common principle learned in chemistry is conservation of mass. In practice, if the mass of the reactants is measured before a reaction and the mass of the products is measured afterward, the results will be the same. In light of special relativity, should we stop teaching the principle of conservation of mass in chemistry classes?

Reasoning Consider a reaction that does not require energy input to occur. This type of reaction occurs because the products represent a lower overall rest energy than the reactants, the difference in rest energy carried away as kinetic energy of ejected particles or radiation. Because the rest energy of the reactants is smaller, according to relativity the mass of the reactants should be smaller than that of the products. Thus the law of conservation of mass is violated. The mass changes are so small, however, that, in practice the law of conservation of mass is still useful.

CONCEPTUAL PROBLEM 5

A photon has zero mass. If a photon is reflected from a surface, does it exert a force on the surface?

Example 9.8   The Energy of a Speedy Proton

The total energy of a proton is three times its rest energy.
(a) Find the proton's rest energy in electron volts.

Solution

$$E_R = m_p c^2 = (1.67 \times 10^{-27} \text{ kg})(3.00 \times 10^8 \text{ m/s})^2$$
$$= (1.50 \times 10^{-10} \text{ J})(1.00 \text{ eV}/1.60 \times 10^{-19} \text{ J})$$
$$= \boxed{938 \text{ MeV}}$$

(b) With what speed is the proton moving?

Solution   Because the total energy E is three times the rest energy, $E = \gamma mc^2$ (Eq. 9.23) gives

$$E = 3m_p c^2 = \frac{m_p c^2}{\sqrt{1 - \dfrac{u^2}{c^2}}}$$

$$3 = \frac{1}{\sqrt{1 - \dfrac{u^2}{c^2}}}$$

Solving for u gives

$$\left(1 - \frac{u^2}{c^2}\right) = \frac{1}{9} \quad \text{or} \quad \frac{u^2}{c^2} = \frac{8}{9}$$

$$u = \frac{\sqrt{8}}{3} c = \boxed{2.83 \times 10^8 \text{ m/s}}$$

(c) Determine the kinetic energy of the proton in electron volts.

Solution

$$K = E - m_p c^2 = 3m_p c^2 - m_p c^2 = 2m_p c^2$$

Because $m_p c^2 = 938$ MeV, $K = \boxed{1876 \text{ MeV}}$

(d) What is the magnitude of the proton's momentum?

Solution   We can use Equation 9.24 to calculate the momentum with $E = 3m_p c^2$:

$$E^2 = p^2 c^2 + (m_p c^2)^2 = (3m_p c^2)^2$$
$$p^2 c^2 = 9(m_p c^2)^2 - (m_p c^2)^2 = 8(m_p c^2)^2$$
$$p = \sqrt{8}\,\frac{m_p c^2}{c} = \sqrt{8}\,\frac{(938 \text{ MeV})}{c} = \boxed{2650\,\frac{\text{MeV}}{c}}$$

The unit of momentum is written MeV/c for convenience.

EXERCISE 7   Find the speed at which the relativistic kinetic energy is two times the non-relativistic value.   Answer $0.786c$

EXERCISE 8   Find the speed of a particle whose total energy is twice its rest energy.
Answer $0.866c$

EXERCISE 9   An electron moves with a speed $u = 0.850c$. Find its (a) total energy and (b) kinetic energy in electron volts.   Answer   (a) 0.970 MeV   (b) 0.459 MeV

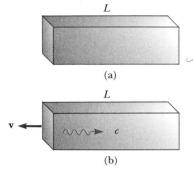

Figure 9.16   (a) A box of length L at rest. (b) When a light pulse is emitted at the left end of the box, the box recoils to the left until the pulse strikes the right end.

9.8 • MASS AS A MEASURE OF ENERGY

To understand the equivalence of mass and energy, consider the following thought experiment proposed by Einstein. Imagine a box of mass M and length L initially at rest, as in Figure 9.16a. Suppose that a pulse of light is emitted from the left side of the box, as in Figure 9.16b. From Equation 9.25, we know that the light of energy E carries momentum $p = E/c$. Hence, the box must recoil to the left with a speed v to conserve momentum. Assuming the box is very massive, the recoil speed is small compared with the speed of light. Conservation of momentum gives $Mv = E/c$, or

$$v = \frac{E}{Mc}$$

The time it takes the light to move the length of the box is approximately $\Delta t = L/c$ (where again, we assume that $v \ll c$). In this time interval, the box moves a small distance Δx to the left, where

$$\Delta x = v\,\Delta t = \frac{EL}{Mc^2}$$

The light then strikes the right end of the box, transfers its momentum to the box, causing the box to stop. With the box in its new position, it appears as if its center of mass has moved to the left. However, its center of mass cannot move because the box is an isolated system. Einstein resolved this perplexing situation by assuming that in addition to energy and momentum, light energy also has kinetic mass. If m_k is the kinetic mass carried by the pulse of light, and the center of mass of the box is to remain fixed, then

$$m_k L = M\,\Delta x$$

Solving for m_k, and using the previous expression for Δx, we get

$$m_k = \frac{M\,\Delta x}{L} = \frac{M}{L}\frac{EL}{Mc^2} = \frac{E}{c^2}$$

or

$$E = m_k c^2$$

Thus, Einstein reached the profound conclusion, "If a body gives off the energy E in the form of radiation, its mass diminishes by E/c^2 . . . The mass of a body is a measure of its energy content."

Although we derived the relationship $E = mc^2$ for light energy, the equivalence of mass and energy is universal. Equation 9.22, $E = \gamma mc^2$, which represents the total energy of any particle, suggests that even when a particle is at rest ($\gamma = 1$) it still possesses enormous energy through its mass. The clearest experimental proof of the equivalence of mass and energy occurs in nuclear and elementary particle interactions in which both the conversion of mass into energy and the conversion of energy into mass take place. Because of this, we can no longer accept the separate classical laws of conservation of mass and conservation of energy; we must instead speak of a unified law of **conservation of mass–energy.** Simply put, this law requires that **the sum of the mass–energy of a system of particles before interaction must equal the sum of the mass–energy of the system after interaction, where the mass–energy of the ith particle is defined as**

• *Conservation of mass–energy*

$$E_i = \frac{m_i c^2}{\sqrt{1 - \dfrac{u_i^2}{c^2}}} \qquad [9.26]$$

The release of enormous energy, accompanied by the change in masses of particles after they have lost their excess energy as they are brought to rest, is the basis of atomic and hydrogen bombs. In a conventional nuclear reactor, the uranium nucleus undergoes fission, a reaction that results in several lighter fragments having considerable kinetic energy. In the case of ^{235}U (the parent nucleus), which undergoes fission, the fragments are two lighter nuclei and two neutrons. The total mass of the fragments is less than that of the parent nucleus by an amount Δm. The corresponding energy Δmc^2 associated with this mass difference is exactly equal to the total kinetic energy of the fragments. This kinetic energy is then used to produce hot water and steam for the generation of electrical power.

Next, consider the basic fusion reaction in which two deuterium atoms combine to form one helium atom. This reaction is of major importance in current research and development of controlled-fusion reactors. The decrease in mass that results from the creation of one helium atom from two deuterium atoms is $\Delta m = 4.25 \times 10^{-29}$ kg. Hence, the corresponding energy that results from one fusion reaction is $\Delta mc^2 = 3.83 \times 10^{-12}$ J $= 23.9$ MeV. To appreciate the magnitude of this result, if 1 g of deuterium is converted to helium, the energy released is about 10^{12} J! At the 1997 cost of electrical energy, this would be worth about \$65 000.

Example 9.9   Binding Energy of the Deuteron

The mass of the deuteron, which is the nucleus of "heavy hydrogen," is not equal to the sum of the masses of its constituents, which are the proton and neutron. Calculate this mass difference and determine its energy equivalence.

Solution   Using atomic mass units (u), we have

$$m_p = \text{mass of proton} = 1.007\ 276\ \text{u}$$

$$m_n = \text{mass of neutron} = 1.008\ 665\ \text{u}$$

$$m_p + m_n = 2.015\ 941\ \text{u}$$

Because the mass of the deuteron is 2.013 553 u (Appen-

dix A), we see that the mass difference Δm is 0.002 388 u. By definition, 1 u $= 1.66 \times 10^{-27}$ kg, and therefore

$$\Delta m = \boxed{0.002\ 388\ \text{u} = 3.96 \times 10^{-30}\ \text{kg}}$$

Using $E = \Delta mc^2$, we find that

$$E = \Delta mc^2 = (3.96 \times 10^{-30}\ \text{kg})(3.00 \times 10^8\ \text{m/s})^2$$

$$= 3.56 \times 10^{-13}\ \text{J} = \boxed{2.23\ \text{MeV}}$$

Therefore, the minimum energy required to separate the proton from the neutron of the deuterium nucleus (the binding energy) is 2.23 MeV.

OPTIONAL

9.9 • GENERAL RELATIVITY

Up to this point, we have sidestepped a curious puzzle. Mass has two seemingly different properties: a *gravitational attraction* for other masses and an *inertial* property that resists acceleration. To designate these two attributes, we use the subscripts g and i and write

Gravitational property $F_g = m_g g$

Inertial property $F_i = m_i a$

The value for the gravitational constant G was chosen to make the magnitudes of m_g and m_i numerically equal. Regardless of how G is chosen, however, the strict proportionality of m_g and m_i has been established experimentally to an extremely high degree: a few parts in 10^{12}. Thus, it appears that gravitational mass and inertial mass may indeed be exactly proportional.

But why? They seem to involve two entirely different concepts: a force of mutual gravitational attraction between two masses and the resistance of a single mass to being accelerated, regardless of what kind of force is producing the acceleration. This question, which puzzled Newton and many other physicists over the years, was answered when Einstein published his theory of gravitation, known as *general relativity*, in 1916. Because it is a mathematically complex theory, we merely offer a hint of its elegance and insight.

In Einstein's view, the remarkable coincidence that m_g and m_i seemed to be exactly proportional was evidence for a very intimate and basic connection between the two concepts. He pointed out that no mechanical experiment (such as dropping a mass) could distinguish between the two situations illustrated in Figures 9.17a and 9.17b. In each case, a mass released by the observer undergoes a downward acceleration of g relative to the floor.

Einstein carried this idea further and proposed that *no* experiment, mechanical or otherwise, could distinguish between the two cases. This extension to include all phenomena (not just mechanical ones) has interesting consequences. For example, suppose that a light pulse is sent horizontally across the box, as in Figure 9.17c. The trajectory of the light pulse bends downward as the box accelerates upward to meet it. Therefore, Einstein proposed that a beam of light should also be bent downward by a gravitational field. (No such bending is predicted in Newton's theory of gravitation.)

The two postulates of Einstein's **general relativity** are as follows:

- All the laws of nature have the same form for observers in any frame of reference, whether accelerated or not.
- In the vicinity of any given point, a gravitational field is equivalent to an accelerated frame of reference in the absence of gravitational effects. (This is the *principle of equivalence.*)

The second postulate implies that gravitational mass and inertial mass are completely equivalent, not just proportional. What were thought to be two different types of mass are actually identical.

One interesting effect predicted by general relativity is that time scales are altered by gravity. A clock in the presence of gravity runs more slowly than one in which gravity is negligible. As a consequence, the frequencies of radiation emitted by atoms in the presence of a strong gravitational field are shifted to lower fre

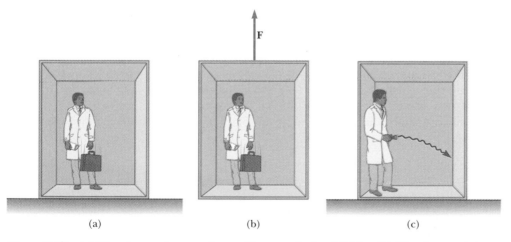

Figure 9.17   (a) The observer is at rest in a uniform gravitational field **g**. (b) The observer is in a region where gravity is negligible, but the frame of reference is accelerated by an external force **F** that produces an acceleration **g**. According to Einstein, the frames of reference in parts (a) and (b) are equivalent in every way. No local experiment could distinguish any difference between the two frames. (c) If parts (a) and (b) are truly equivalent, as Einstein proposed, then a ray of light would bend in a gravitational field.

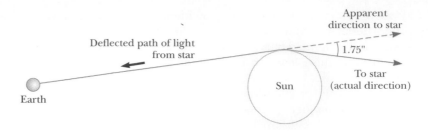

Figure 9.18 Deflection of starlight passing near the Sun. Because of this effect, the Sun and other remote objects can act as a *gravitational lens*. In his general theory of relativity, Einstein calculated that starlight just grazing the Sun's surface should be deflected by an angle of 1.75″.

quencies when compared with the same emissions in a weak field. This gravitational shift has been detected in spectral lines emitted by atoms in massive stars. It has also been verified on the Earth by comparing the frequencies of gamma rays (a high energy form of electromagnetic radiation) emitted from nuclei separated vertically by about 20 m.

The second postulate suggests that a gravitational field may be "transformed away" at any point if we choose an appropriate accelerated frame of reference—a freely falling one. Einstein developed an ingenious method of describing the acceleration necessary to make the gravitational field "disappear." He specified a certain quantity, the *curvature of space–time*, that describes the gravitational effect at every point. In fact, the curvature of space–time completely replaces Newton's gravitational theory. According to Einstein, there is no such thing as a gravitational force. Rather, the presence of a mass causes a curvature of space–time in the vicinity of the mass, and this curvature dictates the space–time path that all freely moving objects must follow. As one physicist said, "Mass one tells space–time how to curve; curved space–time tells mass two how to move." One important test of general relativity is the prediction that a light ray passing near the Sun should be deflected by some angle. This prediction was confirmed by astronomers as bending of starlight during a total solar eclipse shortly following World War I (Fig. 9.18).

If the concentration of mass becomes very great, as is believed to occur when a large star exhausts its nuclear fuel and collapses to a very small volume, a **black hole** may form. Here the curvature of space–time is so extreme that, within a certain distance from the center of the black hole, all matter and light become trapped.

Thinking Physics 4

Atomic clocks are extremely accurate; in fact, an error of 1 second in 3 million years is typical. This error can be described as about 1 part in 10^{14}. However, the atomic clock in Boulder, Colorado, is often 15 ns faster than the one in Washington after only 1 day. This is an error of about 1 part in 6×10^{12}, which is about 17 times larger than the previously expressed error. If atomic clocks are so accurate, why does a clock in Boulder not remain in synchronism with one in Washington? (*Hint:* Denver, near Boulder, is known as the mile-high city.)

Reasoning According to the general theory of relativity, the rate of passage of time depends on gravity—time runs more slowly in strong gravitational fields. Washington is at an elevation very close to sea level, whereas Boulder is about 1 mile higher in altitude. This will result in a weaker gravitational field at Boulder than at Washington. As a result, time runs more rapidly in Boulder than in Washington.

SUMMARY

The two basic postulates of special relativity are

- All the laws of physics are the same in all inertial reference frames.
- The speed of light in vacuum has the same value, $c = 3.00 \times 10^8$ m/s, in all inertial frames, regardless of the velocity of the observer or the velocity of the source emitting the light.

Three consequences of special relativity are

- Events that are simultaneous for one observer are not simultaneous for another observer who is in motion relative to the first.
- Clocks in motion relative to an observer appear to be slowed down by a factor γ. This is known as **time dilation.**
- Lengths of objects in motion appear to be contracted in the direction of motion.

To satisfy the postulates of special relativity, the Galilean transformations must be replaced by the **Lorentz transformations:**

$$x' = \gamma(x - vt)$$
$$y' = y$$
$$z' = z \qquad\qquad \text{[9.7]}$$
$$t' = \gamma\left(t - \frac{v}{c^2}x\right)$$

where $\gamma = (1 - v^2/c^2)^{-1/2}$.

The relativistic form of the **velocity transformation** is

$$u'_x = \frac{u_x - v}{1 - \dfrac{u_x v}{c^2}} \qquad\qquad \text{[9.13]}$$

where u_x is the speed of an object as measured in the S frame and u'_x is its speed measured in the S' frame.

The relativistic expression for the **momentum** of a particle moving with a velocity $\mathbf{u}$ is

$$\mathbf{p} \equiv \frac{m\mathbf{u}}{\sqrt{1 - \dfrac{u^2}{c^2}}} = \gamma m\mathbf{u} \qquad\qquad \text{[9.16]}$$

The relativistic expression for the **kinetic energy** of a particle is

$$K = \gamma mc^2 - mc^2 \qquad\qquad \text{[9.21]}$$

where mc^2 is called the **rest energy** of the particle.

The total energy E of a particle is related to the **mass** through the **energy–mass** equivalence expression:

$$E = \gamma mc^2 = \frac{mc^2}{\sqrt{1 - \dfrac{u^2}{c^2}}} \qquad [9.23]$$

The relativistic momentum is related to the total energy through the equation

$$E^2 = p^2 c^2 + (mc^2)^2 \qquad [9.24]$$

CONCEPTUAL QUESTIONS

1. What two speed measurements do two observers in relative motion *always* agree on?

2. The speed of light in water is 2.3×10^8 m/s. Suppose an electron is moving through water at 2.5×10^8 m/s. Does this violate the principles of relativity?

3. Two identical clocks are synchronized. One is put in orbit directed eastward around the Earth and the other remains on Earth. Which clock runs slower? When the moving clock returns to Earth, are the two still synchronized?

4. A train is approaching you as you stand next to the tracks. Just as an observer on the train passes, you both begin to play the same Beethoven symphony on portable compact disk players. (a) According to you, whose player finishes the symphony first? (b) According to the observer on the train, whose player finishes the symphony first? (c) Whose player *really* finishes the symphony first?

5. A spaceship in the shape of a sphere moves past an observer on Earth with a speed 0.5c. What shape does the observer see as the spaceship moves past?

6. Explain why it is necessary, when defining length, to specify that the positions of the ends of a rod are to be measured simultaneously.

7. When we say that a moving clock runs slower than a stationary one, does this imply that there is something physically unusual about the moving clock?

8. Give a physical argument that shows that it is impossible to accelerate an object of mass m to the speed of light, even with a continuous force acting on it.

9. List some ways our day-to-day lives would change if the speed of light were only 50 m/s.

10. It is said that Einstein, in his teenage years, asked the question, "What would I see in a mirror if I carried it in my hands and ran at the speed of light?" How would you answer this question?

11. Some of the distant star-like objects, called quasars, are receding from us at half the speed of light (or greater). What is the speed of the light we receive from these quasars?

12. How is it possible that photons of light, which have zero rest mass, have momentum?

13. Relativistic quantities must make a smooth transition to their Newtonian counterparts as the speed of a system becomes small compared to the speed of light. Explain.

14. If the speed of a particle is doubled, what effect does it have on its momentum? (Assume that its initial speed is less than $c/2$.)

PROBLEMS

Section 9.1   The Principle of Newtonian Relativity

1. A 2000-kg car moving at 20.0 m/s collides with and sticks to a 1500-kg car at rest at a stop sign. Show that momentum is conserved in a reference frame moving at 10.0 m/s in the direction of the moving car.

2. A ball is thrown at 20.0 m/s inside a boxcar moving along the tracks at 40.0 m/s. What is the speed of the ball relative to the ground if the ball is thrown (a) forward? (b) backward? (c) out the side door?

3. In a laboratory frame of reference, an observer notes that Newton's second law is valid. Show that it is also valid for an observer moving at a constant speed, small compared to the speed of light, relative to the laboratory frame.

4. Show that Newton's second law is *not* valid in a reference frame moving past the laboratory frame of Problem 3 with a constant acceleration.

Section 9.4   Consequences of Special Relativity

5. How fast must a meter stick be moving if its length is observed to shrink to 0.500 m?

6. At what speed does a clock have to move in order to run at a rate that is one half the rate of a clock at rest?

7. An astronaut is traveling in a space vehicle that has a speed of 0.500c relative to the Earth. The astronaut measures his pulse rate at 75.0 per minute. Signals generated by the astronaut's pulse are radioed to Earth when the vehicle is moving perpendicularly to a line that connects the vehicle

with an Earth observer. What pulse rate does the Earth observer measure? What would be the pulse rate if the speed of the space vehicle were increased to 0.990*c*?

8. The proper length of one spaceship is three times that of another. The two spaceships are traveling in the same direction and, while both are passing overhead, an Earth observer measures the two spaceships to have the same length. If the slower spaceship is moving with a speed of 0.350*c*, determine the speed of the faster spaceship.

9. An atomic clock moves at 1000 km/h for 1 hour as measured by an identical clock on Earth. How many nanoseconds slow will the moving clock be at the end of the one-hour interval?

10. If astronauts could travel at *v* = 0.950*c*, we on Earth would say it takes (4.20/0.950) = 4.42 years to reach Alpha Centauri, 4.20 lightyears away. The astronauts disagree. (a) How much time passes on the astronauts' clocks? (b) What distance to Alpha Centauri do the astronauts measure?

11. A spaceship of proper length 300 m takes 0.750 *μ*s to pass an Earth observer. Determine the speed of this spaceship as measured by the Earth observer.

12. A spaceship of proper length L_p takes time *t* to pass an Earth observer. Determine the speed of this spaceship as measured by the Earth observer.

13. A muon formed high in the Earth's atmosphere travels at speed *v* = 0.990*c* for a distance of 4.60 km before it decays into an electron, a neutrino, and an antineutrino ($\mu^- \rightarrow e^- + \nu + \overline{\nu}$). (a) How long does the muon live, as measured in its reference frame? (b) How far does the muon travel, as measured in its frame?

Section 9.5 The Lorentz Transformation Equations

14. For what value of *v* does *γ* = 1.01? Observe that for speeds lower than this value, time dilation and length contraction will be less-than-one-percent effects.

15. A spaceship travels at 0.750*c* relative to Earth. If the spaceship fires a small rocket in the forward direction, how fast (relative to the ship) must it be fired for it to travel at 0.950*c* relative to Earth?

16. A certain quasar recedes from the Earth at *v* = 0.870*c*. A jet of material ejected from the quasar toward the Earth moves at 0.550*c* relative to the quasar. Find the speed of the ejected material relative to the Earth.

17. Two jets of material from the center of a radio galaxy fly away in opposite directions. Both jets move at 0.750*c* relative to the galaxy. Determine the speed of one jet relative to the other.

18. A friend travels by you at a high speed in a spaceship. He tells you that his ship is 20.0 m long and that the identically constructed ship you are sitting in is 19.0 m long. According to your observations, (a) how long is your ship, (b) how long is your friend's ship, and (c) what is the speed of your friend's ship?

19. Observer *A* measures the lengths of two rods, one stationary, the other moving with a speed of 0.955*c*. She finds that the rods have the same length. A second observer, *B*, travels along with the moving rod. What is the ratio of the length of *A*'s rod to the length of *B*'s rod, according to observer *B*?

20. A moving rod is 2.00 m long, and its length is oriented at an angle of 30.0° with respect to the direction of motion. The rod has a speed of 0.995*c*. (a) What is the proper length of the rod? (b) What is the orientation angle in the proper frame?

Section 9.6 Relativistic Momentum and the Relativistic Form of Newton's Laws

21. Calculate the momentum of an electron moving with a speed of (a) 0.0100*c*, (b) 0.500*c*, (c) 0.900*c*.

22. The *nonrelativistic* expression for the momentum of a particle, *p* = *mv*, can be used if *v* ≪ *c*. For what speed does the use of this formula give an error in the momentum of (a) 1.00% and (b) 10.0%?

23. An electron has a momentum that is 90.0% greater than its classical momentum. (a) Find the speed of the electron. (b) How would your result change if the particle were a proton?

24. A golf ball travels with a speed of 90.0 m/s. By what fraction does its relativistic momentum, *p*, differ from its classical value, *mv*? That is, find the ratio (*p* − *mv*)/*mv*.

25. An unstable particle at rest breaks into two fragments of *unequal* mass. The rest mass of the lighter fragment is 2.50×10^{-28} kg, and that of the heavier fragment is 1.67×10^{-27} kg. If the lighter fragment has a speed of 0.893*c* after the breakup, what is the speed of the heavier fragment?

26. Show that the speed of an object having momentum *p* and mass *m* is

$$v = \frac{c}{\sqrt{1 + (mc/p)^2}}$$

Section 9.7 Relativistic Energy

27. A proton moves at 0.950*c*. Calculate its (a) rest energy, (b) total energy, and (c) kinetic energy (in eV).

28. Show that, for any object moving at less than one tenth the speed of light, the relativistic kinetic energy agrees with the result of the classical equation $K = \frac{1}{2}mv^2$ to within less than 1%. Thus for most purposes, the classical equation is good enough to describe these objects, whose motion we call *nonrelativistic*.

29. Determine the energy required to accelerate an electron from (a) 0.500*c* to 0.900*c* and (b) 0.900*c* to 0.990*c*.

30. A spaceship of mass 1.00×10^6 kg is to be accelerated to 0.600*c*. (a) How much energy does this require? (b) How

many kilograms of matter would it take to provide this much energy?

31. Make an order-of-magnitude estimate of the ratio of mass increase to the rest mass of a flag as you run it up a flagpole. In your solution explain what quantities you take as data and the values you estimate or measure for them.

32. When 1.00 g of hydrogen combines with 8.00 g of oxygen, 9.00 g of water is formed. During this chemical reaction, 2.86×10^5 J of energy is released. How much mass do the constituents of this reaction lose? Is the loss of mass likely to be detectable?

33. A cube of steel has a volume of 1.00 cm³ and a mass of 8.00 g when at rest on the Earth. If this cube is now given a speed $v = 0.900c$, what is its density as measured by a stationary observer? Note that relativistic density is $m/V = E/c^2V$.

34. An electron has a kinetic energy five times greater than its rest energy. Find (a) its total energy and (b) its speed.

35. Show that the energy–momentum relationship $E^2 = p^2c^2 + (mc^2)^2$ follows from the expressions $E = \gamma mc^2$ and $p = \gamma mu$.

36. An unstable particle with a mass of 3.34×10^{-27} kg is initially at rest. The particle decays into two fragments that fly off with velocities of $0.987c$ and $-0.868c$. Find the rest masses of the fragments. (*Hint:* Conserve both mass–energy and momentum.)

Additional Problems

37. The net nuclear reaction inside the Sun is $4p \rightarrow \text{He}^4 + \Delta E$. If the rest mass of each proton is 938.2 MeV and the rest mass of the He^4 nucleus is 3727 MeV, calculate the percentage of the starting mass that is released as energy.

38. An electron has a speed of $0.750c$. Find the speed of a proton that has (a) the same kinetic energy as the electron, (b) the same momentum as the electron.

39. The cosmic rays of highest energy are protons, having kinetic energy on the order of 10^{13} MeV. (a) How long would it take a proton of this energy to travel across the Milky Way galaxy, of diameter 10^5 lightyears, as measured in the proton's frame? (b) From the point of view of the proton, how many kilometers across is the galaxy?

40. A spaceship moves away from the Earth at $0.500c$ and fires a shuttle craft in the forward direction at $0.500c$ relative to the ship. The pilot of the shuttle craft launches a probe at forward speed $0.500c$ relative to the shuttle craft. Determine (a) the speed of the shuttle craft relative to the Earth and (b) the speed of the probe relative to the Earth.

41. The average lifetime of a pi meson in its own frame of reference is 2.60×10^{-8} s. If the meson moves with a speed of $0.950c$, what are (a) its mean lifetime as measured by an observer on Earth and (b) the average distance it travels before decaying, as measured by an observer on Earth?

42. An astronaut wishes to visit the Andromeda galaxy (suppose it is 2.00 million lightyears away), making a one-way trip that will take 30.0 years in the spaceship's frame of reference. Assuming that her speed is constant, how fast must she travel relative to the Earth?

43. A physics professor on Earth gives an exam to her students who are on a rocket ship traveling at speed v relative to Earth. The moment the ship passes the professor, she signals the start of the exam. If she wishes her students to have time T_0 (rocket time) to complete the exam, show that she should wait a time (Earth time) of

$$T = T_0 \sqrt{\frac{1 - v/c}{1 + v/c}}$$

before sending a light signal telling them to stop. (*Hint:* Remember that it takes some time for the second light signal to travel from the professor to the students.)

44. Spaceship I, which contains students taking a physics exam, approaches Earth with a speed of $0.600c$ (relative to Earth), and spaceship II, which contains professors proctoring the exam, moves at $0.280c$ (relative to Earth) directly toward the students. If the professors stop the exam after 50.0 min have passed on their clock, how long does the exam last as measured by (a) the students? (b) an observer on Earth?

45. A supertrain (rest length = 100 m) travels at a speed of $0.950c$ as it passes through a tunnel (rest length = 50.0 m). As seen by a trackside observer, is the train ever completely within the tunnel? If so, with how much space to spare?

46. Energy reaches the upper atmosphere of the Earth from the Sun at the rate of 1.79×10^{17} W. If all of this energy were absorbed by the Earth and not re-emitted, how much would the mass of the Earth increase in 1 year?

47. A charged particle moves along a straight line in a uniform electric field **E** with a speed of v. The electric force on a charge q in an electric field is $q\mathbf{E}$. If the motion and the electric field are both in the x direction, (a) show that the acceleration of the charge q in the x direction is given by

$$a = \frac{dv}{dt} = \frac{qE}{m}\left(1 - \frac{v^2}{c^2}\right)^{3/2}$$

(b) Discuss the significance of the dependence of the acceleration on the speed. (c) If the particle starts from rest at $x = 0$ at $t = 0$, how would you proceed to find the speed of the particle and its position after a time t has elapsed?

48. Imagine that the entire Sun collapses to a sphere of radius R_g such that the work required to remove a small mass m from the surface would be equal to its rest energy mc^2. This radius is called the *gravitational radius* for the Sun. Find R_g. (It is believed that the ultimate fate of many stars is to collapse to their gravitational radii or smaller.)

Spreadsheet Problems

S1. Astronomers use the Doppler shift in the Balmer series of the hydrogen spectrum to determine the radial speed of a galaxy. The fractional change in the wavelength of the spectral line is given by

$$Z = \frac{\Delta\lambda}{\lambda_0} = \frac{\lambda - \lambda_0}{\lambda_0} = \sqrt{\frac{1 + v/c}{1 - v/c}} - 1$$

Once this quantity is measured for a particular receding galaxy, the speed of recession can be found by solving for v/c in terms of Z. Spreadsheet 9.1 calculates the speed Z values. (a) What is the speed for galaxies having $Z = 0.2, 0.5, 1.0$, and 2.0? (b) The largest Z values, $Z \approx 3.8$, have been measured for several quasars (quasi-stellar radio sources). How fast are these quasars moving away from us?

S2. Astronauts in a starship traveling at a speed v relative to a starbase are given instructions from mission control to call back in 1 h as measured by the starship clocks. Spreadsheet 9.2 calculates how long Mission Control has to wait for the call for different starship speeds. How long does Mission Control have to wait if the ship is traveling at $v = 0.1c, 0.4c, 0.6c, 0.8c, 0.9c, 0.995c, 0.9995c$?

S3. Most astronomers believe that the Universe began at some instant with an explosion called the Big Bang and that the observed recession of the galaxies is a direct result of this explosion. If the galaxies recede from each other at a constant rate, then we expect that the galaxies moving fastest are now farthest away from Earth. This result, called Hubble's law after Edwin Hubble, can be written $v = Hr$, where H can be determined from observation, v is the speed of recession of the galaxy, and r is its distance from Earth. Current estimates of H range from 15 to 30 km/s/Mly, where Mly is the distance light travels in one million years, and a conservative estimate is $H = 20$ km/s/Mly. Use Spreadsheet 9.1 to calculate the distance from Earth for each galaxy in Problem S1.

S4. Design a spreadsheet program to calculate and plot the relativistic kinetic energy (Eq. 9.20) and the classic kinetic energy ($\frac{1}{2}mu^2$) of a macroscopic object. Plot the relativistic and classical energies versus speed on the same graph. (a) For an object of mass $m = 3$ kg, at what speed does the classical kinetic energy underestimate the relativistic value by 1%? 5%? 50%? What is the relativistic kinetic energy at these speeds? Repeat part (a) for (b) an electron and (c) a proton.

ANSWERS TO CONCEPTUAL PROBLEMS

1. This scenario is not possible with light. Water waves move through a medium—the surface of the water—and therefore have an absolute velocity relative to the reference frame in which this medium is at rest. Thus, the relative velocity of the waves and your boat is determined by Galilean relativity. This makes it possible for you to move faster than the waves. Light waves, which do not require a medium, are described by the principles of special relativity. As you detect the light wave ahead of you and moving away from you (which would be a pretty good trick—think about it!), its velocity relative to you is c. No matter how fast you accelerate in that direction, trying to catch up to it, its velocity relative to you is still c. Thus, you will not be able to catch up to the light wave.

2. The answers to both of these questions is *no*. Both your clothing and your sleeping cabin are at rest in your reference frame, thus, they will have their proper length. There will be no change in measured lengths of objects within your spacecraft. Another observer, on a spacecraft traveling at a high speed relative to yours, will measure you as thinner (if your body is oriented in a direction perpendicular to your velocity vector relative to the other observer) or will claim that you are able to fit into a shorter sleeping cabin (if your body is oriented in a direction parallel to your velocity vector relative to the other observer).

3. Assuming that their on-duty time was kept on Earth, they will be pleasantly surprised with a large paycheck. Less time will have passed for the astronauts in their frame of reference than for their employer back on Earth.

4. The incoming rocket will not appear to have a longer length and a faster clock. Length contraction and time dilation depend only on the magnitude of the relative velocity, not on the direction.

5. A reflected photon does exert a force on a surface. Although a photon has zero mass, a photon does carry momentum. When it reflects from a surface, there is a change in the momentum, just like the change in momentum of a ball bouncing off the floor. According to the momentum interpretation of Newton's second law, a change in momentum results in a force on the surface. This concept is used in theoretical studies of *space sailing*. These studies propose building nonpowered spacecraft with huge reflective sails oriented perpendicularly to the rays from the Sun. The large number of photons from the Sun reflecting from the surface of the sail will exert a force which, although small, will provide a continuous acceleration. This would allow the spacecraft to travel to other planets without fuel.

10

Rotational Motion

In this chapter we investigate the dynamics of particles and systems of particles moving along a circular path and rigid objects rotating about a fixed axis. We shall encounter such terms as angular displacement, angular velocity, angular acceleration, torque, and angular momentum and show how these quantities are useful in describing rotational motion. Next we shall define a vector product, which is a convenient mathematical tool for expressing such physical quantities as torque and angular momentum.

One of the central points of this chapter is to develop the concept of the angular momentum of a system of particles. By analogy with the conservation of linear momentum (Chapter 8), we shall find that the angular momentum of any isolated system is always constant. The results we derive here will enable us to understand the rotational motions of a diverse range of objects in our environment, from an electron orbiting a nucleus to clusters of galaxies orbiting a common center. Although the focus of this chapter is the circular motion of particles, the exten-

Derek Swinson, professor of physics at the University of New Mexico, demonstrating the "gyro-ski" technique. The skier initiates a turn by lifting the axle of the rotating bicycle wheel. The direction of the turn depends on whether the left or right hand is used to lift the axle from the horizontal. Ignoring friction and gravity, the angular momentum of the system (the skier and the bicycle wheel) remains constant.
(Courtesy of Derek Swinson)

sion of this treatment to rigid bodies is reasonably straightforward and is treated in an optional section.

10.1 • ANGULAR VELOCITY AND ANGULAR ACCELERATION

We began our study of linear motion by defining the terms *displacement, velocity,* and *linear acceleration.* We will take the same basic approach now as we turn to the study of rotational motion. Let us begin by considering a circular disk rotating around a fixed axis that is perpendicular to the disk and goes through the point O (Fig. 10.1). A point P on the disk is at a fixed distance r from the origin and rotates about O in a circle of radius r. In fact, **every point on the disk undergoes circular motion about O.**

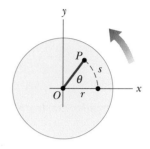

It is convenient to represent the position of the point P with its polar coordinates: (r, θ). The origin of the polar coordinates is chosen to coincide with the center of the circle. In this representation, the only coordinate that changes in time is the angle θ; r remains constant. As a point on the disk moves along the circle of radius r from the positive x axis ($\theta = 0$) to the point P, it moves through an arc of length s, which is related to the angular position θ through the relation $s = r\theta$, or

$$\theta = \frac{s}{r} \qquad [10.1]$$

Figure 10.1   Rotation of a disk about a fixed axis through O perpendicular to the plane of the figure. (In other words, the axis of rotation is the z axis.) Note that a particle at P rotates in a circle of radius r centered at O.

It is important to note the units of θ as expressed by Equation 10.1. The angle θ is the ratio of an arc length and the radius of the circle, and hence is a pure number. However, we commonly refer to the unit of θ as a **radian** (rad). One radian is the angle subtended by an arc length equal to the radius of the arc. Because the circumference of a circle is $2\pi r$, it follows that 360° corresponds to an angle of $2\pi r/r$ rad or 2π rad (one revolution). Hence, 1 rad $= 360°/2\pi \approx 57.3°$. To convert an angle in degrees to an angle in radians, we can use the fact that 2π radians $= 360°$, or π radians $= 180°$; hence,

$$\theta \, (\text{rad}) = \frac{2\pi}{360°} \, \theta \, (\text{deg}) = \frac{\pi}{180°} \, \theta \, (\text{deg})$$

For example, 60° equals $\pi/3$ rad, and 45° equals $\pi/4$ rad.

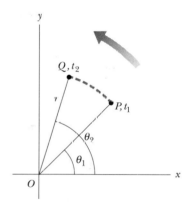

In Figure 10.2, as the particle travels from P to Q in a time Δt, the radius vector sweeps out an angle of $\Delta\theta = \theta_2 - \theta_1$, which equals the **angular displacement** during the time interval Δt. We define the **average angular speed,** $\overline{\omega}$ (omega), as the ratio of this angular displacement to the time interval Δt:

$$\overline{\omega} \equiv \frac{\theta_2 - \theta_1}{t_2 - t_1} = \frac{\Delta\theta}{\Delta t} \qquad [10.2]$$

Figure 10.2   A particle on a rotating rigid object moves from P to Q along the arc of a circle. In the time interval $\Delta t = t_2 - t_1$, the radius vector sweeps out an angle $\Delta\theta = \theta_2 - \theta_1$.

• *Average angular speed*

By analogy with linear speed, the **instantaneous angular speed,** ω, is defined as the limit of the ratio in Equation 10.2 as Δt approaches zero:

$$\omega \equiv \lim_{\Delta t \to 0} \frac{\Delta\theta}{\Delta t} = \frac{d\theta}{dt} \qquad [10.3]$$

• *Instantaneous angular speed*

Angular speed has units of rad/s (or s^{-1} because radians are not dimensional). Let us adopt the convention that the fixed axis of rotation for the disk is the z axis, as in Figure 10.1. We shall take ω to be positive when θ is increasing (counterclockwise motion) and negative when θ is decreasing (clockwise motion).

If the instantaneous angular speed of a particle changes from ω_1 to ω_2 in the time interval Δt, the particle has an angular acceleration. The **average angular acceleration,** $\overline{\alpha}$ (alpha), of a rotating particle is defined as the ratio of the change in the angular speed to the time interval, Δt:

Average angular acceleration •

$$\overline{\alpha} \equiv \frac{\omega_2 - \omega_1}{t_2 - t_1} = \frac{\Delta\omega}{\Delta t} \qquad [10.4]$$

By analogy with linear acceleration, the **instantaneous angular acceleration** is defined as the limit of the ratio $\Delta\omega/\Delta t$ as Δt approaches zero:

Instantaneous angular • *acceleration*

$$\alpha \equiv \lim_{\Delta t \to 0} \frac{\Delta\omega}{\Delta t} = \frac{d\omega}{dt} \qquad [10.5]$$

Angular acceleration has units of rad/s^2 or s^{-2}.

Let us now generalize our argument from the circular disk to any rigid body. A rigid body is any object the elements of which remain fixed with respect to one another. **For rotation about a fixed axis, every particle on a rigid body, such as a circular disk, has the same angular velocity and the same angular acceleration.** That is, the vector quantities $\boldsymbol{\omega}$ and $\boldsymbol{\alpha}$ characterize the rotational motion of the *entire* rigid body. Using these quantities, we can greatly simplify the analysis of rigid-body rotation.

The angular displacement ($\boldsymbol{\theta}$), angular velocity ($\boldsymbol{\omega}$), and angular acceleration ($\boldsymbol{\alpha}$) are analogous to linear displacement ($\mathbf{x}$), linear velocity ($\mathbf{v}$), and linear acceleration ($\mathbf{a}$), respectively, for the corresponding motion discussed in Chapter 2. The variables θ, ω, and α differ dimensionally from the linear variables, x, v, and a, only by a length factor.

We have indicated how the signs for $\boldsymbol{\omega}$ and $\boldsymbol{\alpha}$ are determined, but we have not specified any direction in space associated with these vector quantities.[1] For rotation about a fixed axis, the only direction in space that uniquely specifies the rotational motion is the direction along the axis of rotation. However, we must also decide on the sense of these quantities—that is, whether they point into or out of the plane of Figure 10.1.

The direction of $\boldsymbol{\omega}$ is along the axis of rotation, which is the z axis in Figure 10.1. By convention, we take the direction of $\boldsymbol{\omega}$ to be *out of* the plane of the diagram when the rotation is counterclockwise and *into* the plane of the diagram when the rotation is clockwise. To further illustrate this convention, it is convenient to use the **right-hand rule** illustrated by Figure 10.3a. The four fingers of the right hand are wrapped in the direction of the rotation. The extended right thumb points in

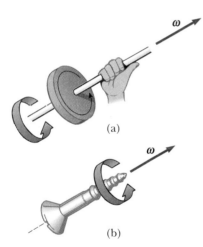

Figure 10.3   (a) The right-hand rule for determining the direction of the angular velocity vector. (b) The direction of $\boldsymbol{\omega}$ is in the direction of advance of a right-handed screw.

---

[1]Although we do not verify it here, the instantaneous angular velocity and instantaneous angular acceleration are vector quantities, but the corresponding average values are not. This is because angular displacement is not a vector quantity for finite rotations.

the direction of $\boldsymbol{\omega}$. Figure 10.3b illustrates that $\boldsymbol{\omega}$ is also in the direction of advance of a similarly rotating right-handed screw.

The sense of $\boldsymbol{\alpha}$ follows from its definition as $d\boldsymbol{\omega}/dt$. It is the same as $\boldsymbol{\omega}$ if the angular speed (the magnitude of $\boldsymbol{\omega}$) is increasing in time and antiparallel to $\boldsymbol{\omega}$ if the angular speed is decreasing in time.

10.2 • ROTATIONAL KINEMATICS

In our study of linear motion, we found that the simplest accelerated motion to analyze is motion under constant linear acceleration (Chapter 2). Likewise, for rotational motion about a fixed axis, the simplest accelerated motion to analyze is motion under constant angular acceleration. Therefore, we shall next develop kinematic relations for rotational motion under constant angular acceleration.

If we write Equation 10.5 in the form $d\omega = \alpha \, dt$ and let $\omega = \omega_0$ at $t_0 = 0$, we can integrate this expression directly:

$$\omega = \omega_0 + \alpha t \qquad (\alpha = \text{constant}) \qquad [10.6]$$

Likewise, substituting Equation 10.6 into Equation 10.3 and integrating once more (with $\theta = \theta_0$ at $t_0 = 0$), we get

$$\theta = \theta_0 + \omega_0 t + \tfrac{1}{2}\alpha t^2 \qquad [10.7]$$

• *Rotational kinematic equations ($\alpha =$ constant)*

If we eliminate t from Equations 10.6 and 10.7, we get

$$\omega^2 = \omega_0{}^2 + 2\alpha(\theta - \theta_0) \qquad [10.8]$$

If we eliminate α, we obtain

$$\theta = \theta_0 + \tfrac{1}{2}(\omega_0 + \omega)t \qquad [10.9]$$

Notice that these kinematic expressions for rotational motion under constant angular acceleration are of the *same form* as those for linear motion under constant linear acceleration, with the substitutions $x \rightarrow \theta$, $v \rightarrow \omega$, and $a \rightarrow \alpha$ (Table 10.1). Furthermore, the expressions are valid for both rigid-body rotation about a *fixed* axis and particle motion about a *fixed* axis.

TABLE 10.1   A Comparison of Kinematic Equations for Rotational and Linear Motion Under Constant Acceleration

Rotational Motion About a Fixed Axis with $\alpha =$ Constant	Linear Motion with $a =$ Constant
Variables: θ and ω	Variables: x and v
$\omega = \omega_0 + \alpha t$	$v = v_0 + at$
$\theta = \theta_0 + \omega_0 t + \tfrac{1}{2}\alpha t^2$	$x = x_0 + v_0 t + \tfrac{1}{2}at^2$
$\theta = \theta_0 + \tfrac{1}{2}(\omega_0 + \omega)t$	$x = x_0 + \tfrac{1}{2}(v_0 + v)t$
$\omega^2 = \omega_0{}^2 + 2\alpha(\theta - \theta_0)$	$v^2 = v_0{}^2 + 2a(x - x_0)$

Example 10.1   Rotating Wheel

A wheel rotates with a constant angular acceleration of 3.50 rad/s². If the angular speed of the wheel is 2.00 rad/s at $t_0 = 0$, (a) what angle does the wheel rotate through in 2.00 s?

Solution

$$\theta - \theta_0 = \omega_0 t + \tfrac{1}{2}\alpha t^2 = \left(2.00 \; \frac{rad}{s}\right)(2.00 \; s)$$

$$+ \tfrac{1}{2}\left(3.50 \; \frac{rad}{s^2}\right)(2.00 \; s)^2$$

$$= \; 11.0 \; rad = 630° = 1.75 \; rev$$

(b) What is the angular speed at $t = 2.00$ s?

Solution

$$\omega = \omega_0 + \alpha t = 2.00 \; rad/s + \left(3.50 \; \frac{rad}{s^2}\right)(2.00 \; s)$$

$$= \; 9.00 \; rad/s$$

We could also obtain this result using Equation 10.8 and the results of part (a). Try it!

EXERCISE 1   Find the angle that the wheel rotates through between $t = 2.00$ s and $t = 3.00$ s.   Answer   10.8 rad

EXERCISE 2   A wheel starts from rest and rotates with constant angular acceleration to an angular speed of 12.0 rad/s in 3.00 s. Find (a) the magnitude of the angular acceleration of the wheel and (b) the angle in radians through which it rotates in this time. Answer   (a) 4.00 rad/s²   (b) 18.0 rad

10.3 • RELATIONS BETWEEN ANGULAR AND LINEAR QUANTITIES

In this section we shall derive some useful relations between the angular speed and angular acceleration of a rotating particle and its linear speed and linear acceleration. Keep in mind that, when a rigid body rotates about a fixed axis, *every* particle of the body moves in a circle whose center is the axis of rotation.

Consider a particle rotating in a circle of radius r about the z axis, as in Figure 10.4. Because the particle moves along a circular path, its linear velocity vector **v** is always tangent to the path; hence, we often call this quantity *tangential velocity*. The magnitude of the tangential velocity of the particle is, by definition, ds/dt, where s is the distance traveled by the particle along the circular path. Recalling from Equation 10.1 that $s = r\theta$, and noting that r is a constant, we get

$$v = \frac{ds}{dt} = r\frac{d\theta}{dt}$$

$$v = r\omega \qquad\qquad [10.10]$$

That is, the magnitude of the tangential velocity of the particle equals the distance of the particle from the axis of rotation multiplied by the particle's angular speed.

We can relate the angular acceleration of the particle to its tangential acceleration, a_t—which is the component of its acceleration tangent to the path of motion—by taking the time derivative of v:

$$a_t = \frac{dv}{dt} = r\frac{d\omega}{dt}$$

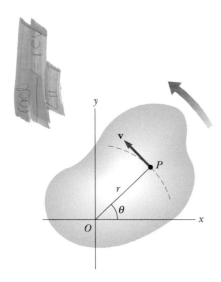

Figure 10.4   As a rigid object rotates about the fixed axis through O, the point P has a linear velocity **v** that is always tangent to the circular path of radius r.

Relationship between linear and angular speed •

$$a_t = r\alpha \qquad\qquad\qquad [10.11]$$

That is, the tangential component of the linear acceleration of a particle undergoing circular motion equals the distance of the particle from the axis of rotation multiplied by the angular acceleration.

In Chapter 3 we found that a particle rotating in a circular path undergoes a centripetal, or radial, acceleration of magnitude v^2/r directed toward the center of rotation (Fig. 10.5). Because $v = r\omega$, we can express the centripetal acceleration of the particle as

$$a_r = \frac{v^2}{r} = r\omega^2 \qquad\qquad\qquad [10.12]$$

The *total linear acceleration* of the particle is $\mathbf{a} = \mathbf{a}_t + \mathbf{a}_r$. Therefore, the magnitude of the total linear acceleration of the particle is

$$a = \sqrt{a_t^2 + a_r^2} = \sqrt{r^2\alpha^2 + r^2\omega^4} = r\sqrt{\alpha^2 + \omega^4} \qquad [10.13]$$

* **Relationship between tangential and angular acceleration**

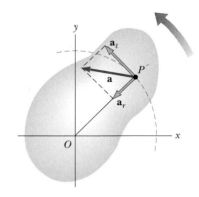

Figure 10.5 As a rigid object rotates about a fixed axis through O, the point P experiences a tangential component of linear acceleration, $\mathbf{a}_t$, and a centripetal component of linear acceleration, $\mathbf{a}_r$. The total linear acceleration of this point is $\mathbf{a} = \mathbf{a}_t + \mathbf{a}_r$.

Thinking Physics 1

A phonograph record (LP—for *long-playing*) is rotated at a constant *angular* speed. A compact disk (CD) is rotated so that the surface sweeps past the laser at a constant *tangential* speed. Consider two circular "grooves" of information on an LP—one near the outer edge and one near the inner edge. Suppose the outer groove "contains" 1.8 seconds of music. Does the inner groove also contain 1.8 seconds of music? How about the CD—do the inner and outer grooves contain the same time interval of music?

Reasoning On the LP, the inner and outer grooves must both rotate once in the same time period. Thus, each groove, regardless of where it is on the record, contains the same time interval of information. Of course, on the inner grooves, this same information must be compressed into a smaller circumference. On a CD, the constant tangential speed requires that there be no such compression—the digital pits representing the information are spaced uniformly everywhere on the surface. Thus, there is more information in an outer groove, due to its larger circumference, and, as a result, a longer time interval of music than the inner groove.

Thinking Physics 2

The launch area for the European Space Agency is not in Europe—it is in French Guiana, near the Equator in South America. Why?

Reasoning Placing a satellite in Earth orbit requires providing a large tangential speed to the satellite. This is the task of the rocket propulsion system. Anything that reduces the requirements on the rocket system is a welcome contribution. The surface of the Earth is already traveling toward the east at a high speed, due to the rotation of the Earth. Thus, if rockets are launched toward the east, the rotation of the Earth provides some initial tangential speed, reducing somewhat the requirements on the rocket. If rockets were launched from Europe, which is at a relatively large angle latitude, the contribution of the Earth's rotation is relatively small, because the distance between Europe and the rotation axis of the Earth is relatively small. The ideal place for launching is at the Equator, which is as far as one can be from the rotation axis of the Earth and still be on the surface of the Earth. This results in the largest possible tangential speed due to the Earth's rotation. The European Space Agency exploits this advantage by launching from French Guiana, which is only a few degrees north of the Equator.

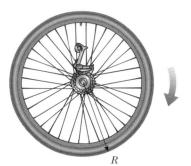

Figure 10.6 (Conceptual Problem 2) A wheel of radius R.

CONCEPTUAL PROBLEM 1

When a tall smokestack falls over, it often breaks somewhere along its length before it hits the ground. Why does this happen?

CONCEPTUAL PROBLEM 2

When a wheel of radius R rotates about a fixed axis as in Figure 10.6, do all points on the wheel have the same angular speed? Do they all have the same linear speed? If the angular speed is constant and equal to ω, describe the linear speeds and linear accelerations of the points located at $r = 0$, $r = R/2$, and $r = R$, where the points are measured from the center of the wheel.

EXERCISE 3 A racing car travels on a circular track of radius 250 m. If the car moves with a constant linear speed of 45.0 m/s, find (a) its angular speed and (b) the magnitude and direction of its acceleration.   Answer   (a) 0.180 rad/s   (b) 8.10 m/s² toward the center of the track

EXERCISE 4 An automobile accelerates from zero to 30 m/s in 6.0 s. The wheels have diameters of 0.40 m. What is the angular acceleration of each wheel?
Answer   25 rad/s²

10.4 • ROTATIONAL KINETIC ENERGY

Let us consider a rigid body as a collection of particles, and let us assume that the body rotates about the fixed z axis with an angular speed of ω (Fig. 10.7). Each particle of the body has some kinetic energy, determined by its mass and speed. If the mass of the ith particle is m_i and its speed is v_i, the kinetic energy of this particle is

$$K_i = \tfrac{1}{2} m_i v_i^2$$

To proceed further, we must recall that, although every particle in the rigid body has the same angular speed ω, the individual linear speeds depend on the distance r_i from the axis of rotation, according to the expression $v_i = r_i \omega$ (Eq. 10.10). The *total* kinetic energy of the rotating rigid body is the sum of the kinetic energies of the individual particles:

$$K_R = \sum K_i = \sum \tfrac{1}{2} m_i v_i^2 = \tfrac{1}{2} \sum m_i r_i^2 \omega^2$$

$$K_R = \tfrac{1}{2} \left(\sum m_i r_i^2 \right) \omega^2$$

where we have factored ω^2 from the sum because it is common to every particle. The quantity in parentheses is called the **moment of inertia**, I:

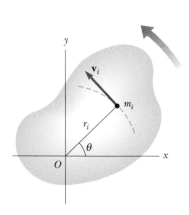

Figure 10.7 A rigid object rotating about the z axis with angular speed ω. The kinetic energy of the particle of mass m_i is $\tfrac{1}{2} m_i v_i^2$. The total energy of the object is $\tfrac{1}{2} I \omega^2$.

*Moment of inertia for a •
system of particles*

$$I = \sum m_i r_i^2 \qquad\qquad [10.14]$$

Therefore, we can express the kinetic energy of the rotating rigid body as

$$K_R = \tfrac{1}{2} I \omega^2 \qquad\qquad [10.15]$$

• *Kinetic energy of a rotating rigid body*

From the definition of moment of inertia, we see that it has dimensions of ML^2 ($\text{kg} \cdot \text{m}^2$ in SI units). The moment of inertia is a measure of a system's resistance to change in its angular velocity. It plays the role of mass in *all* rotational equations. However, the moment of inertia depends on the location and orientation of the axis, and is not a single property of an object the way its mass is. Although we shall commonly refer to the quantity $\tfrac{1}{2} I \omega^2$ as the **rotational kinetic energy,** it is not a new form of energy. It is ordinary kinetic energy, because it was derived from a sum over individual kinetic energies of the particles contained in the rigid body. However, the form of the kinetic energy given by Equation 10.15 is a convenient one for dealing with rotational motion, provided we know how to calculate I. It is important to recognize the analogy between kinetic energy associated with linear motion, $\tfrac{1}{2} mv^2$, and rotational kinetic energy, $\tfrac{1}{2} I \omega^2$. The quantities I and ω in rotational motion are analogous to m and v in linear motion, respectively.

Thinking Physics 3

The temperature of a gas is related to the average *translational* kinetic energy of the molecules of the gas. If we add energy to a gas, the average translational kinetic energy increases and, therefore, the temperature increases. Consider adding the same amount of energy to a monatomic gas, such as helium, and a diatomic gas, such as oxygen (O_2). The temperature of the monatomic gas will rise by more than that of the diatomic gas. Why?

Reasoning In a monatomic gas, the only type of kinetic energy possible is translational kinetic energy. Thus, all of the energy added to the gas appears as translational kinetic energy, with a corresponding large rise in temperature. In the diatomic gas, some of the energy added to the gas can appear as rotational kinetic energy of the molecules. Thus, only part of the energy added appears as translational kinetic energy, and the temperature rises by a lesser amount than for the monatomic gas.

Example 10.2 The Oxygen Molecule

Consider the diatomic molecule oxygen, O_2, which is rotating in the xy plane about the z axis passing through its center, perpendicular to its length. The mass of each oxygen atom is 2.66×10^{-26} kg, and at room temperature, the average separation between the two oxygen atoms is $d = 1.21 \times 10^{-10}$ m (the atoms are treated as point masses). (a) Calculate the moment of inertia of the molecule about the z axis.

Solution Because the distance of each atom from the z axis is $d/2$, the moment of inertia about the z axis is

$$I = \sum m_i r_i^2 = m \left(\frac{d}{2}\right)^2 + m \left(\frac{d}{2}\right)^2 = \frac{md^2}{2}$$

$$= \left(\frac{2.66 \times 10^{-26}}{2}\, \text{kg}\right) (1.21 \times 10^{-10}\ \text{m})^2$$

$$= 1.95 \times 10^{-46}\ \text{kg} \cdot \text{m}^2$$

(b) If the angular speed of the molecule about the z axis is 4.60×10^{12} rad/s, what is its rotational kinetic energy?

Solution

$$K_R = \tfrac{1}{2} I \omega^2$$

$$= \tfrac{1}{2} (1.95 \times 10^{-46}\ \text{kg} \cdot \text{m}^2) \left(4.60 \times 10^{12}\, \frac{\text{rad}}{\text{s}}\right)^2$$

$$= 2.06 \times 10^{-22}\ \text{J}$$

This is approximately equal to the average kinetic energy associated with the linear motion of the molecule at room temperature, which is about 6.2×10^{-21} J.

Example 10.3   Four Rotating Masses

Four point masses are fastened to the corners of a frame of negligible mass lying in the xy plane (Fig. 10.8). (a) If the rotation of the system occurs about the y axis with an angular speed ω, find the moment of inertia about the y axis and the rotational kinetic energy about this axis.

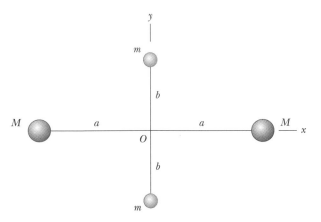

Figure 10.8 (Example 10.3) All four masses are at a fixed separation as shown. The moment of inertia depends on the axis about which it is evaluated.

Solution First, note that the two masses m that lie on the y axis do not contribute to I_y (that is, $r_i = 0$ for these masses about this axis). Applying Equation 10.14, we get

$$I_y = \sum m_i r_i^2 = Ma^2 + Ma^2 = 2Ma^2$$

Therefore, the rotational kinetic energy about the y axis is

$$K_R = \tfrac{1}{2} I_y \omega^2 = \tfrac{1}{2}(2Ma^2)\omega^2 = \boxed{Ma^2\omega^2}$$

The fact that the masses m do not enter into this result makes sense, because they have no motion about the chosen axis of rotation; hence, they have no kinetic energy.

(b) Suppose the system rotates in the xy plane about an axis through O (the z axis). Calculate the moment of inertia about the z axis and the rotational energy about this axis.

Solution Because r_i in Equation 10.14 is the *perpendicular* distance to the axis of rotation, we get

$$I_z = \sum m_i r_i^2 = Ma^2 + Ma^2 + mb^2 + mb^2 = \boxed{2Ma^2 + 2mb^2}$$

$$K_R = \tfrac{1}{2} I_z \omega^2 = \tfrac{1}{2}(2Ma^2 + 2mb^2)\omega^2 = \boxed{(Ma^2 + mb^2)\omega^2}$$

Comparing the results for (a) and (b), we conclude that the moment of inertia and, therefore, the rotational energy associated with a given angular speed depend on the axis of rotation. In (b), we expect the result to include all masses and distances because all four masses are in motion for rotation in the xy plane. Furthermore, the fact that the rotational energy in (a) is smaller than in (b) indicates that it would take less effort (work) to set the system into rotation about the y axis than about the z axis.

10.5 • TORQUE AND THE VECTOR PRODUCT

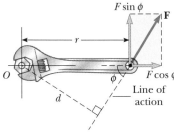

Figure 10.9 The force **F** has a greater rotating tendency about O as F increases and as the moment arm, d, increases. It is the component $F \sin \phi$ that tends to rotate the system about O.

When a force is exerted on a rigid body pivoted about some axis, the body tends to rotate about that axis. The tendency of a force to rotate a body about some axis is measured by a vector quantity called **torque** (τ). Consider the wrench pivoted about the axis through O in Figure 10.9. The applied force **F** generally can act at an angle of ϕ to the horizontal. We define the magnitude of the torque, τ (Greek letter tau), resulting from the force **F** with the expression

$$\tau \equiv rF \sin \phi = Fd \qquad \text{[10.16]}$$

It is very important to recognize that **torque is defined only when a reference axis is specified.** The quantity $d = r \sin \phi$, called the **moment arm** (or *lever arm*) of the force **F**, represents the perpendicular distance from the rotation axis to the line of action of **F**. Note that the only component of **F** that tends to cause a rotation is $F \sin \phi$, the component perpendicular to r. The horizontal component, $F \cos \phi$, passes through O and has no tendency to produce a rotation.

If two or more forces are acting on a rigid body, as in Figure 10.10, then each has a tendency to produce a rotation about the pivot at O. For example, $\mathbf{F}_2$ tends to rotate the body clockwise, and $\mathbf{F}_1$ tends to rotate the body counterclockwise. We shall use the convention that the sign of the torque resulting from a force is positive if its turning tendency is counterclockwise and negative if its turning tendency is clockwise. For example, in Figure 10.10, the torque resulting from $\mathbf{F}_1$, which has a moment arm of d_1, is *positive* and equal to $+F_1d_1$; the torque from $\mathbf{F}_2$ is *negative* and equal to $-F_2d_2$. Hence, the *net* torque acting on the rigid body about O is

$$\tau_{\text{net}} = \tau_1 + \tau_2 = F_1d_1 - F_2d_2$$

From the definition of torque, we see that the rotating tendency increases as F increases and as d increases. For example, it is easier to close a door if we push at the doorknob rather than at a point close to the hinge. **Torque should not be confused with force.** Torque has units of force times length—N·m in SI units.

Now consider a force $\mathbf{F}$ acting on a particle located at the vector position $\mathbf{r}$ (Fig. 10.11). The origin O is assumed to be in an inertial frame, so Newton's second law is valid. The *magnitude* of the torque due to this force relative to the origin is, by definition, equal to $rF\sin\phi$, where ϕ is the angle between $\mathbf{r}$ and $\mathbf{F}$. The axis about which $\mathbf{F}$ would tend to produce rotation is perpendicular to the plane formed by $\mathbf{r}$ and $\mathbf{F}$. If the force lies in the xy plane, as in Figure 10.11, then the torque is represented by a vector parallel to the z axis. The force in Figure 10.11 creates a torque that tends to rotate the body counterclockwise looking down the z axis, and so the sense of τ is toward increasing z, and τ is in the positive z direction. If we reversed the direction of $\mathbf{F}$ in Figure 10.11, τ would then be in the negative z direction. The torque involves two vectors, $\mathbf{r}$ and $\mathbf{F}$, and is in fact defined to be equal to the **vector product,** or **cross product,** of $\mathbf{r}$ and $\mathbf{F}$:

$$\tau \equiv \mathbf{r} \times \mathbf{F} \qquad [10.17]$$

We now give a formal definition of the vector product. Given any two vectors $\mathbf{A}$ and $\mathbf{B}$, the vector product $\mathbf{A} \times \mathbf{B}$ is defined as a third vector, $\mathbf{C}$, the *magnitude* of which is $AB\sin\theta$, where θ is the angle included between $\mathbf{A}$ and $\mathbf{B}$:

$$\mathbf{C} = \mathbf{A} \times \mathbf{B} \qquad [10.18]$$

$$C \equiv |\mathbf{C}| = |AB\sin\theta| \qquad [10.19]$$

Note that the quantity $AB\sin\theta$ is equal to the area of the parallelogram formed by $\mathbf{A}$ and $\mathbf{B}$, as shown in Figure 10.12. The *direction* of $\mathbf{A} \times \mathbf{B}$ is perpendicular to the plane formed by $\mathbf{A}$ and $\mathbf{B}$, and its sense is determined by the advance of a right-handed screw when the screw is turned from $\mathbf{A}$ to $\mathbf{B}$ through the angle θ. A more convenient rule to use for the direction of $\mathbf{A} \times \mathbf{B}$ is the right-hand rule illustrated in Figure 10.12. The four fingers of the right hand are pointed along $\mathbf{A}$ and then

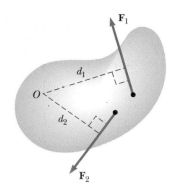

Figure 10.10 The force $\mathbf{F}_1$ tends to rotate the object counterclockwise about O, and $\mathbf{F}_2$ tends to rotate the object clockwise.

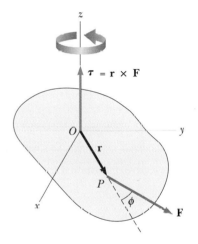

Figure 10.11 The torque vector τ lies in a direction perpendicular to the plane formed by the position vector $\mathbf{r}$ and the applied force $\mathbf{F}$.

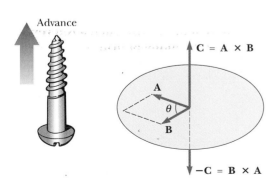 

Figure 10.12 The vector product $\mathbf{A} \times \mathbf{B}$ is a third vector $\mathbf{C}$ having a magnitude $AB\sin\theta$ equal to the area of the parallelogram shown. The direction of $\mathbf{C}$ is perpendicular to the plane formed by $\mathbf{A}$ and $\mathbf{B}$, and its sense is determined by the right-hand rule.

The torque exerted on the nail by the hammer increases as the length of the handle (lever arm) increases.   *(© Richard Megna, 1991, Fundamental Photographs)*

"wrapped" into **B** through the angle θ. The direction of the erect right thumb is the direction of **A** × **B**. Because of the notation, **A** × **B** is often read "**A** cross **B**"; hence the term *cross product*.

Some properties of the vector product follow from its definition:

- Unlike the case of the scalar product, the order in which the two vectors are multiplied in a cross product is important; that is

$$\mathbf{A} \times \mathbf{B} = -(\mathbf{B} \times \mathbf{A}) \qquad [10.20]$$

Therefore, if you change the order of the cross product, you must change the sign. One could easily verify this relation with the right-hand rule (Fig. 10.12).

- If **A** is parallel to **B** ($\theta = 0°$ or $180°$), then **A** × **B** = 0; therefore, it follows that **A** × **A** = 0.
- If **A** is perpendicular to **B**, then $|\mathbf{A} \times \mathbf{B}| = AB$.

From Equations 10.18 and 10.19 and the definition of unit vectors, one finds that the cross products of the rectangular unit vectors **i**, **j**, and **k** obey the following expressions:

Cross products of unit vectors •

$$\mathbf{i} \times \mathbf{i} = \mathbf{j} \times \mathbf{j} = \mathbf{k} \times \mathbf{k} = 0$$

$$\mathbf{i} \times \mathbf{j} = -\mathbf{j} \times \mathbf{i} = \mathbf{k}$$

$$\mathbf{j} \times \mathbf{k} = -\mathbf{k} \times \mathbf{j} = \mathbf{i} \qquad [10.21]$$

$$\mathbf{k} \times \mathbf{i} = -\mathbf{i} \times \mathbf{k} = \mathbf{j}$$

Signs are interchangeable. For example, $\mathbf{i} \times (-\mathbf{j}) = -\mathbf{i} \times \mathbf{j} = -\mathbf{k}$.

CONCEPTUAL PROBLEM 3

Both torque and work are products of force and distance. How are they different? Do they have the same units?

Example 10.4   The Net Torque on a Cylinder

A one-piece cylinder is shaped as in Figure 10.13, with a core section protruding from the larger drum. The cylinder is free to rotate around the central axis shown in the drawing. A rope wrapped around the drum, of radius R_1, exerts a force **F**$_1$ to the right on the cylinder. A rope wrapped around the core, of radius R_2, exerts a force **F**$_2$ downward on the cylinder. (a) What is the net torque acting on the cylinder about the rotation axis (which is the z axis in Fig. 10.13)?

Solution   The torque due to **F**$_1$ is $-R_1F_1$ and is negative because it tends to produce a clockwise rotation. The torque due to **F**$_2$ is $+R_2F_2$ and is positive because it tends to produce a counterclockwise rotation. Therefore, the net torque about the rotation axis is

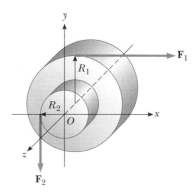

Figure 10.13   (Example 10.4) A solid cylinder pivoted about the z axis through O. The moment arm of **F**$_1$ is R_1, and the moment arm of **F**$_2$ is R_2.

$$\tau_{\text{net}} = \tau_1 + \tau_2 = R_2 F_2 - R_1 F_1$$

(b) Suppose $F_1 = 5.0$ N, $R_1 = 1.0$ m, $F_2 = 6.0$ N, and $R_2 = 0.50$ m. What is the net torque about the rotation axis and which way does the cylinder rotate?

Solution

$$\tau_{\text{net}} = (6.0 \text{ N})(0.50 \text{ m}) - (5.0 \text{ N})(1.0 \text{ m}) = -2.0 \text{ N·m}$$

Because the net torque is negative, the cylinder rotates clockwise.

Example 10.5 The Cross Product

Two vectors lying in the xy plane are given by the equations $\mathbf{A} = 2\mathbf{i} + 3\mathbf{j}$ and $\mathbf{B} = -\mathbf{i} + 2\mathbf{j}$. Find $\mathbf{A} \times \mathbf{B}$, and verify explicitly that $\mathbf{A} \times \mathbf{B} = -\mathbf{B} \times \mathbf{A}$.

Solution Using Equations 10.21 for the cross product of unit vectors gives

$$\mathbf{A} \times \mathbf{B} = (2\mathbf{i} + 3\mathbf{j}) \times (-\mathbf{i} + 2\mathbf{j})$$
$$= 2\mathbf{i} \times 2\mathbf{j} + 3\mathbf{j} \times (-\mathbf{i}) = 4\mathbf{k} + 3\mathbf{k} = 7\mathbf{k}$$

(We have omitted the terms with $\mathbf{i} \times \mathbf{i}$ and $\mathbf{j} \times \mathbf{j}$, because they are zero.)

$$\mathbf{B} \times \mathbf{A} = (-\mathbf{i} + 2\mathbf{j}) \times (2\mathbf{i} + 3\mathbf{j})$$
$$= -\mathbf{i} \times 3\mathbf{j} + 2\mathbf{j} \times 2\mathbf{i} = -3\mathbf{k} - 4\mathbf{k} = -7\mathbf{k}$$

Therefore, $\mathbf{A} \times \mathbf{B} = -\mathbf{B} \times \mathbf{A}$.

EXERCISE 5 Use the results to this example and Equation 10.19 to find the angle between $\mathbf{A}$ and $\mathbf{B}$.   Answer   60.3°

EXERCISE 6 A particle is located at the vector position $\mathbf{r} = (\mathbf{i} + 3\mathbf{j})$ m, and the force acting on it is $\mathbf{F} = (3\mathbf{i} + 2\mathbf{j})$ N. What is the torque about (a) the origin and (b) the point having coordinates $(0,6)$ m?   Answer   (a) $(-7\mathbf{k})$ N·m   (b) $(11\mathbf{k})$ N·m

EXERCISE 7 If $|\mathbf{A} \times \mathbf{B}| = \mathbf{A} \cdot \mathbf{B}$, what is the angle between $\mathbf{A}$ and $\mathbf{B}$?   Answer   45.0°

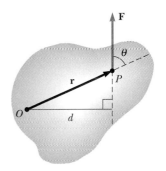

10.6 · EQUILIBRIUM OF A RIGID OBJECT

Consider a single force $\mathbf{F}$ acting on a rigid object as in Figure 10.14. The effect of the force depends on its point of application, P. If $\mathbf{r}$ is the position vector of this point relative to O, the torque associated with the force $\mathbf{F}$ about O is given by $\boldsymbol{\tau} = \mathbf{r} \times \mathbf{F}$.

As you can see from Figure 10.14, the tendency of $\mathbf{F}$ to make the object rotate about an axis through O depends on the moment arm d as well as on the magnitude of $\mathbf{F}$. By definition, the magnitude of $\boldsymbol{\tau}$ is Fd.

Now suppose either of two forces, $\mathbf{F}_1$ and $\mathbf{F}_2$, act on a rigid object. The two forces will have the same effect on the object only if they have the same magnitude, the same direction, and the same line of action. In other words, **two forces $\mathbf{F}_1$ and $\mathbf{F}_2$ have an equivalent effect on a rigid body if and only if $F_1 = F_2$ and if the two have the same torque about any given point**.

Two equal and opposite forces that are *not* equivalent are shown in Figure 10.15. The force directed to the right tends to rotate the object clockwise about an axis perpendicular to the diagram through O, whereas the force directed to the left tends to rotate it counterclockwise about that axis.

When pivoted about an axis through its center of mass, an object undergoes an angular acceleration about this axis if there is a nonzero torque acting. As an example, suppose an object is pivoted about an axis through its center of mass as in Figure 10.16. Two equal and opposite forces act in the directions shown, such

Figure 10.14 A single force $\mathbf{F}$ acts on a rigid object at the point P. The moment arm of $\mathbf{F}$ relative to O is the perpendicular distance d from O to the line of action of $\mathbf{F}$.

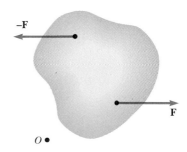

Figure 10.15 The two forces acting on the object are equal in magnitude and opposite in direction, yet the object is not in equilibrium.

that their lines of action do not pass through the center of mass. A pair of forces acting in this manner form what is called a **couple.** (The two forces shown in Figure 10.15 also form a couple.) Because each force produces the same torque, *Fd,* the net torque has a magnitude 2*Fd.* Clearly, the object rotates clockwise and undergoes an angular acceleration about the axis. This is a nonequilibrium situation as far as the rotational motion is concerned.

In general, an object is in rotational equilibrium only if its angular acceleration $\alpha = 0$. Because torque causes angular acceleration, **a necessary condition of equilibrium is that the net torque about any origin must be zero.** We now have **two necessary conditions for equilibrium of an object,** which can be stated as follows:

Two conditions for •
equilibrium of an object

The resultant external force must equal zero.	$\sum \mathbf{F} = 0$	**[10.22]**
The resultant external torque must be zero about *any* origin.	$\sum \boldsymbol{\tau} = 0$	**[10.23]**

The first condition is a statement of translational equilibrium; it tells us that the linear acceleration of the center of mass of the object must be zero when viewed from an inertial reference frame. The second condition is a statement of rotational equilibrium and tells us that the angular acceleration about any axis must be zero. In the special case of **static equilibrium,** the object is at rest so that it has no linear or angular speed (that is, $v_{CM} = 0$ and $\omega = 0$). Note that if the resultant external force is zero, and the resultant external torque is zero about some origin, then the resultant external torque is zero about every other origin.

The two vector expressions given by Equations 10.22 and 10.23 are equivalent, in general, to six scalar equations, three from the first condition of equilibrium and three from the second (corresponding to *x, y,* and *z* components). Hence, in a complex system involving several forces acting in various directions, you would be faced with solving a set of equations with many unknowns. Here, we restrict our discussion to situations in which all the forces lie in the *xy* plane. (When a set of forces lie in the same plane, they are said to be *coplanar.*) With this restriction, we shall have to deal with only three scalar equations. Two of these come from balancing the forces in the *x* and *y* directions. The third comes from the torque equation—namely, that the net torque about *any* point in the *xy* plane must be zero. Hence, the two conditions of equilibrium provide the equations

$$\sum F_x = 0 \qquad \sum F_y = 0 \qquad \sum \tau_z = 0 \qquad \textbf{[10.24]}$$

where the axis of the torque equation is arbitrary, as we show later. In working static equilibrium problems, it is important to recognize external forces acting on the object. Failure to do so will result in an incorrect analysis. The following procedure is recommended when analyzing an object in equilibrium under the action of several external forces:

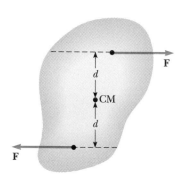

Figure 10.16   Two equal and opposite forces acting on the object form a couple. In this case, the object rotates clockwise. The magnitude of the net torque about the center of mass is 2*Fd.*

PROBLEM-SOLVING STRATEGY • **Objects in Equilibrium**

---

1. Make a sketch of the object under consideration.
2. Draw a free-body diagram and label all external forces acting on the object. Try to guess the correct direction for each force. If you select a direction that leads to a negative sign in your solution for a force, do not

be alarmed; this merely means that the direction of the force is the opposite of what you guessed.

3. Resolve all forces into rectangular components, choosing a convenient coordinate system. Then apply the first condition for equilibrium. Remember to keep track of the signs of the various force components.

4. Choose a convenient axis for calculating the net torque on the object. Remember that the choice of the origin for the torque equation is arbitrary; therefore, choose an origin that will simplify your calculation as much as possible. Usually the most convenient origin for calculating torques is the one at which most of the forces act.

5. The first and second conditions of equilibrium give a set of linear equations with several unknowns. All that is left is to solve the simultaneous equations for the unknowns in terms of the known quantities.

Example 10.6 Standing on a Horizontal Beam

A uniform horizontal beam of length 8.00 m and weight 200 N is attached to a wall by a pin connection. Its far end is supported by a cable that makes an angle of 53.0° with the horizontal (Fig. 10.17a). If a 600-N person stands 2.00 m from the wall, find the tension in the cable and the force exerted by the wall on the beam.

Solution First we must identify all the external forces acting on the beam. These are the gravitational force, the force **T** exerted by the cable, the force **R** exerted by the wall at the pivot (the direction of this force is unknown), and the force exerted by the person on the beam. These are all indicated in the free-body diagram for the beam (Fig. 10.17b). If we resolve **T** and **R** into horizontal and vertical components and apply the first condition for equilibrium, we get

(1) $\sum F_x = R \cos \theta - T \cos 53.0° = 0$

(2) $\sum F_y = R \sin \theta + T \sin 53.0° - 600\ \text{N} - 200\ \text{N} = 0$

Because R, T, and θ are all unknown, we cannot obtain a solution from these expressions alone. (The number of simultaneous equations must equal the number of unknowns in order for us to be able to solve for the unknowns.)

Now let us invoke the condition for rotational equilibrium. A convenient axis to choose for our torque equation is the one that passes through the pivot at O. The feature that makes this point so convenient is that the force **R** and the horizontal component of **T** both have a lever arm of zero, and hence zero torque, about this pivot. Recalling our convention for the sign of the torque about an axis and noting that the lever

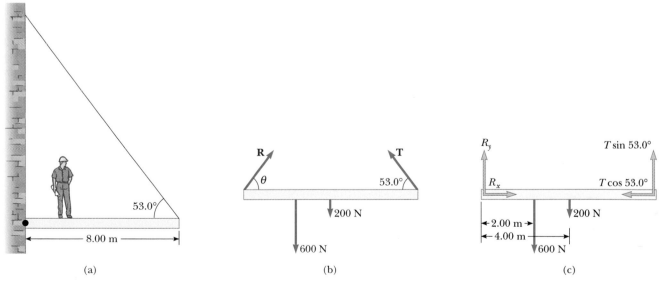

| (a) | (b) | (c) |

Figure 10.17 (Example 10.6) A uniform beam supported by a cable. (b) and (c) The free-body diagram for the beam.

arms of the 600-N, 200-N, and $T \sin 53°$ forces are 2.00 m, 4.00 m, and 8.00 m, respectively, we get

$$\Sigma \tau_O = (T \sin 53.0°)(8.00 \text{ m}) - (600 \text{ N})(2.00 \text{ m})$$
$$- (200 \text{ N})(4.00 \text{ m}) = 0$$

$$T = 313 \text{ N}$$

Thus the torque equation with this axis gives us one of the unknowns directly! This value is substituted into (1) and (2) to give

$$R \cos \theta = 188 \text{ N}$$

$$R \sin \theta = 550 \text{ N}$$

We divide these two equations and recall the trigonometric identity $\sin \theta / \cos \theta = \tan \theta$ to get

$$\tan \theta = \frac{550 \text{ N}}{188 \text{ N}} = 2.93$$

$$\theta = 71.1°$$

Finally,

$$R = \frac{188 \text{ N}}{\cos \theta} = \frac{188 \text{ N}}{\cos 71.1°} = 581 \text{ N}$$

If we had selected some other axis for the torque equation, the solution would have been the same. For example, if we had chosen to have the axis pass through the center of gravity of the beam, the torque equation would involve both T and R. However, this equation, coupled with (1) and (2), could still be solved for the unknowns. Try it!

When many forces are involved in a problem of this nature, it is convenient to set up a table of forces, lever arms, and torques. For instance, in the example just given, we would construct the following table. Setting the sum of the terms in the last column equal to zero represents the condition of rotational equilibrium.

Force Component	Lever Arm Relative to O (m)	Torque About O (N·m)
$T \sin 53.0°$	8.00	$(8.00)\,T \sin 53°$
$T \cos 53.0°$	0	0
200 N	4.00	$-(4.00)(200)$
600 N	2.00	$-(2.00)(600)$
$R \sin \theta$	0	0
$R \cos \theta$	0	0

Example 10.7   The Leaning Ladder

A uniform ladder of length ℓ and weight $w = 50$ N rests against a smooth, vertical wall (Fig. 10.18a). If the coefficient of static friction between ladder and ground is $\mu_s = 0.40$, find the minimum angle θ_{min} such that the ladder does not slip.

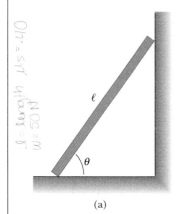

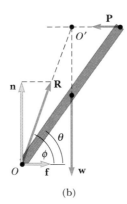

Figure 10.18   (Example 10.7) (a) A uniform ladder at rest, leaning against a frictionless wall. (b) The free-body diagram for the ladder. Note that the forces **R**, **w**, and **P** pass through a common point O'.

Solution   The free-body diagram showing all the external forces acting on the ladder is illustrated in Figure 10.18b. The

reaction **R** exerted by the ground on the ladder is the vector sum of a normal force, **n**, and the force of friction, **f**. The reaction force **P** exerted by the wall on the ladder is horizontal, because the wall is smooth. From the first condition of equilibrium applied to the ladder, we have

$$\sum F_x = f - P = 0$$
$$\sum F_y = n - w = 0$$

Because $w = 50$ N, we see from the second equation here that $n = w = 50$ N. Furthermore, **when the ladder is on the verge of slipping, the force of friction must be a maximum,** given by $f_{max} = \mu_s n = 0.40(50 \text{ N}) = 20$ N. (Recall Eq. 5.9: $f_s \leq \mu_s n$.) Thus, at this angle, $P = 20$ N.

To find θ, we must use the second condition of equilibrium. When the torques are taken about the origin O at the bottom of the ladder, we get

$$\sum \tau_O = P\ell \sin \theta - w\frac{\ell}{2} \cos \theta = 0$$

But $P = 20$ N when the ladder is about to slip and $w = 50$ N, so that this expression gives

$$\tan \theta_{min} = \frac{w}{2P} = \frac{50 \text{ N}}{40 \text{ N}} = 1.25$$

$$\theta_{min} = 51°$$

It is interesting to note that the result does not depend on ℓ or w. The answer depends only on μ_s.

An alternative approach to analyzing this problem is to consider the intersection O' of the lines of action of forces **w** and **P**. Because the torque about any origin must be zero, the torque about O' must be zero. This requires that the line of action of **R** (the resultant of **n** and **f**) pass through O'. That

is, because three forces act on this stationary object, the forces must be concurrent. With this condition, one could then obtain the angle ϕ that **R** makes with the horizontal (where ϕ is greater than θ), assuming the length of the ladder is known.

EXERCISE 8   For the angles labeled in Figure 10.18b, show that $\tan \phi = 2 \tan \theta$.

EXERCISE 9   Two children weighing 500 N and 350 N are on a uniform board weighing 40.0 N supported at its center (a see-saw). If the 500-N child is 1.50 m from the center, determine (a) the upward force exerted on the board by the support and (b) where the 350-N child must sit to balance the system.   Answer   (a) 890 N   (b) 2.14 m from the center

10.7 · RELATION BETWEEN TORQUE AND ANGULAR ACCELERATION

In this section we shall show that the angular acceleration of a particle rotating about a fixed axis is proportional to the net torque acting about that axis. The ideas embodied in this situation can easily be extended to the case of a rigid body rotating about a fixed axis.

Consider a particle of mass m rotating in a circle of radius r under the influence of both a tangential force, $\mathbf{F}_t$, and a radial force, $\mathbf{F}_r$, as in Figure 10.19. (The force $\mathbf{F}_r$, whose nature is not yet specified, *must* be present to keep the particle moving in its circular path.) According to Newton's second law, $\mathbf{F}_t = m\mathbf{a}_t$ and $\mathbf{F}_r = m\mathbf{a}_r$, where $\mathbf{a}_t$ is the tangential acceleration of the particle and $\mathbf{a}_r$ is its centripetal acceleration. The torque due to the force $\mathbf{F}_r$ about the origin is zero because $\mathbf{F}_r$ is antiparallel to $\mathbf{r}$; hence, $\mathbf{r} \times \mathbf{F}_r = 0$. The torque about the origin due to the tangential force is $\mathbf{r} \times \mathbf{F}_t$, but because $\mathbf{r}$ is perpendicular to $\mathbf{F}_t$, the magnitude of the torque is simply $F_t r$. Thus, the magnitude of the net torque on the particle is

$$\sum \tau = F_t r = (ma_t) r$$

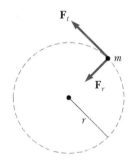

Figure 10.19   A particle rotating in a circle under the influence of a tangential force $\mathbf{F}_t$. A force $\mathbf{F}_r$ must also be present to maintain the circular motion.

Because the tangential acceleration is related to the angular acceleration through Equation 10.11, $a_t = r\alpha$, the torque can be expressed as

$$\sum \tau = (mr\alpha) r = (mr^2) \alpha$$

Recall from Equation 10.14 that the quantity mr^2 is the moment of inertia of the rotating mass about the z axis passing through the origin, so that

$$\sum \tau = I\alpha \qquad\qquad \text{[10.25]}$$

• *Relationship between net torque and angular acceleration*

That is, **the net torque acting on the particle is proportional to its angular acceleration,** and the proportionality constant is the moment of inertia. It is important to note that $\Sigma\tau = I\alpha$ is the rotational analogue of Newton's second law of motion, $\Sigma F = ma$.

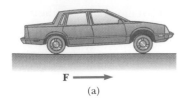

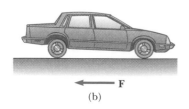

Figure TP10.4
(Thinking Physics 4)

Thinking Physics 4

When an automobile driver steps on the accelerator, the nose of the car moves upward. When the driver brakes, the nose moves downward. Why do these effects occur?

Reasoning When the driver steps on the accelerator, there is an increased force on the tires from the roadway. This force is parallel to the roadway and directed toward the front of the automobile, as suggested in Figure TP10.4a. This force provides a torque about the center of mass that tends to cause the car to rotate in the counterclockwise direction in the diagram. The result of this rotation is a "nosing up" of the car. When the driver steps on the brake, there is an increased force on the tires from the roadway, directed toward the rear of the automobile, as suggested in Figure TP10.4b. This force results in a torque that causes a clockwise rotation and the subsequent "nosing down" of the automobile.

10.8 • ANGULAR MOMENTUM

Again let us consider a particle of mass m, situated at the vector position $\mathbf{r}$ and moving with a momentum $\mathbf{p}$, as shown in Figure 10.20. The **instantaneous angular momentum, L,** of the particle relative to the origin O is defined by the cross product of its instantaneous vector position and the instantaneous linear momentum, $\mathbf{p}$:

Angular momentum of a particle

$$\mathbf{L} = \mathbf{r} \times \mathbf{p} \qquad [10.26]$$

The SI units of angular momentum are $kg \cdot m^2/s$. It is important to note that both the magnitude and the direction of $\mathbf{L}$ depend on the choice of origin. The direction of $\mathbf{L}$ is perpendicular to the plane formed by $\mathbf{r}$ and $\mathbf{p}$, and the sense of $\mathbf{L}$ is governed by the right-hand rule. For example, in Figure 10.20, $\mathbf{r}$ and $\mathbf{p}$ are assumed to be in the xy plane, and $\mathbf{L}$ points in the z direction. Because $\mathbf{p} = m\mathbf{v}$, the magnitude of $\mathbf{L}$ is

$$L = mvr \sin \phi \qquad [10.27]$$

where ϕ is the angle between $\mathbf{r}$ and $\mathbf{p}$. It follows that $\mathbf{L}$ is zero when $\mathbf{r}$ is parallel to $\mathbf{p}$ ($\phi = 0°$ or $180°$). In other words, when the particle moves along a line that passes through the origin, it has zero angular momentum with respect to the origin. This is equivalent to stating that it has no tendency to rotate about the origin. However, if $\mathbf{r}$ is perpendicular to $\mathbf{p}$ ($\phi = 90°$), L is a maximum and equal to mvr. In this case, the particle has maximum tendency to rotate about the origin. In fact, at that instant the particle moves exactly as though it were on the rim of a wheel rotating about the origin in a plane defined by $\mathbf{r}$ and $\mathbf{p}$. A particle has nonzero angular momentum about some point if the position vector of the particle measured from that point rotates about the point.

For linear motion, we found that the resultant force on a particle equals the time rate of change of the particle's linear momentum (Eq. 8.3). We shall now show that Newton's second law implies that the resultant torque acting on a particle equals the time rate of change of the particle's angular momentum. Let us start by writing the torque on the particle in the form

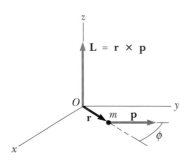

Figure 10.20 The angular momentum $\mathbf{L}$ of a particle of mass m and momentum $\mathbf{p}$ located at the position $\mathbf{r}$ is a vector given by $\mathbf{L} = \mathbf{r} \times \mathbf{p}$. The value of $\mathbf{L}$ depends on the origin and is a vector perpendicular to both $\mathbf{r}$ and $\mathbf{p}$.

$$\boldsymbol{\tau} = \mathbf{r} \times \mathbf{F} = \mathbf{r} \times \frac{d\mathbf{p}}{dt} \qquad \text{[10.28]}$$

where we have used the fact that $\mathbf{F} = d\mathbf{p}/dt$. Now let us differentiate Equation 10.26 with respect to time, using the product rule.

$$\frac{d\mathbf{L}}{dt} = \frac{d}{dt}(\mathbf{r} \times \mathbf{p}) = \mathbf{r} \times \frac{d\mathbf{p}}{dt} + \frac{d\mathbf{r}}{dt} \times \mathbf{p}$$

It is important to adhere to the order of terms, because $\mathbf{A} \times \mathbf{B} = -\mathbf{B} \times \mathbf{A}$.

The last term on the right in the preceding equation is zero because $\mathbf{v} = d\mathbf{r}/dt$ is parallel to $\mathbf{p}$. Therefore,

$$\frac{d\mathbf{L}}{dt} = \mathbf{r} \times \frac{d\mathbf{p}}{dt} \qquad \text{[10.29]}$$

Comparing Equations 10.28 and 10.29, we see that

$$\boldsymbol{\tau} = \frac{d\mathbf{L}}{dt} \qquad \text{[10.30]}$$

• *Torque equals time rate of change of angular momentum*

This result, $\boldsymbol{\tau} = d\mathbf{L}/dt$, is the rotational analog of Newton's second law, $\mathbf{F} = d\mathbf{p}/dt$. Equation 10.30 says that the **torque** acting on a particle is equal to the time rate of change of the particle's angular momentum. It is important to note that Equation 10.30 is valid only if the origins of $\boldsymbol{\tau}$ and $\mathbf{L}$ are the *same*. Equation 10.30 is also valid when several forces are acting on the particle, in which case $\boldsymbol{\tau}$ is the *net* torque on the particle. **Furthermore, the expression is valid for any origin fixed in an inertial frame.** Of course, the same origin must be used in calculating all torques as well as the angular momentum.

A System of Particles

The total angular momentum, $\mathbf{L}$, of a system of particles about some point is defined as the vector sum of the angular momenta of the individual particles:

$$\mathbf{L} = \mathbf{L}_1 + \mathbf{L}_2 + \cdots + \mathbf{L}_n = \sum \mathbf{L}_i$$

where the vector sum is over all of the n particles in the system.

Because the individual momenta of the particles may change in time, the total angular momentum may also vary in time. In fact, from Equations 10.28 and 10.29, we find that the time rate of change of the total angular momentum equals the vector sum of *all* torques, including those associated with internal forces between particles and those associated with external forces. However, the net torque associated with internal forces is zero. To understand this, recall that Newton's third law tells us that the internal forces occur in equal and opposite pairs that lie along the line of separation of each pair of particles. Therefore, the torque due to each action–reaction force pair is zero. By summation, we see that *the net internal torque vanishes.* Finally, we conclude that the total angular momentum can vary with time *only* if there is a net *external* torque on the system, so that we have

$$\sum \boldsymbol{\tau}_{\text{ext}} = \sum \frac{d\mathbf{L}_i}{dt} = \frac{d}{dt} \sum \mathbf{L}_i = \frac{d\mathbf{L}}{dt} \qquad \text{[10.31]}$$

This is an aerial view of Hurricane Fran. The spiral-shaped, nonrigid mass of air undergoes rotation and has angular momentum. *(Courtesy of NASA)*

That is, the time rate of change of the total angular momentum of the system about some origin in an inertial frame equals the net external torque acting on the system about that origin. Note that Equation 10.31 is the rotational analog of $\mathbf{F}_{ext} = d\mathbf{p}/dt$ (Eq. 8.35) for a system of particles.

Thinking Physics 5

Can a particle moving in a straight line have nonzero angular momentum?

Reasoning A particle has angular momentum about an origin when moving in a straight line as long as its line of motion does not pass through the origin. Its angular momentum is zero only if the line of motion passes through the origin, in which case $\mathbf{r}$ is parallel to $\mathbf{v}$, and $\mathbf{L} = \mathbf{r} \times \mathbf{p} = \mathbf{r} \times m\mathbf{v} = 0$.

Example 10.8   Circular Motion

A particle moves in the *xy* plane in a circular path of radius *r*, as in Figure 10.21. (a) Find the magnitude and direction of its angular momentum relative to *O* when its velocity is **v**.

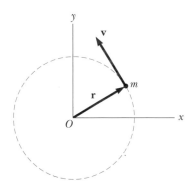

Figure 10.21   (Example 10.8) A particle moving in a circle of radius *r* has an angular momentum equal in magnitude to *mvr* relative to the center. The vector $\mathbf{L} = \mathbf{r} \times \mathbf{p}$ (not shown) points *out* of the diagram in this case.

Solution   Because $\mathbf{r}$ is perpendicular to $\mathbf{v}$, $\phi = 90°$ and the magnitude of $\mathbf{L}$ is simply

$$L = mvr \sin 90° = mvr \qquad \text{(for } \mathbf{r} \text{ perpendicular to } \mathbf{v}\text{)}$$

The direction of $\mathbf{L}$ is perpendicular to the plane of the circle, and its sense depends on the direction of $\mathbf{v}$. If the sense of the rotation is counterclockwise, as in Figure 10.21, then by the right-hand rule, the direction of $\mathbf{L} = \mathbf{r} \times \mathbf{p}$ is *out* of the paper. Hence, we can write the vector expression for the angular momentum as $\mathbf{L} = (mvr)\mathbf{k}$. If the particle were to move clockwise, $\mathbf{L}$ would point into the paper.

(b) Find an alternative expression for *L* in terms of the angular velocity, *ω*.

Solution   Because $v = r\omega$ for a particle rotating in a circle, we can express *L* as

$$L = mvr = m(r\omega)r = mr^2\omega = I\omega$$

where *I* is the moment of inertia of the particle about the *z* axis through *O*. In this case the angular momentum is in the *same* direction as the angular velocity vector, $\boldsymbol{\omega}$ (see Section 10.1), and so we can write $\mathbf{L} = I\boldsymbol{\omega} = I\omega\mathbf{k}$.

Example 10.9   Two Connected Masses

Two masses, m_1 and m_2, are connected by a light cord that passes over a pulley of radius *R* and moment of inertia *I* about its axle, as in Figure 10.22. The mass m_2 slides on a frictionless, horizontal surface. Determine the acceleration of the two masses using the concepts of angular momentum and torque.

Reasoning and Solution   First, calculate the angular momentum of the system, which consists of the two masses plus the pulley. Then calculate the torque about an axis along the axle of the pulley through *O*. At the instant m_1 and m_2 have a speed *v*, the angular momentum of m_1 is m_1vR, and that of m_2 is m_2vR. At the same instant, the angular momentum of the pulley is $I\omega = Iv/R$. Therefore, the total angular momentum of the system is

$$(1) \qquad L = m_1vR + m_2vR + I\frac{v}{R}$$

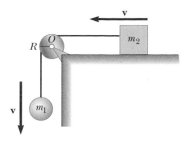

Figure 10.22   (Example 10.9)

about the axle. This is the total external torque about O; that is, $\tau_{\text{ext}} = m_1 g R$. Using this result, together with (1) and Equation 10.30 gives

$$\tau_{\text{ext}} = \frac{dL}{dt}$$

$$m_1 g R = \frac{d}{dt}\left[(m_1 + m_2)Rv + I\frac{v}{R}\right]$$

$$(2)\quad m_1 g R = (m_1 + m_2)R\frac{dv}{dt} + \frac{I}{R}\frac{dv}{dt}$$

Because $dv/dt = a$, we can solve this for a to get

$$a = \frac{m_1 g}{(m_1 + m_2) + I/R^2}$$

Now let us evaluate the total external torque on the system about the axle. Because they have zero moment arms, the forces exerted by the axle on the pulley and by the weight of the pulley do not contribute to the torque. Furthermore, the normal force acting on m_2 is balanced by the force of gravity, $m_2\mathbf{g}$, and so these forces do not contribute to the torque. The external force $m_1\mathbf{g}$ produces a torque about the axle equal in magnitude to $m_1 g R$, where R is the moment arm of the force

You may wonder why we did not include the forces that the cord exerts on the objects in evaluating the net torque about the axle. The reason is that these forces are internal to the system under consideration. Only the external torques contribute to the change in angular momentum.

EXERCISE 10   A car of mass 1500 kg moves on a circular racetrack of radius 50 m with a speed of 40 m/s. What is the magnitude of its angular momentum relative to the center of the track?   Answer 3.0×10^6 kg·m²/s

10.9 · CONSERVATION OF ANGULAR MOMENTUM

In Chapter 8 we found that the total linear momentum of a system of particles remains constant when the net external force acting on the system is zero. In rotational motion, we have an analogous conservation law that states that **the total angular momentum of a system remains constant if the net external torque acting on the system is zero.**

Because the net torque acting on the system equals the time rate of change of the system's angular momentum, we see that if

$$\sum \tau_{\text{ext}} = \frac{d\mathbf{L}}{dt} = 0 \qquad\qquad\qquad [10.32]$$

then

$$\mathbf{L} = \text{constant} \qquad\qquad\qquad [10.33]$$

For a system of particles, we can write this conservation law as $\sum \mathbf{L}_i = $ constant, where $\mathbf{L}_i$ is the angular momentum of the ith particle, given by $\mathbf{L}_i = \mathbf{r}_i \times \mathbf{p}_i$. Hence, we can express the conservation of angular momentum for a system of particles as

$$\sum \mathbf{L}_i = \sum \mathbf{r}_i \times \mathbf{p}_i = \text{constant} \qquad\qquad\qquad [10.34]$$

From Equation 10.34 we derive a third conservation law to add to our list of conserved quantities. We can now state that **the total energy, linear momentum, and angular momentum of an isolated system all remain constant.**

If the system of "particles" under consideration is a rigid body rotating about a fixed axis, then the angular momentum of the body about the axis has a magnitude given by $L = I\omega$, where I is the moment of inertia about the axis. In this case, if the net external torque on the body is zero, we can express the conservation of angular momentum as $I\omega =$ constant.

Although we do not prove it here, an important theorem concerns the angular momentum of a system of particles relative to the center of mass of the system. It can be stated as follows:

> The net torque acting on a system of particles about the center of mass equals the time rate of change of angular momentum, regardless of the motion of the center of mass.

This theorem applies even if the center of mass is accelerating, provided that both τ and **L** are evaluated relative to the center of mass.

Many examples can be used to demonstrate conservation of angular momentum; some of them should be familiar to you. You may have observed a figure skater

Figure 10.23 Nancy Kerrigan. In this spin, her angular momentum increases when she pulls her arms in close to her body, demonstrating that angular momentum is conserved. *(Reuters/Corbis-Bettmann)*

Figure 10.24 The Crab Nebula, in the constellation Taurus. This nebula is the remnant of a supernova explosion, which was seen on Earth in the year A.D. 1054. It is located some 6300 lightyears away and is approximately 6 lightyears in diameter, still expanding outward. *(David Malin, Anglo-Australian Observatory)*

spinning (Fig. 10.23). The angular speed of the skater increases as she pulls her hands and feet close to the trunk of her body. Neglecting friction between skater and ice, we see that there are no external torques on the skater. The moment of inertia of her body decreases as her hands and feet are brought in. The resulting change in angular speed is accounted for as follows. Because angular momentum must be conserved, the product $I\omega$ remains constant, and a decrease in I causes a corresponding increase in ω.

An interesting astrophysical example of conservation of angular momentum occurs when, at the end of its lifetime, a massive star uses up all its fuel and collapses under the influence of gravitational forces, causing a gigantic outburst of energy called a supernova explosion. The best-studied example of a remnant of a supernova explosion is the Crab Nebula, a chaotic, expanding mass of gas (Fig. 10.24). Part of the star's mass is released into space, where it eventually condenses into new stars and planets. Most of what is left behind makes up a **neutron star,** an extremely dense sphere of matter with a diameter of about 10 km in comparison with the 10^6-km diameter of the original star. As the rotational inertia of the system decreases, the star's rotational speed increases. More than 300 rapidly rotating neutron stars have been identified, with periods of rotation ranging from 1.6 ms to 4 s. The neutron star is a most dramatic system—an object with a mass greater than the Sun, rotating about its axis a few times each second!

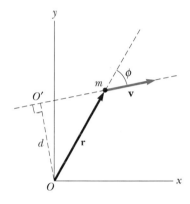

Figure E10.11 (Exercise 11)

EXERCISE 11 A particle of mass m moves in the xy plane with a velocity $\mathbf{v}$ along a straight line, as in Figure E10.11. What are the magnitude and direction of its angular momentum (a) with respect to the origin O and (b) with respect to the origin O'? Answer (a) mvd into the diagram (b) zero

---

Example 10.10 A Revolving Ball on a Horizontal, Frictionless Surface

A ball of mass m on a horizontal, frictionless table is connected to a string that passes through a small hole in the table. The ball is set into circular motion of radius R, at which time its speed is v_0 (Fig. 10.25). If the string is pulled from the bottom so that the radius of the circular path is decreased to r, what is the final speed v of the ball?

Solution Let us take the torque about the center of rotation, O. Note that the force of gravity acting on the ball is balanced by the upward normal force, and so these forces cancel. The force, $\mathbf{F}$, of the string on the ball acts toward the center of rotation, and the vector position $\mathbf{r}$ is directed away from O. Thus, we see that $\boldsymbol{\tau} = \mathbf{r} \times \mathbf{F} = 0$. Because $\boldsymbol{\tau} = d\mathbf{L}/dt = 0$, $\mathbf{L}$ is a constant of the motion. That is, $mv_0R = mvr$, or

$$v = \frac{v_0 R}{r}$$

From this result, we see that as r decreases, the speed v increases.

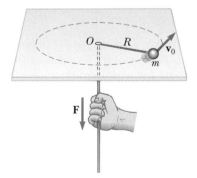

Figure 10.25 (Example 10.10)

10.10 · QUANTIZATION OF ANGULAR MOMENTUM

We have seen that the concept of angular momentum is very useful for describing the motions of macroscopic systems. The concept is also valid and equally useful on a submicroscopic scale, and it has been used extensively in the development of modern theories of atomic, molecular, and nuclear physics. In these developments, it was found that the angular momentum of a system is a *fundamental* quantity. The word "fundamental" in this context implies that angular momentum is an inherent property of atoms, molecules, and their constituents.

In order to explain the results of a variety of experiments on atomic and molecular systems, we must assign discrete values to—in modern terminology, **quantize**—the angular momentum. These discrete values are multiples of a fundamental unit of angular momentum that equals $h/2\pi$, where h is called **Planck's constant.**

$$\text{Fundamental unit of angular momentum} = \frac{h}{2\pi} = 1.054 \times 10^{-34} \text{ kg} \cdot \text{m}^2/\text{s}$$

Example 10.11   Estimating the Rotational Frequency of the Oxygen Molecule

Estimate the rotational frequency of the oxygen molecule using the facts that the "average" separation between the two oxygen atoms is 1.21×10^{-10} m and that the mass of an oxygen atom is 2.66×10^{-26} kg.

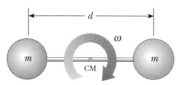

Figure 10.26   (Example 10.11) The rigid-rotor model of the diatomic molecule. The rotation occurs about the center of mass in the plane of the diagram.

Solution   Consider the O_2 molecule as a rigid rotor—that is, two oxygen atoms separated by a fixed distance d and rotating about their center of mass (Fig. 10.26). To find the rotational frequency, first we must evaluate the angular momentum of the molecule about an axis through its center of mass, perpendicular to the plane of rotation.

$$L = L_1 + L_2 = mvr + mvr = 2mvr$$

In this example, $r = d/2 = 0.605 \times 10^{-10}$ m. Also, it is useful to use the relation $v = r\omega$ (Eq. 10.10), giving

$$\begin{aligned}
L &= 2mvr = 2mr^2\omega \\
&= 2(2.66 \times 10^{-26} \text{ kg})(0.605 \times 10^{-10} \text{ m})^2\omega \\
&= (1.95 \times 10^{-46} \text{ kg} \cdot \text{m}^2)\omega
\end{aligned}$$

We can now estimate the lowest rotational frequency of the molecule by equating the rotational angular momentum with the fundamental unit of angular momentum, $h/2\pi$.

$$\begin{aligned}
L &= (1.95 \times 10^{-46} \text{ kg} \cdot \text{m}^2)\omega = \frac{h}{2\pi} \\
&= 1.054 \times 10^{-34} \text{ kg} \cdot \text{m}^2/\text{s}
\end{aligned}$$

Therefore,

$$\omega = \frac{1.054 \times 10^{-34} \text{ kg} \cdot \text{m}^2/\text{s}}{1.95 \times 10^{-46} \text{ kg} \cdot \text{m}^2} = 5.41 \times 10^{11} \text{ rad/s}$$

This result is in good agreement with measured rotational frequencies. The rotational frequencies are much lower than the vibrational frequencies of the molecule, which are typically of the order of 10^{13} rad/s.

Classical concepts and mechanical models can be useful in describing some features of atomic and molecular systems. However, as shown by the preceding example, a wide variety of submicroscopic phenomena can be explained only if one assumes discrete values of the angular momentum associated with a particular type of motion.

The Danish physicist Niels Bohr (1885–1962), in his theory of the hydrogen atom, suggested the radical idea of angular momentum quantization. Strictly classical models were unsuccessful in describing many properties of the hydrogen atom, such as the fact that the atom emits radiation at discrete frequencies. Bohr found that the electron could occupy only those orbits about the proton for which the orbital angular momentum was equal to $nh/2\pi$, where n is an integer, or quantum number. From this simple model, one can estimate the rotational frequencies of the electron in the allowed orbits.

Although Bohr's theory provided some insight concerning the behavior of matter at the atomic level, it is basically incorrect. Subsequent developments in quantum mechanics from 1924 to 1930 provided models and interpretations that are still accepted.

Later developments in atomic physics indicated that an electron possesses another kind of angular momentum, called **spin,** which is an inherent property of the electron itself, unrelated to its motion. The spin angular momentum is also restricted to discrete values. Later we shall return to this important property and discuss its great impact on modern physical science.

10.11 • ROTATION OF RIGID BODIES

O P T I O N A L

In this section we shall investigate the rotational motion of rigid bodies, using as a basis much of what we have learned concerning the circular motion of particles. First we shall discuss the dynamics of a rigid body rotating about a fixed axis. Next we shall show how to determine the kinetic energy of a rotating rigid body and how its change is related to the work done by external forces. Finally we shall use the principle of conservation of energy to describe an object that rolls on a surface without slipping.

Rotational Dynamics

In Section 10.7 we learned that the magnitude of the net torque acting on a particle is given by Equation 10.25:

$$\tau_{net} = I\alpha$$

where I is the moment of inertia and α is the angular acceleration. It is important to note that this result also applies to a rigid body of arbitrary shape rotating about a fixed axis, because the body can be regarded as an infinite number of mass elements of infinitesimal size, each rotating in a circle about the rotation axis. The important and strikingly simple result given by $\tau_{net} = I\alpha$ is in complete agreement with experimental observations. Its simplicity is a result of how the motion is described.

> Although the points on a rigid body rotating about a fixed axis may not experience the same force, linear acceleration, or linear velocity, every point on the body has the same angular acceleration and angular velocity at any instant. Therefore, at any instant the rigid body as a whole is characterized by specific values for angular acceleration, net torque, and angular velocity.

TABLE 10.2 Moments of Inertia of Some Rigid Objects

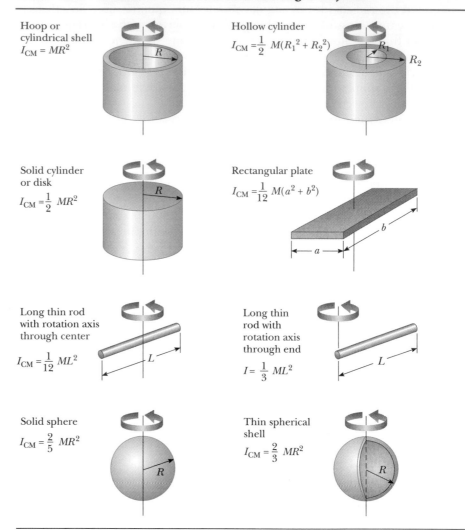

Hoop or cylindrical shell
$I_{CM} = MR^2$

Hollow cylinder
$I_{CM} = \frac{1}{2} M(R_1^2 + R_2^2)$

Solid cylinder or disk
$I_{CM} = \frac{1}{2} MR^2$

Rectangular plate
$I_{CM} = \frac{1}{12} M(a^2 + b^2)$

Long thin rod with rotation axis through center
$I_{CM} = \frac{1}{12} ML^2$

Long thin rod with rotation axis through end
$I = \frac{1}{3} ML^2$

Solid sphere
$I_{CM} = \frac{2}{5} MR^2$

Thin spherical shell
$I_{CM} = \frac{2}{3} MR^2$

The specific form of I depends on the axis of rotation and on the size and shape of the body. The moments of inertia of rigid bodies can be evaluated by methods of integral calculus. Table 10.2 gives the moments of inertia for a number of bodies about specific axes. In all cases, note that I is proportional to the mass of the object and to the square of a geometric factor.

Example 10.12 Angular Acceleration of a Wheel

A wheel of radius R, mass M, and moment of inertia I is mounted on a frictionless, horizontal axle as in Figure 10.27. A light cord wrapped around the wheel supports an object of mass m. Calculate the linear acceleration of the object, the angular acceleration of the wheel, and the tension in the cord.

Reasoning and Solution The torque acting on the wheel about its axis of rotation is $\tau = TR$, where T is the force exerted by the cord on the rim of the wheel. (The weight of the wheel and the normal force exerted by the axle on the wheel pass through the axis of rotation and produce no torque.) Because $\tau = I\alpha$, we get

$$\tau = I\alpha = TR$$

$$(1) \quad \alpha = \frac{TR}{I}$$

Now let us apply Newton's second law to the motion of the suspended object, making use of the free-body diagram in Figure 10.27, taking the upward direction to be positive:

$$\sum F_y = T - mg = -ma$$

$$(2) \quad a = \frac{mg - T}{m}$$

The linear acceleration of the suspended object is equal to the tangential acceleration of a point on the rim of the wheel. Therefore, the angular acceleration of the wheel and this linear acceleration are related by $a = R\alpha$. Using this fact together with (1) and (2), gives

$$(3) \quad a = R\alpha = \frac{TR^2}{I} = \frac{mg - T}{m}$$

$$(4) \quad T = \frac{mg}{1 + mR^2/I}$$

Substituting (4) into (3), and solving for a and α gives

$$a = \frac{g}{1 + I/mR^2}$$

$$\alpha = \frac{a}{R} = \frac{g}{R + I/mR}$$

EXERCISE 12   The wheel in Figure 10.27 is a solid disk of $M = 2.00$ kg, $R = 30.0$ cm, and $I = 0.0900$ kg·m². The sus-

pended object has a mass of $m = 0.500$ kg. Find the tension in the cord and the angular acceleration of the wheel.
Answer   3.27 N; 10.9 rad/s²

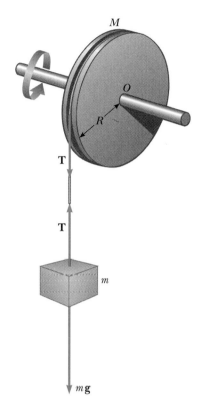

Figure 10.27   (Example 10.12) The cord attached to an object of mass m is wrapped around the wheel, and the tension in the cord produces a torque about the axle through O.

Example 10.13   Rotating Rod Revisited

A uniform rod of length L and mass M is free to rotate on a frictionless pin through one end (Fig. 10.28). The rod is released from rest in the horizontal position. (a) What is the angular speed of the rod at its lowest position?

Reasoning and Solution   The question can be answered by considering the mechanical energy of the system. When the rod is horizontal, it has no rotational energy. The potential energy relative to the lowest position of its center of mass (O') is $MgL/2$. When it reaches its lowest position, the energy is entirely rotational energy, $\frac{1}{2}I\omega^2$, where I is the moment of inertia about the pivot. Because $I = \frac{1}{3}ML^2$ (Table 10.2) and because mechanical energy is constant, we have

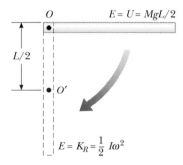

Figure 10.28   (Example 10.13) A uniform rigid rod pivoted at O rotates in a vertical plane under the action of the force of gravity.

$$\tfrac{1}{2}MgL = \tfrac{1}{2}I\omega^2 = \tfrac{1}{2}(\tfrac{1}{3}ML^2)\,\omega^2$$

$$\omega = \sqrt{\frac{3g}{L}}$$

(b) Determine the linear speed of the center of mass and the linear speed of the lowest point on the rod in the vertical position.

Solution

$$v_{CM} = r\omega = \frac{L}{2}\,\omega = \tfrac{1}{2}\sqrt{3gL}$$

The lowest point on the rod has a linear speed equal to $2v_{CM} = \sqrt{3gL}$.

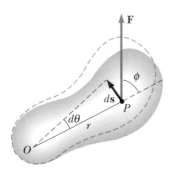

Figure 10.29   A rigid object rotates about an axis through O under the action of an external force **F** applied at P.

Work and Energy in Rotational Motion

Consider a rigid body pivoted at the point O in Figure 10.29. Suppose a single external force, **F**, is applied at the point P and $d\mathbf{s}$ is the displacement of the point of application of the force. The work done by **F** as the body rotates through an infinitesimal distance $ds = r\,d\theta$ in a time dt is

$$dW = \mathbf{F}\cdot d\mathbf{s} = (F\sin\phi)r\,d\theta$$

where $F\sin\phi$ is the tangential component of **F** or the component of the force along the displacement. Note from Figure 10.29 that **the radial component of F does no work because it is perpendicular to the displacement.**

Because the magnitude of the torque due to **F** about the origin was defined as $rF\sin\phi$, we can write the work done for the infinitesimal rotation in the form

$$dW = \tau d\theta \qquad [10.35]$$

The rate at which work is being done by **F** for rotation about the fixed axis is obtained by formally dividing the left and right sides of Equation 10.35 by dt:

$$\frac{dW}{dt} = \tau\frac{d\theta}{dt} \qquad [10.36]$$

But the quantity dW/dt is, by definition, the instantaneous power, P, delivered by the force. Furthermore, because $d\theta/dt = \omega$, Equation 10.36 reduces to

*Power delivered to a rigid •
body*

$$P = \frac{dW}{dt} = \tau\omega \qquad [10.37]$$

This expression is analogous to $P = Fv$ in the case of linear motion, and the expression $dW = \tau d\theta$ is analogous to $dW = F_x dx$.

In linear motion, we found the energy concept, and in particular the work-kinetic energy theorem, to be extremely useful in describing the motion of a system. The energy concept can be equally useful in simplifying the analysis of rotational motion. From what we learned of linear motion, we expect that for rotation of a symmetric object (such as a symmetric wheel) about a fixed axis, the work done by external forces will equal the change in the rotational kinetic energy. To show that this is in fact the case, let us begin with $\tau = I\alpha$. Using the chain rule from the calculus, we can express the torque as

$$\tau = I\alpha = I\frac{d\omega}{dt} = I\frac{d\omega}{d\theta}\frac{d\theta}{dt} = I\frac{d\omega}{d\theta}\,\omega$$

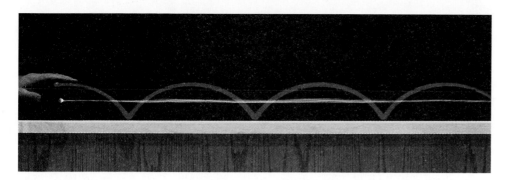

Figure 10.30   Light sources attached at the center and rim of a rolling cylinder illustrate the different paths these points take. The center moves in a straight line, as indicated by the green line, and a point on the rim moves in the path of a cycloid, as indicated by the red curve. *(Henry Leap and Jim Lehman)*

Rearranging this expression and noting that $\tau d\theta = dW$, we get

$$\tau d\theta = dW = I\omega \, d\omega$$

Integrating this expression, we get the total work done:

$$W = \int_{\theta_0}^{\theta} \tau \, d\theta = \int_{\omega_0}^{\omega} I\omega \, d\omega = \tfrac{1}{2} I\omega^2 - \tfrac{1}{2} I\omega_0{}^2 \qquad \textbf{[10.38]}$$

•   *Work-kinetic energy relationship for rotational motion*

Rolling Motion of a Rigid Body

Suppose a cylinder is rolling on a straight path as in Figure 10.30. The center of mass moves in a straight line, and a point on the rim moves in a more complex path corresponding to the shape of a cycloid. Let us further assume that the cylinder of radius R is uniform and rolls on a surface with friction. As the cylinder rotates through an angle of θ, its center of mass moves a distance of $s = r\theta$. Therefore, the speed and acceleration of the center of mass for **pure rolling motion** are

$$v_{CM} = \frac{ds}{dt} = R\frac{d\theta}{dt} = R\omega \qquad \textbf{[10.39]}$$

$$a_{CM} = \frac{dv_{CM}}{dt} = R\frac{d\omega}{dt} = R\alpha \qquad \textbf{[10.40]}$$

The linear velocities of various points on the rolling cylinder are illustrated in Figure 10.31. Note that the linear velocity of any point is in a direction perpendicular to the line from that point to the contact point. At any instant, the point P is at rest relative to the surface, because sliding does not occur.

A general point on the cylinder, such as Q, has both horizontal and vertical components of velocity. However, the points P and P' and the point at the center of mass are unique and of special interest. Relative to the surface on which the cylinder is moving, the center of mass moves with a speed of $v_{CM} = R\omega$, whereas the contact point P has zero velocity. The point P' has a speed equal to $2v_{CM} = 2R\omega$, because all points on the cylinder have the same angular speed.

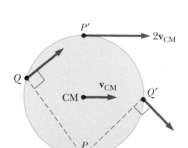

Figure 10.31   All points on a rolling body move in a direction perpendicular to an axis through the contact point P. The center of the body moves with a velocity $\mathbf{v}_{CM}$, and the point P' moves with a velocity $2\mathbf{v}_{CM}$.

We can express the total kinetic energy of the rolling cylinder as

$$K = \tfrac{1}{2} I_p \omega^2 \qquad [10.41]$$

where I_p is the moment of inertia about the axis through P.

A useful theorem called the **parallel axis theorem** enables us to calculate the moment of inertia, I_p, through any axis parallel to the axis through the center of mass of a body. This theorem states that

$$I_p = I_{CM} + MR^2 \qquad [10.42]$$

where R is the distance from the center-of-mass axis to the parallel axis, and M is the total mass of the body. Substitution of this expression into Equation 10.41 gives

$$K = \tfrac{1}{2} I_{CM}\omega^2 + \tfrac{1}{2} MR^2 \omega^2$$

$$K = \tfrac{1}{2} I_{CM}\omega^2 + \tfrac{1}{2} M v_{CM}^2 \qquad [10.43]$$

Total kinetic energy of a rolling body •

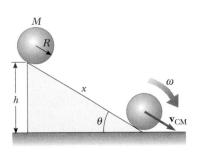

where we have used the fact that $v_{CM} = R\omega$.

We can think of Equation 10.43 as follows: The first term on the right, $\tfrac{1}{2} I_{CM}\omega^2$, represents the rotational kinetic energy about the center of mass, and the term $\tfrac{1}{2} M v_{CM}^2$ represents the kinetic energy the cylinder would have if it were just translating through space without rotating. Thus, we can say that **the total kinetic energy of an object undergoing rolling motion is the sum of a rotational kinetic energy about the center of mass and the translational kinetic energy of the center of mass.**

We can use energy methods to treat a class of problems concerning the rolling motion of a rigid body down a rough incline. We shall assume that the rigid body in Figure 10.32 does not slip and is released from rest at the top of the incline. Note that rolling motion is possible only if a frictional force is present between the object and the incline to produce a net torque about the center of mass. Despite the presence of friction, no loss of mechanical energy takes place because the contact point is at rest relative to the surface at any instant. However, if the rigid body were to slide, mechanical energy would decrease as motion progressed.

Figure 10.32 A round object rolling down an incline. Mechanical energy remains constant if no slipping occurs and there is no rolling friction.

Thinking Physics 6

A common classroom demonstration is to spin on a rotating stool with heavy weights in your outstretched hands and bring the weights close to your body. The result is an increase in the angular velocity, in agreement with the principle of conservation of angular momentum (see Fig. 10.33). Let us imagine we perform such a demonstration, and moving the weights inward results in halving the moment of inertia, and, therefore, doubling the angular velocity. If we consider the rotational kinetic energy, we see that the energy is *doubled* in this situation. Thus, angular momentum is conserved, but kinetic energy is not. Where does this extra energy come from?

Reasoning As you spin with the weights in your hands, there is a tension force in your arms that is causing the weights to move in a circular path. As you bring the weights in, this force from your arms is doing work on the weights, because it is moving them through a radial displacement. As the angular velocity increases in response to the weights moving inward, you must apply more and more force to keep the weights moving in the smaller radius circular path. Thus, the work that you do in bringing the weights in increases as they are brought in. It is this work that you do with your arms

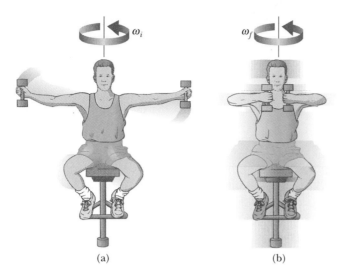

(a)   (b)

Figure 10.33   (Thinking Physics 6) (a) The student is given an initial angular speed while hold-ing two masses as shown. (b) When the masses are pulled in close to the body, the angular speed of the system increases.

that appears as the increase in rotational kinetic energy of the system. You can perform this work due to the energy stored in your body from previous meals, so you are trans-forming potential energy stored in your body into rotational kinetic energy.

CONCEPTUAL PROBLEM 4

Suppose a pencil is balanced on a perfectly frictionless table. If it falls over, what is the path followed by the center of mass of the pencil?

CONCEPTUAL PROBLEM 5

In some motorcycle races, the riders drive over small hills, and the motorcycle becomes airborne for a short time. If the motorcycle racer keeps the throttle open while leaving the hill and going into the air, the motorcycle tends to nose upward. Why does this happen?

CONCEPTUAL PROBLEM 6

Suppose you are designing a car for a coasting race—the cars in this race have no engines, they simply coast down a hill. Do you want large wheels or small wheels? Do you want solid, disk-like wheels or hoop-like wheels? Should the wheels be heavy or light?

(Conceptual Problem 5)
(Tom Raymond/Tony Stone Images)

Example 10.14   Sphere Rolling Down an Incline

If the object in Figure 10.32 is a solid sphere, calculate the speed of its center of mass at the bottom and determine the magnitude of the linear acceleration of the center of mass.

Solution   For a uniform solid sphere, $I_{CM} = \frac{2}{5}MR^2$, and therefore, conservation of mechanical energy gives

$$v_{CM} = \left(\frac{2gh}{1 + \frac{\frac{2}{5}MR^2}{MR^2}} \right)^{1/2} = \left(\frac{10}{7}gh \right)^{1/2}$$

The vertical displacement is related to the displacement x along the incline through the relationship $h = x \sin\theta$.

Hence, after squaring both sides, we can express the equation above as

$$v_{CM}{}^2 = \tfrac{10}{7}gx\sin\theta$$

Comparing this with the familiar expression from kinematics, $v_{CM}{}^2 = 2a_{CM}x$, we see that the acceleration of the center of mass is

$$a_{CM} = \tfrac{5}{7}g\sin\theta$$

These results are quite interesting in that both the speed and the acceleration of the center of mass are *independent* of

the mass and radius of the sphere! That is, **all homogeneous solid spheres experience the same speed and acceleration on a given incline.**

If we repeated the calculations for a hollow sphere, a solid cylinder, or a hoop, we would obtain similar results. The constant factors that appear in the expressions for v_{CM} and a_{CM} depend only on the moment of inertia about the center of mass for the specific body. In all cases, the acceleration of the center of mass is *less* than $g\sin\theta$, the value it would have if the plane were frictionless and no rolling occurred.

EXERCISE 13   The center of mass of a pitched baseball (radius = 3.8 cm) moves at 38 m/s. The ball spins about an axis through its center of mass with an angular speed of 125 rad/s. Calculate the ratio of the rotational energy to the translational kinetic energy. Treat the ball as a uniform sphere.   Answer   1/160

EXERCISE 14   An automobile tire, considered as a solid disk, has a radius of 25 cm and a mass of 6.0 kg. Find its rotational kinetic energy when it is rotating about an axis through its center at an angular speed of 2.0 rev/s.   Answer   15 J

SUMMARY

The **instantaneous angular speed** of a particle rotating in a circle or of a rigid body rotating about a fixed axis is

$$\omega = \frac{d\theta}{dt} \tag{10.3}$$

where ω is in rad/s or in s^{-1}.

The **instantaneous angular acceleration** of a rotating body is

$$\alpha = \frac{d\omega}{dt} \tag{10.5}$$

and has units of rad/s^2 or s^{-2}.

When a rigid body rotates about a fixed axis, every part of the body has the same angular velocity and the same angular acceleration. However, different parts of the body, in general, have different linear velocities and different linear accelerations.

If a particle (or body) undergoes rotational motion about a fixed axis under constant angular acceleration, α, one can apply equations of kinematics by analogy with kinematic equations for linear motion with constant linear acceleration:

$$\omega = \omega_0 + \alpha t \tag{10.6}$$

$$\theta = \theta_0 + \omega_0 t + \tfrac{1}{2}\alpha t^2 \tag{10.7}$$

$$\omega^2 = \omega_0{}^2 + 2\alpha(\theta - \theta_0) \tag{10.8}$$

$$\theta = \theta_0 + \tfrac{1}{2}(\omega_0 + \omega)t \tag{10.9}$$

When a particle rotates about a fixed axis, the angular speed and angular acceleration are related to the linear speed and tangential linear acceleration through the relationships

$$v = r\omega \qquad [10.10]$$

$$a_t = r\alpha \qquad [10.11]$$

The **moment of inertia of a system of particles** is

$$I = \sum m_i r_i^2 \qquad [10.14]$$

If a rigid body rotates about a fixed axis with angular speed ω, its **kinetic energy** can be written

$$K = \tfrac{1}{2} I \omega^2 \qquad [10.15]$$

where I is the moment of inertia about the axis of rotation.

The **torque,** τ, due to a force $\mathbf{F}$ about an origin in an inertial frame is defined to be

$$\tau \equiv \mathbf{r} \times \mathbf{F} \qquad [10.17]$$

Given two vectors, $\mathbf{A}$ and $\mathbf{B}$, their **cross product, $\mathbf{A} \times \mathbf{B}$,** is a vector $\mathbf{C}$ having the magnitude

$$C \equiv |AB \sin \theta| \qquad [10.19]$$

where θ is the angle between $\mathbf{A}$ and $\mathbf{B}$. The direction of $\mathbf{C}$ is perpendicular to the plane formed by $\mathbf{A}$ and $\mathbf{B}$, and its sense is determined by the right-hand rule. Some properties of the cross product include the facts that $\mathbf{A} \times \mathbf{B} = -\mathbf{B} \times \mathbf{A}$ and $\mathbf{A} \times \mathbf{A} = 0$.

The net torque acting on a particle is proportional to the angular acceleration of the particle, and the proportionality constant is the moment of inertia, I:

$$\sum \tau = I\alpha \qquad [10.22]$$

The **angular momentum, L,** of a particle of linear momentum $\mathbf{p} = m\mathbf{v}$ is

$$\mathbf{L} \equiv \mathbf{r} \times \mathbf{p} \qquad [10.23]$$

where $\mathbf{r}$ is the vector position of the particle relative to an origin in an inertial frame. If ϕ is the angle between $\mathbf{r}$ and $\mathbf{p}$, the magnitude of $\mathbf{L}$ is

$$L = mvr \sin \phi = I\omega \qquad [10.24]$$

The **net torque** acting on a particle is equal to the time rate of change of its angular momentum:

$$\sum \tau = \frac{d\mathbf{L}}{dt} \qquad [10.27]$$

The law of conservation of angular momentum states that the total angular momentum of a system remains constant if the net external torque acting on the system is zero:

$$\text{When } \sum \tau_{\text{ext}} = \frac{d\mathbf{L}}{dt} = 0, \text{ then} \qquad [10.29]$$

$$\mathbf{L} = \text{constant} \qquad [10.30]$$

Angular momentum is an inherent property of atoms, molecules, and their constituents. Such systems have discrete (quantized) values of angular momenta that are integer multiples of $h/2\pi$, where h is Planck's constant.

The **total kinetic energy** of a rigid body, such as a cylinder, that is rolling on a rough surface without slipping equals the rotational kinetic energy about the body's center of mass $\tfrac{1}{2} I_{\text{CM}} \omega^2$ plus the translational kinetic energy of the center of mass $\tfrac{1}{2} M v_{\text{CM}}^2$:

$$K = \tfrac{1}{2} I_{\text{CM}} \omega^2 + \tfrac{1}{2} M v_{\text{CM}}^2 \qquad [10.43]$$

In this expression, v_{CM} is the speed of the center of mass and $v_{\text{CM}} = R\omega$ for pure rolling motion.

CONCEPTUAL QUESTIONS

1. In a tape recorder, the tape is pulled past the read-and-write heads at a constant speed by the drive mechanism. Consider the reel from which the tape is pulled—as the tape is pulled off it, the radius of the roll of remaining tape decreases. How does the torque on the reel change with time? How does the angular velocity of the reel change with time? If the tape mechanism is suddenly turned on so that the tape is quickly pulled with a large force, is the tape more likely to break when pulled from a nearly full reel or a nearly empty reel?

2. If a car's wheels are replaced with wheels of a larger diameter, will the reading of the speedometer change? Explain.

3. What is the magnitude of the angular velocity, ω, of the second hand of a clock? What is the direction of ω as you view a clock hanging vertically? What is the magnitude of the angular acceleration, α, of the second hand?

4. If you see an object rotating, is there necessarily a net torque acting on it?

5. In order for a helicopter to be stable as it flies, it must have two propellers. Why?

6. The moment of inertia of an object depends on the choice of rotation axis, as suggested by the parallel axis theorem. Argue that an axis passing through the center of mass of an object must be the axis with the smallest moment of inertia.

7. Suppose you remove two eggs from the refrigerator, one hard-boiled and the other uncooked. You wish to determine which is the hard-boiled egg without breaking the eggs. This can be done by spinning the two eggs on the floor and comparing the rotational motions. Which egg spins faster? Which rotates more uniformly? Explain.

8. A ladder rests inclined against a wall. Would you feel safer climbing up the ladder if you were told that the floor is frictionless but the wall is rough, or that the wall is frictionless but the floor is rough? Justify your answer.

9. Often when a high diver wants to flip in midair, she will draw her legs up against her chest. Why does this make her rotate faster? What should she do when she wants to come out of her flip?

10. If the net force acting on a system is zero, then is it necessarily true that the net torque on it is also zero?

11. Why do tightrope walkers carry a long pole to help balance themselves?

12. Two solid spheres are rolled down a hill—a large, massive sphere and a small sphere with low mass. Which one reaches the bottom of the hill first? Next, we roll a large, low density sphere, and a small high density sphere, and both spheres have the same mass. Which one wins in this case?

13. Consider an object in the shape of a hoop lying in the *xy* plane with all of its mass concentrated on its rim. In two separate experiments, the hoop is rotated by an external agent from rest to an angular speed ω. In one experiment, the rotation occurs about the *z* axis through the center of the hoop. In the other experiment, the rotation occurs about an axis parallel to *z* passing through a point *P* on the rim of the hoop. Which rotation requires more work?

14. Vector **A** is in the negative *y* direction and vector **B** is in the negative *x* direction. What are the directions of (a) **A** × **B** and (b) **B** × **A**?

15. If global warming occurs over the next century, it is likely that the polar ice caps of the Earth will melt and the water will be distributed closer to the Equator. How would this change the moment of inertia of the Earth? Would the length of the day (one revolution) increase or decrease?

PROBLEMS

Section 10.2   Rotational Kinematics

1. What is the angular speed in radians per second of (a) the Earth in its orbit about the Sun and (b) the Moon in its orbit about the Earth?

2. An airliner arrives at the terminal and the engines are shut off. The rotor of one of the engines has an initial clockwise angular speed of 2000 rad/s. The engine's rotation slows with an angular acceleration of magnitude 80.0 rad/s². (a) Determine the angular speed after 10.0 s. (b) How long does it take the rotor to come to rest?

3. An electric motor rotating a grinding wheel at 100 rev/min is switched off. Assuming constant negative acceleration of magnitude 2.00 rad/s², (a) how long does it take the wheel to stop? (b) Through how many radians does it turn during the time found in (a)?

4. A centrifuge in a medical laboratory rotates at an angular speed of 3600 rev/min. When switched off, it rotates 50.0 times before coming to rest. Find the constant angular acceleration of the centrifuge.

5. A dentist's drill starts from rest. After 3.20 s of constant angular acceleration, the drill rotates at a rate of 2.51×10^4 rev/min. (a) Find the drill's angular acceleration. (b) Determine the angle (in radians) through which the drill rotates during this period.

6. The angular position of a swinging door is described by $\theta = 5.00 + 10.0t + 2.00t^2$ rad. Determine the angular po-

sition, angular speed, and angular acceleration of the door at $t = 0$ and $t = 3.00$ s.

7. The tub of a washer goes into its spin cycle, starting from rest and reaching an angular speed of 5.00 rev/s in 8.00 s. At this point the person doing the laundry opens the lid, and a safety switch turns off the washer. The tub slows to rest in 12.0 s. Through how many revolutions does the tub turn? Assume constant angular acceleration while the machine is starting and stopping.

8. A rotating wheel requires 3.00 s to rotate 37.0 revolutions. Its angular velocity at the end of the 3.00 s interval is 98.0 rad/s. What is the constant angular acceleration of the wheel?

Section 10.3   Relations Between Angular and Linear Quantities

9. A discus thrower accelerates a discus from rest to a speed of 25.0 m/s by whirling it through 1.25 rev. Assume the discus moves on the arc of a circle 1.00 m in radius. (a) Calculate the final angular speed of the discus. (b) Determine the magnitude of the angular acceleration of the discus, assuming it to be constant. (c) Calculate the acceleration time.

10. A car accelerates uniformly from rest and reaches a speed of 22.0 m/s in 9.00 s. If the diameter of a tire is 58.0 cm, find (a) the number of revolutions the tire makes during this motion, assuming no slipping. (b) What is the final rotational speed of a tire in revolutions per second?

11. A disk 8.00 cm in radius rotates at a constant rate of 1200 rev/min about its central axis. Determine (a) its angular speed, (b) the linear speed at a point 3.00 cm from its center, (c) the radial acceleration of a point on the rim, and (d) the total distance a point on the rim moves in 2.00 s.

12. A 6.00-kg block is released from A on a frictionless track shown in Figure P10.12. Determine the radial and tangential components of acceleration for the block at P.

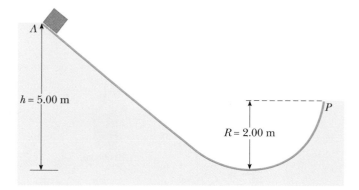

Figure P10.12

13. A car traveling on a flat (unbanked) circular track accelerates uniformly from rest with a tangential acceleration of 1.70 m/s². The car makes it one quarter of the way around the circle before it skids off the track. Determine the coefficient of static friction between the car and track from these data.

Section 10.4   Rotational Kinetic Energy

14. Three particles are connected by rigid rods of negligible mass lying along the y axis (Fig. P10.14). If the system rotates about the x axis with an angular speed of 2.00 rad/s, find (a) the moment of inertia about the x axis and the total rotational energy evaluated from $\frac{1}{2}I\omega^2$ and (b) the linear speed of each particle and the total energy evaluated from $\sum \frac{1}{2}m_i v_i^2$.

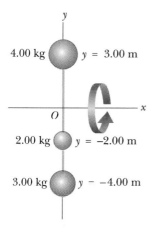

Figure P10.14

15. The four particles in Figure P10.15 are connected by rigid rods of negligible mass. The origin is at the center of the rectangle. If the system rotates in the xy plane about the z axis with an angular speed of 6.00 rad/s, calculate (a) the moment of inertia of the system about the z axis and (b) the rotational energy of the system.

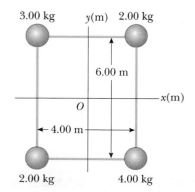

Figure P10.15

Section 10.5   Torque and the Vector Product

16. The fishing pole in Figure P10.16 makes an angle of 20.0°
with the horizontal. What is the torque exerted by the fish
about an axis perpendicular to the page and passing
through the fisher's hand?

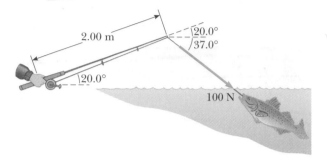

Figure P10.16

17. In Figure P10.17, find the net torque on the wheel about
the axle through O if $a = 10.0$ cm and $b = 25.0$ cm.

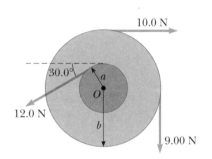

Figure P10.17

18. Given $\mathbf{M} = 6\mathbf{i} + 2\mathbf{j} - \mathbf{k}$ and $\mathbf{N} = 2\mathbf{i} - \mathbf{j} - 3\mathbf{k}$, calculate
the vector product $\mathbf{M} \times \mathbf{N}$.

19. Two vectors are given by $\mathbf{A} = -3\mathbf{i} + 4\mathbf{j}$ and $\mathbf{B} = 2\mathbf{i} + 3\mathbf{j}$.
Find (a) $\mathbf{A} \times \mathbf{B}$ and (b) the angle between $\mathbf{A}$ and $\mathbf{B}$.

20. A force of $\mathbf{F} = 2.00\mathbf{i} + 3.00\mathbf{j}$ N is applied to an object that
is pivoted about a fixed axis aligned along the z coordinate
axis. If the force is applied at the point whose vector position
is $\mathbf{r} = (4.00\mathbf{i} + 5.00\mathbf{j} + 0\mathbf{k})$ m, find (a) the magnitude of
the net torque about the z axis and (b) the direction of the
torque vector, τ.

21. A student claims that she has found a vector $\mathbf{A}$ such that
$(2\mathbf{i} - 3\mathbf{j} + 4\mathbf{k}) \times \mathbf{A} = (4\mathbf{i} + 3\mathbf{j} - \mathbf{k})$. Do you believe this
claim? Explain.

Section 10.6   Equilibrium of a Rigid Object

22. A ladder of weight 400 N and length 10.0 m is placed
against a smooth vertical wall. A person weighing 800 N
stands on the ladder 2.00 m from the bottom as measured
along the ladder. The foot of the ladder is 8.00 m from the

bottom of the wall. Draw a free-body diagram of the ladder.
Calculate the force exerted by the wall and the normal force
exerted by the floor on the ladder.

23. A 1500 kg automobile has a wheel base (the distance between the axles) of 3.00 m. The center of mass of the automobile is on the center line at a point 1.20 m behind the
front axle. Find the force exerted by the ground on each
wheel.

24. A uniform plank of length 6.00 m and mass 30.0 kg rests
horizontally on a scaffold, with 1.50 m of the plank hanging
over one end of the scaffold. Draw a free-body diagram of
the plank. How far can a painter of mass 70.0 kg walk on
the overhanging part of the plank before it tips?

25. A 1200-N uniform boom is supported by a cable as in Figure
P10.25. The boom is pivoted at the bottom, and a 2000-N
object hangs from its top. Find the tension in the cable and
the components of the reaction force on the boom by the
floor.

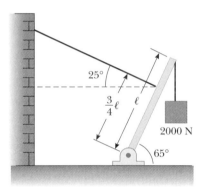

Figure P10.25

26. A crane of mass 3000 kg supports a load of 10 000 kg as in
Figure P10.26. The crane is pivoted with a smooth pin at A
and rests against a smooth support at B. Find the reaction
forces at A and B.

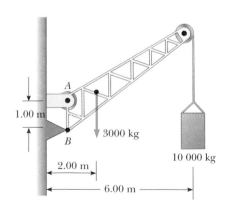

Figure P10.26

Section 10.7 Relation Between Torque and Angular Acceleration

27. A model airplane the mass of which is 0.750 kg is tethered by a wire so that it flies in a circle 30.0 m in radius. The airplane engine provides a net thrust of 0.800 N perpendicular to the tethering wire. (a) Find the torque that the net thrust produces about the center of the circle. (b) Find the angular acceleration of the airplane when it is in level flight. (c) Find the linear acceleration of the airplane tangent to its flight path.

28. The combination of an applied force and a frictional force produces a constant total torque of 36.0 N·m on a wheel rotating about a fixed axis. The applied force acts for 6.00 s, during which time the angular speed of the wheel increases from 0 to 10.0 rad/s. The applied force is then removed, and the wheel comes to rest in 60.0 s. Find (a) the moment of inertia of the wheel, (b) the magnitude of the frictional torque, and (c) the total number of revolutions of the wheel.

Section 10.8 Angular Momentum

29. A light rigid rod 1.00 m in length rotates in the xy plane about a pivot through the rod's center. Two particles of masses of 4.00 kg and 3.00 kg are connected to its ends (Fig. P10.29). Determine the angular momentum of the system about the origin at the instant the speed of each particle is 5.00 m/s.

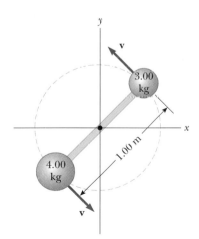

Figure P10.29

30. A 1.50-kg particle moves in the xy plane with a velocity of $\mathbf{v} = (4.20\mathbf{i} - 3.60\mathbf{j})$ m/s. Determine the angular momentum of the particle when its position vector is given by $\mathbf{r} = (1.50\mathbf{i} + 2.20\mathbf{j})$ m.

31. The position vector of a particle of mass 2.00 kg is given as a function of time by $\mathbf{r} = (6.00\,\mathbf{i} + 5.00t\mathbf{j})$ m. Determine the angular momentum of the particle as a function of time.

32. An airplane of mass 12 000 kg flies level to the ground at an altitude of 10.0 km with a constant speed of 175 m/s relative to the Earth. (a) What is the magnitude of the airplane's angular momentum relative to a ground observer directly below the airplane? (b) Does this value change as the airplane continues its motion along a straight line?

Section 10.9 Conservation of Angular Momentum

Section 10.10 Quantization of Angular Momentum (Optional)

33. A ball having mass m is fastened at the end of a flagpole that is connected to the side of a tall building at point P shown in Figure P10.33. The length of the flagpole is ℓ and it makes an angle θ with the horizontal. If the ball become loose and starts to fall, determine the angular momentum (as a function of time) of the ball about point P. Neglect air resistance.

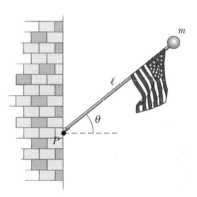

Figure P10.33

34. A student sits on a rotating stool holding two weights, each of mass 3.00 kg. When his arms are extended horizontally, the weights are 1.00 m from the axis of rotation and he rotates with an angular speed of 0.750 rad/s. The moment of inertia of the student plus stool is 3.00 kg·m² and is assumed to be constant. The student pulls the weights horizontally to 0.300 m from the rotation axis. (a) Find the new angular speed of the student. (b) Find the kinetic energy of the student before and after the weights are pulled in.

35. A 60.0-kg woman stands at the rim of a horizontal turntable having a moment of inertia of 500 kg·m² and a radius of

2.00 m. The turntable is initially at rest and is free to rotate about a frictionless, vertical axle through its center. The woman then starts walking around the rim clockwise (as viewed from above the system) at a constant speed of 1.50 m/s relative to the Earth. (a) In what direction and with what angular speed does the turntable rotate? (b) How much work does the woman do to set herself and the turntable into motion?

36. The ball in Figure 10.25 has a mass of 0.120 kg. The distance of the ball from the center of rotation is originally 40.0 cm, and the ball is moving with a speed of 80.0 cm/s. The string is pulled downward 15.0 cm through the hole in the frictionless table. Determine the work done on the ball. (*Hint:* Consider the change of kinetic energy.)

37. A puck of mass 80.0 g and radius 4.00 cm slides along an air table at a speed of 1.50 m/s as shown in Figure P10.37a. It makes a glancing collision with a second puck of radius 6.00 cm and mass 120 g (initially at rest) such that their rims just touch. The pucks stick together and spin after the collision (Fig. P10.37b). (a) What is the angular momentum of the system relative to the center of mass? (b) What is the angular velocity about the center of mass?

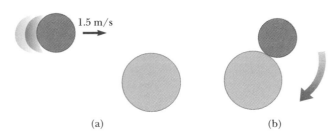

1.5 m/s

(a)   (b)

Figure P10.37

38. In the Bohr model of the hydrogen atom, the electron moves in a circular orbit of radius 0.529×10^{-10} m around the proton. Assuming the orbital angular momentum of the electron is equal to $h/2\pi$, calculate (a) the orbital speed of the electron, (b) the kinetic energy of the electron, and (c) the angular frequency of the electron's motion.

Section 10.11   Rotation of Rigid Bodies (Optional)

39. Attention! About face! Compute an order-of-magnitude estimate for the moment of inertia of your body as you stand tall and turn around an axis through the top of your head and the point half way between your ankles. In your solution state the quantities you measure or estimate and their values.

40. A horizontal 800-N merry-go-round is a solid disk of radius 1.50 m, started from rest by a constant horizontal force of

50.0 N applied tangentially at its edge. Find the kinetic energy of the disk after 3.00 s.

41. A cylinder of mass 10.0 kg rolls without slipping on a horizontal surface. At the instant its center of mass has a speed of 10.0 m/s, determine (a) the translational kinetic energy of its center of mass, (b) the rotational energy about its center of mass, and (c) its total energy.

42. Consider two masses connected by a string that passes over a pulley having a moment of inertia of I about its axis of rotation, as in Figure P10.42. The string does not slip on the pulley, and the system is released from rest. Use the principle of conservation of energy to find the linear speeds of the masses after m_2 descends through a distance h and the angular speed of the pulley at this time.

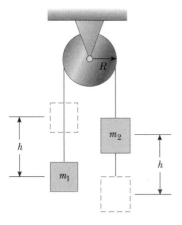

Figure P10.42

43. A uniform rod of length L and mass M is free to rotate about a frictionless pivot at one end, as in Figure P10.43. The rod is released from rest in the horizontal position. What are the *initial* angular acceleration of the rod and the *initial* linear acceleration of the right end of the rod?

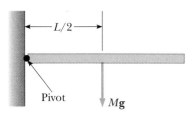

Figure P10.43

44. A bicycle wheel has a diameter of 64.0 cm and a mass of 1.80 kg. The bicycle is placed on a stationary stand on rollers and a resistive force of 120 N is applied to the rim of the tire. (Assume that the wheel is a hoop with all the mass concentrated on the outside radius.) What force must be

applied by a chain passing over a 9.00-cm diameter sprocket in order to give the wheel an acceleration of 4.50 rad/s²? What if you shifted to a 5.60-cm diameter sprocket?

45. (a) A uniform solid disk of radius R and mass M is free to rotate on a frictionless pivot through a point on its rim (Fig. P10.45). If the disk is released from rest in the position shown by the blue circle, what is the speed of its center of mass when the disk reaches the position indicated by the dashed circle? (b) What is the speed of the lowest point on the disk in the dashed position? (c) Repeat part (a) using a uniform hoop.

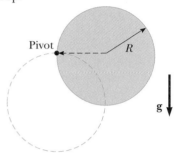

Figure P10.45

46. A block of mass m_1 = 2.00 kg and a block of mass m_2 = 6.00 kg are connected by a massless string over a pulley in the shape of a disk having radius R = 0.25 m, mass M = 10.0 kg. In addition, they are allowed to move on a fixed block-wedge of angle θ = 30.0° as in Figure P10.46. The coefficient of kinetic friction is 0.360 for both blocks. Draw free-body diagrams of both blocks and of the pulley. Determine (a) the acceleration of the two blocks, and (b) the tensions in the string on both sides of the pulley.

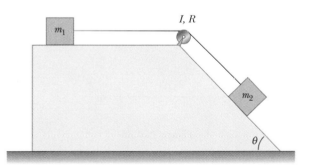

Figure P10.46

47. A can of condensed mushroom soup has mass 215 g, height 10.8 cm, and outer diameter 6.38 cm. It is placed at rest on the top of a 3.00-m long incline, which is at 25.0° to the

horizontal. Assuming energy conservation, calculate the moment of inertia of the can if it takes 1.50 s to reach the bottom of the incline.

Additional Problems

48. An electric motor can accelerate a Ferris wheel of moment of inertia I = 20 000 kg·m² from rest to 10.0 rev/min in 12.0 s. When the motor is turned off, friction causes the wheel to slow down from 10.0 to 8.00 rev/min in 10.0 s. Determine (a) the torque generated by the motor to bring the wheel to 10.0 rev/min and (b) the power needed to maintain this rotational speed.

49. Two blocks, as shown in Figure P10.49, are connected by a string of negligible mass passing over a pulley of radius 0.250 m and moment of inertia I. The block on the frictionless incline is moving up with a constant acceleration of 2.00 m/s². (a) Determine T_1 and T_2, the tensions in the two parts of the string. (b) Find the moment of inertia of the pulley.

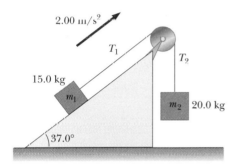

Figure P10.49

50. As a result of friction, the angular speed of a wheel changes with time according to

$$\frac{d\theta}{dt} = \omega_0 e^{-\sigma t}$$

where ω_0 and σ are constants. The angular speed changes from 3.50 rad/s at t = 0 to 2.00 rad/s at t = 9.30 s. Use this information to determine σ and ω_0. Then determine (a) the magnitude of the angular acceleration at t = 3.00 s, (b) the number of revolutions the wheel makes in the first 2.50 s, and (c) the number of revolutions it makes before coming to rest.

51. The pulley shown in Figure P10.51 has radius R and moment of inertia I. One end of the mass m is connected to a spring of force constant k, and the other end is fastened to a cord wrapped around the pulley. The pulley axle and the incline are frictionless. If the pulley is wound counterclock-

wise in order to stretch the spring a distance d from its un-stretched position and then released from rest, find (a) the angular speed of the pulley when the spring is again unstretched and (b) a numerical value for the angular speed at this point if $I = 1.00$ kg · m², $R = 0.300$ m, $k = 50.0$ N/m, $m = 0.500$ kg, $d = 0.200$ m, and $\theta = 37.0°$.

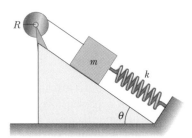

Figure P10.51

52. A solid sphere of mass m and radius r rolls without slipping along the track shown in Figure P10.52. It starts from rest with the lowest point of the sphere at height h above the bottom of the loop of radius R. (a) What is the minimum value of h (in terms of r and R) such that the sphere completes the loop? (b) What are the force components on the sphere at the point P if $h = 3R$?

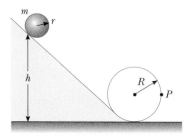

Figure P10.52

53. Two astronauts (Fig. P10.53), each having a mass of 75.0 kg, are connected by a 10.0-m rope of negligible mass. They are isolated in space, orbiting their center of mass at speeds of 5.00 m/s. (a) Calculate the magnitude of the angular momentum of the system by treating the astronauts as particles and (b) the rotational energy of the system. By pulling on the rope, the astronauts shorten the distance between them to 5.00 m. (c) What is the new angular momentum of the system? (d) What are their new speeds? (e) What is the new rotational energy of the system? (f) How much work is done by the astronauts in shortening the rope?

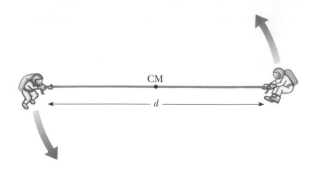

Figure P10.53

54. Two astronauts (Fig. P10.53), each having a mass M, are connected by a rope of length d having negligible mass. They are isolated in space, orbiting their center of mass at speeds v. Calculate (a) the magnitude of the angular momentum of the system by treating the astronauts as particles and (b) the rotational energy of the system. By pulling on the rope, the astronauts shorten the distance between them to $d/2$. (c) What is the new angular momentum of the system? (d) What are their new speeds? (e) What is the new rotational energy of the system? (f) How much work is done by the astronauts in shortening the rope?

55. Figure P10.55 shows a vertical force applied tangentially to a uniform cylinder of weight w. The coefficient of static friction between the cylinder and all surfaces is 0.500. Find, in terms of w, the maximum force **F** that can be applied without causing the cylinder to rotate. (*Hint:* When the cylinder is on the verge of slipping, both friction forces are at their maximum values. Why?)

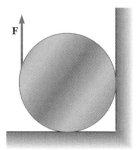

Figure P10.55

56. A common physics demonstration (Fig. P10.56) consists of a ball resting at the end of a board of length ℓ that is elevated at angle θ with the horizontal. A light cup is attached to the board at r_c so that it will catch the ball when the support stick is suddenly removed. (a) Show that the ball will lag behind the falling board when θ is less than 35.3°, and (b) the ball will fall into the cup when the board is supported at this limiting angle and the cup is placed at

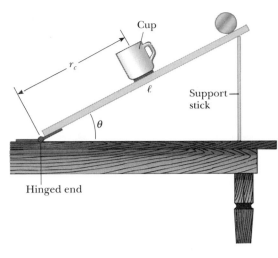

Figure P10.56

$$r_c = \frac{2\ell}{3 \cos \theta}$$

(c) If a ball is at the end of a 1.00-m stick at this critical angle, show that the cup must be 18.4 cm from the moving end.

Spreadsheet Problem

S1. A disk having a moment of inertia of 100 kg·m² is free to rotate about a fixed axis through its center as in Figure 10.1. A tangential force whose magnitude ranges from $F = 0$ to $F = 50.0$ N can be applied at any distance ranging from $R = 0$ to $R = 3.00$ m measured from the axis of rotation. Using Spreadsheet 10.1, find values of F and R that will cause the disk to complete two revolutions in 10.0 s. Are there unique values for F and R?

ANSWERS TO CONCEPTUAL PROBLEMS

1. As the smokestack rotates around its base, each higher portion of the smokestack falls with an increasing tangential acceleration. The tangential acceleration of a given point on the smokestack is proportional to the distance of that portion from the base. As the acceleration increases, eventually higher portions of the smokestack will experience an acceleration higher than that which could result from gravity alone. This can only happen if these portions are being pulled downward by a force in addition to the gravitational force. The force that does this is the shear force from lower portions of the smokestack. Eventually, the shear force to provide this acceleration is larger than the smokestack can withstand, and it breaks.

2. Yes, all points on the wheel have the same angular speed. This is why we use angular quantities to describe rotational motion. Not all points on the wheel have the same linear speed. The point at $r = 0$ has zero linear speed and zero linear acceleration; a point at $r = R/2$ has a linear speed $v = R\omega/2$ and a linear acceleration equal to the centripetal acceleration $v^2/(R/2) = R\omega^2/2$ (the tangential acceleration is zero at all points because ω is constant). A point on the rim at $r = R$ has a linear speed $v = R\omega$ and a linear acceleration $R\omega^2$.

3. There are two major differences between torque and work. The primary difference is that the displacement in the expression for work is directed *along* the force, while the important distance in the torque expression is *perpendicular* to the force. The second difference involves whether there is motion or not—in the case of work, there is work done only if the force succeeds in causing a displacement of the point of application of the force. On the other hand, a force applied at a perpendicular distance from a rotation axis results in a torque whether there is motion or not.

As far as units are concerned, the mathematical expressions for both work and torque result in the product of newtons and meters, but this product is called a joule in the case of work and remains as a newton-meter in the case of torque.

4. On a frictionless table, the only forces on the pencil are the gravitational force, at the center of mass, and the normal force from the table, at the tip. These forces result in a torque on the pencil, causing it to rotate. Both of these forces, however, are vertical—there are no horizontal forces on the pencil. As a result, the center of mass of the pencil must fall straight downward. As the pencil falls, the tip slides to the side, so that the center of mass of the pencil follows a straight line downward to the table.

5. As the motorcycle tire leaves the ground, the friction between the tire and the ground suddenly disappears. If the motorcycle driver keeps the throttle open while leaving the ground, the drive tire will increase its angular velocity, since it no longer experiences the friction force from the ground. The airborne motorcycle is now an isolated system, and its angular momentum must be conserved. The increase in angular momentum of the tire (directed to the left of the motorcycle) must be compensated by an increase in angular momentum of the entire motorcycle (to the right). This ro-

tation results in the nose of the motorcycle rising and the tail dropping.

6. In general, you want the rotational kinetic energy of the system to be as small a fraction as possible of the total energy—you want translation, not rotation! You want the wheels to have as little moment of inertia as possible, so that they represent the lowest resistance to changes in rotational motion. Disk-like wheels would have lower moment of inertia than hoop-like wheels, so disks are preferable. The lower the mass of the wheels, the less is the moment of inertia, so light wheels are preferable. The smaller the radius of the wheels, the less is the moment of inertia, so smaller wheels are preferable, within limits—you want the wheels to be large enough to be able to travel relatively smoothly over irregularities in the road.

11

Orbital Motions and the Hydrogen Atom

We began our study of mechanics with translational motion and the forces that cause it. If we know the forces acting on a system and its initial conditions, we can predict the future of the system. However, describing motion in this manner is often tedious and time consuming. Fortunately, we can often follow the simpler approach of using conservation principles. We easily solved many interesting problems involving motion by recognizing that certain fundamental quantities—such as energy and momentum—were conserved.

In this chapter we return to Newton's universal law of gravitation—one of the fundamental force laws in nature—and show how it, together with Newton's laws of motion, enables us to understand a variety of familiar orbital motions, such as the motions of planets and Earth satellites. Newton's theory of gravitation evolved out of his studies of the motions of the

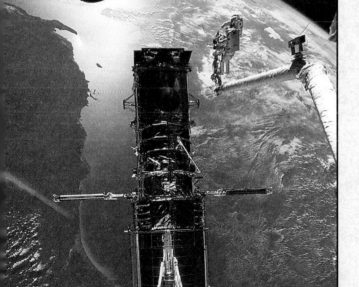

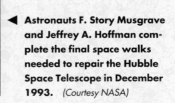

◀ Astronauts F. Story Musgrave and Jeffrey A. Hoffman complete the final space walks needed to repair the Hubble Space Telescope in December 1993. *(Courtesy NASA)*

Moon and planets, for which he used the groundwork provided by Copernicus, Brahe, Kepler, and other astronomers. In examining this evolution, we shall once again make use of conservation of energy and angular momentum.

We conclude this chapter with a discussion of Niels Bohr's famous model of the hydrogen atom, which represents an interesting mixture of classical physics (Newton's laws of motion and the electromagnetic force) and quantum physics (quantization of angular momentum). Although Bohr's theory contains ideas that are contrary to classical physics, his model successfully predicts the observed spectral lines of hydrogen.

11.1 • NEWTON'S UNIVERSAL LAW OF GRAVITY REVISITED

Prior to 1686, many data had been collected on the motions of the Moon and the planets, but a clear understanding of the forces that caused those motions was not yet attainable. In that year, Isaac Newton provided the key that unlocked the secrets of the heavens. He knew, from the first law, that a net force had to be acting on the Moon. If not, the Moon would move in a straight-line path rather than in its almost circular orbit. Newton reasoned that this force between Moon and Earth was an attractive force. He also concluded that there could be nothing special about the Earth–Moon system or the Sun and its planets that would cause gravitational forces to act on them alone. He wrote,

> I deduced that the forces which keep the planets in their orbs must be reciprocally as the squares of their distances from the centers about which they revolve; and thereby compared the force requisite to keep the Moon in her orb with force of gravity at the surface of the Earth; and found them answer pretty nearly.

As you should recall from Chapter 5, every particle in the Universe attracts every other particle with a force that is directly proportional to the product of their masses and inversely proportional to the square of the distance between them. If two particles have masses m_1 and m_2 and are separated by a distance r, the magnitude of the gravitational force between them is

Universal law of gravity •

$$F_g = G\frac{m_1 m_2}{r^2}$$ [11.1]

where G is the *gravitational constant* the value of which in SI units is

$$G = 6.672 \times 10^{-11}\ \frac{\text{N} \cdot \text{m}^2}{\text{kg}^2}$$ [11.2]

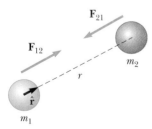

Figure 11.1 The gravitational force between two particles is attractive. The unit vector $\hat{\mathbf{r}}$ is directed from m_1 to m_2. Note that $\mathbf{F}_{21} = -\mathbf{F}_{12}$.

The force law given by Equation 11.1 is often referred to as an **inverse-square law** because the magnitude of the force varies as the inverse square of the separation of the particles. We can express this attractive force in vector form by defining a unit vector, $\hat{\mathbf{r}}$, directed from m_1 to m_2, as shown in Figure 11.1. The force on m_1 due to m_2 is

$$\mathbf{F}_{12} = G\frac{m_1 m_2}{r^2}\hat{\mathbf{r}}$$ [11.3]

Likewise, by Newton's third law, the force on m_2 due to m_1, designated $\mathbf{F}_{21}$, is equal in magnitude to $\mathbf{F}_{12}$ and in the opposite direction. That is, these forces form an action–reaction pair, and $\mathbf{F}_{21} = -\mathbf{F}_{12}$.

The gravitational force exerted by a finite-sized, spherically symmetric mass distribution on a particle outside the sphere is the same as if the entire mass of the sphere were concentrated at its center. For example, the force on a particle of mass m at the Earth's surface has the magnitude

$$F_g = G\,\frac{M_E m}{R_E{}^2}$$

where M_E is the Earth's mass and R_E is the Earth's radius. This force is directed toward the center of the Earth.

Measurement of the Gravitational Constant

The gravitational constant, G, was first measured in an important experiment by Sir Henry Cavendish in 1798. The apparatus he used consisted of two small spheres, each of mass m, fixed to the ends of a light horizontal rod suspended by a thin wire, as in Figure 11.2a. Two large spheres, each of mass M, are then placed near the smaller spheres. The attractive force between the smaller and larger spheres causes the rod to rotate and twist the wire. If the system is oriented as shown in Figure 11.2a, the rod rotates clockwise when viewed from the top. The angle through which it rotates is measured by the deflection of a light beam that is reflected from a mirror attached to the wire. The experiment is carefully repeated with different masses at various separations. In addition to providing a value for G, the results show that the force is attractive, proportional to the product mM, and inversely proportional to the square of the distance r.

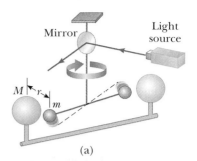

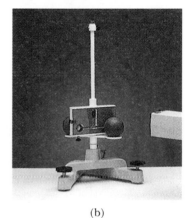

Figure 11.2 (a) Schematic diagram of the Cavendish apparatus for measuring G. The smaller spheres of mass m are attracted to the large spheres of mass M, and the rod rotates through a small angle. A light beam reflected from a mirror on the rotating apparatus measures the angle of rotation. The dashed line represents the original position of the rod. (b) Photograph of a student Cavendish apparatus. *(Courtesy of PASCO Scientific)*

Thinking Physics 1

A student drops a ball and considers why the ball falls to the ground as opposed to the situation in which the ball stays stationary and the Earth moves up to meet it. The student comes up with the following explanation: "The Earth is much more massive than the ball, so the Earth pulls much harder on the ball than the ball pulls on the Earth. Thus, the ball falls while the Earth remains stationary." What do you think about this explanation?

Reasoning According to Newton's universal law of gravity, the force between the ball and the Earth depends on the product of their masses, so both forces—that of the ball on the Earth and that of the Earth on the ball—are equal in magnitude. This follows also, of course, from Newton's third law. The ball has large motion compared to the Earth because, according to Newton's second law, the force of equal magnitude gives a much greater acceleration to the small mass of the ball.

Example 11.1   The Mass of the Earth

Use the gravitational force law to find an approximate value for the mass of the Earth.

Solution   Figure 11.3, obviously not to scale, shows a baseball falling toward the Earth at a location where the acceleration due to gravity is g. We know from Chapter 5 that the magnitude of the gravitational force exerted on the baseball by the Earth is the same as the weight of the ball, $w = m_b g$. That is,

$$m_b g = G \frac{M_E m_b}{R_E^2}$$

We can divide each side of this equation by m_b and solve for the mass of the Earth, M_E:

$$M_E = \frac{g R_E^2}{G}$$

The falling baseball is close enough to the Earth so that the distance of separation between the center of the ball and the center of the Earth can be taken as the radius of the Earth, 6.38×10^6 m. Thus, the mass of the Earth is

$$M_E = \frac{(9.80 \text{ m/s}^2)(6.38 \times 10^6 \text{ m})^2}{6.67 \times 10^{-11} \text{ N} \cdot \text{m}^2/\text{kg}^2} = 5.98 \times 10^{24} \text{ kg}$$

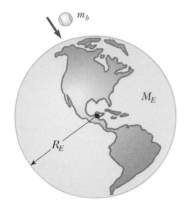

Figure 11.3   (Example 11.2) A baseball falling toward the Earth (not drawn to scale).

Example 11.2   Gravity and Altitude

Derive an expression that shows, for any point above the Earth's surface, how the acceleration due to gravity varies with distance from the center of the Earth.

Solution   The falling baseball of Example 11.1 can be used here also. Now, however, assume that the ball is situated some arbitrary distance, $r > R_E$, from the Earth's center. The first equation in Example 11.1, with r replacing R_E and m_b removed from both sides, becomes

$$g = G \frac{M_E}{r^2}$$

This indicates that the acceleration due to gravity at a point above the Earth's surface decreases as the inverse square of the distance between the point and the center of the Earth. Our assumption of Chapter 2 that objects fall with a constant acceleration is obviously incorrect in light of this example. For short falls, however, the change in g is so small that neglecting the variation does not introduce a significant effect in our results.

Because the weight of an object is mg, a change in the value of g produces a change in weight. For example, if you weigh 800 N at the surface of the Earth, you will weigh only 200 N at a height above the Earth equal to the radius of the Earth.

Also, if the distance of an object from the Earth becomes infinitely large, the weight approaches zero. Values of g at various altitudes are given in Table 11.1.

TABLE 11.1	Free-Fall Acceleration, g, at Various Altitudes Above the Earth's Surface
Altitude h (km)	**g(m/s^2)**
1000	7.33
2000	5.68
3000	4.53
4000	3.70
5000	3.08
6000	2.60
7000	2.23
8000	1.93
9000	1.69
10 000	1.49
50 000	0.13
∞	0

EXERCISE 1   If an object weighs 270 N at the Earth's surface, what will it weigh at an altitude equal to twice the radius of the Earth?   Answer   30 N

EXERCISE 2   Determine the magnitude of the acceleration due to gravity at an altitude of 500 km. By what percentage is the weight of a body reduced at this altitude?
Answer   8.43 m/s²; 14%

11.2 • KEPLER'S LAWS

The movements of the planets, stars, and other celestial bodies have been observed by people for thousands of years. Early in history, scientists regarded the Earth as the center of the Universe. This so-called geocentric model was elaborated and formalized by the Greek astronomer Claudius Ptolemy in the second century A.D. and was accepted for the next 1400 years. In 1543 the Polish astronomer Nicolaus Copernicus (1473–1543) suggested that the Earth and the other planets revolve in circular orbits about the Sun (the heliocentric hypothesis).

The Danish astronomer Tycho Brahe (1546–1601) made accurate astronomical measurements over a period of 20 years and provided the basis for the currently accepted model of the Solar System. It is interesting to note that these precise observations, made on the planets and 777 stars, were carried out with nothing more elaborate than a large sextant and compass; the telescope had not yet been invented.

The German astronomer Johannes Kepler, who was Brahe's assistant, acquired Brahe's astronomical data and spent about 16 years trying to deduce a mathematical model for the motions of the planets. After many laborious calculations, he found that Brahe's precise data on the revolution of Mars about the Sun provided the answer. Kepler's analysis first showed that the concept of circular orbits about the Sun had to be abandoned. He eventually discovered that the orbit of Mars could be accurately described by an ellipse with the Sun at one focus. He then generalized this analysis to include the motions of all planets. The complete analysis is summarized in three statements, known as **Kepler's laws:**

Johannes Kepler (1571–1630)

A German astronomer, Kepler is best known for developing the laws of planetary motion based on the careful observations of Tycho Brahe. After spending several years trying to work out a "regular-solid theory" of the planets, he concluded that the Copernican view of circular planetary orbits had to be abandoned for the view that the planetary orbits are ellipses with the Sun always at one of the foci. *(Art Resource)*

---

1. Every planet moves in an elliptical orbit with the Sun at one of the focal points.
2. The radius vector drawn from the Sun to any planet sweeps out equal areas in equal time intervals.
3. The square of the orbital period of any planet is proportional to the cube of the semimajor axis of the elliptical orbit.

• *Kepler's laws*

Newton demonstrated that these laws were consequences of a simple force that exists between any two masses. Newton's universal law of gravity, together with his laws of motion, provides the basis for a full mathematical solution to the motion of planets and satellites. More important, Newton's universal law of gravity correctly describes the gravitational attractive force between *any* two masses.

11.3 • THE UNIVERSAL LAW OF GRAVITY AND THE MOTIONS OF PLANETS

In formulating his universal law of gravity, Newton built on an observation that suggests that the gravitational force between two bodies is proportional to the inverse square of the separation. Let us compare the acceleration of the Moon in its orbit with the acceleration of an object falling near the Earth's surface, such as an apple (Fig. 11.4). Assume that the two accelerations have the same cause—namely, the gravitational attraction of the Earth. From the inverse-square law, Newton found that the acceleration of the Moon toward the Earth should be proportional to $1/r_M^2$, where r_M is the separation between centers of the Earth and Moon. Furthermore, the acceleration of the apple toward the Earth should be proportional to $1/R_E^2$, where R_E is the radius of the Earth. When the values $r_M = 3.84 \times 10^8$ m and $R_E = 6.37 \times 10^6$ m are used, the ratio of the Moon's acceleration, a_M, to the apple's acceleration, g, is predicted to be

$$\frac{a_M}{g} = \frac{(1/r_M)^2}{(1/R_E)^2} = \left(\frac{R_E}{r_M}\right)^2 = \left(\frac{6.37 \times 10^6 \text{ m}}{3.84 \times 10^8 \text{ m}}\right)^2 = 2.75 \times 10^{-4}$$

Therefore,

$$a_M = (2.75 \times 10^{-4})(9.80 \text{ m/s}^2) = 2.70 \times 10^{-3} \text{ m/s}^2$$

The centripetal acceleration of the Moon can also be calculated kinematically from a knowledge of the Moon's orbital period, T, where $T = 27.32$ days $= 2.36 \times 10^6$ s and its mean distance from the Earth, r_M. In the time T, the Moon travels a distance $2\pi r_M$, which equals the circumference of its orbit. Therefore, its orbital speed is $2\pi r_M/T$, and its centripetal acceleration is

$$a_M = \frac{v^2}{r_M} = \frac{(2\pi r_M/T)^2}{r_M} = \frac{4\pi^2 r_M}{T^2} = \frac{4\pi^2(3.84 \times 10^8 \text{ m})}{(2.36 \times 10^6 \text{ s})^2} = 2.72 \times 10^{-3} \text{ m/s}^2$$

This agreement provides strong evidence that the inverse-square law of force is correct.

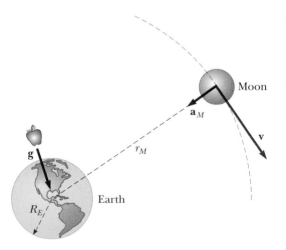

Figure 11.4 As it revolves about the Earth, the Moon experiences a centripetal acceleration $\mathbf{a}_M$ directed toward the Earth. An object near the Earth's surface, such as the apple shown here, experiences an acceleration $\mathbf{g}$. (Dimensions are not to scale.)

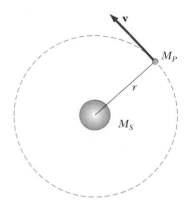

Figure 11.5   A planet of mass M_P moving in a circular orbit about the Sun. The orbits of all planets except Mars, Mercury, and Pluto are nearly circular.

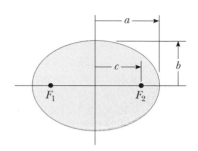

Figure 11.6   Plot of an ellipse. The semimajor axis has a length a, and the semiminor axis has a length b. The focal points are located at a distance c from the center, where $a^2 = b^2 + c^2$, and the eccentricity is defined as $e = c/a$.

Kepler's Third Law

Kepler's third law can be predicted from the inverse-square law for circular orbits.[1] Consider a planet of mass M_P which is assumed to be moving about the Sun (mass M_S) in a circular orbit, as in Figure 11.5. Because the gravitational force on the planet provides the centripetal acceleration as it moves in a circle, we can use Equation 5.3 and write

$$\frac{GM_S M_P}{r^2} = \frac{M_P v^2}{r}$$

But the orbital speed of the planet is simply $2\pi r/T$, where T is the period; therefore, the preceding expression becomes

$$\frac{GM_S}{r^2} = \frac{(2\pi r/T)^2}{r}$$

$$T^2 = \left(\frac{4\pi^2}{GM_S}\right) r^3 = K_S r^3 \qquad \qquad [11.4] \qquad \bullet \textit{ Kepler's third law}$$

where K_S is a constant given by

$$K_S = \frac{4\pi^2}{GM_S} = 2.97 \times 10^{-19} \text{ s}^2/\text{m}^3$$

Equation 11.4 is Kepler's third law. It is also valid for elliptical orbits if we replace r with the length of the semimajor axis, a (Fig. 11.6). Note that the constant of proportionality, K_S, is independent of the mass of the planet. Equation 11.4 is therefore valid for *any* planet. If we were to consider the orbit of a satellite about the Earth, such as the Moon, then the constant would have a different value, with

---

[1]The orbits of all planets except Mars, Mercury, and Pluto are very close to being circular. For example, the ratio of the semiminor axis to the semimajor axis for the Earth is $b/a = 0.999\ 86$. The closeness of the ratio b/a to 1 indicates that the difference between perihelion and aphelion is small as expected.

TABLE 11.2   Useful Planetary Data

Body	Mass (kg)	Mean Radius (m)	Period (s)	Distance from Sun (m)	$\dfrac{T^2}{r^3}\left(\dfrac{s^2}{m^3}\right)$
Mercury	3.18×10^{23}	2.43×10^6	7.60×10^6	5.79×10^{10}	2.97×10^{-19}
Venus	4.88×10^{24}	6.06×10^6	1.94×10^7	1.08×10^{11}	2.99×10^{-19}
Earth	5.98×10^{24}	6.37×10^6	3.156×10^7	1.496×10^{11}	2.97×10^{-19}
Mars	6.42×10^{23}	3.37×10^6	5.94×10^7	2.28×10^{11}	2.98×10^{-19}
Jupiter	1.90×10^{27}	6.99×10^7	3.74×10^8	7.78×10^{11}	2.97×10^{-19}
Saturn	5.68×10^{26}	5.85×10^7	9.35×10^8	1.43×10^{12}	2.99×10^{-19}
Uranus	8.68×10^{25}	2.33×10^7	2.64×10^9	2.87×10^{12}	2.95×10^{-19}
Neptune	1.03×10^{26}	2.21×10^7	5.22×10^9	4.50×10^{12}	2.99×10^{-19}
Pluto	$\approx 1.4 \times 10^{22}$	$\approx 1.5 \times 10^6$	7.82×10^9	5.91×10^{12}	2.96×10^{-19}
Moon	7.36×10^{22}	1.74×10^6	—	—	—
Sun	1.991×10^{30}	6.96×10^8	—	—	—

the Sun's mass replaced by the Earth's mass. In this case, the proportionality constant would equal $4\pi^2/GM_E$.

Table 11.2 is a collection of useful planetary data. The last column verifies that the ratio T^2/r^3 is constant.

Example 11.3   An Earth Satellite

A satellite of mass m moves in a circular orbit about the Earth with a constant speed of v, at a height of $h = 1000$ km above the Earth's surface, as in Figure 11.7. (For clarity, this figure is not drawn to scale.) Find the orbital speed of the satellite. The radius of the Earth is 6.37×10^6 m, and its mass is 5.98×10^{24} kg.

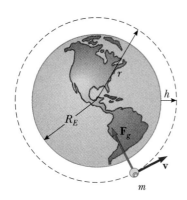

Figure 11.7   (Example 11.3) A satellite of mass m moving around the Earth in a circular orbit of radius r and with constant speed v. The only force acting on the satellite is the gravitational force $\mathbf{F}_g$. (Not drawn to scale.)

Solution   The only external force on the satellite is the gravitational attraction exerted by the Earth. This force is directed toward the center of the satellite's circular path and is the force causing the centripetal acceleration of the satellite. Because the magnitude of the force of gravity is $GM_E m/r^2$, we find that

$$F_g = G\frac{M_E m}{r^2} = m\frac{v^2}{r}$$

$$v^2 = \frac{GM_E}{r}$$

In this expression, the distance r is the Earth's radius plus the height of the satellite; that is, $r = R_E + h = 7.37 \times 10^6$ m, so that

$$v^2 = \frac{(6.67 \times 10^{-11}\ \text{N} \cdot \text{m}^2/\text{kg}^2)(5.98 \times 10^{24}\ \text{kg})}{7.37 \times 10^6\ \text{m}}$$

$$= 5.41 \times 10^7\ \text{m}^2/\text{s}^2$$

Therefore,

$$v = 7.36 \times 10^3\ \text{m/s} \approx \boxed{16\ 400\ \text{mi/h}}$$

Note that v is independent of the mass of the satellite!

EXERCISE 3   Calculate the period of revolution, T, of the satellite.   Answer   105 min

Kepler's Second Law and Conservation of Angular Momentum

Consider a planet of mass M_P moving about the Sun in an elliptical orbit (Fig. 11.8). The gravitational force acting on the planet is a central force, always along the radius vector, directed toward the Sun. The torque on the planet due to this central force is clearly zero, because **F** is parallel to **r**. That is

$$\boldsymbol{\tau} = \mathbf{r} \times \mathbf{F} = \mathbf{r} \times F(r)\hat{\mathbf{r}} = 0$$

But recall that the torque equals the time rate of change of angular momentum; that is, $\boldsymbol{\tau} = d\mathbf{L}/dt$. Therefore,

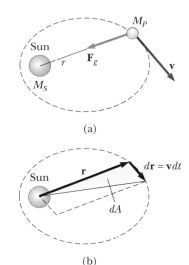

(a)

> because $\boldsymbol{\tau} = 0$, the angular momentum, **L**, of the planet is a constant of the motion:

$$\mathbf{L} = \mathbf{r} \times \mathbf{p} = m\mathbf{r} \times \mathbf{v} = \text{constant}$$

Because **L** is a constant of the motion, the planet's motion at any instant is restricted to the plane formed by **r** and **v**.

We can relate this result to the following geometric consideration. In a time of dt, the radius vector **r** in Figure 11.9b sweeps out the area dA, which equals one half the area $|\mathbf{r} \times d\mathbf{r}|$ of the parallelogram formed by the vectors **r** and $d\mathbf{r}$. Because the displacement of the planet in the time dt is given by $d\mathbf{r} = \mathbf{v}\,dt$, we get

$$dA = \tfrac{1}{2}|\mathbf{r} \times d\mathbf{r}| = \tfrac{1}{2}|\mathbf{r} \times \mathbf{v}\,dt| = \frac{L}{2m}\,dt$$

$$\frac{dA}{dt} = \frac{L}{2m} = \text{constant}$$

[11.5]

(b)

Figure 11.8 (a) The gravitational force acting on a planet acts toward the Sun, along the radius vector. (b) As a planet orbits the Sun, the area swept out by the radius vector in a time dt is equal to one half the area of the parallelogram formed by the vectors **r** and $d\mathbf{r} = \mathbf{v}\,dt$.

• *Kepler's second law*

(Left) This image of Saturn, taken by the *Voyager I* spacecraft on October 18, 1980, was color-enhanced to increase the visibility of large, bright features in Saturn's North Temperate Belt. The distinct color difference between the North Equatorial Belt and other belts may be due to a thicker haze layer covering the northern belt. *(Right)* Jupiter and its four planet-size moons were photographed by *Voyager I* and assembled into this collage. The moons are not to scale but are in their relative positions. Nine other smaller moons circle Jupiter. Not visible is Jupiter's faint ring of particles, seen for the first time by *Voyager I*. *(NASA photos)*

where L and m are both constants of the motion. Thus, we conclude that

> the radius vector from the Sun to any planet sweeps out equal areas in equal times.

It is important to recognize that this result, which is Kepler's second law, is a consequence of the fact that the force of gravity is a central force, which in turn implies that angular momentum of the planet is constant. Therefore, the law applies to *any* situation that involves a central force, whether inverse square or not.

The inverse-square nature of the force of gravity is not revealed by Kepler's second law. Although we do not prove it here, Kepler's first law (as well as his third law) is a direct consequence of the fact that the gravitational force varies as $1/r^2$. That is, under an inverse-square force law, the orbits of the planets can be shown to be ellipses with the Sun at one focus.

Thinking Physics 2

The Earth is closer to the Sun when it is winter in the Northern Hemisphere than when it is summer. July and January both have 31 days. In which month, if either, does the Earth move through a longer distance in its orbit?

Reasoning The Earth is in a slightly elliptical orbit around the Sun. In order to conserve angular momentum, the Earth must move more rapidly when it is close to the Sun and more slowly when it is farther away. Thus, since it is closer to the Sun in January, it is moving faster, and will cover more distance in its orbit than it will in July.

CONCEPTUAL PROBLEM 1

Kepler's first law indicates that planets travel in elliptical orbits. This is intimately related to the fact that the gravitational force varies as the inverse square of the separation between the Sun and the planet. Suppose the gravitational force varied as the inverse *cube*, rather than the inverse square. This opens up some new types of orbits. What about the second and third laws? Will they change if the force depends on the inverse cube of the separation?

CONCEPTUAL PROBLEM 2

A satellite in orbit is not truly traveling through a vacuum—it is moving through very thin air. Does the resulting air friction cause the satellite to slow down?

CONCEPTUAL PROBLEM 3

How would you explain the fact that Saturn and Jupiter have periods much greater than one year?

EXERCISE 4   At its aphelion the planet Mercury is 6.99×10^{10} m from the Sun, and at its perihelion it is 4.60×10^{10} m from the Sun. If its orbital speed is 3.88×10^4 m/s at the aphelion, what is its orbital speed at the perihelion?   Answer 5.90×10^4 m/s

11.4 • ENERGY CONSIDERATIONS IN PLANETARY AND SATELLITE MOTION

Consider a body of mass m moving with a speed of v in the vicinity of a massive body of mass $M \gg m$. The system might be a planet moving around the Sun or a satellite in orbit around the Earth. If we assume that M is at rest in an inertial reference frame,[2] then the total mechanical energy, E, of the two-body system when the bodies are separated by a distance r is the sum of the kinetic energy of the mass m and the potential energy of the system:

$$E = K + U$$

Recall from Chapter 7, Equation 7.22, that the gravitational potential energy, U_g, associated with *any pair* of particles of masses m_1 and m_2 separated by a distance r is given by

$$U_g = -\frac{Gm_1 m_2}{r}$$

Therefore, in our case,

$$E = \tfrac{1}{2}mv^2 - \frac{GMm}{r} \qquad \textbf{[11.6]}$$

Furthermore, the total mechanical energy is constant if we assume that the system is isolated. Therefore, as the mass m moves from P to Q in Figure 11.9, the total energy remains constant and Equation 11.6 gives

$$E = \tfrac{1}{2}mv_i^2 - \frac{GMm}{r_i} = \tfrac{1}{2}mv_f^2 - \frac{GMm}{r_f} \qquad \textbf{[11.7]}$$

Equation 11.6 shows that E may be positive, negative, or zero, depending on the value of v. However, for a bound system, such as the Earth and Sun, E is necessarily *less than zero* if we use the arbitrary convention $U \to 0$ as $r \to \infty$. We can easily establish that $E < 0$ for the system consisting of a mass m moving in a circular orbit about a body of mass M. Newton's second law applied to the body of mass m gives

$$\frac{GMm}{r^2} = \frac{mv^2}{r}$$

Multiplying both sides by r and dividing by 2,

$$\tfrac{1}{2}mv^2 = \frac{GMm}{2r} \qquad \textbf{[11.8]}$$

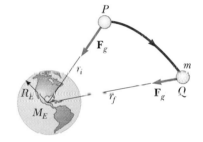

Figure 11.9   As a particle of mass m moves from P to Q above the Earth's surface, the total mechanical energy of the system remains constant.

---

[2]To see that this is reasonable, consider an object of mass m falling toward the Earth. Because the center of mass of the object–Earth system is stationary, it follows that $mv = M_E v_E$. Thus, the Earth acquires a kinetic energy equal to

$$\tfrac{1}{2}M_E v_E^2 = \tfrac{1}{2}\frac{m^2}{M_E}v^2 = \frac{m}{M_E}K$$

where K is the kinetic energy of the object. Because $M_E \gg m$, the kinetic energy of the Earth is negligible.

Substituting this into Equation 11.6, we obtain

$$E = \frac{GMm}{2r} - \frac{GMm}{r}$$

*Total mechanical energy for •
circular orbits*

$$E = -\frac{GMm}{2r} \qquad [11.9]$$

This clearly shows that **the total mechanical energy must be negative in the case of circular orbits.** Furthermore, **the kinetic energy is positive and equal to one half the magnitude of the potential energy** (when the potential energy is chosen to be zero at infinity). The absolute value of E is also equal to the binding energy of the system.

The total mechanical energy is also negative in the case of elliptical orbits. The expression for E for elliptical orbits is the same as Equation 11.9, with r replaced by the semimajor axis length, a. The total energy, the total angular momentum, and the total linear momentum of a planet–Sun system are constants of the motion.

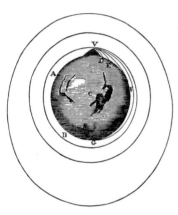

" . . . the greater the velocity . . . with which (a stone) is projected, the farther it goes before it falls to the Earth. We may therefore suppose the velocity to be so increased, that it would describe an arc of 1, 2, 5, 10, 100, 1000 miles before it arrived at the Earth, till at last, exceeding the limits of the Earth, it should pass into space without touching."—Newton, *System of the World.*

Thinking Physics 3

Icebound, by Dean Koontz (Ballantine Books, New York, 1995), is a story of a group of scientists trapped on a floating iceberg near the North Pole. One of the devices that the scientists have with them on their task is a transmitter with which they can fix their position with "the aid of a *geosynchronous polar satellite*." Can a satellite in a *polar* orbit be *geosynchronous*?

Reasoning A geosynchronous satellite is one that stays over one location on the Earth's surface at all times. Thus, an antenna receiving signals from the satellite, such as a television dish, can stay pointed in a fixed direction toward the sky. The satellite must be in an orbit with the correct radius such that its orbital period is the same as that of the Earth's rotation. This would result in the satellite appearing to have no east–west motion relative to the observer at the chosen location. Another requirement is that a geosynchronous satellite *must be in orbit over the Equator.* Otherwise it would appear to undergo a north–south oscillation during one orbit. Thus, it would be impossible to have a geosynchronous satellite in a *polar* orbit. Even if such a satellite were at the proper distance from the Earth, it would be moving rapidly in the north–south direction, resulting in the necessity of accurate tracking equipment. What's more, it would be below the horizon for long periods of time, making it useless for determining one's position.

Thinking Physics 4

A satellite is in a circular orbit in space. Is it necessary that there be a massive object at the center of the orbit?

Reasoning This is the usual situation—satellites are normally in orbit around the Earth, a massive object. Imagine, however, tying a rock somewhere along a length of string and twirling the rock in a circle as shown in Figure 11.10. In this case, there is no entity at the center of the circular path of the rock, nor along its radius, that is exerting a force on the rock. The net force holding the rock in the circular orbit is a

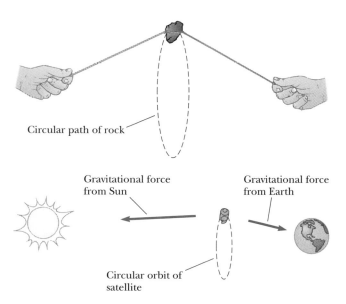

Figure 11.10   (Thinking Physics 4)

combination of the radial components of two forces—one from the string on each side of the rock.

The question posed was not about rocks on strings; it was about satellites. Is there a gravitational counterpart to this situation? There is. Imagine moving from the Earth to the Sun along the line between them. At some point along that line, the gravitational force of attraction to the Earth will balance that toward the Sun. This is similar to the balance along the string of the tension on either side of the rock—the rock is in equilibrium between two balanced forces. Now imagine pulling the satellite perpendicularly off the Earth–Sun line a small amount—as long as the distance is relatively small, the components of the forces from the Earth and Sun along the Sun–Earth centerline will cancel. The components of the two forces perpendicular to the centerline will add, to provide a net force pulling the satellite back toward the centerline. Now, suppose we set the satellite into circular motion with just the right velocity. The satellite will be in orbit around a center point with no mass located there.

The International Sun–Earth Explorer 3 satellite (ISEE-3) was launched in 1978 and placed in just such an orbit in order to study the solar wind. The advantage over a Sun-centered orbit is that the satellite always resides on the side of the Sun toward the Earth, so that it is constantly in communication with Earth. In the absence of a specific mission to Comet Halley, the ISEE-3 satellite was removed from this orbit in 1982 and placed in orbit around the Moon, in anticipation of a flyby of Comet Halley. On the way to this flyby, it passed through the tail of Comet Giacobini-Zinner and successfully gathered important data on both comets.

Example 11.4   A Satellite in an Elliptical Orbit

A satellite moves in an elliptical orbit about the Earth, as in Figure 11.11. The minimum and maximum distances from the surface of the Earth are 400 km and 3000 km. Find the speeds of the satellite at apogee and perigee.

Solution   Because the mass of the satellite is negligible compared with the Earth's mass, we take the center of mass of the Earth to be at rest. Gravity is a *central* force, and so the angular momentum of the satellite about the Earth's center of

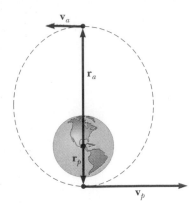

Figure 11.11   (Example 11.4) A satellite in an elliptical orbit about the Earth.

mass remains constant in time. With subscripts a and p for the apogee and perigee positions, conservation of angular momentum gives $L_p = L_a$, or

$$mv_p r_p = mv_a r_a$$

$$v_p r_p = v_a r_a \qquad [1]$$

Applying conservation of energy, we obtain $E_p = E_a$, or

$$U_p + K_p = U_a + K_a$$

$$-G\frac{M_E m}{r_p} + \tfrac{1}{2}mv_p^2 = -G\frac{M_E m}{r_a} + \tfrac{1}{2}mv_a^2$$

$$2GM_E\left(\frac{1}{r_a} - \frac{1}{r_p}\right) = (v_a^2 - v_p^2) \qquad [2]$$

Assuming the Earth's radius is 6.37×10^6 m and using the given data, we get $r_a = 9.37 \times 10^6$ m and $r_p = 6.77 \times 10^6$ m. Because we know the numerical values of G, M_E, r_p, and r_a, we can use Equations (1) and (2) to determine the two unknowns, (v_p and v_a). Solving the equations simultaneously, we obtain

$$v_p = \boxed{8.27 \text{ km/s}} \qquad v_a = \boxed{5.98 \text{ km/s}}$$

Escape Speed

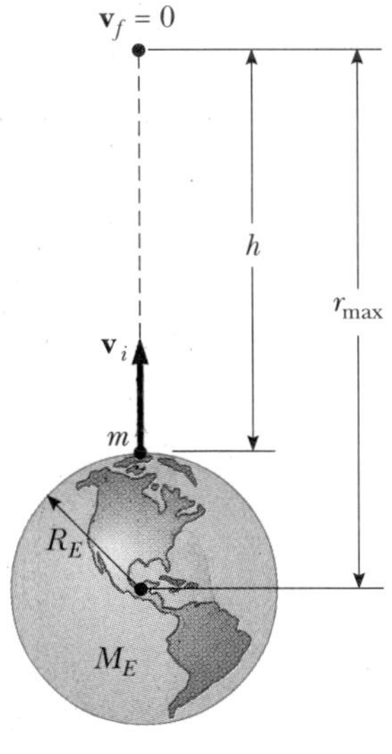

Suppose an object of mass m is projected vertically from the Earth's surface with an initial speed v_i, as in Figure 11.12. We can use energy considerations to find the minimum value of the initial speed such that the object will escape the Earth's gravitational field. Equation 11.6 gives the total energy of the object at any point when its speed and distance from the center of the Earth are known. At the surface of the Earth, where $v_i = v$, $r_i = R_E$. When the object reaches its maximum altitude, $v_f = 0$ and $r_f = r_{max}$. Because the total energy of the system is constant, substitution of these conditions into Equation 11.7 gives

$$\tfrac{1}{2}mv_i^2 - \frac{GM_E m}{R_E} = -\frac{GM_E m}{r_{max}}$$

Solving for v_i^2 gives

$$v_i^2 = 2GM_E\left(\frac{1}{R_E} - \frac{1}{r_{max}}\right) \qquad [11.10]$$

Figure 11.12   An object of mass m projected upward from the Earth's surface with an initial velocity $\mathbf{v}_i$ reaches a maximum altitude h.

Therefore, if the initial speed is known, this expression can be used to calculate the maximum altitude, h, because we know that $h = r_{max} - R_E$.

We are now in a position to calculate the minimum speed the object must have at the Earth's surface in order to escape from the influence of the Earth's gravitational field. This corresponds to the situation in which the object will continue to move away forever, with the speed asymptotically approaching *zero*. Setting $r_{max} = \infty$ in Equation 11.10 and taking $v_i = v_{esc}$ (the escape speed), we get

$$v_{esc} = \sqrt{\frac{2GM_E}{R_E}}$$ [11.11]   • *Escape speed*

Note that this expression for v_{esc} is independent of the mass of the object projected from the Earth. For example, a spacecraft has the same escape speed as a molecule. Furthermore, the result is independent of the *direction* of the velocity, provided the object is not thrown into the ground.

If the object is given an initial speed equal to v_{esc}, its *total* energy is equal to zero. This can be seen by noting that as $r \rightarrow \infty$, the object's kinetic energy and its potential energy are both zero. If v_i is greater than v_{esc}, the *total* energy is greater than zero and the object is still moving as $r \rightarrow \infty$.

Finally, you should note that Equations 11.10 and 11.11 can be applied to objects projected from *any* planet. That is, in general, the escape speed from any planet of mass M and radius R is

$$v_{esc} = \sqrt{\frac{2GM}{R}}$$ [11.12]

A list of escape speeds for the planets, the Moon, and the Sun is given in Table 11.3. Note that the values vary from 1.1 km/s for Pluto to about 618 km/s for the Sun. These results, together with some ideas from the kinetic theory of gases (Chapter 16), explain why some planets have atmospheres and others do not. As we shall see later, a gas molecule has an average kinetic energy that depends on its temperature. Lighter atoms, such as hydrogen and helium, have higher average speeds than do the heavier species. When the average speed of the lighter atoms is not much less than the escape speed, a significant fraction of the molecules have a chance to escape from the planet. This mechanism also explains why the Earth does not retain hydrogen and helium molecules in its atmosphere but does retain much heavier molecules, such as oxygen and nitrogen. By contrast, Jupiter has a very large escape speed (60 km/s), which enables it to retain hydrogen, the primary component of its atmosphere.

TABLE 11.3   Escape Speeds from the Surfaces of the Planets, the Moon, and the Sun

Planet	v_{esc}(km/s)
Mercury	4.3
Venus	10.3
Earth	11.2
Mars	5.0
Jupiter	60.0
Saturn	36.0
Uranus	22.0
Neptune	24.0
Pluto	1.1
Moon	2.3
Sun	618.0

Black Holes

In Chapter 10 we briefly described a rare event called a supernova—the catastrophic explosion of a very massive star. The material that remains in the central core of such an object continues to collapse, but the core's ultimate fate depends on its mass. If the core has a mass less than 1.4 times the mass of our Sun, it gradually cools down and ends its life as a white dwarf star. However, if the core's mass is greater than this, the star may collapse further due to gravitational forces. What remains is a neutron star, a region compressed to a radius of about 10 km. (On Earth, a teaspoon of this material would weigh about 5 billion tons!)

An even more unusual star death may occur when the core has a mass greater than about three solar masses. No known forces in nature are strong enough to prevent the collapse of such a star. The collapse may continue until the star becomes a small object in space, commonly referred to as a **black hole.** In effect, black holes are remains of stars that have collapsed under their own weight. Once an object such as a spaceship comes near a black hole, it experiences an extremely strong gravitational force and is trapped forever (Fig. 11.13).

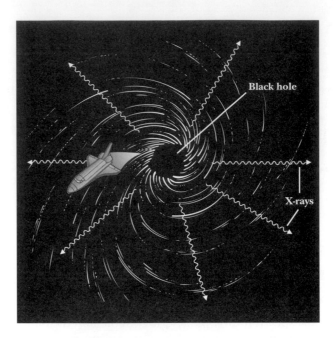

Figure 11.13 The gravitational field in the vicinity of a black hole is so strong that nothing can escape. Any object moving close to a black hole can emit x-rays.

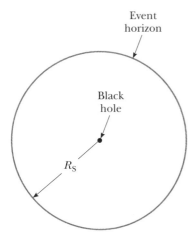

Figure 11.14 A black hole. The distance R_S equals the Schwarzschild radius. Any event occurring within the boundary of radius R_S, called the event horizon, is invisible to an outside observer.

The escape speed from any spherical body can be calculated from Equation 11.12. If the escape speed exceeds the speed of light, c, radiation within the body (such as visible light) cannot escape and the body appears to be black; hence the origin of the terminology "black hole." The critical radius, R_S, for which this occurs is called the **Schwarzschild radius** (Fig. 11.14). Taking $v_{esc} = c$ in Equation 11.12 and solving for R_S, we obtain $R_S = 2GM/c^2$. For example, the value for R_S for a black hole with a mass equal to that of the Sun is calculated to be 3.0 km (about 2 mi); a black hole the mass of which equals that of the Earth has a radius of about 9 mm (about the size of a dime).

Although light from a black hole cannot escape, light from events taking place near the black hole should be visible. A companion star captured by the strong gravitational field of a black hole should emit x-rays (Fig. 11.13). Based on this reasoning, several candidates for black holes have been detected, the most famous being Cygnus X-1, the first x-ray source detected in the constellation Cygnus. There is also evidence that supermassive black holes exist at the centers of galaxies.

EXERCISE 5 The escape speed from the surface of the Earth is 11.2 km/s. Estimate the escape speed for a spacecraft from the surface of the Moon. The Moon has a mass $\frac{1}{81}$ that of Earth and a radius $\frac{1}{4}$ that of Earth.   Answer   2.49 km/s

EXERCISE 6   (a) Calculate the minimum energy required to send a 3000-kg spacecraft from the Earth to a distant point in space where Earth's gravity is negligible. (b) If the journey is to take three weeks, what *average* power will the engines have to supply?
Answer   (a) 1.88×10^{11} J   (b) 103 kW

EXERCISE 7 A satellite moves in an elliptical orbit about the Earth such that, at perigee and apogee positions, the distances from the Earth's center are, respectively, D and $4D$. Find the ratio of the speeds at the two positions, v_p/v_a.   Answer   4

11.5 · ATOMIC SPECTRA AND THE BOHR THEORY OF HYDROGEN

As you may have already learned in chemistry, the hydrogen atom is the simplest known atomic system and an especially important one to understand. Much of what is learned about the hydrogen atom (which consists of one proton and one electron) can be extended to other single-electron ions such as He^+ and Li^{2+}. Furthermore, a thorough understanding of the physics underlying the hydrogen atom can then be used to describe more complex atoms and the periodic table of the elements.

Suppose an evacuated glass tube is filled with hydrogen (or some other gas). If a voltage applied between metal electrodes in the tube is great enough to produce an electric current in the gas, the tube emits light the color of which is characteristic of the gas (this is how a neon sign works). When the emitted light is analyzed with a device called a spectroscope, a series of discrete lines is observed, each line corresponding to a different wavelength, or color, of light. Such a series of spectral lines is commonly referred to as an **emission spectrum.** The wavelengths contained in a given line spectrum are characteristic of the element emitting the light (Fig. 11.15). Because no two elements emit the same line spectrum, this phenomenon represents a marvelous and reliable technique for identifying elements in a substance.

As you shall learn in more detail in Chapter 13, a mechanical wave is a disturbance that transports energy as it moves through a system without transporting matter. A common form of periodic wave is the sinusoidal wave, the shape of which

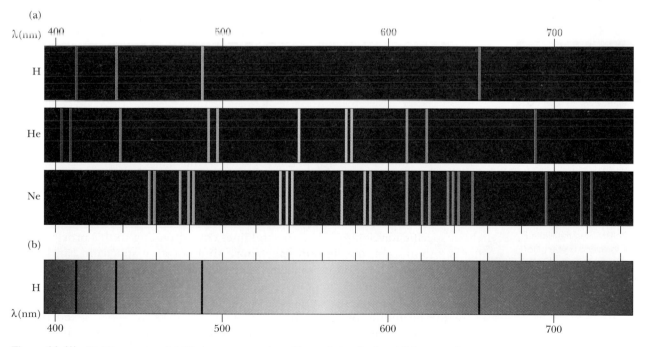

Figure 11.15 Visible spectra. (a) Line spectra produced by emission in the visible range for the elements hydrogen, helium, and neon. (b) The absorption spectrum for hydrogen. The dark absorption lines occur at the same wavelengths as the emission line for hydrogen shown in (a).

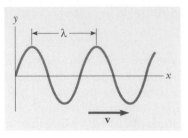

Figure 11.16 A sinusoidal wave traveling to the right. Any point on the wave moves a distance of one wavelength, λ, in a time equal to the period of the wave.

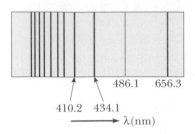

Figure 11.17 A series of spectral lines for atomic hydrogen. The prominent lines labeled are part of the Balmer series.

Niels Bohr (1885–1962)

A Danish physicist, Bohr was an active participant in the early development of quantum mechanics and provided much of its philosophical framework. During the 1920s and 1930s, Bohr headed the Institute for Advanced Studies in Copenhagen. This institute was a magnet for many of the world's best physicists and provided a forum for the exchange of ideas. When Bohr visited the United States in 1939 to attend a scientific conference, he brought news that the fission of uranium had been discovered by Hahn and Strassman in Berlin. The results were the foundations of the atomic bomb developed in the United States during World War II. Bohr was awarded the 1922 Nobel prize for his investigation of the structure of atoms and of the radiation emanating from them.

is depicted in Figure 11.16. The distance between two consecutive crests of the wave is called the **wavelength,** λ. As the wave travels to the right with speed v, any point on the wave travels a distance of one wavelength in a time interval of one period, T (the time for one cycle), so the wave speed is $v = \lambda/T$. The inverse of the period, $1/T$, is called the **frequency,** f, of the wave; it represents the number of cycles per second. Thus, the speed of the wave is often written as $v = \lambda f$. In this section, because we shall deal with electromagnetic waves—which travel at the speed of light, c—the appropriate relation is

$$c = \lambda f \qquad \text{[11.13]}$$

The emission spectrum of hydrogen shown in Figure 11.17 includes four prominent lines that occur at wavelengths of 656.3 nm, 486.1 nm, 434.1 nm, and 410.2 nm. In 1885, Johann Balmer (1825–1898) found that the wavelengths of these and less prominent lines can be described by this simple empirical equation:

$$\frac{1}{\lambda} = R_H \left(\frac{1}{2^2} - \frac{1}{n^2} \right) \qquad \text{Balmer series} \qquad \text{[11.14]}$$

where n may have integral values of 3, 4, 5, . . . , and R_H is a constant, now called the **Rydberg constant.** If the wavelength is in meters, R_H has the value

$$R_H = 1.0973732 \times 10^7 \text{ m}^{-1}$$

The first line in the Balmer series, at 656.3 nm, corresponds to $n = 3$ in Equation 11.14; the line of 486.1 nm corresponds to $n = 4$; and so on.

In addition to emitting light at specific wavelengths, an element can also absorb light at specific wavelengths. The spectral lines corresponding to this process form what is known as an **absorption spectrum.** An absorption spectrum can be obtained by passing a continuous radiation spectrum (one containing all wavelengths) through a vapor of the element being analyzed. The absorption spectrum consists of a series of dark lines superimposed on the otherwise continuous spectrum.

At the beginning of the twentieth century, scientists were perplexed by the failure of classical physics to explain the characteristics of spectra. Why did atoms of a given element emit only certain wavelengths of radiation, so that the emission spectrum displayed discrete lines? Furthermore, why did the atoms absorb only those wavelengths that they emitted? In 1913 Bohr provided an explanation of atomic spectra that included some features of the currently accepted theory. Using the simplest atom, hydrogen, Bohr described a model of what he thought must be the atom's structure. His model of the hydrogen atom contains some classical features as well as some revolutionary postulates that could not be justified within the framework of classical physics.[3] The basic assumptions of the Bohr theory as it applies to the hydrogen atom are as follows:

1. The electron moves in circular orbits about the proton under the influence of the Coulomb force of attraction, as in Figure 11.18.
2. Only certain electron orbits are stable. These are orbits in which the hydrogen atom does not emit energy in the form of radiation. Hence, the total energy of the atom remains constant, and classical mechanics can be used to describe the electron's motion.
3. Radiation is emitted by the hydrogen atom when the electron "jumps" from a more energetic initial state to a lower state. The jump cannot be visualized or treated classically. In particular, the frequency, f, of the radiation emitted in the jump is related to the change in the atom's energy and **is independent of the frequency of the electron's orbital motion.** The frequency of the emitted radiation is found from

$$E_i - E_f = hf \qquad \textbf{[11.15]}$$

where E_i is the energy of the initial state, E_f is the energy of the final state, h is Planck's constant (see Section 10.9), and $E_i > E_f$.
4. The size of the allowed electron orbits is determined by a condition imposed on the electron's orbital angular momentum: the allowed orbits are those for which the electron's orbital angular momentum about the nucleus is an integral multiple of $\hbar = h/2\pi$.

$$m_e v r = n\hbar \qquad n = 1, 2, 3, \ldots \qquad \textbf{[11.16]}$$

Using these four assumptions, we can calculate the allowed energy levels and emission wavelengths of the hydrogen atom. The electrical potential energy of the system shown in Figure 11.18 is given by $U_e = -k_e e^2/r$, where k_e is the Coulomb constant, e is the charge on the electron, and r is the electron–proton separation. Thus, the total energy of the atom, which contains both kinetic and potential energy terms, is

$$E = K + U_e = \tfrac{1}{2} m_e v^2 - k_e \frac{e^2}{r} \qquad \textbf{[11.17]}$$

Applying Newton's second law to this system, we see that the Coulomb attractive force on the electron, $k_e e^2/r^2$ (Eq. 5.15), must equal the mass times the centripetal acceleration ($a = v^2/r$) of the electron:

$$\frac{k_e e^2}{r^2} = \frac{m_e v^2}{r}$$

• *Assumptions of the Bohr theory*

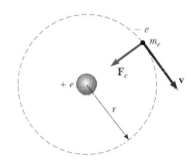

Figure 11.18 Diagram representing Bohr's model of the hydrogen atom in which the orbiting electron is allowed to be only in specific orbits of discrete radii.

---

[3]The Bohr model can be applied successfully to such hydrogen-like ions as singly ionized helium and doubly ionized lithium, but the theory does not properly describe the visible spectra of more complex atoms and ions.

From this expression, we find the kinetic energy to be

$$K = \tfrac{1}{2} m_e v^2 = \frac{k_e e^2}{2r} \qquad \text{[11.18]}$$

Substituting this value of K into Equation 11.17, we find that the total energy of the atom is

Total energy of the hydrogen •
atom

$$E = -\frac{k_e e^2}{2r} \qquad \text{[11.19]}$$

Note that the total energy is negative, indicating a bound electron–proton system. This means that energy in the amount of $k_e e^2/2r$ must be added to the atom just to remove the electron and make the total energy zero. An expression for r, the radius of the allowed orbits, can be obtained by eliminating v by substitution between Equations 11.16 and 11.18:

Radii of Bohr orbits in •
hydrogen

$$r_n = \frac{n^2 \hbar^2}{m_e k_e e^2} \qquad n = 1, 2, 3, \ldots \qquad \text{[11.20]}$$

This result shows that the radii have discrete values, or are *quantized*.

The orbit for which $n = 1$ has the smallest radius; it is called the **Bohr radius,** a_0, and has the value

The Bohr radius •

$$a_0 = \frac{\hbar^2}{m_e k_e e^2} = 0.0529 \text{ nm} \qquad \text{[11.21]}$$

The first three Bohr orbits are shown to scale in Figure 11.19.

The quantization of the orbit radii immediately leads to energy quantization. This can be seen by substituting $r_n = n^2 a_0$ into Equation 11.19. The allowed energy levels are found to be

$$E_n = -\frac{k_e e^2}{2a_0} \left(\frac{1}{n^2} \right) \qquad n = 1, 2, 3, \ldots \qquad \text{[11.22]}$$

Insertion of numerical values into Equation 11.22 gives

$$E_n = -\frac{13.6}{n^2} \text{ eV} \qquad n = 1, 2, 3, \ldots \qquad \text{[11.23]}$$

(Recall from Section 9.7 that 1 eV = 1.6×10^{-19} J.) The lowest stationary state corresponding to $n = 1$, called the **ground state,** has an energy of $E_1 = -13.6$ eV. The next state, the **first excited state,** has $n = 2$ and an energy of $E_2 = E_1/2^2 = -3.4$ eV. Figure 11.20 is an energy-level diagram showing the energies of these discrete energy states and the corresponding quantum numbers. The uppermost level, corresponding to $n = \infty$ (or $r = \infty$) and $E = 0$, represents the state for which the electron is removed from the atom. The minimum energy required to ionize the atom (that is, to completely remove an electron in the ground state from the proton's influence) is called the **ionization energy.** As can be seen from Figure 11.20, the ionization energy for hydrogen, based on Bohr's calculation, is 13.6 eV. This constituted another major achievement for the Bohr theory, because the ionization energy for hydrogen had already been measured to be precisely 13.6 eV.

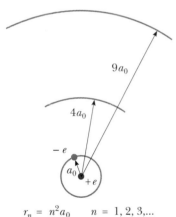

$$r_n = n^2 a_0 \qquad n = 1, 2, 3, \ldots$$

Figure 11.19 The first three Bohr orbits for hydrogen.

Figure 11.20 also shows other spectral series (the Lyman series and the Paschen series) that were found after Balmer's discovery. These spectra obey other empirical formulas, which are reconciled with the Bohr model.

Equation 11.22, together with Bohr's third postulate, can be used to calculate the frequency of the radiation that is emitted when the electron jumps from an outer orbit to an inner orbit:

$$f = \frac{E_i - E_f}{h} = \frac{k_e e^2}{2a_0 h}\left(\frac{1}{n_f^2} - \frac{1}{n_i^2}\right) \qquad [11.24]$$

Because the quantity being measured is wavelength, it is convenient to convert frequency to wavelength, using $c = f\lambda$, to get

$$\frac{1}{\lambda} = \frac{f}{c} = \frac{k_e e^2}{2a_0 hc}\left(\frac{1}{n_f^2} - \frac{1}{n_i^2}\right) \qquad [11.25]$$

The remarkable fact is that the *theoretical* expression, Equation 11.25, is identical to a generalized form of the empirical relations discovered by Balmer and others (see Eq. 11.14),

$$\frac{1}{\lambda} = R_H\left(\frac{1}{n_f^2} - \frac{1}{n_i^2}\right) \qquad [11.26]$$

provided that the combination of constants $k_e e^2/2a_0 hc$ is equal to the experimentally determined Rydberg constant. After Bohr demonstrated the agreement of these two quantities to a precision of about 1%, it was soon recognized as the crowning achievement of his new theory of quantum mechanics. Furthermore, Bohr showed that all of the spectral series for hydrogen have a natural interpretation in his theory. Figure 11.20 shows these spectral series as transitions between energy levels.

Bohr immediately extended his model for hydrogen to other elements in which all but one electron had been removed. Ionized elements such as He^+, Li^{2+}, and Be^{3+} were suspected to exist in hot stellar atmospheres, where frequent atomic collisions occur with enough energy to completely remove one or more atomic electrons. Bohr showed that many mysterious lines observed in the Sun and several stars could not be due to hydrogen but were correctly predicted by his theory if attributed to singly ionized helium.

Although the Bohr model had some success in predicting the spectra of single valence electron atoms, it has some serious drawbacks. For example, it cannot account for the visible spectra of more complex atoms, and it is unable to predict many subtle spectral details of hydrogen and other simple atoms.

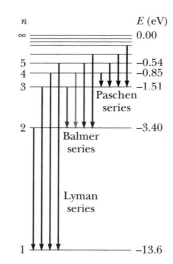

Figure 11.20 An energy level diagram for hydrogen. The discrete allowed energies are plotted on the vertical axis. Nothing is plotted on the horizontal axis, but the horizontal extent of the diagram is made large enough to show allowed transitions. Quantum numbers are given on the left.

CONCEPTUAL PROBLEM 4

Under normal experimental conditions, hydrogen atoms show more lines in emission than in absorption. Why?

CONCEPTUAL PROBLEM 5

Suppose the energy of the electron in a hydrogen atom is $-E$. What is the kinetic energy? What is the potential energy?

Example 11.5 An Electronic Transition in Hydrogen

The electron in the hydrogen atom makes a transition from the $n = 2$ energy state to the ground state (corresponding to $n = 1$). Find the wavelength and frequency of the emitted radiation.

Solution We can use Equation 11.26 directly to obtain λ, with $n_i = 2$ and $n_f = 1$:

$$\frac{1}{\lambda} = R_H\left(\frac{1}{n_f^2} - \frac{1}{n_i^2}\right)$$

$$\frac{1}{\lambda} = R_H\left(\frac{1}{1^2} - \frac{1}{2^2}\right) = \frac{3R_H}{4}$$

$$\lambda = \frac{4}{3R_H} = \frac{4}{3(1.097 \times 10^{-7}\ \text{m}^{-1})}$$

$$= 1.215 \times 10^{-7}\ \text{m} = \boxed{121.5\ \text{nm}} \quad \text{(ultraviolet)}$$

Because $c = f\lambda$, the frequency of the radiation is

$$f = \frac{c}{\lambda} = \frac{3.00 \times 10^8\ \text{m/s}}{1.215 \times 10^{-7}\ \text{m}} = \boxed{2.47 \times 10^{15}\ \text{s}^{-1}}$$

EXERCISE 8 What is the wavelength of the radiation emitted by hydrogen when the electron makes a transition from the $n = 3$ state to the $n = 1$ state?

Answer $\dfrac{9}{8R_H} = 102.6$ nm

Bohr's Correspondence Principle

In our study of relativity in Chapter 9, we found that newtonian mechanics cannot be used to describe phenomena that occur at speeds approaching the speed of light. Newtonian mechanics is a special case of relativistic mechanics and is usable only when v is much less than c. In a similar way, **quantum mechanics is in agreement with classical physics where the energy differences between quantized levels vanish.** This principle, first set forth by Bohr, is called the **correspondence principle.**

For example, consider an electron orbiting the hydrogen atom with $n >$ 10 000. For such large values of n, the energy differences between adjacent levels approach zero and the levels are nearly continuous. As a consequence, the classical model is reasonably accurate in describing the system for large values of n. According to the classical picture, the frequency of the light emitted by the atom is equal to the frequency of revolution of the electron in its orbit about the nucleus. Calculations show that for $n >$ 10 000, this frequency differs from that predicted by quantum mechanics by less than 0.015%.

SUMMARY

Newton's universal law of gravity states that the gravitational force of attraction between any two particles of masses m_1 and m_2 separated by a distance r has the magnitude

$$F_g = G\frac{m_1 m_2}{r^2} \tag{11.1}$$

where G is the gravitational constant, $6.672 \times 10^{-11}\ \text{N}\cdot\text{m}^2/\text{kg}^2$.

Kepler's laws of planetary motion state the following:

1. Every planet moves in an elliptical orbit with the Sun at one of the focal points.
2. The radius vector drawn from the Sun to any planet sweeps out equal areas in equal time intervals.

3. The square of the orbital period of any planet is proportional to the cube of the semimajor axis for the elliptical orbit.

Kepler's second law is a consequence of the fact that the force of gravity is a *central force*. This implies that the angular momentum of the planet–Sun system is a constant of the motion.

Kepler's first and third laws are a consequence of the inverse-square nature of the universal law of gravity. Newton's second law, together with the force law given by Equation 11.1, verifies that the period, T, and radius, r, of the orbit of a planet about the Sun are related by

$$T^2 = \left(\frac{4\pi^2}{GM_S}\right) r^3 \qquad \text{[11.4]}$$

where M_S is the mass of the Sun. Most planets have nearly circular orbits about the Sun. For elliptical orbits, Equation 11.4 is valid if r is replaced by the semimajor axis, a.

If an isolated system consists of a particle of mass m moving with a speed of v in the vicinity of a massive body of mass M, the *total energy* of the system is constant and is

$$E = \tfrac{1}{2}mv^2 - \frac{GMm}{r} \qquad \text{[11.6]}$$

If m moves in a circular orbit of radius r about M, where $M \gg m$, the total energy of the system is

$$E = -\frac{GMm}{2r} \qquad \text{[11.9]}$$

The total energy is negative for any bound system—that is, one in which the orbit is closed, such as a circular or an elliptical orbit.

The minimum speed an object must have to escape the gravitational field of a uniform sphere of mass M and radius R is

$$v_{\text{esc}} = \sqrt{\frac{2GM}{R}} \qquad \text{[11.12]}$$

The Bohr model of the atom is successful in describing the spectra of atomic hydrogen and hydrogen-like ions. One of the basic assumptions of the model is that the electron can exist only in discrete orbits such that the angular momentum mvr is an integral multiple of $h/2\pi = \hbar$. Assuming circular orbits and a simple coulombic attraction between the electron and proton, the energies of the quantum states for hydrogen are calculated to be

$$E_n = -\frac{k_e e^2}{2a_0}\left(\frac{1}{n^2}\right) \qquad \text{[11.22]}$$

where k_e is the Coulomb constant, e is electronic charge, $n = 1, 2, 3, \cdots$ is a positive integer called the **quantum number** of the state, and $a_0 = 0.0529$ nm is the **Bohr radius.**

If the electron in the hydrogen atom makes a transition from an orbit with a quantum number of n_i to one with a quantum number of n_f, where $n_f < n_i$, the frequency of the radiation emitted by the atom is

$$f = \frac{k_e e^2}{2a_0 h}\left(\frac{1}{n_f^2} - \frac{1}{n_i^2}\right) \qquad \text{[11.24]}$$

Using $E = hf = hc/\lambda$, one can calculate the wavelengths of the radiation for various transitions in which there is a change in quantum number, $n_i \rightarrow n_f$. The calculated wavelengths are in excellent agreement with observed line spectra.

CONCEPTUAL QUESTIONS

1. If the gravitational force on an object is directly proportional to its mass, why don't large masses fall with greater acceleration than small ones?

2. The gravitational force that the Sun exerts on the Moon is about twice as great as the gravitational force that the Earth exerts on the Moon. Why doesn't the Sun pull the Moon away from the Earth during a total eclipse of the Sun?

3. Explain why it takes more fuel for a spacecraft to travel from the Earth to the Moon than for the return trip. Estimate the difference.

4. Explain why there is no work done on a planet as it moves in a circular orbit around the Sun, even though a gravitational force is acting on the planet. What is the *net* work done on a planet during each revolution as it moves around the Sun in an elliptical orbit?

5. At what position in its elliptical orbit is the speed of a planet a maximum? At what position is the speed a minimum?

6. Why don't we put a communications satellite in orbit around the 45th parallel? Wouldn't this be more useful in the United States than one in orbit around the Equator?

7. In his 1798 experiment, Cavendish was said to have "weighed the Earth." Explain this statement.

8. If a hole could be dug to the center of the Earth, do you think that the force on a mass *m* would still obey Equation 11.1 there? What do you think the force on *m* would be at the center of the Earth?

9. The *Voyager* spacecraft was accelerated toward escape speed from the Sun by the gravitational force exerted on the spacecraft by Jupiter. How is this possible?

10. The *Apollo 13* spaceship developed trouble in the oxygen system about halfway to the Moon. Why did the mission continue on around the Moon and then return home rather than immediately turn back to Earth?

11. Discuss the similarities and differences between the classical description of planetary motion and the Bohr model of the hydrogen atom.

12. The Bohr theory of the hydrogen atom is based on several assumptions. Discuss those assumptions and their significance. Do any of them contradict classical physics?

13. Explain the significance behind the fact that the total energy of the atom in the Bohr model is negative.

14. Suppose that the electron in the hydrogen atom obeyed classical mechanics rather than quantum mechanics. Why should such a "hypothetical" atom emit a continuous spectrum rather than the observed line spectrum?

PROBLEMS

Section 11.1 Newton's Universal Law of Gravity Revisited

1. Which exerts a greater force of gravitational attraction on objects on the Earth: the Moon or the Sun? Calculate these forces on a 1.00-kg mass.

2. Two ocean liners, each with a mass of 40 000 metric tons, are moving on parallel courses, 100 m apart. What is the magnitude of the acceleration of one of the liners toward the other due to the mutual gravitational attraction? (Treat the ships as spheres.)

3. The gravitational field on the surface of the Moon is about one sixth that on the surface of the Earth. If the radius of the Moon is about one quarter that of the Earth, find the ratio of the average mass density of the Moon to the average mass density of the Earth.

4. A student proposes to measure the gravitational constant *G* by suspending two spherical masses from the ceiling of a tall cathedral and measuring the deflection of the cables from the vertical. Draw a free-body diagram for one of the masses. If two 100.0 kg masses are suspended at the end of 45.00 m long cables and the cables are attached to the ceiling 1.000 m apart, what is the separation of the masses?

5. On the way to the Moon the Apollo astronauts reach a point at which the Moon's gravitational pull becomes equal to that of Earth. Determine the distance of this point from the center of the Earth.

Section 11.2 Kepler's Laws

Section 11.3 The Universal Law of Gravity and the Motions of Planets

6. A satellite is in a circular orbit just above the surface of the Moon. (The radius of the Moon is 1738 km.) (a) What is the acceleration of the satellite? (b) What is the speed of the satellite? (c) What is the period of the satellite orbit?

7. The *Explorer VIII* satellite, placed into orbit November 3, 1960, to investigate the ionosphere, had the following orbit parameters: perigee 459 km and apogee 2289 km (both distances above the Earth's surface); period, 112.7 min. Find the ratio v_p/v_a.

8. Halley's Comet approaches the Sun to within 0.570 A.U., and its orbital period is 75.6 years. (A.U. is the abbreviation for astronomical unit, where 1 A.U. = 1.50×10^{11} m is the

mean Earth–Sun distance.) How far from the Sun will Halley's Comet travel before it starts its return journey? (See Fig. P11.8.)

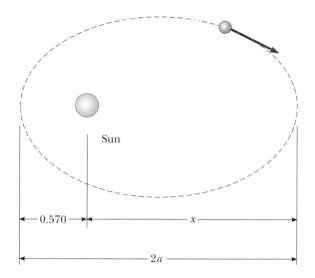

Figure P11.8

9. Io, a satellite of Jupiter, has an orbital period of 1.77 days and an orbital radius of 4.22×10^5 km. From these data, determine the mass of Jupiter.

10. Two planets X and Y travel counterclockwise in circular orbits about a star, as in Figure P11.10. The radii of their orbits is in the ratio $3:1$. At some time, they are aligned as in Figure P11.10a, making a straight line with the star. Five years later, Planet X has rotated through 90° as in Figure P11.10b. Where is planet Y at this time?

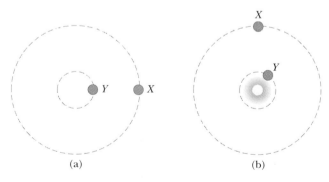

(a)   (b)

Figure P11.10

11. A synchronous satellite, which always remains above the same point on a planet's equator, is put in orbit around Jupiter to study the famous red spot. Jupiter rotates once every 9.84 h. Use the data of Table 11.2 to find the altitude of the satellite.

Section 11.4 Energy Considerations in Planetary and Satellite Motion

12. A satellite of mass 200 kg is placed in Earth orbit at a height of 200 km above the surface. (a) Assuming a circular orbit, how long does the satellite take to complete one orbit? (b) What is the satellite's speed? (c) What is the minimum energy necessary to place this satellite in orbit (assuming no air friction)?

13. A spaceship is fired from the Earth's surface with an initial speed 2.00×10^4 m/s. What will its speed be when it is very far from the Earth? (Neglect friction.)

14. A satellite of the Earth has a mass of 100 kg and is at an altitude of 2.00×10^6 m. (a) What is the potential energy of the satellite–Earth system? (b) What is the magnitude of the gravitational force exerted by the Earth on the satellite?

15. A 1000-kg satellite orbits the Earth at an altitude of 100 km. It is desired to increase the altitude of the circular orbit to 200 km. How much energy must be added to the system to effect this change in altitude?

16. A satellite of mass m orbits the Earth at an altitude h_1. It is desired to increase the altitude of the circular orbit to h_2. How much energy must be added to the system to effect this change in altitude?

17. A satellite moves in a circular orbit just above the surface of a planet. Show that the orbital speed v and escape speed of the satellite are related by the expression $v_{esc} = \sqrt{2}v$.

18. The planet Uranus has a mass about 14.0 times the Earth's mass, and its radius is equal to about 3.70 Earth radii. (a) By setting up ratios with the corresponding Earth values, find the acceleration due to gravity at the cloud tops of Uranus. (b) Ignoring the rotation of the planet, find the minimum escape speed from Uranus.

19. Determine the escape velocity for a rocket on the far side of Ganymede, the largest of Jupiter's moons. Ganymede's radius is 2.64×10^6 m, and its mass is 1.495×10^{23} kg. The mass of Jupiter is 1.90×10^{27} kg, and the distance between Jupiter and Ganymede is 1.071×10^9 m. Be sure to include the gravitational effect due to Jupiter, but you may ignore the motion of Jupiter and Ganymede as they revolve about their center of mass.

20. How much work is done by the Moon's gravitational field as a 1000 kg meteor comes in from outer space and impacts on the Moon's surface?

21. In Robert Heinlein's *The Moon Is a Harsh Mistress*, the colonial inhabitants of the Moon threaten to launch rocks down onto the Earth if they are not given independence (or at least representation). Assume that a rail gun can launch a rock at twice the speed that would be required to escape from an isolated stationary Moon. Calculate the speed of this rock as it enters the Earth's atmosphere.

22. A spacecraft in the shape of a long cylinder has a length of 100 m and its mass with occupants is 1000 kg. It has strayed in too close to a 1.0-m radius black hole having a mass 100 times that of the Sun (Fig. P11.22). If the nose of the space-

craft points toward the center of the black hole, and if distance between the nose of the spaceship and the black hole's center is 10.0 km, (a) determine the average acceleration of the spaceship. (b) What is the difference in the acceleration experienced by the occupants in the nose of the ship and those in the rear of the ship farthest from the black hole?

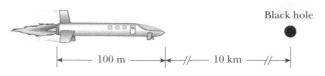

Black hole

|← 100 m →|←//— 10 km —//→|

Figure P11.22

Section 11.5   Atomic Spectra and the Bohr Theory of Hydrogen

23. For a hydrogen atom in its ground state, use the Bohr model to compute (a) the orbital speed of the electron, (b) the kinetic energy of the electron, and (c) the electrical potential energy of the atom.

24. Radiation is emitted from a hydrogen atom, which undergoes a transition from the $n = 3$ state to the $n = 2$ state. Calculate (a) the energy, (b) the wavelength, and (c) the frequency of the emitted light.

25. A hydrogen atom is in its first excited state ($n = 2$). Using the Bohr theory of the atom, calculate (a) the radius of the orbit, (b) the linear momentum of the electron, (c) the angular momentum of the electron, and its (d) kinetic energy, (e) potential energy, and (f) total energy.

26. How much energy is required to ionize hydrogen (a) when it is in the ground state? (b) when it is in the state for which $n = 3$?

27. Show that the speed of the electron in the nth Bohr orbit in hydrogen is given by

$$v_n = \frac{k_e e^2}{n\hbar}$$

28. (a) Calculate the angular momentum of the Moon due to its orbital motion about the Earth. In your calculation, use 3.84×10^8 m as the average Earth–Moon distance and 2.36×10^6 s as the period of the Moon in its orbit. (b) Determine the corresponding quantum number. (c) By what fraction would the Earth–Moon distance have to be increased to increase the quantum number by 1?

Additional Problems

29. *Voyagers 1* and *2* surveyed the surface of Jupiter's moon Io and photographed active volcanoes spewing liquid sulfur to heights of 70 km above the surface of this moon. Estimate the speed with which the liquid sulfur left the volcano. Io's mass is 8.9×10^{22} kg and its radius is 1820 km.

30. You are an astronaut, and you observe a small planet to be spherical. After landing on the planet, you set off, walk straight ahead, and find yourself returning to your spacecraft from the opposite side after completing a lap of 25.0 km. You hold a hammer and a falcon feather at a height of 1.40 m, release them, and observe them to fall together to the surface in 29.2 s. Determine the mass of the planet.

31. A cylindrical habitat in space 6.00 km in diameter and 30 km long has been proposed (by G. K. O'Neill, 1974). Such a habitat would have cities, land, and lakes on the inside surface and air and clouds in the center. This would all be held in place by rotation of the cylinder about its long axis. How fast would the cylinder have to rotate to imitate the Earth's gravitational field at the walls of the cylinder?

32. A "treetop satellite" is a satellite that orbits just above the surface of a spherical object, assumed to offer no air resistance. (a) Find the period of a treetop satellite of Earth. (b) Prove that its speed is given by $v = R\sqrt{4\pi G\rho/3}$, where ρ is the density of the planet.

33. In introductory physics laboratories, a typical Cavendish balance for measuring the gravitational constant G uses lead spheres of masses 1.50 kg and 15.0 g the centers of which are separated by about 4.50 cm. Calculate the gravitational force between these spheres, treating each as a point mass located at the center of the sphere.

34. Show that the escape speed from the surface of a planet of uniform density is directly proportional to the radius of the planet.

35. Two hypothetical planets of masses m_1 and m_2 and radii r_1 and r_2, respectively, are at rest when they are an infinite distance apart. Because of their gravitational attraction, they head toward each other on a collision course. (a) When their center-to-center separation is d, find the speed of each planet and their *relative* velocity. (b) Find the kinetic energy of each planet *just* before they collide if $m_1 = 2.00 \times 10^{24}$ kg, $m_2 = 8.00 \times 10^{24}$ kg, $r_1 = 3.00 \times 10^6$ m, and $r_2 = 5.00 \times 10^6$ m. (*Hint:* Both energy and momentum are conserved.)

36. The maximum distance from the Earth to the Sun (at our aphelion) is 1.521×10^{11} m and the distance of closest approach (at perihelion) is 1.471×10^{11} m. If the Earth's orbital speed at perihelion is 3.027×10^4 m/s, determine (a) the Earth's orbital speed at aphelion, (b) the kinetic and potential energy at perihelion, and (c) the kinetic and potential energy at aphelion. Is the total energy constant? (Neglect the effect of the Moon and other planets.)

37. (a) Determine the amount of work (in joules) that must be done on a 100-kg payload to elevate it to a height of 1000 km above the Earth's surface. (b) Determine the amount of additional work that is required to put the payload into circular orbit at this elevation.

38. After a supernova explosion, a star may undergo a gravitational collapse to an extremely dense state known as a neutron star, in which all the electrons and protons are

squeezed together to form neutrons. A neutron star having a mass about equal to that of the Sun would have a radius of about 10.0 km. Find (a) the free-fall acceleration at its surface, (b) the weight of a 70.0-kg person at its surface, and (c) the energy required to remove a neutron of mass 1.67×10^{-27} kg from its surface to infinity.

39. When it orbited the Moon, the *Apollo 11* spacecraft's mass was 9.979×10^3 kg, its period was 119 min, and its mean distance from the Moon's center was 1.849×10^6 m. Assuming its orbit to be circular and the Moon to be a uniform sphere, find (a) the mass of the Moon, (b) the orbital speed of the spacecraft, and (c) the minimum energy required for the craft to leave the orbit and escape the Moon's gravitational field.

40. Three point objects having masses m, $2m$, and $3m$, are fixed at the corners of a square of side length a such that the lighter object is at the upper left-hand corner, the heavier object is at the lower left-hand corner, and the remaining object is at the upper right-hand corner. Determine the magnitude and direction of the resulting gravitational acceleration at the center of the square.

41. X-ray pulses from Cygnus X-1, a celestial x-ray source, have been recorded during high-altitude rocket flights. The signals can be interpreted as originating when a blob of ionized matter orbits a black hole with a period of 5.00 ms. If the blob were in a circular orbit about a black hole the mass of which is $20.0 M_{\text{Sun}}$, what is the orbital radius?

42. Studies of the relationship of the Sun to its galaxy—the Milky Way—have revealed that the Sun is located near the outer edge of the galactic disc, about 30 000 lightyears from the center. Furthermore, it has been found that the Sun has an orbital speed of approximately 250 km/s around the galactic center. (a) What is the period of the Sun's galactic motion? (b) What is the order of magnitude of the mass of the Milky Way galaxy? Supposing that the galaxy is made mostly of stars of which the Sun is typical, estimate the number of stars in the Milky Way.

43. Four possible transitions for a hydrogen atom are as follows:

(A) $n_i = 2$; $n_f = 5$ (B) $n_i = 5$; $n_f = 3$
(C) $n_i = 7$; $n_f = 4$ (D) $n_i = 4$; $n_f = 7$

(a) Which transition emits the shortest wavelength photon? (b) In which transition does the atom gain the most energy? (c) In which transition(s) does the atom lose energy?

44. *Vanguard I*, launched March 3, 1958, is the oldest human-made satellite still in orbit. Its initial orbit had an apogee of 3970 km and a perigee of 650 km. Its maximum speed was 8.23 km/s and it had a mass of 1.60 kg. (a) Determine the period of the orbit. (Use the semimajor axis.) (b) Determine the speeds at apogee and perigee. (c) Find the total energy of the satellite.

45. Two stars of masses M and m, separated by a distance d, revolve in circular orbits about their center of mass (Fig. P11.45). Show that each star has a period given by

$$T^2 = \frac{4\pi^2}{G(M + m)} d^3$$

(*Hint:* Apply Newton's second law to each star, and note that the center-of-mass condition requires that $Mr_2 = mr_1$, where $r_1 + r_2 = d$)

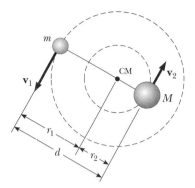

Figure P11.45

46. In 1978, astronomers at the U.S. Naval Observatory discovered that Pluto has a moon, called Charon, that eclipses the planet every 6.39 days. Given that the center-to-center separation between Pluto and Charon is 19 700 km, find the total mass $(M + m)$ of the two bodies. (*Hint:* See Problem 45.)

47. Two hydrogen atoms collide head-on and end up with zero kinetic energy. Each then emits a 121.6-nm photon ($n = 2$ to $n = 1$ transition). At what speed were the atoms moving before the collision?

48. A rocket is given an initial speed of $v_0 = 2\sqrt{Rg}$ vertically upward at the surface of the Earth, which has radius R and surface free-fall acceleration g. The rocket motors are then cut off, and thereafter the rocket coasts under the action of gravitational forces only. (Ignore atmospheric friction and the Earth's rotation.) Derive an expression for the subsequent speed, v, as a function of the distance, r, from the center of the Earth in terms of g, R, and r.

Spreadsheet Problems

S1. Four point masses are fixed as shown in Figure PS11.1. The gravitational potential energy for a test particle of mass m moving along the x axis in the gravitational field of the fixed masses can be written as

$$U(x) = -\frac{GM_1 m}{r_1} - \frac{GM_2 m}{r_2} - \frac{GM_3 m}{r_3} - \frac{GM_4 m}{r_4}$$

where $r_2 = r_4 = [b^2 + (x + a)^2]^{1/2}$ and $r_2 = r_3 = [b^2 + (x - a)^2]^{1/2}$. Spreadsheet 11.1 calculates $U(x)$ and the force $F(x) = -dU(x)/dx$ exerted on the test particle. The quantities M_1, M_2, M_3, M_4, a, and b are input parameters.

The mass of the test particle is 1.00 kg. (This is the case of two particles on the y axis at $y = +1.00$ m and -1.00 m.) (a) Plot $U(x)$ and $F(x)$ versus x for $x = -4.00$ m to $+4.00$ m. (b) If the test particle has an energy of -7.50×10^{-11} J, describe the particle's motion. (c) How does the force on the particle vary as the particle moves along the x axis?

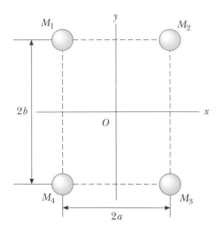

Figure PS11.1

S2. In Spreadsheet 11.1, let $a = 1.00$ m, $b = 0.50$ m, and $M_1 = M_2 = M_3 = M_4 = 1.00$ kg. The mass of the particle is 1.00 kg. (a) Plot $U(x)$ and $F(x)$ versus x. (b) If the test particle has an energy of -2.00×10^{-10} J, what motion is possible?

S3. If a projectile is fired straight up from the surface of the Earth with an initial speed v_0 that is less than the escape speed $v_{esc} = \sqrt{2gR_E} = 11.2$ km/s, the projectile will rise to a maximum height h above the Earth's surface and return. Neglecting air resistance and using conservation of energy, show that

$$h = \frac{R_E v_0^2}{2gR_E - v_0}$$

where R_E is the Earth's radius. Write a spreadsheet to calculate h for $v_0 = 1, 2, 3, \ldots , 11$ km/s. Plot h/R_E versus v_0. What happens as v_0 approaches the escape speed v_{esc}?

S4. The acceleration of an object moving in the gravitational field of the Earth is

$$\mathbf{a} = -GM_E \frac{\mathbf{r}}{r^3}$$

where $\mathbf{r}$ is the position vector directed from the center of the Earth to the object. Choosing the origin at the center of the Earth and assuming the object is moving in the xy plane, the acceleration has rectangular components

$$a_x = -\frac{GM_E x}{(x^2 + y^2)^{3/2}} \qquad a_y = -\frac{GM_E y}{(x^2 + y^2)^{3/2}}$$

Write a spreadsheet (or computer program) to find the position of the object as a function of time. (Use the techniques developed in Spreadsheet 11.1 as a model for your own program. Assume that the initial position of the object is at $x = 0$ and $y = 2R_E$, where R_E is the radius of the Earth, and give the object an initial velocity of 5 km/s in the x direction. The time increment should be made as small as practical. Try 5 s. Plot the x and y coordinates of the object as functions of time. Does the object hit the Earth? Vary the initial velocity until a circular orbit is found.

S5. Modify your spreadsheet (or program) from Problem S4 to calculate the kinetic, potential, and total energies as functions of time. How do they vary with time for the various cases?

S6. Modify your spreadsheet (or program) from Problem S4 to calculate the angular momentum of the object. In rectangular coordinates, $L = m(xv_y - yv_x)$. How does the angular momentum vary with time? Kepler's second law of planetary motion implies that L remains constant. Is your calculated L constant?

ANSWERS TO CONCEPTUAL PROBLEMS

1. Kepler's second law is a consequence of conservation of angular momentum. Angular momentum is conserved as long as there is no torque, which is true for any central force, regardless of the particular dependence on separation. Thus, Kepler's second law would be unchanged. The particular mathematical form of Kepler's third law depends on the inverse square character of the force. If the force were an inverse cube, there would still be a proportional relationship between a power of the period and a power of the semimajor axis, but the powers would be different than in the existing law.

2. Air friction causes a decrease in the mechanical energy of the satellite–Earth system. This reduces the major axis of the orbit, bringing the satellite closer to the surface of the Earth. A satellite in a smaller orbit, however, must travel faster. Thus, the effect of the air friction is to speed up the satellite!

3. Kepler's third law (Eq. 11.4), which applies to all the planets, tells us that the period of a planet is proportional to $r^{3/2}$. Because Saturn and Jupiter are farther from the Sun than Earth, they have longer periods. The Sun's gravitational field (whose magnitude is GM_S/r^2) is much weaker at a distant Jovian planet. Thus, an outer planet experiences much

smaller centripetal acceleration than Earth, and a correspondingly longer period.

4. In an emission spectrum, a gas is initially excited to higher quantum states. The emission spectrum is provided by subsequent downward transitions which can occur to all lower states. If the transition is not to the lowest state, the atom can make a subsequent transition from the intermediate state to the lowest state. For example, a transition can be made from $n = 10$ to $n = 5$, followed by a transition from $n = 5$ to $n = 1$. Thus, there is a rich mixture of spectral lines in the spectrum, corresponding to this large number of possible transitions. An absorption spectrum is normally performed with the gas at room temperature, at which almost all of the atoms are in the ground state. Thus, when radiation is absorbed, upward transitions involving *only the ground state* are possible. Thus, there are fewer transitions than for the emission spectrum and fewer lines.

5. In the discussion of the hydrogen atom, it can be seen that the kinetic energy is equal to half of the magnitude of the potential energy. Thus, we must have $K = E$ and $U_e = -2E$.

B.C.   by John Hart

By permission of John Hart and Field Enterprises, Inc.

12

Oscillatory Motion

I f a force acting on a body varies in time, the acceleration of the body also changes with time. If this force always acts toward the equilibrium position of the body, a repetitive back-and-forth motion about that position results. The motion is an example of what is called *periodic* or *oscillatory* motion.

You are most likely familiar with several examples of periodic motion, such as the oscillations of a mass on a spring, the motion of a pendulum, and the vibrations of a stringed musical instrument. Numerous systems exhibit oscillatory motion. For example, the molecules in a solid oscillate about their equilibrium positions; electromagnetic waves, such as light waves, radar, and radio waves, are characterized by oscillating electric and

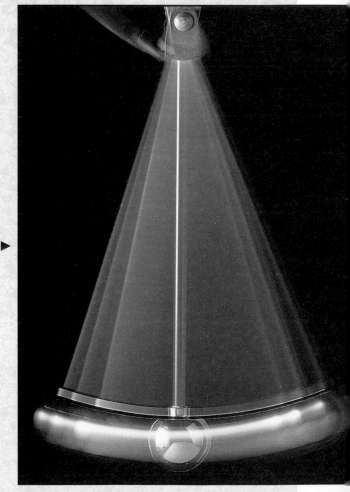

Time exposure of a pendulum, ▶ **which consists of a small sphere suspended by a light string. The time it takes the pendulum to undergo one complete oscillation is called the period of its motion, and the maximum displacement of the pendulum from the vertical position is called the** *amplitude.* **As we shall see in this chapter, the period of its motion for small amplitudes depends only on the length of the pendulum and the value of the free-fall acceleration.** *(James Stevenson/ SPL/Photo Researchers)*

magnetic field vectors; in alternating-current circuits, voltage, current, and electrical charge vary periodically with time.

Much of the material in this chapter deals with **simple harmonic motion.** This special kind of motion occurs when a restoring force acts on a body. In simple harmonic motion, an object oscillates between two spatial positions for an indefinite period of time. In real mechanical systems, retarding (frictional) forces are always present. Such forces reduce the mechanical energy of the system with time, and the oscillations are said to be *damped.* If an external driving force is applied, we call the motion a *forced oscillation.*

12.1 · SIMPLE HARMONIC MOTION

A particle moving along the x axis is said to exhibit **simple harmonic motion** when x, its displacement from equilibrium, varies in time according to the relationship

$$x = A \cos(\omega t + \phi) \qquad \text{[12.1]}$$

• *Displacement versus time for simple harmonic motion*

where A, ω, and ϕ are constants of the motion. In order to give physical significance to these constants, it is convenient to plot x as a function of t, as in Figure 12.1. First, we note that A, called the **amplitude** of the motion, is simply **the maximum displacement of the particle in either the positive or negative x direction.** The constant ω is called the **angular frequency** (defined in Eq. 12.4). The constant angle ϕ is called the **phase constant** (or phase angle) and, along with the amplitude A, is determined uniquely by the initial displacement and velocity of the particle. The constants ϕ and A tell us what the displacement and velocity were at time $t = 0$. The quantity $(\omega t + \phi)$ is called the **phase** of the motion and is useful for comparing the motions of two particles. Note that the function x is periodic and repeats itself each time ωt increases by 2π radians.

The **period,** T, of the motion is the time required for the particle to go through one full cycle of its motion. That is, the value of x at time t equals the value of x at time $t + T$. We can show that the period is given by $T = 2\pi/\omega$ by using the fact that the phase increases by 2π radians in a time of T:

$$\omega t + \phi + 2\pi = \omega(t + T) + \phi$$

Simplifying this expression, we see that $\omega T = 2\pi$, or

$$T = \frac{2\pi}{\omega} \qquad \text{[12.2]}$$

• *Period*

The inverse of the period is called the **frequency** of the motion, f. The frequency represents the **number of oscillations the particle makes per unit time:**

$$f = \frac{1}{T} = \frac{\omega}{2\pi} \qquad \text{[12.3]}$$

• *Frequency*

The units of f are cycles per second, or hertz (Hz). Rearranging Equation 12.3 gives

$$\omega = 2\pi f = \frac{2\pi}{T} \qquad \text{[12.4]}$$

• *Angular frequency*

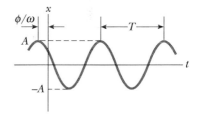

Figure 12.1 Displacement versus time for a particle undergoing simple harmonic motion. The amplitude of the motion is A and the period is T.

The angular frequency ω has the units radians per second. We shall discuss the geometric significance of ω in Section 12.4.

We can obtain the velocity of a particle undergoing simple harmonic motion by differentiating Equation 12.1 with respect to time:

Velocity in simple harmonic motion •

$$v = \frac{dx}{dt} = -\omega A \sin(\omega t + \phi) \qquad \text{[12.5]}$$

The acceleration of the particle is dv/dt:

Acceleration in simple harmonic motion •

$$a = \frac{dv}{dt} = -\omega^2 A \cos(\omega t + \phi) \qquad \text{[12.6]}$$

Because $x = A \cos(\omega t + \phi)$, we can express Equation 12.6 in the form

$$a = -\omega^2 x \qquad \text{[12.7]}$$

From Equation 12.5 we see that, because the sine and cosine functions oscillate between ± 1, the extreme values of v are $\pm \omega A$. Likewise, Equation 12.6 tells us that the extreme values of the acceleration are $\pm \omega^2 A$. Therefore, the *maximum* values of the velocity and acceleration are

Maximum values of velocity and acceleration in simple harmonic motion •

$$v_{\text{max}} = \omega A \qquad \text{[12.8]}$$

$$a_{\text{max}} = \omega^2 A \qquad \text{[12.9]}$$

Figure 12.2a plots displacement versus time for an arbitrary value of the phase constant. The velocity and acceleration-versus-time curves are illustrated in Figures 12.2b and 12.2c. They show that the phase of the velocity differs from the phase of the displacement by $\pi/2$ rad, or 90°. That is, when x is a maximum or a minimum,

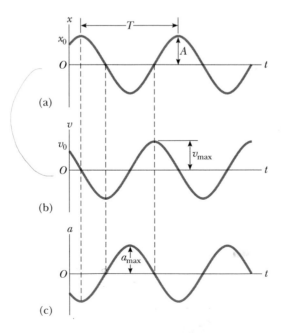

(a)

(b)

(c)

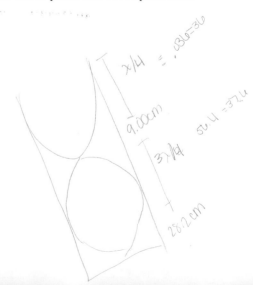

Figure 12.2 Graphical representation of simple harmonic motion: (a) displacement versus time, (b) velocity versus time, and (c) acceleration versus time. Note that at any specified time the velocity is 90° out of phase with the displacement and the acceleration is 180° out of phase with the displacement.

the velocity is zero. Likewise, when x is zero, the speed is a maximum. Furthermore, note that the phase of the acceleration differs from the phase of the displacement by π radians, or 180°. That is, when x is a maximum, a is a maximum in the opposite direction.

As we stated earlier, $x = A \cos(\omega t + \phi)$ is a general expression for the displacement of the particle from equilibrium, where the phase constant ϕ and the amplitude A must be chosen to meet the initial conditions of the motion. Suppose that the initial position, x_0, and initial velocity, v_0, of a single oscillator of known angular frequency are given; that is, at $t = 0$, $x = x_0$ and $v = v_0$. Under these conditions, Equations 12.1 and 12.5 give

$$x_0 = A \cos \phi \qquad \text{and} \qquad v_0 = -\omega A \sin \phi \qquad \text{[12.10]}$$

Dividing these two equations eliminates A, giving $\dfrac{v_0}{x_0} = -\omega \tan \phi$, or

$$\tan \phi = -\frac{v_0}{\omega x_0} \qquad \text{[12.11]}$$

Furthermore, if we take the sum $x_0{}^2 + (v_0/\omega)^2 = A^2 \cos^2 \phi + A^2 \sin^2 \phi$ (where we have used Eq. 12.10) and solve for A, we find that

$$A = \sqrt{x_0{}^2 + \left(\frac{v_0}{\omega}\right)^2} \qquad \text{[12.12]}$$

- *The phase angle ϕ and amplitude A can be obtained from the initial conditions.*

Thus, we see that ϕ and A are known if x_0, ω, and v_0 are specified. We shall treat a few specific cases in the next section.

We conclude this section by pointing out the following important properties of a particle moving in simple harmonic motion:

- The displacement, velocity, and acceleration all vary sinusoidally with time but are not in phase, as shown in Figure 12.2.
- The acceleration is proportional to the displacement, but in the opposite direction.
- The frequency and the period of motion are independent of the amplitude.

- *Properties of simple harmonic motion*

Example 12.1 An Oscillating Body

A body oscillates with simple harmonic motion along the x axis. Its displacement varies with time according to the equation

$$x = (4.00 \text{ m}) \cos \left(\pi t + \frac{\pi}{4} \right)$$

where t is in seconds and the angles in the parentheses are in radians.

(a) Determine the amplitude, frequency, and period of the motion.

Solution By comparing this equation with the general equa-

tion for simple harmonic motion, $x = A \cos(\omega t + \phi)$, we see that $A = 4.00$ m and $\omega = \pi$ rad/s; therefore we find $f = \omega/2\pi = \pi/2\pi = 0.500$ s^{-1} and $T = 1/f = 2.00$ s.

(b) Calculate the velocity and acceleration of the body at any time t.

Solution

$$v = \frac{dx}{dt} = -4.00 \sin \left(\pi t + \frac{\pi}{4} \right) \frac{d}{dt} (\pi t)$$

$$= -(4.00\pi \text{ m/s}) \sin \left(\pi t + \frac{\pi}{4} \right)$$

$$a = \frac{dv}{dt} = -4.00\pi \cos\left(\pi t + \frac{\pi}{4}\right)\frac{d}{dt}(\pi t)$$

$$= -(4.00\pi^2 \text{ m/s}^2)\cos\left(\pi t + \frac{\pi}{4}\right)$$

EXERCISE 1   Using the results to part (b), determine the position, velocity, and acceleration of the body at $t = 1.00$ s.

Answer $x = -2.83$ m; $v = 8.89$ m/s; $a = 27.9$ m/s^2

EXERCISE 2   Determine the maximum speed and the maximum acceleration of the body.   Answer   From the general expressions for v and a found in part (b), we see that the maximum values of the sine and cosine functions are unity. Therefore, v varies between $\pm 4.00\pi$ m/s, and a varies between $\pm 4.00\pi^2$ m/s^2. Thus, $v_{max} = 4.00\pi$ m/s and $a_{max} = 4.00\pi^2$ m/s^2.

EXERCISE 3   A particle moving with simple harmonic motion travels a total distance of 20.0 cm in each cycle of its motion, and its maximum acceleration is 50.0 m/s^2. Find (a) the angular frequency of the motion and (b) the maximum speed of the particle.

Answer   (a) 31.6 rad/s   (b) 1.58 m/s

12.2 • MOTION OF A MASS ATTACHED TO A SPRING

Consider a physical system consisting of a mass, m, attached to the end of a spring where the mass is free to move on a horizontal track (Fig. 12.3). We know from experience that such a system oscillates back and forth if disturbed from the equilibrium position $x = 0$, where the spring is unstretched. If the surface is frictionless, the mass can exhibit simple harmonic motion. One experimental arrangement that clearly demonstrates that such a system exhibits simple harmonic motion is illustrated in Figure 12.4. A mass oscillating vertically on a spring has a marking pen attached to it. While the mass is in motion, a sheet of paper is moved horizontally as shown, and the marking pen traces out a sinusoidal pattern. We can understand this qualitatively by first recalling that when the mass is displaced a small distance, x, from equilibrium, the spring exerts a force on it, given by **Hooke's law,**

$$F_s = -kx \qquad [12.13]$$

where k is the force constant of the spring. We call this a **linear restoring force** because it is linearly proportional to the displacement and always directed toward the equilibrium position *opposite* the displacement. That is, when the mass is displaced to the right in Figure 12.3, x is positive and the restoring force is to the left. When the mass is displaced to the left of $x = 0$, then x is negative and the restoring force is to the right.

If we apply Newton's second law to the motion of the mass in the x direction, we get

$$F_s = -kx = ma$$

$$a = -\frac{k}{m}x \qquad [12.14]$$

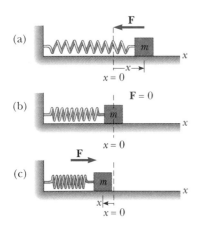

Figure 12.3   A mass attached to a spring on a frictionless track moves in simple harmonic motion. (a) When the mass is displaced to the right of equilibrium, the displacement is positive and the acceleration is negative. (b) At the equilibrium position, $x = 0$, the acceleration is zero but the speed is a maximum. (c) When the displacement is negative, the acceleration is positive.

That is, just as we learned in Section 12.1, **the acceleration is proportional to the displacement of the mass from equilibrium and is in the opposite direction.** If the mass is displaced a maximum distance, $x = A$, at some initial time and released from rest, its *initial* acceleration is $-kA/m$ (that is, the acceleration has its extreme negative value). When the mass passes through the equilibrium position,

$x = 0$ and its acceleration is zero. At this instant, its speed is a maximum. It then continues to travel to the left of equilibrium and finally reaches $x = -A$, at which time its acceleration is kA/m (maximum positive) and its speed is again zero. Thus, we see that the mass oscillates between the turning points $x = \pm A$. In one full cycle of its motion it travels a distance of $4A$.

Let us now describe this motion in a quantitative fashion. Recall that, by definition, $a = dv/dt = d^2x/dt^2$, and so we can express Equation 12.14 as

$$\frac{d^2x}{dt^2} = -\frac{k}{m}x \qquad [12.15]$$

If we denote the ratio k/m with the symbol ω^2,

$$\omega^2 = \frac{k}{m} \qquad [12.16]$$

then Equation 12.15 can be written in the form

$$\frac{d^2x}{dt^2} = -\omega^2 x \qquad [12.17]$$

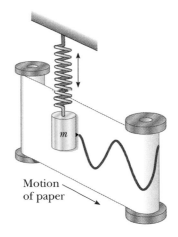

Figure 12.4 An experimental apparatus for demonstrating simple harmonic motion. A pen attached to the oscillating mass traces out a sine wave on the moving chart paper.

What we now require is a solution to Equation 12.17—that is, a function $x(t)$ that satisfies this second-order differential equation. Because Equations 12.17 and 12.7 are equivalent, we see that the solution must be that of simple harmonic motion:

$$x(t) = A\cos(\omega t + \phi)$$

To see this explicitly, note that if $x = A\cos(\omega t + \phi)$, then

$$\frac{dx}{dt} = A\frac{d}{dt}\cos(\omega t + \phi) = -\omega A\sin(\omega t + \phi)$$

$$\frac{d^2x}{dt^2} = -\omega A\frac{d}{dt}\sin(\omega t + \phi) = -\omega^2 A\cos(\omega t + \phi)$$

Comparing the expressions for x and d^2x/dt^2, we see that $d^2x/dt^2 = -\omega^2 x$ and Equation 12.17 is satisfied.

The following general statement can be made based on the foregoing discussion:

> Whenever the force acting on a particle is linearly proportional to the displacement and in the opposite direction, the particle exhibits simple harmonic motion.

We shall give additional physical examples in subsequent sections.

Because the period of simple harmonic motion is $T = 2\pi/\omega$ and the frequency is the inverse of the period, we can express the period and frequency of the motion for this mass–spring system as

$$T = \frac{2\pi}{\omega} = 2\pi\sqrt{\frac{m}{k}} \qquad [12.18]$$

• *Period of a mass–spring system*

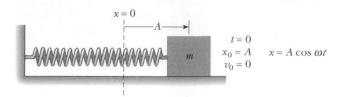

Figure 12.5   A mass–spring system that starts from rest at $x_0 = A$. In this case, $\phi = 0$, and so $x = A \cos \omega t$.

Frequency of a mass–spring •
system

$$f = \frac{1}{T} = \frac{1}{2\pi} \sqrt{\frac{k}{m}} = \frac{\omega}{2\pi}$$ [12.19]

That is, the period and frequency depend *only* on the mass and on the force constant of the spring. As we might expect, the frequency is larger for a stiffer spring (larger value of k) and decreases with increasing mass.

Special Case I   In order to better understand the physical significance of our solution of the equation of motion, let us consider the following special case. Suppose we pull the mass from equilibrium by a distance of A and release it from rest in this stretched position, as in Figure 12.5. We must then require that our solution for $x(t)$ obey the initial conditions that at $t = 0$, $x_0 = A$ and $v_0 = 0$. These conditions are met if we choose $\phi = 0$, giving $x = A \cos \omega t$ as our solution. To check this solution, we note that it satisfies the condition that $x_0 = A$ at $t = 0$, because $\cos 0 = 1$. Thus, we see that A and ϕ contain the information on initial conditions.

Now let us investigate the behavior of the velocity and acceleration in this special case. Because $x = A \cos \omega t$, we have

$$v = \frac{dx}{dt} = -\omega A \sin \omega t \qquad \text{and} \qquad a = \frac{dv}{dt} = -\omega^2 A \cos \omega t$$

From the preceding velocity expression we see that at $t = 0$, $v_0 = 0$, as we require. The expression for the acceleration tells us that at $t = 0$, $a = -\omega^2 A$. Physically this makes sense, because the force on the mass is to the left when the displacement is positive. In fact, at this position $F_s = -kA$ (to the left), and the initial acceleration is $-kA/m$.

We could also use a more formal approach to show that $x = A \cos \omega t$ is the correct solution by using the relation $\tan \phi = -v_0/\omega x_0$ (Eq. 12.10). Because $v_0 = 0$ at $t = 0$, $\tan \phi = 0$ and so $\phi = 0$.

The displacement, velocity, and acceleration versus time are plotted in Figure 12.6 for this special case. Note that the acceleration reaches extreme values of $\pm \omega^2 A$ when the displacement has extreme values of $\mp A$. Furthermore, the velocity has extreme values of $\pm \omega A$, which both occur at $x = 0$. Hence, the quantitative solution agrees with our qualitative description of this system.

Special Case II   Now suppose that the mass is given an initial velocity of $\mathbf{v}_0$ to the *right* at the unstretched position of the spring, so that at $t = 0$, $x_0 = 0$ and $v = v_0$ (Fig. 12.7). Our particular solution must now satisfy these initial conditions.

Applying Equation 12.11, $\tan \phi = -v_0/\omega x_0$, and the initial condition that $x_0 = 0$ at $t = 0$ gives $\tan \phi = -\infty$ or $\phi = -\pi/2$. Hence, the solution is given by $x = A \cos(\omega t - \pi/2)$, which can be written $x = A \sin \omega t$. Furthermore, from Equation 12.11 we see that $A = v_0/\omega$; therefore, we can express our solution as

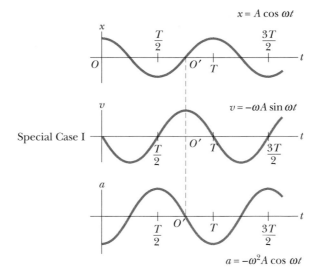

$$x = A \cos \omega t$$

$$v = -\omega A \sin \omega t$$

Special Case I

$$a = -\omega^2 A \cos \omega t$$

Figure 12.6   Displacement, velocity, and acceleration versus time for a particle undergoing simple harmonic motion under the initial conditions that at $t = 0$, $x = A$, and $v = 0$.

$$x = \frac{v_0}{\omega} \sin \omega t$$

The velocity and acceleration in this case are

$$v = \frac{dx}{dt} = v_0 \cos \omega t \quad \text{and} \quad a = \frac{dv}{dt} = -\omega v_0 \sin \omega t$$

This is consistent with the fact that the mass always has its maximum speed at $x = 0$, and the force and acceleration are zero at that position. The graphs of these functions versus time in Figure 12.6 correspond to the origin at O'. What would be the solution for x if the mass were initially moving to the left in Figure 12.7?

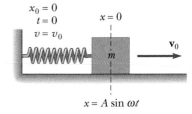

Figure 12.7   The mass–spring system starts its motion at the equilibrium position, $x = 0$ at $t = 0$. If its initial velocity is v_0 to the right, its x coordinate varies as

$$x = \frac{v_0}{\omega} \sin \omega t.$$

Thinking Physics 1

We know that the period of oscillation of a mass hung on a spring is proportional to the square root of the mass. Thus, if we perform an experiment in which we hang a range of masses on the end of a spring and measure the period of oscillation, a graph of the square of the period against the mass will result in a straight line, as suggested in Figure TP12.1. But we find that the line does not go through the origin. Why not?

Reasoning   The reason that the line does not go through the origin is that the spring itself has mass. Thus, the resistance to changes in motion of the system is a combination of the mass hung on the end of the spring and the mass of the oscillating coils of the spring. The entire mass of the spring is not oscillating, however. The bottom-most coil is oscillating over the same amplitude as the mass, and the top-most coil is not oscillating at all. For a cylindrical spring, energy arguments can be used to show that the effective additional mass representing the oscillations of the spring is one third of the mass of the spring. The square of the period is proportional to the total oscillating mass, but the graph in the diagram shows the square of the period against only the mass hung on the spring. A graph of period squared against total mass (mass hung on the spring plus the effective oscillating mass of the spring) would pass through the origin.

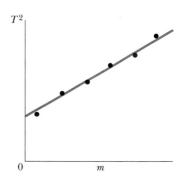

Figure TP12.1   (Thinking Physics 1)

Thinking Physics 2

If you go bungee jumping, you will bounce up and down at the end of the cord after your daring dive off a bridge. Suppose you perform this dive and measure the frequency of your bouncing. You then move to another bridge. You discover that the bungee cord is too long for dives off this bridge—you will hit the ground. Thus, you fold the bungee cord in half and make the dive from the doubled bungee cord. How does the frequency of your bouncing at the end of this dive compare to the frequency after the dive from the first bridge?

Reasoning Let us model the bungee cord as a spring. The force exerted by a spring is proportional to the separation of the coils as the spring is extended. Imagine that we extend a spring by a given distance and measure the distance between coils. We then cut the spring in half. If one of the half-springs is now extended by the same distance, the coils will be twice as far apart as they were for the complete spring. Thus, it takes twice as much force to stretch the half-spring, from which we conclude that the half-spring has a spring constant that is twice that of the complete spring. Now consider the folded bungee cord that we model as two half-springs in parallel. Each half has a spring constant that is twice the original spring constant of the bungee cord. In addition, a given mass hanging on the folded bungee cord will experience two forces—one from each half-spring. As a result, the required force for a given extension will be four times as much as for the original bungee cord. The effective spring constant of the folded bungee cord is, therefore, four times as large as the original spring constant. Because the frequency of oscillation is proportional to the square root of the spring constant, your bouncing frequency on the folded cord will be twice that of the original cord.

(Thinking Physics 2) Bungee jumping.   *(Gamma)*

CONCEPTUAL PROBLEM 1

Is a bouncing ball an example of simple harmonic motion? Is the daily movement of a student from home to school and back simple harmonic motion?

CONCEPTUAL PROBLEM 2

A mass is hung on a spring and the frequency of oscillation of the system, f, is measured. The mass, a second identical mass, and the spring are carried in the Space Shuttle to space. The two masses are attached to the ends of the spring and the system is taken out into space on a space walk. The spring is extended and the system is released to oscillate while floating in space. What is the frequency of oscillation for this system, in terms of f?

Example 12.2   A Mass–Spring System

A mass of 200 g is connected to a light spring of force constant 5.00 N/m and is free to oscillate on a horizontal, frictionless surface. If the mass is displaced 5.00 cm from equilibrium and released from rest, as in Figure 12.5, (a) find the period of its motion.

Solution This situation corresponds to Special Case I, in which $x = A \cos \omega t$ and $A = 5.00 \times 10^{-2}$ m. Therefore,

$$\omega = \sqrt{\frac{k}{m}} = \sqrt{\frac{5.00 \text{ N/m}}{200 \times 10^{-3} \text{ kg}}} = 5.00 \text{ rad/s}$$

and

$$T = \frac{2\pi}{\omega} = \frac{2\pi}{5.00} = 1.26 \text{ s}$$

(b) Determine the maximum speed and maximum acceleration of the mass.

Solution

$$v_{max} = \omega A = (5.00 \text{ rad/s})(5.00 \times 10^{-2} \text{ m}) = 0.250 \text{ m/s}$$

$$a_{max} = \omega^2 A = (5.00 \text{ rad/s})^2(5.00 \times 10^{-2} \text{ m}) = 1.25 \text{ m/s}^2$$

(c) Express the displacement, speed, and acceleration as functions of time.

Solution   The expression $x = A \cos \omega t$ is our solution for Special Case I, and so we can use the results from (a), (b), and (c) to get

$$x = A \cos \omega t = \boxed{(0.0500 \text{ m}) \cos 5.00t}$$

$$v = -\omega A \sin \omega t = \boxed{-(0.250 \text{ m/s}) \sin 5.00t}$$

$$a = -\omega^2 A \cos \omega t = \boxed{-(1.25 \text{ m/s}^2) \cos 5.00t}$$

EXERCISE 4   (a) A 400-g mass is suspended from a spring hanging vertically, and the spring is found to stretch 8.0 cm. Find the spring constant. (b) How much will the spring stretch if the suspended mass is 575 g?   Answer   (a) 49 N/m   (b) 12 cm

EXERCISE 5   A 3.0-kg mass is attached to a spring and pulled out horizontally to a maximum displacement from equilibrium of 0.50 m. What spring constant must the spring have if the mass is to achieve an acceleration equal to the free-fall acceleration?   Answer   59 N/m

12.3 • ENERGY OF THE SIMPLE HARMONIC OSCILLATOR

Let us examine the mechanical energy of the mass–spring described in Figure 12.5. Because the surface is frictionless, we expect that the total mechanical energy is constant. We can use Equation 12.5 to express the kinetic energy as

$$K = \tfrac{1}{2}mv^2 = \tfrac{1}{2}m\omega^2 A^2 \sin^2(\omega t + \phi) \qquad \text{[12.20]}$$

• *Kinetic energy of a simple harmonic oscillator*

The elastic potential energy stored in the spring for any elongation x is $\tfrac{1}{2}kx^2$. Using Equation 12.1, we get

$$U = \tfrac{1}{2}kx^2 = \tfrac{1}{2}kA^2 \cos^2(\omega t + \phi) \qquad \text{[12.21]}$$

• *Potential energy of a simple harmonic oscillator*

We see that K and U are always positive quantities. Because $\omega^2 = k/m$, we can express the *total energy* of the simple harmonic oscillator as

$$E = K + U = \tfrac{1}{2}kA^2[\sin^2(\omega t + \phi) + \cos^2(\omega t + \phi)]$$

But $\sin^2 \theta + \cos^2 \theta = 1$; therefore, this equation reduces to

$$E = \tfrac{1}{2}kA^2 \qquad \text{[12.22]}$$

• *Total energy of a simple harmonic oscillator*

That is, the energy of a simple harmonic oscillator is a constant of the motion and proportional to the square of the amplitude. In fact, the total mechanical energy is just equal to the maximum potential energy stored in the spring when $x = \pm A$. At these points, $v = 0$ and there is no kinetic energy. At the equilibrium position, $x = 0$ and $U = 0$, so that the total energy is all in the form of kinetic energy. That is, at $x = 0$, $E = \tfrac{1}{2}mv_{max}^2 = \tfrac{1}{2}m\omega^2 A^2$.

Plots of the kinetic and potential energies versus time are shown in Figure 12.8a, where $\phi = 0$. In this situation, both K and U are always positive, and their sum at all times is a constant equal to $\tfrac{1}{2}kA^2$, the total energy of the system. The variations of K and U with displacement are plotted in Figure 12.8b. Energy is continuously being transferred between potential energy stored in the spring and the kinetic energy of the mass. Figure 12.9 illustrates the position, velocity, acceleration, kinetic energy, and potential energy of the mass–spring system for one full period of the motion. Most of the ideas discussed so far are incorporated in this important figure. We suggest that you study it carefully.

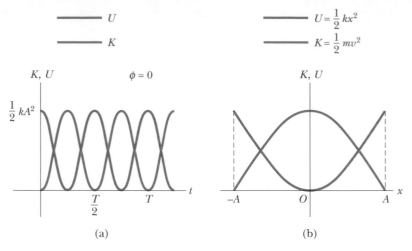

Figure 12.8   (a) Kinetic energy and potential energy versus time for a simple harmonic oscillator with $\phi = 0$. (b) Kinetic energy and potential energy versus displacement for a simple harmonic oscillator. In either plot, note that $K + U =$ constant.

Finally, we can use energy to obtain the velocity for an arbitrary displacement, x, expressing the total energy at some arbitrary position as

$$E = K + U = \tfrac{1}{2}mv^2 + \tfrac{1}{2}kx^2 = \tfrac{1}{2}kA^2$$

Velocity as a function of • position for a simple harmonic oscillator

$$v = \pm\sqrt{\frac{k}{m}(A^2 - x^2)} = \pm\omega\sqrt{A^2 - x^2} \qquad \textbf{[12.23]}$$

Again, this expression substantiates the fact that the speed is a maximum at $x = 0$ and zero at the turning points, $x = \pm A$.

Thinking Physics 3

A mass oscillating on the end of a horizontal spring slides back and forth over a frictionless surface. During one oscillation, you set an identical mass with Velcro (a sticky fabric) attached at the maximum displacement point. Just as the oscillating mass reaches its largest displacement and is momentarily at rest, it adheres to the new mass by means of the Velcro, and the two masses continue the oscillation together. Does the period of the oscillation change? Does the amplitude of oscillation change? Does the energy of the oscillation change?

Reasoning   The period of oscillation *does* change, because the period depends on the mass that is oscillating. The amplitude does not change. Because the new mass was added while the original mass was at rest, the combined masses are at rest at this point, also, defining the amplitude as the same as in the original oscillation. The energy does not change, either. At the maximum displacement point, the energy is all potential energy stored in the spring, which depends only on the spring constant and the amplitude, not the mass. The increased mass will pass through the equilibrium point with less velocity than in the original oscillation, but with the same kinetic energy. Another approach is to think about how energy could be transferred into the oscillating system—no work was done (nor was there any other form of energy transfer), so the energy in the system cannot change.

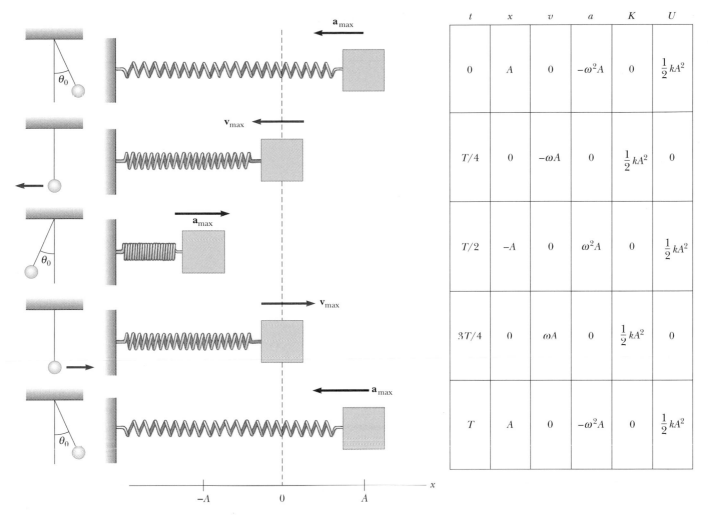

t	x	v	a	K	U
0	A	0	$-\omega^2 A$	0	$\frac{1}{2}kA^2$
$T/4$	0	$-\omega A$	0	$\frac{1}{2}kA^2$	0
$T/2$	$-A$	0	$\omega^2 A$	0	$\frac{1}{2}kA^2$
$3T/4$	0	ωA	0	$\frac{1}{2}kA^2$	0
T	A	0	$-\omega^2 A$	0	$\frac{1}{2}kA^2$

Figure 12.9   Simple harmonic motion for a mass–spring system and its analogy to the motion of a simple pendulum. The parameters in the table at the right refer to the mass–spring system, assuming that at $t = 0$, $x = A$ so that $x = A \cos \omega t$ (Special Case I).

CONCEPTUAL PROBLEM 3

A mass–spring system undergoes simple harmonic motion with an amplitude A. Does the total energy change if the mass is doubled but the amplitude is not changed? Are the kinetic and potential energies at a given point in its motion affected by the change in mass? Explain.

Example 12.3   Oscillations on a Horizontal Surface

A 0.50-kg mass connected to a light spring of force constant 20.0 N/m oscillates on a horizontal, frictionless track. (a) Calculate the total energy of the system and the maximum speed of the mass if the amplitude of the motion is 3.0 cm.

Solution   Using Equation 12.22, we get

$$E = \tfrac{1}{2}kA^2 = \tfrac{1}{2}\left(20.0 \,\frac{\text{N}}{\text{m}}\right)(3.0 \times 10^{-2}\,\text{m})^2 = 9.0 \times 10^{-3}\,\text{J}$$

When the mass is at $x = 0$, $U = 0$ and $E = \frac{1}{2}mv^2_{max}$; therefore

$$\frac{1}{2}mv^2_{max} = 9.0 \times 10^{-3}\,J$$

$$v_{max} = \sqrt{\frac{18.0 \times 10^{-3}\,J}{0.50\,kg}} = 0.19\,m/s$$

(b) What is the velocity of the mass when the displacement is equal to 2.0 cm?

Solution   We can apply Equation 12.23 directly:

$$v = \pm\sqrt{\frac{k}{m}(A^2 - x^2)}$$

$$= \pm\sqrt{\frac{20.0}{0.50}(3.0^2 - 2.0^2) \times 10^{-4}}$$

$$= \pm 0.14\,m/s$$

The positive and negative signs indicate that the mass could be moving to the right or left at this instant.

(c) Compute the kinetic and potential energies of the system when the displacement equals 2.0 cm.

Solution   Using the result to part (b), we get

$$K = \frac{1}{2}mv^2 = \frac{1}{2}(0.50\,kg)(0.14\,m/s)^2 = 5.0 \times 10^{-3}\,J$$

$$U = \frac{1}{2}kx^2 = \frac{1}{2}\left(20.0\,\frac{N}{m}\right)(2.0 \times 10^{-2}\,m)^2$$

$$= 4.0 \times 10^{-3}\,J$$

Note that the sum $K + U$ equals the total mechanical energy, E.

EXERCISE 6   For what values of x does the speed of the mass equal 0.10 m/s?   Answer ± 2.6 cm

12.4 • MOTION OF A PENDULUM

The **simple pendulum** is another mechanical system that exhibits periodic motion. It consists of a point mass, m, suspended by a light string of length L, where the upper end of the string is fixed as in Figure 12.10. The motion occurs in a vertical plane and is driven by the force of gravity. We shall show that the motion is that of a simple harmonic oscillator, provided the angle, θ, that the pendulum makes with the vertical is small.

The forces acting on the mass are the tension force, **T**, acting along the string, and the force of gravity, $m\mathbf{g}$. The tangential component of the force of gravity, $mg \sin \theta$, always acts toward $\theta = 0$, opposite the displacement. Therefore, the tangential force is a restoring force, and we can use Newton's second law to write the equation of motion in the tangential direction as

$$F_t = -mg \sin \theta = m\frac{d^2s}{dt^2}$$

where s is the displacement measured along the arc in Figure 12.10 and the minus sign indicates that F_t acts toward the equilibrium position. Because $s = L\theta$ and L is constant, this equation reduces to

$$\frac{d^2\theta}{dt^2} = -\frac{g}{L}\sin \theta$$

The right side is proportional to $\sin \theta$ rather than to θ; hence, we conclude that the motion is not simple harmonic motion, because it is not of the form of Equation 12.17. However, if we assume that θ is *small*, we can use the approxi-

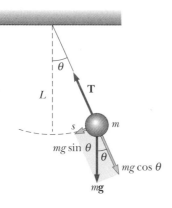

Figure 12.10   When θ is small, the simple pendulum oscillates in simple harmonic motion about the equilibrium position ($\theta = 0$). The restoring force is $mg \sin \theta$, the component of the force of gravity tangent to the circle.

mation $\sin \theta \approx \theta$, where θ is measured in radians,[1] and the equation of motion becomes

$$\frac{d^2\theta}{dt^2} = -\frac{g}{L}\theta \qquad [12.24]$$

• *Equation of motion for the simple pendulum (small θ)*

Now we have an expression with exactly the same form as Equation 12.17, and so we conclude that the motion is approximately simple harmonic motion for small amplitudes. Therefore, θ can be written as $\theta = \theta_0 \cos(\omega t + \phi)$, where θ_0 is the **maximum angular displacement** and the angular frequency ω is

$$\omega = \sqrt{\frac{g}{L}} \qquad [12.25]$$

• *Angular frequency of motion for a simple pendulum*

The period of the motion is

$$T = \frac{2\pi}{\omega} = 2\pi\sqrt{\frac{L}{g}} \qquad [12.26]$$

• *Period of motion for a simple pendulum*

In other words, **the period and frequency of a simple pendulum oscillating at small angles depend only on the length of the string and the free-fall acceleration.** Because the period is *independent* of the mass, we conclude that *all* simple pendula of equal length at the same location oscillate with equal periods. Experiment shows that this is correct. The analogy between the motion of a simple pendulum and the mass–spring system is illustrated in Figure 12.9.

The simple pendulum can be used as a timekeeper. It is also a convenient device for making precise measurements of the free-fall acceleration. Such measurements are important, because variations in local values of **g** can provide information on the locations of oil and other valuable underground resources.

CONCEPTUAL PROBLEM 4

A simple pendulum is suspended from the ceiling of a stationary elevator, and the period is determined. Describe the changes, if any, in the period when the elevator (a) accelerates upward, (b) accelerates downward, and (c) moves with constant velocity.

CONCEPTUAL PROBLEM 5

Imagine that a pendulum is hanging from the ceiling of a car. As the car coasts freely down a hill, is the equilibrium position of the pendulum vertical? Does the period of oscillation change from that in a stationary car?

CONCEPTUAL PROBLEM 6

A grandfather clock depends on the period of a pendulum to keep correct time. Suppose a grandfather clock is calibrated correctly and then the temperature of the room in which it resides increases. Does the grandfather clock run slow, fast, or correctly? *Hint:* A metal expands when its temperature is raised.

[1]See Appendix A for more details on the small-angle approximation.

Example 12.4   A Measure of Height

A man enters a tall tower, needing to know its height. He notes that a long pendulum extends from the ceiling almost to the floor and that its period is 12.0 s. How tall is the tower?

Solution   If we use $T = 2\pi\sqrt{L/g}$ and solve for L, we get

$$L = \frac{gT^2}{4\pi^2} = \frac{(9.80 \text{ m/s}^2)(12.0 \text{ s})^2}{4\pi^2} = 35.7 \text{ m}$$

EXERCISE 7   If the pendulum described in this example is taken to the moon, where the free-fall acceleration is 1.67 m/s^2, what is the period there?   Answer   29.1 s

O P T I O N A L

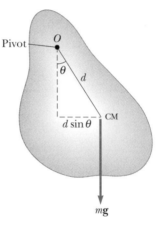

Figure 12.11   The physical pendulum consists of a rigid body pivoted at the point O, which is not at the center of mass. At equilibrium, the weight vector passes through O, corresponding to $\theta = 0$. The restoring torque about O when the system is displaced through an angle θ is $mgd \sin \theta$.

The Physical Pendulum

If a hanging object oscillates about a fixed axis that does not pass through its center of mass, and the object cannot be accurately approximated as a point mass, then it must be treated as a physical, or compound, pendulum. Consider a rigid body pivoted at a point O that is a distance of d from the center of mass (Fig. 12.11). The torque about O is provided by the force of gravity, and its magnitude is $mgd \sin \theta$. Using the fact that $\tau = I\alpha$, where I is the moment of inertia about the axis through O, we get

$$-mgd \sin \theta = I \frac{d^2\theta}{dt^2}$$

The minus sign on the left indicates that the torque about O tends to decrease θ. That is, the force of gravity produces a restoring torque.

If we again assume that θ is small, then the approximation $\sin \theta \approx \theta$ is valid and the equation of motion reduces to

$$\frac{d^2\theta}{dt^2} = -\left(\frac{mgd}{I}\right)\theta = -\omega^2\theta \qquad \text{[12.27]}$$

Note that the equation has the same form as Equation 12.17, and so the motion is approximately simple harmonic motion for small amplitudes. That is, the solution of Equation 12.27 is $\theta = \theta_0 \cos(\omega t + \phi)$, where θ_0 is the maximum angular displacement and

$$\omega = \sqrt{\frac{mgd}{I}}$$

The period is

Period of motion for a   •
physical pendulum

$$T = \frac{2\pi}{\omega} = 2\pi\sqrt{\frac{I}{mgd}} \qquad \text{[12.28]}$$

One can use this result to measure the moment of inertia of a planar rigid body. If the location of the center of mass and, hence, the distance d are known, the moment of inertia can be obtained through a measurement of the period. Finally, note that Equation 12.28 reduces to the period of a simple pendulum (Eq. 12.26) when $I = md^2$—that is, when all the mass is concentrated at the center of mass.

CONCEPTUAL PROBLEM 7

Two students are watching the swinging of a Foucault pendulum with a large bob in a museum. One student says, "I'm going to sneak past the fence and stick some chewing gum on the top of the pendulum bob, to change its frequency of oscillation." The other student says, "That won't change the frequency—the frequency of a pendulum is independent of mass." Which student is correct?

Example 12.5   A Swinging Rod

A uniform rod of mass M and length L is pivoted about one end and oscillates in a vertical plane (Fig. 12.12). Find the period of oscillation if the amplitude of the motion is small.

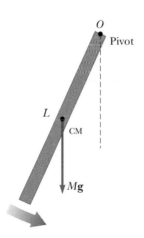

Figure 12.12   (Example 12.5) A rigid rod oscillating about a pivot through one end is a physical pendulum with $d = L/2$ and from Table 10.2, $I_0 = \frac{1}{3}ML^2$.

Solution   The moment of inertia of a uniform rod about an axis through one end is $\frac{1}{3}ML^2$. The distance d from the pivot to the center of mass is $L/2$. Substituting these quantities into Equation 12.28 gives

$$T = 2\pi\sqrt{\frac{\frac{1}{3}ML^2}{Mg\frac{L}{2}}} = 2\pi\sqrt{\frac{2L}{3g}}$$

Comment   In one of the Moon landings, an astronaut walking on the Moon's surface had a belt hanging from his space suit, and the belt oscillated as a compound pendulum. A scientist on Earth observed this motion on TV and from it was able to estimate the free-fall acceleration on the Moon. How do you suppose this calculation was done?

EXERCISE 8   Calculate the period of a meter stick pivoted about one end and oscillating in a vertical plane as in Figure 12.12.   Answer   1.64 s

EXERCISE 9   A simple pendulum has a length of 3.00 m. Determine the change in its period if it is taken from a point where $g = 9.80$ m/s² to an elevation where the free-fall acceleration decreases to 9.79 m/s².   Answer 1.78×10^{-3} s

12.5 • DAMPED OSCILLATIONS

The oscillatory motions we have considered so far have occurred in the context of an ideal system—that is, one that oscillates indefinitely under the action of a linear restoring force. In realistic systems, dissipative forces, such as friction, are present and retard the motion of the system. As a consequence, the mechanical energy of the system diminishes in time, and the motion is said to be *damped*.

One common type of drag force, which we discussed in Chapter 5, is proportional to the velocity and acts in the direction opposite the motion. This type of drag is often observed when an object is oscillating in air, for instance. Because the drag force can be expressed as $\mathbf{R} = -b\mathbf{v}$, where b is a constant, and the restoring force exerted on the system is $-kx$, we can write Newton's second law as

$$\sum F_x = -kx - bv = ma_x \qquad [12.29]$$

$$-kx - b\frac{dx}{dt} = m\frac{d^2x}{dt^2}$$

The solution of this equation requires mathematics that may not yet be familiar to you, and so it will simply be stated without proof. When the parameters of the system are such that $b < \sqrt{4mk}$, the solution to Equation 12.29 is

$$x = Ae^{-(b/2m)t}\cos(\omega t + \phi) \qquad [12.30]$$

where the angular frequency of motion is

$$\omega = \sqrt{\frac{k}{m} - \left(\frac{b}{2m}\right)^2} \qquad [12.31]$$

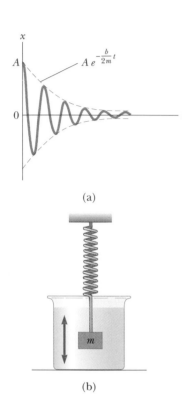

(a)

(b)

Figure 12.13 (a) Graph of the displacement versus time for a damped oscillator. Note the decrease in amplitude with time. (b) One example of a damped oscillator is a mass on a spring submersed in a liquid.

This result can be verified by substituting Equation 12.30 into Equation 12.29. Figure 12.13a shows the displacement as a function of time in this case. We see that **when the drag force is relatively small, the oscillatory character of the motion is preserved but the amplitude of vibration decreases in time** and the motion ultimately ceases. This is known as an **underdamped oscillator.** The dashed blue lines in Figure 12.13a, which outline the *envelope* of the oscillatory curve, represent the exponential factor that appears in Equation 12.30. The exponential factor shows that *the amplitude decays exponentially with time.* For motion with a given spring constant and particle mass, the oscillations dampen more rapidly as the maximum value of the drag force approaches the maximum value of the restoring force. One example of a damped harmonic oscillator is a mass immersed in a fluid, as in Figure 12.13b.

It is convenient to express the angular frequency of vibration of a damped system (Eq. 12.31) in the form

$$\omega = \sqrt{\omega_0{}^2 - \left(\frac{b}{2m}\right)^2}$$

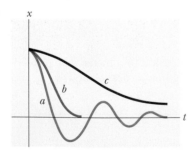

Figure 12.14 Plots of displacement versus time for (a) an underdamped oscillator, (b) a critically damped oscillator, and (c) an overdamped oscillator.

where $\omega_0 = \sqrt{k/m}$ represents the angular frequency of oscillation in the absence of a drag force (the undamped oscillator). In other words, when $b = 0$, the drag force is zero and the system oscillates with its natural frequency, ω_0. As the magnitude of the drag force increases, the oscillations dampen more rapidly. When b reaches a critical value, b_c, so that $b_c/2m = \omega_0$, the system does not oscillate and is said to be

critically damped. In this case it returns to equilibrium in an exponential manner with time, as in Figure 12.14.

If the medium is highly viscous and the parameters meet the condition that $b/2m > \omega_0$, the system is **overdamped.** Again, the displaced system does not oscillate but simply returns to its equilibrium position. As the damping increases, the time it takes the displacement to reach equilibrium also increases, as indicated in Figure 12.14. In any case, when a drag force is present, the energy of the oscillator eventually falls to zero. The lost mechanical energy dissipates into thermal energy in the resistive medium.

12.6 · FORCED OSCILLATIONS

O P T I O N A L

We have seen that the energy of a damped oscillator decreases in time as a result of the drag force. It is possible to compensate for this energy loss by applying an external force that does positive work on the system. At any instant, energy can be put into the system by an applied force that acts in the direction of motion of the oscillator. For example, a child on a swing can be kept in motion by appropriately timed "pushes." The amplitude of motion remains constant if the energy input per cycle of motion exactly equals the energy lost as a result of air drag.

A common example of a forced oscillator is a damped oscillator driven by an external force that varies periodically, such as $F = F_0 \cos \omega t$, where ω is the angular frequency of the force and F_0 is a constant. Adding this driving force to the left side of Equation 12.29 gives

$$F_0 \cos \omega t - b\frac{dx}{dt} - kx = m\frac{d^2x}{dt^2} \qquad [12.32]$$

Again, the solution of this equation is rather lengthy and will not be presented. However, after a sufficiently long period of time, when the energy input per cycle equals the energy lost per cycle, a steady-state condition is reached in which the oscillations proceed with constant amplitude. At this time, Equation 12.32 has the solution

$$x = A \cos(\omega t + \phi) \qquad [12.33]$$

where

$$A = \frac{F_0/m}{\sqrt{(\omega^2 - \omega_0{}^2)^2 + \left(\dfrac{b\omega}{m}\right)^2}} \qquad [12.34]$$

and where $\omega_0 = \sqrt{k/m}$ is the angular frequency of the undamped oscillator $(b = 0)$. From a physical point of view, one can argue that in a steady state the oscillator must have the same angular frequency as the driving force, and so the solution given by Equation 12.33 is expected.

Equation 12.34 shows that the amplitude of the forced oscillator is constant, because it is being driven by an external force. That is, the external agent provides the energy necessary to overcome the losses due to the drag force. Note that the mass oscillates at the angular frequency of the driving force, ω. For small damping, the amplitude becomes large when the frequency of the driving force is near the

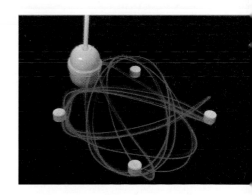

This is a computer-generated simulation of the chaotic pattern of a driven pendulum. The nonlinearity in the motion is provided by the gravitational torque, which is proportional to the sine of the angle. An important research problem is to predict the conditions under which different physical systems exhibit chaotic behavior. *(© Yoav Levy/Phototake)*

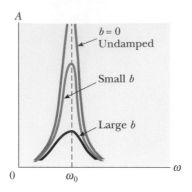

Figure 12.15 Graph of amplitude versus frequency for a damped oscillator when a periodic driving force is present. When the frequency of the driving force equals the natural frequency, ω_0, resonance occurs. Note that the shape of the resonance curve depends on the size of the damping coefficient, b.

Figure 12.16 If pendulum P is set into oscillation, pendulum Q will eventually oscillate with the greatest amplitude because its length is equal to that of P and so they have the same natural frequency of vibration. The pendula are oscillating in a direction perpendicular to the plane formed by the stationary strings.

natural frequency of oscillation or when $\omega \approx \omega_0$. The dramatic increase in amplitude near the natural frequency is called **resonance,** and the angular frequency, ω_0, is called the **resonance frequency** of the system.

The reason for large-amplitude oscillations at the resonance frequency is that energy is being transferred to the system under the most favorable conditions. This can be better understood by taking the first time derivative of x, which gives an expression for the velocity of the oscillator. In doing so, one finds that v is proportional to $\sin(\omega t + \phi)$. When the applied force is in phase with **v**, the rate at which work is done on the oscillator by the force **F** (in other words, the power) equals Fv. Because the quantity Fv is always positive when **F** and **v** are in phase, we conclude that **at resonance the applied force is in phase with the velocity, and the power transferred to the oscillator is a maximum.**

Figure 12.15 is a graph of amplitude as a function of frequency for the forced oscillator, with and without a resistive force. Note that the amplitude increases with decreasing damping ($b \rightarrow 0$) and that the resonance curve broadens as the damping increases. Under steady-state conditions, and at any driving frequency, the energy transferred into the system equals the energy lost because of the damping force; hence, the average total energy of the oscillator remains constant. In the absence of a damping force ($b = 0$), we see from Equation 12.34 that the steady-state amplitude approaches infinity as $\omega \rightarrow \omega_0$. In other words, if there are no losses in the system and we continue to drive an initially motionless oscillator with a sinusoidal force that is in phase with the velocity, the amplitude of motion will build up without limit (Fig. 12.15). This does not occur in practice because some damping is always present. That is, at resonance the amplitude is large but finite for small damping.

One experiment that demonstrates a resonance phenomenon is illustrated in Figure 12.16. Several pendula of different lengths are suspended from a stretched string. If one of them, such as P, is set swinging in the plane perpendicular to the plane of the string, the others will begin to oscillate, because they are coupled by the stretched string. Of those that are forced into oscillation by this coupling, pendulum Q, which is the same length as P (hence, the two pendula have the same natural frequency), will oscillate with the greatest amplitude.

Resonance appears in other areas of physics. For example, certain electrical circuits have natural (resonant) frequencies. A structure such as a bridge has natural frequencies and can be set into resonance by an appropriate driving force. A striking example of such a structural resonance occurred in 1940, when the Tacoma Narrows Bridge in Washington was destroyed by resonant vibrations (Fig. 12.17). The winds were not particularly strong on that occasion, but the bridge still collapsed because vortices (turbulences) generated by the wind blowing through the bridge occurred at a frequency that matched the natural frequency of the bridge.

Many other examples of resonant vibrations can be cited. A resonant vibration you may have experienced is the "singing" of telephone wires in the wind. Machines often break if one vibrating part is at resonance with some other moving part.

Thinking Physics 4

In an earthquake region, should damping in a building be large or small?

Reasoning The damping in a building should be large, so that the resonance curve is broad and flat. The building should respond to a wide range of earthquake frequen-

(a)   (b)

Figure 12.17 (a) High winds set up vibrations in the bridge, causing it to oscillate at a frequency near one of the natural frequencies of the bridge structure. (b) Once established, this resonance condition led to the bridge's collapse. *(UPI/Bettmann Newsphotos)*

cies, with a relatively small response. If the damping were small, so that the resonance curve were tall and skinny, the building would reject many earthquake frequencies but would have a disastrously large response to a narrow range of frequencies.

CONCEPTUAL PROBLEM 8

You stand on the end of a diving board and bounce to set it into oscillation. You find a maximum response, in terms of the amplitude of oscillation of the end of the board, when you bounce at frequency f. You now move to the middle of the board and repeat the experiment. Is the resonance frequency for forced oscillations at this point higher than f, lower than f, or the same as f? Why?

CONCEPTUAL PROBLEM 9

Some parachutes have holes in them to allow air to move smoothly through the chute. Without the holes, the air gathered under the chute as the parachutist falls is sometimes released from under the edges of the chute alternately and periodically from one side and then the other. Why might this periodic release of air cause a problem?

SUMMARY

The position of a simple harmonic oscillator varies periodically in time according to the relation

$$x = A \cos(\omega t + \phi) \tag{12.1}$$

where A is the amplitude of the motion, ω is the angular frequency, and ϕ is the phase constant. The values of A and ϕ depend on the initial position and velocity of the oscillator.

The time for one complete oscillation is called the **period** of the motion, defined by

$$T = \frac{2\pi}{\omega} \tag{12.2}$$

The inverse of the period is the **frequency** of the motion, which equals the number of oscillations per second.

The **velocity** and **acceleration** of a simple harmonic oscillator are

$$v = \frac{dx}{dt} = -\omega A \sin(\omega t + \phi) \tag{12.5}$$

$$a = \frac{dv}{dt} = -\omega^2 A \cos(\omega t + \phi)$$ [12.6]

Thus, the maximum velocity is ωA, and the maximum acceleration is $\omega^2 A$. The velocity is zero when the oscillator is at its turning points, $x = \pm A$, and the speed is a maximum at the equilibrium position, $x = 0$. The magnitude of the acceleration is a maximum at the turning points and is zero at the equilibrium position.

A mass–spring system moving on a frictionless track exhibits simple harmonic motion with the period

$$T = \frac{2\pi}{\omega} = 2\pi \sqrt{\frac{m}{k}}$$ [12.18]

where k is the force constant of the spring and m is the mass attached to the spring.

The kinetic energy and potential energy of a simple harmonic oscillator vary with time and are

$$K = \tfrac{1}{2}mv^2 = \tfrac{1}{2}m\omega^2 A^2 \sin^2(\omega t + \phi)$$ [12.20]

$$U = \tfrac{1}{2}kx^2 = \tfrac{1}{2}kA^2 \cos^2(\omega t + \phi)$$ [12.21]

The **total energy** of a simple harmonic oscillator is a constant of the motion and is

$$E = \tfrac{1}{2}kA^2$$ [12.22]

The potential energy of a simple harmonic oscillator is a maximum when the particle is at its turning points (maximum displacement from equilibrium) and is zero at the equilibrium position. The kinetic energy is zero at the turning points and is a maximum at the equilibrium position.

A **simple pendulum** of length L exhibits simple harmonic motion for small angular displacements from the vertical, with a period of

$$T = 2\pi \sqrt{\frac{L}{g}}$$ [12.26]

That is, the period is independent of the suspended mass.

A **physical pendulum** exhibits simple harmonic motion about a pivot that does not go through the center of mass. The period of this motion is

$$T = 2\pi \sqrt{\frac{I}{mgd}}$$ [12.28]

where I is the moment of inertia about an axis through the pivot and d is the distance from the pivot to the center of mass.

Damped oscillations occur in a system in which a drag force opposes the linear restoring force. If such a system is set in motion and then left to itself, its mechanical energy decreases in time because of the presence of the nonconservative drag force. It is possible to compensate for this loss in energy by driving the system with an external periodic force that is in phase with the motion of the system. When the frequency of the driving force matches the natural frequency of the *undamped* oscillator that starts its motion from rest, energy is continuously transferred to the oscillator and its amplitude increases without limit.

CONCEPTUAL QUESTIONS

1. Does the acceleration of a simple harmonic oscillator remain constant during its motion? Is the acceleration ever zero? Explain.

2. Explain why the kinetic and potential energies of a mass–spring system can never be negative.

3. What is the total distance traveled by a body executing simple harmonic motion in a time equal to its period if its amplitude is *A*?

4. If a mass–spring system is hung vertically and set into oscillation, why does the motion eventually stop?

5. If a pendulum clock keeps perfect time at the base of a mountain, will it also keep perfect time when moved to the top of the mountain? Explain.

6. If a grandfather clock were running slow, how could we adjust the length of the pendulum to correct the time?

7. A pendulum bob is made from a ball filled with water. What happens to the frequency of vibration of this pendulum if the ball has a hole that allows water to slowly leak out?

8. A platoon of soldiers marches in step along a road. Why are they ordered to break step when crossing a bridge?

9. If the length of a simple pendulum is quadrupled, what happens to (a) its frequency and (b) its period?

10. The amplitude of a system moving in simple harmonic motion is doubled. Determine the change in (a) the total energy, (b) the maximum speed, (c) the maximum acceleration, and (d) the period.

11. A simple harmonic oscillator has a total energy of *E*. (a) Determine the kinetic and potential energies when the displacement equals one-half the amplitude. (b) For what value of the displacement does the kinetic energy equal the potential energy?

12. If the coordinate of a particle varies as $x = -A \cos \omega t$, what is the phase constant in Equation 12.1? At what position does the particle begin its motion?

13. Does the displacement of an oscillating particle between $t = 0$ and a later time *t* necessarily equal the position of the particle at time *t*? Explain.

14. Determine whether or not the following quantities can be in the same direction for a simple harmonic oscillator: (a) displacement and velocity, (b) velocity and acceleration, (c) displacement and acceleration.

15. Can the amplitude *A* and phase constant ϕ be determined for an oscillator if only the position is specified at $t = 0$? Explain.

16. A simple pendulum undergoes simple harmonic motion when θ is small. Is the motion periodic when θ is large? How does the period of motion change as θ increases?

17. Will damped oscillations occur for any values of *b* and *k*? Explain.

18. Is it possible to have damped oscillations when a system is at resonance? Explain.

PROBLEMS

Section 12.1 Simple Harmonic Motion

1. The displacement of a particle is given by the expression $x = (4.00 \text{ m}) \cos(3.00 \pi t + \pi)$, where *x* is in meters and *t* is in seconds. Determine (a) the frequency and period of the motion, (b) the amplitude of the motion, (c) the phase constant, and (d) the displacement of the particle at $t = 0.250$ s.

2. A particle oscillates with simple harmonic motion so that its displacement varies according to the expression $x = (5.00 \text{ cm}) \cos(2t + \pi/6)$, where *x* is in centimeters and *t* is in seconds. At $t = 0$, find (a) the displacement of the particle, (b) its velocity, and (c) its acceleration. (d) Find the period and amplitude of the motion.

3. A piston in an automobile engine is in simple harmonic motion. If its amplitude of oscillation from the centerline is ± 5.00 cm and its mass is 2.00 kg, find the maximum velocity and acceleration of the piston when the auto engine is running at the rate of 3600 rev/min.

4. A 20.0-g particle moves in simple harmonic motion with a frequency of 3.00 oscillations/s and an amplitude of 5.00 cm. (a) Through what total distance does the particle move during one cycle of its motion? (b) What is its maximum speed? Where does this occur? (c) Find the maximum acceleration of the particle. Where in the motion does the maximum acceleration occur?

5. A particle moving along the *x* axis in simple harmonic motion starts from the origin at $t = 0$ and moves to the right. If the amplitude of its motion is 2.00 cm and the frequency is 1.50 Hz, (a) show that its displacement is given by the expression $x = (2.00 \text{ cm}) \sin(3.00\pi t)$. Determine (b) the maximum speed and the earliest time $(t > 0)$ at which the particle has this speed, (c) the maximum acceleration and the earliest time $(t > 0)$ at which the particle has this acceleration to the right, and (d) the total distance traveled between $t = 0$ and $t = 1.00$ s.

6. If the initial position, velocity, and acceleration of an object moving in simple harmonic motion are x_0, v_0, and a_0, and

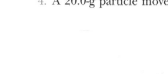

if the angular frequency of oscillation is ω, (a) show that the position and velocity of the object for all time can be written as

$$x(t) = x_0 \cos \omega t + \left(\frac{v_0}{\omega}\right) \sin \omega t$$

$$v(t) = -x_0\omega \sin \omega t + v_0 \cos \omega t$$

(b) If the amplitude of the motion is A, show that

$$v^2 - ax = v_0{}^2 - a_0 x_0 = A^2\omega^2$$

Section 12.2   Motion of a Mass Attached to a Spring

(*Note:* Neglect the mass of the spring in all of these problems.)

7. An archer pulls her bow string back 0.400 m by exerting a force that increases uniformly from zero to 230 N. (a) What is the equivalent spring constant of the bow? (b) How much work is done in pulling the bow?

8. A 1.00-kg mass attached to a spring of force constant 25.0 N/m oscillates on a horizontal, frictionless track. At $t = 0$, the mass is released from rest at $x = -3.00$ cm. (That is, the spring is compressed by 3.00 cm.) Find (a) the period of its motion, (b) the maximum values of its speed and acceleration, and (c) the displacement, velocity, and acceleration as functions of time.

9. A 0.500-kg mass attached to a spring of force constant 8.00 N/m vibrates in simple harmonic motion with an amplitude of 10.0 cm. Calculate (a) the maximum value of its speed and acceleration, (b) the speed and acceleration when the mass is 6.00 cm from the equilibrium position, and (c) the time it takes the mass to move from $x = 0$ to $x = 8.00$ cm.

10. A block of unknown mass is attached to a spring of force constant 6.50 N/m and undergoes simple harmonic motion with an amplitude of 10.0 cm. When the mass is halfway between its equilibrium position and the endpoint, its speed is measured to be +30.0 cm/s. Calculate (a) the mass of the block, (b) the period of the motion, and (c) the maximum acceleration of the block.

11. A 1.00-kg mass is attached to a horizontal spring. The spring is initially stretched by 0.100 m and the mass is released from rest there. After 0.500 s, the speed of the mass is zero. What is the maximum speed of the mass?

12. A particle that hangs from a spring oscillates with an angular frequency of 2.00 rad/s. The spring is suspended from the ceiling of an elevator car and hangs motionless (relative to the elevator car) as the car descends at a constant speed of 1.50 m/s. The car then stops suddenly. (a) With what amplitude does the particle oscillate? (b) What is the equation of motion for the particle? (Choose the upward direction to be positive.)

Section 12.3   Energy of the Simple Harmonic Oscillator

(Neglect spring masses.)

13. An automobile having a mass of 1000 kg is driven into a brick wall in a safety test. The bumper behaves like a spring of constant 5.00×10^6 N/m and compresses 3.16 cm as the car is brought to rest. What was the speed of the car before impact, assuming no energy is lost during impact with the wall?

14. An ore car of mass 4000 kg starts from rest and rolls downhill from the mine on tracks. At the end of the tracks, 10.0 m lower in elevation, is a spring with force constant $k = 400\ 000$ N/m. How much is the spring compressed in stopping the ore car? Ignore friction.

15. A 200-g mass is attached to a spring and executes simple harmonic motion with a period of 0.250 s. If the total energy of the system is 2.00 J, find (a) the force constant of the spring and (b) the amplitude of the motion.

16. A mass–spring system oscillates with an amplitude of 3.50 cm. If the force constant of the spring is 250 N/m and the mass is 0.500 kg, determine (a) the mechanical energy of the system, (b) the maximum speed of the mass, and (c) the maximum acceleration.

17. A 50.0-g mass connected to a spring of force constant 35.0 N/m oscillates on a horizontal, frictionless surface with an amplitude of 4.00 cm. Find (a) the total energy of the system and (b) the speed of the mass when the displacement is 1.00 cm. When the displacement is 3.00 cm, find (c) the kinetic energy and (d) the potential energy.

18. A 2.00-kg mass is attached to a spring and placed on a horizontal, smooth surface. A horizontal force of 20.0 N is required to hold the mass at rest when it is pulled 0.200 m from its equilibrium position (the origin of the x axis). The mass is now released from rest with an initial displacement of $x_0 = 0.200$ m, and it subsequently undergoes simple harmonic oscillations. Find (a) the force constant of the spring, (b) the frequency of the oscillations, and (c) the maximum speed of the mass. Where does this maximum speed occur? (d) Find the maximum acceleration of the mass. Where does it occur? (e) Find the total energy of the oscillating system. When the displacement equals one third the maximum value, find (f) the speed and (g) the acceleration.

Section 12.4   Motion of a Pendulum

19. A simple pendulum has a period of 2.50 s. (a) What is its length? (b) What would its period be on the Moon, where $g_M = 1.67$ m/s²?

20. A "seconds" pendulum is one that moves through its equilibrium position once each second. (The period of the pendulum is 2.000 s.) The length of a seconds pendulum is 0.9927 m at Tokyo and 0.9942 m at Cambridge, England.

What is the ratio of the free-fall accelerations at these two locations?

21. A simple pendulum has a mass of 0.250 kg and a length of 1.00 m. It is displaced through an angle of 15.0° and then released. What are (a) the maximum speed? (b) the maximum angular acceleration? (c) the maximum restoring force?

22. The angular displacement of a pendulum is represented by the equation $\theta = 0.320 \cos \omega t$, where θ is in radians and $\omega = 4.43$ rad/s. Determine the period and length of the pendulum.

23. A particle of mass m slides inside a frictionless hemispherical bowl of radius R. Show that if it starts from rest with a small displacement from equilibrium, the particle moves in simple harmonic motion with an angular frequency equal to that of a simple pendulum of length R. That is, $\omega = \sqrt{g/R}$.

24. A simple pendulum is 5.00 m long. (a) What is the period of simple harmonic motion for this pendulum if it is located in an elevator accelerating upward at 5.00 m/s²? (b) What is its period if the elevator is accelerating downward at 5.00 m/s²? (c) What is the period of simple harmonic motion for this pendulum if it is placed in a truck that is accelerating horizontally at 5.00 m/s²?

25. A very light rod extends rigidly 0.500 m out from one end of a meter stick. The stick is suspended from a pivot at the far end of the rod and set into oscillation. (a) Determine the period of oscillation. (b) By what percentage does this differ from a 1.00–m-long simple pendulum?

26. Consider the physical pendulum of Figure 12.11. (a) If its moment of inertia about an axis passing through its center of mass and parallel to the axis passing through its pivot point is I_{CM}, show that its period is

$$T = 2\pi \sqrt{\frac{I_{CM} + md^2}{mgd}}$$

where d is the distance between the pivot point and center of mass. (b) Show that the period has a minimum value when d satisfies $md^2 = I_{CM}$.

Section 12.5 Damped Oscillations (Optional)

27. A pendulum of length 1.00 m is released from an initial angle of 15.0°. After 1000 s, its amplitude is reduced by friction to 5.50°. What is the value of $b/2m$?

28. Show that Equation 12.30 is a solution of Equation 12.29 provided that $b^2 < 4mk$.

29. Show that the time rate of change of mechanical energy for a damped, undriven oscillator is given by $dE/dt = -bv^2$ and hence is always negative. (*Hint:* Differentiate the expression for the mechanical energy of an oscillator, $E = \frac{1}{2}mv^2 + \frac{1}{2}kx^2$, and use Eq. 12.29.)

Section 12.6 Forced Oscillations (Optional)

30. Calculate the resonant frequencies of (a) a 3.00-kg mass attached to a spring of force constant 240 N/m and (b) a simple pendulum 1.50 m in length.

31. A 2.00-kg mass attached to a spring is driven by an external force $F = (3.00 \text{ N}) \cos(2\pi t)$. If the force constant of the spring is 20.0 N/m, determine (a) the period and (b) the amplitude of the motion. (*Hint:* Assume there is no damping—that is, $b = 0$, and use Eq. 12.34.)

32. Consider an *undamped*, forced oscillator ($b = 0$), and show that Equation 12.33 is a solution of Equation 12.32, with an amplitude given by Equation 12.34.

Additional Problems

33. A mass m is oscillating freely on a vertical spring (Fig. P12.33). When $m = 0.810$ kg, the period is 0.910 s. An unknown mass on the same spring has a period of 1.16 s. Determine (a) the spring constant k and (b) the unknown mass.

Figure P12.33

34. A mass m is oscillating freely on a vertical spring with a period T (Fig. P12.33). An unknown mass m' on the same spring oscillates with a period T'. Determine (a) the force constant k and (b) the unknown mass m'.

35. A pendulum of length L and mass M has a spring of force constant k connected to it at a distance h below its point of suspension (Fig. P12.35). Find the frequency of vibration of the system for small values of the amplitude (small θ). (Assume the vertical suspension of length L is rigid but neglect its mass.)

36. A horizontal plank of mass m and length L is pivoted at one end, and the opposite end is attached to a spring of force constant k (Fig. P12.36). The moment of inertia of the plank about the pivot is $\frac{1}{3}mL^2$. When the plank is displaced a small angle θ from its horizontal equilibrium position and released, show that it moves with simple harmonic motion with an angular frequency $\omega = \sqrt{3k/m}$.

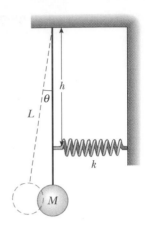

Figure P12.35

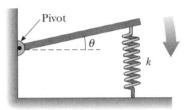

Figure P12.36

37. A large block *P* executes horizontal simple harmonic motion by sliding across a frictionless surface with a frequency $f = 1.50$ Hz. Block *B* rests on it, as shown in Figure P12.37, and the coefficient of static friction between the two is $\mu_s = 0.600$. What maximum amplitude of oscillation can the system have if the block is not to slip?

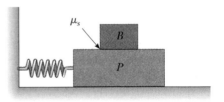

Figure P12.37

38. A 2.00-kg block hangs without vibrating at the end of a spring ($k = 500$ N/m) that is attached to the ceiling of an elevator car. The car is rising with an upward acceleration of $g/3$ when the acceleration suddenly ceases (at $t = 0$). (a) What is the angular frequency of oscillation of the block after the acceleration ceases? (b) By what amount is the spring stretched during the time that the elevator car is ac-

celerating? (c) What are the amplitude of the oscillation and the initial phase angle observed by a rider in the car? Take the upward direction to be positive.

39. A simple pendulum having a length of 2.23 m and a mass of 6.74 kg is given an initial speed of 2.06 m/s at its equilibrium position. Assume it undergoes simple harmonic motion and determine its (a) period, (b) total energy, and (c) maximum angular displacement.

40. People who ride motorcycles and bicycles learn to look out for bumps in the road—and especially for washboarding, a condition of many equally spaced ridges or bumps a certain distance apart. What is so bad about washboarding? A motorcycle has several springs and shock absorbers in its suspension, but you can model it as a single spring supporting a mass. You can estimate the spring constant by thinking about how far the spring compresses when a big biker sits down on the seat. Compute, on the basis of reasonable estimates, the distance between the washboard bumps of which a motorcyclist must be particularly careful when he or she is traveling at highway speed.

41. A mass *M* is connected to a spring of mass *m* and oscillates in simple harmonic motion on a horizontal, frictionless track (Fig. P12.41). The force constant of the spring is *k* and the equilibrium length is ℓ. Find (a) the kinetic energy of the system when the mass has a speed of *v* and (b) the period of oscillation. [*Hint:* Assume that all portions of the spring oscillate in phase and that the velocity of a segment *dx* is proportional to the distance from the fixed end; that is, $v_x = (x/\ell)v$. Also, note that the mass of a segment of the spring is $dm = (m/\ell)\, dx$.]

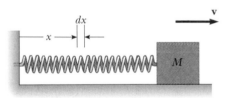

Figure P12.41

42. When a mass *M*, connected to the end of a spring of mass $m_s = 7.40$ g and force constant *k* is set into simple harmonic motion, the period of its motion is

$$T = 2\pi \sqrt{\frac{M + (m_s/3)}{k}}$$

A two-part experiment is conducted using various masses suspended vertically from the spring, as in Figure P12.33. (a) Displacements of 17.0, 29.3, 35.3, 41.3, 47.1, and 49.3 cm are measured for *M* values of 20.0, 40.0, 50.0, 60.0, 70.0, and 80.0 g, respectively. Construct a graph of *Mg* versus *x*, and perform a linear least-squares fit to the data. From the slope of your graph, determine a value for *k* for this spring. (b) The system is now set into simple harmonic motion, and periods are measured with a stopwatch. With $M = 80.0$ g, the total time for 10 oscillations is measured to be 13.41 s. The experiment is repeated with M values of 70.0, 60.0, 50.0, 40.0, and 20.0 g, with corresponding times for 10 oscillations of 12.52, 11.67, 10.67, 9.62, and 7.03 s. Obtain experimental values for T for each of these M values. Plot a graph of T^2 versus M and determine a value for k from the slope of the linear least-squares fit through the data points. Compare this value of k with that obtained in part (a). (c) Obtain a value for m_s from your graph and compare it with the given value of 7.40 g.

43. A mass m is connected to two rubber bands of length L, each under tension F, as in Figure P12.43. The mass is displaced vertically by a small distance, y. Assuming the tension does not change, show that (a) the restoring force is $-(2F/L)y$ and (b) the system exhibits simple harmonic motion with an angular frequency $\omega = \sqrt{2F/mL}$.

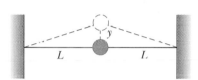

Figure P12.43

44. Consider a damped oscillator as illustrated in Figure 12.13. Assume the mass is 375 g, the spring constant is 100 N/m, and $b = 0.100$ kg/s. (a) How long does it take for the amplitude to drop to half its initial value? (b) How long does it take for the mechanical energy to drop to half its initial value? (c) Show that, in general, the fractional rate at which the amplitude decreases in a damped harmonic oscillator is one half the fractional rate at which the mechanical energy decreases.

45. A small, thin disk of radius r and mass m is attached rigidly to the face of a second thin disk of radius R and mass M as shown in Figure P12.45. The center of the small disk is located at the edge of the large disk. The large disk is mounted at its center on a frictionless axle. The assembly is rotated through a small angle θ from its equilibrium position and released. (a) Show that the speed of the center of the small disk as it passes through the equilibrium position is

$$v = 2\left[\frac{Rg(1 - \cos\theta)}{(M/m) + (r/R)^2 + 2}\right]^{1/2}$$

(b) Show that the period of the motion is

$$T = 2\pi\left[\frac{(M + 2m)R^2 + mr^2}{2mgR}\right]^{1/2}$$

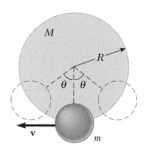

Figure P12.45

46. A mass m is connected to two springs of force constants k_1 and k_2, as in Figures P12.46a and P12.46b. In each case, the mass moves on a frictionless table and is displaced from equilibrium and released. Show that in each case the mass exhibits simple harmonic motion with periods

(a) $T = 2\pi\sqrt{\dfrac{m(k_1 + k_2)}{k_1 k_2}}$

(b) $T = 2\pi\sqrt{\dfrac{m}{k_1 + k_2}}$

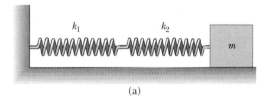

(a)

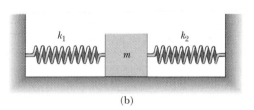

(b)

Figure P12.46

47. Imagine that a hole is drilled through the center of the Earth to its other side. It can be shown that an object of mass m at a distance r from the center of the Earth is pulled toward the center of the Earth only by the mass within the sphere of radius r (the reddish region in Fig. P12.47). Write Newton's law of gravitation for an object at the distance r from the center of the Earth, and show that the force on it is of Hooke's law form, $F = -kr$, where the effective force constant is $k = (4/3)\pi\rho Gm$, where ρ is the density of the Earth (assumed uniform) and G is the gravitational constant. Show that a sack of mail dropped into the hole will execute simple harmonic motion. When will it arrive at the other side of the Earth?

Earth

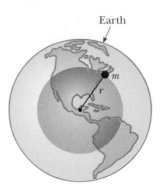

m

r

Figure P12.47

▣ Spreadsheet Problems

S1. Use a spreadsheet to plot the position x of an object undergoing simple harmonic motion as a function of time. (a) Use $A = 2.00$ m, $\omega = 5.00$ rad/s, and $\phi = 0$. On the same graph, plot the function for $\phi = \pi/4$, $\pi/2$, and π. Explain how the different phase constants affect the position of the object. (b) Repeat part (a) with $\omega = 5.00$ rad/s. How does the change in ω affect the plots?

S2. The potential and kinetic energies of a particle of mass m oscillating on a spring are given by Equations 12.20 and 12.21. Use a spreadsheet to plot $U(t)$, $K(t)$, and their sum as functions of time on the same graph. The input variables should be k, A, m, and ϕ. What happens to the sum of $U(t)$ and $K(t)$ as the input variables are changed?

S3. Consider a simple pendulum of length L that is displaced from the vertical by an angle θ_0 and released. In this problem, you are to determine the angular displacement for any value of the initial angular displacement. Prepare a spreadsheet to integrate the equation of motion of a simple pendulum:

$$\frac{d^2\theta}{dt^2} = -\frac{g}{L}\sin\theta$$

Take the initial condition to be $\theta = \theta_0$ and $d\theta/dt = 0$ at $t = 0$. Choose values for θ_0 and L, and find the angular displacement θ as a function of time. Using the same values of θ_0, compare your results for θ with those obtained from $\theta(t) = \theta_0 \cos \omega_0 t$ where $\omega_0 = \sqrt{g/L}$. Repeat for different values of θ_0. Be sure to include large initial angular displace-

ments. Does the period change when the initial angular displacement is changed? How do the periods for large values of θ_0 compare to those for small values of θ_0? (*Note:* Using the modified Euler's method to solve this differential equation you may find that the amplitude tends to increase with time. The fourth-order Runge-Kutta method discussed in the *Spreadsheet Investigations* Supplement would be a better choice to solve the differential equation. However, if you choose dt small enough, the solution using the modified Euler's method will still be good.)

S4. In the real world friction is always present. Modify your spreadsheet for Problem S3 to include the force of air resistance: $F_{air} = -bv$. In this situation, the equation of motion is

$$\frac{d^2\theta}{dt^2} = -\frac{g}{L}\sin\theta - \frac{b}{m}\frac{d\theta}{dt}$$

where m is the mass of the pendulum and b is the drag coefficient. Integrate this equation and calculate the sum of the kinetic and potential energies as functions of time using your calculated values of θ and $d\theta/dt$. Because energy is not constant, when in the pendulum's cycle is energy dissipated the fastest?

S5. The differential equation for the position x of a driven, damped harmonic oscillator of mass m is

$$m\frac{d^2x}{dt^2} + b\frac{dx}{dt} + m\omega_0^2 x = F_0 \cos \omega t$$

where ω_0 and b are constants, F_0 is the amplitude of the driving force, and ω is the driving angular frequency. For the case of a mass on a spring, $\omega_0 = \sqrt{k/m}$, where k is the force constant. Write a computer program or spreadsheet to integrate this equation of motion. Your input variables are the damping coefficient, b, the mass m, ω (the natural frequency of the system), and F_0. Choose the values for the initial velocity and position and calculate the position x as a function of time. Vary the driving angular frequency ω. Pick some values near ω_0. Does resonance occur? (*Note:* Your program may crash or blow up near resonance. The amplitudes of the motion may have large values when $\omega = \omega_0$. Also, the Euler method usually tends to overestimate the solution. Other methods such as the fourth-order Runge-Kutta method discussed in the *Spreadsheet Investigations* Supplement give a more accurate solution.)

ANSWERS TO CONCEPTUAL PROBLEMS

1. The bouncing ball is not an example of simple harmonic motion. The ball does not follow a sinusoidal function for its position as a function of time. The daily movement of a stu-

dent is also not simple harmonic motion, since the student stays at a fixed position—school—for a long period of time. If this motion were sinusoidal, the student would move more and more slowly as she approached her desk, and, as soon as she sat down at the desk, she would start to move back toward home again!

2. When the spring with two masses is set into oscillation in space, the coil in the exact center of the spring does not move. Thus, we can imagine clamping the center coil in place without affecting the motion. If we do this, we have two separate oscillating systems, one on each side of the clamp. The half-spring on each side of the clamp has twice the spring constant of the full spring, as shown by the following argument. The force exerted by a spring is proportional to the separation of the coils as the spring is extended. Imagine that we extend a spring by a given distance and measure the distance between coils. We then cut the spring in half. If one of the half-springs is now extended by the same distance, the coils will be twice as far apart as they were for the complete spring. Thus, it takes twice as much force to stretch the half-spring, from which we conclude that the half-spring has a spring constant which is twice that of the complete spring. Thus, our clamped system of masses on two half-springs will vibrate with a frequency that is higher than f by a factor of $\sqrt{2}$.

3. No. Because $E = \frac{1}{2}kA^2$, changing the mass while keeping A constant has no effect on the total energy. The potential energy $U = \frac{1}{2}kx^2$ depends only on position, not mass. Because $K = E - U$, the kinetic energy will also depend on position and not mass.

4. If it accelerates upward, the effective "g" is greater than the free-fall acceleration, so the period decreases. If it accelerates downward, the effective "g" is less than the free-fall acceleration, so the period increases. If it moves with constant velocity, the period does not change. (If the pendulum is in free-fall, it does not oscillate.)

5. Since the car is accelerating downhill, the principle of equivalence tells us that there is an effective gravitational field component uphill. Combining this with the actual gravitational field, the net effective gravitational field is no longer vertical. Thus, the equilibrium position of the pendulum will not be vertical. In addition, the net effective gravitational field has changed in value from the actual field, so the period of the pendulum will change.

6. As the temperature increases, the length of the pendulum will increase, due to thermal expansion. With a longer length, the period of the pendulum will increase. Thus, it will take longer to execute each swing, so that each second according to the clock will take longer than an actual second. Thus, the clock will run *slow*.

7. The first student is correct. While changing the mass of a simple pendulum does not change the frequency, the Foucault pendulum is a *physical* pendulum—the bob has a size. Assuming that the cord is very light, when the gum is placed on top of the bob, the center of mass of the physical pendulum is shifted very slightly upward. This shortens the effective length of the pendulum, which alters the frequency.

8. The diving board is acting as a physical pendulum. With your mass at the end of the board, you and the board have the maximum possible moment of inertia. When you move to the center of the board, the moment of inertia of the system decreases, and the natural frequency of oscillation increases above f.

9. If the period of the alternating release of air matched the swinging period of the parachutist, there could be a resonance response. The swinging of the parachutist could become so energetic that the parachute could become fouled and its braking action could disappear.

13

Wave Motion

Most of us experienced waves as children, when we dropped pebbles into a pond. The disturbance created by a pebble manifests itself as ripple waves that move outward until they finally reach the shore. If you were to carefully examine the motion of a leaf floating near the point where the pebble entered the water, you would see that the leaf moves up and down and sideways about its original position, but does not undergo any net displacement away from or toward the source of the disturbance. The water wave (the disturbance) moves from one place to another, *yet the water is not carried with it.*

The world is full of other kinds of waves, including sound waves, waves on strings, earthquake waves, radio waves, and x-rays. Most waves can be placed in one of two categories. **Mechanical waves** are waves that disturb and propagate through a medium; sound waves, for which air is the medium, and earthquake waves, for which the Earth's crust is the medium, are two examples of mechanical waves. **Electromagnetic waves** are a special class of waves that do not require a medium in order to propagate; light waves and radio waves are two familiar examples. In this chapter we shall confine our attention to the study of mechanical waves, deferring our study of electromagnetic waves to Chapter 24.

The wave concept is abstract. When we observe a water wave, what we see is a

Large waves such as this travel ▶ **great distances over the surface of the ocean, yet the water does not flow with the wave. The crests and troughs of the wave often form repetitive patterns.** *(Superstock)*

rearrangement of the water's surface. Without the water, there would be no wave. In the case of this or any other mechanical wave, what we interpret as a wave corresponds to the disturbance of a medium. Therefore, we can consider a mechanical wave to be **the motion of a disturbance.** This motion is not to be confused with the motion of the particles that make up the medium. In general, we describe mechanical wave motion by specifying the positions of all particles of the disturbed medium as a function of time.

13.1 • THREE WAVE CHARACTERISTICS

All mechanical waves require (1) some source of disturbance, (2) a medium that can be disturbed, and (3) some physical mechanism through which particles of the medium can influence each other. All waves carry energy and momentum, but the amount of energy transmitted through a medium and the mechanism responsible for the energy transport differ from case to case. For instance, the power of ocean waves during a storm is much greater than the power of sound waves generated by a musical instrument.

Three physical characteristics are important in describing waves: wavelength, frequency, and wave speed. **One wavelength is the minimum distance between any two points on a wave that behave identically**—for example, adjacent crests or adjacent troughs. Figure 13.1a shows displacement versus position for a sinusoidal wave at a specific time. The symbol λ (Greek lambda) is used to denote wavelength.

Most waves are periodic. **The frequency of such waves is the rate at which the disturbance repeats itself.** The period of the wave is the minimum time it takes the disturbance to repeat itself and is equal to the inverse of the frequency. Figure 13.1b shows displacement versus time for a sinusoidal wave at a fixed position, where the period is the time between identical displacements of a point on the wave.

Waves travel, or *propagate*, with a specific speed, which depends on the properties of the medium being disturbed. For instance, sound waves travel through air at 20°C with a speed of about 343 m/s, whereas the speed of sound in most solids is higher than 343 m/s.

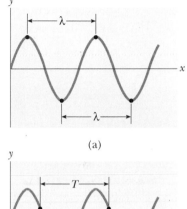

(a)

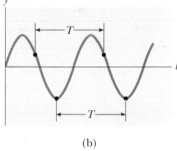

(b)

Figure 13.1   (a) The wavelength λ of a wave is the distance between adjacent crests or adjacent troughs. (b) The period T of a wave is the time it takes the wave to travel one wavelength.

Thinking Physics 1

Why is it important to be quiet in avalanche country? In the movie *On Her Majesty's Secret Service* (United Artists, 1969), the bad guys try to stop James Bond, who is escaping on skis, by firing a gun and causing an avalanche. Why did this happen?

Reasoning   The essence of a wave is the propagation of a *disturbance* through a medium. An impulsive sound, like the gunshot in the James Bond movie, can cause an acoustical disturbance that propagates through the air and can impact a ledge of snow that is just ready to break free to begin an avalanche. Such a disastrous event occurred in 1916 during World War I, when Austrian soldiers in the Alps were smothered by an avalanche caused by cannon fire.

13.2 · TYPES OF WAVES

One way to demonstrate wave motion is to flip the free end of a long rope that is under tension and has its opposite end fixed, as in Figure 13.2. In this manner, a single wave bump (or pulse) is formed and travels (to the right in Fig. 13.2) with a definite speed. This type of disturbance is called a **traveling wave,** and the rope is the medium through which it travels. Figure 13.2 represents four consecutive "snapshots" of the traveling wave. The shape of the wave pulse changes very little as it travels along the rope.[1]

Note that **as the wave pulse travels, each rope segment that is disturbed moves in a direction perpendicular to the wave motion.** Figure 13.3 illustrates this point for a particular segment, labeled *P.* Note that there is no motion of any part of the rope in the direction of the wave. A traveling wave such as this, in which the particles of the disturbed medium move perpendicularly to the wave velocity, is called a **transverse wave.**

In another class of waves, called **longitudinal waves,** the particles of the medium undergo displacements *parallel* to the direction of wave motion. Sound waves in air, for instance, are longitudinal. Their disturbance corresponds to a series of high- and low-pressure regions that may travel through air or through any material medium with a certain speed. A longitudinal pulse can be easily produced in a stretched spring, as in Figure 13.4. The free end is pumped back and forth along the length of the spring. This action produces compressed and stretched regions of the coil that travel along the spring, parallel to the wave motion.[2]

Some waves are neither transverse nor longitudinal but a combination of the two. Surface water waves are a good example. Figure 13.5 shows the motion of water particles at the surface as a wave moves to the right. Each particle moves in a circular path, and hence the disturbance has both transverse and longitudinal components. As the wave passes, water particles at the crests are moving in the direction of the wave, and particles at the troughs move in the opposite direction. Hence, no *net* displacement of water particles takes place. A cork bobbing on a pond surface as a wave passes exhibits this circular motion.

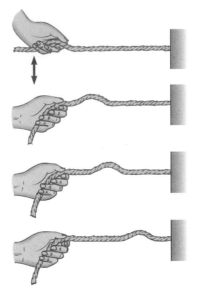

Figure 13.2 A wave pulse traveling down a stretched rope. The shape of the pulse is approximately unchanged as it travels along the rope.

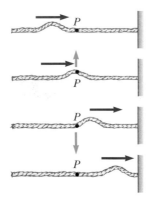

Figure 13.3 A pulse traveling on a stretched rope is a transverse wave. That is, any element *P* on the rope moves (blue arrows) in a direction perpendicular to the wave motion (red arrows).

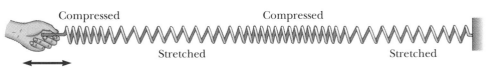

Compressed   Compressed

Stretched   Stretched

Longitudinal wave

Figure 13.4 A longitudinal pulse along a stretched spring. The displacement of the coils is in the direction of the wave motion. For the starting motion described in the text, the compressed region is followed by a stretched region.

---

[1]Strictly speaking, the pulse changes its shape and gradually spreads out during the motion. This effect is called *dispersion* and is common to many mechanical waves.

[2]In the case of longitudinal waves in a gas, each compressed area is a region of higher-than-average pressure and density, and each stretched region is a region of lower-than-average pressure and density.

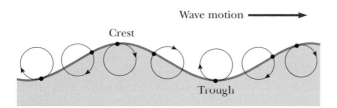

Figure 13.5   Wave motion on the surface of water. The molecules at the water's surface move in nearly circular paths. Each molecule is displaced horizontally and vertically from its equilibrium position, represented by circles.

CONCEPTUAL PROBLEM 1

In a long line of people waiting to buy tickets at a movie theater, when the first person leaves, a pulse of motion occurs, as people step forward to fill in the gap. The gap moves through the line of people. What determines the speed of this pulse? Is it transverse or longitudinal? How about "the wave" at a baseball game, where people in the stands stand up and shout as the wave arrives at their location, and this pulse moves around the stadium—what determines the speed of this pulse? Is it transverse or longitudinal?

13.3 • ONE-DIMENSIONAL TRANSVERSE TRAVELING WAVES

So far we have provided only verbal and graphical descriptions of a traveling wave. Let us now develop a mathematical description of a one-dimensional traveling wave. Consider a wave pulse traveling to the right with constant speed, v, along a long, stretched string, as in Figure 13.6. The pulse moves along the x axis (the axis of the string), and the transverse (up-and-down) displacement of the string is measured with the coordinate y.

Figure 13.6a represents the shape and position of the pulse at time $t = 0$. At this time, the shape of the pulse, whatever it may be, can be represented as $y = f(x)$. That is, y is some definite function of x. The *maximum displacement* of the string, y_m, is called the **amplitude** of the wave. Because the speed of the wave pulse is v, the pulse travels to the right a distance of vt in the time t (Fig. 13.6b). If the shape

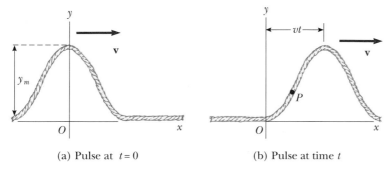

(a) Pulse at $t = 0$ (b) Pulse at time t

Figure 13.6   A one-dimensional wave pulse traveling to the right with a speed v. (a) At $t = 0$, the shape of the pulse is given by $y = f(x)$. (b) At some later time t, the shape remains unchanged and the vertical displacement of any point P of the medium is given by $y = f(x - vt)$.

of the wave pulse doesn't change with time, we can represent the displacement y for all later times, measured in a stationary frame with the origin at O, as

Pulse traveling to the right •

$$y = f(x - vt) \qquad [13.1]$$

In the same way, if the wave pulse travels to the left, the displacement of the string is

Pulse traveling to the left •

$$y = f(x + vt) \qquad [13.2]$$

The displacement y, sometimes called the **wave function,** depends on the two variables x and t. For this reason, it is often written $y(x, t)$, which is read "y as a function of x and t." It is important to understand the meaning of y.

Consider a point, P, on the string, identified by a particular value of its coordinates. As the wave pulse passes through P, the y coordinate of this point increases, reaches a maximum, and then decreases to zero. **The wave function $y(x, t)$ represents the y coordinate of any point P at any time t.** Furthermore, if t is fixed, then the wave function y as a function of x, sometimes called the *waveform*, defines a curve representing the actual shape of the pulse at that time. This is equivalent to a snapshot of the pulse at time t.

For a pulse that moves without changing its shape, the velocity of the pulse is the same as the velocity of any point along the pulse, such as the crest. To find the velocity of the pulse, we can calculate how far the crest moves horizontally in a short time and then divide that distance by the time interval. In order to follow the motion of the crest, some particular value, say x_0, must be substituted for $x - vt$ in Equation 13.1. (The value x_0 is called the *argument* of the function y.) Regardless of how x and t change individually, we must require that $x - vt = x_0$ in order to stay with the crest. This equation, therefore, represents the motion of the crest. At $t = 0$, the crest is at $x = x_0$; after an interval of dt, the crest is at $x = x_0 + v\,dt$. Therefore, in the time dt the crest has moved a distance of $dx = (x_0 + v\,dt) - x_0 = v\,dt$. Clearly, the wave speed, often called the **phase speed,** is

Phase speed •

$$v = \frac{dx}{dt} \qquad [13.3]$$

The *wave velocity*, or *phase velocity*, must not be confused with the *transverse velocity* of a particle in the medium (which is always perpendicular to the wave velocity). The phase velocity is the same at any point along the pulse.

The following example illustrates how a specific wave function is used to describe the motion of a traveling wave pulse.

Example 13.1   A Pulse Moving to the Right

A wave pulse moving to the right along the x axis is represented by the wave function

$$y(x, t) = \frac{2}{(x - 3.0t)^2 + 1}$$

where x and y are measured in centimeters and t is in seconds. Let us plot the waveform at $t = 0$, $t = 1.0$ s, and $t = 2.0$ s.

Solution   First, note that this function is of the form $y = f(x - vt)$. By inspection, we see that the speed of the wave is $v = 3.0$ cm/s. Furthermore, the wave amplitude (the maximum value of y) is given by $A = 2.0$ cm. The location of the peak of the pulse occurs at the value of x for which the denominator is a minimum, that is, where $(x - 3.0t) = 0$. At times $t = 0$, $t = 1.0$ s, and $t = 2.0$ s, the wave function expressions are

$$y(x, 0) = \frac{2}{x^2 + 1} \quad \text{at } t = 0$$

$$y(x, 1.0) = \frac{2}{(x - 3.0)^2 + 1} \quad \text{at } t = 1.0 \text{ s}$$

$$y(x, 2.0) = \frac{2}{(x - 6.0)^2 + 1} \quad \text{at } t = 2.0 \text{ s}$$

We can now use these expressions to plot the wave function versus x at these times. For example, let us evaluate $y(x, 0)$ at $x = 0.50$ cm:

$$y(0.50, 0) = \frac{2}{(0.50)^2 + 1} = 1.6 \text{ cm}$$

Likewise, $y(1.0, 0) = 1.0$ cm, $y(2.0, 0) = 0.40$ cm, and so on. A continuation of this procedure for other values of x yields the waveform shown in Figure 13.7a. In a similar manner, one obtains the graphs of $y(x, 1.0)$ and $y(x, 2.0)$, shown in Figures 13.7b and 13.7c, respectively. These snapshots show that the wave pulse moves to the right without changing its shape and has a constant speed of 3.0 cm/s.

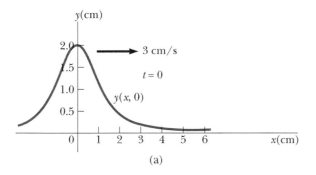

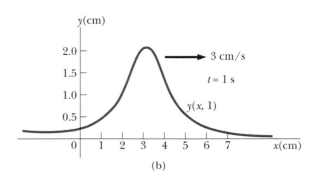

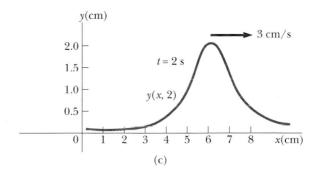

Figure 13.7 (Example 13.1) Graphs of the function $y(x, t) = 2/[(x - 3t)^2 + 1]$. (a) $t = 0$, (b) $t = 1$ s, and (c) $t = 2$ s.

13.4 · SINUSOIDAL TRAVELING WAVES

In this section we introduce an important periodic waveform known as a **sinusoidal wave** (Fig. 13.8). The red curve represents a snapshot of a sinusoidal traveling wave at $t = 0$, and the blue curve represents a snapshot of the wave at some later t. At $t = 0$, the displacement of the curve can be written

$$y = A \sin \left(\frac{2\pi}{\lambda} x \right) \qquad \text{[13.4]}$$

where the amplitude, A, as usual, represents the maximum value of the displacement, and λ is the wavelength as defined in Figure 13.1a. Thus, we see that the displacement repeats itself when x is increased by an integral multiple of λ. If the wave moves to the right with a phase speed of v, the wave function at some later t is

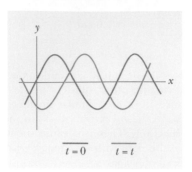

Figure 13.8   A one-dimensional sinusoidal wave traveling to the right with a speed v. The red curve represents a snapshot of the wave at $t = 0$, and the blue curve represents a snapshot at some later time t.

$$y = A \sin \left[\frac{2\pi}{\lambda} (x - vt) \right] \qquad [13.5]$$

That is, the sinusoidal wave moves to the right a distance of vt in the time t, as in Figure 13.8. Note that the wave function has the form $f(x - vt)$ and represents a wave traveling to the right. If the wave were traveling to the left, the quantity $x - vt$ would be replaced by $x + vt$, just as in the case of the traveling pulse described by Equations 13.1 and 13.2.

Because the period T is the time it takes the wave to travel a distance of one wavelength, the phase speed, wavelength, and period are related by

$$v = \frac{\lambda}{T} \qquad [13.6]$$

or

$$\lambda = vT \qquad [13.7]$$

Substituting Equation 13.6 into Equation 13.5, we find that

$$y = A \sin \left[2\pi \left(\frac{x}{\lambda} - \frac{t}{T} \right) \right] \qquad [13.8]$$

This form of the wave function clearly shows the periodic nature of y. That is, at any given time t (a snapshot of the wave), y has the same value at the positions x, $x + \lambda$, $x + 2\lambda$, and so on. Furthermore, at any given position x, the value of y at times t, $t + T$, $t + 2T$, and so on, is the same.

We can express the sinusoidal wave function in a convenient form by defining two other quantities: **angular wave number,** k, and **angular frequency,** ω:

Angular wave number •

$$k \equiv \frac{2\pi}{\lambda} \qquad [13.9]$$

Angular frequency •

$$\omega \equiv \frac{2\pi}{T} = 2\pi f \qquad [13.10]$$

Note that in Equation 13.10, we used the definition of frequency, $f = 1/T$. Using these definitions, we see that Equation 13.8 can be written in the more compact form

$$y = A \sin(kx - \omega t) \qquad [13.11]$$

We shall use this form most frequently.

Using Equations 13.9 and 13.10, we can express the phase speed, v, in the alternative forms

$$v = \frac{\omega}{k} \qquad [13.12]$$

Speed of a traveling •
sinusoidal wave

$$v = f\lambda \qquad [13.13]$$

The wave function given by Equation 13.11 assumes that the displacement, y, is zero at $x = 0$ and $t = 0$. This need not be the case. If the transverse displacement is not zero at $x = 0$ and $t = 0$, we generally express the wave function in the form

$$y = A \sin(kx - \omega t - \phi) \qquad \text{[13.14]}$$

• ***Wave function for a traveling sinusoidal wave***

where ϕ is called the **phase constant** and can be determined from the initial conditions.

Thinking Physics 2

The quantity that we call the angular wave number k is sometimes called *spatial frequency*. Why?

Reasoning The angular frequency ω is 2π divided by the *time* interval for one cycle of oscillation for the wave, the period, T. The angular wave number k is 2π divided by the *space* interval for one cycle of oscillation for the wave, the wavelength, λ (the spatial interval of one cycle of the wave). Due to the similarity in these parameters, it is reasonable to consider ω the *temporal* frequency and k the *spatial* frequency.

CONCEPTUAL PROBLEM 2

Sound waves from a musical performance are encoded onto radio waves for transmission from the radio studio to your home receiver. A radio wave with a wavelength of 3 meters carries a sound wave that has a wavelength of 3 meters in air. Which has the higher frequency, the radio wave or the sound wave?

Example 13.2   A Traveling Sinusoidal Wave

A sinusoidal wave traveling in the positive x direction has an amplitude of 15.0 cm, a wavelength of 40.0 cm, and a frequency of 8.00 Hz. The vertical displacement of the medium at $t = 0$ and $x = 0$ is also 15.0 cm, as shown in Figure 13.9. (a) Find the angular wave number, period, angular frequency, and speed of the wave.

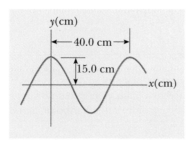

Figure 13.9   (Example 13.2) A sinusoidal wave of wavelength $\lambda = 40.0$ cm and amplitude $A = 15.0$ cm. The wave function can be written in the form $y = A \cos(kx - \omega t)$.

Solution   Using Equations 13.9, 13.10, and 13.13, we find the following:

$$k = \frac{2\pi}{\lambda} = \frac{2\pi \,\text{rad}}{40.0 \,\text{cm}} = 0.157 \,\text{rad/cm}$$

$$T = \frac{1}{f} = \frac{1}{8.00 \,\text{s}^{-1}} = 0.125 \,\text{s}$$

$$\omega = 2\pi f = 2\pi(8.00 \,\text{s}^{-1}) = 50.3 \,\text{rad/s}$$

$$v = f\lambda = (8.00 \,\text{s}^{-1})(40.0 \,\text{cm}) = 320 \,\text{cm/s}$$

(b) Determine the phase constant ϕ, and write a general expression for the wave function.

Solution   Because $A = 15.0$ cm and because it is given that $y = 15.0$ cm at $x = 0$ and $t = 0$, substitution into Equation 13.14 gives

$$15 = 15 \sin(-\phi) \qquad \text{or} \qquad \sin(-\phi) = 1$$

Because $\sin(-\phi) = -\sin\phi$, we see that $\phi = -\pi/2$ rad (or $-90°$). Hence, the wave function is of the form

$$y = A \sin\left(kx - \omega t + \frac{\pi}{2}\right) = A \cos(kx - \omega t)$$

That the wave function must have this form can be seen by inspection, noting that the cosine argument is displaced by $90°$ from the sine function. Substituting the values for A, k, and ω into this expression gives

$$y = (15.0 \text{ cm})\cos(0.157x - 50.3t)$$

Sinusoidal Waves on Strings

One method of producing a traveling sinusoidal wave on a very long string is shown in Figure 13.10. One end of the string is connected to a blade that is set in vibration. As the blade oscillates vertically with simple harmonic motion, a traveling wave moving to the right is set up on the string. Figure 13.10 represents snapshots of the wave at intervals of one quarter of a period. Note that **each particle of the string, such as P, oscillates vertically in the y direction with simple harmonic motion.** This must be the case because each particle follows the simple harmonic motion of the blade. Therefore, every segment of the string can be treated as a simple harmonic oscillator vibrating with a frequency equal to the frequency of vibration of the blade that drives the string.[3] Note that although each segment oscillates in the y direction, the wave (or disturbance) travels in the x direction with speed v.

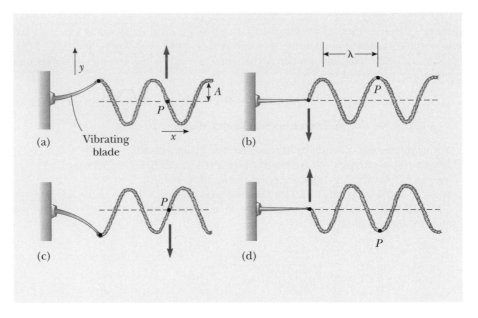

Figure 13.10 One method for producing a train of sinusoidal wave pulses on a continuous string. The left end of the string is connected to a blade that is set into vibration. Every segment of the string, such as the point P, oscillates with simple harmonic motion in the vertical direction.

[3]In this arrangement, we are assuming that the segment always oscillates in a vertical line. The tension in the string would vary if the segment were allowed to move sideways. Such a motion would make the analysis very complex.

Of course, this is the definition of a transverse wave. In this case, the energy carried by the traveling wave is supplied by the vibrating blade.

If the waveform at $t = 0$ is as described in Figure 13.10b, then the wave function can be written

$$y = A \sin(kx - \omega t)$$

We can use this expression to describe the motion of any point on the string. The point P (or any other point on the string) moves vertically, and so its x coordinate *remains constant.* Therefore, the **transverse speed,** v_y, of the point P (not to be confused with the wave speed, v) and its **transverse acceleration,** a_y, are

$$v_y = \frac{dy}{dt}\bigg]_{x=\text{constant}} = \frac{\partial y}{\partial t} = -\omega A \cos(kx - \omega t) \qquad \textbf{[13.15]}$$

$$a_y = \frac{dv_y}{dt}\bigg]_{x=\text{constant}} = \frac{\partial v_y}{\partial t} = -\omega^2 A \sin(kx - \omega t) \qquad \textbf{[13.16]}$$

The maximum values of these quantities are simply the absolute values of the coefficients of the cosine and sine functions:

$$(v_y)_{\text{max}} = \omega A \qquad \textbf{[13.17]}$$

$$(a_y)_{\text{max}} = \omega^2 A \qquad \textbf{[13.18]}$$

You should recognize that the transverse speed and transverse acceleration of any point on the string do not reach their maximum values simultaneously. In fact, the transverse speed reaches its maximum value (ωA) when the displacement $y = 0$, whereas the transverse acceleration reaches its maximum value $(\omega^2 A)$ when $y = -A$. Finally, Equations 13.17 and 13.18 are identical to the corresponding equations for simple harmonic motion.

Example 13.3 A Sinusoidally Driven String

The string shown in Figure 13.10 is driven at a frequency of 5.00 Hz. The amplitude of the motion is 12.0 cm, and the wave speed is 20.0 m/s. Furthermore, the wave is such that $y = 0$ at $x = 0$ and $t = 0$. Determine the angular frequency and wave number for this wave, and write an expression for the wave function.

Solution Using Equations 13.10 and 13.12 gives

$$\omega = \frac{2\pi}{T} = 2\pi f = 2\pi(5.00 \text{ Hz}) = 31.4 \text{ rad/s}$$

$$k = \frac{\omega}{v} = \frac{31.4 \text{ rad/s}}{20.0 \text{ m/s}} = 1.57 \text{ rad/m}$$

Because $A = 12.0$ cm $= 0.120$ m, we have

$$y = A \sin(kx - \omega t) = (0.120 \text{ m})\sin(1.57x - 31.4t)$$

EXERCISE 1 Calculate the maximum values for the transverse speed and transverse acceleration of any point on the string.   Answer   3.77 m/s; 118 m/s²

EXERCISE 2 When a particular wire is vibrating with a frequency of 4.00 Hz, a transverse wave of wavelength 60.0 cm is produced. Determine the speed of wave pulses along the wire.   Answer   2.40 m/s

EXERCISE 3 For a certain transverse wave, the distance between two successive maxima is 1.2 m and eight maxima pass a given point along the direction of travel every 12 s. Calculate the wave speed.   Answer   0.80 m/s

13.5 • SUPERPOSITION AND INTERFERENCE OF WAVES

Many interesting wave phenomena in nature cannot be described by a single moving pulse. Instead, one must analyze complex waveforms in terms of a combination of many traveling waves. To analyze such wave combinations, one can make use of the **superposition principle,** which states that

Linear waves obey the •
superposition principle.

> if two or more traveling waves are moving through a medium, the resultant wave function at any point is the sum of the wave functions of the individual waves.

This rather striking property is exhibited by many waves in nature. Waves that obey this principle are called *linear waves,* and they are generally characterized by small amplitudes. Waves that violate the superposition principle are called *nonlinear waves* and are often characterized by large amplitudes. In this book, we shall deal only with linear waves.

One consequence of the superposition principle is that **two traveling waves can pass through each other without being destroyed or even altered.** For instance, when two pebbles are thrown into a pond, the expanding circular surface waves do not destroy each other. In fact, the ripples pass through each other. The complex pattern that is observed can be viewed as two independent sets of expanding circles. Likewise, when sound waves from two sources move through air, they

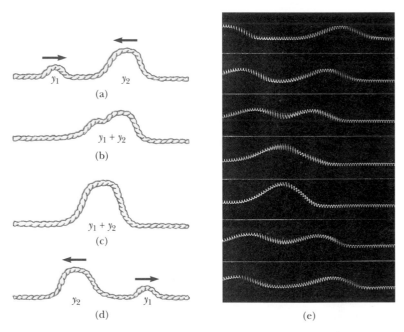

Figure 13.11   *(Left)* Two wave pulses traveling on a stretched string in opposite directions pass through each other. When the pulses overlap, as in (b) and (c), the net displacement of the string equals the sum of the displacements produced by each pulse. Because each pulse produces positive displacements of the string, we refer to their superposition as *constructive interference.*   *(Right)* Photograph of superposition of two equal and symmetric pulses traveling in opposite directions on a stretched spring.   *(Photo, Education Development Center, Newton, Mass.)*

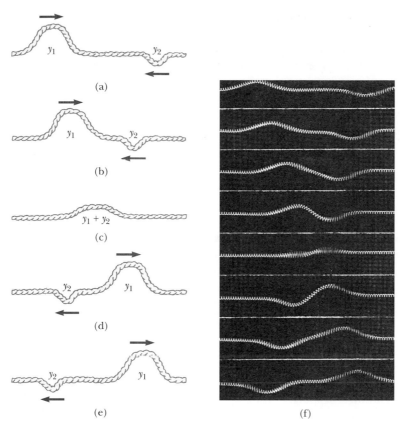

(a)

(b)

(c)

(d)

(e)

(f)

Figure 13.12 *(Left)* Two wave pulses traveling in opposite directions with displacements that are inverted relative to each other. When the two overlap as in (c), their displacements subtract from each other. *(Right)* Photograph of superposition of two symmetric pulses traveling in opposite directions, where one is inverted relative to the other. *(Photo, Education Development Center, Newton, Mass.)*

also can pass through each other. The sound one hears at a given point is the resultant of both disturbances.

A simple pictorial representation of the superposition principle is obtained by considering two pulses traveling in opposite directions on a stretched string, as in Figure 13.11. The wave function for the pulse moving to the right is y_1, and the wave function for the pulse moving to the left is y_2. The pulses have the same speed but different shapes. Each pulse is assumed to be symmetric (although this is not a necessary condition), and in both cases the vertical displacements of the string are taken to be positive. When the waves begin to overlap (Fig. 13.11b), the resulting complex waveform is given by $y_1 + y_2$. When the crests of the pulses exactly coincide (Fig. 13.11c), the resulting waveform, $y_1 + y_2$, is symmetric. The two pulses finally separate and continue moving in their original directions (Fig. 13.11d). Note that the final waveforms remain unchanged, as if the two pulses had never met! The combination of separate waves in the same region of space to produce a resultant wave is called **interference.**

For the two pulses shown in Figure 13.11, the vertical displacements caused by the pulses are in the same direction, and so the resultant waveform (when the pulses overlap) exhibits a displacement greater than those of the individual pulses. Now

Interference patterns produced by outward spreading waves from several drops of water falling into a pond. *(Martin Dohrn/SPL/Photo Researchers)*

consider two identical pulses, again traveling in opposite directions on a stretched string, but this time one pulse is inverted relative to the other, as in Figure 13.12. In this case, when the pulses begin to overlap, the resultant waveform is the sum of the two separate displacements, but one displacement is negative. Again, the two pulses pass through each other.

EXERCISE 4   Two waves are traveling in the same direction along a stretched string. Each has an amplitude of 4.0 cm and they are 90° out of phase. Find the amplitude of the resultant wave.   Answer   5.7 cm

13.6 · THE SPEED OF TRANSVERSE WAVES ON STRINGS

For linear mechanical waves, **the speed depends only on the properties of the medium through which the disturbance travels.** In this section we shall focus our attention on determining the speed of a transverse pulse traveling on a stretched string. We shall show that if the tension in the string is F and its mass per unit length is μ, then the wave speed, v, is

Speed of a wave on a •
stretched string

$$v = \sqrt{\frac{F}{\mu}} \qquad\qquad [13.19]$$

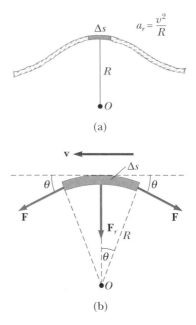

Figure 13.13   (a) To obtain the speed v of a wave on a stretched string, it is convenient to describe the motion of a small segment of the string in a moving frame of reference. (b) The net force on a small segment of length Δs is in the radial direction. The horizontal components of the tension force cancel.

Now let us use a mechanical analysis to derive the preceding expression for the speed of a pulse traveling on a stretched string. Consider a pulse moving to the right with a uniform speed of v, measured relative to a stationary frame of reference. Instead of using a stationary frame, however, it is more convenient to choose as our reference frame one that moves along with the pulse at the same speed, so that the pulse appears to be at rest in the frame, as in Figure 13.13a. This is permitted because Newton's laws are valid in either a stationary frame or one that moves with constant velocity. A small segment of the string, of length Δs, forms the approximate arc of a circle of radius R, as shown in Figure 13.13a and magnified in Figure 13.13b. In the pulse's frame of reference, the segment moves to the left with the speed v. This segment has a centripetal acceleration of v^2/R, which is supplied by the tension, F, in the string. The force **F** acts on each side of the segment, tangent to the arc, as in Figure 13.13b. The horizontal components of **F** cancel, and each vertical component $F \sin \theta$ acts radially inward toward the center of the arc. Hence, the total radial force is $2F \sin \theta$. Because the segment is small, θ is small, and we can use the small-angle approximation $\sin \theta \approx \theta$. Therefore, the total radial force can be expressed as

$$F_r = 2F \sin \theta \approx 2F\theta$$

The segment has the mass $m = \mu \Delta s$, where μ is the mass per unit length of the string. Because the segment forms part of a circle and subtends an angle of 2θ at the center, $\Delta s = R(2\theta)$, and hence

$$m = \mu \Delta s = 2\mu R\theta$$

If we apply Newton's second law to this segment, the radial component of motion gives

$$F_r = \frac{mv^2}{R} \qquad \text{or} \qquad 2F\theta = \frac{2\mu R\theta v^2}{R}$$

where F_r is the force that supplies the centripetal acceleration of the segment and maintains the curvature of this segment.

Solving for v gives

$$v = \sqrt{\frac{F}{\mu}}$$

Notice that this derivation is based on the assumption that the pulse height is small relative to the length of the string. Using this assumption, we were able to use the approximation $\sin \theta \approx \theta$. Furthermore, the model assumes that the tension, F, is not affected by the presence of the pulse, so that F is the same at all points on the string. Finally, this proof does *not* assume any particular shape for the pulse. Therefore, we conclude that a pulse of *any shape* will travel on the string with speed $v = \sqrt{F/\mu}$, without changing its shape.

Thinking Physics 3

A secret agent is trapped in a building on top of the elevator car at a lower floor. She attempts to signal a fellow agent on the roof by tapping a message on the elevator cable. As the pulses from this tapping move up the cable to the accomplice, does the speed with which they move stay the same, increase, or decrease? If the pulses are sent one second apart, are they received one second apart by her partner?

Reasoning   The elevator cable can be modeled as a heavy string. The speed of waves on the cable is a function of the tension in the cable. As the waves move higher on the cable, they encounter increased tension, because each higher point on the cable must support the weight of all the cable below it (and the elevator). Thus, the speed of the pulses increases as they move higher on the cable. The frequency of the pulses will not be affected, because each pulse takes the same total time to reach the top—they will still arrive at the top of the cable at intervals of one second.

CONCEPTUAL PROBLEM 3

In mechanics, massless strings are often assumed. Why is this not a good assumption when discussing waves on strings?

Example 13.4   The Speed of a Pulse on a Cord

A uniform cord has a mass of 0.300 kg and a total length of 6.00 m. Tension is maintained in the cord by suspending a 2.00-kg mass from one end (Fig. 13.14). Find the speed of a pulse on this cord. Assume the tension is not affected by the mass of the cord.

Solution   The tension F in the cord is equal to the weight of the suspended 2.00-kg mass:

$$F = mg = (2.00 \text{ kg})(9.80 \text{ m/s}^2) = 19.6 \text{ N}$$

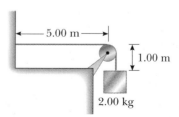

Figure 13.14   (Example 13.4) The tension F in the cord is maintained by the suspended mass. The wave speed is given by the expression $v = \sqrt{F/\mu}$.

(This calculation of the tension neglects the small mass of the cord. Strictly speaking, the cord can never be exactly horizontal, and therefore the tension is not uniform.)

The mass per unit length μ is

$$\mu = \frac{m}{\ell} = \frac{0.300 \text{ kg}}{6.00 \text{ m}} = 0.0500 \text{ kg/m}$$

Therefore, the wave speed is

$$v = \sqrt{\frac{F}{\mu}} = \sqrt{\frac{19.6 \text{ N}}{0.0500 \text{ kg/m}}} = 19.8 \text{ m/s}$$

EXERCISE 5   Show that Equation 13.19 is dimensionally correct using the known dimensions of F and μ.

EXERCISE 6   Transverse waves travel with a speed of 20.0 m/s in a string under a tension of 6.00 N. What tension is required for a wave speed of 30.0 m/s in the same string?
Answer   13.5 N

13.7 • REFLECTION AND TRANSMISSION OF WAVES

Whenever a traveling wave pulse reaches a boundary, part or all of the pulse is *reflected*. Any part not reflected is said to be *transmitted* through the boundary. Consider a pulse traveling on a string that is fixed at one end (Fig. 13.15). When the pulse reaches the fixed boundary, it is reflected. Because the support attaching the string to the wall is rigid, none of the pulse is transmitted through the wall.

Note that the reflected pulse has exactly the same amplitude as the incoming pulse but is inverted. The inversion can be explained as follows. When the pulse meets the end of the string that is fixed at the support, the string produces an upward force on the support. By Newton's third law, the support must exert an equal and opposite reaction force on the string. This downward force causes the pulse to invert on reflection.

Now consider another situation in which there is total reflection and zero transmission. In this case the pulse arrives at the end of a string that is free to move vertically, as in Figure 13.16. The tension at the free end is maintained by tying the string to a ring of negligible mass that is free to slide vertically on a frictionless post. Again, the pulse is reflected, but this time it is not inverted. As the pulse reaches the post, it exerts a force on the free end, causing the ring to accelerate upward. In the process, the ring has upward momentum as it reaches the top of its motion and is then returned to its original position by the downward component of the tension force. This produces a reflected pulse that is not inverted, the amplitude of which is the same as that of the incoming pulse.

Finally, we may have a situation in which the boundary is intermediate between these two extreme cases; that is, it is neither completely rigid nor completely free. In this case, part of the wave is transmitted and part is reflected. For instance, suppose a light string is attached to a heavier string, as in Figure 13.17. When a pulse traveling on the light string reaches the boundary between the two strings, part of the pulse is reflected and inverted and part is transmitted to the heavier string. As one would expect, both the reflected pulse and the transmitted one have a smaller amplitude than the incident pulse. The inversion in the reflected wave is similar to the behavior of a pulse meeting a rigid boundary.

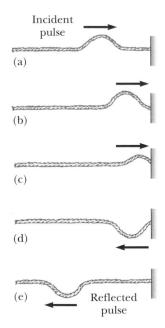

Incident pulse

(a)

(b)

(c)

(d)

(e)   Reflected pulse

Figure 13.15   The reflection of a traveling wave pulse at the fixed end of a stretched string. The reflected pulse is inverted, but its shape remains the same.

When a pulse traveling on a heavy string strikes the boundary of a lighter string, as in Figure 13.18, again part is reflected and part transmitted. This time, however, the reflected pulse is not inverted. In both the lighter-to-heavier case and the heavier-to-lighter case, the relative heights of the reflected and transmitted pulses depend on the relative densities of the two strings. In an extreme case, if the second string were infinitely more dense, it would behave like a rigid wall.

In the preceding section, we found that the speed of a wave on a string increases as the mass per unit length of the string decreases. In other words, a pulse travels more slowly on a heavy string than on a light string if both are under the same tension. The following general rules apply to reflected waves:

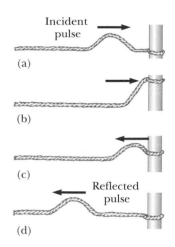

Figure 13.16   The reflection of a traveling wave pulse at the free end of a stretched string. In this case, the reflected pulse is not inverted.

> When a wave pulse travels from medium A to medium B and $v_A > v_B$ (that is, when B is more massive than A), the reflected part of the pulse is inverted on reflection. When a wave pulse travels from medium A to medium B and $v_A < v_B$ (A is more massive than B), the reflected pulse is not inverted.

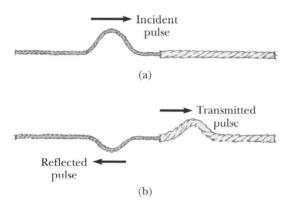

Figure 13.17   (a) A pulse traveling to the right on a light string attached to a heavier string. (b) Part of the incident pulse is reflected (and inverted), and part is transmitted to the heavier string. (Note that the change in pulse width is not shown.)

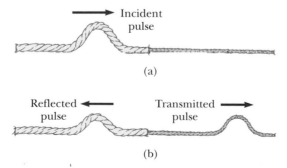

Figure 13.18   (a) A pulse traveling to the right on a heavy string attached to a lighter string. (b) The incident pulse is partially reflected and partially transmitted. In this case, the reflected pulse is not inverted. (Note that the change in pulse width is not shown.)

13.8 • ENERGY TRANSMITTED BY SINUSOIDAL WAVES ON STRINGS

As waves propagate through a medium, they transport energy. This is easily demonstrated by hanging a mass on a stretched string and then sending a pulse down the string, as in Figure 13.19. When the pulse meets the suspended mass, the mass is momentarily displaced, as in Figure 13.19b. In the process, energy is transferred to the mass, because work must be done in moving it upward.

This section examines the rate at which energy is transported along a string. We shall assume that this one-dimensional wave is sinusoidal when we calculate the power transferred. Later we shall extend the same ideas to three-dimensional waves.

Consider a sinusoidal wave traveling on a string (Fig. 13.20). The source of the energy is some external agent at the left end of the string, which does work in producing the oscillations. Let us focus our attention on an element of the string of length Δx and mass Δm. Each such segment moves vertically with simple harmonic motion. Furthermore, each segment has the same angular frequency, ω, and the same amplitude, A. As we found in Chapter 12, the total energy, E, associated with a particle moving with simple harmonic motion is $E = \frac{1}{2}kA^2 = \frac{1}{2}m\omega^2A^2$ (Eq. 12.22), where k is the effective force constant of the restoring force. If we apply this equation to the element of length Δx, we see that the total energy of this element is

$$\Delta E = \tfrac{1}{2}(\Delta m)\omega^2 A^2$$

If μ is the mass per unit length of the string, then the element of length Δx has a mass, Δm, that is equal to $\mu\,\Delta x$. Hence, we can express the energy as

$$\Delta E = \tfrac{1}{2}(\mu\,\Delta x)\omega^2 A^2 \qquad\qquad \textbf{[13.20]}$$

If the wave travels from left to right, as in Figure 13.20, the energy ΔE arises from the work done on the element Δm by the string element to the left of Δm. Similarly, the element Δm does work on the element to its right, so we see that energy is transmitted to the right. The rate at which energy is transmitted along the string, or the power (P), is dE/dt. If we let Δx approach 0, Equation 13.20 gives

$$P = \frac{dE}{dt} = \tfrac{1}{2}\left(\mu\,\frac{dx}{dt}\right)\omega^2 A^2$$

Figure 13.19   (a) A pulse traveling to the right on a stretched string on which a mass has been suspended. (b) Energy is transmitted to the suspended mass when the pulse arrives.

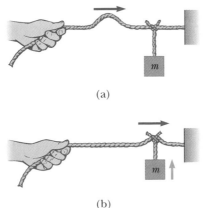

(a)

(b)

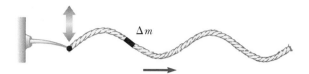

Figure 13.20   A sinusoidal wave traveling along the *x* axis on a stretched string. Every segment moves vertically, and each segment has the same total energy. The power transmitted by the wave equals the energy contained in one wavelength divided by the period of the wave.

Because dx/dt is equal to the wave speed, v, we have

$$P = \tfrac{1}{2}\mu\omega^2 A^2 v \qquad\qquad\qquad \textbf{[13.21]} \qquad \bullet \ \textit{\textbf{Power}}$$

This shows that the power transmitted by a sinusoidal wave on a string is proportional to (1) the wave speed, (2) the square of the frequency, and (3) the square of the amplitude. In fact, *all* sinusoidal waves have the following general property: **The power transmitted by any sinusoidal wave is proportional to the square of the angular frequency and to the square of the amplitude.**

Thus, we see that a wave traveling through a medium corresponds to energy transport through the medium, with no net transfer of matter. An oscillating source provides the energy and produces a harmonic disturbance of the medium. The disturbance is able to propagate through the medium as the result of the interaction between adjacent particles. In order to verify Equation 13.21 by direct experiment, one would have to design some device at the far end of the string to extract the energy of the wave without producing any reflections.

Thinking Physics 4

We have shown that waves carry energy—not only waves on strings, but all waves. Devise a demonstration that shows that waves transfer momentum as well as energy.

Reasoning   Imagine that we cover the ends of a cardboard tube with rubber sheets, secured with rubber bands, to form a crude drum with two heads. We now lay the tube horizontally on the table and suspend a Ping-Pong ball from a string so that the Ping-Pong ball is just touching one of the rubber sheets. If we now thump the other rubber sheet with a finger, the sound will travel down the tube, strike the first rubber sheet, and the Ping-Pong ball will bounce away from the rubber sheet. This demonstrates that the sound wave is carrying momentum to the Ping-Pong ball.

Example 13.5   Power Supplied to a Vibrating String

A string of mass per unit length $\mu = 5.00 \times 10^{-2}$ kg/m is under a tension of 80.0 N. How much power must be supplied to the rope to generate sinusoidal waves at a frequency of 60.0 Hz and an amplitude of 6.00 cm?

Solution   The wave speed on the string is

$$v = \sqrt{\frac{F}{\mu}} = \left(\frac{80.0 \text{ N}}{5.00 \times 10^{-2} \text{ kg/m}}\right)^{1/2} = 40.0 \text{ m/s}$$

Because $f = 60.0$ Hz, the angular frequency ω of the sinusoidal waves on the string has the value

$$\omega = 2\pi f = 2\pi(60.0 \text{ Hz}) = 377 \text{ s}^{-1}$$

Using these values in Equation 13.21 for the power, with $A = 6.00 \times 10^{-2}$ m, gives

$$\text{Power} = \tfrac{1}{2}\mu\omega^2 A^2 v$$

$$= \tfrac{1}{2}(5.00 \times 10^{-2} \text{ kg/m})(377 \text{ s}^{-1})^2$$

$$\times (6.00 \times 10^{-2} \text{ m})^2(40.0 \text{ m/s})$$

$$= 512 \text{ W}$$

13.9 • SOUND WAVES

Let us turn our attention from transverse waves to longitudinal ones. As stated in Section 13.2, longitudinal waves are waves in which the particles of the medium undergo displacements parallel to the direction of wave motion. Sound waves in air are the most important example of longitudinal waves. They can travel through any material medium, their speed depending on the properties of that medium.

The displacements accompanying a sound wave in air are longitudinal displacements of individual molecules from their equilibrium positions. Such displacements result if the source of the waves, such as the diaphragm of a loudspeaker, oscillates with simple harmonic motion. For instance, you can produce a one-dimensional harmonic sound wave in a long, narrow tube containing a gas by means of a vibrating piston at one end, as in Figure 13.21. The darker regions in the figure represent regions in which the gas is compressed and, consequently, the density and pressure are *above* their equilibrium values. Such a compressed layer of gas, called a **condensation,** is formed when the piston is being pushed into the tube. The condensation moves down the tube as a pulse, continuously compressing the layers in front of it. When the piston is withdrawn from the tube, the gas in front of it expands and consequently the pressure and density in this region fall below their equilibrium values. These low-pressure regions, called **rarefactions,** are represented by the lighter areas in Figure 13.21. The rarefactions also propagate along the tube, following the condensations. Both regions move with a speed equal to the speed of sound in that medium. The speed of sound waves in air at 20°C is about 343 m/s.

As the piston oscillates back and forth in a sinusoidal fashion, regions of condensation and rarefaction are continuously set up. The distance between two successive condensations (or two successive rarefactions) equals the wavelength, λ. As these regions travel down the tube, any small volume of the medium moves with simple harmonic motion parallel to the direction of the wave (in other words, longitudinally). If $s(x, t)$ is the displacement of a small volume element measured from its equilibrium position, we can express this displacement function as

$$s(x, t) = s_{max} \cos(kx - \omega t) \qquad [13.22]$$

where s_{max} is the **maximum displacement from equilibrium** (the displacement amplitude), k is the angular wave number, and ω is the angular frequency of the piston. Note that the displacement of the volume element is along x, the direction of motion of the sound wave, which of course means we are describing a longitudinal wave. The variation in the pressure of the gas, ΔP, measured from its equilibrium value is also sinusoidal; it is given by

$$\Delta P = \Delta P_{max} \sin(kx - \omega t) \qquad [13.23]$$

The **pressure amplitude,** ΔP_{max}, is the **maximum change in pressure from the equilibrium value.** It is proportional to the displacement amplitude, s_{max}:

Pressure amplitude •

$$\Delta P_{max} = \rho v \omega s_{max} \qquad [13.24]$$

where ρ is the density of the medium, v is the wave speed, and ωs_{max} is the maximum longitudinal speed of the medium in front of the piston.

Thus, we see that a sound wave may be considered as either a displacement wave or a pressure wave. A comparison of Equations 13.22 and 13.23 shows that **the**

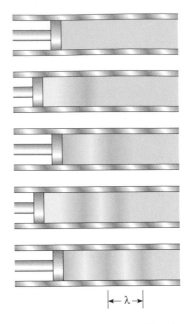

Figure 13.21 A sinusoidal longitudinal wave propagating down a tube filled with a compressible gas. The source of the wave is a vibrating piston at the left. The high- and low-pressure regions are dark and light, respectively.

pressure wave is 90° out of phase with the displacement wave. Graphs of these functions are shown in Figure 13.22. Note that the pressure variation is a maximum when the displacement is zero, whereas the displacement is a maximum when the pressure variation is zero. Because the pressure is proportional to the density of the medium, the expression that describes how that density varies from its equilibrium value is similar to Equation 13.23.

Thinking Physics 5

Garth Brooks had a hit song with "The Thunder Rolls" (Caged Panther Music, Inc., 1990). Why *does* thunder produce an extended "rolling" sound? And how does lightning produce thunder in the first place?

Reasoning Let us assume we are at ground level and neglect ground reflections. When lightning strikes, a channel of ionized air carries a very large electric current from a cloud to the ground. This results in a very rapid temperature increase of this channel of air as the current flows through it. The temperature increase causes a sudden expansion of the air. This expansion is so sudden and so intense that a tremendous disturbance is produced in the air—thunder. The thunder rolls due to the fact that the lightning channel is a long extended source—the entire length of the channel produces the sound at essentially the same instant of time. Sound produced at the bottom of the channel reaches you first, if you are on the ground, because that is the point closest to you, and then sounds from progressively higher portions of the channel reach you. If the lightning channel were a perfectly straight line, the resulting sound might be a steady roar, but the zig-zagged shape of the path results in the rolling variation in loudness, with sound from some portions of the channel arriving at your location simultaneously. This results in periods of loud sounds, interspersed with other instants when the net sound reaching you is low in intensity.

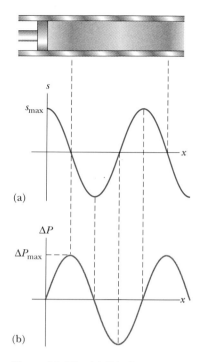

(a)

(b)

Figure 13.22 (a) Displacement amplitude versus position and (b) pressure amplitude versus position for a sinusoidal longitudinal wave. The displacement wave is 90° out of phase with the pressure wave.

13.10 · THE DOPPLER EFFECT

When a vehicle sounds its horn as it travels along a highway, the frequency of the sound you hear is higher as the vehicle approaches you than it is as the vehicle moves away from you. This is one example of the **Doppler effect,** named after the Austrian physicist Christian Johann Doppler (1803–1853).

> In general, a Doppler effect is experienced whenever there is relative motion between the source of sound and the observer. When the source and observer are moving toward each other, the observer hears a frequency that is higher than the true frequency of the source. When the source and observer are moving away from each other, the observer hears a frequency that is lower than the true frequency of the source.[4]

[4]Although the Doppler effect is most commonly experienced with sound waves, it is a phenomenon common to all harmonic waves. For example, a shift in frequencies of light waves (electromagnetic waves) is produced by the relative motion of source and observer. The relativistic Doppler effect demands that there can be no distinction between observer moving toward source and source moving toward observer. In 1842 Doppler first reported the frequency shift in connection with light emitted by stars revolving about each other in double-star systems.

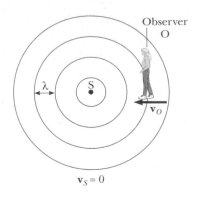

Figure 13.23   An observer O moving with a speed v_O toward a stationary point source S hears a frequency f' that is greater than the source frequency.

"I love hearing that lonesome wail of the train whistle as the magnitude of the frequency of the wave changes due to the Doppler effect."

Frequency heard with •
observer and source in motion

The Doppler effect for electromagnetic waves is used in police radar systems to measure the speeds of motor vehicles. Likewise, astronomers use the effect to determine the relative motions of stars, galaxies, and other celestial objects.

First, let us consider the case where the observer O is moving with a speed of v_O and the sound source S is stationary, as in Figure 13.23. For simplicity, we shall assume that the air is stationary and that the observer moves directly toward the source (considered as a point source). We shall take the frequency of the source to be f, the wavelength to be λ, and the speed of sound to be v. A stationary observer would detect f vibrations each second, where $f = v/\lambda$. (That is, when the source and observer are both at rest, the observed frequency must equal the true frequency of the source.) However, the observer moving toward the source with the speed v_O receives an additional number of vibrations per second, equal to v_O/λ. Hence, the frequency the observer hears, f', is *increased*: $f' = v/\lambda + v_O/\lambda$. Because $\lambda = v/f$, we can express f' as

$$f' = f\left(\frac{v + v_O}{v}\right) \qquad \text{(Observer moving toward source)} \qquad \textbf{[13.25]}$$

Now consider the situation in which the source moves with a speed of v_S relative to the medium, and the observer is at rest. If the source moves directly toward observer A in Figure 13.24a, the wave fronts seen by the observer are closer to each other than they would be if the source were at rest. As a result, the wavelength λ' measured by observer A is shorter than the true wavelength λ of the source. During each vibration, with a duration of T (the period), the source moves a distance of $v_S T = v_S/f$. Therefore, the wavelength is *shortened* by this amount, and the observed wavelength has the shorter value $\lambda' = \lambda - v_S/f$. Because $\lambda = v/f$, the frequency heard by observer A is

$$f' = \frac{v}{\lambda'} = f\left(\frac{v}{v - v_S}\right) \qquad \text{(source moving toward observer)} \qquad \textbf{[13.26]}$$

That is, the frequency is *increased* when the source moves toward the observer. In a similar manner, if the source moves away from observer B at rest, the sign of v_S is reversed in Equation 13.26.

Finally, if both the source and the observer are in motion, one finds the following general formula for the observed frequency:

$$f' = f\left(\frac{v \pm v_O}{v \mp v_S}\right) \qquad \textbf{[13.27]}$$

In this expression, the upper signs ($+ v_O$ and $- v_S$) refer to motion of the one *toward* the other, and the lower signs ($- v_O$ and $+ v_S$) refer to motion of one *away from* the other.

When working with any Doppler effect problem, a convenient rule to remember concerning signs is the following: The word *toward* is associated with an *increase* in the observed frequency. The words *away from* are associated with a *decrease* in the observed frequency.

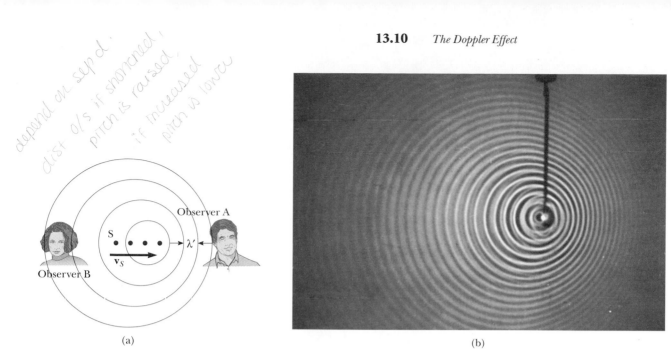

*depend on sep'd.
dist o/s if shortened
pitch is raised
if increased
pitch is lower*

(a)

(b)

Figure 13.24   (a) A source S moving with a speed v_S toward a stationary observer A and away from a stationary observer B. Observer A hears an increased frequency, and observer B hears a decreased frequency. (b) The Doppler effect in water observed in a ripple tank   *(Courtesy Educational Development Center, Newton, Mass.)*

Thinking Physics 6

Suppose you place your stereo speakers far apart and run past them from right to left, or left to right. If you run rapidly enough and have excellent pitch discrimination, you may notice that the music that is playing seems to be out of tune when you are between the speakers. Why?

Reasoning   When you are between the speakers, you are running away from one of them and toward the other. Thus, there is a Doppler shift downward for the sound from the speaker behind you and a Doppler shift upward for the sound from the speaker ahead of you. As a result, the sound from the two speakers will not be in tune. A calculation shows that a world-class sprint runner could run fast enough to generate about a semitone difference in the sound from the two speakers.

CONCEPTUAL PROBLEM 4

Suppose the wind blows. Does this cause a Doppler effect for an observer listening to a sound moving through the air? Is it like a moving source or a moving observer?

CONCEPTUAL PROBLEM 5

You are driving toward a cliff and you honk your horn. Is there a Doppler shift of the sound when you hear the echo? Is it like a moving source or moving observer? What if the reflection occurs not from a cliff but from the forward edge of a huge alien spacecraft which is moving toward you as you drive?

Example 13.6   The Noisy Siren

An ambulance travels down a highway at a speed of 33.5 m/s (75 mi/h). Its siren emits sound at a frequency of 400 Hz. What is the frequency heard by a passenger in a car traveling at 24.6 m/s (55 mi/h) in the opposite direction as the car approaches the ambulance and as the car moves away from the ambulance?

Solution   Let us take the velocity of sound in air to be $v = 343$ m/s. We can use Equation 13.27 in both cases. As the ambulance and car approach each other, the observed apparent frequency is

$$f' = f\left(\frac{v + v_O}{v - v_S}\right) = (400 \text{ Hz})\left(\frac{343 \text{ m/s} + 24.6 \text{ m/s}}{343 \text{ m/s} - 33.5 \text{ m/s}}\right)$$

$$= 475 \text{ Hz}$$

Likewise, as they recede from each other, a passenger in the car hears a frequency

$$f' = f\left(\frac{v - v_O}{v + v_S}\right) = (400 \text{ Hz})\left(\frac{343 \text{ m/s} - 24.6 \text{ m/s}}{343 \text{ m/s} + 33.5 \text{ m/s}}\right)$$

$$= 338 \text{ Hz}$$

The *change* in frequency as detected by the passenger in the car is $475 - 338 = 137$ Hz, which is more than 30% of the actual frequency emitted.

EXERCISE 7   Suppose that the passenger car is parked on the side of the highway as the ambulance travels down the highway at the speed of 33.5 m/s. What frequency will the passenger in the car hear as the ambulance (a) approaches the parked car and (b) recedes from the parked car?   Answer   (a) 443 Hz   (b) 364 Hz

EXERCISE 8   A band is playing on a moving truck. The band strikes the note middle C (262 Hz), but it is heard by spectators ahead of the truck as C# (277 Hz). How fast is the truck moving?   Answer   18.6 m/s

SUMMARY

A **transverse wave** is a wave in which the particles of the medium move in a direction perpendicular to the direction of the wave velocity. An example is a wave on a stretched string.

   Longitudinal waves are waves in which the particles of the medium move parallel to the direction of the wave speed. Sound waves in air are longitudinal.

   Any one-dimensional wave traveling with a speed of v in the positive x direction can be represented by a wave function of the form $y = f(x - vt)$. Likewise, the wave function for a wave traveling in the negative x direction has the form $y = f(x + vt)$. The shape of the wave at any instant (a snapshot of the wave) is obtained by holding t constant.

   The **superposition principle** says that when two or more waves move through a medium, the resultant wave function equals the algebraic sum of the individual wave functions. Waves that obey this principle are said to be *linear*. When two waves combine in space, they interfere to produce a resultant wave.

   The wave function for a one-dimensional harmonic wave traveling to the right can be expressed as

$$y = A \sin\left[\frac{2\pi}{\lambda}(x - vt)\right] = A \sin(kx - \omega t) \qquad \text{[13.6, 13.12]}$$

where A is the amplitude, λ is the wavelength, k is the angular wave number, and ω is the angular frequency. If T is the period and f is the frequency, then v, k, and ω can be written

$$v = \frac{\lambda}{T} = f\lambda \qquad\qquad \text{[13.7, 13.14]}$$

$$k \equiv \frac{2\pi}{\lambda} \qquad\qquad \text{[13.10]}$$

$$\omega \equiv \frac{2\pi}{T} = 2\pi f \qquad\qquad \text{[13.11]}$$

[handwritten: $\lambda = \sigma T$]

[handwritten: $v = \frac{\omega}{k}$]

[handwritten: $v = f\lambda$]

[handwritten: Traveling Sinusoidal Waves]

The speed of a wave traveling on a stretched string of mass per unit length μ and tension F is

$$v = \sqrt{\frac{F}{\mu}} \qquad\qquad \text{[13.19]}$$

When a pulse traveling on a string meets a fixed end, the pulse is reflected and inverted. If the pulse reaches a free end, it is reflected but not inverted.

The **power** transmitted by a harmonic wave on a stretched string is

$$P = \tfrac{1}{2}\mu\omega^2 A^2 v \qquad\qquad \text{[13.21]}$$

[handwritten: transverse speed, v_y
$v_y = \frac{dy}{dt} = -\omega A \cos(kx - \omega t)$
transverse acceleration, a_y
$a_y = \frac{da_y}{dt} = -\omega^2 A \sin(kx - \omega t)$
antinodes spaced by $\frac{\lambda}{2}$ so are nodes]

The change in frequency heard by an observer whenever there is relative motion between a wave source and the observer is called the **Doppler effect.** When the source and observer are moving toward each other, the observer hears a frequency that is higher than the true frequency of the source. When the source and observer are moving away from each other, the observer hears a frequency that is lower than the true frequency of the source.

CONCEPTUAL QUESTIONS

1. How would you set up a longitudinal wave in a stretched spring? Would it be possible to set up a transverse wave in a spring?

2. By what factor would you have to increase the tension in a taut string in order to double the wave speed?

3. When traveling on a taut string, does a wave pulse always invert on reflection? Explain.

4. Does the vertical speed of a segment of a horizontal taut string through which a wave is traveling depend on the wave speed?

5. A vibrating source generates a sinusoidal wave on a string under constant tension. If the power delivered to the string is doubled, by what factor does the amplitude change? Does the wave speed change under these circumstances?

6. Consider a wave traveling on a taut rope. What is the difference, if any, between the speed of the wave and the speed of a small section of the rope?

7. What happens to the wavelength of a wave on a string when the frequency is doubled? What happens to its speed? Assume the tension in the string remains the same.

8. When all the strings on a guitar are stretched to the same tension, will the speed of a wave along the more massive bass strings be faster or slower than the speed of a wave on the lighter strings?

9. If you stretch a rubber hose and pluck it, you can observe a pulse traveling up and down the hose. What happens to the speed if you stretch the hose tighter? If you fill the hose with water?

10. In a longitudinal wave in a spring, the coils move back and forth in the direction of wave motion. Does the speed of the wave depend on the maximum speed of each coil?

11. When two waves interfere, can the amplitude of the resultant wave be larger than either of the two original waves? Under what conditions?

12. A solid can transport both longitudinal waves and transverse waves, but a fluid can transport only longitudinal waves. Why?

13. In an earthquake both S (transverse) and P (longitudinal) waves are sent out. The S waves travel through the Earth more slowly than the P waves (4.5 km/s versus 7.8 km/s). By detecting the time of arrival of the waves, how can one determine how far away the epicenter of the quake was? How many detection centers are necessary to pinpoint the location of the epicenter? *[handwritten: (3)]*

[handwritten: 2-D problem]

PROBLEMS

Section 13.3   One-Dimensional Transverse Traveling Waves

1. At $t = 0$, a transverse wave pulse in a wire is described by the function

$$y = \frac{6}{x^2 + 3}$$

where x and y are in meters. Write the function $y(x, t)$ that describes this wave if it is traveling in the positive x direction with a speed of 4.50 m/s.

2. Ocean waves with a crest-to-crest distance of 10.0 m can be described by

$$y(x, t) = (0.800 \text{ m})\sin [0.628(x - vt)]$$

where $v = 1.20$ m/s. (a) Sketch $y(x, t)$ at $t = 0$. (b) Sketch $y(x, t)$ at $t = 2.00$ s. Note how the entire waveform has shifted 2.40 m in the positive x direction in this time interval.

3. A wave moving along the x axis is described by

$$y(x, t) = 5.0e^{-(x+5.0t)^2}$$

where x is in meters and t is in seconds. Determine (a) the direction of the wave motion and (b) the speed of the wave.

4. Two wave pulses A and B are moving in opposite directions along a taut string with a speed of 2.00 cm/s. The amplitude of A is twice the amplitude of B. The pulses are shown in Figure P13.4 at $t = 0$. Sketch the shape of the string at $t = 1, 1.5, 2, 2.5,$ and 3 s.

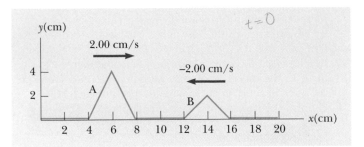

Figure P13.4

Section 13.4   Sinusoidal Traveling Waves

5. A sinusoidal wave is traveling along a rope. The oscillator that generates the wave completes 40.0 vibrations in 30.0 s. Also, a given peak travels 425 cm along the rope in 10.0 s. What is the wavelength?

6. A sinusoidal wave of wavelength 2.00 m and amplitude 0.100 m travels with a speed of 1.00 m/s on a string. Initially, the left end of the string is at the origin and the wave moves from left to right. Find (a) the frequency and angular frequency, (b) the angular wave number, and (c) the wave function for this wave. Determine the equation of motion for (d) the left end of the string, and (e) the point on the string at $x = 1.50$ m to the right of the left end. (f) What is the maximum speed of any point on the string?

7. (a) Write the expression for the function $y(x, t)$ for a sinusoidal wave traveling along a rope in the *negative x* direction with the following characteristics: $A = 8.00$ cm, $\lambda = 80.0$ cm, $f = 3.00$ Hz, and $y(0, 0) = 0$. (b) Write the expression for y as a function of x and t for the wave in part (a) assuming that $y(10 \text{ cm}, 0) = 0$.

8. A transverse wave on a string is described by

$$y = (0.120 \text{ m})\sin \pi(x/8 + 4t)$$

(a) Determine the transverse speed and acceleration of the string at $t = 0.200$ s for the point on the string located at $x = 1.60$ m. (b) What are the wavelength, period, and speed of propagation of this wave?

9. A wave is described by $y = (2.00 \text{ cm}) \sin(kx - \omega t)$, where $k = 2.11$ rad/m, $\omega = 3.62$ rad/s, x is in meters, and t is in seconds. Determine the amplitude, wavelength, frequency, and speed of the wave.

10. A sinusoidal wave traveling in the $-x$ direction (to the left) has an amplitude of 20.0 cm, a wavelength of 35.0 cm, and a frequency of 12.0 Hz. The displacement of the wave at $t = 0, x = 0$ is $y = -3.00$ cm and has a positive velocity here. (a) Sketch the wave at $t = 0$. (b) Find the angular wave number, period, angular frequency, and phase velocity of the wave. (c) Write an expression for the wave function $y(x, t)$.

11. A transverse sinusoidal wave on a string has a period $T = 25.0$ ms and travels in the negative x direction with a speed of 30.0 m/s. At $t = 0$, a particle on the string at $x = 0$ has a displacement of 2.00 cm and is moving in the negative y direction with a speed of 2.00 m/s. (a) What is the amplitude of the wave? (b) What is the initial phase angle? (c) What is the maximum transverse speed of the string? (d) Write the wave function for the wave.

Section 13.5   Superposition and Interference of Waves

12. Two pulses traveling on the same string are described by

$$y_1 = \frac{5}{(3x - 4t)^2 + 2} \quad \text{and} \quad y_2 = \frac{-5}{(3x + 4t - 6)^2 + 2}$$

(a) In which direction does each pulse travel? (b) At what time do the two cancel? (c) At what point do the two waves always cancel?

13. Two sinusoidal waves in a string are defined by the functions

$$y_1 = (2.00 \text{ cm})\sin(20.0x - 30.0t)$$

$$y_2 = (2.00 \text{ cm})\sin(25.0x - 40.0t)$$

where y and x are in centimeters and t is in seconds. (a) What is the phase difference between these two waves at the point $x = 5.00$ cm at $t = 2.00$ s? (b) What is the positive x value closest to the origin for which the two phases differ by π at $t = 2.00$ s? (This is where the two waves add to zero.)

Section 13.6 The Speed of Transverse Waves on Strings

14. A piano string of mass-per-length 5.00×10^{-3} kg/m is under a tension of 1350 N. Find the speed with which a wave travels on this string.

15. A phone cord is 4.00 m long. The cord has a mass of 0.200 kg. If a transverse wave pulse travels from the receiver to the phone box in 0.100 s, what is the tension in the cord?

16. A 30.0-m steel wire and a 20.0-m copper wire, both with 1.00-mm diameters, are connected end to end and stretched to a tension of 150 N. How long does it take a transverse wave to travel the entire length of the two wires?

17. Transverse pulses travel with a speed of 200 m/s along a taut copper wire the diameter of which is 1.50 mm. What is the tension in the wire? (The density of copper is 8.92 g/cm³.)

18. A light string of mass-per-unit length 8.00 g/m has its ends tied to two walls separated by a distance equal to three fourths the length of the string (Fig. P13.18). A mass m is suspended from the center of the string, putting a tension in the string. (a) Find an expression for the transverse wave speed in the string as a function of the hanging mass.

$$a = \quad v = \sqrt{mg/2\mu\cos2}$$

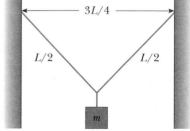

Figure P13.18

(b) How much mass should be suspended from the string to have a wave speed of 60.0 m/s? $= 3.89$ kg

19. An astronaut on the Moon wishes to measure the local value of the free-fall acceleration g_{Moon} by timing pulses traveling down a wire that has a large mass suspended from it. Assume a wire of mass 4.00 g is 1.60 m long and has a 3.00 kg mass suspended from it. A pulse requires 36.1 ms to traverse the length of the wire. Calculate g_{Moon} from these data. (You may neglect the mass of the wire when calculating the tension in it.)

Section 13.8 Energy Transmitted by Sinusoidal Waves on Strings

20. A taut rope has a mass of 0.180 kg and a length of 3.60 m. What power must be supplied to the rope in order to generate sinusoidal waves having an amplitude of 0.100 m and a wavelength of 0.500 m and traveling with a speed of 30.0 m/s? $= 1070$ N

21. Sinusoidal waves 5.00 cm in amplitude are to be transmitted along a string of linear density 4.00×10^{-2} kg/m. If the maximum power delivered by the source is 300 W and the string is under a tension of 100 N, what is the highest vibrational frequency at which the source can operate?

22. It is found that a 6.00-m segment of a long string contains four complete waves and has a mass of 180 g. The string is vibrating sinusoidally with a frequency of 50.0 Hz and a peak-to-valley displacement of 15.0 cm. (*Peak-to-valley* means the vertical distance from the farthest positive displacement to the farthest negative displacement.) (a) Write down the function that describes this wave traveling in the positive x direction. (b) Determine the power being supplied to the string. $y = .075 \sin(4.19 \, x - 314t)$
$P = 1624$ W

23. A horizontal string can transmit a maximum power of P_0 (without breaking) if a wave with amplitude A and angular frequency ω is traveling along it. In order to increase this maximum power, a student folds the string and uses this double string as a transmitter. Determine the maximum power that can be transmitted along the double string, assuming that the tension remains the same.

24. A two-dimensional water wave spreads in circular wave fronts. Show that the amplitude A at a distance r from the initial disturbance is proportional to $1/\sqrt{r}$. (*Hint:* Consider the energy concentrated in the outward-moving ripple.)

Section 13.9 Sound Waves

(In this section, use the following values as needed unless otherwise specified: the equilibrium density of air, $\rho = 1.20$ kg/m³; the speed of sound in air, $v = 343$ m/s. Also, pressure variations, ΔP, are measured relative to atmospheric pressure.)

25. A flowerpot is knocked off a balcony 20.0 m above the sidewalk heading for a man of height 1.75 m standing below. How high from the ground can the flowerpot be after which it would be too late for a shouted warning to reach the man in time? Assume that the man below requires 0.300 s to respond to the warning.

26. A rescue plane flies horizontally at a constant speed searching for a disabled boat. When the plane is directly above the boat, the boat's crew blows a loud horn. By the time the plane's sound detector perceives the horn's sound, the plane has traveled a distance equal to one-half its altitude above the ocean. If it takes the sound 2.00 s to reach the plane, determine (a) the speed of the plane, and (b) its altitude. Take the speed of sound to be 343 m/s.

27. A sound wave in a cylinder is described by Equations 13.22 through 13.24. Show that $\Delta P = \pm \rho v \omega \sqrt{s_{max}^2 - s^2}$.

28. Calculate the pressure amplitude of a 2.00-kHz sound wave in air if the displacement amplitude is 2.00×10^{-8} m.

29. An experimenter wishes to generate in air a sound wave that has a displacement amplitude of 5.50×10^{-6} m. The pressure amplitude is to be limited to 8.40×10^{-1} N/m². What is the minimum wavelength the sound wave can have? $\lambda = 5.81\,m$

30. A sound wave in air has a pressure amplitude of 4.00 N/m² and a frequency of 5.00 kHz. Take $\Delta P = 0$ at the point $x = 0$ when $t = 0$. (a) What is ΔP at $x = 0$ when $t = 2.00 \times 10^{-4}$ s, and (b) what is ΔP at $x = 0.0200$ m when $t = 0$?

a.) $0 N/m^2$, 2π
b.) $3.84\ N/m^2$

Section 13.10   The Doppler Effect

31. A commuter train passes a passenger platform at a constant speed of 40.0 m/s. The train horn is sounded at its characteristic frequency of 320 Hz. (a) What total change in frequency is detected by a person on the platform as the train goes from approaching to receding? (b) What wavelength is detected by a person on the platform as the train approaches?

32. A driver traveling northbound on a highway is driving at a speed of 25.0 m/s. A police car driving southbound at a speed of 40.0 m/s approaches with its siren sounding at a base frequency of 2500 Hz. (a) What frequency is observed by the driver as the police car approaches? (b) What frequency is detected by the driver after the police car passes him? (c) Repeat parts (a) and (b) for the case when the police car is traveling northbound.

33. Standing at a crosswalk, you hear a frequency of 560 Hz from the siren of an approaching police car. After the police car passes, the observed frequency of the siren is 480 Hz. Determine the car's speed from these observations.

34. A tuning fork vibrating at 512 Hz falls from rest and accelerates at 9.80 m/s². How far below the point of release is

the tuning fork when waves of frequency 485 Hz reach the release point? Take the speed of sound in air to be 340 m/s.

35. A block with a speaker bolted to it is connected to a spring having spring constant $k = 20.0$ N/m as in Figure P13.35. The total mass of the block and speaker is 5.00 kg, and the amplitude of this unit's motion is 0.500 m. If the speaker emits sound waves of frequency 440 Hz, determine the range in frequencies heard by the person to the right of the speaker.

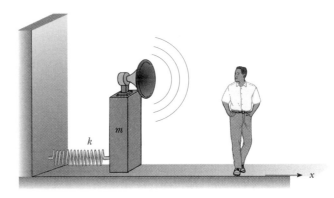

Figure P13.35

Additional Problems

36. Unoccupied by spectators, a large set of football bleachers has solid seats and risers. You stand on the field in front of it and sharply clap two wooden boards together once. You hear from the bleachers a sound with definite pitch, which may remind you of a short toot on a trumpet. Account for this sound. Compute order-of-magnitude estimates for its frequency, wavelength, and duration on the basis of data you specify.

37. A 2.00-kg block hangs from a rubber string, being supported so that the string is not stretched. The unstretched length of the string is 0.500 m and its mass is 5.00 g. The spring constant for the string is 100 N/m. The block is released and stops at the lowest point. (a) Determine the tension in the string when the block is at this lowest point. (b) What is the length of the string in this "stretched" position? (c) Find the speed of a transverse wave in the string if the block is held in this lowest position.

38. A block of mass M hangs from a rubber string, being supported so that the string is not stretched. The unstretched length of the string is L_0 and its mass is m, much less than M. The spring constant for the string is k. The block is released and stops at the lowest point. (a) Determine the

tension in the string when the block is at this lowest point. (b) What is the length of the string in this "stretched" position? (c) Find the speed of a transverse wave in the string if the block is held in this lowest position.

39. The wave function for a traveling wave on a taut string is (in SI units)

$$y(x, t) = (0.350 \text{ m})\sin(10\pi t - 3\pi x + \pi/4)$$

(a) What is the speed and direction of travel of the wave? (b) What is the vertical displacement of the string at $t = 0$, $x = 0.100$ m? (c) What are the wavelength and frequency of the wave? (d) What is the maximum magnitude of the transverse speed of the string?

40. A stone is dropped into a deep canyon and is heard to strike the bottom 10.2 s after release. The speed of sound waves in air is 343 m/s. How deep is the canyon? What would be the percentage error in the depth if the time required for the sound to reach the canyon rim were ignored?

41. The ocean floor is underlain by a layer of basalt that constitutes the crust, or uppermost layer of the Earth, in this region. Below this crust is found denser peridotite rock, which forms the Earth's mantle. The boundary between these two layers is called the *Mohorovičić discontinuity* ("Moho" for short). If an explosive charge is set off at the surface of the basalt, it generates a seismic wave that is reflected back at the Moho. If the speed of this wave in basalt is 6.50 km/s, and the two-way travel time is 1.85 s, what is the thickness of this oceanic crust?

42. A worker strikes a steel pipeline with a hammer, generating both longitudinal and transverse waves. Reflected waves return 2.40 s apart. How far away is the reflection point? (For steel, $v_{\text{long}} = 6.20$ km/s and $v_{\text{trans}} = 3.20$ km/s.)

43. A rope of total mass m and length L is suspended vertically. Show that a transverse wave pulse will travel the length of the rope in a time $t = 2\sqrt{L/g}$. (*Hint:* First find an expression for the wave speed at any point a distance x from the lower end by considering the tension in the rope as resulting from the weight of the segment below that point.)

44. A train whistle ($f = 400$ Hz) sounds higher or lower in pitch depending on whether it approaches or recedes. (a) Prove that the difference in frequency between the approaching and receding train whistle is

$$\Delta f = \frac{2f\left(\dfrac{u}{v}\right)}{1 - \dfrac{u^2}{v^2}} \qquad \begin{array}{l} u = \text{speed of train} \\ v = \text{speed of sound} \end{array}$$

(b) Calculate this difference for a train moving at a speed of 130 km/h. Take the speed of sound in air to be 340 m/s.

45. In order to be able to determine her speed, a skydiver carries a tone generator. A friend on the ground at the landing site has equipment for receiving and analyzing sound waves. While the skydiver is falling at terminal speed, her tone generator emits a steady tone of 1800 Hz. (Assume that the air is calm and that the sound speed is 343 m/s, independent of altitude.) (a) If her friend on the ground (directly beneath the skydiver) receives waves of frequency 2150 Hz, what is the skydiver's speed of descent? (b) If the skydiver were also carrying sound-receiving equipment sensitive enough to detect waves reflected from the ground, what frequency would she receive?

46. A wave pulse traveling along a string of linear mass density is described by the relationship

$$y = [A_0 e^{-bx}]\sin(kx - \omega t)$$

where the factors in brackets before the sine are said to be the amplitude. (a) What is the power $P(x)$ carried by this wave at a point x? (b) What is the power $P(0)$ carried by this wave at the origin? (c) Compute the ratio $P(x)/P(0)$.

47. A bat, moving at 5.00 m/s, is chasing a flying insect. If the bat emits a 40.0-kHz chirp and receives back an echo at 40.4 kHz, at what speed is the insect moving toward or away from the bat? (Take the speed of sound in air to be $v = 340$ m/s.)

48. An earthquake on the ocean floor in the Gulf of Alaska induces a *tsunami* (sometimes called a *tidal wave*) that reaches Hilo, Hawaii, 4450 km distant, in a time of 9 h 30 min. Tsunamis have enormous wavelengths (100 km to 200 km), and for such waves the propagation speed is $v \approx \sqrt{gd}$, where d is the average depth of the water. From the information given, find the average wave speed and the average ocean depth between Alaska and Hawaii. (This method was used in 1856 to estimate the average depth of the Pacific Ocean long before soundings were made to give a direct determination.)

🖳 Spreadsheet Problem

S1. Two transverse wave pulses traveling in opposite directions along the x axis are represented by the following wave functions:

$$y_1(x, t) = \frac{6}{(x - 3t)^2} \qquad y_2(x, t) = -\frac{3}{(x + 3t)^2}$$

where x and y are measured in centimeters and t is in seconds. Write a spreadsheet or program to add the two pulses and obtain the shape of the composite waveform $y_{\text{tot}} = y_1 + y_2$ as a function of time. Plot y_{tot} versus x. Make separate plots for $t = 0, 0.5, 1, 1.5, 2, 2.5,$ and 3.0 s.

ANSWERS TO CONCEPTUAL PROBLEMS

1. A pulse in a long line of people is longitudinal, since the movement of people is parallel to the direction of propagation of the pulse. The speed is determined by the reaction time of the people and the speed with which they can move, once a space opens up. There is also a psychological factor, in that people will not want to fill a space that opens up in front of them too quickly, so as not to intimidate the person in front of them. The "wave" at a stadium is transverse, since the fans stand up vertically, as the wave sweeps past them horizontally. The speed of this pulse depends on the limits of the fans' abilities to rise and sit rapidly, and on psychological factors associated with the anticipation of seeing the pulse approach the observer's location.

2. Both the radio wave and the sound wave have the same wavelength. The speed of propagation for electromagnetic waves, however, is much higher than the speed of sound, so the frequency of the radio wave is much higher than that of the sound wave. Audible sound frequencies range from about 15 Hz to 15 000 Hz, while radio frequencies are in kilohertz for AM radio and megahertz for FM.

3. A wave on a massless string would have an infinite speed of propagation because its linear mass density is zero.

4. Wind can change a Doppler shift. Both v_O and v_S in our equation must be interpreted as speeds of observer and source relative to air. If source and observer are moving relative to each other, the observer will hear one shifted frequency in still air and a different shifted frequency if wind is blowing. However, if source and observer are at rest relative to each other, there is no shift when wind blows past.

5. The echo *is* Doppler shifted, and the shift is like both a moving source and a moving observer. The sound which leaves your horn in the forward direction is Doppler shifted to a higher frequency, because it is coming from a moving source. As the sound reflects back and comes toward you, you are a moving observer, so there is a second Doppler shift to an even higher frequency. If the sound reflects from the spacecraft moving toward you, there is another moving source shift to an even higher frequency. The reflecting surface of the spacecraft acts as a moving source.

14

Superposition and Standing Waves

An important aspect of waves is their combined effect when two or more travel in the same medium. For instance, what happens to a string when a wave traveling toward a fixed end is reflected back on itself?

In a linear medium—that is, one in which the restoring force of the medium is proportional to the displacement of the medium—one can apply the **principle of superposition** to obtain the resultant disturbance. This principle can be applied to many types of waves under certain conditions, including waves on strings, sound waves, surface water waves, and electromagnetic waves. The superposition principle states that the net displacement of any part of the disturbed medium equals the vector sum of the displacements caused by the individual waves. In Chapter 13, the term **interference** was used to describe the effect produced by combining two waves that are moving simultaneously through a medium.

This chapter is concerned with the superposition principle as it applies to sinusoidal waves. If the sinusoidal waves that com-

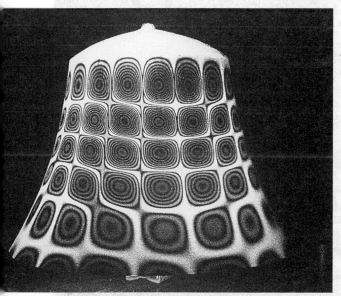

◀ Photograph of standing waves on a vibrating handbell. The photograph was taken using a technique called *time-average holographic interferometry.* Such a hologram consists of millions of superimposed holograms, but this hologram emphasizes the two positions of maximum deflection of the vibrating bell. The pattern recorded here occurred at a frequency of 2684 Hz, but other patterns are observed at lower and higher frequencies.
(Courtesy of Professor Thomas D. Rossing, Northern Illinois University)

bine in a given medium have the same frequency and wavelength, one finds that a stationary pattern, called a **standing wave,** can be produced at certain frequencies under certain circumstances. For example, a stretched string fixed at both ends has a discrete set of standing wave patterns, called **modes of vibration,** that depend on the tension and the mass-per-unit-length of the string. These modes of vibration are found in the strings of stringed musical instruments. Other musical instruments, such as the organ and flute, make use of the natural frequencies of sound waves in hollow pipes. Such frequencies depend on the length and shape of the pipe and on whether the ends are open or closed.

In this chapter we also consider the superposition and interference of waves with different frequencies and wavelengths. When two sound waves with nearly the same frequency interfere, one hears variations in loudness called *beats.* The beat frequency corresponds to the rate of alternation between constructive and destructive interference. Finally, we describe how any complex periodic waveform can, in general, be described by a sum of sine and cosine functions.

14.1 • SUPERPOSITION AND INTERFERENCE OF SINUSOIDAL WAVES

Let us apply the superposition principle to two sinusoidal waves traveling in the same direction in a medium. If the two waves are traveling to the right and have the same frequency, wavelength, and amplitude but differ in phase, we can express their individual wave functions as

$$y_1 = A \sin(kx - \omega t) \qquad \text{and} \qquad y_2 = A \sin(kx - \omega t - \phi)$$

Hence, the resultant wave function, y, is

$$y = y_1 + y_2 = A[\sin(kx - \omega t) + \sin(kx - \omega t - \phi)]$$

In order to simplify this expression, it is convenient to use the trigonometric identity

$$\sin a + \sin b = 2 \cos\left(\frac{a - b}{2}\right) \sin\left(\frac{a + b}{2}\right)$$

If we let $a = kx - \omega t$ and $b = kx - \omega t - \phi$, the resultant wave, y, reduces to

Resultant of two traveling •
sinusoidal waves

$$y = \left(2A \cos\frac{\phi}{2}\right) \sin\left(kx - \omega t - \frac{\phi}{2}\right) \qquad \text{[14.1]}$$

This result has several important features. The resultant wave function, y, is also harmonic and has the *same* frequency and wavelength as the individual waves. The amplitude of the resultant wave is $2A \cos(\phi/2)$ and the phase is $\phi/2$. If the phase constant, ϕ, equals 0, then $\cos(\phi/2) = \cos 0 = 1$ and the amplitude of the resultant wave is $2A$. In other words, the amplitude of the resultant wave is twice the amplitude of either individual wave. In this case, the waves are said to be everywhere *in phase* and to thus **interfere constructively.** That is, the crests and troughs of the individual waves occur at the same positions, as is shown by the blue lines in Figure 14.1a. In general, constructive interference occurs when $\cos(\phi/2) = \pm 1$, or when $\phi = 0, 2\pi, 4\pi, \ldots$. However, if ϕ is equal to π radians or to any *odd* multiple of π then $\cos(\phi/2) = \cos(\pi/2) = 0$ and the resultant wave has *zero* amplitude everywhere. In this case, the two waves **interfere destructively.** That is, the crest of one

Constructive interference •

Destructive interference •

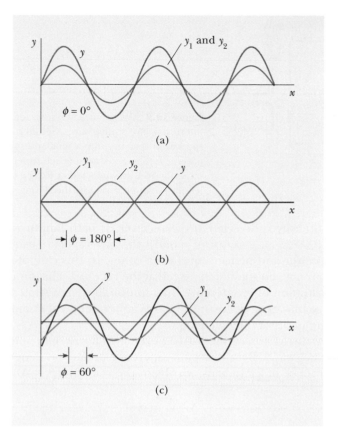

Figure 14.1   The superposition of two waves with amplitudes y_1 and y_2, where $y_1 = y_2$.
(a) When the two waves are in phase, the result is constructive interference. (b) When the two waves are 180° out of phase, the result is destructive interference. (c) When the phase angle lies in the range $0 < \phi < 180°$, the resultant y falls somewhere between that shown in (a) and that shown in (b).

wave coincides with the trough of the second (Fig. 14.1b) and their displacements cancel at every point. Finally, when the phase constant has an arbitrary value between 0 and π, as in Figure 14.1c, the resultant wave has an amplitude the value of which is somewhere between 0 and $2A$.

Interference of Waves

One simple device for demonstrating interference of sound waves is illustrated in Figure 14.2. Sound from speaker S is sent into a tube at P, at which point there is a T-shaped junction. Half the sound power travels in one direction, and half in the opposite direction. Thus, the sound waves that reach receiver R at the other side can travel along either of two paths. The total distance from speaker to receiver is called the **path length,** r. The length of the lower path is fixed at r_1. The upper path length, r_2, can be varied by sliding the U-shaped tube (similar to that on a slide trombone). When the difference in the path lengths, $\Delta r = |r_2 - r_1|$, is either zero or some integral multiple of the wavelength λ, the two waves reaching the receiver are in phase and interfere constructively, as in Figure 14.1a. In this case, a maximum

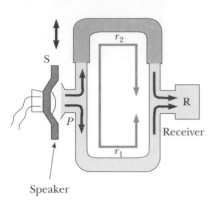

Speaker

Figure 14.2 An acoustical system for demonstrating interference of sound waves. Sound from the speaker propagates into a tube and splits into two parts at *P*. The two waves, which superimpose at the opposite side, are detected at R. The upper path length r_2 can be varied by the sliding section.

in the sound intensity is detected at the receiver. If path length r_2 is adjusted so that Δr is $\lambda/2, 3\lambda/2, \ldots . n\lambda/2$ (for n odd), the two waves are exactly 180° out of phase at the receiver and hence cancel each other. In this case of completely destructive interference, no sound is detected at the receiver. This simple experiment is a striking illustration of interference. In addition, it demonstrates the fact that a phase difference may arise between two waves generated by the same source when they travel along paths of unequal lengths.

It is often useful to express the path difference in terms of the phase difference, ϕ, between the two waves. Because a path difference of one wavelength corresponds to a phase difference of 2π radians, we obtain the ratio $\lambda/2\pi = \Delta r/\phi$, or

Relationship between path •
difference and phase angle

$$\Delta r = \frac{\lambda}{2\pi}\, \phi \qquad\qquad [14.2]$$

Nature provides many other examples of interference phenomena. Later in the book we shall describe several interesting interference effects involving light waves.

Thinking Physics 1

If stereo speakers are connected to the amplifier "out of phase," one speaker is moving outward when the other is moving inward. This results in a weakness in the bass notes, which can be corrected by reversing the wires on one of the speaker connections. Why is there only an effect on the bass notes in this case and not the treble notes?

Reasoning Imagine that you are sitting in front of the speakers, midway between them. Then, the sound from each speaker travels the same distance to you, and there is no phase difference in the sound that is due to a path difference. Because the speakers are connected out of phase, the sound waves are half a wavelength out of phase on leaving the speaker and, consequently, on arriving at your ear. As a result, there will be cancellation of the sound for all frequencies in the ideal case of a zero-size head located exactly on the midpoint between the speakers. If the ideal head were moved off the centerline, there will be an additional phase difference introduced by the path-length difference for the sound from the two speakers. In the case of low frequency, long wavelength bass notes, the path-length differences will be a small fraction of a wavelength, so there will still be significant cancellation. For the high-frequency, short-wavelength treble notes, a small movement of the ideal head will result in a much larger fraction of a wavelength in path length difference or even multiple wavelengths. Thus, the treble notes could be in phase with this head movement. If we now add the fact that the head is not of zero size and the fact that there are two ears, we can see that the complete cancellation is not possible, and, with even small movements of the head, one or both ears will be at or near maxima for the treble notes. The size of the head is much smaller than bass wavelengths, however, so there is significant weakening of the bass over much of the region in front of the speakers.

CONCEPTUAL PROBLEM 1

As oppositely moving pulses of the same shape (one upward, one downward) on a string pass through each other, there is one instant at which the string shows no displacement from the equilibrium position at any point. Has the energy carried by the pulses disappeared at this instant in time? If not, where is it?

Example 14.1   Two Speakers Driven by the Same Source

A pair of speakers placed 3.00 m apart are driven by the same oscillator (Fig. 14.3). A listener is originally at point O, which is located 8.00 m from the center of the line connecting the two speakers. The listener then walks to point P, which is a perpendicular distance 0.350 m from O before reaching the *first minimum* in sound intensity. What is the frequency of the oscillator?

Solution   The first minimum occurs when the two waves reaching the listener at P are 180° out of phase—in other words, when their path difference equals $\lambda/2$. In order to calculate the path difference, we must first find the path lengths r_1 and r_2. Making use of the two shaded triangles in Figure 14.3, we find the path lengths to be

$$r_1 = \sqrt{(8.00 \text{ m})^2 + (1.15 \text{ m})^2} - 8.08 \text{ m}$$

$$r_2 = \sqrt{(8.00 \text{ m})^2 + (1.85 \text{ m})^2} = 8.21 \text{ m}$$

Hence, the path difference is $r_2 - r_1 = 0.13$ m. Because we require that this path difference be equal to $\lambda/2$ for the first minimum, we find that $\lambda = 0.26$ m.

To obtain the oscillator frequency, we can use $v = \lambda f$, where v is the speed of sound in air, 343 m/s:

$$f = \frac{v}{\lambda} = \frac{343 \text{ m/s}}{0.26 \text{ m}} - 1.3 \text{ kHz}$$

EXERCISE 1   If the oscillator frequency is adjusted such that the listener hears the first minimum at a distance of 0.75 m from O, what is the new frequency?   Answer   0.63 kHz

Figure 14.3   (Example 14.1)

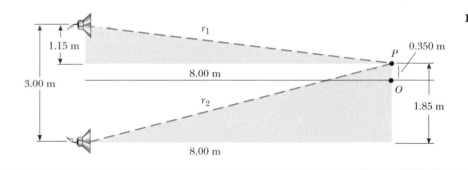

14.2 • STANDING WAVES

If a stretched string is clamped at both ends, waves traveling in both directions are reflected from the ends. The incident and reflected waves combine according to the superposition principle.

Consider two sinusoidal waves, in the same medium, that have the same amplitude, frequency, and wavelength but are traveling in *opposite* directions. If we neglect any phase angle ϕ for simplicity, their wave functions can be written

$$y_1 = A \sin(kx - \omega t) \qquad \text{and} \qquad y_2 = A \sin(kx + \omega t)$$

where y_1 represents a wave traveling to the right and y_2 represents a wave traveling to the left. Adding these two functions gives the resultant wave function, y:

$$y = y_1 + y_2 = A \sin(kx - \omega t) + A \sin(kx + \omega t)$$

where $k = 2\pi/\lambda$ and $\omega = 2\pi f$, as usual. Using the trigonometric identity $\sin(a \pm b) = \sin a \cos b \pm \cos a \sin b$, this reduces to

Wave function for a standing •
wave

$$y = (2A \sin kx)\cos \omega t \qquad\qquad [14.3]$$

This expression represents the wave function of a **standing wave.** From this result, we see that a standing wave has an angular frequency of ω and an amplitude of $2A \sin kx$ (the quantity in the parentheses of Eq. 14.3). That is, every particle of the string vibrates in simple harmonic motion with the same frequency. However, the amplitude of motion of a given particle depends on x. This is in contrast to the situation involving a traveling sinusoidal wave, in which all particles oscillate with the same amplitude as well as the same frequency.

Because the amplitude of the standing wave at any value of x is equal to $2A \sin kx$, we see that the *maximum* amplitude has the value $2A$. This occurs when the coordinate x satisfies the condition $\sin kx = 1$, or when

$$kx = \frac{\pi}{2}, \frac{3\pi}{2}, \frac{5\pi}{2}, \cdots$$

Because $k = 2\pi/\lambda$, the positions of maximum amplitude, called **antinodes,** are

Position of antinodes •

$$x = \frac{\lambda}{4}, \frac{3\lambda}{4}, \frac{5\lambda}{4}, \cdots = \frac{n\lambda}{4} \qquad\qquad [14.4]$$

where $n = 1, 3, 5, \ldots$. Note that **adjacent antinodes are separated by a distance of $\lambda/2$.**

In a similar way, the standing wave has a *minimum* amplitude of zero when x satisfies the condition $\sin kx = 0$ or when

$$kx = \pi, 2\pi, 3\pi, \ldots$$

giving

Position of nodes •

$$x = \frac{\lambda}{2}, \lambda, \frac{3\lambda}{2}, \cdots = \frac{n\lambda}{2} \qquad\qquad [14.5]$$

where $n = 1, 2, 3, \ldots$. **These points of zero amplitude, called nodes, are also spaced by $\lambda/2$.** The distance between a node and an adjacent antinode is $\lambda/4$.

The standing wave patterns produced at various times by two waves traveling in opposite directions are depicted graphically in Figure 14.4. The upper part of each figure represents the individual traveling waves, and the lower part represents the standing wave patterns. The nodes of the standing wave are labeled N, and the antinodes are labeled A. At $t = 0$ (Fig. 14.4a), the two waves are spatially identical,

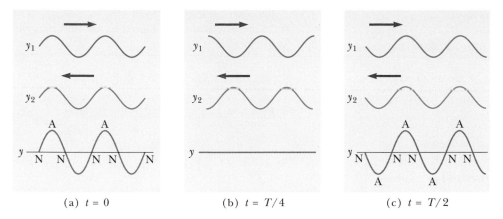

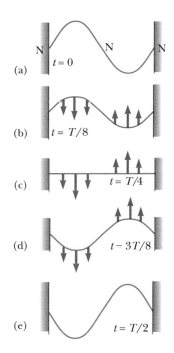

(a) $t = 0$ (b) $t = T/4$ (c) $t = T/2$

Figure 14.4 Standing wave patterns at various times produced by two waves of equal amplitude traveling in opposite directions. For the resultant wave y, the nodes (N) are points of zero displacement and the antinodes (A) are points of maximum displacement.

Figure 14.5 A standing wave pattern in a taut string showing snapshots during one half-cycle. (a) At $t = 0$, the string is momentarily at rest, and so $K = 0$ and all of the energy is potential energy U associated with the vertical displacements of the string segments. (b) At $t = T/8$, the string is in motion, and the energy is half kinetic and half potential. (c) At $t = T/4$, the string is horizontal (undeformed) and, therefore, $U = 0$; all of the energy is kinetic. The motion continues as indicated in (d) and (e), and ultimately the initial configuration in part (a) is repeated.

giving a standing wave of maximum amplitude, $2A$. One quarter of a period later, at $t = T/4$ (Fig. 14.4b), the individual waves have moved one quarter of a wavelength (one to the right and the other to the left). At this time, the individual displacements are equal and opposite for all values of x, and hence the resultant wave has zero displacement everywhere. At $t = T/2$ (Fig. 14.4c), the individual waves are again identical spatially, producing a standing wave pattern that is inverted relative to the $t = 0$ pattern.

It is instructive to describe the energy associated with the motion of a standing wave. To illustrate this point, consider a standing wave formed on a stretched string that is fixed at each end, as in Figure 14.5. Except for the nodes, which are stationary, all points on the string oscillate vertically with the same frequency. Furthermore, different points have different amplitudes of motion. Figure 14.5 represents snapshots of the standing wave at various times over one half-cycle. The nodes represent points on the string that are always at rest, and the locations of the maxima and minima never change. Each point on the string executes simple harmonic motion in the vertical direction. That is, one can view the standing wave as a large number of oscillators vibrating parallel to each other. The energy of the vibrating string continually alternates between elastic potential energy, at which time the string is momentarily stationary (Fig. 14.5a), and kinetic energy, at which time the string is horizontal and the particles have their maximum speed (Fig. 14.5c). The string particles have both potential energy and kinetic energy at intermediate times (Figs. 14.5b and 14.5d).

Example 14.2 Formation of a Standing Wave

Two waves traveling in opposite directions produce a standing wave. The individual wave functions are

$$y_1 = (4.0 \text{ cm}) \sin(3.0x - 2.0t)$$
$$y_2 = (4.0 \text{ cm}) \sin(3.0x + 2.0t)$$

where x and y are in centimeters. (a) Find the maximum displacement of the motion at $x = 2.3$ cm.

Solution When the two waves are summed, the result is a standing wave the function of which is given by Equation 14.3, with $A_0 = 4.0$ cm and $k = 3.0$ rad/cm:

$$y = (2A_0 \sin kx)\cos \omega t = [(8.0 \text{ cm})\sin 3.0x]\cos \omega t$$

Thus, the maximum displacement of the motion at the position $x = 2.3$ cm is

$$y_{\text{max}} = (8.0 \text{ cm})\sin 3.0x]_{x=2.3}$$
$$= (8.0 \text{ cm})\sin(6.9 \text{ rad}) = 4.6 \text{ cm}$$

(b) Find the positions of the nodes and antinodes.

Solution   Because $k = 2\pi/\lambda = 3$ rad/cm, we see that $\lambda = 2\pi/3$ cm. Therefore, from Equation 14.4 we find that the antinodes are located at

$$x = n\left(\frac{\pi}{6}\right) \text{ cm} \qquad (n = 1, 3, 5, \ldots)$$

and from Equation 14.5 we find that the nodes are located at

$$x = n\frac{\lambda}{2} = n\left(\frac{\pi}{3}\right) \text{ cm} \qquad (n = 1, 2, 3, \ldots)$$

14.3 • NATURAL FREQUENCIES IN A STRETCHED STRING

Consider a string that has a length of L and is fixed at both ends, as in Figure 14.6. Standing waves are set up in the string by a continuous superposition of waves incident on and reflected from the ends. The string has a number of natural patterns of vibration, called **normal modes.** Each of these modes has a characteristic frequency; the frequencies are easily calculated.

First, note that the ends of the string must be nodes, because these points are fixed. If the string is displaced at its midpoint and released, all modes that do not have a node at the center point are excited. Consider the normal mode in which the center of the string is an antinode (Fig. 14.6b). For this normal mode, the length of the string equals $\lambda/2$ (the distance between nodes):

$$L = \lambda_1/2 \qquad \text{or} \qquad \lambda_1 = 2L$$

The next normal mode, of wavelength λ_2 (Fig. 14.6c), occurs when the length of the string equals one wavelength, that is, when $\lambda_2 = L$. The third normal mode (Fig. 14.6d) corresponds to the case in which the length equals $3\lambda/2$; therefore,

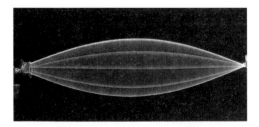

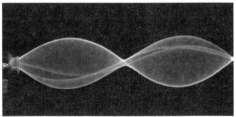

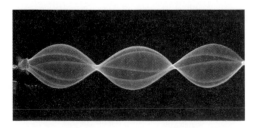

Multiflash photographs of standing wave patterns in a cord driven by a vibrator at the left end. The single loop pattern at the top left represents the fundamental ($n = 1$), the two-loop pattern at the right represents the second harmonic ($n = 2$), and the three-loop pattern at the lower left represents the third harmonic ($n = 3$). *(Richard Megna, Fundamental Photographs, NYC)*

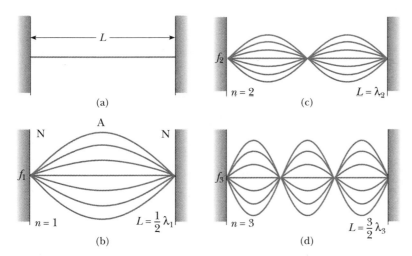

Figure 14.6   (a) A string of length L fixed at both ends. The normal modes of vibration form a harmonic series: (b) the fundamental frequency, or first harmonic; (c) the second harmonic; and (d) the third harmonic.

$\lambda_3 = 2L/3$. In general, the wavelengths of the various normal modes can be conveniently expressed as

$$\lambda_n = \frac{2L}{n} \qquad (n = 1, 2, 3, \dots) \qquad \text{[14.6]}$$

• *Wavelengths of normal modes*

where the index n refers to the nth mode of vibration. The natural frequencies associated with these modes are obtained from the relationship $f = v/\lambda$, where **the wave speed, v, is the same for all frequencies.** Using Equation 14.6, we find that the frequencies of the normal modes are

$$f_n = \frac{v}{\lambda_n} = \frac{n}{2L} v \qquad (n = 1, 2, 3, \dots) \qquad \text{[14.7]}$$

• *Frequencies of normal modes as functions of wave speed and length of string*

Because $v = \sqrt{F/\mu}$ (Eq. 13.19), where F is the tension in the string and μ is its mass per unit length, we can express the natural frequencies of a stretched string (sometimes called *harmonics*) as

$$f_n = \frac{n}{2L} \sqrt{\frac{F}{\mu}} \qquad (n = 1, 2, 3, \dots) \qquad \text{[14.8]}$$

• *Frequencies of normal modes as functions of string tension and linear mass density*

The lowest frequency, corresponding to $n = 1$, is called the **fundamental frequency,** f_1, and is

$$f_1 = \frac{1}{2L} \sqrt{\frac{F}{\mu}} \qquad \text{[14.9]}$$

• *Fundamental frequency of a taut string*

It is clear that the frequencies of the remaining modes are integral multiples of the fundamental frequency—that is, $2f_1$, $3f_1$, $4f_1$, and so on. These higher natural frequencies, together with the fundamental frequency, form a **harmonic series.** The fundamental, f_1, is the first harmonic; the frequency $f_2 = 2f_1$ is the second harmonic; the frequency f_n is the nth harmonic.

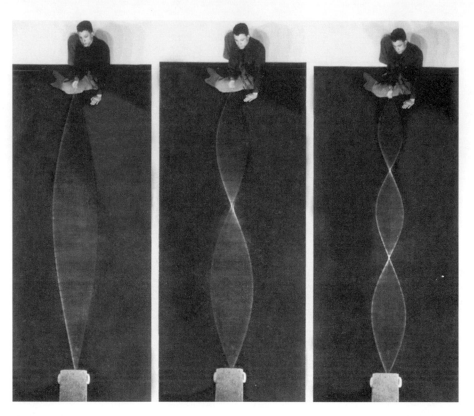

Photographs of standing waves. As one end of the rope is moved from side to side with increasing frequency, patterns with more and more loops are formed; only certain definite frequencies will produce fixed patterns.   *(Photos, Education Development Center, Newton, Mass.)*

We can obtain the foregoing results in an alternative manner. Because we require that the string be fixed at $x = 0$ and $x = L$, the wave function $y(x, t)$ given by Equation 14.3 must be *zero* at these points for *all* times. That is, the boundary conditions require that $y(0, t) = 0$ and $y(L, t) = 0$ for all values of t. Because $y = (2A \sin kx)\cos \omega t$, the first condition, $y(0, t) = 0$, is automatically satisfied because $\sin kx = 0$ at $x = 0$. To meet the second condition, $y(L, t) = 0$, we require that $\sin kL = 0$. This condition is satisfied when the angle kL equals an integral multiple of π (180°). Therefore, the allowed values of k are[1]

$$k_n L = n\pi \qquad (n = 1, 2, 3, \dots)$$ **[14.10]**

Because $k_n = 2\pi/\lambda_n$, we find that

$$\left(\frac{2\pi}{\lambda_n}\right) L = n\pi \qquad \text{or} \qquad \lambda_n = \frac{2L}{n}$$

which is identical to Equation 14.6.

When a stretched string is distorted to a shape that corresponds to any one of its harmonics, after being released it will vibrate at the frequency of that harmonic. However, if the string is plucked or bowed, the resulting vibration will include frequencies of various harmonics, including the fundamental. In effect, the string

---

[1] We exclude $n = 0$, because this corresponds to the trivial case in which no wave exists ($k = 0$).

"selects" the normal-mode frequencies when disturbed by a nonharmonic disturbance (for example, a finger plucking it).

Figure 14.7 shows a stretched string vibrating with its first and second harmonics simultaneously. In this figure, the combined vibration is the superposition of
the two vibrations shown in Figures 14.6b and 14.6c. The larger loop corresponds
to the fundamental frequency of vibration, f_1, and the smaller loops correspond to
the second harmonic, f_2. In general, the resulting motion, or displacement, of the
string can be described by a superposition of the various harmonic wave functions,
with different frequencies and amplitudes. Hence, the sound that one hears corresponds to a complex waveform associated with these modes of vibration. (We
shall return to this point in Section 14.6.)

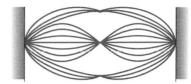

Figure 14.7   Multiple exposures
of a string vibrating simultaneously
in its first and second harmonics.

The frequency of a stringed instrument can be changed either by varying the
strings' tension, F, or by changing their length, L. For example, the tension in the
strings of guitars and violins is adjusted by a screw mechanism or by turning pegs
on the neck of the instrument. As the tension increases, the frequencies of the
normal modes increase according to Equation 14.8. Once the instrument is
"tuned," the player varies the frequency by moving his or her fingers along the
neck, thereby changing the length of the vibrating portion of the string. As this
length is reduced, the frequency increases, because the normal-mode frequencies
are inversely proportional to (vibrating) string length.

CONCEPTUAL PROBLEM 2

Guitarists sometimes play a "harmonic," by lightly touching a string at the exact center and
plucking the string. The result is a clear note one octave higher than the fundamental of the
string, even though the string is not pressed to the fingerboard. Why does this happen?

CONCEPTUAL PROBLEM 3

An archer shoots an arrow from a bow. Does the string of the bow exhibit standing waves
after the arrow leaves? If so, and if the bow is perfectly symmetric so that the arrow leaves
from the center of the string, what harmonics are excited?

Example 14.3   Give Me a C Note

A middle C string of the C-major scale on a piano has a fundamental frequency of 262 Hz, and the A note has a fundamental frequency of 440 Hz. (a) Calculate the frequencies of
the next two harmonics of the C string.

Solution   Because f_1 = 262 Hz, we can use Equations 14.8
and 14.9 to find the frequencies f_2 and f_3:

$$f_2 = 2f_1 = \boxed{524 \text{ Hz}}$$

$$f_3 = 3f_1 = \boxed{786 \text{ Hz}}$$

(b) If the strings for the A and C notes are assumed to
have the same mass per unit length and the same length,
determine the ratio of tensions in the two strings.

Solution   Using Equation 14.8 for the two strings vibrating at
their fundamental frequencies gives

$$f_{1A} = \frac{1}{2L}\sqrt{F_A/\mu} \quad \text{and} \quad f_{1C} = \frac{1}{2L}\sqrt{F_C/\mu}$$

$$f_{1A}/f_{1C} = \sqrt{F_A/F_C}$$

$$F_A/F_C = (f_{1A}/f_{1C})^2 = (440/262)^2 = \boxed{2.82}$$

(c) In a real piano, the assumption we made in (b) is only
half true. The string densities are equal, but the A string is
64% as long as the C string. What is the ratio of their tensions?

$$f_{1A}/f_{1C} = (L_C/L_A)\sqrt{F_A/F_C} = (100/64)\sqrt{F_A/F_C}$$

$$F_A/F_C = (0.64)^2(440/262)^2 = \boxed{1.16}$$

EXERCISE 2   A stretched string is 160 cm long and has a linear density of 0.0150 g/cm. What tension in the string will result in a second harmonic of 460 Hz?   Answer   813 N

EXERCISE 3   A string of linear density 1.0×10^{-3} kg/m and length 3.0 m is stretched between two points. One end is vibrated transversely at 200 Hz. What tension in the string will establish a standing-wave pattern with three loops along the string's length? Answer   160 N

14.4 • STANDING WAVES IN AIR COLUMNS

Standing longitudinal waves can be set up in a tube of air, such as an organ pipe, as the result of interference between longitudinal waves traveling in opposite directions. The phase relationship between the incident wave and the wave reflected from one end depends on whether that end is open or closed. This is analogous to the phase relationships between incident and reflected transverse waves at the ends of a string. **The closed end of an air column is a displacement node,** just as the fixed end of a vibrating string is a displacement node. As a result, the reflected wave at a closed end of a tube of air is 180° out of phase with the incident wave. Furthermore, because the pressure wave is 90° out of phase with the displacement wave (Section 13.9), **the closed end of an air column corresponds to a pressure antinode** (that is, a point of maximum pressure variation).

If the end of an air column is open to the atmosphere, the air molecules have complete freedom of motion. The wave reflected from an open end is nearly in phase with the incident wave when the tube's diameter is small relative to the wavelength of the sound. As a consequence, **the open end of an air column is approximately a displacement antinode and a pressure node.**

Strictly speaking, the open end of an air column is not exactly an antinode. A condensation does not reach full expansion until it passes somewhat beyond an open end. For a thin-walled tube of circular cross-section, this correction that must be added for each open end is about 0.6R, where R is the tube's radius. Hence, the effective length of the tube is somewhat greater than the true length, L.

The first three modes of vibration of a pipe that is open at both ends are shown in Figure 14.8a. Note that in this figure, the standing longitudinal waves are drawn as transverse waves because it is difficult to draw longitudinal displacements. When air is directed into the pipe from the left, longitudinal standing waves are formed and the pipe resonates at its natural frequencies. All modes of vibration are excited simultaneously (although not with the same amplitude). Note that the ends are displacement antinodes (approximately). In the fundamental mode, the wavelength is twice the length of the pipe, and hence the frequency of the fundamental, f_1, is $v/2L$. In a similar way, the frequencies of the higher harmonics are $2f_1$, $3f_1$, ... Thus, in a pipe that is open at both ends, the natural frequencies of vibration form a harmonic series; that is, the higher harmonics are integral multiples of the fundamental frequency. Because all harmonics are present, we can express the natural frequencies of vibration as

Natural frequencies of a pipe •
open at both ends

$$f_n = n \frac{v}{2L} \qquad (n = 1, 2, 3, \ldots) \qquad \textbf{[14.11]}$$

where v is the speed of sound in air.

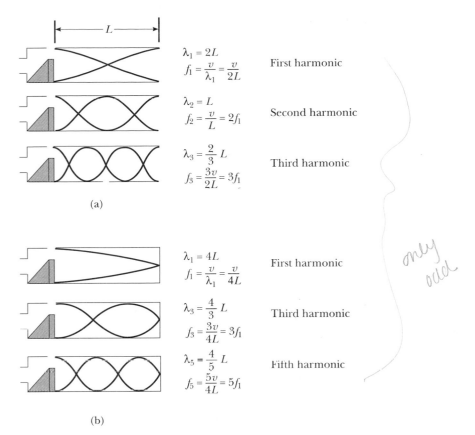

Figure 14.8   (a) Standing longitudinal waves in an organ pipe open at both ends. The natural frequencies that form a harmonic series are $f_1, 2f_1, 3f_1, \ldots$. (b) Standing longitudinal waves in an organ pipe closed at one end. Only the *odd* harmonics are present, and so the natural frequencies are $f_1, 3f_1, 5f_1, \ldots$.

If a pipe is closed at one end and open at the other, the closed end is a displacement node (Fig. 14.8b). In this case, the wavelength for the fundamental mode is four times the length of the tube. Hence, the fundamental, f_1, is equal to $v/4L$, and the frequencies of the higher harmonics are equal to $3f_1, 5f_1, \ldots$. That is, **in a pipe that is closed at one end, only odd harmonics are present,** and these are

$$f_n = n\frac{v}{4L} \qquad (n = 1, 3, 5, \ldots) \qquad \textbf{[14.12]}$$

• *Natural frequencies of a pipe closed at one end and open at the other*

Standing waves in air columns are the primary sources of the sounds produced by wind instruments.

Thinking Physics 2

Passing ocean waves sometimes cause the water in a harbor to undergo very large oscillations, called *seiches*. Why would this happen?

Reasoning Water in a harbor is enclosed and possesses a natural frequency based on the size of the harbor. This is similar to the natural frequency of the enclosed air in a bottle, which can be excited by blowing across the edge of the opening. Ocean waves pass by the opening of the harbor at a certain frequency. If this frequency matches that of the enclosed harbor, then a large standing wave can be set up in the water by resonance. This can be simulated by carrying a fish tank with water. If your walking frequency matches the natural frequency of the water as it sloshes back and forth, a large standing wave in the fish tank can be established.

Thinking Physics 3

A bugle has no valves, keys, slides, or finger holes—how can it play a song?

Reasoning Songs for the bugle are limited to harmonics of the fundamental frequency, because there is no control over frequencies without valves, keys, slides, or finger holes. The player obtains different notes by changing the tension in the lips as the bugle is played, in order to excite different harmonics. The normal playing range of a bugle is among the third, fourth, fifth, and sixth harmonics of the fundamental. For example, "Reveille" is played with just the three notes G, C, and E, and "Taps" is played with these three notes and the G one octave above the lower G.

Thinking Physics 4

If an orchestra doesn't warm up before a performance, the strings go flat and the wind instruments go sharp during the performance. Why?

Reasoning Without warming up, all the instruments will be at room temperature at the beginning of the concert. As the wind instruments are played, they fill with warm air from the player's exhalation. The increase in temperature of the air in the instrument causes an increase in the speed of sound, which raises the resonance frequencies of the air columns. As a result, the instruments go sharp. The strings on the stringed instruments also increase in temperature due to the friction of rubbing with the bow. This results in thermal expansion, which causes a decrease in the tension in the strings. With a decrease in tension, the wave speed on the strings drops, and the fundamental frequencies decrease. Thus, the stringed instruments go flat.

CONCEPTUAL PROBLEM 4

In Balboa Park in San Diego, California, there is an outdoor organ. Does the fundamental frequency of a particular pipe on this organ change on hot and cold days? How about on days with high and low atmospheric pressure?

CONCEPTUAL PROBLEM 5

If you have a series of identical glass bottles, with varying amounts of water in them, you can play musical notes by either striking the bottles with a spoon, or blowing across the open tops of the bottles. When hitting the bottles, the frequency of the note decreases as the water level rises. When blowing on the bottles, the frequency of the note increases as the water level rises. Why is the behavior of the frequency different in these two cases?

Example 14.4   Resonance in a Pipe

A pipe has a length of 1.23 m. (a) Determine the frequencies of the first three harmonics if the pipe is open at each end. Take $v = 343$ m/s as the speed of sound in air.

Solution   The first harmonic of a pipe open at both ends is

$$f_1 = \frac{v}{2L} = \frac{343 \text{ m/s}}{2(1.23 \text{ m})} = \boxed{139 \text{ Hz}}$$

Because all harmonics are present, the second and third harmonics are $f_2 = 2f_1 = 278$ Hz and $f_3 = 3f_1 = 417$ Hz.

(b) What are the three frequencies determined in part (a) if the pipe is closed at one end?

Solution   The fundamental frequency of a pipe closed at one end is

$$f_1 = \frac{v}{4L} = \frac{343 \text{ m/s}}{4(1.23 \text{ m})} = 69.7 \text{ Hz}$$

In this case, only odd harmonics are present, and so the next two resonances have frequencies $f_3 = 3f_1 = 209$ Hz and $f_5 = 5f_1 = 349$ Hz.

(c) For the pipe open at both ends, how many harmonics are present in the normal human hearing range (20 to 20 000 Hz)?

Solution   Because all harmonics are present, $f_n = nf_1$. For $f_n = 20\ 000$ Hz, we have $n = 20\ 000/139 = 144$, so that 144 harmonics are present in the audible range. Actually, only the first few harmonics have sufficient amplitude to be heard.

Example 14.5   Measuring the Frequency of a Tuning Fork

A simple apparatus for demonstrating resonance in a tube is described in Figure 14.9a. A long, vertical tube open at both ends is partially submerged in a beaker of water, and a vibrating tuning fork of unknown frequency is placed near the top. The length of the air column, L, is adjusted by moving the tube vertically. The sound waves generated by the fork are reinforced when the length of the air column corresponds to one of the resonant frequencies of the tube.

For a certain tube, the smallest value of L for which a peak occurs in the sound intensity is 9.00 cm. From this measurement, determine the frequency of the tuning fork and the value of L for the next two resonant modes.

Reasoning   Although the tube is open at both ends to allow the water in, the water surface acts like a wall at one end of the length L. Therefore this setup represents a pipe closed at one end and the fundamental has a frequency of $v/4L$ (Fig. 14.9b).

Solution   Taking $v = 343$ m/s for the speed of sound in air and $L = 0.0900$ m, we get

$$f_1 = \frac{v}{4L} = \frac{343 \text{ m/s}}{4(0.0900 \text{ m})} = 953 \text{ Hz}$$

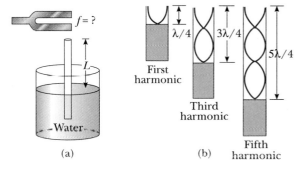

Figure 14.9   (Example 14.5) (a) Apparatus for demonstrating the resonance of sound waves in a tube closed at one end. The length L of the air column is varied by moving the tube vertically while it is partially submerged in water. (b) The first three normal modes of the system shown in part (a).

From this information about the fundamental mode, we see that the wavelength is $\lambda = 4L = 0.360$ m. Because the frequency of the source is constant, the next two resonance modes (Fig. 14.9b) correspond to lengths of $3\lambda/4 = 0.270$ m and $5\lambda/4 = 0.450$ m.

EXERCISE 4   The longest pipe on a certain organ is 4.88 m. What is the fundamental frequency (at 0.0°C) if the nondriven end of the pipe is (a) closed and (b) open?
Answer   (a) 17.0 Hz   (b) 34.0 Hz

EXERCISE 5   An air column 2.00 m in length is open at both ends. The frequency of a certain harmonic is 410 Hz, and the frequency of the next higher harmonic is 492 Hz. Determine the speed of sound in the air column.   Answer   328 m/s

14.5 • BEATS: INTERFERENCE IN TIME

The interference phenomena with which we have been dealing so far involve the superposition of two or more waves with the same frequency, traveling in opposite directions. Because the resultant waveform in this case depends on the coordinates of the disturbed medium, we can refer to the phenomenon as *spatial interference*. Standing waves in strings and pipes are common examples of spatial interference.

We now consider another type of interference effect, one that results from the superposition of two waves with slightly *different frequencies*. In this case, when the two waves are observed at a given point, they are periodically in and out of phase. That is, there occurs a temporal alternation between constructive and destructive interference. We refer to this phenomenon as **interference in time** or **temporal interference.**

For example, if two tuning forks of slightly different frequencies are struck, one hears a sound of pulsating intensity, called a **beat.**

Definition of beats •

> A **beat** is the periodic variation in intensity at a given point due to the superposition of two waves with slightly different frequencies.

The number of beats one hears per second, or **beat frequency,** equals the difference in frequency between the two sources. The maximum beat frequency that the human ear can detect is about 20 beats/s. When the beat frequency exceeds this value, it blends indistinguishably with the compound sounds producing the beats.

One can use beats to tune a stringed instrument, such as a piano, by beating a note against a reference tone of known frequency. The string can then be adjusted to equal the frequency of the reference by tightening or loosening it until the beats become too infrequent to notice.

Consider two waves with equal amplitudes, traveling through a medium with slightly different frequencies, f_1 and f_2. We can represent the displacement that each wave would produce at a point as

$$y_1 = A \cos 2\pi f_1 t \qquad \text{and} \qquad y_2 = A \cos 2\pi f_2 t$$

Using the superposition principle, we find that the resultant displacement at that point is given by

$$y = y_1 + y_2 = A(\cos 2\pi f_1 t + \cos 2\pi f_2 t)$$

It is convenient to write this in a form that uses the trigonometric identity

$$\cos a + \cos b = 2 \cos \left(\frac{a-b}{2}\right) \cos \left(\frac{a+b}{2}\right)$$

Letting $a = 2\pi f_1 t$ and $b = 2\pi f_2 t$, we find that

Resultant of two waves of •
different frequencies but equal
amplitude

$$y = 2A \cos 2\pi \left(\frac{f_1 - f_2}{2}\right) t \cos 2\pi \left(\frac{f_1 + f_2}{2}\right) t \qquad \text{[14.13]}$$

Graphs demonstrating the individual waveforms as well as the resultant wave are shown in Figure 14.10. From the factors in Equation 14.13, we see that the resultant

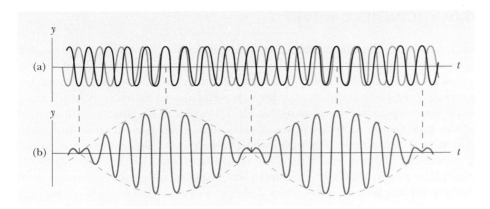

Figure 14.10   Beats are formed by the combination of two waves of slightly different frequencies traveling in the same direction. (a) The individual waves. (b) The combined wave has an amplitude (broken line) that oscillates in time.

vibration at a point has an effective frequency equal to the average frequency, $(f_1 + f_2)/2$, and an amplitude of

$$A - 2A \cos 2\pi \left(\frac{f_1 - f_2}{2} \right) t \qquad [14.14]$$

That is, the *amplitude varies in time* with a frequency of $(f_1 - f_2)/2$. When f_1 is close to f_2, this amplitude variation is slow, as illustrated by the envelope (broken line) of the resultant waveform in Figure 14.10b.

Note that a beat, or a maximum in amplitude, will be detected whenever

$$\cos 2\pi \left(\frac{f_1 - f_2}{2} \right) t = \pm 1$$

That is, there are *two* maxima in each cycle. Because the amplitude varies with frequency as $(f_1 - f_2)/2$, the number of beats per second, or the beat frequency, f_b, is twice this value:

$$f_b = f_1 - f_2 \qquad [14.15] \qquad \bullet \ \textit{Beat frequency}$$

For instance, if two tuning forks vibrate individually at frequencies of 438 Hz and 442 Hz, respectively, the resultant sound wave of the combination has a frequency of 440 Hz (the musical note A) and a beat frequency of 4 Hz. That is, the listener hears the 440-Hz sound wave go through an intensity maximum four times every second.

CONCEPTUAL PROBLEM 6

You have a standard tuning fork whose frequency you know and a second tuning fork whose frequency you don't know. When you play them together, you hear a beat frequency of 4 Hz. You know that the frequency of the mystery tuning fork is different from that of the standard fork by 4 Hz, but you can't tell whether it is higher or lower. What could you do to determine this? (*Hint:* You are chewing gum at the time that you perform these measurements.)

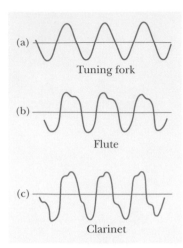

Figure 14.11 Waveform produced by (a) a tuning fork, (b) a flute, and (c) a clarinet, each at approximately the same frequency. *(Adapted from C. A. Culver, Musical Acoustics, 4th ed., New York, McGraw-Hill, 1956, p. 128.)*

14.6 · COMPLEX WAVES

The sound wave patterns produced by most instruments are very complex. Some characteristic waveforms produced by a tuning fork, a harmonic flute, and a clarinet are shown in Figure 14.11. Although each instrument has its own characteristic pattern, Figure 14.11 shows that all three waveforms are periodic. A struck tuning fork produces primarily one harmonic (the fundamental), whereas the flute and clarinet produce many frequencies, which include the fundamental and various harmonics. Thus, the complex waveforms produced by a violin or clarinet, and the corresponding richness of musical tones, are the result of the superposition of various harmonics. This is in contrast to the drum, in which the overtones do not form a harmonic series.

Analysis of complex waveforms appears at first sight to be a formidable task. However, if the waveform is periodic, it can be represented with arbitrary precision by the combination of a sufficiently large number of sinusoidal waves that form a harmonic series. In fact, one can represent any periodic function or any function over a finite interval as a series of sine and cosine terms by using a mathematical technique based on *Fourier's theorem*. The corresponding sum of terms that represents the periodic waveform is called a **Fourier series.**

Let $y(t)$ be any function that is periodic in time, with a period of T, so that $y(t + T) = y(t)$. **Fourier's theorem** states that this function can be written

$$y(t) = \sum_n (A_n \sin 2\pi f_n t + B_n \cos 2\pi f_n t) \qquad [14.16]$$

where the lowest frequency is $f_1 = 1/T$.

The higher frequencies are integral multiples of the fundamental, and so $f_n = nf_1$. The coefficients A_n and B_n represent the amplitudes of the various waves. The amplitude of the nth harmonic is proportional to $\sqrt{A_n^2 + B_n^2}$, and its intensity is proportional to $A_n^2 + B_n^2$.

Figure 14.12 represents a harmonic analysis of the waveforms shown in Figure 14.11. Note the variation of relative intensity with harmonic content for the flute

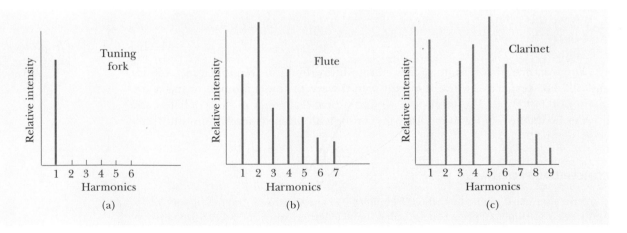

Figure 14.12 Harmonics of the waveforms shown in Figure 14.11. Note the variations in intensity of the various harmonics. *(Adapted from C. A. Culver, Musical Acoustics, 4th ed., New York, McGraw-Hill, 1956.)*

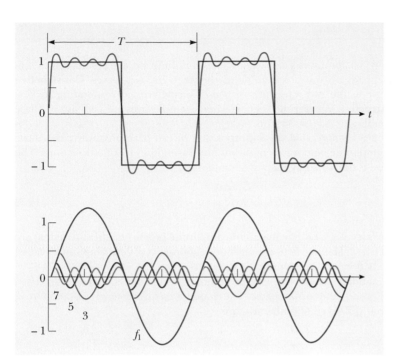

Figure 14.13   Harmonic synthesis of a square wave, which can be represented by the sum of odd harmonics of the fundamental.   *(From M. L. Warren, Introductory Physics, New York, W. H. Freeman, 1979, p. 178; by permission of the publisher.)*

and clarinet. In general, any musical sound contains components that are members of a harmonic set with varying relative intensities.

As an example of Fourier synthesis, consider the periodic square wave shown in Figure 14.13. The square wave is synthesized by a series of *odd* harmonics of the fundamental. The series contains only sine functions (that is, $B_n = 0$ for all n). Only the first four odd harmonics and their respective amplitudes are shown. One obtains a better fit to the true waveform by adding more harmonics.

Using modern technology, one can produce a variety of musical tones by mixing harmonics with varying amplitudes. Electronic music synthesizers that do this are now widely used.

Thinking Physics 5

A professor performs a demonstration in which he breathes helium and then speaks with a comical voice. One student explains, "The velocity of sound in helium is higher than in air, so the fundamental frequency of the standing waves in the mouth is increased." Another student says, "No, the fundamental frequency is determined by the vocal folds and cannot be changed. Only the *quality* of the voice has changed." Which student is correct?

Reasoning   The second student is correct. The fundamental frequency of the complex tone from the voice is determined by the vibration of the vocal folds and is not changed by substituting a different gas in the mouth. The introduction of the helium into the mouth results in higher harmonics being excited more than in the normal voice, but the fundamental frequency of the voice is the same—only the quality has changed. The unusual inclusion of the higher frequency harmonics results in a common description of this effect as a "high-pitched" voice, but this is incorrect.

SUMMARY

[Handwritten margin notes:]

$\Delta r = \frac{\lambda}{2\pi} \phi$ relationship btwn path diffe & ϕ

Length of string $L = \lambda/2$

$\lambda = 2L$

open at both ends

$f_n = n \frac{v}{2L}$

closed at one end

$f_n = n \frac{v}{4L}$

$v = \lambda f$ speed of traveling sinusoidal wave

$v = \sqrt{F/\mu}$ ← speed of a wave on a stretched string

When two waves with equal amplitudes and frequencies superimpose, the resultant wave has an amplitude that depends on the phase angle, ϕ, between the two waves. **Constructive interference** occurs when the two waves are *in phase* everywhere, corresponding to $\phi = 0, 2\pi, 4\pi, \ldots$. **Destructive interference** occurs when the two waves are 180° out of phase everywhere, corresponding to $\phi = \pi, 3\pi, 5\pi, \ldots$.

Standing waves are formed from the superposition of two harmonic waves that have the same frequency, amplitude, and wavelength but are traveling in *opposite* directions. The resultant standing wave is described by the wave function

$$y = (2A \sin kx)\cos \omega t \qquad [14.3]$$

Hence, its amplitude varies as $\sin kx$. The maximum amplitude points (called **antinodes**) are separated by a distance $\lambda/2$. Halfway between antinodes are points of zero amplitude (called **nodes**). The distance between adjacent nodes is $\lambda/2$.

One can set up standing waves with specific frequencies in such systems as stretched strings, hollow pipes, rods, and drumheads. The natural frequencies of vibration of a stretched string of length, L, fixed at both ends, are

$$f_n = \frac{n}{2L}\sqrt{\frac{F}{\mu}} \qquad (n = 1, 2, 3, \ldots) \qquad [14.8]$$

where F is the tension in the string and μ is its mass-per-unit length. The natural frequencies of vibration form a **harmonic series**—that is, $f_1, 2f_1, 3f_1, \ldots$.

The standing wave patterns for longitudinal waves in a hollow pipe depend on whether the ends of the pipe are open or closed. If the pipe is open at both ends, the natural frequencies of vibration form a harmonic series. If one end is closed, only odd harmonics of the fundamental are present.

The phenomenon of **beats** occurs as a result of the superposition of two traveling waves of slightly different frequencies. For sound waves at a given point, one hears an alternation in sound intensity with time. Thus, beats correspond to *interference as time passes*.

Any periodic waveform can be represented by the combination of the sinusoidal waves that form a harmonic series. The process is called *Fourier synthesis* and is based on *Fourier's theorem.*

[Handwritten notes:]

SHM

$T = 2\pi/\omega$

$f = 1/T = \omega/2\pi$

$\omega = 2\pi f = 2\pi/T$

$v = -\omega A \sin(\omega t + \phi)$

$a = \omega^2 A \cos(\omega t + \phi)$

$k = 2\pi/\lambda$ angular wave #

CONCEPTUAL QUESTIONS

1. When two waves interfere constructively or destructively, is there any gain or loss in energy? Explain.
2. Does the phenomenon of wave interference apply only to sinusoidal waves?
3. Some singers claim to be able to shatter a wine glass by maintaining a certain vocal pitch over a period of several seconds (Fig. Q14.3). What mechanism causes the glass to break? (The glass must be very clean in order to break.)
4. What limits the amplitude of motion of a real vibrating system that is driven at one of its natural frequencies?
5. Explain why all harmonics are present in an organ pipe open at both ends, but only the odd harmonics are present in a pipe closed at one end.
6. An airplane mechanic notices that the sound from a twin-engine aircraft rapidly varies in loudness when both engines are running. What could be causing this variation from loud to soft?
7. Why does a vibrating guitar string sound louder when placed on the instrument than it would if allowed to vibrate in the air while off the instrument?

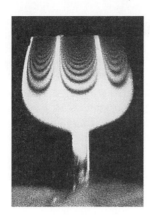 

Figure Q14.3   (Question 3) *(Left)* Standing wave pattern in a vibrating wine glass. The wine glass will shatter if the amplitude of vibration becomes too large.   *(Courtesy of Professor Thomas D. Rossing, Northern Illinois University)*   *(Right)* A wine glass shattered by the amplified sound of a human voice.   *(© Ben Rose 1992/The IMAGE Bank)*

8. When the base of a vibrating tuning fork is placed against a chalkboard, the sound becomes louder. How does this affect the length of time for which the fork vibrates? Does this agree with conservation of energy?

9. To keep animals away from their cars, some people mount short, thin pipes on the fenders. The pipes give out a high-pitched wail when the cars are moving. How do they create the sound?

10. If you wet your fingers and lightly run them around the rim of a fine wine glass, a high-pitched sound is heard. Why? How could you produce various musical notes with a set of wine glasses?

11. When a bell is rung, standing waves are set up around the bell's circumference. What boundary conditions must be satisfied by the resonant wavelengths? How does a crack in the bell, such as in the Liberty Bell, affect the satisfying of the boundary conditions and the sound emanating from the bell?

12. Explain why your voice seems to sound better than usual when you sing in the shower.

13. What is the purpose of the slide on a trombone or the valves on a trumpet?

14. Despite a reasonably steady hand, a certain person often spills his coffee when carrying it to his seat. Discuss resonance as a possible cause of this difficulty and devise a means for solving the problem.

PROBLEMS

Section 14.1   Superposition and Interference of Sinusoidal Waves

1. Two harmonic waves are described by

$$y_1 = (5.00 \text{ m}) \sin[\pi(4.00x - 1200t)]$$

$$y_2 = (5.00 \text{ m}) \sin[\pi(4.00x - 1200t - 0.250)]$$

where x, y_1, and y_2 are in meters and t is in seconds. (a) What is the amplitude of the resultant wave? (b) What is the frequency of the resultant wave?

2. A sinusoidal wave is described by

$$y_1 = (0.0800 \text{ m}) \sin[2\pi(0.100x - 80.0t)]$$

where y_1 and x are in meters and t is in seconds. Write an expression for a wave that has the same frequency, amplitude, and wavelength as y_1 but when added to y_1 gives a resultant with an amplitude of $8\sqrt{3}$ cm.

3. Two waves are traveling in the same direction along a stretched string. Each has an amplitude of 4.00 cm and they are 90.0° out of phase. Find the amplitude of the resultant wave.

4. Two identical sinusoidal waves with wavelengths of 3.00 m travel in the same direction at a speed of 2.00 m/s. The second wave originates from the same point as the first but at a later time. Determine the minimum possible time interval between the starting moments of the two waves if the amplitude of the resultant wave is the same as that of the two initial waves.

5. Two speakers are driven by a common oscillator at 800 Hz and face each other at a distance of 1.25 m. Locate the points along a line joining the two speakers at which relative minima would be expected. (Use v = 343 m/s.)

6. A tuning fork generates sound waves with a frequency of 246 Hz. The waves travel in opposite directions along a hall-

way, are reflected by walls, and return. What is the phase difference between the reflected waves when they meet? The corridor is 47.0 m long and the tuning fork is located 14.0 m from one end. The speed of sound in air is 343 m/s.

Section 14.2   Standing Waves

7. Use the trigonometric identity

$$\sin(a \pm b) = \sin a \cos b \pm \cos a \sin b$$

to show that the resultant of two wave functions each of amplitude A, angular frequency ω, and propagation number k and traveling in opposite directions can be written

$$y = (2A \sin kx) \cos \omega t$$

For simplicity, set all phase angles ϕ to zero.

8. Two sinusoidal waves traveling in opposite directions interfere to produce a standing wave described by

$$y = (1.50 \text{ m}) \sin(0.400x) \cos(200t)$$

where x is in meters and t is in seconds. Determine the wavelength, frequency, and speed of the interfering waves.

9. Two waves in a long string are given by

$$y_1 = (0.0150 \text{ m}) \cos\left(\frac{x}{2} - 40t\right)$$

$$y_2 = (0.0150 \text{ m}) \cos\left(\frac{x}{2} + 40t\right)$$

where the y's and x are in meters and t is in seconds. (a) Determine the positions of the nodes of the resulting standing wave. (b) What is the maximum displacement at the position $x = 0.400$ m?

10. Two waves that set up a standing wave in a long string are given by

$$y_1 = A \sin(kx - \omega t + \phi)$$

$$y_2 = A \sin(kx + \omega t)$$

Show (a) that the addition of the arbitrary phase angle will change only the position of the nodes and (b) that the distance between nodes remains constant.

Section 14.3   Natural Frequencies in a Stretched String

11. A student wants to establish a standing wave on a wire that is 1.80 m long and clamped at both ends. The wave speed is 540 m/s. What is the minimum frequency the student should apply to set up standing waves?

12. A standing wave is established in a 120–cm-long string fixed at both ends. The string vibrates in four segments when driven at 120 Hz. (a) Determine the wavelength. (b) What is the fundamental frequency?

13. A cello A-string vibrates in its fundamental mode with a frequency of 220 vibrations/s. The vibrating segment is 70.0 cm long and has a mass of 1.20 g. (a) Find the tension in the string. (b) Determine the frequency of the harmonic that causes the string to vibrate in three segments.

14. A string of length L, mass-per-unit length μ, and tension F is vibrating at its fundamental frequency. What effect will the following have on the fundamental frequency? (a) The length of the string is doubled, with all other factors held constant. (b) The mass per unit length is doubled, with all other factors held constant. (c) The tension is doubled, with all other factors held constant.

15. A 60.0-cm guitar string under a tension of 50.0 N has a mass per unit length of 0.100 g/cm. What is the highest resonant frequency that can be heard by a person capable of hearing frequencies up to 20 000 Hz?

16. A stretched wire vibrates in its fundamental mode at a frequency of 400 vibrations/s. What would be the fundamental frequency if the wire were half as long, with twice the diameter and with four times the tension?

17. A 2.00–m-long wire having a mass of 0.100 kg is fixed at both ends. The tension in the wire is maintained at 20.0 N. What are the frequencies of the first three allowed modes of vibration? If a node is observed at a point 0.400 m from one end, in what mode and with what frequency is it vibrating?

18. A violin string has a length of 0.350 m and is tuned to concert G, having a frequency $f_G = 392$ Hz. Where must the violinist place her finger to play concert A, having a frequency $f_A = 440$ Hz? If this position is to remain correct to one half the width of a finger (i.e., to within 0.600 cm), the string tension cannot be allowed to slip by more than what fraction?

19. Find the fundamental frequency and the next three frequencies that could cause a standing wave pattern on a string that is 30.0 m long, has a mass per unit length 9.00×10^{-3} kg/m, and is stretched to a tension of 20.0 N.

20. Standing-wave vibrations are set up in a crystal goblet with two nodes and two antinodes equally spaced around the 20.0-cm circumference of its rim. If transverse waves move around the glass at 900 m/s, an opera singer would have to produce a high harmonic with what frequency in order to shatter the glass with a resonant vibration?

Section 14.4   Standing Waves in Air Columns

(In this section, unless otherwise indicated, assume that the speed of sound in air is 343 m/s.)

21. Calculate the length for a pipe that has a fundamental frequency of 240 Hz if the pipe is (a) closed at one end and (b) open at both ends.

22. A glass tube (open at both ends) of length L is positioned near an audio speaker of frequency $f = 0.680$ kHz. For what values of L will the tube resonate with the speaker?

23. The overall length of a piccolo is 32.0 cm. The resonating air column vibrates as a pipe open at both ends. (a) Find the frequency of the lowest note a piccolo can play, assuming the speed of sound in air is 340 m/s. (b) Opening holes in the side effectively shortens the length of the resonant column. If the highest note a piccolo can sound is 4000 Hz, find the distance between adjacent nodes for this mode of vibration.

24. The fundamental frequency of an open organ pipe corresponds to middle C (261.6 Hz on the chromatic musical scale). The third resonance of a closed organ pipe has the same frequency. What are the lengths of the two pipes?

25. A tuning fork the frequency of which is f is used to set up a resonance condition in a pipe. Write an expression for the length that will cause the pipe to resonate in its nth mode if it is (a) open at both ends; (b) closed at one end. (Assume that the speed of sound is v.)

26. Do not stick anything into your ear! Estimate the length of your ear canal, from its opening at the external ear in to the eardrum. If you regard the canal as a tube that is open at one end and closed at the other, at approximately what fundamental frequency would you expect your hearing to be most sensitive? Explain why you can hear especially soft sounds just around this frequency.

27. A shower stall measures 86.0 cm × 86.0 cm × 210 cm. When you sing in the shower, which frequencies will sound the richest (resonate), assuming the shower acts as a pipe closed at both ends (nodes at opposite sides)? Assume also that the human voice ranges from 130 Hz to 2000 Hz (not necessarily one person's voice, however). Let the speed of sound in the hot shower stall be 355 m/s.

28. An open pipe 0.400 m in length is placed vertically in a cylindrical bucket, nearly touching the bottom of the bucket, which has area 0.100 m². Water is slowly poured into the bucket until a sounding tuning fork of frequency 440 Hz, held over the pipe, produces resonance. Find the mass of water in the bucket at this moment.

29. Water is pumped into a long cylinder at a rate of 18.0 cm³/s. The radius of the cylinder is 4.00 cm, and at the open top of the cylinder there is a tuning fork vibrating with a frequency of 200 Hz. As the water rises, how much time elapses between successive resonances?

30. Water is pumped into a long cylinder at a volume flow rate R. The radius of the cylinder is r (cm), and at the open top of the cylinder there is a tuning fork vibrating with a frequency f. As the water rises, how much time elapses between successive resonances?

31. A piece of metal pipe is just the right length so that when it is cut into two pieces, their lowest resonance frequencies are 256 Hz for one and 440 Hz for the other. (a) What resonant frequency would have been produced by the original length of pipe, and (b) how long was the original piece?

32. A piece of cardboard tubing, closed at one end, is just the right length so that when it is cut into two (unequal) pieces, their lowest resonant frequencies are 256 Hz for the piece with the closed end and 440 Hz for the piece with both ends open. (a) What resonant frequency would have been produced by the original cardboard tubing, and (b) how long was the original piece?

Section 14.5 Beats: Interference in Time (Optional)

33. In certain ranges of a piano keyboard, more than one string is tuned to the same note to provide extra loudness. For example, the note at 110 Hz has two strings at this pitch. If one string slips from its normal tension of 600 N to 540 N, what beat frequency will be heard when the two strings are struck simultaneously?

34. A student holds a tuning fork oscillating at 256 Hz. He walks toward a wall at a constant speed of 1.33 m/s. (a) What beat frequency does he observe between the tuning fork and its echo? (b) How fast must he walk away from the wall to observe a beat frequency of 5.00 Hz?

35. A flute is designed so that it plays a frequency of 261.6 Hz, middle C, when all the holes are covered and the temperature is 20.0°C. (a) Consider the flute as a pipe that is open at both ends; find the length of the flute, assuming that middle C is the fundamental. (b) A second player, nearby in a colder room, also attempts to play middle C on an identical flute. A beat frequency of 3.00 Hz is heard. What is the temperature of the room? The speed of sound in air is described by

$$v = (331 \text{ m/s}) \sqrt{1 + \frac{T}{273°}}$$

where T is the Celsius temperature.

Additional Problems

36. Two loudspeakers are placed on a wall, 2.00 m apart. A listener stands directly in front of one of the speakers, 3.00 m from the wall. The speakers are being driven by a single oscillator at a frequency of 300 Hz. (a) What is the phase difference between the two waves when they reach the observer? (b) What is the frequency closest to 300 Hz to which the oscillator may be adjusted such that the observer will hear minimal sound?

37. On a marimba, the wooden bar that sounds a tone when struck vibrates as a transverse standing wave with three antinodes and two nodes. The lowest frequency note is 87.0 Hz, produced by a bar 40.0 cm long. (a) Find the speed of transverse waves on the bar. (b) The loudness and duration of the emitted sound are enhanced by a resonant pipe suspended vertically below the center of the bar. If the pipe is open at the top end only and the speed of sound in air is 340 m/s, what is the length of the pipe required to resonate with the bar in part (a)?

(Problem 37) Marimba players in Mexico City. *(Murray Greenberg)*

38. A speaker at the front of a room and an identical speaker at the rear of the room are being driven by the same oscillator at 456 Hz. A student walks at a uniform rate of 1.50 m/s along the length of the room. How many beats does the student hear per second?

39. Two train whistles have identical frequencies of 180 Hz. When one train is at rest in the station sounding its whistle, a beat frequency of 2.00 Hz is heard from a moving train. What two possible speeds and directions can the moving train have?

40. Jane waits on a railroad platform, while two trains approach from the same direction at equal speeds of 8.00 m/s. Both trains are blowing their whistles (which have the same frequency), and one train is some distance behind the other. After the first train passes Jane, but before the second train passes her, she hears beats of frequency 4.00 Hz. What is the frequency of the train whistles?

41. A string (mass = 4.80 g, length = 2.00 m, and tension = 48.0 N), fixed at both ends, vibrates in its second ($n = 2$) natural mode. What is the wavelength in air of the sound generated by this vibrating string?

42. A pipe that is open at both ends has a fundamental frequency of 300 Hz when the speed of sound in air is 333 m/s. (a) What is the length of the pipe? (b) What is the frequency of the second harmonic when the temperature of the air is increased so that the speed of sound in the pipe is 344 m/s?

43. A string with a mass of 8.00 g and a length of 5.00 m has one end attached to a wall; the other end is draped over a pulley and attached to a hanging mass of 4.00 kg. If the string is plucked, what is the fundamental frequency of vibration?

44. In a major chord on the physical-pitch musical scale, the frequencies are in the ratios $4:5:6:8$. A set of pipes, closed at one end, are to be cut so that when sounded in their fundamental mode, they will sound out a major chord. (a) What is the ratio of the lengths of the pipes? (b) What length pipes are needed if the lowest frequency of the chord is 256 Hz? (c) What are the frequencies of this chord?

45. Two wires are welded together. The wires are of the same material, but one is twice the diameter of the other one. They are subjected to a tension of 4.60 N. The thin wire has a length of 40.0 cm and a linear mass density of 2.00 g/m. The combination is fixed at both ends and vibrated in such a way that two antinodes are present with the node between them being right at the weld. (a) What is the frequency of vibration? (b) How long is the thick wire?

46. Two identical strings, each fixed at both ends, are arranged near each other. If string *A* starts oscillating in its fundamental mode, it is observed that string *B* will begin vibrating in its third ($n = 3$) natural mode. Determine the ratio of the tension of string *B* to the tension of string *A*.

47. A standing wave is set up in a string of variable length and tension by a vibrator of variable frequency. When the vibrator has a frequency *f*, using a string of length *L* and tension *F*, there are *n* antinodes set up in the string. (a) If the length of the string is doubled, by what factor should the frequency be changed to get the same number of antinodes? (b) If the frequency and length are held constant, what tension will produce $n + 1$ antinodes? (c) If the frequency is tripled and the length halved, by what factor should the tension be changed to get twice as many antinodes?

48. Radar detects the speed of a car, using the Doppler shift of microwaves that are reflected off the moving car, by beating the received wave with the transmitted wave and measuring the difference. The Doppler shift for microwaves is

$$f = f_0 \sqrt{\frac{c + v}{c - v}}$$

where f_0 is the transmitted frequency, *c* is the speed of light (3×10^8 m/s), and *v* is the relative speed of the two objects. (a) Show that the wave that reflects back to the source has a frequency

$$f = f_0 \frac{(c + v)}{(c - v)}$$

(b) Show that the expression for the beat frequency of the microwaves may be written as $f_b = 2v/\lambda$. (Because the beat frequency is much smaller than the transmitted frequency, use the approximation $f + f_0 = 2f_0$.) (c) What beat frequency is measured for a speed of 30.0 m/s (67 mph) if the microwaves have a frequency of 10.0 GHz? (1 GHz = 10^9 Hz.) (d) If the beat frequency measurement is accurate to ± 5 Hz, how accurate is the velocity measurement?

49. If two adjacent natural frequencies of an organ pipe are determined to be 0.550 kHz and 0.650 kHz, calculate the fundamental frequency and length of this pipe. (Use $v = 340$ m/s.)

50. A 0.0100 kg and 2.00–m-long wire is fixed at both ends and vibrates in its simplest mode under tension 200 N. When a tuning fork is placed near the wire, a beat frequency of 5.00 Hz is heard. (a) What is the frequency (frequencies) of the tuning fork? (b) What should the tension in the wire be to make the beats disappear?

51. The wave function for a standing wave is given in Equation 14.3 as $y = 2A \sin kx \cos \omega t$. (a) Rewrite this wave function in terms of the wavelength λ and wavespeed v of the wave. (b) Write the wave function of the simplest standing-wave vibration of a stretched string of length L. (c) Write the wave function for the second harmonic. (d) Generalize these results and write the wave function for the nth resonance vibration.

Spreadsheet Problems

S1. Spreadsheet 14.1 adds two traveling waves at some fixed time t. The resultant wave function is

$$y = y_1 + y_2 = A_1 \sin(k_1 x - \omega_1 t + \phi_1)$$
$$+ A_2 \sin(k_2 x - \omega_2 t + \phi_2)$$

Add two waves traveling in opposite directions having the same wavelengths, the same phases, and the same speeds. (a) Use $A_1 = A_2 = 0.10$ m, $\omega_1 = -\omega_2 = 3.0$ rad/s, $\phi_1 = \phi_2 = 0$, and $k_1 = k_2 = 2.0$ rad/m. View the associated graph. Choose different values of t to see the time evolution of the resultant wave function. Do you get standing waves? (b) Repeat part (a) using $A_1 = 0.10$ m, $A_2 = 0.20$ m.

S2. Use Spreadsheet 14.1 to add two traveling waves that differ only in phase. (a) Choose $A_1 = A_2 = 0.10$ m, $\omega_1 = \omega_2 = 2.5$ rad/s, $k_1 = k_2 = 1.0$ rad/m, $\phi_1 = 0$, and $\phi_2 = 0, \pi/8, \pi/4, \pi/2$, and π. View the associated graphs. For which values of ϕ_2 do you get constructive interference? Destructive interference? (b) Repeat part (a) using $A_2 = 0.20$ m.

S3. Use Spreadsheet 14.1 to add two traveling waves with different wavelengths. (a) Choose $A_1 = A_2 = 0.10$ m, $\omega_1 = \omega_2 = 3.0$ rad/s, $k_1 = 2.0$ rad/m, $k_2 = 1.0$ rad/m, and $\phi_1 = \phi_2 = 0$. View the associated graph. (b) Repeat part (a) with $k_2 = 3, 4, 5, 6, 7, 8, 9$, and 10. Examine the graph for each case. Explain the appearance of the graphs.

S4. Three waves with the same frequency, wavelength, and amplitude are traveling in the same direction. Each differs in phase such that $(\phi_2 - \phi_1) = (\phi_3 - \phi_2) = \Delta\phi$. Modify Spreadsheet 14.1 to add the three waves at some fixed time t. Calculate the resultant wave function at $t = 0.0$, 1.0 s, and 2.0 s. Use $A_1 = A_2 = A_3 = 0.050$ m, $\omega_1 = \omega_2 = \omega_3 = 2.5$ rad/s, and $k_1 = k_2 = k_3 = 1.0$ rad/m. Choose $\Delta\phi = \pi/6$. View the associated graph. Repeat for different values of $\Delta\phi$.

S5. Write a spreadsheet or computer program to add two traveling waves of different frequencies:

$$y_1 = y_0 \cos 2\pi f_1 t \qquad y_2 = y_0 \cos 2\pi f_2 t$$

If the two frequencies are close to each other, beats will be produced. Does your program show the beats? From your numerical results, how does the beat frequency relate to the frequencies of the two waves?

S6. The Fourier theorem states that any periodic wave of frequency f, no matter how complicated, can be expressed as a sum of even or odd harmonic functions. That is,

$$y(t) = \sum_{n=0}^{\infty} A_n \sin(n\omega t + \phi_n)$$

where $\omega = 2\pi f$ is the fundamental angular frequency. (a) Use $A_1 = 1$, $A_2 = 1/2$, $A_3 = 1/3$, and so on. All the ϕ_n's are zero. (b) Use $A_1 = 1$, $A_3 = 1/3$, $A_5 = 1/5$, and so on. All the even A_n's and all the ϕ_n's are zero. Write a spreadsheet or computer program that will calculate this sum. You should design your spreadsheet or program so that the number of terms in the sum can be easily specified. Start with just the first term ($n = 1$). Then rerun the calculation with more terms in the sum ($n = 2, 3, 4, \ldots$ 20). Plot the sum so that one or two periods are displayed. Note how the waveform builds up as more terms are added.

ANSWERS TO CONCEPTUAL PROBLEMS

1. At the instant at which there is no displacement of the string, the string is still moving. Thus, the energy is stored at that instant completely as kinetic energy of the string.

2. At the center of the string, there is a node for the second harmonic, as well as for every even-numbered harmonic. By placing the finger at the center and plucking, the guitarist is eliminating any harmonic that does not have a node at that point, which is all of the odd harmonics. The even harmonics can vibrate relatively freely with the finger at the center because they exhibit no displacement at that point. The result is a sound with a mixture of frequencies that are integer multiples of the second harmonic, which is one octave higher than the fundamental.

3. The bow string is pulled away from equilibrium and released, similar to the way that a guitar string is pulled and released when it is plucked. Thus, standing waves will be excited in the bow string. If the arrow leaves from the exact center of the string, then a series of odd harmonics will be excited. Even harmonics will not be excited because they have a node at the point where the string exhibits its maximum displacement.

4. A change in temperature will result in a change in the speed

of sound and, therefore, in the fundamental frequency. The speed of sound for audible frequencies in the open atmosphere is only a function of temperature, and does not depend on the pressure. Thus, there will be no effect on the fundamental frequency of variations in atmospheric pressure.

5. When the bottles are struck, standing wave vibrations are established in the glass material of the bottles. The frequencies of these vibrations are determined by the tension in the glass and the mass loading of the glass material. As the water level rises, there is more mass loading, since the glass is in contact with the water. This increased mass loading decreases the frequency. On the other hand, blowing into a bottle establishes a standing wave vibration in the air cavity above the water. As the water level rises, the length of this cavity decreases, and the frequency rises.

6. If you take the gum out of your mouth and stick it to one of the tines of the mystery tuning fork, it will lower its frequency, due to the mass loading of the gum. Now, when the two forks are played together, the beat frequency will be different. If the beat frequency is lower, then the forks are closer together in frequency, so the mystery fork must be higher in frequency than the standard. If the beat frequency is higher, the frequencies are farther apart, and the mystery fork must have had a lower frequency. This argument assumes that the piece of gum is small enough that the frequency shift for the fork is less than 4 Hz.

15

Fluid Mechanics

atter is normally classified as being in one of three states: solid, liquid, or gaseous. Everyday experience tells us that a solid has a definite volume and shape. A brick maintains its familiar shape and size day in and day out. We also know that a liquid has a definite volume but no definite shape. Finally, an unconfined gas has neither definite volume nor definite shape. These definitions help us to picture the states of matter, but they are somewhat artificial. For example, asphalt and plastics are normally considered solids, but over long periods of time they tend to flow like liquids. Likewise, most substances can be a solid, liquid, or gas (or combinations of these), depending on the temperature and pressure. In general, the time it takes a particular substance to change its shape in response to an external force determines whether we treat the substance as a solid, liquid, or gas.

A **fluid** is a collection of molecules that are randomly arranged and held together by weak cohesive forces and forces exerted by the walls of a container. Both liquids and gases are fluids. In our treatment of the mechanics of fluids, we shall see that no new physical principles are needed to

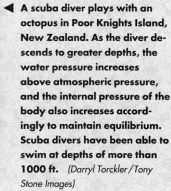

◀ A scuba diver plays with an octopus in Poor Knights Island, New Zealand. As the diver descends to greater depths, the water pressure increases above atmospheric pressure, and the internal pressure of the body also increases accordingly to maintain equilibrium. Scuba divers have been able to swim at depths of more than 1000 ft. *(Darryl Torckler/Tony Stone Images)*

415

explain such effects as the buoyant force on a submerged object and the dynamic lift on an airplane wing. First, we consider a fluid at rest and derive an expression for the pressure exerted by the fluid as a function of its density and depth. We then treat fluids in motion, an area of study called fluid dynamics. A fluid in motion can be described by a model in which certain simplifying assumptions are made. We use this model to analyze some situations of practical importance. An underlying principle known as the **Bernoulli principle** enables us to determine relationships between the pressure, density, and velocity at every point in a fluid.

15.1 · PRESSURE

The study of fluid mechanics involves the density of a substance, defined as its mass per unit volume. For this reason, Table 15.1 lists the densities of various substances. These values vary slightly with temperature, because the volume of a substance is temperature dependent (as we shall see in Chapter 16). Note that under standard conditions (0°C and atmospheric pressure) the densities of gases are about 1/1000 the densities of solids and liquids. This difference implies that the average molecular spacing in a gas under these conditions is about ten times greater in each dimension than in a solid or liquid.

Fluids do not sustain shearing stresses, and thus the only stress that can exist on an object submerged in a fluid is one that tends to compress the object. The force exerted by the fluid on the object is always perpendicular to the surfaces of the object, as shown in Figure 15.1.

The pressure at a specific point in a fluid can be measured with the device pictured in Figure 15.2. The device consists of an evacuated cylinder enclosing a light piston connected to a spring. As the device is submerged in a fluid, the fluid presses down on the top of the piston and compresses the spring until the inward force of the fluid is balanced by the outward force of the spring. The fluid pressure can be measured directly if the spring is calibrated in advance. This is accomplished by applying a known force to the spring to compress it a given distance.

Figure 15.1   The force of the fluid on a submerged object at any point is perpendicular to the surface of the object. The force of the fluid on the walls of the container is perpendicular to the walls at all points.

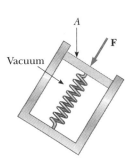

Figure 15.2   A simple device for measuring pressure in a fluid.

TABLE 15.1   Densities of Some Common Substances

Substance	ρ (kg/m³)[a]	Substance	ρ (kg/m³)[a]
Ice	0.917×10^3	Water	1.00×10^3
Aluminum	2.70×10^3	Sea water	1.03×10^3
Iron	7.86×10^3	Ethyl alcohol	0.806×10^3
Copper	8.92×10^3	Benzene	0.879×10^3
Silver	10.5×10^3	Mercury	13.6×10^3
Lead	11.3×10^3	Air	1.29
Gold	19.3×10^3	Oxygen	1.43
Platinum	21.4×10^3	Hydrogen	8.99×10^{-2}
Glycerine	1.26×10^3	Helium	1.79×10^{-1}

[a] All values are at standard atmospheric pressure and temperature (STP) — that is, atmospheric pressure and 0°C. To convert to grams per cubic centimeter, multiply by 10^{-3}.

If *F* is the magnitude of the normal force on the piston and *A* is the surface area of the piston, then the pressure, *P*, of the fluid at the level to which the device has been submerged is defined as the ratio of force to area:

$$P \equiv \frac{F}{A}$$

[15.1]   • *Definition of pressure*

To define the pressure at a specific point, consider a fluid acting on the device shown in Figure 15.2. If the normal force exerted by the fluid is *F* over a surface element of area δA that contains the point in question, then the pressure at that point is

$$P = \lim_{\delta A \to 0} \frac{F}{\delta A} = \frac{dF}{dA}$$

[15.2]

As we see in the next section, the pressure in a fluid varies with depth. Therefore, to get the total force on a flat wall of a container, we have to integrate Equation 15.2 over the surface.

Because pressure is force per unit area, it has units of N/m² in the SI system. Another name for the SI unit of pressure is **pascal** (Pa).

$$1 \text{ Pa} = 1 \text{ N/m}^2$$

[15.3]

Snowshoes prevent the person from sinking into the soft snow because the person's weight is spread over a larger area, which reduces the pressure on the snow's surface. *(Earl Young/FPG)*

Thinking Physics 1

The daring physics professor, after a long lecture, stretches out for a nap on a bed of nails, as in the photograph. How is this possible?

Reasoning If you try to support your entire weight on a single nail, the pressure on your body is your weight divided by the very small area of the nail. This pressure is sufficiently large to penetrate the skin. However, if you distribute your weight over several hundred nails, as the professor is doing, the pressure is considerably reduced because the area that supports your weight is the total area of all nails in contact with your body. (Note that lying on a bed of nails is much more comfortable than sitting on a bed of nails. Standing on the bed of nails without shoes is not recommended.)

(Thinking Physics 1)   *(Jim Lehman)*

SERWAY
PUNCTURE-PEDIC

Suction cups can be used to hold objects onto surfaces. Why don't astronauts use suction cups to hold onto the outside surface of the space shuttle?

Reasoning   The action of a suction cup depends on the fact that air is pushed out from under the cup when it is pressed against a surface. When released, it tends to spring back a bit, causing the trapped gas under the cup to expand and have a reduced pressure. Thus, the difference between the atmospheric pressure on the outside of the cup and the reduced pressure inside provides a net force pushing the cup against the surface. For astronauts in orbit around the Earth, there is no air outside the surface of the spacecraft. Thus, if a suction cup were to be pressed against the outside surface of the spacecraft, the pressure differential could not occur.

CONCEPTUAL PROBLEM 1

A woman wearing high-heeled shoes is invited into a home in which the kitchen has vinyl floor covering. Why should the homeowner be concerned?

EXERCISE 1   Estimate the density of the *nucleus* of an atom. What does this result suggest concerning the structure of matter? (Use the fact that the mass of a proton is equal to 1.67×10^{-27} kg and its radius is approximately 10^{-15} m.)   Answer 4.0×10^{17} kg/m^3; matter is mostly empty space.

15.2 • VARIATION OF PRESSURE WITH DEPTH

As divers know well, the pressure in the sea or a lake increases as they dive to greater depths. Likewise, atmospheric pressure decreases with increasing altitude. For this reason, aircraft flying at high altitudes must have pressurized cabins.

   We now show how the pressure in a liquid increases linearly with depth. Consider a liquid of density ρ at rest and open to the atmosphere, as in Figure 15.3. Let us select a sample of the liquid contained within an imaginary cylinder of cross-sectional area A extending from the surface of the liquid to a depth h. The pressure exerted by the fluid on the bottom face is P, and the pressure on the top face of the cylinder is atmospheric pressure, P_0. Therefore, the upward force exerted by the liquid on the bottom of the cylinder is PA, and the downward force exerted by the atmosphere on the top is P_0A. Because the mass of liquid in the cylinder is $\rho V = \rho Ah$, the weight of the fluid in the cylinder is $w = \rho gV = \rho gAh$. Because the cylinder is in equilibrium, the upward force at the bottom must be greater than the downward force at the top of the sample to support its weight:

$$PA - P_0A = \rho gAh$$

or

$$P = P_0 + \rho gh \qquad [15.4]$$

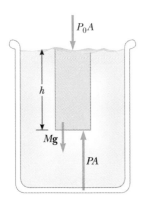

Figure 15.3   The variation of pressure with depth in a fluid. The net force on the volume of water within the darker region must be zero.

*Variation of pressure with •
depth*

That is, the absolute pressure P at a depth h below the surface of a liquid open to the atmosphere is *greater* than atmospheric pressure by an amount $\rho g h$.

In our calculations and end-of-chapter problems, we usually take atmospheric pressure to be

$$P_0 = 1.00 \text{ atm} \approx 1.013 \times 10^5 \text{ Pa}$$

Equation 15.4 verifies that the pressure is the same at all points having the same depth, independent of the shape of the container.

In view of the fact that the pressure in a liquid depends only on depth, any increase in pressure at the surface must be transmitted to every point in the fluid. This was first recognized by the French scientist Blaise Pascal (1623–1662) and is called **Pascal's law: A change in the pressure applied to an enclosed liquid is transmitted undiminished to every point of the liquid and to the walls of the container.**

• *Pascal's law*

An important application of Pascal's law is the hydraulic press, illustrated by Figure 15.4. A force F_1 is applied to a small piston of area A_1. The pressure is transmitted through a liquid to a larger piston of area A_2. Because the pressure is the same on both sides, we see that $P = F_1/A_1 = F_2/A_2$. Therefore, the force F_2 is larger than F_1 by the multiplying factor A_2/A_1. Hydraulic brakes, car lifts, hydraulic jacks, and forklifts all make use of this principle.

Thinking Physics 3

Blood pressure is normally measured with the cuff of the sphygmomanometer around the arm. Suppose that the blood pressure were measured with the cuff around the calf of the leg of a standing person. Would the reading of the blood pressure be the same here as it was for the arm?

Reasoning The blood pressure measured at the calf would be larger than that measured at the arm. If we imagine the vascular system of the body to be a vessel containing a liquid (the blood), the pressure in the liquid will increase with depth. The blood at the calf is deeper in the liquid than that at the arm and is at a higher pressure.

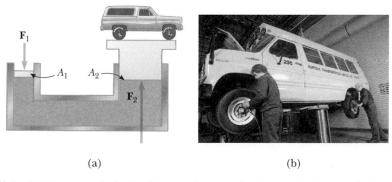

(a)   (b)

Figure 15.4   (a) Diagram of a hydraulic press. Because the increase in pressure is the same at the left and right sides, a small force $\mathbf{F}_1$ at the left produces a much larger force $\mathbf{F}_2$ at the right. (b) A bus under repair is supported by a hydraulic lift in a garage.   *(Superstock)*

(Conceptual Problem 2)   *(Henry Leap)*

CONCEPTUAL PROBLEM 2

A typical silo on a farm has many bands wrapped around its perimeter as shown in the photograph. Why is the spacing between successive bands smaller at the lower portions of the silo?

Example 15.1   The Car Lift

In a car lift used in a service station, compressed air exerts a force on a small piston of circular cross-section having a radius of 5.00 cm. This pressure is transmitted by a liquid to a second piston of radius 15.0 cm. What force must the compressed air exert in order to lift a car weighing 13 300 N? What air pressure will produce this force?

Solution   Because the pressure exerted by the compressed air is transmitted undiminished throughout the fluid, we have

$$F_1 = \left(\frac{A_1}{A_2}\right)F_2 = \frac{\pi(5.00 \times 10^{-2}\ \text{m})^2}{\pi(15.0 \times 10^{-2}\ \text{m})^2}\ (1.33 \times 10^4\ \text{N})$$

$$= 1.48 \times 10^3\ \text{N}$$

The air pressure that will produce this force is

$$P = \frac{F_1}{A_1} = \frac{1.48 \times 10^3\ \text{N}}{\pi(5.00 \times 10^{-2}\ \text{m})^2} = 1.88 \times 10^5\ \text{Pa}$$

This pressure is approximately twice atmospheric pressure.

The input work (the work done by $\mathbf{F}_1$) is equal to the output work (the work done by $\mathbf{F}_2$), so that energy is conserved.

Example 15.2   The Force on a Dam

Water is filled to a height H behind a dam of width w (Fig. 15.5). Determine the resultant force on the dam.

Reasoning   We cannot calculate the force on the dam by simply multiplying the area times the pressure, because the pressure varies with depth. The problem can be solved by finding the force dF on a narrow horizontal strip at depth h and then integrating the expression to find the total force on the dam.

Solution   The pressure at the depth h beneath the surface at the shaded portion is

$$P = \rho g h = \rho g (H - y)$$

(We have left out atmospheric pressure because it acts on both sides of the dam.) Using Equation 15.2, we find the force on the shaded strip of area $dA = w\,dy$ to be

$$dF = P\,dA = \rho g(H - y)\,w\,dy$$

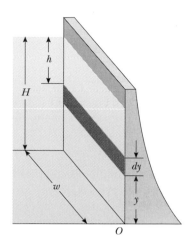

Figure 15.5 (Example 15.2) The total force on a dam must be obtained from the expression $F = \int P\, dA$, where dA is the area of the dark strip.

Therefore, the total force on the dam is

$$F = \int P\, dA = \int_0^H \rho g(H - y)\, w\, dy = \tfrac{1}{2}\rho g w H^2$$

Note that because the pressure increases with depth, the dam is designed so that its thickness increases with depth, as in Figure 15.5.

EXERCISE 2 Use the fact that the pressure increases linearly with depth to find the average pressure on the dam.
Answer $\tfrac{1}{2}\rho g H$

EXERCISE 3 What is the hydrostatic force on the back of Grand Coulee Dam if the water in the reservoir is 150 m deep and the width of the dam is 1200 m?
Answer 1.32×10^{11} N

EXERCISE 4 In some places, the Greenland ice sheet is 1.0 km thick. Estimate the pressure on the ground underneath the ice ($\rho_{\text{ice}} = 920 \text{ kg/m}^3$). Answer 9×10^6 Pa

15.3 · PRESSURE MEASUREMENTS

One simple device for measuring pressure is the open-tube manometer illustrated in Figure 15.6a. One end of a U-shaped tube containing a liquid is open to the atmosphere, and the other end is connected to a system of unknown pressure P. The difference in pressure $P - P_0$ is equal to $\rho g h$. Therefore, we see that $P = P_0 + \rho g h$. The pressure P is called the **absolute pressure,** and the difference $P - P_0$ is called the **gauge pressure.** For example, the pressure you measure in your bicycle tire is gauge pressure.

Another instrument used to measure pressure is the common barometer, invented by Evangelista Torricelli (1608–1647). A long tube closed at one end is filled with mercury and then inverted into a dish of mercury (Fig. 15.6b). The closed end of the tube is nearly a vacuum, so its pressure can be taken as zero. Therefore, it follows that $P_0 = \rho_{\text{Hg}} g h$, where ρ_{Hg} is the density of the mercury and h is the height of the mercury column. One atmosphere ($P_0 = 1$ atm) of pressure is defined to be the pressure equivalent of a column of mercury that is exactly 0.7600 m in height at 0°C, with $g = 9.80665$ m/s². At this temperature, mercury has a density of 13.595×10^3 kg/s³; therefore

$$P_0 = \rho_{\text{Hg}} g h = (13.595 \times 10^3 \text{ kg/m}^3)(9.80665 \text{ m/s}^2)(0.7600 \text{ m})$$

$$= 1.013 \times 10^5 \text{ Pa}$$

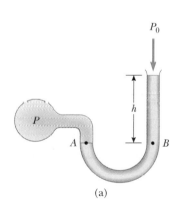

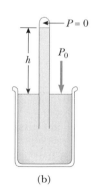

Figure 15.6 Two devices for measuring pressure: (a) an open-tube manometer and (b) a mercury barometer.

15.4 · BUOYANT FORCES AND ARCHIMEDES' PRINCIPLE

Archimedes' principle •

Figure 15.7 The external forces on the cube of water are the force of gravity **w** and the buoyant force **B**. Under equilibrium conditions, $B = w = mg$.

Archimedes (287–212 B.C.)

A Greek mathematician, physicist, and engineer, Archimedes was perhaps the greatest scientist of antiquity. According to legend, Archimedes was asked by King Hieron to determine whether the king's crown was made of pure gold or had been alloyed with some other metal. The task was to be performed without damaging the crown. Archimedes presumably arrived at a solution while taking a bath, noting a partial loss of weight after submerging his arms and legs in the water. As the story goes, he was so excited about his great discovery that he ran through the streets of Syracuse naked shouting, "Eureka!" which is Greek for "I have found it."

Archimedes' principle can be stated as follows:

> Any body completely or partially submerged in a fluid is buoyed up by a force the magnitude of which is equal to the weight of the fluid displaced by the body.

Everyone has experienced Archimedes' principle. Recall that it is relatively easy to lift someone if the person is in a swimming pool, whereas lifting that same individual on dry land is much harder. Evidently, water provides partial support to any object placed in it. The upward force that the fluid exerts on an object submerged in it is called the **buoyant force.** According to Archimedes' principle, **the magnitude of the buoyant force always equals the weight of the fluid displaced by the object.**

Archimedes' principle can be verified in the following manner. Suppose we focus our attention on the indicated cube of fluid in the container of Figure 15.7. This cube of fluid is in equilibrium under the action of the forces on it. One of these forces is the force of gravity. What cancels this downward force? Apparently, the rest of the fluid inside the container is holding it in equilibrium. Thus, the magnitude of the buoyant force is exactly equal to the weight of the fluid inside the cube:

$$B = w$$

Now, imagine that the cube of fluid is replaced by a cube of steel of the same dimensions. What is the buoyant force on the steel? The fluid surrounding a cube behaves in the same way whether a cube of fluid or a cube of steel is being buoyed up; therefore, **the buoyant force acting on the steel is the same as the buoyant force acting on a cube of fluid of the same dimensions.** This result applies for a submerged object of any shape, size, or density.

Let us show explicitly that the magnitude of the buoyant force is equal to the weight of the displaced fluid. The pressure at the bottom of the cube in Figure 15.7 is greater than the pressure at the top by an amount $\rho_f gh$, where ρ_f is the density of the fluid and h is the height of the cube. The pressure difference, ΔP, is equal to the buoyant force per unit area—that is, $\Delta P = B/A$. Thus, we see that $B = (\Delta P)A = (\rho_f gh)A = \rho_f gV$, where V is the volume of the cube. Because the mass of the fluid in the cube is $M = \rho_f V$, we see that

$$B = w = \rho_f Vg = Mg \qquad \text{[15.5]}$$

where w is the weight of the displaced fluid.

Before proceeding with a few examples, it is instructive to compare the forces acting on a totally submerged object with those acting on a floating object.

Case I: A Totally Submerged Object   When an object is totally submerged in a fluid of density ρ_f, the magnitude of the upward buoyant force is $B = \rho_f V_0 g$, where V_0 is the volume of the object. If the object has a density ρ_0, its weight is $w = Mg = \rho_0 V_0 g$, and the net force on it is $B - w = (\rho_f - \rho_0)V_0 g$. Hence, if the

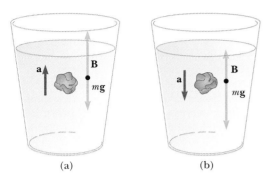

(a)   (b)

Figure 15.8   (a) A totally submerged object that is less dense than the fluid in which it is submerged will experience a net upward force. (b) A totally submerged object that is denser than the fluid sinks.

density of the object is less than the density of the fluid, as in Figure 15.8a, the unsupported object will accelerate upward. If the density of the object is greater than the density of the fluid as in Figure 15.8b, the unsupported object will sink.

Case II: A Floating Object   Now consider an object in static equilibrium floating on a fluid—that is, an object that is only partially submerged. In this case, the upward buoyant force is balanced by the downward force of gravity exerted on the object. If V is the volume of the fluid displaced by the object (which corresponds to that volume of the object beneath the fluid level), then the buoyant force has a magnitude $B = \rho_f V g$. Because the weight of the object is $w = Mg = \rho_0 V_0 g$, and $w = B$, we see that $\rho_f V g = \rho_0 V_0 g$, or

$$\frac{\rho_0}{\rho_f} = \frac{V}{V_0}$$ [15.6]

Under normal conditions, the average density of a fish is slightly greater than the density of water. This being the case, a fish would sink if it did not have some mechanism for adjusting its density. The fish accomplishes this by internally regulating the size of its swim bladder. In this manner, fish are able to swim to various depths.

Hot-air balloons over Albuquerque, New Mexico. Because hot air is less dense than cold air, there is a net upward buoyant force on the ballons.   *(William Moriarity/Rainbow)*

Thinking Physics 4

Suppose an office party is taking place on the top floor of a tall building. Carrying an iced soft drink, you step on the elevator, which begins to accelerate downward. What happens to the ice in the drink? Does it rise farther out of the liquid? Sink deeper into the liquid? Or is it unaffected by the motion?

Reasoning   The level of the ice in the liquid is unaffected by the motion. The acceleration of the elevator is equivalent to a change in the gravitational field, according to the principle of equivalence. If the elevator accelerates downward, one might be tempted to say that the effect is the same as if gravity decreases—the weight of the ice cube decreases, causing it to float higher in the liquid. Recall, however, that the magnitude of the buoyant force is equal to the weight of the liquid displaced by the ice cube. The weight of the liquid also decreases with the effectively decreased gravity. With both the weight of the ice cube and the buoyant force decreasing by the same factor, the level of the ice cube in the liquid is unaffected.

Thinking Physics 5

A florist delivery person is delivering a flower basket to a home. The basket includes an attached helium-filled balloon, which suddenly comes loose from the basket and begins to accelerate upward toward the sky. Startled by the release of the balloon, the delivery person drops the flower basket. As the basket falls, the basket–Earth system experiences an increase in kinetic energy and a decrease in gravitational potential energy, consistent with the conservation of mechanical energy. The balloon–Earth system, however, experiences an increase in *both* gravitational potential energy and kinetic energy. Is this inconsistent with the conservation of mechanical energy principle? If not, from where is the extra energy coming?

Reasoning   In the case of the system of the flower basket and the Earth, a good approximation to the motion of the basket can be made by ignoring the effects of the air. Thus, the conservation of mechanical energy principle is obeyed, with the exception of a small amount of transformation to internal energy of the air and the basket due to air friction. For the balloon–Earth system, we cannot ignore the effects of the air—it is the buoyant force of the air that causes the balloon to rise.

CONCEPTUAL PROBLEM 3

Atmospheric pressure varies from day to day. Does a ship float higher in the water on a high pressure day compared to a low pressure day?

CONCEPTUAL PROBLEM 4

Suppose a damaged ship just barely floats in the ocean after a hole in its hull has been plugged. It is pulled toward shore and into a river, heading toward a dry dock for repair. As it is pulled up the river, it sinks. Why?

CONCEPTUAL PROBLEM 5

A pound of styrofoam and a pound of lead have the same weight. If they are placed on an equal arm balance, will it balance?

CONCEPTUAL PROBLEM 6

A person in a boat floating in a small pond throws an anchor overboard. Does the level of the pond rise, fall, or remain the same?

Example 15.3   A Submerged Object

A piece of aluminum is suspended from a string and then completely immersed in a container of water (Fig. 15.9). The mass of the aluminum is 1.0 kg, and its density is 2.7×10^3 kg/m³. Calculate the tension in the string before and after the aluminum is immersed.

Solution   When the aluminum is suspended in air, as in Figure 15.9a, the tension in the string, T_1 (the reading on the scale), is equal to the weight, Mg, of the aluminum, assuming that the buoyant force of air can be neglected:

$$T_1 = Mg = (1.0 \text{ kg})(9.80 \text{ m/s}^2) = 9.8 \text{ N}$$

When immersed in water, the aluminum experiences an upward buoyant force **B**, as in Figure 15.9b, which reduces the tension in the string. Because the system is in equilibrium,

$$T_2 + B - Mg = 0$$
$$T_2 = Mg - B = 9.8 \text{ N} - B$$

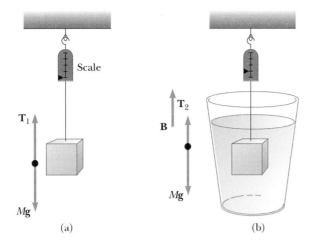

Figure 15.9   (Example 15.3) (a) When the aluminum is suspended in air, the scale reads the weight, Mg (neglecting the buoyancy of air). (b) When the aluminum is immersed in water, the buoyant force **B** reduces the scale reading to $T_2 = Mg - B$.

In order to calculate B, we must first calculate the volume of the aluminum:

$$V_{Al} = \frac{M}{\rho_{Al}} = \frac{1.0 \text{ kg}}{2.7 \times 10^3 \text{ kg/m}^3} = 3.7 \times 10^{-4} \text{ m}^3$$

Because the buoyant force equals the weight of the water displaced, we have

$$B = M_w g = \rho_w V_{Al} g$$
$$= (1.0 \times 10^3 \text{ kg/m}^3)(3.7 \times 10^{-4} \text{ m}^3)(9.80 \text{ m/s}^2)$$
$$= 3.6 \text{ N}$$

Therefore,

$$T_2 = 9.8 \text{ N} - B = 9.8 \text{ N} - 3.6 \text{ N} = \boxed{6.2 \text{ N}}$$

EXERCISE 5   What is the true weight of one cubic meter of a light plastic that has a specific gravity of 0.15? Note the true weight of an object is its weight in vacuum.
Answer   1.47 kN

15.5 · FLUID DYNAMICS

Thus far, our study of fluids has been restricted to fluids at rest. We now turn our attention to fluid dynamics—that is, fluids in motion. Instead of trying to study the motion of each particle of the fluid as a function of time, we describe the properties of the fluid at each point as a function of time.

Flow Characteristics

When fluid is in motion, its flow can be characterized as being one of two main types. The flow is said to be **steady** or **laminar** if each particle of the fluid follows a smooth path, so that the paths of different particles never cross each other, as in Figure 15.10. Thus, in steady flow, the velocity of the fluid at any point remains constant in time.

Above a certain critical speed, fluid flow becomes **turbulent.** Turbulent flow is an irregular flow characterized by small whirlpool-like regions, as in Figure 15.11. As an example, the flow of water in a stream becomes turbulent in regions in which rocks and other obstructions are encountered, often forming white-water rapids.

The term **viscosity** is commonly used in fluid flow to characterize the degree of internal friction in the fluid. This internal friction or viscous force is associated with the resistance of two adjacent layers of the fluid against moving relative to each other. Because of viscosity, part of the kinetic energy of a fluid is converted to thermal energy. This is similar to the mechanism by which an object sliding on a rough horizontal surface loses kinetic energy.

Figure 15.10   An illustration of streamline flow around an automobile in a test wind tunnel. The streamlines in the airflow are made visible by smoke particles.   *(Andy Sacks/Tony Stone Images)*

Figure 15.11   Turbulent flow: The tip of a rotating blade (the dark region at the top) forms a vortex in air that is being heated by an alcohol lamp (the wick is at the bottom). Note the air turbulence on both sides of the rotating blade.   *(Kim Vandiver and Harold Edgerton, Palm Press, Inc.)*

Because the motion of a real fluid is complex and not yet fully understood, we make some simplifying assumptions in our approach. As we shall see, many features of real fluids in motion can be understood by considering the behavior of an ideal fluid. In our model, we make the following four assumptions:

Properties of an ideal fluid •

1. **Nonviscous fluid.** In a nonviscous fluid, internal friction is neglected. An object moving through the fluid experiences no viscous force.
2. **Incompressible fluid.** The density of an incompressible fluid is assumed to remain constant because it does not change due to a change in pressure on the fluid.
3. **Steady flow.** In steady flow, we assume that the velocity of the fluid at each point remains constant in time.
4. **Irrotational flow.** Fluid flow is irrotational if there is no angular momentum of the fluid about any point. If a small wheel placed anywhere in the fluid does not rotate about the wheel's center of mass, the flow is irrotational. (If the wheel were to rotate, as it would if turbulence were present, the flow would be rotational.)

The first two assumptions are properties of our ideal fluid. The last two are descriptions of the way that the fluid flows.

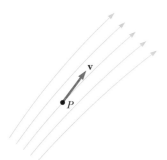

Figure 15.12   This diagram represents a set of streamlines (blue lines). A particle at *P* follows one of these streamlines, and its velocity is tangent to the streamline at each point along its path.

15.6 • STREAMLINES AND THE EQUATION OF CONTINUITY

The path taken by a fluid particle under steady flow is called a **streamline.** The velocity of the fluid particle is always tangent to the streamline, as shown in Figure 15.12. No two streamlines can cross each other, for if they did, a fluid particle could move either way at the crossover point and then the flow would not be steady. A set of streamlines, as shown in Figure 15.12, forms what is called a *tube of flow*. Note that fluid particles cannot flow into or out of the sides of this tube, because if they did, the streamlines would be crossing each other.

Consider an ideal fluid flowing through a pipe of nonuniform size, as in Figure 15.13. The particles in the fluid move along the streamlines in steady flow. At all points the velocity of any particle is tangent to the streamline along which it moves.

In a small time interval, Δt, the fluid at the bottom end of the pipe moves a distance $\Delta x_1 = v_1 \Delta t$. If A_1 is the cross-sectional area in this region, then the mass contained in the shaded region is $\Delta m_1 = \rho A_1 \Delta x_1 = \rho A_1 v_1 \Delta t$. In a similar way, the fluid that moves through the upper end of the pipe in the time Δt has a mass $\Delta m_2 = \rho A_2 v_2 \Delta t$. However, because *mass is conserved* and because the flow is steady, the mass that crosses A_1 in a time Δt must equal the mass that crosses A_2 in a time Δt. That is, $\Delta m_1 = \Delta m_2$, or $\rho A_1 v_1 = \rho A_2 v_2$. Because the density is common to both sides of this expression, we get

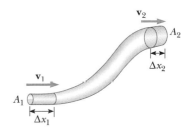

Figure 15.13 A fluid moving with streamline flow through a pipe of varying cross-sectional area. The volume of fluid flowing through A_1 in a time interval Δt must equal the volume flowing through A_2 in the same time interval. Therefore, $A_1 v_1 = A_2 v_2$.

$$A_1 v_1 = A_2 v_2 = \text{constant} \qquad [15.7]$$

This expression, called the **equation of continuity,** says that **the product of the area and the fluid speed at all points along the pipe is a constant.** Therefore, the speed is high where the tube is constricted and low where the tube is wide. The product Av, which has the dimensions of volume/time, is called the **volume flow rate.** The condition $Av = $ constant is equivalent to the fact that **the amount of fluid that enters one end of the tube in a given time interval equals the amount leaving in the same time interval, assuming no leaks.**

• *Equation of continuity*

Example 15.4   Filling a Water Bucket

A water hose 2.00 cm in diameter is used to fill a 20.0-liter bucket. If it takes 1.00 min to fill the bucket, what is the speed v at which the water leaves the hose? ($1 \text{ L} = 10^3 \text{ cm}^3$)

Solution   The cross-sectional area of the hose is

$$A = \pi r^2 = \pi \frac{d^2}{4} = \pi \left(\frac{2.00^2}{4} \right) \text{cm}^2 = \pi \text{ cm}^2$$

According to the data given, the flow rate is equal to 20.0 liters/min. Equating this to the product Av gives

$$Av = 20.0 \, \frac{\text{L}}{\text{min}} = \frac{20.0 \times 10^3 \text{ cm}^3}{60.0 \text{ s}}$$

$$v = \frac{20.0 \times 10^3 \text{ cm}^3}{(\pi \text{ cm}^2)(60.0 \text{ s})} = 106 \text{ cm/s}$$

EXERCISE 6   If the diameter of the hose is reduced to 1.00 cm, what will the speed of the water be as it leaves the hose, assuming the same flow rate?   Answer   424 cm/s

---

15.7 • BERNOULLI'S PRINCIPLE

As a fluid moves through a pipe of varying cross-section and elevation, the pressure changes along the pipe. In 1738 the Swiss physicist Daniel Bernoulli first derived an expression that relates the pressure to fluid speed and elevation.

Consider the flow of an ideal fluid through a nonuniform pipe in a time, Δt, as illustrated in Figure 15.14. The force on the lower end of the fluid is $P_1 A_1$, where P_1 is the pressure in section 1. The work done by this force is $W_1 = F_1 \Delta x_1 = P_1 A_1 \Delta x_1 = P_1 \Delta V$, where ΔV is the volume of section 1. In a similar manner, the

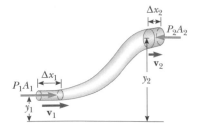

Figure 15.14 A fluid flowing through a constricted pipe with streamline flow. The fluid in the section of length Δx_1 moves to the section of length Δx_2. The volumes of fluid in the two sections are equal.

work done on the fluid at the upper end in the time Δt is $W_2 = -P_2 A_2 \, \Delta x_2 = -P_2 \, \Delta V$. (The volume that passes through section 1 in a time Δt equals the volume that passes through section 2 in the same time interval.) This work is negative because the fluid force opposes the displacement. Thus the net work done by these forces in the time Δt is

$$W_F = (P_1 - P_2)\Delta V$$

If a fluid element of mass Δm enters the tube with speed v_1 and leaves with speed v_2, then the change in its kinetic energy is

$$\Delta K = \tfrac{1}{2}(\Delta m) v_2{}^2 - \tfrac{1}{2}(\Delta m) v_1{}^2$$

As the fluid element rises a vertical distance $y_2 - y_1$, the work done by gravity is negative and given by

$$W_g = -\Delta mg(y_2 - y_1)$$

Because the net work done on the fluid is $W_F + W_g$, we can apply the work-kinetic energy theorem to this volume of fluid to give

$$(P_1 - P_2)\Delta V - \Delta mg(y_2 - y_1) = \tfrac{1}{2}(\Delta m) v_2{}^2 - \tfrac{1}{2}(\Delta m) v_1{}^2$$

If we divide each term by ΔV, noting that $\rho = \Delta m/\Delta V$, and rearrange terms, the expression reduces to

$$P_1 + \tfrac{1}{2}\rho v_1{}^2 + \rho g y_1 = P_2 + \tfrac{1}{2}\rho v_2{}^2 + \rho g y_2 \qquad \textbf{[15.8]}$$

This is **Bernoulli's equation** as applied to an ideal fluid. It is often expressed as

$$P + \tfrac{1}{2}\rho v^2 + \rho g y = \text{constant} \qquad \textbf{[15.9]}$$

> Bernoulli's equation says that the sum of the pressure, (P), the kinetic energy per unit volume $(\tfrac{1}{2}\rho v^2)$, and gravitational potential energy per unit volume $(\rho g y)$ has the same value at all points along a streamline.

Note the Bernoulli's equation is *not* a sum of energy density terms because P is pressure, not energy density. When the fluid is at rest, $v_1 = v_2 = 0$ and Equation 15.8 becomes

$$P_1 - P_2 = \rho g(y_2 - y_1) = \rho g h$$

which agrees with Equation 15.4.

Daniel Bernoulli (1700–1782)

Bernoulli was a Swiss physicist and mathematician who made important discoveries in hydrodynamics. His most famous work, *Hydrodynamica*, published in 1738, is both a theoretical and a practical study of equilibrium, pressure, and velocity of fluids. In this publication, Bernoulli also attempted the first explanation of the behavior of gases with changing pressure and temperature; this was the beginning of the kinetic theory of gases. *(Corbis-Bettmann)*

Thinking Physics 6

In tornadoes, it sometimes occurs that the windows of a house explode outward. Sometimes, the roof of a house is lifted off the house. What causes these events to happen? How does opening the windows in a tornado (which may seem counterintuitive!) help in theory?

Reasoning The wind speed in a tornado is very high. According to Bernoulli's principle, such high wind speeds will result in low pressures. When these winds pass by a window, the pressure outside the window is reduced well below that of the pressure on the inside surface, which is due to the still air inside the house. As a result, the net force outward can burst the window. When these high-speed winds pass over the roof of a house, the pressure above the roof is much lower than the pressure below the roof, again from the still air inside the house. The pressure difference can be enough to lift the roof off the house. By opening the windows, the pressure differential between the outside and inside air can be reduced by allowing a flow of air through the open windows. This will reduce the possibility of exploding windows or a flying roof.

CONCEPTUAL PROBLEM 7

When you are driving a small car on the freeway and a truck passes you at a high speed, you feel pulled *toward* the truck. Why?

Example 15.5 The Venturi Tube

The horizontal constricted pipe illustrated in Figure 15.15, known as a *Venturi tube,* can be used to measure flow speeds such as air speed in flight. Let us determine the flow speed at point 2 if the pressure difference $P_1 - P_2$ is known.

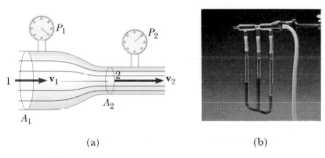

(a)   (b)

Figure 15.15   (a) (Example 15.5) The pressure P_1 is greater than the pressure P_2, because $v_1 < v_2$. This device can be used to measure the speed of fluid flow. (b) A Venturi tube. *(Courtesy of Central Scientific Co.)*

Solution Because the pipe is horizontal, $y_1 = y_2$ and Equation 15.8 applied to points 1 and 2 gives

$$P_1 + \tfrac{1}{2}\rho v_1{}^2 = P_2 + \tfrac{1}{2}\rho v_2{}^2$$

From the equation of continuity (Eq. 15.7), we see that $A_1 v_1 = A_2 v_2$, or

$$v_1 = \frac{A_2}{A_1}\, v_2$$

Substituting this expression into the previous equation gives

$$P_1 + \tfrac{1}{2}\rho\left(\frac{A_2}{A_1}\right)^2 v_2{}^2 = P_2 + \tfrac{1}{2}\rho v_2{}^2$$

$$v_2 = A_1 \sqrt{\frac{2(P_1 - P_2)}{\rho(A_1{}^2 - A_2{}^2)}}$$

We can also obtain an expression for v_1 using this result and the continuity equation. Note that because $A_2 < A_1$, it follows that $P_1 > P_2$. In other words, the pressure is reduced in the constricted part of the pipe. This result is somewhat analogous to the following situation: Consider a very crowded room in which people are squeezed together. As soon as a door is opened and people begin to exit, the squeezing (pressure) is least near the door where the motion (flow) is greatest.

EXERCISE 7 The Garfield Thomas water tunnel at Pennsylvania State University has a circular cross-section that constricts from a diameter of 3.6 m to the test section, which is 1.2 m in diameter. If the speed of flow is 3.0 m/s in the larger-diameter pipe, determine the speed of flow in the test section.   Answer   27 m/s

15.8 • OTHER APPLICATIONS OF BERNOULLI'S PRINCIPLE   O P T I O N A L

Consider the streamlines that flow around an airplane wing as shown in Figure 15.16. Let us assume that the airstream approaches the wing horizontally from the right with a velocity $\mathbf{v}_1$. The tilt of the wing causes the airstream to be deflected downward with a velocity $\mathbf{v}_2$. Because the airstream is deflected by the wing, the wing must exert a force on the airstream. According to Newton's third law, the airstream must exert an equal and opposite force $\mathbf{F}$ on the wing. This force has a

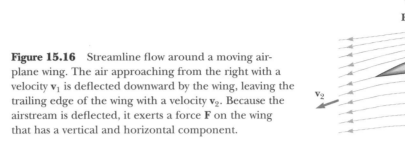

Figure 15.16 Streamline flow around a moving airplane wing. The air approaching from the right with a velocity $\mathbf{v}_1$ is deflected downward by the wing, leaving the trailing edge of the wing with a velocity $\mathbf{v}_2$. Because the airstream is deflected, it exerts a force $\mathbf{F}$ on the wing that has a vertical and horizontal component.

Figure 15.17 A stream of air passing over a tube dipped into a liquid will cause the liquid to rise in the tube as shown.

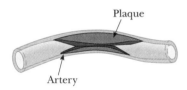

Figure 15.18 Blood must travel faster than normal through a constricted artery.

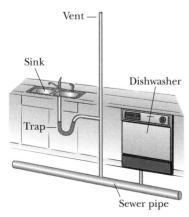

Figure 15.19 (Thinking Physics 7)

vertical component called the **lift** (or aerodynamic lift) and a horizontal component called **drag.** The lift depends on several factors, such as the speed of the airplane, the area of the wing, its curvature, and the angle between the wing and the horizontal. As this angle increases, turbulent flow can set in above the wing to reduce the lift.

The lift on the wing is consistent with Bernoulli's equation. The speed of the airstream is greater above the wing, hence the air pressure above the wing is less than the pressure below the wing, which results in a net upward force.

In general, an object experiences lift by any effect that causes the fluid to change its direction as it flows past the object. Some factors that influence the lift are the shape of the object, its orientation with respect to the fluid flow, spinning motion (for example, a spinning baseball), and the texture of the object's surface.

A number of devices operate in the manner described in Figure 15.17. A stream of air passing over an open tube reduces the pressure above the tube. This reduction in pressure causes the liquid to rise into the air stream. The liquid is then dispersed into a fine spray of droplets. You might recognize that this so-called atomizer is used in perfume bottles and paint sprayers.

Bernoulli's equation explains one symptom called vascular flutter in a person with advanced arteriosclerosis. The artery is constricted as a result of an accumulation of plaque on its inner walls (Fig. 15.18). If the blood speed is sufficiently high in the constricted region, the artery may collapse under external pressure, causing a momentary interruption in blood flow. At this point, Bernoulli's equation does not apply, and the vessel reopens under arterial pressure. As the blood rushes through the constricted artery, the internal pressure drops and again the artery closes. Such variations in blood flow can be heard with a stethoscope. If the plaque becomes dislodged and ends up in a smaller vessel that delivers blood to the heart, the person can suffer a heart attack.

Thinking Physics 7

Consider the portion of a home plumbing system shown in Figure 15.19. The water trap in the pipe below the sink captures a plug of water that prevents sewer gas from finding its way from the sewer pipe, up the sink drain, and into the home. Suppose the dishwasher is draining, so that water is moving to the left in the sewer pipe. What is the purpose of the vent, which is open to the air above the roof of the house? In which direction is air moving at the opening of the vent, upward or downward?

Reasoning Let us imagine that the vent is not present, so that the drain pipe for the sink is simply connected through the trap to the sewer pipe. As water from the dishwasher moves to the left in the sewer pipe, the pressure in the sewer pipe is reduced below atmospheric pressure, according to Bernoulli's principle. The pressure at the drain in the sink is still at atmospheric pressure. Thus, this pressure differential can push the plug of water in the water trap of the sink down the drain pipe into the sewer pipe, removing it as a barrier to sewer gas. With the addition of the vent to the roof, the reduced pressure of the dishwasher water will result in air entering the vent pipe at the roof. This will keep the pressure in the vent pipe and the right-hand side of the sink drain pipe at a pressure close to atmospheric, so that the plug of water in the water trap will remain.

SUMMARY

The **pressure,** P, in a fluid is the force per unit area that the fluid exerts on any surface:

$$P \equiv \frac{F}{A} \qquad \text{[15.1]}$$

In the SI system, pressure has units of N/m^2, and $1 \ N/m^2 = 1$ pascal (Pa).

The pressure in a fluid varies with depth h according to the expression

$$P = P_0 + \rho g h \qquad \text{[15.4]}$$

where P_0 is atmospheric pressure ($= 1.013 \times 10^5 \ N/m^2$) and ρ is the density of the fluid, assumed uniform.

Pascal's law states that when pressure is applied to an enclosed fluid, the pressure is transmitted undiminished to every point in the fluid and to every point on the walls of the container.

When an object is partially or fully submerged in a fluid, the fluid exerts an upward force on the object called the **buoyant force.** According to **Archimedes' principle,** the buoyant force is equal to the weight of the fluid displaced by the body.

Various aspects of fluid dynamics can be understood by assuming that the fluid is nonviscous and incompressible and that the fluid motion is a steady flow with no turbulence.

Using these assumptions, two important results regarding fluid flowing through a pipe of nonuniform size can be obtained:

- The flow rate through the pipe is a constant, which is equivalent to stating that the product of the cross-sectional area, A, and the speed, v, at any point is a constant. This gives the **equation of continuity:**

$$A_1 v_1 = A_2 v_2 = \text{constant} \qquad \text{[15.7]}$$

- The sum of the pressure, kinetic energy per unit volume, and gravitational potential energy per unit volume has the same value at all points along a streamline. This corresponds to **Bernoulli's equation:**

$$P + \tfrac{1}{2}\rho v^2 + \rho g y = \text{constant} \qquad \text{[15.9]}$$

CONCEPTUAL QUESTIONS

1. Figure Q15.1 shows aerial views from directly above two dams. Both dams are equally long (the vertical dimension in the diagram) and equally deep (into the page in the diagram). The dam on the left holds back a very large lake, and the dam on the right holds back a narrow river. Which dam has to be built more strongly?

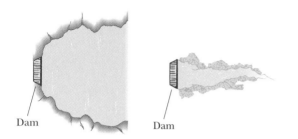

Dam   Dam

Figure Q15.1   (Conceptual Question 1)

2. Some physics students attach a long tube to the opening of a balloon. Leaving the balloon on the ground, the other end of the tube is hoisted to the roof of a multistory campus building. Students at the top of the building start pouring water into the tube. The students on the ground watch the balloon fill with water. On the roof, the students are surprised to see that the tube never seems to fill up—more and more water can continue to be poured into the tube. On the ground, the balloon bursts and drenches the students there. Why did the balloon-tube system never fill up, and why did the balloon burst?

3. A barge is carrying a load of gravel along a river. It approaches a low bridge, and the captain realizes that the top of the pile of gravel is not going to make it under the bridge. The captain orders the crew to quickly shovel gravel from the pile into the water. Is this a good decision?

4. Two drinking glasses having equal weights but different shapes and different cross-sectional areas are filled to the same level with water. According to the expression $P = P_0 + \rho g h$, the pressure is the same at the bottom of both glasses. In view of this, why does one weigh more than the other?

5. Pascal used a barometer with water as the working fluid. Why is it impractical to use water for a typical barometer?

6. A helium-filled balloon rises until its density becomes the same as that of the air. If a sealed submarine begins to sink, will it go all the way to the bottom of the ocean or will it stop when its density becomes the same as that of the surrounding water?

7. A fish rests on the bottom of a bucket of water while the bucket is being weighed. When the fish begins to swim around, does the weight change?

8. Lead has a greater density than iron, and both are denser than water. Is the buoyant force on a lead object greater than, less than, or equal to the buoyant force on an iron object of the same volume?

9. An ice cube is placed in a glass of water. What happens to the level of the water as the ice melts?

10. The water supply for a city is often provided from reservoirs built on high ground. Water flows from the reservoir, through pipes, and into your home when you turn the tap on your faucet. Why is the water flow more rapid out of a faucet on the first floor of a building than in an apartment on a higher floor?

11. If air from a hair dryer is blown over the top of a Ping-Pong ball, the ball can be suspended in air. Explain.

12. When ski-jumpers are airborne, why do they bend their bodies forward and keep their hands at their sides?

13. Explain why a sealed bottle partially filled with a liquid can float.

14. When is the buoyant force on a swimmer greater: after exhaling or after inhaling?

15. A piece of unpainted wood is partially submerged in a container filled with water. If the container is sealed and pressurized above atmospheric pressure, does the wood rise, fall, or remain at the same level? (*Hint:* Wood is porous.)

16. Because atmospheric pressure is about 10^5 N/m^2 and the area of a person's chest is about 0.13 m^2, the force of the atmosphere on one's chest is approximately 13 000 N. In view of this enormous force, why don't our bodies collapse?

17. If you release a ball while inside a freely falling elevator, the ball remains in front of you rather than falling to the floor, because the ball, the elevator, and you all experience the same downward acceleration, **g**. What happens if you repeat this experiment with a helium-filled balloon? (This one is tricky.)

18. Two identical ships set out to sea. One is loaded with a cargo of Styrofoam, and the other is empty. Which ship is more submerged?

19. A small piece of steel is tied to a block of wood. When the wood is placed in a tub of water with the steel on top, half of the block is submerged. If the block is inverted so that the steel is under water, does the amount of the block submerged increase, decrease, or remain the same? What happens to the water level in the tub when the block is inverted?

20. Prairie dogs ventilate their burrows by building a mound over one entrance, which is open to a stream of air. A second entrance at ground level is open to almost stagnant air. How does this construction create an air flow through the burrow?

21. An unopened can of diet cola floats when placed in a tank of water, whereas a can of regular cola of the same brand sinks in the tank. What do you suppose could explain this behavior?

PROBLEMS

Section 15.1 Pressure

1. A 50.0-kg woman balances on one heel of a high-heel shoe. If the heel is circular with radius 0.500 cm, what pressure does she exert on the floor?

2. A king orders a gold crown having a mass of 0.500 kg. When it arrives from the metalsmith, the volume of the crown is found to be 185 cm^3. Is the crown made of solid gold?

3. What is the total mass of the earth's atmosphere? (The radius of the Earth is 6.37×10^6 m, and atmospheric pressure at the surface is 1.013×10^5 N/m^2.)

Section 15.2 Variation of Pressure with Depth

4. Determine the absolute pressure at the bottom of a lake that is 30.0 m deep.

5. The spring of the pressure gauge shown in Figure 15.2 has a force constant of 1000 N/m, and the piston has a diameter of 2.00 cm. Find the depth in water for which the spring compresses by 0.500 cm.

6. The small piston of a hydraulic lift has a cross-sectional area of 3.00 cm^2, and the large piston has an area of 200 cm^2 (Fig. 15.4). What force must be applied to the small piston to raise a load of 15.0 kN? (In service stations this is usually accomplished with compressed air.)

7. What must be the contact area between a suction cup (completely exhausted) and a ceiling in order to support the weight of an 80.0-kg student?

8. A swimming pool has dimensions 30.0 m × 10.0 m and a flat bottom. When the pool is filled to a depth of 2.00 m with fresh water, what is the total force due to the water on the bottom? On each end? On each side?

9. Normal atmospheric pressure is 1.013×10^5 Pa. The approach of a storm causes the height of a mercury barometer to drop by 20.0 mm from the normal height. What is the atmospheric pressure? (The density of mercury is 13.59 g/cm^3.)

10. Mercury is poured into a U-shaped tube, as in Figure P15.10a. The left arm of the tube has a cross-sectional area of $A_1 = 10.0$ cm^2, and the right arm has a cross-sectional area of $A_2 = 5.00$ cm^2. One hundred grams of water are then poured into the right arm, as in Figure P15.10b. (a) Determine the length of the water column in the right arm of the U-shaped tube. (b) Given that the density of mercury is 13.6 g/cm^3, what distance, h, does the mercury rise in the left arm?

11. Blaise Pascal duplicated Torricelli's barometer using (as a Frenchman would) a red Bordeaux wine as the working liquid (Fig. P15.11). The density of the wine he used was 984 kg/m^3. What was the height h of the wine column for normal atmospheric pressure? Would you expect the vacuum above the column to be as good as for mercury?

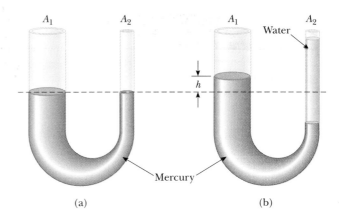

Figure P15.10

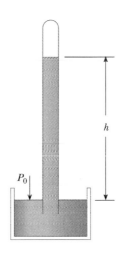

Figure P15.11

Section 15.4 Buoyant Forces and Archimedes' Principle

12. A Styrofoam slab has a thickness of 10.0 cm and a density of 300 kg/m^3. What is the area of the slab if it floats just awash in fresh water when a 75.0-kg swimmer is aboard?

13. A Styrofoam slab has thickness h and density ρ_s. What is the area of the slab if it floats with its top surface just awash in fresh water when a swimmer of mass m is on top?

14. A balloon is filled with 400 m^3 of helium. How big a payload can the balloon lift? (The density of air is 1.29 kg/m^3; the density of helium is 0.180 kg/m^3.)

15. A cube of wood 20.0 cm on a side and having a density of 650 kg/m^3 floats on water. (a) What is the distance from the horizontal top of the cube to the water level? (b) How

much lead weight has to be placed on top of the cube so that its top is just level with the water?

16. About how many helium-filled toy balloons would be required to lift you? Because helium is an irreplaceable resource, develop a theoretical answer rather than an experimental answer. In your solution state what physical quantities you take as data and the values you measure or estimate for them.

17. A plastic sphere floats in water with 50.0 percent of its volume submerged. This same sphere floats in oil with 40.0 percent of its volume submerged. Determine the densities of the oil and the sphere.

18. A frog in a hemispherical pod finds that he just floats without sinking into a sea of blue-green ooze with density 1.35 g/cm³ (Fig. P15.18). If the pod has radius 6.00 cm and negligible mass, what is the mass of the frog?

Figure P15.18

19. How many cubic meters of helium are required to lift a balloon with a 400-kg payload to a height of 8000 m? ($\rho_{He} = 0.18$ kg/m³.) Assume the balloon maintains a constant volume and that the density of air decreases with altitude z according to the expression $\rho_{air} = \rho_0 e^{-z/8000}$, where z is in meters, and ρ_0 (= 1.25 kg/m³) is the density of air at sea level.

20. A 10.0-kg block of metal measuring 15.0 cm × 10.0 cm × 10.0 cm is suspended from a scale and immersed in water, as in Figure 15.9b. The 15.0-cm dimension is vertical, and the top of the block is 5.00 cm from the surface of the water. (a) What are the forces on the top and bottom of the block? (Take $P_0 = 1.0130 \times 10^5$ N/m².) (b) What is the reading of the spring scale? (c) Show that the buoyant force equals the difference between the forces at the top and bottom of the block.

Section 15.5   Fluid Dynamics

Section 15.6   Streamlines and the Equation of Continuity

Section 15.7   Bernoulli's Principle

21. The volume flow rate of water through a horizontal pipe is 2.00 m³/min. Determine the speed of flow at a point

at which the diameter of the pipe is (a) 10.0 cm, (b) 5.00 cm.

22. The legendary Dutch boy who saved Holland by placing his finger in the hole of a dike had a finger 1.20 cm in diameter. Assuming the hole was 2.00-m below the surface of the sea (density 1030 kg/m³), (a) what was the force on his finger? (b) If he removed his finger from the hole, how long would it take the released water to fill 1 acre of land to a depth of 1 foot, assuming the hole remained constant in size? (A typical U.S. family of four uses 1 acre-foot of water, 1234 m³, in 1 year.)

23. A large storage tank, open at the top and filled with water, develops a small hole in its side at a point 16.0 m below the water level. If the rate of flow from the leak is 2.50×10^{-3} m³/min, determine (a) the speed at which the water leaves the hole and (b) the diameter of the hole.

24. A horizontal pipe 10.0 cm in diameter has a smooth reduction to a pipe 5.00 cm in diameter. If the pressure of the water in the larger pipe is 8.00×10^4 Pa and the pressure in the smaller pipe is 6.00×10^4 Pa, at what rate does water flow through the pipes?

25. Water is pumped from the Colorado River up to Grand Canyon Village through a 15.0-cm diameter pipe. The river is at 564 m elevation and the village is at 2096 m. (a) What is the minimum pressure the water must be pumped at to arrive at the village? (b) If 4500 m³ are pumped per day, what is the speed of the water in the pipe? (c) What additional pressure is necessary to deliver this flow? *Note:* You may assume that the acceleration due to gravity and the density of air are constant over this range of elevations.

26. Old Faithful geyser in Yellowstone Park erupts at approximately 1-hour intervals, and the height of the fountain reaches 40.0 m. (a) With what speed does the water leave the ground? (b) What is the pressure (above atmospheric) in the heated underground chamber?

Section 15.8   Other Applications of Bernoulli's Principle (Optional)

27. An airplane has a mass of 1.60×10^4 kg, and each wing has an area of 40.0 m². During level flight, the pressure on the lower wing surface is 7.00×10^4 Pa. Determine the pressure on the upper wing surface.

28. A Venturi tube may be used as a fluid flow meter (Fig. 15.15). If the difference in pressure $P_1 - P_2 = 21.0$ kPa, find the fluid flow rate in m³/s given that the radius of the outlet tube is 1.00 cm, the radius of the inlet tube is 2.00 cm, and the fluid is gasoline ($\rho = 700$ kg/m³).

29. A Pitot tube can be used to determine the velocity of air flow by measuring the difference between the total pressure and the static pressure (Fig. P15.29). If the fluid in the tube is mercury, density $\rho_{Hg} = 13\ 600$ kg/m³, and $\Delta h = 5.00$ cm, find the speed of air flow. (Assume that the air is stagnant at point A and take $\rho_{air} = 1.25$ kg/m³.)

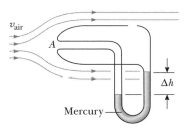

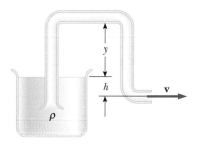

Figure P15.29

30. A siphon is used to drain water from a tank, as indicated in Figure P15.30. The siphon has a uniform diameter. Assume steady flow. (a) If the distance $h = 1.00$ m, find the speed of outflow at the end of the siphon. (b) What is the limitation on the height of the top of the siphon above the water surface? (In order to have a continuous flow of liquid, the pressure in the liquid cannot go below the vapor pressure of the liquid.)

Figure P15.30

31. A large storage tank is filled to a height h_0. The tank is punctured at a height h from the bottom of the tank (Fig. P15.31). (a) Prove that the speed at which the water comes out is $\sqrt{2g(h_0 - h)}$ if the flow is steady and frictionless. (b) How far from the tank will the stream land?

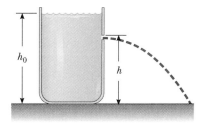

Figure P15.31

32. A hole is punched at height h in the side of a container of height h_0, full of water as in Figure P15.31. If the water is to shoot as far as possible horizontally, (a) how far from the bottom of the container should the hole be punched?

(b) Neglecting friction losses, how far (initially) from the side of the container will the water land?

Additional Problems

33. A Ping-Pong ball has a diameter of 3.80 cm and average density of 0.0840 g/cm^3. What force would be required to hold it completely submerged under water?

34. Figure P15.34 shows a water tank with a valve at the bottom. If this valve is opened, what is the maximum height attained by the water stream coming out of the right side of the tank? Assume that $h = 10.0$ m, $L = 2.00$ m, and $\theta = 30.0°$ and that the cross-sectional area at A is very large compared with that at B.

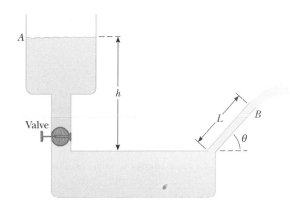

Figure P15.34

35. A helium-filled balloon is tied to a 2.00–m-long, 0.0500-kg uniform string. The balloon is spherical with a radius of 0.400 m. When released, it lifts a length, h, of string and then remains in equilibrium, as in Figure P15.35. Determine the value of h. The envelope of the balloon has mass 0.250 kg.

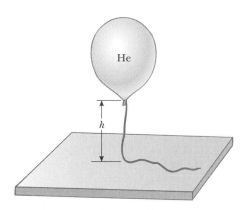

Figure P15.35

36. Water is forced out of a fire extinguisher by air pressure, as shown in Figure P15.36. How much gauge air pressure (above atmospheric) is required for a water jet to have a speed of 30.0 m/s when the water level is 0.500 m below the nozzle?

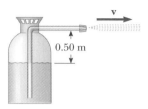

Figure P15.36

37. Torricelli was the first to realize that we live at the bottom of an ocean of air. He correctly surmised that the pressure of our atmosphere is due to the weight of the air. The density of air at 0°C at the Earth's surface is 1.29 kg/m³. The density actually decreases with increasing altitude as the atmosphere thins out. But *if we assume* that the density is constant (1.29 kg/m³) up to some altitude h and zero above, then h would represent the thickness of our atmosphere. Use this model to determine the value of h that gives a pressure of 1.00 atm at the surface of the Earth. Would the peak of Mt. Everest rise above the surface of such an atmosphere?

38. The true weight of a body is its weight when measured in a vacuum in which there are no buoyant forces. A body of volume V is weighed in air on a balance using weights of density ρ. If the density of air is ρ_a and the balance reads w', show that the true weight w is

$$w = w' + \left(V - \frac{w'}{\rho g}\right)\rho_a g$$

39. A wooden dowel has a diameter of 1.20 cm. It floats in water with 0.400 cm of its diameter above water (Fig. P15.39). Determine the density of the dowel.

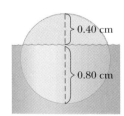

Figure P15.39

40. A 1.00-kg beaker containing 2.00 kg of oil (density = 916.0 kg/m³) rests on a scale. A 2.00-kg block of iron is suspended from a spring scale and completely submerged in the oil, as in Figure P15.40. Determine the equilibrium readings of both scales.

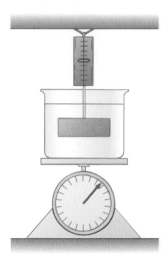

Figure P15.40

41. If a 1-megaton nuclear weapon is exploded at ground level, the peak overpressure (that is, the pressure increase above normal atmospheric pressure) will be 0.200 atm at a distance of 6.00 km. What force due to such an explosion will be exerted on the side of a house with dimensions 4.50 m × 22.0 m?

42. A light spring of constant $k = 90.0$ N/m rests vertically on a table (Fig. P15.42a). A 2.00-g balloon is filled with helium (density = 0.180 kg/m³) to a volume of 5.00 m³ and connected to the spring, causing it to expand as in Figure P15.42b. Determine the expansion length L when the balloon is in equilibrium.

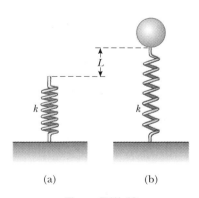

(a)   (b)

Figure P15.42

43. With reference to Figure 15.5, show that the total torque exerted by the water behind the dam about an axis through O is $\frac{1}{6}\rho gwH^3$. Show that the effective line of action of the total force exerted by the water is at a distance $\frac{1}{3}H$ above O.

44. About 1654 Otto von Guericke, inventor of the air pump, evacuated a sphere made of two brass hemispheres. Two teams of eight horses each could pull the hemispheres apart only on some trials, and then "with greatest difficulty," making a sound like a cannon firing (Fig. P15.44). (a) Show that the force F required to pull the evacuated hemispheres apart is $\pi R^2(P_0 - P)$, where R is the radius of the hemispheres and P is the pressure inside the hemispheres, which is much less than P_0. (b) Determine the force if $P = 0.100P_0$ and $R = 0.300$ m.

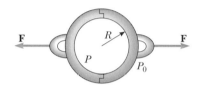

Figure P15.44   *(Henry Leap and Jim Lehman)*

45. In 1983, the United States began coining the cent piece out of copper-clad zinc rather than pure copper. If the mass of the old copper cent is 3.083 g and that of the new cent is 2.517 g, calculate the percent of zinc (by volume) in the new cent. The density of copper is 8.960 g/cm³ and that of zinc is 7.133 g/cm³. The new and old coins have the same volume.

46. A thin spherical shell of mass 4.00 kg and diameter 0.200 m is filled with helium (density = 0.180 kg/m³). It is then released from rest on the bottom of a pool of water that is 4.00 m deep. (a) Neglecting frictional effects, show that the shell rises with constant acceleration, and determine the value of that acceleration. (b) How long will it take for the top of the shell to reach the water surface?

47. An incompressible, nonviscous fluid initially rests in the vertical portion of the pipe shown in Figure P15.47a, where $L = 2.00$ m. When the valve is opened, the fluid flows into the horizontal section of the pipe. What is the speed of the

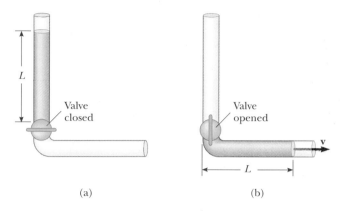

Figure P15.47

fluid when it is entirely in the horizontal section as in Figure P15.47b? Assume the cross-sectional area of the entire pipe is constant.

48. Water falls over a dam of height h meters at a rate of R kg/s. (a) Show that the power available from the water is

$$P = Rgh$$

where g is the acceleration of gravity. (b) Each hydroelectric unit at the Grand Coulee Dam discharges water at a rate of 8.50×10^5 kg/s from a height of 87.0 m. The power developed by the falling water is converted to electric power with an efficiency of 85.0%. How much electric power is produced by each hydroelectric unit?

49. Show that the variation of atmospheric pressure with altitude is given by $P = P_0e^{-\alpha h}$, where $\alpha = \rho_0 g/P_0$, P_0 is atmospheric pressure at some reference level, and ρ_0 is the atmospheric density at this level. Assume that the decrease in atmospheric pressure with increasing altitude is given by Equation 15.4, so that $dP/dh = -\rho g$, and that the density of air is proportional to the pressure.

50. A cube of ice the edge of which is 20.0 mm is floating in a glass of ice-cold water with one of its faces parallel to the water surface. (a) How far below the water surface is the bottom face of the block? (b) Ice-cold ethyl alcohol is gently poured onto the water surface to form a layer 5.00 mm thick above the water. When the ice cube attains hydrostatic equilibrium again, what will be the distance from the top of the water to the bottom face of the block? (c) Additional cold ethyl alcohol is poured onto the water surface until the top surface of the alcohol coincides with the top surface of the ice cube (in hydrostatic equilibrium). How thick is the required layer of ethyl alcohol?

51. A light balloon filled with helium of density 0.180 kg/m³ is tied to a light string of length $L = 3.00$ m. The string is tied to the ground, forming an "inverted" simple pendulum as in Figure P15.51a. If the balloon is displaced slightly from equilibrium, as in Figure P15.51b, (a) show that the motion

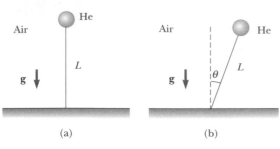

(a)   (b)

Figure P15.51

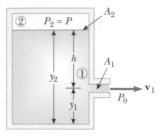

Figure P15.52

is simple harmonic, and (b) determine the period of the motion. Take the density of air to be 1.29 kg/m³, and ignore any energy lost due to air friction.

52. *Torricelli's Law:* A tank, with a cover, containing a liquid of density ρ has a hole in its side at a distance y_1 from the bottom (Fig. P15.52). The diameter of the hole is small compared to the diameter of the tank. The air above the liquid

is maintained at a pressure P. Assume steady, frictionless flow. Show that the speed at which the fluid leaves the hole when the liquid level is a distance h above the hole is

$$v_1 = \sqrt{\frac{2(P - P_0)}{\rho} + 2gh}$$

ANSWERS TO CONCEPTUAL PROBLEMS

1. She can exert enough pressure on the floor to dent or puncture the floor covering. The large pressure is caused by the fact that her weight is distributed over the very small cross-sectional area of her high heels. If you are the homeowner, you might want to suggest that she remove her high heels and put on some slippers.

2. If you think of the grain stored in the silo as a fluid, the pressure the grain exerts on the walls of the silo increases with increasing depth just as water pressure in a lake increases with increasing depth. Thus, the spacing between bands is made smaller at the lower portions to overcome the larger outward forces on the walls in these regions.

3. The level of floating of a ship in the water is unaffected by the atmospheric pressure. Air exerts negligible buoyant force compared to water. The buoyant force of the water results from the pressure *differential* in the fluid. On a high-pressure day, the pressure at all points in the water is higher than on a low-pressure day. Since water is almost incompressible, however, the rate of change of pressure with depth is the same, resulting in no change in the buoyant force.

4. In the ocean, the ship floats due to the buoyant force from *salt water*. Salt water is denser than fresh water. As the ship is pulled up the river, the buoyant force from the fresh water in the river is not sufficient to support the weight of the ship, and it sinks.

5. The balance will not be in equilibrium—the lead side will be lower. Despite the fact that the weights on both sides of the balance are the same, the styrofoam, due to its larger volume, will experience a larger buoyant force from the surrounding air. Thus, the net force of the weight and buoyant force is larger, in the downward direction, for the lead than for the styrofoam.

6. The level of the pond falls. This is because the anchor displaces more water while in the boat. A floating object displaces a volume of water whose weight is equal to the weight of the object. A submerged object displaces a volume of water equal to the volume of the object. Because the density of the anchor is greater than that of water, a volume of water that weighs the same as the anchor will be greater than the volume of the anchor.

7. As the truck passes, the air between your car and the truck is compressed into the channel between you and the truck and moves at a higher speed than when your car is in the open. According to Bernoulli's principle, this high-speed air has a lower pressure than the air on the outer side of your car. The difference in pressure provides a net force on your car toward the truck.

16

Temperature and the Kinetic Theory of Gases

Our study thus far has focused mainly on Newtonian mechanics and relativity, which explain a wide range of phenomena, such as the motion of baseballs, rockets, and planets. We now turn to the study of thermodynamics, which is concerned with the concepts of heat and temperature. As we shall see, thermodynamics is very successful in explaining the bulk properties of matter and the correlation between these properties and the mechanics of atoms and molecules.

Have you ever wondered how a refrigerator cools, or what types of transformations occur in an automobile engine, or what happens to the kinetic energy of an object once the object comes to rest? The laws of thermodynamics and the concepts of heat and temperature enable us to answer such practical questions.

Many things can happen to an object when its temperature is raised. Its size changes slightly, but it may also melt, boil, ignite,

◄ The glass vessel contains dry ice (solid carbon dioxide). The white cloud is carbon dioxide vapor, which is denser than air and hence falls from the vessel. *(R. Folwell/Science Photo Library/Photo Researchers)*

or even explode. The outcome depends on the composition of the object, the degree to which its temperature is raised, and its environment. In general, thermodynamics deals with the physical and chemical transformations of matter in all of its forms: solid, liquid, gas, and plasma.

This chapter concludes with a study of ideal gases. We shall approach this study on two levels. The first will examine ideal gases on the macroscopic scale. Here we shall be concerned with the relationships among such quantities as pressure, volume, and temperature. On the second level, we shall examine gases on a microscopic scale, using a model that pictures the components of a gas as small particles. The latter approach, called the *kinetic theory of gases*, will help us to understand what happens on the atomic level to affect such macroscopic properties as pressure and temperature.

16.1 • TEMPERATURE AND THE ZEROTH LAW OF THERMODYNAMICS

We often associate the concept of temperature with how hot or cold an object feels when we touch it. Thus, our senses provide us with a qualitative indication of temperature. However, our senses are unreliable and often misleading. For example, if we remove a metal ice tray and a cardboard package of frozen vegetables from the freezer, the ice tray feels colder to the hand than the vegetables even though the two are at the same temperature. This is because metal is a better thermal conductor than cardboard, and so the ice tray transfers energy from our hand more efficiently than does the cardboard package. What we need is a reliable and reproducible method for establishing the relative "hotness" or "coldness" of objects. Scientists have developed a variety of thermometers for making such quantitative measurements.

We are all probably familiar with the fact that two objects at different initial temperatures may eventually reach some intermediate temperature when placed in contact with each other. For example, if two soft drinks, one hot and the other cold, are placed in an insulated container, the two eventually reach an equilibrium temperature, with the cold one warming up and the hot one cooling off. Likewise, if a cup of hot coffee is cooled with an ice cube, the ice eventually melts and the coffee's temperature decreases.

In order to understand the concept of temperature, it is useful to first define two often used phrases, **thermal contact** and **thermal equilibrium.** To grasp the meaning of thermal contact, imagine two objects placed in an insulated container so that they interact with each other but not with the rest of the world. If the objects are at different temperatures, energy is exchanged between them. **The energy exchanged between objects because of a temperature difference is called heat.** We shall examine the concept of heat in more detail in Chapter 17. For purposes of the current discussion, we shall assume that two objects are in **thermal contact** with each other if heat can be exchanged between them. **Thermal equilibrium** is the situation in which two objects in thermal contact with each other cease to have any exchange of heat.

Now consider two objects, A and B, that are not in thermal contact with each other, and a third object, C, that will be our thermometer. We wish to determine whether or not A and B would be in thermal equilibrium with each other once placed in thermal contact. The thermometer (object C) is first placed in thermal contact with A until thermal equilibrium is reached. At that point, the thermometer's reading remains constant, and we record it. The thermometer is then placed in thermal contact with B, and its reading is recorded after thermal equilibrium is reached. If the two readings are the same, then A and B are in thermal equilibrium with each other.

We can summarize these results in a statement known as the **zeroth law of thermodynamics** (the law of equilibrium):

• *Zeroth law of thermodynamics*

> If objects A and B are separately in thermal equilibrium with a third object, C, then A and B are in thermal equilibrium with each other.

This statement, insignificant and obvious as it may seem, is easily proved experimentally and is very important because it can be used to define temperature. We can think of temperature as the property that determines whether or not an object is in thermal equilibrium with other objects. **Two objects in thermal equilibrium with each other are at the same temperature.**

16.2 • THERMOMETERS AND TEMPERATURE SCALES

Thermometers are devices used to measure the temperature of a system. All thermometers make use of a change in some physical property with temperature. Some of the physical properties used are (1) the change in volume of a liquid, (2) the change in length of a solid, (3) the change in pressure of a gas held at constant volume, (4) the change in volume of a gas held at constant pressure, (5) the change in electric resistance of a conductor, and (6) the change in color of a very hot object. For a given substance, a temperature scale can be established based on any one of these physical quantities.

The most common thermometer in everyday use consists of a mass of liquid—usually mercury or alcohol—that expands into a glass capillary tube when its temperature rises (Fig. 16.1). In this case the physical property is the change in volume of a liquid. Note that the glass tube also expands as its temperature rises. One can define any temperature change to be proportional to the change in length of the liquid column. The thermometer can be calibrated by placing it in thermal contact with some natural systems that remain at constant temperature. One such system is a mixture of water and ice in thermal equilibrium at atmospheric pressure, which is defined to have a temperature of zero degrees Celsius, written 0°C; this temperature is called the *ice point of water*. Another commonly used system is a mixture of water and steam in thermal equilibrium at atmospheric pressure; its temperature is 100°C, the *steam point of water*. Once the liquid levels in the thermometer have been established at these two points, the column is divided into 100 equal segments, each denoting a change in temperature of one Celsius degree.

Thermometers calibrated in this way present problems when extremely accurate readings are needed. For instance, an alcohol thermometer calibrated at the ice and steam points of water might agree with a mercury thermometer only at the calibration points. Because mercury and alcohol have different thermal expansion properties, when one indicates a temperature of 50°C, say, the other may indicate a slightly different value. The discrepancies between different types of thermome-

Figure 16.1 Schematic diagram of a mercury thermometer. As a result of thermal expansion, the level of the mercury rises as the mercury is heated from 0°C (the ice point) to 100°C (the steam point).

ters are especially large when the temperatures to be measured are far from the calibration points.

An additional practical problem of any thermometer is its limited temperature range. A mercury thermometer, for example, cannot be used below the freezing point of mercury, which is $-39°C$. To surmount these problems, we need a universal thermometer the readings of which are independent of the substances used. The gas thermometer approaches this requirement.

The Constant-Volume Gas Thermometer and the Kelvin Scale

In a **gas thermometer**, the temperature readings are nearly independent of the substance used in the thermometer. One type of gas thermometer is the constant-volume unit, shown in Figure 16.2. The physical property used in this device is the pressure variation with temperature of a fixed volume of gas. When the constant-volume gas thermometer was developed, it was calibrated using the ice and steam points of water, as follows. (A different calibration procedure, to be discussed shortly, is now used.) The gas flask was inserted into an ice bath, and mercury reservoir B was raised or lowered until the volume of the confined gas was at some value, indicated by the zero point on the scale. The height h, the difference between the levels in the reservoir and column A, indicated the pressure in the flask at $0°C$. The flask was inserted into water at the steam point, and reservoir B was readjusted until the height in column A was again brought to zero on the scale, ensuring that the gas volume was the same as it had been in the ice bath (hence the designation "constant-volume"). This gave a value for the pressure at $100°C$. These pressure and temperature values were then plotted on a graph, as in Figure 16.3. The line connecting the two points serves as a calibration curve for measuring unknown temperatures. If we wanted to measure the temperature of a substance, we would place the gas flask in thermal contact with the substance and adjust the column of mercury until the level in column A again returns to zero. The height of the mercury column would tell us the pressure of the gas, and we could then find the temperature of the substance from the graph.

Now suppose that temperatures are measured with various gas thermometers containing different gases. Experiments show that the thermometer readings are nearly independent of the type of gas used, so long as the gas pressure is low and the temperature is well above the point at which the gas liquefies (Fig. 16.4). The agreement among thermometers using different gases improves as the pressure is reduced.

If you extend the curves in Figure 16.4 back toward negative temperatures, you find, in every case, that the pressure is zero when the temperature is $-273.15°C$. This significant temperature is used as the basis for the Kelvin temperature scale, which sets $-273.15°C$ as its zero point (0 K). The size of a degree on the Kelvin scale is identical to the size of a degree on the Celsius scale. Thus, the relationship that enables one to convert between these temperatures is

$$T_C = T - 273.15 \qquad \textbf{[16.1]}$$

where T_C is the **Celsius temperature** and T is the **Kelvin temperature** (sometimes called the **absolute temperature).**

Early gas thermometers made use of ice and steam points according to the procedure just described. However, these points are experimentally difficult to duplicate. For this reason, a new procedure based on a single fixed point was adopted

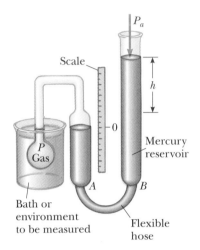

Figure 16.2 A constant-volume gas thermometer measures the pressure of the gas contained in the flask immersed in the bath. The volume of gas in the flask is kept constant by raising or lowering reservoir B such that the mercury level in column A remains constant.

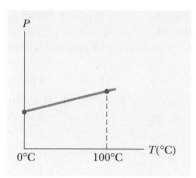

Figure 16.3 A typical graph of pressure versus temperature taken with a constant-volume gas thermometer. The dots represent known reference temperatures (the ice point and the steam point).

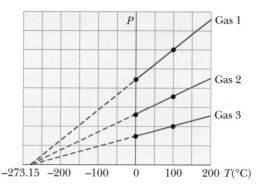

Figure 16.4 Pressure versus temperature for dilute gases. Note that, for all gases, the pressure extrapolates to zero at the unique temperature of −273.15°C.

in 1954 by the International Committee on Weights and Measures. The **triple point of water, which corresponds to the single temperature and pressure at which water, water vapor, and ice can co-exist in equilibrium,** was chosen as a convenient and reproducible reference temperature for the Kelvin scale. It occurs at a temperature of 0.01°C and a pressure of 4.58 mm of mercury. The temperature at the triple point of water on the Kelvin scale has been assigned a value of 273.16 kelvin (K). Thus, the SI unit of temperature, **the kelvin, is defined as 1/273.16 of the temperature of the triple point of water.**

• *The kelvin*

Figure 16.5 shows the Kelvin temperature for various physical processes and structures. The temperature 0 K is often referred to as **absolute zero,** and as Figure 16.5 shows, this temperature has never been achieved, although laboratory experiments have come close.

What would happen to a substance if its temperature could reach 0 K? As Figure 16.4 indicates, the pressure it exerted on the walls of its container would be zero. In Section 16.5 we shall show that the pressure of a gas is proportional to the kinetic energy of the molecules of that gas. Thus, according to classical physics, the kinetic energy of the gas would go to zero, and there would be no motion at all of the individual components of the gas; hence, the molecules would settle out on the bottom of the container. Quantum theory, to be discussed in Chapter 28, modifies this statement to indicate that there would be some residual energy, called the *zero-point energy*, at this low temperature.

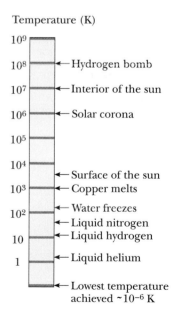

Temperature (K)

Figure 16.5 Absolute temperatures at which various selected physical processes occur. Note that the scale is logarithmic.

The Celsius, Fahrenheit, and Kelvin Temperature Scales

Equation 16.1 shows that the Celsius temperature, T_C, is shifted from the absolute (Kelvin) temperature, T, by 273.15°. Because the size of a degree is the same on the two scales, a temperature difference of 5°C is equal to a temperature difference of 5 K. The two scales differ only in the choice of the zero point. Thus, the ice point (273.15 K) corresponds to 0.00°C, and the steam point (373.15 K) is equivalent to 100.00°C.

The most common temperature scale in everyday use in the United States is the **Fahrenheit scale.** This scale sets the temperature of the ice point at 32°F and the temperature of the steam point at 212°F. The relationship between the Celsius and Fahrenheit temperature scales is

$$T_F = \tfrac{9}{5}T_C + 32°F \qquad \qquad \text{[16.2]}$$

Equation 16.2 can easily be used to find a relationship between changes in temperature on the Celsius and Fahrenheit scales. It is left as a problem for you to show that if the Celsius temperature changes by ΔT_C, the Fahrenheit temperature changes by an amount ΔT_F, given by

$$\Delta T_F = \tfrac{9}{5}\Delta T_C \qquad \qquad \text{[16.3]}$$

Thinking Physics 1

A group of future astronauts lands on an inhabited planet. They strike up a conversation with the life forms about temperature scales. It turns out that the inhabitants of this planet have a temperature scale based on the freezing and boiling points of water separated by 100 of the inhabitants' degrees. Would these two temperatures on this planet be the same as those on Earth? Would the size of the planet inhabitants' degrees be the same as ours? Suppose that the inhabitants have also devised a scale similar to the Kelvin scale. Would their absolute zero be the same as ours?

Reasoning The values of 0°C and 100°C for the freezing and boiling points of water are defined at atmospheric pressure. On another planet, it is unlikely that atmospheric pressure would be exactly the same as that on Earth. Thus, water would freeze and boil at different temperatures on the planet. The inhabitants may *call* these temperatures 0° and 100°, but they would not be the same temperatures as our 0° and 100°. If the inhabitants did assign values of 0° and 100° for these temperatures, then their degrees would not be the same size as our Celsius degrees (unless their atmospheric pressure was the same as ours). For a version of the Kelvin scale from this other planet, the absolute zero would be the same as ours, because it is based on a natural, universal definition, rather than being associated with a particular substance or a given atmospheric pressure.

CONCEPTUAL PROBLEM 1

In an astronomy class, the temperature at the core of a star is given by the professor as 1.5×10^7 degrees. A student asks if this is Kelvin or Celsius. How would you respond?

Example 16.1   Heating a Pan of Water

A pan of water is heated from 25°C to 80°C. What is the change in its temperature on the Kelvin scale and on the Fahrenheit scale?

Solution   From Equation 16.1, we see that the change in temperature on the Celsius scale equals the change on the Kelvin scale. Therefore,

$$\Delta T = \Delta T_C = 80 - 25 = 55°C = 55 \text{ K}$$

From Equation 16.3, we find

$$\Delta T_F = \tfrac{9}{5}\Delta T_C = \tfrac{9}{5}(80 - 25) = 99°F$$

EXERCISE 1   The melting point of gold is 1064°C, and the boiling point is 2660°C. (a) Express these temperatures in kelvin. (b) Compute the difference between these temperatures in Celsius degrees and kelvin.
Answer   (a) 1337 K; 2933 K   (b) 1596°C; 1596 K

16.3 • THERMAL EXPANSION OF SOLIDS AND LIQUIDS

Our discussion of the liquid thermometer made use of one of the best-known changes that occurs in a substance: As its temperature increases, its volume increases. (As we shall see shortly, in some materials the volume decreases when the temperature increases.) This phenomenon, known as **thermal expansion,** plays an important role in numerous applications. For example, thermal expansion joints must be included in buildings, concrete highways, and bridges to compensate for changes in dimensions with temperature variations.

The overall thermal expansion of an object is a consequence of the change in the average separation between its constituent atoms or molecules. Consider how the atoms in a solid substance behave. At ordinary temperatures, the atoms vibrate about their equilibrium positions with an amplitude of about 10^{-11} m, and the average spacing between the atoms is about 10^{-10} m. As the temperature of the solid increases, the average separation between atoms increases. The increase in average separation with increasing temperature (and subsequent thermal expansion) is due to the asymmetry of the potential energy between atoms. If the thermal expansion of an object is sufficiently small compared with the object's initial dimensions, then the change in any dimension is, to a good approximation, dependent on the first power of the temperature change.

Suppose an object has an initial length of L_0 along some direction at some temperature. The length increases by ΔL for the change in temperature ΔT. Experiments show that, when ΔT is small enough, ΔL is proportional to ΔT and to L_0:

$$\Delta L = \alpha L_0 \, \Delta T \qquad [16.4]$$

or

$$L - L_0 = \alpha L_0 (T - T_0) \qquad [16.5]$$

where L is the final length, T is the final temperature, and the proportionality constant α is called the **average coefficient of linear expansion** for a given material and has units of $(°C)^{-1}$.

It may be helpful to think of the process of thermal expansion as a magnification or a photographic enlargement. For example, as a metal washer is heated (Fig. 16.6), all dimensions, including the radius of the hole, increase according to Equation 16.4. Table 16.1 lists the average coefficient of linear expansion for various materials. Note that for these materials α is positive, indicating an increase in length with increasing temperature. This is not always the case. For example, some substances, such as calcite ($CaCO_3$), expand along one dimension (positive α) and contract along another (negative α) with increasing temperature.

Because the linear dimensions of an object change with temperature, it follows that volume and surface area also change with temperature. Consider a cube having an initial edge length, L_0, therefore an initial volume of $V_0 = L_0{}^3$. As the temperature is increased, the length of each side increases to

$$L = L_0 + \alpha L_0 \, \Delta T$$

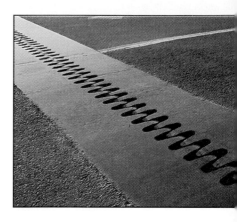

Thermal expansion joints are used to separate sections of roadways on bridges. Without these joints, the surfaces would buckle due to thermal expansion on very hot days or crack due to contraction on very cold days.   *(Frank Siteman, Stock/Boston)*

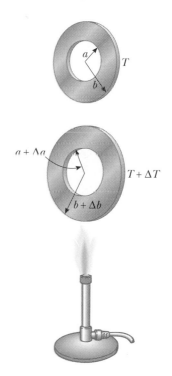

Figure 16.6   Thermal expansion of a homogeneous metal washer. Note that as the washer is heated, all dimensions increase. (The expansion is exaggerated in this figure.)

TABLE **16.1**   **Expansion Coefficients for Some Materials Near Room Temperature**

Material	Linear Expansion Coefficient $\alpha(°C)^{-1}$	Material	Volume Expansion Coefficient $\beta(°C)^{-1}$
Aluminum	24×10^{-6}	Alcohol, ethyl	1.12×10^{-4}
Brass and bronze	19×10^{-6}	Benzene	1.24×10^{-4}
Copper	17×10^{-6}	Acetone	1.5×10^{-4}
Glass (ordinary)	9×10^{-6}	Glycerin	4.85×10^{-4}
Glass (Pyrex)	3.2×10^{-6}	Mercury	1.82×10^{-4}
Lead	29×10^{-6}	Turpentine	9.0×10^{-4}
Steel	11×10^{-6}	Gasoline	9.6×10^{-4}
Invar (Ni–Fe alloy)	0.9×10^{-6}	Air at 0°C	3.67×10^{-3}
Concrete	12×10^{-6}	Helium at 0°C	3.665×10^{-3}

Thermal expansion: The extreme heat of a July day in Asbury Park, New Jersey, caused these railroad tracks to buckle. *(Wide World Photos)*

The new volume, $V = L^3$, is

$$L^3 = (L_0 + \alpha L_0 \, \Delta T)^3 = L_0{}^3 + 3\alpha L_0{}^3 \, \Delta T + 3\alpha^2 L_0{}^3 (\Delta T)^2 + \alpha^3 L_0{}^3 (\Delta T)^3$$

The last two terms in this expression contain the quantity $\alpha \, \Delta T$ raised to the second and third powers. Because $\alpha \, \Delta T$ is much less than one, squaring it makes it even smaller. Therefore, we can neglect these two terms to get a simpler expression:

$$V = L^3 = L_0{}^3 + 3\alpha L_0{}^3 \, \Delta T = V_0 + 3\alpha V_0 \, \Delta T$$

or

$$\Delta V = V - V_0 = \beta V_0 \, \Delta T \qquad \qquad \textbf{[16.6]}$$

where $\beta = 3\alpha$. The quantity β is called the **average coefficient of volume expansion.** We considered a solid cube in deriving it, but Equation 16.6 describes a sample of any shape of solid, liquid, or gas at constant pressure.

By a similar procedure, we can show that the *increase in area* of an object accompanying an increase in temperature is

$$\Delta A = \gamma A_0 \, \Delta T \qquad \qquad \textbf{[16.7]}$$

where γ, the **average coefficient of area expansion,** is given by $\gamma = 2\alpha$.

As Table 16.1 indicates, each substance has its own characteristic coefficients of expansion. For example, when the temperatures of a brass rod and a steel rod of equal length are raised by the same amount from some common initial value, the brass rod expands more than the steel rod because brass has a larger coefficient of expansion than steel. A simple device called a bimetallic strip that uses this principle is found in practical devices such as thermostats. The strip is made by securely bonding two different metals together. As the temperature of the strip increases, the two metals expand by different amounts, and the strip bends as in Figure 16.7.

Figure 16.7   The two metals that form this bimetallic strip are bonded along their longest dimension. The strip in this photograph was straight before being heated and bends when heated. Which way would it bend if it were cooled? *(Courtesy of Central Scientific Company)*

Thinking Physics 2

Suppose a solid object slightly denser than water is sitting on the bottom of a container of water. Can the temperature of this system be changed in any way to make the object float to the surface?

Reasoning Let us imagine lowering the temperature. Liquids tend to have larger co-efficients of linear expansion than solids, so the density of the water will increase more rapidly with the dropping temperature than that of the solid. If the density of the solid were initially only slightly larger than that of water, it would in principle be possible to make it float by lowering the temperature, although one faces the possible problem of reaching the freezing point of the water before this happens.

Thinking Physics 3

A homeowner is painting the ceiling, and a drop of paint falls from the brush onto an operating incandescent light bulb. The bulb breaks. Why?

Reasoning The glass envelope of an incandescent light bulb receives energy on the inside surface by radiation from the very hot filament and movement of the gas filling the bulb. Thus, the glass can become very hot. If a drop of paint falls onto the glass, that portion of the glass envelope will suddenly become cold, and the contraction of this region could cause thermal stresses that would break the glass.

CONCEPTUAL PROBLEM 2

Common thermometers are made of a mercury column in a glass tube. Based on the operation of these common thermometers, which has the larger coefficient of linear expansion—glass or mercury? (Don't answer this by looking in a table!)

CONCEPTUAL PROBLEM 3

Two spheres are made of the same metal and have the same radius, but one is hollow and the other is solid. The spheres are taken through the same temperature increase. Which sphere expands more?

Example 16.2   Does the Hole Get Bigger or Smaller?

A hole of cross-sectional area 100 cm² is cut in a piece of steel at 20°C. What is the change in area of the hole if the steel is heated from 20°C to 100°C?

Solution A hole in a substance expands in exactly the same way as would a piece of the substance having the same shape as the hole. The change in the area of the hole can be found by using Equation 16.7.

$$\Delta A = \gamma A_0 \, \Delta T$$
$$= [22 \times 10^{-6}(°C)^{-1}](100 \text{ cm}^2)(80°C)$$
$$= 0.18 \text{ cm}^2$$

The Unusual Behavior of Water

Liquids generally increase in volume with increasing temperature and have volume expansion coefficients about ten times greater than those of solids. Water is an exception to this rule, as we can see from its density-versus-temperature curve in Figure 16.8. As the temperature increases from 0°C to 4°C, water contracts and thus its density increases. Above 4°C, water expands with increasing temperature. In other words, the density of water reaches a maximum value of 1000 kg/m³ at 4°C.

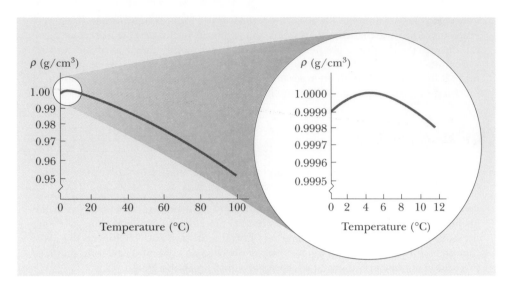

Figure 16.8 The variation of density with temperature for water at atmospheric pressure. The inset at the right shows that the maximum density of water occurs at 4°C.

We can use this unusual thermal expansion behavior of water to explain why a pond freezes at the surface. When the atmospheric temperature drops from, say, 7°C to 6°C, the water at the surface of the pond also cools and consequently decreases in volume. This means that the surface water is denser than the water below it, which has not cooled and decreased in volume. As a result, the surface water sinks and warmer water from below is forced to the surface to be cooled. When the atmospheric temperature is between 4°C and 0°C, however, the surface water expands as it cools, becoming less dense than the water below it. The mixing process stops, and eventually the surface water freezes. As the water freezes, the ice remains on the surface because ice is less dense than water. The ice continues to build up on the surface, and water near the bottom of the pool remains at 4°C. If this did not happen, fish and other forms of marine life would not survive.

EXERCISE 2 An automobile fuel tank is filled to the brim with 45 L of gasoline at 10°C. Immediately afterward, the vehicle is parked in a location where the temperature is 35°C. How much gasoline overflows from the tank as a result of expansion? (Neglect the expansion of the tank.) Answer 1.08 L

Figure 16.9 An ideal gas confined to a cylinder the volume of which can be varied with a movable piston. The state of the gas is defined by any two properties, such as pressure, volume, and temperature.

16.4 • MACROSCOPIC DESCRIPTION OF AN IDEAL GAS

In this section we shall be concerned with the properties of a gas of mass m confined to a container of volume V, pressure P, and temperature T. It is useful to know how these quantities are related. In general, the equation that interrelates these quantities, called the *equation of state*, is very complicated. However, if the gas is maintained at a very low pressure (or low density), the equation of state is found experimentally to be quite simple. Such a low-density gas is commonly referred to as an **ideal gas.** Most gases at room temperature and atmospheric pressure behave approximately as ideal gases.

It is convenient to express the amount of gas in a given volume in terms of the number of moles, n. As we learned in Section 1.2, **one mole** of any substance is that mass of the substance that contains Avogadro's number, $N_A = 6.022 \times 10^{23}$, of molecules. The number of moles, n, of a substance is related to its mass, m, through the expression

$$n = \frac{m}{M} \qquad [16.8]$$

where M is the **molar mass** of the substance, usually expressed in grams per mole. For example, the molar mass of molecular oxygen, O_2, is 32.0 g/mol. Therefore, the mass of one mole of oxygen is 32.0 g.

Now suppose an ideal gas is confined to a cylindrical container the volume of which can be varied by means of a movable piston, as in Figure 16.9. We shall assume that the cylinder does not leak, and so the mass (or the number of moles) remains constant. For such a system, experiments provide the following information. First, when the gas is kept at a constant temperature, its pressure is inversely proportional to the volume (Boyle's law). Second, when the pressure of the gas is kept constant, the volume is directly proportional to the temperature (the law of Charles and Gay-Lussac). These observations can be summarized by the following **equation of state for an ideal gas:**

$$PV = nRT \qquad [16.9]$$

In this expression, called the **ideal gas law,** R is a constant for a specific gas that can be determined from experiments, and T is the absolute temperature in kelvin. Experiments on several gases show that as the pressure approaches zero, the quantity PV/nT approaches the same value of R for all gases. For this reason, R is called the **universal gas constant.** In SI units, where pressure is expressed in pascals and volume in cubic meters, the product PV has units of newton-meters, or joules, and R has the value

$$R = 8.31 \text{ J/mol} \cdot \text{K} \qquad [16.10]$$

• *The universal gas constant*

If the pressure is expressed in atmospheres and the volume in liters (1 L = 10^3 cm^3 = 10^{-3} m^3), then R has the value

$$R = 0.0821 \text{ L} \cdot \text{atm/mol} \cdot \text{K}$$

Using this value of R and Equation 16.9, one finds that the volume occupied by 1 mol of any gas at atmospheric pressure and 0°C (273 K) is 22.4 L.

The ideal gas law is often expressed in terms of the total number of molecules, N. Because the total number of molecules equals the product of the number of moles and Avogadro's number, N_A, we can write Equation 16.9 as

$$PV = nRT = \frac{N}{N_A} RT$$

$$PV = Nk_B T \qquad [16.11]$$

where k_B is called **Boltzmann's constant** and has the value

$$k_B = \frac{R}{N_A} = 1.38 \times 10^{-23} \text{ J/K} \qquad [16.12]$$

• *Boltzmann's constant*

These colorful balloons rise as a gas burner heats the air inside them. Because warm air is less dense than cool air, the buoyant force upward can exceed the total force downward, causing the balloon to rise. The vertical motion can also be controlled by venting hot air at the top of the balloon. *(Richard Megna, Fundamental Photographs, NYC)*

CONCEPTUAL PROBLEM 4

Some picnickers stop at a convenience store to buy food, including bags of snack chips. They drive up into the mountains to their picnic site. When they unload the food, they notice that the chip bags are puffed up like balloons. Why did this happen?

CONCEPTUAL PROBLEM 5

A common material for cushioning objects in packages is made by trapping small bubbles of air between sheets of plastic. Many individuals find satisfaction in popping the bubbles after receiving a package (other individuals who are nearby and listening to the sound may not be so satisfied). Is this material more effective at cushioning the package contents on a hot day or a cold day?

Example 16.3   Squeezing a Tank of Gas

Pure helium gas is admitted into a tank containing a movable piston. The initial volume, pressure, and temperature of the gas are 15×10^{-3} m^3, 200 kPa, and 300 K. If the volume is decreased to 12×10^{-3} m^3 and the pressure increased to 350 kPA, find the final temperature of the gas. (Assume that helium behaves like an ideal gas.)

Solution   If no gas escapes from the tank, the number of moles remains constant; therefore, using $PV = nRT$ at the initial and final points gives

$$\frac{P_i V_i}{T_i} = \frac{P_f V_f}{T_f}$$

where i and f refer to the initial and final values. Solving for T_f, we get

$$T_f = \left(\frac{P_f V_f}{P_i V_i}\right) T_i = \frac{(350 \text{ kPa})(12 \times 10^{-3} \text{ m}^3)}{(200 \text{ kPa})(15 \times 10^{-3} \text{ m}^3)} (300 \text{ K})$$

$$= \boxed{420 \text{ K}}$$

Example 16.4   Heating a Bottle of Air

A sealed glass bottle containing air at atmospheric pressure (101 kPa) and having a volume of 30 cm^3 is at 27°C. It is then tossed into an open fire. When the temperature of the air in the bottle reaches 200°C, what is the pressure inside the bottle? Assume any volume changes of the bottle are negligible.

Solution   This example is approached in the same fashion as that used in Example 16.3. We start with the expression

$$\frac{P_i V_i}{T_i} = \frac{P_f V_f}{T_f}$$

Because the initial and final volumes of the gas are assumed equal, this expression reduces to

$$\frac{P_i}{T_i} = \frac{P_f}{T_f}$$

This gives

$$P_f = \left(\frac{T_f}{T_i}\right) P_i = \left(\frac{473 \text{ K}}{300 \text{ K}}\right)(101 \text{ kPa}) = \boxed{159 \text{ kPa}}$$

It is obvious that the higher the temperature, the higher the pressure exerted by the trapped air. Of course, if the pressure rises high enough, the bottle will shatter.

EXERCISE 3   In this example we neglected the change in volume of the bottle. If the coefficient of volume expansion for glass is 27×10^{-6} (°C)$^{-1}$, find the magnitude of this volume change.   Answer   0.14 cm^3

EXERCISE 4   An ideal gas is held in a container at constant volume. Initially, its temperature is 10.0°C and its pressure is 2.50 atm. What is its pressure when its temperature is 80.0°C?   Answer   3.12 atm

EXERCISE 5   A full tank of oxygen (O_2) contains 12.0 kg of oxygen under a gauge pressure of 40.0 atm. Determine the mass of oxygen that has been withdrawn from the tank when the pressure reading is 25.0 atm. Assume the temperature of the tank remains constant.   Answer   4.39 kg

16.5 • THE KINETIC THEORY OF GASES

In the preceding section we discussed the properties of an ideal gas, using such quantities as pressure, volume, number of moles, and temperature. In this section we shall show that these macroscopic properties can be understood on the basis of what is happening on the atomic (microscopic) scale. In addition, we shall re-examine the ideal gas law in terms of the behavior of the individual molecules that make up the gas.

Because the molecular interactions in a gas are much weaker than those in solids and liquids, our present discussion will be restricted to the molecular behavior of gases.

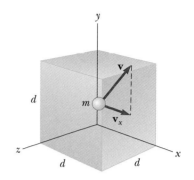

Figure 16.10 A cubical box with sides of length *d* containing an ideal gas. The molecule shown moves with velocity **v**.

Molecular Model for the Pressure of an Ideal Gas

From the kinetic theory of gases, we shall find that the pressure a gas exerts on the walls of its container is a consequence of the collisions of the gas molecules with the walls. We make the following assumptions:

- The number of molecules is large, and the average separation between them is large compared with their dimensions. This means that the molecules occupy a negligible volume in the container.
- The molecules obey Newton's laws of motion, but as a whole they move randomly. By "randomly" we mean that any molecule can move in any direction with any speed. We also assume that the distribution of speeds does not change in time, despite the collisions between molecules. That is, at any given moment, a certain percentage of molecules move at high speeds, a certain constant percentage move at low speeds, and so on.
- We ignore the collisions of molecules with each other in our model.
- The forces between molecules are negligible except during a collision. The forces between molecules are short-range, so the molecules interact with each other only during collisions.
- The gas under consideration is a pure substance; that is, all molecules are identical.

• *Assumptions of kinetic theory for an ideal gas*

Although we often picture an ideal gas as consisting of single atoms, molecular gases exhibit equally good approximations to ideal behavior at low pressures. Effects of molecular rotations or vibrations have no effect, on the average, on the motions considered here.

Now let us derive an expression for the pressure of *N* molecules of an ideal gas in a container of volume *V*. The container is a cube with edges of length *d* (Fig. 16.10). We shall focus our attention on one of these molecules, of mass *m* and assumed to be moving so that its component of velocity in the *x* direction is v_x (Fig. 16.11). As the molecule collides elastically with any wall, its velocity is reversed. Because the *x*-component of the momentum, p_x, of the molecule is mv_x before the collision and $-mv_x$ after the collision, the **change in momentum of the molecule is**

$$\Delta p_x = mv_f - mv_i = -mv_x - mv_x = -2mv_x$$

Applying the impulse-momentum theorem (Eq. 8.12) to the molecule gives

$$F_1 \Delta t = \Delta p_x = -2mv_x$$

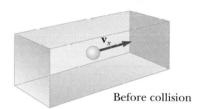

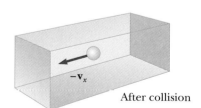

Figure 16.11 A molecule makes an elastic collision with the wall of the container. Its *x* component of momentum is reversed, thereby imparting momentum to the wall, and its *y* component remains unchanged. In this construction, we assume that the molecule moves in the *xy* plane.

In order for the molecule to make two collisions with the same wall, it must travel a distance of $2d$ in the x direction. Therefore, the time interval between two collisions with the same wall is

$$\Delta t = \frac{2d}{v_x}$$

The substitution of this result into the impulse-momentum equation enables us to express the force of the wall on the molecule:

$$F_1 = \frac{-2mv_x}{\Delta t} = \frac{-2mv_x^2}{2d} = \frac{-mv_x^2}{d}$$

Now, by Newton's third law, the force of the molecule on the wall is equal in magnitude and opposite in direction:

$$F_{1,\text{on wall}} = -F_1 = -\left(\frac{-mv_x^2}{d}\right) = \frac{mv_x^2}{d}$$

The total force exerted on the wall by all the molecules is found by adding the forces exerted by the individual molecules:

$$F = \frac{m}{d}\left(v_{x1}^2 + v_{x2}^2 + \cdots + v_{xN}^2\right)$$

In this equation, v_{x1} is the x component of velocity of molecule 1, v_{x2} is the x component of velocity of molecule 2, and so on. The summation terminates when we reach molecule N, because there are N molecules in the container.

To proceed further, note that the average value of the square of the velocity in the x direction for the N molecules is

$$\overline{v_x^2} = \frac{v_{x1}^2 + v_{x2}^2 + \cdots + v_{xN}^2}{N}$$

Thus, the total force on the wall can be written

$$F = \frac{m}{d}\, N\overline{v_x^2}$$

Now let us focus on one molecule in the container and say that this molecule has velocity components v_x, v_y, and v_z. The Pythagorean theorem relates the square of the velocity to the squares of these components:

$$v^2 = v_x^2 + v_y^2 + v_z^2$$

Hence, the average value of v^2 for all the molecules in the container is related to the average values of v_x^2, v_y^2, and v_z^2 according to the expression

$$\overline{v^2} = \overline{v_x^2} + \overline{v_y^2} + \overline{v_z^2}$$

However, the average of the square of the velocity component is the same in any direction because the motion is completely random. Therefore,

$$\overline{v_x^2} = \overline{v_y^2} = \overline{v_z^2}$$

and we have

$$\overline{v^2} = 3\overline{v_x^2}$$

Thus, the total force on the wall is

$$F = \frac{N}{3}\left(\frac{m\overline{v^2}}{d}\right)$$

This expression allows us to find the pressure exerted on the wall:

$$P = \frac{F}{A} = \frac{F}{d^2} = \frac{1}{3}\left(\frac{N}{d^3}\,m\overline{v^2}\right) = \frac{1}{3}\left(\frac{N}{V}\right)(m\overline{v^2})$$

$$P = \frac{2}{3}\left(\frac{N}{V}\right)\left(\frac{1}{2}m\overline{v^2}\right) \qquad\qquad \textbf{[16.13]}$$

• *Pressure of an ideal gas*

This result shows that **the pressure is proportional to the number of molecules per unit volume and to the average translational kinetic energy of the molecules, $\frac{1}{2}m\overline{v^2}$.** With this simplified model of an ideal gas, we have arrived at an important result that relates the large-scale quantity of pressure to an atomic quantity, the average value of the square of the molecular speed. Thus, we have a key link between the atomic world and the large-scale world.

You should note that Equation 16.13 verifies some features of pressure that are probably familiar to you. One way to increase the pressure inside a container is to increase the number of molecules per unit volume in the container. You do this when you add air to a tire. The pressure in the tire can also be increased by increasing the average translational kinetic energy of the molecules in the tire. As we shall see shortly, this can be accomplished by increasing the temperature of the gas inside the tire. That is why the pressure inside a tire increases as the tire heats up during long trips. The continuous flexing of the tires as they move along the road surface results in work done as parts of the tire distort and an increase in internal energy of the rubber. The increased temperature of the rubber results in transfer of energy into the air by conduction, increasing the air's temperature, which in turn produces an increase in pressure.

Molecular Interpretation of Temperature

We can obtain some insight into the meaning of temperature by first writing Equation 16.13 in the form

$$PV = \frac{2}{3}N\left(\frac{1}{2}m\overline{v^2}\right)$$

Let us now compare this with the equation of state for an ideal gas:

$$PV = Nk_\text{B}T$$

Recall that the equation of state is based on experimental facts concerning the macroscopic behavior of gases. Equating the right sides of these expressions, we find that

$$T = \frac{2}{3k_\text{B}}\left(\frac{1}{2}m\overline{v^2}\right) \qquad\qquad \textbf{[16.14]}$$

• *Temperature is proportional to average kinetic energy*

That is, **temperature is a direct measure of average translational molecular kinetic energy.**

By rearranging Equation 16.14, we can relate the average translational molecular kinetic energy to the temperature:

Average kinetic energy per molecule •

$$\tfrac{1}{2} m \overline{v^2} = \tfrac{3}{2} k_B T \qquad\qquad [16.15]$$

That is, average translational kinetic energy per molecule is $\tfrac{3}{2} k_B T$. Because $\overline{v_x^2} = \tfrac{1}{3}\overline{v^2}$, it follows that

$$\tfrac{1}{2} m \overline{v_x^2} = \tfrac{1}{2} k_B T \qquad\qquad [16.16]$$

In a similar manner, for the y and z motions we find

$$\tfrac{1}{2} m \overline{v_y^2} = \tfrac{1}{2} k_B T \qquad \text{and} \qquad \tfrac{1}{2} m \overline{v_z^2} = \tfrac{1}{2} k_B T$$

Thus, each translational degree of freedom contributes an equal amount of energy to the gas, namely, $\tfrac{1}{2} k_B T$. (In general, the phrase "degrees of freedom" refers to the number of independent means by which a molecule can possess energy.) A generalization of this result, known as the **theorem of equipartition of energy,** says that **the energy of a system in thermal equilibrium is equally divided among all degrees of freedom.**

Theorem of equipartition of energy •

The total translational kinetic energy of N molecules of gas is simply N times the average energy per molecule, which is given by Equation 16.15:

Total kinetic energy of N molecules •

$$E = N \left(\tfrac{1}{2} m \overline{v^2} \right) = \tfrac{3}{2} N k_B T = \tfrac{3}{2} n R T \qquad\qquad [16.17]$$

where we have used $k_B = R/N_A$ for Boltzmann's constant and $n = N/N_A$ for the number of moles of gas. From this result, we see that **the total kinetic energy of a system of molecules is proportional to the absolute temperature of the system.**

The square root of $\overline{v^2}$ is called the *root-mean-square* (rms) *speed* of the molecules. From Equation 16.15 we get, for the rms speed,

Root-mean-square speed •

$$v_{rms} = \sqrt{\overline{v^2}} = \sqrt{\frac{3 k_B T}{m}} = \sqrt{\frac{3 R T}{M}} \qquad\qquad [16.18]$$

where M is the molar mass in kg/mol. This expression shows that, at a given temperature, lighter molecules move faster, on the average, than heavier molecules. For example, hydrogen, with a molar mass of 2×10^{-3} kg/mol, moves four times as fast as oxygen, the molar mass of which is 32×10^{-3} kg/mol. Table 16.2 lists the rms speeds for various molecules at 20°C.

TABLE 16.2   Some rms Speeds

Gas	Molar Mass (g/mol)	v_{rms} at 20°C (m/s)
H_2	2.02	1902
He	4.0	1352
H_2O	18	637
Ne	20.1	603
N_2 or CO	28	511
NO	30	494
CO_2	44	408
SO_2	64	338

Thinking Physics 4

Imagine a gas in an insulated cylinder with a movable piston. The piston is pushed inward, compressing the gas, and then released. As the molecules of the gas strike the piston, they move it outward. From the points of view of (a) energy principles, and (b) kinetic theory, explain how the expansion of this gas causes its temperature to drop.

Reasoning (a) From the point of view of energy principles, the molecules strike the piston and move it through a distance. Thus, the molecules do work on the piston, which represents a transfer of energy out of the gas. As a result, the internal energy of the gas drops. Because temperature is related to internal energy, the temperature of the gas drops.

(b) From the point of view of kinetic theory, a molecule colliding with the piston causes the piston to rebound with some velocity. According to conservation of momentum, then, the molecule must rebound with less velocity than it had before the collision. Thus, as these collisions occur, the average velocity of the collection of molecules is reduced. Because temperature is related to the average velocity of molecules, the temperature of the gas drops.

CONCEPTUAL PROBLEM 6

Small planets tend to have little or no atmosphere. Why is this?

Example 16.5   A Tank of Helium

A tank of volume 0.300 m^3 contains 2.00 mol of helium gas at 20.0°C. Assuming the helium behaves like an ideal gas, (a) find the total thermal energy of the system.

Solution   Using Equation 16.17 with $n = 2.00$ and $T = 293$ K, we get

$$E = \tfrac{3}{2}nRT = \tfrac{3}{2}(2.00 \text{ mol})(8.31 \text{ J/mol·K})(293 \text{ K})$$

$$= 7.30 \times 10^3 \text{ J}$$

(b) What is the average kinetic energy per molecule?

Solution   From Equation 16.15, we see that the average kinetic energy per molecule is

$$\tfrac{1}{2}m\overline{v^2} = \tfrac{3}{2}k_B T = \tfrac{3}{2}(1.38 \times 10^{-23} \text{ J/K})(293 \text{ K})$$

$$= 6.07 \times 10^{-21} \text{ J}$$

EXERCISE 6   Using the fact that the molar mass of helium is 1.00×10^{-3} kg/mol, determine the rms speed of the atoms at 20.0°C.   Answer 1.35×10^3 m/s

---

SUMMARY

The **zeroth law of thermodynamics** states that if two objects, A and B, are separately in thermal equilibrium with a third object, then A and B are in thermal equilibrium with each other.

The relationship between T_C, the **Celsius temperature,** and T, the **Kelvin (absolute) temperature,** is

$$T_C = T - 273.15 \tag{16.1}$$

The relationship between the **Fahrenheit and Celsius** temperatures is

$$T_F = \tfrac{9}{5}T_C + 32°\text{F} \tag{16.2}$$

When a substance is heated, it generally expands. If an object has an initial length of L_0 at some temperature and undergoes a change in temperature of ΔT, its length changes by the amount ΔL, which is proportional to the object's initial length and the temperature change:

$$\Delta L = \alpha L_0 \Delta T \tag{16.4}$$

The parameter α is called the **average coefficient of linear expansion.**

The change in area of a substance is given by

$$\Delta A = \gamma A_0 \Delta T \tag{16.7}$$

where γ is the **average coefficient of area expansion** and is equal to 2α.

The change in volume of most substances is proportional to the initial volume, V_0, and the temperature change, ΔT:

$$\Delta V = \beta V_0 \, \Delta T$$

where β is the **average coefficient of volume expansion** and is equal to 3α.

An **ideal gas** is one that obeys the equation

$$PV = nRT \qquad [16.9]$$

where P is the pressure of the gas, V is its volume, n is the number of moles of gas, R is the universal gas constant (8.31 J/mol·K), and T is the absolute temperature in kelvin. A real gas at very low pressures behaves approximately as an ideal gas.

The **pressure** of N molecules of an ideal gas contained in a volume V is given by

$$P = \tfrac{2}{3}\left(\frac{N}{V}\right)\left(\tfrac{1}{2}m\overline{v^2}\right) \qquad [16.13]$$

where $\tfrac{1}{2}m\overline{v^2}$ is the **average kinetic energy per molecule.**

The average kinetic energy of the molecules of a gas is directly proportional to the absolute temperature of the gas:

$$\tfrac{1}{2}m\overline{v^2} = \tfrac{3}{2}k_B T \qquad [16.15]$$

where k_B is **Boltzmann's constant** (1.38×10^{-23} J/K).

The **root-mean-square** (rms) speed of the molecules of gas is

$$v_{\text{rms}} = \sqrt{\overline{v^2}} = \sqrt{\frac{3k_B T}{m}} = \sqrt{\frac{3RT}{M}} \qquad [16.18]$$

CONCEPTUAL QUESTIONS

1. A piece of copper is dropped into a beaker of water. If the water's temperature rises, what happens to the temperature of the copper? Under what conditions are the water and copper in thermal equilibrium?

2. In the movie *Apollo 13* (Universal, 1995), the character who depicts astronaut Jim Lovell says, in describing his upcoming trip to the Moon, "I'll be walking in a place where there's 400° difference between sunlight and shadow." What exactly is it that is hot in sunlight and cold in shadow? Suppose an astronaut standing on the Moon holds a thermometer in his or her gloved hand. Is it reading the temperature of the vacuum at the Moon's surface? Does it read a temperature? If so, what object or substance has that temperature?

3. When the metal ring and metal sphere in Figure Q16.3 are both at room temperature, the sphere does not fit through the ring. After the ring is heated, the sphere can be passed through the ring. Explain.

4. Explain why a column of mercury in a thermometer first descends slightly and then rises when placed in hot water.

5. Markings to indicate length are placed on a steel tape in a room that has a temperature of 22°C. Are measurements made with the tape on a day when the temperature is 27°C too long, too short, or accurate? Defend your answer.

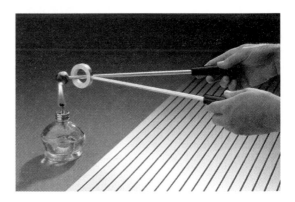

Figure Q16.3   *(Courtesy of Central Scientific Company)*

6. If a helium-filled balloon initially at room temperature is placed in a freezer, will its volume increase, decrease, or remain the same?

7. What does the ideal gas law predict about the volume of a gas at absolute zero? Why is this prediction incorrect?

8. What happens to a helium-filled balloon released into the air? Will it expand or contract? Will it stop rising at some height?

9. An inflated rubber balloon filled with air is immersed in a flask of liquid nitrogen that is at 77 K. Describe what happens to the balloon, assuming that it remains flexible while being cooled.

10. Two cylinders at the same temperature each contain the same kind and quantity of gas. If the volume of cylinder *A* is three times greater than the volume of cylinder *B*, what can you say about the relative pressures in the cylinders?

11. After food is cooked in a pressure cooker, why is it very important to cool the container with cold water before attempting to remove the lid?

12. The shore at a particular point of the ocean is very rocky. At one location, the rock formation is such that a cave is located with an opening that is underwater, as shown in Figure Q16.12a. (a) Inside the cave, there is a pocket of trapped air. As the level of the ocean rises and falls due to the tides, will the level of the water in the cave rise and fall? If so, will it have the same amplitude as that of the ocean? (b) Suppose the cave is deeper in the water, so that it is completely submerged and filled with water at high tide, as shown in Figure Q16.12b. At low tide, will the level of the water in the cave be the same as that of the ocean?

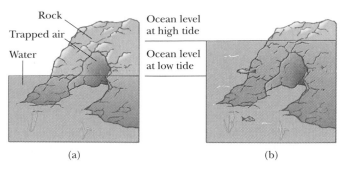

Figure Q16.12

13. An alcohol rub is sometimes used to help lower the temperature of a sick patient. Why does this help?

14. Ideal gas is contained in a vessel at 300 K. If the temperature is increased to 900 K, by what factor does each one of the following change? (a) The average kinetic energy of one molecule; (b) the rms molecular speed; (c) the average momentum change in one collision of a molecule with a wall; (d) the rate of collisions of molecules with walls; (e) the pressure of the gas.

PROBLEMS

Section 16.2   Thermometers and Temperature Scales

1. Convert the following temperatures to Celsius and kelvin: (a) the normal human body temperature, 98.6°F; (b) the air temperature on a cold day, −5.00°F.

2. In a constant-volume gas thermometer, the pressure at 20.0°C is 0.980 atm. (a) What is the pressure at 45.0°C? (b) What is the temperature if the pressure is 0.500 atm?

3. A constant-volume gas thermometer is calibrated in dry ice (which is carbon dioxide in the solid state and has a temperature of −80.0°C) and in boiling ethyl alcohol (78.0°C). The two pressures are 0.900 atm and 1.635 atm. (a) What value of absolute zero does the calibration yield? What is the pressure at (b) the freezing point of water and (c) the boiling point of water?

4. Show that the temperature −40.0° is unique in that it has the same numerical value on the Celsius and Fahrenheit scales.

5. Liquid nitrogen has a boiling point of −195.81°C at atmospheric pressure. Express this temperature in (a) degrees Fahrenheit and (b) kelvin.

6. On a Strange temperature scale, the freezing point of water is −15.0°S and the boiling point is +60.0°S. Develop a *linear* conversion equation between this temperature scale and the Celsius scale.

7. Show that if the temperature on the Celsius scale changes by ΔT_C, the Fahrenheit temperature changes by ΔT_F, which equals $\frac{9}{5}\Delta T_C$.

8. The temperature difference between the inside and the outside of an automobile engine is 450°C. Express this temperature difference on the (a) Fahrenheit scale, (b) Kelvin scale.

9. The temperature of one northeastern state varies from 105°F in the summer to −25.0°F in winter. Express this range of temperature in degrees Celsius.

Section 16.3   Thermal Expansion of Solids and Liquids

(Use Table 16.1.)

10. The concrete sections of a certain superhighway are designed to have a length of 25.0 m. The sections are poured and cured at 10.0°C. What minimum spacing should the engineer leave between the sections to eliminate buckling if the concrete is to reach a temperature of 50.0°C?

11. A copper telephone wire has essentially no sag between poles 35.0 m apart on a winter day when the temperature is −20.0°C. How much longer is the wire on a summer day when $T_C = 35.0$°C?

12. A brass ring of diameter 10.00 cm at 20.0°C is heated and slipped over an aluminum rod of diameter 10.01 cm at 20.0°C. Assuming the average coefficients of linear expansion are constant, (a) to what temperature must this combination be cooled to separate them? Is this attainable? (b) What if the aluminum rod were 10.02 cm in diameter?

13. A pair of eyeglass frames is made of epoxy plastic. At room temperature (assume 20.0°C), the frames have circular lens holes 2.20 cm in radius. To what temperature must the frames be heated in order to insert lenses 2.21 cm in radius? The average coefficient of linear expansion for epoxy is 1.30×10^{-4} (°C)$^{-1}$.

14. A copper rod and steel rod are heated. At 0°C the copper rod has a length of L_C, the steel one has a length L_S. When the rods are being heated or cooled, a difference of 5.00 cm is maintained between their lengths. Determine the values of L_C and L_S.

15. A square hole 8.00 cm along each side is cut in a sheet of copper. Calculate the change in the area of this hole if the temperature of the sheet is increased by 50.0 K.

16. The average volume coefficient of expansion for carbon tetrachloride is 5.81×10^{-4}(°C)$^{-1}$. If a 50.0-gal steel container is filled completely with carbon tetrachloride when the temperature is 10.0°C, how much will spill over when the temperature rises to 30.0°C?

17. The active element of a certain laser is made of a glass rod 30.0 cm long by 1.50 cm in diameter. If the temperature of the rod increases by 65.0°C, find the increase in (a) its length, (b) its diameter, and (c) its volume. (Take $\alpha = 9.00 \times 10^{-6}$(°C)$^{-1}$.)

18. A volumetric glass flask made of Pyrex is calibrated at 20.0°C. It is filled to the 100-mL mark with 35.0°C acetone. (a) What is the volume of the acetone when it cools to 20.0°C? (b) How significant is the change in volume of the flask?

Section 16.4   Macroscopic Description of an Ideal Gas

19. Gas is contained in an 8.00-liter vessel at a temperature of 20.0°C and a pressure of 9.00 atm. (a) Determine the number of moles of gas in the vessel. (b) How many molecules are there in the vessel?

20. A tank having a volume of 0.100 m^3 contains helium gas at 150 atm. How many balloons can the tank blow up if each filled balloon is a sphere 0.300 m in diameter at an absolute pressure of 1.20 atm?

21. An auditorium at 20.0°C and 101 kPa pressure has dimensions 10.0 m $\times$ 20.0 m $\times$ 30.0 m. How many molecules of air fill the auditorium?

22. Nine grams of water are placed in a 2.00-L pressure cooker and heated to 500°C. What is the pressure inside the container?

23. The mass of a hot-air balloon and its cargo (not including the air inside) is 200 kg. The air outside is at 10.0°C and 101 kPa. The volume of the balloon is 400 m^3. To what temperature must the air in the balloon be heated before the balloon will lift off? (Air density at 10.0°C is 1.25 kg/m^3.)

24. One mole of oxygen gas is at a pressure of 6.00 atm and a temperature of 27.0°C. (a) If the gas is heated at constant volume until the pressure triples, what is the final temperature? (b) If the gas is heated until both the pressure and volume are doubled, what is the final temperature?

25. An automobile tire is inflated using air originally at 10.0°C and normal atmospheric pressure. During the process, the air is compressed to 28.0 percent of its original volume and the temperature is increased to 40.0°C. (a) What is the tire pressure? (b) After the car is driven at high speed, the tire air temperature rises to 85.0°C and the interior volume of the tire increases by 2.00%. What is the new tire pressure in pascals (absolute)?

26. A weather balloon is designed to expand to a maximum radius of 20.0 m when in flight at its working altitude, where the air pressure is 0.0300 atm and the temperature is 200 K. If the balloon is filled at atmospheric pressure and 300 K, what is its radius at lift-off?

27. Estimate the mass of the air in your bedroom. State the quantities you take as data and the value you measure or estimate for each.

28. A room of volume 80.0 m^3 contains air having an average molar mass of 29.0 g/mol. If the temperature of the room is raised from 18.0°C to 25.0°C, what mass of air (in kg) will leave the room? Assume that the air pressure in the room is maintained at 101 kPa.

29. A room of volume V contains air having an average molar mass of M g/mol. If the temperature of the room is raised from T_1 to T_2, what mass of air will leave the room? Assume that the air pressure in the room is maintained at P_0.

30. At 25.0 m below the surface of the sea (density = 1025 kg/m^3), at which point the temperature is 5.00°C, a diver exhales an air bubble having a volume of 1.00 cm^3. If the surface temperature of the sea is 20.0°C, what is the volume of the bubble right before it breaks the surface?

Section 16.5   The Kinetic Theory of Gases

31. Use the definition of Avogadro's number to find the mass of a helium atom.

32. A sealed cubical container 20.0 cm on a side contains three times Avogadro's number of molecules at a temperature of 20.0°C. Find the force exerted by the gas on one of the walls of the container.

33. A cylinder contains a mixture of helium and argon gas in equilibrium at 150°C. (a) What is the average kinetic energy of each gas molecule? (b) What is the root-mean-square speed of each type of molecule?

34. At what temperature would the rms speed of helium atoms (mass = 6.66×10^{-27} kg) equal (a) the escape speed from Earth, 1.12×10^4 m/s, and (b) the escape speed from the Moon, 2.37×10^3 m/s?

35. In a period of 1.00 s, 5.00×10^{23} nitrogen molecules strike a wall of area 8.00 cm^2. If the molecules move with a speed

of 300 m/s and strike the wall head-on in perfectly elastic collisions, find the pressure exerted on the wall. (The mass of one N_2 molecule is 4.68×10^{-26} kg.)

36. Suppose that the temperature at the center of the Sun is 2.00×10^7 K. Find (a) the average translational kinetic energy of a proton in the Sun's interior and (b) the proton's root-mean-square speed.

37. A 5.00-liter vessel contains 0.125 moles of an ideal gas at a pressure of 1.50 atm. What is the average translational kinetic energy of a single molecule?

38. A mixture of two gases will diffuse through a filter at rates proportional to their rms speeds. (a) Find the ratio of speeds for the two isotopes of chlorine, ^{35}Cl and ^{37}Cl, as they diffuse through the air. (b) Which isotope moves faster?

Additional Problems

39. At what Fahrenheit temperature are the Kelvin and Fahrenheit temperatures numerically equal?

40. A student measures the length of a brass rod with a steel tape at 20.0°C. The reading is 95.00 cm. What will the tape indicate for the length of the rod when the rod and the tape are at (a) -15.0°C, (b) 55.0°C?

41. The density of gasoline is 730 kg/m^3 at 0°C. Its volume expansion coefficient is 9.60×10^{-4}/C°. If 1.00 gallon of gasoline occupies 0.00380 m^3, how many extra kilograms of gasoline would you get if you bought 10.0 gallons of gasoline at 0°C rather than at 20.0°C from a pump not temperature-compensated?

42. A steel ball bearing is 4.000 cm in diameter at 20.0°C. A bronze plate has a hole in it that is 3.994 cm in diameter at 20.0°C. What common temperature must they have in order that the ball just squeeze through the hole?

43. A mercury thermometer is constructed as in Figure P16.43. The capillary tube has a diameter of 0.00400 cm, and the bulb has a diameter of 0.250 cm. Neglecting the expansion of the glass, find the change in height of the mercury column for a temperature change of 30.0°C.

44. A liquid with a coefficient of volume expansion β just fills a spherical shell of volume V at a temperature of T (Fig. P16.43). The shell is made of a material that has a coefficient of linear expansion of α. The liquid is free to expand into a capillary of cross-sectional area A at the top. (a) If the temperature increases by ΔT, show that the liquid rises in the capillary by the amount Δh given by $\Delta h = (V/A)(\beta - 3\alpha)\Delta T$. (b) For a typical system, such as a mercury thermometer, why is it a good approximation to neglect the expansion of the shell?

45. The rectangular plate shown in Figure P16.45 has an area A equal to ℓw. If the temperature increases by ΔT, show that the increase in area is $\Delta A = 2\alpha A \Delta T$, where α is the average coefficient of linear expansion. What approximation does this expression assume? (*Hint:* Note that each dimension increases according to $\Delta \ell = \alpha \ell \Delta T$.)

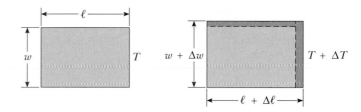

Figure P16.45

46. Show that the equation of state of an ideal gas can be written as $PM = \rho RT$, where M is its molar mass.

47. A liquid has a density ρ. (a) Show that the fractional change in density for a change in temperature ΔT is $\Delta \rho / \rho = -\beta \Delta T$. What does the negative sign signify? (b) Fresh water has a maximum density of 1.000 g/cm^3 at 4.0°C. At 10.0°C, its density is 0.9997 g/cm^3. What is β for water over this temperature interval?

48. An expandable cylinder has its top connected to a spring of constant 2.00×10^3 N/m (see Fig. P16.48). The cylinder is filled with 5.00 L of gas with the spring relaxed at a pressure

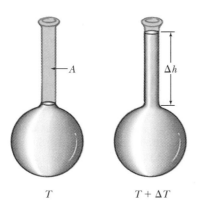

Figure P16.43

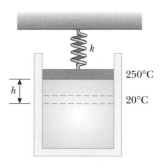

Figure P16.48

of 1.00 atm and a temperature of 20.0°C. (a) If the lid has a cross-sectional area of 0.0100 m^2 and negligible mass, how high will the lid rise when the temperature is raised to 250°C? (b) What is the pressure of the gas at 250°C?

49. A vertical cylinder of cross-sectional area A is fitted with a tight-fitting, frictionless piston of mass m (Fig. P16.49). (a) If there are n mol of an ideal gas in the cylinder at a temperature of T, determine the height h at which the pis-

ton is in equilibrium under its own weight. (b) What is the value for h if $n = 0.200$ mol, $T = 400$ K, $A = 0.00800$ m^2, and $m = 20.0$ kg?

50. A bimetallic bar is made of two thin strips of dissimilar metals bonded together. As they are heated, the one with the larger average coefficient of expansion expands more than the other, forcing the bar into an arc, with the outer radius having a larger circumference (see Fig. P16.50). (a) Derive an expression for the angle of bending θ as a function of the initial length of the strips, their average coefficients of linear expansion, the change in temperature, and the separation of the centers of the strips ($\Delta r = r_2 - r_1$). (b) Show that the angle of bending goes to zero when ΔT goes to zero or when the two coefficients of expansion become equal. (c) What happens if the bar is cooled?

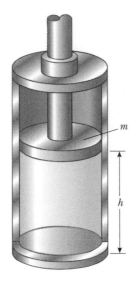

Figure P16.49

Figure P16.50

ANSWERS TO CONCEPTUAL PROBLEMS

1. The accurate response would be that it does not matter! Temperatures on the Kelvin and Celsius scales differ by only 273 degrees. This difference is insignificant for temperatures on the order of 10^7. If we imagine that the temperature given is in kelvin, and ignore any problems with significant figures, then the Celsius temperature is 1.4999727×10^7 °C.

2. Mercury must have the larger coefficient of expansion. As the temperature of a thermometer rises, both the mercury and the glass expand. If they both had the same coefficient of linear expansion, the mercury and the cavity in the glass would both expand by the same amount, and there would be no apparent movement of the end of the mercury column relative to the calibration scale on the glass. If the glass expanded more than the mercury, the reading would go down as the temperature went up! Now that we have argued this conceptually, we can look in a table and find that the coefficient for mercury is about 20 times as large as for

glass, so that the expansion of the glass can effectively be ignored.

3. A cavity in a material expands in exactly the same way as if the cavity were filled with material. Thus, both spheres will expand by the same amount.

4. The chip bags contain a sealed sample of air. When the bags are taken up the mountain, the external atmospheric pressure on the bags is reduced. As a result, the difference between the pressure of the air inside the bags and the reduced pressure outside results in a net force pushing the plastic of the bag outward.

5. On a cold day, the trapped air in the bubbles would be reduced in pressure, according to the ideal gas law. Thus, the volume of the bubbles may be smaller than on a hot day, and the material would not be as effective in cushioning the package contents.

6. The retention of an atmosphere on a planet is due to the gravitational force holding the gas of the atmosphere to the

planet. On a small planet, the gravitational force is very small, and the escape speed is correspondingly small. If a small planet starts its existence with an atmosphere, the molecules of the gas will have a distribution of speeds, according to kinetic theory. Some of these molecules will have speeds higher than the escape speed of the planet and will leave the atmosphere. As the remaining atmosphere is warmed by radiation from the Sun, more molecules will attain speeds high enough to escape. As a result, the atmosphere bleeds off into space.

17

Heat and the First Law of Thermodynamics

Until about 1850, the fields of heat and mechanics were considered to be two distinct branches of science, and the law of conservation of energy seemed to describe only certain kinds of mechanical systems. Mid–nineteenth-century experiments performed by the Englishman James Joule (1818–1889) and others showed that energy may be added to (or removed from) a system either as heat or as work done on (or by) the system. Now thermal energy is treated as a form of energy that can be transformed into mechanical energy. Once the concept of energy was broadened to include thermal energy, the law of conservation of energy emerged as a universal law of nature.

This chapter focuses on the concept of heat, the first law of thermodynamics, and some important applications. The first law of thermodynamics is merely the law of conservation of energy. It tells us only that an increase in one form of energy must be accompanied by a decrease in some other form of energy. The first law places no restrictions on the types of energy conversions that can occur. Furthermore, the first law makes no distinction between the results of heat

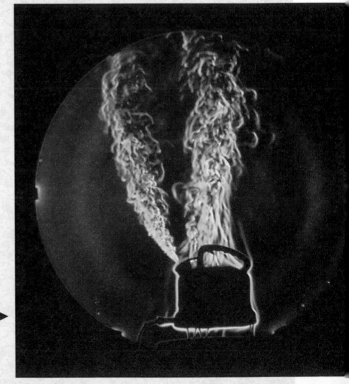

Schlieren photograph of a tea-kettle showing steam and turbulent convection currents. ▶
(Gary Settles/Science Source/Photo Researchers)

and work—a change in internal energy. According to the first law, a system's internal energy can be increased either by transfer of thermal energy to the system or by work done on the system. An important difference between thermal energy and mechanical energy is not evident from the first law: Energy transferred to a system as work can be stored completely as internal energy, but it is impossible for internal energy to completely leave a system only by the mechanism of work.

17.1 · HEAT, THERMAL ENERGY, AND INTERNAL ENERGY

A major distinction must be made between internal energy and heat. **Internal energy** is all of the energy belonging to a system while it is stationary (neither translating nor rotating), including nuclear energy, chemical energy, and strain energy (as in a compressed or stretched spring), as well as thermal energy associated with the random motion of atoms or molecules. **Heat is the energy that is transferred between the system and its surroundings because of a temperature difference between them.**

In the previous chapter we showed that the thermal energy of a monatomic ideal gas is associated with the internal motion of its atoms. In this special case, the thermal energy is simply the kinetic energy on a microscopic scale; the higher the temperature of the gas, the greater the kinetic energy of the atoms and the greater the thermal energy of the gas. More generally, however, thermal energy includes other forms of molecular energy, such as rotational energy and vibrational kinetic and potential energy.

As an analogy, consider the distinction between work and energy that we discussed in Chapter 7. The work done on (or by) a system is a measure of the energy transferred between the system and its surroundings, whereas the mechanical energy of the system (kinetic or potential) is a consequence of its motion and coordinates. Thus, when a person does work on a system, energy is transferred from the person to the system. It makes no sense to talk about the work *of* a system—one should instead refer only to the **work done on or by a system** when some process has occurred in which energy has been transferred to or from the system. Likewise, it makes no sense to use the term **heat** unless energy has been transferred as a result of a temperature difference.

It is also important to recognize that energy can be transferred between two systems even when no thermal energy transfer occurs. For example, when a gas is compressed by a piston, the gas is warmed and its thermal energy increases, but there is no transfer of thermal energy from the surroundings; if the gas then expands rapidly, it cools and its thermal energy decreases, but there is no transfer of thermal energy to the surroundings. In each case, energy is transferred to or from the system as work but appears within the system as an increase or decrease of thermal energy. The changes in internal energy in these examples are equal to the changes in thermal energy and are measured by corresponding changes in temperature.

Units of Heat

Early in the development of thermodynamics, before scientists realized that heat is transferred energy, heat was defined in terms of the temperature changes it produced in a body. Hence, the **calorie** (cal) was defined as **the amount of heat**

James Prescott Joule
(1818–1889)

A British physicist, Joule received some formal education in mathematics, philosophy, and chemistry from John Dalton but was in large part self-educated. Joule's most active research period, from 1837 through 1847, led to the establishment of the principle of conservation of energy and the equivalence of heat and other forms of energy. His study of the quantitative relationship among electrical, mechanical, and chemical effects of heat culminated in his announcement in 1843 of the amount of work required to produce a unit of heat, called the mechanical equivalent of heat.   *(By kind permission of the President and Council of the Royal Society)*

· *Definition of the calorie*

necessary to raise the temperature of 1 g of water from 14.5°C to 15.5°C.[1] (Note that the "Calorie," with a capital C, used in describing the chemical energy content of foods, is actually a kilocalorie.) Likewise, the unit of heat in the British system, the **British thermal unit** (Btu), was defined as **the heat required to raise the temperature of 1 lb of water from 63°F to 64°F.**

In 1948, scientists decided that because heat (like work) is a measure of the transfer of energy, its SI unit should be the joule. The calorie is now defined to be exactly 4.186 J:

Mechanical equivalent of heat •

$$1 \text{ cal} \equiv 4.186 \text{ J} \qquad\qquad [17.1]$$

Note that this definition makes no reference to the heating of water. This relation is known, for purely historical reasons, as the **mechanical equivalent of heat.**

Example 17.1   Losing Weight the Hard Way

A student eats a dinner rated at 2000 (food) Calories. He wishes to do an equivalent amount of work in the gymnasium by lifting a 50.0-kg mass. How many times must he raise the mass to expend this much energy? Assume that he raises it a distance of 2.00 m each time and that he gains no work energy when it is dropped to the floor.

Solution   Because 1 Calorie $= 1.00 \times 10^3$ cal, the work required is 2.00×10^6 cal. Converting this to J, we have for the total work required

$$W = (2.00 \times 10^6 \text{ cal})(4.186 \text{ J/cal}) = 8.37 \times 10^6 \text{ J}$$

The work done in lifting the mass a distance h is equal to mgh, and the work done in lifting it n times is $nmgh$. We equate this to the total work required:

$$W = nmgh = 8.37 \times 10^6 \text{ J}$$

$$n = \frac{8.37 \times 10^6 \text{ J}}{(50.0 \text{ kg})(9.80 \text{ m/s}^2)(2.00 \text{ m})} = 8.54 \times 10^3 \text{ times}$$

If the student is in good shape and lifts the weight once every 5 s, it will take him about 12 h to perform this feat.

17.2 • SPECIFIC HEAT

The quantity of heat required to raise the temperature of a given mass of a substance by some amount varies with the substance. For example, the heat required to raise the temperature of 1 kg of water by 1.0°C is 4186 J, but the heat required to raise the temperature of 1 kg of copper by 1.0°C is only 387 J. Every substance requires a unique amount of heat to change the temperature of 1 kg of it by 1.0°C, and this number is a measure of the **specific heat** of the substance. Table 17.1 lists specific heats for several substances.

Suppose that a quantity Q of energy is transferred to mass m of a substance, thereby changing its temperature by ΔT. The **specific heat,** c, of the substance is defined as

Specific heat •

$$c \equiv \frac{Q}{m \, \Delta T} \qquad\qquad [17.2]$$

From this definition, we can express the energy transferred, Q, between a system of mass m and its surroundings for the temperature change ΔT as

---

[1]Originally, the calorie was defined as the heat necessary to raise the temperature of 1 g of water by 1°C. However, careful measurements showed that energy depends somewhat on temperature; hence, a more precise definition evolved.

TABLE 17.1 Specific Heats of Some Substances at 25.0°C and Atmospheric Pressure

Substance	Specific heat, c	
	J/kg·°C	cal/g·°C
Elemental Solids		
Aluminum	900	0.215
Beryllium	1830	0.436
Cadmium	230	0.055
Copper	387	0.0924
Germanium	322	0.077
Gold	129	0.0308
Iron	448	0.107
Lead	128	0.0305
Silicon	703	0.168
Silver	234	0.056
Other Solids		
Brass	380	0.092
Wood	1700	0.41
Glass	837	0.200
Ice (−5.0°C)	2090	0.50
Marble	860	0.21
Liquids		
Alcohol (ethyl)	2400	0.58
Mercury	140	0.033
Water (15.0°C)	4186	1.00

$$Q = mc \, \Delta T \qquad\qquad [17.3]$$

For example, the energy required to raise the temperature of 0.5 kg of water by 3.0°C is $(0.5 \text{ kg})(4186 \text{ J/kg·°C})(3.0°C) = 6280$ J. Note that when the temperature increases, ΔT and Q are taken to be *positive*, corresponding to energy flowing *into* the system. When the temperature decreases, ΔT and Q are *negative* and energy flows *out* of the system.

When specific heats are measured, the values obtained are also found to depend on the conditions of the experiment. In general, measurements made at constant pressure are different from those made at constant volume. For solids and liquids, the difference between the two values is usually no more than a few percent and is often neglected. For gases, the difference between the two values is significant.

Note from Table 17.1 that water has the highest specific heat of any substance with which we are likely to come in routine contact. The high specific heat of water is responsible for the moderate temperatures found in regions near large bodies of water. As the temperature of a body of water decreases during the winter, the

Figure 17.1 Circulation of air at the beach. (a) On a hot day, the air above the warm sand warms faster than the air above the cooler water. The cooler air over the water moving toward the beach displaces the rising warmer air. (b) At night, the sand cools more rapidly than the water and hence the air currents reverse their directions.

water transfers energy to the air, which carries the energy landward when prevailing winds are favorable. For example, the prevailing winds off the western coast of the United States are toward the land, and the energy liberated by the Pacific Ocean as it cools keeps coastal areas much warmer than they would be otherwise. This explains why the western coastal states generally have more favorable winter weather than the eastern coastal states, where the winds do not transfer energy toward land.

The fact that the specific heat of water is higher than that of sand is responsible for the pattern of air flow at a beach. During the day, the Sun adds roughly equal amounts of energy to beach and water, but the lower specific heat of sand causes the beach to reach a higher temperature than the water. The air above the land reaches a higher temperature than that above the water and the cold air pushes the hot air upward (due to Archimedes' principle). This results in a breeze from ocean to land during the day. Because the hot air gradually cools as it rises, it subsequently sinks, setting up the circulating pattern shown in Figure 17.1a. During the night, the sand cools more quickly than the water, and the circulating pattern reverses itself because the hotter air is now over the water (Fig. 17.1b). (The offshore and onshore breezes are certainly well known to sailors.)

Conservation of Energy: Calorimetry

Situations in which mechanical energy is converted to thermal energy occur frequently. We shall look at some in the examples following this section and in the problems at the end of the chapter, but most of our attention will be directed toward a particular kind of conservation-of-energy situation. In problems using the procedure we shall describe, called **calorimetry** problems, only the thermal energy transfer between the system and its surroundings is considered.

One technique for measuring the specific heat of a solid or liquid is simply to heat the substance to some known temperature, place it in a vessel containing water of known mass and temperature, and measure the temperature of the water after equilibrium is reached. Because a negligible amount of mechanical work is done in the process, the law of conservation of energy requires that the energy that leaves the warmer substance (of unknown specific heat) equals the energy that enters the water.[2] Devices in which this thermal energy transfer occurs are called **calorimeters.**

---

[2]For precise measurements, the container holding the water should be included in our calculations, because it also transfers thermal energy. This would require a knowledge of its mass and composition. However, if the mass of the water is large compared with that of the container, we can neglect the thermal energy gained by the container. Furthermore, precautions must be taken in such measurements to minimize thermal energy transfer between the system and the surroundings.

For example, suppose that m_x is the mass of a substance the specific heat of which we wish to determine, c_x its specific heat, and T_x its initial temperature. Let m_w, c_w, and T_w represent corresponding values for the water. If T is the final equilibrium temperature after everything is mixed, then from Equation 17.3 we find that the thermal energy gained by the water is $m_w c_w (T - T_w)$ and the thermal energy lost by the substance of unknown c is $- m_x c_x (T - T_x)$. Assuming that the combined system (water + unknown) does not lose or gain any thermal energy, it follows that the thermal energy gained by the water must equal the thermal energy lost by the unknown:

$$m_w c_w (T - T_w) = - m_x c_x (T - T_x)$$

Solving for c_x gives

$$c_x = \frac{m_w c_w (T - T_w)}{m_x (T_x - T)} \qquad \text{[17.4]}$$

Thinking Physics 1

The equation $Q = mc\,\Delta T$ indicates the relationship between energy, Q, transferred to an object of mass, m, and specific heat, c, by means of heat and the resulting temperature change, ΔT. In reality, the energy on the left-hand side of the equation could be transferred by any method, not just heat. Give a few examples in which this equation could be used to calculate a temperature change of an object but in which there is no heat involved.

Reasoning There are a number of examples, a few of which follow:

1. During the first few seconds after turning on a toaster, the temperature of the electrical coils rises. The transfer mechanism here is *electrical transmission* of energy through the power cord.
2. The temperature of a potato in a microwave oven increases due to the absorption of microwaves. In this case, the energy transfer mechanism is by *electromagnetic radiation*—the microwaves.
3. A bag of lead pellets is dropped to the floor many times, resulting in an increase in temperature of the pellets. The transfer mechanism here is *work* done during the deformation of the pellets as they strike the floor.
4. A carpenter attempts to use a dull drill bit to form a hole in a piece of wood. The bit fails to make much headway but becomes very warm. The increase in temperature in this case is due to *work* done on the bit by the wood.
5. The temperature of the fuel–air mixture in an automobile engine cylinder increases when the spark plug fires. In this case, the energy is already in the fuel–air mixture as electrical potential energy and is transformed to thermal energy in the mixture by the *chemical reaction* of combustion.

In each of these cases, as well as many others that could be generated, the Q on the left of the equation of interest is not a measure of heat but is replaced with the energy transferred or transformed by other means. Regardless of the fact that heat is not involved, the equation can still be used to calculate the temperature change.

Example 17.2   Cooling a Hot Ingot

A 0.0500-kg ingot of metal is heated to 200.0°C and then dropped into a beaker containing 0.400 kg of water initially at 20.0°C. If the final equilibrium temperature of the mixed system is 22.4°C, find the specific heat of the metal.

Solution   Because the thermal energy lost by the ingot equals the thermal energy gained by the water, we can write

$$m_w c_w (T_{wf} - T_{wi}) = -m_x c_x (T_{xf} - T_{xi})$$

$$(0.400 \text{ kg})(4186 \text{ J/kg} \cdot °\text{C})(22.4°\text{C} - 20.0°\text{C})$$

$$= (0.0500 \text{ kg})(c_x)(200.0°\text{C} - 22.4°\text{C})$$

from which we find that

$$c_x = \boxed{453 \text{ J/kg} \cdot °\text{C}}$$

The ingot is most likely iron, as can be seen by comparing this result with the data in Table 17.1.

EXERCISE 1   What is the total thermal energy transferred to the water as the ingot is cooled?   Answer   4020 J

EXERCISE 2   A cowboy fires a silver bullet of mass 2.00 g with a muzzle speed of 200 m/s into the pine wall of a saloon. Assume that all the internal energy generated by the impact remains with the bullet. What is the temperature change of the bullet?   Answer   85.5°C

EXERCISE 3   How many calories of heat are required to raise the temperature of 3.0 kg of aluminum from 20°C to 50°C?   Answer   19.4 kcal

17.3 • LATENT HEAT AND PHASE CHANGES

A substance usually undergoes a change in temperature when energy is transferred between the substance and its surroundings. There are situations, however, in which the transfer of energy does not result in a change in temperature. This is the case whenever the physical characteristics of the substance change from one form to another, commonly referred to as a **phase change.** Some common phase changes are solid to liquid (melting), liquid to gas (boiling), and a change in crystalline structure of a solid. All such phase changes involve a change in internal energy.

The thermal energy transfer required to change the phase of a given mass, m, of a pure substance is

$$Q = mL \qquad \qquad \textbf{[17.5]}$$

Latent heat •

where L is called the **latent heat** ("hidden" heat) of the substance and depends on the nature of the phase change as well as on the properties of the substance. **Latent heat of fusion,** L_f, is the term used when the phase change occurs during melting or freezing, and **latent heat of vaporization,** L_v, is the term used when the phase change occurs during boiling or condensing.[3] For example, the latent heat of fusion for water at atmospheric pressure is 3.33×10^5 J/kg, and the latent heat of vaporization of water is 2.26×10^6 J/kg. The latent heats of different substances vary considerably, as Table 17.2 illustrates.

[3]When a gas cools, it eventually returns to the liquid phase, or *condenses*. The thermal energy transfer per unit mass during the process is called the **latent heat of condensation,** and it equals in magnitude the latent heat of vaporization. When a liquid cools, it eventually solidifies, and the thermal energy transfer per unit mass during the process is called the **latent heat of fusion.**

TABLE 17.2 Latent Heats of Fusion and Vaporization

Substance	Melting Point (°C)	Latent Heat of Fusion (J/kg)	Boiling Point (°C)	Latent Heat of Vaporization (J/kg)
Helium	−269.65	5.23×10^3	−268.93	2.09×10^4
Nitrogen	−209.97	2.55×10^4	−195.81	2.01×10^5
Oxygen	−218.79	1.38×10^4	−182.97	2.13×10^5
Ethyl alcohol	−114	1.04×10^5	78	8.54×10^5
Water	0.00	3.33×10^5	100.00	2.26×10^6
Sulfur	119	3.81×10^4	444.60	3.26×10^5
Lead	327.3	2.45×10^4	1750	8.70×10^5
Aluminum	660	3.97×10^5	2450	1.14×10^7
Silver	960.80	8.82×10^4	2193	2.33×10^6
Gold	1063.00	6.44×10^4	2660	1.58×10^6
Copper	1083	1.34×10^5	1187	5.06×10^6

Consider, for example, the thermal energy required to convert a 1-g block of ice at −30.0°C to steam (water vapor) at 120.0°C. Figure 17.2 indicates the experimental results obtained when energy is gradually added to the ice. Let us examine each portion of the curve separately.

Part A. During this portion of the curve, we are changing the temperature of the ice from −30.0°C to 0.0°C. Because the specific heat of ice is 2090 J/kg·°C, we can calculate the amount of energy added from Equation 17.3:

$$Q = m_i c_i \Delta T = (1.00 \times 10^{-3} \text{ kg})(2090 \text{ J/kg·°C})(30.0°C) - 62.7 \text{ J}$$

Part B. When the ice reaches 0.0°C, the ice–water mixture remains at this temperature—even though energy is being added—until all the ice melts to become water at 0.0°C. The energy required to melt 1.00 g of ice at 0.0°C is, from Equation 17.5,

$$Q = mL_f = (1.00 \times 10^{-3} \text{ kg})(3.33 \times 10^5 \text{ J/kg}) = 333 \text{ J}$$

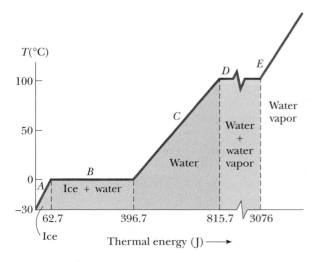

Figure 17.2 A plot of temperature versus thermal energy added when 1 g of ice initially at −30.0°C is converted to steam.

Part C.   Between 0.0°C and 100.0°C, nothing surprising happens. No phase change occurs in this region. The energy added to the water is being used to increase its temperature. The amount of energy necessary to increase the temperature from 0.0°C to 100.0°C is

$$Q = m_w c_w \, \Delta T = (1.00 \times 10^{-3} \text{ kg})(4.19 \times 10^3 \text{ J/kg} \cdot {}^\circ\text{C})(100.0°\text{C})$$
$$= 4.19 \times 10^2 \text{ J}$$

Part D.   At 100.0°C, another phase change occurs as the water changes from water at 100.0°C to steam at 100.0°C. Just as in Part B, the water–steam mixture remains at 100.0°C—even though energy is being added—until all the liquid has been converted to steam. The energy required to convert 1.00 g of water to steam at 100.0°C is

$$Q = mL_v = (1.00 \times 10^{-3} \text{ kg})(2.26 \times 10^6 \text{ J/kg}) = 2.26 \times 10^3 \text{ J}$$

Part E.   On this portion of the curve, energy is being added to the steam with no phase change occurring. The energy that must be added to raise the temperature of the steam to 120.0°C is

$$Q = m_s c_s \, \Delta T = (1.00 \times 10^{-3} \text{ kg})(2.01 \times 10^3 \text{ J/kg} \cdot {}^\circ\text{C})(20.0°\text{C}) = 40.2 \text{ J}$$

The total amount of energy that must be added to change 1 gram of ice at $-30.0°\text{C}$ to steam at 120.0°C is therefore about 3.11×10^3 J. Conversely, to cool 1 gram of steam at 120.0°C down to the point at which we have ice at $-30.0°\text{C}$, we must remove 3.11×10^3 J of energy.

Phase changes can be described in terms of a rearrangement of molecules when energy is added to or removed from a substance. Consider first the liquid-to-gas phase change. The molecules in a liquid are close together, and the forces between them are stronger than those between the molecules of a gas, which are far apart. Therefore, work must be done on the liquid, at the molecular level, against these attractive molecular forces in order to separate the molecules. The latent heat of vaporization is the amount of energy that must be added to 1 kg of the liquid to accomplish this.

Similarly, at the melting point of a solid, we imagine that the amplitude of vibration of the atoms about their equilibrium position becomes large enough to allow the atoms to pass the barriers of adjacent atoms and move to their new positions. The new locations are, on the average, less symmetrical and therefore have higher energy. The latent heat of fusion is equal to the work required, at the molecular level, to transform 1 kg of the mass from the ordered solid phase to the disordered liquid phase.

The average distance between atoms is much greater in the gas phase than in either the liquid or the solid phase. Each atom or molecule is removed from its neighbors, without the compensation of attractive forces to new neighbors. Therefore, it is not surprising that more work is required, at the molecular level, to vaporize a given mass of substance than to melt it, and that the latent heat of vaporization is much greater than the latent heat of fusion (Table 17.2).

Thinking Physics 2

The intuitive result of adding energy to a substance is to observe an increase in temperature. But if energy is added to ice at 0°C, its temperature does not increase. Where does the energy go?

Reasoning When energy is added to ice at, say, −10°C, much of the energy goes into increased molecular vibration, which has a physical manifestation of an increased temperature. When energy is added to ice at 0°C, the temperature of the ice remains constant and the ice melts. Much of the energy goes into increasing the electrical potential energy and breaking the bonds between molecules. This is similar to adding energy to a rock–Earth system by throwing the rock so fast that it escapes the Earth—the gravitational bond between the rock and the Earth has been broken. Breaking the bonds between the water molecules in the ice allows the molecules to move with respect to each other, which we recognize as the behavior of the liquid that results from the melting.

PROBLEM-SOLVING STRATEGY • Calorimetry Problems

If you are having difficulty with calorimetry problems, consider the following factors.

1. Be sure your units are consistent throughout. For instance, if you are using specific heats in cal/g·°C, be sure that masses are in grams and temperatures are Celsius throughout.

2. Losses and gains in energy are found by using $Q = mc\,\Delta T$ only for those intervals in which no phase changes are occurring. The equations $Q = mL_f$ and $Q = mL_v$ are to be used only when phase changes *are* taking place.

3. Often sign errors occur in heat loss = heat gain equations. One way to check your equation is to examine the signs of all ΔT's that appear in it.

Example 17.3 Boiling Liquid Helium

Liquid helium has a very low boiling point, 4.2 K, and a very low latent heat of vaporization, 2.09×10^4 J/kg (Table 17.2). A constant power of 10.0 W is transferred to a container of liquid helium from an immersed electric heater. At this rate, how long does it take to boil away 1.00 kg of liquid helium at 4.2 K?

Reasoning and Solution Because $L_v = 2.09 \times 10^4$ J/kg for liquid helium, we must supply 2.09×10^4 J of energy to boil away 1.00 kg. The power supplied to the helium is 10.0 W = 10.0 J/s. That is, in 1.00 s, 10.0 J of energy is trans-

ferred to the helium. Therefore, the time it takes to transfer an energy of 2.09×10^4 J is

$$t = \frac{2.09 \times 10^4 \text{ J}}{10.0 \text{ J/s}} = 2.09 \times 10^3 \text{ s} \approx 35 \text{ min}$$

EXERCISE 4 If 10.0 W of power is supplied to 1.00 kg of water at 100.0°C, how long will it take for the water to completely boil away? Answer 62.8 h

17.4 · WORK AND THERMAL ENERGY IN THERMODYNAMIC PROCESSES

In the macroscopic approach to thermodynamics, we describe the *state* of a system with such variables as pressure, volume, temperature, and internal energy. The number of macroscopic variables needed to characterize a system depends on the system's nature. For a homogeneous system, such as a gas containing only one type of molecule, usually only two variables are needed. However, it is important to note that a *macroscopic state* of an isolated system can be specified only if the system is in thermal equilibrium internally. In the case of a gas in a container, internal thermal equilibrium requires that every part of the container be at the same pressure and temperature.

Consider gas contained in a cylinder fitted with a movable piston (Fig. 17.3). In equilibrium, the gas occupies a volume V and exerts a uniform pressure P on the cylinder walls and piston. If the piston has a cross-sectional area of A, the force exerted by the gas on the piston is $F = PA$. Now let us assume that the gas expands **quasi-statically**—that is, slowly enough to allow the system to remain essentially in thermodynamic equilibrium at all times. As the piston moves up a distance of dy, the work done by the gas on the piston is

$$dW = F\,dy = PA\,dy$$

Because $A\,dy$ is the increase in volume of the gas dV, we can express the work done as

$$dW = P\,dV \qquad \qquad \textbf{[17.6]}$$

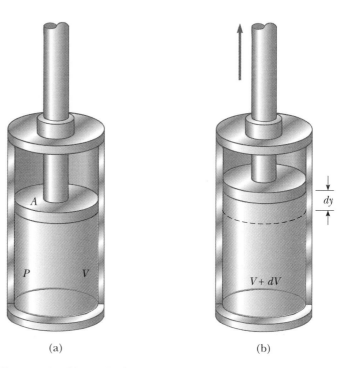

(a)   (b)

Figure 17.3 Gas contained in a cylinder at a pressure P does work on a moving piston as the system expands from a volume V to a volume $V + dV$.

Because the gas expands, dV is positive and the work done by the gas is positive, whereas if the gas is compressed, dV is negative, indicating that the work done by the gas is negative. (In the latter case, negative work can be interpreted as being work done *on* the gas.) Clearly, the work done by the gas is zero when the volume remains constant. The total work done by the gas as its volume changes from V_i to V_f is given by the integral of Equation 17.6:

$$W = \int_{V_i}^{V_f} P \, dV \qquad [17.7]$$

To evaluate this integral, one must know how the pressure varies during the expansion process. (Note that a *process* is *not* specified merely by the initial and final states. Rather, a process is a *fully specified* change in state of a system.) In general, the pressure is not constant throughout the expansion process but depends on the volume and temperature. If the pressure and volume are known at each step of the process, the states of the gas can be represented as a curve on a *PV* diagram, as in Figure 17.4. The work done in the expansion from the initial state to the final state is the area under the curve.

- **Work equals area under the curve in a PV diagram**

As Figure 17.4 shows, the work done in the expansion from the initial state to the final state depends on the path taken between the two states. To illustrate this important point, consider several different paths connecting i and f (Fig. 17.5). In the process depicted in Figure 17.5a, the pressure of the gas is first reduced from P_i to P_f by cooling at constant volume, V_i, and the gas then expands from V_i to V_f at constant pressure, P_f. The work done along this path is $P_f(V_f - V_i)$. In Figure 17.5b, the gas first expands from V_i to V_f at constant pressure, P_i, and then its pressure is reduced to P_f at constant volume, V_f. The work done along this path is $P_i(V_f - V_i)$, which is greater than that for the process described in Figure 17.5a. Finally, for the process described in Figure 17.5c, where both P and V change continuously, the work done has some value intermediate between the values obtained in the first two processes. To evaluate the work in this case, the shape of the *PV* curve must be known. Therefore, we see that the work done by a system depends on the process by which the system goes from the initial to the final state. In other words, the work done depends on the initial, final, and intermediate states of the system.

In a similar manner, the thermal energy transferred into or out of the system is also found to depend on the process. This can be demonstrated by the situations

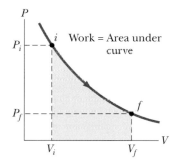

Figure 17.4 A gas expands reversibly (slowly) from state i to state f. The work done by the gas equals the area under the *PV* curve.

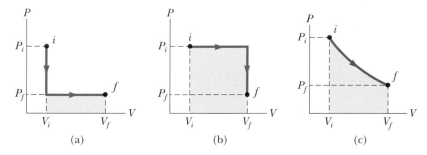

Figure 17.5 The work done by a gas as it is taken from an initial state to a final state depends on the path between these states.

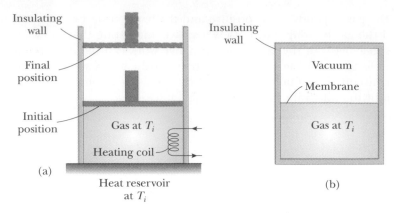

(a)

Heat reservoir
at T_i

(b)

Figure 17.6   (a) A gas at temperature T_i expands slowly by absorbing thermal energy from a reservoir at the same temperature. (b) A gas expands rapidly into an evacuated region after a membrane is broken.

This device, called Hero's engine, was invented around 150 B.C. by Hero in Alexandria. When water is boiled in the flask, which is suspended by a cord, steam exits through two tubes at the sides of the flask (in opposite directions), creating a torque that rotates the flask.   *(Courtesy of Central Scientific Co.)*

depicted in Figure 17.6. In each case, the gas has the same initial volume, temperature, and pressure and is assumed to be ideal. In Figure 17.6a, the gas is in thermal contact with a heat reservoir, and a small heating coil allows energy to be transferred to the gas by heat from the surface of the coil. As an infinitesimal amount of energy enters the gas from the coil, the pressure of the gas becomes infinitesimally greater than atmospheric pressure, and the gas expands, causing the piston to rise. During this expansion to some final volume, V_f, and final pressure P_f, sufficient thermal energy to maintain a constant temperature of T_i is transferred from the coil to the gas.

Now consider the thermally insulated system shown in Figure 17.6b. When the membrane is broken, the gas expands rapidly into the vacuum until it occupies a volume of V_f and is at a pressure of P_f. In this case, the gas does no work, because there is no movable piston. Furthermore, no energy is transferred through the thermally insulated wall, which we call an *adiabatic wall*, and so the temperature remains at T_i. This process is often referred to as **adiabatic free expansion** or simply **free expansion.** In general, **an adiabatic process is one in which no thermal energy is transferred between the system and its surroundings.**

The initial and final states of the ideal gas in Figure 17.6a are identical to the initial and final states in Figure 17.6b, but the paths are different. In the first case, thermal energy is transferred slowly to the gas, and the gas does work on the piston. In the second case, no thermal energy is transferred and the work done is zero. Therefore, we conclude that **heat, like work, depends on the initial, final, and intermediate states of the system.** Furthermore, because heat and work depend on the path, neither quantity is determined by the end points of a thermodynamic process.

Free expansion of a gas •

Heat depends on the path •
between the initial and final
states

Thinking Physics 3

Why is air at the top of a mountain cold? And why is the windward side of a mountain rainier than the leeward side?

Reasoning   Imagine a parcel of air being pushed by prevailing winds from a valley to the top of a mountain. Air is a very poor thermal conductor, so the thermodynamic process that occurs in this parcel of air is close to adiabatic. As the parcel rises, it moves into regions of lower atmospheric pressure. As a result, the parcel expands. By expanding, it is doing work on the surrounding air. This represents a transfer of energy out of the system, so that the internal energy drops. Thus, the parcel of air becomes cold as it rises up the mountain. If the parcel of air contains water vapor, the drop in temperature as it rises up the mountain can result in the condensation into raindrops. Thus, the rain tends to fall more on the windward side as the temperature drops. On the leeward side, much of the water vapor has already condensed out as rain, and the air parcel is warming up again as it descends the mountain. Thus, there is little rain on the leeward side of the mountain.

An interesting example of this occurs in Hawaii. The prevailing winds are often from the northeast. The island of Kauai has a very tall mountain (Mount Waialeale, 1548 m), and it is one of the rainiest places on Earth, with an average annual rainfall of 1170 cm. Just to the southwest of Kauai is the island of Niihau, which is very dry and requires a system of irrigation for its farming. The leeward position of Niihau relative to Kauai is the reason for this.

17.5 • THE FIRST LAW OF THERMODYNAMICS

When the law of conservation of energy was first introduced in Chapter 7, it was stated that the mechanical energy of a system is constant in the absence of non conservative forces, such as friction. That is, the changes in the internal energy of the system were not included in this mechanical model. The first law of thermodynamics is a generalization of the law of conservation of energy and encompasses possible changes in internal energy. It is a universally valid law that can be applied to all kinds of processes. Furthermore, it provides us with a connection between the microscopic and macroscopic worlds.

We have seen that energy can be transferred between a system and its surroundings in two ways. One is work done by (or on) the system that requires that there be a macroscopic displacement of the point of application of a force (or pressure). The other mode is heat, which occurs through random molecular collisions. Each of these represents a change of energy of the system, and therefore usually results in measurable changes in the macroscopic variables of the system, such as pressure, temperature, and volume of a gas.

To put these ideas on a more quantitative basis, suppose a system undergoes a change from an initial state to a final state. During this change, positive Q is the heat transferred *to* the system, and positive W is the work done *by* the system. As an example, suppose the system is a gas the pressure and volume of which change from P_i, V_i to P_f, V_f. If the quantity $Q - W$ is measured for various paths connecting the initial and final equilibrium states (that is, for various *processes*), one finds that it is the same for *all* paths connecting the initial and final states. We conclude that the quantity $Q - W$ is determined completely by the initial and final states of the system, and we call it the **change in the internal energy of the system.** Although Q and W both depend on the path, **$Q - W$ is independent of the path.** If we represent the internal energy function with the letter U, then the *change* in internal energy, ΔU, can be expressed as

$$\Delta U = Q - W \qquad\qquad\qquad \text{[17.8]}$$ • *First-law equation*

where all quantities must have the same energy units. Equation 17.8 is known as the **first-law equation** and is a key equation to many applications.

When a system undergoes an infinitesimal change in state, in which a small amount of heat dQ is transferred and a small amount of work dW is done, the energy also changes by a small amount dU. Thus, for infinitesimal processes we can express the first-law equation as[4]

First-law equation for •
infinitesimal changes

$$dU = dQ - dW \qquad\qquad \textbf{[17.9]}$$

In thermodynamics, we do not concern ourselves with the specific form of the internal energy. An analogy can be made between the internal energy of a system and the potential energy function associated with a body moving under the influence of gravity without friction. The potential energy function is independent of the path, and it is only the changes in it that are of concern. Likewise, the change in internal energy of a thermodynamic system is what matters, because only differences are defined. Because absolute values are not defined, any reference state can be chosen for the internal energy.

Let us look at some special cases in which the only changes in energy will be changes in internal energy. First consider an **isolated system**—that is, one that does not interact with its surroundings. In this case, no heat is added or removed from the system and the work done is zero; hence, the internal energy remains constant. That is, because $Q = W = 0$, then $\Delta U = 0$, and we conclude that **the internal energy of an isolated system remains constant.**

For isolated systems, U •
remains constant

Next consider the case in which a system (one not isolated from its surroundings) is taken through a **cyclic process**—that is, one that originates and ends at the same state. In this case, the change in the internal energy must again be *zero* and therefore the heat added to the system must equal the work done during the cycle. That is, in a cyclic process,

Cyclic process •

$$\Delta U = 0 \qquad \text{and} \qquad Q = W$$

Note that **the net work done per cycle equals the area enclosed by the path representing the process on a *PV* diagram.** As we shall see in Chapter 18, cyclic processes are very important in describing the thermodynamics of **heat engines**—devices in which some part of the heat added to the system is extracted as mechanical work.

If a process occurs in which the work done is zero, then the change in internal energy equals the heat entering or leaving the system. If heat enters the system, Q is positive and the internal energy increases. For a gas, we can associate this increase in internal energy with an increase in the kinetic energy of the molecules. However, if a process occurs in which the heat transferred is zero and work is done by the system, then the magnitude of the change in internal energy equals the negative of the work done by the system. That is, the internal energy of the system decreases. For example, if a gas is compressed (by a moving piston, say), with no transfer of heat, the work done by the gas is negative and the internal energy again increases.

---

[4]It should be noted that dQ and dW are not true differential quantities, although dU is a true differential. In fact, dQ and dW are not differentials of any definable quantity. For further details on this point, see R. P. Bauman, *Modern Thermodynamics and Statistical Mechanics*, New York, Macmillan, 1992.

No practical distinction exists between heat and work on a microscopic scale. Each can produce a change in the internal energy of a system. Although the macroscopic quantities Q and W are *not* properties of a system, they are related to changes of the internal energy of a stationary system through the first-law equation. Once a process, or path, is defined, Q and W can be either calculated or measured, and the change in internal energy can be found from the first-law equation. One of the important consequences of the first law is the existence of a quantity we called internal energy, the value of which is determined by the state of the system. The internal energy function is therefore called a **state function.**

Thinking Physics 4

If an ideal gas undergoes a free expansion (an expansion into a vacuum) in an insulated container, it does no work, because it does not have to push anything out of the way. There is also no heat exchanged with its surroundings, because the container is insulated. Thus, the first law tells us that the internal energy and the temperature remain constant. What about a gas of positively charged ions? If it undergoes a free expansion in an insulated container, what happens to the temperature of this gas?

Reasoning   In an ideal gas, it is assumed that there are no interactions between molecules (except during collisions), so that the energy of the gas does not depend on the intermolecular separation. For a gas of charged particles, however, the energy does depend on intermolecular separation, because there is electrical potential energy associated with the electrical forces between the ions. If the charged gas undergoes a free expansion, no work is done (by pushing anything out of the way), and there is no heat exchange. The overall electrical potential energy of the gas has decreased, because the particles have moved farther apart. As a result, the average kinetic energy of the particles must increase, and, in turn, the temperature of the gas will rise.

Thinking Physics 5

In the late 1970s, casino gambling was approved in Atlantic City, New Jersey, which can become quite cold in the winter. Energy projections that were performed for the design of the casinos showed that the air conditioning would need to operate in the casino even in the middle of a very cold January. Why?

Reasoning   If we consider the air in the casino to be the gas to which we apply the first law, imagine that there is no air conditioning and no ventilation, so that this gas simply stays in the room. There is no work being done by the gas, so we focus on the heat input. A casino contains a large number of people, many of whom are active (throwing dice, cheering, etc.), and many of whom are in excited states (celebration, frustration, panic, etc.). As a result, these people have large rates of energy flow by heat from their bodies. This energy goes into the air of the casino, resulting in an increase in internal energy of the air. With the large number of excited people in a casino (along with the large number of incandescent lights and machines), the temperature of the gas can rise quickly and to a very high value. In order to keep the temperature at a reasonable value, there must be an energy transfer out of the air. Calculations show that conduction through the walls even in a 10°F January is not sufficient to maintain the required energy transfer, so the air conditioning system must be in almost continuous use throughout the year.

17.6 · SOME APPLICATIONS OF THE FIRST LAW OF THERMODYNAMICS

In order to apply the first law of thermodynamics to specific systems, it is useful to first define some common thermodynamic processes. As we learned in Section 17.4, an **adiabatic process** is defined as one in which no heat enters or leaves the system—that is, $Q = 0$. An adiabatic process can be achieved either by thermally insulating the system from its surroundings (as in Fig. 17.6b) or by performing the process rapidly. Applying the first-law equation in this case, we see that

Adiabatic process •

First-law equation applied to an adiabatic process •

$$\Delta U = -W \qquad\qquad [17.10]$$

From this result, we see that when a gas expands adiabatically, W is positive and ΔU is negative. Conversely, when the gas is compressed adiabatically, ΔU is positive.

Adiabatic processes are very important in engineering practice. Common applications include the expansion of hot gases in an internal combustion engine, the liquefaction of gases in a cooling system, and the compression stroke in a diesel engine.

The **free expansion** depicted in Figure 17.6b is an adiabatic process in which no work is done on or by the gas. Because $Q = 0$ and $W = 0$, we see from the first law that $\Delta U = 0$ for this process. That is, **the initial and final internal energies of a gas are equal in an adiabatic free expansion.** As we shall see in Chapter 18, the internal energy of an ideal gas depends only on its temperature. Thus, we would expect no change in temperature during an adiabatic free expansion. This is in accord with experiments performed at low pressures. Careful experiments with real gases at high pressures show a slight increase or decrease in temperature after the expansion.

Isobaric process (constant pressure) •

A process that occurs at constant pressure is called an **isobaric process.** When such a process occurs, energy is transferred by heat and the work done is simply the pressure multiplied by the change in volume, or $P(V_f - V_i)$.

Isovolumetric process (constant volume) •

A process that takes place at constant volume is called an **isovolumetric process.** In such a process, the work done is clearly zero. Hence, from the first-law equation we see that, in an isovolumetric process,

First-law equation applied to an isovolumetric process •

$$\Delta U = Q \qquad\qquad [17.11]$$

This equation tells us that **if heat is added to a system kept at constant volume, all of the heat goes into increasing the internal energy of the system.** For example, when a can of deodorant is thrown into a fire, heat enters the system (the gas in the can) by conduction through the metal walls of the can. As a consequence, the temperature and pressure in the can rise until the can possibly explodes.

Isothermal process (constant temperature) •

A process that occurs at constant temperature is called an **isothermal process.** The internal energy of an ideal gas is a function of temperature only. Hence, in an isothermal process of a stationary ideal gas, $\Delta U = 0$.

Isothermal Expansion of an Ideal Gas

Suppose an ideal gas is allowed to expand quasi-statically at constant temperature as described by the *PV* diagram in Figure 17.7. (Isothermal expansion can be achieved by placing the gas in good thermal contact with a reservoir at the same temperature and carrying out the expansion very slowly, as in Figure 17.6a.) The curve is a hyperbola with the equation $PV =$ constant. Let us calculate the work done by the gas in the expansion from state i to state f.

This work is given by Equation 17.7. Because the gas is ideal and the process is quasi-static, we can apply $PV = nRT$ for each point on the path. Therefore, we have

$$W = \int_{V_i}^{V_f} P\, dV = \int_{V_i}^{V_f} \frac{nRT}{V}\, dV$$

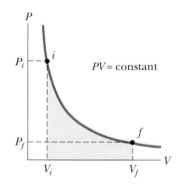

Figure 17.7   The PV diagram for an isothermal expansion of an ideal gas from an initial state to a final state. The curve is a hyperbola.

Because T is constant in this case, it can be removed from the integral, along with n and R:

$$W = nRT \int_{V_i}^{V_f} \frac{dV}{V}$$

To evaluate the integral, we use $\int \dfrac{dx}{x} = \ln x$. Thus, we find

$$W = nRT \ln \left(\frac{V_f}{V_i}\right) \qquad\qquad \textbf{[17.12]}$$

• **Work done in an isothermal process**

Numerically, this work equals the shaded area under the PV curve in Figure 17.7. Because the gas expands, $V_f > V_i$, and the work done by the gas is positive, as we would expect. If the gas is compressed, then $V_f < V_i$, and the work done by the gas is negative. In the next chapter we shall find that the internal energy of an ideal gas depends only on temperature. Hence, for an isothermal process $\Delta U = 0$, and from the first law we conclude that the heat transferred from the reservoir to the gas equals the work done by the gas, or $Q = W$.

The Boiling Process

Suppose that a liquid of mass m vaporizes at constant pressure P. Its volume in the liquid state is V_ℓ, and its volume in the vapor state is V_v. Let us find the work done in the expansion and the change in internal energy of the system.

Because the expansion takes place at constant pressure, the work done by the system is

$$W = \int_{V_\ell}^{V_v} P\, dV = P \int_{V_\ell}^{V_v} dV = P(V_v - V_\ell)$$

The heat that must be transferred to the liquid to vaporize all of it is equal to $Q = mL_v$, where L_v is the heat of vaporization of the liquid. Using the first-law equation and the preceding result, we get

$$\Delta U = Q - W = mL_v - P(V_v - V_\ell) \qquad\qquad \textbf{[17.13]}$$

CONCEPTUAL PROBLEM 1

Suppose an isothermal process occurs for a dilute (or low density) gas, during which work W is done. How much energy was transferred by heat?

CONCEPTUAL PROBLEM 2

Consider the human body performing a strenuous exercise, such as lifting weights or riding a bicycle. Work is being done by the body, and energy is leaving by conduction from the skin into the surrounding air. According to the first law of thermodynamics, the temperature of the body should be decreasing steadily during the exercise. This is not what happens, however. Is the first law invalid in this situation? Explain.

Example 17.4   Boiling Water

One gram of water occupies a volume of 1.00 cm³ at atmospheric pressure. When this amount of water is boiled, it becomes 1671 cm³ of steam. Calculate the change in internal energy for this process.

Solution   Because the latent heat of vaporization of water is 2.26×10^6 J/kg at atmospheric pressure, the heat required to boil 1.00 g is

$$Q = mL_v = (1.00 \times 10^{-3} \text{ kg})(2.26 \times 10^6 \text{ J/kg}) = 2260 \text{ J}$$

The work done by the system is positive and equal to

$$W = P(V_v - V_\ell)$$
$$= (1.013 \times 10^5 \text{ N/m}^2)[(1671 - 1.00) \times 10^{-6} \text{ m}^3]$$
$$= 169 \text{ J}$$

Hence, the change in internal energy is

$$\Delta U = Q - W = 2260 \text{ J} - 169 \text{ J} = 2.09 \text{ kJ}$$

The internal energy of the system increases because ΔU is positive. We see that most (93%) of the thermal energy transferred to the liquid goes into increasing the internal energy. Only 7% goes into external work.

OPTIONAL

17.7 · HEAT TRANSFER

In practice, it is important to understand the rate at which energy is transferred between a system and its surroundings and the mechanisms responsible for the transfer. You may have used a Thermos bottle or some other thermally insulated vessel to store hot coffee for a length of time. The vessel reduces energy transfer between the outside air and the hot coffee. Ultimately, of course, the liquid will reach air temperature because the vessel is not a perfect insulator. There is no energy transfer between a system and its surroundings when they are at the same temperature.

Heat Conduction

The easiest heat transfer process to describe quantitatively is called **conduction.** In this process, the transfer mechanism can be viewed on an atomic scale as an exchange of kinetic energy between molecules, where the less energetic particles gain energy by colliding with the more energetic particles. For example, if you insert a metallic bar into a flame while holding one end, you will find that the temperature of the metal in your hand increases. The energy reaches your hand through conduction. The manner in which energy is transferred from the flame, through the bar, and to your hand can be understood by examining what is happening to the atoms and electrons of the metal. Initially, before the rod is inserted into the flame, the metal atoms and electrons are vibrating about their equilibrium positions. At the surface of the rod, rapidly moving gas molecules in the flame collide with atoms at the surface of the metal, transferring energy into the metal by conduction. Those metal atoms and electrons near the flame begin to vibrate with larger and larger amplitudes. These, in turn, collide with their neighbors and transfer some of their

Melted snow patterns on a parking lot indicate the presence of underground hot water pipes used to aid snow removal. Heat from the water is conducted to the pavement from the pipes, causing the snow to melt.   *(Courtesy of Dr. Albert A. Bartlett, University of Colorado, Boulder)*

energy in the collisions. Slowly, metal atoms and electrons farther down the rod increase their amplitude of vibration, until the large-amplitude vibrations arrive at the end being held. The effect of this increased vibration is an increase in temperature of the metal and possibly a burned hand.

Although the transfer of energy through a metal can be partially explained by atomic vibrations and electron motion, the rate of conduction also depends on the properties of the substance being heated. For example, it is possible to hold a piece of asbestos in a flame indefinitely. This implies that very little energy is being conducted through the asbestos. In general, metals are good thermal conductors, and materials such as asbestos, cork, paper, and fiberglass are poor conductors. Gases also are poor thermal conductors because of their dilute nature. Metals are good thermal conductors because they contain large numbers of electrons that are relatively free to move through the metal and can transport energy from one region to another. Thus, in a good conductor, such as copper, conduction takes place via the vibration of atoms and via the motion of free electrons.

Conduction occurs only if there is a difference in temperature between two parts of the conducting medium. Consider a slab of material of thickness Δx and cross-sectional area A with its opposite faces at different temperatures T_1 and T_2, where $T_2 > T_1$ (Fig. 17.8). It is found from experiment that the energy Q transferred in a time Δt flows from the hotter end to the colder end. The rate of heat transfer, $Q/\Delta t$, is found to be proportional to the cross-sectional area and the temperature difference, and inversely proportional to the thickness:

$$\frac{Q}{\Delta t} \propto A \frac{\Delta T}{\Delta x}$$

It is convenient to use the symbol H to represent the rate of heat transfer. That is, we take $H = Q/\Delta t$. Note that H has units of watts when Q is in joules and Δt is in seconds. For a slab of infinitesimal thickness dx and temperature difference dT, we can write the **law of heat conduction**

$$H = -kA \frac{dT}{dx}$$

[17.14]

where the proportionality constant k is called the **thermal conductivity** of the material, and dT/dx is the **temperature gradient** (the variation of temperature with position). The minus sign in Equation 17.14 denotes the fact that energy transfers as heat in the direction of decreasing temperature.

Suppose a substance is in the shape of a long uniform rod of length L, as in Figure 17.9, and is insulated so that heat cannot escape from its surface except at the ends, which are in thermal contact with reservoirs having temperatures T_1 and T_2. When a steady state has been reached, the temperature at each point along the rod is constant in time. In this case, the temperature gradient is the same everywhere along the rod and is

$$\frac{dT}{dx} = \frac{T_1 - T_2}{L}$$

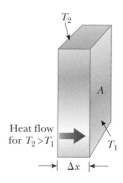

Figure 17.8 Heat transfer through a conducting slab of cross-sectional area A and thickness Δx. The opposite faces are at different temperatures T_1 and T_2.

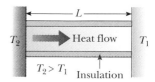

Figure 17.9 Conduction of heat through a uniform, insulated rod of length L. The opposite ends are in thermal contact with reservoirs at different temperatures.

Thus the rate of heat transfer is

$$H = kA \frac{(T_2 - T_1)}{L} \qquad \text{[17.15]}$$

Substances that are good thermal conductors have large thermal conductivity values, whereas good thermal insulators have low thermal conductivity values. Table 17.3 lists thermal conductivities for various substances. We see that metals are generally better thermal conductors than nonmetals.

For a compound slab containing several materials of thicknesses L_1, L_2, . . . and thermal conductivities k_1, k_2, . . . , the rate of heat transfer through the slab at steady state is

$$H = \frac{A(T_2 - T_1)}{\Sigma_i (L_i / k_i)} \qquad \text{[17.16]}$$

where T_1 and T_2 are the temperatures of the outer extremities of the slab (which are held constant) and the summation is over all slabs.

TABLE 17.3   Thermal Conductivities of Some Substances

Substance	Thermal Conductivity (W/m·°C)
Metals (at 25°C)	
Aluminum	238
Copper	397
Gold	314
Iron	79.5
Lead	34.7
Silver	427
Gases (at 20°C)	
Air	0.0234
Helium	0.138
Hydrogen	0.172
Nitrogen	0.0234
Oxygen	0.0238
Nonmetals (approximate values)	
Asbestos	0.08
Concrete	0.8
Diamond	2300
Glass	0.8
Ice	2
Water	0.6
Wood	0.08

Convection

At one time or another you probably have warmed your hands by holding them over an open flame. In this situation, the air directly above the flame is heated and expands. As a result, the density of the air decreases and the air rises. This warmed mass of air heats your hands as it flows by. **Thermal energy transferred by the movement of a heated substance is said to have been transferred by convection.** When the movement results from differences in density, as in the example of air around a fire, it is referred to as **natural convection.** When the heated substance is forced to move by a fan or pump, as in some hot-air and hot-water heating systems, the process is called **forced convection.**

The circulating pattern of air flow at a beach is an example of convection. Likewise, the mixing that occurs as water is cooled and eventually freezes at its surface (Chapter 16) is an example of convection in nature. Recall that the mixing by convection currents ceases when the water temperature reaches 4°C. Because the water in a lake cannot be cooled by convection below 4°C, and because water is a relatively poor thermal conductor, the water near the bottom remains near 4°C for a long time. As a result, fish have a comfortable temperature in which to live even in periods of prolonged cold weather.

If it were not for convection currents, it would be difficult to boil water. As water is heated in a tea kettle, the lower layers are warmed first. These heated regions expand and rise to the top because their density is lowered. At the same time, the denser cool water falls to the bottom of the kettle so that it can be heated.

The same process occurs when a room is heated by a stove. The stove warms the air in the lower regions of the room. The warm air expands and rises to the ceiling because of its lower density. The denser regions of cooler air from above replace the warm air, setting up the continuous air current pattern shown in Figure 17.10.

Figure 17.10 Convection currents are set up in a room heated by a wood-burning stove.

Radiation

The third way of transferring energy is through **radiation.** All objects radiate energy continuously in the form of electromagnetic waves. Through electromagnetic radiation, approximately 1340 J of energy from the Sun strikes 1 m² of the top of the Earth's atmosphere every second. Some of this energy is reflected back into space and some is absorbed by the atmosphere, but enough arrives at the surface of the Earth each day to supply all of our energy needs on this planet hundreds of times over—if it could be captured and used efficiently. The growth in the number of solar houses in this country is one example of an attempt to make use of this free energy.

Radiant energy from the Sun affects our day-to-day existence in a number of ways. It influences the Earth's average temperature, ocean currents, agriculture, rain patterns, and so on. For example, consider what happens to the atmospheric temperature at night. If there is a cloud cover above the Earth, the water vapor in the clouds reflects back a part of the infrared radiation emitted by the Earth and consequently the temperature remains at moderate levels. In the absence of this cloud cover, however, there is nothing to prevent this radiation from escaping into space, and thus the temperature drops more on a clear night than when it is cloudy.

The rate at which an object emits radiant energy is proportional to the fourth power of its absolute temperature. This is known as **Stefan's law** and is expressed in equation form as

$$P = \sigma A e T^4 \qquad \textbf{[17.17]}$$

where P is the power radiated by the body in watts, σ is a constant equal to 5.6696×10^{-8} W/m$^2 \cdot$K^4, A is the surface area of the object in square meters, e is a constant called the **emissivity,** and T is temperature in kelvin. The value of e can vary between zero and unity, depending on the properties of the surface.

An object radiates energy at a rate given by Equation 17.17. At the same time, the object also absorbs electromagnetic radiation. If the latter process did not occur, an object would eventually radiate all of its energy and its temperature would reach absolute zero. The energy that a body absorbs comes from its surroundings, which consists of other objects that radiate energy. If an object is at a temperature T and its surroundings are at a temperature T_0, the net energy gained or lost each second by the object as a result of radiation is

$$P_{\text{net}} = \sigma A e (T^4 - T_0^{\,4}) \qquad \textbf{[17.18]}$$

When an object is in equilibrium with its surroundings, it radiates and absorbs energy at the same rate, and so its temperature remains constant. When an object is hotter than its surroundings, it radiates more energy than it absorbs and so it cools. An **ideal absorber** is defined as an object that absorbs all of the energy incident on it. The emissivity of an ideal absorber is equal to unity. Such an object is often referred to as a **black body**. An ideal absorber is also an ideal radiator of energy. In contrast, an object with an emissivity equal to zero absorbs none of the energy incident on it. Such an object reflects all the incident energy and so is a perfect reflector.

The Dewar Flask

The Thermos bottle, called a **Dewar flask** in the scientific community, is a practical example of a container designed to minimize energy losses by conduction, convection, and radiation. Such a container is used to store either cold or hot liquids for long periods of time. The standard construction (Fig. 17.11) consists of a double-walled Pyrex vessel with silvered inner walls. The space between the walls is evacuated to minimize heat transfer by conduction and convection. The silvered surfaces minimize heat transfer by radiation by reflecting most of the radiated energy. Very little thermal energy is lost over the neck of the flask, because glass is not a very good thermal conductor, and the cross-section of the glass in the direction of conduction is small. A further reduction in thermal energy loss is obtained by reducing the size of the neck. Dewar flasks are commonly used to store liquid nitrogen (boiling point 77 K) and liquid oxygen (boiling point 90 K).

In order to confine liquid helium, which has a very low heat of vaporization (boiling point 4.2 K), it is often necessary to use a double Dewar system in which the Dewar flask containing the liquid is surrounded by a second Dewar flask. The space between the two flasks is filled with liquid nitrogen.

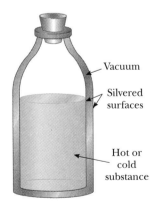

Vacuum

Silvered surfaces

Hot or cold substance

Figure 17.11   A cross-sectional view of a Dewar vessel, used to store hot or cold liquids or other substances.

Thinking Physics 6

If you sit in front of a fire in a fireplace with your eyes closed, you can feel significant warmth in your eyelids. If you now put on a pair of eyeglasses and repeat this activity, your eyelids will not feel nearly so warm. Why?

Reasoning Much of the warm feeling that one has while sitting in front of a fireplace is due to radiation from the fire. A large fraction of this radiation is in the infrared part of the spectrum. Your eyelids are particularly sensitive to infrared radiation. However, glass is opaque to infrared radiation. Thus, when you put on the glasses, you block much of the radiation from reaching your eyelids, and thus they feel cooler.

Thinking Physics 7

If you inspect an old or burned-out light bulb, you will notice a dark region on the inner surface. What's more, this region will be located on that part of the glass envelope that was the highest when the bulb was operating. What is the origin of this dark region, and why is it located at the high point?

Reasoning The dark region is tungsten, which has vaporized from the filament of the light bulb and collected on the inner surface of the glass. Many light bulbs contain gas, which allows convection to occur within the bulb. The gas near the filament is at a very high temperature, causing it to expand and float upward, due to Archimedes' principle. As it floats upward, it carries the vaporized tungsten with it, so that the tungsten collects on the surface at the top of the light bulb.

CONCEPTUAL PROBLEM 3

Rub the palm of your hand on a metal surface for 30 to 45 seconds. Place the palm of your other hand on an unrubbed portion of the surface and then the rubbed portion. The rubbed portion will feel warmer. Now repeat this process on a wooden surface. You should notice that the difference in temperature between the rubbed and unrubbed portions of the wood surface seems larger than that of the metal surface. Why is this?

CONCEPTUAL PROBLEM 4

In usually warm climates that experience an occasional hard freeze, fruit growers will spray the fruit trees with water, hoping that a layer of ice will form on the fruit. Why is this advantageous?

CONCEPTUAL PROBLEM 5

On a clear, cold night, why does frost tend to form on the tops of mailboxes and cars rather than the sides?

CONCEPTUAL PROBLEM 6

Cups of water for coffee or tea can be warmed with an immersion coil, which is immersed in the water and raised to a high temperature by means of electricity. The instructions warn the user not to operate the coils when they are out of water. Why? Should the immersion coil be used to warm up a cup of stew?

SUMMARY

Thermal energy transfer is a form of energy transfer that takes place as a consequence of a temperature difference. The **internal energy** of a substance is a function of the state of the substance and generally increases with increasing temperature.

The **calorie** is the amount of heat necessary to raise the temperature of 1 g of water by 1.0°C. The calorie is related to the joule by experiment, yielding 4.186 J/cal. This relationship was historically called the mechanical equivalent of heat.

The thermal energy required to change the temperature of a substance by ΔT is

$$Q = mc\,\Delta T \qquad\qquad \textbf{[17.3]}$$

where m is the mass of the substance and c is its **specific heat.**

The thermal energy required to change the phase of a pure substance of mass m is

$$Q = mL \qquad\qquad \textbf{[17.5]}$$

The parameter L is called the **latent heat** of the substance and depends on the nature of the phase change and the properties of the substance.

A **quasi-static process** is one that proceeds slowly enough to allow the system to always be in a state of equilibrium.

The **work done** by a gas as its volume changes from some initial value, V_i, to some final value, V_f, is

$$W = \int_{V_i}^{V_f} P\,dV \qquad\qquad \textbf{[17.7]}$$

where P is the pressure, which may vary during the process. In order to evaluate W, the nature of the process must be specified—that is, P and V must be known during each step of the process. The work done depends on the initial, final, and intermediate states. In other words, W depends on the path taken between the initial and final states.

The **first law of thermodynamics** is the law of conservation of energy. It includes all forms of energy, including thermal energy.

When a stationary system undergoes a change from one state to another, the change in its internal energy, ΔU, is

$$\Delta U = Q - W \qquad\qquad \textbf{[17.8]}$$

where Q is the thermal energy transferred into (or out of) the system and W is the work done by (or on) the system. Although Q and W both depend on the path taken from the initial state to the final state, the quantity ΔU is independent of the path.

An isolated system is one that does not interact with its surroundings. The internal energy of an isolated system remains constant. In a **cyclic process** (one that originates and terminates at the same state), $\Delta U = 0$, and therefore $Q = W$. An **adiabatic process** is one in which no thermal energy is transferred between the system and its surroundings ($Q = 0$). In this case, the first-law equation gives $\Delta U = -W$. In the **adiabatic free expansion** of a gas, $Q = 0$ and $W = 0$, and so $\Delta U = 0$. An **isobaric process** is one that occurs at constant pressure. The work done in such a process is $P\,\Delta V$. An **isovolumetric process** is one that occurs at constant volume. No work (of expansion) is done in such a process. An **isothermal process** is one that occurs at constant temperature. The work done by an ideal gas during an isothermal process is

$$W = nRT \ln\left(\frac{V_f}{V_i}\right) \qquad\qquad \textbf{[17.12]}$$

CONCEPTUAL QUESTIONS

1. Clearly distinguish among temperature, thermal energy, heat, and internal energy.
2. When a sealed Thermos bottle full of hot coffee is shaken, what are the changes, if any, in (a) the temperature of the coffee and (b) the internal energy of the coffee?
3. Using the first law of thermodynamics, explain why the *total* energy of an isolated system is always constant.
4. Ethyl alcohol has about one half the specific heat of water. If equal masses of alcohol and water in separate beakers are supplied with the same amount of heat, compare the temperature increases of the two liquids.
5. A small crucible is taken from a 200°C oven and immersed in a tub full of water at room temperature (this process is often referred to as *quenching*). What is the approximate final equilibrium temperature?
6. The U.S. penny is now made of copper-coated zinc. Can a calorimetric experiment be devised to test for the metal content in a collection of pennies? If so, describe the procedure you would use.
7. What is wrong with the statement: "Given any two bodies, the one with the higher temperature contains more heat"?
8. What is a major problem that arises in measuring specific heats if a sample with a temperature above 100°C is placed in water?
9. The air temperature above coastal areas is profoundly influenced by the large specific heat of water. One reason is that the heat released when 1 cubic meter of water cools by 1.0°C will raise the temperature of an enormously larger volume of air by 1.0°C. Estimate this volume of air. The specific heat of air is approximately 1.0 kJ/kg·°C. Take the density of air to be 1.3 kg/m³.
10. You need to pick up a very hot cooking pot in your kitchen. You have a pair of hot pads. Should you soak them in cold water or keep them dry in order to pick up the pot most comfortably?
11. Pioneers stored fruits and vegetables in underground cellars. Discuss fully this choice for a storage site.
12. In winter, the pioneers mentioned in Question 11 stored an open barrel of water alongside their produce. Why?

PROBLEMS

Section 17.1 Heat, Thermal Energy, and Internal Energy

Section 17.2 Specific Heat

1. Consider Joule's apparatus described in Figure P17.1. The two masses are 1.50 kg each, and the tank is filled with 200 g of water. What is the increase in the temperature of the water after the masses fall through a distance of 3.00 m?
2. A 50.0-g sample of copper is at 25.0°C. If 1200 J of thermal energy is added to it, what is the final temperature of the copper?
3. The temperature of a silver bar rises by 10.0°C when it absorbs 1.23 kJ of heat. The mass of the bar is 525 g. Determine the specific heat of silver.
4. What is the order of magnitude of the heat that a water heater puts into the water for a hot bath? How does this energy compare with your daily food–energy intake? In your solution state the quantities you estimate and the values you estimate for them.
5. A 1.50-kg iron horseshoe initially at 600°C is dropped into a bucket containing 20.0 kg of water at 25.0°C. What is the final temperature of the water and horseshoe? (Neglect the heat capacity of the container and assume that a negligible amount of water boils away.)
6. An aluminum cup of mass 200 g contains 800 g of water in

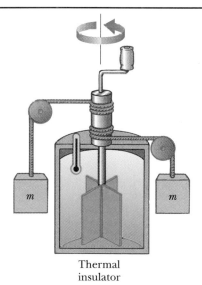

Figure P17.1 The falling weights rotate the paddles, causing the temperature of the water to increase.

thermal equilibrium at 80.0°C. The combination of cup and water is cooled uniformly so that the temperature decreases by 1.50 °C per minute. At what rate is thermal energy being removed? Express your answer in watts.

7. An aluminum calorimeter of mass 100 g contains 250 g of

water. The calorimeter and water are in thermal equilibrium at 10.0°C. Two metallic blocks are placed in the water. One is a 50.0-g piece of copper at 80.0°C; the other has a mass of 70.0 g and is originally at a temperature of 100°C. The entire system stabilizes at a final temperature of 20.0°C. (a) Determine the specific heat of the unknown sample. (b) Guess the material of the unknown using Table 17.1.

8. Lake Erie contains roughly 4.00×10^{11} m³ of water. (a) How much heat is required to raise the temperature of that volume of water from 11.0°C to 12.0°C? (b) Approximately how many years would it take to supply this amount of heat by using the full output of a 1000-MW electric power plant?

9. A 3.00-g copper penny at 25.0°C drops 50.0 m to the ground. (a) If 60.0% of its initial potential energy goes into increasing the internal energy, determine its final temperature. (b) Does the result depend on the mass of the penny? Explain.

Section 17.3   Latent Heat and Phase Changes

10. How much thermal energy is required to change a 40.0-g ice cube from a solid at -10.0°C to steam at 110°C?

11. If 90.0 g of molten lead at 327.3°C is poured into a 300-g casting form made of iron and initially at 20.0°C, what is the final temperature of the system? (Assume no energy loss to the environment occurs.)

12. Steam at 100°C is added to ice at 0°C. (a) Find the amount of ice melted and the final temperature when the mass of steam is 10.0 g and the mass of ice is 50.0 g. (b) Repeat when the mass of steam is 1.00 g and the mass of ice is 50.0 g.

13. A 1.00-kg block of copper at 20.0°C is dropped into a large vessel of liquid nitrogen at 77.3 K. How many kilograms of nitrogen boil away by the time the copper reaches 77.3 K? (The specific heat of copper is 0.0920 cal/g·°C. The latent heat of vaporization of nitrogen is 48.0 cal/g.)

14. A 50.0-g copper calorimeter contains 250 g of water at 20.0°C. How much steam must be condensed into the water to make the final temperature of the system 50.0°C?

15. In an insulated vessel, 250 g of ice at 0°C is added to 600 g of water at 18.0°C. (a) What is the final temperature of the system? (b) How much ice remains when the system reaches equilibrium?

16. A 3.00-g lead bullet at 30.0°C is fired at a speed of 240 m/s into a large block of ice at 0°C, in which it embeds itself. What quantity of ice melts?

17. Two speeding lead bullets, each of mass 5.00 g and at temperature 20.0°C, collide head-on at speeds of 500 m/s each. Assuming a perfectly inelastic collision and no loss of heat to the atmosphere, describe the final state of the two-bullet system.

Section 17.4   Work and Thermal Energy in Thermodynamic Processes

18. Gas in a container is at a pressure of 1.50 atm and a volume of 4.00 m³. What is the work done by the gas if (a) it expands at constant pressure to twice its initial volume? (b) It is compressed at constant pressure to one quarter of its initial volume?

19. A sample of ideal gas is expanded to twice its original volume of 1.00 m³ in a quasi-static process for which $P = \alpha V^2$, with $\alpha = 5.00$ atm/m⁶, as shown in Figure P17.19. How much work was done by the expanding gas?

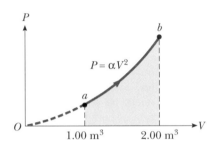

Figure P17.19

20. (a) Determine the work done by a fluid that expands from i to f as indicated in Figure P17.20. (b) How much work is performed by the fluid if it is compressed from f to i along the same path?

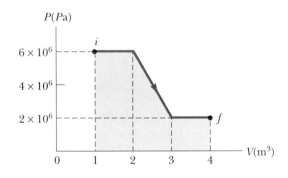

Figure P17.20

21. One mole of an ideal gas is heated slowly so that it goes from the PV state (P_0, V_0) to $(3P_0, 3V_0)$ in such a way that the pressure of the gas is directly proportional to the volume. (a) How much work is done in the process? (b) How is the temperature of the gas related to its volume during this process?

22. A sample of helium behaves as an ideal gas as it is heated at constant pressure from 273 K to 373 K. If the gas does 20.0 J of work, what is the mass of helium present?

Section 17.5 The First Law of Thermodynamics

23. A gas is compressed at a constant pressure of 0.800 atm from 9.00 L to 2.00 L. In the process, 400 J of thermal energy leaves the gas. (a) What is the work done by the gas? (b) What is the change in its internal energy?

24. A thermodynamic system undergoes a process in which its internal energy decreases by 500 J. If at the same time, 220 J of work is done on the system, find the thermal energy transferred to or from it.

25. A gas is taken through the cyclic process described in Figure P17.25. (a) Find the net thermal energy transferred to the system during one complete cycle. (b) If the cycle is reversed—that is, the process goes along *ACBA*—what is the net thermal energy transferred per cycle?

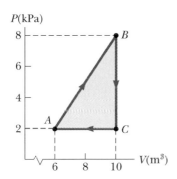

Figure P17.25

26. Consider the cyclic process depicted in Figure P17.25. If Q is negative for the process *BC*, and ΔU is negative for the process *CA*, determine the signs of Q, W, and ΔU that are associated with each process.

Section 17.6 Some Applications of the First Law of Thermodynamics

27. An ideal gas initially at 300 K undergoes an isobaric expansion at 2.50 kPa. If the volume increases from 1.00 m³ to 3.00 m³ and 12.5 kJ of thermal energy is transferred to the gas, find (a) the change in its internal energy and (b) its final temperature.

28. One mole of an ideal gas does 3000 J of work on its surroundings as it expands isothermally to a final pressure of 1.00 atm and volume of 25.0 L. Determine (a) the initial volume and (b) the temperature of the gas.

29. How much work is done by the steam when 1.00 mol of water at 100°C boils and becomes 1.00 mol of steam at 100°C at 1.00 atm pressure? Determine the change in internal energy of the steam as it vaporizes. Consider the steam to be an ideal gas.

30. One mole of gas initially at a pressure of 2.00 atm and a volume of 0.300 L has an internal energy that may be set equal to 91.0 J. In its final state, the pressure is 1.50 atm, the volume is 0.800 L, and the internal energy equals 182 J. For the three paths *IAF*, *IBF*, and *IF* in Figure P17.30, calculate (a) the work done by the gas and (b) the net thermal energy transferred in the process.

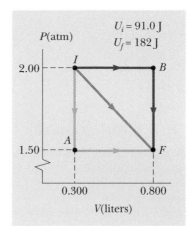

Figure P17.30

31. Two moles of helium gas initially at 300 K and 0.400 atm are compressed isothermally to 1.20 atm. Find (a) the final volume of the gas, (b) the work done by the gas, and (c) the thermal energy transferred. Consider the helium to behave as an ideal gas.

32. A 1.00-kg block of aluminum is heated at atmospheric pressure such that its temperature increases from 22.0°C to 40.0°C. Find (a) the work done by the aluminum, (b) the thermal energy added to it, and (c) the change in its internal energy.

33. One mole of water vapor at a temperature of 373 K cools down to 283 K. The heat given off by the cooling liquid is absorbed by 10.0 mol of an ideal gas, causing it to expand at a constant temperature of 273 K. If the final volume of the ideal gas is 20.0 L, determine the initial volume of the ideal gas.

34. One mole of steam at temperature T_h cools down to liquid water at T_c. The heat it gives off is absorbed by n mol of an ideal gas, causing it to expand at a constant temperature of T_0. If the final volume of the ideal gas is V_f, determine the initial volume of the ideal gas.

35. An ideal gas is carried through a thermodynamic cycle consisting of two isobaric and two isothermal processes as shown in Figure P17.35. Show that the net work done in the entire cycle is given by

$$W_{net} = P_1(V_2 - V_1) \ln\left(\frac{P_2}{P_1}\right)$$

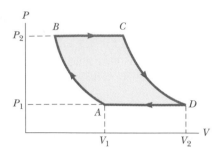

Figure P17.35

Section 17.7   Heat Transfer (Optional)

36. A steam pipe is covered with 1.50–cm-thick insulating material of heat conductivity 0.200 cal/cm·°C·s. How much heat is lost every second when the steam is at 200°C and the surrounding air is at 20.0°C? The pipe has a circumference of 20.0 cm and a length of 50.0 m. Neglect losses through the ends of the pipe.

37. A bar of gold is in thermal contact with a bar of silver of the same length and area (Fig. P17.37). One end of the compound bar is maintained at 80.0°C, and the opposite end is at 30.0°C. When the heat flow reaches steady state, find the temperature at the junction.

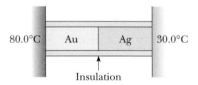

Figure P17.37

38. A box with a total surface area of 1.20 m² and a wall thickness of 4.00 cm is made of an insulating material. A 10.0-W electric heater inside the box maintains the inside temperature at 15.0°C above the outside temperature. Find the thermal conductivity k of the insulating material.

39. A Thermopane window of area 6.00 m² is constructed of two layers of glass, each 4.00 mm thick, separated by an air space of 5.00 mm. If the inside is at 20.0°C and the outside is at −30.0°C, what is the rate of heat loss through the window?

40. The surface of the Sun has a temperature of about 5800 K. Taking the radius of the Sun to be 6.96×10^8 m, calculate the total energy radiated by the sun each day. (Assume emissivity $e = 1.00$.)

Additional Problems

41. One hundred grams of liquid nitrogen at 77.3 K is stirred into a beaker containing 200 g of 5.00°C water. If the nitrogen leaves the solution as soon as it turns to gas, how much water freezes? (The heat of vaporization of nitrogen is 48.0 cal/g, and the heat of fusion of water is 80.0 cal/g.)

42. A 75.0-kg cross-country skier moves across snow so that the coefficient of friction between skis and snow is 0.200. Assume that all the snow beneath his skis is at 0°C and that all the thermal energy generated by friction is added to the snow, which sticks to his skis until melted. How far would he have to ski to melt 1.00 kg of snow?

43. An aluminum rod, 0.500 m in length and of cross-sectional area 2.50 cm², is inserted into a thermally insulated vessel containing liquid helium at 4.22 K. The rod is initially at 300 K. (a) If one half of the rod is inserted into the helium, how many liters of helium boil off by the time the inserted half cools to 4.22 K? (Assume the upper half does not yet cool.) (b) If the upper end of the rod is maintained at 300 K, what is the approximate boil-off rate of liquid helium after the lower half has reached 4.22 K? (Note that aluminum has a thermal conductivity of 31.0 J/s·cm·K at 4.2 K, a specific heat of 0.210 cal/g·°C, and a density of 2.70 g/cm³. The density of liquid helium is 0.125 g/cm³.)

44. A *flow calorimeter* is an apparatus used to measure the specific heat of a liquid. The technique is to measure the temperature difference between the input and output points of a flowing stream of the liquid while adding heat at a known rate. In one particular experiment, a liquid of density 0.780 g/cm³ flows through the calorimeter at the rate of 4.00 cm³/s. At steady state, a temperature difference of 4.80°C is established between the input and output points when heat is supplied at the rate of 30.0 J/s. What is the specific heat of the liquid?

45. One mole of an ideal gas is contained in a cylinder with a movable piston. The initial pressure, volume, and temperature are P_0, V_0, and T_0, respectively. Find the work done by the gas for the following processes and show each process on a PV diagram: (a) an isobaric compression in which the final volume is one half the initial volume, (b) an isothermal compression in which the final pressure is four times the initial pressure, (c) an isovolumetric process in which the final pressure is triple the initial pressure.

46. One mole of an ideal gas, initially at 300 K, is cooled at constant volume so that the final pressure is one fourth the initial pressure. Then the gas expands at constant pressure until it reaches the initial temperature. Determine the work done by the gas.

47. An ideal gas initially at P_0, V_0, and T_0 is taken through a cycle as in Figure P17.47. (a) Find the net work done by the gas per cycle. (b) What is the net heat added to the system per cycle? (c) Obtain a numerical value for the net work done per cycle for 1.00 mol of gas initially at 0°C.

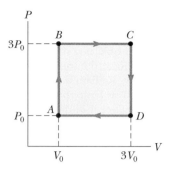

P
B ——— C
$3P_0$

P_0 A ——— D

V_0 $3V_0$ V

Figure P17.47

48. An iron plate is held against an iron wheel so that there is a sliding frictional force of 50.0 N acting between the two pieces of metal. The relative speed at which the two surfaces slide over each other is 40.0 m/s. (a) Calculate the rate at which mechanical energy is converted to thermal energy. (b) The plate and the wheel have a mass of 5.00 kg each, and each receives 50.0% of the thermal energy. If the system is run as described for 10.0 s and each object is then allowed to reach a uniform internal temperature, what is the resultant temperature increase?

49. A "solar cooker" consists of a curved reflecting mirror that focuses sunlight onto the object to be heated (Fig. P17.49). The solar power-per-unit area reaching the Earth at the location is 600 W/m², and the cooker has a diameter of 0.600 m. Assuming that 40.0% of the incident energy is transferred to the water, how long would it take to completely boil off 0.500 liter of water initially at 20.0°C? (Neglect the heat capacity of the container.)

Figure P17.49

50. An electric teakettle is boiling. The electric power consumed by the water is 1.00 kW. Assuming that the pressure of vapor in the kettle equals atmospheric pressure, determine the speed of effusion of vapor from the kettle's spout if the spout has a cross-sectional area of 2.00 cm².

51. An ideal gas is enclosed in a cylinder with a movable piston on top of it. The piston has a mass of 8000 g and an area of 5.00 cm² and is free to slide up and down, keeping the pressure of the gas constant. How much work is done as the temperature of 0.200 mol of the gas is raised from 20.0°C to 300°C?

52. A chamber contains 1.00 kg of water at 0°C under a piston, just touching the water surface. The piston is then raised quickly, so that part of the water is vaporized and the other part is frozen (no liquid left). Assuming the temperature remains constant at 0°C, determine the mass of ice formed in the chamber.

Spreadsheet Problems

S1. The rate at which an object with an initial temperature T_i cools when the surrounding temperature is T_0 is given by Newton's law of cooling:

$$\frac{dQ}{dt} = hA(T - T_0)$$

where A is the surface area of the object, and h is a constant that characterizes the rate of cooling, called the surface coefficient of heat transfer. The object's temperature T at any time t is

$$T = T_0 + (T_i - T_0)e^{-hAt/mc}$$

where m is the object's mass and c is its specific heat.

Spreadsheet 17.1 evaluates and plots the temperature of an object as a function of time. Consider the cooling of a cup of coffee the initial temperature of which is 75°C. After 3 min, it cools to a drinkable 45°C. Room temperature is 22°C, the effective area of the coffee is 125 cm², and its mass is 200 g. Take the specific heat c to be that of water. Use Spreadsheet 17.1 to find h for this cup of coffee. That is, find the value of h for which $T = 45°C$ at $t = 180$ s.

S2. In the Einstein model of a crystalline solid the molar specific heat at constant volume is

$$C_V = 3R \left(\frac{T_E}{T}\right)^2 \frac{e^{T_E/T}}{(e^{T_E/T} - 1)^2}$$

where T_E is a characteristic temperature called the Einstein temperature and T is the temperature in Kelvin. (a) Verify that $C_V \approx 3R$, the Dulong-Petit law, for $T \gg T_E$, by evaluating $C_V - 3R$. (b) For diamond, T_E is approximately 1060 K. If 1 mol of diamond is heated from 300 K to 600 K, numerically integrate

$$\Delta U = \int C_V \, dT$$

to find the increase in the internal energy of the material. Use your spreadsheet to carry out a simple rectangular numeric integration. (You may also want to try to use the trapezoid and Simpson's methods to carry out the integration.)

ANSWERS TO CONCEPTUAL PROBLEMS

1. If the process is isothermal, there is no change in internal energy. According to the first law, then, the heat is equal to the work done by the gas, $Q = W$.

2. The energy that is leaving the body by work and heat is replaced by means of biological processes that transform the potential energy in the food that the individual ate into internal energy. Thus, the temperature of the body can be maintained.

3. When you rub the surface, you are doing work on the surface, and the internal energy of the surface increases. The increase in internal energy is detected as an increase in temperature. With the metal surface, some of this energy is transferred away from the rubbing site by *conduction*. Thus, the internal energy remaining in the rubbed area is not as high for the metal as it is for the wood, and it feels relatively cooler than the wood.

4. If the fruit were open to the cold air, it is possible that the temperature of the fruit could drop to a point several degrees below freezing, causing damage. By spraying with water, a film of ice can form on the fruit. This provides some insulation to the fruit, so that its temperature will remain somewhat higher than that of the air.

5. One of the ways that objects transfer energy is by radiation. If we consider a mailbox, the top of the mailbox is oriented toward the clear sky. Radiation emitted by the top of the mailbox goes upward and into space. There is little radiation coming down from space into the top of the mailbox. Radiation leaving the sides of the mailbox is absorbed by the environment. Radiation *from* the environment (trees, houses, cars, etc.), however, can enter the sides of the mailbox, keeping it warmer than the top. As a result, the top is the coldest portion and frost forms there first.

6. The operation of the immersion coil depends on the convection of water to maintain a safe temperature. As the water near the coils warms up, it floats to the top due to Archimedes' principle. The temperature of the coils cannot go higher than the boiling temperature of water, 100°C. If the coils are operated in air, the convection process is reduced, and the upper limit of 100°C is removed. As a result, the coils can become hot enough to be damaged. If the coils are used in an attempt to warm a thick liquid like stew, the convection process cannot occur fast enough to carry energy away from the coils, so that they again become hot enough to be damaged.

18

Heat Engines, Entropy, and the Second Law of Thermodynamics

The first law of thermodynamics, studied in the previous chapter, is a statement of conservation of energy, generalized to include heat as a form of energy transfer. This law places no restrictions on the types of energy conversions that can occur. Furthermore, it makes no fundamental distinction between the results of energy input to a system by heat and work: According to the first law, the internal (thermal) energy of a body can be increased either by adding heat to it or by doing work on it. An important difference exists between heat and work, however, that is not evident from the first law. One manifestation of this difference is the fact that it is impossible to convert thermal energy into mechanical energy in an isothermal process.

◄ This steam-driven locomotive runs from Durango to Silverton, Colorado. Early steam-driven locomotives obtained their energy by burning wood or coal. The generated heat produces the steam, which powers the locomotive. Modern trains use electricity or diesel fuel to power their locomotives. All heat engines extract heat from a burning fuel and convert only a fraction of this energy to mechanical energy. *(Lois Moulton, Tony Stone Images)*

Lord Kelvin, British physicist and mathematician (1824–1907)

Born William Thomson in Belfast, Kelvin was the first to propose the use of an absolute scale of temperature. The Kelvin scale, named in his honor, is discussed in Section 16.2. Kelvin's study of Carnot's theory led to the idea that heat cannot pass spontaneously from a colder body to a hotter body. *(J. L. Charmet/SPL/Photo Researchers)*

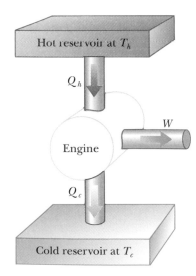

Figure 18.1 Schematic representation of a heat engine. The engine absorbs thermal energy, Q_h, from the hot reservoir, expels thermal energy Q_c to the cold reservoir, and does work W.

Contrary to what the first law implies, only certain types of energy conversions can take place. The second law of thermodynamics establishes which natural processes can and which cannot occur. The following are examples of processes that are consistent with the first law of thermodynamics in either direction, but proceed only in a particular direction governed by the second law.

1. When two objects at different temperatures are placed in thermal contact with each other, thermal energy always flows from the warmer to the cooler object—never from the cooler to the warmer.
2. A rubber ball dropped to the ground bounces several times and eventually comes to rest, but a ball lying on the ground never begins bouncing on its own.
3. An oscillating pendulum eventually comes to rest because of collisions with air molecules and friction at the point of suspension, and the initial mechanical energy is converted to thermal energy; the reverse conversion of energy never occurs.

These are all *irreversible processes*—that is, processes that occur naturally in only one direction. No irreversible process ever runs backward, because if it did, it would violate the second law of thermodynamics.[1]

From an engineering viewpoint, perhaps the most important application of the second law of thermodynamics is the limited efficiency of heat engines. The second law says that a machine capable of continuously converting thermal energy in a cyclic process *completely* to other forms of energy cannot be constructed.

18.1 • HEAT ENGINES AND THE SECOND LAW OF THERMODYNAMICS

A **heat engine** is a device that converts thermal energy to other useful forms, such as mechanical energy. In a typical process for producing electricity in a power plant, for instance, coal or some other fuel is burned, and the thermal energy produced is used to convert water to steam. This steam is directed at the blades of a turbine, setting it into rotation. Finally, the mechanical energy associated with this rotation is used to drive an electric generator. In another heat engine, the internal combustion engine in an automobile, energy enters the engine by mass transfer as the fuel is injected into the cylinder and a fraction of this energy is converted to mechanical energy.

A heat engine carries some working substance through a cyclic process during which (1) thermal energy is absorbed from a source at a high temperature, (2) work is done by the engine, and (3) thermal energy is expelled by the engine to a source at a lower temperature. As an example, consider the operation of a steam engine in which the working substance is water. The water is carried through a cycle in which it first evaporates into steam in a boiler and then expands against a piston. After the steam is condensed with cooling water, it is returned to the boiler, and the process is repeated.

---

[1]To be more precise, we should say that the set of events in the time-reversed sense is highly improbable. From this viewpoint, events in one direction are vastly more probable than those in the opposite direction.

It is useful to represent a heat engine schematically, as in Figure 18.1. The engine absorbs a quantity of energy that we will model as heat, Q_h, from the hot reservoir, does work W, and then gives up heat, Q_c, to the cold reservoir. Because the working substance goes through a cycle, its initial and final internal energies are equal, so $\Delta U = 0$. Hence, from the first-law equation we see that **the net work, W, done by a heat engine equals the net thermal energy flowing into it.** As we can see from Figure 18.1, $Q_{net} = Q_h - Q_c$; therefore,

$$W = Q_h - Q_c \qquad [18.1]$$

where Q_h and Q_c are taken to be positive quantities.

If the working substance is a gas, **the net work done for a cyclic process is the area enclosed by the curve representing the process on a *PV* diagram.** This is shown for an arbitrary cyclic process in Figure 18.2.

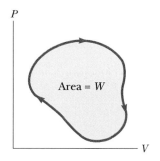

Figure 18.2 The *PV* diagram for an arbitrary cyclic process. The net work done equals the area enclosed by the curve.

> The **thermal efficiency,** e, of a heat engine is defined as the ratio of the net work done to the thermal energy absorbed at the higher temperature during one cycle:

• *Thermal efficiency*

$$e = \frac{W}{Q_h} = \frac{Q_h - Q_c}{Q_h} = 1 - \frac{Q_c}{Q_h} \qquad [18.2]$$

We can think of the efficiency as the ratio of what you get (mechanical energy) to what you give (thermal energy at the higher temperature). Equation 18.2 shows that a heat engine has 100% efficiency ($e = 1$) only if $Q_c = 0$—that is, if no thermal energy is expelled to the cold reservoir. In other words, a heat engine with perfect efficiency would have to expel all the input energy by mechanical work. One of the consequences of the second law of thermodynamics is that this is impossible.

The **second law of thermodynamics** can be stated as follows: **It is impossible to construct a heat engine that, operating in a cycle, produces no other effect than the absorption of thermal energy from a reservoir and the performance of an equal amount of work.**

• *Second law of thermodynamics*

This form of the second law is useful in understanding the operation of heat engines. With reference to Equation 18.2, the second law says that, during the operation of a heat engine, W can never be equal to Q_h or, alternatively, that some heat, Q_c, must be rejected to the environment. As a result, it is theoretically impossible to construct an engine that works with 100% efficiency.

Our assessment of the first two laws of thermodynamics can be summed up as follows: the first law says **we cannot get more energy out of a cyclic process than the amount of thermal energy we put in,** and the second law says **we cannot break even because we must put more thermal energy in, at the higher temperature, than the net amount of work output.**

Thinking Physics 1

Does a heat engine have to take in and expel energy *by heat*?

Reasoning The typical schematic model of a heat engine shows the energy entering the engine by heat and leaving by heat and work. This is only a schematic, however. The energy could enter and leave by other means. In an automobile engine, for example, the energy enters the cylinder by *mass transfer*—the energy is carried in with the fuel, and then transformed to thermal energy by the ignition process. The exhaust is also by mass transfer—of the expended fuel–air mixture out the tailpipe. The atmosphere can be considered to be a heat engine—radiation from the ground is absorbed in the atmosphere. About 0.5% of this energy does work by driving the zonal wind systems in the general atmospheric circulation. The remainder is radiated out into space. Here, the input and output energy is transferred by *electromagnetic radiation*.

Example 18.1   The Efficiency of an Engine

Find the efficiency of an engine that introduces 2000 J of heat during the combustion phase and loses 1500 J at exhaust.

Solution   The efficiency of the engine is given by Equation 18.2 as

$$e = 1 - \frac{Q_c}{Q_h} = 1 - \frac{1500 \text{ J}}{2000 \text{ J}} = 0.25, \text{ or } 25\%$$

EXERCISE 1   If an engine has an efficiency of 20% and loses 3000 J to the cooling water, how much work is done by the engine?   Answer   750 J

EXERCISE 2   The heat absorbed by an engine is three times greater than the work it performs. (a) What is its thermal efficiency? (b) What fraction of the heat absorbed is expelled to the cold reservoir?   Answer   (a) 0.333   (b) 2/3

18.2 • REVERSIBLE AND IRREVERSIBLE PROCESSES

In the next section we shall discuss a theoretical heat engine that is the most efficient engine possible. In order to understand its nature, we must first examine the meaning of reversible and irreversible processes. A **reversible** process is one in which the system can be returned to its initial conditions along the same path and in which every point along the path is an equilibrium state. A process that does not satisfy these requirements is **irreversible.**

All natural processes are known to be irreversible. From the endless number of examples that could be selected, let us examine the free expansion of a gas, already discussed in Section 17.4, and show that it cannot be reversible. The gas is contained in an insulated container, as shown in Figure 18.3, with a membrane separating the gas from a vacuum. If the membrane is punctured, the gas expands freely into the vacuum. Because the gas does not exert a force through a distance on the surroundings, it does no work as it expands. In addition, no thermal energy is transferred to or from the gas, because the container is insulated from its surroundings. Thus, in this process, the system has changed but the surroundings have not. Now imagine that we try to reverse the process by first compressing the gas to its original volume. Let's say an engine is being used to force the piston inward. Note, however, that this action is changing both the system and surroundings. The surroundings are changing because work is being done by an outside agent on the system, and the system is changing because the compression is increasing the temperature of the gas. We can lower the temperature of the gas by allowing it to come into contact with an external heat reservoir. Although this second procedure returns the gas to its original state, the surroundings are again affected because ther-

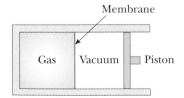

Figure 18.3   Free expansion of a gas.

mal energy is added to the surroundings. If this thermal energy could somehow be used to drive the engine and compress the gas, the system and its surroundings could be returned to their initial states. However, our statement of the second law says that this extracted thermal energy cannot be completely converted into mechanical energy isothermally. We must conclude that a reversible process has not occurred.

Although real processes are always irreversible, some are *almost* reversible. If a real process occurs very slowly so that the system is virtually always in equilibrium, the process can be considered reversible. For example, imagine compressing a gas very slowly by dropping some grains of sand onto a frictionless piston, as in Figure 18.4. The compression process can be reversed by the placement of the gas in thermal contact with a heat reservoir. The pressure, volume, and temperature of the gas are well defined during this isothermal compression. Each added grain of sand represents a change to a new equilibrium state. The process can be reversed by the slow removal of grains of sand from the piston.

A general characteristic of a reversible process is that there can be no dissipative effects present, such as turbulence or friction, that convert mechanical energy into thermal energy. In reality, such effects are impossible to eliminate completely, and hence it is not surprising that real processes in nature are irreversible.

Figure 18.4 A gas in thermal contact with a heat reservoir is compressed slowly by dropping grains of sand onto the piston. The compression is isothermal and reversible.

18.3 • THE CARNOT ENGINE

In 1824 a French engineer named Sadi Carnot (1796–1832) described a theoretical engine, now called a **Carnot engine,** that is of great importance from both practical and theoretical viewpoints. He showed that a heat engine operating in an ideal, reversible cycle—called a **Carnot cycle**—between two heat reservoirs is the most efficient engine possible. Such an ideal engine establishes an upper limit on the efficiencies of all engines. That is, **the net work done by a working substance taken through the Carnot cycle is the greatest amount of work possible for a given amount of thermal energy supplied to the substance at the upper temperature.**

To describe the Carnot cycle, we shall assume that the substance working between temperatures T_c and T_h is an ideal gas contained in a cylinder with a movable piston at one end. The cylinder walls and the piston are thermally nonconducting. Four stages of the Carnot cycle are shown in Figure 18.5.

- The process $A \rightarrow B$ is an isothermal expansion at temperature T_h, in which the gas is placed in thermal contact with a heat reservoir at temperature T_h (Fig. 18.5a). During the process, the gas absorbs heat Q_h from the reservoir and does work W_{AB} in raising the piston.
- In the process $B \rightarrow C$, the base of the cylinder is replaced by a thermally nonconducting wall and the gas expands adiabatically—that is, no thermal energy enters or leaves the system (Fig. 18.5b). During the process, the temperature falls from T_h to T_c, and the gas does work W_{BC} in raising the piston.
- In the process $C \rightarrow D$, the gas is placed in thermal contact with a heat reservoir at temperature T_c (Fig. 18.5c) and is compressed isothermally at temperature T_c. During this time, the gas expels heat Q_c to the reservoir, and the work done on the gas is W_{CD}.

Sadi Carnot (1796–1832)

A French physicist, Carnot is considered to be the founder of the science of thermodynamics. Some of his notes found after his death indicate that he was the first to recognize the relationship between work and heat.

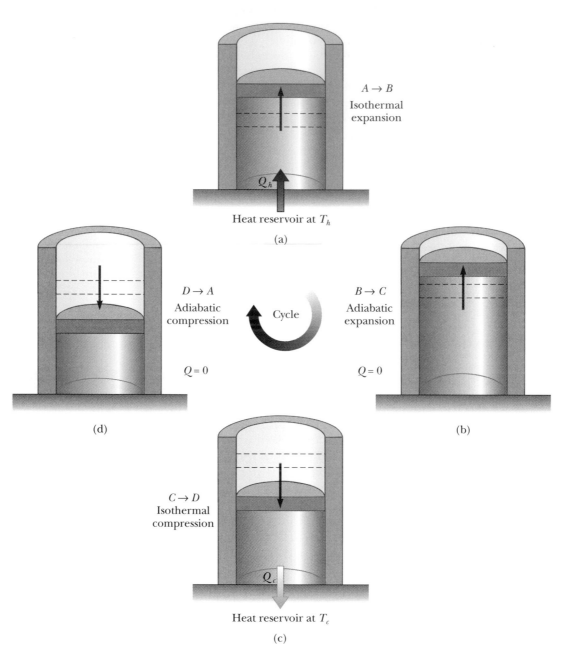

Figure 18.5   The Carnot cycle. (a) In process $A \rightarrow B$, the gas expands isothermally while in contact with a reservoir at T_h. (b) In process $B \rightarrow C$, the gas expands adiabatically ($Q = 0$). (c) In process $C \rightarrow D$, the gas is compressed isothermally while in contact with a reservoir at $T_c < T_h$. (d) In process $D \rightarrow A$, the gas is compressed adiabatically. The upward arrows on the piston in (a) and (b) indicate weights being removed during the expansions, and the downward arrows in (c) and (d) indicate the addition of weights during the compressions.

• In the final stage, $D \rightarrow A$, the base of the cylinder is again replaced by a thermally nonconducting wall (Fig. 18.5d) and the gas is compressed adiabatically. The temperature of the gas increases to T_h, and the work done on the gas is W_{DA}.

Figure 18.6 is the *PV* diagram for the cycle. The cycle consists of two adiabatic and two isothermal processes, all reversible.

Carnot showed that the thermal efficiency of a Carnot engine is

$$e_c = \frac{T_h - T_c}{T_h} = 1 - \frac{T_c}{T_h} \qquad \text{[18.3]}$$

• *Carnot efficiency*

It is important to note that in the equation, T must be in kelvin. From this result, we see that all Carnot engines operating between the same two temperatures have the same efficiency. Furthermore, the efficiency of a Carnot engine operating between two temperatures is greater than the efficiency of any real engine operating between the same two temperatures.

Equation 18.3 can be applied to any working substance operating in a Carnot cycle between two heat reservoirs. According to this result, the efficiency is zero if $T_c = T_h$, as one would expect. The efficiency increases as T_c is lowered and as T_h is increased. However, the efficiency can be unity (100%) only if $T_c = 0$ K. Such reservoirs are not available, and so the maximum efficiency is always less than unity. In most practical cases, the cold reservoir is near room temperature, about 300 K. Therefore, one usually strives to increase the efficiency by raising the temperature of the hot reservoir. **All real engines are less efficient than the Carnot engine because they are all subject to practical difficulties, including friction, but especially the need to operate irreversibly to complete a cycle in a brief time period.**

• *An alternative form of the second law of thermodynamics*

CONCEPTUAL PROBLEM 1

Which is more effective in increasing the efficiency of a Carnot engine—increasing T_h by ΔT, or decreasing T_c by ΔT?

CONCEPTUAL PROBLEM 2

An automobile manufacturer makes a claim that their cars obtain the same gasoline mileage with or without the air conditioner running. Do you believe this? Does playing the radio change the gas mileage? Turning on the headlights? Operating the windshield washer system? Rolling down a mechanical window?

CONCEPTUAL PROBLEM 3

If friction could be totally removed from a Carnot engine, would it then be 100% efficient?

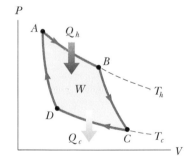

Figure 18.6 The *PV* diagram for the Carnot cycle. The net work done, *W*, equals the net heat received in one cycle, $Q_h - Q_c$. Note that $\Delta U = 0$ for the cycle.

Example 18.2   The Steam Engine

A steam engine has a boiler that operates at 500 K. The heat changes water to steam, and this steam then drives the piston. The exhaust temperature is that of the outside air, approximately 300 K. What is the maximum thermal efficiency of this steam engine?

A model steam engine equipped with a built-in horizontal boiler. The water is heated electrically, which generates steam that is used to power the electric generator at the right. *(Courtesy of Central Scientific Co.)*

Solution   From the expression for the efficiency of a Carnot engine, we find the maximum thermal efficiency for any engine operating between these temperatures:

$$e_c = 1 - \frac{T_c}{T_h} = 1 - \frac{300 \text{ K}}{500 \text{ K}} = 0.4, \text{ or } 40\%$$

You should note that this is the highest theoretical efficiency of the engine. In practice, the efficiency is considerably lower.

EXERCISE 3   Determine the maximum work the engine can perform in each cycle of operation if it absorbs 200 J of thermal energy from the hot reservoir during each cycle.
Answer   80 J

EXERCISE 4   A heat engine operates between two reservoirs at 20°C and 300°C. What is the maximum efficiency possible for this engine?   Answer   49%

18.4 • HEAT PUMPS AND REFRIGERATORS

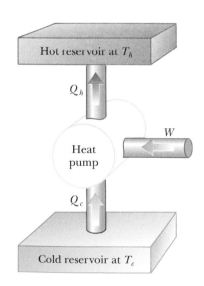

Figure 18.7   Schematic diagram of a heat pump, which absorbs thermal energy, Q_c, from the cold reservoir and expels thermal energy, Q_h, to the hot reservoir. The work done on the heat pump is W.

A heat pump is a mechanical device that moves thermal energy from a region at lower temperature to a region at higher temperature. Heat pumps have long been popular for cooling and are now becoming increasingly popular for heating purposes as well. In the heating mode, a circulating coolant fluid absorbs thermal energy from the outside and releases energy to the interior of the structure. The fluid is usually in the form of a low-pressure vapor when in the coils of the exterior part of the unit, where it absorbs thermal energy from either the air or the ground. This gas is then compressed into a hot, high-pressure vapor and enters the interior part of the unit where it condenses to a liquid and releases its stored thermal energy. An air conditioner is simply a heat pump installed backward, with "exterior" and "interior" interchanged.

Figure 18.7 is a schematic representation of a heat pump. The outside temperature is T_c, the inside temperature is T_h, and the thermal energy absorbed by the circulating fluid is Q_c. The compressor does work W on the fluid, and the thermal energy transferred from the pump into the structure is Q_h.

The effectiveness of a heat pump, in its heating mode, is described in terms of a number called the **coefficient of performance,** COP. This is defined as the ratio of the energy transferred as heat into the hot reservoir and the work required to transfer that heat:

$$\text{COP (heat pump)} = \frac{\text{energy transferred as heat}}{\text{work done by pump}} = \frac{Q_h}{W} \qquad [18.4]$$

If the outside temperature is 25°F or higher, the COP for a heat pump is about 4. That is, the energy transferred as heat into the house is about four times greater than the work done by the compressor in the heat pump. However, as the outside temperature decreases, it becomes more difficult for the heat pump to extract sufficient thermal energy from the air, and the COP drops. In fact, the COP can fall below unity for temperatures below the mid-teens.

A Carnot-cycle heat engine run in reverse constitutes a heat pump—in fact, the heat pump with the highest possible coefficient of performance for the temperatures between which it operates. The maximum coefficient of performance is

$$COP_c \text{ (heat pump)} = \frac{T_h}{T_h - T_c}$$

Although heat pumps are relatively new products in heating, the refrigerator has been a standard appliance in homes for years. The refrigerator works much like a heat pump, except that it cools its interior by pumping thermal energy from the food storage compartments into the warmer air outside. During its operation, a refrigerator removes thermal energy, Q_c, from the interior of the refrigerator, and in the process its motor does work W. The coefficient of performance of a refrigerator or of a heat pump used in its cooling cycle is

$$COP \text{ (refrigerator)} = \frac{Q_c}{W} \qquad \text{[18.5]}$$

An efficient refrigerator is one that removes the greatest amount of thermal energy from the cold reservoir with the least amount of work. Thus, a good refrigerator should have a high coefficient of performance, typically 5 or 6.

The highest possible coefficient of performance is again that of a refrigerator the working substance of which is carried through the Carnot heat-engine cycle in reverse:

$$COP_c \text{ (refrigerator)} = \frac{T_c}{T_h - T_c}$$

As the difference between temperatures of the two reservoirs approaches zero, the theoretical coefficient of performance of a Carnot heat pump approaches infinity. In practice, the low temperature of the cooling coils and the high temperature at the compressor limit the COP to values below 10.

Thinking Physics 2

Suppose the temperature inside a house is held constant at 72°F. Assume that it is 92°F outside in the summer, and the air conditioner (heat pump) is operated to keep the house at 72°F. Later, during the winter, it is 52°F outside and the heat pump is used to keep the house at 72°F again. In both cases, there was a 20°F difference between the inside and outside temperatures. The electric bill for the heat pump will be higher in the summer than in the winter. Why?

Reasoning In the summer, the interior of the house serves as the low temperature reservoir for the heat engine, and the outside air is the high temperature reservoir. Energy drawn from the interior air combines with energy entering the heat pump by electrical transmission, and the combination is transferred to the external air. Thus, the only advantage of the energy entering by electrical transmission is to pump the energy from the interior to the exterior. In the winter, energy drawn from the exterior air combines with energy entering the heat pump by electrical transmission, and the combination is transferred to the internal air. Thus, the energy entering by electricity does two things—it operates the heat pump *and* it adds to the internal energy of the interior air. Thus, less energy must be drawn from the external air in winter than from the internal air in the summer, and it costs less to run the heat pump in the winter.

Thinking Physics 3

It is a sweltering summer day, and your air conditioning system is not operating. In your kitchen, you have a working refrigerator and an ice chest full of ice. Which should you open, and leave open, to cool the room more efficiently?

Reasoning   The high temperature reservoir for your kitchen refrigerator is the air in the kitchen. If the refrigerator door were left open, energy would be drawn from the air in the kitchen, passed through the refrigeration system and transferred right back into the air. The result would be that the kitchen would become *warmer* due to the addition of the energy coming in by electricity to run the refrigeration system. If the ice chest were opened, energy in the air would enter the ice, raising its temperature and causing it to melt. The transfer of energy from the air would cause its temperature to drop. Thus, it would be more effective to open the ice chest.

CONCEPTUAL PROBLEM 4

Why are room air conditioners placed in windows or through holes in walls?

CONCEPTUAL PROBLEM 5

Can a heat pump deliver more energy than it consumes?

18.5 • AN ALTERNATIVE STATEMENT OF THE SECOND LAW

The first law of thermodynamics says that energy is conserved in every process. There is more to the story than that, however, and the second law adds the finishing touches. To illustrate what we mean, suppose you wish to cool off a hot piece of pizza by placing it on a block of ice. You will certainly be successful, because in every like situation that has ever been encountered, heat transfer has always taken place from a hot object to a cooler one. If you wait a short while, some amount of thermal energy—500 J, say—is transferred from the pizza to the ice. Energy is conserved, and the first law is satisfied. Yet, nothing in the first law says that this thermal energy transfer could not proceed in the opposite direction. (Imagine your astonishment if someday you place a piece of hot pizza on ice and 500 J of thermal energy moves from the ice to the pizza!) It is the second law that determines the directions of such natural phenomena.

An analogy can be made with the impossible sequence of events seen in a movie film running backward—such as a person diving out of a swimming pool, an apple rising from the ground and latching onto the branch of a tree, or a pot of hot water becoming colder as it rests over an open flame. Such events occurring backward in time are impossible because they violate the second law of thermodynamics. **Real processes have a preferred direction of time, often called the arrow of time.**

The second law has been stated in many different ways, but all the statements can be shown to be equivalent. Which form you use depends on the application you have in mind. For example, if you were concerned about the thermal energy transfer between pizza and ice, you might choose to concentrate on the second law in this form: **Thermal energy will not flow spontaneously from a cold object to a hot object.** At first glance, this statement of the second law seems to be radically different from that in Section 18.1. The two are, in fact, equivalent in all respects. Although we shall not prove it here, it can be shown that if either statement is false, so is the other.

18.6 · ENTROPY

The zeroth law of thermodynamics involves the concept of temperature, and the first law involves the concept of internal (thermal) energy. Temperature and internal energy are both state functions; that is, they can be used to describe the thermodynamic state of a system. Another state function, this one related to the second law of thermodynamics, is the **entropy,** *S*. In this section we define entropy on a macroscopic scale as it was first expressed by the German physicist Rudolf Clausius (1822–1888) in 1865.

Consider any infinitesimal process for a system between two equilibrium states. If dQ_r is the amount of thermal energy that would be transferred if the system had followed a reversible path, then the change in entropy, regardless of the actual path followed, is equal to this amount of thermal energy transferred along the reversible path divided by the absolute temperature of the system:

$$dS = \frac{dQ_r}{T} \qquad \text{[18.6]}$$

The subscript *r* on the term dQ_r is a reminder that the thermal energy transfer is to be measured along a reversible path, even though the system may actually have followed some irreversible path. When thermal energy is absorbed by the system, dQ_r is positive and hence the entropy increases. When thermal energy is expelled by the system, dQ_r is negative and the entropy decreases. Note that Equation 18.6 defines not entropy but rather the *change* in entropy. Hence, the meaningful quantity in a description of a process is the *change* in entropy.

Entropy originally found its place in thermodynamics, but its importance grew tremendously as the field of physics called statistical mechanics developed, because this method of analysis provided an alternative way of interpreting entropy. In statistical mechanics, a substance's behavior is described in terms of the statistical behavior of its atoms and molecules. One of the main outcomes of this treatment is the principle that **isolated systems tend toward disorder, and entropy is a measure of that disorder.**

For example, consider the molecules of a gas in the air in your room. If all the gas molecules moved together like soldiers marching in step, they would be in a very ordered state. It is also an unlikely state. If you could see the molecules, you would see that they actually move haphazardly in all directions, bumping into one another, changing speed on collision, some going fast, some slowly. This is a highly

Rudolf Clausius (1822–1888)

"I propose . . . to call *S* the entropy of a body, after the Greek word 'transformation.' I have designedly coined the word 'entropy' to be similar to energy, for these two quantities are so analogous in their physical significance, that an analogy of denominations seems to be helpful." *(AIP Niels Bohr Library, Lande Collection)*

• *Clausius's definition of change in entropy*

disordered state. It is also highly possible that the actual state of the system is such a highly disordered state.

The cause of this perpetual drive toward disorder is easily seen. For any given energy of the system, only certain states are possible or accessible. Among those states, it is assumed that all are equally probable. However, when such possible states are examined, it is found that far more of them are disordered states than ordered states. Because each of the states is equally probable, it is highly probable that the actual state will be one of the highly disordered states, or more precisely, that the actual state will be one in which the system moves between states of equivalent amounts of disorder. In the following discussion, therefore, the term *state* will indicate a collection of states of equivalent amounts of disorder.

All physical processes tend toward more probable states for the system and its surroundings. The more probable state is always one of higher disorder. Because *The second law again* • entropy is a measure of disorder, an alternative way of saying this is that **the entropy of the Universe increases in all natural processes.** This statement is yet another way of stating the second law of thermodynamics.

To calculate the change in entropy for a finite process, we must recognize that T is generally not constant. If dQ_r is the thermal energy transferred reversibly when the system is at a temperature of T, then the change in entropy in an arbitrary reversible process between an initial state and a final state is

Changes in entropy for a •
finite process

$$\Delta S = \int_i^f dS = \int_i^f \frac{dQ_r}{T} \qquad \text{(reversible path)} \qquad \text{[18.7]}$$

Although we do not prove it here, the change in entropy of a system that goes from one state to another has the same value for *all* paths connecting the two states. That is, the change in entropy of a system depends only on the properties of the initial and final equilibrium states.

In the case of a **reversible, adiabatic process,** no thermal energy is transferred between the system and its surroundings, and therefore $\Delta S = 0$. Because no change in entropy occurs, such a process is often referred to as an **isentropic process.**

Consider the changes in entropy that occur in a Carnot heat engine operating between the temperatures T_c and T_h. In one cycle, the engine absorbs thermal energy Q_h from the hot reservoir and rejects thermal energy Q_c to the cold reservoir. Thus, the total change in entropy for one cycle is

$$\Delta S = \frac{Q_h}{T_h} - \frac{Q_c}{T_c}$$

where the negative sign in the second term represents the fact that thermal energy Q_c is expelled by the system. For a Carnot cycle,

$$\frac{Q_c}{Q_h} = \frac{T_c}{T_h}$$

Change in entropy for a •
Carnot cycle is zero.

Using this result in the preceding expression for ΔS, we find that the total change in entropy for a Carnot engine operating in a cycle is *zero.*

Now consider a system taken through an arbitrary reversible cycle. Because the entropy function is a state function and hence depends only on the properties of a given equilibrium state, we conclude that $\Delta S = 0$ for *any* cycle. In general, we can write this condition in the mathematical form

$$\oint \frac{dQ_r}{T} = 0 \qquad\qquad \text{[18.8]}$$

- $\Delta S = 0$ *for any reversible cycle*

where the symbol $\oint$ indicates that the integration is over a *closed* path.

Another important property of entropy is the fact that the entropy of the Universe is not changed by a reversible process. This can be understood by noting that two bodies, A and B, that interact with each other reversibly must always be in thermal equilibrium with each other. That is, their temperatures must always be equal. Therefore, when a small amount of thermal energy, dQ, is transferred from A to B, the increase in entropy of B is dQ/T, whereas the corresponding change in entropy of A is $- dQ/T$. Thus, the total change in entropy of the system $(A + B)$ is zero, and the entropy of the Universe is unaffected by the reversible process.

As a special case, the following example shows how to calculate the change in entropy of a substance that undergoes a phase change.

Example 18.3   Change in Entropy: Melting Process

A solid substance that has a latent heat of fusion L_f melts at a temperature T_m. Calculate the change in entropy that occurs when m grams of this substance is melted.

Solution   Let us assume that the melting occurs so slowly that it can be considered a reversible process. In that case the temperature can be regarded as constant and equal to T_m. Making use of Equation 18.7 and the fact that the latent heat of fusion $Q = mL_f$, (Eq. 17.5), we find that

$$\Delta S = \int \frac{dQ_r}{T} = \frac{1}{T_m} \int dQ = \frac{Q}{T_m} = \frac{mL_f}{T_m}$$

Note that we were able to remove T_m from the integral because the process is isothermal.

EXERCISE 5   Calculate the change in entropy when 0.30 kg of lead melts at 327°C. Lead has a latent heat of fusion equal to 24.5 kJ/kg.   Answer $\Delta S = 12.3$ J/K

EXERCISE 6   What is the entropy decrease in 1 mol of helium gas that is cooled at 1 atm from room temperature (20°C) to 4 K? (C of helium = 21 J/mol K.)
Answer -90.2 J/K

18.7 • ENTROPY CHANGES IN IRREVERSIBLE PROCESSES

By definition, calculation of the change in entropy requires information about a reversible path connecting the initial and final equilibrium states. In order to calculate changes in entropy for real (irreversible) processes, we must first recognize that the entropy function (like internal energy) depends only on the *state* of the system. That is, entropy is a state function. Hence, the change in entropy when a system moves between any two equilibrium states depends only on the initial and final states. Experimentally one finds that the entropy change is the same for all processes that can occur between a given set of initial and final states. It is possible to show that if this were not the case, the second law of thermodynamics would be violated.

In view of the fact that the entropy of a system depends only on the state of the system, we can now calculate entropy changes for irreversible processes between two equilibrium states. We can do this by devising a reversible process (or series of reversible processes) between the same two equilibrium states and computing

$\int dQ_r/T$ for the reversible process. The entropy change for the irreversible process is the same as that for the reversible process between the same two equilibrium states. In irreversible processes, it is critically important to distinguish between Q, the actual thermal energy transfer in the process, and Q_r, the thermal energy that would have been transferred along a reversible path. It is only the latter that gives the correct value for entropy change. For example, as we will see, if an ideal gas expands adiabatically into a vacuum, $Q = 0$, but $\Delta S \neq 0$ because $Q_r \neq 0$. The reversible path between the same two states is the reversible, isothermal expansion that gives $\Delta S > 0$.

As we shall see in the following examples, change in entropy for the system plus its surroundings is always positive for an irreversible process. In general, the total entropy (and disorder) always increases in irreversible processes. From these considerations, the second law of thermodynamics can be stated as follows: **The total entropy of an isolated system that undergoes a change cannot decrease.** Furthermore, if the process is *irreversible,* the total entropy of an isolated system always *increases.* However, in a reversible process, the total entropy of an isolated system remains constant.

When dealing with interacting bodies that are not isolated from the environment, remember that the increase of entropy applies for the system *and* its surroundings. When two bodies interact in an irreversible process, the increase in entropy of one part of the system is greater than the decrease in entropy of the other part. Hence, we conclude that the change in entropy of the Universe must be greater than zero for an irreversible process and equal to zero for a reversible process. Ultimately, the entropy of the Universe should reach a maximum value. At this point the Universe will be in a state of uniform temperature and density. All physical, chemical, and biological processes will cease, because a state of perfect disorder implies that no energy is available for doing work. This gloomy state of affairs is sometimes referred to as the *heat death of the Universe.*

Heat Conduction

Consider the reversible transfer of thermal energy Q from a hot reservoir at temperature T_h to a cold reservoir at temperature T_c. Because the cold reservoir absorbs thermal energy Q, its entropy increases by Q/T_c. At the same time, the hot reservoir loses thermal energy Q and its entropy decreases by Q/T_h. The increase in entropy of the cold reservoir is greater than the decrease in entropy of the hot reservoir, because T_c is less than T_h. Therefore, the total change in entropy of the system (and of the Universe) is greater than zero:

$$\Delta S_u = \frac{Q}{T_c} - \frac{Q}{T_h} > 0$$

Thinking Physics 4

The entropy statement of the second law states that the entropy of the Universe increases in irreversible processes. This sounds very different from the previous two forms of the second law: "1. It is impossible to construct a heat engine that, operating in a cycle, produces no other effect than the absorption of thermal energy from a reservoir and the performance of an equal amount of work. 2. Energy will not flow by heat spontaneously from a cold object to a hot object." Can these two statements be made consistent with the entropy interpretation of the second law?

Reasoning There is no problem with consistency. In the first statement, the thermal energy in the reservoir is disordered energy—the random motion of molecules. Performing work results in ordered energy—pushing a piston through a displacement, for example. In this case, the motion of all molecules of the piston is in the same direction. If a heat engine absorbs thermal energy and performs an equal amount of work, then we have converted disorder into order, in violation of the entropy statement. In the second statement, we start with an ordered system—higher temperature in the hot object, lower in the cold object. This separation of temperatures is an example of order. If energy were to flow from the cold object to the hot object, so that the temperatures became even farther apart, this would be an increase in order, in violation of the entropy statement.

Example 18.4   Which Way Does the Heat Flow?

A large cold object is at 273 K, and a large hot object is at 373 K. Show that it is impossible for a small amount of thermal energy, say 8.00 J, to be transferred from the cold object to the hot one without decreasing the entropy of the Universe.

Reasoning We assume that, during the thermal energy transfer, the two objects do not undergo a temperature change. This is not a necessary assumption; it is used to avoid reliance on the techniques of integral calculus. The process as described is irreversible, so we must find an equivalent reversible process. It is sufficient to assume that the hot and cold objects are connected by a poor thermal conductor the temperature of which spans the range from 273 K to 373 K, which transfers thermal energy but the state of which does not change during the process. Then, the thermal energy transfer to or from each object is reversible, and we may set $Q = Q_r$.

Solution The entropy change of the hot object is

$$\Delta S_h = \frac{Q}{T_h} = \frac{8.00 \text{ J}}{373 \text{ K}} = 0.0214 \text{ J/K}$$

The cold object loses thermal energy, and its entropy change is

$$\Delta S_c = \frac{Q}{T_c} = \frac{-8.00 \text{ J}}{273 \text{ K}} = -0.0293 \text{ J/K}$$

The net entropy change of the Universe is

$$\Delta S_u = \Delta S_c + \Delta S_h = -0.0079 \text{ J/K}$$

This is in violation of the concept that the entropy of the Universe always increases in natural processes. That is, *the spontaneous transfer of thermal energy from a cold to a hot object cannot occur.*

EXERCISE 7 In the preceding example, suppose that 8.00 J of thermal energy is transferred from the hot to the cold object. What is the net entropy change of the Universe?

Answer   +0.0079 J/K

Free Expansion

An ideal gas in an insulated container initially occupies a volume of V_i (Fig. 18.8). A partition separating the gas from an evacuated region is suddenly broken so that the gas expands (irreversibly) to the volume V_f. Let us find the change in entropy of the gas and the Universe.

The process is clearly neither reversible nor quasi-static. The work done by the gas against the vacuum is zero, and because the walls are insulating, no thermal energy is transferred during the expansion. That is, $W = 0$ and $Q = 0$. Using the

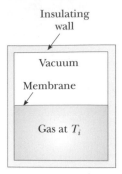

Figure 18.8 Free expansion of a gas. When the partition separating the gas from the evacuated region is ruptured, the gas expands freely and irreversibly so that it occupies a greater final volume. The container is thermally insulated from its surroundings, so $Q = 0$.

first law, we see that the change in internal energy is zero; therefore, $U_i = U_f$. Because the gas is ideal, U depends on temperature only, so we conclude that $T_i = T_f$.

To apply Equation 18.7, we must find Q_r—that is, we must find an equivalent reversible path that shares the same initial and final states. A simple choice is an isothermal, reversible expansion in which the gas pushes slowly against a piston. Because T is constant in this process, Equation 18.7 gives

$$\Delta S = \int \frac{dQ_r}{T} = \frac{1}{T} \int_i^f dQ_r$$

But $\int dQ_r$ is simply the work done by the gas during the reversible isothermal expansion from V_i to V_f, which is given by Equation 17.12. Because this is an isothermal process, $\Delta U = 0$, so the first law tells us that the energy input by heat is equal to the work done by the gas. Using this result, we find that

$$\Delta S = nR \ln \left(\frac{V_f}{V_i} \right) \tag{18.9}$$

Because $V_f > V_i$, we conclude that ΔS is positive, and so both the entropy and the disorder of the gas (and the Universe) increase as a result of the irreversible, adiabatic expansion.

Irreversible Heat Transfer

A substance of mass m_1, specific heat c_1, and initial temperature T_1 is placed in thermal contact with a second substance of mass m_2, specific heat c_2, and initial temperature T_2, where $T_2 > T_1$. The two substances are contained in an insulated box so that no thermal energy is lost to the surroundings. The system is allowed to reach thermal equilibrium. What is the total entropy change for the system?

First, let us calculate the final equilibrium temperature, T_f. Energy conservation requires that the thermal energy lost by one substance equal the thermal energy gained by the other. Because, by definition, $Q = mc\,\Delta T$ for each substance, we get $Q_1 = -Q_2$, or

$$m_1 c_1 \,\Delta T_1 = -m_2 c_2 \,\Delta T_2$$

$$m_1 c_1 (T_f - T_1) = -m_2 c_2 (T_f - T_2)$$

Solving for T_f gives

$$T_f = \frac{m_1 c_1 T_1 + m_2 c_2 T_2}{m_1 c_1 + m_2 c_2} \tag{18.10}$$

Note that $T_1 < T_f < T_2$, as expected.

The process is irreversible, because the system goes through a series of nonequilibrium states. During such a transformation, the temperature at any time is not well defined. However, we can imagine that the hot body, initially at temperature T_i, is slowly cooled to temperature T_f by contact with a series of reservoirs differing infinitesimally in temperature, the first reservoir being at T_i and the last at T_f. Such a series of very small changes in temperature would approximate a reversible process. Applying Equation 18.7 and noting that $dQ = mc\,dT$ for an infinitesimal change, we get

$$\Delta S = \int_1 \frac{dQ_1}{T} + \int_2 \frac{dQ_2}{T} = m_1 c_1 \int_{T_1}^{T_f} \frac{dT}{T} + m_2 c_2 \int_{T_2}^{T_f} \frac{dT}{T}$$

where we have assumed that the specific heats remain constant. Integrating, we find

$$\Delta S = m_1 c_1 \ln\left(\frac{T_f}{T_1}\right) + m_2 c_2 \ln\left(\frac{T_f}{T_2}\right) \qquad \textbf{[18.11]}$$

• *Change in entropy for a heat transfer process*

where T_f is given by Equation 18.10. If Equation 18.10 is substituted into Equation 18.11, it can be shown that one of the terms in Equation 18.11 will always be positive and the other negative. (You may want to verify this for yourself.) However, the positive term will always be larger than the negative term, resulting in a positive value for ΔS. Thus, we conclude that the entropy of the system and of the Universe increases in this irreversible process.

Finally, you should note that Equation 18.11 is valid only when no mixing occurs between the two substances in thermal contact. If the substances are liquids or gases and mixing occurs, Equation 18.11 applies only if the two fluids are identical.

18.8 • ENTROPY ON A MICROSCOPIC SCALE[2]

As we have seen, entropy can be approached via macroscopic concepts, using parameters such as pressure and temperature. Entropy can also be treated from a microscopic viewpoint through statistical analysis of molecular motions. We shall use a microscopic model to investigate the free expansion of an ideal gas discussed in the preceding section.

In the kinetic theory of gases, gas molecules are taken to be particles moving in a random fashion. Suppose all the gas is confined to the volume V_i (Fig. 18.9a). When the partition separating V_i from the larger container is removed, the molecules eventually are distributed in some fashion throughout the greater volume, V_f (Fig. 18.9b). The exact nature of the distribution is a matter of probability.

Such a probability can be determined by first finding the probabilities for the variety of molecular locations involved in the free expansion process. The instant after the partition is removed (and before the molecules have had a chance to rush into the other half of the container), all the molecules are in the initial volume. Now let us estimate the probability of the molecules' arriving at a particular configuration through natural random motions into some larger volume V_f. Assume each molecule occupies some microscopic volume, V_m. Then the total number of possible locations of that molecule, in a macroscopic initial volume, V_i, is the ratio, $W_i = V_i/V_m$, which is a huge number. We assume each of these locations is equally probable.

As more molecules are added to the system, the number of possible states multiply together. Neglecting the small probability of two molecules trying to occupy the same location, each molecule may go into any of the (V_i/V_m) locations, so the number of possibilities is $W_i^N = (V_i/V_m)^N$. Although a large number of the possible states would be considered unusual, the total number of states is so enormous that these unusual states have a negligible probability of occurring. Hence,

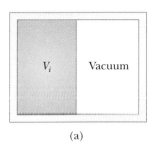

(a)

(b)

Figure 18.9 In a free expansion, the gas is allowed to expand into a larger volume that was previously a vacuum.

[2]This section was adapted from A. Hudson and R. Nelson, *University Physics*, Philadelphia, Saunders College Publishing, 1990, used with permission of the publisher.

A full house (a pair and three of a kind) is a good hand in the game of poker. Can you calculate the probability of being dealt this hand in a standard deck of 52 cards assuming that 5 cards are dealt at one time?   *(Tom Mareschel, The IMAGE Bank)*

the number of equivalent states is proportional to $(V_i/V_m)^N$. In a similar way, if the volume is increased to V_f, the number of equivalent states increases to $W_f^N = (V_f/V_m)^N$. Hence, the probabilities are $P_i = cW_i^N$ and $P_f = cW_f^N$, where the constant, c, has been left undetermined. The ratio of these probabilities is

$$\frac{P_f}{P_i} = \frac{cW_f^N}{cW_i^N} = \frac{\left(\dfrac{V_f}{V_m}\right)^N}{\left(\dfrac{V_i}{V_m}\right)^N} = \left(\frac{V_f}{V_i}\right)^N$$

If we now take the natural logarithm of this equation and multiply by Boltzmann's constant, k_B, we find

$$Nk_B \ln\left(\frac{W_f}{W_i}\right) = nN_A k_B \ln\left(\frac{V_f}{V_i}\right)$$

where we write the number of molecules, N, as nN_A, the number of moles times Avogadro's number. From Section 13.7 we know that $N_A k_B$ is the universal gas constant, R, so this equation may be written as

$$Nk_B \ln W_f - Nk_B \ln W_i = nR \ln\left(\frac{V_f}{V_i}\right) \qquad \text{[18.12]}$$

From thermodynamic considerations (specifically, Eq. 18.9) we know that when n mol of a gas undergoes a free expansion from V_i to V_f, the change in entropy, ΔS, is

$$S_f - S_i = nR \ln\left(\frac{V_f}{V_i}\right) \qquad \text{[18.13]}$$

Note that the right-hand sides of Equations 18.12 and 18.13 are identical. We thus make the following connection between entropy and the number of (microscopic) states available to the system in its (macroscopic) state:

Entropy (microscopic definition) •

$$S \equiv Nk_B \ln W \qquad \text{[18.14]}$$

Although our discussion used the specific example of the free expansion of an ideal gas, a more rigorous development of the statistical interpretation of entropy leads to the same conclusion: **Entropy is a measure of microscopic disorder.**

Imagine that a container of gas has molecules with speeds above the mean value on the left side and molecules with speeds lower than the mean value on the right side (an *ordered* arrangement), as in Figure 18.10a. Compare with an even mixture of fast- and slow-moving molecules like that in Figure 18.10b (a *disordered* arrangement). You might expect the ordered arrangement to be unlikely because random motions tend to mix the slow- and fast-moving molecules uniformly. Yet each of these arrangements *individually* is equally probable. However, there are far more disordered arrangements than ordered arrangements, so *collectively* the ordered arrangements are much less probable.

Let us estimate this probability for 100 molecules. The chance of any one molecule being in the left part of the container as a result of random motion is $1/2$. If the molecules move independently, the probability of 50 faster molecules being found in the left part at any instant is $(1/2)^{50}$. Likewise, the probability of the remaining 50 slower molecules being found in the right part at any instant is $(1/2)^{50}$. Therefore, the probability of finding this fast–slow separation through

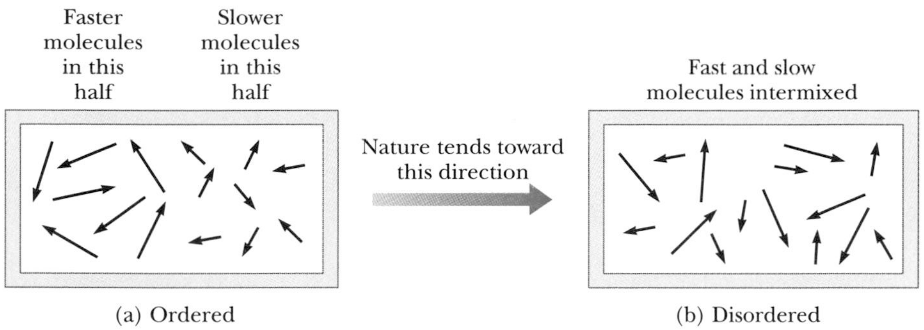

Faster molecules in this half

Slower molecules in this half

Nature tends toward this direction

Fast and slow molecules intermixed

(a) Ordered

(b) Disordered

Figure 18.10 A box of gas in two equally probable states of molecular motion that may exist. (a) An *ordered* arrangement; one of a *few*, and therefore *collectively unlikely*, set. (b) A *disordered* arrangement; one of *many*, and therefore a *collectively likely set*.

random motion is the product $(1/2)^{50}(1/2)^{50} = (1/2)^{100}$, which corresponds to about 1 chance in 10^{30}. When this calculation is extrapolated from 100 molecules to a mole of gas (about 10^{23} molecules), the ordered arrangement is found to be *extremely* improbable!

Thinking Physics 5

A box contains 5 gas molecules, spread throughout the box. At some point in time, all 5 are in one half of the box. Does this violate the second law of thermodynamics? Is the second law valid for this system?

Reasoning Strictly speaking, this situation does violate the second law of thermodynamics. In response to the second question, however, the second law is not valid for small numbers of particles. The second law is based on collections of huge numbers of particles, so that disordered states have astronomically higher probabilities than ordered states. Because the macroscopic world is built from these huge numbers of particles, we see the second law as valid, as processes proceed from order to disorder. In the 5-molecule system, the general idea of the second law is valid in that there are more disordered states than ordered states, but the relatively high probability of the ordered states results in their existence from time to time.

CONCEPTUAL PROBLEM 6

A variety of flowers grow in a garden. The flowers are picked and arranged into a bouquet. After the flowers wilt, they are thrown into a fireplace and burned. Arrange these three states of the flowers (growing in the garden, arranged in a bouquet, and burned in a fireplace) in order of increasing entropy.

CONCEPTUAL PROBLEM 7

Imagine an adiabatic expansion of a gas. According to the definition of entropy, there is no entropy change, because there was no energy transferred by heat. But the molecules of the gas now have more volume in which to move around, so there seems to be an increase in disorder. An increase in disorder means an increase in entropy. How can we reconcile these notions—one that suggests no entropy change and the other that suggests an increase in entropy?

Example 18.5   Free Expansion of an Ideal Gas: Macroscopic vs. Microscopic

Again consider the free expansion of an ideal gas. Let us verify that the macroscopic and microscopic approaches lead to the same conclusion. Suppose that 1 mol of gas undergoes a free expansion to four times its initial volume. The initial and final temperatures are, of course, the same. (a) Using a macroscopic approach, calculate the entropy change. (b) Find the probability, P_i, that all molecules will, through random motions, be found simultaneously in the original volume. (c) Calculate the change in entropy for the free expansion and show that it agrees with part (a).

Solution   (a) From Equation 18.9 we obtain

$$\Delta S = nR \ln\left(\frac{V_f}{V_i}\right) = (1)R \ln\left(\frac{4V_i}{V_i}\right) = R \ln 4$$

(b) The number of states available to a single molecule in the initial volume V_i is $W_i = (V_i/V_m)$. For one mole (N_A molecules), the number of available states is $W_i^{N_A}$. Thus, the probability that all molecules are found in the original volume is

$$P_i = cW_i^{N_A} = c\left(\frac{V_i}{V_m}\right)^{N_A}$$

(c) The number of states for all N_A molecules in the volume $V_f = 4V_i$ is

$$W_f^{N_A} = \left(\frac{V_f}{V_m}\right)^{N_A} = \left(\frac{4V_i}{V_m}\right)^{N_A}$$

From Equation 18.14 we obtain

$$\Delta S = k_B \ln W_f^{N_A} - k_B \ln W_i^{N_A} = k_B \ln\left(\frac{W_f^{N_A}}{W_i^{N_A}}\right) = k_B \ln (4^{N_A})$$

$$= N_A k_B \ln 4 = R \ln 4$$

The answer is the same as that in part (a), which dealt with macroscopic parameters.

OPTIONAL

18.9 • ENTROPY AND DISORDER

As you look around at the beauty in nature, it is easy to recognize that the events taking place within natural processes involve a large element of chance. For example, the spacing between trees in a natural forest is quite random. If you were to discover a forest in which all the trees were equally spaced, you would conclude that the forest had probably been planted by humans. We can express the results of such an observation by saying that **a disorderly arrangement is much more probable than an orderly one if the laws of nature are allowed to act without interference.**

We already found that entropy is a measure of the disorder in a system through the relation $S = k_B \ln W$, where W is the number of available states over which the molecules can be distributed and is therefore a measure of the disorder, or randomness, of the system. We can further explore the meaning of this relation by considering another specific example. Imagine that you have a bag of 100 marbles, of which 50 are red and 50 are green. You are allowed to draw four marbles from the bag according to the following rules. Draw one marble, record its color, return it to the bag, and draw another. Continue this process until four marbles have been drawn. Because each marble is returned to the bag before another is drawn, the probability of drawing a red marble is always the same as that of drawing a green one. All the possible outcomes are shown in Table 18.1. For example, the result RRGR means that you drew a red marble on the first draw, a red one on the second, a green one on the third, and a red one on the fourth. As this table indicates, there is only one way to draw four red marbles. However, there are four possible sequences that could produce one green and three red marbles, six sequences that could produce two green and two red, four sequences that could produce three

TABLE 18.1 Possible Results of Drawing Four Marbles from a Bag

End Result	Possible Draws	Total Number of Same Results
All R	RRRR	1
1G, 3R	RRRG, RRGR, RGRR, GRRR	4
2G, 2R	RRGG, RGRG, GRRG, RGGR, GRGR, GGRR	6
3G, 1R	GGGR, GGRG, GRGG, RGGG	4
All G	GGGG	1

green and one red, and one sequence that could produce all green. The most likely outcome—two red and two green marbles—corresponds to the most disordered state. This is considered disorder simply because we cannot distinguish between individual molecules of the same color or between individual molecules. The probability that you would draw four red or four green marbles, these being the most ordered states, is much lower. From Equation 18.14 we see that the state with the greatest disorder has the highest entropy because it is the most probable. However, the most ordered states (all red or all green) are the least likely to occur and are the states of lowest entropy. The outcome of the draw can range from a highly ordered state, which has the lowest entropy, to a highly disordered state, which has the highest entropy.

Thinking Physics 6

Touch a rubber band to your forehead and note the sensation of temperature. Remove the rubber band from your forehead, stretch it and touch it to your forehead again. It should now feel warmer. Remove the rubber band again from your forehead, let it return to its original length and touch it to your forehead again. It now feels cooler.

  (a) Why do these temperature changes occur?
  (b) Someone proposes that the answer to part (a) is *internal friction,* as portions of the rubber band rub against each other during the size changes. How would you respond to this proposal?

Reasoning (a) The answer lies in the second law of thermodynamics. Rubber consists of long, coiled molecules, the axes of which lie in random directions relative to the length of the rubber band. The force of the stretch tends to cause the coiled molecules to line up parallel with the rubber band. This represents an increase in the degree of *order* of these molecules, which represents a *decrease* in entropy. If the rubber band is stretched quickly, however, then the stretching process is very close to adiabatic, and the net entropy change is close to zero. Thus, there must be some change that occurs to compensate for the decrease in entropy due to the coils lining up. We can compensate for the decrease if the molecules increase their vibrational motion, so that their positions in the rubber represent a lower degree of order. Increased vibrational motion is represented by more internal energy that, in turn, means that the temperature increases—this is the warmth that is felt on the forehead.

  When the rubber band is allowed to return to its original length, the coiled molecules can become more disordered, resulting in an increase in entropy. The increase in entropy of the directions of the molecules is accompanied by a decrease in entropy of the vibrations of the molecules. This decreased vibration represents a lower internal energy, and you feel a cooler temperature in this situation.

(b) The internal friction argument could be claimed to be consistent with the increase in temperature associated with stretching the rubber band. When the rubber band is returned to its original length, however, there would be *additional rubbing* of portions of the rubber band, and the internal friction would result in a further temperature *increase*. This is inconsistent with the observation that the temperature decreases on returning the rubber band to its original length. Internal friction can be argued as the reason that a wire coat hanger increases in temperature when it is bent back and forth, but the rubber band phenomenon has the second law of thermodynamics as its basis.

SUMMARY

Real processes proceed in directions governed by the second law of thermodynamics. A **heat engine** is a device that converts thermal energy to other useful forms of energy. The net work done by a heat engine in carrying a substance through a cyclic process ($\Delta U = 0$) is

$$W = Q_h - Q_c \qquad [18.1]$$

where Q_h is the thermal energy absorbed from a hot reservoir and Q_c is the thermal energy rejected to a cold reservoir.

The **thermal efficiency,** e, of a heat engine is defined as the ratio of the net work done to the thermal energy absorbed per cycle from the higher temperature reservoir:

$$e \equiv \frac{W}{Q_h} = 1 - \frac{Q_c}{Q_h} \qquad [18.2]$$

The **second law of thermodynamics** can be stated as follows: **It is impossible to construct a heat engine that, operating in a cycle, produces no effect other than the absorption of thermal energy from a reservoir and the performance of an equal amount of work.**

A **reversible process** is one in which the system can be returned to its initial conditions along the same path and in which every point along the path is an equilibrium state. A process that does not satisfy these requirements is **irreversible.**

The **efficiency of a heat engine** operating in the **Carnot cycle** is given by

$$e_c = 1 - \frac{T_c}{T_h} \qquad [18.3]$$

where T_c is the absolute temperature of the cold reservoir and T_h is the absolute temperature of the hot reservoir.

No real heat engine operating between the temperatures T_c and T_h can be more efficient than an engine operating reversibly in a Carnot cycle between the same two temperatures.

The second law of thermodynamics states that when real (irreversible) processes occur, the degree of disorder in the system plus the surroundings increases. When a process occurs in an isolated system, ordered energy is converted into disordered energy. The measure of disorder in a system is called **entropy,** S.

The **change in entropy,** dS, of a system moving through an infinitesimal process between two equilibrium states is

$$dS = \frac{dQ_r}{T} \qquad [18.6]$$

The change in entropy of a system moving between two equilibrium states is

$$\Delta S = \int_i^f \frac{dQ_r}{T} \qquad [18.7]$$

The value of ΔS is the same for all reversible paths connecting the initial and final states.

The change in entropy for any reversible, cyclic process is zero, and when such a process occurs, the entropy of the Universe remains constant. Entropy is a state function—that is, it depends on the state of the system. The change in entropy for a system undergoing a real (irreversible) process between two equilibrium states is the same as that for a reversible process between the same states. In an irreversible process, the total entropy of an isolated system always increases. In general, the total entropy (and disorder) always increases in any irreversible process. Furthermore, the change in entropy of the Universe is greater than zero for an irreversible process.

From a microscopic viewpoint, **entropy,** S, is defined as

$$S \equiv k_B \ln W \qquad [18.14]$$

where k_B is Boltzmann's constant and W is the number of (microscopic) states available to the system in its (macroscopic) state. Because of the statistical tendency of systems to proceed toward states of greater probability and greater disorder, all natural processes are irreversible and increase in entropy. Thus, **entropy is a measure of microscopic disorder.**

CONCEPTUAL QUESTIONS

1. What are some factors that affect the efficiency of automobile engines?
2. A steam-driven turbine is one major component of an electric power plant. Why is it advantageous to have the temperature of the steam as high as possible?
3. Is it possible to construct a heat engine that creates no thermal pollution?
4. Discuss the change in entropy of a gas that expands (a) at constant temperature and (b) adiabatically.
5. In solar ponds constructed in Israel, the Sun's energy is concentrated near the bottom of a salty pond. With the proper layering of salt in the water, convection is prevented, and temperatures of 100°C may be reached. Can you guess the maximum efficiency with which useful energy can be extracted from the pond?
6. Why does your automobile burn more gas in winter than in summer?
7. Can a heat pump have a coefficient of performance less than unity? Explain.
8. The Carnot cycle was illustrated in the chapter on a *PV* diagram. What would the Carnot cycle look like on a *TS* diagram?
9. Give some examples of irreversible processes that occur in nature. Give an example of a process in nature that is nearly reversible.
10. A thermodynamic process occurs in which the entropy of a system changes by −8.0 J/K. According to the second law of thermodynamics, what can you conclude about the entropy change of the environment?
11. If a supersaturated sugar solution is allowed to slowly evaporate, sugar crystals form in the container. Hence, sugar molecules go from a disordered form (in solution) to a highly ordered, crystalline form. Does this process violate the second law of thermodynamics? Explain.
12. How could you increase the entropy of 1 mol of a metal that is at room temperature? How could you decrease its entropy?
13. A heat pump is to be installed in a region where the average outdoor temperature in the winter months is −20°C. In view of this, why would it be advisable to place the outdoor compressor unit deep in the ground? Why are heat pumps not commonly used for heating in cold climates?
14. Suppose your roommate is "Mr. Clean" and tidies up your messy room after a big party. Because more order is being created by your roommate, does this represent a violation of the second law of thermodynamics?
15. Discuss the entropy changes that occur when you (a) bake a loaf of bread and (b) consume the bread.
16. A diesel engine operates with a compression ratio of about 15. A gasoline engine has a compression ratio of about 6. Which engine runs hotter? Which engine (at least theoretically) can operate more efficiently?
17. If you shake a jar full of jelly beans of different sizes, the larger beans tend to appear near the top, and the

smaller ones tend to fall to the bottom. Why does this occur? Does this process violate the second law of thermodynamics?

18. The device in Figure Q18.18, called a thermoelectric converter, uses a series of semiconductor cells to convert thermal energy to electrical energy. In the photograph at the left, both legs of the device are at the same temperature, and no electrical energy is produced. However, when one leg is at a higher temperature than the other, as in the photograph on the right, electrical energy is produced as the device extracts energy from the hot reservoir and drives a small electric motor. (a) Why does the temperature differential produce electrical energy in this demonstration? (b) In what sense does this intriguing experiment demonstrate the second law of thermodynamics?

Figure Q18.18   (*Courtesy of PASCO Scientific Company*)

PROBLEMS

Section 18.1   Heat Engines and the Second Law of Thermodynamics

1. A heat engine absorbs 360 J of thermal energy and performs 25.0 J of work in each cycle. Find (a) the efficiency of the engine and (b) the thermal energy expelled in each cycle.

2. A heat engine performs 200 J of work in each cycle and has an efficiency of 30.0%. For each cycle, how much thermal energy is (a) absorbed and (b) expelled?

3. A particular engine has a power output of 5.00 kW and an efficiency of 25.0%. If the engine expels 8000 J of thermal energy in each cycle, find (a) the heat absorbed in each cycle and (b) the time for each cycle.

Section 18.3   The Carnot Engine

4. A heat engine operating between 200°C and 80.0°C achieves 20.0% of the maximum possible efficiency. What energy input will enable the engine to perform 10.0 kJ of work?

5. One of the most efficient engines ever built (its actual efficiency is 42.0%) operates between 430°C and 1870°C. (a) What is its maximum theoretical efficiency? (b) How much power does the engine deliver if it absorbs 1.40×10^5 J of thermal energy each second?

6. A Carnot engine has a power output of 150 kW. The engine operates between two reservoirs at 20.0°C and 500°C. (a) How much thermal energy is absorbed per hour? (b) How much thermal energy is lost per hour?

7. A steam engine is operated in a cold climate in which the exhaust temperature is 0°C. (a) Calculate the theoretical maximum efficiency of the engine using an intake steam temperature of 100°C. (b) If, instead, superheated steam at 200°C is used, find the maximum possible efficiency.

8. The exhaust temperature of a Carnot heat engine is 300°C. What is the intake temperature if the efficiency of the engine is 30.0%?

9. An ideal gas is taken through a Carnot cycle. The isothermal expansion occurs at 250°C, and the isothermal compression takes place at 50.0°C. If the gas absorbs 1200 J of heat during the isothermal expansion, find (a) the heat expelled to the cold reservoir in each cycle and (b) the net work done by the gas in each cycle.

10. A power plant operates at a 32.0% efficiency during the summer when the sea water for cooling is at 20.0°C. The plant uses 350°C steam to drive turbines. Assuming that the plant's efficiency changes in the same proportion as the ideal efficiency, what would be the plant's efficiency in the winter when the sea water is at 10.0°C?

11. A power plant has been proposed that would make use of the temperature gradient in the ocean. The system is to operate between 20.0°C (surface water temperature) and 5.00°C (water temperature at a depth of about 1 km). (a) What is the maximum efficiency of such a system? (b) If the power output of the plant is 75.0 MW, how much thermal energy is absorbed per hour? (c) In view of your answer to (a), do you think such a system is worthwhile (considering that there is no charge for fuel)?

12. Suppose a heat engine is connected to two heat reservoirs, one a pool of molten aluminum (660°C) and the other a block of solid mercury (-38.9°C). The engine runs by freezing 1.00 g of aluminum and melting 15.0 g of mercury during each cycle. The heat of fusion of aluminum is 3.97×10^5 J/kg; the heat of fusion of mercury is 1.18×10^4 J/kg. What is the efficiency of this engine?

13. A 20.0%-efficient real engine is used to speed up a train from rest to 5.00 m/s. It is known that an ideal (Carnot) engine using the same cold and hot reservoirs would accelerate the same train from rest to a speed of 6.50 m/s using the same amount of fuel. If the engines use air at 300 K as a cold reservoir, find the temperature of the steam serving as the hot reservoir.

Section 18.4   Heat Pumps and Refrigerators

14. What is the coefficient of performance of a refrigerator that operates with Carnot efficiency between temperatures $-3.00°C$ and $+27.0°C$?

15. An ideal refrigerator or ideal heat pump is equivalent to a Carnot engine running in reverse. That is, heat Q_c is absorbed from a cold reservoir and heat Q_h is rejected to a hot reservoir. (a) Show that the work that must be supplied to run the refrigerator or pump is

$$W = \frac{T_h - T_c}{T_c} Q_c$$

(b) Show that the coefficient of performance of the ideal refrigerator is

$$COP = \frac{T_c}{T_h - T_c}$$

16. How much work is required, using an ideal Carnot refrigerator, to remove 1.00 J of thermal energy from helium gas at 4.00 K and reject this thermal energy to a room-temperature (293 K) environment?

17. A refrigerator has a coefficient of performance equal to 5.00. If the refrigerator absorbs 120 J of thermal energy from a cold reservoir in each cycle, find (a) the work done in each cycle and (b) the thermal energy expelled to the hot reservoir.

18. A heat pump, shown in Figure P18.18, is essentially a heat engine run backward. It extracts thermal energy from colder air outside and deposits it in a warmer room. Suppose that the ratio of the actual thermal energy entering the room to the work done by the device's motor is 10.0% of the theoretical maximum. Determine the thermal energy entering the room per joule of work done by the motor if the inside temperature is 20.0°C and the outside temperature is $-5.00°C$.

Section 18.6   Entropy

19. An ice tray contains 500 g of water at 0°C. Calculate the change in entropy of the water as it freezes completely and slowly at 0°C.

20. At a pressure of 1 atm, liquid helium boils at 4.20 K. The latent heat of vaporization is 20.5 kJ/kg. Determine the entropy change (per kilogram) resulting from vaporization.

21. Calculate the change in entropy of 250 g of water heated slowly from 20.0°C to 80.0°C. (*Hint:* Note that $dQ = mc\,dT$.)

Section 18.7   Entropy Changes in Irreversible Processes

22. The surface of the Sun is approximately at 5700 K, and the temperature of the Earth's surface is approximately 290 K. What entropy change occurs when 1000 J of thermal energy is transferred from the Sun to the Earth?

23. A 1500-kg car is moving at 20.0 m/s. The driver brakes to a stop. The brakes cool off to the temperature of the surrounding air, which is nearly constant at 20.0°C. What is the total entropy change?

24. An avalanche of ice and snow, of mass 1000 kg, slides downhill a vertical distance of 200 m. What is the total entropy change if the mountain air temperature is nearly constant at $-3.00°C$?

25. A 1.00-kg iron horseshoe is taken from a furnace at 900°C and dropped into 4.00 kg of water at 10.0°C. If no heat is lost to the surroundings, determine the total entropy change.

26. How fast are you personally making the entropy of the Universe increase right now? Compute an order-of-magnitude estimate, stating what quantities you take as data and the values you measure or estimate for them.

27. One mole of H_2 gas is contained in the left-hand side of the container shown in Figure P18.27, which has equal volumes

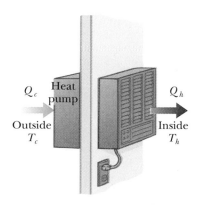

Figure P18.18

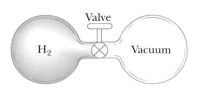

Figure P18.27

left and right. The right-hand side is evacuated. When the valve is opened, the gas streams into the right side. What is the final entropy change? Does the temperature of the gas change?

28. A 2.00-liter container has a center partition that divides it into two equal parts, as shown in Figure P18.28. The left side contains H_2 gas, and the right side contains O_2 gas. Both gases are at room temperature and at atmospheric pressure. The partition is removed and the gases are allowed to mix. What is the entropy increase?

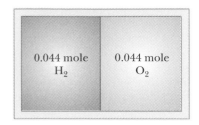

| 0.044 mole
H_2 | 0.044 mole
O_2 |

Figure P18.28

18.8   Entropy on a Microscopic Scale

18.9   Entropy and Disorder (Optional)

29. If you toss two dice, what is the total number of ways in which you can obtain (a) a 12 and (b) a 7?

30. Prepare a table like Table 18.1 for the following occurrence. You toss four coins into the air simultaneously. Record the results of your tosses in terms of the numbers of heads and tails that result. For example, HHTH and HTHH are two possible ways in which three heads and one tail can be achieved. (a) On the basis of your table, what is the most probable result of a toss? In terms of entropy, (b) what is the most ordered state, and (c) what is the most disordered?

31. Repeat the procedure used to construct Table 18.1 (a) for the case in which you draw three marbles from your bag rather than four and (b) for the case in which you draw five rather than four.

Additional Problems

32. On a hot day, a Florida home is kept cool by an air conditioner. The outside temperature is 36.0°C and the interior air is at 18.0°C. If 50 000 kJ/h of thermal energy is removed from the house, what is the minimum power that must be provided to the air conditioner?

33. Every second at Niagara Falls, some 5000 m³ of water falls a distance of 50.0 m. What is the increase in entropy per second due to the falling water? (Assume that the mass of the surroundings is so large that its temperature and that

of the water stays nearly constant at 20.0°C. Suppose that negligible water evaporates.)

34. If a 35.0%-efficient heat engine (Fig. 18.1) is run in reverse so as to form a refrigerator (Fig. 18.7), what would be this refrigerator's coefficient of performance?

35. How much work is required, using an ideal Carnot refrigerator, to change 0.500 kg of tap water at 10.0°C into ice at −20.0°C? Assume the freezer compartment is held at −20.0°C and the refrigerator exhausts heat into a room at 20.0°C.

36. A heat engine operates between two reservoirs at T_2 = 600 K and T_1 = 350 K. It absorbs 1000 J of thermal energy from the higher temperature reservoir and performs 250 J of work. Find (a) the entropy change of the Universe, ΔS_u, for this process and (b) the work, W, that could have been done by an ideal Carnot engine operating between these two reservoirs. (c) Show that the difference between the work done in parts (a) and (b) is $T_1 \Delta S_u$.

37. A house loses heat through the exterior walls and roof at a rate of 5000 J/s = 5.00 kW when the interior temperature is 22.0°C and the outside temperature is −5.00°C. Calculate the electric power required to maintain the interior temperature at 22.0°C for the following two cases. (a) The electric power is used in electric resistance heaters (which convert all of the electricity supplied into heat). (b) The electric power is used to drive an electric motor that operates the compressor of a heat pump (which has a coefficient of performance equal to 60.0% of the Carnot cycle value).

38. An athlete whose mass is 70.0 kg drinks 16 ounces (453.6 g) of refrigerated water. The water is at a temperature of 35.0°F. (a) Neglecting the temperature change of the body that results from the water intake (so that the body is regarded as a reservoir always at 98.6°F), find the entropy increase of the entire system. (b) Assume that the entire body is cooled by the drink and that the average specific heat of a human is equal to the specific heat of liquid water. Neglecting any other heat transfers and any metabolic heat release, find the athlete's temperature after she drinks the cold water, given an initial body temperature of 98.6°F. Under *these* assumptions, what is the entropy increase of the entire system? Compare this result with (a).

39. A refrigerator has a coefficient of performance of 3.00. The ice tray compartment is at −20.0°C, and the room temperature is 22.0°C. The refrigerator can convert 30.0 g of water at 22.0°C to 30.0 g of ice at −20.0°C each minute. What input power is required? Give your answer in watts.

40. An ideal (Carnot) freezer in a kitchen has a constant temperature of 260 K, whereas the air in the kitchen has a constant temperature of 300 K. Suppose the insulation for the freezer is not perfect so that some heat flows into the freezer at a rate of 0.150 W. Determine the average power of the freezer's motor which is needed to maintain the constant temperature in the freezer.

41. A power plant, having a Carnot efficiency, produces

1000 MW of electrical power from turbines that take in steam at 500 K and reject water at 300 K into a flowing river. If the water downstream is 6.00 K warmer due to the output of the power plant, determine the flow rate of the river.

42. A power plant, having a Carnot efficiency, produces electric power P from turbines that take in heat from steam at temperature T_h and discharge heat at temperature T_c through a heat exchanger into a flowing river. If the water downstream is warmer by ΔT due to the output of the power plant, determine the flow rate of the river.

43. What is the minimum amount of work that must be done to extract 400 J of heat from a massive object at 0°C while rejecting heat to a hot reservoir at a temperature of 20.0°C?

44. A gas follows the path 123 on the PV diagram in Figure P18.44 and 418 J of heat flows into the system. Also, 167 J of work is done. (a) What is the internal energy change of the system? (b) How much thermal energy flows into the system if the process follows the path 143? The work done by the gas along this path is 63.0 J. What net work would be done on or by the system if the system followed (c) the path 12341? (d) the path 14321? (e) What is the change in internal energy of the system in the processes described in parts (c) and (d)?

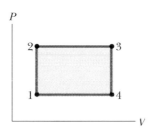

Figure P18.44

45. Figure P18.45 represents n mol of an ideal monatomic gas being taken through a cycle that consists of two isothermal processes at temperatures $3T_0$ and T_0 and two constant-volume processes. For each cycle, determine, in terms of n, R, and T_0, (a) the net heat transferred to the gas and (b) the efficiency of an engine operating in this cycle.

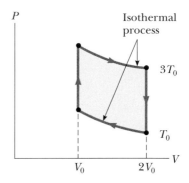

Figure P18.45

46. An electric power plant has an overall efficiency of 15.0%. The plant is to deliver 150 MW of power to a city, and its turbines use coal as the fuel. The burning coal produces steam that drives the turbines. This steam is then condensed to water at 25.0°C by passing it through cooling coils in contact with river water. (a) How many metric tons of coal does the plant consume each day (1 metric ton = 10^3 kg)? (b) What is the total cost of the fuel per year if the delivered price is $8.00/metric ton? (c) If the river water is delivered at 20.0°C, at what minimum rate must it flow over the cooling coils in order that its temperature not exceed 25.0°C? (*Note:* The heat of combustion of coal is 33.0 kJ/g.)

ANSWERS TO CONCEPTUAL PROBLEMS

1. The efficiency of a Carnot engine is given by the ratio of the difference in operating temperatures $(T_h - T_c)$ to the temperature of the hot reservoir (T_h). Increasing T_h by ΔT or decreasing T_c by ΔT both have the same effect on the difference in operating temperatures. But only increasing T_h has an effect on the denominator of the fraction, increasing the denominator, and, thus, reducing the value of the fraction. Thus, it would be more effective to decrease T_c by ΔT.

2. An automobile engine has a certain efficiency, which determines how much energy is available to operate the car. With no accessories operating, this energy is completely available for changing the kinetic energy of the car (during acceleration) and overcoming friction as the car moves. If accessories are operating, the energy for their operation must come from the output energy of the engine, leaving less available for moving the car, and thus decreasing the gas mileage. Therefore, the air conditioner, the radio, the headlights, and the windshield washer system will all reduce the gas mileage when they are operating. Rolling down a manual window does not—the energy for this comes from the driver's body.

3. If friction were removed, the engine would be as efficient as possible, but even a Carnot engine would not be 100% efficient—this would violate the second law. The maximum efficiency depends on the operating temperatures.

4. The role of an air conditioner is that of a heat pump—it pumps energy from a cold reservoir (the interior air) to a warm reservoir (the exterior air). Thus, the "exhaust" of the air conditioner must be to the outside air, which is the case

if the air conditioner is installed in a window or through a hole in the wall. If an air conditioner were operated in the middle of a room, air would enter the unit, have energy added to it (the energy entering the unit by electricity) and be expelled back into the room, warming the room up.

5. Yes, a heat pump can deliver significantly more energy than it consumes. In the winter, the energy delivered to the interior of the home is a combination of the energy drawn from the exterior air and the energy coming into the unit by electricity. If the COP of the heat pump is 5, then it is delivering 5 times as much energy as it consumes.

6. In order of increasing entropy: arranged in a bouquet, growing in the garden, burned in a fireplace.

7. There is no increase in entropy from either point of view. In an adiabatic expansion, the temperature must drop. Thus, although the molecules have more volume in which to move around, they move with lower velocities. The decrease in velocity represents more order, because the random jostling of the molecules is reduced. A calculation will show that the decrease in entropy due to the drop in jostling exactly compensates for the increase due to the increased volume. Thus, the net entropy change is zero.

19

Electric Forces and Electric Fields

This chapter begins with a review of some of the basic properties of the electrostatic force, as well as some properties of the electric field associated with stationary charged particles. Our study of electrostatics then continues with the concept of an electric field that is associated with a continuous charge distribution and the effect of this field on other charged particles.

Two methods for calculating the electric fields associated with continuous charge distributions are discussed. The first method uses Coulomb's law and integral calculus, and the second uses Gauss's law. An outgrowth of Coulomb's law, Gauss's law is much more convenient for calculating electric fields of highly symmetric charge distributions. Furthermore, Gauss's law serves as a guide for understanding more complicated problems.

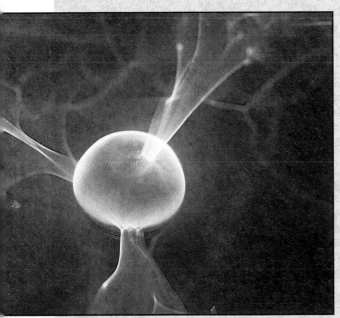

◄ This metallic sphere is charged by a generator to a high voltage. The high concentration of charge formed on the sphere creates a strong electric field around the sphere. The charges then leak through the surrounding gas, producing the pink glow. *(E. R. Degginger, H. Armstrong Roberts)*

19.1 • HISTORICAL OVERVIEW

In this chapter we continue our study of electric and magnetic phenomena that we introduced in Chapter 5. The laws of electricity and magnetism play a central role in the operation of devices such as radios, televisions, electric motors, computers, high-energy particle accelerators, and a host of electronic devices used in medicine. More fundamental, however, is the fact that the interatomic and intermolecular forces that are responsible for the formation of solids and liquids are electric in origin. Furthermore, such forces as the pushes and pulls between objects and the elastic force in a spring arise from electric forces at the atomic level.

Chinese documents suggest that magnetism was recognized as early as about 2000 B.C. The ancient Greeks observed electric and magnetic phenomena possibly as early as 700 B.C. They found that a piece of amber, when rubbed, became electrified and attracted pieces of straw or feathers. The existence of magnetic forces was known from observations that pieces of a naturally occurring stone called *magnetite* (Fe_3O_4) were attracted to iron. (The word *electric* comes from the Greek word for amber, *elektron*. The word *magnetic* comes from *Magnesia*, a city on the coast of Turkey where magnetite was found.)

In 1600, the English scientist William Gilbert discovered that electrification was not limited to amber but was a general phenomenon. Scientists went on to electrify a variety of objects, including chickens and people! Experiments by Charles Coulomb in 1785 confirmed the inverse-square force law for electricity.

It was not until the early part of the 19th century that scientists established that electricity and magnetism are related phenomena. In 1820, the Danish scientist Hans Oersted discovered that a compass needle, which is magnetic, is deflected when placed near a circuit carrying an electric current. In 1831, Michael Faraday in England and, almost simultaneously, Joseph Henry in the United States, showed that when a wire is moved near a magnet (or when a magnet is moved near a wire), an electric current is observed in the wire. In 1873, James Clerk Maxwell used these observations and other experimental facts as a basis for formulating the laws of electromagnetism as we know them today. Shortly thereafter (around 1888), Heinrich Hertz verified Maxwell's predictions by producing electromagnetic waves in the laboratory. This achievement was followed by such practical developments as radio and television.

Maxwell's contributions to the science of electromagnetism were especially significant because the laws he formulated are basic to *all* forms of electromagnetic phenomena. His work is comparable in importance to Newton's discovery of the laws of motion and the theory of gravitation.

19.2 • PROPERTIES OF ELECTRIC CHARGES

A number of simple experiments demonstrate the existence of electrostatic forces. For example, after running a comb through your hair, you will find that the comb attracts bits of paper. The attractive electrostatic force is often strong enough to suspend the bits. The same effect occurs with other rubbed materials, such as glass or rubber.

Another simple experiment is to rub an inflated balloon with wool (or across your hair). On a dry day, the rubbed balloon will stick to the wall of a room, often

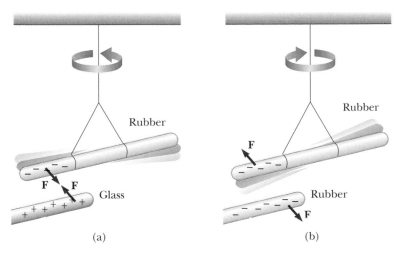

(a)   (b)

Figure 19.1   (a) A negatively charged rubber rod, suspended by a thread, is attracted to a positively charged glass rod. (b) A negatively charged rubber rod is repelled by another negatively charged rubber rod.

Charles Coulomb (1736–1806), the great French physicist for whom the unit of electric charge called the *coulomb* was named.

for hours. When materials behave this way, they are said to have become **electrically charged.** You can give your body an electric charge by walking across a wool rug or by sliding across a car seat. You can then feel, and remove, the charge on your body by lightly touching another person. Under the right conditions, a visible spark is seen when you touch, and a slight tingle is felt by both parties. (Such an experiment works best on a dry day because excessive moisture can provide a pathway for charge to leak off a charged object.)

Experiments also demonstrate that there are two kinds of electric charge, which were given the names **positive** and **negative** by Benjamin Franklin (1706–1790). Figure 19.1 illustrates the interactions of the two charges. A hard rubber (or plastic) rod that has been rubbed with fur (or an acrylic material) is suspended by a piece of string. When a glass rod that has been rubbed with silk is brought near the rubber rod, the rubber rod is attracted toward the glass rod (Fig. 19.1a). If two charged rubber rods (or two charged glass rods) are brought near each other, as in Figure 19.1b, the force between them is repulsive. This observation demonstrates that the rubber and glass have different kinds of charge. We use the convention suggested by Franklin, wherein the electric charge on the glass rod is called positive and that on the rubber rod is called negative. On the basis of observations such as these, we conclude that **like charges repel one another, and unlike charges attract one another.**

A natural tendency exists for charge to be transferred between unlike materials. Rubbing the two materials together only increases the area of contact and thus enhances the charge-transfer process.

Another important characteristic of electric charge is that **it is always conserved.** That is, when two initially neutral objects are charged by being rubbed together, charge is not created in the process. The objects become charged because *negative charge is transferred* from one object to the other. One object gains some amount of negative charge while the other loses an equal amount of negative charge and hence is left with a positive charge. For example, when a glass rod is rubbed with silk, as in Figure 19.2, the silk obtains a negative charge that is equal in mag-

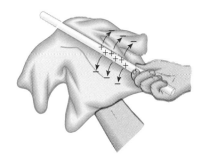

Figure 19.2   When a glass rod is rubbed with silk, electrons are transferred from the glass to the silk. Because of conservation of charge, each electron adds negative charge to the silk, and an equal positive charge is left behind on the rod. Also, because the charges are transferred in discrete bundles, the charges on the two objects are $\pm e$, $\pm 2e$, or $\pm 3e$, and so on.

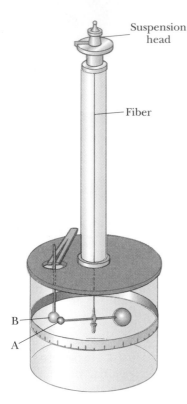

Figure 19.3   Coulomb's torsion balance, which was used to establish the inverse-square law for the electrostatic force between two charges.

Figure 19.4   (Conceptual Problem 1)

nitude to the positive charge on the glass rod as negatively charged electrons are transferred from the glass to the silk. Likewise, when rubber is rubbed with fur, electrons are transferred from the fur to the rubber. An *uncharged object* contains an enormous number of electrons (on the order of 10^{23}). However, for every negative electron, a positively charged proton is also present; hence, an uncharged object has no net charge of either sign.

Electric forces between charged objects were measured by Coulomb with a torsion balance (Fig. 19.3). The apparatus consists of two small spheres fixed to the ends of a light horizontal rod made of an insulating material and suspended by a silk thread. Sphere A is given a charge, and charged object B is brought near sphere A. The attractive or repulsive force between the two charged objects causes the rod to rotate and to twist the suspension. The angle through which the rod rotates is measured via the deflection of a light beam by a mirror attached to the suspension. The rod rotates through some angle against the restoring force of the twisted thread before reaching equilibrium. The value of the angle of rotation increases as the charge on the objects increases. Thus, the angle of rotation provides a quantitative measure of the electric force of attraction or repulsion.

From our discussion thus far, and from what we learned in Chapter 5, we conclude that electric charge has the following important properties:

- Two kinds of charges exist in nature, with the property that unlike charges attract one another and like charges repel one another.
- The force between charges varies as the inverse square of their separation.
- Charge is conserved.
- Charge is quantized.

CONCEPTUAL PROBLEM 1

Two insulated rods are oppositely charged on the ends. They are mounted at their centers so that they are free to rotate, and then held in the position shown in a view from above (Fig. 19.4). The plane of the rotation of the rods is in the plane of the paper. Will the rods stay in these positions when released? If not, into what position(s) will they move? Will the final configuration(s) be stable?

19.3 • INSULATORS AND CONDUCTORS

It is convenient to classify substances in terms of their ability to conduct electrical charge.

Conductors are materials in which electric charges move freely, and **insulators** are materials in which electric charges do not move freely.

Materials such as glass, rubber, and lucite are insulators. When such materials are charged by rubbing, only the rubbed area becomes charged, and there is no tendency for charge to move to other regions of the material. In contrast, materials such as copper, aluminum, and silver are good conductors. When such materials are charged in some small region, the charge readily distributes itself over the entire surface of the material. If you hold a copper rod in your hand and rub it with wool or fur, it will not attract a small piece of paper. This might suggest that a metal cannot be charged. However, if you hold the copper rod by an insulating handle and then rub, the rod will remain charged and attract the piece of paper. In the first case, the electric charges produced by rubbing readily move from the copper through your body and finally to Earth. In the second case, the insulating handle prevents the flow of charge to Earth.

Semiconductors are a third class of materials, and their electrical properties are somewhere between those of insulators and those of conductors. Silicon and germanium are well-known examples of semiconductors that are widely used to fabricate a variety of electronic devices. The electrical properties of semiconductors can be changed over many orders of magnitude by adding controlled amounts of certain foreign atoms to the materials.

• *Metals are good conductors.*

Charging by Induction

When a conductor is connected to Earth by means of a conducting wire or pipe, it is said to be **grounded.** For our purposes, the Earth can be considered an infinite reservoir for electrons; this means that it can accept or supply an unlimited number of electrons. With this in mind, we can understand how to charge a conductor by a process known as **induction.**

Consider a negatively charged rubber rod brought near a neutral (uncharged) conducting sphere that is insulated so that there is no conducting path to ground (Fig. 19.5). The repulsive force between the electrons in the rod and those in the sphere causes a redistribution of charge on the sphere so that some electrons move to the side of the sphere farthest away from the rod. The region of the sphere nearest the rod has an excess of positive charge because of the migration of electrons away from this location. If a grounded conducting wire is then connected to the sphere, as in Figure 19.5b, some of the electrons leave the sphere and travel to the Earth. If the wire to ground is then removed (Fig. 19.5c), the conducting sphere is left with an excess of induced positive charge. Finally, when the rubber rod is removed from the vicinity of the sphere (Fig. 19.5d), the induced positive charge remains on the ungrounded sphere. This excess positive charge becomes uniformly distributed over the surface of the ungrounded sphere because of the repulsive forces among the like charges and the high mobility of charge carriers in a metal.

In the process of inducing a charge on the sphere, the charged rubber rod loses none of its negative charge, because it never came in contact with the sphere. **Charging an object by induction requires no contact with the object inducing the charge.** This is in contrast to charging an object by rubbing, which does require contact between the two objects.

A process very similar to charging by induction in conductors also takes place in insulators. In most neutral atoms and molecules, the center of positive charge

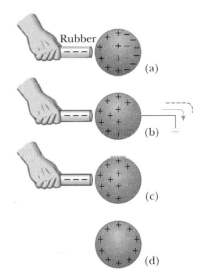

Figure 19.5 Charging a metallic object by induction. (a) The charge on a neutral metallic sphere is redistributed when a charged rubber rod is placed near the sphere. (b) The sphere is grounded, and some of the electrons leave. (c) The ground connection is removed, and the sphere has a net nonuniform positive charge. (d) When the rubber rod is removed, the excess positive charge becomes uniformly distributed over the surface of the sphere.

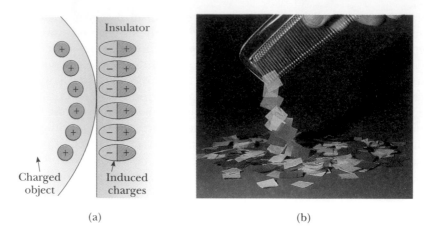

(a)   (b)

Figure 19.6   (a) A charged object induces charges on the surface of an insulator. (b) A charged comb attracts bits of paper because charges are displaced in the paper. The paper is neutral but polarized.   *(Fundamental Photographs, NYC)*

coincides with the center of negative charge. However, in the presence of a charged object, these centers may shift slightly, resulting in more positive charge on one side of the molecule than on the other. This effect is known as **polarization.** The realignment of charge within individual molecules produces an induced charge on the surface of the insulator, as shown in Figure 19.6a. With these concepts in mind, you should be able to explain why a comb that has been rubbed through hair attracts bits of neutral paper or why a balloon that has been rubbed against your clothing can stick to a neutral wall.

Thinking Physics 1

A positively charged ball, hanging from a string, is brought near a nonconducting object. Based on the behavior of the ball–string combination, the ball is seen to be attracted to the object. From this experiment, it is not possible to determine whether the object is negatively charged or neutral. Why not? What additional experiment would help you decide between these two possibilities?

Reasoning   The attraction between the ball and the object could be an attraction of unlike charges, *or* it could be an attraction between a charged object and a neutral object due to polarization of the molecules of the neutral object. Two possible additional experiments would help to determine if the object is charged. First, a known neutral ball could be brought near the object—if there is an attraction, the object is negatively charged. Another possibility is to bring a known negatively charged ball near the object—if there is a repulsion, then the object is negatively charged. If there is attraction, then the object is neutral.

CONCEPTUAL PROBLEM 2

If a suspended object A is attracted to object B, which is charged, can we conclude that object A is charged?

19.4 • COULOMB'S LAW

In Chapter 5 we introduced Coulomb's law, which describes the electrostatic force between two stationary charged particles. When the charges are of the same sign, the force on either charge is repulsive, and when the charges are of opposite sign, the force is attractive (Fig. 19.7). The magnitude of the electrostatic force between two charges separated by a distance of r is

$$F = k_e \frac{|q_1| |q_2|}{r^2}$$ [19.1] • *Coulomb's law*

where $k_e (= 8.99 \times 10^9 \text{ N} \cdot \text{m}^2/\text{C}^2)$ is the Coulomb constant. The constant k_e is also written

$$k_e = \frac{1}{4\pi\epsilon_0}$$

where the constant ϵ_0 is known as the *permittivity of free space* and has the value

$$\epsilon_0 = 8.8542 \times 10^{-12} \text{ C}^2/\text{N} \cdot \text{m}^2$$ *Permittivity of free space*

The charge of an electron or proton has the magnitude $|e| = 1.60 \times 10^{-19}$ C. Therefore, 1 C of charge is equal to the charge of 6.25×10^{18} electrons (that is, $1/e$). This can be compared with the number of free electrons in 1 cm^3 of copper, which is on the order of 10^{23}. Note that 1 C is a substantial amount of charge. In typical electrostatic experiments, where a rubber or glass rod is charged by friction, a net charge on the order of 10^{-6} C $(= 1 \mu\text{C})$ is obtained. In other words, only a very small fraction of the total available charge is transferred between the rod and the rubbing material. The charges and masses of the electron, proton, and neutron are given in Table 19.1.

When dealing with Coulomb's force law, you must remember that force is a *vector* quantity and must be treated accordingly. Furthermore, note that **Coulomb's law applies exactly only to point charges or particles.** The electrostatic force on q_2 due to q_1, written $\mathbf{F}_{21}$, can be expressed in vector form as

$$\mathbf{F}_{21} = k_e \frac{q_1 q_2}{r^2} \hat{\mathbf{r}}$$ [19.2]

where $\hat{\mathbf{r}}$ is a unit vector directed from q_1 to q_2 as in Figure 19.7a. From Newton's third law, we see that the electric force on q_1 due to q_2 is equal in magnitude to the

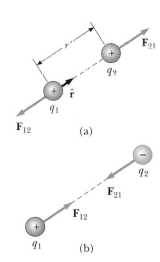

Figure 19.7 Two point charges separated by a distance r exert a force on each other given by Coulomb's law. Note that the force on q_1 is equal to and opposite the force on q_2. (a) When the charges are of the same sign, the force is repulsive. (b) When the charges are of the opposite sign, the force is attractive.

TABLE 19.1 Charge and Mass of the Electron, Proton, and Neutron

Particle	Charge (C)	Mass (kg)
Electron (e)	$-1.6021917 \times 10^{-19}$	9.1095×10^{-31}
Proton (p)	$+1.6021917 \times 10^{-19}$	1.67261×10^{-27}
Neutron (n)	0	1.67492×10^{-27}

force on q_2 due to q_1 and in the opposite direction; that is, $\mathbf{F}_{12} = -\mathbf{F}_{21}$. Finally, from Equation 19.2 we see that if q_1 and q_2 have the same sign, the product $q_1 q_2$ is positive and the force is repulsive, as in Figure 19.7a. If q_1 and q_2 are of opposite sign, as in Figure 19.7b, the product $q_1 q_2$ is negative and the force is attractive.

When more than two charges are present, the force between any pair is given by Equation 19.2. Therefore, **the resultant force on any one charge equals the vector sum of the forces due to the various individual charges.** This **principle of superposition** as applied to electrostatic forces is an experimentally observed fact. For example, if there are four charges, then the resultant force on particle 1 due to particles 2, 3, and 4 is given by the vector sum

$$\mathbf{F}_1 = \mathbf{F}_{12} + \mathbf{F}_{13} + \mathbf{F}_{14}$$

Example 19.1   Where Is the Resultant Force Zero?

Three charges lie along the x axis as in Figure 19.8. The positive charge $q_1 = 15.0\ \mu\text{C}$ is at $x = 2.00$ m, and the positive charge $q_2 = 6.00\ \mu\text{C}$ is at the origin. Where must a negative charge q_3 be placed on the x axis such that the resultant force on it is zero?

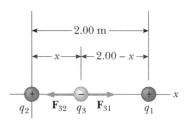

Figure 19.8 (Example 19.1) Three point charges are placed along the x axis. The charge q_3 is negative, whereas q_1 and q_2 are positive. If the net force on q_3 is zero, then the force on q_3 due to q_1 must be equal and opposite to the force on q_3 due to q_2.

Solution Because q_3 is negative and both q_1 and q_2 are positive, the forces $\mathbf{F}_{31}$ and $\mathbf{F}_{32}$ are both attractive, as indicated in Figure 19.8. If we let x be the coordinate of q_3, then the forces $\mathbf{F}_{31}$ and $\mathbf{F}_{32}$ have magnitudes

$$F_{31} = k_e \frac{|q_3|\,|q_1|}{(2.00 - x)^2} \quad \text{and} \quad F_{32} = k_e \frac{|q_3|\,|q_2|}{x^2}$$

In order for the resultant force on q_3 to be zero, $\mathbf{F}_{32}$ must be equal to and opposite $\mathbf{F}_{31}$, or

$$k_e \frac{|q_3|\,|q_2|}{x^2} = k_e \frac{|q_3|\,|q_1|}{(2.00 - x)^2}$$

Because k_e and q_3 are common to both sides, we solve for x and find that

$$(2.00 - x)^2 |q_2| = x^2 |q_1|$$

$$(4.00 - 4.00x + x^2)(6.00 \times 10^{-6}\ \text{C}) = x^2 (15.0 \times 10^{-6}\ \text{C})$$

Solving this quadratic equation for x, we find that

$x = 0.775$ m. Why is the negative root not acceptable?

Example 19.2   The Hydrogen Atom

The electron and proton of a hydrogen atom are separated (on the average) by a distance of 5.3×10^{-11} m. Find the magnitude of the electrostatic force and the gravitational force between the two particles.

Solution From Coulomb's law, we find that the attractive electrical force has the magnitude

$$F_e = k_e \frac{|e|^2}{r^2} = 8.99 \times 10^9\ \frac{\text{N} \cdot \text{m}^2}{\text{C}^2}\ \frac{(1.60 \times 10^{-19}\ \text{C})^2}{(5.3 \times 10^{-11}\ \text{m})^2}$$

$$= 8.2 \times 10^{-8}\ \text{N}$$

Using Newton's law of gravity and Table 19.1 for the particle masses, we find that the gravitational force has the magnitude

$$F_g = G \frac{m_e m_p}{r^2} =$$

$$\left(6.7 \times 10^{-11}\ \frac{\text{N} \cdot \text{m}^2}{\text{kg}^2}\right) \frac{(9.11 \times 10^{-31}\ \text{kg})(1.67 \times 10^{-27}\ \text{kg})}{(5.3 \times 10^{-11}\ \text{m})^2}$$

$$= 3.6 \times 10^{-47}\ \text{N}$$

The ratio $F_e/F_g \approx 2 \times 10^{39}$. Thus the gravitational force between charged atomic particles is negligible compared with the electric force.

EXERCISE 1 Two point charges of magnitudes 3.0 nC and 6.0 nC are separated by a distance of 0.30 m. Find the electric force of repulsion between them.
Answer 1.8×10^{-6} N

19.5 • ELECTRIC FIELDS

The gravitational field **g** at a point in space was defined in Chapter 5 to be equal to the gravitational force **F** acting on a test mass m_0 divided by the test mass: $\mathbf{g} = \mathbf{F}/m_0$. In similar manner, an electric field at a point in space can be defined in terms of the electric force acting on a test charge q_0 placed at that point. To be more precise, **the electric field vector E at a point in space is defined as the electric force F acting on a positive test charge placed at that point divided by the magnitude of the test charge** q_0:

$$\mathbf{E} \equiv \frac{\mathbf{F}}{q_0} \qquad [19.3]$$

Note that **E** is the field produced by some charge *external* to the test charge—not the field produced by the test charge. This is analogous to the gravitational field set up by some body such as the Earth. The vector **E** has the SI units of newtons per coulomb (N/C). The direction of **E** is in the direction of **F** because we have assumed that **F** acts on a positive test charge. Thus, we can say that **an electric field exists at a point if a test charge at rest placed at that point experiences an electric force.** Once the electric field is known at some point, the force on *any* charged particle placed at that point can be calculated from Equation 19.3. Furthermore, the electric field is said to exist at some point (even empty space) regardless of whether or not a test charge is located at that point.

Consider a point charge q located a distance r from a small test charge q_0. According to Coulomb's law, the force exerted on the test charge by q is

$$\mathbf{F} = k_e \frac{q q_0}{r^2} \hat{\mathbf{r}}$$

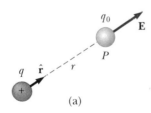

Because the electric field at the position of the test charge is defined by $\mathbf{E} = \mathbf{F}/q_0$, we find that, at the position of q_0, the electric field created by q is

$$\mathbf{E} = k_e \frac{q}{r^2} \hat{\mathbf{r}} \qquad [19.4]$$

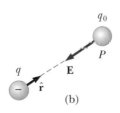

where $\hat{\mathbf{r}}$ is a unit vector directed from q toward q_0 (Fig. 19.9). If q is positive, as in Figure 19.9a, the electric field is directed radially outward from it. If q is negative, as in Figure 19.9b, the field is directed toward it.

In order to calculate the electric field at a point P due to a group of point charges, we first calculate the electric field vectors at P individually using Equation 19.4 and then add them vectorially. In other words, **the total electric field due to a group of charges equals the vector sum of the electric fields of all the charges.** This superposition principle applied to fields follows directly from the superposi-

Figure 19.9 A test charge q_0 at point P is a distance r from a point charge q. (a) If q is positive, the electric field at P points radially *outward* from q. (b) If q is negative, the electric field at P points radially *inward* toward q.

tion property of electric forces. Thus, the electric field of a group of charges (excluding the test charge q_0) can be expressed as

$$\mathbf{E} = k_e \sum_i \frac{q_i}{r_i^2} \hat{\mathbf{r}}_i$$ [19.5]

where r_i is the distance from the ith charge, q_i, to the point P (the location of the test charge) and $\hat{\mathbf{r}}_i$ is a unit vector directed from q_i toward P.

Thinking Physics 2

As an electron moving horizontally passes through deflection plates in an oscilloscope, a uniform electric field in the vertical direction exerts an electric force on the electron. What is the shape of the path followed by the electron while it is between the plates?

Reasoning The magnitude of the electrical force on the electron having charge e due to a uniform electric field of magnitude E is $F_e = eE$. Thus, the force is constant. Compare this to the force on a projectile of mass m moving in two dimensions in a uniform gravitational field g near the surface of the Earth. The magnitude of the gravitational force is $F_g = mg$. In both cases, the particle is subject to a constant force in the vertical direction and has an initial velocity in the horizontal direction. Thus, the path will be the same—the electron will follow a *parabolic* path, just like the projectile. Once the electron leaves the region between the plates, the electric field disappears and the electron continues moving in a straight line according to Newton's first law.

CONCEPTUAL PROBLEM 3

Why do TV screens and computer monitors become so dusty?

Example 19.3   Electric Field of a Dipole

An **electric dipole** consists of a positive charge q and a negative charge $-q$ separated by a distance of $2a$, as in Figure 19.10. Find the electric field $\mathbf{E}$ due to these charges along the y axis at the point P, which is a distance y from the origin. Assume that $y \gg a$.

Solution   At P, the fields $\mathbf{E}_1$ and $\mathbf{E}_2$ due to the two charges are equal in magnitude, because P is equidistant from the two equal and opposite charges. The total field is given by $\mathbf{E} = \mathbf{E}_1 + \mathbf{E}_2$, where

$$E_1 = E_2 = k_e \frac{q}{r^2} = k_e \frac{q}{y^2 + a^2}$$

The y components of $\mathbf{E}_1$ and $\mathbf{E}_2$ cancel each other. The x components are equal, because they are both along the x axis. Therefore, $\mathbf{E}$ is parallel to the x axis and has a magnitude equal to $2E_1 \cos \theta$. From Figure 19.10 we see that $\cos \theta = a/r = a/(y^2 + a^2)^{1/2}$. Therefore,

$$E = 2E_1 \cos \theta = 2k_e \frac{q}{(y^2 + a^2)} \frac{a}{(y^2 + a^2)^{1/2}}$$

$$= k_e \frac{2qa}{(y^2 + a^2)^{3/2}}$$

Using the approximation $y \gg a$, we can neglect a^2 in the denominator and write

$$E \approx k_e \frac{2qa}{y^3}$$

Thus, we see that along the y axis the field of a dipole at a distant point varies as $1/r^3$, whereas the more slowly varying field of a point charge goes as $1/r^2$. (Note that in the geometry of this example, $r = y$.) This is because at distant points, the fields of the two equal and opposite charges almost cancel each other. The $1/r^3$ variation in E for the dipole

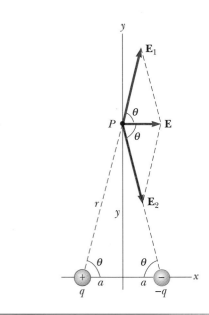

is also obtained for a distant point along the *x* axis (Problem 8) and for a general distant point. The dipole is a good model of many molecules, such as HCl.

As we shall see in later chapters, neutral atoms and molecules behave as dipoles when placed in an external electric field. Furthermore, many molecules, such as HCl, are permanent dipoles. (HCl is partially described as an H$^+$ ion combined with a Cl$^-$ ion.) The effect of such dipoles on the behavior of materials subjected to electric fields is discussed in Chapter 20.

Figure 19.10 (Example 19.3) The total electric field **E** at *P* due to two equal and opposite charges (an electric dipole) equals the vector sum **E**$_1$ + **E**$_2$. The field **E**$_1$ is due to the positive charge *q*, and **E**$_2$ is the field due to the negative charge $-q$.

EXERCISE 2 A piece of aluminum foil of mass 5.0×10^{-2} kg is suspended by a string in an electric field directed vertically upward. If the charge on the foil is 3.0 μC, find the strength of the field that will reduce the tension in the string to zero.   Answer 1.6×10^5 N/C

EXERCISE 3 The nucleus of a hydrogen atom, a proton, sets up an electric field. The average distance between the proton and the electron of a hydrogen atom is approximately 5.3×10^{-11} m. What is the magnitude of the electric field at this distance from the proton? Answer 5.1×10^{11} N/C

Electric Field Due to Continuous Charge Distributions

In most practical situations (such as an object charged by rubbing), the average separation between charges is small compared with their distances from the field point. In such cases, the system of charges can be considered *continuous*. That is, we imagine that the system of closely spaced charges is equivalent to a total charge that is continuously distributed through some volume or over some surface.

To evaluate the electric field of a continuous charge distribution, the following procedure is used. First, we divide the charge distribution into small elements, each of which contains a small charge Δq as in Figure 19.11. Next, we use Coulomb's law to calculate the electric field due to one of these elements at a point *P*. Finally, we evaluate the total field at *P* due to the charge distribution by summing the contributions of all the charge elements (that is, by applying the superposition principle).

The electric field at *P* due to one element of charge Δq is given by

$$\Delta \mathbf{E} = k_e \frac{\Delta q}{r^2} \hat{\mathbf{r}}$$

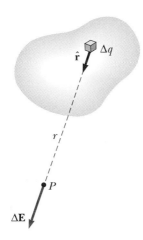

Figure 19.11 The electric field at *P* due to a continuous charge distribution is the vector sum of the fields due to all the elements Δq of the charge distribution.

where r is the distance from the element to point P and $\hat{\mathbf{r}}$ is a unit vector directed from the element toward P. The total electric field at P due to all elements in the charge distribution is approximately

$$\mathbf{E} \approx k_e \sum_i \frac{\Delta q_i}{r_i^2} \hat{\mathbf{r}}_i$$

where the index i refers to the ith element in the distribution. If the separation between elements in the charge distribution is small compared with the distance to P, the charge distribution can be approximated as continuous. Therefore, the total field at P in the limit $\Delta q_i \rightarrow 0$ becomes

Electric field of a continuous •
charge distribution

$$\mathbf{E} = k_e \lim_{\Delta q_i \to 0} \sum_i \frac{\Delta q_i}{r_i^2} \hat{\mathbf{r}}_i = k_e \int \frac{dq}{r^2} \hat{\mathbf{r}} \qquad \text{[19.6]}$$

where the integration is a *vector* operation and must be treated with caution. We shall illustrate this type of calculation with several examples in which we shall assume that the charge is *uniformly* distributed on a line or a surface or throughout some volume. When performing such calculations, it is convenient to use the concept of a charge density along with the following notations.

If a charge Q is uniformly distributed throughout a volume V the **charge per unit volume,** ρ, is defined by

Volume charge density •

$$\rho \equiv \frac{Q}{V} \qquad \text{[19.7]}$$

where ρ has units of coulombs per cubic meter.

If a charge Q is uniformly distributed on a surface of area A, the **surface charge density,** σ, is defined by

Surface charge density •

$$\sigma \equiv \frac{Q}{A} \qquad \text{[19.8]}$$

where σ has units of coulombs per square meter.

Finally, if a charge Q is uniformly distributed along a line of length ℓ, the **linear charge density,** λ, is defined by

Linear charge density •

$$\lambda \equiv \frac{Q}{\ell} \qquad \text{[19.9]}$$

where λ has units of coulombs per meter.

Example 19.4 The Electric Field Due to a Charged Rod

A rod of length ℓ has a uniform positive charge per unit length λ and a total charge Q. Calculate the electric field at a point P along the axis of the rod, a distance d from one end (Fig. 19.12).

Reasoning and Solution For this calculation, the rod is taken to be along the x axis. Let us use Δx to represent the length of one small segment of the rod and let Δq be the charge on the segment. The ratio of Δq to Δx is equal to the ratio of the total charge to the total length of the rod. That is, $\Delta q/\Delta x = Q/\ell = \lambda$. Therefore, the charge Δq on the small segment is $\Delta q = \lambda \, \Delta x$.

The field $\Delta \mathbf{E}$ due to this segment at the point P is in the negative x direction, and its magnitude is[1]

$$\Delta E = k_e \frac{\Delta q}{x^2} = k_e \frac{\lambda \, \Delta x}{x^2}$$

Note that each element produces a field in the negative x direction, and so the problem of summing their contributions is particularly simple in this case. The total field at P due to all segments of the rod, which are at different distances

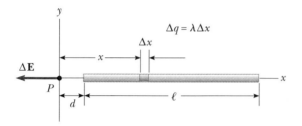

Figure 19.12   (Example 19.4) The electric field at P due to a uniformly charged rod lying along the x axis. The field at P due to the segment of charge Δq is $k_e \, \Delta q / x^2$. The total field at P is the vector sum over all segments of the rod.

from P, is given by Equation 19.6, which in this case becomes

$$E = \int_d^{\ell + d} k_e \lambda \, \frac{dx}{x^2}$$

where the limits on the integral extend from one end of the rod ($x = d$) to the other ($x = \ell + d$). Because k_e and λ are constants, they can be removed from the integral. Thus, we find that

$$E = k_e \lambda \int_d^{\ell + d} \frac{dx}{x^2} = k_e \lambda \left[-\frac{1}{x} \right]_d^{\ell + d}$$

$$= k_e \lambda \left(\frac{1}{d} - \frac{1}{\ell + d} \right) = \frac{k_e Q}{d(\ell + d)}$$

where we have used the fact that the total charge $Q = \lambda \ell$.

From this result we see that if the point P is far from the rod ($d \gg \ell$), then ℓ in the denominator can be neglected, and $E \approx k_e Q / d^2$. This is just the form you would expect for a point charge. Therefore, at large values of d/ℓ the charge distribution appears to be a point charge of magnitude Q. The use of the limiting technique (in this case, $d/\ell \rightarrow \infty$) is often a good method for checking a theoretical formula.

Example 19.5   The Electric Field of a Uniform Ring of Charge

A ring of radius a has a uniform positive charge per unit length, with a total charge Q. Calculate the electric field along the axis of the ring at a point P lying a distance x from the center of the ring (Fig. 19.13a).

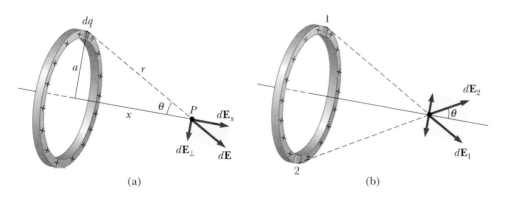

(a)   (b)

Figure 19.13   (Example 19.5) A uniformly charged ring of radius a. (a) The field at P on the x axis due to an element of charge dq. (b) The total electric field at P is along the x axis. Note that the perpendicular component of the electric field at P due to segment 1 is canceled by the perpendicular component due to segment 2, which is located on the ring opposite segment 1.

[1] It is important that you understand the procedure being used to carry out integrations such as this. First, choose an element the parts of which are all equidistant from the point at which the field is being calculated. Next, express the charge element Δq in terms of the other variables within the integral (in this example, there is one variable, x). The integral must be over scalar quantities and therefore must be expressed in terms of components. Then reduce the form to an integral over a single variable (or multiple integrals, each over a single variable). In examples that have spherical or cylindrical symmetry, the variables will be radial coordinates.

Reasoning and Solution   The magnitude of the electric field at P due to the segment of charge dq is

$$dE = k_e \frac{dq}{r^2}$$

This field has an x component $dE_x = dE \cos \theta$ along the axis of the ring and a component $dE_\perp$ perpendicular to the axis. But as we see in Figure 19.13b, the resultant field at P must lie along the x axis because the perpendicular components sum to zero. That is, the perpendicular component of any element is canceled by the perpendicular component of an element on the opposite side of the ring. Because $r = (x^2 + a^2)^{1/2}$ and $\cos \theta = x/r$, we find that

$$dE_x = dE \cos \theta = \left(k_e \frac{dq}{r^2} \right) \frac{x}{r} = \frac{k_e x}{(x^2 + a^2)^{3/2}} \, dq$$

In this case, all segments of the ring give the same contribution to the field at P because they are all equidistant from this point. Thus, we can integrate the expression above to get the total field at P:

$$E_x = \int \frac{k_e x}{(x^2 + a^2)^{3/2}} \, dq = \frac{k_e x}{(x^2 + a^2)^{3/2}} \int dq$$

$$= \frac{k_e x}{(x^2 + a^2)^{3/2}} Q$$

This result shows that the field is zero at $x = 0$. Does this surprise you?

EXERCISE 4   Show that at large distances from the ring ($x \gg a$) the electric field along the axis approaches that of a point charge of magnitude Q.

EXERCISE 5   A sphere of radius 4 cm has a net charge of $+39 \ \mu C$. (a) If this charge is uniformly distributed throughout the volume of the sphere, what is the volume charge density? (b) If this charge is uniformly distributed on the sphere's surface, what is the surface charge density?   Answer   (a) $0.15 \ C/m^3$ (b) $1.9 \times 10^{-3} \ C/m^2$

PROBLEM-SOLVING STRATEGY AND HINTS

1. Units: When performing calculations that involve the Coulomb constant, $k_e (= 1/4\pi\epsilon_0)$, charges must be in coulombs and distances in meters. If they appear in other units, you must convert them.

2. Applying Coulomb's law to point charges: It is important to use the superposition principle properly when dealing with a collection of interacting charges. When several charges are present, the resultant force on any one is the *vector sum* of the forces due to the individual charges. Be very careful in the algebraic manipulation of vector quantities. It may be useful to review the material on vector addition in Chapter 1.

3. Calculating the electric field of point charges: Remember that the superposition principle can be applied to electric fields, which are also vector quantities. To find the total electric field at a given point, first calculate the electric field at that point due to each individual charge. The resultant field at the point is the vector sum of the fields due to the individual charges.

4. Continuous charge distributions: When you are confronted with problems that involve a continuous distribution of charge, the vector sums for evaluating the total electric field at some point must be replaced by vector integrals. The charge distribution is divided into infinitesimal pieces, and the vector sum is carried out by integrating over the entire charge distribution. Examples 19.4 and 19.5 demonstrate such procedures.

5. Symmetry: Whenever dealing with either a distribution of point charges or a continuous charge distribution, take advantage of any symmetry in the system to simplify your calculations.

19.6 • ELECTRIC FIELD LINES

A convenient aid for visualizing electric field patterns is to draw lines pointing in the same direction as the electric field vector at any point. These lines, called **electric field lines,** are related to the electric field in any region of space in the following manner:

- The electric field vector **E** is *tangent* to the electric field line at each point.
- The number of lines per unit area through a surface that is perpendicular to the lines is proportional to the strength of the electric field in that region. Thus, **E** is large where the field lines are close together and small where they are far apart.

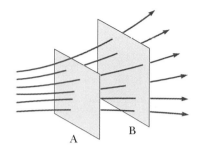

Figure 19.14 Electric field lines penetrating two surfaces. The magnitude of the field is greater on surface A than on surface B.

These properties are illustrated in Figure 19.14. The density of lines through surface A is greater than the density of lines through surface B. Therefore, the electric field is more intense on surface A than on surface B. Furthermore, the field drawn in Figure 19.14 is nonuniform, because the lines at different locations point in different directions.

Some representative electric field lines for a single positive point charge are shown in Figure 19.15a. Note that in this two-dimensional drawing we show only the field lines that lie in the plane containing the point charge. The lines are actually directed radially outward in *all* directions from the charge, somewhat like the needles of a porcupine. Because a positive test charge placed in this field would be repelled by the charge q, the lines are directed radially away from q. In a similar way, the electric field lines for a single negative point charge are directed toward the charge (Fig. 19.15b). In either case, the lines are radial and extend to infinity. Note that the lines are closer together as they get nearer the charge, indicating that the strength of the field is increasing.

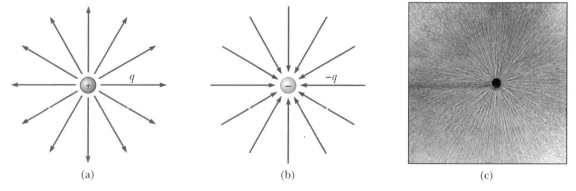

(a)   (b)   (c)

Figure 19.15 The electric field lines for a point charge. (a) For a positive point charge, the lines are radially outward. (b) For a negative point charge, the lines are radially inward. Note that the figures show only those field lines that lie in the plane containing the charge. (c) The dark areas are small pieces of thread suspended in oil, which align with the electric field produced by a small charged conductor at the center. *(Photo courtesy of Harold M. Waage, Princeton University)*

The rules for drawing electric field lines for any charge distribution follow:

- The lines must begin on positive charges (or at infinity) and must terminate either on negative charges or, in the case of an excess of positive charge, at infinity.
- The number of lines drawn leaving a positive charge or approaching a negative charge is proportional to the magnitude of the charge.
- No two field lines can cross each other.

Is this visualization of the electric field in terms of field lines consistent with Coulomb's law? To answer this question, consider an imaginary spherical surface of radius r, concentric with the charge. Using the principle of symmetry, we see that the magnitude of the electric field is the same everywhere on the surface of the sphere. The number of lines, N, that emerge from the charge is equal to the number that penetrate the spherical surface. Hence, the number of lines per unit area on the sphere is $N/4\pi r^2$ (where the surface area of the sphere is $4\pi r^2$). Because E is proportional to the number of lines per unit area, we see that E varies as $1/r^2$. This is consistent with the result obtained from Coulomb's law—that is, $E = k_e q/r^2$.

It is important to note that electric field lines are not material objects. They are used only to provide a qualitative description of the electric field. One problem with this model is the fact that one always draws a finite number of lines from each charge, which makes it appear as if the field were quantized and acted only in a certain direction. The field, in fact, is continuous—existing at every point. Another problem with this model is the danger of getting the wrong impression from a two-dimensional drawing of field lines used to describe a three-dimensional situation.

Because charge is quantized, the number of lines leaving any material object must be 0, $\pm C'e$, $\pm 2C'e$, . . . , where C' is an arbitrary (but fixed) proportionality constant. Once C' is chosen, the number of lines is not arbitrary. For example, if object 1 has charge Q_1 and object 2 has charge Q_2, then the ratio of numbers of lines is $N_2/N_1 = Q_2/Q_1$.

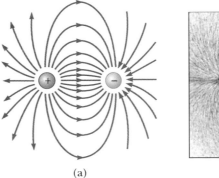
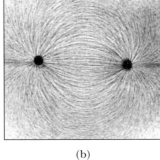

(a)   (b)

Figure 19.16   (a) The electric field lines for two equal and opposite point charges (an electric dipole). Note that the number of lines leaving the positive charge equals the number terminating at the negative charge. (b) The photograph was taken using small pieces of thread suspended in oil, which align with the electric field.   *(Photo courtesy of Harold M. Waage, Princeton University)*

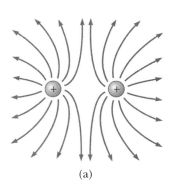 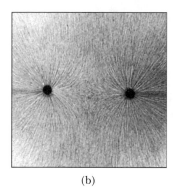

(a)   (b)

Figure 19.17   (a) The electric field lines for two positive point charges. (b) The photograph was taken using small pieces of thread suspended in oil, which align with the electric field. *(Photo courtesy of Harold M. Waage, Princeton University)*

The electric field lines for two point charges of equal magnitude but opposite signs (the electric dipole) are shown in Figure 19.16. In this case, the number of lines that begin at the positive charge must equal the number that terminate at the negative charge. At points very near the charges, the lines are nearly radial. The high density of lines between the charges indicates a region of strong electric field. The attractive nature of the force between the charges can also be seen from Figure 19.16.

Figure 19.17 shows the electric field lines in the vicinity of two equal positive point charges. Again, close to either charge the lines are nearly radial. The same number of lines emerges from each charge because the charges are equal in magnitude. At great distances from the charges, the field is approximately equal to that of a single point charge of magnitude $2q$. The bulging out of the electric field lines between the charges indicates the repulsive nature of the electric force between like charges.

Finally, in Figure 19.18 we sketch the electric field lines associated with a positive charge $+2q$ and a negative charge $-q$. In this case, the number of lines leaving the charge $+2q$ is twice the number terminating on the charge $-q$. Hence, only half of the lines that leave the positive charge end at the negative charge. The remaining half terminate on a negative charge we assume to be located at infinity. At great distances from the charges (great compared with the charge separation), the electric field lines are equivalent to those of a single charge $+q$.

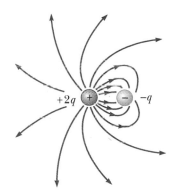

Figure 19.18   The electric field lines for a point charge $+2q$ and a second point charge $-q$. Note that two lines leave the charge $+2q$ for every one that terminates on $-q$.

CONCEPTUAL PROBLEM 4

In fair weather, there is an electric field at the surface of the Earth, pointing down into the ground. What is the sign of the electric charge on the ground in fair weather?

CONCEPTUAL PROBLEM 5

An uncharged, metallic-coated styrofoam ball is suspended in the region between two vertical metal plates. If the two plates are charged, one positive and one negative, describe the motion of the ball after it is brought into contact with one of the plates.

19.7 • ELECTRIC FLUX

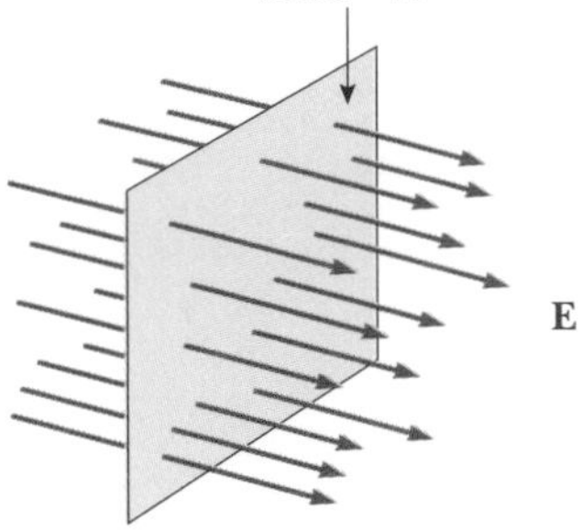

Now that we have described the concept of electric field lines qualitatively, let us use a new concept, electric flux, to approach electric field lines on a quantitative basis. **Electric flux is a quantity proportional to the number of electric field lines penetrating some surface.** When the surface being penetrated is a closed surface (such as a sphere) that encloses some net charge, the number of lines that go through the surface is proportional to the net charge within the surface. The number of lines through the surface is independent of the shape of the surface enclosing the charge. This is essentially a statement of Gauss's law, described in the next section.

First consider an electric field that is uniform in both magnitude and direction, as in Figure 19.19. The field lines penetrate a plane rectangular surface of area A, which is perpendicular to the field. Recall that the number of lines per unit area is proportional to the magnitude of the electric field. Therefore, the number of lines penetrating the surface of area A is proportional to the product EA. The product of the electric field strength, E, and a surface area, A, perpendicular to the field is called the **electric flux,** Φ:

$$\Phi = EA \qquad [19.10]$$

From the SI units of E and A, we see that electric flux has the units $N \cdot m^2/C$.

If the surface under consideration is not perpendicular to the field, the number of lines (the flux) through it must be less than that given by Equation 19.10. This can be easily understood by considering Figure 19.20 where the normal to the surface of area A is at an angle of θ to the uniform electric field. Note that the number of lines that cross this area is equal to the number that cross the projected area A', which is perpendicular to the field. From Figure 19.20 we see that the two areas are related by $A' = A \cos \theta$. Because the flux through area A equals the flux through A', we conclude that the desired flux is

$$\Phi = EA \cos \theta \qquad [19.11]$$

From this result, we see that the flux through a surface of fixed area has the maximum value EA when the surface is perpendicular to the field (in other words, when the *normal* to the surface is parallel to the field—that is, $\theta = 0°$); the flux is zero when the surface is parallel to the field (when the normal to the surface is perpendicular to the field—that is, $\theta = 90°$).

In more general situations, the electric field may vary over the surface in question. Therefore, our definition of flux given by Equation 19.11 has meaning only over a small element of area. Consider a general surface divided up into a large

Figure 19.19 Field lines of a uniform electric field penetrating a plane of area A perpendicular to the field. The electric flux Φ through this area is equal to EA.

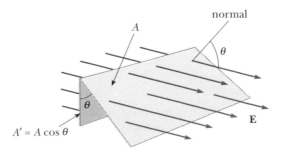

Figure 19.20 Field lines for a uniform electric field through an area A that is at an angle θ to the field. Because the number of lines that go through the shaded area A' is the same as the number that go through A, we conclude that the flux through A' is equal to the flux through A and is given by the expression $\Phi = EA \cos \theta$.

number of small elements, each of area ΔA. The variation in the electric field over the element can be neglected if the element is small enough. It is convenient to define a vector $\Delta \mathbf{A}_i$ the magnitude of which represents the area of the ith element and the direction of which is *defined to be perpendicular* to the surface, as in Figure 19.21. The electric flux $\Delta \Phi_i$ through this small element is

$$\Delta \Phi_i = E_i \Delta A_i \cos \theta = \mathbf{E}_i \cdot \Delta \mathbf{A}_i$$

where we have used the definition of the scalar product of two vectors ($\mathbf{A} \cdot \mathbf{B} = AB \cos \theta$). By summing the contributions of all elements, we obtain the total flux through the surface.[2] If we let the area of each element approach zero, then the number of elements approaches infinity and the sum is replaced by an integral. Therefore, the general definition of electric flux is

$$\Phi \equiv \lim_{\Delta A_i \to 0} \sum \mathbf{E}_i \cdot \Delta \mathbf{A}_i = \int_{\text{surface}} \mathbf{E} \cdot d\mathbf{A} \qquad \text{[19.12]}$$

Equation 19.12 is a surface integral, which must be evaluated over the surface in question. In general, the value of Φ depends both on the field pattern and on the specified surface.

We shall usually be interested in evaluating the flux through a *closed surface*. (A closed surface is defined as one that divides space into an inside region and an outside region, so that movement cannot take place from one region to the other without penetrating the surface. The surface of a sphere, for example, is a closed surface.) Consider the closed surface in Figure 19.22. Note that the vectors $\Delta \mathbf{A}_i$ point in different directions for the various surface elements. At each point, these vectors are *perpendicular* to the surface and, by convention, always point *outward*. At the elements labeled ① and ②, $\mathbf{E}$ is outward and $\theta < 90°$; hence, the flux $\Delta \Phi = \mathbf{E} \cdot \Delta \mathbf{A}$ through these elements is positive. For elements such as ③, where the field lines are directed into the surface, $\theta > 90°$ and the flux becomes negative because of the cos θ factor. (Element ③ is on the side of the surface we cannot see; hence the dashed lines representing the vectors $\mathbf{E}_3$ and $\Delta \mathbf{A}_3$.) The net flux through the surface is proportional to the net number of lines penetrating the surface, where

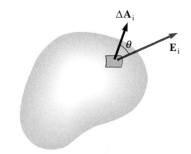

Figure 19.21   A small element of a surface of area ΔA_i. The electric field makes an angle θ with the normal to the surface (the direction of ΔA_i), and the flux through the element is equal to $E_i \Delta A_i \cos \theta$.

Figure 19.22   A closed surface in an electric field. The area vectors $\Delta \mathbf{A}_i$ are, by convention, normal to the surface and point outward. The flux through an area element can be positive (elements ① and ②) or negative (element ③). Element ③ is on the back side of the surface.

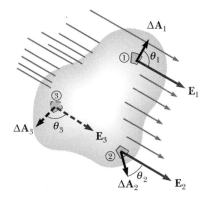

[2]It is important to note that drawings with field lines can have inaccuracies, because a small area (depending on its location) may happen to have too many or too few penetrating lines. At any rate, it is stressed that the basic definition of electric flux is $\int \mathbf{E} \cdot d\mathbf{A}$. The use of lines is only an aid for visualizing the concept.

the *net number* means **the number leaving the volume surrounded by the surface minus the number entering the volume.** If more lines are leaving the surface than entering, the net flux is positive. If more lines enter than leave the surface, the net flux is negative. (The net flux can be nonzero only when some net charge is contained within the closed surface.) Using the symbol $\oint$ to represent an integral over a closed surface, we can write the net flux, Φ_c, through a closed surface

$$\Phi_c = \oint \mathbf{E} \cdot d\mathbf{A} = \oint E_n \, dA \qquad \text{[19.13]}$$

where E_n represents the component of the electric field perpendicular to the surface and the subscript c denotes a closed surface.

Evaluating the net flux through a closed surface can be very cumbersome. However, if the field is perpendicular to the surface at each point and constant in magnitude, the calculation is straightforward. The following example illustrates this point.

Example 19.6   Flux Through a Cube

Consider a uniform electric field $\mathbf{E}$ oriented in the x direction. Find the net electric flux through the surface of a cube of edges ℓ oriented as shown in Figure 19.23.

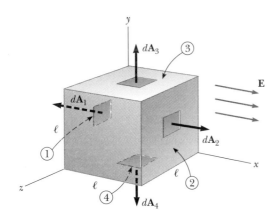

Figure 19.23   (Example 19.6) A hypothetical surface in the shape of a cube in a uniform electric field parallel to the x axis. The net flux through the surface is zero.

Solution   The net flux can be evaluated by summing up the fluxes through each face of the cube. First, note that the flux through four of the faces is zero, because $\mathbf{E}$ is perpendicular

to $d\mathbf{A}$ on these faces. In particular, the orientation of $d\mathbf{A}$ is perpendicular to $\mathbf{E}$ for the faces labeled ③ and ④ in Figure 19.23. Therefore, $\theta = 90°$, so that $\mathbf{E} \cdot d\mathbf{A} = E \, dA \cos 90° = 0$. The flux through each of the planes parallel to the yx plane is also zero for the same reason.

Now consider the faces labeled ① and ②. The net flux through these faces is

$$\Phi_c = \int_1 \mathbf{E} \cdot d\mathbf{A} + \int_2 \mathbf{E} \cdot d\mathbf{A}$$

For face ①, $\mathbf{E}$ is constant and inward and $d\mathbf{A}$ is outward ($\theta = 180°$), so that we find the flux through this face is

$$\int_1 \mathbf{E} \cdot d\mathbf{A} = \int_1 E \, dA \cos 180° = -E \int_1 dA = -EA = -E\ell^2$$

because the area of each face is $A = \ell^2$.

Likewise, for ②, $\mathbf{E}$ is constant and outward and in the same direction as $d\mathbf{A}$ ($\theta = 0°$), so that the flux through this face is

$$\int_2 \mathbf{E} \cdot d\mathbf{A} = \int_2 E \, dA \cos 0° = E \int_2 dA = +EA = E\ell^2$$

Hence, the net flux over all faces is zero, because

$$\Phi_c = -E\ell^2 + E\ell^2 = 0$$

EXERCISE 6   A flat surface having an area of 3.2 m² can be oriented in any direction in a uniform electric field of magnitude $E = 6.2 \times 10^5$ N/C. Calculate the electric flux through this area when the electric field (a) is perpendicular to the surface; (b) is parallel to the surface; (c) makes an angle of 75° with the plane of the surface.
Answer   (a) 1.98×10^6 N·m²/C   (b) zero   (c) 1.92×10^6 N·m²/C

19.8 · GAUSS'S LAW

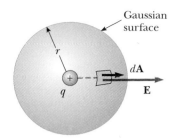

In this section we describe a general relation between the net electric flux through a closed surface (often called a *gaussian surface*) and the charge *enclosed* by the surface. This relation, known as **Gauss's law,** is of fundamental importance in the study of electrostatic fields.

First, let us consider a positive point charge, q, located at the center of a sphere of radius, r, as in Figure 19.24. From Equation 19.4, we know that the magnitude of the electric field everywhere on the surface of the sphere is $E = k_e q/r^2$. Furthermore, the field lines radiate outward and hence are perpendicular (or normal) to the surface at each point. That is, at each point on the surface, **E** is parallel to the vector $\Delta \mathbf{A}_i$ representing the local element of area ΔA_i. Therefore,

$$\mathbf{E} \cdot \Delta \mathbf{A}_i = E_n \, \Delta A_i = E \, \Delta A_i$$

and from Equation 19.13 we find that the net flux through the gaussian surface is

Figure 19.24 A spherical gaussian surface of radius r surrounding a point charge q. When the charge is at the center of the sphere, the electric field is normal to the surface and constant in magnitude everywhere on the surface.

$$\Phi_c = \oint E_n \, dA = \oint E \, dA = E \oint dA$$

because E is constant over the surface and is given by $E = k_e q/r^2$. Furthermore, for a spherical gaussian surface, $\oint dA = A = 4\pi r^2$ (the surface area of a sphere). Hence, the net flux through the gaussian surface is

$$\Phi_c = \frac{k_e q}{r^2} (4\pi r^2) = 4\pi k_e q$$

Recalling that $k_e = 1/4\pi\epsilon_0$, we can write this in the form

$$\Phi_c = \frac{q}{\epsilon_0} \qquad\qquad \textbf{[19.14]}$$

Note that this result, which is independent of r, says that the net flux through a spherical gaussian surface is proportional to the charge q *inside* the surface. The fact that the flux is independent of the radius is a consequence of the inverse-square dependence of the electric field given by Equation 19.4. That is, E varies as $1/r^2$, but the area of the sphere varies as r^2. Their combined effect produces a flux that is independent of r.

Now consider several closed surfaces surrounding a charge, q, as in Figure 19.25. Surface S_1 is spherical, whereas surfaces S_2 and S_3 are nonspherical. The flux that passes through surface S_1 has the value q/ϵ_0. As we discussed in the preceding section, the flux is proportional to the number of electric field lines passing through that surface. The construction in Figure 19.25 shows that the number of electric field lines through the spherical surface S_1 is equal to the number of electric field lines through the nonspherical surfaces S_2 and S_3. Therefore, it is reasonable to conclude that the net flux through any closed surface is independent of the shape of that surface. (One can prove that this is the case if $E \propto 1/r^2$.) In fact, **the net flux through any closed surface surrounding the point charge q is q/ϵ_0.**

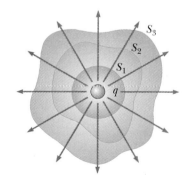

Figure 19.25 Closed surfaces of various shapes surrounding a charge q. Note that the net electric flux through each surface is the same.

Now consider a point charge located *outside* a closed surface of arbitrary shape, as in Figure 19.26. As you can see from this construction, electric field lines enter the surface, and field lines leave it. However, **the number of electric field lines entering the surface equals the number leaving the surface.** Therefore, we conclude that **the net electric flux through a closed surface that surrounds no**

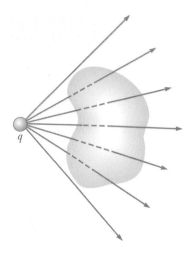

Figure 19.26 A point charge located *outside* a closed surface. In this case, note that the number of lines entering the surface equals the number leaving the surface.

charge is zero. If we apply this result to Example 19.6, we can easily see that the net flux through the cube is zero, because it was assumed there was no charge inside the cube.

Let us extend these arguments to the generalized case of many point charges, which we consider to be a continuous distribution of charge. We shall again make use of the superposition principle. That is, we can express the flux through any closed surface as

$$\oint \mathbf{E} \cdot d\mathbf{A} = \oint (\mathbf{E}_1 + \mathbf{E}_2 + \mathbf{E}_3) \cdot d\mathbf{A}$$

where $\mathbf{E}$ is the total electric field at any point on the surface and $\mathbf{E}_1$, $\mathbf{E}_2$, and $\mathbf{E}_3$ are the fields produced by the individual charges at that point. Consider the system of charges shown in Figure 19.27. The surface S surrounds only one charge, q_1; hence, the net flux through S is q_1/ϵ_0. The flux through S due to the charges outside it is zero, because each electric field line that enters S at one point leaves it at another. The surface S' surrounds charges q_2 and q_3; hence, the net flux through S' is $(q_2 + q_3)/\epsilon_0$. Finally, the net flux through surface S'' is zero, because there is no charge inside this surface. That is, *all* lines that enter S'' at one point leave S'' at another.

Gauss's law, which is a generalization of the foregoing discussion, states that the net flux through *any* closed surface is

Gauss's law •

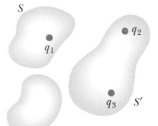

Figure 19.27 The net electric flux through any closed surface depends only on the charge *inside* that surface. The net flux through surface S is q_1/ϵ_0, the net flux through surface S' is $(q_2 + q_3)/\epsilon_0$, and the net flux through surface S'' is zero.

Gauss's law is useful for • *evaluating* $\mathbf{E}$ *when the charge distribution has symmetry.*

$$\Phi_c = \oint \mathbf{E} \cdot d\mathbf{A} = \frac{q_{in}}{\epsilon_0} \qquad \text{[19.15]}$$

where q_{in} represents the *net charge inside the gaussian surface* and $\mathbf{E}$ represents the electric field at any point on the gaussian surface. In other words,

> **Gauss's law** states that the net electric flux through any closed surface is equal to the net charge inside the surface divided by ϵ_0.

When using Equation 19.15, note that, although the charge q_{in} is the net charge inside the gaussian surface, the $\mathbf{E}$ that appears in Gauss's law represents the *total electric field,* which includes contributions from charges both inside and outside the gaussian surface. This point is often neglected or misunderstood.

In principle, Gauss's law can always be used to calculate the electric field of a system of charges or a continuous distribution of charge. In practice, however, **the technique is useful only for calculating the electric field in situations in which there is a high degree of symmetry.** As we shall see in the next section, **Gauss's law can be used to evaluate the electric field for charge distributions that have spherical, cylindrical, or plane symmetry.** If the gaussian surface surrounding the charge distribution is chosen carefully, the integral in Equation 19.15 will be easy to evaluate. Also note that a gaussian surface is a mathematical surface and need not coincide with any real physical surface.

Thinking Physics 3

Imagine a cubical conducting shell with a net charge placed in it. A cube has a high degree of symmetry. Why can't we use Gauss's law to evaluate the electric field around the cube?

Reasoning Although Gauss's law certainly *applies* to the electric field of the cube, it cannot be used to *evaluate* the field. Using Gauss's law to evaluate the electric field requires bringing the electric field outside of the integral in the equation for Gauss's law. The only way that the electric field can be brought outside the integral is if the electric field is a constant. There is no surface around a cube over which one can use symmetry arguments to convince oneself that the electric field is constant. Thus, the attempt is futile—the electric field must remain within the integral, so we need to know the field to do the integral.

However, the other geometries to which Gauss's law is applied in an effort to evaluate the field—spheres, cylinders, flat sheets—have sufficient symmetry that we can find a surface over which a valid argument can be made that the field is constant.

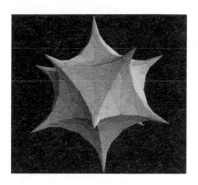

This beautiful closed surface generated by a computer is one example of a mathematical construct in hyperbolic space. *(Courtesy of Wolfram Research, Inc.)*

Thinking Physics 4

If the net flux through a gaussian surface is zero, which of the following statements are true? (a) There are no charges inside the surface. (b) The net charge inside the surface is zero. (c) The electric field is zero everywhere on the surface. (d) The number of electric field lines entering the surface equals the number leaving the surface.

Reasoning Statements (b) and (d) are true and follow from Gauss's law. Statement (a) is not necessarily true because Gauss's law says that the net flux through any closed surface equals the net charge inside the surface divided by ϵ_0. For example, an electric dipole (whose net charge is zero) might be inside the surface. Statement (c) is not necessarily true. Although the net flux through the surface is zero, the electric field in that region may not be zero (Fig. 19.8).

Thinking Physics 5

A spherical gaussian surface surrounds a point charge q. Describe what happens to the total flux through the surface if (a) the charge is tripled, (b) the volume of the sphere is doubled, (c) the surface is changed to a cube, and (d) the charge is moved to another location *inside* the surface.

Reasoning (a) If the charge is tripled, the flux through the surface is also tripled, because the net flux is proportional to the charge inside the surface. (b) The flux remains constant when the volume changes, because the surface surrounds the same amount of charge, regardless of its volume. (c) The total flux does not change when the shape of the closed surface changes. (d) The total flux through the closed surface remains unchanged as the charge inside the surface is moved to another location inside that surface. All of these conclusions are arrived at through an understanding of Gauss's law.

CONCEPTUAL PROBLEM 6

Explain why Gauss's law cannot be used to calculate the electric field of (a) an electric dipole, (b) a charged disk, and (c) three point charges at the corners of a triangle.

EXERCISE 7 A point charge of $+5.0\ \mu C$ is located at the center of a sphere with a radius of 12 cm. What is the electric flux through the surface of this sphere?
Answer $5.7 \times 10^5\ N \cdot m^2/C$

EXERCISE 8 The electric field in the Earth's atmosphere is $E = 100\ N/C$, pointing downward. Determine the electric charge on the Earth. Answer $-4.6 \times 10^5\ C$

19.9 · APPLICATION OF GAUSS'S LAW TO CHARGED INSULATORS

In this section we give some examples of how to use Gauss's law to calculate **E** for a given charge distribution. It is important to recognize that **Gauss's law is useful when there is a high degree of symmetry in the charge distribution, as in the cases of uniformly charged spheres, long cylinders, and plane sheets.** In such cases, it is possible to find a simple gaussian surface over which the surface integral given by Equation 19.15 is easily evaluated.

> A surface should always be chosen to take advantage of the symmetry of the charge distribution.

Example 19.7 The Electric Field Due to a Point Charge

Starting with Gauss's law, calculate the electric field due to an isolated point charge q and show that Coulomb's law follows from this result.

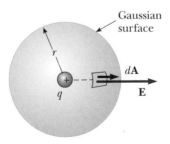

Figure 19.28 (Example 19.7) The point charge q is at the center of the spherical gaussian surface, and **E** is parallel to $d\mathbf{A}$ at every point on the surface.

Solution For this situation we choose a spherical gaussian surface of radius r and centered on the point charge, as in Figure 19.28. The electric field of a positive point charge is radial outward by symmetry and is therefore normal to the surface at every point. That is, **E** is parallel to $d\mathbf{A}$ at each point on the surface, and so $\mathbf{E} \cdot d\mathbf{A} = E\, dA$ and Gauss's law gives

$$\Phi_c = \oint \mathbf{E} \cdot d\mathbf{A} = \oint E\, dA = \frac{q}{\epsilon_0}$$

By symmetry, E is constant everywhere on the surface, and so it can be removed from the integral. Therefore,

$$\oint E\, dA = E \int dA = E(4\pi r^2) = \frac{q}{\epsilon_0}$$

where we have used the fact that the surface area of a sphere is $4\pi r^2$. Hence, the magnitude of the field a distance r from q is

$$E = \frac{q}{4\pi\epsilon_0 r^2} = k_e \frac{q}{r^2}$$

If a second point charge q_0 is placed at a point where the field is E, the electric force on this charge has a magnitude

$$F = q_0 E = k_e \frac{q q_0}{r^2}$$

Previously we obtained Gauss's law from Coulomb's law. Here we show that Coulomb's law follows from Gauss's law: They are equivalent.

Example 19.8   A Spherically Symmetric Charge Distribution

An insulating sphere of radius a has a uniform charge density ρ and a total positive charge Q (Fig. 19.29). (a) Calculate the magnitude of the electric field at a point outside the sphere.

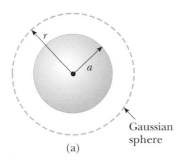

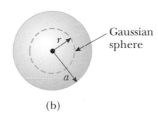

Figure 19.29   (Example 19.8) A uniformly charged insulating sphere of radius a and total charge Q. (a) The field at a point exterior to the sphere is k_eQ/r^2. (b) The field inside the sphere is due only to the charge *within* the gaussian surface and is given by $(k_eQ/a^3)r$.

Solution   Because the charge distribution is spherically symmetric, we again select a spherical gaussian surface of radius r, concentric with the sphere, as in Figure 19.29a. Following the line of reasoning given in Example 19.7, we find that

$$E = k_e \frac{Q}{r^2} \quad \text{(for } r > a\text{)}$$

Note that this result is identical to that obtained for a point charge. Therefore, we conclude that, for a uniformly charged sphere, the field in the region external to the sphere is *equivalent* to that of a point charge located at the center of the sphere.

(b) Find the magnitude of the electric field at a point inside the sphere.

Reasoning and Solution   In this case we select a spherical gaussian surface with radius $r < a$, concentric with the charge distribution (Fig. 19.29b). Let us denote the volume of this smaller sphere by V'. To apply Gauss's law in this situation, it is important to recognize that the charge q_{in} within the gaussian surface of volume V' is a quantity less than the total charge Q. To calculate the charge q_{in}, we use the fact that

$q_{in} = \rho V'$, where ρ is the charge per unit volume and V' is the volume enclosed by the gaussian surface, given by $V' = \frac{4}{3}\pi r^3$ for a sphere. Therefore,

$$q_{in} = \rho V' = \rho(\tfrac{4}{3}\pi r^3)$$

As in Example 19.7, the magnitude of the electric field is constant everywhere on the spherical gaussian surface and is normal to the surface at each point. Therefore, Gauss's law in the region $r < a$ gives

$$\oint E \, dA = E \oint dA = E(4\pi r^2) = \frac{q_{in}}{\epsilon_0}$$

Solving for E gives

$$E = \frac{q_{in}}{4\pi\epsilon_0 r^2} = \frac{\rho\frac{4}{3}\pi r^3}{4\pi\epsilon_0 r^2} = \frac{\rho}{3\epsilon_0}r$$

Because by definition $\rho = Q/\frac{4}{3}\pi a^3$, this can be written

$$E = \frac{Qr}{4\pi\epsilon_0 a^3} = \frac{k_eQ}{a^3}r \quad \text{(for } r < a\text{)}$$

Note that this result for E differs from that obtained in part (a). It shows that $E \rightarrow 0$ as $r \rightarrow 0$, as you might have guessed based on the spherical symmetry of the charge distribution. Therefore, the result fortunately eliminates the singularity that would exist at $r = 0$ if E varied as $1/r^2$ inside the sphere. That is, if $E \propto 1/r^2$, the field would be infinite at $r = 0$, which is clearly a physically impossible situation. A plot of E versus r is shown in Figure 19.30.

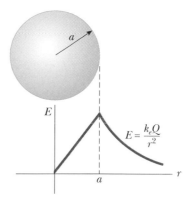

Figure 19.30   (Example 19.8) A plot of E versus r for a uniformly charged insulating sphere. The field inside the sphere ($r < a$) varies linearly with r. The field outside the sphere ($r > a$) is the same as that of a point charge Q located at the origin.

Example 19.9   A Cylindrically Symmetric Charge Distribution

Find the electric field a distance r from a uniform positive line charge of infinite length the charge per unit length of which is λ = constant (Fig. 19.31).

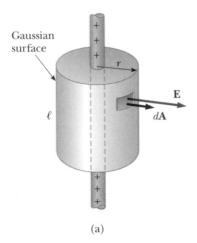

Gaussian surface

r

$\mathbf{E}$

ℓ

$d\mathbf{A}$

(a)

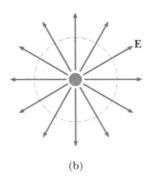

$\mathbf{E}$

(b)

Figure 19.31   (Example 19.9) (a) An infinite line of charge surrounded by a cylindrical gaussian surface concentric with the line charge. (b) An end view shows that the field on the cylindrical surface is constant in magnitude and perpendicular to the surface.

Reasoning   The symmetry of the charge distribution shows that E must be perpendicular to the line charge and directed outward as in Figure 19.31a. The end view of the line charge

shown in Figure 19.31b should help visualize the directions of the electric field lines. In this situation, we select a cylindrical gaussian surface of radius r and length ℓ that is coaxial with the line charge. For the curved part of this surface, $\mathbf{E}$ is constant in magnitude and perpendicular to the surface at each point. Furthermore, the flux through the ends of the gaussian cylinder is zero because $\mathbf{E}$ is parallel to these surfaces.

Solution   The total charge inside our gaussian surface is $\lambda\ell$. Applying Gauss's law and noting that $\mathbf{E}$ is parallel to $d\mathbf{A}$ everywhere on the cylindrical surface, we find that

$$\Phi_c = \oint \mathbf{E} \cdot d\mathbf{A} = E \oint dA = \frac{q_{in}}{\epsilon_0} = \frac{\lambda\ell}{\epsilon_0}$$

But the area of the curved surface is $A = 2\pi r\ell$; therefore,

$$E(2\pi r\ell) = \frac{\lambda\ell}{\epsilon_0}$$

$$E = \frac{\lambda}{2\pi\epsilon_0 r} = 2k_e\frac{\lambda}{r} \qquad [19.16]$$

Thus, we see that the field of a cylindrically symmetric charge distribution varies as $1/r$, whereas the field external to a spherically symmetric charge distribution varies as $1/r^2$. Equation 19.16 can also be obtained using Coulomb's law and integration; however, the mathematical techniques necessary for this calculation are more cumbersome.

If the line charge has a finite length, the result for E is not that given by Equation 19.16. For points close to the line charge and far from the ends, Equation 19.16 gives a good approximation for the value of the field. It turns out that Gauss's law is not useful for calculating E for a finite line charge. This is because the magnitude of the electric field is no longer constant over the surface of the gaussian cylinder. Furthermore, $\mathbf{E}$ is not perpendicular to the cylindrical surface at all points. When there is little symmetry in the charge distribution, as in this situation, it is necessary to calculate $\mathbf{E}$ using Coulomb's law.

It is left as a problem (Problem 29) to show that the $\mathbf{E}$ field inside a long uniformly charged rod of finite thickness is proportional to r.

Example 19.10   A Nonconducting Plane Sheet of Charge

Find the electric field due to a nonconducting, infinite plane with uniform charge per unit area σ.

Reasoning and Solution   The symmetry of the situation shows that $\mathbf{E}$ must be perpendicular to the plane and that the

direction of $\mathbf{E}$ on one side of the plane must be opposite its direction on the other side, as in Figure 19.32. It is convenient to choose for our gaussian surface a small cylinder the axis of which is perpendicular to the plane and the ends of which each have an area A and are equidistant from the plane. Here we see that because **E** is parallel to the cylindrical surface, there is no flux through this surface. The flux out of each end of the cylinder is *EA* (because **E** is perpendicular to the ends); hence, the total flux through our gaussian surface is 2*EA*.

Solution   Noting that the total charge inside the surface is σA, we use Gauss's law to get

$$\Phi_c = 2EA = \frac{q_{in}}{\epsilon_0} = \frac{\sigma A}{\epsilon_0}$$

$$E = \frac{\sigma}{2\epsilon_0} \qquad\qquad \textbf{[19.17]}$$

Because the distance of the surfaces from the plane does not appear in Equation 19.17, we conclude that $E = \sigma/2\epsilon_0$ at any distance from the plane. That is, the field is uniform everywhere.

An important configuration related to this example is the case of two parallel planes of charge, with charge densities σ

and $-\sigma$, respectively (Problem 46). In this situation, the electric field is σ/ϵ_0 between the planes and approximately zero elsewhere.

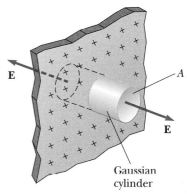

Figure 19.32   (Example 19.10) A cylindrical gaussian surface penetrating an infinite sheet of charge. The flux through each end of the gaussian surface is *EA*. There is no flux through the cylinder's curved surface.

EXERCISE 9   A conducting spherical shell of radius 15 cm carries a net charge of $-6.4\ \mu C$ uniformly distributed on its surface. Find the electric field at points (a) just outside the shell and (b) inside the shell.   Answer   (a) 2.6×10^6 N/C   (b) 0

PROBLEM-SOLVING STRATEGY AND HINTS • Applying Gauss's Law

Gauss's law may seem mysterious to you, and it is usually one of the most challenging concepts in introductory physics. However, as we have seen, Gauss's law is very powerful in solving problems with a high degree of symmetry. In this section, we have used Gauss's law to solve problems with three kinds of symmetry: plane, cylindrical, and spherical. It is important to review Examples 19.7 through 19.10 and to use the following procedure:

1. Select a gaussian surface that *has a symmetry appropriate for the charge distribution.* For point charges or spherically symmetric charge distributions, the gaussian surface should be a sphere centered on the charge, as in Examples 19.7 and 19.8. For uniform line charges or uniformly charged cylinders, the gaussian surface should be a cylindrical surface that is co-axial with the line charge or cylinder, as in Example 19.9. For sheets of charge having plane symmetry, the gaussian surface should be a small cylinder that straddles the sheet, as in Example 19.10. Note that in all cases, the gaussian surface is selected so that the electric field has the same magnitude everywhere on the surface and is directed perpendicular to the surface or is parallel to the surface. This enables you to easily evaluate the surface integral that appears on the left side of Gauss's law, and this integral represents the total electric flux through that surface.
2. Evaluate the right side of Gauss's law, which amounts to calculating the total electric charge, q_{in}, *inside* the gaussian surface. If the charge density

is uniform, as is usually the case (that is, if λ, σ, or ρ is constant), simply multiply that charge density by the length, area, or volume enclosed by the gaussian surface. However, if the charge distribution is *nonuniform*, you must integrate the charge density over the region enclosed by the gaussian surface. For example, if the charge were distributed along a line, you would integrate the expression $dq = \lambda \, dx$, where dq is the charge on an infinitesimal element dx and λ is the charge per unit length. For a plane of charge, you would integrate $dq = \sigma \, dA$, where σ is the charge per unit area and dA is an infinitesimal element of area. For a volume of charge, you would integrate $dq = \rho \, dV$, where ρ is the charge per unit volume and dV is an infinitesimal element of volume.

3. Once the left and right sides of Gauss's law have been evaluated, you can calculate the electric field on the gaussian surface, assuming the charge distribution is given in the problem. Conversely, if the electric field is known, you can calculate the charge distribution that produces the field.

19.10 • CONDUCTORS IN ELECTROSTATIC EQUILIBRIUM

A good electrical conductor, such as copper, contains charges (electrons) that are not bound to any atom and are free to move about within the material. When no *net* motion of charge occurs within the conductor, the conductor is in **electrostatic equilibrium.** As we shall see, an isolated conductor (one that is insulated from ground) in electrostatic equilibrium has the following properties:

> - The electric field is zero everywhere inside the conductor.
> - Any excess charge on the conductor must reside entirely on its surface.
> - The electric field just outside the charged conductor is perpendicular to the conductor's surface and has the magnitude σ/ϵ_0, where σ is the charge per unit area at that point.
> - On an irregularly shaped conductor, charge tends to accumulate at locations where the radius of curvature of the surface is the smallest—that is, at sharp points. This will be discussed further in Chapter 20.

The first property can be understood by considering a conducting slab placed in an external field **E** (Fig. 19.33). In electrostatic equilibrium, the electric field *inside* the conductor must be zero. If this were not the case, the free charges would accelerate under the action of an electric field. Before the external field is applied, the electrons are uniformly distributed throughout the conductor. When the external field is applied, the free electrons accelerate to the left, causing a buildup of negative charge on the left surface (excess electrons) and of positive charge on the right (where electrons have been removed). These charges create their own electric field, which *opposes* the external field. The surface-charge density increases until the magnitude of the electric field set up by these charges equals that of the external field, giving a net field of zero *inside* the conductor. In a good conductor, the time it takes the conductor to reach equilibrium is on the order of 10^{-16} s, which for most purposes can be considered instantaneous.

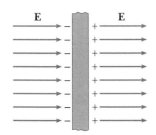

Figure 19.33 A conducting slab in an external electric field **E**. The charges induced on the surfaces of the slab produce an electric field that opposes the external field, giving a resultant field of zero inside the conductor.

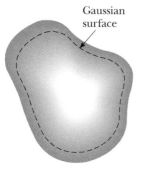

Gaussian surface

Figure 19.34 An insulated conductor of arbitrary shape. The broken line represents a gaussian surface just inside the conductor.

We can use Gauss's law to verify the second and third properties. Figure 19.34 shows an arbitrarily shaped insulated conductor. A gaussian surface is drawn inside the conductor as close to the surface as we wish. As we have just shown, the electric field everywhere inside the conductor is zero when it is in electrostatic equilibrium. Because the electric field is therefore *zero* at every point on the gaussian surface, we see that the net flux through this surface is zero. From this result and Gauss's law, we conclude that the net charge inside the gaussian surface is zero. Because there can be no net charge inside the gaussian surface (which is arbitrarily close to the conductor's surface), **any net charge on the conductor must reside on its surface.** Gauss's law does *not* tell us how this excess charge is distributed on the surface.

Examining the third property, we can use Gauss's law to relate the electric field just outside the surface of a charged conductor in equilibrium to the charge distribution on the conductor. To do this, it is convenient to draw a gaussian surface in the shape of a small cylinder with end faces parallel to the surface (Fig. 19.35). Part of the cylinder is just outside the conductor and part is inside. There is no flux through the face on the inside of the cylinder, because **E** = 0 inside the conductor. Furthermore, the field outside the conductor is normal to the surface. If **E** had a tangential component, the free charges would move along the surface, creating surface currents, and the conductor would not be in equilibrium. There is no flux through the cylindrical face of the gaussian surface, because **E** is tangent to this surface. Hence, the net flux through the gaussian surface is $E_n A$, where E_n is the electric field just outside the conductor. Applying Gauss's law to this surface gives

$$\Phi_c = \oint E_n \, dA = E_n A = \frac{q_{\text{in}}}{\epsilon_0} = \frac{\sigma A}{\epsilon_0}$$

We have used the fact that the charge inside the gaussian surface is $q_{\text{in}} = \sigma A$, where A is the area of the cylinder's face and σ is the (local) charge per unit area. Solving for E_n gives

$$E_n = \frac{\sigma}{\epsilon_0} \qquad\qquad \textbf{[19.18]}$$

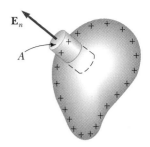

Figure 19.35   A gaussian surface in the shape of a small cylinder is used to calculate the electric field just outside a charged conductor. The flux through the gaussian surface is $E_n A$. Note that **E** is zero inside the conductor.

- *Electric field just outside a charged conductor*

Thinking Physics 6

Suppose a point charge $+Q$ is in empty space. We sneak up and surround the charge with a spherical conducting shell. What effect does this have on the field lines from the charge?

Reasoning   When the spherical shell is placed around the charge, the charges in the shell will adjust to satisfy the rules for a conductor in equilibrium and Gauss's law. A net charge of $-Q$ will move to the interior surface of the conductor, so that the electric field within the conductor will be zero (a spherical gaussian surface totally within the shell will enclose no net charge). A net charge of $+Q$ will reside on the outer surface. Thus, a gaussian surface outside the sphere will enclose a net charge of $+Q$, just as if the shell were not there. Thus, the only change in the field lines from the initial situation will be the absence of field lines within the conductor.

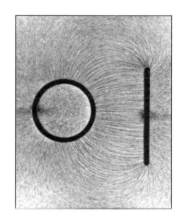

Electric field pattern of a charged conducting plate near an oppositely charged conducting cylinder. Small pieces of thread suspended in oil align with the electric field lines. Note that (1) the electric field lines are perpendicular to the conductors and (2) there are no lines inside the cylinder ($E = 0$). *(Courtesy of Harold M. Waage, Princeton University)*

CONCEPTUAL PROBLEM 7

There are great similarities between electric and gravitational fields. A room can be electrically shielded so that there are no electric fields in the room by surrounding it with a conductor. Can a room be *gravitationally* shielded? Why or why not?

SUMMARY

Resultant Force

$F_1 = F_{12} + F_{13} + F_{14}$

Electric field Vector

$E = \dfrac{F}{q_0}$ F = electric force
             q_0 = magnitude of
                 test charge

Volume charge density

$\rho = \dfrac{Q}{V}$ Q = charge
          V = volume

Surface charge density

$\sigma = \dfrac{Q}{A}$

Linear charge density

$\lambda = \dfrac{Q}{\ell}$

Electric charges have the following important properties:

- Unlike charges attract one another, and like charges repel one another.
- Electric charge is always conserved.
- Charge is quantized; that is, the charge on any object is an integral multiple of the electronic charge.
- The force between charged particles varies as the inverse square of their separation.

Conductors are materials in which charges move freely. **Insulators** are materials that do not readily transport charge.

Coulomb's law states that the electrostatic force between two stationary, charged particles separated by a distance of r has the magnitude

$$F = k_e \frac{|q_1|\,|q_2|}{r^2} \qquad \text{[19.1]}$$

where the Coulomb constant $k_e = 8.99 \times 10^9 \ \text{N·m}^2/\text{C}^2$.

The *electric field* due to the point charge q at a distance of r from the charge is

$$\mathbf{E} = k_e \frac{q}{r^2}\hat{\mathbf{r}} \qquad \text{[19.4]}$$

where $\hat{\mathbf{r}}$ is a unit vector directed from the charge to the point in question. The electric field is directed radially outward from a positive charge and is directed *toward* a negative charge.

The electric field due to a group of charges can be obtained using the superposition principle. That is, the total electric field equals the *vector sum* of the electric fields of all the charges at some point:

$$\mathbf{E} = k_e \sum_i \frac{q_i}{r_i^2}\hat{\mathbf{r}}_i \qquad \text{[19.5]}$$

In a similar way, the electric field of a continuous charge distribution at some point is

$$\mathbf{E} = k_e \int \frac{dq}{r^2}\hat{\mathbf{r}} \qquad \text{[19.6]}$$

where dq is the charge on one element of the charge distribution and r is the distance from the element to the point in question.

Electric field lines are useful for describing the electric field in any region of space. The electric field vector $\mathbf{E}$ is always tangent to the electric field lines at every point. Furthermore, the number of lines per unit area through a surface perpendicular to the lines is proportional to the magnitude of $\mathbf{E}$ in that region.

Electric flux is proportional to the number of electric field lines that penetrate a surface. If the electric field is uniform and makes an angle of θ with the normal to the surface, the electric flux through the surface is

$$\Phi = EA \cos \theta \qquad \text{[19.11]}$$

In general, the electric flux through a surface is defined by the expression

$$\Phi = \int_{\text{surface}} \mathbf{E} \cdot d\mathbf{A} \qquad \text{[19.12]}$$

use proportion for
unequal charges.   →

$T = mg$

Gauss's law says that the net electric flux Φ_c through any closed gaussian surface is equal to the *net* charge *inside* the surface divided by ϵ_0:

$$\Phi_c = \oint \mathbf{E} \cdot d\mathbf{A} = \frac{q_{in}}{\epsilon_0} \qquad \text{[19.15]}$$

Using Gauss's law, one can calculate the electric field due to various symmetric charge distributions.

A **conductor in electrostatic equilibrium** has the following properties:

- The electric field is zero everywhere inside the conductor.
- Any excess charge on an isolated conductor must reside entirely on its surface.
- The electric field just outside the conductor is perpendicular to its surface and has the magnitude σ/ϵ_0, where σ is the charge per unit area at that point.
- On an irregularly shaped conductor, charge tends to accumulate where the radius of curvature of the surface is the smallest—that is, at sharp points.

CONCEPTUAL QUESTIONS

1. Explain from an atomic viewpoint why charge is usually transferred by electrons.

2. A balloon is negatively charged by rubbing and then clings to a wall. Does this mean that the wall is positively charged? Why does the balloon eventually fall?

3. A charged comb often attracts small bits of dry paper that then fly away when they touch the comb. Explain.

4. Operating room personnel must wear special conducting shoes while working around oxygen. Why? Contrast this procedure with what might happen if personnel wore rubber shoes.

5. Would life be different if the electron were positively charged and the proton were negatively charged? Does the choice of signs have any bearing on physical and chemical interactions? Explain.

6. When defining the electric field, why is it necessary to specify that the magnitude of the test charge be very small (that is, take the limit of $\mathbf{F}/q$ as $q \to 0$)?

7. Consider two equal point charges separated by some distance d. At what point (other than ∞) would a third test charge experience no net force?

8. A negative point charge $-q$ is placed at the point P near the positively charged ring shown in Figure 19.13 (Example 19.5). If $x \ll a$, describe the motion of the point charge if it is released from rest.

9. A very large, thin, flat plate of aluminum of area A has a total charge of Q uniformly distributed over its surfaces. The same charge is spread uniformly over only the *upper* surface of an otherwise identical glass plate. Compare the electric fields just above the centers of the plates' upper surfaces.

10. If there are more electric field lines leaving a gaussian surface than entering, what can you conclude about the net charge enclosed by that surface?

11. A uniform electric field exists in a region of space in which there are no charges. What can you conclude about the *net* electric flux through a gaussian surface placed in this region of space?

12. A common demonstration involves charging a balloon, which is an insulator, by rubbing it on your head, and touching the balloon to a ceiling or wall, which is also an insulator. The electrical attraction between the charged balloon and the neutral wall results in the balloon sticking to the wall. Imagine now that we have two infinitely large sheets of insulating material. One is charged and the other is neutral. If these are brought into contact, will there be an attractive force between them, as there was for the balloon and the wall?

13. If the total charge inside a closed surface is known but the distribution of the charge is unspecified, can you use Gauss's law to find the electric field? Explain.

14. Explain why excess charge on an isolated conductor must reside on its surface, using the repulsive nature of the force between like charges and the freedom of motion of charge within the conductor.

15. A person is placed in a large, hollow, metallic sphere that is insulated from ground. If a large charge is placed on the sphere, will the person be harmed on touching the inside of the sphere? Explain what will happen if the person also has an initial charge the sign of which is opposite that of the charge on the sphere.

16. Imagine two electric dipoles in empty space. Each dipole has zero net charge. Is there an electric force between the dipoles—that is, can two objects with *zero net charge* exert electric forces on each other? If so, is the force attractive or repulsive?

PROBLEMS

Section 19.4   Coulomb's Law

1. Two identical conducting small spheres are placed with their centers 0.300 m apart. One is given a charge of 12.0 nC and the other a charge of -18.0 nC. (a) Find the electrostatic force exerted on one sphere by the other. (b) The spheres are connected by a conducting wire. After equilibrium has occurred, find the electrostatic force between the two.

2. (a) Two protons in a molecule are separated by 3.80×10^{-10} m. Find the electrostatic force exerted by one proton on the other. (b) How does the magnitude of this force compare to the magnitude of the gravitational force between the two protons? (c) What must be the charge-to-mass ratio of a particle if the magnitude of the gravitational force between two of these particles equals the magnitude of electrostatic force between them?

3. Richard Feynman once said that if two persons stood at arm's length from each other and each person had 1% more electrons than protons, the force of repulsion between them would be enough to lift a "weight" equal to that of the entire Earth. Carry out an order-of-magnitude calculation to substantiate this assertion.

4. In the Bohr theory of the hydrogen atom, an electron moves in a circular orbit about a proton, where the radius of the orbit is 0.529×10^{-10} m. (a) Find the electrostatic force between the two. (b) If this force provides the centripetal acceleration of the electron, what is the speed of the electron?

5. Two small silver spheres, each with a mass of 10.0 g, are separated by 1.00 m. Calculate the fraction of the electrons in one sphere that must be transferred to the other in order to produce an attractive force of 1.00×10^4 N (about 1 ton) between the spheres. (The number of electrons per atom of silver is 47, and the number of atoms per gram is Avogadro's number divided by the molar mass of silver, 107.87.)

Section 19.5   Electric Fields

6. The electrons in a particle beam each have a kinetic energy of 1.60×10^{-17} J. What are the magnitude and direction of the electric field that will stop these electrons in a distance of 10.0 cm?

7. The electrons in a particle beam each have a kinetic energy K. What are the magnitude and direction of the electric field that will stop these electrons in a distance d?

8. Consider the electric dipole shown in Figure P19.8. Show that the electric field at a *distant* point along the x axis is $E_x \cong 4k_e qa/x^3$.

9. In Figure P19.9, determine the point (other than infinity) at which the electric field is zero.

10. A positively charged bead having a mass of 1.00 g falls from

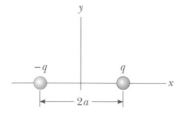

Figure P19.8

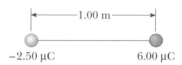

Figure P19.9

rest in a vacuum at a height of 5.00 m in a uniform vertical electric field of magnitude 1.00×10^4 N/C. The bead hits the ground at a speed of 21.0 m/s. Determine (a) the direction of the electric field (up or down), and (b) the charge on the bead.

11. A continuous line of charge lies along the x axis, extending from $x = +x_0$ to positive infinity. The line carries a uniform linear charge density λ_0. What are the magnitude and direction of the electric field at the origin?

12. Consider n equal positive point charges each of magnitude q/n placed symmetrically around a circle of radius R. (a) Calculate the magnitude of the electric field E at a point a distance x on the line passing through the center of the circle and perpendicular to the plane of the circle. (b) Explain why this result is identical to the calculation done in Example 19.5.

13. Three equal positive charges q are at the corners of an equilateral triangle of sides a, as in Figure P19.13. (a) At what point in the plane of the charges (other than infinity) is the electric field zero? (b) What are the magnitude and direction of the electric field at P due to the two charges at the base?

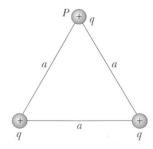

Figure P19.13

14. Three point charges, q, $2q$, and $3q$, are arranged on the vertices of an equilateral triangle with sides of length a. Determine the magnitude of electric field at the geometric center of the triangle.

15. A rod 14.0 cm long is uniformly charged and has a total charge of $-22.0\ \mu C$. Determine the magnitude and direction of the electric field along the axis of the rod at a point 36.0 cm from its center.

16. A uniformly charged ring of radius 10.0 cm has a total charge of $75.0\ \mu C$. Find the electric field on the axis of the ring at (a) 1.00 cm, (b) 5.00 cm, (c) 30.0 cm, and (d) 100 cm from the center of the ring.

17. A uniformly charged insulating rod of length 14.0 cm is bent into the shape of a semicircle as in Figure P19.17. If the rod has a total charge of 7.50 μC, find the magnitude and direction of the electric field at O, the center of the semicircle.

• O

Figure P19.17

Section 19.7 Electric Flux

18. An electric field of intensity 3.50 kN/C is applied along the x axis. Calculate the electric flux through a rectangular plane 0.350 m wide and 0.700 m long if (a) the plane is parallel to the yz plane; (b) the plane is parallel to the xy plane; (c) the plane contains the y axis, and its normal makes an angle of 40.0° with the x axis.

19. A 40.0-cm-diameter loop is rotated in a uniform electric field until the position of maximum electric flux is found. The flux in this position is 5.20×10^5 N·m²/C. What is the electric field strength?

20. An electric field of magnitude 2.00×10^4 N/C exists perpendicular to the Earth's surface on a day when a thunderstorm is brewing. A car of approximately rectangular size 6.00 m by 3.00 m is traveling along a roadway sloping down at 10.0°. Determine the electric flux through the bottom of the car.

Section 19.8 Gauss's Law

21. A charge of 12.0 μC is at the geometric center of a cube. What is the electric flux through one of the faces?

22. The following charges are located inside a submarine: 5.00 μC, $-9.00\ \mu C$, 27.0 μC, and $-84.0\ \mu C$. Calculate the net electric flux through the submarine. Compare the number of electric field lines leaving the submarine with the number entering it.

23. A point charge Q is located just above the center of the flat face of a hemisphere of radius R as in Figure P19.23. What is the electric flux (a) through the curved surface and (b) through the flat face?

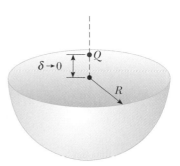

$\delta \to 0$ $\bullet Q$

R

mary

Figure P19.23

24. The electric field everywhere on the surface of a thin spherical shell of radius 0.750 m is measured to be equal to 890 N/C and points radially toward the center of the sphere. (a) What is the net charge within the sphere's surface? (b) What can you conclude about the nature and distribution of the charge inside the spherical shell?

Section 19.9 Application of Gauss's Law to Charged Insulators

25. Determine the magnitude of the electric field at the surface of a lead-208 nucleus, which contains 82 protons and 126 neutrons. Assume the lead nucleus has a volume 208 times that of 1 proton, and consider a proton to be a sphere of radius 1.20×10^{-15} m.

26. A solid sphere of radius 40.0 cm has a total positive charge of 26.0 μC uniformly distributed throughout its volume. Calculate the magnitude of the electric field (a) 0 cm, (b) 10.0 cm, (c) 40.0 cm, and (d) 60.0 cm from the center of the sphere.

27. A very thin, spherical shell of radius a has a total charge of Q distributed uniformly over its surface (Fig. P19.27). Find the electric field at points inside and outside the shell.

28. A cylindrical shell of radius 7.00 cm and length 240 cm has its charge uniformly distributed on its surface. The electric field at a point 19.0 cm radially outward from its axis (measured from the midpoint of the shell) is 36.0 kN/C. Use approximate relations to find (a) the net charge on the shell and (b) the electric field at a point 4.00 cm from the

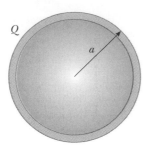

Figure P19.27

axis, measured radially outward from the midpoint of the shell.

29. Consider a long cylindrical charge distribution of radius R with a uniform charge density ρ. Find the electric field at distance r from the axis where $r < R$.

30. A 10.0-g piece of Styrofoam carries a net charge of $-0.700\ \mu C$ and floats above the center of a very large horizontal sheet of plastic that has a uniform charge density on its surface. What is the charge per unit area on the plastic sheet?

Section 19.10 Conductors in Electrostatic Equilibrium

31. A long, straight metal rod has a radius of 5.00 cm and a charge per unit length of 30.0 nC/m. Find the electric field (a) 3.00 cm, (b) 10.0 cm, and (c) 100 cm from the axis of the rod, where distances are measured perpendicular to the rod.

32. A charge q is placed on a conducting sheet that has a finite but very small thickness. What is the approximate magnitude of the electric field at an external point very near the surface of the sheet at its center if the area of the sheet is A?

33. A square plate of copper with 50.0-cm sides has no net charge and is placed in a region of uniform electric field of 80.0 kN/C directed *perpendicularly* to the plate. Find (a) the charge density of each face of the plate and (b) the total charge on each face.

34. A solid conducting sphere of radius 2.00 cm has a charge 8.00 μC. A conducting spherical shell of inner radius 4.00 cm and outer radius 5.00 cm is concentric with the solid sphere and has a charge $-4.00\ \mu C$. Find the electric field at (a) $r = 1.00$ cm, (b) $r = 3.00$ cm, (c) $r = 4.50$ cm, and (d) $r = 7.00$ cm from the center of this charge configuration.

35. Two identical conducting spheres each having a radius of 0.500 cm are connected by a light 2.00-m-long conducting wire. Determine the tension in the wire if 60.0 μC is placed on one of the conductors. (*Hint:* Assume that the surface distribution of charge on each sphere is uniform.)

36. A long, straight wire is surrounded by a hollow metal cylinder the axis of which coincides with that of the wire. The wire has a charge per unit length of λ, and the cylinder has a net charge per unit length of 2λ. From this information, use Gauss's law to find (a) the charge per unit length on the inner and outer surfaces of the cylinder and (b) the electric field outside the cylinder a distance r from the axis.

37. The electric field on the surface of an irregularly shaped conductor varies from 56.0 kN/C to 28.0 kN/C. Calculate the local surface charge density at the point on the surface at which the radius of curvature of the surface is (a) smallest and (b) greatest.

Additional Problems

38. Three point charges are aligned along the x axis as shown in Figure P19.38. Find the electric field at (a) the position (2.00, 0) and (b) the position (0, 2.00).

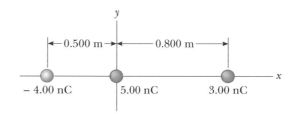

Figure P19.38

39. A small, 2.00-g plastic ball is suspended by a 20.0-cm-long string in a uniform electric field, as in Figure P19.39. If the ball is in equilibrium when the string makes a 15.0° angle with the vertical, what is the net charge on the ball?

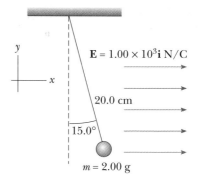

Figure P19.39

40. A uniform electric field of magnitude 640 N/C exists between two parallel plates that are 4.00 cm apart. A proton is released from the positive plate at the same instant that an

electron is released from the negative plate. (a) Determine the distance from the positive plate that the two pass each other. (Ignore the electrostatic attraction between the proton and electron.) (b) Repeat part (a) for a sodium ion (Na^+) and a chlorine ion (Cl^-).

41. A charged cork ball of mass 1.00 g is suspended on a light string in the presence of a uniform electric field, as in Figure P19.41. When $\mathbf{E} = (3.00\mathbf{i} + 5.00\mathbf{j}) \times 10^5$ N/C, the ball is in equilibrium at $\theta = 37.0°$. Find (a) the charge on the ball and (b) the tension in the string.

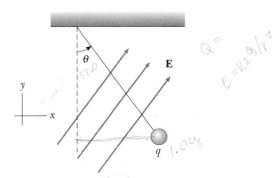

Figure P19.41

42. Four identical point charges ($q = +10.0 \mu C$) are located on the corners of a rectangle, as shown in Figure P19.42. The dimensions of the rectangle are $L = 60.0$ cm and $W = 15.0$ cm. Calculate the magnitude and direction of the net electric force exerted on the charge at the lower left corner by the other three charges.

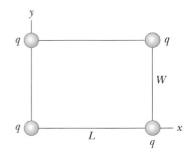

Figure P19.42

43. A solid, *insulating* sphere of radius a has a uniform charge density of ρ and a total charge of Q. Concentric with this sphere is an *uncharged, conducting* hollow sphere, the inner and outer radii of which are b and c, as in Figure P19.43. (a) Find the electric field intensity in the regions $r < a$, $a < r < b$, $b < r < c$, and $r > c$. (b) Determine the induced charge per unit area on the inner and outer surfaces of the hollow sphere.

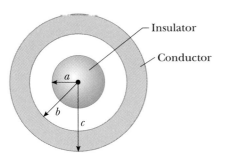

Figure P19.43

44. In nuclear fission, a nucleus of uranium-238, which contains 92 protons, divides into two smaller spheres, each having 46 protons and a radius of 5.90×10^{-15} m. What is the magnitude of the repulsive electric force pushing the two spheres apart?

45. An infinitely long cylindrical insulating shell of inner radius a and outer radius b has a uniform volume charge density ρ. A line of charge density λ is placed along the axis of the shell. Determine the electric field intensity everywhere.

46. Two infinite, nonconducting sheets of charge are parallel to each other as in Figure P19.46. The sheet on the left has a uniform surface charge density σ, and the one on the right has a uniform charge density $-\sigma$. Calculate the value of the electric field at points (a) to the left of, (b) in between, and (c) to the right of the two sheets. (*Hint:* See Example 19.10.)

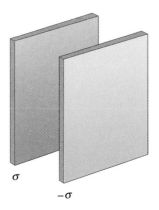

Figure P19.46

47. Repeat the calculations for Problem 46 when both sheets have *positive* uniform charge densities of value σ.

48. A sphere of radius $2a$ is made of a nonconducting material that has a uniform volume charge density ρ. (Assume that the material does not affect the electric field.) A spherical cavity of radius a is now removed from the sphere, as shown in Figure P19.48. Show that the electric field within the cavity is uniform and is given by $E_x = 0$ and $E_y = \rho a/3\epsilon_0$. (*Hint:* The field within the cavity is the superposition of the field due to the original uncut sphere plus the field due to a sphere the size of the cavity with a uniform negative charge density $-\rho$.)

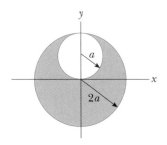

Figure P19.48

49. Three identical small Styrofoam balls ($m = 2.00$ g) are suspended from a fixed point by three nonconducting threads, each with a length of 50.0 cm and with negligible mass. At equilibrium the three balls form an equilateral triangle with sides of 30.0 cm. What is the common charge q carried by each ball?

50. Identical thin rods of length $2a$ carry equal charges, $+Q$, uniformly distributed along their lengths. The rods lie along the x axis with their centers separated by a distance of $b > 2a$ (Fig. P19.50). Show that the force exerted on the right rod is given by

$$F = \frac{k_e Q^2}{4a^2} \ln\left(\frac{b^2}{b^2 - 4a^2}\right)$$

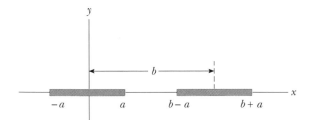

Figure P19.50

51. A line of positive charge is formed into a semicircle of radius $R = 60.0$ cm, as shown in Figure P19.51. The charge per unit length along the semicircle is described by the expres-

sion $\lambda = \lambda_0 \cos\theta$. The total charge on the semicircle is 12.0 μC. Calculate the total force on a charge of 3.00 μC placed at the center of curvature.

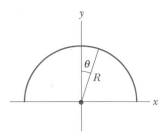

Figure P19.51

52. A particle is said to be nonrelativistic as long as its speed is less than one tenth the speed of light, or less than 3.00×10^7 m/s. (a) How long will an electron remain nonrelativistic if it starts from rest in a region of electric field 1.00 N/C? (b) How long will a proton remain nonrelativistic in the same electric field? (c) Electric fields are commonly much larger than 1 N/C. Will the charged particle remain nonrelativistic for a shorter or a longer time in a much larger electric field?

▣ Spreadsheet Problems

S1. Spreadsheet 19.1 calculates the x and y components of the electric field **E** along the x axis due to any two point charges Q_1 and Q_2. Enter the values of the charges and their x and y coordinates. Choose $Q_1 = 4$ μC at the origin, and set $Q_2 = 0$. Plot the electric field along the x axis. *Note:* You must choose a step size and adjust the scaling for the graph because the electric field at the position of the point charge is infinite.

S2. (a) Using Spreadsheet 19.1, place $Q_1 = 4$ μC at the origin and $Q_2 = 4$ μC at $x = 0.08$ m, $y = 0$. Plot the components of the electric field **E** as a function of x. (b) Change Q_2 to -4 μC, and plot the components of the electric field as a function of x. (This is a dipole.) You will need to adjust the scales of your graphs.

S3. (a) Using Spreadsheet 19.1, take $Q_1 = 6$ nC at $x = 0$, $y = 0.03$ m, and $Q_2 = 6$ nC at $x = 0$, $y = -0.03$ m. Plot the components of the electric field **E** as a function of x. (b) Change Q_2 to -6 nC, and repeat part (a).

S4. Consider an electric dipole in which a charge $+q$ is on the x axis at $x = a$ and a charge $-q$ is on the x axis at $x = -a$. Take $q = 6$ nC and $a = 0.02$ m. Use Spreadsheet 19.1 to plot the components of the electric field along the x axis for values of x greater than 4 cm. Modify Spreadsheet 19.1 to add a plot $2k_e p/x^3$ (the **E** field in the approximation $x \gg a$), where $p = 2aq$ is the dipole moment of the charge distribution. For what range of values of x is the dipole approximation within 20% of the actual value? Within 5%?

ANSWERS TO CONCEPTUAL PROBLEMS

1. The configuration shown is inherently unstable. The negative charges are repelling each other. If there is any slight rotation of one of the rods, the repulsion can result in further rotation away from this configuration. There are three possible final configurations shown in Figure CP19.1: Configuration (a) is stable—if the positive upper ends are pushed toward each other, they will repel and move the system back to the original configuration. Configuration (b) is an equilibrium configuration, but is unstable—if the upper ends are moved toward each other, the attraction of the upper ends will be larger than that of the lower ends and the configuration will shift to (c). Configuration (c) is stable.

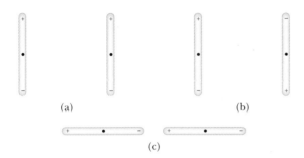

(a)   (b)

(c)

Figure CP19.1

2. No. Object A might have a charge opposite in sign to that of B, but it also might be a neutral conductor. In this latter case, object B causes A to be polarized, pulling charge of one sign to the near face of A and pushing an equal amount of charge of the opposite sign to the far face, as in Figure CP19.2. Then the force of attraction exerted on B by the induced charge

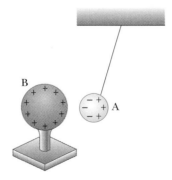

Figure CP19.2

on the near side of A is slightly larger than the force of repulsion exerted on B by the induced charge on the far side of A. Therefore, the net force on A is toward B.

3. A television screen is constantly bombarded with electrons from the electron gun. As a result, it becomes negatively charged, and there is an electric field established in the region in front of the screen. This electric field will polarize dust particles in the air in front of the screen, just like a charged object polarizes molecules in a neutral object. What's more, the electric field is not uniform, because the television screen is not an infinite sheet. Thus, the positively charged end of the dust particles will be nearer the screen and in a stronger electric field than the negative end. As a result, there is a net attractive force, pulling the dust particle to the screen, where it adheres.

4. Electric field lines start on positive charges and end on negative charges. Thus, if the fair weather field is directed into the ground, the ground must have a negative charge.

5. The two charged plates create a region of uniform electric field between them, directed from the positive toward the negative plate. Once the ball is disturbed so as to touch one plate, say the negative one, some negative charge will be transferred to the ball and it will experience an electric force that will accelerate it to the positive plate. Once the charge touches the positive plate, it will release its negative charge, acquire a positive charge, and accelerate back to the negative plate. The ball will continue to move back and forth between the plates until it has transferred all their net charge, thereby making both plates neutral.

6. The electric field patterns of each of these three configurations do not have sufficient symmetry to make the calculations practical. (Gauss's law is only useful for calculating the electric field of highly symmetric charge distributions, such as uniformly charged spheres, cylinders, and sheets.) In order to apply Gauss's law, you must be able to find a closed surface surrounding the charge distribution, which can be subdivided so that the field over the separate regions of the surface is constant. Such a surface cannot be found for these cases.

7. The electric shielding effect of conductors depends on the fact that there are two kinds of charge—positive and negative. As a result, charges can move within the conductor so that the combination of positive and negative charges establishes an electric field that exactly cancels the external field within the conductor and any cavities inside the conductor. There is only one type of gravitational charge, however—there is no "negative mass." As a result, gravitational shielding is not possible.

20

Electric Potential and Capacitance

The concept of potential energy was introduced in Chapter 7 in connection with such conservative forces as the force of gravity, the elastic force of a spring, and the electrostatic force. By using the principle of conservation of energy, we were often able to avoid working directly with forces when solving mechanical problems. In this chapter we shall use the energy concept in our study of electricity. Because the electrostatic force (given by Coulomb's law) is conservative, electrostatic phenomena can conveniently be described in terms of an *electric* potential energy function. This concept enables us to define a quantity called *electric potential,* which is a scalar function of position and thus leads to a simpler method of describing some electrostatic phenomena than the electric field method. Although the potential is clearly an easier path for many problems, one often requires **E**. In some situations, a knowledge of **E** provides the simplest path to calculation of the potential. As we shall see in subsequent chapters,

Jennifer is holding on to a charged sphere that reaches a potential of about 100 000 volts. The device that generates this high potential is called a Van de Graaff generator. Why do you suppose Jennifer's hair stands on end like the needles of a porcupine? Why is it important that she stand on a pedestal insulated from ground? *(Courtesy of Henry Leap and Jim Lehman)*

the concept of electric potential is of great practical value. For example, the measured voltage between any two points in an electrical circuit is simply the difference in electric potential between the points.

This chapter is also concerned with the properties of capacitors, devices that store charge. Capacitors are commonly used in a variety of electrical circuits. For instance, they are used to tune the frequency of radio receivers, as filters in power supplies, to eliminate unwanted sparking in automobile ignition systems, and as energy-storing devices in electronic flash units.

A capacitor basically consists of two conductors separated by an insulator. We shall see that the capacitance of a given device depends on its geometry and on the material called a *dielectric,* separating the charged conductors.

20.1 · POTENTIAL DIFFERENCE AND ELECTRIC POTENTIAL

When a test charge q_0 is placed in an electrostatic field $\mathbf{E}$, the electric force on the charge is $q_0\mathbf{E}$. This force is the vector sum of the individual forces exerted on q_0 by the various charges producing the field $\mathbf{E}$. It follows that the force $q_0\mathbf{E}$ is conservative because the individual forces governed by Coulomb's law are conservative. When a charge is moved within an electric field at constant velocity, the work done on q_0 by the electric field is equal to the negative of the work done by the external agent causing the displacement. For an infinitesimal displacement $d\mathbf{s}$, the work done by the electric field is $\mathbf{F} \cdot d\mathbf{s} = q_0\mathbf{E} \cdot d\mathbf{s}$. This decreases the potential energy of the charge–field system by an amount $dU = -q_0\mathbf{E} \cdot d\mathbf{s}$. For a finite displacement of the test charge between points A and B, the **change in potential energy** is

$$\Delta U = U_B - U_A = -q_0 \int_A^B \mathbf{E} \cdot d\mathbf{s} \qquad \text{[20.1]}$$

• *Change in potential energy*

The integral in Equation 20.1 is performed along the path by which q_0 moves from A to B and is called either a **path integral** or a **line integral.** Because the force $q_0\mathbf{E}$ is conservative, **this integral does not depend on the path taken between A and B.**

The potential energy per unit charge, U/q_0, is independent of the value of q_0 and has a unique value at every point in an electric field. The quantity U/q_0 is called the **electric potential** (or simply the **potential**), V. Thus, the electric potential at any point in an electric field is

$$V = \frac{U}{q_0} \qquad \text{[20.2]}$$

Because potential energy is a scalar, electric potential is also a scalar quantity.

The **potential difference,** $\Delta V = V_B - V_A$, between the points A and B is defined as the change in potential energy divided by the test charge q_0:

$$\Delta V = \frac{\Delta U}{q_0} = -\int_A^B \mathbf{E} \cdot d\mathbf{s} \qquad \text{[20.3]}$$

• *Potential difference between two points*

Potential difference should not be confused with potential energy. The potential difference is *proportional* to the potential energy difference, and we see from

Equation 20.3 that the two are related by $\Delta U = q_0 \, \Delta V$. The potential is characteristic of the field, independent of the charges that may be placed in the field. However, when we speak of potential energy, we are referring to the charge–field system.

Because, as already noted, the change in the potential energy is the negative of the work done by the electric force, the potential difference ΔV equals the work per unit charge that an external agent must perform to move a test charge from A to B without a change in kinetic energy.

Equation 20.3 defines potential difference only. That is, only *differences* in V are meaningful. The electric potential function is often taken to be zero at some convenient point. We shall usually set the potential at zero for a point at infinity (that is, a point infinitely remote from the charges producing the electric field). With this choice, we can say that **the electric potential at an arbitrary point equals the work required per unit charge to bring a positive test charge from infinity to that point.** Thus, if we take $V_A = 0$ at infinity in Equation 20.3, then the potential at any point P is

$$V_P = -\int_{\infty}^{P} \mathbf{E} \cdot d\mathbf{s} \qquad \text{[20.4]}$$

In reality, V_P represents the potential difference between the point P and a point at infinity. (Equation 20.4 is a special case of Equation 20.3.)

Because potential is a measure of energy per unit charge, the SI units of potential are joules per coulomb, called the **volt** (V):

Definition of a volt •
$$1 \text{ V} \equiv 1 \text{ J/C}$$

That is, 1 J of work must be done to take a 1-C charge through a potential difference of 1 V. Equation 20.3 shows that the potential difference also has the same units as electric field times distance. From this, it follows that the SI units of electric field, newtons per coulomb, can also be expressed as volts per meter.

$$1 \text{ N/C} = 1 \text{ V/m}$$

As we learned in Chapter 9, Section 9.7, a unit of energy commonly used in atomic and nuclear physics is the **electron volt** (eV):

The electron volt •
$$1 \text{ eV} = 1.60 \times 10^{-19} \text{ C} \cdot \text{V} = 1.60 \times 10^{-19} \text{ J} \qquad \text{[20.5]}$$

For instance, an electron in the beam of a typical TV picture tube has a speed of 5.0×10^7 m/s. This corresponds to a kinetic energy of 1.1×10^{-15} J, which is equivalent to 7.1×10^3 eV. Such an electron has to be accelerated from rest through a potential difference of 7.1 kV to reach this speed.

Thinking Physics 1

Suppose scientists had chosen to measure small energies in *proton volts* rather than electron volts. What difference would this make?

Reasoning There would be no change at all. An electron volt is the kinetic energy gained by an electron in being accelerated through a potential difference of one volt. A proton accelerated through one volt would have the same kinetic energy, because it carries the same charge as the electron (except for the sign). The proton would be moving more slowly after accelerating through one volt, due to its larger mass, but it would still gain one electron volt, or one proton volt, of kinetic energy.

20.2 • POTENTIAL DIFFERENCES IN A UNIFORM ELECTRIC FIELD

In this section we shall describe the potential difference between any two points in a *uniform* electric field. The potential difference is independent of the path between the two points; that is, the work done in taking a test charge from point A to point B is the same along all paths. This confirms that a static, uniform electric field is conservative.

First, consider a uniform electric field directed along the negative y axis, as in Figure 20.1a. Let us calculate the potential difference between two points, A and B, separated by the distance d, where d is measured parallel to the field lines. If we apply Equation 20.3 to this situation, we get

$$V_B - V_A = \Delta V = -\int_A^B \mathbf{E} \cdot d\mathbf{s} = -\int_A^B E \cos 0° \, d\mathbf{s} = -\int_A^B E \, ds$$

Because E is constant, it can be removed from the integral sign, giving

$$\Delta V = -E \int_A^B ds = -Ed \qquad\qquad \text{[20.6]}$$

• *Potential difference between two points in a uniform $\mathbf{E}$ field*

The minus sign results from the fact that point B is at a lower potential than point A; that is, $V_B < V_A$. In general, **electric field lines always point in the direction of decreasing electric potential,** as shown in Figure 20.1a.

Now suppose that a test charge q_0 moves from A to B. The change in its electric potential energy can be found from Equations 20.3 and 20.6:

$$\Delta U = q_0 \, \Delta V = -q_0 Ed \qquad\qquad \text{[20.7]}$$

From this result, we see that if q_0 is positive, ΔU is negative. This means that **when a positive charge moves in the direction of the electric field, its electric potential energy decreases.** (This is analogous to the decrease in gravitational potential energy of a mass falling in a gravitational field as suggested in Fig. 20.1b.) If a positive test charge is released from rest in the electric field, it experiences an electric force $q_0\mathbf{E}$ in the direction of $\mathbf{E}$ (downward in Fig. 20.1a). Therefore, it accelerates downward, gaining kinetic energy. **As the charged particle gains kinetic energy, it loses an equal amount of potential energy.**

Figure 20.1 (a) When the electric field $\mathbf{E}$ is directed downward, point B is at a lower electric potential than point A. A positive test charge that moves from A to B loses electric potential energy. (b) A mass m moving downward in the direction of the gravitational field $\mathbf{g}$ loses gravitational potential energy.

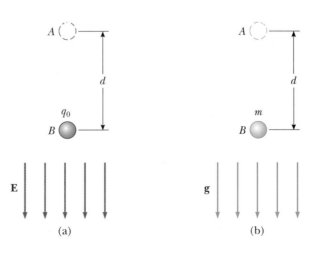

(a)   (b)

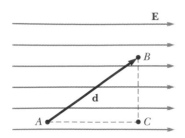

Figure 20.2   A uniform electric field directed along the positive *x* axis. Point *B* is at a lower potential than point *A*. Points *B* and *C* are at the *same* potential.

An equipotential surface •

If q_0 is negative, then ΔU is positive and the situation is reversed. **A negative charge gains electric potential energy when it moves in the direction of the electric field.** If a negative charge is released from rest in the field **E**, it accelerates in a direction opposite the electric field.

Now consider the more general case of a charged particle moving between any two points in a uniform electric field directed along the *x* axis, as in Figure 20.2. If **d** represents the displacement vector between points *A* and *B*, Equation 20.3 gives

$$\Delta V = -\int_A^B \mathbf{E} \cdot d\mathbf{s} = -\mathbf{E} \cdot \int_A^B d\mathbf{s} = -\mathbf{E} \cdot \mathbf{d} \qquad [20.8]$$

where again we are able to remove **E** from the integral, because it is constant. Furthermore, the change in electric potential energy of the charge is

$$\Delta U = q_0 \, \Delta V = -q_0 \mathbf{E} \cdot \mathbf{d} \qquad [20.9]$$

Finally, our results show that all points in a plane *perpendicular* to a uniform electric field are at the same potential. This can be seen in Figure 20.2, where the potential difference $V_B - V_A$ is equal to $V_C - V_A$. Therefore, $V_B = V_C$. **The name equipotential surface is given to any surface consisting of a continuous distribution of points having the same electrical potential.** Note that because $\Delta U = q_0 \, \Delta V$, no work is done in moving a test charge between any two points on an equipotential surface. The equipotential surfaces of a uniform electric field consist of a family of planes, all perpendicular to the field. Equipotential surfaces for fields with other symmetries will be described in later sections.

CONCEPTUAL PROBLEM 1

If a proton is released from rest in a uniform electric field, does its electric potential energy increase or decrease?

Example 20.1   The Electric Field Between Two Parallel Plates of Opposite Charge

A 12-V battery is connected between two parallel plates, as in Figure 20.3. The separation between the plates is 0.30 cm, and the electric field is assumed to be uniform. (This assumption is reasonable if the plate separation is small relative to the plate size and if we do not consider points near the edges of the plates.) Find the magnitude of the electric field between the plates.

Solution   The electric field is directed from the positive plate toward the negative plate. We see that the positive plate is at a higher potential than the negative plate. Note that the potential difference between plates must equal the potential dif-

ference between the battery terminals. This can be understood by noting that all points on a conductor in equilibrium are at the same potential,[1] and hence there is no potential difference between a terminal of the battery and any portion of the plate to which it is connected. Therefore, the magnitude of the electric field between the plates is

$$E = \frac{|V_B - V_A|}{d} = \frac{12 \text{ V}}{0.30 \times 10^{-2} \text{ m}} = 4.0 \times 10^3 \text{ V/m}$$

This configuration, which is called a *parallel-plate capacitor*, is examined in more detail later in this chapter.

---

[1]The electric field vanishes within a conductor in electrostatic equilibrium, and so the path integral $\int \mathbf{E} \cdot d\mathbf{s}$ between any two points within the conductor must be zero. A fuller discussion of this point is given in Section 20.6.

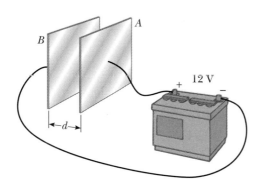

Figure 20.3 (Example 20.1) A 12-V battery connected to two parallel plates. The electric field between the plates has a magnitude given by the potential difference divided by the plate separation d.

Example 20.2 Motion of a Proton in a Uniform Electric Field

A proton is released from rest in a uniform electric field of magnitude 8.0×10^4 V/m directed along the positive x axis (Fig. 20.4). The proton undergoes a displacement of 0.50 m in the direction of **E**. (a) Find the change in the electric potential between the points A and B.

Solution   From Equation 20.6 we have

$$\Delta V = -Ed = -\left(8.0 \times 10^4 \, \frac{V}{m}\right)(0.50 \text{ m}) - -4.0 \times 10^4 \text{ V}$$

This negative result tells us that the electric potential decreases as the proton moves from A to B.

(b) Find the change in potential energy for this displacement.

Solution

$$\Delta U = q_0 \, \Delta V = e \, \Delta V$$
$$= (1.6 \times 10^{-19} \text{ C})(-4.0 \times 10^4 \text{ V})$$
$$= -6.4 \times 10^{-15} \text{ J}$$

The negative sign here means that the potential energy decreases as the proton moves in the direction of the electric field. This makes sense, because as the proton accelerates in

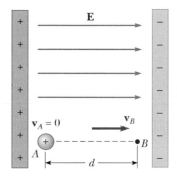

Figure 20.4 (Example 20.2) A proton accelerates from A to B in the direction of the electric field.

the direction of the field, it gains kinetic energy and at the same time (the proton–field system) loses electrical potential energy (energy is conserved).

EXERCISE 1   Apply the principle of energy conservation to find the speed of the proton after it has moved 0.50 m, starting from rest.   Answer 2.77×10^6 m/s

EXERCISE 2   The difference in potential between the deflection plates of a cathode ray tube is 300 V. If the distance between these plates is 2.0 cm, find the magnitude of the uniform electric field in this region.   Answer 1.5×10^4 N/C

20.3 • ELECTRIC POTENTIAL AND ELECTRIC POTENTIAL ENERGY DUE TO POINT CHARGES

Consider an isolated positive point charge, q (Fig. 20.5). Recall that such a charge produces an electric field that is directed radially outward from the charge. In order to find the electric potential at a distance of r from the charge, we begin with the general expression for potential difference:

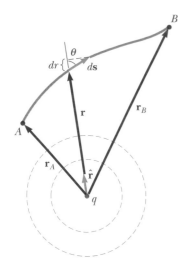

Figure 20.5 The potential difference between points A and B due to a point charge q depends *only* on the initial and final radial coordinates, r_A and r_B, respectively. The two dashed circles represent cross-sections of spherical equipotential surfaces.

$$V_B - V_A = -\int_A^B \mathbf{E} \cdot d\mathbf{s}$$

Because the electric field due to the point charge is given by $\mathbf{E} = kq\hat{\mathbf{r}}/r^2$, where $\hat{\mathbf{r}}$ is a unit vector directed from the charge to the field point, the quantity $\mathbf{E} \cdot d\mathbf{s}$ can be expressed as

$$\mathbf{E} \cdot d\mathbf{s} = k_e \frac{q}{r^2} \hat{\mathbf{r}} \cdot d\mathbf{s}$$

The dot product $\hat{\mathbf{r}} \cdot d\mathbf{s} = ds \cos\theta$, where θ is the angle between $\hat{\mathbf{r}}$ and $d\mathbf{s}$, as in Figure 20.5. Furthermore, note that $ds \cos\theta$ is the projection of $d\mathbf{s}$ onto r, so that $ds \cos\theta = dr$. That is, any displacement $d\mathbf{s}$ produces the change dr in the magnitude of $\mathbf{r}$. With these substitutions, we find that $\mathbf{E} \cdot d\mathbf{s} = (k_e q/r^2) dr$, so that the expression for the potential difference becomes

$$V_B - V_A = -\int E_r \, dr = -k_e q \int_{r_A}^{r_B} \frac{dr}{r^2} = \frac{k_e q}{r} \Big]_{r_A}^{r_B}$$

$$V_B - V_A = k_e q \left[\frac{1}{r_B} - \frac{1}{r_A} \right] \qquad [20.10]$$

The line integral of $\mathbf{E} \cdot d\mathbf{s}$ is *independent* of the path between A and B—as it must be, because the electric field of a point charge is conservative. Furthermore, Equation 20.10 expresses the important result that the potential difference between any two points A and B depends *only* on the *radial* coordinates r_A and r_B. As we learned in Section 20.1, it is customary to choose the reference of potential to be zero at $r_A = \infty$. With this choice, the electric potential due to a point charge at any distance r from the charge is

Potential of a point charge •

$$V = k_e \frac{q}{r} \qquad [20.11]$$

From this we see that V is constant on a spherical surface of radius r. Hence, we conclude that **the equipotential surfaces for an isolated point charge consist of a family of spheres concentric with the charge,** as shown in Figure 20.5. Note that the equipotential surfaces are perpendicular to the lines of electric force, as was the case for a uniform electric field.

The electric potential of two or more point charges is obtained by applying the superposition principle. That is, the total potential at some point P due to several point charges is the sum of the potentials due to the individual charges. For a group of charges, we can write the total potential at P in the form

The potential of several point •
charges

$$V = k_e \sum_i \frac{q_i}{r_i} \qquad [20.12]$$

where the potential is again taken to be zero at infinity and r_i is the distance from the point P to the charge q_i. Note that the sum in Equation 20.12 is an *algebraic sum* of scalars rather than a vector sum (which is used to calculate the electric field of a group of charges). Thus, it is much easier to evaluate V than to evaluate $\mathbf{E}$.

We now consider the electric potential energy of interaction of a system of charged particles. If V_1 is the electric potential due to charge q_1 at point P, then the work required to bring a second charge q_2 from infinity to point P without acceleration is $q_2 V_1$. By definition, this work equals the potential energy U of the two-particle system when the particles are separated by a distance of r_{12} (Fig. 20.6). We can therefore express the potential energy as

$$U = q_2 V_1 = k_e \frac{q_1 q_2}{r_{12}}$$ [20.13]   • *Electric potential energy of two charges*

Note that if the charges are of the same sign, U is positive. This is consistent with the fact that like charges repel, and so positive work must be done *on* the system to bring the two charges near one another. If the charges are of opposite sign, the force is attractive and U is negative. This means that negative work must be done to bring the unlike charges near one another.

If the system consists of more than two charged particles, the total electric potential energy can be obtained by calculating U for every pair of charges and summing the terms algebraically. The total electric potential energy of a system of point charges is equal to the work required to bring the charges, one at a time, from an infinite separation to their final positions.

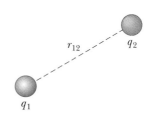

Figure 20.6 If two point charges are separated by a distance r_{12}, the potential energy of the pair of charges is given by $k_e q_1 q_2 / r_{12}$.

Thinking Physics 2

A spherical rubber balloon contains a charged object, which is mounted so that it always stays at the center of the balloon. As the balloon is inflated to a larger volume, what happens to the potential at the surface of the balloon? To the electric field at the surface of the balloon? To the total electric flux through the balloon? Answer the same questions for a *metallic* balloon being inflated.

Reasoning The potential at the surface of the balloon will decrease in inverse proportion to the radius. The electric field will decrease as the reciprocal of the squared radius of the balloon. The electric flux will remain constant, because all electric field lines must pass through the balloon, regardless of the size of the balloon. The answers will be exactly the same for the metallic balloon—the metal makes no difference, except for the fact that there will be no electric field within the skin of the balloon.

CONCEPTUAL PROBLEM 2

If the electric potential at some point is zero, can you conclude that there are no charges in the vicinity of that point?

Example 20.3   The Potential Due to Two Point Charges

A 2.00-μC point charge is located at the origin, and a second point charge of -6.00 μC is located on the y axis at the position (0, 3.00) m, as in Figure 20.7a. (a) Find the total electric potential due to these charges at the point P, the coordinates of which are (4.00, 0) m.

Solution For two charges, the sum in Equation 20.12 gives

$$V_P = k_e \left(\frac{q_1}{r_1} + \frac{q_2}{r_2} \right)$$

In this example, $q_1 = 2.00$ μC, $r_1 = 4.00$ m, $q_2 = -6.00$ μC, and $r_2 = 5.00$ m. Therefore, V_P reduces to

$$V_P = 8.99 \times 10^9 \ \frac{\text{N} \cdot \text{m}^2}{\text{C}^2} \left(\frac{2.00 \times 10^{-6} \ \text{C}}{4.00 \ \text{m}} - \frac{6.00 \times 10^{-6} \ \text{C}}{5.00 \ \text{m}} \right)$$

$$= -6.29 \times 10^3 \ \text{V}$$

(b) How much work is required to bring a 3.00-μC point charge from infinity to the point P?

Solution

$$W = q_3 V_P = (3.00 \times 10^{-6}\ \text{C})(-6.29 \times 10^3\ \text{V})$$

$$= -18.9 \times 10^{-3}\ \text{J}$$

The negative sign means that work is done by the field on the charge as it is displaced from infinity to P. Therefore, positive

work would have to be done by an external agent to remove the charge from P back to infinity.

EXERCISE 3   Find the total potential energy of the system of three charges in the configuration shown in Figure 20.7b.
Answer $-5.48 \times 10^{-2}\ \text{J}$

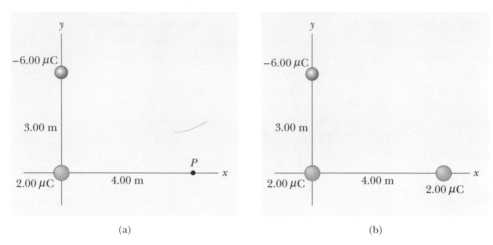

(a)   (b)

Figure 20.7   (Example 20.3) (a) The electric potential at the point P due to the two point charges q_1 and q_2 is the algebraic sum of the potentials due to the individual charges. (b) What is the potential energy of the system of three charges?

EXERCISE 4   (a) Find the potential at a distance of 1.0 cm from a proton. (b) What is the potential difference between two points that are 1.0 cm and 2.0 cm from a proton? (Take $V = 0$ at $r = \infty$)   Answer   (a) $1.4 \times 10^{-7}\ \text{V}$ (b) $-7.2 \times 10^{-8}\ \text{V}$

20.4 • OBTAINING E FROM THE ELECTRIC POTENTIAL

The electric field **E** and the electric potential V are related by Equation 20.3. Both quantities are determined by a specific charge distribution. We now show how to calculate the electric field if the electric potential is known in a certain region.

From Equation 20.3 we can express the potential difference dV between two points a distance ds apart as

$$dV = -\mathbf{E} \cdot d\mathbf{s} \qquad [20.14]$$

If the electric field has only *one* component, E_x, then $\mathbf{E} \cdot d\mathbf{s} = E_x\ dx$. Therefore, Equation 20.14 becomes $dV = -E_x\ dx$, or

$$E_x = -\frac{dV}{dx} \qquad [20.15]$$

That is, the electric field is equal to the negative of the derivative of the electric potential with respect to some coordinate. Note that the potential change is zero

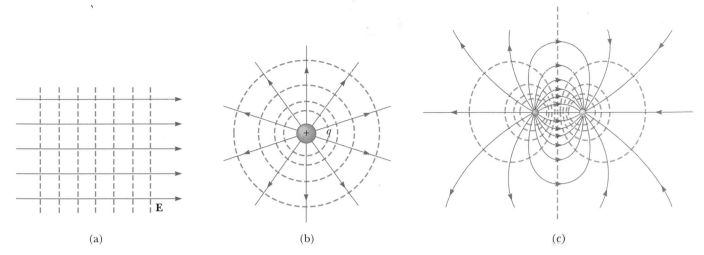

Figure 20.8   Equipotential surfaces (dashed blue lines) and electric field lines (red lines) for (a) a uniform electric field produced by an infinite sheet of charge, (b) a point charge, and (c) an electric dipole. In all cases, the equipotential surfaces are *perpendicular* to the electric field lines at every point.

for any displacement perpendicular to the electric field. This is consistent with the notion of equipotential surfaces being perpendicular to the field, as in Figure 20.8a.

If the charge distribution has spherical symmetry, where the charge density depends only on the radial distance, r, then the electric field is radial. In this case, $\mathbf{E} \cdot d\mathbf{s} = E_r\, dr$, and so we can express dV in the form $dV = -F_r\, dr$. Therefore,

$$E_r = -\frac{dV}{dr} \qquad\qquad [20.16]$$

For example, the potential of a point charge is $V = k_e q / r$. Because V is a function of r only, the potential function has spherical symmetry. Applying Equation 20.16, we find that the electric field due to the point charge is $E_r = k_e q / r^2$, a familiar result. Note that the potential changes only in the radial direction, not in a direction perpendicular to r. Thus, V (like E_r) is a function only of r. Again, this is consistent with the idea that **equipotential surfaces are perpendicular to field lines.** In this case the equipotential surfaces are a family of spheres concentric with the spherically symmetric charge distribution (Fig. 20.8b). The equipotential surfaces for the electric dipole are sketched in Figure 20.8c.

CONCEPTUAL PROBLEM 3

Can electric field lines ever cross? Why or why not? Can equipotential surfaces ever cross? Why or why not?

CONCEPTUAL PROBLEM 4

Suppose you know the value of the electric potential at one point. Can you find the electric field at that point from this information?

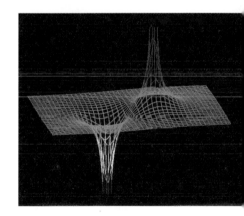

Computer-generated plot of the electric potential associated with an electric dipole. The charges lie in the horizontal plane, at the centers of the potential spikes. The contour lines help visualize the size of the potential the values of which are plotted vertically.   *(Richard Megna, Fundamental Photographs, NYC)*

CONCEPTUAL PROBLEM 5

If the electric potential is constant in some region, what can you conclude about the electric field in that region? If the electric field is zero in some region, what can you say about the electric potential in that region?

Example 20.4   The Electric Potential of a Dipole

An electric dipole consists of two equal and opposite charges separated by a distance $2a$, as in Figure 20.9. The dipole is along the x axis and is centered at the origin. Calculate the electric potential and the electric field at P.

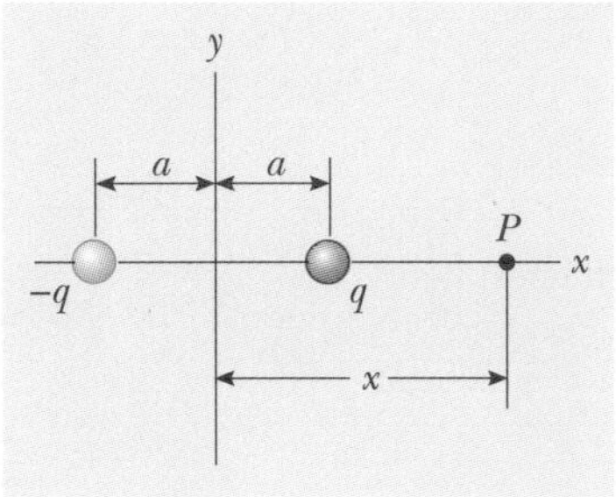

Solution

$$V = k_e \sum \frac{q_i}{r_i} = k_e \left(\frac{q}{x-a} - \frac{q}{x+a} \right) = \frac{2k_e qa}{x^2 - a^2}$$

If P is far from the dipole, so that $x \gg a$, then a^2 can be neglected in the term $x^2 - a^2$ and V becomes

$$V \approx \frac{2k_e qa}{x^2} \qquad (x \gg a)$$

Using Equation 20.15 and this result, we calculate the electric field at P:

$$E = -\frac{dV}{dx} = \frac{4k_e qa}{x^3} \qquad \text{for } x \gg a$$

Figure 20.9   (Example 20.4) An electric dipole located on the x axis.

20.5 • ELECTRIC POTENTIAL DUE TO CONTINUOUS CHARGE DISTRIBUTIONS

The electric potential due to a continuous charge distribution can be calculated in two ways. If the charge distribution is known, we can start with Equation 20.11 for the potential of a point charge. We then consider the potential due to a small charge element dq, treating this element as a point charge (Fig. 20.10). The potential dV at some point P due to the charge element dq is

$$dV = k_e \frac{dq}{r} \qquad\qquad \textbf{[20.17]}$$

where r is the distance from the charge element to P. To get the total potential at P, we integrate Equation 20.17 to include contributions from all elements of the charge distribution. Because each element is, in general, at a different distance from P and because k_e is a constant, we can express V as

$$V = k_e \int \frac{dq}{r} \qquad\qquad \textbf{[20.18]}$$

In effect, we have replaced the sum in Equation 20.12 with an integral. Note that this expression for V uses a particular reference: The potential is taken to be zero when P is infinitely far from the charge distribution.

The second method for calculating the potential of a continuous charge distribution makes use of Equation 20.3. This procedure is useful when the electric

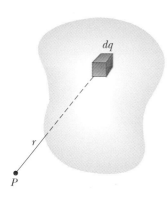

Figure 20.10   The electric potential at the point P due to a continuous charge distribution can be calculated by dividing the charged body into segments of charge dq and summing the potential contributions over all segments.

field is already known from other considerations, such as Gauss's law. If the charge distribution is highly symmetric, we first evaluate **E** at any point using Gauss's law and then substitute the obtained value into Equation 20.3 to determine the potential difference between any two points. We then choose V to be zero at some convenient point. Let us illustrate both methods with two examples.

Example 20.5   Potential Due to a Uniformly Charged Ring

Find the electric potential at a point P located on the axis of a uniformly charged ring of radius a and total charge Q. The plane of the ring is chosen perpendicular to the x axis (Fig. 20.11).

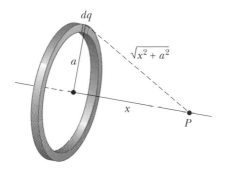

Figure 20.11   (Example 20.5) A uniformly charged ring of radius a, the plane of which is perpendicular to the x axis. All segments of the ring are at the same distance from any axial point P.

Reasoning and Solution   Let us take P to be at a distance x from the center of the ring, as in Figure 20.11. The charge element dq is at a distance equal to $\sqrt{x^2 + a^2}$ from the point P. Hence, we can express V as

$$V = k_e \int \frac{dq}{r} = k_e \int \frac{dq}{\sqrt{x^2 + a^2}}$$

In this case, each element dq is at the same distance from P.

Therefore, the term $\sqrt{x^2 + a^2}$ can be removed from the integral, and V reduces to

$$V = \frac{k_e}{\sqrt{x^2 + a^2}} \int dq = \frac{k_e Q}{\sqrt{x^2 + a^2}}$$

The only variable in this expression for V is x. This is not surprising, because our calculation is valid only for points along the x axis, where y and z are both zero. From the symmetry, we see that along the x axis **E** can have only an x component. Therefore, we can use Equation 20.15 to find the electric field at P:

$$E_x = -\frac{dV}{dx} = -k_e Q \frac{d}{dx} (x^2 + a^2)^{-1/2}$$

$$= -k_e Q \left(-\frac{1}{2}\right)(x^2 + a^2)^{-3/2}(2x)$$

$$= \frac{k_e Q x}{(x^2 + a^2)^{3/2}}$$

This result agrees with that obtained by direct integration (see Example 19.5). Note that $E_x = 0$ at $x = 0$ (the center of the ring). Could you have guessed this from Coulomb's law?

EXERCISE 5   What is the electric potential at the center of the uniformly charged ring? What does the field at the center imply about this result?   Answer $V = k_e Q/a$ at $x = 0$. Because $E = 0$, V must have a maximum or minimum value; it is in fact a maximum.

Example 20.6   Potential of a Uniformly Charged Sphere

An insulating solid sphere of radius R has a uniform positive charge density with a total charge of Q (Fig. 20.12a). (a) Find the electric potential at a point outside the sphere—that is, for $r > R$. Take the potential to be zero at $r = \infty$.

Solution   In Example 19.8 we found from Gauss's law that the magnitude of the electric field outside a uniformly charged sphere is

$$E_r = k_e \frac{Q}{r^2} \qquad \text{(for } r > R)$$

where the field is directed radially outward when Q is positive. To obtain the potential at an exterior point, such as B in Figure 20.12a, we substitute this expression for E into Equation 20.4. Because $\mathbf{E} \cdot d\mathbf{s} = E_r \, dr$ in this case, we get

$$V_B = -\int_\infty^r E_r \, dr = -k_e Q \int_\infty^r \frac{dr}{r^2}$$

$$V_B = k_e \frac{Q}{r} \qquad \text{(for } r > R)$$

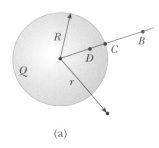

(a)

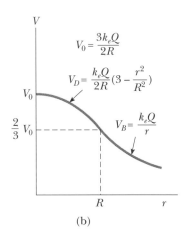

(b)

Figure 20.12   (Example 20.6) (a) A uniformly charged insulating sphere of radius R and total charge Q. The electric potential at points B and C is equivalent to that of a point charge Q located at the center of the sphere. (b) A plot of the electric potential V versus the distance r from the center of a uniformly charged, insulating sphere of radius R. The curve for V_D inside the sphere is parabolic and joins smoothly with the curve for V_B outside the sphere, which is a hyperbola. The potential has a maximum value V_0 at the center of the sphere.

Note that the result is identical to that for the electric potential due to a point charge. Because the potential must be continuous at $r = R$, we can use this expression to obtain the potential at the surface of the sphere. That is, the potential at a point such as C in Figure 20.12a is

$$V_C = k_e \frac{Q}{R} \qquad \text{(for } r = R)$$

(b) Find the potential at a point inside the charged sphere—that is, for $r < R$.

Solution   In Example 19.8 we found that the electric field inside a uniformly charged sphere is

$$E_r = \frac{k_e Q}{R^3} r \qquad \text{(for } r < R)$$

We can use this result and Equation 20.3 to evaluate the potential difference $V_D - V_C$, where D is an interior point:

$$V_D - V_C = -\int_R^r E_r \, dr = -\frac{k_e Q}{R^3} \int_R^r r \, dr = \frac{k_e Q}{2R^3} (R^2 - r^2)$$

Substituting $V_C = k_e Q/R$ into this expression and solving for V_D, we get

$$V_D = \frac{k_e Q}{2R} \left(3 - \frac{r^2}{R^2} \right) \qquad \text{(for } r < R)$$

At $r = R$, this expression gives a result for the potential that agrees with that for the potential at the surface—that is, V_C. A plot of V versus r for this charge distribution is given in Figure 20.12b.

EXERCISE 6   What are the electric field and electric potential at the center of a uniformly charged sphere?

Answer: $E = 0$ and $V_0 = 3k_e Q/2R$

PROBLEM-SOLVING STRATEGY AND HINTS

1. When working problems involving electric potential, remember that it is a *scalar quantity,* and so there are no components to worry about. Therefore, when using the superposition principle to evaluate the electric potential at a point due to a system of point charges, simply take the algebraic sum of the potentials due to each charge. You must keep track of signs, however. The potential for each positive charge ($V = k_e q/r$) is positive, whereas the potential for each negative charge is negative.

2. Just as with potential energy in mechanics, only *changes* in electric potential are significant; hence, the point at which the potential is set at zero is arbitrary. When dealing with point charges or a finite-sized charge dis-

tribution, we usually define $V = 0$ to be at a point infinitely far from the charges. However, if the charge distribution itself extends to infinity, some other nearby point must be selected as the reference point.

3. The electric potential at some point P due to a continuous distribution of charge can be evaluated by dividing the charge distribution into infinitesimal elements of charge dq located at a distance of r from the point P. This element is then treated as a point charge, and so the potential at P due to the element is $dV = k_e \, dq/r$. The total potential at P is obtained by integrating dV over the entire charge distribution. For most problems, it is necessary in performing the integration to express dq and r in terms of a single variable. In order to simplify the integration, it is important to give careful consideration to the geometry involved in the problem. Review Example 20.5 as a guide for using this method.

4. Another method that can be used to obtain the potential due to a finite continuous charge distribution is to start with the definition of the potential difference given by Equation 20.3. If $\mathbf{E}$ is known or can be obtained easily (say, from Gauss's law), then the line integral of $\mathbf{E} \cdot d\mathbf{s}$ can be evaluated. Example 20.6 uses this method.

5. Once you know the electric potential at a point, it is possible to obtain the electric field at that point by remembering that **the electric field is equal to the negative of the derivative of the potential with respect to some coordinate.** Example 20.5 illustrates how to use this procedure.

20.6 · ELECTRIC POTENTIAL OF A CHARGED CONDUCTOR

In Chapter 19 we found that when a solid conductor in electrostatic equilibrium carries a net charge, the charge resides on the outer surface of the conductor. Furthermore, we showed that the electric field just outside the surface of a conductor in equilibrium is perpendicular to the surface, whereas the field *inside* the conductor is zero. If the electric field had a component parallel to the surface, this would cause surface charges to move, creating a current and nonequilibrium.

We shall now show that **every point on the surface of a charged conductor in electrostatic equilibrium is at the same electric potential.** Consider two points A and B on the surface of a charged conductor, as in Figure 20.13. Along a surface path connecting these points, $\mathbf{E}$ is always perpendicular to the displacement, $d\mathbf{s}$; therefore, $\mathbf{E} \cdot d\mathbf{s} = 0$. Using this result and Equation 20.3, we conclude that the potential difference between A and B is necessarily zero. That is,

$$V_B - V_A = -\int_A^B \mathbf{E} \cdot d\mathbf{s} = 0$$

This result applies to *any* two points on the surface. Therefore, V is constant everywhere on the surface of a charged conductor in equilibrium, and so such a surface is an equipotential surface. Furthermore, because the electric field is zero inside the conductor, we conclude that the potential is constant everywhere inside the conductor and equal to its value at the surface. It follows that no work is required

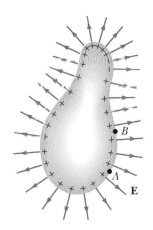

Figure 20.13 An arbitrarily shaped conductor with an excess positive charge. When the conductor is in electrostatic equilibrium, all of the charge resides at the surface, $\mathbf{E} = 0$ inside the conductor, and the electric field just outside the conductor is perpendicular to the surface. The potential is constant inside the conductor and is equal to the potential at the surface. The surface charge density is nonuniform.

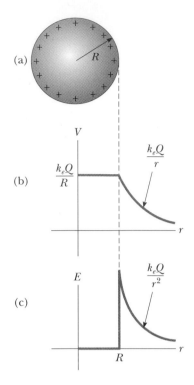

Figure 20.14 (a) The excess charge on a conducting sphere of radius R is uniformly distributed on its surface. (b) The electric potential versus the distance r from the center of the charged conducting sphere. (c) The electric field intensity versus the distance r from the center of the charged conducting sphere.

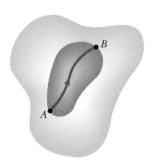

Figure 20.15 A conductor in electrostatic equilibrium containing an empty cavity. The electric field in the cavity is *zero* regardless of the charge of the conductor.

to move a test charge from the interior of a charged conductor to its surface. (Note that the potential is not zero inside the conductor, even though the electric field is zero.)

For example, consider a solid metal sphere of radius R and total positive charge Q, as shown in Figure 20.14a. The electric field outside the sphere is $k_e Q / r^2$ and points radially outward. Following Example 20.6, we see that the potential at the interior and surface of the sphere must be $k_e Q / R$ relative to infinity. The potential outside the sphere is $k_e Q / r$. Figure 20.14b is a plot of the potential as a function of r, and Figure 20.14c shows the variations of the electric field with r.

When a net charge is placed on a spherical conductor, the surface charge density is uniform, as indicated in Figure 20.14a. However, if the conductor is nonspherical, as in Figure 20.13, the surface charge density is high where the radius of curvature is small and convex and low where the radius of curvature is large and concave. Because the electric field just outside a charged conductor is proportional to the surface charge density, σ, we see that **the electric field is large near points having small convex radii of curvature and reaches very high values at sharp points.**

A Cavity Within a Conductor

Now consider a conductor of arbitrary shape containing a cavity, as in Figure 20.15. Let us assume there are no charges inside the cavity. We shall show that **the electric field inside the cavity must be zero,** regardless of the charge distribution on the outside surface of the conductor. Furthermore, the field in the cavity is zero even if an electric field exists outside the conductor.

In order to prove this point, we shall use the fact that every point on the conductor is at the same potential, and therefore any two points A and B on the surface of the cavity must be at the same potential. Now imagine that a field $\mathbf{E}$ exists in the cavity, and evaluate the potential difference $V_B - V_A$, defined by the expression

$$V_B - V_A = -\int_A^B \mathbf{E} \cdot d\mathbf{s}$$

If $\mathbf{E}$ is nonzero, we can invariably find a path between A and B for which $\mathbf{E} \cdot d\mathbf{s}$ is always a positive number (a path along the direction of $\mathbf{E}$), and so the integral must be positive. However, because $V_B - V_A = 0$, the integral must also be zero. This contradiction can be reconciled only if $\mathbf{E} = 0$ inside the cavity. Thus, we conclude that a cavity surrounded by conducting walls is a field-free region as long as there are no charges inside the cavity.

This result has some interesting applications. For example, it is possible to shield an electronic circuit or even an entire laboratory from external fields by surrounding it with conducting walls. Shielding is often necessary during highly sensitive electrical measurements.

Thinking Physics 3

Why is the end of a lightning rod pointed?

Reasoning The role of a lightning rod is to serve as a point at which the lightning strikes, so that the charge delivered by the lightning will pass safely to the ground. A lightning strike begins when a channel of ionized air, called a *stepped leader,* starts to drop from the cloud. As it nears the ground, an upward-moving channel of ionized air, called a *return stroke,* jumps up to meet the stepped leader. The desire is for the return stroke to begin on a lightning rod, rather than on the rooftop of a house. If the lightning rod is pointed, then the electric field is very strong near the point, because the radius of curvature of the conductor is very small. This large electric field will cause large accelerations of stray electrons in the air and greatly increase the likelihood that ionization of the air to begin the return stroke will occur near the tip of the lightning rod rather than elsewhere.

CONCEPTUAL PROBLEM 6

Suppose you are sitting in a car and a 20 000-volt power line drops across the car. Should you stay in the car or get out? The power line potential is 20 000 volts compared to the potential of the ground.

20.7 · CAPACITANCE

Consider two conductors having a potential difference of ΔV between them. Let us assume that the conductors have equal and opposite charges, as in Figure 20.16. This can be accomplished by connecting two uncharged conductors to the terminals of a battery. As we learned in Example 20.1, such a combination of two conductors is called a **capacitor.** The potential difference ΔV is proportional to the magnitude of the charge Q on the capacitor. The **capacitance,** C, of a capacitor is defined as the ratio of the magnitude of the charge on either conductor to the magnitude of the potential difference between them:

$$C \equiv \frac{Q}{\Delta V} \qquad [20.19]$$

Note that by definition **capacitance is always a positive quantity.** Furthermore, because the potential difference increases as the stored charge increases (see Eq. 20.7), the ratio $Q/\Delta V$ is constant for a given capacitor. Therefore, the capacitance of a device is a measure of its ability to store charge and electrical potential energy.

From Equation 20.19, we see that capacitance has the SI units coulombs per volt, called the **farad** (F) in honor of Michael Faraday. That is,

$$[\text{Capacitance}] = 1 \text{ F} = 1 \text{ C/V}$$

The farad is a very large unit of capacitance. In practice, typical devices have capacitances ranging from microfarads to picofarads. Capacitors are often labeled "mF" for microfarads and "mmF" for micromicrofarads (picofarads).

As you will soon see, the capacitance of a device depends on the geometric arrangement of the conductors. To illustrate this point, let us calculate the capacitance of an isolated spherical conductor of radius R and charge Q. (The second conductor can be taken as a concentric hollow, conducting sphere of infinite radius.) Because the sphere's potential is simply $k_e Q/R$ (where $V = 0$ at infinity), its capacitance is

$$C = \frac{Q}{\Delta V} = \frac{Q}{k_e Q/R} = \frac{R}{k_e} = 4\pi\epsilon_0 R \qquad [20.20]$$

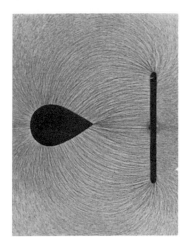

The electric field pattern of a charged conducting plate near an oppositely charged pointed conductor. Small pieces of thread suspended in oil align with the electric field lines. Note that the electric field is most intense near the pointed part of the conductor and at other points at which the radius of curvature is small.
(Courtesy of Harold M. Waage, Princeton University)

• *Definition of capacitance*

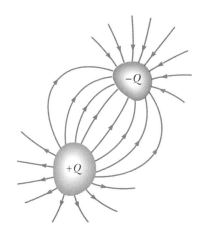

Figure 20.16 A capacitor consists of two conductors electrically isolated from each other and their surroundings. Once the capacitor is charged, the two conductors carry equal but opposite charges.

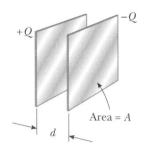

Figure 20.17 A parallel-plate capacitor consists of two parallel plates each of area A, separated by a distance d. When the capacitor is charged, the plates carry equal charges of opposite sign.

(Remember from Chapter 19, Section 19.4, that the Coulomb constant $k_e = 1/4\pi\epsilon_0$.) Equation 20.20 shows that the capacitance of an isolated charged sphere is proportional to the sphere's radius and is independent of both the charge and the potential difference.

Calculation of Capacitance

The capacitance of a pair of oppositely charged conductors can be calculated in the following manner. A convenient charge of magnitude Q is assumed, and the potential difference is calculated using the techniques described in Section 20.5. One then simply uses $C = Q/\Delta V$ to evaluate the capacitance. As you might expect, the calculation is relatively easy if the geometry of the capacitor is simple.

Let us illustrate this with two geometries with which we are all familiar: parallel plates and concentric cylinders. In these examples, we shall assume that the charged conductors are separated by a vacuum. (The effect of a dielectric material between the conductors will be treated in Section 20.10.)

The Parallel-Plate Capacitor

Two parallel plates of equal area A are separated by a distance of d, as in Figure 20.17. One plate has charge Q; the other, charge $-Q$. The charge per unit area on either plate is $\sigma = Q/A$. If the plates are very close together (compared with their length and width), we assume that the electric field is uniform between the plates and zero elsewhere. According to Example 19.10, the electric field between the plates is

$$E = \frac{\sigma}{\epsilon_0} = \frac{Q}{\epsilon_0 A}$$

The potential difference between the plates equals Ed; therefore,

$$\Delta V = Ed = \frac{Qd}{\epsilon_0 A}$$

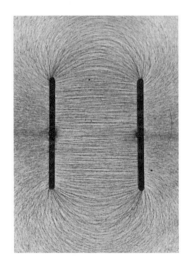

Figure 20.18 Electric field pattern of two oppositely charged conducting parallel plates. Small pieces of thread on an oil surface align with the electric field. Note the nonuniform nature of the electric field at the ends of the plates. Such end effects can be neglected if the plate separation is small compared to the length of the plates. *(Courtesy of Harold M. Waage, Princeton University)*

Substituting this result into Equation 20.19, we find that the capacitance is

$$C = \frac{Q}{\Delta V} = \frac{Q}{Qd/\epsilon_0 A}$$

$$C = \frac{\epsilon_0 A}{d} \qquad \text{[20.21]}$$

That is **the capacitance of a parallel-plate capacitor is proportional to the area of its plates and inversely proportional to the plate separation.**

As you can see from the definition of capacitance, $C = Q/\Delta V$, the amount of charge a given capacitor can store for a given potential difference across its plates increases as the capacitance increases. Therefore, it seems reasonable that a capacitor constructed from plates having large areas should be able to store a large charge. The amount of charge needed to produce a given potential difference increases with decreasing plate separation.

A careful inspection of the electric field lines for a parallel-plate capacitor reveals that the field is uniform in the central region between the plates. However, the field is nonuniform at the edges of the plates. Figure 20.18 is a photograph of the electric field pattern of a parallel-plate capacitor showing the nonuniform field lines at the plates' edges.

PROBLEM-SOLVING STRATEGY AND HINTS

In practice, a variety of phrases are used to describe the potential difference between two points, the most common being *voltage*. A voltage *applied* to a device or *across* a device has the same meaning as the potential difference across the device. For example, if we say that the voltage across a certain capacitor is 12 volts, we mean that the potential difference between the capacitor's plates is 12 volts.

CONCEPTUAL PROBLEM 7

Why is it dangerous to touch the terminals of a high-voltage capacitor even after the voltage source that charged the battery is disconnected from the capacitor? What can be done to make the capacitor safe to handle after the voltage source has been removed?

Example 20.7 Parallel-Plate Capacitor

A parallel-plate capacitor has an area $A = 2.00 \times 10^{-4}$ m^2 and a plate separation $d = 1.00$ mm. Find its capacitance.

Solution From Equation 20.21, we find

$$C = \epsilon_0 \frac{A}{d} = \left(8.85 \times 10^{-12} \, \frac{\text{C}^2}{\text{N} \cdot \text{m}^2} \right) \left(\frac{2.00 \times 10^{-4} \, \text{m}^2}{1.00 \times 10^{-3} \, \text{m}} \right)$$

$= 1.77 \times 10^{-12}$ F $= 1.77$ pF

EXERCISE 7 If the plate separation is increased to 3.00 mm, find the capacitance. Answer 0.590 pF

Example 20.8 The Cylindrical Capacitor

A cylindrical conductor of radius a and charge Q is coaxial with a larger cylindrical shell of radius b and charge $-Q$ (Fig. 20.19a). Find the capacitance of this cylindrical capacitor if its length is ℓ.

Reasoning and Solution If we assume that ℓ is long compared with a and b, we can neglect end effects. In this case, the field is perpendicular to the axis of the cylinders and is confined to the region between them (Fig. 20.19b). We must first calculate the potential difference between the two cylinders, which is given in general by

$$V_b - V_a = - \int_a^b \mathbf{E} \cdot d\mathbf{s}$$

where $\mathbf{E}$ is the electric field in the region $a < r < b$. In Chapter 19, using Gauss's law, we showed that the electric field of a

cylinder of charge per unit length λ is $E = 2k_e\lambda/r$. The same result applies here, because the outer cylinder does not contribute to the electric field inside it. Using this result and noting that $\mathbf{E}$ is along r in Figure 20.19b, we find that

$$V_b - V_a = - \int_a^b E_r \, dr = -2k_e\lambda \int_a^b \frac{dr}{r} = -2k_e\lambda \ln\left(\frac{b}{a}\right)$$

Substituting this into Equation 20.19 and using the fact that $\lambda = Q/\ell$, we get

$$C = \frac{Q}{|\Delta V|} = \frac{Q}{\dfrac{2k_e Q}{\ell} \ln\left(\dfrac{b}{a}\right)} = \frac{\ell}{2k_e \ln\left(\dfrac{b}{a}\right)} \quad \textbf{[20.22]}$$

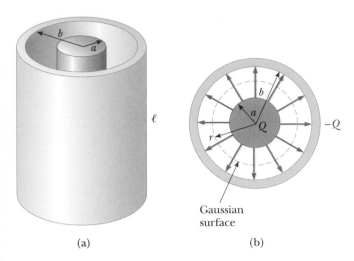

(a)   (b)

Figure 20.19 (Example 20.8) (a) A cylindrical capacitor consists of a cylindrical conductor of radius a and length ℓ surrounded by a coaxial cylindrical shell of radius b. (b) The end view of a cylindrical capacitor. The dashed line represents the end of the cylindrical gaussian surface of radius r and length ℓ.

where ΔV is the magnitude of the potential difference, given by $2k_e\lambda \ln (b/a)$, a positive quantity. That is, $\Delta V = V_a - V_b$ is positive because the inner cylinder is at the higher potential. Our result for C makes sense because it shows that the capacitance is proportional to the length of the cylinders. As you might expect, the capacitance also depends on the radii of the two cylindrical conductors. As an example, a coaxial cable consists of two concentric cylindrical conductors of radii a and b separated by an insulator. The cable carries currents in opposite directions in the inner and outer conductors. Such a geometry is especially useful for shielding an electrical signal from external influences. From Equation 20.22, we see that the capacitance per unit length of a coaxial cable is

$$\frac{C}{\ell} = \frac{1}{2k_e \ln\left(\dfrac{b}{a}\right)}$$

20.8 • COMBINATIONS OF CAPACITORS

Two or more capacitors are often combined in circuits in several ways. The equivalent capacitance of certain combinations can be calculated using methods described in this section. Figure 20.20 shows the circuit symbols for capacitors and batteries, together with their color codes. The positive terminal of the battery is at the higher potential and is represented by the longer vertical line in the battery symbol.

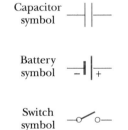

Figure 20.20 Circuit symbols for capacitors, batteries, and switches. Note that capacitors are in blue, and batteries and switches are in red.

Parallel Combination

Two capacitors connected as shown in Figure 20.21a are known as a **parallel combination** of capacitors. The left plates of the capacitors are connected by a conducting wire to the positive terminal of the battery, and therefore both are at the same potential. Likewise, the right plates are connected to the negative terminal of the battery. When the capacitors are first connected in the circuit, electrons are transferred through the battery from the left plates to the right plates, leaving the left plates positively charged and the right plates negatively charged. The energy source for this charge transfer is the internal chemical energy stored in the battery, which is converted to electric potential energy. The flow of charge ceases when the voltage across the capacitors is equal to that of the battery. The capacitors reach their maximum charge when the flow of charge ceases. Let us call the maximum charges on the two capacitors Q_1 and Q_2. Then the *total charge, Q,* stored by the two capacitors is

$$Q = Q_1 + Q_2 \qquad\qquad [20.23]$$

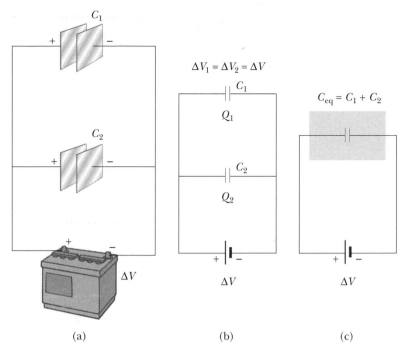

$\Delta V_1 = \Delta V_2 = \Delta V$

$C_{eq} = C_1 + C_2$

(a)   (b)   (c)

Figure 20.21 (a) A parallel combination of two capacitors. (b) The circuit diagram for the parallel combination. (c) The potential difference is the same across each capacitor, and the equivalent capacitance is $C_{eq} = C_1 + C_2$.

Suppose we wish to replace these two capacitors with one equivalent capacitor having the capacitance C_{eq}. This equivalent capacitor must have exactly the same external effect on the circuit as the original two. That is, it must store Q units of charge. We see from Figure 20.21b that **in the parallel circuit the potential differences across the two capacitors are the same,** and each is equal to the voltage of the battery, ΔV. From Figure 20.21c, we see that the voltage across the equivalent capacitor is also ΔV. Thus, we have

$$Q_1 = C_1 \Delta V \qquad Q_2 = C_2 \Delta V$$

and, for the equivalent capacitor,

$$Q = C_{eq} \Delta V$$

Substitution of these relations into Equation 20.23 gives

$$C_{eq} \Delta V = C_1 \Delta V + C_2 \Delta V$$

or

$$C_{eq} = C_1 + C_2 \qquad \text{(parallel combination)} \qquad [20.24]$$

If we extend this treatment to three or more capacitors connected in parallel, the equivalent capacitance is

$$C_{eq} = C_1 + C_2 + C_3 + \cdots \qquad \text{(parallel combination)} \qquad [20.25]$$

Thus, we see that **the equivalent capacitance of a parallel combination of capacitors is larger than any of the individual capacitances and is the algebraic sum of the individual capacitances.**

EXERCISE 8   Two capacitors, $C_1 = 5.0\ \mu F$ and $C_2 = 12\ \mu F$, are connected in parallel, and the resulting combination is connected to a 9.0-V battery. (a) What is the value of the equivalent capacitance of the combination? What is (b) the potential difference across each capacitor and (c) the charge stored on each capacitor?   Answer   (a) 17 μF (b) 9.0 V   (c) 45 μC and 108 μC

Series Combination

Now consider two capacitors connected in **series,** as illustrated in Figure 20.22a. For this series combination of capacitors, **the magnitude of the charge must be the same on all the plates.**

To see why this must be true, let us consider the charge transfer process in some detail. We start with uncharged capacitors and follow what happens just after a battery is connected to the circuit. When the connection is made, the right plate of C_1 and the left plate of C_2 form an isolated conductor. Thus, whatever negative charge enters one plate must be equal to the positive charge of the other plate, to maintain neutrality of the isolated conductor. As a result, both capacitors must have the same charge.

Suppose an equivalent capacitor performs the same function as the series combination. After it is fully charged, **the equivalent capacitor must end up with a charge of $-Q$ on its right plate and $+Q$ on its left plate.** By applying the definition of capacitance to the circuit shown in Figure 20.22b, we have

$$\Delta V = \frac{Q}{C_{eq}}$$

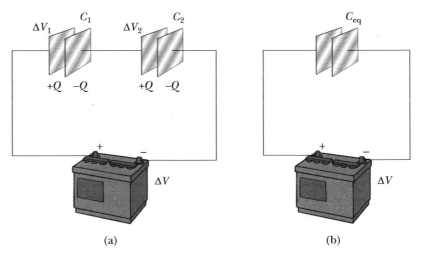

(a)   (b)

Figure 20.22   A series combination of two capacitors. The charge on each capacitor is the same, and the equivalent capacitance can be calculated from the relationship $\dfrac{1}{C_{eq}} = \dfrac{1}{C_1} + \dfrac{1}{C_2}$.

where ΔV is the potential difference between the terminals of the battery and C_{eq} is the equivalent capacitance. From Figure 20.22a, we see that

$$\Delta V = \Delta V_1 + \Delta V_2 \qquad \text{[20.26]}$$

where ΔV_1 and ΔV_2 are the potential differences across capacitors C_1 and C_2. In general the potential difference across any number of capacitors in series is equal to the sum of the potential differences across the individual capacitors. Because $Q = C \Delta V$ can be applied to each capacitor, the potential difference across each is

$$\Delta V_1 = \frac{Q}{C_1} \qquad \Delta V_2 = \frac{Q}{C_2}$$

$\Delta V = \dfrac{Q}{C_1 + C_2}$

Substituting these expressions into Equation 20.26, and noting that $\Delta V = Q/C_{eq}$, we have

$$\frac{Q}{C_{eq}} = \frac{Q}{C_1} + \frac{Q}{C_2}$$

$\Delta V_1 = Q/C_1$

Canceling Q, we arrive at the relationship

$$\frac{1}{C_{eq}} = \frac{1}{C_1} + \frac{1}{C_2} \qquad \text{(series combination)} \qquad \text{[20.27]}$$

If this analysis is applied to three or more capacitors connected in series, the equivalent capacitance is found to be

$U = 1/2 C \Delta V^2$

$$\frac{1}{C_{eq}} = \frac{1}{C_1} + \frac{1}{C_2} + \frac{1}{C_3} + \cdots \qquad \text{(series combination)} \qquad \text{[20.28]}$$

$1/C_{eq}$

This shows that **the equivalent capacitance of a series combination is always less than any individual capacitance in the combination and the inverse of the equivalent capacitance is the algebraic sum of the inverses of the individual capacitances.**

Example 20.9 Equivalent Capacitance

Find the equivalent capacitance between a and b for the combination of capacitors shown in Figure 20.23a. All capacitances are in microfarads.

Solution Using Equations 20.25 and 20.28, we reduce the combination step by step as indicated in the figure. The 1.0-μF and 3.0-μF capacitors are in parallel and combine according to $C_{eq} = C_1 + C_2$. Their equivalent capacitance is 4.0 μF. Likewise, the 2.0-μF and 6.0-μF capacitors are also in parallel and have an equivalent capacitance of 8.0 μF. The upper branch in Figure 20.23b now consists of two 4.0-μF capacitors in series, which combine according to

$$\frac{1}{C_{eq}} = \frac{1}{C_1} + \frac{1}{C_2} = \frac{1}{4.0 \ \mu\text{F}} + \frac{1}{4.0 \ \mu\text{F}} = \frac{1}{2.0 \ \mu\text{F}}$$

$$C_{eq} = 2.0 \ \mu\text{F}$$

Likewise, the lower branch in Figure 20.23b consists of two 8.0-μF capacitors in *series*, which give an equivalent of 4.0 μF. Finally, the 2.0-μF and 4.0-μF capacitors in Figure 20.23c are in parallel and have an equivalent capacitance of 6.0 μF. Hence, the equivalent capacitance of the circuit is 6.0 μF, as shown in Figure 20.23d.

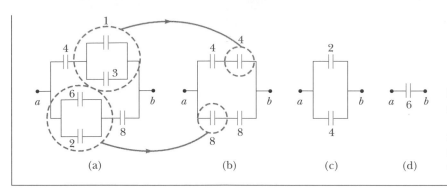

Figure 20.23 (Example 20.9) To find the equivalent combination of the capacitors in (a), the various combinations are reduced in steps as indicated in (b), (c), and (d), using the series and parallel rules described in the text.

EXERCISE 9 Consider three capacitors having capacitance of 3.0 μF, 6.0 μF, and 12 μF. Find their equivalent capacitance if they are connected (a) in parallel and (b) in series.
Answer   (a) 21 μF   (b) 1.7 μF

20.9 • ENERGY STORED IN A CHARGED CAPACITOR

Almost everyone who works with electronic equipment has at some time verified that a capacitor can store energy. If the plates of a charged capacitor are connected by a conductor, such as wire, charge transfers from one plate to the other until the two are uncharged. The discharge can often be observed as a visible spark. If you accidentally touched the opposite plates of a charged capacitor, your fingers would act as pathways by which the capacitor could discharge, resulting in an electric shock. The degree of shock would depend on the capacitance and voltage applied to the capacitor. Where high voltages are present, such as in the power supply of a television set, the shock could be fatal.

Consider a parallel-plate capacitor that is initially uncharged, so that the initial potential difference across the plates is zero. Now imagine that the capacitor is connected to a battery and develops a maximum charge of Q. We shall assume that the capacitor is charged *slowly* so that the problem can be treated as an electrostatic system. The final potential difference across the capacitor is $\Delta V = Q/C$. Because the initial potential difference is zero, the *average* potential difference during the charging process is $\Delta V/2 = Q/2C$. From this we might conclude that the work needed to charge the capacitor is $W = Q\,\Delta V/2 = Q^2/2C$. Although this result is correct, a more detailed proof is desirable.

Suppose that q is the charge on the capacitor at some instant during the charging process. At the same instant, the potential difference across the capacitor is $\Delta V = q/C$. The work that is necessary[2] to transfer an increment of charge dq from the plate of charge $-q$ to the plate of charge q (which is at the higher potential) is

$$dW = \Delta V\,dq = \frac{q}{C}\,dq$$

Thus, the total work required to charge the capacitor from $q = 0$ to some final charge $q = Q$ is

---

[2]One mechanical analog of this process is the work required to stretch or compress a spring through some small distance.

$$W = \int_0^Q \frac{q}{C} \, dq = \frac{Q^2}{2C}$$

But the work done in charging the capacitor can be considered as potential energy U stored in the capacitor. Using $Q = C\,\Delta V$, we can express the electrostatic energy stored in a charged capacitor in the following alternative forms:

$$U = \frac{Q^2}{2C} = \tfrac{1}{2}Q\,\Delta V = \tfrac{1}{2}C(\Delta V)^2 \qquad \textbf{[20.29]}$$

• *Energy stored in a charged capacitor*

This result applies to *any* capacitor, regardless of its geometry. We see that the stored energy increases as C increases and as the potential difference increases. In practice, there is a limit to the maximum energy (or charge) that can be stored. This is because electrical discharge ultimately occurs between the plates of the capacitor at a sufficiently large value of ΔV. For this reason, capacitors are usually labeled with a maximum operating voltage.

The energy stored in a capacitor can be thought of as stored in the electric field created between the plates as the capacitor is charged. This is reasonable in view of the fact that the magnitude of the electric field is proportional to the charge on the capacitor. For a parallel-plate capacitor, the potential difference is related to the electric field through the relationship $\Delta V = Ed$. Furthermore, the capacitance is $C = \epsilon_0 A/d$. Substituting these expressions into Equation 20.29 gives

$$U = \tfrac{1}{2}\frac{\epsilon_0 A}{d}(E^2 d^2) = \tfrac{1}{2}(\epsilon_0 Ad)E^2 \qquad \textbf{[20.30]}$$

• *Energy stored in a parallel-plate capacitor*

Because the volume of a parallel-plate capacitor that is occupied by the electric field is Ad, the energy per unit volume $u = U/Ad$, called the **energy density,** is

$$u = \tfrac{1}{2}\epsilon_0 E^2 \qquad \textbf{[20.31]}$$

• *Energy density in an electric field*

Although Equation 20.31 was derived for a parallel-plate capacitor, the expression is generally valid. That is, **the energy density in any electrostatic field is proportional to the square of the electric field intensity at a given point.**

Thinking Physics 4

You have three capacitors and two batteries. How should you connect the capacitors and batteries so that the capacitors will store the maximum possible energy?

Reasoning The energy stored in a capacitor is proportional to the capacitance and the square of the potential difference. Thus, we would like to maximize each of these quantities. We can do this by connecting the three capacitors in *parallel* (so that the capacitances add) across the two batteries in *series* (so that the potential differences add).

Thinking Physics 5

You charge a capacitor and then remove it from the battery. The capacitor consists of large movable plates with air between them. You pull the plates a bit farther apart. What happens to the charge on the capacitor? To the potential difference? To the energy? To the capacitance? To the electric field between the plates? Was work done in pulling the plates apart?

Reasoning Because the capacitor is removed from the battery, charges on the plates have nowhere to go. Thus, the charge on the capacitor remains the same as the plates are pulled apart. Because the electric field of large plates is independent of distance, the electric field remains constant. Because the electric field is a measure of the rate of change of potential with distance, the potential difference between the plates must increase as the separation distance increases. Because the same charge is stored at a higher potential difference, the capacitance has decreased. Because energy stored is proportional to both charge and potential difference, the energy stored in the capacitor must increase. This energy must have been transferred from somewhere, so work was done. This is consistent with the fact that the plates attract one another, so work must be done to pull them apart.

Example 20.10   Rewiring Two Charged Capacitors

Two capacitors C_1 and C_2 (where $C_1 > C_2$) are charged to the same potential difference ΔV_0 but with opposite polarity. The charged capacitors are removed from the battery, and their plates are connected as shown in Figure 20.24a. The switches S_1 and S_2 are then closed, as in Figure 20.24b. (a) Find the final potential difference between a and b after the switches are closed.

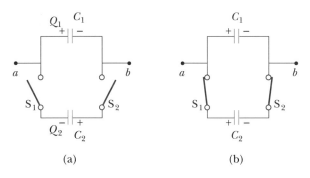

(a)   (b)

Figure 20.24   (Example 20.10)

Solution   The charges on the left-hand plates of the capacitors before the switches are closed are

$$Q_1 = C_1 \Delta V_0 \quad \text{and} \quad Q_2 = - C_2 \Delta V_0$$

The negative sign for Q_2 is necessary, because this capacitor's polarity is opposite that of capacitor C_1. After the switches are closed, the charges on the plates redistribute until the total charge Q shared by the capacitors is

$$Q = Q_1 + Q_2 = (C_1 - C_2)\Delta V_0$$

The two capacitors are now in parallel, and so the final potential differences across each is the same:

$$\Delta V = \frac{Q}{C_1 + C_2} = \left(\frac{C_1 - C_2}{C_1 + C_2}\right)\Delta V_0$$

(b) Find the total energy stored in the capacitors before and after the switches are closed.

Solution   Before the switches are closed, the total energy stored in the capacitors is

$$U_i = \tfrac{1}{2}C_1(\Delta V_0)^2 + \tfrac{1}{2}C_2(\Delta V_0)^2 = \tfrac{1}{2}(C_1 + C_2)(\Delta V_0)^2$$

After the switches are closed and the capacitors have reached an equilibrium charge, the total energy stored in them is

$$U_f = \tfrac{1}{2}C_1(\Delta V)^2 + \tfrac{1}{2}C_2(\Delta V)^2 = \tfrac{1}{2}(C_1 + C_2)(\Delta V)^2$$

$$= \tfrac{1}{2}(C_1 + C_2)\left(\frac{C_1 - C_2}{C_1 + C_2}\right)^2(\Delta V_0)^2 = \left(\frac{C_1 - C_2}{C_1 + C_2}\right)^2 U_i$$

Therefore, the ratio of the final to the initial energy stored is

$$\frac{U_f}{U_i} = \left(\frac{C_1 - C_2}{C_1 + C_2}\right)^2$$

This shows that the final energy is less than the initial energy. At first, you might think that energy conservation has been violated, but that is not the case. Part of the missing energy appears as thermal energy in the connecting wires, and part is radiated away in the form of electromagnetic waves (see Chapter 24).

EXERCISE 10   A 3.0-μF capacitor is connected to a 12-V battery. How much energy is stored in the capacitor?   Answer 2.2×10^{-4} J

EXERCISE 11 A parallel-plate capacitor is charged and then disconnected from a battery. By what fraction does the stored energy change (increase or decrease) when the plate separation is doubled?   Answer   The stored energy doubles.

20.10 • CAPACITORS WITH DIELECTRICS

A **dielectric** is a nonconducting material such as rubber, glass, or waxed paper. When a dielectric material is inserted between the plates of a capacitor, the capacitance increases. If the dielectric completely fills the space between the plates, the capacitance increases by the dimensionless factor κ, called the **dielectric constant.**

The following experiment can be performed to illustrate the effect of a dielectric in a capacitor. Consider a parallel-plate capacitor of charge Q_0 and capacitance C_0 in the absence of a dielectric. The potential difference across the capacitor as measured by a voltmeter is $\Delta V_0 = Q_0/C_0$ (Fig. 20.25a). Notice that the capacitor circuit is *open;* that is, the plates of the capacitor are *not* connected to a battery and charge cannot flow through an ideal voltmeter. Hence, there is *no* path by which charge can flow and alter the charge on the capacitor. If a dielectric is now inserted between the plates as in Figure 20.25b, it is found that the voltmeter reading *decreases* by a factor of κ to the value ΔV, where

$$\Delta V = \frac{\Delta V_0}{\kappa}$$

Because $\Delta V < \Delta V_0$, we see that $\kappa > 1$.

Because the charge Q_0 on the capacitor *does not change,* we conclude that the capacitance must change to the value

$$C = \frac{Q_0}{\Delta V} = \frac{Q_0}{\Delta V_0/\kappa} = \kappa \frac{Q_0}{\Delta V_0}$$

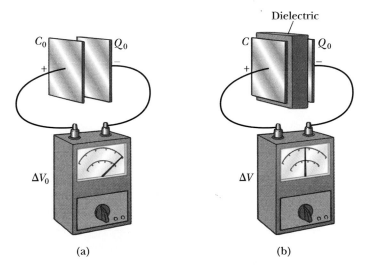

(a)   (b)

Figure 20.25 When a dielectric is inserted between the plates of a charged capacitor, the charge on the plates remains unchanged, but the potential difference as recorded by an electrostatic voltmeter is reduced from ΔV_0 to $\Delta V = \Delta V_0/\kappa$. Thus, the capacitance *increases* in the process by the factor κ.

$$C = \kappa C_0 \qquad [20.32]$$

where C_0 is the capacitance in the absence of the dielectric. That is, the capacitance *increases* by the factor κ when the dielectric completely fills the region between the plates.[3] For a parallel-plate capacitor, where $C_0 = \epsilon_0 A/d$, we can express the capacitance when the capacitor is filled with a dielectric as

$$C = \kappa \frac{\epsilon_0 A}{d} \qquad [20.33]$$

From this result, it would appear that the capacitance could be made very large by decreasing d, the distance between the plates. In practice, the lowest value of d is limited by the electrical discharge that could occur through the dielectric medium separating the plates. For any given separation d, the maximum voltage that can be applied to a capacitor without causing a discharge depends on the **dielectric strength** (maximum electric field intensity) of the dielectric, which for air is equal to 3×10^6 V/m. If the field strength in the medium exceeds the dielectric strength, the insulating properties break down and the medium begins to conduct. Most insulating materials have dielectric strengths and dielectric constants greater than those of air, as Table 20.1 indicates. Thus, we see that a dielectric provides the following advantages:

- It increases the capacitance of a capacitor.
- It increases the maximum operating voltage of a capacitor.
- It may provide mechanical support between the conducting plates.

TABLE 20.1   Dielectric Constants and Dielectric Strengths of Various Materials at Room Temperature

Material	Dielectric Constant κ	Dielectric Strength[a] (V/m)
Vacuum	1.00000	—
Air (dry)	1.00059	3×10^6
Bakelite	4.9	24×10^6
Fused quartz	3.78	8×10^6
Pyrex glass	5.6	14×10^6
Polystyrene	2.56	24×10^6
Teflon	2.1	60×10^6
Neoprene rubber	6.7	12×10^6
Nylon	3.4	14×10^6
Paper	3.7	16×10^6
Strontium titanate	233	8×10^6
Water	80	—
Silicone oil	2.5	15×10^6

[a] The dielectric strength equals the maximum electric field that can exist in a dielectric without electrical breakdown.

---

[3]If another experiment is performed in which the dielectric is introduced while the potential difference is held constant by means of a battery, the charge increases to the value $Q = \kappa Q_0$. The additional charge is supplied by the battery, and the capacitance still increases by the factor κ.

Types of Capacitors

Commerical capacitors are often made using metal foil interlaced with a dielectric such as thin sheets of paraffin-impregnated paper. These alternate layers of metal foil and dielectric are then rolled into the shape of a cylinder to form a small package. High-voltage capacitors commonly consist of interwoven metal plates immersed in silicone oil. Small capacitors are often constructed from ceramic materials. Variable capacitors (typically 10 pF to 500 pF) usually consist of two interwoven sets of metal plates, one fixed and the other movable, with air as the dielectric.

An *electrolytic capacitor* is often used to store large amounts of charge at relatively low voltages. This device consists of a metal foil in contact with an electrolyte—a solution that conducts electricity by virtue of the motion of ions contained in the solution. When a voltage is applied between the foil and the electrolyte, a thin layer of metal oxide (an insulator) is formed on the foil, and this layer serves as the dielectric. Very large capacitance values can be attained because the dielectric layer is very thin.

When electrolytic capacitors are used in circuits, the polarity must be installed properly. If the polarity of the applied voltage is opposite what is intended, the oxide layer will be removed and the capacitor will not be able to store charge.

Thinking Physics 6

Consider a parallel-plate capacitor with a dielectric material between the plates. Is the capacitance higher on a cold day or a hot day?

Reasoning It is the polarization of the molecules in the dielectric that increases the capacitance when the dielectric is added. As the temperature increases, there is more vibrational motion of the polarized molecules. This disturbs the orderly arrangement of the polarized molecules, and the net polarization decreases. Thus, the capacitance must decrease as the temperature increases.

CONCEPTUAL PROBLEM 8

An electric stud finder will locate the studs in a wall in a home. It consists of a parallel-plate capacitor, with the plates next to each other, as shown in Figure 20.26. How does this device detect the presence of a wooden stud in the wall?

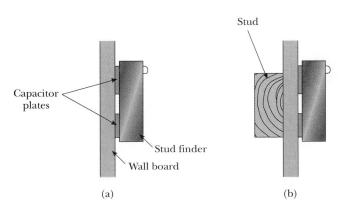

(a)   (b)

Figure 20.26 (Conceptual Problem 8) An electric stud finder positioned (a) on a piece of wall board and (b) over a wooden stud.

Example 20.11   A Paper-Filled Capacitor

A parallel-plate capacitor has plates of dimensions 2.0 cm × 3.0 cm separated by a 1.0-mm thickness of paper. (a) Find the capacitance of this device.

Solution   Because $\kappa = 3.7$ for paper (Table 20.1), we get

$$C = \kappa \frac{\epsilon_0 A}{d} = 3.7 \left(8.85 \times 10^{-12} \, \frac{C^2}{N \cdot m^2} \right) \left(\frac{6.0 \times 10^{-4} \, m^2}{1.0 \times 10^{-3} \, m} \right)$$

$$= 20 \times 10^{-12} \, F = \boxed{20 \text{ pF}}$$

(b) What is the maximum charge that can be placed on the capacitor?

Solution   From Table 20.1 we see that the dielectric strength of paper is 16×10^6 V/m. Because the thickness of the paper

is 1.0 mm, the maximum voltage that can be applied before breakdown is

$$\Delta V_{max} = E_{max} d = \left(16 \times 10^6 \, \frac{V}{m} \right) (1.0 \times 10^{-3} \, m)$$

$$= 16 \times 10^3 \, V$$

Hence, the maximum charge is

$$Q_{max} = C \Delta V_{max} = (20 \times 10^{-12} \, F) (16 \times 10^3 \, V)$$

$$= 0.32 \, \mu C$$

EXERCISE 12   What is the maximum energy that can be stored in the capacitor?   Answer 2.5×10^{-3} J

Example 20.12   Energy Stored Before and After Inserting a Dielectric

A parallel-plate capacitor is charged with a battery to a charge, Q_0, as in Figure 20.27a. The battery is then removed, and a slab of material that has a dielectric constant κ is inserted between the plates, as in Figure 20.27b. Find the energy stored in the capacitor before and after the dielectric is inserted.

Solution   The energy stored in the capacitor in the absence of the dielectric is

$$U_0 = \tfrac{1}{2} C_0 (\Delta V_0)^2$$

Because $\Delta V_0 = Q_0 / C_0$, this can be expressed as

$$U_0 = \frac{Q_0{}^2}{2 C_0}$$

After the battery is removed and the dielectric is inserted between the plates, the *charge on the capacitor remains the same.* Hence, the energy stored in the presence of the dielectric is

$$U = \frac{Q_0{}^2}{2C}$$

But the capacitance in the presence of the dielectric is $C = \kappa C_0$, and so U becomes

$$U = \frac{Q_0{}^2}{2 \kappa C_0} = \frac{U_0}{\kappa}$$

Because $\kappa > 1$, we see that the final energy is less than the initial energy by the factor $1/\kappa$. This missing energy can be

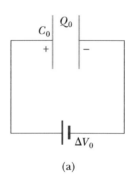

(a)

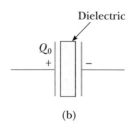

(b)

Figure 20.27   (Example 20.12)

accounted for by noting that when the dielectric is inserted into the capacitor, it gets pulled into the device. An external agent must do negative work to keep the slab from accelerating. This work is simply the difference $U - U_0$. (Alternatively, the positive work done by the system on the external agent is $U_0 - U$.)

EXERCISE 13 Suppose that the capacitance in the absence of a dielectric is 8.50 pF, and the capacitor is charged to a potential difference of 12.0 V. If the battery is disconnected and a slab of polystyrene ($\kappa = 2.56$) is inserted between the plates, calculate the energy difference $U - U_0$. Answer 373 pJ

EXERCISE 14 (a) How much charge can be placed on a capacitor with air between the plates before it breaks down if the area of each of the plates is 5.00 cm²? (b) Find the maximum charge if polystyrene is used between the plates instead of air. Answer (a) 13.3 nC (b) 272 nC

PROBLEM-SOLVING STRATEGY AND HINTS

1. Be careful with your choice of units. To calculate capacitance in farads, make sure that distances are in meters and use the SI value of ϵ_0. When checking consistency of units, remember that the units for electric fields can be either newtons per coulomb or volts per meter.
2. When two or more unequal capacitors are connected in series, they carry the same charge, but their potential differences are not the same. The capacitances add as reciprocals, and the equivalent capacitance of the combination is always less than the smallest individual capacitor.
3. When two or more capacitors are connected in parallel, the potential differences across them are the same. The charge on each capacitor is proportional to its capacitance; hence, the capacitances add directly to give the equivalent capacitance of the parallel combination.
4. A dielectric increases capacitance by the factor κ (the dielectric constant) because induced surface charges on the dielectric reduce the electric field inside the material from E_0 to E_0/κ.
5. Be careful about problems in which you may be connecting or disconnecting a battery to a capacitor. It is important to note whether modifications to the capacitor are being made while the capacitor is connected to the battery or after it is disconnected. If the capacitor remains connected to the battery, the voltage across the capacitor necessarily remains the same (equal to the battery voltage), and the charge is proportional to the capacitance, *regardless of how it may be modified* (say, by insertion of a dielectric). However, if you disconnect the capacitor from the battery before making any modifications to the capacitor, then its charge remains the same. In this case, as you vary the capacitance, the voltage across the plates changes in inverse proportion to capacitance, according to $\Delta V = Q/C$.

This photograph illustrates dielectric breakdown in air. Sparks are produced when a large alternating voltage is applied across the electrodes by way of a high-voltage induction coil power supply. *(Courtesy of Central Scientific Co.)*

SUMMARY

When a positive test charge q_0 is moved between points A and B in an electrostatic field **E,** the **change in potential energy** is

$$\Delta U = -q_0 \int_A^B \mathbf{E} \cdot d\mathbf{s} \qquad [20.1]$$

The **potential difference,** ΔV, between points A and B in an electrostatic field $\mathbf{E}$ is defined as the change in potential energy divided by the test charge q_0.

$$\Delta V = \frac{\Delta U}{q_0} = -\int_A^B \mathbf{E} \cdot d\mathbf{s}$$ [20.3]

where the electric potential V is a scalar and has the units joules per coulomb, defined as 1 volt (V).

The potential difference between two points, A and B, in a uniform electric field, $\mathbf{E}$, is

$$\Delta V = -\mathbf{E} \cdot \mathbf{d}$$ [20.8]

where $\mathbf{d}$ is the displacement vector between A and B.

Equipotential surfaces are surfaces on which the electric potential remains constant. Equipotential surfaces are *perpendicular* to the electric field lines.

The electric potential due to a point charge q at any distance r from the charge is

$$V = k_e \frac{q}{r}$$ [20.11]

The electric potential due to a group of point charges is obtained by summing the potentials due to the individual charges. Because V is a scalar, the sum is a simple algebraic operation.

The **electric potential energy of a pair of point charges** separated by a distance r is

$$U = k_e \frac{q_1 q_2}{r}$$ [20.13]

This represents the work required to bring the charges from an infinite separation to the separation r. The potential energy of a distribution of point charges is obtained by summing terms like Equation 20.13 over *all* pairs of particles.

If the electric potential is known as a function of coordinates x, y, z, the components of the electric field can be obtained by taking the negative derivative of the potential with respect to the coordinates. For example, the x component of the electric field is

$$E_x = -\frac{dV}{dx}$$ [20.15]

The **electric potential due to a continuous charge distribution** is

$$V = k_e \int \frac{dq}{r}$$ [20.18]

Every point on the surface of a charged conductor in electrostatic equilibrium is at the same potential. Furthermore, the potential is constant everywhere inside the conductor and equal to its value at the surface.

A *capacitor* is a charge storage device. It consists of two equal and oppositely charged conductors spaced very close together compared to their size, with a potential difference of ΔV between them. The **capacitance, C,** of any capacitor is defined to be the ratio of the magnitude of the charge Q on either conductor to the magnitude of the potential difference, ΔV:

$$C \equiv \frac{Q}{\Delta V}$$ [20.19]

The SI units of capacitance are coulombs per volt, or the farad (F), and $1 \text{ F} = 1 \text{ C/V}$.

If two or more capacitors are connected in parallel, the potential differences across them must be the same. The equivalent capacitance of a parallel combination of capacitors is

$$C_{eq} = C_1 + C_2 + C_3 + \ldots \qquad \text{[20.25]}$$

If two or more capacitors are connected in series, the charges on them are the same, and the equivalent capacitance of the series combination is

$$\frac{1}{C_{eq}} = \frac{1}{C_1} + \frac{1}{C_2} + \frac{1}{C_3} + \ldots \qquad \text{[20.28]}$$

Work is required to charge a capacitor, because the charging process consists of transferring charges from one conductor at a lower potential to another conductor at a higher potential. The work done in charging the capacitor to the charge Q equals the electrostatic potential energy U stored in the capacitor, where

$$U = \frac{Q^2}{2C} = \tfrac{1}{2}Q\,\Delta V = \tfrac{1}{2}C(\Delta V)^2 \qquad \text{[20.29]}$$

When a dielectric material is inserted between the plates of a capacitor, the capacitance generally increases by the dimensionless factor κ, called the **dielectric constant.** That is,

$$C = \kappa C_0 \qquad \text{[20.32]}$$

where C_0 is the capacitance in the absence of the dielectric.

CONCEPTUAL QUESTIONS

1. Distinguish between electric potential and electrical potential energy.

2. Give a physical explanation of the fact that the potential energy of a pair of like charges is positive whereas the potential energy of a pair of unlike charges is negative.

3. Explain why, under static conditions, all points in a conductor must be at the same electric potential.

4. Why is it important to avoid sharp edges or points on conductors used in high-voltage equipment?

5. In what type of weather would a car battery be more likely to discharge and why?

6. How would you shield an electronic circuit or laboratory from stray electric fields? Why does this work?

7. If you are given three different capacitors, C_1, C_2, C_3, how many different combinations of capacitance can you produce?

8. The plates of a capacitor are connected to a battery. What happens to the charge on the plates if the connecting wires are removed from the battery? What happens to the charge if the wires are removed from the battery and connected to each other?

9. If you want to increase the maximum operating voltage of a parallel-plate capacitor, describe how you can do this for a fixed plate separation.

10. If the potential difference across a capacitor is doubled, by what factor does the energy stored change?

11. Because the charges on the plates of a parallel-plate capacitor are equal and opposite, they attract each other. Hence, it takes positive work to increase the plate separation. What happens to the external work done in this process? (Assume the capacitor plates are removed from the charging battery.)

12. If you were asked to design a capacitor where small size and large capacitance were required, what factors would be important in your design?

13. Explain why a dielectric increases the maximum operating voltage of a capacitor although the physical size of the capacitor does not change.

14. A pair of capacitors are connected in parallel and an identical pair are connected in series. Which pair would be more dangerous to handle after being connected to the same voltage source? Explain.

15. The energy stored in a particular capacitor is increased fourfold. What is the accompanying change in (a) the charge and (b) the potential difference across the capacitor?

PROBLEMS

Section 20.1   Potential Difference and Electric Potential

1. A uniform electric field of magnitude 250 V/m is directed in the positive x direction. A $+12.0\text{-}\mu C$ charge moves from the origin to the point $(x, y) = (20.0 \text{ cm}, 50.0 \text{ cm})$. (a) What was the change in the potential energy of this charge? (b) Through what potential difference did the charge move?

2. How much work is done (by a battery, generator, or some other source of electrical energy) in moving Avogadro's number of electrons from an initial point where the electric potential is 9.00 V to a point where the potential is -5.00 V? (The potential in each case is measured relative to a common reference point.)

3. (a) Calculate the speed of a proton that is accelerated from rest through a potential difference of 120 V. (b) Calculate the speed of an electron that is accelerated through the same potential difference.

4. Through what potential difference would an electron need to be accelerated for it to achieve a speed of 40.0% of the speed of light ($c = 3.00 \times 10^8$ m/s), starting from rest? Note that this problem must be solved using a relativistic expression for the energy.

5. An electron starts from rest and is accelerated through a potential difference of 20.0 kV. (a) Find its actual relativistic speed. (b) Find the speed that classical physics would predict for it. By how much is this in error?

Section 20.2   Potential Differences in a Uniform Electric Field

6. The difference in potential between the accelerating plates of a TV set is about 25 000 V. If the distance between these plates is 1.50 cm, find the magnitude of the uniform electric field in this region.

7. Suppose an electron is released from rest in a uniform electric field the strength of which is 5.90×10^3 V/m. (a) Through what potential difference will it have passed after moving 1.00 cm? (b) How fast will the electron be moving after it has traveled 1.00 cm?

8. An electron moving parallel to the x axis has an initial speed of 3.70×10^6 m/s at the origin. Its speed is reduced to 1.40×10^5 m/s at the point $x = 2.00$ cm. Calculate the potential difference between the origin and the point. Which point is at the higher potential?

9. A uniform electric field of magnitude 325 V/m is directed in the *negative y* direction in Figure P20.9. The coordinates of point A are $(-0.200, -0.300)$ m, and those of point B are $(0.400, 0.500)$ m. Calculate the potential $V_B - V_A$ using the path shown in blue.

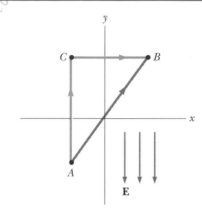

Figure P20.9

10. On planet Tehar, the acceleration of gravity is the same as that on Earth, but there is also a strong downward electric field with the field being uniform close to the planet's surface. A 2.00-kg ball having a charge of 5.00 μC is thrown upward at a speed of 20.1 m/s and it hits the ground after an interval of 4.10 s. What is the potential difference between the starting point and the top point of the trajectory?

Section 20.3   Electric Potential and Electric Potential Energy Due to Point Charges

(*Note:* Assume a reference level for potential as $V = 0$ at $r = \infty$ unless the statement of the problem requires otherwise.)

11. (a) Find the potential at a distance of 1.00 cm from a proton. (b) What is the potential difference between two points that are 1.00 cm and 2.00 cm from a proton? (c) Repeat parts (a) and (b) for an electron.

12. Given two 2.00-μC charges, as in Figure P20.12, and a positive test charge $q = 1.28 \times 10^{-18}$ C at the origin, (a) what is the net force exerted on q by the two 2.00-μC charges? (b) What is the electric field at the origin due to the two 2.00-μC charges? (c) What is the electrical potential at the origin due to the two 2.00-μC charges?

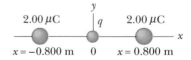

Figure P20.12

13. The three charges in Figure P20.13 are at the vertices of an isosceles triangle. Calculate the electric potential at the midpoint of the base, taking $q = 7.00$ μC.

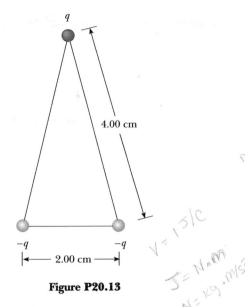

Figure P20.13

Handwritten notes in margin:
$V = J/C$
$J = N \cdot m$
$N = kg \cdot m/s^2$

14. A charge $+q$ is at the origin. A charge $-2q$ is at $x = 2.00$ m on the x axis. For what finite value(s) of x is (a) the electric field zero? (b) the electric potential zero?

15. The Bohr model of the hydrogen atom holds that the electron can exist only in certain allowed orbits. The radius of each Bohr orbit is $r = n^2(0.0529 \text{ nm})$ where $n = 1, 2, 3, \ldots$. Calculate the electric potential energy of a hydrogen atom when the electron is in the (a) first allowed orbit, $n = 1$; (b) second allowed orbit, $n = 2$; and (c) when the electron has escaped from the atom, $r = \infty$. Express your answers in electron volts.

16. Two point charges, $Q_1 = +5.00$ nC and $Q_2 = -3.00$ nC, are separated by 35.0 cm. (a) What is the potential energy of the pair? What is the significance of the algebraic sign of your answer? (b) What is the electric potential at a point midway between the charges?

17. Show that the amount of work required to assemble four identical point charges of magnitude Q at the corners of a square of side s is $5.41 k_e Q^2/s$.

18. Compare this problem with Problem 42 in Chapter 19. Four identical point charges ($q = +10.0$ μC) are located on the corners of a rectangle, as shown in Figure P19.42. The dimensions of the rectangle are $L = 60.0$ cm and $W = 15.0$ cm. Calculate the electric potential energy of the charge at the lower left corner due to the other three charges.

19. Two insulating spheres having radii 0.300 cm and 0.500 cm, masses 0.100 kg and 0.700 kg, and charges -2.00 μC and 3.00 μC, are released from rest when their centers are separated by 1.00 m. (a) How fast will each be moving when they collide? (*Hint:* Consider conservation of energy and linear momentum.) (b) If the spheres were conductors, would the speeds be larger or smaller than those calculated in part (a)? Explain.

Section 20.4 Obtaining E from the Electric Potential

20. The potential in a region between $x = 0$ and $x = 6.00$ m is $V = a + bx$ where $a = 10.0$ V and $b = -7.00$ V/m. Determine (a) the potential at $x = 0$, 3.00 m, and 6.00 m, and (b) the magnitude and direction of the electric field at $x = 0$, 3.00 m and $x = 6.00$ m.

21. Over a certain region of space, the electric potential is $V = 5x - 3x^2y + 2yz^2$. Find the expressions for the x, y, and z components of the electric field over this region. What is the magnitude of the field at the point P, which has coordinates $(1.00, 0, -2.00)$ m?

22. The electric potential inside a uniformly charged spherical insulator of radius R is

$$V = \frac{k_e Q}{2R}\left(3 - \frac{r^2}{R^2}\right)$$

Outside it is

$$V = \frac{k_e Q}{r}$$

Use $E_r = -\dfrac{dV}{dr}$ to derive the electric field both (a) inside ($r < R$) and (b) outside ($r > R$) this charge distribution.

Section 20.5 Electric Potential Due to Continuous Charge Distributions

23. Consider a ring of radius R with the total charge Q spread uniformly over its perimeter. What is the potential difference between the point at the center of the ring and a point on its axis a distance $2R$ from the center?

24. Compare this problem with Problem 17 in Chapter 19. A uniformly charged insulating rod of length 14.0 cm is bent into the shape of a semicircle, as in Figure P19.17. If the rod has a total charge of -7.50 μC, find the electric potential at O, the center of the semicircle.

25. A rod of length L (Fig. P20.25) lies along the x axis with its left end at the origin and has a nonuniform charge density

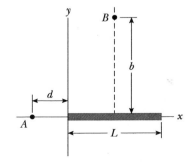

Figure P20.25

$\lambda = \alpha x$ (where α is a positive constant). (a) What are the units of α? (b) Calculate the electric potential at A.

26. For the arrangement described in the previous problem, calculate the electric potential at point B that lies on the perpendicular bisector of the rod a distance b above the x axis.

Section 20.6   Electric Potential of a Charged Conductor

27. Two charged spherical conductors are connected by a long conducting wire, and a charge of 20.0 μC is placed on the combination. (a) If one sphere has a radius of 4.00 cm and the other has a radius of 6.00 cm, what is the electric field near the surface of each sphere? (b) What is the electrical potential of each sphere?

Section 20.7   Capacitance

28. (a) If a drop of liquid has capacitance 1.00 pF, what is its radius? (b) If another drop has radius 2.00 mm, what is its capacitance? (c) What is the charge on the smaller drop if its potential is 100 V?

29. Two conductors having net charges of $+10.0$ μC and -10.0 μC have a potential difference of 10.0 V. Determine (a) the capacitance of the system and (b) the potential difference between the two conductors if the charges on each are increased to $+100$ μC and -100 μC.

30. (a) How much charge is on each plate of a 4.00-μF capacitor when it is connected to a 12.0-V battery? (b) If this same capacitor is connected to a 1.50-V battery, what charge is stored?

31. A **spherical capacitor** consists of a spherical conducting shell of radius b and charge $-Q$ that is concentric with a smaller conducting sphere of radius a and charge $+Q$ (Fig. P20.31). (a) Show that its capacitance is

$$C = \frac{ab}{k_e(b - a)}$$

(b) Show that as b approaches infinity, the capacitance approaches the value $a/k_e = 4\pi\epsilon_0 a$.

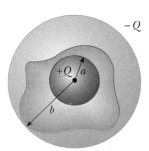

Figure P20.31

32. Regarding the Earth and a cloud layer 800 m above the Earth as the "plates" of a capacitor, calculate the capacitance if the cloud layer has an area of 1.00 km². If an electric field of 3.00×10^6 N/C makes the air break down and conduct electricity (that is, causes lightning), what is the maximum charge the cloud can hold?

33. An air-filled capacitor consists of two parallel plates, each with area 7.60 cm², separated by a distance of 1.80 mm. If a 20.0-V potential difference is applied to these plates, calculate (a) the electric field between the plates, (b) the surface charge density, (c) the capacitance, and (d) the charge on each plate.

34. A 50.0-m length of coaxial cable has an inner conductor that has a diameter of 2.58 mm and carries a charge of 8.10 μC. The surrounding conductor has an inner diameter of 7.27 mm and a charge of -8.10 μC. (a) What is the capacitance of this cable? (b) What is the potential difference between the two conductors? Assume the region between the conductors is air.

Section 20.8   Combinations of Capacitors

35. Two capacitors, $C_1 = 5.00$ μF and $C_2 = 12.0$ μF, are connected in parallel, and the resulting combination is connected to a 9.00-V battery. (a) What is the value of the equivalent capacitance of the combination? What is (b) the potential difference across each capacitor and (c) the charge stored on each capacitor?

36. The two capacitors of Problem 35 are now connected in series and to a 9.00-V battery. Find (a) the value of the equivalent capacitance of the combination, (b) the voltage across each capacitor, and (c) the charge on each capacitor.

37. Two capacitors when connected in parallel give an equivalent capacitance of 9.00 pF and an equivalent capacitance of 2.00 pF when connected in series. What is the capacitance of each capacitor?

38. Two capacitors when connected in parallel give an equivalent capacitance of C_p and an equivalent capacitance of C_s when connected in series. What is the capacitance of each capacitor?

39. Four capacitors are connected as shown in Figure P20.39. (a) Find the equivalent capacitance between points a and b. (b) Calculate the charge on each capacitor if $\Delta V_{ab} = 15.0$ V.

40. Evaluate the equivalent capacitance of the configuration shown in Figure P20.40. All the capacitors are identical and each has capacitance C.

41. Consider the circuit shown in Figure P20.41, where $C_1 = 6.00$ μF, $C_2 = 3.00$ μF, and $\Delta V = 20.0$ V. Capacitor C_1 is first charged by the closing of switch S_1. Switch S_1 is then opened, and the charged capacitor is connected to the uncharged capacitor by the closing of S_2. Calculate the initial charge acquired by C_1 and the final charge on each.

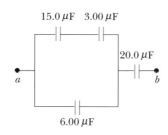

15.0 μF 3.00 μF

20.0 μF

a   b

6.00 μF

Figure P20.39

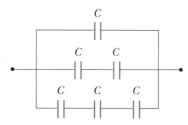

C

C   C

C   C   C

Figure P20.40

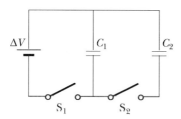

ΔV C_1 C_2

S_1 S_2

Figure P20.41

42. How should four 2.00-μF capacitors be connected to have a total capacitance of (a) 8.00 μF, (b) 2.00 μF, (c) 1.50 μF, and (d) 0.500 μF?

Section 20.9   Energy Stored in a Charged Capacitor

43. (a) A 3.00-μF capacitor is connected to a 12.0-V battery. How much energy is stored in the capacitor? (b) If the capacitor had been connected to a 6.00-V battery, how much energy would have been stored?

44. A parallel-plate capacitor is charged and then disconnected from a battery. By what fraction does the stored energy change (increase or decrease) when the plate separation is doubled?

45. Two capacitors, $C_1 = 25.0$ μF and $C_2 = 5.00$ μF, are connected in parallel and charged with a 100-V power supply. (a) Draw a circuit diagram and calculate the total energy stored in the two capacitors. (b) What potential difference would be required across the same two capacitors connected in series in order that the combination store the

same energy as in (a)? Draw a circuit diagram of this circuit.

46. A uniform electric field $E = 3000$ V/m exists within a certain region. What volume of space contains an energy equal to 1.00×10^{-7} J? Express your answer in cubic meters and in liters.

47. A parallel-plate capacitor has a charge Q and plates of area A. Show that the force exerted on each plate by the other is $F = Q^2/2\epsilon_0 A$. (*Hint:* Let $C = \epsilon_0 A/x$ for an arbitrary plate separation x; then require that the work done in separating the two charged plates be $W = \int F \, dx$.)

Section 20.10   Capacitors with Dielectrics

48. Find the capacitance of a parallel-plate capacitor that uses Bakelite as a dielectric if each of the plates has an area of 5.00 cm² and the plate separation is 2.00 mm. (Refer to Table 20.1.)

49. A capacitor that has air between its plates is connected across a potential difference of 12.0 V and stores 48.0 μC of charge. It is then disconnected from the source while still charged. (a) Find the capacitance of the capacitor. (b) A piece of Teflon is inserted between the plates. Find the new capacitance. (c) Find the voltage and charge now on the capacitor.

50. Determine (a) the capacitance and (b) the maximum voltage that can be applied to a Teflon-filled parallel-plate capacitor having a plate area of 1.75 cm² and insulation thickness of 0.0400 mm.

51. The supermarket sells rolls of aluminum foil, of plastic wrap, and of waxed paper. Describe a capacitor made from supermarket materials. Compute order-of-magnitude estimates for its capacitance and its breakdown voltage.

52. (a) How much charge can be placed on a capacitor with air between the plates before it breaks down if the area of each of the plates is 5.00 cm²? (b) Find the maximum charge if polystyrene is used between the plates instead of air.

Additional Problems

53. The liquid-drop model of the nucleus suggests that high-energy oscillations of cetain nuclei can split the nucleus into two unequal fragments plus a few neutrons. The fragments acquire kinetic energy from their mutual Coulomb repulsion. Calculate the electric potential energy (in electron volts) of two spherical fragments from a uranium nucleus having the following charges and radii: $38e$ and 5.50×10^{-15} m; $54e$ and 6.20×10^{-15} m. Assume that the charge is distributed uniformly throughout the volume of each sperical fragment and that their surfaces are initially in contact at rest. (The electrons surrounding the nucleus can be neglected.)

54. The charge distribution shown in Figure P20.54 is referred to as a linear quadrupole. (a) Show that the potential at a

point on the x axis where $x > d$ is

$$V = \frac{2k_e Q d^2}{x^3 - xd^2}$$

(b) Show that the expression obtained in (a) when $x \gg d$ reduces to

$$V = \frac{2k_e Q d^2}{x^3}$$

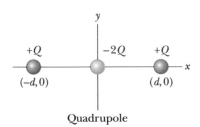

Quadrupole

Figure P20.54

55. Two parallel plates having equal but opposite charge are separated by 12.0 cm. Each plate has a surface-charge density of 36.0 nC/m². A proton is released from rest at the positive plate. Determine (a) the potential difference between the plates, (b) the energy of the proton when it reaches the negative plate, (c) the speed of the proton just before it strikes the negative plate, (d) the acceleration of the proton, and (e) the force on the proton. (f) From the force, find the electric field intensity and show that it is equal to the electric field intensity found from the charge densities on the plates.

56. For the system of capacitors shown in Figure P20.56, find (a) the equivalent capacitance of the system, (b) the potential across each capacitor, (c) the charge on each capacitor, and (d) the total energy stored by the group.

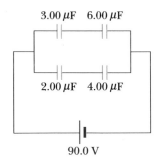

Figure P20.56

57. A certain spherical object can be raised to a maximum potential of 600 kV; then additional charge leaks off in sparks by producing dielectric breakdown of the surrounding dry air. Determine (a) the charge on the object and (b) the radius of the object.

58. The energy stored in a 52.0-μF capacitor is used to melt a 6.00-mg sample of lead. To what voltage must the capacitor be initially charged assuming that the initial temperature of the lead is 20.0°C? Lead has a specific heat of 128 J/kg·°C, a melting point of 327.3°C, and a latent heat of fusion of 24.5 kJ/kg.

59. A 2.00-nF parallel-plate capacitor is charged to an initial potential difference $\Delta V_i = 100$ V and then isolated. The dielectric material between the plates is mica ($\kappa = 5.00$). (a) How much work is required to withdraw the mica sheet? (b) What is the potential difference of the capacitor after the mica is withdrawn?

60. An electron is released from rest on the axis of a uniform positively charged ring 0.100 m from the ring's center. If the linear charge density of the ring is $+0.100$ μC/m and the radius of the ring is 0.200 m, how fast will the electron be moving when it reaches the center of the ring?

61. A parallel-plate capacitor is constructed using a dielectric material the dielectric constant of which is 3.00 and the dielectric strength of which is 2.00×10^8 V/m. The desired capacitance is 0.250 μF, and the capacitor must withstand a maximum potential difference of 4000 V. Find the minimum area of the capacitor plates.

62. A parallel-plate capacitor is constructed using three dielectric materials, as in Figure P20.62. (a) Find an expression for the capacitance of the device in terms of the plate area A and d, κ_1, κ_2, and κ_3. (b) Calculate the capacitance using the values $A = 1.00$ cm², $d = 2.00$ mm, $\kappa_1 = 4.90$, $\kappa_2 = 5.60$, and $\kappa_3 = 2.10$.

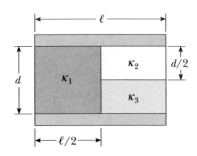

Figure P20.62

63. A conducting slab of thickness d and area A is inserted into the space between the plates of a parallel-plate capacitor with spacing s and surface area A, as in Figure P20.63. What is the capacitance of the system?

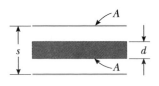

Figure P20.63

64. (a) Two spheres have radii a and b and their centers are a distance d apart. Show that the capacitance of this system is

$$C = \frac{4\pi\epsilon_0}{\dfrac{1}{a} + \dfrac{1}{b} - \dfrac{2}{d}}$$

provided that d is large compared with a and b. (*Hint:* Because the spheres are far apart, assume that the potential of each equals the sum of the potentials due to each sphere, and when calculating those potentials assume that $V = k_e Q/r$ applies.) (b) Show that as d approaches infinity the above result reduces to that of two spherical capacitors in series.

Spreadsheet Problems

S1. Spreadsheet 20.1 calculates the electric potential in the two-dimensional region around two charges. The region is divided into a 10×10 grid. The upper left corner of the grid corresponds to $x = 1.0$ m, $y = 1.0$ m, and the lower-right corner corresponds to $x = 10$ m, $y = 10$ m. The two charges can be placed anywhere in this grid. The first two tables calculate the distances from each charge to each grid point. The spreadsheet then calculates the electric potential at each grid point. Place charges of $+2.0$ μC at $x = 3.0$ m, $y = 5.0$ m and $x = 7.0$ m, $y = 5.0$ m. Print out the electric potential matrix. The location of the point

charges will appear on the grid in the cells denoted in *ERR* (in Lotus1-2-3) or *#DIV/0* (in Excel) because the potential is infinite at the location of the point charges. Sketch by hand the equipotential lines for $V = 20$ kV, 15 kV, and 10 kV. Sketch in a representative set of electric field lines. (*Hint:* By adjusting the cell width on the spreadsheet and the printer line spacing, you can get the grid to print out approximately as a square.)

S2. Repeat Problem S1 with one charge replaced by -2.0 μC. Print out the electric potential matrix. Sketch the equipotential lines for $V = \pm 20$ kV, ± 15 kV, and ± 10 kV. Sketch in a representative set of electric field lines.

S3. Place two charges anywhere in the grid in Problem S1. Choose any values for the charges. Print out the electric potential matrix. Sketch representative equipotentials and electric field lines.

S4. Modify Spreadsheet 20.1 to include additional charges. Choose locations and values for each charge. Sketch the equipotentials and the electric field lines. Place a row of charges together. Are the field lines for this case uniform? If not, why not?

S5. Spreadsheet 11.1 calculates the gravitational potential energy of a mass moving along the x axis in the gravitational field of four fixed-point masses. Rework this spreadsheet to calculate the electrical potential along the x axis for four fixed-point charges at the same locations as the four-point masses. Find the electric field on the x axis as well. (*Hint:* $GM_1 m$ must be replaced by $k_e Q_1$, and so forth.) Investigate various choices for the charges and their locations.

ANSWERS TO CONCEPTUAL PROBLEMS

1. The proton is displaced in the direction of the electric field, so the electric potential and electric potential energy will *decrease*. The decrease in potential energy is accompanied by an equal increase in kinetic energy, as required by the law of conservation of energy. An electron would be displaced opposite to the electric field so the electric potential would increase but the electric potential energy would decrease.

2. No. Suppose there are several charges in the vicinity of the point in question. If some charges are positive and some negative, their contributions to the potential at the point may cancel. For example, the electric potential at the midpoint of two equal but opposite charges is zero.

3. Field lines represent the direction of the electric force on a positive test charge. If electric field lines were to cross, then, at the point of crossing, there would be an ambiguity regarding the direction of the force on the test charge, because there would be two possible forces. Thus, electric field lines cannot cross. It is possible for equipotential surfaces to cross.

For example, suppose two identical positive charges are at diagonally opposite corners of a square, and two negative charges of equal magnitude are at the other two corners. Then the planes perpendicular to the sides of the square at their midpoints are equipotential surfaces. These two planes cross each other, at the line perpendicular to the square at its center.

4. The value of the electric potential at one point is not sufficient to determine the electric field. The electric field is the negative of the gradient of the potential, so the nature in which the potential varies in space must be known.

5. If V is constant in a certain region, the electric field in that region must be zero because the electric field is equal to the negative gradient of E. (In one dimension, $E_x = -dV/dx$, so if $V = $ constant, then $E = 0$.) Likewise, if $E = 0$ in some region, one can only conclude that V is a constant in that region.

6. The power line, if it makes electrical contact with the metal of the car, will raise the potential of the car to 20 000 volts.

It will also raise the potential of your body to 20 000 volts, because you are in contact with the car. This is not a problem. If you step out of the car, your body at 20 000 volts will make contact with the ground, which is at ground potential of zero volts. As a result, a current will flow through your body, and you will likely be injured. Thus, it is best to stay in the car until help arrives.

7. The capacitor often remains charged long after the voltage source is disconnected. This residual charge can be lethal. The capacitor can be safely handled after discharging the plates by short circuiting the device with a conductor, such as a screwdriver with an insulating handle.

8. The capacitance of the device is measured by the electronics in the stud finder. When a (dielectric) wooden stud is brought into the electric field region, it has the same effect as introducing a dielectric between the plates of a parallel-plate capacitor—the capacitance changes. The stud finder detects this change in capacitance and signals the presence of the stud.

21

Current and Direct Current Circuits

Thus far, our discussion of electrical phenomena has been confined to charges at rest, or electrostatics. We shall now consider situations involving electric charges in motion. The term *electric current*, or simply *current*, is used to describe the rate of flow of charge through some region of space. Most practical applications of electricity involve electric currents. For example, the battery of a flashlight supplies current to the filament of the bulb when the switch is turned on. In these common situations, the flow of charge takes place in a conductor, such as a copper wire. It is also possible for currents to exist outside a conductor. For instance, a beam of electrons in a television picture tube constitutes a current.

In this chapter we shall first define current and current density. A microscopic description of current will be given, and some of the factors that contribute to resistance to the flow of charge in conductors will be discussed. Mechanisms responsible for the electrical resistances of various

◀ **Photograph of a carbon filament incandescent lamp. The resistance of such a lamp is typically 10 Ω, but its value changes with temperature. Most modern lightbulbs use tungsten filaments, the resistance of which increases with increasing temperature.**
(Courtesy of Central Scientific Co.)

materials depend on the materials' compositions and on temperature. A classical model is used to describe electrical conduction in metals; we shall point out some of the limitations of this model.

This chapter is also concerned with the analysis of some simple circuits the elements of which include batteries, resistors, and capacitors in varied combinations. The analysis of these circuits is simplified by the use of two rules known as *Kirchhoff's rules,* which follow from the laws of conservation of energy and conservation of charge. Most of the circuits analyzed are assumed to be in *steady state,* which means that the currents are constant in magnitude and direction. We shall close with a discussion of circuits containing resistors and capacitors, in which the current varies with time.

21.1 · ELECTRIC CURRENT

Whenever there is a net flow of charge, a **current** is said to exist. To define current more precisely, suppose the charges are moving perpendicular to a surface of area A, as in Figure 21.1. (This area could be the cross-sectional area of a wire, for example.) **The current is the rate at which charge flows through this surface.** If ΔQ is the amount of charge that passes through this area in a time interval Δt, the average current, I_{av}, is the ratio of the charge to the time interval:

$$I_{av} = \frac{\Delta Q}{\Delta t} \qquad [21.1]$$

If the rate at which charge flows varies in time, the current also varies in time. We define the **instantaneous current** I as the differential limit of the preceding expression:

$$I = \frac{dQ}{dt} \qquad [21.2]$$

The SI unit of current is the **ampere** (A):

$$1\ A = 1\ C/s \qquad [21.3]$$

That is, 1 A of current is equivalent to 1 C of charge passing through a surface in 1 s.

When charges flow through a surface as in Figure 21.1, they can be positive, negative, or both. **It is conventional to give the current the same direction as the flow of positive charge.** In a common conductor such as copper, the current is due to the motion of the negatively charged electrons. Therefore, when we speak of current in such a conductor, **the direction of the current is opposite the direction of flow of electrons.** However, if one considers a beam of positively charged protons in an accelerator, the current is in the direction of motion of the protons. In some cases—gases and electrolytes, for example—the current is the result of the flow of both positive and negative charges. It is common to refer to a moving charge (whether it is positive or negative) as a mobile **charge carrier.** For example, the charge carriers in a metal are electrons.

It is instructive to relate current to the motions of the charged particles. To illustrate this point, consider the current in a conductor of cross-sectional area A

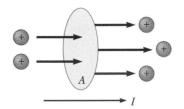

Figure 21.1 Charges in motion through an area A. The time rate of flow of charge through the area is defined as the current I. The direction of the current is the direction in which positive charge would flow if free to do so.

Electric current •

The direction of the current •

(Fig. 21.2). The volume of an element of the conductor of length Δx is $A \Delta x$. If n represents the number of mobile charge carriers per unit volume, then the number of carriers in the volume element is $nA \Delta x$. Therefore, the charge ΔQ in this element is

$$\Delta Q = \text{number of carriers} \times \text{charge per carrier} = (nA \Delta x)q$$

where q is the charge on each carrier. If the carriers move with a speed of v_d, the distance they move in the time Δt is $\Delta x = v_d \Delta t$. Therefore, we can write ΔQ in the form

$$\Delta Q = (nAv_d \Delta t)q$$

If we divide both sides of this equation by Δt, we see that the current in the conductor is

$$I = \frac{\Delta Q}{\Delta t} = nqv_d A \qquad [21.4]$$

The speed of the charge carriers, v_d, is an average speed called the **drift speed.** To understand its meaning, consider a conductor in which the charge carriers are free electrons. If the conductor is isolated, these electrons undergo random motion similar to that of gas molecules. When a potential difference is applied across the conductor (say, by means of a battery), an electric field is set up in the conductor, which creates an electric force on the electrons, accelerating them, and hence producing a current. In reality, the electrons do not simply move in straight lines along the conductor. Instead, they undergo repeated collisions with the metal atoms, and the result is a complicated zigzag motion (Fig. 21.3). The energy transferred from the electrons to the metal atoms during collision causes an increase in the vibrational energy of the atoms and a corresponding increase in the temperature of the conductor. However, despite the collisions, the electrons move slowly along the conductor (in a direction opposite $\mathbf{E}$) with the drift velocity, $\mathbf{v}_d$. One can think of the collisions within a conductor as being an effective internal friction (or drag force) similar to that experienced by the molecules of a liquid flowing through a pipe stuffed with steel wool.

The **current density** J in the conductor is defined to be the current per unit area. Because $I = nqv_d A$, the current density is

$$J \equiv \frac{I}{A} = nqv_d \qquad [21.5]$$

where J has the SI units amperes per square meter. In general, the current density is a *vector quantity*. That is,

$$\mathbf{J} \equiv nq\mathbf{v}_d \qquad [21.6]$$

From this definition, we see that the current density vector is in the direction of motion of positive charge carriers and opposite the direction of motion of negative charge carriers. Because the drift velocity is proportional to the electric field $\mathbf{E}$ in the conductor, we conclude that the current density is also proportional to $\mathbf{E}$.

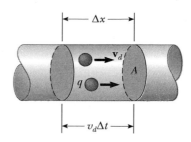

Figure 21.2 A section of a uniform conductor of cross-sectional area A. The charge carriers move with a speed v_d, and the distance they travel in a time Δt is given by $\Delta x = v_d \Delta t$. The number of mobile charge carriers in the section of length Δx is given by $nAv_d \Delta t$, where n is the number of mobile carriers per unit volume.

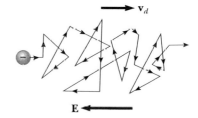

Figure 21.3 A schematic representation of the zigzag motion of a charge carrier in a conductor. The changes in direction are due to collisions with atoms in the conductor. Note that the net motion of electrons is opposite the direction of the electric field. The zigzag paths are actually parabolic segments.

• *Current density*

> ### Thinking Physics 1
>
> Suppose a current-carrying wire has a cross-sectional area that gradually becomes smaller along the wire, so that the wire has the shape of a very long cone. How does the drift velocity of electrons vary along the wire?
>
> **Reasoning** Every portion of the wire is carrying the same amount of current. Thus, as the cross-sectional area decreases, the drift velocity must increase to maintain the constant value of the current. This increased drift velocity is a result of the electric field lines in the wire being compressed into a smaller area, thus increasing the strength of the field.

Example 21.1   Drift Speed in a Copper Wire

A copper wire of cross-sectional area 3.00×10^{-6} m^2 carries a current of 10.0 A. Find the drift speed of the electrons in this wire. The density of copper is 8.95 g/cm^3.

Solution From the periodic table of the elements in Appendix C, we find that the atomic mass of copper is 63.5 g/mol. Recall that one atomic mass of any substance contains Avogadro's number of atoms, 6.02×10^{23} atoms. Knowing the density of copper enables us to calculate the volume occupied by 63.5 g of copper:

$$V = \frac{m}{\rho} = \frac{63.5 \text{ g}}{8.95 \text{ g/cm}^3} = 7.09 \text{ cm}^3$$

If we now assume that each copper atom contributes one free electron to the body of the material, we have

$$n = \frac{6.02 \times 10^{23} \text{ electrons}}{7.09 \text{ cm}^3}$$

$$= 8.48 \times 10^{22} \text{ electrons/cm}^3$$

$$= \left(8.48 \times 10^{22} \, \frac{\text{electrons}}{\text{cm}^3} \right) \left(10^6 \, \frac{\text{cm}^3}{\text{m}^3} \right)$$

$$= 8.48 \times 10^{28} \text{ electrons/m}^3$$

From Equation 21.4, we find that the drift speed is

$$v_d = \frac{I}{nqA}$$

$$= \frac{10.0 \text{ C/s}}{(8.48 \times 10^{28} \text{ m}^{-3})(1.60 \times 10^{-19} \text{ C})(3.00 \times 10^{-6} \text{ m}^2)}$$

$$= 2.46 \times 10^{-4} \text{ m/s}$$

Example 21.1 shows that typical drift speeds are very small. In fact, the drift speed is much smaller than the average speed between collisions. For instance, electrons traveling with this speed would take about 68 min to travel 1 m! In view of this low speed, you might wonder why a light turns on almost instantaneously when a switch is thrown. In a conductor, the electric field that drives the free electrons travels through the conductor with a speed close to that of light. Thus, when you flip a light switch, the message for the electrons to start moving through the wire (the electric field) reaches them at a speed on the order of 10^7 m/s.

EXERCISE 1 If a current of 80.0 mA exists in a metal wire, how many electrons flow past a given cross section of the wire in 10.0 min?   Answer 3.0×10^{20} electrons

21.2 • RESISTANCE AND OHM'S LAW

When a voltage (potential difference) ΔV is applied across the ends of a metallic conductor, as in Figure 21.4, the current in the conductor is found to be proportional to the applied voltage; that is, $I \propto \Delta V$. If the proportionality is exact, we can write $\Delta V = IR$, where the proportionality constant R is called the resistance of the

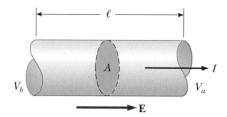

Figure 21.4 A uniform conductor of length ℓ and cross-sectional area A. A potential difference $V_b - V_a$ maintained across the conductor sets up an electric field **E** in the conductor, and this field produces a current I that is proportional to the potential difference.

conductor. In fact, we define this **resistance** as the ratio of the voltage across the conductor to the current it carries:

$$R \equiv \frac{\Delta V}{I}$$

[21.7]   • *Resistance*

Resistance has the SI units volts per ampere, called **ohms (Ω).** Thus, if a potential difference of 1 V across a conductor produces a current of 1 A, the resistance of the conductor is 1 Ω. For example, if an electrical appliance connected to a 120-V source carries a current of 6 A, its resistance is 20 Ω.

It is useful to compare the concepts of electric current, voltage, and resistance with the flow of water in a river. As water flows downhill in a river of constant width and depth, the flow rate (water current) depends on the angle of flow and the effects of rocks, the river bank, and other obstructions. Likewise, electric current in a uniform conductor depends on the applied voltage and the resistance of the conductor caused by collisions of the electrons with atoms in the conductor.

For many materials, including most metals, experiments show that **the resistance is constant over a wide range of applied voltages.** This statement is known as **Ohm's law** after Georg Simon Ohm (1787–1854), who was the first to conduct a systematic study of electrical resistance.

Ohm's law is *not* a fundamental law of nature, but an empirical relationship that is valid only for certain materials. Materials that obey Ohm's law, and hence have a constant resistance over a wide range of voltages, are said to be **ohmic.** Materials that do not obey Ohm's law are **nonohmic.** Ohmic materials have a linear current-voltage relationship over a large range of applied voltages (Fig. 21.5a). Nonohmic materials have a nonlinear current-voltage relationship (Fig. 21.5b). One

Georg Simon Ohm (1787–1854). *(Courtesy of North Wind Picture Archives)*

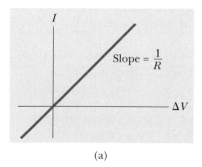

(a)

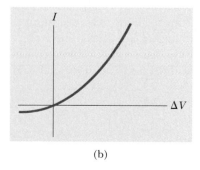

(b)

Figure 21.5 (a) The current-voltage curve for an ohmic material. The curve is linear, and the slope gives the resistance of the conductor. (b) A nonlinear current-voltage curve for a semiconducting diode. This device does not obey Ohm's law.

TABLE 21.1   Resistivities and Temperature Coefficients of Resistivity for Various Materials

Material	Resistivity[a] $(\Omega \cdot m)$	Temperature Coefficient $\alpha[(°C)^{-1}]$
Silver	1.59×10^{-8}	3.8×10^{-3}
Copper	1.7×10^{-8}	3.9×10^{-3}
Gold	2.44×10^{-8}	3.4×10^{-3}
Aluminum	2.82×10^{-8}	3.9×10^{-3}
Tungsten	5.6×10^{-8}	4.5×10^{-3}
Iron	10×10^{-8}	5.0×10^{-3}
Platinum	11×10^{-8}	3.92×10^{-3}
Lead	22×10^{-8}	3.9×10^{-3}
Nichrome[b]	1.50×10^{-6}	0.4×10^{-3}
Carbon	3.5×10^{-5}	-0.5×10^{-3}
Germanium	0.46	-48×10^{-3}
Silicon	640	-75×10^{-3}
Glass	$10^{10} - 10^{14}$	—
Hard rubber	$\sim 10^{13}$	—
Sulfur	10^{15}	—
Quartz (fused)	75×10^{16}	—

[a] All values at 20°C.

[b] A nickel–chromium alloy commonly used in heating elements.

common semiconducting device that is nonohmic is the diode. Its resistance is small for currents in one direction (positive ΔV) and large for currents in the reverse direction (negative ΔV). Most modern electronic devices, such as transistors, have nonlinear current-voltage relationships; their operation depends on the particular ways in which they violate Ohm's law.

We can express Equation 21.7 in the form

$$\Delta V = IR \qquad \text{[21.8]}$$

where R is understood to be independent of ΔV. We shall use this expression later in the discussion of electrical circuits. A **resistor** is a simple circuit element that provides a specified resistance in an electrical circuit. The symbol for a resistor in circuit diagrams is a zigzag line (–⋀⋀⋀–).

The resistance of an ohmic conducting wire is found to be proportional to its length and inversely proportional to its cross-sectional area. That is,

$$R = \rho \frac{\ell}{A} \qquad \text{[21.9]}$$

where the constant of proportionality ρ is called[1] the **resistivity** of the material, which has the unit ohm-meter $(\Omega \cdot m)$. To understand this relationship between

[1]The symbol ρ used for resistivity should not be confused with the same symbol used earlier in the text for mass density and charge density.

resistance and resistivity, note that every ohmic material has a characteristic resistivity, a parameter that depends on the properties of the material and on temperature. However, as you can see from Equation 21.7, the resistance of a conductor depends on size and shape as well as on resistivity. Table 21.1 provides a list of resistivities for various materials measured at 20°C.

The inverse of the resistivity is defined[2] as the **conductivity, σ.** Hence, the resistance of an ohmic conductor can also be expressed in terms of its conductivity as

$$R = \frac{\ell}{\sigma A} \qquad\qquad [21.10]$$

where $\sigma(= 1/\rho)$ has the unit $(\Omega \cdot m)^{-1}$.

Equation 21.10 shows that the resistance of a cylindrical conductor is proportional to its length and inversely proportional to its cross-sectional area. This is analogous to the flow of liquid through a pipe. As the length of the pipe is increased, the resistance to liquid flow increases because of a gain in friction between the fluid and the walls of the pipe. As its cross-sectional area is increased, the pipe can transport more fluid in a given time interval, so its resistance drops.

Thinking Physics 2

We have seen that an electric field must exist inside a conductor that carries a current. How is this possible in view of the fact that in electrostatics we concluded that the electric field is zero inside a conductor?

Reasoning   In the electrostatic case in which charges are stationary, the internal electric field must be zero because a nonzero field would produce a current (by interacting with the free electrons in the conductor), which would violate the condition of static equilibrium. In this chapter we deal with conductors that carry current, a nonelectrostatic situation. The current arises because of a potential difference applied between the ends of the conductor, which produces an internal electric field. So there is no paradox.

CONCEPTUAL PROBLEM 1

Newspaper articles often have statements such as, "10 000 volts of electricity surged through the victim's body." What is wrong with this statement?

Example 21.2   The Resistance of Nichrome Wire

(a) Calculate the resistance per unit length of a 22-gauge Nichrome wire, which has a radius of 0.321 mm.

Solution   The cross-sectional area of this wire is

$$A = \pi r^2 = \pi(0.321 \times 10^{-3} \text{ m})^2 = 3.24 \times 10^{-7} \text{ m}^2$$

The resistivity of Nichrome is $1.5 \times 10^{-6} \ \Omega \cdot m$ (Table 21.1). Thus, we can use Equation 21.9 to find the resistance per unit length:

$$\frac{R}{\ell} = \frac{\rho}{A} = \frac{1.5 \times 10^{-6} \ \Omega \cdot m}{3.24 \times 10^{-7} \text{ m}^2} = 4.6 \ \Omega/m$$

(b) If a potential difference of 10 V is maintained across a 1.0-m length of the Nichrome wire, what is the current in the wire?

Solution   Because a 1.0-m length of this wire has a resistance of 4.6 Ω, Equation 21.7 gives

---

[2]Again, do not confuse the symbol σ for conductivity with the same symbol used for surface charge density.

$$I = \frac{\Delta V}{R} = \frac{10\text{ V}}{4.6\text{ }\Omega} = 2.2\text{ A}$$

Note from Table 21.1 that the resistivity of Nichrome wire is about 100 times that of copper. Therefore, a copper wire of the same radius would have a resistance per unit length of only 0.052 Ω/m. A 1.0-m length of copper wire of the same radius would carry the same current (2.2 A) with an applied voltage of only 0.11 V.

Because of its high resistivity and its resistance to oxida-tion, Nichrome is often used for heating elements in toasters, irons, and electric heaters.

EXERCISE 2   What is the resistance of a 6.0-m length of 22-gauge Nichrome wire? How much current does it carry when connected to a 120-V source?   Answer   28 Ω; 4.3 A

EXERCISE 3   Calculate the current density and electric field in the wire assuming that it carries a current of 2.2 A. Answer 6.7×10^6 A/m^2; 10 N/C

Change in Resistivity with Temperature

Resistivity depends on a number of factors, one of which is temperature. For most metals, resistivity increases approximately linearly with increasing temperature over a limited temperature range, according to the expression

Variation of ρ with temperature •

$$\rho = \rho_0[1 + \alpha(T - T_0)] \qquad [21.11]$$

where ρ is the resistivity at some temperature T (in degrees Celsius), ρ_0 is the resistivity at some reference temperature T_0 (usually 20°C), and α is called the **temperature coefficient of resistivity.** From Equation 21.11, we see that α can also be expressed as

Temperature coefficient of resistivity •

$$\alpha = \frac{1}{\rho_0}\frac{\Delta\rho}{\Delta T} \qquad [21.12]$$

where $\Delta\rho = \rho - \rho_0$ is the change in resistivity in the temperature interval $\Delta T = T - T_0$.

The resistivities and temperature coefficients of certain materials are listed in Table 21.1. Note the enormous range in resistivities, from very low values for good conductors, such as copper and silver, to very high values for good insulators, such as glass and rubber. An ideal, or "perfect," conductor would have zero resistivity, and an ideal insulator would have infinite resistivity.

Because resistance is proportional to resistivity according to Equation 21.9, the temperature variation of the resistance can be written

$$R = R_0[1 + \alpha(T - T_0)] \qquad [21.13]$$

Precise temperature measurements are often made using this property, as shown in Example 21.3.

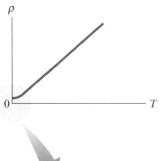

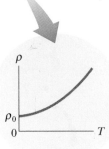

Figure 21.6   Resistivity versus temperature for a normal metal, such as copper. The curve is linear over a wide range of temperatures, and ρ increases with increasing temperature. As T approaches absolute zero (insert), the resistivity approaches a finite value ρ_0.

CONCEPTUAL PROBLEM 2

Aliens with strange powers visit Earth and double every linear dimension of every object on the surface of the earth. Does the electrical cord from the wall socket to your floor lamp now have more resistance than before, less resistance, or the same resistance? Does the lightbulb filament glow more brightly than before, less brightly, or the same? (Assume the resistivities of materials remain the same.)

CONCEPTUAL PROBLEM 3

When incandescent bulbs burn out, they usually do so just after they are switched on. Why?

Example 21.3 A Platinum Resistance Thermometer

A resistance thermometer, which measures temperature by measuring the change in resistance of a conductor, is made from platinum and has a resistance of 50.0 Ω at 20.0°C. When immersed in a vessel containing melting indium, its resistance increases to 76.8 Ω. What is the melting point of indium?

Solution Solving Equation 21.13 for ΔT and obtaining α from Table 21.1, we get

$$\Delta T = \frac{R - R_0}{\alpha R_0} = \frac{76.8\ \Omega - 50.0\ \Omega}{[3.92 \times 10^{-3}\ (\text{°C})^{-1}](50.0\ \Omega)} = 137\text{°C}$$

Because $T_0 = 20.0$°C, we find that $T = \boxed{157\text{°C}}$.

For several metals, resistivity is nearly proportional to absolute temperature, as shown in Figure 21.6. In reality, however, there is always a nonlinear region at very low temperatures, and the resistivity usually approaches some finite value near absolute zero (see the magnified inset in Fig. 21.6). This residual resistivity near absolute zero is due primarily to collisions of electrons with impurities and to imperfections in the metal. In contrast, the high-temperature resistivity (the linear region) is dominated by collisions of electrons with the metal atoms. We shall describe this process in more detail in Section 21.4.

Semiconductors, such as silicon and germanium, have intermediate resistivity values. Their resistivity generally decreases with increasing temperature, corresponding to a negative temperature coefficient of resistivity (Fig. 21.7). This is due to the increase in the density of charge carriers at the higher temperatures. Because the charge carriers in a semiconductor are often associated with impurity atoms, the resistivity is very sensitive to the type and concentration of such impurities. A **thermistor** is a semiconducting thermometer that makes use of the large changes in its resistivity with temperature.

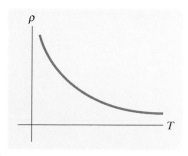

Figure 21.7 Resistivity versus temperature for a pure semiconductor, such as silicon or germanium.

EXERCISE 4 If a silver wire has a resistance of 10 Ω at 20°C, what resistance does it have at 40°C? Neglect any change in length or cross-sectional area due to the change in temperature.
Answer 10.8 Ω

21.3 • SUPERCONDUCTORS

There is a class of metals and compounds the resistances of which go to zero below certain *critical temperatures*, T_c. These materials are known as **superconductors.** The resistance–temperature graph for a superconductor follows that of a normal metal at temperatures greater than T_c (Fig. 21.8). When the temperature is equal to or less than T_c, the resistivity drops suddenly to zero. This phenomenon was discovered by the Dutch physicist Heike Kamerlingh Onnes in 1911 as he worked with mercury, which is a superconductor below 4.2 K. Recent measurements have shown that the resistivities of superconductors below T_c are less than 4×10^{-25} $\Omega \cdot$m, which is around 10^{17} times smaller than the resistivity of copper and considered to be zero in practice.

Today thousands of superconductors are known. Such common metals as aluminum, tin, lead, zinc, and indium are superconductors. Table 21.2 lists the critical temperatures of several superconductors. The value of T_c is sensitive to chemical composition, pressure, and crystalline structure. It is interesting to note that copper, silver, and gold, which are excellent conductors, do not exhibit superconductivity.

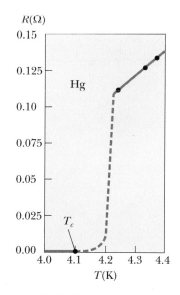

Figure 21.8 Resistance versus temperature for mercury. The graph follows that of a normal metal above the critical temperature, T_c. The resistance drops to zero at the critical temperature, which is 4.2 K for mercury.

TABLE 21.2
Critical Temperatures for Various Superconductors

Material	T_C (K)
$YBa_2Cu_3O_7$	92
Bi-Sr-Ca-Cu-O	105
Tl-Ba-Ca-Cu-O	125
$HgBa_2Ca_2Cu_3O_8$	134
Nb_3Ge	23.2
Nb_3Sn	21.05
Nb	9.46
Pb	7.18
Hg	4.15
Sn	3.72
Al	1.19
Zn	0.88

One of the truly remarkable features of superconductors is the fact that once a current is set up in them, it persists *without any applied voltage* (because $R = 0$). In fact, steady currents have been observed to persist in superconducting loops for several years with no apparent decay!

An important recent development in physics that has created much excitement in the scientific community has been the discovery of high-temperature copper-oxide-based superconductors. The excitement began with a 1986 publication by Georg Bednorz and K. Alex Müller, scientists at the IBM Zurich Research Laboratory in Switzerland, in which they reported evidence for superconductivity at a temperature near 30 K in an oxide of barium, lanthanum, and copper. Bednorz and Müller were awarded the Nobel Prize in 1987 for their remarkable discovery. Shortly thereafter, a new family of compounds was open for investigation, and research activity in the field of superconductivity proceeded vigorously. In early 1987, groups at the University of Alabama at Huntsville and the University of Houston announced the discovery of superconductivity at about 92 K in an oxide of yttrium, barium, and copper ($YBa_2Cu_3O_7$). Late in 1987, teams of scientists from Japan and the United States reported superconductivity at 105 K in an oxide of bismuth, strontium, calcium, and copper. More recently, scientists have reported superconductivity at temperatures as high as 125 K in an oxide containing thallium. At this point one cannot rule out the possibility of room-temperature superconductivity, and the search for novel superconducting materials continues. It is an important search both for scientific reasons and because practical applications become more probable and widespread as the critical temperature is raised.

An important and useful application is superconducting magnets in which the magnetic field strengths are about ten times greater than those of the best normal electromagnets. Such superconducting magnets are being considered as a means of storing energy. The idea of using superconducting power lines for transmitting power efficiently is also receiving some consideration. Modern superconducting electronic devices consisting of two thin-film superconductors separated by a thin insulator have been constructed. They include magnetometers (a magnetic-field measuring device) and various microwave devices.

Photograph of a small permanent magnet levitated above a disk of the superconductor $YBa_2Cu_3O_7$, which is at 77 K. This superconductor has zero electric resistance at temperatures below 92 K and expels any applied magnetic field.
(Courtesy of IBM Research Laboratory)

21.4 • A MODEL FOR ELECTRICAL CONDUCTION

The classical model of electrical conduction in metals leads to Ohm's law and shows that resistivity can be related to the motion of electrons in metals.

Consider a conductor as a regular array of atoms containing free electrons (sometimes called *conduction* electrons). Such electrons are free to move through the conductor (as we learned in our discussion of drift speed in Section 21.1) and are approximately equal in number to the atoms. In the absence of an electric field, the free electrons move in random directions with average speeds on the order of 10^6 m/s. The situation is similar to the motion of gas molecules confined in a vessel. In fact, some scientists refer to conduction electrons in a metal as an *electron gas*.

The conduction electrons are not totally free, because they are confined to the interior of the conductor and undergo frequent collisions with the array of atoms. The collisions are the predominant mechanism for the resistivity of a metal at normal temperatures. Note that there is no current through a conductor in the absence of an electric field, because the average velocity of the free electrons is zero. In

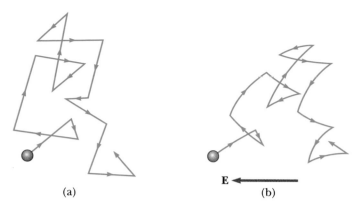

(a)   E ◄——————   (b)

Figure 21.9 (a) A schematic diagram of the random motion of a charge carrier in a conductor in the absence of an electric field. The drift velocity is zero. (b) The motion of a charge carrier in a conductor in the presence of an electric field. Note that the random motion is modified by the field, and the charge carrier has a drift velocity.

other words, just as many electrons move in one direction as in the opposite direction, on the average, and so there is no net flow of charge.

The situation is modified when an electric field is applied to the metal. In addition to random thermal motion, the free electrons drift slowly in a direction opposite that of the electric field, with an average drift speed of v_d, which is much less (typically 10^{-4} m/s; see Example 21.1) than the average speed between collisions (typically 10^6 m/s). Figure 21.9 provides a crude depiction of the motion of free electrons in a conductor. In the absence of an electric field, there is no net displacement after many collisions (Fig. 21.9a). An electric field **E** modifies the random motion and causes the electrons to drift in a direction opposite that of **E** (Fig. 21.9b). The slight curvature in the paths in Figure 21.9b results from the acceleration of the electrons between collisions, caused by the applied field. One mechanical system somewhat analogous to this situation is a ball rolling down a slightly inclined plane through an array of closely spaced, fixed pegs (Fig. 21.10). The ball represents a conduction electron, the pegs represent defects in the crystal lattice, and the component of the gravitational force along the incline represents the electric force, $e\mathbf{E}$.

In our model, we shall assume that the excess energy acquired by the electrons in the electric field is lost to the conductor in the collision process. The energy given up to the atoms in the collisions increases the vibrational energy of the atoms, causing the conductor to warm up. The model also assumes that an electron's motion after a collision is independent of its motion *before* the collision.[3]

We are now in a position to obtain an expression for the drift velocity. When a mobile, charged particle of mass m and charge q is subjected to an electric field

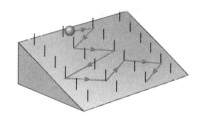

Figure 21.10 A mechanical system somewhat analogous to the motion of charge carriers in the presence of an electric field. The collisions of the ball with the pegs represent the resistance to the ball's motion down the incline.

[3]Because the collision process is random, each collision event is *independent* of what happened earlier. This is analogous to the random process of throwing a die. The probability of rolling a particular number on one throw is independent of the result of the previous throw. On the average, it would take six throws to come up with that number, starting at any arbitrary time.

E, it experiences a force of $q\mathbf{E}$. Because $\mathbf{F} = m\mathbf{a}$, we conclude that the acceleration of the particle is

$$\mathbf{a} = \frac{q\mathbf{E}}{m} \qquad \text{[21.14]}$$

This acceleration, which occurs for only a short time between collisions, enables the electrons to acquire a small drift velocity. If t is the time since the last collision (at $t = 0$), and $\mathbf{v}_0$ is the initial velocity, then the velocity of the electron after the time t is

$$\mathbf{v} = \mathbf{v}_0 + \mathbf{a}t = \mathbf{v}_0 + \frac{q\mathbf{E}}{m}\,t \qquad \text{[21.15]}$$

We now take the average value of $\mathbf{v}$ over all possible times t and all possible values of $\mathbf{v}_0$. If the initial velocities are assumed to be randomly distributed in space, we see that the average value of $\mathbf{v}_0$ is zero. The term $(qE/m)\,t$ is the velocity added by the field at the end of one trip between atoms. If the electron starts with zero velocity, the average value of the second term of Equation 21.15 is $(qE/m)\tau$, where τ is the *average time between collisions*. Because the average of $\mathbf{v}$ is equal to the drift velocity, we have

Drift velocity •
$$\mathbf{v}_d = \frac{q\mathbf{E}}{m}\,\tau \qquad \text{[21.16]}$$

Substituting this result into Equation 21.6, we find that the magnitude of the current density is

Current density •
$$J = nqv_d = \frac{nq^2E}{m}\,\tau \qquad \text{[21.17]}$$

Comparing this expression with an alternative form of Equation 21.7,[4] $J = \sigma E$, we obtain the following relationships for the conductivity and resistivity:

Conductivity •
$$\sigma = \frac{nq^2\tau}{m} \qquad \text{[21.18]}$$

Resistivity •
$$\rho = \frac{1}{\sigma} = \frac{m}{nq^2\tau} \qquad \text{[21.19]}$$

According to this classical model, conductivity and resistivity do not depend on the electric field. This feature is characteristic of a conductor obeying Ohm's law. The model shows that the conductivity can be calculated from a knowledge of the density of the charge carriers, their charge and mass, and the average time between

---

[4]The relation $J = \sigma E$ can be derived as follows: The potential difference across a conductor of length ℓ is $\Delta V = E\ell$, and, from the definition of resistance, $\Delta V = IR$. Using these relations, together with Equations 21.5 and 21.10, we find that the magnitude of the current density is $J = I/A = \Delta V/RA = E\ell/RA = \sigma E$.

collisions, which is related to the average distance between collisions ℓ (the mean free path) and the average thermal speed $\bar{v}$ through the expression[5]

$$\tau = \frac{\ell}{\bar{v}} \qquad\qquad [21.20]$$

Example 21.4   Electron Collisions in Copper

(a) Using the data and results from Example 21.1 and the classical model of electron conduction, estimate the average time between collisions for electrons in copper at 20°C.

Solution   From Equation 21.19 we see that

$$\tau = \frac{m}{nq^2\rho}$$

where $\rho = 1.7 \times 10^{-8}\ \Omega\cdot m$ for copper and the carrier density $n = 8.48 \times 10^{28}$ electrons/m^3 for the wire described in Example 21.1. Substitution of these values into the previous expression gives

$$\tau = \frac{(9.11 \times 10^{-31}\ \text{kg})}{(8.48 \times 10^{28}\ \text{m}^{-3})(1.6 \times 10^{-19}\ \text{C})^2(1.7 \times 10^{-8}\ \Omega\cdot\text{m})}$$
$$= 2.5 \times 10^{-14}\ \text{s}$$

(b) Assuming the average speed for free electrons in copper to be 1.6×10^6 m/s and using the result from part (a), calculate the mean free path for electrons in copper.

Solution

$$\ell = \bar{v}\tau = (1.6 \times 10^6\ \text{m/s})(2.5 \times 10^{-14}\ \text{s}) = 4.0 \times 10^{-8}\ \text{m}$$

which is equivalent to 40 nm (compared with atomic spacings of about 0.2 nm). Thus, although the time between collisions is very short, the electrons travel about 200 atomic distances before colliding with an atom.

Although this classical model of conduction is consistent with Ohm's law, it is not satisfactory for explaining some important phenomena. For example, classical calculations for $\bar{v}$ using the ideal-gas model are about a factor of 10 smaller than the true values. Furthermore, according to Equations 21.19 and 21.20, the temperature variation of the resistivity is predicted to vary as $\bar{v}$, which, according to an ideal-gas model (Chap. 16, Eq. 16.15), is proportional to $\sqrt{T}$. This is in disagreement with the linear dependence of resistivity with temperature for pure metals (Fig. 21.5a). It is possible to account for such observations only by using a quantum mechanical model, which we shall describe briefly.

According to quantum mechanics, electrons have wave-like properties. If the array of atoms in a conductor is regularly spaced (that is, periodic), the wave-like character of the electrons makes it possible for them to move freely through the conductor, and a collision with an atom is unlikely. For an idealized conductor there would be no collisions, the mean free path would be infinite, and the resistivity would be zero. Electron waves are scattered only if the atomic arrangement is irregular (not periodic)—for example, as a result of structural defects or impurities. At low temperatures, the resistivity of metals is dominated by scattering caused by collisions between the electrons and impurities. At high temperatures, the resistivity is dominated by scattering caused by collisions between the electrons and the atoms of the conductor, which are continuously displaced as a result of thermal agitation. The thermal motion of the atoms makes the structure irregular (compared with an atomic array at rest), thereby reducing the electron's mean free path.

---

[5] Recall that the average speed is the average of the speeds that particles have as a consequence of the temperature of the system of particles (Chap. 16).

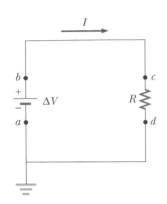

Figure 21.11 A circuit consisting of a battery of emf $\mathcal{E}$ and resistance R. Positive charge flows in the clockwise direction, from the negative to the positive terminal of the battery. Points a and d are grounded.

This versatile circuit enables the experimenter to examine the properties of circuit elements such as capacitors and resistors and their effects on circuit behavior. *(Courtesy of Central Scientific Company)*

CONCEPTUAL PROBLEM 4

Why don't the free electrons in a metal fall to the bottom of the metal due to gravity? And charges in a conductor are supposed to reside on the surface—why don't the free electrons all go to the surface?

21.5 • ELECTRICAL ENERGY AND POWER

If a battery is used to establish an electric current in a conductor, there occurs a continuous transformation of chemical energy stored in the battery to kinetic energy of the charge carriers. This kinetic energy is quickly lost as a result of collisions between the charge carriers and the lattice ions, resulting in an increase in the temperature of the conductor. Thus, the chemical energy stored in the battery is continuously transformed into thermal energy.

In order to understand the process of energy transfer in a simple circuit, consider a battery the terminals of which are connected to a resistor R, as shown in Figure 21.11. (Remember that the positive terminal of the battery is always at the higher potential.) Now imagine following a positive quantity of charge ΔQ around the circuit from point a through the battery and resistor and back to a. Point a is a reference point that is grounded (the ground symbol is $\perp$), and its potential is taken to be zero. As the charge moves from a to b through the battery the potential difference of which is ΔV, its electrical potential energy increases by the amount $\Delta Q \Delta V$, and the chemical potential energy in the battery decreases by the same amount. (Recall from Chap. 20 that $\Delta U = q \Delta V$.) However, as the charge moves from c to d through the resistor, it loses this electrical potential energy during collisions with atoms in the resistor, thereby producing thermal energy. Note that, if we neglect the resistance of the interconnecting wires, no loss in energy occurs for paths bc and da. When the charge returns to point a, it must have the same potential energy (zero) as it had at the start.[6]

The rate at which the charge ΔQ loses potential energy in going through the resistor is

$$\frac{\Delta U}{\Delta t} = \frac{\Delta Q}{\Delta t} \Delta V = I \Delta V$$

where I is the current in the circuit. Of course, the charge regains this energy when it passes through the battery. Because the rate at which the charge loses energy equals the power P dissipated in the resistor, we have

$$P = I \Delta V \qquad \text{[21.21]}$$

In this case, the power is supplied to a resistor by a battery. However, Equation 21.21 can be used to determine the power transferred from a battery to *any* device carrying a current I and having a potential difference ΔV between its terminals.

Using Equation 21.21 and the fact that $\Delta V = IR$ for a resistor, we can express the power dissipated by the resistor in the alternative forms

---

[6]Note that when the current reaches its steady-state value, there is *no* change with time in the kinetic energy associated with the current.

$$P = I^2R = \frac{(\Delta V)^2}{R}$$ [21.22] • *Power dissipated by a resistor*

The SI unit of power is the watt, introduced in Chapter 6. The power dissipated in a conductor of resistance R is also often referred to as an I^2R *loss.*

As we learned in Chapter 6, Section 6.5, the unit of energy the electric company uses to calculate energy consumption, the kilowatt-hour, is the energy consumed in 1 h at the constant rate of 1 kW. Because 1 W = 1 J/s, we have

$$1 \text{ kWh} = (10^3 \text{ W})(3600 \text{ s}) = 3.6 \times 10^6 \text{ J}$$ [21.23]

Thinking Physics 3

When is more power being delivered to a lightbulb—just after it is turned on and the glow of the filament is increasing or after it has been on for a few seconds and the glow is steady?

Reasoning Once the switch is closed, the line voltage is applied across the lightbulb. As the voltage is applied across the cold filament when first turned on, the resistance of the filament is low. Thus, the current is high, and a relatively large amount of power is delivered to the bulb. As the filament warms up, its resistance rises, and the current falls. As a result, the power delivered to the bulb falls. The large current spike at the beginning of operation is the reason that lightbulbs often fail just as they are turned on as noted in Conceptual Problem 3.

Thinking Physics 4

Two lightbulbs A and B are connected across the same potential difference, as in Figure 21.12. The resistance of A is twice that of B. Which lightbulb dissipates more power? Which carries the greater current?

Reasoning Because the voltage across each lightbulb is the same, and the power dissipated by a conductor is $P = (\Delta V)^2/R$, the conductor with the lower resistance will dissipate more power. In this case, the power dissipated by B is twice that of A and provides twice as much illumination. Furthermore, because $P = I\,\Delta V$, we see that the current carried by B is twice that of A.

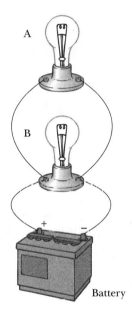

Figure 21.12 (Thinking Physics 4)

Example 21.5 **Electrical Rating of a Lightbulb**

A lightbulb is rated at 120 V/75 W, which means its operating voltage is 120 V and it has a power rating of 75.0 W. The bulb is powered by a 120-V direct-current power supply. Find the current in the bulb and its resistance.

Solution Because the power rating of the bulb is 75.0 W and the operating voltage is 120 V, we can use $P = I\,\Delta V$ to find the current:

$$I = \frac{P}{\Delta V} = \frac{75.0 \text{ W}}{120 \text{ V}} = 0.625 \text{ A}$$

Using $\Delta V = IR$, the resistance is calculated to be

$$R = \frac{\Delta V}{I} = \frac{120 \text{ V}}{0.625 \text{ A}} = 192 \text{ }\Omega$$

EXERCISE 5 What would the resistance be in a lamp rated at 120 V and 100 W? Answer 144 Ω

Example 21.6   The Cost of Operating a Lightbulb

How much does it cost to burn a 100-W lightbulb for 24 h if electricity costs eight cents per kilowatt hour?

Solution   Because the energy consumed equals power × time, the amount of energy you must pay for, expressed in kWh, is

$$\text{Energy} = (0.10 \text{ kW})(24 \text{ h}) = 2.4 \text{ kWh}$$

If energy is purchased at 8¢ per kWh, the cost is

$$\text{Cost} = (2.4 \text{ kWh})(\$0.080/\text{kWh}) = \boxed{\$0.19}$$

That is, it costs 19¢ to operate the lightbulb for one day. This is a small amount, but when larger and more complex devices are being used, the costs go up rapidly.

Demands on our energy supplies have made it necessary to be aware of the energy requirements of our electric devices. This is true not only because they are becoming more expensive to operate but also because, with the dwindling of the coal and oil resources that ultimately supply us with electrical energy, increased awareness of conservation becomes necessary. On every electric appliance is a label that contains the information you need to calculate the power requirements of the appliance. The power consumption in watts is often stated directly, as on a lightbulb. In other cases, the amount of current used by the device and the voltage at which it operates are given. This information and Equation 21.21 are sufficient to calculate the operating cost of any electric device.

EXERCISE 6   If electricity costs 8¢ per kilowatt hour, what does it cost to operate an electric oven, which operates at 20.0 A and 220 V, for 5.00 h?   Answer $1.76

EXERCISE 7   A 12-V battery is connected to a 60-Ω resistor. Neglecting the internal resistance of the battery, calculate the power dissipated in the resistor.   Answer   2.4 W

21.6 · SOURCES OF emf

The source that maintains the constant voltage in Figure 21.13 is called an "emf."[7] Sources of emf are any devices (such as batteries and generators) that increase the potential energy of charges circulating in circuits. One can think of a source of emf as a charge pump that forces electrons to move in a direction opposite the electrostatic field inside the source. The emf, $\mathcal{E}$, of a source describes the work done per unit charge, and hence the SI unit of emf is the volt.

Consider the circuit shown in Figure 21.13, consisting of a battery connected to a resistor. We shall assume that the connecting wires have no resistance. If we neglect the internal resistance of the battery, the potential difference across the battery (the terminal voltage) equals the emf of the battery. However, because a real battery always has some internal resistance, r, the terminal voltage is not equal to the emf. The circuit shown in Figure 21.13 can be described by the circuit diagram in Figure 21.14a. The battery within the dashed rectangle is represented by a source of emf, $\mathcal{E}$, in series with the internal resistance r. Now imagine a positive charge moving from a to b in Figure 21.14a. As the charge passes from the negative to the positive terminal within the battery, the potential of the charge increases by $\mathcal{E}$. However, as it moves through the resistance r, its potential decreases by an amount Ir, where I is the current in the circuit. Thus, the terminal voltage of the battery, $\Delta V = V_b - V_a$, is[8]

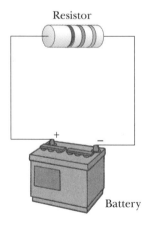

Figure 21.13   A circuit consisting of a resistor connected to the terminals of a battery.

Resistor

+   −

Battery

---

[7]The term was originally an abbreviation for *electromotive force*, but it is not a force, so the long form is discouraged.

[8]The terminal voltage in this case is less than the emf by the amount Ir. In some situations, the terminal voltage may *exceed* the emf by the amount Ir. This happens when the direction of the current is *opposite* that of the emf, as when a battery is charged with another source of emf.

$$\Delta V = \mathcal{E} - Ir \qquad [21.24]$$

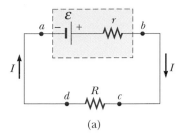

Note from this expression that $\mathcal{E}$ is equivalent to the **open-circuit voltage—that is, the terminal voltage when the current is zero.** Figure 21.14b is a graphical representation of the changes in potential as the circuit is traversed clockwise. By inspecting Figure 21.14a, we see that the terminal voltage ΔV must also equal the potential difference across the external resistance R, often called the **load resistance;** that is, $\Delta V = IR$. Combining this with Equation 21.24, we see that

$$\mathcal{E} = IR + Ir \qquad [21.25]$$

Solving for the current gives

$$I = \frac{\mathcal{E}}{R + r}$$

This shows that the current in this simple circuit depends on both the resistance external to the battery and the internal resistance. If R is much greater than r, we can neglect r in our analysis. In many circuits we shall ignore this internal resistance.

If we multiply Equation 21.25 by the current I, we get

$$I\mathcal{E} = I^2R + I^2r$$

This equation tells us that the total power output of the source of emf, $I\mathcal{E}$, is equal to the power that is dissipated in the load resistance, I^2R, *plus* power that is dissipated in the internal resistance, I^2r. Again, if $r \ll R$, most of the power delivered by the battery is dissipated in the load resistance.

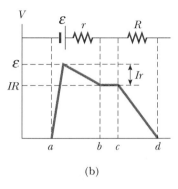

Figure 21.14 (a) Circuit diagram of an emf $\mathcal{E}$ of internal resistance r connected to an external resistor R. (b) Graphical representation showing how the potential changes as the series circuit in part (a) is traversed clockwise.

CONCEPTUAL PROBLEM 5

If the energy transferred to a dead battery during charging is E, is the total energy transferred out of the battery to an electrical load during use in which it completely discharges also E?

CONCEPTUAL PROBLEM 6

If you have your headlights on while you start your car, why do they dim while the car is starting?

CONCEPTUAL PROBLEM 7

Electrical devices are often rated with a voltage and a current—for example, 120 volts, 5 amperes. Batteries, however, are only rated with a voltage—for example, 1.5 volts. Why?

EXERCISE 8 A battery with an emf of 12 V and an internal resistance of 0.90 Ω is connected across a load resistor R. If the current in the circuit is 1.4 A, what is the value of R?
Answer 7.7 Ω

21.7 • RESISTORS IN SERIES AND IN PARALLEL

When two or more resistors are connected together so that they have only one common point per pair, they are said to be in *series*. Figure 21.15 shows two resistors connected in series. Note that the current is the same through the two resistors,

- *For a series connection of resistors, the current is the same in all the resistors.*

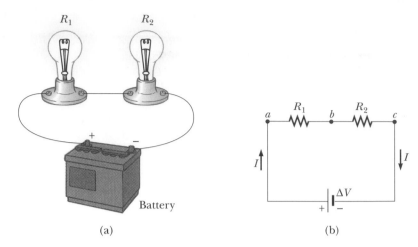

Figure 21.15   Series connection of two resistors, R_1 and R_2. The current is the same in each resistor.

because any charge that flows through R_1 must also flow through R_2. Because the potential difference between a and b in Figure 21.15b equals IR_1 and the potential difference between b and c equals IR_2, the potential difference between a and c is

$$\Delta V = IR_1 + IR_2 = I(R_1 + R_2)$$

Therefore, we can replace the two resistors in series with a single **equivalent resistance,** R_{eq}, the value of which is the *sum* of the individual resistances:

$$R_{eq} = R_1 + R_2 \qquad \text{[21.26]}$$

The resistance R_{eq} is equivalent to the series combination $R_1 + R_2$ in the sense that the circuit current is unchanged when R_{eq} replaces $R_1 + R_2$. The equivalent resistance of three or more resistors connected in series is simply

Equivalent resistance of •
several resistors in series

$$R_{eq} = R_1 + R_2 + R_3 + \cdots \qquad \text{[21.27]}$$

Therefore, **the equivalent resistance of a series connection of resistors is always greater than any individual resistance and is the algebraic sum of the individual resistances.**

Note that if the filament of one lightbulb in Figure 21.15 were to break, or "burn out," the circuit would no longer be complete (an open-circuit condition would exist) and the second bulb would also go out. Some Christmas-tree-light sets (especially older ones) are connected in this way, and the tedious task of determining which bulb is burned out is familiar to many people.

In many circuits, fuses are used in series with other circuit elements for safety purposes. The conductor in the fuse is designed to melt and open the circuit at some maximum current, the value of which depends on the nature of the circuit. If a fuse were not used, excessive currents could damage circuit elements, overheat wires, and perhaps cause a fire. In modern home construction, circuit breakers are used in place of fuses. When the current in a circuit exceeds some value (typically 15 A), the circuit breaker acts as a switch and opens the circuit.

Now consider two resistors connected in *parallel,* as shown in Figure 21.16. In this case, **the potential differences across the resistors are equal.** However, the currents are generally not the same. When the current I reaches point a (called a *junction*) in Figure 21.16b, it splits into two parts, I_1 going through R_1 and I_2 going through R_2. If R_1 is greater than R_2, then I_1 is less than I_2. Clearly, because charge must be conserved, the current I that enters point a must equal the total current leaving point b:

$$I = I_1 + I_2$$

The potential drops across the resistors must be the *same,* and applying $I = \Delta V/R$ gives

$$I = I_1 + I_2 = \frac{\Delta V}{R_1} + \frac{\Delta V}{R_2} = \Delta V\left(\frac{1}{R_1} + \frac{1}{R_2}\right) = \frac{\Delta V}{R_{eq}}$$

From this result, we see that the equivalent resistance of two resistors in parallel is

$$\frac{1}{R_{eq}} = \frac{1}{R_1} + \frac{1}{R_2} \qquad\qquad \textbf{[21.28]}$$

This can be rearranged to become

$$R_{eq} = \frac{R_1 R_2}{R_1 + R_2}$$

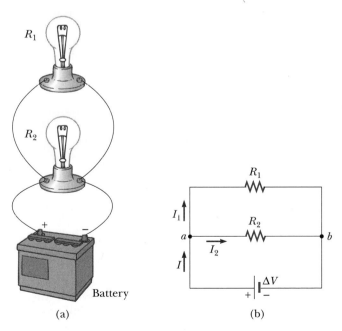

(a)   (b)

Figure 21.16   Parallel connection of two resistors, R_1 and R_2. The potential difference across each resistor is the same, and the equivalent resistance of the combination is given by $R_{eq} = R_1 R_2/(R_1 + R_2)$.

An extension of this analysis to three or more resistors in parallel yields the following general expression:

Equivalent resistance of • several resistors in parallel

$$\frac{1}{R_{eq}} = \frac{1}{R_1} + \frac{1}{R_2} + \frac{1}{R_3} + \cdots$$

[21.29]

From this expression it can be seen that **the equivalent resistance of two or more resistors connected in parallel is always less than the smallest resistance in the group and the inverse of the equivalent resistance is the algebraic sum of the inverses of the individual resistances.**

Household circuits are always wired so that the lightbulbs (or appliances, or whatever) are connected in parallel, as in Figure 21.16a. In this manner, each device operates independently of the others, so that if one is switched off, the others remain on. Equally important, each device operates on the same voltage.

Finally, it is interesting to note that parallel resistors combine in the same way that series capacitors combine, and vice versa.

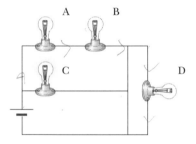

Figure 21.17   (Thinking Physics 5)

Thinking Physics 5

Predict the relative brightnesses of the four identical bulbs in Figure 21.17. What happens if bulb A "burns out," so that it cannot conduct current? What if C "burns out"? What if D "burns out"?

Reasoning   Bulbs A and B are connected in series across the emf of the battery, whereas bulb C is connected by itself across this emf. Thus, the emf is split between bulbs A and B. As a result, bulb C will be brighter than bulbs A and B, which should be equally as bright as each other. Bulb D has an equipotential (the vertical wire) connected across it. Thus, there is no potential difference across D and it does not glow at all. If bulb A "burns out," B goes out but C stays lighted. If C "burns out," there is no effect on the other bulbs. If D "burns out," the event is undetectable, because D was not glowing anyway.

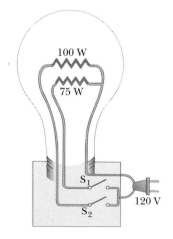

Figure 21.18   (Thinking Physics 6)

Thinking Physics 6

Figure 21.18 illustrates how a three-way lightbulb is constructed to provide three levels of light intensity. The socket of the lamp is equipped with a three-way switch for selecting different light intensities. The bulb contains two filaments. Why are the filaments connected in parallel? Explain how the two filaments are used to provide three different light intensities.

Reasoning   If the filaments were connected in series and one of them were to burn out, no current could pass through the bulb, and the bulb would give no illumination, regardless of the switch position. However, when the filaments are connected in parallel and one of them (say the 75-W filament) burns out, the bulb will still operate in one of the switch positions as current passes through the other (100 W) filament. The three light intensities are made possible by selecting one of three values of filament resistance, using a single value of 120 V for the applied voltage. The 75-W filament offers one value of resistance, the 100-W filament offers a second value, and the third resistance is obtained by combining the two filaments in parallel. When switch 1 is closed and switch 2 is opened, current passes only through the 75-W filament. When switch 1 is open and switch 2 is closed, current passes only through the 100-W filament. When both switches are closed, current passes through both filaments, and a total illumination of 175 W is obtained.

CONCEPTUAL PROBLEM 8

Connecting batteries in series increases the emf. What advantage might there be in connecting them in parallel?

CONCEPTUAL PROBLEM 9

You have a large supply of lightbulbs and a battery. You start with one lightbulb connected to the battery and notice its brightness. You then add one lightbulb at a time, each new bulb being added in series to the previous bulbs. As you add the lightbulbs, what happens to the brightness of the bulbs? To the current through the bulbs? To the power transferred from the battery? To the lifetime of the battery? To the terminal voltage of the battery? Answer the same questions if the lightbulbs are added one by one in parallel with the first.

PROBLEM-SOLVING STRATEGY • Resistors

1. When two or more unequal resistors are connected in *series*, they carry the same current, but the potential differences across them are not the same. The resistors add directly to give the equivalent resistance of the series combination.
2. When two or more unequal resistors are connected in *parallel*, the potential differences across them are the same. Because the current is inversely proportional to the resistance, the currents through them are not the same. The equivalent resistance of a parallel combination of resistors is found through reciprocal addition, and the equivalent resistor is always *less* than the smallest individual resistor.
3. A complicated circuit consisting of resistors can often be reduced to a simple circuit containing only one resistor. To do so, examine the initial circuit and replace any resistors in series or any in parallel using the procedures outlined in Steps 1 and 2. Draw a sketch of the new circuit after these changes have been made. Examine the new circuit and replace any series or parallel combinations. Continue this process until a single equivalent resistance is found.
4. If the current through or the potential difference across a resistor in the complicated circuit is to be found, start with the final circuit found in Step 3 and gradually work your way back through the circuits, using $\Delta V = IR$ and the rules of Steps 1 and 2.

Example 21.7   Find the Equivalent Resistance

Four resistors are connected as shown in Figure 21.19a. (a) Find the equivalent resistance between *a* and *c*.

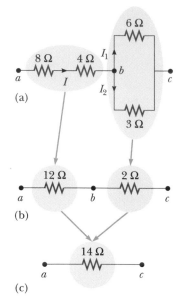

(a)

(b)

(c)

Figure 21.19   (Example 21.17) The resistances of the four resistors shown in (a) can be reduced in steps to an equivalent 14-Ω resistor.

Solution   The circuit can be reduced in steps as shown in Figure 21.19. The 8.0-Ω and 4.0-Ω resistors are in series, and

so the equivalent resistance between *a* and *b* is 12 Ω (Eq. 21.26). The 6.0-Ω and 3.0-Ω resistors are in parallel, and so from Equation 21.28 we find that the equivalent resistance from *b* to *c* is 2.0 Ω. Hence, the equivalent resistance from *a* to *c* is   14 Ω.

(b) What is the current in each resistor if a potential difference of 42 V is maintained between *a* and *c*?

Solution   The current *I* in the 8.0-Ω and 4.0-Ω resistors is the same because they are in series. Using Equation 21.7 and the results from part (a), we get

$$I = \frac{\Delta V_{ac}}{R_{eq}} = \frac{42 \text{ V}}{14 \text{ } \Omega} = \ 3.0 \text{ A}$$

When this current enters the junction at *b*, it splits. Part of it passes through the 6.0-Ω resistor (I_1) and part goes through the 3.0-Ω resistor (I_2). Because the potential difference across these resistors, ΔV_{bc}, is the same (they are in parallel), we see that $6I_1 = 3I_2$, or $I_2 = 2I_1$. Using this result and the fact that $I_1 + I_2 = 3.0$ A, we find that $I_1 = 1.0$ A and $I_2 = 2.0$ A.   We could have guessed this from the start by noting that the current through the 3.0-Ω resistor has to be twice the current through the 6.0-Ω resistor in view of their relative resistances and the fact that the same voltage is applied to each of them.

As a final check, note that $\Delta V_{bc} = 6I_1 = 3I_2 = 6.0$ V and $\Delta V_{ab} = 12I = 36$ V; therefore, $\Delta V_{ac} = \Delta V_{ab} + \Delta V_{bc} = 42$ V, as it must.

Example 21.8   Three Resistors in Parallel

Three resistors are connected in parallel, as in Figure 21.20. A potential difference of 18 V is maintained between points *a* and *b*. (a) Find the current in each resistor.

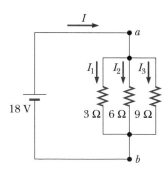

Figure 21.20   (Example 21.8) Three resistors connected in parallel. The voltage across each resistor is 18 V.

Solution   The resistors are in parallel, and the potential difference across each is 18 V. Applying $\Delta V = IR$ to each resistor gives

$$I_1 = \frac{\Delta V}{R_1} = \frac{18 \text{ V}}{3.0 \text{ } \Omega} = \ 6.0 \text{ A}$$

$$I_2 = \frac{\Delta V}{R_2} = \frac{18 \text{ V}}{6.0 \text{ } \Omega} = \ 3.0 \text{ A}$$

$$I_3 = \frac{\Delta V}{R_3} = \frac{18 \text{ V}}{9.0 \text{ } \Omega} = \ 2.0 \text{ A}$$

(b) Calculate the power dissipated by each resistor and the total power dissipated by the three resistors.

Solution   Applying $P = I^2R$ to each resistor gives

$$3.0\text{-}\Omega: \ P_1 = I_1{}^2R_1 = (6.0 \text{ A})^2(3.0 \text{ } \Omega) = \ 110 \text{ W}$$

6.0-Ω: $P_2 = I_2{}^2 R_2 = (3.0\text{ A})^2 (6.0\ \Omega) = \boxed{54\text{ W}}$

9.0-Ω: $P_3 = I_3{}^2 R_3 = (2.0\text{ A})^2 (9.0\ \Omega) = \boxed{36\text{ W}}$

This shows that the smallest resistor dissipates the most power because it carries the most current. (Note that you can also use $P = (\Delta V)^2 / R$ to find the power dissipated by each resistor.) Summing the three quantities gives a total power of $\boxed{200\text{ W}}$.

(c) Calculate the equivalent resistance of the three resistors. We can use Equation 21.28 to find R_{eq}:

Solution

$$\frac{1}{R_{eq}} = \frac{1}{3.0} + \frac{1}{6.0} + \frac{1}{9.0}$$

$$R_{eq} = \frac{18}{11}\ \Omega = \boxed{1.6\ \Omega}$$

EXERCISE 9   Use R_{eq} to calculate the total power dissipated in the circuit.   Answer   200 W

21.8 • KIRCHHOFF'S RULES AND SIMPLE DC CIRCUITS

As indicated in the preceding section, we can analyze simple circuits using $\Delta V = IR$ and the rules for series and parallel combinations of resistors. However, there are many ways in which resistors can be connected so that the circuits formed cannot be reduced to a single equivalent resistor. The procedure for analyzing such complex circuits is greatly simplified by the use of two simple rules called **Kirchhoff's rules:**

1. The sum of the currents entering any junction must equal the sum of the currents leaving that junction. (This rule is often referred to as the **junction rule.**)
2. The sum of the potential differences across each element around any closed circuit loop must be zero. (This rule is usually called the **loop rule.**)

The junction rule is a statement of **conservation of charge.** Whatever current enters a given point in a circuit must leave that point, because charge cannot build up or disappear at a point. If we apply this rule to the junction in Figure 21.21a, we get

$$I_1 = I_2 + I_3$$

Figure 21.21b represents a mechanical analog to this situation in which water flows through a branched pipe with no leaks. The flow rate into the pipe equals the total flow rate out of the two branches.

The second rule is equivalent to the law of **conservation of energy.** A charge that moves around any closed loop in a circuit (the charge starts and ends at the same point) must gain as much energy as it loses if a potential is defined for each point in the circuit. Its energy may decrease in the form of a potential drop, $-IR$, across a resistor or as a result of having the charge move in the reverse direction through an emf. In the latter case, electric potential energy is converted to chemical energy as the battery is charged. In a similar way, electrical energy may be converted to mechanical energy for operating a motor.

As an aid in applying the loop rule, the following points should be noted. They are summarized in Figure 21.22, where it is assumed that movement is from point a toward point b:

• If a resistor is traversed in the direction of the current, the change in potential across the resistor is $-IR$ (Fig. 21.22a).

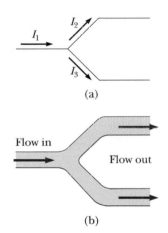

Figure 21.21   (a) A schematic diagram illustrating Kirchhoff's junction rule. Conservation of charge requires that whatever current enters a junction must leave that junction. Therefore, in this case, $I_1 = I_2 + I_3$. (b) A mechanical analog of the junction rule: The flow out must equal the flow in.

(a)
$$\Delta V = V_b - V_a = -IR$$

(b)
$$\Delta V = V_b - V_a = +IR$$

(c)
$$\Delta V = V_b - V_a = +\mathcal{E}$$

(d)
$$\Delta V = V_b - V_a = -\mathcal{E}$$

Figure 21.22 Rules for determining the potential changes across a resistor and a battery, assuming the battery has no internal resistance.

- If a resistor is traversed in the direction *opposite* the current, the change in potential across the resistor is $+IR$ (Figure 21.22b).
- If a source of emf is traversed in the direction of the emf (from $-$ to $+$ on the terminals), the change in potential is $+\mathcal{E}$ (Fig. 21.22c).
- If a source of emf is traversed in the direction opposite the emf (from $+$ to $-$ on the terminals), the change in potential is $-\mathcal{E}$ (Fig. 21.22d).

There are limitations on the use of the junction rule and the loop rule. You may use the junction rule as often as needed so long as each time you write an equation, you include in it a current that has not been used in a previous junction rule equation. In general, the number of times the junction rule must be used is one fewer than the number of junction points in the circuit. The loop rule can be used as often as needed so long as a new circuit element (a resistor or battery) or a new current appears in each new equation. In general, **the number of independent equations you need must equal the number of unknowns in order to solve a particular circuit problem.**

The following examples illustrate the use of Kirchhoff's rules in analyzing circuits. In all cases, it is assumed that the circuits have reached steady-state conditions—that is, the currents in the various branches are constant. If a capacitor is included as an element in one of the branches, **it acts as an open circuit: The current in the branch containing the capacitor is zero under steady-state conditions.**

Gustav Robert Kirchhoff (1824–1887). *(Courtesy of North Wind Picture Archives)*

PROBLEM-SOLVING STRATEGY AND HINTS • Kirchhoff's Rules

1. First, draw the circuit diagram and assign labels to all the known quantities and symbols to all the unknown quantities. You must assign *directions* to the currents in each part of the circuit. Do not be alarmed if you guess the direction of a current incorrectly; the result will have a negative value, but *its magnitude will be correct*. Although the assignment of current directions is arbitrary, you must adhere *rigorously* to the directions you assigned when you apply Kirchhoff's rules.
2. Apply the junction rule (Kirchhoff's first rule) to all but one of the junctions in the circuit; doing so provides independent equations relating the currents. (This step is easy!)
3. Now apply the loop rule (Kirchhoff's second rule) to as many loops in the circuit as are needed to solve for the unknowns. In order to apply this rule, you must correctly identify the change in potential as you cross each element in traversing the closed loop (either clockwise or counterclockwise). Watch out for signs!
4. Solve the equations simultaneously for the unknown quantities. Be careful in your algebraic steps, and check your numerical answers for consistency.

Example 21.9   Applying Kirchhoff's Rules

Find the currents I_1, I_2, and I_3 in the circuit shown in Figure 21.23.

Reasoning   We choose the directions of the currents as in Figure 21.23. Applying Kirchhoff's first rule to junction c gives

(1) $I_1 + I_2 = I_3$

There are three loops in the circuit, *abcda*, *befcb*, and *aefda* (the outer loop). Therefore, we need only two loop equations to determine the unknown currents. The third loop equation

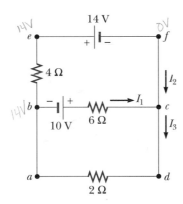

Figure 21.23 (Example 21.9) A circuit containing three loops.

would give no new information. Applying Kirchhoff's second rule to loops *abcda* and *befcb* and traversing these loops in the clockwise direction, we obtain the expressions

(2) Loop *abcda*: $10 \text{ V} - (6 \,\Omega)I_1 - (2 \,\Omega)I_3 = 0$

(3) Loop *befcb*: $-(4 \,\Omega)I_2 - 14 \text{ V} + (6 \,\Omega)I_1 - 10 \text{ V} = 0$

Note that in loop *befcb*, a positive sign is obtained when traversing the 6-Ω resistor, because the direction of the path is opposite the direction of I_1. A third loop equation for *aefda* gives $-14 = 2I_3 + 4I_2$, which is just the sum of (2) and (3).

Solution Expressions (1), (2), and (3) represent three independent equations with three unknowns. We can solve the problem as follows: Substituting (1) into (2) gives

$$10 - 6I_1 - 2(I_1 + I_2) = 0$$

(4) $10 = 8I_1 + 2I_2$

Dividing each term in (3) by 2 and rearranging the equation gives

(5) $-12 = -3I_1 + 2I_2$

Subtracting (5) from (4) eliminates I_2, giving

$$22 = 11I_1$$

$$I_1 = 2 \text{ A}$$

Using this value of I_1 in (5) gives a value for I_2:

$$2I_2 = 3I_1 - 12 = 3(2) - 12 = -6$$

$$I_2 = -3 \text{ A}$$

Finally, $I_3 = I_1 + I_2 = -1$ A. Hence, the currents have the values

$$I_1 = 2 \text{ A} \qquad I_2 = -3 \text{ A} \qquad I_3 = -1 \text{ A}$$

The fact that I_2 and I_3 are both negative indicates only that we chose the wrong direction for these currents. However, the numerical values are correct.

EXERCISE 10 Find the potential difference between points *b* and *c*. Answer $V_b - V_c = 2$ V

Example 21.10 A Multiloop Circuit

(a) Under steady-state conditions, find the unknown currents in the multiloop circuit shown in Figure 21.24.

Reasoning First note that the capacitor represents an open circuit, and hence there is no current along path *ghab* under steady-state conditions. Therefore, $I_{fg} = I_{gb} = I_{bc} \equiv I_1$. Labeling the currents as shown in Figure 21.24 and applying Kirchhoff's first rule to junction *c*, we get

(1) $I_1 + I_2 = I_3$

Kirchhoff's second rule applied to loops *defcd* and *cfgbc* gives

(2) Loop *defcd*: $4.00 \text{ V} - (3.00 \,\Omega)I_2 - (5.00 \,\Omega)I_3 = 0$

(3) Loop *cfgbc*: $(3.00 \,\Omega)I_2 - (5.00 \,\Omega)I_1 + 8.00 \text{ V} = 0$

Solution From (1) we see that $I_1 = I_3 - I_2$, which when substituted into (3) gives

(4) $8.00 \text{ V} - (5.00 \,\Omega)I_3 + (8.00 \,\Omega)I_2 = 0$

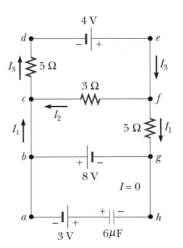

Figure 21.24 (Example 21.10) A multiloop circuit. Note that Kirchhoff's loop equation can be applied to *any* closed loop, including one containing the capacitor.

Subtracting (4) from (2), we eliminate I_3 and find

$$I_2 = -\frac{4.00}{11.0}\,A = -0.364\,A$$

Because I_2 is negative, we conclude that the direction of I_2 is from c to f through the 3.00-Ω resistor. Using this value of I_2 in (3) and (1) gives the following values for I_1 and I_3:

$$I_1 = 1.38\,A \qquad I_3 = 1.02\,A$$

Under steady-state conditions, the capacitor represents an *open* circuit, and so there is no current in the branch *ghab*.

(b) What is the charge on the capacitor?

Solution   We can apply Kirchhoff's second rule to loop *abgha* (or any other loop that contains the capacitor) to find the potential difference ΔV_c across the capacitor:

$$-8.00\,V + \Delta V_c - 3.00\,V = 0$$

$$\Delta V_c = 11.0\,V$$

Because $Q = C\,\Delta V_c$, the charge on the capacitor is

$$Q = (6.00\ \mu F)(11.0\,V) = 66.0\ \mu C$$

Why is the left side of the capacitor positively charged?

EXERCISE 11   Find the voltage across the capacitor by traversing any other loop.   Answer   11.0 V

21.9 • *RC* CIRCUITS

So far we have discussed circuits with constant currents, or so-called *steady-state circuits*. We shall now consider circuits containing capacitors, in which the currents may vary in time.

Charging a Capacitor

Consider the series circuit shown in Figure 21.25. Let us assume that the capacitor is initially uncharged. There is no current when switch S is open (Fig. 21.25b). If the switch is closed at $t = 0$, charges begin to flow, setting up a current in the circuit, and the capacitor begins to charge (Fig. 21.25c). Note that during the charging, charges do not jump across the plates of the capacitor, because the gap between the plates represents an open circuit. Instead, electrons move from the top plate to the bottom plate only by moving through the resistor, switch, and battery until the capacitor is fully charged. The value of the maximum charge depends on the

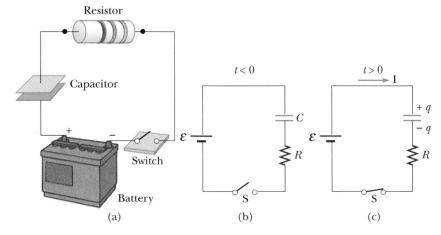

Figure 21.25   (a) A capacitor in series with a resistor, battery, and switch. (b) Circuit diagram representing this system before the switch is closed, $t < 0$. (c) Circuit diagram after the switch is closed, $t > 0$.

emf of the battery. Once the maximum charge is reached, the current in the circuit is zero.

To put this discussion on a quantitative basis, let us apply Kirchhoff's second rule to the circuit after the switch is closed. Choosing clockwise as our direction around the circuit, we get

$$\mathcal{E} - \frac{q}{C} - IR = 0 \qquad\qquad [21.30]$$

where q/C is the potential difference across the capacitor and IR is the potential difference across the resistor. Note that q and I are *instantaneous* values of the charge and current, respectively, as the capacitor is charged.

We can use Equation 21.30 to find the initial current in the circuit and the maximum charge on the capacitor. At $t = 0$, when the switch is closed, the charge on the capacitor is zero, and from Equation 21.30 we find that the initial current in the circuit I_0 is a maximum and equal to

$$I_0 = \frac{\mathcal{E}}{R} \qquad \text{(current at } t = 0) \qquad [21.31]$$

• *Maximum current*

At this time, **the potential difference is entirely across the resistor.** Later, when the capacitor is charged to its maximum value Q, charges cease to flow, the current in the circuit is zero, and **the potential difference is entirely across the capacitor.** Substituting $I = 0$ into Equation 21.30 yields the following expression for Q:

$$Q = C\mathcal{E} \qquad \text{(maximum charge)} \qquad [21.32]$$

• *Maximum charge on the capacitor*

To determine analytical expressions for the time dependence of the charge and current, we must solve Equation 21.30, a single equation containing two variables, q and I. In order to do this, let us substitute $I = dq/dt$ and rearrange the equation:

$$\frac{dq}{dt} = \frac{\mathcal{E}}{R} - \frac{q}{RC}$$

An expression for q may be found in the following way. Rearrange the equation by placing terms involving q on the left side and those involving t on the right side. Then integrate both sides:

$$\frac{dq}{(q - C\mathcal{E})} = -\frac{1}{RC} \, dt$$

$$\int_0^q \frac{dq}{(q - C\mathcal{E})} = -\frac{1}{RC} \int_0^t dt$$

$$\ln\left(\frac{q - C\mathcal{E}}{-C\mathcal{E}}\right) = -\frac{t}{RC}$$

From the definition of the natural logarithm, we can write this expression as

$$q(t) = C\mathcal{E}\,[1 - e^{-t/RC}] = Q[1 - e^{-t/RC}] \qquad [21.33]$$

• *Charge versus time for a capacitor being charged through a resistor*

where e is the base of the natural logarithm and $Q = C\mathcal{E}$ is the *maximum* charge on the capacitor.

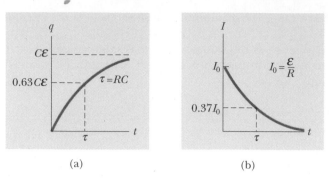

(a)   (b)

Figure 21.26   (a) Plot of capacitor charge versus time for the circuit shown in Figure 21.25. After one time constant, τ, the charge is 63.2% of the maximum value, $C\mathcal{E}$. The charge approaches its maximum value as t approaches infinity. (b) Plot of current versus time for the RC circuit shown in Figure 21.25. The current has its maximum value, $I_0 = \mathcal{E}/R$, at $t = 0$ and decays to zero exponentially as t approaches infinity. After one time constant, τ, the current decreases to 36.8% of its initial value.

An expression for the charging current may be found by differentiating Equation 21.33 with respect to time. Using $I = dq/dt$, we obtain

Current versus time •

$$I(t) = \frac{\mathcal{E}}{R}\, e^{-t/RC} \qquad\qquad [21.34]$$

where $\mathcal{E}/R$ is the initial current in the circuit.

Plots of charge and current versus time are shown in Figure 21.26. Note that the charge is zero at $t = 0$ and approaches the maximum value of $C\mathcal{E}$ as $t \to \infty$ (Fig. 21.26a). Furthermore, the current has its maximum value of $I_0 = \mathcal{E}/R$ at $t = 0$ and decays exponentially to zero as $t \to \infty$ (Fig. 21.26b). The quantity RC, which appears in the exponential of Equations 21.33 and 21.34, is called the **time constant, τ,** of the circuit. It represents the time it takes the current to decrease to $1/e$ of its initial value; that is, in the time τ, $I = e^{-1}\, I_0 = 0.37I_0$. In a time of 2τ, $I = e^{-2}I_0 = 0.135I_0$, and so forth. Likewise, in a time τ the charge increases from zero to $C\mathcal{E}[1 - e^{-1}] = 0.63C\mathcal{E}$.

The following dimensional analysis shows that τ has units of time:

$$[\tau] = [RC] = \left[\frac{\Delta V}{I} \times \frac{Q}{\Delta V}\right] = \left[\frac{Q}{Q/t}\right] = \mathrm{T}$$

The energy decrease of the battery during the charging process is $Q\mathcal{E} = C\mathcal{E}^2$. After the capacitor is fully charged, the energy stored in it is $\frac{1}{2}Q\mathcal{E} = \frac{1}{2}C\mathcal{E}^2$, which is just half the energy decrease of the battery. It is left to an end-of-chapter problem to show that the remaining half of the energy supplied by the battery goes into thermal energy dissipated in the resistor (Problem 52).

Discharging a Capacitor

Now consider the circuit in Figure 21.27, consisting of a capacitor with an initial charge of Q, a resistor, and a switch. When the switch is open (Fig. 21.27a), there is a potential difference of Q/C across the capacitor and zero potential difference across the resistor, because $I = 0$. If the switch is closed at $t = 0$, the capacitor begins

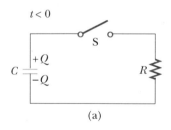

(a)

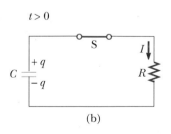

(b)

Figure 21.27   (a) A charged capacitor connected to a resistor and a switch, which is open at $t < 0$. (b) After the switch is closed, a nonsteady current is set up in the direction shown and the charge on the capacitor decreases exponentially with time.

to discharge through the resistor. At some time during the discharge, the current in the circuit is I and the charge on the capacitor is q (Fig. 21.27b). From Kirchhoff's second rule, we see that the potential difference across the resistor, IR, must equal the potential difference across the capacitor, q/C:

$$IR = \frac{q}{C} \qquad [21.35]$$

However, **the current in the circuit must equal the rate of decrease of charge on the capacitor.** That is, $I = -dq/dt$, and so Equation 21.35 becomes

$$-R\frac{dq}{dt} = \frac{q}{C}$$

$$\frac{dq}{q} = -\frac{1}{RC}dt$$

Integrating this expression, using the fact that $q = Q$ at $t = 0$, gives

$$\int_Q^q \frac{dq}{q} = -\frac{1}{RC}\int_0^t dt$$

$$\ln\left(\frac{q}{Q}\right) = -\frac{t}{RC}$$

$$q(t) = Qe^{-t/RC} \qquad [21.36]$$

• *Charge versus time for a discharging capacitor*

Differentiating Equation 21.36 with respect to time gives the current as a function of time:

$$I(t) = -\frac{dq}{dt} - I_0 e^{-t/RC} \qquad [21.37]$$

• *Current versus time for a discharging capacitor*

where the initial current is $I_0 = Q/RC$. Thus we see that both the charge on the capacitor and the current decay exponentially at a rate characterized by the time constant $\tau = RC$.

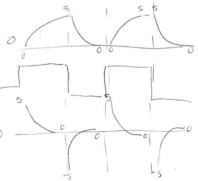

Thinking Physics 7

Many automobiles are equipped with windshield wipers that can be used intermittently during a light rainfall. How does the operation of this feature depend on the charging and discharging of a capacitor?

Reasoning The wipers are part of an *RC* circuit the time constant of which can be varied by selecting different values of R through a multipositioned switch. The brief time that the wipers remain on and the time they are off are determined by the value of the time constant of the circuit.

Example 21.11 Charging a Capacitor in an *RC* Circuit

An uncharged capacitor and a resistor are connected in series to a battery as in Figure 21.25. If $\varepsilon = 12.0$ V, $C = 5.00$ μF, and $R = 8.00 \times 10^5$ Ω, find the time constant of the circuit, the maximum charge on the capacitor, the maximum current in the circuit, and the charge and current as a function of time.

Solution The time constant of the circuit is $\tau = RC = (8.00 \times 10^5 \ \Omega)(5.00 \times 10^{-6} \ \text{F}) = 4.00$ s. The maximum charge on the capacitor is $Q = C\mathcal{E} = (5.00 \times 10^{-6} \ \text{F})(12.0 \ \text{V}) = 60.0 \ \mu\text{C}$. The maximum current in the circuit is $I_0 = \mathcal{E}/R = (12.0 \ \text{V})/(8.00 \times 10^5 \ \Omega) = 15.0 \ \mu\text{A}$. Using these values and Equations 21.33 and 21.34, we find that

$$q(t) = 60.0[1 - e^{-t/4}] \ \mu\text{C}$$

$$I(t) = 15.0 e^{-t/4} \ \mu\text{A}$$

EXERCISE 12 Calculate the charge on the capacitor and the current in the circuit after one time constant has elapsed.
Answer   37.9 μC, 5.52 μA

Example 21.12   Discharging a Capacitor in an *RC* Circuit

Consider a capacitor C being discharged through a resistor R as in Figure 21.27. (a) After how many time constants is the charge on the capacitor one fourth of its initial value?

Solution The charge on the capacitor varies with time according to Equation 21.36, $q(t) = Qe^{-t/RC}$. To find the time it takes the charge q to drop to one fourth of its initial value, we substitute $q(t) = Q/4$ into this expression and solve for t:

$$\tfrac{1}{4}Q = Qe^{-t/RC}$$

$$\tfrac{1}{4} = e^{-t/RC}$$

Taking logarithms of both sides, we find

$$-\ln 4 = -\frac{t}{RC}$$

$$t = RC \ln 4 = 1.39 \ RC$$

(b) The energy stored in the capacitor decreases with time as it discharges. After how many time constants is this stored energy one fourth of its initial value?

Solution Using Equations 20.29 and 21.36, we can express the energy stored in the capacitor at any time t as

$$U = \frac{q^2}{2C} = \frac{Q^2}{2C} e^{-2t/RC} = U_0 e^{-2t/RC}$$

where U_0 is the initial energy stored in the capacitor. As in part (a), we now set $U = U_0/4$ and solve for t:

$$\tfrac{1}{4}U_0 = U_0 e^{-2t/RC}$$

$$\tfrac{1}{4} = e^{-2t/RC}$$

Again, taking logarithms of both sides and solving for t gives

$$t = \tfrac{1}{2} RC \ln 4 = 0.693 \ RC$$

EXERCISE 13 After how many time constants is the current in the *RC* circuit one half of its initial value?
Answer   0.693 RC

EXERCISE 14 An uncharged capacitor and a resistor are connected in series to a source of emf. If $\mathcal{E} = 9.0$ V, $C = 20 \ \mu\text{F}$, and $R = 100 \ \Omega$, find (a) the time constant of the circuit, (b) the maximum charge on the capacitor, and (c) the maximum current in the circuit.
Answer   (a) 2.0 ms   (b) 180 μC   (c) 90 mA

SUMMARY

The **electric current** I in a conductor is defined as

$$I \equiv \frac{dQ}{dt} \qquad \text{[21.2]}$$

where dQ is the charge that passes through a cross-section of the conductor in the time dt. The SI unit of current is the ampere (A); 1 A = 1 C/s.

The current in a conductor is related to the motion of the charge carriers through the relationship

$$I = nqv_d A \qquad \text{[21.4]}$$

where n is the density of charge carriers, q is their charge, v_d is the drift speed, and A is the cross-sectional area of the conductor.

The **current density J** in a conductor is defined as the current per unit area:

$$J \equiv \frac{I}{A} = nqv_d \qquad \text{[21.5]}$$

The **resistance R** of a conductor is defined as the ratio of the potential difference across the conductor to the current:

$$R \equiv \frac{\Delta V}{I} \qquad \text{[21.7]}$$

The SI units of resistance are volts per ampere, defined as ohms (Ω). That is, $1 \, \Omega = 1 \, V/A$.

If the resistance is independent of the applied voltage, the conductor obeys Ohm's law, and conductors that have a constant resistance over a wide range of voltages are said to be *ohmic*.

If a conductor has a uniform cross-sectional area of A and a length of ℓ, its resistance is

$$R = \rho \frac{\ell}{A} \qquad \text{[21.9]}$$

where ρ is called the **resistivity** of the conductor. The inverse of the resistivity is defined as the **conductivity, σ.** That is, $\sigma = 1/\rho$.

The resistivity of a conductor varies with temperature in an approximately linear fashion; that is,

$$\rho = \rho_0[1 + \alpha(T - T_0)] \qquad \text{[21.11]}$$

where α is the temperature coefficient of resistivity and ρ_0 is the resistivity at some reference temperature T_0.

In a classical model of electronic conduction in a metal, the electrons are treated as molecules of a gas. In the absence of an electric field, the average velocity of the electrons is zero. When an electric field is applied, the electrons move (on the average) with a **drift velocity $\mathbf{v}_d$**, which is opposite the electric field:

$$\mathbf{v}_d = \frac{q\mathbf{E}}{m} \tau \qquad \text{[21.16]}$$

where τ is the average time between collisions with the atoms of the metal. The resistivity of the material according to this model is

$$\rho = \frac{m}{nq^2\tau} \qquad \text{[21.19]}$$

where n is the number of free electrons per unit volume.

If a potential difference ΔV is maintained across a resistor, the **power,** or rate at which energy is supplied to the resistor, is

$$P = I\Delta V \qquad \text{[21.21]}$$

Because the potential difference across a resistor is $\Delta V = IR$, we can express the power dissipated in a resistor in the form

$$P = I^2R = \frac{(\Delta V)^2}{R} \qquad \text{[21.22]}$$

The electrical energy supplied to a resistor appears in the form of thermal energy in the resistor.

The **emf** of a battery is the voltage across its terminals when the current is zero. That is, the emf is equivalent to the open-circuit voltage of the battery.

The **equivalent resistance** of a set of resistors connected in **series** is

$$R_{eq} = R_1 + R_2 + R_3 + \cdots \qquad [21.27]$$

The **equivalent resistance** of a set of resistors connected in *parallel* is given by

$$\frac{1}{R_{eq}} = \frac{1}{R_1} + \frac{1}{R_2} + \frac{1}{R_3} + \cdots \qquad [21.29]$$

Complex circuits involving more than one loop are conveniently analyzed using two simple rules called **Kirchhoff's rules:**

1. The sum of the currents entering any junction must equal the sum of the currents leaving that junction.
2. The sum of the potential differences across the elements around any closed-circuit loop must be *zero.*

The first rule is a statement of **conservation of charge;** the second rule is equivalent to a statement of **conservation of energy.**

When a resistor is traversed in the direction of the current, the change in potential, ΔV, across the resistor is $-IR$. If a resistor is traversed in the direction opposite the current, $\Delta V = +IR$.

If a source of emf is traversed in the direction of the emf (negative to positive) the change in potential is $+\mathcal{E}$. If it is traversed opposite the emf (positive to negative), the change in potential is $-\mathcal{E}$.

If a capacitor is charged with a battery of emf $\mathcal{E}$ through a resistance R, the charge on the capacitor and the current in the circuit vary in time according to the expressions

$$q(t) = Q[1 - e^{-t/RC}] \qquad [21.33]$$

$$I(t) = \frac{\mathcal{E}}{R} e^{-t/RC} \qquad [21.34]$$

where $Q = C\mathcal{E}$ is the *maximum* charge on the capacitor. The product RC is called the **time constant** of the circuit.

If a charged capacitor is discharged through a resistance R, the charge and current decrease exponentially in time according to the expressions

$$q(t) = Qe^{-t/RC} \qquad [21.36]$$

$$I(t) = I_0 e^{-t/RC} \qquad [21.37]$$

where $I_0 = Q/RC$ is the initial current in the circuit and Q is the initial charge on the capacitor.

CONCEPTUAL QUESTIONS

1. In an analogy between automobile traffic flow and electrical current, what would correspond to the charge Q? What would correspond to the current I?

2. What factors affect the resistance of a conductor?

3. Two wires A and B of circular cross section are made of the same metal and have equal lengths, but the resistance of wire A is three times greater than that of wire B. What is the ratio of their cross-sectional areas? How do their radii compare?

4. Use the atomic theory of matter to explain why the resistance of a material should increase as its temperature increases.

5. Explain how a current can persist in a superconductor without any applied voltage.

6. What would happen to the drift velocity of the electrons in a wire and to the current in the wire if the electrons could move freely without resistance through the wire?

7. If charges flow very slowly through a metal, why does it not

require several hours for a light to come on when you throw a switch?

8. If you were to design an electric heater using Nichrome wire as the heating element, what parameters of the wire could you vary to meet a specific power output, such as 1000 W?

9. Car batteries are often rated in ampere-hours. Does this designate the amount of current, power, energy, or charge that can be drawn from the battery?

10. How would you connect resistors so that the equivalent resistance is larger than the individual resistances? Give an example involving two or three resistors.

11. How would you connect resistors so that the equivalent resistance is smaller than the individual resistances? Give an example involving two or three resistors.

12. Why is it possible for a bird to sit on a high-voltage wire without being electrocuted?

13. A "short circuit" is a circuit containing a path of very low resistance in parallel with some other part of the circuit. Discuss the effect of a short circuit on the portion of the circuit it parallels. Use a lamp with a frayed line cord as an example.

14. A series circuit consists of three identical lamps connected to a battery, as in Figure Q21.14. When the switch S is closed, what happens (a) to the intensities of lamps A and B; (b) to the intensity of lamp C; (c) to the current in the circuit; and (d) to the voltage drop across the three lamps? (e) Does the power dissipated in the circuit increase, decrease, or remain the same?

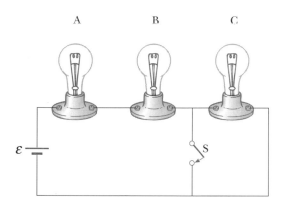

Figure Q21.14

15. Two lightbulbs both operate from 110 V, but one has a power rating of 25 W and the other of 100 W. Which bulb has the higher resistance? Which bulb carries the greater current?

16. If electrical power is transmitted over long distances, the resistance of the wires becomes significant. Why? Which mode of transmission would result in less energy loss—high current and low voltage or low current and high voltage? Discuss.

17. Two sets of Christmas tree lights are available. For set A, when one bulb is removed, the remaining bulbs remain illuminated. For set B, when one bulb is removed, the remaining bulbs do not operate. Explain the difference in wiring for the two sets.

18. Are the two headlights on a car wired in series or in parallel? How can you tell?

19. A ski resort consists of a few chair lifts and several interconnected downhill runs on the side of a mountain, with a lodge at the bottom. The lifts are analogous to batteries and the runs are analogous to resistors. Sketch how two runs can be in series. Sketch how three runs can be in parallel. Sketch a junction of one lift and two runs. One of the skiers is carrying an altimeter. State Kirchhoff's junction rule and Kirchhoff's loop rule for ski resorts.

20. In Figure Q21.20, describe what happens to the lightbulb after the switch is closed. Assume the capacitor has a large capacitance and is initially uncharged, and assume that the light illuminates when connected directly across the battery terminals.

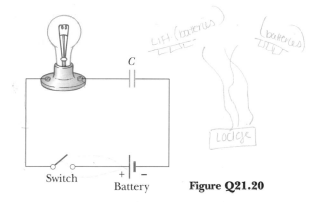

Figure Q21.20

21. Figure Q21.21 shows a series connection of three lamps, all rated at 120 V, with power ratings of 60 W, 75 W, and

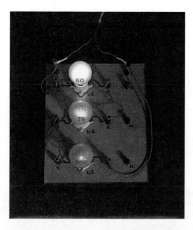

Figure Q21.21 *(Henry Leap and Jim Lehman)*

200 W. Why do the intensities of the lamps differ? Which lamp has the greatest resistance? How would their intensities differ if they were connected in parallel?

22. A student claims that a second lightbulb in series is less bright than the first, because the first bulb uses up some of the current. How would you respond to this statement?

PROBLEMS

Section 21.1   Electric Current

1. In a particular cathode ray tube, the measured beam current is 30.0 μA. How many electrons strike the tube screen every 40.0 s?

2. A teapot with a surface area of 700 cm^2 is to be silver plated. It is attached to the negative electrode of an electrolytic cell containing silver nitrate ($Ag^+NO_3^-$). If the cell is powered by a 12.0-V battery and has a resistance of 1.80 Ω, how long does it take to build up a 0.133-mm layer of silver on the teapot? (Density of silver = 10.5×10^3 kg/m^3.)

3. Suppose that the current through a conductor decreases exponentially with time according to $I(t) = I_0 e^{-t/\tau}$, where I_0 is the initial current (at $t = 0$), and τ is a constant having dimensions of time. Consider a fixed observation point within the conductor. (a) How much charge passes this point between $t = 0$ and $t = \tau$? (b) How much charge passes this point between $t = 0$ and $t = 10\tau$? (c) How much charge passes this point between $t = 0$ and $t = \infty$?

4. A Van de Graaff generator produces a beam of 2.00-MeV deuterons, which are heavy hydrogen nuclei containing a proton and a neutron. (a) If the beam current is 10.0 μA, how far apart are the deuterons? (b) Is their electrostatic repulsion a factor of beam stability? Explain.

5. An aluminum wire has cross-sectional area 4.00×10^{-6} m^2 and carries a current of 5.00 A. Find the drift speed of the electrons in the wire. The density of aluminum is 2.70 g/cm^3. (Assume one electron is supplied by each atom.)

Section 21.2   Resistance and Ohm's Law

6. A lightbulb has a resistance of 240 Ω when operating at a voltage of 120 V. What is the current through the lightbulb?

7. A 0.900-V potential difference is maintained across a 1.50-m length of tungsten wire that has a cross-sectional area of 0.600 mm^2. What is the current in the wire?

8. Suppose that you wish to fabricate a uniform wire out of 1.00 g of copper. If the wire is to have a resistance of $R = 0.500$ Ω, and all of the copper is to be used, what will be (a) the length and (b) the diameter of this wire?

9. A 12.0-Ω metal wire is cut into three equal pieces that are then connected side by side to form a new wire the length of which is equal to one third the original length. What is the resistance of this new wire?

10. (a) Make an order-of-magnitude estimate of the resistance between the ends of a rubber band. (b) Make an order-of-magnitude estimate of the resistance between the "heads" and "tails" sides of a penny. In each case state what quantities you take as data and the values you measure or estimate for them. (c) *Don't* try this at home, but each would carry current of what order of magnitude if it were connected across a 120-V power supply?

11. While traveling through Death Valley on a day when the temperature is 58.0°C, Bill Hiker finds that a certain voltage applied to a copper wire produces a current of 1.00 A. Bill then travels to Antarctica and applies the same voltage to the same wire. What current does he register there if the temperature is −88.0°C? Assume no change in the wire's shape and size.

12. An aluminum rod has a resistance of 1.234 Ω at 20.0°C. Calculate the resistance of the rod at 120°C by accounting for the changes in both the resistivity and the dimensions of the rod.

13. A certain lightbulb has a tungsten filament with a resistance of 19.0 Ω when cold and 140 Ω when hot. Assume that Equation 21.13 can be used over the large temperature range involved here, and find the temperature of the filament when hot. Assume an initial temperature of 20.0°C.

14. A carbon wire and a Nichrome wire are connected in series. If the combination has a resistance of 10.0 kΩ at 0°C, what is the resistance of each wire at 0°C so that the resistance of the combination does not change with temperature?

Section 21.4   A Model for Electrical Conduction

15. If the drift velocity of free electrons in a copper wire is 7.84×10^{-4} m/s, calculate the electric field in the conductor.

16. If the current carried by a conductor is doubled, what happens to the (a) charge carrier density? (b) current density? (c) electron drift velocity? (d) average time between collisions?

Section 21.5   Electrical Energy and Power

17. A toaster is rated at 600 W when connected to a 120-V source. What current does the toaster carry, and what is its resistance?

18. In a hydroelectric installation, a turbine delivers 1500 hp to a generator, which in turn converts 80.0% of the mechanical energy into electrical energy. Under these conditions, what current will the generator deliver at a terminal potential difference of 2000 V?

19. What is the required resistance of an immersion heater that will increase the temperature of 1.50 kg of water from 10.0°C to 50.0°C in 10.0 min while operating at 110 V?

20. What is the required resistance of an immersion heater that will increase the temperature of a mass m of water from T_1 to T_2 in a time interval Δt while operating at a voltage ΔV?

21. Suppose that a voltage surge produces 140 V for a moment. By what percentage will the power output of a 120-V, 100-W lightbulb increase assuming its resistance does not change?

22. A coil of Nichrome wire is 25.0 m long. The wire has a diameter of 0.400 mm and is at 20.0°C. If it carries a current of 0.500 A, what are (a) the electric field intensity in the wire, and (b) the power dissipated in it? (c) If the temperature is increased to 340°C and the voltage across the wire remains constant, what is the power dissipated?

23. Batteries are rated in terms of ampere hours (A·h), where a battery that can produce a current of 2.00 A for 3.00 h is rated at 6.00 A·h. (a) What is the total energy, in kilowatt hours, stored in a 12.0-V battery rated at 55.0 A·h? (b) At $0.0600 per kilowatt hour, what is the value of the electricity produced by this battery?

Section 21.6   Sources of emf

24. (a) What is the current in a 5.60-Ω resistor connected to a battery that has a 0.200-Ω internal resistance if the terminal voltage of the battery is 10.0 V? (b) What is the emf of the battery?

25. A battery has an emf of 15.0 V. The terminal voltage of the battery is 11.6 V when it is delivering 20.0 W of power to an external load resistor R. (a) What is the value of R? (b) What is the internal resistance of the battery?

26. Two 1.50-V batteries—with their positive terminals in the same direction—are inserted in series into the barrel of a flashlight. One battery has an internal resistance of 0.255 Ω, the other an internal resistance of 0.153 Ω. When the switch is closed, a current of 600 mA occurs in the lamp. (a) What is the lamp's resistance? (b) What fraction of the power dissipated is dissipated in the batteries?

Section 21.7   Resistors in Series and in Parallel

27. A television repairperson needs a 100-Ω resistor to repair a malfunctioning set. She is temporarily out of resistors of this value. All she has in her toolbox are a 500-Ω resistor and two 250-Ω resistors. How can the desired resistance be obtained from the resistors on hand?

28. (a) Find the equivalent resistance between points a and b in Figure P21.28. (b) If a potential difference of 34.0 V is

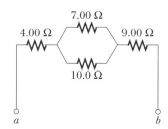

Figure P21.28

applied between points a and b, calculate the current in each resistor.

29. Consider the circuit shown in Figure P21.29. Find (a) the current in the 20.0-Ω resistor and (b) the potential difference between points a and b.

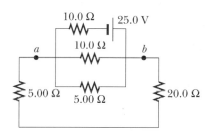

Figure P21.29

30. A lightbulb marked "75 W [at] 120 V" is screwed into a socket at one end of a long extension cord in which each of the two conductors has resistance 0.800 Ω. The other end of the extension cord is plugged into a 120-V outlet. Draw a circuit diagram and find the actual power of the bulb in this circuit.

31. Three 100-Ω resistors are connected, as shown in Figure P21.31. The maximum power that can safely be dissipated in any one resistor is 25.0 W. (a) What is the maximum voltage that can be applied to the terminals a and b? (b) For the voltage determined in part (a), what is the power dissipation in each resistor? What is the total power dissipation?

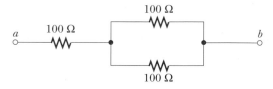

Figure P21.31

32. Four copper wires of equal length are connected in series. Their cross-sectional areas are 1.00 cm², 2.00 cm²,

3.00 cm^2, and 5.00 cm^2. If a voltage of 120 V is applied to the arrangement, determine the voltage across the 2.00-cm^2 wire.

Section 21.8 Kirchhoff's Rules and Simple DC Circuits

(*Note:* The currents are not necessarily in the directions shown for some circuits.)

33. Determine the current in each branch of the circuit in Figure P21.33.

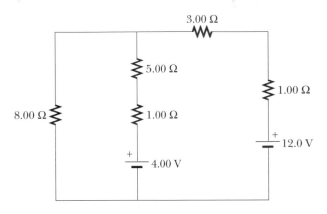

Figure P21.33

34. In Figure P21.33, show how to add just enough ammeters to measure every different current that is flowing. Show how to add just enough voltmeters to measure the potential difference across each resistor and across each battery.

35. The circuit considered in problem 33 and drawn in Figure P21.33 is connected for two minutes. (a) Find the energy converted by each battery. (b) Find the energy converted by each resistor. (c) Find the total energy converted by the circuit.

36. The ammeter in Figure 21.36 reads 2.00 A. Find I_1, I_2, and $\mathcal{E}$.

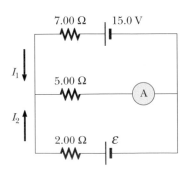

Figure P21.36

37. Using Kirchhoff's rules, (a) find the current in each resistor in Figure P21.37. (b) Find the potential difference between points *c* and *f*. Which point is at the higher potential?

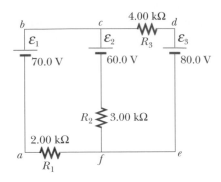

Figure P21.37

Section 21.9 RC Circuits

38. A $2.00 \times 10^{-3} \, \mu\text{F}$ capacitor with an initial charge of $5.10 \, \mu\text{C}$ is discharged through a $1.30\text{-k}\Omega$ resistor. (a) Calculate the current through the resistor $9.00 \, \mu\text{s}$ after the resistor is connected across the terminals of the capacitor. (b) What charge remains on the capacitor after $8.00 \, \mu\text{s}$? (c) What is the maximum current in the resistor?

39. Consider a series *RC* circuit (Fig. 21.25) for which $R = 1.00 \, \text{M}\Omega$, $C = 5.00 \, \mu\text{F}$, and $\mathcal{E} = 30.0 \, \text{V}$. Find (a) the time constant of the circuit and (b) the maximum charge on the capacitor after the switch is closed. (c) If the switch is closed at $t = 0$, find the current in the resistor 10.0 s later.

40. In the circuit of Figure P21.40, the switch S has been open for a long time. It is then suddenly closed. Determine the time constant (a) before the switch is closed and (b) after the switch is closed. (c) If the switch is closed at $t = 0$ s, determine the current through it as a function of time.

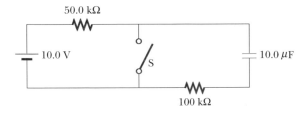

Figure P21.40

41. The circuit in Figure P21.41 has been connected for a long time. (a) What is the voltage across the capacitor? (b) If the battery is disconnected, how long does it take the capacitor to discharge to one tenth of its initial voltage?

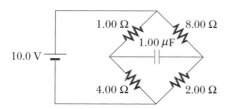

Figure P21.41

Additional Problems

42. One lightbulb is marked "25 W 120 V" and another "100 W 120 V" to mean that each converts that respective power when plugged into a constant 120-V potential difference. (a) Find the resistance of each. (b) In what time will 1.00 C pass through the dim bulb? How is this charge different on its exit versus its entry? (c) In what time will 1.00 J pass through the dim bulb? How is this energy different on its exit versus its entry? (d) Find the cost of running the dim bulb continuously for 30.0 days if the electric company sells its product at $0.0700 per kWh. What *physical quantity* does the electric company sell? What is its price for one SI unit of this quantity?

43. A high-voltage transmission line of diameter 2.00 cm and length 200 km carries a steady current of 1000 A. If the conductor is a wire made of copper with a free charge density of 8.00×10^{28} electrons/m³, how long does it take one electron to travel the full length of the cable?

44. A high-voltage transmission line carries 1000 A starting at 700 kV for a distance of 100 miles. If the resistance in the wire is 0.500 Ω/mi, what is the power loss due to resistive losses?

45. A copper cable is to be designed to carry a current of 300 A with a power loss of only 2.00 W/m. What is the required radius of the copper cable?

46. Four 1.50-V AA batteries in series are used to power a transistor radio. If the batteries can move a charge of 240 C before being depleted, how long will they last if the radio has a resistance of 200 Ω?

47. A battery has emf 9.20 V and internal resistance 1.20 Ω. (a) What resistance across the battery will dissipate heat energy from it at a rate of 12.8 W? (b) 21.2 W?

48. A 10.0-μF capacitor is charged by a 10.0-V battery through a resistance R. The capacitor reaches a potential difference of 4.00 V in a time 3.00 s after charging begins. Find R.

49. An electric heater is rated at 1500 W, a toaster at 750 W, and an electric grill at 1000 W. The three appliances are connected to a common 120-V circuit. (a) How much current does each draw? (b) Is a circuit fused at 25.0 A sufficient in this situation? Explain.

50. A more general definition of the temperature coefficient of resistivity is

$$\alpha = \frac{1}{\rho} \frac{d\rho}{dT}$$

where ρ is the resistivity at temperature T. (a) Assuming that α is constant, show that

$$\rho = \rho_0 e^{\alpha(T-T_0)}$$

where ρ_0 is the resistivity at temperature T_0. (b) Using the series expansion ($e^x \approx 1 + x$; $x \ll 1$), show that the resistivity is given approximately by the expression $\rho = \rho_0[1 + \alpha(T - T_0)]$ for $\alpha(T - T_0) \ll 1$.

51. An experiment is conducted to measure the electrical resistivity of Nichrome in the form of wires with different lengths and cross-sectional areas. For one set of measurements, a student uses #30-gauge wire, which has a cross-sectional area of 7.3×10^{-8} m². The voltage across the wire and the current in the wire are measured with a voltmeter and ammeter, respectively. For each of the measurements given in the table below taken on wires of three different lengths, calculate the resistance of the wires and the corresponding values of the resistivity. What is the average value of the resistivity, and how does it compare with the value given in Table 21.1?

ℓ(m)	ΔV(V)	I(A)	$R(\Omega)$	$\rho(\Omega \cdot m)$
0.54	5.22	0.500		
1.028	5.82	0.276		
1.543	5.94	0.187		

52. A battery is used to charge a capacitor through a resistor, as in Figure 21.25. Show that in the process of charging the capacitor, half of the energy supplied by the battery is dissipated as heat in the resistor and half is stored in the capacitor.

53. A straight cylindrical wire lying along the x axis has length ℓ and diameter d. It is made of a material described by Ohm's law with resistivity ρ. Assume that potential V_0 is maintained at $x = 0$, and $V = 0$ at $x = \ell$. In terms of ℓ, d, V_0, ρ, and physical constants, derive expressions for: (a) the electric field in the wire; (b) the resistance of the wire; (c) the electric current in the wire; and (d) the current density in the wire. Express vectors in vector notation. (e) Prove that $\mathbf{E} = \rho\mathbf{J}$.

Spreadsheet Problems

S1. Spreadsheet 21.1 calculates the average annual lighting cost per bulb for fluorescent and incandescent bulbs and the average yearly savings realized with fluorescent bulbs. It also graphs the average annual lighting cost per bulb versus the cost of electrical energy. (a) Suppose that a fluorescent bulb costs $5, lasts for 5000 h, consumes 40 W of power, but provides the light intensity of a 100-W incandescent bulb. Assume that a 100-W incandescent bulb is on at all times and that energy costs 8.3 cents per kWh. How much does a consumer save each year by switching to fluorescent bulbs? (b) Check with your local electric company for their current rates, and find the cost of bulbs in your area. Would it pay you to switch to fluorescent bulbs? (c) Vary the parameters for bulbs of different wattages and reexamine the annual savings.

S2. The current-voltage characteristic curve for a semiconductor diode as a function of temperature T is given by

$$I = I_0(e^{e\Delta V/k_B T} - 1)$$

where e is the charge on the electron, k_B is Boltzmann's constant, ΔV is the applied voltage, and T is the absolute temperature. Set up a spreadsheet to calculate I and $R = \Delta V/I$ for $\Delta V = 0.40$ V to $\Delta V = 0.60$ V in increments of 0.01 V. Assume $I_0 = 1.0$ nA. Plot R versus ΔV for $T = 280$ K, 300 K, and 320 K.

S3. The application of Kirchhoff's rules to a dc circuit leads to a set of n linear equations in n unknowns. It is very tedious to solve these algebraically if $n > 3$. The purpose of this problem is to solve for the currents in a moderately complex circuit using matrix operations on a spreadsheet. You can solve equations very easily this way, and you can also readily explore the consequences of changing the values of the circuit parameters. (a) Consider the circuit in Figure S21.3. Assume the four unknown currents are in the directions shown.

- Apply Kirchhoff's rules to get four independent equations for the four unknown currents I_i, $i = 1, 2, 3$, and 4.
- Write these equations in matrix form $\mathbf{AI} = \mathbf{B}$, that is,

$$\sum_{j=1}^{4} A_{ij}I_j = B_i \qquad i = 1, 2, 3, 4$$

The solution is $\mathbf{I} = \mathbf{A}^{-1}\mathbf{B}$, where $\mathbf{A}^{-1}$ is the inverse matrix of $\mathbf{A}$.

- Set $R_1 = 2\ \Omega$, $R_2 = 4\ \Omega$, $R_3 = 6\ \Omega$, $R_4 = 8\ \Omega$, $\mathcal{E}_1 = 3$ V, $\mathcal{E}_2 = 9$ V, and $\mathcal{E}_3 = 12$ V.
- Enter the matrix $\mathbf{A}$ into your spreadsheet, one value per cell. Use the matrix inversion operation of the spreadsheet to calculate $\mathbf{A}^{-1}$.
- Find the currents by using the matrix multiplication operation of the spreadsheet to calculate $\mathbf{I} = \mathbf{A}^{-1}\mathbf{B}$.

(b) Change the sign of $\mathcal{E}_3$, and repeat the calculations in part (a). This is equivalent to changing the polarity of $\mathcal{E}_3$. (c) Set $\mathcal{E}_1 = \mathcal{E}_2 = 0$ and repeat the calculations in part (a). For these values, the circuit can be solved using simple series-parallel rules. Compare your results using both methods. (d) Investigate any other cases of interest. For example, see how the currents change if you vary R_4.

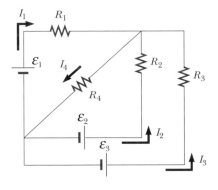

Figure S21.3

ANSWERS TO CONCEPTUAL PROBLEMS

1. A voltage is not something that "surges" through a completed circuit. A voltage is a potential difference that is applied *across* a device or a circuit. What goes *through* the circuit is *current*. Thus, it would be more correct to say, "1 ampere of electricity surges through the victim's body." Although this current would have disastrous results on the human body, a value of 1 (ampere) doesn't sound as exciting for a newspaper article as 10 000 (volts). Another possibility is to write, "10 000 volts of electricity were applied across the victim's body," which still doesn't sound quite as exciting!

2. The length of the line cord will double in this event. This would tend to increase the resistance of the line cord. But the doubling of the radius of the line cord results in the cross-sectional area increasing by a factor of 4. This would

reduce the resistance more than the doubling of length increases it. The net result is a decrease in resistance. The same effect will occur for the lightbulb filament. The lowered resistance will result in more current flowing through the filament, causing it to glow more brightly.

3. The bulb filaments are cold when the lamp is first switched on, hence they have a lower resistance and draw more current than when they are hot. The increased current can overheat the filament and destroy it.

4. The gravitational force pulling the electrons to the bottom of a piece of metal is much smaller than the electrical repulsion pushing the electrons apart. Thus, they stay distributed throughout the metal. The concept of charges residing on the surface of a metal is true for a metal with an excess charge. The number of free electrons in a piece of metal is

the same as the number of positive crystal lattice ions—the metal has zero net charge.

5. The total amount of energy delivered by the battery will be less than *E*. Recall that a battery can be considered to be an ideal, resistanceless battery in series with the internal resistance. When charging, the energy delivered to the battery includes the energy necessary to charge the ideal battery, plus the energy that goes into raising the temperature of the battery due to "joule heating" in the internal resistance. This latter energy is not available during the discharge of the battery. During discharge, part of the reduced available energy again transforms into internal energy in the internal resistance, further reducing the available energy below *E*.

6. The starter in the automobile draws a relatively large current from the battery. This large current causes a significant voltage drop across the internal resistance of the battery. As a result, the terminal voltage of the battery is reduced, and the headlights dim accordingly.

7. An electrical device has a given resistance. Thus, when it is attached to a power source with a known potential difference, a definite current will be drawn. The device can be labeled with both the voltage and the current. Batteries, however, can be applied to a number of devices. Each device will have a different resistance, so the current from the battery will vary with the device. As a result, only the voltage of the battery can be specified.

8. Connecting batteries in parallel does not increase the emf. A high-current device connected to batteries in parallel can draw current from both batteries. Thus, connecting the batteries in parallel does increase the possible current output, and, therefore, the possible power output.

9. As you add more lightbulbs in *series,* the overall resistance of the circuit is increasing. Thus, the current through the bulbs will decrease. This decrease in current will result in a decrease in power transferred from the battery. As a result, the battery lifetime will increase. The whole string will be dimmer than the original lightbulb. As the current drops, the terminal voltage across the battery will become closer and closer to the battery emf.

As you add more lightbulbs in *parallel,* the overall resistance of the circuit is decreasing. The current through each bulb remains nearly the same (until the battery starts to get hot). Each new bulb will be nearly as bright as the original lightbulb. The current leaving the battery will increase with the addition of each bulb. This increase in current will result in an increase in power transferred from the battery. As a result, the battery lifetime will decrease. As the current rises, the terminal voltage across the battery will drop further below the battery emf.

22

Magnetism

The list of important technological applications of magnetism is very long. For instance, large electromagnets are used to pick up heavy loads. Magnets are also used in such devices as meters, motors, and loudspeakers. Magnetic tapes are routinely used in sound- and video-recording equipment and for computer memory, and magnetic recording material is used on computer disks. Intense magnetic fields generated by superconducting magnets are currently being used as a means of containing the plasmas (heated to temperatures on the order of 10^8 K) used in controlled nuclear fusion research.

As we investigate magnetism in this chapter, you will find that the subject cannot be divorced from electricity. For example, magnetic fields affect moving charges, and moving charges produce magnetic fields. Ultimately we shall find that the source of all magnetic fields is electric

The white arc in this photograph indicates the circular path followed by an electron beam moving in a magnetic field. The vessel contains gas at very low pressure, and the beam is made visible as the electrons collide with the gas atoms, which in turn emit visible light. The magnetic field is produced by two coils (not shown). The apparatus can be used to measure the charge : mass ratio for the electron. *(Courtesy of CENCO)*

current, whether it be the current in a wire or the current produced by the motion of charges within atoms or molecules.

22.1 • HISTORICAL OVERVIEW: MAGNETS

Many historians of science believe that the compass, which uses a magnetic needle, was used in China as early as the thirteenth century B.C., its invention being of Arab or Indian origin. The phenomenon of magnetism was known to the Greeks as early as about 800 B.C. They discovered that certain stones, now called magnetite (Fe_3O_4), attracted pieces of iron. Legend ascribes the name *magnetite* to the shepherd Magnes, "the nails of whose shoes and the tip of whose staff stuck fast in a magnetic field while he pastured his flocks." In 1269 Pierre de Maricourt mapped out the directions taken by a needle when it was placed at a variety of points on the surface of a spherical natural magnet. He found that the directions formed lines that encircled the sphere and passed through two points diametrically opposite each other, which he called the *poles* of the magnet. Subsequent experiments have shown that every magnet, regardless of its shape, has two poles, called *north poles* and *south poles,* which exhibit forces on each other in a manner analogous to electrical charges. That is, like poles repel each other and unlike poles attract each other. The poles received their names because of the behavior of a magnet in the presence of the Earth's magnetic field. If a bar magnet is suspended from its midpoint by a piece of string so that it can swing freely in a horizontal plane, it will rotate until its "north" pole points to the north of the Earth and its "south" pole points to the south. (The same idea is used to construct a simple compass.)

In 1600 William Gilbert extended these experiments to a variety of materials. Using the fact that a compass needle orients in preferred directions, he suggested that magnets are attracted to land masses. In 1750 John Michell (1724–1793) used a torsion balance to show that magnetic poles exert attractive or repulsive forces on each other and that these forces vary as the inverse square of their separation. Although the force between two magnetic poles is similar to the force between two electric charges, an important difference exists. Electric charges can be isolated (witness the electron and proton), whereas **magnetic poles cannot be isolated.** That is, **magnetic poles are always found in pairs.** No matter how many times a permanent magnet is cut, each piece always has a north pole and a south pole. (There is some theoretical basis for speculating that magnetic monopoles— isolated north or south poles—may exist in nature, and attempts to detect them currently make up an active experimental field of investigation. However, none of these attempts has proved successful.)

The relationship between magnetism and electricity was discovered in 1819 when, while preparing for a lecture demonstration, the Danish scientist Hans Oersted found that an electric current in a wire deflected a nearby compass needle. Shortly thereafter, André Ampère (1775–1836) deduced quantitative laws of magnetic force between current-carrying conductors. He also suggested that electric current loops of molecular size are responsible for *all* magnetic phenomena.

In the 1820s, further connections between electricity and magnetism were identified by Faraday and, independently, Joseph Henry (1797–1878). They showed that an electric current could be produced in a circuit either by moving a magnet near the circuit or by changing the current in another, nearby circuit. Their observations demonstrated that a changing magnetic field produces an electric field.

Hans Christian Oersted (1777– 1851), Danish physicist. *(The Bettmann Archive)*

An assortment of commercially available magnets. Some of the magnets are made of metallic alloys and others are ceramic compounds. *(Courtesy of Central Scientific Co.)*

Years later, theoretical work by Maxwell showed that the reverse is also true: A changing electric field gives rise to a magnetic field.

There is a similarity between electric and magnetic effects that has given rise to methods of making permanent magnets. In Chapter 16 we learned that when rubber and wool are rubbed together, both become charged, one positively and the other negatively. In a somewhat analogous fashion, an unmagnetized piece of iron can be magnetized by stroking it with a magnet. Magnetism can also be induced in iron (and other materials) by other means. For example, if a piece of unmagnetized iron is placed near a strong permanent magnet, the piece of iron eventually becomes magnetized. The process of magnetizing the piece of iron in the presence of a strong external field can be accelerated either by heating and cooling the iron or by hammering.

Thinking Physics 1

On a business trip to Australia, you take along your American-made compass that you used in your Boy Scout days. Does this compass work correctly in Australia?

Reasoning   There is no problem with using the compass in Australia. The north pole of the magnet in the compass will be attracted to the south magnetic pole near the north geographic pole, just as it was in the United States. The only difference in the magnetic field lines is that they have an upward component in Australia, whereas they have a downward component in the United States. Your compass cannot detect this, however—it only displays the direction of the *horizontal* component of the magnetic field.

22.2 • THE MAGNETIC FIELD

In earlier chapters we found it convenient to describe the interaction between charged objects in terms of electric fields. Recall that an electric field surrounds any electric charge. The region of space surrounding a *moving* charge includes a

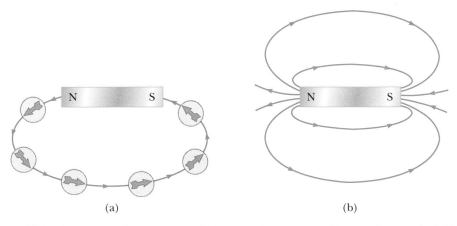

(a)   (b)

Figure 22.1   (a) Tracing the magnetic field lines of a bar magnet. (b) Several magnetic field lines of a bar magnet.

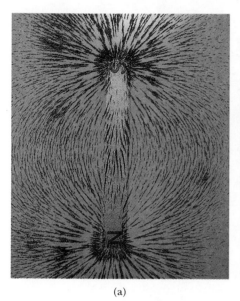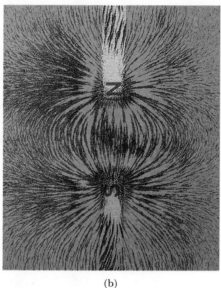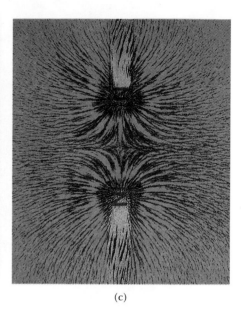

(a) (b) (c)

Figure 22.2 (a) Magnetic field patterns surrounding a bar magnet as displayed with iron filings. (b) Magnetic field patterns between *unlike* poles of two bar magnets. (c) Magnetic field pattern between *like* poles of two bar magnets. *(Courtesy of Henry Leap and Jim Lehman)*

magnetic field in addition to the electric field. A magnetic field also surrounds any material with permanent magnetism.

In order to describe any type of field, we must define its magnitude, or strength, and its direction. The direction of the magnetic field, **B**, at any location is the direction in which the north pole of a compass needle points at that location. Figure 22.1a shows how the magnetic field of a bar magnet can be traced with the aid of a compass. Several magnetic field lines of a bar magnet traced out in this manner are shown in Figure 22.1b. Magnetic field patterns can be displayed by small iron filings, as shown in Figure 22.2.

We can define a magnetic field vector, **B** (sometimes called the *magnetic induction*), at some point in space in terms of the magnetic force exerted on an appropriate test object. Our test object is taken to be a charged particle moving with a velocity of **v**. For the time being, let us assume that no electric or gravitational fields are present in the region of the charge. Experiments on the motions of various charged particles in a magnetic field give the following results:

- The magnetic force is proportional to the charge, q, and speed, v, of the particle.
- The magnitude and direction of the magnetic force depend on the velocity of the particle and on the magnitude and direction of the magnetic field.
- When a charged particle moves parallel to the magnetic field vector, the magnetic force, **F**, on the particle is zero.
- When the velocity vector makes an angle of θ with the magnetic field, the magnetic force acts in a direction perpendicular to both **v** and **B**; that is, **F** is perpendicular to the plane formed by **v** and **B** (Fig. 22.3a).
- The magnetic force on a positive charge is directed opposite the force on a negative charge moving in the same direction (Fig. 22.3b).

• *Properties of the magnetic force on a charge moving in a **B** field*

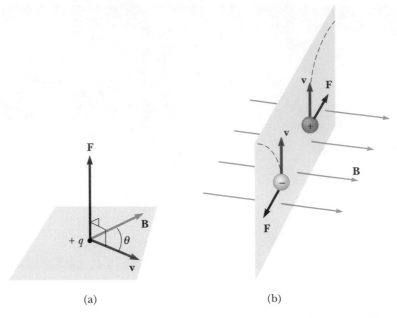

(a)   (b)

Figure 22.3   The direction of the magnetic force on a charged particle moving with a velocity **v** in the presence of a magnetic field **B**. (a) When **v** is at an angle θ to **B**, the magnetic force is perpendicular to both **v** and **B**. (b) In the presence of a magnetic field, the moving charged particles are deflected as indicated by the dotted lines.

• If the velocity vector makes an angle of θ with the magnetic field, the magnitude of the magnetic force is proportional to sin θ.

These observations can be summarized by writing the magnetic force in the form

Magnetic force on a charged • *particle moving in a magnetic field*

$$\mathbf{F} = q\mathbf{v} \times \mathbf{B} \qquad [22.1]$$

where the direction of the magnetic force is that of **v** × **B**, which, by definition of the cross-product, is perpendicular to both **v** and **B**. We can regard this equation as an operational definition of the magnetic field at a point in space. That is, the magnetic field is defined in terms of a sideways force acting on a moving charged particle. The SI unit of magnetic field is the **tesla** (T), where

$$1\ \text{T} = 1\ \text{N·s/C·m}$$

Figure 22.4 reviews the right-hand rule for determining the direction of the cross product **v** × **B**. Point the four fingers of your right hand along the direction of **v**, then turn them until they point along the direction of **B**. The thumb then points in the direction of **v** × **B**. Because $\mathbf{F} = q\mathbf{v} \times \mathbf{B}$, **F** is in the direction of **v** × **B** if q is positive (Fig. 22.4a) and *opposite* the direction of **v** × **B** if q is negative (Fig. 22.4b). The magnitude of the magnetic force is

$$F = qvB \sin \theta \qquad [22.2]$$

where θ is the angle between **v** and **B**. From this expression, we see that F is zero when **v** is either parallel or antiparallel to **B** ($\theta = 0$ or 180°). Furthermore, the force has its maximum value, $F = qvB$, when **v** is perpendicular to **B** ($\theta = 90°$).

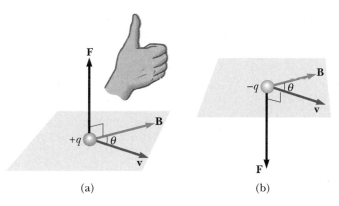

Figure 22.4 The right-hand rule for determining the direction of the magnetic force **F** acting on a charge q moving with a velocity **v** in a magnetic field **B**. If q is positive, **F** is upward in the direction of the thumb. If q is negative, **F** is downward.

There are several important differences between electric and magnetic forces on charged particles:

- The electric force is always parallel to the direction of the electric field, whereas the magnetic force is perpendicular to the magnetic field.
- The electric force acts on a charged particle independent of the particle's velocity, whereas the magnetic force acts on a charged particle only when the particle is in motion.
- The electric force does work in displacing a charged particle, whereas the magnetic force associated with a steady magnetic field does *no* work when a charged particle is displaced.

• *Differences between electric and magnetic fields*

This last statement is a consequence of the fact that when a charge moves in a steady magnetic field, the magnetic force is always *perpendicular* to the displacement. That is, $\mathbf{F} \cdot d\mathbf{s} = (\mathbf{F} \cdot \mathbf{v}) \, dt = 0$, because the magnetic force is a vector perpendicular to **v**. From this property and the work–kinetic energy theorem, we conclude that the kinetic energy of a charged particle *cannot* be altered by a magnetic field alone. In other words, when a charge moves with a velocity of **v**, an applied magnetic field can alter the direction of the velocity vector, but it cannot change the speed of the particle.

• *A magnetic field cannot change the speed of a particle.*

Thinking Physics 2

Suppose a uniform magnetic field exists in a finite region of space. Can you inject a charged particle into this region and have it stay trapped in the region by the magnetic force?

Reasoning Let us consider separately the components of the particle velocity parallel and perpendicular to the field lines in the region. For the component parallel to the field lines, there will be no force on the particle—it will continue to move with the parallel component until it leaves the region of the magnetic field. Now consider the component perpendicular to the field lines. This component will result in a magnetic force that is perpendicular to both the field lines and the velocity component. The

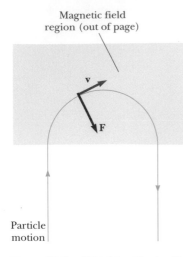

Magnetic field region (out of page)

v

F

Particle motion

Figure 22.5   (Thinking Physics 2)

path of a particle for which the force is always perpendicular to the velocity is a circle. Thus, the particle will follow a circular arc and exit the field on the other side of the circle, as shown in Figure 22.5.

CONCEPTUAL PROBLEM 1

A charged particle moves in a circular path in the presence of a magnetic field applied perpendicular to the particle's velocity. Does the particle gain energy from the magnetic field?

CONCEPTUAL PROBLEM 2

If a charged particle moves in a straight line through some region of space, can you say that the magnetic field in that region is zero?

CONCEPTUAL PROBLEM 3

How can the motion of a charged particle be used to distinguish between a magnetic field and an electric field in a certain region?

CONCEPTUAL PROBLEM 4

Why does the picture on a television screen become distorted when a magnet is brought near the screen? *Caution:* You should not do this at home, because it may permanently affect the television set.

Example 22.1   A Proton Moving in a Magnetic Field

A proton moves with a speed of 8.0×10^6 m/s along the x axis. It enters a region where there is a magnetic field of magnitude 2.5 T, directed at an angle of 60° to the x axis and lying in the xy plane (Fig. 22.6). Calculate the initial magnetic force on and acceleration of the proton.

Solution   From Equation 22.2, we get

$$F = qvB \sin \theta$$
$$= (1.60 \times 10^{-19} \text{ C})(8.0 \times 10^6 \text{ m/s})(2.5 \text{ T})(\sin 60°)$$
$$= 2.8 \times 10^{-12} \text{ N}$$

Because $\mathbf{v} \times \mathbf{B}$ is in the positive z direction (the right-hand rule), and the charge is positive, $\mathbf{F}$ is in the positive z direction.

The mass of the proton is 1.67×10^{-27} kg, and so its initial acceleration is

$$a = \frac{F}{m} = \frac{2.8 \times 10^{-12} \text{ N}}{1.67 \times 10^{-27} \text{ kg}} = 1.7 \times 10^{15} \text{ m/s}^2$$

in the positive z direction.

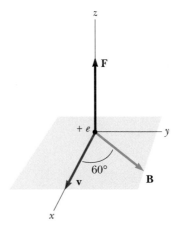

Figure 22.6   (Example 22.1) The magnetic force $\mathbf{F}$ on a proton is in the positive z direction when $\mathbf{v}$ and $\mathbf{B}$ lie in the xy plane.

EXERCISE 1   Calculate the acceleration of an electron that moves through the same magnetic field at the same speed as the proton.   Answer 3.0×10^{18} m/s^2

EXERCISE 2   A proton moves with a speed of 2.5×10^6 m/s horizontally at right angles to a magnetic field. What magnetic field is required to just balance the weight of the proton and keep it moving horizontally?   Answer 4.09×10^{-14} T in the horizontal plane

22.3 • MAGNETIC FORCE ON A CURRENT-CARRYING CONDUCTOR

If a force is exerted on a single charged particle when it moves through an external magnetic field, it should not surprise you to find that a current-carrying wire also experiences a force when placed in an external magnetic field. This follows from the fact that the current represents a collection of many charged particles in motion; hence, the resultant force on the wire is due to the sum of the individual forces on the charged particles. The force on the particles is transmitted to the "bulk" of the wire through collisions with the atoms making up the wire.

Before we continue our discussion, some explanation is in order concerning notation in many of our figures. To indicate the direction of **B**, we use the following convention. If **B** is directed into the page, as in Figure 22.7, we use a series of blue crosses, which represent the tails of arrows. If **B** is directed out of the page, we use a series of blue dots, which represent the tips of arrows. If **B** lies in the plane of the page, we use a series of blue field lines with arrowheads.

The force on a current-carrying conductor can be demonstrated by hanging a wire between the faces of a magnet as in Figure 22.7, where the magnetic field is directed into the page. The wire deflects to the left or right when a current is passed through it.

Let us quantify this discussion by considering a straight segment of wire of length ℓ and cross-sectional area A, carrying a current, I, in a uniform external magnetic field, **B**, as in Figure 22.8. The magnetic force on the charge q moving with the drift velocity $\mathbf{v}_d$ is $q\mathbf{v}_d \times \mathbf{B}$. To find the total force on the wire segment,

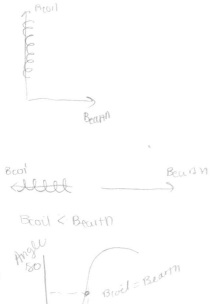

This apparatus demonstrates the force on a current-carrying conductor in an external magnetic field. Why does the bar swing *into* the magnet after the switch is closed?   *(Courtesy of Henry Leap and Jim Lehman)*

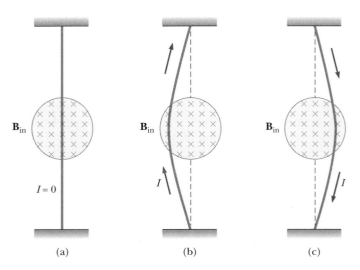

Figure 22.7   A segment of a flexible vertical wire is stretched between the poles of a magnet with the magnetic field (blue crosses) directed into the paper. (a) When there is no current in the wire, it remains vertical. (b) When the current is upward, the wire deflects to the left. (c) When the current is downward, the wire deflects to the right.

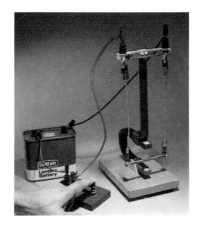

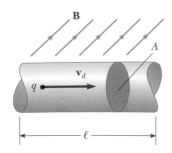

Figure 22.8 A section of a wire containing moving charges in a magnetic field **B**. The magnetic force on each charge is $q\mathbf{v}_d \times \mathbf{B}$, and the net force on a straight wire of length ℓ is $I\ell \times \mathbf{B}$.

we multiply the force on one charge by the number of charges in the segment. Because the volume of the segment is $A\ell$, the number of charges in the segment is $nA\ell$, where n is the number of charges per unit volume. Hence, the total magnetic force on the wire of length ℓ is

$$\mathbf{F} = (q\mathbf{v}_d \times \mathbf{B})\, nA\ell$$

This can be written in a more convenient form by noting that, from Equation 21.4, the current in the wire is $I = nqv_d A$. Therefore, **F** can be expressed as

$$\mathbf{F} = I\ell \times \mathbf{B} \qquad [22.3]$$

where ℓ is a vector in the direction of the current I; the magnitude of ℓ equals the length of the segment. Note that this expression applies only to a straight segment of wire in a uniform external magnetic field. Furthermore, we have neglected the internal magnetic field produced by the current.

Now consider an arbitrarily shaped wire of uniform cross-section in an external magnetic field, as in Figure 22.9. It follows from Equation 22.3 that the magnetic force on a very small segment, $d\mathbf{s}$, in the presence of an external field, **B**, is

$$d\mathbf{F} = I\, d\mathbf{s} \times \mathbf{B} \qquad [22.4]$$

where $d\mathbf{F}$ is directed out of the page for the directions assumed in Figure 22.9. We can consider Equation 22.4 as an alternative definition of **B**. That is, the field **B** can be defined in terms of a measurable force on a current element, where the force is a maximum when **B** is perpendicular to the element and zero when **B** is parallel to the element.

To get the total force **F** on the wire, we integrate Equation 22.4 over the length of the wire:

$$\mathbf{F} = I\int_a^b d\mathbf{s} \times \mathbf{B} \qquad [22.5]$$

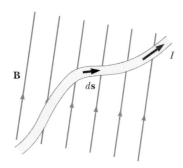

Figure 22.9 A wire of arbitrary shape carrying a current I in a magnetic field **B** experiences a magnetic force. The force on any segment $d\mathbf{s}$ is $I\, d\mathbf{s} \times \mathbf{B}$ and is directed out of the page.

In this expression, a and b represent the end points of the wire. When this integration is carried out, the magnitude of the magnetic field and the direction the field makes with the vector $d\mathbf{s}$ (that is, the element orientation) may vary from point to point.

Thinking Physics 3

In a lightning stroke, there is rapid movement of negative charge from a cloud to the ground. In what direction is a lightning stroke deflected by the Earth's magnetic field?

Reasoning The downward flow of negative charge in a lightning stroke is equivalent to an upward-moving current. Thus, the length element vector representing the current is upward, and the magnetic field vector has a northward component. According to the cross product of the length element and magnetic field vectors, then, the lightning stroke would be deflected to the *west*.

Example 22.2   Force on a Semicircular Conductor

A wire bent into the shape of a semicircle of radius R forms a closed circuit and carries a current I. The circuit lies in the xy plane, and a uniform magnetic field is present along the positive y axis, as in Figure 22.10. Find the magnetic force on the straight portion of the wire and on the curved portion.

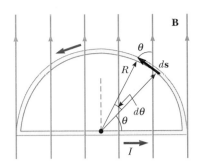

Figure 22.10   (Example 22.2) The net force on a closed current loop in a uniform magnetic field is zero. In this case, the force on the straight portion is $2IRB$ and outward, whereas the force on the curved portion is also $2IRB$ and inward.

Reasoning and Solution   The force on the straight portion of the wire has a magnitude $F_1 = I\ell B = 2IRB$, because $\ell = 2R$ and the wire is perpendicular to **B**. The direction of $\mathbf{F}_1$ is out of the paper because $\boldsymbol{\ell} \times \mathbf{B}$ is outward. (That is, $\boldsymbol{\ell}$ is to the right, in the direction of the current, and so by the rule of cross products, $\boldsymbol{\ell} \times \mathbf{B}$ is outward.)

To find the force on the curved part, we must first write an expression for the force $d\mathbf{F}_2$ on the element $d\mathbf{s}$. If θ is the

angle between **B** and $d\mathbf{s}$ in Figure 22.10, then the magnitude of $d\mathbf{F}_2$ is

$$dF_2 = I|d\mathbf{s} \times \mathbf{B}| = IB \sin \theta \, ds$$

where ds is the length of the small element measured along the circular arc. In order to integrate this expression, we must express ds in terms of θ. Because $s = R\theta$, $ds = R \, d\theta$, and the expression for dF_2 can be written

$$dF_2 = IRB \sin \theta \, d\theta$$

To get the total force F_2 on the curved portion, we can integrate this expression to account for contributions from all elements. Note that the direction of the force on every element is the same: into the paper (because $d\mathbf{s} \times \mathbf{B}$ is inward). Therefore, the resultant force $\mathbf{F}_2$ on the curved wire must also be into the paper. Integrating dF_2 over the limits $\theta = 0$ to $\theta = \pi$ (that is, the entire semicircle) gives

$$F_2 = IRB \int_0^\pi \sin \theta \, d\theta = IRB\left[-\cos \theta \right]_0^\pi$$

$$= -IRB(\cos \pi - \cos 0) = -IRB(-1 - 1) - \; 2IRB$$

Because $F_2 = 2IRB$ and is directed into the paper and the force on the straight wire $F_1 = 2IRB$ is out of the paper, we see that the net force on the closed loop is zero.

EXERCISE 3   Calculate the magnitude of the force per unit length exerted on a conductor carrying a current of 22.0 A in a region in which a uniform magnetic field has a magnitude of 0.770 T and is directed perpendicular to the conductor.   Answer   16.9 N/m

22.4 • TORQUE ON A CURRENT LOOP IN A UNIFORM MAGNETIC FIELD

In the preceding section we showed how a force is exerted on a current-carrying conductor when the conductor is placed in an external magnetic field. With this as a starting point, we shall show that a torque is exerted on a current loop placed in a magnetic field. The results of this analysis will be of great practical value when we discuss generators in Chapter 23.

Consider a rectangular loop carrying a current, I, in the presence of a uniform external magnetic field *in the plane of the loop*, as in Figure 22.11a. The forces on the sides of length a are zero, because these wires are parallel to the field; hence, $d\mathbf{s} \times \mathbf{B} = 0$ for these sides. The magnitude of the forces on the sides of length b, however, is

$$F_1 = F_2 = IbB$$

The direction of $\mathbf{F}_1$, the force on the left side of the loop, is out of the paper, and that of $\mathbf{F}_2$, the force on the right side of the loop, is into the paper. If we view the loop from an end, as in Figure 22.11b, we see the forces directed as shown. If we

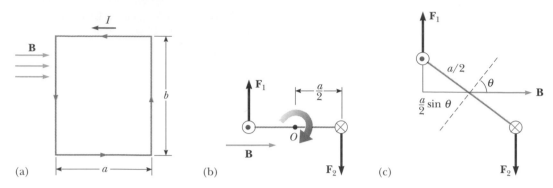

Figure 22.11   (a) Front view of a rectangular loop in a uniform magnetic field. There are no forces acting on the sides of width a parallel to **B**, but there are forces acting on the sides of length b. (b) Bottom view of the rectangular loop shows that the forces $\mathbf{F}_1$ and $\mathbf{F}_2$ on the sides of length b create a torque that tends to twist the loop clockwise as shown. (c) If **B** is at an angle θ with respect to a line perpendicular to the plane of the loop, the torque is $IAB \sin \theta$.

assume that the loop is pivoted so that it can rotate about point O, we see that these two forces produce a torque about O that rotates the loop clockwise. The magnitude of this torque, $\tau_{\max}$, is

$$\tau_{\max} = F_1 \frac{a}{2} + F_2 \frac{a}{2} = (IbB) \frac{a}{2} + (IbB) \frac{a}{2} = IabB$$

where the moment arm about O is $a/2$ for each force. Because the area of the loop is $A = ab$, the magnitude of the torque can be expressed as

$$\tau_{\max} = IAB \qquad \text{[22.6]}$$

Remember that this result is valid only when the field **B** is parallel to the plane of the loop. The sense of the rotation is clockwise when the loop is viewed from the bottom end, as indicated in Figure 22.11b. If the current were reversed, the forces would reverse their directions and the rotational tendency would be counterclockwise.

Now suppose the uniform magnetic field makes an angle of θ with a line perpendicular to the plane of the loop, as in Figure 22.11c. For convenience, we shall assume that **B** is perpendicular to the sides of length b. (The end view of these sides is shown in Fig. 22.11c.) In this case, the forces on the sides of length a cancel each other and produce no torque, because they pass through a common origin. However, the forces acting on the sides of length b, $\mathbf{F}_1$ and $\mathbf{F}_2$, form a couple and hence produce a torque about *any point*. Referring to the end view in Figure 22.11c, we note that the moment arm of $\mathbf{F}_1$ about O is $(a/2)\sin \theta$. Likewise, the moment arm of $\mathbf{F}_2$ about O is also $(a/2)\sin \theta$. Because $F_1 = F_2 = IbB$, the net torque about O has the magnitude

$$\tau = F_1 \frac{a}{2} \sin \theta + F_2 \frac{a}{2} \sin \theta$$

$$= IbB \left(\frac{a}{2} \sin \theta \right) + IbB \left(\frac{a}{2} \sin \theta \right) = IabB \sin \theta$$

$$= IAB \sin \theta$$

where $A = ab$ is the area of the loop. This result shows that the torque has its maximum value, IAB, when the field is parallel to the plane of the loop ($\theta = 90°$) and is zero when the field is perpendicular to the plane of the loop ($\theta = 0$). As we see in Figure 22.11c, the loop tends to rotate in the direction of decreasing values of θ (that is, so that the normal to the plane of the loop rotates toward the direction of the magnetic field).

A convenient vector expression for the torque is

$$\tau = I\mathbf{A} \times \mathbf{B} \qquad [22.7]$$

where **A**, a vector perpendicular to the plane of the loop, has a magnitude equal to the area of the loop. The sense of **A** is determined by the right-hand rule illustrated in Figure 22.12. When the four fingers of the right hand are rotated in the direction of the current in the loop, the thumb points in the direction of **A**. The product $I\mathbf{A}$ is defined to be the **magnetic moment, $\boldsymbol{\mu}$,** of the loop:

$$\boldsymbol{\mu} \equiv I\mathbf{A} \qquad [22.8]$$

Figure 22.12   A right-hand rule for determining the direction of the vector **A**. The magnetic moment μ is also in the direction of **A**.

The SI unit of magnetic moment is the ampere-meter² ($A \cdot m^2$). Using this definition, the torque can be expressed as

$$\tau = \boldsymbol{\mu} \times \mathbf{B} \qquad [22.9]$$

• *Torque on a current loop*

Although the torque was obtained for a particular orientation of **B** with respect to the loop, Equation 22.9 is valid for any orientation. Furthermore, although the torque expression was derived for a rectangular loop, the result is valid for a loop of any shape.

If a coil consists of N turns of wire, each carrying the same current and having the same area, the total magnetic moment of the coil is the product of the number of turns and the magnetic moment for one turn. The torque on an N-turn coil is N times greater than that on a one-turn coil.

Example 22.3   The Magnetic Moment of a Coil

A rectangular coil of dimensions 5.40 cm × 8.50 cm consists of 25 turns of wire. The coil carries a current of 15.0 mA. (a) Calculate the magnitude of its magnetic moment.

Solution   The magnitude of the magnetic moment of a current loop is $\mu = IA$ (see Eq. 22.8). In this case, $A = (0.0540 \text{ m})(0.0850 \text{ m}) = 4.59 \times 10^{-3} \text{ m}^2$. Because the coil has 25 turns, and assuming that each turn has the same area A, we have

$$\mu_{coil} = NIA = (25)(15.0 \times 10^{-3} \text{ A})(4.59 \times 10^{-3} \text{ m}^2)$$
$$= 1.72 \times 10^{-3} \text{ A} \cdot m^2$$

(b) Suppose a magnetic field of magnitude 0.350 T is ap-

plied parallel to the plane of the loop. What is the magnitude of the torque acting on the loop?

Solution   The torque is given by Equation 22.9, $\tau = \boldsymbol{\mu} \times \mathbf{B}$. In this case, **B** is *perpendicular* to $\boldsymbol{\mu}_{coil}$, so that

$$\tau = \mu_{coil}B = (1.72 \times 10^{-3} \text{ A} \cdot m^2)(0.350 \text{ T})$$
$$= 6.02 \times 10^{-4} \text{ N} \cdot m$$

EXERCISE 4   Show that the units of τ, $A \cdot m^2 \cdot T$, reduce to $N \cdot m$.

EXERCISE 5   Calculate the magnitude of the torque on the coil when the 0.350-T magnetic field makes angles of (a) 60° and (b) 0° with $\boldsymbol{\mu}$.

Answer   (a) $5.21 \times 10^{-4} \text{ N} \cdot m$ (b) 0

22.5 • THE BIOT–SAVART LAW

From their investigations on the force between a current-carrying conductor and a magnet, Jean-Baptiste Biot and Félix Savart arrived at an expression for the magnetic field at some point in space in terms of the current that produces the field. The **Biot–Savart law** says that if a wire carries a steady current, I, then at point P the magnetic field $d\mathbf{B}$ associated with a wire of length ds (Fig. 22.13) has the following properties:

Properties of the magnetic field due to a current element •

- The vector $d\mathbf{B}$ is perpendicular both to $d\mathbf{s}$ (which is in the direction of the current) and to the unit vector, $\hat{\mathbf{r}}$, directed from the element to P.
- The magnitude of $d\mathbf{B}$ is inversely proportional to r^2, where r is the distance from the element to P.
- The magnitude of $d\mathbf{B}$ is proportional to the current and to the length, ds, of the element.
- The magnitude of $d\mathbf{B}$ is proportional to $\sin\theta$, where θ is the angle between $d\mathbf{s}$ and $\hat{\mathbf{r}}$.

The **Biot–Savart law** can be summarized in the following convenient form:

Biot–Savart law •

$$d\mathbf{B} = k_m \frac{I \, d\mathbf{s} \times \hat{\mathbf{r}}}{r^2} \qquad [22.10]$$

where k_m is a constant that in SI units is exactly 10^{-7} T·m/A. The constant k_m is usually written $\mu_0/4\pi$, where μ_0 is another constant, called the **permeability of free space:**

$$\frac{\mu_0}{4\pi} = k_m = 10^{-7} \text{ T·m/A} \qquad [22.11]$$

Permeability of free space •

$$\mu_0 = 4\pi k_m = 4\pi \times 10^{-7} \text{ T·m/A} \qquad [22.12]$$

Hence, the Biot–Savart law, Equation 22.10, can also be written

$$d\mathbf{B} = \frac{\mu_0}{4\pi} \frac{I \, d\mathbf{s} \times \hat{\mathbf{r}}}{r^2} \qquad [22.13]$$

It is important to note that the Biot–Savart law gives the magnetic field at a point only for a small element of the conductor. To find the total magnetic field **B** at some point due to a conductor of finite size, we must sum up contributions from all current elements making up the conductor. That is, we must evaluate **B** by integrating Equation 22.13.

There are two similarities between the Biot–Savart law of magnetism and Coulomb's law of electrostatics, and one important difference. The current element $I \, d\mathbf{s}$ produces a magnetic field, whereas the point charge q produces an electric field. Furthermore, the magnitude of the magnetic field varies as the inverse square of the distance from the current element, as does the electric field due to a point charge. However, the directions of the two fields are quite different. The electric field due to a point charge is radial; in the case of a positive point charge, **E** is directed from the charge to the field point. The magnetic field due to a current element is perpendicular to both the current element and the radius vector. Hence, if the conductor lies in the plane of the page, as in Figure 22.13, $d\mathbf{B}$ points out of the page at the point P and into the page at P'.

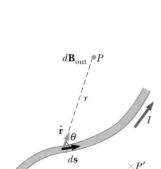

Figure 22.13 The magnetic field $d\mathbf{B}$ at a point P due to a current element $d\mathbf{s}$ is given by the Biot–Savart law. The field is out of the page at P and into the page at P'.

Figure 22.14 gives a convenient rule for determining the direction of the magnetic field due to a wire. Note that the field lines generally encircle the current. In the case of a long, straight wire, the field lines form circles that are concentric with the wire and are in a plane perpendicular to the wire. Although the magnetic field due to a long, current-carrying wire can be calculated using the Biot–Savart law (Problem 51), in Section 22.7 we use a different method to show that the magnitude of this field at a distance *r* from the wire is

$$B = \frac{\mu_0 I}{2\pi r}$$

[22.14]

• *Magnetic field due to a long, straight wire*

When applying the Biot–Savart law, it is important that you recognize that **the magnetic field described in these calculations is the field due to a given current-carrying conductor.** This is not to be confused with any *external* field that may be applied to the conductor.

Thinking Physics 4

In electrical circuits, it is often the case that wires carrying currents in opposite directions are twisted together. What is the advantage of this decision?

Reasoning If the wires are not twisted together, the combination of the two wires forms a current loop. Although it is not a circle, it is a loop of current surrounding an open area. Thus, there will be a magnetic field generated by the loop that might affect adjacent circuits or components.

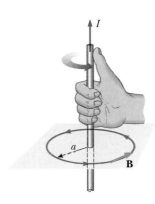

Figure 22.14 The right-hand rule for determining the direction of the magnetic field surrounding a long, straight wire carrying a current. Note that the magnetic field lines form circles around the wire.

CONCEPTUAL PROBLEM 5

Suppose you move along a wire at the same speed as the drift electrons in the current. Do you now measure a magnetic field of zero?

CONCEPTUAL PROBLEM 6

Can you use a compass to detect the currents in wires in the walls near light switches in your home?

Example 22.4   Magnetic Field on the Axis of a Circular Current Loop

Consider a circular loop of wire of radius *R* located in the *yz* plane and carrying a steady current *I*, as in Figure 22.15. Calculate the magnetic field at an axial point *P* a distance *x* from the center of the loop.

Reasoning In this situation, note that any element *d***s** is perpendicular to *r̂*. Furthermore, all elements around the loop are at the same distance *r* from *P*, where $r^2 = x^2 + R^2$. Hence, the magnitude of *d***B** due to the element *d***s** is

$$dB = \frac{\mu_0 I}{4\pi} \frac{|d\mathbf{s} \times \hat{\mathbf{r}}|}{r^2} = \frac{\mu_0 I}{4\pi} \frac{ds}{(x^2 + R^2)}$$

[22.15]

The direction of the magnetic field *d***B** due to the element *d***s** is perpendicular to the plane formed by *r̂* and *d***s**, as in Figure 22.15. The vector *d***B** can be resolved into a component *dB_x*, along the *x* axis, and a component *dB_y*, which is perpendicular to the *x* axis. When the components perpendicular to the *x* axis are summed over the whole loop, the result is zero. That is, by symmetry any element on one side of the loop sets up a perpendicular component that cancels the component set up by an element diametrically opposite it.

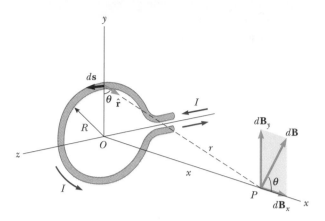

Figure 22.15 (Example 22.4) The geometry for calculating the magnetic field at an axial point *P* for a current loop. Note that by symmetry the total field **B** is along the *x* axis.

Solution For the reasons given previously, **the resultant field at *P* must be along the *x* axis** and can be found by integrating the components $dB_x = dB \cos \theta$, where this expression is obtained from resolving the vector $d\mathbf{B}$ into its components as shown in Figure 22.15. That is, $\mathbf{B} = B_x\mathbf{i}$, where

$$B_x = \oint dB \cos \theta = \frac{\mu_0 I}{4\pi} \oint \frac{ds \cos \theta}{x^2 + R^2}$$

and the integral must be taken over the entire loop. Because θ, x, and R are constants for all elements of the loop and because $\cos \theta = R/(x^2 + R^2)^{1/2}$, we get

$$B_x = \frac{\mu_0 I R}{4\pi(x^2 + R^2)^{3/2}} \oint ds = \frac{\mu_0 I R^2}{2(x^2 + R^2)^{3/2}} \quad \textbf{[22.16]}$$

where we have used the fact that $\oint ds = 2\pi R$ (the circumference of the loop).

To find the magnetic field at the center of the loop, we set $x = 0$ in Equation 22.16. At this special point, this gives

$$B = \frac{\mu_0 I}{2R} \quad \text{(at } x = 0\text{)} \quad \textbf{[22.17]}$$

It is also interesting to determine the behavior of the magnetic field far from the loop—that is, when x is large compared with R. In this case, we can neglect the term R^2 in the denominator of Equation 22.16 and get

$$B \approx \frac{\mu_0 I R^2}{2x^3} \quad \text{(for } x \gg R\text{)} \quad \textbf{[22.18]}$$

Because the magnitude of the magnetic dipole moment μ of the loop is defined as the product of the current and the area (Eq. 22.8), $\mu = I(\pi R^2)$ and we can express Equation 22.18 in the form

$$B = \frac{\mu_0}{2\pi} \frac{\mu}{x^3} \quad \textbf{[22.19]}$$

This result is similar in form to the expression for the electric field due to an electric dipole, $E = k_e p/y^3$ (Example 19.3), where p is the electric dipole moment. The pattern of the magnetic field lines for a circular loop is shown in Figure 22.16. For clarity, the lines are drawn only for one plane that contains the axis of the loop. The field pattern is axially symmetric.

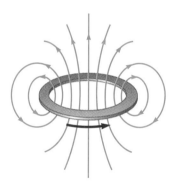

Figure 22.16 Magnetic field lines for a current loop. Far from the loop, the field lines are identical in form to those of an electric dipole.

EXERCISE 6 A wire in which there is a current of 5.00 A is to be formed into a circular loop of one turn. If the required value of the magnetic field at the center of the loop is 10.0 μT, what is the required radius? Answer 31.4 cm

22.6 • THE MAGNETIC FORCE BETWEEN TWO PARALLEL CONDUCTORS

In Section 22.3 we described the magnetic force that acts on a current-carrying conductor when the conductor is placed in an external magnetic field. Because a current in a conductor sets up its own magnetic field, it is easy to understand that two current-carrying conductors exert magnetic forces on each other. As we shall see, such forces can be used as the basis for defining the ampere and the coulomb.

Consider two long, straight, parallel wires separated by the distance a and carrying currents I_1 and I_2 in the same direction, as in Figure 22.17. We can easily determine the force on one wire due to a magnetic field set up by the other wire. Wire 2, which carries the current I_2, sets up a magnetic field, $\mathbf{B}_2$, at the position of wire 1. The direction of $\mathbf{B}_2$ is perpendicular to the wire, as shown in Figure 22.17. According to Equation 22.3, the magnetic force on a length, ℓ, of wire 1 is $\mathbf{F}_1 = I_1\boldsymbol{\ell} \times \mathbf{B}_2$. Because $\boldsymbol{\ell}$ is perpendicular to $\mathbf{B}_2$, the magnitude of $\mathbf{F}_1$ is $F_1 = I_1\ell B_2$. Because the field due to wire 2 is given by Equation 22.14,

$$B_2 = \frac{\mu_0 I_2}{2\pi a}$$

we see that

$$F_1 = I_1\ell B_2 = I_1\ell\left(\frac{\mu_0 I_2}{2\pi a}\right) = \frac{\ell\mu_0 I_1 I_2}{2\pi a}$$

We can rewrite this in terms of the force per unit length, as

$$\frac{F_1}{\ell} = \frac{\mu_0 I_1 I_2}{2\pi a} \qquad\qquad \textbf{[22.20]}$$

The direction of $\mathbf{F}_1$ is downward, toward wire 2, because $\boldsymbol{\ell} \times \mathbf{B}_2$ is downward. If one considers the field set up at wire 2 due to wire 1, the force $\mathbf{F}_2$ on wire 2 is found to be equal to and opposite $\mathbf{F}_1$. That is what one would expect, because Newton's third law of action–reaction must be obeyed.

When the currents are in opposite directions, the forces are reversed and the wires repel each other. Hence, we find that **parallel conductors carrying currents in the same direction attract each other,** whereas **parallel conductors carrying currents in opposite directions repel each other.**

The force between two parallel wires, each carrying a current, is used to define the **ampere:** If two long, parallel wires 1 m apart carry the same current and the force per unit length on each wire is 2×10^{-7} N/m, then the current is defined to be 1 A. The numerical value of 2×10^{-7} N/m is obtained from Equation 22.20, with $I_1 = I_2 = 1$ A and $a = 1$ m.

The SI unit of charge, the **coulomb,** can now be defined in terms of the ampere: If a conductor carries a steady current of 1 A, then the quantity of charge that flows through a cross-section of the conductor in 1 s is 1 C.

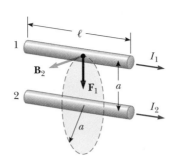

Figure 22.17   Two parallel wires each carrying a steady current exert a force on each other. The field $\mathbf{B}_2$ at wire 1 due to the current in wire 2 produces a force on wire 1 given by $F_1 = I_1\ell B_2$. The force is attractive if the currents are parallel as shown and repulsive if the currents are antiparallel.

CONCEPTUAL PROBLEM 7

Two wires carry current in opposite directions and are oriented parallel, with one above the other. The wires repel each other. Is the upper wire in a stable levitation over the lower wire? Suppose the current in one wire is reversed, so that the wires now attract. Is the lower wire hanging in a stable attraction to the upper wire?

EXERCISE 7   At what distance from a long, straight wire carrying a current of 5 A is the magnetic field due to the wire equal to the strength of the Earth's field—approximately 5×10^{-5} T?   Answer   2 cm

André-Marie Ampère (1775–
1836), a French mathematician,
chemist, and philosopher.   *(Photo
courtesy of AIP Niels Bohr Library)*

22.7 • AMPÈRE'S LAW

A simple experiment first carried out by Oersted in 1820 clearly demonstrates that
a current-carrying conductor produces a magnetic field. In this experiment, several
compass needles are placed in a horizontal plane near a long vertical wire, as in
Figure 22.18a. When there is no current in the wire, all needles point in the same
direction (that of the Earth's field), as one would expect. However, when the wire
carries a strong, steady current, the needles all deflect in a direction tangent to the
circle, as in Figure 22.18b. These observations show that the direction of **B** is consistent with the right-hand rule described in Section 22.6.

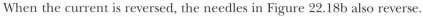

> If the wire is grasped in the right hand with the thumb in the direction of
> the current, the fingers will curl in the direction of **B**.

When the current is reversed, the needles in Figure 22.18b also reverse.

Because the needles point in the direction of **B**, we conclude that the lines of
B form circles about the wire, as discussed in Section 22.5. By symmetry, the magnitude of **B** is the same everywhere on a circular path that is centered on the wire
and lies in a plane perpendicular to the wire. By varying the current and distance
from the wire, one finds that **B** is proportional to the current and inversely proportional to the distance from the wire.

Now let us evaluate the product **B** · d**s** and sum the products $B \, ds$ over the closed
circular path centered on the wire. Along this path, the vectors d**s** and **B** are parallel
at each point (Fig. 22.18b), so that **B** · d**s** = $B \, ds$. Furthermore, **B** is constant in
magnitude on this circle and is given by Equation 22.14. Therefore, the sum of the
products $B \, ds$ over the closed path, which is equivalent to the line integral of
B · d**s**, is

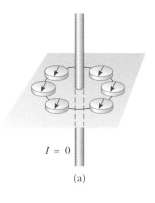

(a)

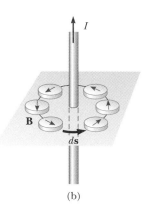

(b)

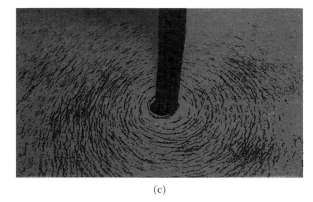

(c)

Figure 22.18   (a) When there is no current in the vertical wire, all compass needles point in
the same direction. (b) When the wire carries a strong current, the compass needles deflect in a
direction tangent to the circle, which is the direction of **B** due to the current. (c) Circular magnetic field lines surrounding a current-carrying conductor as displayed with iron filings. The
photograph was taken using 30 parallel wires each carrying a current of 0.50 A.   *(Henry Leap and
Jim Lehman)*

$$\oint \mathbf{B} \cdot d\mathbf{s} = B \oint ds = \frac{\mu_0 I}{2\pi r}(2\pi r) = \mu_0 I \qquad \text{[22.21]}$$

where $\oint ds = 2\pi r$ is the circumference of the circle.

This result, known as **Ampère's law,** was calculated for the special case of a circular path surrounding a wire. However, it can also be applied in the general case in which a steady current passes through the area surrounded by an arbitrary closed path. That is, Ampère's law says that the line integral of $\mathbf{B} \cdot d\mathbf{s}$ around any closed path equals $\mu_0 I$, where I is the total steady current passing through any surface bounded by the closed path.

$$\oint \mathbf{B} \cdot d\mathbf{s} = \mu_0 I \qquad \text{[22.22]} \qquad \bullet \quad \textit{Ampère's law}$$

Ampère's law is valid only for steady currents. Furthermore, even though Ampère's law is true for all current configurations, it is useful only for calculating the magnetic fields of configurations with high degrees of symmetry. (Recall that Gauss's law is useful only for calculating the electric fields of highly symmetric charge distributions.) The following examples illustrate some symmetric current configurations for which Ampère's law is useful.

Example 22.5   The Magnetic Field Created by a Long Current-Carrying Wire

A long, straight wire of radius R carries a steady current I_0 that is uniformly distributed through the cross section of the wire (Fig. 22.19). Calculate the magnetic field a distance r from the center of the wire in the regions $r \geq R$ and $r < R$.

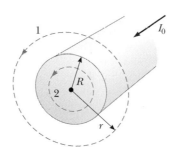

Figure 22.19   (Example 22.5) A long, straight wire of radius R carrying a steady current I_0 uniformly distributed across the wire. The magnetic field at any point can be calculated from Ampère's law using a circular path of radius r, concentric with the wire.

Solution   In Region 1, where $r \geq R$, let us choose for our path of integration a circle of radius r centered at the wire. From symmetry, we see that $\mathbf{B}$ must be constant in magnitude

and parallel to $d\mathbf{s}$ at every point on this circle. Because the total current passing through the plane of the circle is I_0, Ampère's law applied to the circle gives

$$\oint \mathbf{B} \cdot d\mathbf{s} = B \oint ds = B(2\pi r) = \mu_0 I_0$$

$$B = \frac{\mu_0 I_0}{2\pi r} \qquad \text{(for } r \geq R)$$

Now consider the interior of the wire—that is, Region 2—where $r < R$. Here the current I passing through the plane of the circle of radius $r < R$ is less than the total current I_0. Because the current is uniform over the cross-section of the wire, the fraction of the current enclosed by the circle of radius $r < R$ must equal the ratio of the area πr^2 enclosed by Circle 2 and the cross-sectional area πR^2 of the wire.

$$\frac{I}{I_0} = \frac{\pi r^2}{\pi R^2}$$

$$I = \frac{r^2}{R^2} I_0$$

Following the same procedure as for Circle 1, we apply Ampère's law to Circle 2:

$$\oint \mathbf{B} \cdot d\mathbf{s} = B(2\pi r) = \mu_0 I = \mu_0 \left(\frac{r^2}{R^2} I_0 \right)$$

$$B = \left(\frac{\mu_0 I_0}{2\pi R^2}\right) r \quad \text{(for } r < R) \qquad \textbf{[22.23]}$$

The magnetic field strength versus r for this configuration is sketched in Figure 22.20. Note that inside the wire, $B \to 0$ as $r \to 0$. This result is similar in form to that of the electric field inside a uniformly charged rod.

Figure 22.20   A sketch of the magnetic field versus r for the wire described in Example 22.5. The field is proportional to r inside the wire and varies as $1/r$ outside the wire.

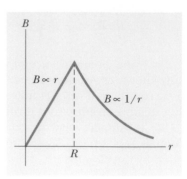

Example 22.6   The Magnetic Field Created by a Toroid

A toroid consists of N turns of wire wrapped around a ring-shaped structure, as in Figure 22.21. You can think of a toroid as a solenoid bent into the shape of a doughnut. Assuming that the turns are closely spaced, calculate the magnetic field inside the toroid, a distance r from the center.

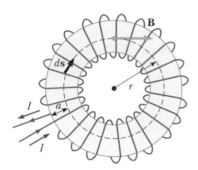

Figure 22.21   (Example 22.6) A toroid consists of many turns of wire wrapped around a doughnut-shaped structure (called a torus). If the coils are closely spaced, the field in the interior of the toroid is tangent to the dashed circle and varies as $1/r$, and the exterior field is zero.

Reasoning   To calculate the field inside the toroid, we evaluate the line integral of $\mathbf{B} \cdot d\mathbf{s}$ over a circle of radius r. By

symmetry, we see that the magnetic field is constant in magnitude on this circle and tangent to it, so that $\mathbf{B} \cdot d\mathbf{s} = B\,ds$.

Furthermore, note that the closed path threads N loops of wire, each of which carries a current I. Therefore, the right side of Equation 22.22 is $\mu_0 NI$ in this case.

Solution   Ampère's law applied to the circle gives

$$\oint \mathbf{B} \cdot d\mathbf{s} = B\oint ds = B(2\pi r) = \mu_0 NI$$

$$B = \frac{\mu_0 NI}{2\pi r} \qquad \textbf{[22.24]}$$

This result shows that B varies as $1/r$ and hence is non-uniform within the coil. However, if r is large compared with the cross-sectional radius of the toroid, then the field is approximately uniform inside the coil. Furthermore, for an ideal toroid, where the turns are closely spaced, the external magnitude field is zero. This can be seen by noting that the net current threaded by any circular path lying outside the toroid is zero (including the region of the "hole in the doughnut"). Therefore, from Ampère's law we find that $B = 0$ in the regions exterior to the coil. In reality, the turns of a toroid form a helix rather than circular loops. As a result, there is always a small field external to the coil.

Thinking Physics 5

Consider a plastic annular ring encircling a long, straight wire, which is coming out of the page in Figure 22.22. On the plastic ring are fastened two bar magnets, as shown. If a current flows in the wire in a direction out of the page, is there a net torque on the ring–magnet combination? If so, in which direction?

Reasoning   The small magnetic dipoles that form the bar magnets will experience a magnetic torque due to the magnetic field set up by the wire. The torque on each magnet can be represented by a force at the south pole acting perpendicular to its body, and an oppositely directed force at the north pole. Since the circular magnetic field generated by the wire will fall off inversely with distance, the force on the north poles will be smaller than the force on the south poles. However, because the moment arms of these forces increase linearly with distance from the wire, the effects of decreases in field strength with distance is balanced by increasing moment arms. Consequently, there is no net torque on the ring–magnet combination.

22.8 • THE MAGNETIC FIELD OF A SOLENOID

A solenoid is a long wire wound in the form of a helix. If the turns are closely spaced, this configuration can produce a reasonably uniform magnetic field within a small volume of the solenoid's interior region. Each of the turns can be regarded as a circular loop, and the net magnetic field is the vector sum of the fields due to all the turns.

Figure 22.23a shows the magnetic field lines of a loosely wound solenoid. Note that the interior field lines are nearly parallel, uniformly distributed, and close together. This indicates that the field in the solenoid's interior is uniform. The field lines between turns tend to cancel each other. The field outside the solenoid is both nonuniform and weak. To see why, consider the point *P* in Figure 22.23a. The field is nonuniform because of the finite length of the solenoid and a resultant field distribution that resembles that of a bar magnet. The field is weak because the field due to current elements on the upper portions tends to cancel the field due to current elements on the lower portions.

If the turns are closely spaced and the solenoid is of finite length, the field lines are as shown in Figure 22.23b. In this case, the field lines diverge from one end

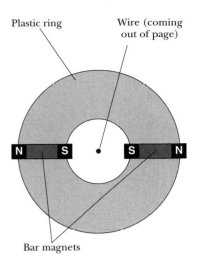

Figure 22.22 (Thinking Physics 5)

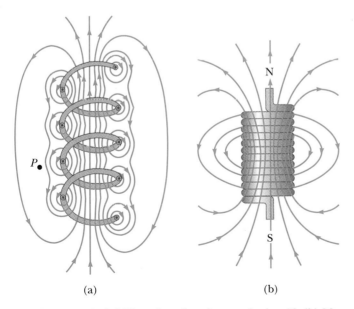

(a)   (b)

Figure 22.23 (a) The magnetic field lines for a loosely wound solenoid. (b) Magnetic field lines for a tightly wound solenoid of finite length carrying a steady current. The field in the space enclosed by the solenoid is nearly uniform and strong. Note that the field lines resemble those of a bar magnet, so that the solenoid effectively has north and south poles.

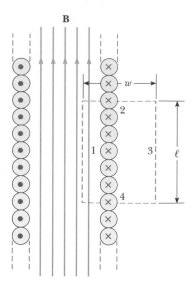

Figure 22.24 A cross-sectional view of a tightly wound solenoid. If the solenoid is long relative to its radius, we can assume that the magnetic field inside is uniform and the field outside is zero. Ampère's law applied to the red dashed rectangular path can then be used to calculate the field inside the solenoid.

and converge at the opposite end. An inspection of this field distribution exterior to the solenoid shows a similarity to the field of a bar magnet (see Fig. 22.1b). Hence, one end of the solenoid behaves like the north pole of a magnet and the opposite end behaves like the south pole. As the length of the solenoid increases, the field within it becomes more and more uniform. When the solenoid's turns are closely spaced and its length is great compared with its radius, it approaches the case of an *ideal solenoid*. The field outside the solenoid is weak compared with the field inside, and the field inside is uniform over a large volume.

We can use Ampère's law to obtain an expression for the magnetic field inside an ideal solenoid. A longitudinal cross-section of part of our ideal solenoid (Fig. 22.24) carries current I. **B** inside the ideal solenoid is uniform and parallel to the axis, and **B** outside is zero. Consider a rectangular path of length ℓ and width w, as shown in Figure 22.24. We can apply Ampère's law to this path by evaluating the integral of $\mathbf{B} \cdot d\mathbf{s}$ over each of the four sides of the rectangle. The contribution along side 3 is clearly zero, because $\mathbf{B} = 0$ in this region. The contributions from sides 2 and 4 are both zero, because **B** is perpendicular to $d\mathbf{s}$ along these paths. Side 1, the length of which is ℓ, gives a contribution of $B\ell$ to the integral, because **B** along this path is uniform and parallel to $d\mathbf{s}$. Therefore, the integral over the closed rectangular path has the value

$$\oint \mathbf{B} \cdot d\mathbf{s} = \int_{\text{path } 1} \mathbf{B} \cdot d\mathbf{s} = B \int_{\text{path } 1} ds = B\ell$$

The right side of Ampère's law involves the *total* current that passes through the area bound by the path of integration. In our case, the total current through the rectangular path equals the current through each turn of the solenoid multiplied by the number of turns. If N is the number of turns in the length ℓ, then the total current through the rectangle equals NI. Therefore, Ampère's law applied to this path gives

$$\oint \mathbf{B} \cdot d\mathbf{s} = B\ell = \mu_0 NI$$

Magnetic field inside a long • *solenoid*

$$B = \mu_0 \frac{N}{\ell} I = \mu_0 nI \qquad \text{[22.25]}$$

where $n = N/\ell$ is the number of turns *per unit length* (not to be confused with N).

We also could obtain this result in a simpler manner by reconsidering the magnetic field of a toroidal coil (Ex. 22.6). If the radius, r, of the toroidal coil containing N turns is large compared with its cross-sectional radius, a, then a short section of the toroidal coil approximates a solenoid, with $n = N/2\pi r$. In this limit, we see that Equation 22.24 derived for the toroidal coil agrees with Equation 22.25.

Equation 22.25 is valid only for points near the center of a very long solenoid. As you might expect, the field near each end is smaller than the value given by Equation 22.25. At the very end of a long solenoid, the magnitude of the field is about one half that of the field at the center (see Problem 22.32).

CONCEPTUAL PROBLEM 8

A hanging Slinky toy is attached to a powerful battery and a switch. When the switch is closed so that current suddenly flows through the Slinky, does the Slinky compress or expand?

EXERCISE 8 A closely wound, long solenoid of overall length 30.0 cm has a magnetic field of magnitude 5.00×10^{-4} T at its center produced by a current of 1.00 A through its windings. How many turns of wire are on the solenoid? Answer 120 turns

22.9 • MAGNETISM IN MATTER

O P T I O N A L

The magnetic field produced by a current in a coil of wire gives us a hint regarding what causes certain materials to exhibit strong magnetic properties. In order to understand why some materials are magnetic, it is instructive to begin our discussion with the classical model of the atom, in which electrons are assumed to move in circular orbits around the much more massive nucleus. In this model, each electron, with its charge of 1.6×10^{-19} C, circles the atom once in about 10^{-16} s. If we divide the electronic charge by this time interval, we find that the orbiting electron is equivalent to a current of 1.6×10^{-3} A. Therefore, each orbiting electron is viewed as a tiny current loop with a corresponding magnetic moment. Such a current loop produces a magnetic field of about 12.5 T at the center of the circular path (see Problem 16).

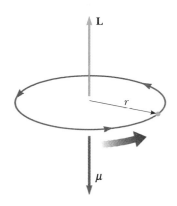

Figure 22.25 An electron moving in a circular orbit of radius r has an angular momentum **L** in one direction and a magnetic moment **μ** in the opposite direction.

Consider an electron moving with a constant speed of v in a circular orbit of radius r about the nucleus, as in Figure 22.25. Because the electron travels a distance of $2\pi r$ (the circumference of the circle) in the time T, which is the time required for one revolution, the electron's orbital speed is $v = 2\pi r/T$. The current associated with this orbiting electron equals its charge divided by the time for one revolution:

$$I = \frac{e}{T} = \frac{e}{2\pi r/v} = \frac{ev}{2\pi r}$$

The magnitude of the magnetic moment associated with this current loop is given by $\mu = IA$, where $A = \pi r^2$ is the area of the orbit. Therefore,

$$\mu = IA = \left(\frac{ev}{2\pi r}\right)\pi r^2 = \tfrac{1}{2}evr$$

Because the magnitude of the orbital angular momentum of the electron is $L = m_e v r$ (Eq. 11.24), the magnitude of the magnetic moment can be written as

$$\mu = \left(\frac{e}{2m_e}\right)L \qquad\qquad [22.26]$$

• *Orbital magnetic moment*

This result says that **the magnetic moment of an orbiting electron is proportional to its orbital angular momentum.** Note that, because the electron is negatively charged, the vectors **μ** and **L** point in opposite directions. Both vectors are perpendicular to the plane of the orbit, as indicated in Figure 22.25.

As we learned in Chapter 11, Section 11.9, the Bohr theory of the atom includes the notion that the orbital angular momentum must be quantized and is always some integer multiple of $\hbar = h/2\pi = 1.06 \times 10^{-34}$ J·s, where h is Planck's constant. That is,

$$L = 0, \hbar, 2\hbar, 3\hbar, \ldots$$

• *Angular momentum is quantized.*

Hence, the smallest nonzero value of the magnetic moment of an orbiting electron is

$$\mu = \frac{e}{2m_e}\hbar \qquad\qquad [22.27]$$

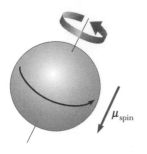

μ_{spin}

Figure 22.26 Classical model of a spinning electron.

In most substances, the magnetic moment of one electron in an atom is canceled by that of another electron in the atom, orbiting in the opposite direction. The net result is that **the magnetic effect produced by the orbital motion of the electrons is either zero or very small for most materials.**

As mentioned earlier, an electron has another intrinsic property called **spin,** which also contributes to its magnetic moment. The spin of an electron is a separate angular momentum from its orbital angular momentum, just as the spin of the Earth is separate from its orbital motion around the Sun. One can model the electron as a sphere of charge spinning around its axis as it orbits the nucleus, as in Figure 22.26. (This classical description of a spinning electron should not be taken literally. The property of spin can be understood only through a relativistic model.) This spinning motion produces an effective current loop and hence a magnetic moment of the same order of magnitude as that due to the orbital motion. The magnitude of the spin angular momentum predicted by quantum theory is

$$S = \frac{\hbar}{2} = 5.2729 \times 10^{-35} \, \text{J} \cdot \text{s}$$

Every electron has an intrinsic magnetic moment associated with its spin with the value

Bohr magneton •

$$\mu_{\text{B}} = \frac{e}{2m_e} \hbar = 9.27 \times 10^{-24} \, \text{J/T} \qquad \text{[22.28]}$$

which is called the **Bohr magneton.**

In atoms or ions containing multiple electrons, the electrons usually pair up with their spins opposite each other, an arrangement that results in a cancellation of the spin magnetic moments. However, an atom with an odd number of electrons must have at least one "unpaired" electron and a corresponding spin magnetic moment. The magnetic moments of several atoms and ions are listed in Table 22.1.

TABLE 22.1
Magnetic Moments of Some Atoms and Ions (in Units of Bohr Magnetons)

Atom (or ion)	Magnet Moment per Atom or Ion (μ/μ_{B})
H	1
He	0
Ne	0
Fe	2.22
Co	1.72
Ni	0.606
Gd	7.1
Dy	10.0
Co^{2+}	4.8
Ni^{2+}	3.2
Fe^{2+}	5.4
Ce^{3+}	2.14
Yb^{3+}	4.00

Ferromagnetic Materials

Iron, cobalt, nickel, gadolinium, and dysprosium are strongly magnetic materials and are said to be **ferromagnetic.** Ferromagnetic substances, used to fabricate permanent magnets, contain atoms with spin magnetic moments that tend to align parallel to each other even in a weak external magnetic field. Once the moments are aligned, the substance remains magnetized after the external field is removed. This permanent alignment is due to strong coupling between neighboring atoms, which can only be understood using quantum physics.

All ferromagnetic materials contain microscopic regions called **domains,** within which all magnetic moments are aligned. The domains range from about 10^{-12} to $10^{-8} \, \text{m}^3$ in volume and contain 10^{-17} to 10^{-21} atoms. The boundaries between domains having different orientations are called **domain walls.** In an unmagnetized sample, the domains are randomly oriented so that the net magnetic moment is zero, as in Figure 22.27a. When the sample is placed in an external magnetic field, the domains tend to align with the field, which results in a magnetized sample, as in Figure 22.27b. Observations show that domains initially oriented along the external field grow in size at the expense of the less favorably oriented domains. When the external field is removed, the sample may retain most of its magnetism.

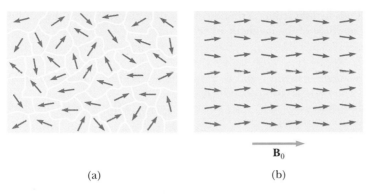

$\mathbf{B}_0$

(a)   (b)

Figure 22.27   (a) Random orientation of domains in an unmagnetized substance. (b) When an external magnetic field, $\mathbf{B}_0$, is applied, the domains tend to align with the magnetic field.

The extent to which a ferromagnetic substance retains its magnetism depends on whether it is classified as being magnetically **hard** or **soft.** *Soft magnetic materials,* such as iron, are easily magnetized but also tend to lose their magnetism easily. When a soft magnetic material is magnetized and the external magnetic field is removed, thermal agitation produces domain motion and the material quickly returns to an unmagnetized state. In contrast, *hard magnetic materials,* such as cobalt and nickel, are difficult to magnetize but tend to retain their magnetism, and domain alignment persists in them after the external magnetic field is removed. Such hard magnetic materials are referred to as **permanent magnets.** Rare-earth permanent magnets, such as samarium-cobalt, are now regularly used in industry.

Thinking Physics 6

Suppose an aluminum tube is filled with very fine iron filings. Each filing is in the form of a sliver—long and thin. If the tube of filings is shaken, what is the likelihood that all of them will wind up aligned in the same direction? Suppose the shaking is repeated while the tube is in a strong magnetic field. What is the likelihood now that the filings will be aligned? Is this a macroscopic analog to anything discussed in this section?

Reasoning   When the tube is shaken outside of a magnetic field, the orientation of the filings will be random—it is extremely unlikely that they will all be aligned in the same direction. If the shaking occurs in a magnetic field, the filings, which become magnetized, can align with the field while they are momentarily airborne and free to rotate. Thus, it is likely that a large number of the filings will be aligned with the field after the shaking. The filings are an analog to magnetic domains in a ferromagnetic material. When the material is placed in a magnetic field, more domains end up aligned with the field than misaligned, similar to the iron filings.

Oxygen, a paramagnetic substance, is attracted to a magnetic field. The liquid oxygen in this photograph is suspended between the poles of a permanent magnet. Paramagnetic substances contain atoms (or ions) that have permanent magnetic dipole moments. These dipoles interact weakly with each other and are randomly oriented in the absence of an external magnetic field. When the substance is placed in an external magnetic field, its atomic dipoles tend to align with the field. *(Courtesy of Leon Lewandowski)*

SUMMARY

The **magnetic force** that acts on a charge, q, moving with a velocity of $\mathbf{v}$ in an external magnetic field, $\mathbf{B}$, is

$$\mathbf{F} = q\mathbf{v} \times \mathbf{B} \qquad\qquad [22.1]$$

This force is in a direction perpendicular both to the velocity of the particle and to the field. The magnitude of the magnetic force is

$$F = qvB \sin \theta \qquad [22.2]$$

where θ is the angle between **v** and **B**. $F = 0$ when **v** is either parallel or antiparallel to **B**, and $F = qvB$ when **v** is perpendicular to **B**.

If a straight conductor of length ℓ carries a current, I, the force on that conductor when placed in a uniform *external* magnetic field, **B**, is

$$\mathbf{F} = I\boldsymbol{\ell} \times \mathbf{B} \qquad [22.3]$$

where $\boldsymbol{\ell}$ is in the direction of the current and $|\boldsymbol{\ell}| = \ell$.

If an arbitrarily shaped wire carrying current I is placed in an *external* magnetic field, the force on a very small segment, $d\mathbf{s}$, is

$$d\mathbf{F} = I \, d\mathbf{s} \times \mathbf{B} \qquad [22.4]$$

To determine the total force on the wire, one has to integrate Equation 22.4.

The **magnetic moment, $\boldsymbol{\mu}$,** of a current loop carrying current I is

$$\boldsymbol{\mu} = I\mathbf{A} \qquad [22.8]$$

where **A** is perpendicular to the plane of the loop and $|\mathbf{A}|$ is equal to the area of the loop. The SI unit of $\boldsymbol{\mu}$ is the ampere-meter^2.

The torque, $\boldsymbol{\tau}$, on a current loop when the loop is placed in a uniform *external* magnetic field, **B**, is

$$\boldsymbol{\tau} = \boldsymbol{\mu} \times \mathbf{B} \qquad [22.9]$$

The **Biot–Savart law** says that the magnetic field $d\mathbf{B}$ at the point P, due to the wire element $d\mathbf{s}$ carrying a steady current, I, is

$$d\mathbf{B} = k_m \frac{I \, d\mathbf{s} \times \hat{\mathbf{r}}}{r^2} \qquad [22.10]$$

where $k_m = 10^{-7} \text{ T} \cdot \text{m/A}$ and r is the distance from the element to the point P. To find the total field at P due to a current-carrying conductor, one must integrate this vector expression over the entire conductor.

The magnitude of the **magnetic field** at a distance of r from a long, straight wire carrying current I is

$$B = \frac{\mu_0 I}{2\pi r} \qquad [22.15]$$

where $\mu_0 = 4\pi \times 10^{-7} \text{ T} \cdot \text{m/A}$ is the **permeability of free space.** The field lines are circles concentric with the wire.

The force per unit length between two parallel wires separated by a distance of a and carrying currents I_1 and I_2 has the magnitude

$$\frac{F_1}{\ell} = \frac{\mu_0 I_1 I_2}{2\pi a} \qquad [22.20]$$

The force is attractive if the currents are in the same direction and repulsive if they are in opposite directions.

Ampère's law says that the line integral of $\mathbf{B} \cdot d\mathbf{s}$ around any closed path equals $\mu_0 I$, where I is the total steady current passing through any surface bounded by the closed path:

$$\oint \mathbf{B} \cdot d\mathbf{s} = \mu_0 I \qquad [22.22]$$

Using Ampère's law, one finds that the fields inside a toroidal coil and solenoid are

$$B = \frac{\mu_0 NI}{2\pi r} \quad \text{(toroid)} \qquad\qquad \textbf{[22.24]}$$

$$B = \mu_0 \frac{N}{\ell} I = \mu_0 n I \quad \text{(solenoid)} \qquad \textbf{[22.25]}$$

where N is the total number of turns and n is the number of turns per unit length.

The fundamental sources of all magnetic fields are the magnetic dipole moments associated with atoms. The atomic dipole moments can arise both from the orbital motions of the electrons and from an intrinsic property of electrons known as *spin*.

CONCEPTUAL QUESTIONS

1. Two charged particles are projected into a region in which there is a magnetic field perpendicular to their velocities. If the charges are deflected in opposite directions, what can you say about them?

2. A current-carrying conductor experiences no magnetic force when placed in a certain manner in a uniform magnetic field. Explain.

3. Is it possible to orient a current loop in a uniform magnetic field so that the loop does not tend to rotate? Explain.

4. The north seeking pole of a magnet is attracted toward the geographic North Pole of the Earth. Yet, like poles repel. What is the way out of this dilemma?

5. Which way would a compass point if you were at the north magnetic pole?

6. A magnet attracts a piece of iron. The iron can then attract another piece of iron. On the basis of domain alignment, explain what happens in each piece of iron.

7. Will a nail be attracted to either pole of a magnet? Explain what is happening inside the nail when placed near the magnet.

8. Why does hitting a magnet with a hammer cause the magnetism to be reduced?

9. Parallel wires exert magnetic forces on each other. What about perpendicular wires? Imagine two wires oriented perpendicular to each other and almost touching. Each wire carries a current. Is there a force between the wires?

10. Explain why it is not possible to determine the charge and mass of a charged particle separately via electric and magnetic forces.

11. How can a current loop be used to determine the presence of a magnetic field in a given region of space?

12. It is found that charged particles from outer space, called cosmic rays, strike the Earth more frequently at the poles than at the Equator. Why?

13. Explain why two parallel wires carrying currents in opposite directions repel each other.

14. A hollow copper tube carries a current. Why is **B** = 0 inside the tube? Is **B** nonzero outside the tube?

15. Describe the change in the magnetic field in the space enclosed by a solenoid carrying a steady current I if (a) the length of the solenoid is doubled but the number of turns remains the same and (b) the number of turns is doubled but the length remains the same.

16. Consider an electron near the magnetic equator. In which direction will it tend to be deflected if its velocity is directed (a) downward? (b) northward? (c) westward? (d) southeastward?

17. Figure Q22.17 shows two permanent magnets, each having a hole through its center. Note that the upper magnet is levitated above the lower one. (a) How does this occur? (b) What purpose does the pencil serve? (c) What can you say about the poles of the magnets from this observation? (d) If the upper magnet were inverted, what do you suppose would happen?

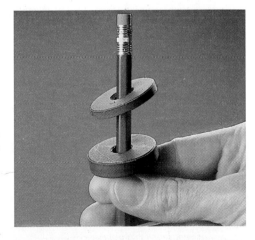

Figure Q22.17 Magnetic levitation using two ceramic magnets.
(Courtesy of Central Scientific Company)

18. You are an astronaut stranded on a planet with no test equipment for minerals. The planet does not have a mag-

netic field. You have two bars of iron in your possession: One is magnetized, one is not. How could you determine which is which?

19. A Hindu ruler once suggested that he be entombed in a magnetic coffin with the polarity arranged so that he would be forever suspended between heaven and Earth. Is such magnetic levitation possible? Discuss.

20. Should the surface of a computer disk be made from a hard or a soft ferromagnetic substance?

21. The *bubble chamber* is a device used for observing tracks of particles that pass through the chamber, which is immersed in a magnetic field. If some tracks are spirals and others are straight lines, what can you say about the particles?

22. The electron beam in Figure Q22.22 is projected to the right. The beam deflects downward in the presence of a magnetic field produced by a pair of current-carrying coils. (a) What is the direction of the magnetic field? (b) What would happen to the beam if the current in the coils were reversed?

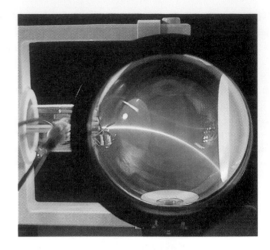

Figure Q22.22   Bending of a beam of electrons in a magnetic field.   *(Courtesy of Central Scientific Company)*

PROBLEMS

Section 22.2   The Magnetic Field

1. Determine the initial direction of the deflection of charged particles as they enter the magnetic fields, as shown in Figure P22.1.

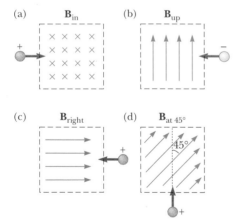

Figure P22.1

2. A proton travels with a speed of 3.00×10^6 m/s at an angle of 37.0° with the direction of a magnetic field of 0.300 T in the positive y direction. What are (a) the magnitude of the magnetic force on the proton and (b) its acceleration?

3. A proton moves perpendicular to a uniform magnetic field **B** at 1.00×10^7 m/s and experiences an acceleration of 2.00×10^{13} m/s² in the positive x direction when its

velocity is in the positive z direction. Determine the magnitude and direction of the field.

4. An electron is accelerated through 2400 V from rest and then enters a region in which there is a uniform 1.70-T magnetic field. What are (a) the maximum and (b) the minimum values of the magnetic force this charge can experience?

5. At the Equator, near the surface of the Earth, the magnetic field is approximately 50.0 μT northward, and the electric field is about 100 N/C downward in fair weather. Find the gravitational, electric, and magnetic forces on an electron moving eastward in a straight line at 6.00×10^6 m/s in this environment.

6. A metal ball having net charge $Q = 5.00$ μC is thrown out of a window horizontally at a speed $v = 20.0$ m/s. The window is at a height $h = 20.0$ m above the ground. A uniform horizontal magnetic field of magnitude $B = 0.0100$ T is perpendicular to the plane of the ball's trajectory. Find the magnetic force acting on the ball just before it hits the ground.

7. At the Fermilab accelerator in Batavia, Illinois, protons having momentum 4.80×10^{-16} kg·m/s are held in a circular orbit of radius 1.00 km by an upward magnetic field. What is the magnitude of this field?

Section 22.3   Magnetic Force on a Current-Carrying Conductor

8. A wire carries a steady current of 2.40 A. A straight section of the wire is 0.750 m long and lies along the x axis within a uniform magnetic field, **B** = (1.60**k**) T. If the current is

in the positive *x* direction, what is the magnetic force on the section of wire?

9. A wire having a mass per unit length of 0.500 g/cm carries a 2.00-A current horizontally to the south. What are the direction and magnitude of the minimum magnetic field needed to lift this wire vertically upward?

10. In Figure P22.10, the cube is 40.0 cm on each edge. Four straight segments of wire—*ab, bc, cd,* and *da*—form a closed loop that carries a current $I = 5.00$ A, in the direction shown. A uniform magnetic field, $B = 0.0200$ T, is in the positive *y* direction. Determine the magnitude and direction of the magnetic force on each segment.

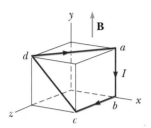

Figure P22.10

11. A strong magnet is placed under a horizontal conducting ring of radius *r* that carries current *I*, as in Figure P22.11. If, at the ring's location, the magnetic field **B** makes an angle θ with the vertical, what are the magnitude and direction of the resultant force on the ring?

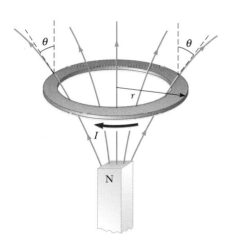

Figure P22.11

12. An arbitrarily shaped, closed loop carrying current *I* is placed in a uniform external magnetic field, **B**, as in Figure P22.12. Show that the net magnetic force on the loop is zero.

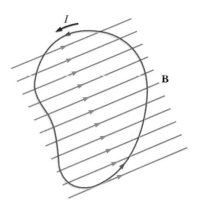

Figure P22.12

Section 22.4 Torque on a Current Loop in a Uniform Magnetic Field

13. A rectangular loop consists of $N = 100$ closely wrapped turns and has dimensions $a = 0.400$ m and $b = 0.300$ m. The loop is hinged along the *y* axis, and its plane makes an angle $\theta = 30.0°$ with the *x* axis (Fig. P22.13). What is the magnitude of the torque exerted on the loop by a uniform magnetic field $B = 0.800$ T directed along the *x* axis when the current is $I = 1.20$ A in the direction shown? What is the expected direction of rotation of the loop?

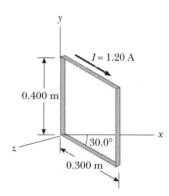

Figure P22.13

14. A small bar magnet is suspended in a uniform 0.250-T magnetic field. The maximum torque experienced by the bar magnet is 4.60×10^{-3} N·m. Calculate the magnetic moment of the bar magnet.

15. A long piece of wire of mass 0.100 kg and total length of 4.00 m is used to make a square coil with side 0.100 m. The coil is hinged along a horizontal side, carries a 3.40-A current, and is placed in a vertical magnetic field of magnitude 0.0100 T. (a) Determine the angle that the plane of the coil

makes with the vertical when the coil is in equilibrium. (b) Find the torque acting on the coil due to the magnetic force at equilibrium.

Section 22.5   The Biot–Savart Law

16. In Niels Bohr's 1913 model of the hydrogen atom, an electron circles the proton at a distance of 5.29×10^{-11} m with a speed of 2.19×10^6 m/s. Compute the magnetic field strength that this motion produces at the location of the proton.

17. (a) A conductor in the shape of a square of edge length $\ell = 0.400$ m carries a current $I = 10.0$ A (Fig. P22.17). Calculate the magnitude and direction of the magnetic field at the center of the square. (*Hint:* Use the result to Problem 22.51a to find the field due to each straight section.) (b) If this conductor is formed into a single circular turn and carries the same current, what is the value of the magnetic field at the center?

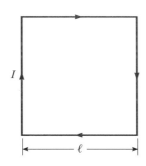

Figure P22.17

18. A current path shaped as shown in Figure P22.18 produces a magnetic field at P, the center of the arc. If the arc subtends an angle of 30.0° and the radius of the arc is 0.600 m, what are the magnitude and direction of the field produced at P if the current is 3.00 A?

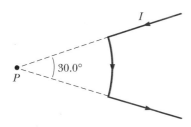

Figure P22.18

19. Determine the magnetic field at a point P located a distance x from the corner of an infinitely long wire bent at a right angle, as in Figure P22.19. The wire carries a steady current I.

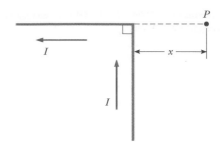

Figure P22.19

Section 22.6   The Magnetic Force between Two Parallel Conductors

20. Two long, parallel conductors, separated by 10.0 cm, carry currents in the same direction. If $I_1 = 5.00$ A and $I_2 = 8.00$ A, what is the force per unit length exerted on each conductor by the other?

21. In Figure P22.21, the current in the long, straight wire is $I_1 = 5.00$ A and the wire lies in the plane of the rectangular loop, which carries 10.0 A. The dimensions are $c = 0.100$ m, $a = 0.150$ m, and $\ell = 0.450$ m. Find the magnitude and direction of the net force exerted on the loop by the magnetic field created by the wire.

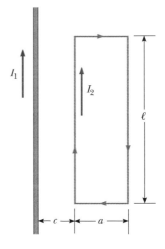

Figure P22.21

22. Two long, parallel wires, each having a mass per unit length of 40.0 g/m, are supported in a horizontal plane by strings 6.00 cm long, as in Figure P22.22. Each wire carries the same current I, causing the wires to repel each other so that the angle θ between the supporting strings is 16.0°. (a) Are the currents in the same or opposite directions? (b) Find the magnitude of each current.

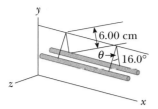

Figure P22.22

Section 22.7 Ampère's Law

23. Four long, parallel conductors carry equal currents of $I = 5.00$ A. Figure P22.23 is an end view of the conductors. The current direction is into the page at points A and B (indicated by the crosses) and out of the page at C and D (indicated by the dots). Calculate the magnitude and direction of the magnetic field at point P, located at the center of the square of edge length 0.200 m.

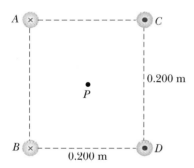

Figure P22.23

24. The magnetic field 40.0 cm away from a long, straight wire carrying current 2.00 A is 1.00 μT. (a) At what distance is it 0.100 μT? (b) At one instant, the two conductors in a long household extension cord carry equal 2.00-A currents in opposite directions. The two wires are 3.00 mm apart. Find the magnetic field 40.0 cm away from the middle of the cord, in the plane of the two wires. (c) At what distance is it one tenth as large? (d) The center wire in a coaxial cable carries current 2.00 A in one direction and the sheath around it carries current 2.00 A in the opposite direction. What magnetic field does the cable create at points outside?

25. The magnetic coils of a tokamak fusion reactor are in the shape of a toroid having an inner radius of 0.700 m and an outer radius of 1.30 m. If the toroid has 900 turns of large-diameter wire, each of which carries a current of 14.0 kA, find the magnetic field strength along (a) the inner radius and (b) the outer radius.

26. A cylindrical conductor of radius $R = 2.50$ cm carries a current of $I = 2.50$ A along its length; this current is uniformly distributed throughout the cross section of the conductor. (a) Calculate the magnetic field midway along the radius of the wire (that is, at $r = R/2$). (b) Find the distance beyond the surface of the conductor at which the magnitude of the magnetic field has the same value as the magnitude of the field at $r = R/2$.

27. A packed bundle of 100 long, straight, insulated wires forms a cylinder of radius $R = 0.500$ cm. (a) If each wire carries 2.00 A, what are the magnitude and direction of the magnetic force per unit length acting on a wire located 0.200 cm from the center of the bundle? (b) Would a wire on the outer edge of the bundle experience a force greater or smaller than the value calculated in part (a)?

28. Niobium metal becomes a superconductor when cooled below 9 K. If superconductivity is destroyed when the surface magnetic field exceeds 0.100 T, determine the maximum current a 2.00-mm-diameter niobium wire can carry and remain superconducting.

Section 22.8 The Magnetic Field of a Solenoid

29. What current is required in the windings of a long solenoid that has 1000 turns uniformly distributed over a length of 0.400 m in order to produce at the center of the solenoid a magnetic field of magnitude 1.00×10^{-4} T?

30. A superconducting solenoid is to generate a magnetic field of 10.0 T. (a) If the solenoid winding has 2000 turns/m, what is the required current? (b) What force per unit length is exerted on the windings by this magnetic field?

31. A solenoid of radius $R = 5.00$ cm is made of a long piece of wire of radius $r = 2.00$ mm, length $\ell = 10.0$ m ($\ell \gg R$), and resistivity $\rho = 1.70 \times 10^{-8}$ $\Omega \cdot$m. Find the magnetic field at the center of the solenoid if the wire is connected to a battery having an emf $\mathcal{E} = 20.0$ V.

32. Consider a solenoid of length ℓ and radius R, containing N closely spaced turns and carrying a steady current I. (a) Using these parameters, find the magnetic field along the axis as a function of distance a from the end of the solenoid. (b) Show that as ℓ becomes very long, B approaches $\mu_0 NI/2\ell$ at each end of the solenoid.

Section 22.9 Magnetism in Matter (Optional)

33. The magnetic moment of the Earth is approximately 8.00×10^{22} A$\cdot$m^2. (a) If this were caused by the complete magnetization of a huge iron deposit, how many unpaired electrons would this correspond to? (b) At two unpaired electrons per iron atom, how many kilograms of iron would this correspond to? (The density of iron is 7900 kg/m^3, and there are approximately 8.50×10^{28} iron atoms/m^3.)

34. In Bohr's 1913 model of the hydrogen atom, the electron is in a circular orbit of radius 5.29×10^{-11} m, and its speed is 2.19×10^6 m/s. (a) What is the magnitude of the mag-

netic moment due to the electron's motion? (b) If the electron orbits counterclockwise in a horizontal circle, what is the direction of this magnetic moment vector?

Additional Problems

35. A lightning bolt may carry a current of 1.00×10^4 A for a short period of time. What is the resulting magnetic field 100 m from the bolt? Assume that the bolt extends far above and below the point of observation.

36. A long, straight wire lies on a horizontal table and carries a current of 1.20 μA. At a distance d above the wire, a proton moves parallel to the wire (opposite the current) with a constant velocity of 2.30×10^4 m/s. Determine the value of d. You may ignore the magnetic field due to the Earth.

37. The magnitude of the Earth's magnetic field at either pole is approximately 7.00×10^{-5} T. Using a model in which you assume that this field is produced by a current loop around the Equator, determine the current that would generate such a field. (Note that the radius of the Earth is $R_E = 6.37 \times 10^6$ m.)

38. An electron enters a region of magnetic field of magnitude 0.100 T, traveling perpendicular to the linear boundary of the region. The direction of the field is perpendicular to the velocity of the electron. (a) Determine the time it takes for the electron to leave the "field-filled" region, noting that its path is a semicircle. (b) Find the kinetic energy of the electron if the maximum depth of penetration in the field is 2.00 cm.

39. Two parallel conductors carry current in opposite directions, as shown in Figure P22.39. One conductor carries a current of 10.0 A. Point A is at the midpoint between the wires and point C is a distance $d/2$ to the right of the 10.0-A current. If $d = 18.0$ cm and I is adjusted so that the magnetic field at C is zero, find (a) the value of the current I and (b) the value of the magnetic field at A.

40. Suppose you install a compass on the center of the dashboard of a car. Compute an order-of-magnitude estimate for the magnetic field at this location produced by the current when you switch on the headlights. How does it compare with the Earth's magnetic field?

41. A very long, thin strip of metal of width w carries a current I along its length, as in Figure P22.41. Find the magnetic field in the plane of the strip (at an external point P) a distance b from one edge.

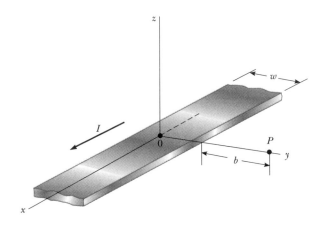

Figure P22.41

42. A 0.200-kg metal rod carrying a current of 10.0 A glides on two horizontal rails 0.500 m apart. What vertical magnetic field is required to keep the rod moving at a constant speed if the coefficient of kinetic friction between the rod and rails is 0.100?

43. A nonconducting ring of radius 10.0 cm is uniformly charged with a total positive charge 10.0 μC. The ring rotates at a constant angular speed 20.0 rad/s about an axis through its center, perpendicular to the plane of the ring. What is the magnitude of the magnetic field on the axis of the ring 5.00 cm from its center?

44. A nonconducting ring of radius R is uniformly charged with a total positive charge q. The ring rotates at a constant angular speed ω about an axis through its center, perpendicular to the plane of the ring. What is the magnitude of the magnetic field on the axis of the ring a distance $R/2$ from its center?

45. Sodium melts at 99°C. Liquid sodium, an excellent thermal conductor, is used in some nuclear reactors to remove thermal energy from the reactor core. The liquid sodium is moved through pipes by pumps that exploit the force on a moving charge in a magnetic field. The principle is as follows: Imagine the liquid metal to be in an electrically insulating pipe having a rectangular cross-section of width w and height h. A uniform magnetic field perpendicular to the pipe affects a section of length L (Fig. P22.45). An electric

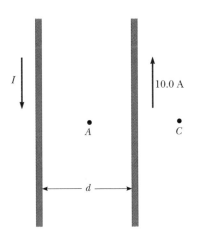

Figure P22.39

current directed perpendicular to the pipe and to the magnetic field produces a current density of magnitude J in the liquid sodium. (a) Explain why this arrangement produces on the liquid a force that is directed along the length of the pipe. (b) Show that the section of liquid in the magnetic field experiences a pressure increase JLB.

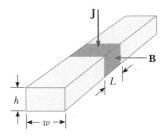

Figure P22.45

46. Two circular coils of radius R are each perpendicular to a common axis. The coil centers are a distance R apart and a steady current I flows in the same direction around each coil, as shown in Figure P22.46. (a) Show that the magnetic field on the axis at a distance x from the center of one coil is

$$B = \frac{\mu_0 I R^2}{2} \left[\frac{1}{(R^2 + x^2)^{3/2}} + \frac{1}{(2R^2 + x^2 - 2Rx)^{3/2}} \right]$$

(b) Show that $\dfrac{dB}{dx}$ and $\dfrac{d^2B}{dx^2}$ are both zero at a point midway between the coils. This means the magnetic field in the region midway between the coils is uniform. Coils in this configuration are called **Helmholtz coils.**

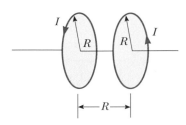

Figure P22.46

47. For a research project, a student needs a solenoid that produces an interior magnetic field of 0.0300 T. She decides to use a current of 1.00 A and a wire 0.500 mm in diameter. She winds the solenoid as layers on an insulating form

1.00 cm in diameter and 10.0 cm long. Determine the number of layers of wire needed and the total length of the wire.

48. Two circular loops are parallel, coaxial, and almost in contact, 1.00 mm apart (Fig. P22.48). Each loop is 10.0 cm in radius. The top loop carries a clockwise current of 140 A. The bottom loop carries a counterclockwise current of 140 A. (a) Calculate the magnetic force that the bottom loop exerts on the top loop. (b) The upper loop has a mass of 0.0210 kg. Calculate its acceleration, assuming that the only forces acting on it are the force in part (a) and its weight. (*Hint:* Think about how one loop looks to a bug perched on the other loop.)

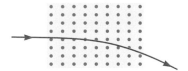

Figure P22.48

49. Protons having a kinetic energy of 5.00 MeV are moving in the positive x direction and enter a magnetic field **B** = $(0.0500 \text{ T})\mathbf{k}$ directed out of the plane of the page and extending from $x = 0$ to $x = 1.00$ m, as in Figure P22.49. (a) Calculate the y component of the protons' momentum as they leave the magnetic field. (b) Find the angle α between the initial velocity vector of the proton beam and the velocity vector after the beam emerges from the field. (*Hint:* Neglect relativistic effects and note that 1 eV = 1.60 × 10^{-19} J.)

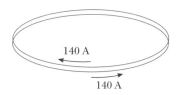

Figure P22.49

50. An infinite sheet of current lying in the yz plane carries a surface current of density $\mathbf{J}_s$. The current is in the y direction, and $\mathbf{J}_s$ represents the current per unit length measured along the z axis. Figure P22.50 is an edge view of the sheet. Find the magnetic field near the sheet. (*Hint:* Use Ampère's law and evaluate the line integral that takes a rectangular path around the sheet, represented by the dashed line in Fig. P22.50.)

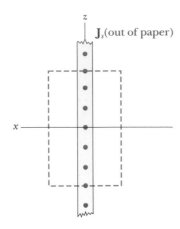

Figure P22.50   Top view of an infinite current sheet lying in the yz plane, where the current is in the y direction (out of the paper).

51. Consider a thin, straight wire segment carrying a constant current, I, and placed along the x axis, as in Figure P22.51. (a) Use the Biot–Savart law to show that the total magnetic field at the point P, located a distance a from the wire, is

$$B = \frac{\mu_0 I}{4\pi a} (\cos \theta_1 - \cos \theta_2)$$

(b) If the wire is infinitely long, show that the result in (a) gives a magnetic field that agrees with that obtained using Ampère's law (see Ex. 22.5).

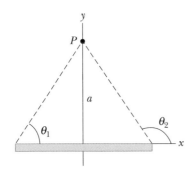

Figure P22.51

52. A long, straight wire—wire 1—oriented along the y axis carries a steady current I_1, as in Figure P22.52. A rectangular circuit located to the right of the wire carries current I_2. Consider wire 2, which is the upper horizontal segment of the circuit that runs from $x = a$ to $x = a + b$. Show that the magnetic force exerted on wire 2 is upward and has a magnitude given by

$$F = \frac{\mu_0 I_1 I_2}{2\pi} \ln \left(1 + \frac{b}{a} \right)$$

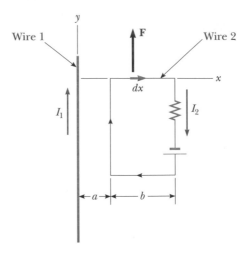

Figure P22.52

Spreadsheet Problems

S1. Spreadsheet 22.1 calculates the magnetic field along the x axis of a solenoid of length ℓ, radius R, and n turns per unit length carrying a current I. The solenoid is centered at the origin with its axis along the x axis. The field at any point x is

$$B = \frac{1}{2} \mu_0 n I [f_1(x) - f_2(x)]$$

where

$$f_1(x) = \frac{x + \frac{1}{2}\ell}{\left[R^2 + \left(x + \frac{1}{2}\ell \right)^2 \right]^{1/2}}$$

$$f_2(x) = \frac{x - \frac{1}{2}\ell}{\left[R^2 + \left(x - \frac{1}{2}\ell \right)^2 \right]^{1/2}}$$

(a) Let $\ell = 10$ cm, $R = 0.50$ cm, $I = 4.0$ A, and $n = 1500$ turns/m. Plot B versus x for positive values of x ranging from $x = 0$ to $x = 20$ cm. (b) Change R to the following values and observe the effects on the graph in each case: $R = 0.10$ cm, 1.0 cm, 2.0 cm, 5.0 cm, and 10.0 cm. For small values of R (small compared to ℓ), the field inside the solenoid should look like that of an ideal solenoid; for large values of R, the field should look like that of a loop. Does it?

S2. If $x \gg \ell$ (where x is measured from the center of the solenoid), the magnetic field of a finite solenoid along the x axis approaches

$$\frac{\mu_0}{4\pi} \frac{2\mu}{x^3}$$

where $\mu = NI(\pi R^2)$ is the magnetic dipole moment of the solenoid. Modify Spreadsheet 22.1 to calculate and plot $x^3 B(x)$ as a function of x. Verify that $x^3 B(x)$ approaches a constant for $x \gg \ell$.

S3. An electron has an initial velocity $\mathbf{v}_0$ at the origin of a coordinate system at $t = 0$. A uniform electric field $\mathbf{E} = (1.00 \times 10^3 \text{ V/m})\mathbf{i}$ and a uniform magnetic field $\mathbf{B} = (0.500 \times 10^{-4} \text{ T})\mathbf{j}$ are turned on at $t = 0$. Write a spreadsheet or computer program to integrate the equations of motion. Because v_0 can be in any direction, you must consider the motion along x, y, and z directions. Describe the general motion of the electron for various values of $\mathbf{v}_0$.

ANSWERS TO CONCEPTUAL PROBLEMS

1. No. The magnetic field produces a magnetic force that is directed toward the center of the circular path of the particle, and this causes a change in direction of the particle's velocity vector. However, because the magnetic force is always perpendicular to the displacement of the particle, no work is done on the particle, and its speed remains constant. Hence, its kinetic energy also remains constant.

2. No. There may be another field such as an electric field or gravitational field that produces another force on the charged particle that has the same magnitude but is opposite the magnetic force, resulting in a net force of zero. Also, a charged particle moving parallel to a magnetic field would experience no magnetic force.

3. The magnetic force on a moving charged particle is always perpendicular to the direction of motion. There is no magnetic force on the charge when it moves parallel to the direction of the magnetic field. However, the force on a charged particle moving in an electric field is never zero and is always parallel to the direction of the electric field. Therefore, by projecting the charged particle in different directions, it is possible to determine the nature of the field. Note that the constant force of the electric field produces a parabolic trajectory, while the magnetic field produces a circular path.

4. The magnetic field of the magnet produces a magnetic force on the electrons moving toward the screen that produce the image. This magnetic force deflects the electrons to regions on the screen other than the ones to which they are supposed to go. The result is a distorted image.

5. If you are moving along with the electrons, you will measure zero current for the electrons, so the electrons would not produce a magnetic field according to your observations. However, the positive crystal lattice ions in the metal are now moving backward relative to you and creating a current equivalent to the forward movement of electrons when you were stationary. Thus, you will measure the same magnetic field as when you were stationary, but it will be due to the moving positive crystal lattice ions in your reference frame.

6. A compass would not detect currents in wires near light switches for two reasons. Because the cable to the light switch contains two wires, with one carrying current to the switch and the other away from the switch, the net magnetic field would be very small and fall off rapidly. The second reason is that the current is alternating at 60 Hz. As a result, the magnetic field is oscillating at 60 Hz. This frequency would be too fast for the compass to follow, so the effect on the compass reading would average to zero.

7. The levitating wire is stable with respect to vertical motion— if it is displaced upward, the repulsive force weakens, and the wire drops back down. Conversely, if it drops lower, the repulsive force increases, and it moves back up. The wire is not stable, however, with respect to lateral movement. If it moves away from the vertical position directly over the lower wire, the repulsive force will have a sideways component, which will push the wire away.

   In the case of the attracting wires, the hanging wire is not stable for vertical movement. If it rises, the attractive force increases, and the wire moves even closer to the upper wire. If the hanging wire falls, the attractive force weakens, and the wire falls further. The hanging wire is also unstable to lateral movement. If the wire moves to the right, it moves farther from the upper wire and the attractive force decreases. Although there is a restoring-force component pulling it back to the left, the vertical-force component is not strong enough to hold it up and it falls.

8. Each coil of the Slinky will generate a small magnetic dipole at its center, because a coil will act as a current loop. Because the sense of rotation of the current is the same in all coils, each coil becomes a magnet with the same orientation of poles. Thus, all of the coils attract, and the Slinky will compress.

23

Faraday's Law and Inductance

Our studies so far have been concerned with the electric fields due to stationary charges and the magnetic fields produced by moving charges. This chapter deals with electric fields that originate in changing magnetic fields.

As we learned in Chapter 19, Section 19.1, experiments conducted by Michael Faraday in England in the early 1800s and—at the same time, independently—by Joseph Henry in the United States showed that an electric current could be induced in a circuit by a changing magnetic field. The results of those experiments led to a very basic and important law of electromagnetism known as *Faraday's law of induction*. This law holds that the magnitude of the emf induced in a circuit is proportional to the time rate of change of the magnetic field through the circuit.

With the exploration of Faraday's law, we complete our introduction to the fundamental laws of electromagnetism. These laws can be summarized in a set of four equations called *Max-*

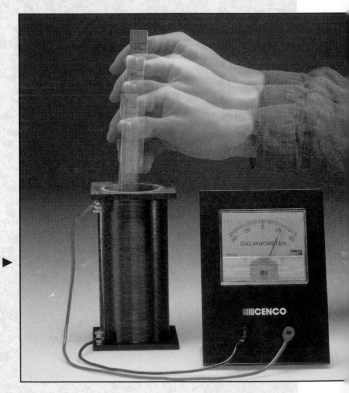

When a strong magnet is ▶ moved toward or away from the coil attached to a galvanometer, an electric current is induced, indicated by the momentary deflection of the galvanometer during the movement of the magnet. *(Richard Megna, Fundamental Photographs)*

well's equations. Together with the Lorentz force law, which we shall discuss briefly, they represent a complete theory for describing the interaction of charged objects. Maxwell's equations relate electric and magnetic fields to each other and to their ultimate source: electric charges.

The phenomenon of electromagnetic induction has some practical consequences. One of the most important practical applications of electromagnetic induction is the field of power generation. We shall describe an effect known as *self-induction,* in which a time-varying current in a conductor induces an emf that opposes the external emf that set up the current. This phenomenon is the basis of a circuit element known as an *inductor,* which plays an important role in circuits with time-varying currents. Finally, we shall discuss the energy stored in the magnetic field of an inductor and the energy density associated with a magnetic field.

23.1 • FARADAY'S LAW OF INDUCTION

We begin by describing two simple experiments that demonstrate that an electric current can be produced by a changing magnetic field. First, consider a loop of wire connected to a galvanometer, a device that measures current, as in Figure 23.1. If a magnet is moved toward the loop, the galvanometer needle deflects in one direction, as in Figure 23.1a. If the magnet is moved away from the loop, the galvanometer needle deflects in the opposite direction, as in Figure 23.1b. If the magnet is held stationary relative to the loop, no deflection is observed. Finally, if the magnet is held stationary and the coil is moved either toward or away from it, the

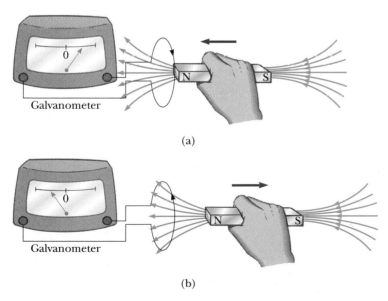

(a)

(b)

Figure 23.1   (a) When a magnet is moved toward a loop of wire connected to a galvanometer, the galvanometer deflects as shown. This shows that a current is induced in the loop. (b) When the magnet is moved away from the loop, the galvanometer deflects in the opposite direction, indicating that the induced current is opposite that shown in part (a).

Michael Faraday (1791–1867)

Faraday was a British physicist and chemist who is often regarded as the greatest experimental scientist of the 1800s. His many contributions to the study of electricity include the invention of the electric motor, electric generator, and transformer, as well as the discovery of electromagnetic induction and the laws of electrolysis.   *(By kind permission of the President and Council of the Royal Society)*

needle deflects. From these observations comes the conclusion that **an electric current is set up in the galvanometer circuit as long as there is relative motion between the magnet and the coil.**[1]

These results are quite remarkable in view of the fact that **a current is set up in the circuit even though there are no batteries in the circuit!** We call such a current an **induced current,** and it is produced by an **induced emf** or generated emf.

Now let us describe an experiment, first conducted by Faraday, that is illustrated in Figure 23.2. Part of the apparatus consists of a coil connected to a switch and a battery. We shall refer to this coil as the *primary coil* of wire and to the corresponding circuit as the primary circuit. The primary coil is wrapped around an iron ring to intensify the magnetic field produced by the current through the coil. A second coil, at the right, is also wrapped around the iron ring and is connected to a galvanometer. We shall refer to this as the *secondary coil* and to the corresponding circuit as the secondary circuit. There is no battery in the secondary circuit, and the secondary coil is not physically connected to the primary coil. The only purpose of this circuit is to detect any current that might be generated by a change in the magnetic field produced by the primary circuit.

At first you might guess that no current would ever be detected in the secondary circuit. However, something quite amazing happens when the switch in the primary circuit is suddenly closed or opened. At the instant the switch is closed, the galvanometer deflects in one direction and then returns to zero. When the switch is opened, the galvanometer deflects in the opposite direction and again returns to zero. Finally, the galvanometer reads zero when a steady current exists in the primary circuit.

As a result of these observations, Faraday concluded that **an electric current can be produced by a changing magnetic field.** A current cannot be produced by a steady magnetic field. The current produced in the secondary circuit occurs for only an instant while the magnetic field acting on the secondary coil is changing. In effect, the secondary circuit behaves as though a source of emf were connected to it for an instant. It is customary to say that an emf is induced in the secondary

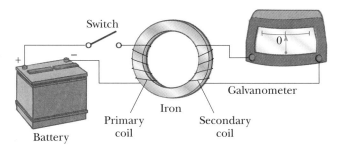

Figure 23.2   Faraday's experiment. When the switch in the primary circuit at the left is closed, the galvanometer in the secondary circuit at the right deflects momentarily. The emf induced in the secondary circuit is caused by the changing magnetic field through the coil in this circuit.

---

[1]The magnitude of the current depends on the particular resistance of the circuit, but the existence of the current (or its algebraic sign) does *not*.

circuit by the changing magnetic field produced by the current in the primary circuit.

In order to quantify such observations, it is necessary to define a quantity called **magnetic flux.** The flux associated with a magnetic field is defined in a similar manner to the electric flux (Chapter 19, Section 19.8). Consider an element of area dA on an arbitrarily shaped surface that is not closed, as in Figure 23.3. If the magnetic field at this element is **B**, then the magnetic flux through the element is **B** $\cdot$ d**A**, where d**A** is a vector perpendicular to the surface the magnitude of which equals the area dA. Hence, the total magnetic flux, Φ_B, through the surface is

$$\Phi_B = \int \mathbf{B} \cdot d\mathbf{A} \qquad \text{[23.1]}$$

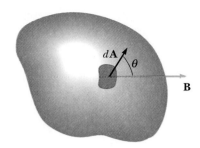

Figure 23.3 The magnetic flux through an area element $d\mathbf{A}$ is given by $\mathbf{B} \cdot d\mathbf{A} = B\,dA \cos \theta$. Note that $d\mathbf{A}$ is perpendicular to the surface.

The SI unit of magnetic flux is the tesla-meter squared, which is named the *weber* (Wb): 1 Wb = 1 T·m^2.

The two experiments illustrated in Figures 23.1 and 23.2 have one thing in common. In both cases, **an emf is induced in a circuit when the magnetic flux through the circuit changes with time.** In fact, a general statement summarizes such experiments involving induced emfs:

> The emf induced in a circuit is directly proportional to the time rate of change of magnetic flux through the circuit.

This statement, known as **Faraday's law of induction,** can be written

$$\mathcal{E} = -\frac{d\Phi_B}{dt} \qquad \text{[23.2]}$$

• *Faraday's law*

where Φ_B is the magnetic flux through the circuit, given by Equation 23.1. The negative sign in Equation 23.2 is a consequence of Lenz's law and will be discussed in Section 23.3. If the circuit is a coil consisting of N loops all of the same area and if the flux threads all loops, the induced emf is

$$\mathcal{E} = -N\frac{d\Phi_B}{dt} \qquad \text{[23.3]}$$

Suppose the magnetic field is uniform over a loop of area A lying in a plane as in Figure 23.4. In this case, the flux through the loop is equal to $BA \cos \theta$; hence, the induced emf is

$$\mathcal{E} = -\frac{d}{dt}(BA \cos \theta) \qquad \text{[23.4]}$$

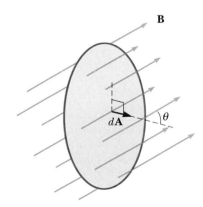

Figure 23.4 A conducting loop of area A in the presence of a uniform magnetic field **B,** which is at an angle θ with the normal to the loop.

From this expression, we see that an emf can be induced in a circuit in several ways: (1) the magnitude of **B** can vary with time; (2) the area of the circuit can change with time; (3) the angle θ between **B** and the normal to the plane can change with time; and (4) any combination of these can occur.

Examples 23.1 and 23.2 illustrate cases in which an emf is induced in a circuit as a result of a time variation of the magnetic field.

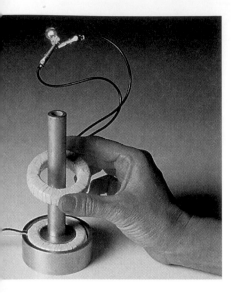

A demonstration of electromagnetic induction. An ac voltage is applied to the lower coil. A voltage is induced in the upper coil, as indicated by the illuminated lamp connected to the upper coil. What do you think happens to the lamp's intensity as the upper coil is moved over the vertical tube?
(Courtesy of Central Scientific Company)

Thinking Physics 1

Argentina has more land area (2.8×10^6 km^2) than Greenland (2.2×10^6 km^2). Yet the magnetic flux of the Earth's magnetic field is larger through Greenland than through Argentina. Why?

Reasoning There are two reasons for the larger magnetic flux through Greenland. Greenland (latitude 60° north to 80° north) is closer to a magnetic pole than is Argentina (latitude 20° south to 50° south). As a result, the angle that the magnetic field lines make with the vertical is smaller in Greenland than in Argentina. The dot product in the definition of the magnetic flux through an area incorporates this angle. The second reason is also a result of Greenland's proximity to a magnetic pole. The field strength is larger in the vicinity of a magnetic pole than farther from the pole. Thus, $|\mathbf{B}|$ in the definition of flux is larger for Greenland than for Argentina. These two influences will dominate over the slightly larger area of Argentina.

CONCEPTUAL PROBLEM 1

A circular loop is located in a uniform and constant magnetic field. Describe how an emf can be induced in the loop in this situation.

CONCEPTUAL PROBLEM 2

A spacecraft orbiting the Earth has a coil of wire in it. An astronaut measures a small current in the coil, although there is no battery connected to it and there are no magnets on the spacecraft. What is causing the current?

CONCEPTUAL PROBLEM 3

Suppose you would like to steal power for your home from the electric company by placing a loop of wire near a transmission cable in order to induce an emf in the loop (an illegal procedure). Should you orient your loop so that the transmission cable passes through your loop or simply place your loop near the transmission cable?

Example 23.1   One Way to Induce an emf in a Coil

A coil is wrapped with 200 turns of wire around the perimeter of a square frame of sides 18 cm. Each turn has the same area, equal to that of the frame, and the total resistance of the coil is 2.0 Ω. A uniform magnetic field is turned on perpendicular to the plane of the coil. If the field changes linearly from 0 to 0.50 T in a time of 0.80 s, find the magnitude of the induced emf in the coil while the field is changing.

Solution The area of the loop is $(0.18 \text{ m})^2 = 0.0324 \text{ m}^2$. The magnetic flux through the loop at $t = 0$ is zero, because $B = 0$. At $t = 0.80$ s, the magnetic flux through the loop is $\Phi_B = BA = (0.50 \text{ T})(0.0324 \text{ m}^2) = 0.0162 \text{ Wb}$. Therefore, the magnitude of the induced emf is

$$|\mathcal{E}| = \frac{N \Delta \Phi_B}{\Delta t} = \frac{200(0.0162 \text{ Wb} - 0 \text{ Wb})}{0.80 \text{ s}} = 4.1 \text{ V}$$

(Note that 1 Wb/s = 1 V.)

EXERCISE 1 What is the magnitude of the induced current in the coil while the field is changing?   Answer   2.0 A

Example 23.2   An Exponentially Decaying B Field

A plane loop of wire of area A is placed in a region in which the magnetic field is perpendicular to the plane. The magnitude of **B** varies in time according to the expression $B = B_0e^{-at}$. That is, at $t = 0$ the field is B_0, and for $t > 0$, the field decreases exponentially in time (Fig. 23.5). Find the induced emf in the loop as a function of time.

Solution   Because **B** is perpendicular to the plane of the loop, the magnetic flux through the loop at time $t > 0$ is

$$\Phi_B = BA = AB_0e^{-at}$$

The induced emf can be calculated from Faraday's Law, noting that AB_0 and the parameter a are constants.

$$\mathcal{E} = -\frac{d\Phi_B}{dt} = -AB_0\frac{d}{dt}e^{-at} = aAB_0e^{-at}$$

That is, the induced emf decays exponentially in time. Note that the maximum emf occurs at $t = 0$, where $\mathcal{E}_{max} = aAB_0$.

Why is this true? The plot of $\mathcal{E}$ versus t is similar to the B versus t curve shown in Figure 23.5.

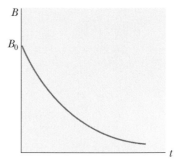

Figure 23.5   (Example 23.2) Exponential decrease of the magnetic field with time. The induced emf and induced current have similar time variations.

EXERCISE 2   A flat loop of wire consisting of a single turn of cross-sectional area 8.0 cm^2 is perpendicular to a magnetic field that increases uniformly in magnitude from 0.50 T to 2.50 T in 1.0 s. What is the resulting induced current if the loop has a resistance of 2.0 Ω?
Answer   800 μA

23.2 • MOTIONAL emf

In Examples 23.1 and 23.2, we considered cases in which an emf is produced in a circuit when the magnetic field changes with time. In this section we describe **motional emf,** which is the emf induced in a conductor moving through a magnetic field.

First, consider a straight conductor of length ℓ moving with constant velocity through a uniform magnetic field directed into the page, as in Figure 23.6. For simplicity, we shall assume that the conductor is moving perpendicularly to the field. The electrons in the conductor experience a force along the conductor of $\mathbf{F}_m = q\mathbf{v} \times \mathbf{B}$. Under the influence of this force, the electrons move to the lower end and accumulate there, leaving a net positive charge at the upper end. As a result of this charge separation, an electric field is produced within the conductor. The charge at the ends builds up until the magnetic force qvB is balanced by the electric force qE. At this point, charge stops flowing, and the condition for equilibrium requires that the electric force must balance the magnetic force:

$$qE = qvB \qquad \text{or} \qquad E = vB$$

Because the electric field produced in the conductor is constant, it is related to the potential difference across the ends according to the relation $\Delta V = E\ell$. Thus,

$$\Delta V = E\ell = B\ell v$$

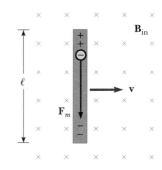

Figure 23.6   A straight conducting bar of length ℓ moving with a velocity **v** through a uniform magnetic field **B** directed perpendicular to **v**. An emf equal to $B\ell v$ is induced between the ends of the bar.

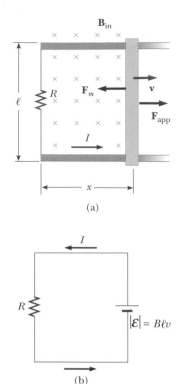

(a)

(b)

Figure 23.7   (a) A conducting bar sliding with a velocity **v** along two conducting rails under the action of an applied force $\mathbf{F}_{app}$. The magnetic force $\mathbf{F}_m$ opposes the motion, and a counterclockwise current is induced in the loop. (b) The equivalent circuit of (a).

where the upper end is at a higher potential than the lower end. Thus, **a potential difference is maintained as long as the conductor is moving through the magnetic field. If the motion is reversed, the polarity of ΔV is also reversed.**

Another interesting situation occurs if we now consider what happens when the moving conductor is part of a closed circuit. This circumstance is particularly useful for illustrating how a changing magnetic flux can cause an induced current in a closed circuit. Consider a circuit consisting of a conducting bar of length ℓ sliding along two fixed parallel conducting rails, as in Figure 23.7a. For simplicity, we assume that the moving bar has zero electrical resistance and that the stationary part of the circuit has a resistance of R. A uniform and constant magnetic field, **B**, is applied perpendicular to the plane of the circuit.

As the bar is pulled to the right with a velocity of **v**, under the influence of an applied force, $\mathbf{F}_{app}$, free charges in the bar experience a magnetic force along the length of the bar. This magnetic force sets up an induced current, because the charges on which the force is acting are free to move in the circuit. In this case, the rate of change of magnetic flux through the loop and the accompanying induced emf across the moving bar are proportional to the change in loop area as the bar moves through the magnetic field. As we shall see, if the bar is pulled to the right with a constant velocity, the work done by the applied force is dissipated in the form of thermal energy in the circuit's resistive element.

Because the area of the circuit at any instant is ℓx, the external magnetic flux through the circuit is

$$\Phi_B = B\ell x$$

where x is the width of the circuit, which changes with time. Using Faraday's law, we find that the induced emf is

$$\mathcal{E} = -\frac{d\Phi_B}{dt} = -\frac{d}{dt}(B\ell x) = -B\ell\frac{dx}{dt}$$

$$\mathcal{E} = -B\ell v \qquad\qquad [23.5]$$

Because the resistance of the circuit is R, the magnitude of the induced current is

$$I = \frac{|\mathcal{E}|}{R} = \frac{B\ell v}{R} \qquad\qquad [23.6]$$

The equivalent circuit diagram for this example is shown in Figure 23.7b.

Let us examine the system using energy considerations. Because there is no battery in the circuit, one might wonder about the origin of the induced current and the electrical energy in the system. We can understand this by noting that the external force does work on the conductor, thereby moving charges through a magnetic field. This causes the charges to move along the conductor with some average drift velocity, and hence a current is established. From the viewpoint of energy conservation, the total work done by the applied force during some time interval should equal the electrical energy that the induced emf supplies in that same period. Furthermore, if the bar moves with constant speed, the work done must equal the energy delivered to the resistor during this time interval.

As the conductor of length ℓ moves through the uniform magnetic field **B**, it experiences a magnetic force, $\mathbf{F}_m$, of magnitude $I\ell B$ (Eq. 22.3 in Chapter 22). The direction of this force is opposite the motion of the bar, or to the left in Figure 23.7a.

If the bar is to move with a *constant* velocity, the applied force must be equal to and opposite the magnetic force, or to the right in Figure 23.7a. (If the magnetic force acted in the direction of motion, it would cause the bar to accelerate once it was in motion, thereby increasing its velocity. This state of affairs would represent a violation of the principle of energy conservation.) Using Equation 23.6 and the fact that $F_{app} = I\ell B$, we find that the power delivered by the applied force is

$$P = F_{app}v = (I\ell B)v = \frac{B^2\ell^2 v^2}{R} \qquad [23.7]$$

This power is equal to the rate at which energy is delivered to the resistor, I^2R, as we would expect. It is also equal to the power $I\mathcal{E}$ supplied by the induced emf. This example is a clear demonstration of the conversion of mechanical energy to electrical energy and finally to thermal energy (joule heating).

Thinking Physics 2

As an airplane flies from Los Angeles to Seattle, it cuts through the lines of the Earth's magnetic field. As a result, an emf is developed between the wingtips of the airplane. Which wingtip is positively charged and which is negatively charged? Would the wingtips be charged in the same way on an airplane traveling from Antarctica to Australia? Could this emf be a source of energy to operate the lighting system of the airplane?

Reasoning The magnetic field lines of the Earth have a downward component in the northern hemisphere. As the airplane flies northward, the right-hand rule indicates that positive charges experience a force to the left side of the airplane. Thus, the left wingtip becomes positively charged and the right wingtip negatively charged. In the southern hemisphere, the field lines have an upward component, so the charging of the wingtips would be in the opposite sense. Based on the relative weakness of the Earth's magnetic field, it is not likely that the emf induced could operate a lighting system. An order of magnitude calculation shows that the emf between the wingtips is only a fraction of a volt.

CONCEPTUAL PROBLEM 4

Is it possible to induce a constant emf for a long continuous time?

CONCEPTUAL PROBLEM 5

An electric guitar pickup consists of a permanent magnet around which is wrapped a coil of wire (Fig. 23.8). How does this detect the movement of a steel guitar string?

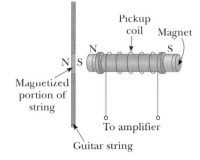

Figure 23.8 (Conceptual Problem 5) In an electric guitar, a vibrating metallic string induces a voltage in a pickup coil.

Example 23.3 emf Induced in a Rotating Bar

A conducting bar of length ℓ rotates with a constant angular speed ω about a pivot at one end. A uniform magnetic field **B** is directed perpendicular to the plane of rotation, as in Figure 23.9. Find the emf induced between the ends of the bar.

Solution Consider a segment of the bar of length dr, the velocity of which is **v**. According to Equation 23.5, the emf induced in a conductor of this length moving perpendicular to a field **B** is

$$(1) \quad d\mathcal{E} = Bv\,dr$$

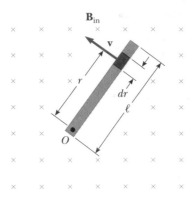

Figure 23.9 (Example 23.3) A conducting bar rotating around a pivot at one end in a uniform magnetic field that is perpendicular to the plane of rotation. An emf is induced across the ends of the bar.

Each segment of the bar is moving perpendicular to **B**, so there is an emf generated across each segment; the value of this emf is given by (1). Summing up the emfs induced across all elements, which are in series, gives the total emf between the ends of the bar. That is,

$$\mathcal{E} = \int Bv \, dr$$

In order to integrate this expression, note that the linear speed of an element is related to the angular speed ω through the relationship $v = r\omega$. Therefore, because B and ω are constants, we find that

$$\mathcal{E} = B \int v \, dr = B\omega \int_0^\ell r \, dr = \tfrac{1}{2}B\omega\ell^2$$

Example 23.4   Magnetic Force on a Sliding Bar

A bar of mass m and length ℓ moves on two frictionless parallel rails in the presence of a uniform magnetic field directed into the paper (Fig. 23.10). The bar is given an initial velocity $\mathbf{v}_0$ to the right and is released. Find the velocity of the bar as a function of time.

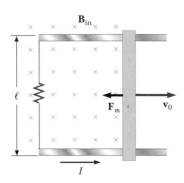

Figure 23.10 (Example 23.4) A conducting bar of length ℓ sliding on two fixed conducting rails is given an initial velocity $\mathbf{v}_0$ to the right.

Solution First note that the induced current is counterclockwise and the magnetic force on the bar is $F_m = -I\ell B$, where the negative sign denotes that the force is to the left and retards the motion. This is the only horizontal force acting on the bar, and hence Newton's second law applied to motion in the horizontal direction gives

$$F_x = ma = m\frac{dv}{dt} = -I\ell B$$

Because the induced current is given by Equation 23.6, $I = B\ell v/R$, we can write this expression as

$$m\frac{dv}{dt} = -\frac{B^2\ell^2}{R} v$$

$$\frac{dv}{v} = -\left(\frac{B^2\ell^2}{mR}\right) dt$$

Integrating this last equation using the initial condition that $v = v_0$ at $t = 0$, we find that

$$\int_{v_0}^v \frac{dv}{v} = -\left(\frac{B^2\ell^2}{mR}\right) \int_0^t dt$$

$$\ln\left(\frac{v}{v_0}\right) = -\left(\frac{B^2\ell^2}{mR}\right) t = -\frac{t}{\tau}$$

where the constant $\tau = mR/B^2\ell^2$. From this, we see that the velocity can be expressed in the exponential form

$$v = v_0 e^{-t/\tau}$$

Therefore, the velocity of the bar decreases exponentially with time under the action of the magnetic retarding force. Furthermore, if we substitute this result into Equation 23.6, we find that the induced emf and induced current also decrease exponentially with time. That is,

$$I = \frac{B\ell v}{R} = \frac{B\ell v_0}{R} e^{-t/\tau}$$

$$\mathcal{E} = IR = B\ell v_0 e^{-t/\tau}$$

The Alternating-Current Generator

The alternating current generator (or ac generator), a device that converts mechanical energy to electrical energy, consists of a coil of wire rotated by some external means in an external magnetic field. In commercial power plants, the energy required to rotate the loop can be derived from a variety of sources. For example, in a hydroelectric plant, falling water directed against the blades of a turbine produces the rotary motion; in a coil-fired plant, the energy from burning coal is used to convert water to steam, and this steam is directed against the turbine blades. As the loop rotates, the magnetic flux through it changes with time, inducing an emf and a current in an external circuit.

Suppose that the coil has N turns, all of the same area A, and suppose that the coil rotates with a constant angular speed ω about an axis perpendicular to the magnetic field. If θ is the angle between the magnetic field and the direction perpendicular to the plane of the coil, then the magnetic flux through the loop at any time t is given by

$$\Phi_B = BA \cos \theta = BA \cos \omega t$$

where we have used the relationship between angular displacement and angular speed, $\theta = \omega t$. (We have set the clock so that $t = 0$ when $\theta = 0$.) Hence, the induced emf in the coil is given by

$$\mathcal{E} = -N\frac{d\Phi_B}{dt} = -NAB\frac{d}{dt}(\cos \omega t) = NAB\omega \sin \omega t \qquad [23.8]$$

This result shows that the emf varies sinusoidally with time. From Equation 23.8 we see that the maximum emf has the value $\mathcal{E}_{max} = NAB\omega$, which occurs when $\omega t = 90°$ or $270°$. In other words, $\mathcal{E} = \mathcal{E}_{max}$ when the plane of the coil is parallel to the magnetic field and the time rate of change of flux is a maximum. Furthermore, the emf is *zero* when $\omega t = 0$ or $180°$—that is, when the plane of the coil is perpendicular to **B** and the time rate of change of flux is zero.

23.3 • LENZ'S LAW

The direction of an induced emf and induced current can be found from Lenz's law:

> The polarity of an induced emf is such that it produces a current the magnetic field of which opposes the change in magnetic flux through the loop. That is, the induced current is in a direction such that the induced magnetic field attempts to maintain the original flux through the circuit.

The induced current tends to keep the original flux through the circuit from changing. Interpretation of this statement depends on the circumstances. As we shall see, this law is a consequence of the law of conservation of energy.

In order to attain a better understanding of Lenz's law, let us return to the example of a bar moving to the right on two parallel rails in the presence of a uniform magnetic field directed into the page (Fig. 23.11a). As the bar moves to the right, the magnetic flux through the circuit increases with time, because the

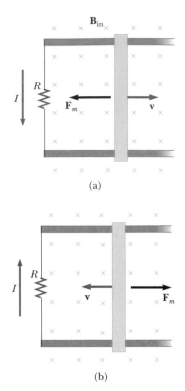

(a)

(b)

Figure 23.11 (a) As the conducting bar slides on the two fixed conducting rails, the magnetic flux through the loop increases in time. By Lenz's law, the induced current must be *counterclockwise* in order to produce a counteracting flux *out* of the paper. (b) When the bar moves to the left, the induced current must be *clockwise*. Why?

area of the loop increases. Lenz's law says that the induced current must be in such a direction that the magnetic flux *it* produces opposes the change in the magnetic flux caused by the external magnetic field. Because the magnetic flux due to the external field is increasing into the page, the induced current, if it is to oppose the change, must produce a flux out of the page. Hence, the induced current must be counterclockwise when the bar moves to the right to give a counteracting flux out of the page in the region inside the loop. (Use the right-hand rule to verify this direction.) If the bar is moving to the left, as in Figure 23.11b, the magnetic flux through the loop decreases with time. Because the flux is into the page, the induced current has to be clockwise to produce a flux into the page inside the loop. In either case, the induced current tends to maintain the original flux through the circuit.

Let us examine this situation from the viewpoint of energy considerations. Suppose that the bar is given a slight push to the right. In the preceding analysis, we found that this motion leads to a counterclockwise current in the loop. Let us see what happens if we assume that the current is clockwise. For a clockwise current, *I*, the direction of the magnetic force, *BIℓ*, on the sliding bar would be to the right. This force would accelerate the rod and increase its speed, which in turn would cause the area of the loop to increase more rapidly. This would increase the induced current, which would increase the force, which would increase the current, which would. . . . In effect, the system would acquire energy with no additional energy input. This result is clearly inconsistent with all experience and with the law of conservation of energy. Thus, we are forced to conclude that the current must be counterclockwise.

Consider another situation, one in which a bar magnet is moved to the right toward a stationary loop of wire, as in Figure 23.12a. As the magnet moves toward the loop, the magnetic flux through the loop increases with time. To counteract this increase in flux to the right, the induced current produces a flux to the left, as in Figure 23.12b; hence, the induced current is in the direction shown. Note that the magnetic field lines associated with the induced current oppose the motion of the magnet. Therefore, the left face of the current loop is a north pole and the right face is a south pole.

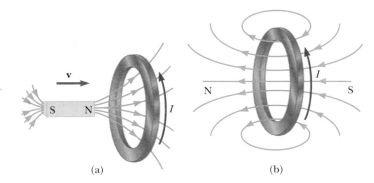

(a)   (b)

Figure 23.12 (a) When the magnet is moved toward the stationary conducting loop, a current is induced in the direction shown. (b) This induced current produces its own flux to the left to counteract the increasing external flux to the right.

If the magnet moved to the left, its flux through the loop, which is toward the right, would decrease in time. Under these circumstances, the induced current in the loop would be in a direction to set up a field through the loop from left to right, in an effort to maintain a constant number of flux lines. Hence, the direction of the induced current in the loop would be opposite that shown in Figure 23.12b. In this case, the left face of the loop would be a south pole and the right face would be a north pole.

Thinking Physics 3

In some circuits, a spark occurs between the poles of a switch when the switch is opened. Yet, when the switch for this circuit is closed, there is no spark. Why is there this difference?

Reasoning According to Lenz's law, induced emfs are in a direction such as to attempt to maintain the original magnetic flux when a change occurs. When the switch is opened, the sudden drop in the magnetic field in the circuit induces an emf in a direction which attempts to keep the original current flowing. This can cause a spark as the current bridges the air gap between the poles of the switch. The spark does not occur when the switch is closed, because the original current is zero, and the induced emf attempts to maintain it at zero.

CONCEPTUAL PROBLEM 6

A transformer is a pair of coils wrapped around an iron form, as in Figure 23.2. When ac voltage is applied to one coil, the *primary*, the magnetic field lines cutting through the other coil, the *secondary*, induce an emf. By varying the number of turns of wire on each coil, the applied ac voltage can be stepped up or stepped down. It is clear that this device will not work with dc voltage. What's more, if dc voltage of the same magnitude is applied, the primary coil sometimes will overheat and burn. Why?

CONCEPTUAL PROBLEM 7

In equal arm balances from the early twentieth century, it is sometimes observed that an aluminum sheet hangs from one of the arms and passes between the poles of a magnet. Why?

Example 23.5   Application of Lenz's Law

A coil of wire is placed near an electromagnet, as shown in Figure 23.13a. Find the direction of the induced current in the coil (a) at the instant the switch is closed, (b) after the switch has been closed for several seconds, and (c) when the switch is opened.

Reasoning (a) When the switch is closed, the situation changes from a condition in which no lines of flux pass through the coil to one in which lines of flux pass through in the direction shown in Figure 23.13b. To counteract this change in the number of lines, the coil must set up a field

from left to right in the figure. This requires a current directed as shown in Figure 23.13b.

(b) After the switch has been closed for several seconds, there is no change in the number of lines through the loop; hence the induced current is zero.

(c) Opening the switch causes the magnetic field to change from a condition in which flux lines thread through the coil from right to left to a condition of zero flux. The induced current must then be as shown in Figure 23.13c, so as to set up its own field from right to left.

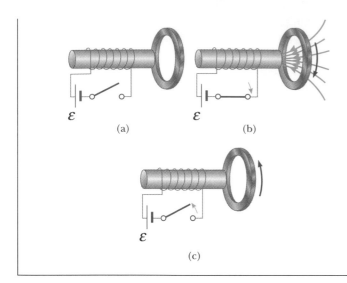

(a)

(b)

(c)

Figure 23.13   (Example 23.5)

EXERCISE 3   A small airplane with a wing span of 14 m flies due north at a speed of 70 m/s over a region in which the vertical component of the Earth's magnetic field is 1.2 μT. (a) What potential difference is developed between the wingtips? (b) How would the answer to (a) change if the plane were flying due east?   Answer   (a) 1.2 mV   (b) same emf with the north end positive

EXERCISE 4   Coils rotating in a magnetic field are often used to measure unknown magnetic fields. As an example, consider a coil of radius 1.0 cm with 50 turns that is rotated about an axis perpendicular to the field at a rate of 20 Hz. If the maximum induced emf in the coil is 3.0 V, find the strength of the magnetic field.   Answer   1.5 T

23.4 • INDUCED emfs AND ELECTRIC FIELDS

We have seen that a changing magnetic flux induces an emf and a current in a conducting loop. We therefore must conclude that **an electric field is created in the conductor as a result of changing magnetic flux.** In fact, the law of electromagnetic induction demonstrates that **an electric field is always generated by a changing magnetic flux,** even in free space in which no charges are present. However, this induced electric field has properties that are quite different from those of the electrostatic field produced by stationary charges.

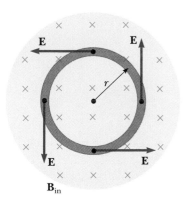

Figure 23.14 A loop of radius r in a uniform magnetic field perpendicular to the plane of the loop. If **B** changes in time, an electric field is induced in a direction tangent to the loop.

We can illustrate this point by considering a conducting loop of radius r, situated in a uniform magnetic field that is perpendicular to the plane of the loop, as in Figure 23.14. If the magnetic field changes with time, then Faraday's law tells us that an emf of $\mathcal{E} = - d\Phi_B/dt$ is induced in the loop. The induced current that is produced implies the presence of an induced electric field, **E**, which must be tangent to the loop, because all points on the loop are equivalent. The work done in moving a test charge, q, once around the loop is equal to $q\mathcal{E}$. Because the electric force on the charge is qE, the work done by that force in moving the charge once around the loop is $qE(2\pi r)$, where $2\pi r$ is the circumference of the loop. These two expressions for the work must be equal; therefore, we see that

$$q\mathcal{E} = qE(2\pi r)$$

$$E = \frac{\mathcal{E}}{2\pi r}$$

Using this result along with Faraday's law and the fact that $\Phi_B = BA = \pi r^2 B$ for a circular loop, we find that the induced electric field can be expressed as

$$E = -\frac{1}{2\pi r}\frac{d\Phi_B}{dt} = -\frac{r}{2}\frac{dB}{dt}$$

This expression can be used to calculate the induced electric field if the time variation of the magnetic field is specified. The negative sign indicates that the induced electric field, **E**, opposes the change in the magnetic field. It is important to understand that **this result is also valid in the absence of a conductor.** That is, a free charge placed in a changing magnetic field experiences the same electric field.

In general, the emf for any closed path can be expressed as the line integral of **E**· *d***s** over that path. Hence, the general form of Faraday's law of induction is

$$\mathcal{E} = \oint \mathbf{E}\cdot d\mathbf{s} = -\frac{d\Phi_B}{dt} \qquad \text{[23.9]}$$

• *Faraday's law in general form*

It is important to recognize that **the induced electric field, E, that appears in Equation 23.9 is a nonconservative field that is generated by a changing magnetic field.** The field **E** that satisfies Equation 23.9 could not possibly be an electrostatic field, for the following reason. If the field were electrostatic, and hence conservative, the line integral of **E**· *d***s** over a closed loop would be zero, contrary to Equation 23.9.

Thinking Physics 4

In studying electric fields, the point is made that electric field lines begin on positive charges and end on negative charges. Do *all* electric field lines begin and end on charges?

Reasoning The statement that electric field lines begin and end on charges is true only for *electrostatic* fields—electric fields due to stationary charges. Electric field lines due to changing magnetic fields form closed loops, with no beginning and no end, and are independent of the presence of charges.

EXERCISE 5 The current in a solenoid is increasing at a rate of 10 A/s. The cross-sectional area of the solenoid is π cm^2, and there are 300 turns on its 15-cm length. What is the induced emf opposing the increasing current?   Answer   2.4 mV

23.5 • SELF-INDUCTANCE

Consider an isolated circuit consisting of a switch, a resistor, and a source of emf, as in Figure 23.15. When the switch is closed, the current doesn't immediately jump from zero to its maximum value, $\mathcal{E}/R$; the law of electromagnetic induction (Faraday's law) describes this behavior. What happens is that, as the current increases

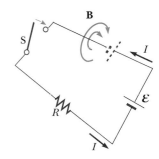

Figure 23.15 After the switch is closed, the current produces a magnetic flux through the loop. As the current increases toward its equilibrium value, the flux changes in time and induces an emf in the loop. The battery drawn with dashed lines is a symbol for the self-induced emf.

with time, the magnetic flux through the loop due to the current also increases with time. This increasing flux induces an emf in the circuit that opposes the change in the net magnetic flux through the loop. By Lenz's law, the induced electric field in the wires must therefore be opposite the direction of the current, and this opposing emf results in a *gradual* increase in the current. This effect is called *self-induction,* because the changing flux through the circuit arises from the circuit itself. The emf that is set up in this case is called a **self-induced emf.**

To obtain a quantitative description of self-induction, we recall from Faraday's law that the induced emf is the negative time rate of change of the magnetic flux. The magnetic flux is proportional to the magnetic field, which in turn is proportional to the current in the circuit. Therefore, **the self-induced emf is always proportional to the time rate of change of the current.** For a closely spaced coil of N turns of fixed geometry (a toroidal coil or the ideal solenoid), we find that

Self-induced emf •

$$\mathcal{E}_L = -N\frac{d\Phi_B}{dt} = -L\frac{dI}{dt} \qquad [23.10]$$

where L is a proportionality constant, called the **inductance** of the device, that depends on the geometric features of the circuit and other physical characteristics. From this expression, we see that the inductance of a coil containing N turns is

Inductance of an N-turn coil •

$$L = \frac{N\Phi_B}{I} \qquad [23.11]$$

where it is assumed that the same flux passes through each turn. Later we shall use this equation to calculate the inductance of some special current geometries.

From Equation 23.10 we can also write the inductance as the ratio

$$L = -\frac{\mathcal{E}_L}{dI/dt} \qquad [23.12]$$

This is usually taken to be the defining equation for the inductance of any coil, regardless of its shape, size, or material characteristics. Just as resistance is a measure of opposition to current, inductance is a measure of opposition to the *change* in current.

The SI unit of inductance is the **henry (H),** which, from Equation 23.12, is seen to be equal to 1 volt-second per ampere:

$$1\ \text{H} = 1\ \frac{\text{V}\cdot\text{s}}{\text{A}} = 1\ \frac{\text{Wb}}{\text{A}} = 1\ \frac{\text{T}\cdot\text{m}^2}{\text{A}}$$

As we shall see, **the inductance of a device depends on its geometry.** Because inductance calculations can be quite difficult for complicated geometries, Examples 23.6 and 23.7 involve simple situations for which inductances are easily evaluated.

Joseph Henry (1797–1878)

An American physicist, Henry became the first director of the Smithsonian Institution and first president of the Academy of Natural Science. He improved the design of the electromagnet and constructed one of the first motors. He also discovered the phenomenon of self-induction but failed to publish his findings. The unit of inductance, the henry, is named in his honor.   *(North Wind Picture Archives)*

Thinking Physics 5

If inductors are connected in series and/or parallel, one expects the inductances combine like resistors—they add in series, and the inverses add in parallel. The previous statement holds true for toroidal inductors in tight quarters in an electrical circuit but not for inductors formed from solenoids. Why?

Reasoning The inductance of a coil is *self*-inductance—an effect due to the magnetic field of a toroid or solenoid cutting through its own coils. In an ideal toroidal inductor, the magnetic field is contained within the interior of the toroid. Thus, if toroidal inductors are placed close to each other, there is no magnetic interaction between the inductors. Finite solenoids, however, have magnetic fields that exist outside the solenoid. If solenoids are placed in close proximity, the field of one solenoid can interact with the coils of the other solenoid. Thus, there is *mutual* inductance in addition to self-inductance, and the simple inductance combination rules do not apply.

Thinking Physics 6

The ground fault interrupter (GFI) is a safety device that protects users of electrical power against electric shock when they touch appliances. Its essential parts are shown in Figure 23.16. How does its operation make use of Faraday's law?

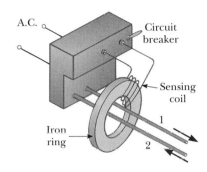

Figure 23.16 (Thinking Physics 6) Essential components of a ground fault interrupter.

Reasoning Wire 1 leads from the wall outlet to the appliance being protected, and wire 2 leads from the appliance back to the wall outlet. An iron ring surrounding the two wires confines the magnetic field set up by each wire. A sensing coil, that can activate a circuit breaker when changes in magnetic flux occur, is wrapped around part of the iron ring. (This can happen, for example, if one of the wires loses its insulation and accidentally touches the metal case of the appliance, providing a direct path to ground.) Because the currents in the two wires are in opposite directions, the net magnetic field through the sensing coil due to the currents is zero. If a short to ground occurs, the current in one of the wires will suddenly increase, producing a rapid change in flux through the coil and generating an induced voltage in it. This induced voltage is used to trigger a circuit breaker, stopping the current before it reaches a level that might be harmful to the person using the appliance.

Example 23.6 Inductance of a Solenoid

Find the inductance of a uniformly wound solenoid having N turns and length ℓ. Assume that ℓ is long compared with the radius and that the core of the solenoid is air.

Solution In this case, we can take the interior magnetic field to be uniform and given by Equation 22.25:

$$B = \mu_0 nI = \mu_0 \frac{N}{\ell} I$$

where n is the number of turns per unit length, N/ℓ. The magnetic flux through each turn is

$$\Phi_B = BA = \mu_0 \frac{NA}{\ell} I$$

where A is the cross-sectional area of the solenoid. Using this expression and Equation 23.11 we find that

$$L = \frac{N\Phi_B}{I} = \frac{\mu_0 N^2 A}{\ell}$$

This shows that L depends on geometry and is proportional to the square of the number of turns. Because $N = n\ell$, we can also express the result in the form

$$L = \mu_0 \frac{(n\ell)^2}{\ell} A = \mu_0 n^2 A\ell = \mu_0 n^2 \text{ (volume)}$$

where $A\ell$ is the volume of the solenoid.

Example 23.7 Calculating Inductance and emf

(a) Calculate the inductance of a solenoid containing 300 turns if the length of the solenoid is 25.0 cm and its cross-sectional area is $4.00 \text{ cm}^2 = 4.00 \times 10^{-4} \text{ m}^2$.

Solution Using the first expression for L in Example 23.6 we get

$$L = \frac{\mu_0 N^2 A}{\ell} = (4\pi \times 10^{-7} \text{ T} \cdot \text{m/A}) \frac{(300)^2 (4.00 \times 10^{-4} \text{ m}^2)}{25.0 \times 10^{-2} \text{ m}}$$

$$= 1.81 \times 10^{-4} \text{ T} \cdot \text{m}^2/\text{A} = \boxed{0.181 \text{ mH}}$$

(b) Calculate the self-induced emf in the solenoid described in (a) if the current through it is decreasing at the rate of 50.0 A/s.

Solution Using Equation 23.10 and given that $dI/dt = -50.0$ A/s, we get

$$\mathcal{E}_L = -L\frac{dI}{dt} = -(1.81 \times 10^{-4} \text{ H})(-50.0 \text{ A/s}) = \boxed{9.05 \text{ mV}}$$

EXERCISE 6 A 0.388-mH inductor has a length that is four times its diameter. If it is wound with 22 turns per centimeter, what is its length?   Answer   0.109 m

23.6 · *RL* CIRCUITS

A circuit that contains a coil, such as a solenoid, has a self-inductance that prevents the current from increasing or decreasing instantaneously. A circuit element that has a large inductance is called an **inductor.** The circuit symbol for an inductor is ◦◦◦. We shall always assume that the self-inductance of the remainder of the circuit is negligible compared with that of the inductor.

Consider the circuit—consisting of a resistor, inductor, and battery—shown in Figure 23.17. The internal resistance of the battery will be neglected. Suppose the switch S is closed at $t = 0$. The current will begin to increase, and due to the increasing current, the inductor will produce an emf (sometimes referred to as a *back emf*) that opposes the increasing current. In other words, the inductor acts like a battery the polarity of which is opposite that of the real battery in the circuit. The back emf produced by the inductor is

$$\mathcal{E}_L = -L\frac{dI}{dt}$$

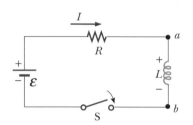

Figure 23.17 A series *RL* circuit. As the current increases toward its maximum value, the inductor produces an emf that opposes the increasing current.

Because the current is increasing, dI/dt is positive; therefore $\mathcal{E}_L$ is negative. This corresponds to the fact that there is a potential drop from a to b across the inductor. For this reason, point a is at a higher potential than point b, as illustrated in Figure 23.17.

With this in mind, we can apply Kirchhoff's loop rule to this circuit. If we begin at the battery and travel clockwise, we have

$$\mathcal{E} - IR - L\frac{dI}{dt} = 0 \qquad \text{[23.13]}$$

where IR is the voltage drop across the resistor. We must now look for a solution to this differential equation, which is similar to Equation 21.30 (Chapter 21) for the *RC* circuit.

To obtain a mathematical solution of Equation 23.13, it is convenient to change variables by letting $x = (\mathcal{E}/R) - I$, so that $dx = -dI$. With these substitutions, Equation 23.13 can be written

$$x + \frac{L}{R}\frac{dx}{dt} = 0$$

$$\frac{dx}{x} = -\frac{R}{L} dt$$

Integrating this last expression gives

$$\ln \frac{x}{x_0} = -\frac{R}{L} t$$

where the integrating constant is taken to be $-\ln x_0$. Taking the antilog of this result gives

$$x = x_0 e^{-Rt/L}$$

where the current is zero at $t = 0$ and $x_0 = \mathcal{E}/R$. Hence, the last expression is equivalent to

$$\frac{\mathcal{E}}{R} - I = \frac{\mathcal{E}}{R} e^{-Rt/L}$$

$$I = \frac{\mathcal{E}}{R} (1 - e^{-Rt/L})$$

which represents the solution of Equation 23.13.

This mathematical solution of Equation 23.13, which represents the current as a function of time, can also be written

$$I(t) = \frac{\mathcal{E}}{R} (1 - e^{-Rt/L}) = \frac{\mathcal{E}}{R} (1 - e^{-t/\tau}) \qquad \textbf{[23.14]}$$

where the constant τ is the **time constant** of the *RL* circuit:

$$\tau = L/R \qquad \textbf{[23.15]}$$

It is left to an exercise to show that the dimension of τ is time. Physically, τ is the time it takes the current to reach $(1 - e^{-1}) = 0.63$ of its final value, $\mathcal{E}/R$.

Figure 23.18 plots the current versus time, where $I = 0$ at $t = 0$. Note that the final equilibrium value of the current, which occurs at $t = \infty$, is $\mathcal{E}/R$. This can be seen by setting dI/dt equal to zero in Equation 23.13 (at equilibrium, the change in the current is zero) and solving for the current. Thus, we see that the current rises very fast initially and then gradually approaches the equilibrium value $\mathcal{E}/R$ as $t \to \infty$.

Taking the first time derivative of Equation 23.14, we get

$$\frac{dI}{dt} = \frac{\mathcal{E}}{L} e^{-t/\tau} \qquad \textbf{[23.16]}$$

From Equation 23.16 we see that the rate of increase of current, dI/dt, is a *maximum* (equal to $\mathcal{E}/L$) at $t = 0$ and falls off exponentially to zero as $t \to \infty$ (Fig. 23.19).

Now consider the *RL* circuit arranged as shown in Figure 23.20. The circuit contains two switches that operate so that when one is closed, the other is opened. Now suppose that S_1 is closed for a long enough time to allow the current to reach its equilibrium value, $\mathcal{E}/R$. If S_1 is now opened and S_2 is closed at $t = 0$, we

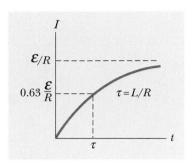

Figure 23.18 Plot of the current versus time for the *RL* circuit shown in Figure 23.17. The switch is closed at $t = 0$, and the current increases toward its maximum value, $\mathcal{E}/R$. The time constant τ is the time it takes I to reach 63% of its maximum value.

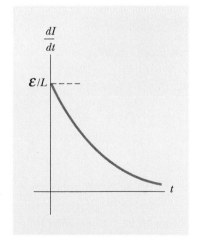

Figure 23.19 Plot of dI/dt versus time for the *RL* circuit shown in Figure 23.17. The time rate of change of current is a maximum at $t = 0$ when the switch is closed. The rate decreases exponentially with time as I increases toward its maximum value.

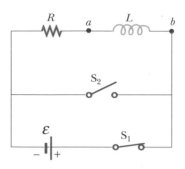

Figure 23.20   An *RL* circuit containing two switches. When S_1 is closed and S_2 is open as shown, the battery is in the circuit. At the instant S_2 is closed, S_1 is opened, and the battery is removed from the circuit.

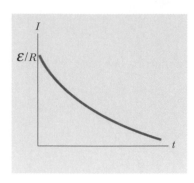

Figure 23.21   Current versus time for the circuit shown in Figure 23.20. At $t < 0$, S_1 is closed and S_2 is open. At $t = 0$, S_2 is closed, S_1 is open, and the current has its maximum value $\mathcal{E}/R$.

have a circuit with no battery ($\mathcal{E} = 0$). If we apply Kirchhoff's loop rule to this circuit, we obtain

$$IR + L\frac{dI}{dt} = 0 \qquad\qquad \textbf{[23.17]}$$

It is left to Problem 27 to show that the solution of this differential equation is

$$I(t) = \frac{\mathcal{E}}{R}\, e^{-t/\tau} = I_0 e^{-t/\tau} \qquad\qquad \textbf{[23.18]}$$

where the current at $t = 0$ is $I_0 = \mathcal{E}/R$ and $\tau = L/R$.

The graph of current versus time for the circuit of Figure 23.20 (Fig. 23.21) shows that the current is continuously decreasing with time, as one would expect. Furthermore, note that the slope, dI/dt, is always negative and has its maximum value at $t = 0$. The negative slope signifies that $\mathcal{E}_L = -L(dI/dt)$ is not *positive*—that is, point *a* in Figure 23.20 is at a lower potential than point *b*.

CONCEPTUAL PROBLEM 8

The switch in the circuit in Figure 23.22 is closed and the lightbulb glows steadily. The inductor is a simple air-core solenoid. What happens to the brightness of the lightbulb as an iron rod is being inserted into the interior of the solenoid?

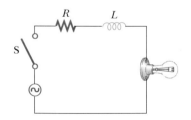

Figure 23.22   (Conceptual Problem 8)

Example 23.8   Time Constant of an *RL* Circuit

The switch in Figure 23.23a is closed at $t = 0$. (a) Find the time constant of the circuit.

Solution   The time constant is given by Equation 23.15

$$\tau = \frac{L}{R} = \frac{30.0 \times 10^{-3}\ \text{H}}{6.00\ \Omega} = 5.00\ \text{ms}$$

(b) Calculate the current in the circuit at $t = 2.00$ ms.

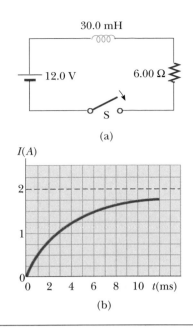

(a)

(b)

Figure 23.23   (Example 23.8) (a) The switch in this *RL* circuit is closed at *t* = 0. (b) A graph of the current versus time for the circuit in part (a).

Solution   Using Equation 23.14 for the current as a function of time (with *t* and τ in milliseconds), we find that at *t* = 2.00 ms

$$I = \frac{\varepsilon}{R}\,(1 - e^{-t/\tau}) = \frac{12.0\text{ V}}{6.00\ \Omega}\,(1 - e^{-0.400}) = 0.659\text{ A}$$

A plot of Equation 23.14 for this circuit is given in Figure 23.23b.

EXERCISE 7   Calculate the current in the circuit and the voltage across the resistor after one time constant has elapsed.
Answer   1.26 A, 7.59 V

23.7 • ENERGY STORED IN A MAGNETIC FIELD

In the preceding section we found that the induced emf set up by an inductor prevents a battery from establishing an instantaneous current. Hence, a battery has to do work against an inductor to create a current. Part of the energy supplied by the battery goes into thermal energy dissipated in the resistor, and the remaining energy is stored in the inductor. If we multiply each term in Equation 23.13 by the current *I* and rearrange the expression, we get

$$I\varepsilon = I^2 R + LI\,\frac{dI}{dt} \qquad \textbf{[23.19]}$$

This expression tells us that the rate at which energy is supplied by the battery, $I\varepsilon$, equals the sum of the rate at which thermal energy is dissipated in the resistor, I^2R, and the rate at which energy is stored in the inductor, $LI(dI/dt)$. Thus, Equation 23.19 is simply an expression of energy conservation. If we let U_B denote the energy stored in the inductor at any time, then the rate dU_B/dt at which energy is stored in the inductor can be written

$$\frac{dU_B}{dt} = LI\,\frac{dI}{dt}$$

To find the total energy stored in the inductor, we can rewrite this expression as $dU_B = LI\,dI$ and integrate:

$$U_B = \int_0^{U_B} dU_B = \int_0^I LI\,dI$$

$$U_B = \tfrac{1}{2}LI^2 \qquad \textbf{[23.20]}$$

• *Energy stored in an inductor*

where L is constant and so has been removed from the integral. Equation 23.20 represents the energy stored as magnetic energy in the field of the inductor when the current is I. Note that it is similar to the equation for the energy stored in the electric field of a capacitor, $U_E = Q^2/2C$ (Eq. 20.29 in Chapter 20). In either case, we see that it takes work to establish a field.

We can also determine the energy per unit volume, or energy density, stored in a magnetic field. For simplicity, consider a solenoid the inductance of which is $L = \mu_0 n^2 A\ell$ (see Example 23.6). The magnetic field of the solenoid is $B = \mu_0 nI$. Substituting the expression for L and $I = B/\mu_0 n$ into Equation 23.20 gives

$$U_B = \tfrac{1}{2}LI^2 = \tfrac{1}{2}\mu_0 n^2 A\ell \left(\frac{B}{\mu_0 n}\right)^2 = \frac{B^2}{2\mu_0}(A\ell) \qquad \textbf{[23.21]}$$

Because $A\ell$ is the volume of the solenoid, the energy stored per unit volume in a magnetic field—in other words, the *energy density*—is

Magnetic energy density •

$$u_B = \frac{U_B}{A\ell} = \frac{B^2}{2\mu_0} \qquad \textbf{[23.22]}$$

Although Equation 23.22 was derived for the special case of a solenoid, **it is valid for any region of space in which a magnetic field exists.** Note that it is similar to the equation for the energy per unit volume stored in an electric field, given by $\tfrac{1}{2}\epsilon_0 E^2$ (Eq. 20.31). In both cases, the energy density is proportional to the square of the field strength.

Example 23.9   What Happens to the Energy in the Inductor?

Consider once again the RL circuit shown in Figure 23.20, in which switch S_2 is closed at the instant S_1 is opened (at $t = 0$). Recall that the current in the upper loop decays exponentially with time according to the expression $I = I_0 e^{-t/\tau}$, where $I_0 = \mathcal{E}/R$ is the initial current in the circuit and $\tau = L/R$ is the time constant. Let us show explicitly that all the energy stored in the magnetic field of the inductor is delivered to the resistor.

Solution   The rate at which energy is dissipated in the resistor, dU/dt, (or the power) is equal to I^2R, where I is the instantaneous current:

$$\frac{dU}{dt} = I^2R = (I_0 e^{-Rt/L})^2 R = I_0^2 R e^{-2Rt/L}$$

To find the total energy dissipated in the resistor, we integrate this expression over the limits $t = 0$ to $t = \infty$ (∞ because it takes an infinite time for the current to reach zero):

(1) $$U = \int_0^\infty I_0^2 R e^{-2Rt/L}\, dt = I_0^2 R \int_0^\infty e^{-2Rt/L}\, dt$$

The value of the definite integral is $L/2R$, and so U becomes

$$U = I_0^2 R \left(\frac{L}{2R}\right) = \tfrac{1}{2}LI_0^2$$

Note that this is equal to the initial energy stored in the magnetic field of the inductor, given by Equation 23.20, as we set out to prove.

EXERCISE 8   Show that the integral on the right-hand side of Equation (1) has the value $L/2R$.

Example 23.10   The Coaxial Cable

A long coaxial cable consists of two concentric cylindrical conductors of radii a and b and length ℓ, as in Figure 23.24. The inner conductor is assumed to be a thin cylindrical shell. Each conductor carries a current I (the outer one being a return path). (a) Calculate the self-inductance, L, of this cable.

Solution   To obtain L, we must know the magnetic flux through any cross-section between the two conductors. From Ampère's law (Section 22.7), it is easy to see that the magnetic field between the conductors is $B = \mu_0 I/2\pi r$. The field is zero outside the conductors ($r > b$) because the net current

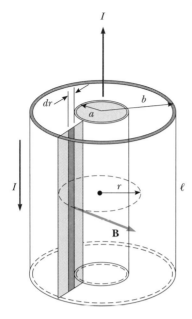

Figure 23.24 (Example 23.10) Section of a long coaxial cable. The inner and outer conductors carry equal and opposite currents.

through a circular path surrounding both wires is zero, and hence from Ampère's law, $\oint \mathbf{B} \cdot d\mathbf{s} = 0$. The field is zero inside the inner conductor because it is hollow and there is no current within a radius $r < a$.

The magnetic field is perpendicular to the outlined rectangular strip of length ℓ and width $(b - a)$, the cross section of interest because it is the surface one would choose for applying Lenz's law. Dividing this rectangle into strips of width dr, we see that the area of each strip is $\ell \, dr$ and the flux through the shaded strip is $B \, dA = B\ell \, dr$. Hence, the total flux through the region between the conductors is

$$\Phi_B = \int B \, dA = \int_a^b \frac{\mu_0 I}{2\pi r} \ell \, dr = \frac{\mu_0 I \ell}{2\pi} \int_a^b \frac{dr}{r} = \frac{\mu_0 I \ell}{2\pi} \ln\left(\frac{b}{a}\right)$$

Using this result, we find that the self-inductance of the cable is

$$L = \frac{\Phi_B}{I} = \frac{\mu_0 \ell}{2\pi} \ln\left(\frac{b}{a}\right)$$

(b) Calculate the total energy stored in the magnetic field of the cable.

Solution Using Equation 23.20 and the results to part (a) gives

$$U_B = \tfrac{1}{2} L I^2 = \frac{\mu_0 \ell I^2}{4\pi} \ln\left(\frac{b}{a}\right)$$

EXERCISE 9 A 10.0-V battery, a 5.00-Ω resistor, and a 10.0-H inductor are connected in series. After the current in the circuit has reached its maximum value, calculate (a) the power supplied by the battery, (b) the power dissipated in the resistor, (c) the power dissipated in the inductor, and (d) the energy stored in the magnetic field of the inductor.
Answer (a) 20 W (b) 20 W (c) 0 (d) 20 J

SUMMARY

Faraday's law of induction states that the emf induced in a circuit is directly proportional to the time rate of change of magnetic flux through the circuit:

$$\mathcal{E} = -N\frac{d\Phi_B}{dt} \qquad \text{[23.3]}$$

where N is the number of turns and Φ_B is the magnetic flux through each, given by

$$\Phi_B = \int \mathbf{B} \cdot d\mathbf{A} \qquad \text{[23.1]}$$

When a conducting bar of length ℓ moves through a magnetic field, **B**, with a velocity of **v** so that **v** is perpendicular to **B**, the emf induced in the bar (called the **motional emf**) is

$$\mathcal{E} = -B\ell v \qquad \text{[23.5]}$$

Lenz's law states that the induced current and induced emf in a conductor are in such a direction as to oppose the change that produced them.

A general form of **Faraday's law of induction** is

$$\mathcal{E} = \oint \mathbf{E} \cdot d\mathbf{s} = -\frac{d\Phi_B}{dt} \qquad \text{[23.9]}$$

where **E** is a nonconservative electric field that is produced by the changing magnetic flux.

When the current in a coil changes with time, an emf is induced in the coil according to Faraday's law. The **self-induced emf** is defined by the expression

$$\mathcal{E} = -L\frac{dI}{dt} \qquad \text{[23.10]}$$

where L is the *inductance* of the coil. Inductance is a measure of the opposition of a device to a change in current. Inductance has the SI unit the **henry** (H), where $1\ \text{H} = 1\ \text{V}\cdot\text{s/A}$.

The **inductance** of a coil is

$$L = \frac{N\Phi_B}{I} \qquad \text{[23.11]}$$

where Φ_B is the magnetic flux through the coil and N is the total number of turns.

If a resistor and inductor are connected in series to a battery of emf $\mathcal{E}$, as shown in Figure 23.17, and a switch in the circuit is closed at $t = 0$, the current in the circuit varies in time according to the expression

$$I(t) = \frac{\mathcal{E}}{R}(1 - e^{-t/\tau}) \qquad \text{[23.14]}$$

where $\tau = L/R$ is the **time constant** of the *RL* circuit. That is, the current rises to an equilibrium value of $\mathcal{E}/R$ after a time interval that is long compared with τ.

If the battery is removed from an *RL* circuit, as in Figure 23.20 with S_1 open and S_2 closed, the current decays exponentially with time according to the expression

$$I(t) = \frac{\mathcal{E}}{R}e^{-t/\tau} \qquad \text{[23.18]}$$

where $\mathcal{E}/R$ is the initial current in the circuit.

The **energy** stored in the magnetic field of an inductor carrying a current I is

$$U_B = \tfrac{1}{2}LI^2 \qquad \text{[23.20]}$$

The **energy per unit volume** (or energy density) at a point where the magnetic field is B is

$$u_B = \frac{B^2}{2\mu_0} \qquad \text{[23.22]}$$

CONCEPTUAL QUESTIONS

1. A loop of wire is placed in a uniform magnetic field. For what orientation of the loop is the magnetic flux a maximum? For what orientation is the flux zero?

2. As the conducting bar in Figure Q23.2 moves to the right, an electric field is set up directed downward. If the bar were moving to the left, explain why the electric field would be upward.

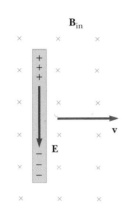

Figure Q23.2

3. As the bar in Figure Q23.2 moves perpendicular to the field, is an external force required to keep it moving with constant speed?

4. Magnetic storms on the Sun can cause difficulties with communications here on Earth. Why do these sunspots affect us in this way?

5. Wearing a metal bracelet in a region of strong magnetic field could be hazardous. Discuss.

6. How is electrical energy produced in dams (that is, how is the energy of motion of the water converted to ac electricity)?

7. A piece of aluminum is dropped vertically downward between the poles of an electromagnet. Does the magnetic field affect the velocity of the aluminum?

8. The bar in Figure Q23.8 moves on rails to the right with a velocity **v**, and the uniform, constant magnetic field is directed out of the page. Why is the induced current clockwise? If the bar were moving to the left, what would be the direction of the induced current?

9. In a beam-balance scale, an aluminum plate is sometimes used to slow the oscillations of the beam near equilibrium. The plate is mounted at the end of the beam and moves between the poles of a small horseshoe magnet attached to the frame. Why are the oscillations of the beam strongly damped near equilibrium?

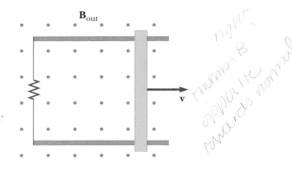

Figure Q23.8

10. When the switch in Figure Q23.10a is closed, a current is set up in the coil and the metal ring springs upward (Fig. Q23.10b). Explain this behavior.

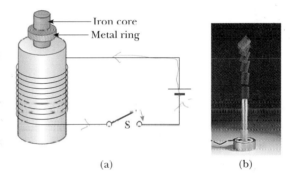

(a)   (b)

Figure Q23.10   (Questions 10 and 11)   *(Photo Courtesy of Central Scientific Company)*

11. Assume that the battery in Figure Q23.10a is replaced by an ac source and the switch is held closed. If held down, the metal ring on top of the solenoid becomes hot. Why?

12. A bar magnet is held above a loop of wire in a horizontal plane, as in Figure Q23.12. The south end of the magnet is toward the loop of wire. The magnet is dropped toward the loop. Find the direction of the current through the resistor (a) while the magnet is falling toward the loop and (b) after the magnet has passed through the loop and moves away from it.

13. Find the direction of the current through the resistor in Figure Q23.13 (a) at the instant the switch is closed, (b) after the switch has been closed for several minutes, and (c) at the instant the switch is opened.

14. Discuss the similarities between the energy stored in the electric field of a charged capacitor and the energy stored in the magnetic field of a current-carrying coil.

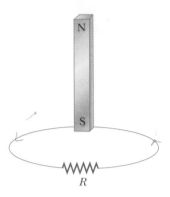

Figure Q23.12

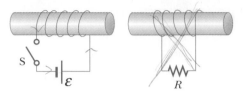

Figure Q23.13

15. What is the inductance of two inductors connected in series?

16. If the current in an inductor is doubled, by what factor does the stored energy change?

17. Suppose the switch in Figure 23.17 has been closed for a long time and is suddenly opened. Does the current instantaneously drop to zero? Why does a spark appear at the switch contacts at the instant the switch is thrown open?

PROBLEMS

Section 23.1   Faraday's Law of Induction

1. A strong electromagnet has a field of 1.60 T and a cross-sectional area of 0.200 m². If we place a coil having 200 turns and a total resistance of 20.0 Ω around the electromagnet and then turn off the power to the electromagnet in 20.0 ms, what is the current induced in the coil?

2. A rectangular loop of area A is placed in a region in which the magnetic field is perpendicular to the plane of the loop. The magnitude of the field is allowed to vary in time according to $B = B_0 e^{-t/\tau}$, where B_0 and τ are constants. The field has a value of B_0 at $t \le 0$. (a) Use Faraday's law to show that the emf induced in the loop is

$$\mathcal{E} = \frac{AB_0}{\tau} e^{-t/\tau}$$

(b) Obtain a numerical value for $\mathcal{E}$ at $t = 4.00$ s when $A = 0.160$ m², $B_0 = 0.350$ T, and $\tau = 2.00$ s. (c) For the values of A, B_0, and τ given in part (b), what is the maximum value of $\mathcal{E}$?

3. An aluminum ring of radius 5.00 cm and resistance 3.00×10^{-4} Ω is placed on top of a long air-core solenoid with 1000 turns per meter and radius 3.00 cm, as in Figure P23.3. Suppose that the magnetic field at the end of the solenoid due to the current in the solenoid is one half that at the center of the solenoid, and nearly parallel to its axis. Assume that the magnetic field is negligible outside of the solenoid. If the current in the solenoid is increasing at a rate of 270 A/s, (a) what is the induced current in the ring? (b) At the center of the ring, what is the magnetic field produced by the induced current in the ring? (c) What is the direction of this field?

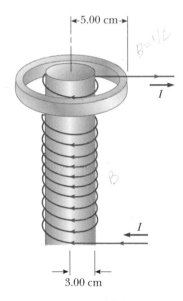

Figure P23.3

4. An aluminum ring of radius r_1 and resistance R is placed on top of a long air-core solenoid with n turns per meter and radius r_2, as in Figure P23.3. Suppose that the magnetic field at the end of the solenoid due to the current in the solenoid is one half that at the center of the solenoid, and nearly parallel to its axis. Assume that the magnetic field is negligible outside of the solenoid. If the current in the solenoid is increasing at a rate $\Delta I/\Delta t$, (a) what is the induced current in the ring? (b) At the center of the ring, what is the magnetic field produced by the induced current in the ring? (c) What is the direction of this field?

5. A long, straight wire carries a current $I = I_0 \sin(\omega t + \phi)$ and lies in the plane of a rectangular coil of N turns of wire, as shown in Figure P23.5. The quantities I_0, ω, and ϕ are all constants. Determine the emf induced in the coil by the magnetic field created by the current in the straight wire. Assume $I_0 = 50.0$ A, $\omega = 200\pi$ s^{-1}, $N = 100$, $a = b = 5.00$ cm, and $\ell = 20.0$ cm.

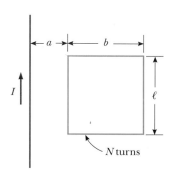

Figure P23.5

6. A magnetic field of 0.200 T exists within a solenoid of 500 turns and a diameter of 10.0 cm. How rapidly (that is, within what period of time) must the field be reduced to zero if the average magnitude of the induced emf within the coil during this time interval is to be 10.0 kV?

Section 23.3 Lenz's Law

7. Consider the arrangement shown in Figure P23.7. Assume that $R = 6.00$ Ω, $\ell = 1.20$ m, and a uniform 2.50-T magnetic field is directed into the page. At what speed should the bar be moved to produce a current of 0.500 A in the resistor?

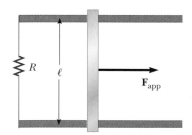

Figure P23.7

8. A conducting rod of length ℓ moves on two horizontal, frictionless rails, as in Figure P23.7. If a constant force of magnitude 1.00 N moves the bar at 2.00 m/s through a magnetic field **B** that is into the page, (a) what is the current through an 8.00-Ω resistor R? (b) What is the rate of energy dissipation in the resistor? (c) What is the mechanical power delivered by the 1.00-N force?

9. A Boeing 747 jet with a wingspan of 60.0 m is flying horizontally at 300 m/s over Phoenix, where the direction of the Earth's magnetic field is 58.0° below the horizontal. If the

magnitude of the field is 50.0 μT, what is the voltage generated between the wing tips?

10. The 10.0-Ω square loop in Figure P23.10 is placed in a uniform 0.100-T magnetic field directed perpendicular to the plane of the loop. The loop, which is hinged at each vertex, is pulled as shown until the separation between points A and B is 3.00 m. If the process takes 0.100 s, what is the average current generated in the loop?

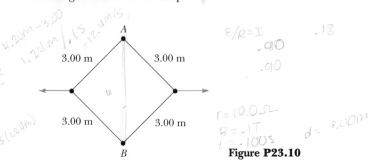

Figure P23.10

11. A helicopter has blades of length 3.00 m, extending out from a central hub and rotating at 2.00 rev/s. If the vertical component of the Earth's magnetic field is 0.500×10^{-4} T, what is the emf induced between the blade tip and the center hub?

12. Use Lenz's law to answer the following questions concerning the direction of induced currents. (a) What is the direction of the induced current in resistor R in Figure P23.12a when the bar magnet is moved to the left? (b) What

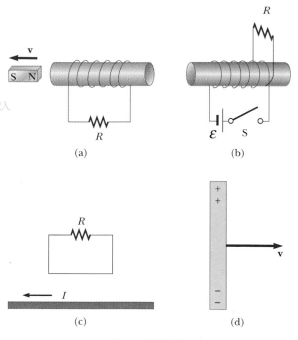

Figure P23.12

is the direction of the current induced in the resistor R right after the switch S in Figure P23.12b is closed? (c) What is the direction of the induced current in R when the current I in Figure P23.12c decreases rapidly to zero? (d) A copper bar is moved to the right, as in Figure P23.12d, while its axis is maintained perpendicularly to a magnetic field. If the top of the bar becomes positive relative to the bottom, what is the direction of the magnetic field?

13. A 25-turn circular coil of wire has a diameter of 1.00 m. It is placed with its axis along the direction of the Earth's magnetic field (magnitude 50.0 μT), and then in 0.200 s it is flipped 180°. What average emf is generated?

14. In a 250-turn automotive alternator, the magnetic flux in each turn is $\Phi_B = (2.50 \times 10^{-4} \text{ Wb})\cos(\omega t)$, where ω is the angular frequency of the alternator. The alternator rotates three times for each engine rotation. When the engine is running at 1000 rpm, determine (a) the induced emf in the alternator as a function of time and (b) the maximum emf in the alternator.

15. A conducting rectangular loop of mass M, resistance R, and dimensions w by ℓ falls from rest into a magnetic field **B**, as in Figure P23.15. The loop accelerates until it reaches a terminal speed v_t. (a) Show that

$$v_t = \frac{MgR}{B^2 w^2}$$

(b) Why is v_t proportional to R? (c) Why is it inversely proportional to B^2?

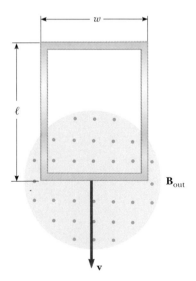

Figure P23.15

16. In 1832 Faraday proposed that the apparatus shown in Figure P23.16 could be used to generate electric current from the flowing water in the Thames River. Two conducting plates of lengths a and widths b are placed facing one another on opposite sides of the river, a distance w apart, and immersed entirely. The flow velocity of the river is **v**, and the vertical component of the Earth's magnetic field is B. (a) Show that the current in the load resistor R is

$$I = \frac{abvB}{\rho + abR/w}$$

where ρ is the resistivity of the water. (b) Calculate the short-circuit current ($R = 0$) if $a = 100$ m, $b = 5.00$ m, $v = 3.00$ m/s, $B = 50.0$ μT, and $\rho = 100$ $\Omega \cdot$m.

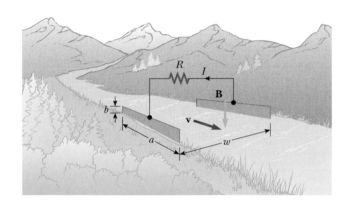

Figure P23.16

Section 23.4   Induced emfs and Electric Fields

17. A magnetic field directed into the page changes with time according to $B = (0.0300t^2 + 1.40)$ T, where t is in seconds. The field has a circular cross-section of radius $R = 2.50$ cm (Fig. P23.17). What are the magnitude and direction of the electric field at point P_1 when $t = 3.00$ s and $r_1 = 0.0200$ m?

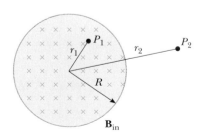

Figure P23.17

18. For the situation described in Figure P23.17, the magnetic field varies as $B = (2.00t^3 - 4.00t^2 + 0.800)$ T, and $r_2 = 2R = 5.00$ cm. (a) Calculate the magnitude and direction of the force exerted on an electron located at point P_2 when $t = 2.00$ s. (b) At what time is this force equal to zero?

19. A solenoid has a radius of 2.00 cm and 1000 turns/m. The current varies with time according to the expression $I = 3e^{0.2t}$, where I is in amperes and t is in seconds. Calculate the electric field 5.00 cm from the axis of the solenoid at $t = 10.0$ s.

20. A coil of 15 turns and radius 10.0 cm surrounds a long solenoid of radius 2.00 cm and 1.00×10^3 turns/meter (Fig. P23.20). If the current in the solenoid changes as $I = (5.00$ A) $\sin(120t)$, what is the induced emf in the 15-turn coil?

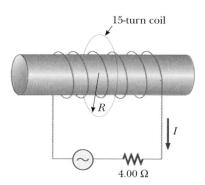

15-turn coil

R

I

4.00 Ω

Figure P23.20

Section 23.5   Self-Inductance

21. A coil has an inductance of 3.00 mH, and a current through it changes from 0.200 A to 1.50 A in a time of 0.200 s. Find the magnitude of the average induced emf in the coil during this time.

22. An emf of 24.0 mV is induced in a 500-turn coil at an instant when the current is 4.00 A and is changing at the rate of 10.0 A/s. What is the magnetic flux through each turn of the coil?

23. A 10.0-mH inductor carries a current $I = I_{max} \sin \omega t$, with $I_{max} = 5.00$ A and $\omega/2\pi = 60.0$ Hz. What is the back emf as a function of time?

24. The current in a 90.0-mH inductor changes with time as $I = t^2 - 6t$ (in SI units). Find the magnitude of the induced emf at (a) $t = 1.00$ s and (b) $t = 4.00$ s. (c) At what time is the emf zero?

25. A toroid has a major radius R and a minor radius r and is tightly wound with N turns of wire, as shown in Figure P23.25. If $R \gg r$, the magnetic field within the region of the toroid of cross-sectional area $A = \pi r^2$ is essentially that of a long solenoid that has been bent into a large circle of radius

R. Using the uniform field of a long solenoid, show that the self-inductance of such a toroid is given (approximately) by

$$L \cong \frac{\mu_0 N^2 A}{2\pi R}$$

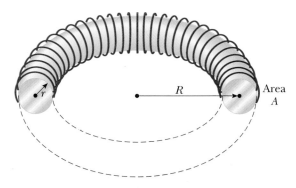

r

R

Area A

Figure P23.25

26. An inductor in the form of a solenoid contains 420 turns, is 16.0 cm in length, and has a cross-sectional area of 3.00 cm². What uniform rate of decrease of current through the inductor induces an emf of 175 μV?

Section 23.6   RL Circuits

27. Show that $I = I_0 e^{-t/\tau}$ is a solution of the differential equation

$$IR + L\frac{dI}{dt} = 0$$

where $\tau = L/R$ and $I_0 = \mathcal{E}/R$ is the current at $t = 0$.

28. Calculate the inductance in an RL circuit in which $R = 0.500$ Ω and the current increases to one fourth its final value in 1.50 s.

29. A 12.0-V battery is about to be connected to a series circuit containing a 10.0-Ω resistor and a 2.00-H inductor. How long will it take the current to reach (a) 50.0% and (b) 90.0% of its final value?

30. Consider the circuit in Figure P23.30, taking $\mathcal{E} = 6.00$ V, $L = 8.00$ mH, and $R = 4.00$ Ω. (a) What is the inductive time constant of the circuit? (b) Calculate the current in the circuit 250 μs after the switch is closed. (c) What is the value of the final steady-state current? (d) How long does it take the current to reach 80.0% of its maximum value?

31. For the RL circuit shown in Figure P23.30, let $L = 3.00$ H, $R = 8.00$ Ω, and $\mathcal{E} = 36.0$ V. (a) Calculate the ratio of the potential difference across the resistor to that across the inductor when $I = 2.00$ A after the switch is closed. (b) Calculate the voltage across the inductor when $I = 4.50$ A.

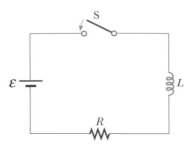

Figure P23.30

32. In the circuit shown in Figure P23.30, let $L = 7.00$ H, $R = 9.00$ Ω, and $\mathcal{E} = 120$ V. What is the self-induced emf 0.200 s after the switch is closed?

33. A 12.0-V battery is connected in series with a resistor and an inductor. The circuit has a time constant of 500 μs and the maximum current is 200 mA. What is the value of the inductance?

34. An inductor that has an inductance of 15.0 H and a resistance of 30.0 Ω is connected across a 100-V battery. What is the rate of increase of the current (a) at $t = 0$ and (b) at $t = 1.50$ s?

35. One application of an RL circuit is the generation of time-varying high-voltage from a low-voltage source, as shown in Figure P23.35. (a) What is the current in the circuit a long time after the switch has been in position A? (b) Now the switch is thrown quickly from A to B. Compute the initial voltage across each resistor and the inductor. (c) How much time elapses before the voltage across the inductor drops to 12.0 V?

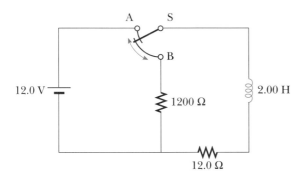

Figure P23.35

Section 23.7   Energy Stored in a Magnetic Field

36. Calculate the energy associated with the magnetic field of a 200-turn solenoid in which a current of 1.75 A produces a flux of 3.70×10^{-4} Wb in each turn.

37. The magnetic field inside a superconducting solenoid is 4.50 T. The solenoid has an inner diameter of 6.20 cm and

a length of 26.0 cm. Determine (a) the magnetic energy density in the field and (b) the energy stored in the magnetic field within the solenoid.

38. At $t = 0$, an emf of 500 V is applied to a coil that has an inductance of 0.800 H and a resistance of 30.0 Ω. (a) Find the energy stored in the magnetic field when the current reaches half its maximum value. (b) After the emf is connected, how long does it take the current to reach this value?

39. On a clear day there is a 100-V/m vertical electric field near the Earth's surface. At the same time, the Earth's magnetic field has a magnitude of 0.500×10^{-4} T. Compute the energy densities of the two fields.

40. An RL circuit in which $L = 4.00$ H and $R = 5.00$ Ω is connected to a 22.0-V battery at $t = 0$. (a) What energy is stored in the inductor when the current is 0.500 A? (b) At what rate is energy being stored in the inductor when $I = 1.00$ A? (c) What power is being delivered to the circuit by the battery when $I = 0.500$ A?

41. The magnitude of the magnetic field outside a sphere of radius R is $B = B_0(R/r)^2$, where B_0 is a constant. Determine the total energy stored in the magnetic field outside the sphere and evaluate your result for $B_0 = 5.00 \times 10^{-5}$ T and $R = 6.00 \times 10^6$ m. These values are appropriate for the Earth's magnetic field.

Additional Problems

42. A solenoid wound with 2 000 turns/m is supplied with current that varies in time according to $I = 4 \sin(120\pi t)$, where I is in A and t is in s. A small coaxial circular coil of 40 turns and radius $r = 5.00$ cm is located inside the solenoid near its center. (a) Derive an expression that describes the manner in which the emf in the small coil varies in time. (b) At what average rate is energy dissipated in the small coil if the windings have a total resistance of 8.00 Ω?

43. A flat compact coil of area 0.100 m² is rotating at 60.0 rev/s with the axis of rotation perpendicular to a 0.200-T magnetic field. (a) If there are 1 000 turns on the coil, what is the maximum voltage induced in it? (b) When the maximum induced voltage occurs, what is the orientation of the coil with respect to the magnetic field?

44. Figure P23.44 is a graph of the induced emf versus time for a coil of N turns rotating with angular velocity ω in a uniform magnetic field directed perpendicular to the axis of rotation of the coil. Copy this sketch (on a larger scale), and on the same set of axes show the graph of emf versus t when (a) the number of turns in the coil is doubled, (b) the angular velocity is doubled, (c) the angular velocity is doubled and the number of turns in the coil is halved.

45. A bar of mass m, length d, and resistance R slides without friction on parallel rails, as shown in Figure P23.45. A battery

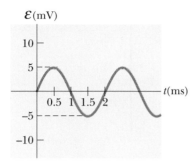

$\mathcal{E}$(mV)

Figure P23.44

that maintains a constant emf $\mathcal{E}$ is connected between the rails, and a constant magnetic field **B** is directed perpendicular to the plane of the page. If the bar starts from rest, show that at time t it moves with a speed

$$v = \frac{\mathcal{E}}{Bd}\left(1 - e^{-B^2d^2t/mR}\right)$$

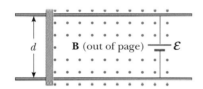

Figure P23.45

46. An automobile has a vertical radio antenna 1.20 m long. The automobile travels at 65.0 km/h on a horizontal road where the Earth's magnetic field is 50.0 μT directed downward (toward the north) at an angle of 65.0° below the horizontal. (a) Specify the direction that the automobile should move in order to generate the maximum motional emf in the antenna, with the top of the antenna positive relative to the bottom. (b) Calculate the magnitude of this induced emf.

47. The magnetic flux threading a metal ring varies with time t according to $\Phi_B = 3(at^3 - bt^2)$ T·m², with $a = 2.00$ s⁻³, and $b = 6.00$ s⁻². The resistance of the ring is 3.00 Ω. Determine the maximum current induced in the ring during the interval from $t = 0$ to $t = 2.00$ s.

48. Magnetic field values are often determined by using a device known as a *search coil*. This technique depends on the measurement of the total charge passing through a coil in a time interval during which the magnetic flux linking the windings changes either because of the motion of the coil or because of a change in the value of B. (a) Show that if the flux through the coil changes from Φ_1 to Φ_2, the charge

transferred through the coil between t_1 and t_2 will be given by $Q = N(\Phi_2 - \Phi_1)/R$, where R is the resistance of the coil and associated circuitry (galvanometer) and N is the number of turns. (b) As a specific example, calculate B when a 100-turn coil of resistance 200 Ω and cross-sectional area 40.0 cm² produces the following results. A total charge of 5.00×10^{-4} C passes through the coil when it is rotated in a uniform field from a position where the plane of the coil is perpendicular to the field to a position where the coil's plane is parallel to the field.

49. Assume that the switch in Figure P23.49 is initially in position 1. Show that if the switch is thrown from position 1 to position 2, all the energy stored in the magnetic field of the inductor is dissipated as thermal energy in the resistor.

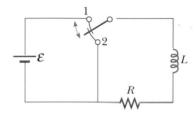

Figure P23.49

50. The inductor in Figure P23.50 has negligible resistance. When the switch is thrown open after having been closed for a long time, the current in the inductor drops to 0.250 A in 0.150 s. What is the inductance of the inductor?

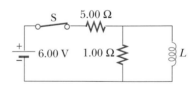

Figure P23.50

51. A novel method of storing electrical energy has been proposed. A huge underground superconducting coil, 1.00 km in diameter, would be fabricated. It would carry a maximum current of 50.0 kA through each winding of a 150-turn Nb₃Sn solenoid. (a) If the inductance of this huge coil were 50.0 H, what would be the total energy stored? (b) What would be the compressive force per meter length acting between two adjacent windings 0.250 m apart?

52. In Figure P23.52, the rolling axle, 1.50 m long, is pushed along horizontal rails at a constant speed $v = 3.00$ m/s. A resistor $R = 0.400$ Ω is connected to the rails at points a

and *b*, directly opposite each other. (The wheels make good electrical contact with the rails, and so the axle, rails, and *R* form a closed-loop circuit. The only significant resistance in the circuit is *R*.) There is a uniform magnetic field, *B* = 0.0800 T, vertically downward. (a) Find the induced current *I* in the resistor. (b) What horizontal force **F** is required to keep the axle rolling at constant speed? (c) Which end of the resistor, *a* or *b*, is at the higher electric potential? (d) After the axle rolls past the resistor, does the current in *R* reverse direction?

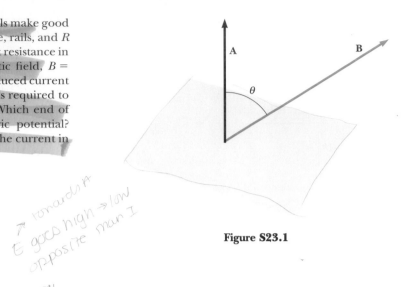

Figure S23.1

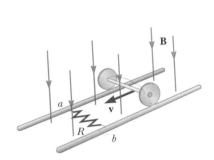

Figure P23.52

53. (a) A flat circular coil does not really produce a uniform magnetic field in the area it encloses, but estimate the self-inductance of a flat circular coil, with radius *R* and *N* turns, by supposing the field at its center is uniform over its area. (b) A circuit on a laboratory table consists of a 1.5-V battery, a 270-Ω resistor, a knife switch, and three 30-cm-long patch cords connecting them. Suppose the circuit is arranged to be circular. Think of it as a flat coil with one turn. Compute the order of magnitude of its self-inductance and (c) of the time constant describing how fast the current increases when you close the switch.

Spreadsheet Problems

S1. The size and orientation of a flat surface of area *A* can be described by a vector **A** = *A***n̂**, where **n̂** is a unit vector perpendicular to the surface. Suppose that a magnetic field **B** exists in the region of this surface. If the field is constant over the area *A*, then the magnetic flux, Φ_B, through the surface is $\Phi_B = \mathbf{B} \cdot \mathbf{A} = BA \cos\theta = B_x A_x + B_y A_y + B_z A_z$, where $\mathbf{A} = A_x\mathbf{i} + A_y\mathbf{j} + A_z\mathbf{k}$ and $\mathbf{B} = B_x\mathbf{i} + B_y\mathbf{j} + B_z\mathbf{k}$ (Fig. S23.1). Spreadsheet 23.1 will calculate the flux as a function of time. It will also numerically differentiate the flux to find the induced emf, $\mathcal{E} = -\Delta\Phi_B/\Delta t$. (a) Set $A_x = 0$, $A_y = 0$, $A_z = 0.2$ m², $B_x = 0$, $B_y = 0.5$ T, and $B_z = 0.6$ T at $t = 0$. Copy these values down through $t = 4.5$ s.

What is the induced emf? (b) Modify the spreadsheet so that B_z increases with time, $B_z = 0.2t$, where *t* is in seconds and B_z is in teslas. What is the induced emf?

S2. Modify Spreadsheet 23.1 so that $\mathbf{A} = 0.02 \sin(\omega t)\mathbf{i} + 0.02 \cos(\omega t)\mathbf{k}$, where *A* is in square meters. Set $\mathbf{B} = 0.5$ T **k**. (a) What physical situation does this variation in **A** correspond to? (b) Let $\omega = 1$ rad/s. View the included graph and describe the induced emf. (c) Increase ω to 2 rad/s. How does the induced emf change from part (b)?

S3. (a) With **A** as given in Problem S2, modify Spreadsheet 23.1 so that $B_z = 0.2t$, where *t* is in seconds and B_z is in teslas. View the included graph, and explain your numerical results. (b) Choose any other time dependence for **B**. Explore the consequences of your choice.

S4. A coil of self-inductance *L* carries a current given by $I = I_{max} \sin 2\pi ft$. The self-induced emf in the coil is $\mathcal{E}_L = -L\, dI/dt$. Develop a spreadsheet to calculate *I* as a function of time. Numerically differentiate the current and calculate $\mathcal{E}_L$. Choose $I_{max} = 2.00$ A, $f = 60.0$ Hz, and $L = 10.0$ mH. Plot $\mathcal{E}_L$ versus *t*.

S5. Spreadsheet 23.2 calculates the current and the energy stored in the magnetic field of an *RL* circuit when the circuit is charging and when it is discharging. Use $R = 1000$ Ω, $L = 0.35$ H, and $\mathcal{E} = 10$ V. (a) What is the time constant of the circuit? From the graph of the current versus time when the circuit is charging, how much time has elapsed when the current is 50% of its maximum value? 90%? 99%? Give your answers both in seconds and in multiples of the time constant. (b) Repeat for the case when the circuit is discharging. (c) How much time has elapsed when the energy is 50% of its maximum value for the two cases? 90%? 99%?

ANSWERS TO CONCEPTUAL PROBLEMS

1. According to Faraday's law, an emf is induced in a wire loop if the magnetic flux through the loop changes with time. In this situation, an emf can be induced by either rotating the loop around an arbitrary axis or by changing the shape of the loop.

2. As the spacecraft moves through space, it is apparently moving from a region of one magnetic field strength to a region of a different magnetic field strength. The changing magnetic field through the coil induces an emf and a corresponding induced current in the coil.

3. The magnetic field lines around the transmission cable will be circular. If you place your loop around the cable, there will be no field lines passing through the loop, so no emf will be induced. The loop needs to be placed next to the cable, with the plane of the loop parallel to the cable, to maximize the changing flux through the area as the alternating current produces an oscillating magnetic field.

4. A constant induced emf requires a magnetic field that is changing at a constant rate in one direction—for example, always increasing or always decreasing. It is impossible for a magnetic field to increase forever, both in terms of energy considerations and technological concerns. In the case of a decreasing field, once it reaches zero and then reverses direction, we again face the problem with the field increasing without bound in the opposite direction.

5. The pickup coil is placed near the vibrating guitar string, which is made of a metal that can be magnetized. The permanent magnet inside the coil magnetizes the portion of the string nearest the coil. When the guitar string vibrates at some frequency, its magnetized segment produces a changing magnetic flux through the pickup coil. The changing flux induces a voltage in the coil, and this voltage is fed to an amplifier. The output of the amplifier is sent to the loudspeakers, producing the sound waves that we hear.

6. The primary coil of the transformer is an inductor. When an ac voltage is applied, the back emf due to the inductance will limit the current flow through the coil. If dc voltage is applied, there is no back emf, and the current can rise to a higher value. It is possible that this increased current will deliver so much energy to the resistance in the coil that its temperature rises to the point at which the insulation on the wire can burn.

7. The aluminum sheet and magnet supply *magnetic damping* to the system, so that the oscillations of the arms damp out rapidly. When the sheet moves in the magnetic field, eddy currents are induced in the aluminum. According to Lenz's law, these currents are in a direction such as to oppose the movement of the aluminum sheet, providing magnetic friction.

8. As the iron rod is being inserted into the solenoid, the inductance of the inductor increases. As a result, more potential difference appears across the inductor than before. As a consequence, less potential difference appears across the bulb, and its brightness decreases.

24

Electromagnetic Waves

Although we are not always aware of their presence, electromagnetic waves permeate our environment. In the form of visible light, they enable us to view the world around us; infrared waves warm our environment; radio-frequency waves carry our favorite television and radio programs; microwaves cook our food and are used in radar communication systems. The list goes on and on. The waves we described in Chapters 13 and 14 were mechanical waves, which, by definition, can exist only if a medium is present. Electromagnetic waves, in contrast, can propagate through a vacuum.

The purpose of this chapter is to explore the properties of electromagnetic waves. The fundamental laws of electricity and magnetism—Maxwell's equations—form the basis of all electromagnetic phenomena. One of these equations predicts that a time-varying electric field produces a magnetic field just as a time-varying magnetic field produces an electric field. From this generalization, Maxwell provided the final important link between electric and magnetic fields. The most dramatic prediction of his equations is the existence of electromagnetic waves that propagate through empty space with the speed of light. This discovery led to many practical applications, such as radio and television, and to the realization that light is one form of electromagnetic radiation.

An old-style rancher in St. Augustine Plains, New Mexico, rides past one of the 27 radio telescopes that comprise the Very Large Array (VLA). Arranged in a Y-shaped configuration on a system of railroad tracks, the radio telescopes of the VLA capture and focus electromagnetic waves from space. *(© Danny Lehman)*

24.1 · DISPLACEMENT CURRENT AND THE GENERALIZED AMPÈRE'S LAW

We have seen that charges in motion, or currents, produce magnetic fields. When a current-carrying conductor has high symmetry, we can calculate the magnetic field using Ampère's law, given by Equation 22.22:

$$\oint \mathbf{B} \cdot d\mathbf{s} = \mu_0 I \qquad [22.22]$$

where **the line integral is over any closed path through which the conduction current passes,** and the conduction current is defined by $I = dQ/dt$. (In this section, we use the term *conduction current* to refer to the current carried by charged particles such as electrons in a wire or ions in an electrolytic solution.) We now show that **Ampère's law in this form is valid only if the conduction current is continuous in space.** Maxwell recognized this limitation and modified Ampère's law to include all possible situations.

We can understand this problem by considering a capacitor being charged as in Figure 24.1. When the conduction current flows, the charge on the plate changes, but **no conduction current passes between the plates.** Now consider the two surfaces S_1 and S_2 in Figure 24.1 bounded by the same path P. Ampère's law says that the line integral of $\mathbf{B} \cdot d\mathbf{s}$ around this path must equal $\mu_0 I$, where I is the total current through any surface bounded by the path P.

When the path P is considered as bounding S_1, the result of the integral is $\mu_0 I$ because the conduction current passes through S_1. When the path bounds S_2, however, the result is zero, because no conduction current passes through S_2. Thus, we have a contradictory situation that arises from the discontinuity of the conduction current! Maxwell solved this problem by postulating an additional term on the right side of Equation 22.22, called the **displacement current,** I_d, defined as

$$I_d \equiv \epsilon_0 \frac{d\Phi_E}{dt} \qquad [24.1]$$

• *Displacement current*

Recall that Φ_E is the flux of the electric field, defined as $\Phi_E = \int \mathbf{E} \cdot d\mathbf{A}$.

As the capacitor is being charged (or discharged), the changing electric field between the plates may be modeled as a current that bridges the discontinuity in the conduction current in the wire. When the expression for the displacement current given by Equation 24.1 is added to the right side of Ampère's law, the difficulty represented by Figure 24.1 is resolved. No matter what surface bounded by the path P is chosen, some combination of conduction and displacement current will pass through it. With this new term I_d, we can express the generalized form of Ampère's law (sometimes called the **Ampère–Maxwell law**) as[1]

$$\oint \mathbf{B} \cdot d\mathbf{s} = \mu_0 (I + I_d) = \mu_0 I + \mu_0 \epsilon_0 \frac{d\Phi_E}{dt} \qquad [24.2]$$

• *Ampère-Maxwell law*

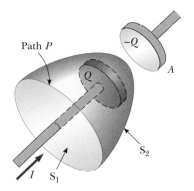

Figure 24.1 The surfaces S_1 and S_2 are bounded by the same path P. The conduction current in the wire passes only through S_1. This leads to a contradiction in Ampère's law that is resolved only if one postulates a displacement current through S_2.

---

[1]Strictly speaking, this expression is valid only in a vacuum. If a magnetic material is present, a magnetizing current must also be included on the right side of Equation 24.2 to make Ampère's law fully general.

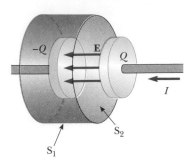

Figure 24.2 The conduction current $I = dQ/dt$ passes through S_1. The displacement current $I_d = \epsilon_0 d\Phi_E/dt$ passes through S_2. The two currents must be equal for continuity. In general, the total current through any surface bounded by some path is $I + I_d$.

The meaning of this expression can be understood by referring to Figure 24.2. The electric flux through S_2 is $\Phi_E = \int \mathbf{E} \cdot d\mathbf{A} = EA$, where A is the area of the capacitor plates and E is the uniform electric field strength between the plates. If Q is the charge on the plates at any instant, then $E = Q/\epsilon_0 A$ (Section 20.7). Therefore, the electric flux through S_2 is simply

$$\Phi_E = EA = \frac{Q}{\epsilon_0}$$

Hence, the displacement current I_d through S_2 is

$$I_d = \epsilon_0 \frac{d\Phi_E}{dt} = \frac{dQ}{dt} \tag{24.3}$$

That is, the displacement current is precisely equal to the conduction current I through S_1!

The central point of this formalism is the fact that **magnetic fields are produced both by conduction currents and by changing electric fields.**

24.2 • MAXWELL'S WONDERFUL EQUATIONS

In this section we present four equations that can be regarded as the basis of all electrical and magnetic phenomena. These relationships, known as Maxwell's equations after James Clerk Maxwell, are as fundamental to electromagnetic phenomena as Newton's laws are to mechanical phenomena. In fact, the theory developed by Maxwell was more far-reaching than even he imagined, because it was shown by Einstein in 1905 to be in agreement with the special theory of relativity. As we shall see, Maxwell's equations represent laws of electricity and magnetism that have already been discussed. However, the equations have additional important consequences, in that they predict the existence of electromagnetic waves (traveling patterns of electric and magnetic fields), which travel with a speed of $c = 1/\sqrt{\mu_0 \epsilon_0} \approx 3 \times 10^8$ m/s, the speed of light. Furthermore, Maxwell's equations show that such waves are radiated by accelerating charges.

For simplicity, we present **Maxwell's equations** as applied to free space—that is, in the absence of any dielectric or magnetic material. The four equations are:

Maxwell's equations •

$$\oint \mathbf{E} \cdot d\mathbf{A} = \frac{Q}{\epsilon_0} \tag{24.4}$$

$$\oint \mathbf{B} \cdot d\mathbf{A} = 0 \tag{24.5}$$

$$\oint \mathbf{E} \cdot d\mathbf{s} = -\frac{d\Phi_B}{dt} \tag{24.6}$$

$$\oint \mathbf{B} \cdot d\mathbf{s} = \mu_0 I + \epsilon_0 \mu_0 \frac{d\Phi_E}{dt} \tag{24.7}$$

Equation 24.4 is **Gauss's law,** which states that **the total electric flux through any closed surface equals the net charge inside that surface divided by ϵ_0.** This law describes how charges create electric fields, where electric field lines originate on positive charges and terminate on negative charges.

Equation 24.5, which can be considered **Gauss's law in magnetism, says that the net magnetic flux through a closed surface is zero.** That is, the number of magnetic field lines that enter a closed volume must equal the number that leave that volume. This implies that magnetic field lines cannot begin or end at any point. If they did, this would mean that isolated magnetic monopoles existed at those points. The fact that isolated magnetic monopoles have not been observed in nature can be taken as a basis of Equation 24.5.

Equation 24.6 is **Faraday's law of induction,** which describes how a changing magnetic field creates an electric field. This law states that **the line integral of the electric field around any closed path (which equals the emf) equals the rate of change of magnetic flux through any surface area bounded by that path.** One consequence of Faraday's law is the current induced in a conducting loop placed in a time-varying magnetic field.

Equation 24.7, which is the **Ampère–Maxwell law,** describes how both an electric current and a changing electric field will create a magnetic field. That is, **the line integral of the magnetic field around any closed path is determined by the sum of the net current through that path and the rate of change of electric flux through any surface bounded by that path.**

Once the electric and magnetic fields are known at some point in space, the force those fields exert on a particle of charge q can be calculated from the expression

$$\mathbf{F} = q\mathbf{E} + q\mathbf{v} \times \mathbf{B} \qquad [24.8]$$

This is called the **Lorentz force.** Maxwell's equations, together with this force law, give a complete description of all classical electromagnetic interactions.

It is interesting to note the symmetry of Maxwell's equations. Equations 24.4 and 24.5 are symmetric, apart from the absence of a magnetic monopole term in Equation 24.5. Furthermore, Equations 24.6 and 24.7 are symmetric in that the line integrals of **E** and **B** around a closed path are related to the rate of change of magnetic flux and electric flux, respectively.

24.3 • ELECTROMAGNETIC WAVES

The fundamental laws governing the behavior of electric and magnetic fields are Maxwell's equations, discussed in Section 24.2. In his unified theory of electromagnetism, Maxwell showed that time-dependent electric and magnetic fields satisfy a wave equation. The most significant outcome of this theory is the prediction of the existence of electromagnetic waves.

First let us assume that the electromagnetic wave is a **plane wave**—that is, it travels in one direction only. The plane wave we are describing has the following properties. It travels in the x direction (the direction of propagation), the electric field **E** is in the y direction, and the magnetic field **B** is in the z direction, as in Figure 24.3. Waves in which the electric and magnetic fields are restricted to being parallel to certain lines in the yz plane are said to be **linearly polarized waves.** Furthermore, we assume that at any point, P, E and B depend on x and t only and not on the y or z coordinate.

The properties of electromagnetic waves can be deduced from Maxwell's equations. We can relate E and B to each other by Equations 24.6 and 24.7. In empty

James Clerk Maxwell (1839–1879)

James Clerk Maxwell is generally regarded as the greatest theoretical physicist of the 19th century. Born in Edinburgh to a well-known Scottish family, he entered the University of Edinburgh at age 15. He was appointed to his first professorship, at Aberdeen, in 1856. This was the beginning of a career during which Maxwell would develop the electromagnetic theory of light, the kinetic theory of gases, and explanations of the nature of Saturn's rings and of color vision. *(North Wind Picture Archives)*

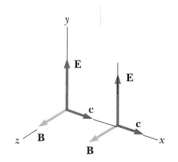

Figure 24.3 A plane-polarized electromagnetic wave traveling in the positive x direction. The electric field is along the y direction, and the magnetic field is along the z direction. These fields depend only on x and t.

space, where $Q = 0$ and $I = 0$, Equation 24.6 remains unchanged and Equation 24.7 becomes

$$\oint \mathbf{B} \cdot d\mathbf{s} = \epsilon_0 \mu_0 \frac{d\Phi_E}{dt} \qquad [24.9]$$

Using Equations 24.6 and 24.9 and the plane wave assumption, the following differential equations relating E and B are obtained. For simplicity of notation, we drop the subscripts on the components E_y and B_z:

$$\frac{\partial E}{\partial x} = -\frac{\partial B}{\partial t} \qquad [24.10]$$

$$\frac{\partial B}{\partial x} = -\mu_0 \epsilon_0 \frac{\partial E}{\partial t} \qquad [24.11]$$

Note that the derivatives here are partial derivatives. For example, when $\partial E/\partial x$ is evaluated, we assume that t is constant. Likewise, when evaluating $\partial B/\partial t$, x is held constant. Taking the derivative of Equation 24.10 with respect to x and combining this with Equation 24.11 we get

$$\frac{\partial^2 E}{\partial x^2} = -\frac{\partial}{\partial x}\left(\frac{\partial B}{\partial t}\right) = -\frac{\partial}{\partial t}\left(\frac{\partial B}{\partial x}\right) = -\frac{\partial}{\partial t}\left(-\mu_0\epsilon_0\frac{\partial E}{\partial t}\right)$$

$$\frac{\partial^2 E}{\partial x^2} = \mu_0\epsilon_0 \frac{\partial^2 E}{\partial t^2} \qquad [24.12]$$

• *Electric field wave equation for electromagnetic waves in free space*

In the same manner, taking a derivative of Equation 24.11 and combining it with Equation 24.12, we get

$$\frac{\partial^2 B}{\partial x^2} = \mu_0\epsilon_0 \frac{\partial^2 B}{\partial t^2} \qquad [24.13]$$

• *Magnetic field wave equation for electromagnetic waves in free space*

Equations 24.12 and 24.13 both have the form of the general wave equation,[2] with a speed, c, of

$$c = \frac{1}{\sqrt{\mu_0\epsilon_0}} \qquad [24.14]$$

• *The speed of electromagnetic waves*

Taking $\mu_0 = 4\pi \times 10^{-7}$ T·m/A and $\epsilon_0 = 8.854\,18 \times 10^{-12}$ C^2/N·m^2 in Equation 24.14, we find that $c = 2.997\,92 \times 10^8$ m/s. Because this speed is precisely the same as the speed of light in empty space,[3] one is led to believe (correctly) that light is an electromagnetic wave.

The simplest plane wave solution to Equations 24.12 and 24.13 are those for which the field amplitudes E and B vary with x and t according to the expressions

$$E = E_{\text{max}} \cos(kx - \omega t) \qquad [24.15]$$

• *Sinusoidal electric and magnetic fields*

$$B = B_{\text{max}} \cos(kx - \omega t) \qquad [24.16]$$

[2]The general wave equation is of the form $(\partial^2 g/\partial x^2) = (1/v^2)(\partial^2 g/\partial t^2)$, where v is the speed of the wave and g is the wave variable.

[3]Because of the redefinition of the meter in 1983, the speed of light is now a *defined* quantity with an *exact* value of $c = 2.997\,924\,58 \times 10^8$ m/s.

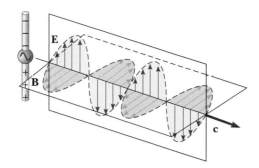

Figure 24.4   An electromagnetic wave sent out by oscillating charges in an antenna. This represents the wave at one instant of time. Note that the electric field is perpendicular to the magnetic field, and both are perpendicular to the direction of wave propagation.

where E_{max} and B_{max} are the maximum values of the fields. The angular wave number $k = 2\pi/\lambda$, where λ is the wavelength, and the angular frequency $\omega = 2\pi f$, where f is the number of cycles per second. Figure 24.4 represents one instant of a sinusoidal, linearly polarized plane wave moving in the positive x direction. Note that the electric and magnetic fields of a plane electromagnetic wave are perpendicular to each other and to the direction of propagation. Thus, **electromagnetic waves are transverse waves.**

Taking partial derivatives of Equations 24.15 and 24.16, and substituting into Equations 24.10 and 24.11, we find that

$$\frac{E}{B} = c \qquad\qquad\qquad \textbf{[24.17]}$$

That is, **at every instant the ratio of the electric field to the magnetic field of an electromagnetic wave equals the speed of light.**

Finally, electromagnetic waves obey the **superposition principle,** because the differential equations involving E and B are *linear* equations. For example, two waves traveling in opposite directions with the same frequency could be added by simply adding the wave fields algebraically.

Let us summarize the properties of electromagnetic waves as we have described them:

- The solutions of Maxwell's third and fourth equations are wavelike.
- Electromagnetic waves travel through empty space with the speed of light, $c = 1/\sqrt{\epsilon_0\mu_0}$.
- The electric and magnetic field components of plane electromagnetic waves are perpendicular to each other and also perpendicular to the direction of wave propagation. The latter property can be summarized by saying that electromagnetic waves are transverse waves.
- The relative magnitudes of **E** and **B** in empty space are related by $E/B = c$.
- Electromagnetic waves obey the superposition principle.

• *Properties of electromagnetic waves*

Thinking Physics 1

There is a Doppler effect for light, which is demonstrated in astronomical observations by the shift of spectral lines from receding galaxies toward the red end of the visible spectrum. The equation for the Doppler effect for light is not the same equation as that for sound. Why would the equation be different?

Reasoning In general, for waves requiring a medium, the velocities of the source and observer can be separately measured with respect to a third entity, the medium. In the Doppler effect for sound, these two velocities are the velocities of the source and of the observer relative to the air. Light as an electromagnetic wave does not require a medium—there is no third entity. Thus, we cannot identify separate velocities for the source and observer—only their relative velocity can be identified. As a result, a different equation must be used, one that contains only this single velocity. This equation can be generated from the laws of relativity.

Example 24.1 An Electromagnetic Wave

A plane electromagnetic sinusoidal wave of frequency 40.0 MHz travels in free space in the x direction, as in Figure 24.5. At some point and at some instant, the electric field has its maximum value of 750 N/C and is along the y axis.

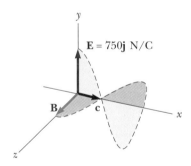

Figure 24.5 (Example 24.1) At some instant, a plane electromagnetic wave moving in the x direction has a maximum electric field of 750 N/C in the positive y direction. The corresponding magnetic field at that point has a magnitude E/c and is in the z direction.

(a) Determine the wavelength and period of the wave.

Solution Because $c = \lambda f$ and the wave has a frequency $f = 40.0$ MHz $= 4.00 \times 10^7$ s^{-1}, we get

$$\lambda = \frac{c}{f} = \frac{3.00 \times 10^8 \text{ m/s}}{4.00 \times 10^7 \text{ s}^{-1}} = \boxed{7.50 \text{ m}}$$

The period T of the wave equals the inverse of the frequency, and so

$$T = \frac{1}{f} = \frac{1}{4.00 \times 10^7 \text{ s}^{-1}} = 2.50 \times 10^{-8} \text{ s}$$

(b) Calculate the magnitude and direction of the magnetic field when $\mathbf{E} = 750\mathbf{j}$ N/C.

Solution From Equation 24.17 we see that

$$B_{\text{max}} = \frac{E_{\text{max}}}{c} = \frac{750 \text{ N/C}}{3.00 \times 10^8 \text{ m/s}} = 2.50 \times 10^{-6} \text{ T}$$

Because $\mathbf{E}$ and $\mathbf{B}$ must be perpendicular to each other and both must be perpendicular to the direction of wave propagation (x in this case), we conclude that $\mathbf{B}$ is in the z direction.

(c) Write expressions for the space–time variation of the electric and magnetic field components for this wave.

Solution We can apply Equations 24.15 and 24.16 directly:

$$E = E_{\text{max}} \cos(kx - \omega t) = (750 \text{ N/C}) \cos(kx - \omega t)$$

$$B = B_{\text{max}} \cos(kx - \omega t) = (2.50 \times 10^{-6} \text{ T}) \cos(kx - \omega t)$$

where

$$\omega = 2\pi f = 2\pi(4.00 \times 10^7 \text{ s}^{-1}) = 8\pi \times 10^7 \text{ rad/s}$$

$$k = \frac{2\pi}{\lambda} = \frac{2\pi}{7.50 \text{ m}} = 0.838 \text{ rad/m}$$

24.4 · HERTZ'S DISCOVERIES

In 1888, Heinrich Hertz (1857–1894) was the first to generate and detect electromagnetic waves in a laboratory setting. In order to appreciate the details of his experiment, let us first examine the properties of an *LC* circuit. In such a circuit, a charged capacitor is connected to an inductor, as in Figure 24.6. When the switch is closed, both the current in the circuit and the charge on the capacitor oscillate.

Let us assume that the capacitor has an initial charge of Q_{max} and that the switch is closed at $t = 0$. It is convenient to describe what happens from an energy viewpoint. When the capacitor is fully charged, the total energy in the circuit is stored in the electric field of the capacitor and is equal to $Q_{max}^2/2C$. At this time, the current is zero, and so no energy is stored in the inductor. As the capacitor begins to discharge, the energy stored in its electric field decreases. At the same time, the current increases and an amount of energy equal to $LI^2/2$ is now stored in the magnetic field of the inductor. Thus, we see that energy is transferred from the electric field of the capacitor to the magnetic field of the inductor. When the capacitor is fully discharged, it stores no energy. At this time, the current reaches its maximum value and all of the energy is stored in the inductor. The process then repeats in the reverse direction. The energy continues to transfer between the inductor and the capacitor, corresponding to oscillations of both current and charge.

A representation of this energy transfer is shown in Figure 24.7. The behavior of the circuit is analogous to that of the oscillating mass–spring system studied in Chapter 12. The potential energy stored in a stretched spring, $kx^2/2$, is analogous to the potential energy stored in the capacitor, $Q_{max}^2/2C$. The kinetic energy of the moving mass, $mv^2/2$, is analogous to the energy stored in the inductor, $LI^2/2$, which requires the presence of moving charges. In Figure 24.7a, all of the energy is stored as electric potential energy in the capacitor at $t = 0$ (because $I = 0$). In Figure 24.7b, all of the energy is stored as magnetic potential energy in the inductor, $LI_{max}^2/2$, where I_{max} is the maximum current. At intermediate points, part of the energy is electric and part is magnetic.

The frequency of oscillation of an *LC* circuit, called the *resonance frequency*, is

$$f_0 = \frac{1}{2\pi\sqrt{LC}} \qquad \text{[24.18]}$$

The circuit Hertz used in his investigations of electromagnetic waves is similar to that already discussed and is shown schematically in Figure 24.8. An induction coil (a large coil of wire) is connected to two metal spheres with a narrow gap between them to form a capacitor. Oscillations are initiated in the circuit by sending short voltage pulses via the coil to the spheres, charging one positive, the other negative. Because *L* and *C* are quite small in this circuit, the frequency of oscillation is quite high, $f \approx 100$ MHz. This circuit is called a *transmitter* because it produces electromagnetic waves.

Hertz placed a second circuit, the receiver, several meters from the transmitter circuit. This receiver circuit, which consisted of a single loop of wire connected to two spheres, had its own effective inductance, capacitance, and natural frequency of oscillation. Hertz found that energy was being sent from the transmitter to the

Heinrich Rudolf Hertz
(1857–1894)

Hertz was born in Hamburg, Germany. He studied physics under Helmholtz and Kirchhoff at the University of Berlin. In 1885, Hertz accepted the position of professor of physics at Karlsruhe. Discovering radio waves (in 1887), demonstrating their generation, and determining their speed are among Hertz's many achievements. After finding that the speed of a radio wave was the same as that of light, Hertz showed that radio waves, like light waves, could be reflected, refracted, and diffracted. Hertz died of blood poisoning at the age of 36. During his short life, he made many contributions to science. The hertz, equal to one complete vibration or cycle per second, is named after him. *(The Bettman Archive)*

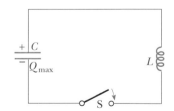

Figure 24.6 A simple *LC* circuit. The capacitor has an initial charge Q_{max} and the switch is closed at $t = 0$.

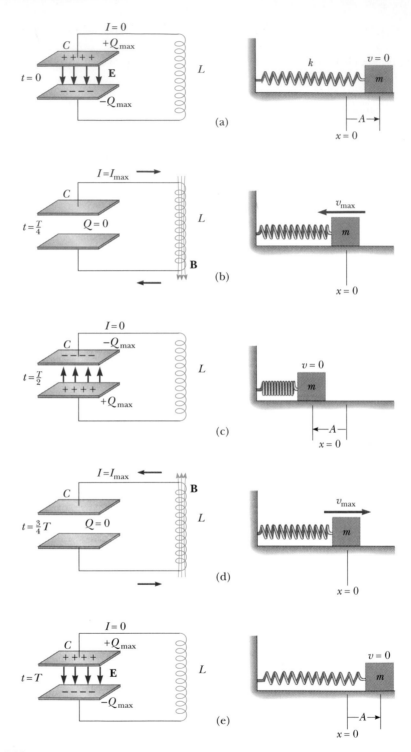

Figure 24.7 Energy transfer in a resistanceless LC circuit. The capacitor has a charge Q_{max} at $t = 0$ when the switch is closed. The mechanical analog of this circuit, the mass–spring system, is shown at the right.

receiver when the resonance frequency of the receiver was adjusted to match that of the transmitter.[4] The energy transfer was detected when the voltage across the spheres in the receiver circuit became high enough to ionize air molecules, which caused sparks to appear in the air gap separating the spheres. Hertz's experiment is analogous to the mechanical phenomenon in which one tuning fork picks up the vibrations from an identical vibrating fork.

Hertz assumed that the energy transferred from the transmitter to the receiver was carried in the form of waves, which are now known to have been electromagnetic waves. In a series of experiments, he also showed that the radiation generated by the transmitter exhibited the wave properties of interference, diffraction, reflection, refraction, and polarization. As we shall see shortly, all of these properties are exhibited by light. Thus, it became evident that the waves observed by Hertz had properties similar to those of light waves and differed only in frequency and wavelength.

Perhaps Hertz's most convincing experiment was his measurement of the speed of the waves from the transmitter. Waves of known frequency from the transmitter were reflected from a metal sheet so that an interference pattern was set up, much like the standing wave pattern on a stretched string. As we saw in our discussion of standing waves, the distance between nodes is $\lambda/2$, and so Hertz was able to determine the wavelength, λ. Using the relationship $v = f\lambda$, Hertz found that v was close to 3×10^8 m/s, the known speed of visible light. Thus, Hertz's experiments provided the first evidence in support of Maxwell's theory.

24.5 • PRODUCTION OF ELECTROMAGNETIC WAVES BY AN ANTENNA

Electromagnetic waves are radiated by *any* circuit carrying an alternating current. The fundamental mechanism responsible for this radiation is the acceleration of a charged particle. **Whenever a charged particle undergoes an acceleration, it must radiate energy by electromagnetic radiation.**

An alternating voltage applied to the wires of an antenna forces an electric charge in the antenna to oscillate. This is a common technique for accelerating charged particles and is the source of the radio waves emitted by the antenna of a radio station.

Figure 24.9 illustrates how an electromagnetic wave is produced by oscillating electric charges in an antenna. Two metal rods are connected to an ac generator, which causes charges to oscillate between the two rods. The output voltage of the generator is sinusoidal. At $t = 0$, the upper rod is given a maximum positive charge and the bottom rod an equal negative charge, as in Figure 24.9a. The electric field near the antenna at this instant is also shown in Figure 24.9a. As the charges oscillate, the rods become less charged, the field near the rods decreases in strength, and the downward-directed maximum electric field produced at $t = 0$ moves away from the rod. When the charges are neutralized, as in Figure 24.9b, the electric field has dropped to zero. This occurs at a time equal to one quarter of the period of oscillation. Continuing in this fashion, the upper rod soon obtains a maximum negative charge and the lower rod becomes positive, as in Figure 24.9c, resulting

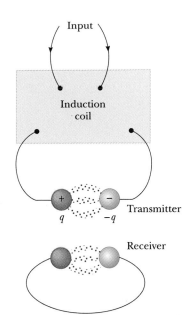

Figure 24.8 Schematic diagram of Hertz's apparatus for generating and detecting electromagnetic waves. The transmitter consists of two spherical electrodes connected to an induction coil, which provides short voltage surges to the spheres, setting up oscillations in the discharge. The receiver is a nearby loop containing a second spark gap.

---

[4]Following Hertz's discoveries, Guglielmo Marconi succeeded in developing a practical, long-range radio communication system.

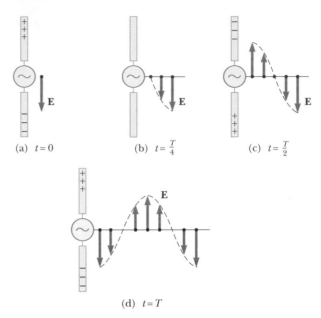

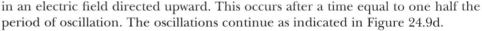

(d) $t = T$

Figure 24.9   The electric field set up by charges oscillating in an antenna. The field moves away from the antenna with the speed of light.

Figure 24.10   Magnetic field lines around an antenna carrying a changing current. Why do the circles have different radii?

in an electric field directed upward. This occurs after a time equal to one half the period of oscillation. The oscillations continue as indicated in Figure 24.9d.

The electric field near the antenna oscillates in phase with the charge distribution. That is, the field points down when the upper rod is positive and up when the upper rod is negative. Furthermore, the magnitude of the field at any instant depends on the amount of charge on the rods at that instant. As the charges continue to oscillate (and accelerate) between the rods, the electric field set up by the charges moves away from the antenna at the speed of light. One cycle of charge oscillation produces one full wavelength in the electric field pattern.

Because the oscillating charges create a current in the rods, a magnetic field is also generated when the current in the rods is upward, as shown in Figure 24.10. The magnetic field lines circle the antenna and are *perpendicular to the electric field at all points*. As the current changes with time, the magnetic field lines spread out from the antenna. At great distances from the antenna, the electric and magnetic fields become very weak. However, at these great distances, it is necessary to take into account the facts that (1) a changing magnetic field produces a changing electric field and (2) a changing electric field produces a changing magnetic field, as predicted by Maxwell. These induced electric and magnetic fields are in phase: At any point, the two fields reach their maximum values at the same instant. This is illustrated for one instant of time in Figure 24.11.

CONCEPTUAL PROBLEM 1

Radio antennas can be in the form of conducting lines or loops. What should the orientation of each of these types of antennas be, relative to a linear broadcasting antenna?

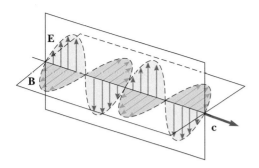

Figure 24.11 An electromagnetic wave sent out by oscillating charges in an antenna. This drawing represents the wave at one instant of time.

CONCEPTUAL PROBLEM 2

In radio transmission, a radio wave serves as a carrier wave, and the sound wave is superimposed on the carrier wave. In amplitude modulation (AM radio), the amplitude of the carrier wave varies according to the sound wave. In frequency modulation (FM radio), the frequency of the carrier wave varies according to the sound wave. The navy sometimes uses flashing lights to send Morse code to neighboring ships, a process that has similarities to radio broadcasting. Is this AM or FM? What is the carrier frequency? What is the signal frequency? What is the broadcasting antenna? What is the receiving antenna?

24.6 • ENERGY CARRIED BY ELECTROMAGNETIC WAVES

Electromagnetic waves carry energy, and as they propagate through space they can transfer energy to objects placed in their path. The intensity of an electromagnetic wave is described by a vector **S**, called the **Poynting vector,** defined by the expression

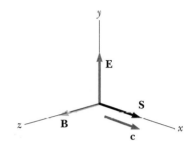

Figure 24.12 The Poynting vector **S** for a plane electromagnetic wave is along the direction of propagation.

$$\mathbf{S} \equiv \frac{1}{\mu_0} \mathbf{E} \times \mathbf{B} \qquad [24.19]$$

• *Poynting vector*

The magnitude of the Poynting vector represents the rate at which energy flows through a unit surface area perpendicular to the flow and its direction is along the direction of wave propagation (Fig. 24.12). The SI units of the Poynting vector are $J/s \cdot m^2 = W/m^2$. (These are the units **S** must have because it represents the power per unit area, where the unit area is oriented at right angles to the direction of wave propagation.)

As an example, let us evaluate the magnitude of **S** for a plane electromagnetic wave where $|\mathbf{E} \times \mathbf{B}| = EB$. In this case

$$S = \frac{EB}{\mu_0} \qquad [24.20]$$

• *Poynting vector for a plane wave*

Because $B = E/c$, we can also express this as

$$S = \frac{E^2}{\mu_0 c} = \frac{c}{\mu_0} B^2$$

These equations for S apply at any instant of time.

What is of more interest for a sinusoidal plane electromagnetic wave is the time average of S over one or more cycles, called the wave intensity, I. When this average is taken, we obtain an expression involving the time average of $\cos^2(kx - \omega t)$, which equals $\frac{1}{2}$. Hence, the average value of S (or the intensity of the wave) is

Average power per unit area •

$$I = S_{\mathrm{av}} = \frac{E_{\max} B_{\max}}{2\mu_0} = \frac{E_{\max}^2}{2\mu_0 c} = \frac{c}{2\mu_0} B_{\max}^2 \qquad \textbf{[24.21]}$$

Recall that the energy per unit volume, u_E, which is the instantaneous energy density associated with an electric field (Section 20.9), is given by Equation 20.31:

$$u_E = \tfrac{1}{2}\epsilon_0 E^2 \qquad \textbf{[24.22]}$$

and that the instantaneous energy density u_B associated with a magnetic field (Section 23.7) is given by Equation 23.22:

$$u_B = \frac{B^2}{2\mu_0} \qquad \textbf{[24.23]}$$

Because E and B vary with time for an electromagnetic wave, the energy densities also vary with time. Using the relationships $B = E/c$ and $c = 1/\sqrt{\epsilon_0\mu_0}$, Equation 24.23 becomes

$$u_B = \frac{(E/c)^2}{2\mu_0} = \frac{\epsilon_0\mu_0}{2\mu_0} E^2 = \tfrac{1}{2}\epsilon_0 E^2$$

Comparing this result with the expression for u_E, we see that

$$u_B = u_E = \tfrac{1}{2}\epsilon_0 E^2 = \frac{B^2}{2\mu_0}$$

That is, **for an electromagnetic wave, the instantaneous energy density associated with the magnetic field equals the instantaneous energy density associated with the electric field.** Hence, in a given volume the energy is equally shared by the two fields.

The **total instantaneous energy density,** u, is equal to the sum of the energy densities associated with the electric and magnetic fields:

Total energy density of an •
electromagnetic wave

$$u = u_E + u_B = \epsilon_0 E^2 = \frac{B^2}{\mu_0}$$

When this is averaged over one or more cycles of an electromagnetic wave, we again get a factor of $\frac{1}{2}$. Hence, the total average energy per unit volume of an electromagnetic wave is

Average energy density of an •
electromagnetic wave

$$u_{\mathrm{av}} = \epsilon_0 (E^2)_{\mathrm{av}} = \tfrac{1}{2}\epsilon_0 E_{\max}^2 = \frac{B_{\max}^2}{2\mu_0} \qquad \textbf{[24.24]}$$

Comparing this result with Equation 24.21 for the average value of S, we see that

$$I = S_{\mathrm{av}} = c u_{\mathrm{av}} \qquad \textbf{[24.25]}$$

In other words, the intensity of an electromagnetic wave equals the average energy density multiplied by the speed of light.

Example 24.2 Fields Due to a Point Source

A point source of electromagnetic radiation has an average power output of 800 W. Calculate the maximum values of the electric and magnetic fields at a point 3.50 m from the source.

Solution Recall from Chapter 13 that the wave intensity, I, a distance r from a point source is

$$I = \frac{P_{av}}{4\pi r^2}$$

where P_{av} is the average power output of the source and $4\pi r^2$ is the area of a sphere of radius r centered on the source. Because the intensity of an electromagnetic wave is also given by Equation 24.21, we have

$$I = \frac{P_{av}}{4\pi r^2} = \frac{E_{max}^2}{2\mu_0 c}$$

Solving for E_{max} gives

$$E_{max} = \sqrt{\frac{\mu_0 c P_{av}}{2\pi r^2}}$$

$$= \sqrt{\frac{(4\pi \times 10^{-7} \text{ T·m/A})(3.00 \times 10^8 \text{ m/s})(800 \text{ W})}{2\pi(3.50 \text{ m})^2}}$$

$$= 62.6 \text{ V/m}$$

We calculate the maximum value of the magnetic field using this result and the relationship $B_{max} = E_{max}/c$ (Eq. 24.17):

$$B_{max} = \frac{E_{max}}{c} = \frac{62.6 \text{ V/m}}{3.00 \times 10^8 \text{ m/s}} = 2.09 \times 10^{-7} \text{ T}$$

EXERCISE 1 Calculate the energy density 3.50 m from the point source. *Answer* $1.73 \times 10^{-8} \text{ J/m}^3$

24.7 • MOMENTUM AND RADIATION PRESSURE

Electromagnetic waves transport linear momentum as well as energy. Hence, it follows that pressure is exerted on a surface when an electromagnetic wave impinges on it. In what follows, we assume that the electromagnetic wave transports a total energy U to a surface in a time t. If the surface absorbs all the incident energy U in this time, Maxwell showed that the total momentum **p** delivered to this surface has a magnitude

$$p = \frac{U}{c} \quad \text{(complete absorption)} \qquad \text{[24.26]}$$

• *Momentum delivered to an absorbing surface*

Furthermore, if the Poynting vector of the wave is **S**, the radiation pressure P (force per unit area) exerted on the perfect absorbing surface is

$$P = \frac{S}{c} \qquad \text{[24.27]}$$

• *Radiation pressure exerted on a perfect absorbing surface*

A black body is such a perfectly absorbing surface, for which all of the incident energy is abosrbed (none is reflected). (A more detailed discussion of a black body will be presented in Chapter 28.)

If the surface is a perfect reflector, then the momentum delivered in a time t for normal incidence is twice that given by Equation 24.26, or $p = 2U/c$. That is, a momentum U/c is delivered first by the incident wave and then again by the reflected wave, in analogy with a ball colliding elastically with a wall.[5] Finally, the

• *Momentum delivered to a perfectly reflecting surface*

[5]For *oblique* incidence, the momentum transferred is $2U \cos\theta/c$ and the pressure is given by $P = 2S \cos\theta/c$, where θ is the angle between the normal to the surface and the direction of propagation.

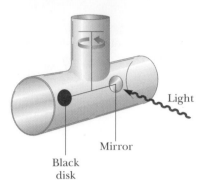

Light

Mirror

Black
disk

Figure 24.13   An apparatus for measuring the pressure exerted by light. In practice, the system is contained in a high vacuum.

radiation pressure exerted on a perfect reflecting surface for normal incidence of the wave is twice that given by Equation 24.27, or $P = 2S/c$.

Although radiation pressures are very small (about 5×10^{-6} N/m^2 for direct sunlight), they have been measured using torsion balances such as the one shown in Figure 24.13. Light is allowed to strike either a mirror or a black disk, both of which are suspended from a fine fiber. Light striking the black disk is completely absorbed, and so all of its momentum is transferred to the disk. Light striking the mirror (normal incidence) is totally reflected, hence the momentum transfer is twice as great as that transferred to the disk. The radiation pressure is determined by measuring the angle through which the horizontal connecting rod rotates. The apparatus must be placed in a high vacuum to eliminate the effects of air currents.

Thinking Physics 2

In the interplanetary space in the solar system, there is a large amount of dust. Although interplanetary dust can in theory have a variety of sizes, from molecular size upward, there is very little dust smaller than about 0.2 μm in our solar system. Why? (*Hint:* The solar system originally contained dust particles of all sizes.)

Reasoning   Dust particles in the solar system are subject to two forces—the gravitational force toward the Sun, and the force from radiation pressure, which is away from the Sun. The gravitational force is proportional to the cube of the radius of a spherical dust particle, because it is proportional to the mass of the particle. The radiation pressure is proportional to the square of the radius, because it depends on the planar cross-section of the particle. For large particles, the gravitational force is larger than the force from radiation pressure. For small particles, less than about 0.2 μm, the larger force from radiation pressure sweeps these particles out of the solar system.

CONCEPTUAL PROBLEM 3

In *space sailing*, which is a proposed alternative method to reach the planets, a spacecraft carries a very large sail that experiences a force due to radiation pressure from the Sun. Should the sail be absorptive or reflective to be most effective?

Example 24.3   Solar Energy

The Sun delivers about 1000 W/m² of electromagnetic flux to the Earth's surface. (a) Calculate the total power incident on a roof of dimensions 8.00 m × 20.0 m.

Solution   The magnitude of the Poynting vector is $S = 1000$ W/m², which represents the power per unit area, or the light intensity. Assuming the radiation is incident normal to the roof (Sun directly overhead), we get

$$\text{Power} = SA = (1000 \text{ W/m}^2)(8.00 \times 20.0 \text{ m}^2)$$
$$= 1.60 \times 10^5 \text{ W}$$

If this power could all be converted to electrical energy, it would provide more than enough power for the average home. However, solar energy is not easily harnessed, and the prospects for large-scale conversion are not as bright as they may appear from this simple calculation. For example, the conversion efficiency from solar to electrical energy is typically 10% for photovoltaic cells. Roof systems for converting solar energy to thermal energy are approximately 50% effi-

cient; however, there are other practical problems with solar energy that must be considered, such as overcast days, geographic location, and energy storage.

(b) Determine the radiation pressure and radiation force on the roof assuming the roof covering is a perfect absorber.

Solution   Using Equation 24.27 with $S = 1000$ W/m², we find that the radiation pressure is

$$P = \frac{S}{c} = \frac{1000 \text{ W/m}^2}{3.00 \times 10^8 \text{ m/s}} = 3.33 \times 10^{-6} \text{ N/m}^2$$

Because pressure equals force per unit area, this corresponds to a radiation force of

$$F = PA = (3.33 \times 10^{-6} \text{ N/m}^2)(160 \text{ m}^2) = 5.33 \times 10^{-4} \text{ N}$$

EXERCISE 2   How much solar energy (in joules) is incident on the roof in 1 hour?   Answer 5.76×10^8 J

Example 24.4   Poynting Vector for a Wire

A long, straight wire of resistance R, radius a, and length ℓ carries a constant current I, as in Figure 24.14. Calculate the Poynting vector for this wire.

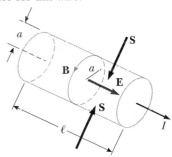

Figure 24.14   (Example 24.4) A wire of length ℓ, resistance R, and radius a carrying a current I. The Poynting vector **S** is directed radially *inward*.

Solution   First, let us find the electric field **E** along the wire. If ΔV is the potential difference across its ends, then $\Delta V = IR$ and

$$E = \frac{\Delta V}{\ell} = \frac{IR}{\ell}$$

Recall that the magnetic field at the surface of the wire (Example 22.5) is

$$B = \frac{\mu_0 I}{2\pi a}$$

The vectors **E** and **B** are mutually perpendicular, as shown in Figure 24.14, and therefore $|\mathbf{E} \times \mathbf{B}| = EB$. Hence, the Poynting vector **S** is directed radially inward and has a magnitude

$$S = \frac{EB}{\mu_0} = \frac{1}{\mu_0} \frac{IR}{\ell} \frac{\mu_0 I}{2\pi a} = \frac{I^2 R}{2\pi a \ell} = \frac{I^2 R}{A}$$

where $A = 2\pi a \ell$ is the surface area of the wire and the total area through which S passes. From this result, we see that

$$SA = I^2 R$$

where SA has units of power (J/s = W). That is, **the rate at which electromagnetic energy flows into the wire, SA, equals the rate of energy (or power) dissipated as joule heat, I^2R.**

EXERCISE 3   A heater wire of radius 0.30 mm, length 1.0 m, and resistance 5.0 Ω carries a current of 2.0 A. Determine the magnitude and direction of the Poynting vector for this wire.   Answer 1.1×10^4 W/m² directed radially inward

24.8 • THE SPECTRUM OF ELECTROMAGNETIC WAVES

We have seen that all electromagnetic (EM) waves travel in a vacuum with the speed of light, *c*. These waves transport energy and momentum from some source to a receiver. In 1888, Hertz successfully generated and detected the radio-frequency electromagnetic waves predicted by Maxwell. Maxwell himself had recognized as EM waves both visible light and the near infrared radiation discovered in 1800 by William Herschel. It is now known that other forms of electromagnetic waves exist; they are distinguished by their frequencies and wavelengths.

Because all electromagnetic waves travel through vacuum with the speed *c*, their frequency, *f*, and wavelength, λ, are related by the important expression

$$c = f\lambda \qquad \text{[24.28]}$$

The various types of electromagnetic waves, all produced by accelerating charges, are listed in Figure 24.15. Note the wide range of frequencies and wavelengths. For instance, a radio wave of frequency 5 MHz (a typical value) has the wavelength

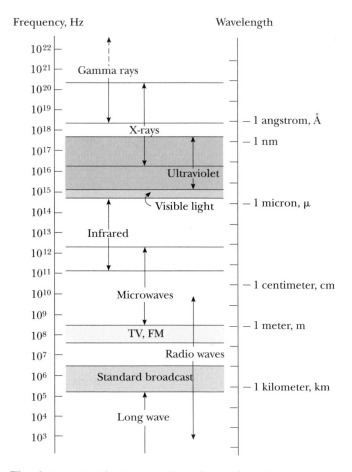

Figure 24.15   The electromagnetic spectrum. Note the overlap between adjacent wave types.

$$\lambda = \frac{c}{f} = \frac{3 \times 10^8 \text{ m/s}}{5 \times 10^6 \text{ s}^{-1}} = 60 \text{ m}$$

Let us briefly describe the wave types shown in Figure 24.15.

Radio waves, which were discussed in the preceding section, are the result of charges accelerating through conducting wires. They are generated by such electronic devices as *LC* oscillators and are used in radio and television communication systems.

• *Radio waves*

Microwaves (short-wavelength radio waves) have wavelengths ranging between about 1 mm and 30 cm and are also generated by electronic devices. Because of their short wavelengths, they are well suited for radar systems used in aircraft navigation and for studying the atomic and molecular properties of matter. Microwave ovens are an interesting domestic application of these waves.

• *Microwaves*

Infrared waves have wavelengths ranging from about 1 mm to the longest wavelength of visible light, 7×10^{-7} m. These waves, produced by hot bodies and molecules, are readily absorbed by most materials. The infrared energy absorbed by a substance appears as thermal energy because the energy agitates the atoms of the body, increasing their vibrational or translational motion, which results in a temperature rise. Infrared radiation has many practical and scientific applications, including physical therapy, infrared photography, and vibrational spectroscopy.

• *Infrared waves*

Visible light, the most familiar form of electromagnetic waves, is that part of the spectrum the human eye can detect. Light is produced by the rearrangement of electrons in atoms and molecules and hot objects like lightbulb filaments. The wavelengths of visible light are classified by color, ranging from violet ($\lambda \approx 4 \times 10^{-7}$ m) to red ($\lambda \approx 7 \times 10^{-7}$ m). The eye's sensitivity is a function of wavelength and is a maximum at a wavelength of about 5.6×10^{-7} m (yellow–green). Light is the basis of the science of optics and optical instruments, which will be discussed later.

• *Visible light*

Ultraviolet light covers wavelengths ranging from about 3.8×10^{-7} m (380 nm) down to 6×10^{-10} m (0.6 nm). The Sun is an important source of ultraviolet waves, which is the main cause of suntans and sunburns. Most of the ultraviolet waves from the Sun are absorbed by atoms in the stratosphere. (This is fortunate, because ultraviolet waves in large quantities have harmful effects on humans.) One important constituent of the stratosphere is ozone (O_3), which results from reactions of oxygen with ultraviolet radiation. This ozone shield converts lethal high-energy ultraviolet radiation to thermal energy, which raises the temperature of the stratosphere. A great deal of controversy has arisen concerning the depletion of the protective ozone layer by the use of Freon in aerosol spray cans and as refrigerants.

• *Ultraviolet waves*

X-rays are electromagnetic waves with wavelengths in the range of about 10^{-8} m (10 nm) down to 10^{-13} m (10^{-4} nm). The most common source of x-rays is the acceleration of high-energy electrons bombarding a metal target. X-rays are used as a diagnostic tool in medicine and as a treatment for certain forms of cancer. Because x-rays damage or destroy living tissues and organisms, care must be taken to avoid unnecessary exposure and overexposure. X-rays are also used in the study of crystal structure; x-ray wavelengths are comparable to the atomic separation distances (≈ 0.1 nm) in solids.

• *X-rays*

Gamma rays are electromagnetic waves emitted by radioactive nuclei (such as ^{60}Co and ^{137}Cs) and during certain nuclear reactions. They have wavelengths ranging from about 10^{-10} m to less than 10^{-14} m. Gamma rays are highly penetrating

• *Gamma rays*

and produce serious damage when absorbed by living tissues. As a consequence, those working near such dangerous radiation must be protected with heavily absorbing materials, such as layers of lead.

Thinking Physics 3

The center of sensitivity of our eyes coincides with the center of the wavelength distribution of the Sun. Is this an amazing coincidence?

Reasoning This is not a coincidence—it is the result of biological evolution. Humans have evolved with vision most sensitive to the wavelengths that are strongest from the Sun. It is an interesting conjecture to imagine aliens from another planet, with a Sun with a different temperature, arriving at Earth. Their eyes would have the center of sensitivity at different wavelengths than ours. How would their vision of the Earth compare to ours?

CONCEPTUAL PROBLEM 4

The color of an object is said to depend on wavelength. So if you view colored objects under water, in which the wavelength of the light will be different, does the color change?

24.9 • POLARIZATION

As we learned in Section 24.3, the electric and magnetic vectors associated with an electromagnetic wave are at right angles to each other and also to the direction of wave propagation, as shown in Figure 24.16. The phenomenon of polarization described in this section is a property that specifies the directions of the electric and magnetic fields associated with an electromagnetic wave.

An ordinary beam of light consists of a large number of waves emitted by the atoms or molecules of the light source. Each atom produces a wave with its own orientation, **E**, corresponding to the direction of atomic vibration. The direction of polarization of the electromagnetic wave is defined to be the direction in which **E** is vibrating. However, because all directions of vibration are possible, the resultant electromagnetic wave is a superposition of waves produced by the individual atomic sources. The result is an **unpolarized** light wave, represented schematically in Figure 24.17a. The direction of wave propagation in this figure is perpendicular to the page. Note that *all* directions of the electric field vector, lying in a plane perpendicular to the direction of propagation, are equally probable. At any given point and at some instant of time, there is only one resultant electric field; hence, you should not be misled by the meaning of Figure 24.17a.

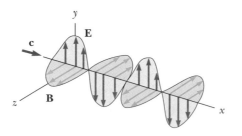

Figure 24.16 Schematic diagram of an electromagnetic wave propagating in the *x* direction. The electric field vector, **E**, vibrates in the *xy* plane, and the magnetic field vector, **B**, vibrates in the *xz* plane. This is an example of a linearly polarized wave.

A wave is said to be **linearly polarized** if **E** vibrates in the same direction *at all times* at a particular point, as in Figure 24.17b. (Sometimes such a wave is described as **plane-polarized.**) The wave described in Figure 24.16 is an example of a wave linearly polarized in the *y* direction. As the field propagates in the *x* direction, **E** is always in the *y* direction. The plane formed by **E** and the direction of propagation is called the **plane of polarization** of the wave. In Figure 24.16, the plane of polarization is the *xy* plane. It is possible to obtain a linearly polarized wave from an unpolarized wave by removing from the unpolarized wave all components of electric field vectors except those that lie in a single plane.

The most common technique for polarizing light is to send it through a material that passes only components of electric field vectors that are parallel to a characteristic direction of the material called the **polarizing direction.**

In 1938 E. H. Land discovered a material, which he called **polaroid,** that polarizes light through selective absorption by oriented molecules. This material is fabricated in thin sheets of long-chain hydrocarbons, which are stretched during manufacture so that the molecules align. After a sheet is dipped into a solution containing iodine, the molecules become good electrical conductors. However, the conduction takes place primarily along the hydrocarbon chains, because the valence electrons of the molecules can move easily only along the chains (recall that valence electrons are "free" electrons that can readily move through the conductor). As a result, the molecules readily *absorb* light the electric field vector of which is parallel to their length and *transmit* light the electric field vector of which is perpendicular to their length. It is common to refer to the direction perpendicular to the molecular chains as the **transmission axis.** An ideal polarizer passes the components of electric vectors that are parallel to the transmission axis. Components that are perpendicular to the transmission axis are absorbed. If light passes through several polarizers, whatever is transmitted has the plane of polarization parallel to the polarizing direction of the last polarizer it passed.

Let us now obtain an expression for the intensity of light that passes through a polarizing material. In Figure 24.18, an unpolarized light beam is incident on the

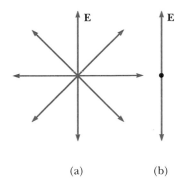

Figure 24.17   (a) An unpolarized light beam viewed along the direction of propagation (perpendicular to the page). The electric field vector can vibrate in any direction with equal probability. (b) A linearly polarized light beam with the electric field vector vibrating in the vertical direction.

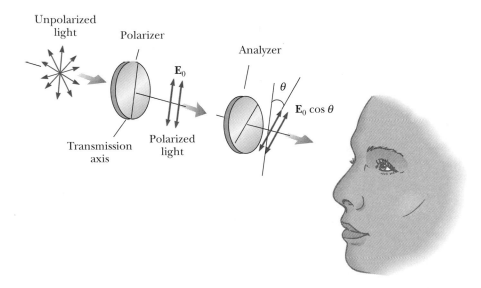

Figure 24.18   Two polarizing sheets the transmission axes of which make an angle θ with each other. Only a fraction of the polarized light incident on the analyzer is transmitted.

first polarizing sheet, called the **polarizer,** where the transmission axis is as indicated. The light that is passing through this sheet is polarized vertically, and the transmitted electric field vector is $\mathbf{E}_0$. A second polarizing sheet, called the **analyzer,** intercepts this beam with its transmission axis at an angle of θ to the axis of the polarizer. The component of $\mathbf{E}_0$ that is perpendicular to the axis of the analyzer is completely absorbed, and the component parallel to that axis is $E_0 \cos \theta$. We know from Equation 24.18 that the transmitted intensity varies as the *square* of the transmitted amplitude, and so we conclude that the intensity of the transmitted (polarized) light varies as

Malus's law •

$$I = I_0 \cos^2 \theta \qquad \qquad [24.29]$$

where I_0 is the intensity of the polarized wave incident on the analyzer. This expression, known as **Malus's law,** applies to any two polarizing materials the transmission axes of which are at an angle of θ to each other. From this expression, note that the transmitted intensity is a maximum when the transmission axes are parallel ($\theta = 0$ or $180°$), and zero (complete absorption by the analyzer) when the transmission axes are perpendicular to each other. This variation in transmitted intensity through a pair of polarizing sheets is illustrated in Figure 24.19.

Optical Activity

Many important applications of polarized light involve materials that display **optical activity.** A substance is said to be optically active if it rotates the plane of polarization of transmitted light. The angle through which the light is rotated by a specific material depends on the length of the sample and on the concentration if the substance is in solution. One optically active material is a solution of the common sugar dextrose. A standard method for determining the concentration of sugar solutions is to measure the rotation produced by a fixed length of the solution.

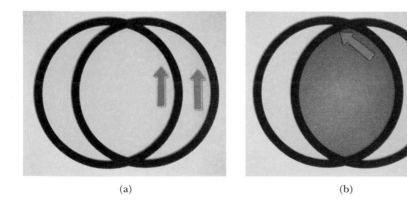
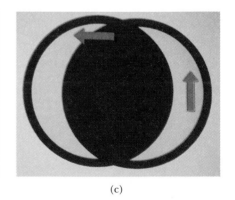

(a)   (b)   (c)

Figure 24.19   The intensity of light transmitted through two polarizers depends on the relative orientation of their transmission axes. (a) The transmitted light has maximum intensity when the transmission axes are aligned with each other. (b) The transmitted light intensity diminishes when the transmission axes are at an angle of $45°$ with each other. (c) The transmitted light intensity is a minimum when the transmission axes are at right angles to each other.   *(Henry Leap and Jim Lehman)*

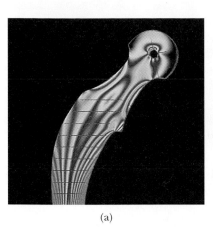 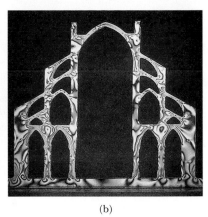

(a)   (b)

Figure 24.20 (a) Photograph showing strain distribution in a plastic model of a hip replacement used in a medical research laboratory. The pattern is produced when the plastic model is viewed between a polarizer and an analyzer oriented perpendicular to the polarizer *(Sepp Seitz 1981)* (b) A plastic model of an arch structure under load conditions observed between two crossed polarizers. Such patterns are useful in the optimum design of architectural components. *(Peter Aprahamian/Science Photo Library)*

A material is optically active because of an asymmetry in the shape of its constituent molecules. For example, some proteins are optically active because of their spiral shape. Other materials, such as glass and plastic, become optically active when stressed. Suppose an unstressed piece of plastic is placed between a polarizer and an analyzer so that light passes from polarizer to plastic to analyzer. When the analyzer axis is perpendicular to the polarizer axis, none of the polarized light passes through the analyzer. In other words, the unstressed plastic has no effect on the light passing through it. If the plastic is stressed, however, the regions of greatest stress rotate the polarized light through the largest angles. Hence, a series of bright and dark bands is observed in the transmitted light, with the bright bands corresponding to regions of greatest stress.

Engineers often use this technique, called **optical stress analysis,** to assist in designing structures ranging from bridges to small tools. A plastic model is built and analyzed under different load conditions to determine regions of potential weakness and failure under stress. Some examples of a plastic model under stress are shown in Figure 24.20.

Thinking Physics 4

A polarizer for microwaves can be made as a grid of parallel metal wires about a centimeter apart. Is the electric field vector for microwaves transmitted through this polarizer parallel to, or perpendicular to, the metal wires?

Reasoning Electric field vectors parallel to the metal wires will cause electrons in the metal to oscillate parallel to the wires. Thus, the energy from the waves with these electric field vectors will be transferred to the metal by accelerating these electrons and will eventually be transformed to internal energy through the resistance of the metal. Waves with electric field vectors perpendicular to the metal wires will not be able to accelerate electrons and will pass through. Thus, the electric field polarization will be perpendicular to the metal wires.

CONCEPTUAL PROBLEM 5

In everyday experience, radio waves are polarized, but light is not—why?

SUMMARY

Electromagnetic waves, which are predicted by Maxwell's equations, have the following properties:

- The electric and magnetic fields satisfy the following wave equations, which can be obtained from Maxwell's third and fourth equations:

$$\frac{\partial^2 E}{\partial x^2} = \mu_0 \epsilon_0 \frac{\partial^2 E}{\partial t^2} \qquad \text{[24.12]}$$

$$\frac{\partial^2 B}{\partial x^2} = \mu_0 \epsilon_0 \frac{\partial^2 B}{\partial t^2} \qquad \text{[24.13]}$$

- Electromagnetic waves travel through a vacuum with the speed of light, *c*, where

$$c = \frac{1}{\sqrt{\mu_0 \epsilon_0}} = 3.00 \times 10^8 \text{ m/s} \qquad \text{[24.14]}$$

- The electric and magnetic fields of an electromagnetic wave are perpendicular to each other and perpendicular to the direction of wave propagation. (Hence electromagnetic waves are transverse waves.)
- The instantaneous magnitudes of **E** and **B** in an electromagnetic wave are related by the expression

$$\frac{E}{B} = c \qquad \text{[24.17]}$$

- Electromagnetic waves carry energy. The rate of flow of energy crossing a unit area is described by the **Poynting vector S,** where

$$\mathbf{S} \equiv \frac{1}{\mu_0} \mathbf{E} \times \mathbf{B} \qquad \text{[24.19]}$$

- Electromagnetic waves carry momentum and hence can exert pressure on surfaces. If an electromagnetic wave the Poynting vector of which is **S** is completely absorbed by a surface on which it is normally incident, the radiation pressure on that surface is

$$P = \frac{S}{c} \qquad \text{(complete absorption)} \qquad \text{[24.27]}$$

If the surface totally reflects a normally incident wave, the pressure is doubled.

The electric and magnetic fields of a sinusoidal plane electromagnetic wave propagating in the positive *x* direction can be written

$$E = E_{max} \cos(kx - \omega t) \qquad \text{[24.15]}$$

$$B = B_{max} \cos(kx - \omega t) \qquad \text{[24.16]}$$

where ω is the angular frequency of the wave and k is the angular wave number. These equations represent special solutions to the wave equations for **E** and **B**.

The average value of the Poynting vector for a plane electromagnetic wave has the magnitude

$$S_{av} = \frac{E_{max}B_{max}}{2\mu_0} = \frac{E_{max}^2}{2\mu_0 c} = \frac{c}{2\mu_0} B_{max}^2 \qquad [24.21]$$

The average power per unit area (intensity) of a sinusoidal plane electromagnetic wave equals the average value of the Poynting vector taken over one or more cycles.

The electromagnetic spectrum includes waves covering a broad range of frequencies and wavelengths. The frequency, f, and wavelength, λ, of a given wave are related by

$$c = f\lambda \qquad [24.28]$$

When polarized light of intensity I_0 is incident on a polarizing film, the light transmitted through the film has an intensity equal to $I_0 \cos^2 \theta$, where θ is the angle between the transmission axis of the polarizer and the electric field vector of the incident light.

CONCEPTUAL QUESTIONS

1. What is the fundamental source of electromagnetic radiation?

2. Electrical engineers often speak of the radiation resistance of an antenna. What do you suppose they mean by this phrase?

3. Describe the physical significance of the Poynting vector.

4. If a high-frequency current is passed through a solenoid containing a metallic core, the core heats up by induction. Explain why the materials heat up in these situations.

5. Certain orientations of the receiving antenna on a television set give better reception than others. Furthermore, the best orientation varies from station to station. Explain.

6. If you charge a comb by running it through your hair and then hold the comb next to a bar magnet, do the electric and magnetic fields produced constitute an electromagnetic wave?

7. An empty plastic or glass dish removed from a microwave oven is cool to the touch right after it is removed from the oven. How can this be possible? (You can assume that your electric bill has been paid.)

8. What does a radio wave do to the charges in the receiving antenna to provide a signal for your car radio?

9. When light (or other electromagnetic radiation) travels across a given region, what is it that moves?

10. Suppose a creature from another planet had eyes that were sensitive to infrared radiation. Describe what he would see if he looked around the room you are now in. That is, what would be bright and what would be dim?

11. Why should an infrared photograph of a person look different from a photograph taken with visible light?

12. A welder must wear protective glasses and clothing to prevent eye damage and sunburn. What does this imply about the light produced by the welding?

13. Radio stations often advertise "instant news." If what they mean is that you hear the news at the instant they speak it, is their claim true? About how long would it take for a message to travel across this country by radio waves, assuming that these waves could travel this great distance and still be detected?

14. Light from the Sun takes approximately $8\frac{1}{3}$ minutes to reach the Earth. During this time the Earth has continued to move in its orbit around the Sun. How far is the actual location of the Sun from its image in the sky?

15. Do Maxwell's equations allow for the existence of magnetic monopoles?

PROBLEMS

Section 24.1 Displacement Current and the Generalized Ampère's Law

Section 24.2 Maxwell's Wonderful Equations

1. Consider the situation shown in Figure P24.1. An electric field of 300 V/m is confined to a circular area 10.0 cm in diameter and directed outward from the plane of the figure. If the field is increasing at a rate of 20.0 V/m·s, what are the direction and magnitude of the magnetic field at the point P, 15.0 cm from the center of the circle?

Section 24.3 Electromagnetic Waves

2. The speed of an electromagnetic wave traveling in a nonmagnetic transparent substance is $v = 1/\sqrt{\kappa\mu_0\epsilon_0}$, where κ is the dielectric constant of the substance. Determine the speed of light in water, which has a dielectric constant of 1.78.

3. An electromagnetic wave in vacuum has an electric field amplitude of 220 V/m. Calculate the amplitude of the corresponding magnetic field.

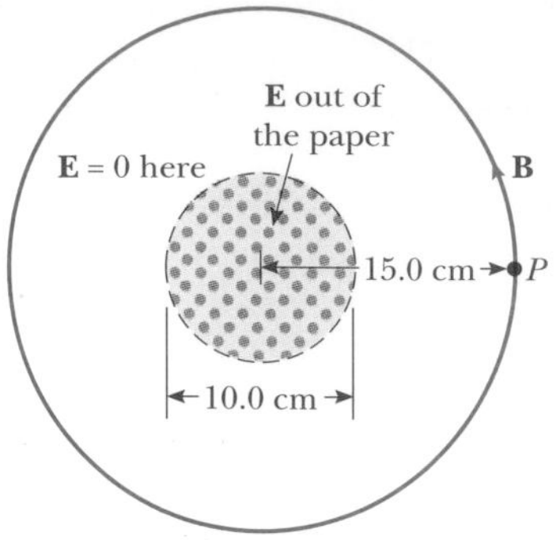

Figure P24.1

4. Calculate the maximum value of the magnetic field in a medium in which the speed of light is two thirds the speed of light in vacuum and the amplitude of the electric field is 7.60 mV/m.

5. Figure 24.4 shows a plane electromagnetic sinusoidal wave propagating in what is assumed to be the x direction. The wavelength is 50.0 m, and the electric field vibrates in the xy plane with an amplitude of 22.0 V/m. Calculate (a) the sinusoidal frequency and (b) the magnitude and direction of **B** when the electric field has its maximum value in the negative y direction. (c) Write an expression for B in the form

$$B = B_{max} \cos(kx - \omega t)$$

with numerical values for B_{max}, k, and ω.

6. Write down expressions for the electric and magnetic fields of a sinusoidal plane electromagnetic wave having a frequency of 3.00 GHz and traveling in the positive x direction. The amplitude of the electric field is 300 V/m.

7. Verify that the following equations are solutions to Equations 24.12 and 24.13, respectively:

$$E = E_{max} \cos(kx - \omega t) \qquad B = B_{max} \cos(kx - \omega t)$$

8. In SI units, the electric field in an electromagnetic wave is described by

$$E_y = 100 \sin(1.00 \times 10^7 x - \omega t)$$

Find (a) the amplitude of the corresponding magnetic wave, (b) the wavelength λ, and (c) the frequency f.

Section 24.4   Hertz's Discoveries

9. If the coil in the resonant circuit of a radio has an inductance of 2.00 μH, what range of values must the tuning capacitor have in order to cover the complete range of FM frequencies? (The FM range of frequencies is 88.0 MHz to 108 MHz.)

10. A standing wave interference pattern is set up by radio waves between two metal sheets 2.00 m apart. This is the shortest

distance between the plates that will produce a standing wave pattern. What is the fundamental frequency?

Section 24.5   Production of Electromagnetic Waves by an Antenna

11. A television set uses a dipole receiving antenna for VHF channels and a loop antenna for UHF channels. The UHF antenna produces a voltage from the changing magnetic flux through the loop. (a) Using Faraday's law, derive an expression for the amplitude of the voltage that appears in a single-turn circular loop antenna with a radius r, small compared to the wavelength of the wave. The TV station broadcasts a signal with a frequency f, and the signal has an electric field amplitude E_{max} and magnetic field amplitude B_{max} at the receiving antenna's location. (b) If the electric field in the signal points vertically, what should be the orientation of the loop for best reception?

12. Two hand-held radio transceivers with dipole antennas are separated by a large fixed distance. Assuming a vertical transmitting antenna, what fraction of the maximum received power will occur in the receiving antenna when it is inclined from the vertical by (a) 15.0°? (b) 45.0°? (c) 90.0°?

13. Two radio-transmitting antennas are separated by half the broadcast wavelength and are driven in phase with each other. In which directions are (a) the strongest and (b) the weakest signals radiated?

Section 24.6   Energy Carried by Electromagnetic Waves

14. How much electromagnetic energy per cubic meter is contained in sunlight if the intensity of sunlight at the Earth's surface under a fairly clear sky is 1000 W/m²?

15. What is the average magnitude of the Poynting vector 5.00 miles from a radio transmitter broadcasting isotropically with an average power of 250 kW?

16. An AM radio station broadcasts isotropically with an average power of 4.00 kW. A dipole receiving antenna 65.0 cm long is located 4.00 miles from the transmitter. Compute the amplitude of the emf induced by this signal between the ends of the receiving antenna.

17. A community plans to build a facility to convert solar radiation to electrical power. They require 1.00 MW of power, and the system to be installed has an efficiency of 30.0% (that is, 30.0% of the solar energy incident on the surface is converted to electrical energy). What must be the effective area of a perfectly absorbing surface used in such an installation, assuming a constant energy flux of 1000 W/m²?

18. A monochromatic light source emits 100 W of electromagnetic power uniformly in all directions. (a) Calculate the average electric-field energy density 1.00 m from the source. (b) Calculate the average magnetic-field energy density at

the same distance from the source. (c) Find the wave intensity at this location.

19. The filament of an incandescent lamp has a 150-Ω resistance and carries a direct current of 1.00 A. The filament is 8.00 cm long and 0.900 mm in radius. (a) Calculate the magnitude of the Poynting vector at the surface of the filament. (b) Find the magnitude of the electric and magnetic fields at the surface of the filament.

20. Assuming that the antenna of a 10.0-kW radio station radiates spherical electromagnetic waves, compute the maximum value of the magnetic field 5.00 km from the antenna, and compare this value with the magnetic field of the Earth.

21. In a certain region of empty space, the electric field intensity at a given instant is $\mathbf{E} = (80.0\mathbf{i} + 32.0\mathbf{j} - 64.0\mathbf{k})$ N/C and the magnetic field intensity is $\mathbf{B} = (0.200\mathbf{i} + 0.0800\mathbf{j} + 0.290\mathbf{k})$ μT. (a) Show that the two fields are perpendicular to each other. (b) Determine their Poynting vector.

22. A lightbulb's filament has a resistance of 110 Ω. The bulb is plugged into a standard 120-V (rms) outlet and emits 1.00% of the input energy as electromagnetic radiation of frequency f. Assuming that the bulb is covered with a filter, absorbing all other frequencies, find the amplitude of the magnetic field 1.00 m from the bulb.

23. High-power lasers in factories are used to cut through cloth and metal. One such laser has a beam diameter of 1.00 mm and generates an electric field at the target having an amplitude 0.700 MV/m. Find (a) the amplitude of the magnetic field produced, (b) the intensity of the laser, and (c) the power dissipated.

27. A 15.0-mW helium-neon laser ($\lambda = 632.8$ nm) emits a beam of circular cross section the diameter of which is 2.00 mm. (a) Find the maximum electric field in the beam. (b) What total energy is contained in a 1.00-m length of the beam? (c) Find the momentum carried by a 1.00-m length of the beam.

Section 24.8   The Spectrum of Electromagnetic Waves

28. Compute an order-of-magnitude estimate for the frequency of an electromagnetic wave with wavelength equal to (a) your height, (b) the thickness of this sheet of paper. How is each wave classified on the electromagnetic spectrum?

29. The eye is most sensitive to light having a wavelength of 5.50×10^{-7} m, which is in the green–yellow region of the electromagnetic spectrum. What is the frequency of this light?

30. A "rabbit ears" television antenna consists of a pair of adjustable rods. If we set each rod to a quarter-wavelength for channel 3, which has a central frequency of 63.0 MHz, how long are the rods?

31. Suppose you are located 180 m from a radio transmitter. (a) How many wavelengths are you from the transmitter if the station calls itself 1150 AM? (The AM band frequencies are in kHz.) (b) What if this station were 98.1 FM? (The FM band frequencies are in MHz.)

32. A radar pulse returns to the receiver after a total travel time of 4.00×10^{-4} s. How far away is the object that reflected the wave?

Section 24.7   Momentum and Radiation Pressure

24. A plane electromagnetic wave of intensity 6.00 W/m² strikes a small pocket mirror, of area 40.0 cm², held perpendicular to the approaching wave. (a) What momentum does the wave transfer to the mirror each second? (b) Find the force that the wave exerts on the mirror.

25. A radio wave transmits 25.0 W/m² of power per unit area. A flat surface of area A is perpendicular to the direction of propagation of the wave. Calculate the radiation pressure on it if the surface is a perfect absorber.

26. A possible means of space flight is to place a perfectly reflecting aluminized sheet into Earth's orbit and use the light from the Sun to push this "solar sail." Suppose a sail of area 6.00×10^5 m² and mass 6000 kg is placed in orbit facing the Sun. (a) What force is exerted on the sail? (b) What is the sail's acceleration? (c) How long does it take the sail to reach the Moon, 3.84×10^8 m away? Ignore all gravitational effects, assume that the acceleration calculated in part (b) remains constant, and assume a solar intensity of 1340 W/m². (*Hint:* The radiation pressure exerted by a reflected wave is given as twice the average power per area divided by the speed of light.)

Section 24.9   Polarization

33. Unpolarized light passes through two polaroid sheets. The axis of the first is vertical and that of the second is at 30.0° to the vertical. What fraction of the initial light is transmitted?

34. Three polarizing disks, the planes of which are parallel, are centered on a common axis. The direction of the transmission axis in each case is shown in Figure P24.34 relative to the common vertical direction. A plane-polarized beam of light with E_0 parallel to the vertical reference direction is

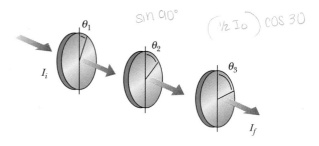

Figure P24.34

incident from the left on the first disk with intensity I_i = 10.0 units (arbitrary). Calculate the transmitted intensity I_f when (a) $\theta_1 = 20.0°$, $\theta_2 = 40.0°$, and $\theta_3 = 60.0°$; (b) $\theta_1 = 0°$, $\theta_2 = 30.0°$, and $\theta_3 = 60.0°$.

35. Plane-polarized light is incident on a single polarizing disk with the direction of $\mathbf{E}_0$ parallel to the direction of the transmission axis. Through what angle should the disk be rotated so that the intensity in the transmitted beam is reduced by a factor of (a) 3.00, (b) 5.00, (c) 10.0?

36. In Figure P24.34, suppose that the left and right polarizing disks have their transmission axes perpendicular to each other. Also, let the center disk be rotated on the common axis with an angular speed ω. Show that if unpolarized light is incident on the left disk with an intensity I_0, the intensity of the beam emerging from the right disk is

$$I = \frac{1}{16} I_0 (1 - \cos 4\,\omega t)$$

This means that the intensity of the emerging beam is modulated at a rate that is four times the rate of rotation of the center disk. [*Hint:* Use the trigonometric identities $\cos^2 \theta = (1 + \cos 2\theta)/2$ and $\sin^2 \theta = (1 - \cos 2\theta)/2$, and recall that $\theta = \omega t$.]

Additional Problems

37. Assume that the intensity of solar radiation incident on the cloudtops of Earth is 1340 W/m². (a) Calculate the total power radiated by the Sun, taking the average Earth–Sun separation to be 1.49×10^{11} m. (b) Determine the maximum values of the electric and magnetic fields at the Earth's location due to solar radiation.

38. Show that the instantaneous value of the Poynting vector has the magnitude

$$S = \frac{c}{2} \left(\epsilon_0 E^2 + \frac{B^2}{\mu_0} \right)$$

39. A dish antenna having a diameter of 20.0 m receives (at normal incidence) a radio signal from a distant source, as shown in Figure P24.39. The radio signal is a continuous sinusoidal wave with amplitude $E_{max} = 0.200\ \mu V/m$. Assume the antenna absorbs all the radiation that falls on the dish. (a) What is the amplitude of the magnetic field in this wave? (b) What is the intensity of the radiation received by this antenna? (c) What is the power received by the antenna? (d) What force is exerted on the antenna by the radio waves?

40. The intensity of solar radiation at the top of the Earth's atmosphere is 1340 W/m². Assuming that 60% of the incoming solar energy reaches the Earth's surface and assuming that you absorb 50% of the incident energy, make an order-of-magnitude estimate of the amount of solar energy you absorb in a 60-minute sunbath.

Figure P24.39

41. In 1965, Penzias and Wilson discovered the cosmic microwave radiation left over from the Big Bang explosion of the Universe. Suppose the energy density of this radiation is 4.00×10^{-14} J/m³. Determine the corresponding electric field amplitude.

42. Consider a small, spherical particle of radius r located in space a distance R from the Sun. (a) Show that the ratio $F_{rad}/F_{grav} \propto 1/r$, where F_{rad} = the force exerted by solar radiation and F_{grav} = the force of gravitational attraction. (b) The result of part (a) means that, for a sufficiently small value of r, the force exerted on the particle by solar radiation exceeds the force of gravitational attraction. Calculate the value of r for which the particle is in equilibrium under the two forces. (Assume that the particle has a perfectly absorbing surface and a mass density of 1.50 g/cm³. Let the particle be located 3.75×10^{11} m from the Sun and use 214 W/m² as the value of the solar flux at that point.)

43. A linearly polarized microwave of wavelength 1.50 cm is directed along the positive x axis. The electric field vector has a maximum value of 175 V/m and vibrates in the xy plane. (a) Assume that the magnetic field component of the wave can be written in the form $B = B_{max} \sin(kx - \omega t)$ and give values for B_{max}, k, and ω. Also, determine in which plane the magnetic field vector vibrates. (b) Calculate the magnitude of the Poynting vector for this wave. (c) What maximum radiation pressure would this wave exert if directed at normal incidence onto a perfectly reflecting sheet? (d) What maximum acceleration would be imparted to a 500-g sheet (perfectly reflecting and at normal incidence) of dimensions 1.00 m × 0.750 m?

44. An astronaut, stranded in space 10.0 m from her spacecraft and at rest relative to it, has a mass (including equipment) of 110 kg. Having a 100-W light source that forms a directed beam, she decides to use the beam as a photon rocket to propel herself continuously toward the spacecraft. (a) Calculate how long it takes her to reach the spacecraft by this method. (b) Suppose she instead decides to throw

the light source away in a direction opposite the spacecraft. If the mass of the light source is 3.00 kg and, after being thrown, moves at 12.0 m/s relative to the recoiling astronaut, how long does she take to reach the spacecraft?

45. You want to rotate the plane of polarization of a polarized light beam by 45.0° with a maximum intensity reduction of 10.0%. (a) How many sheets of perfect polarizers do you need to achieve your goal? (b) What is the angle between adjacent polarizers?

46. Lasers have been used to suspend spherical glass beads in the Earth's gravitational field. (a) If a bead has a mass of 1.00 μg and a density of 0.200 g/cm^3, determine the radiation intensity needed to support the bead. (b) If the laser beam has a radius of 0.200 cm, what is the power required for this laser?

47. Lasers have been used to suspend spherical glass beads in the Earth's gravitational field. (a) If a bead has mass m and density ρ, determine the radiation intensity needed to support the bead. (b) If the beam has a radius r, what is the power required for this laser?

48. A singer's voice is transmitted by a radio wave to a person 100 km away. (a) How much time does it take for the radio wave to travel through the air to the person? (b) By the time the radio message reaches a listener, how far from the singer has the sound wave moved in the auditorium? Assume that the speed of sound is 345 m/s.

49. The Earth reflects approximately 38.0% of the incident sunlight by reflection from its clouds and oceans. (a) Given that the intensity of solar radiation is 1340 W/m^2, what is the radiation pressure on the Earth in Pa when the Sun is straight overhead? (b) Compare this to normal atmospheric pressure, which is 1.01×10^5 Pa, at the Earth's surface.

50. A 1.00 m diameter mirror focuses the Sun's rays onto an absorbing plate 2.00 cm in radius, which holds a can containing 1.00 liter of water at 20.0°C. (a) If the solar intensity is 1.00 kW/m^2, what is the intensity on the absorbing plate? (b) What are the **E** and **B** field amplitudes? (c) If 40.0% of the energy is absorbed, how long would it take to bring the water to its boiling point?

ANSWERS TO CONCEPTUAL PROBLEMS

1. An antenna that is a conducting line responds to the electric field of the electromagnetic wave—the oscillating electric field causes an electric force on electrons in the wire along its length. The movement of electrons along the wire is detected as a current by the radio, demodulated, and amplified. Thus, a line antenna must have the same orientation as the broadcasting antenna. A loop antenna responds to the magnetic field in the radio wave. The varying magnetic field induces a varying current in the loop, and this signal is demodulated and amplified. The axis of the loop antenna must be perpendicular to the broadcasting antenna, so that the magnetic field lines cut through the area of the loop.

2. The flashing of the light according to Morse code is a drastic amplitude modulation—the amplitude is changing from a maximum to zero. In this sense, it is similar to the on-and-off binary code used in computers and compact disks. The carrier frequency is that of the light, on the order of 10^{14} Hz. The signal frequency depends on the skill of the signal operator, but it is on the order of single hertz, as the light is flashed on and off. The broadcasting antenna for this modulated signal is the filament of the lightbulb in the signal source. The receiving antenna is the eye.

3. The sail should be as reflective as possible, so that the maximum momentum is transferred to the sail from the reflection of sunlight.

4. Assuming water does not preferentially absorb one wavelength over another, the color will not change for two reasons. First, despite the popular statement that color depends on wavelength, it actually depends on the *frequency* of the light, which does not change under water. Second, when the light enters the eye, it travels through the fluid within the eye. Thus, even if color did depend on wavelength, the important wavelength is that of the light in the ocular fluid, which does not depend on the medium through which the light traveled to reach the eye.

5. Radio waves originate from antennas, which tend to be long vertical towers. This results in a natural polarization of the radio waves in the vertical direction. Light originates from the vibrations of atoms, or electronic transitions within atoms, which represent oscillations in all possible directions. Thus, light is not generally polarized.

25

Reflection and Refraction
of Light

25.1 • THE NATURE OF LIGHT

Until the beginning of the 19th century, light was thought to be a stream of particles that was emitted by a light source and stimulated the sense of sight on entering the eye. The chief architect of this particle theory of light was Isaac Newton. The theory provided a simple explanation of some known experimental facts concerning the nature of light—namely, the laws of reflection and refraction.

Most scientists accepted the particle theory of light. However, during Newton's lifetime, another theory was proposed—that light might be some sort of wave motion. In 1678, a Dutch physicist and astronomer, Christian Huygens (1629–1695), showed that a wave theory of light could also explain the laws of reflection and refraction. The wave theory did not receive immediate acceptance, for several reasons. All the waves known at the time (sound, water, and so on) traveled through some sort of medium, but light from the Sun could travel to us through empty space. Furthermore, it was argued that if light were some form of wave motion, the waves would be able to bend around obstacles; hence, we should be able to see around corners. It is now known that light does indeed bend around the edges of objects. This phenomenon, known as **diffraction,** is not easy to observe, because

Photograph of a rainbow over Niagara Falls in Ontario, Canada. Rainbows are often seen by an observer located between the Sun and a rain shower. The colors of the visible spectrum that we see in a rainbow are due to the dispersion and refraction of light by raindrops. *(John Edwards/Tony Stone Images)*

light waves have such short wavelengths. Even though experimental evidence for the diffraction of light was discovered by Francesco Grimaldi (1618–1663) around 1660, most scientists for more than a century rejected the wave theory and adhered to Newton's particle theory. This was, for the most part, due to Newton's great reputation as a scientist.

The first clear demonstration of the wave nature of light was provided in 1801 by Thomas Young (1773–1829), who showed that, under appropriate conditions, light exhibits interference behavior. That is, light waves coming from a single source arriving at a point by two different paths can combine and cancel each other by destructive interference, with the energy appearing at some other point. Such behavior could not be explained at that time by a particle theory, because scientists thought that there was no conceivable way in which two or more particles could come together and cancel one another. Several years later, a French physicist, Augustin Fresnel (1788–1829), performed a number of experiments on interference and diffraction. In 1850 Jean Foucault (1791–1868) provided further evidence of the inadequacy of the particle theory by showing that the speed of light in liquids is less than that in air; according to the particle model, the speed of light would be *higher* in glasses and liquids than in air. Additional developments during the 19th century led to the general acceptance of the wave theory of light.

The most important development concerning the theory of light was the work of James Clerk Maxwell, who in 1865 predicted that light was a form of high-frequency electromagnetic wave. As discussed in Chapter 24, Hertz in 1887 provided experimental confirmation of Maxwell's theory by producing and detecting electromagnetic waves. Furthermore, Hertz and other investigators showed that **these waves exhibited reflection, refraction, and all the other characteristic properties of waves.**

Although the classical theory of electricity and magnetism could explain most known properties of light, some experiments could not be accounted for by the assumption that light was a wave. The most striking of these was the photoelectric effect, discovered by Hertz, in which electrons are ejected from a metal when the surface of the metal is exposed to light. As one example of the difficulties that arose, experiments showed that the kinetic energy of an ejected electron is independent of the light intensity. This finding contradicted the wave theory, which held that a more intense beam of light should add more energy to the electron. An explanation proposed by Einstein in 1905 used the concept of quantization developed in 1900 by Max Planck (1858–1947). Einstein's prediction that the kinetic energy of the ejected electrons would depend on the frequency of the radiation, but not on the intensity, was initially rejected. However, experiments performed between 1905 and 1912 eventually led to the acceptance of Einstein's theory. The quantization model assumes that the energy of a light wave is present in bundles of energy called *photons;* hence, the energy is said to be *quantized.* More specifically, Einstein showed that the energy of a photon is proportional to the frequency of the electromagnetic wave. That is,

$$E = hf \qquad\qquad [25.1] \qquad \bullet \ \textit{Energy of a photon}$$

where $h = 6.63 \times 10^{-34}$ J·s is Planck's constant. This theory retains some features of both the wave theory and the particle theory of light. As we shall discuss later, the photoelectric effect is the result of energy transfer from a single photon to an electron in the metal. That is, the electron interacts with one photon of light as if it (the electron) had been struck by a particle. Yet this photon has wave-like characteristics, because its energy is determined by the frequency (wave-like quantity).

In view of these developments, light must be regarded as having a dual nature: **In some cases light acts like a wave and in others it acts like a particle.** Classical electromagnetic wave theory provides an adequate explanation of light propagation and interference, whereas the photoelectric effect and other experiments involving the interaction of light with matter are best explained by assuming that light is a particle. Light is light, to be sure. The question, "Is light a wave or a particle?" is inappropriate; in some experiments, we measure its wave properties; in other experiments we measure its particle properties.

25.2 • THE RAY APPROXIMATION IN GEOMETRIC OPTICS

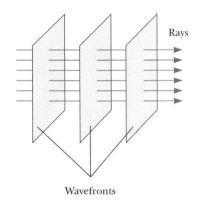

Figure 25.1 A plane wave propagating to the right. Note that the rays, which always point in the direction of wave motion, are straight lines perpendicular to the wave fronts.

A wavefront is a surface over which the wave displacement has the same value. For example, a point source of light waves generates spherical waves moving in all directions. The wavefronts in this case are concentric spherical surfaces. One wavefront is the surface where the wave displacement has its maximum value, corresponding to the crest of the wave. In studying geometric optics here and in Chapter 26, we shall use what is called the **ray approximation.** To understand it, first note that the rays of a given wave are straight lines that are perpendicular to the wavefronts, as illustrated in Figure 25.1 for a plane wave. In the ray approximation, we assume that a wave moving through a medium travels in a straight line in the direction of its rays.

If the wave meets a barrier containing a circular opening the diameter of which is large relative to the wavelength, as in Figure 25.2a, the wave emerging from the opening continues to move in a straight line (apart from some small edge effects); hence, the ray approximation continues to be valid. If the diameter of the opening is on the order of the wavelength, as in Figure 25.2b, the waves (and, as a consequence, the rays we draw) spread out from the opening in all directions. We say that the outgoing wave is noticeably diffracted. If the opening is small relative to the wavelength, the opening can be approximated as a point source of waves (Fig. 25.2c). Thus, diffraction is more pronounced as the ratio d/λ approaches zero.

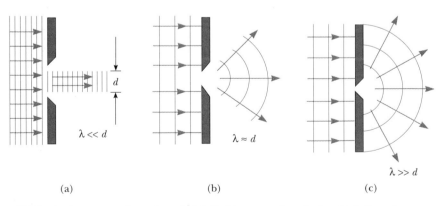

|(a)|(b)|(c)|

Figure 25.2 A plane wave of wavelength λ is incident on a barrier in which there is an opening of diameter d. (a) When $\lambda \ll d$, there is almost no observable diffraction and the ray approximation remains valid. (b) When $\lambda \approx d$, diffraction becomes significant. (c) When $\lambda \gg d$, the opening behaves as a point source emitting spherical waves.

Similar effects are seen when waves encounter an opaque circular object. In this case, when $\lambda \ll d$, the object casts a sharp shadow.

The ray approximation and the assumption that $\lambda \ll d$ are used here and in Chapter 26, which also deals with geometric optics. This approximation is very good for the study of mirrors, lenses, prisms, and associated optical instruments, such as telescopes, cameras, and eyeglasses. We return to the subject of diffraction (where $\lambda \approx d$) in Chapter 27.

25.3 • REFLECTION AND REFRACTION

Reflection of Light

Figure 25.3a shows several rays of a beam of light incident on a smooth, reflecting surface. The reflected rays are parallel to each other, as indicated in the figure. Reflection of light from such a smooth surface is called **specular reflection.** If the reflecting surface is rough, as in Figure 25.3b, it will reflect the rays in various directions. Reflection from any rough surface is known as **diffuse reflection.** A surface behaves as a smooth surface as long as the surface variations are small compared with the wavelength of the incident light. Figure 25.3c and 25.3d are photographs of specular reflection and diffuse reflection using laser light.

Consider the two types of reflection from a road surface that you observe when you drive at night. On a dry night, light from oncoming vehicles is scattered off the road in different directions (diffuse reflection) and the road is quite visible. On a

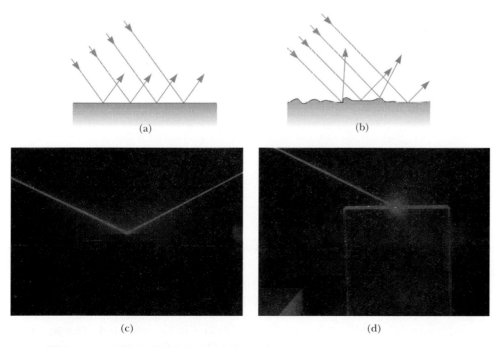

Figure 25.3   Schematic representation of (a) specular reflection, where the reflected rays are all parallel to each other, and (b) diffuse reflection, where the reflected rays travel in random directions. (c) and (d) Photographs of specular and diffuse reflection using laser light.   *(Photographs courtesy of Henry Leap and Jim Lehman)*

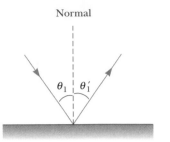

Normal

θ_1 | θ_1'

Figure 25.4   According to the law of reflection, $\theta_1 = \theta_1'$. The incident ray, the reflected ray, and the normal all lie in the same plane.

rainy night, all the little irregularities in the road surface are filled with water. Because the water surface is quite smooth, the light undergoes specular reflection and the glare from reflected light makes the road less visible. We shall concern ourselves only with specular reflection, and we shall use the term *reflection* to mean specular reflection.

Consider a light ray that is traveling in air and incident at an angle on a flat, smooth surface, as in Figure 25.4. The incident and reflected rays make angles of θ_1 and θ_1', respectively, with a line drawn normal to the surface at the point where the incident ray strikes the surface. Experiments show that **the angle of reflection equals the angle of incidence:**

$$\theta_1' = \theta_1 \qquad\qquad [25.2]$$

By convention, the angles of incidence and reflection are measured from the normal to the surface rather than from the surface itself.

Example 25.1   The Double Reflected Light Ray

Two mirrors make an angle of 120° with each other, as in Figure 25.5. A ray is incident on mirror M_1 at an angle of 65° to the normal. Find the direction of the ray after it is reflected from mirror M_2.

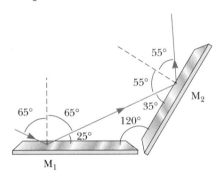

Figure 25.5   (Example 25.1) Mirrors M_1 and M_2 make an angle of 120° with each other.

Reasoning and Solution   From the law of reflection, we know that the first reflected ray also makes an angle of 65° with the normal. Thus, this ray makes an angle of 90° − 65°, or 25°, with the horizontal. From the triangle made by the first reflected ray and the two mirrors, we see that the first reflected ray makes an angle of 35° with M_2 (because the sum of the interior angles of any triangle is 180°). This means that this ray makes an angle of 55° with the normal to M_2. Hence, from the law of reflection, the second reflected ray makes an angle of 55° with the normal to M_2. By comparing the direction of the ray incident on M_1 with its direction after reflecting from M_2, we see that the ray is reflected through 120°, which corresponds to the angle between the mirrors.

Refraction of Light

When a ray of light traveling through a transparent medium is obliquely incident on a boundary leading into another transparent medium, as in Figure 25.6a, part of the ray is reflected but part enters the second medium. The ray that enters the second medium is bent at the boundary and is said to be **refracted. The incident ray, the reflected ray, and the refracted ray all lie in the same plane.** The **angle of refraction,** θ_2 in Figure 25.6, depends on the properties of the two media and on the angle of incidence, through the relationship

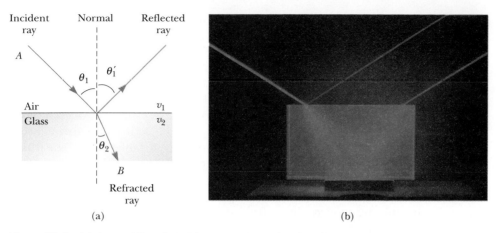

(a)   (b)

Figure 25.6   (a) A ray obliquely incident on an air–glass interface. The refracted ray is bent toward the normal because $v_2 < v_1$. All rays and the normal lie in the same plane. (b) Light incident on the Lucite block bends both when it enters the block and when it leaves the block. *(Henry Leap and Jim Lehman)*

$$\frac{\sin\theta_2}{\sin\theta_1} = \frac{v_2}{v_1} = \text{constant}$$ [25.3]

where v_1 is the speed of light in medium 1 and v_2 is the speed of light in medium 2. The experimental discovery of this relationship is usually credited to Willebrord Snell (1591–1627) and is therefore known as **Snell's law.**

   The path of a light ray through a refracting surface is reversible. For example, the ray in Figure 25.6 travels from point A to point B. If the ray originated at B, it would follow the same path to reach point A. In the latter case, however, the reflected ray would be in the glass.

   When light moves from a material in which its speed is high to a material in which its speed is lower, the angle of refraction, θ_2, is less than the angle of incidence, and so the refracted ray is bent toward the normal, as shown in Figure 25.7a. If the ray moves from a material in which it travels slowly to a material in which it travels more rapidly, θ_2 is greater than θ_1, and so the ray is bent away from the normal, as shown in Figure 25.7b.

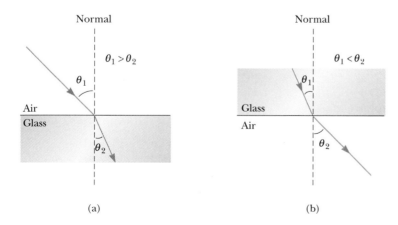

(a)   (b)

Figure 25.7   (a) When the light beam moves from air into glass, its path is bent toward the normal. (b) When the beam moves from glass into air, its path is bent away from the normal.

The pencil partially immersed in water appears bent because light is refracted as it travels across the boundary between water and air. (*Jim Lehman*)

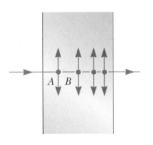

Figure 25.8 Light passing from one atom to another in a medium. The dots are electrons, and the vertical arrows represent their oscillations.

The behavior of light as it passes from air into another substance and then re-emerges into air is often a source of confusion to students. Let us take a look at why this behavior is so different from other occurrences in our daily lives. When light travels in air, its speed is equal to 3×10^8 m/s; on entry into a block of glass, its speed is reduced to approximately 2×10^8 m/s. When the light re-emerges into air, its speed instantaneously increases to its original value, 3×10^8 m/s. This is far different from what happens, for example, when a bullet is fired through a block of wood. In this case, the speed of the bullet is reduced as it moves through the wood because some of its original energy is used to tear apart the fibers of the wood. When the bullet enters the air once again, it emerges at the speed it had just before leaving the block of wood.

To see why light behaves as it does, consider Figure 25.8, which represents a beam of light entering a piece of glass from the left. Once inside the glass, the light may encounter an electron bound to an atom, represented by point *A* in the figure. Let us assume that light is absorbed by the atom, which causes the electron to oscillate (a detail represented by the double-headed arrow in the drawing). The oscillating electron then acts as an antenna and radiates (emits) the beam of light toward an atom at *B*, at which point the light is again absorbed by an atom. The details of these absorptions and emissions are best explained in terms of quantum mechanics, a subject we shall study in Chapter 28. For now, it is sufficient to think of the process as one in which the light passes from one atom to another through the glass. (The situation is somewhat analogous to a relay race in which a baton is passed between runners on the same team.) Although light travels from one atom to another with a speed of 3×10^8 m/s, the absorptions and emissions of light by the atoms cause the overall speed of light in the glass to be lowered. Once the light emerges into the air, the absorptions and emissions cease and the light's speed returns to its original value.

TABLE 25.1   Index of Refraction for Various Substances Measured with Light of Vacuum Wavelength $\lambda_0 = 589$ nm

Substance	Index of Refraction	Substance	Index of Refraction
Solids at 20°C		**Liquids at 20° C**	
Cubic zirconia	2.21	Benzene	1.501
Diamond (C)	2.419	Carbon disulfide	1.628
Fluorite (CaF$_2$)	1.434	Carbon tetrachloride	1.461
Silica (SiO$_2$)	1.458	Ethyl alcohol	1.361
Glass, crown	1.52	Glycerine	1.473
Glass, flint	1.66	Corn syrup	2.21
Ice (H$_2$O)	1.309	Water	1.333
Polystyrene	1.49		
Sodium chloride (NaCl)	1.544	**Gases at 0°C, 1 atm**	
Zircon	1.923	Air	1.000293
		Carbon dioxide	1.00045

The Law of Refraction

Light passing from one medium to another is refracted because the speed of light is different in the two media. In fact, light travels at its maximum speed in vacuum. It is convenient to define the **index of refraction, n,** of a medium to be the ratio

$$n \equiv \frac{\text{speed of light in vacuum}}{\text{speed of light in a medium}} = \frac{c}{v}$$ [25.4]

• *Index of refraction*

From this definition we see that the index of refraction is a dimensionless number greater than or equal to unity, because v in a medium is less than c. Furthermore, n is equal to unity for vacuum. The indices of refraction for various substances are listed in Table 25.1.

   As light travels from one medium to another, its frequency does not change. To see why this is so, consider Figure 25.9. Wavefronts pass an observer at point A in medium 1 with a certain frequency and are incident on the boundary between medium 1 and medium 2. The frequency with which the wavefronts pass an observer at point B in medium 2 must equal the frequency at which they arrive at point A. If this were not the case, either wavefronts would pile up at the boundary or they would be destroyed or created at the boundary. Because there is no mechanism for this to occur, the frequency must be a constant as a light ray passes from one medium into another.

   Therefore, because the relation $v = f\lambda$ must be valid in both media and because $f_1 = f_2 = f$, we see that

$$v_1 = f\lambda_1 \quad \text{and} \quad v_2 = f\lambda_2$$

Because $v_1 \neq v_2$, it follows that $\lambda_1 \neq \lambda_2$. A relationship between index of refraction and wavelength can be obtained by dividing these two equations and making use of the definition of the index of refraction given by Equation 25.4:

$$\frac{\lambda_1}{\lambda_2} = \frac{v_1}{v_2} = \frac{c/n_1}{c/n_2} = \frac{n_2}{n_1}$$ [25.5]

which gives

$$\lambda_1 n_1 = \lambda_2 n_2$$ [25.6]

If medium 1 is vacuum or, for all practical purposes, air, then $n_1 = 1$. Hence, it follows from Equation 25.5 that the index of refraction of any medium can be expressed as the ratio

$$n = \frac{\lambda_0}{\lambda_n}$$ [25.7]

where λ_0 is the wavelength of light in vacuum and λ_n is the wavelength in the medium the index of refraction of which is n. A schematic representation of this reduction in wavelength is shown in Figure 25.10.

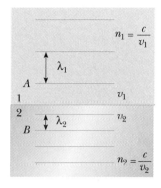

Figure 25.9   As a wave front moves from medium 1 to medium 2, its wavelength changes but its frequency remains constant.

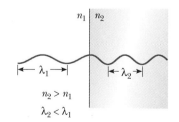

Figure 25.10   Schematic diagram of the *reduction* in wavelength when light travels from a medium of low index of refraction to one of higher index of refraction.

We are now in a position to express Snell's law (Eq. 25.3) in an alternative form. If we combine Equations 25.5 and 25.3, we get

Snell's law of refraction •

$$n_1 \sin \theta_1 = n_2 \sin \theta_2 \qquad\qquad [25.8]$$

This form of Snell's law is practical because it is expressed in terms of n values.

Thinking Physics 1

An observer on the west-facing beach of a large lake is watching the beginning of a sunset. The water is very smooth except for some areas with small ripples. The observer notices that some areas of the water are blue and some are pink. Why does the water appear to be different colors in different areas?

Reasoning The different colors arise from specular and diffuse reflection. The smooth areas of the water will specularly reflect the light from the west, which is the pink light from the sunset. The areas with small ripples will reflect the light diffusely. Thus, light from all parts of the sky will be reflected into the observers' eyes. Because most of the sky is still blue at the beginning of the sunset, these areas will appear to be blue.

Thinking Physics 2

When looking through a glass window to the outdoors at night, you sometimes see a *double* image of yourself. Why?

Reasoning Reflection occurs whenever there is an interface between two optical media. For the glass in the window, there are two such interfaces. The first is the inner surface of the glass and the second is the outer surface. Each of these interfaces results in an image.

CONCEPTUAL PROBLEM 1

Why do face masks make observing clearer under water? A face mask includes a flat piece of glass—there are no lenses in the mask.

CONCEPTUAL PROBLEM 2

As light from the Sun passes through the atmosphere, it refracts, due to the small difference between the speeds of light in air and in vacuum. The *optical* length of the day is defined as the time interval between the instant when the top of the Sun is just visibly observed above the horizon to the instant at which the top of the Sun just disappears below the horizon. The *geometric* length of the day is defined as the time interval between the instant when a geometric straight line drawn from the observer to the actual top of the Sun just clears the horizon to the instant at which this line just dips below the horizon. Which is longer, the optical day or the geometric day?

CONCEPTUAL PROBLEM 3

If a beam of light with a given cross section enters a new medium, the cross section of the refracted beam is different from the incident beam. Is it larger or smaller, or is there no definite direction to the change?

Example 25.2 An Index of Refraction Measurement

A beam of light of wavelength 550 nm traveling in air is incident on a slab of transparent material. The incident beam makes an angle of 40.0° with the normal, and the refracted beam makes an angle of 26.0° with the normal. Find the index of refraction of the material.

Solution Snell's law of refraction (Eq. 25.8) with these data gives

$$n_1 \sin \theta_1 = n_2 \sin \theta_2$$

$$n_2 = \frac{n_1 \sin \theta_1}{\sin \theta_2} = (1.00) \frac{\sin 40.0°}{\sin 26.0°} = \frac{0.643}{0.438} = 1.47$$

If we compare this value with the data in Table 25.1, we see that the material may be silica.

EXERCISE 1 What is the wavelength of light in the material? *Answer* 374 nm

Example 25.3 The Speed of Light in Silica

Light of wavelength 589 nm in vacuum passes through a piece of silica ($n = 1.458$). (a) Find the speed of light in silica.

Solution The speed of light in silica can be easily obtained from Equation 25.4:

$$v = \frac{c}{n} = \frac{3.00 \times 10^8 \text{ m/s}}{1.458} = 2.06 \times 10^8 \text{ m/s}$$

(b) What is the wavelength of this light in silica?

Solution We use Equation 25.7 to calculate the wavelength in silica, noting that we are given the wavelength in vacuum to be $\lambda_0 = 589$ nm:

$$\lambda_n = \frac{\lambda_0}{n} = \frac{589 \text{ nm}}{1.458} - 404 \text{ nm}$$

EXERCISE 2 Find the frequency of the light. *Answer* 5.09×10^{14} Hz

Example 25.4 Light Passing Through a Slab

A light beam passes from medium 1 to medium 2 with the latter being a thick slab of material the index of refraction of which is n_2 (Fig. 25.11). Show that the emerging beam is parallel to the incident beam.

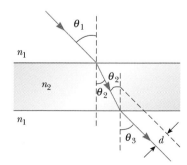

Figure 25.11 (Example 25.4) When light passes through a flat slab of material, the emerging beam is parallel to the incident beam, and therefore $\theta_1 = \theta_3$. The dashed line represents the path the light would take if the slab were not there.

Solution First, let us apply Snell's law to the upper surface:

$$(1) \quad \sin \theta_2 = \frac{n_1}{n_2} \sin \theta_1$$

Applying Snell's law to the lower surface gives

$$(2) \quad \sin \theta_3 = \frac{n_2}{n_1} \sin \theta_2$$

Substituting (1) into (2) gives

$$\sin \theta_3 = \frac{n_2}{n_1} \left(\frac{n_1}{n_2} \sin \theta_1 \right) = \sin \theta_1$$

That is, $\theta_3 = \theta_1$, and so the layer does not alter the direction of the beam as shown by the dashed line. It does, however, produce a displacement of the beam. The same result is obtained when light passes through multiple layers of materials.

25.4 · DISPERSION AND PRISMS

An important property of the index of refraction is that its value in anything but vacuum depends on the wavelength of light. This phenomenon is called **dispersion** (Fig. 25.12). Because n is a function of wavelength, Snell's law indicates that **light rays of different wavelengths are bent at different angles when incident on a refracting material.** As we see from Figure 25.12, the index of refraction for a material decreases with increasing wavelength. This means that blue light ($\lambda \cong 470$ nm) bends more than red light ($\lambda \cong 650$ nm) when passing into a refracting material.

To understand how dispersion can affect light, let us consider what happens when light strikes a prism, as in Figure 25.13a. A single ray of light that is incident on the prism from the left emerges bent away from its original direction of travel by an angle, δ, called the **angle of deviation.** Now suppose a beam of white light (a combination of all visible wavelengths) is incident on a prism, as in Figure 25.13b. Because of dispersion, the blue component of the incident beam is bent more than the red component, and the rays that emerge from the second face of the prism spread out in a series of colors known as a **visible spectrum,** as shown in Figure 25.14. These colors, in order of decreasing wavelength, are red, orange, yellow, green, blue, indigo, and violet. Clearly, the angle of deviation, δ, depends on wavelength. Violet light ($\lambda \cong 400$ nm) deviates the most, red light ($\lambda \cong 650$ nm) deviates the least, and the remaining colors in the visible spectrum fall between these extremes.

The dispersion of light into a spectrum is demonstrated most vividly in nature through the formation of a rainbow, often seen by an observer positioned between the Sun and a rain shower (see page 730). To understand how a rainbow is formed, consider Figure 25.15. A ray of light passing overhead strikes a drop of water in the atmosphere and is refracted and reflected as follows. It is first refracted at the front surface of the drop, with the violet light deviating the most and the red light the least. At the back surface of the drop, the light is reflected and returns to the front

Figure 25.12 Variation of index of refraction with vacuum wavelength for three materials.

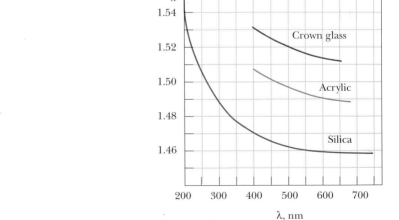

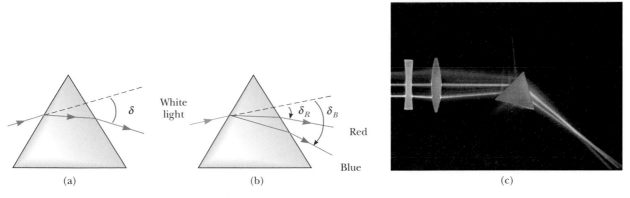

Figure 25.13 (a) A prism refracts single-wavelength light and deviates the light through an angle δ. (b) When light is incident on a prism, the blue light is bent more than the red. (c) Light of different colors passes through a prism and two lenses. As the light passes through the prism, different wavelengths are refracted at different angles. *(David Parker, SPL/Photo Researchers)*

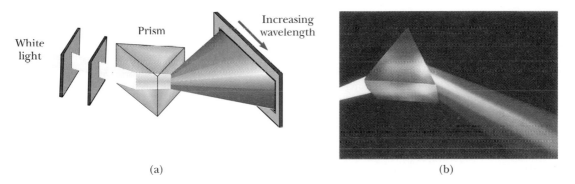

Figure 25.14 (a) Dispersion of white light by a prism. Because *n* varies with wavelength, the prism disperses the white light into its various spectral components. (b) Different colors are refracted at different angles because the index of refraction of the glass depends on wavelength. Violet light deviates the most; red light deviates the least. *(Photograph courtesy of Bausch and Lomb)*

surface, where it again undergoes refraction as it moves from water into air. The rays leave the drop so that the angle between the incident white light and the most intense returning violet light is 40°, and the angle between the white light and the most intense returning red light is 42°. This small angular difference between the returning rays causes us to see the rainbow.

Thinking Physics 3

When a beam of light enters a glass prism, which has nonparallel sides, the rainbow of color exiting the prism is a testimonial to the dispersion occurring in the glass. Suppose a beam of light enters a slab of material with *parallel* sides. When the beam exits the other side, traveling in the same direction as the original beam, is there any evidence of dispersion?

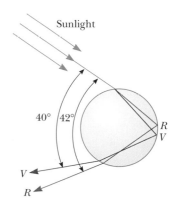

Figure 25.15 Refraction of sunlight by a spherical raindrop.

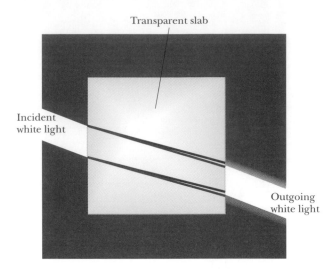

Figure 25.16   (Thinking Physics 3)

Reasoning Due to dispersion, light at the violet end of the spectrum will exhibit a larger angle of refraction on entering the glass than light at the red end. All colors of light will return to the original *direction* of propagation on refracting back out into the air. Thus, the outgoing beam will be predominantly white. But the net shift in the *position* of the violet light along the edge of the slab will be larger than the red light. Thus, one edge of the outgoing beam will have a bluish tinge to it (it will appear blue rather than violet, because the eye is not very sensitive to violet light), whereas the other edge will have a reddish tinge. This effect is indicated in Figure 25.16. It is the colored edges of the outgoing beam of white light that are the evidence for dispersion.

Christian Huygens (1629–1695).
(Courtesy of Rijksmuseum voor de Geschiedenis der Natuurwetenschappen. Courtesy AIP Niels Bohr Library)

CONCEPTUAL PROBLEM 4

In dispersive materials, the angle of refraction for a light ray depends on the wavelength of the light. Does the angle of reflection from the surface of the material depend on the wavelength?

25.5 • HUYGENS'S PRINCIPLE

In this section, we shall discuss a geometric method proposed by Huygens in 1678, which is useful for describing the wave properties of light. Huygens assumed that light was some form of wave motion rather than a stream of particles. He had no knowledge of the nature of light or of its electromagnetic character. Nevertheless, his simplified wave model is adequate for understanding many practical aspects of the propagation of light.

Huygens's principle is a geometric construction for determining the position of a wavefront from a knowledge of an earlier wavefront. In Huygens's construction,

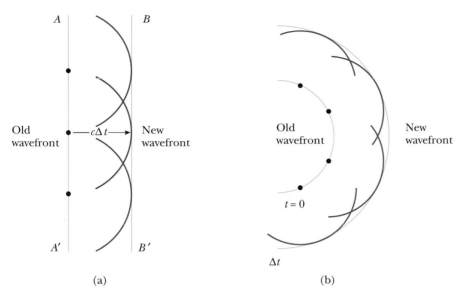

Figure 25.17 Huygens's construction for (a) a plane wave propagating to the right and (b) a spherical wave.

all points on a given wavefront are taken as point sources for the production of spherical secondary waves, called *wavelets*, that propagate outward with speeds characteristic of waves in that medium. After some time has elapsed, the new position of the wavefront is the surface tangent to the wavelets.

Figure 25.17 illustrates two simple examples of Huygens's construction. First, consider a plane wave moving through free space, as in Figure 25.17a. At $t = 0$, the wavefront is indicated by the plane labeled AA'. Each point on this wavefront is a point source for a wavelet. Showing three of these points, we draw circles, each of radius $c \Delta t$, where c is the speed of light in free space and Δt is the time of propagation from one wavefront to the next. The surface drawn tangent to the wavelets is the plane BB', which is parallel to AA'. In a similar manner, Figure 25.17b shows Huygens's construction for an outgoing spherical wave.

A convincing demonstration of the existence of Huygens's wavelets is obtained with water waves in a shallow tank (called a *ripple tank*), as in Figure 25.18. Plane waves produced below the slits emerge above the slits as two-dimensional circular waves propagating outward.

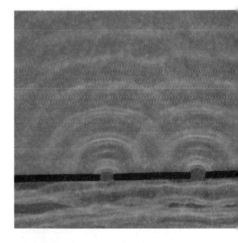

Figure 25.18 Water waves in a ripple tank, which demonstrates Huygens's wavelets. A plane wave is incident on a barrier with two small openings. Each opening acts as a source of circular wavelets. *(Erich Schrempp/Photo Researchers)*

25.6 · TOTAL INTERNAL REFLECTION

An interesting effect called **total internal reflection** can occur when light travels from a medium with a high index of refraction to one with a lower index of refraction. Consider a light beam traveling in medium 1 and meeting the boundary

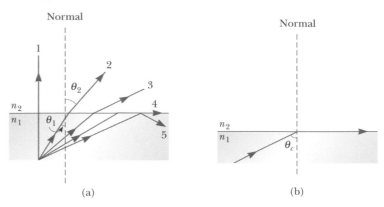

Normal

Normal

(a)   (b)

Figure 25.19   (a) Rays from a medium of index of refraction n_1 travel into a medium of index of refraction n_2, where $n_2 < n_1$. As the angle of incidence increases, the angle of refraction θ_2 increases until θ_2 is 90° (ray 4). For even larger angles of incidence, total internal reflection occurs (ray 5). (b) The angle of incidence producing an angle of refraction equal to 90° is the critical angle, θ_c.

between mediums 1 and 2, where $n_1 > n_2$ (Fig. 25.19). Various possible directions of the beam are indicated by rays 1 through 5. The refracted rays are bent away from the normal because $n_1 > n_2$. (Remember that when light refracts at the interface between the two media, it is also partially reflected. For simplicity, we ignore these reflected rays.) At some particular angle of incidence, θ_c, called the **critical angle,** the refracted light ray will move parallel to the boundary so that $\theta_2 = 90°$ (Fig. 25.19b). For angles of incidence greater than or equal to θ_c, the beam is entirely reflected at the boundary, as is ray 5 in Figure 25.19a. This ray is reflected at the boundary as though it had struck a perfectly reflecting surface. It and all rays like it obey the law of reflection—that is, the angle of incidence equals the angle of reflection.

Critical angle •

We can use Snell's law to find the critical angle. When $\theta_1 = \theta_c$, $\theta_2 = 90°$ and Snell's law (Eq. 25.8) gives

$$n_1 \sin \theta_c = n_2 \sin 90° = n_2$$

$$\sin \theta_c = \frac{n_2}{n_1} \qquad \text{(for } n_1 > n_2\text{)} \qquad \text{[25.9]}$$

This equation can be used only when n_1 is greater than n_2. That is, **total internal reflection occurs only when light travels from a medium of high index of refraction to a medium of lower index of refraction.** If n_1 were less than n_2, Equation 25.9 would give $\sin \theta_c > 1$, which is meaningless because the sine of an angle can never be greater than unity.

The critical angle for total internal reflection is small when n_1 is considerably larger than n_2. Examples of this combination are diamond ($n = 2.42$ and $\theta_c = 24°$) and crown glass ($n = 1.52$ and $\theta_c = 41°$). The angles given are only valid when the material is in air. This property combined with proper faceting causes diamonds and crystal glass to sparkle.

This photograph shows nonparallel light rays entering a glass prism. The bottom two rays undergo total internal reflection at the longest side of the prism. The top three rays are refracted at the longest side as they leave the prism. *(Henry Leap and Jim Lehman)*

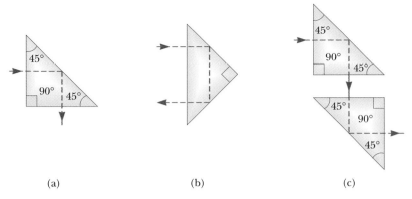

(a)   (b)   (c)

Figure 25.20 Internal reflection in a prism. (a) The ray is deviated by 90°. (b) The direction of the ray is reversed. (c) Two prisms used as a periscope.

A prism and the phenomenon of total internal reflection can be used to alter the direction of travel of a light beam. Two such possibilities are illustrated in Figure 25.20. In one case the light beam is deflected by 90° (Fig. 25.20a), and in the second case the path of the beam is reversed (Fig. 25.20b). A common application of total internal reflection is in a submarine periscope. In this device, two prisms are arranged as in Figure 25.20c so that an incident beam of light follows the path shown and the periscope user is able to "see around corners."

Example 25.5 A View from the Fish's Eye

(a) Find the critical angle for a water–air boundary if the index of refraction of water is 1.33.

Solution Applying Equation 25.9, we find the critical angle to be

$$\sin \theta_c = \frac{n_2}{n_1} = \frac{1}{1.33} = 0.752$$

$$\theta_c = 48.8°$$

(b) Use the results of part (a) to predict what a fish sees when it looks upward toward the water surface at an angle of 40°, 49°, and 60° (Fig. 25.21).

Reasoning Examine Figure 25.19a. Note that because the path of a light ray is reversible, light traveling from medium 2 into medium 1 follows the paths shown in Figure 25.19a, but in the *opposite* direction. This situation is illustrated in Figure 25.21 for the case of a fish looking upward toward the water surface. A fish can see out of the water if it looks toward the surface at an angle less than the critical angle. Thus, at 40°, the fish can see into the air above the water. At an angle of 49°, the critical angle for water, the light that reaches the

fish has to skim along the water surface before being refracted to the fish's eye. At angles greater than the critical angle, the light reaching the fish comes via internal reflection at the surface. Thus, at 60°, the fish sees a reflection of some object on the bottom of the pool.

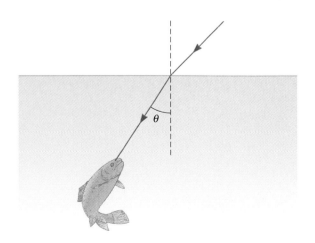

Figure 25.21 (Example 25.5)

Figure 25.22 Light travels in a curved transparent rod by multiple internal reflections.

Strands of glass optical fibers are used to carry voice, video, and data signals in telecommunication networks. Typical fibers have diameters of 60 μm. (*© Richard Megna 1983, Fundamental Photographs*)

Fiber Optics

Another interesting application of total internal reflection is the use of glass or transparent plastic rods to "pipe" light from one place to another. As indicated in Figure 25.22, light is limited to traveling within the rods, even around gentle curves, as the result of successive internal reflections. Such a light pipe can be flexible if thin fibers are used rather than thick rods. If a bundle of parallel fibers is used to construct an optical transmission line, images can be transferred from one point to another.

This technique is used in the industry known as **fiber optics.** Very little light intensity is lost in the fibers as a result of reflections on the sides. Any loss in intensity is due essentially to reflections from the two ends and absorption by the fiber material. Fiber optic devices are particularly useful for viewing an image produced at an inaccessible location. Physicians often use fiber optic cables to aid in the diagnosis and repair of certain medical conditions, without the intrusive effects of major surgery. For example, a fiber optic cable can be threaded through the esophagus and into the stomach to enable the physician to look for ulcers. In this application, the cable actually consists of two fiber optic lines, one to transmit a beam of light into the stomach for illumination and the other to allow this light to be transmitted out of the stomach. The resulting image can be viewed directly by the physician in some cases but most often is displayed on a television monitor or captured on film. The field of fiber optics is also finding increasing use in telecommunications, because the fibers can carry a much higher volume of telephone calls, or other forms of communication, than electrical wires.

Thinking Physics 4

A beam of white light is incident on the curved edge of a semicircular piece of glass, as shown in Figure 25.23. The light enters the curved surface along the normal, so it shows no refraction. It encounters the straight side at the center of curvature of the curved side and refracts into the air. The incoming beam is moved clockwise (so that the angle θ increases) such that the beam always enters along the normal to the curved side and encounters the straight side at the center of curvature of the curved side. As the refracted beam approaches a direction parallel to the straight side, it becomes redder. Why?

Figure 25.23   (Thinking Physics 4)

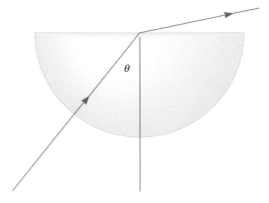

Reasoning When the outgoing beam approaches the direction parallel to the straight side, the incident angle is approaching the critical angle for total internal reflection. But there is dispersion occurring as the light passes through the glass. The index of refraction for light at the violet end of the visible spectrum is larger than at the red end. Thus, as the outgoing beam approaches the straight side, the violet light experiences total internal reflection first, followed by the other colors. The red light is the last to experience total internal reflection, so, just before the outgoing beam disappears, it is composed of light from the red end of the visible spectrum.

Thinking Physics 5

An optical fiber consists of a transparent core, surrounded by a *cladding,* which is a material with a lower index of refraction than the core. There is a cone of angles, called the *acceptance cone,* at the entrance to the fiber. Incoming light at angles within this cone will be transmitted through the fiber, whereas light entering the core from angles outside the cone will not be transmitted. Figure 25.24 shows a light ray entering the fiber just within the acceptance cone and experiencing total internal reflection at the interface between the core and the cladding. If it is technologically difficult to produce light entering the fiber from a small range of angles, how could you adjust the indices of refraction of the core and cladding to increase the size of the acceptance cone would you design them to be farther apart or closer together?

Reasoning The acceptance cone would become larger if the critical angle (θ_c in the diagram) could be made smaller. This can be done by making the index of refraction of the cladding smaller, so that the indices of refraction of the core and cladding would be farther apart.

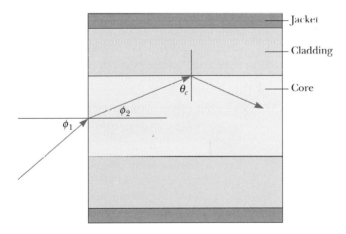

Figure 25.24 (Thinking Physics 5)

SUMMARY

In geometric optics, we use the **ray approximation,** in which we assume that a wave travels through a medium in straight lines in the direction of the rays of that wave. We neglect diffraction effects, which is a good approximation as long as the wavelength is short compared with any aperture dimensions.

The **index of refraction** of a material, n, is defined as

$$n \equiv \frac{c}{v}$$ [25.4]

where c is the speed of light in a vacuum and v is the speed of light in the material. In general, n is related to wavelength as

$$n = \frac{\lambda_0}{\lambda_n}$$ [25.7]

where λ_0 is the wavelength of the light in vacuum and λ_n is its wavelength in the material.

The **law of reflection** states that a wave reflects from a surface so that the angle of reflection, θ_1', equals the angle of incidence, θ_1.

The **law of refraction,** or **Snell's law,** states that

$$n_1 \sin \theta_1 = n_2 \sin \theta_2$$ [25.8]

Huygens's principle states that all points on a wavefront can be taken as point sources for the production of secondary wavelets. At some later time, the new position of the wavefront is the surface tangent to these secondary wavelets.

Total internal reflection can occur when light travels from a medium of high index of refraction to one of lower index of refraction. The minimum angle of incidence, θ_c, for which total reflection occurs at an interface is

$$\sin \theta_c = \frac{n_2}{n_1}$$ (where $n_1 > n_2$)   [25.9]

CONCEPTUAL QUESTIONS

1. You can make a corner reflector by placing three flat mirrors in the corner of a room where the ceiling meets the walls. Show that no matter where you are in the room, you can see yourself reflected in the mirrors—upside down.

2. Several corner reflectors were left on the Moon's Sea of Tranquility by the astronauts of Apollo 11. How can scientists use a laser beam sent from Earth even today to determine the precise distance from the Earth to the Moon?

3. The rectangular aquarium sketched in Figure Q25.3 contains only one goldfish. When the fish is near a corner of the aquarium and is viewed along a direction that makes an equal angle with two adjacent faces, the observer sees two fish mirroring each other, as shown. Explain this observation.

4. As light travels from vacuum ($n = 1$) to a medium such as glass ($n > 1$), does the wavelength of the light change? Does the frequency change? Does the speed change? Explain.

5. Explain why an oar in the water appears bent.

6. Explain why a diamond loses most of its sparkle when submerged in carbon disulfide and why an imitation diamond of cubic zirconia loses all of its sparkle in corn syrup.

7. A laser beam passing through a nonhomogeneous sugar solution follows a curved path. Explain.

8. Light of wavelength λ is incident on a slit of width d. Under what conditions is the ray approximation valid? Under what circumstances does the slit produce enough diffraction to make the ray approximation invalid?

9. A laser beam ($\lambda = 632.8$ nm) is incident on a piece of Lucite, as in Figure Q25.9. Part of the beam is reflected and part is refracted. What information can you get from this photograph?

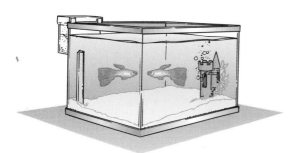

Figure Q25.3

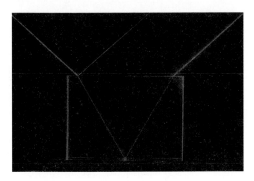

Figure Q25.9 Light from a helium–neon laser beam ($\lambda = 632.8$ nm, red light) is incident on a block of Lucite. The photograph shows both reflected and refracted rays. Can you identify the incident, reflected, and refracted rays? *(Henry Leap and Jim Lehman)*

Figure Q25.18 *(Ron Chapple/FPG)*

10. The level of water in a clear, colorless glass is easily observed with the naked eye. The level of liquid helium in a clear glass vessel is extremely difficult to see with the naked eye. Explain.

11. Describe an experiment in which internal reflection is used to determine the index of refraction of a medium.

12. Why does a diamond show flashes of color when observed under white light?

13. Explain why a diamond sparkles more than a glass crystal of the same shape and size.

14. Why do astronomers looking at distant galaxies talk about looking backward in time?

15. A solar eclipse occurs when the Moon gets between the Earth and the Sun. Use a diagram to show why some areas of the Earth see a total eclipse, other areas see a partial eclipse, and most areas see no eclipse.

16. Figure 25.15 represents sunlight striking a drop of water in the atmosphere. Use the laws of refraction and reflection and the fact that sunlight consists of a wide range of wavelengths to discuss the formation of rainbows.

17. Why does the arc of a rainbow appear with red on top and violet on the bottom?

18. How is it possible that a complete circle of rainbow can sometimes be seen from an airplane (Fig. Q25.18)?

19. Under what conditions is a mirage formed? On a hot day, what are we seeing when we observe "water on the road"?

20. Suppose you are told that only two colors of light (X and Y) are sent through a glass prism and that X is bent more than Y. Which color travels more slowly in the prism?

PROBLEMS

Section 25.3 Reflection and Refraction

(*Note:* In this section if an index of refraction value is not given, refer to Table 25.1.)

1. A narrow beam of sodium yellow light with wavelength 589 nm in vacuum is incident from air onto a smooth water surface at an angle $\theta_1 = 35.0°$. Determine the angle of refraction, θ_2, and the wavelength of the light in water.

2. The wavelength of red helium-neon laser light in air is 632.8 nm. (a) What is its frequency? (b) What is its wavelength in glass that has an index of refraction of 1.50? (c) What is its speed in the glass?

3. An underwater scuba diver sees the Sun at an apparent angle of 45.0° from the vertical. Where is the Sun actually?

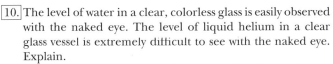

4. A laser beam is incident at an angle of 30.0° to the vertical onto a solution of corn syrup in water. If the beam is refracted to 19.24° to the vertical, (a) what is the index of refraction of the syrup solution? Suppose the light is red, with vacuum wavelength 632.8 nm. Find its (b) wavelength, (c) frequency, and (d) speed in the solution.

5. A ray of light strikes a flat block of glass ($n = 1.50$) of thickness 2.00 cm at an angle of 30.0° with the normal. Trace the light beam through the glass and find the angles of incidence and refraction at each surface.

6. Light of wavelength 436 nm in air enters a fishbowl filled with water, then exits through the crown glass wall of the container. What is the wavelength of the light (a) in the water and (b) in the glass?

7. A cylindrical tank with an open top has a diameter of 3.00 m and is completely filled with water. When the afternoon Sun reaches an angle of 28.0° above the horizon, sunlight ceases to illuminate the bottom of the tank. How deep is the tank?

8. The angle between the two mirrors in Figure P25.8 is a right angle. The beam of light in the vertical plane *P* strikes mirror 1 as shown. (a) Determine the distance the reflected light beam travels before striking mirror 2. (b) In what direction does the light beam travel after being reflected from mirror 2?

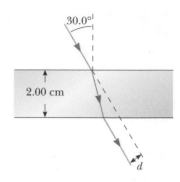

Figure P25.11

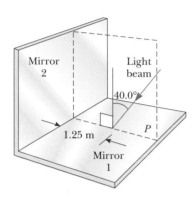

Figure P25.8

9. How many times will the incident beam shown in Figure P25.9 be reflected by each of the parallel mirrors?

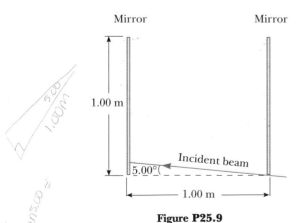

Figure P25.9

10. A beam of light strikes the surface of mineral oil at an angle of 23.1° with the normal to the surface. If the light travels with a speed of 2.17×10^8 m/s through the oil, what is the angle of refraction?

11. When the light in Figure P25.11 passes through the glass block, it is shifted laterally by the distance *d*. If $n = 1.50$, find the value of *d*.

12. Find the time required for the light to pass through the glass block described in Problem 11.

13. The light beam shown in Figure P25.13 makes an angle of 20.0° with the normal line *NN′* in the linseed oil. Determine the angles θ and θ'. (The index of refraction for linseed oil is 1.48.)

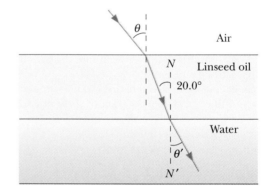

Figure P25.13

14. Two light pulses are emitted simultaneously from a source. Both pulses travel to a detector, but one first passes through 6.20 m of ice. Determine the difference in the pulses' times of arrival at the detector.

15. A piece of wire is bent through an angle θ. The bent wire is partially submerged in benzene (index of refraction = 1.50) so that looking along the dry part, the wire appears to be straight and makes an angle of 30.0° with the horizontal. Determine the value of θ.

16. When you look through a window, by how much time is the light you see delayed by having to go through glass instead of air? Make an order-of-magnitude estimate on the basis of data you specify. By how many wavelengths is it delayed?

Section 25.4 Dispersion and Prisms

17. A narrow white light beam is incident on a block of silica at an angle of 30.0°. Find the angular width of the light beam inside the silica.

18. A ray of light strikes the midpoint of one face of an equiangular glass prism ($n = 1.50$) at an angle of incidence of 30.0°. Trace the path of the light ray through the glass and find the angles of incidence and refraction at each surface.

19. The index of refraction for violet light in silica flint glass is 1.66, and that for red light is 1.62. What is the angular dispersion of visible light passing through a prism of apex angle 60.0° if the angle of incidence is 50.0°? (See Fig. P25.19.)

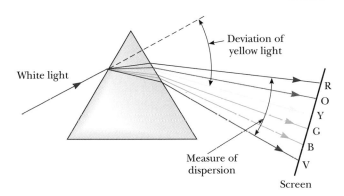

Figure P25.19

20. Light of wavelength 700 nm is incident on the face of a silica prism at an angle of 75.0° (with respect to the normal to the surface). The apex angle of the prism is 60.0°. Use the value of n from Figure 25.12 and calculate the angle (a) of refraction at this first surface, (b) of incidence at the second surface, (c) of refraction at the second surface, and (d) between the incident and emerging rays.

21. A light ray is incident on a prism and refracted at its first surface, as shown in Figure P25.21. Suppose the apex angle of the prism is $\Phi = 60.0°$ and its index of refraction is 1.50. (a) Find the smallest value of the angle of incidence at the first surface for which the refracted ray does not undergo total internal reflection at the second surface. (b) For what angle of incidence θ_1 does the light ray pass symmetrically through the prism and leave at the same angle θ_1?

22. A light ray is incident on a prism and refracted at the first surface, as shown in Figure P25.21. Let Φ represent the apex angle of the prism and n its index of refraction. Find in terms of n and Φ the smallest allowed value of the angle of incidence at the first surface for which the refracted ray does not undergo total internal reflection at the second surface.

23. A 1.00-cm high by 4.00-cm long glass plate is made up of two cemented prisms. The top prism has an index of refraction of 1.486 for blue light and 1.472 for red light. The bottom prism has an index of refraction of 1.878 for blue light and 1.862 for red light. A ray consisting of red and blue light is incident at 50.0° on the top face, as shown in Figure P25.23. Determine the exit angles for both rays that pass through the plate and the angle between them.

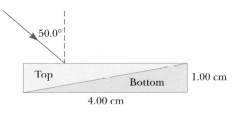

Figure P25.23

Section 25.6 Total Internal Reflection

24. For 589 nm light, calculate the critical angle for the following materials surrounded by air: (a) diamond, (b) flint glass, and (c) ice.

25. Consider a common mirage formed by superheated air just above the roadway. If an observer viewing from 2.00 m above the road (where $n = 1.0003$) has the illusion of seeing water up the road at $\theta_1 = 88.8°$, find the index of refraction of the air just above the road surface. (*Hint:* Treat this as a problem in total internal reflection.)

26. An optical fiber is made of a clear plastic for which the index of refraction is 1.50. For what angles with the surface does light remain contained within the fiber?

27. Determine the maximum angle θ for which the light rays incident on the end of the pipe in Figure P25.27 are subject

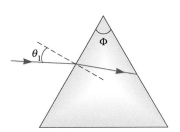

Figure P25.21

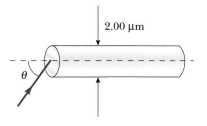

Figure P25.27

to total internal reflection along the walls of the pipe. Assume that the pipe has an index of refraction of 1.36 and the outside medium is air.

28. A glass cube is placed on a newspaper, which rests on a table. A person reads the news through the vertical side of the cube. Determine the maximum index of refraction of the cube.

Additional Problems

29. A glass block having $n = 1.52$ and surrounded by air measures 10.0 cm × 10.0 cm. For an angle of incidence of 45.0°, what is the maximum distance x shown in Figure P25.29 so that the ray emerges from the opposite side?

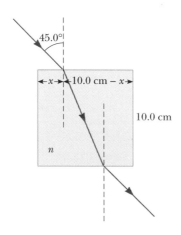

Figure P25.29

30. A light ray enters the atmosphere of a planet and propagates vertically down to the surface 20.0 km below. The index of refraction where the light enters the atmosphere is 1.000 and it increases uniformly to the surface where it has a value of 1.005. (a) How long does it take the ray to traverse this path? (b) Compare this to the time with no atmosphere.

31. A layer of ice floats on water. If light is incident on the upper surface of the ice at an angle of 30.0°, what is the angle of refraction in the water?

32. (a) Consider a horizontal interface between air above and glass of index 1.55 below. Draw a light ray incident from the air at angle of incidence 30.0°. Determine the angles of the reflected and refracted rays and show them on the diagram. (b) Suppose instead that the light ray is incident from the glass at angle of incidence 30.0°. Determine the angles of the reflected and refracted rays and show all three rays on a new diagram. (c) For rays incident from the air onto the air–glass surface, determine and tabulate the angles of reflection and refraction for all the angles of incidence at 10.0° intervals from 0° and 90.0°. (d) Do the

same for light rays coming up to the interface through the glass.

33. A small light source is on the bottom of a swimming pool 1.00 m deep. The light escaping from the water forms a circle on the water surface. What is the diameter of this circle?

34. A light beam is incident on a water surface from air. What is the maximum possible value for the angle of refraction?

35. A narrow beam of light is incident from air onto a glass surface of index of refraction 1.56. Find the angle of incidence for which the corresponding angle of refraction is one half the angle of incidence. (*Hint:* You might want to use the trigonometric identity $\sin 2\theta = 2 \sin \theta \cos \theta$.)

36. Light is incident normally on a 1.00-cm layer of water that lies on top of a flat Lucite plate with a thickness of 0.500 cm. How much longer does light take to pass through this double layer than to traverse the same distance in air? ($n_{\text{Lucite}} = 1.59$.)

37. One technique to measure the apex angle A of a prism is shown in Figure P25.37. A parallel beam of light is directed on the angle so that the beam reflects from opposite sides.

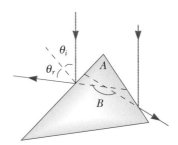

Figure P25.37

Show that the angular separation of the two beams is given by $B = 2A$.

38. The laws of refraction and reflection are the same for sound as for light. The speed of sound in air is 340 m/s and that in water is 1510 m/s. If a sound wave approaches a plane water surface at an angle of incidence of 12.0°, what is the angle of refraction?

39. A layer of kerosene ($n = 1.45$) is floating on water. For what angles of incidence at the kerosene–water interface is light totally internally reflected within the kerosene?

40. When the Sun is directly overhead, a narrow shaft of light enters a cathedral through a small hole in the ceiling and forms a spot on the floor 10.0 m below. (a) At what speed (in centimeters per minute) does the spot move across the (flat) floor? (b) If a mirror is placed on the floor to intercept the light, at what speed does the reflected spot move across the ceiling?

41. A hiker stands on a mountain peak near sunset and observes a rainbow caused by water droplets in the air about 8.00 km away. The valley is 2.00 km below the mountain peak and entirely flat. What fraction of the complete circular arc of the rainbow is visible to the hiker?

42. A fish is at a depth d under water. Show that when viewed from an angle of incidence θ_1, the apparent depth z of the fish is

$$z = \frac{3d \cos \theta_1}{\sqrt{7 + 9 \cos^2 \theta_1}}$$

43. A laser beam strikes one end of a slab of material, as shown in Figure P25.43. The index of refraction of the slab is 1.48. Determine the number of internal reflections of the beam before it emerges from the opposite end of the slab.

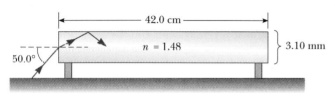

Figure P25.43

44. The prism shown in Figure P25.44 has an index of refraction of 1.55. Light is incident at an angle of 20.0°. Determine the angle θ at which the light emerges.

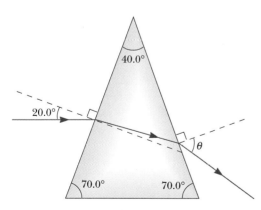

Figure P25.44

45. The light beam in Figure P25.45 strikes surface 2 at the critical angle. Determine the angle of incidence, θ_1.

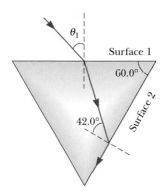

Figure P25.45

46. Figure P25.46 shows a top view of a square enclosure. The inner surfaces are plane mirrors. A ray of light enters a small hole in the center of one mirror. (a) At what angle θ must the ray enter in order to exit through the hole after being reflected once by each of the other three mirrors? (b) Are there other values of θ for which the ray can exit after multiple reflections? If so, make a sketch of one of the ray's paths.

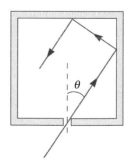

Figure P25.46

47. Students allow a narrow beam of laser light to strike a water surface. They arrange to measure the angle of refraction for selected angles of incidence and record the data shown in the accompanying table. Use the data to verify Snell's law of refraction by plotting the sine of the angle of incidence versus the sine of the angle of refraction. Use the resulting plot to deduce the index of refraction of water.

Angle of Incidence (degrees)	Angle of Refraction (degrees)
10.0	7.5
20.0	15.1
30.0	22.3
40.0	28.7
50.0	35.2
60.0	40.3
70.0	45.3
80.0	47.7

48. A 4.00-m-long pole stands vertically in a lake having a depth of 2.00 m. When the Sun is 40.0° above the horizontal, determine the length of the pole's shadow on the bottom of the lake. Take the index of refraction for water to be 1.33.

49. A light ray of wavelength 589 nm is incident at an angle θ on the top surface of a block of polystyrene, as shown in Figure P25.49. (a) Find the maximum value of θ for which the refracted ray undergoes total internal reflection at the left vertical face of the block. Repeat the calculation for the case in which the polystyrene block is immersed in (b) water and (c) carbon disulfide.

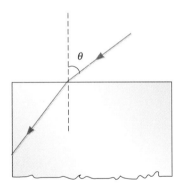

50. A ray of light passes from air into water. In order for its deviation angle $\delta = |\theta_1 - \theta_2|$ to be 10.0°, what must be its angle of incidence? You may need to use a calculator.

51. A drinking glass is 4.00 cm wide, as shown in Figure P25.51. When an observer's eye is placed as shown, the observer sees the edge of the bottom of the glass. When this glass is filled with water, an observer at the same position relative to the glass sees the center of the bottom of the glass. Find the height of the glass.

Spreadsheet Problem

S1. Spreadsheet 25.1 calculates the angle of refraction and angle of deviation as light beams of different incidence angles travel through a prism of apex angle Φ. Input parameters are the index of refraction n and the apex angle Φ (Fig. S25.1). As the angle of incidence θ_1 increases from zero, the angle of deviation δ decreases, reaches a minimum value, and then increases. At minimum deviation

$$n \sin \frac{\Phi}{2} = \sin \frac{\Phi + \delta_{min}}{2} = \sin \theta_1$$

This equation is often used to experimentally find n of any transparent prism. (a) For crown glass ($n = 1.52$ and $\Phi = 60°$), find the minimum angle of deviation from the graph. Using the spreadsheet, verify these equations. At the angle of minimum deviation, how is angle ρ_1 related to ρ_2? (b) Because the index of refraction depends slightly on wavelength, vary n between 1.50 and 1.55, and note how the minimum angle of deviation changes with wavelength. (c) Choose several different values of Φ and note the minimum angles of deviation for each.

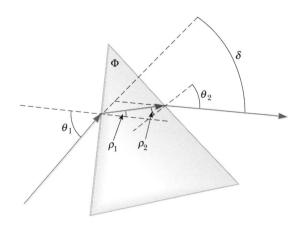

ANSWERS TO CONCEPTUAL PROBLEMS

1. The refraction necessary for focused viewing in the eye occurs at the air–cornea interface. The lens of the eye only performs some fine tuning of this image, allowing for accommodation for objects at various distances. When the eye is opened underwater, the interface is water–cornea, rather than the normal air–cornea for which we are designed. Thus, the image of the scene is not focused on the retina, and the scene is blurry. The face mask simply provides a layer of air in front of the eyes, so that air–cornea interface is reestablished, and the refraction is correct to form the image on the retina.

2. The optical day is longer than the geometric day. Due to the refraction of light by air, light rays from the Sun deviate slightly downward toward the surface of the Earth as the light enters the atmosphere. Thus, in the morning, light rays from the upper edge of the Sun will arrive at your eyes before the geometric line from your eyes to the top of the Sun clears the horizon. In the evening, light rays from the top of the Sun will continue to arrive at your eyes even after the geometric line from your eyes to the top of the Sun dips below the horizon.

3. The cross section can be visualized by considering just the two rays of light on the edges of the beam. If the beam of

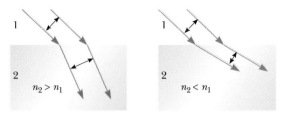

Change in cross section of a light beam on entering a new medium.

light enters a new medium with a higher index of refraction, the rays bend toward the normal, and the cross-section of the refracted beam will be larger than that of the incident beam, as suggested by the diagram on the left. If the new index of refraction is lower, the rays bend away from the normal, and the cross section of the beam is reduced, as shown on the diagram on the right.

4. There is no dependence of the angle of reflection on wavelength, because the light does not enter deeply into the material during reflection—it reflects from the surface.

26

Mirrors and Lenses

This chapter is concerned with the images formed when spherical waves fall on flat and spherical surfaces. We find that images can be formed by reflection or by refraction and that mirrors and lenses work because of this reflection and refraction. Such devices, commonly used in optical instruments and systems, are discussed in detail. In this chapter, we continue to use the ray approximation and to assume that light travels in straight lines, both steps valid because we are studying the field called **geometric optics.** The field of **wave optics** is discussed in Chapter 27.

26.1 · IMAGES FORMED BY FLAT MIRRORS

The main objective of this chapter is to discuss the manner in which optical elements such as lenses and mirrors form images. We begin our investigation by considering the simplest possible mirror, the flat mirror.

Consider a point source of light placed at O in Figure 26.1, a distance p in front of a flat mirror. The distance p is called the **object distance.** Light rays leave the source and are reflected from the mirror. After reflection, the rays diverge (spread apart), but they appear to the viewer to come from a point I located behind the mirror. Point I is called the **image** of the object at O. Regardless of the system under study, images are al-

Some common navigational instruments. *(left)* A magnifying glass is used to view fine print and small objects. *(middle)* A compass is used to determine directions. *(right)* Binoculars are used to view distant objects.
(Murray Alcosser/The Image Bank)

ways formed in the same way. **Images are formed either at the point where rays of light intersect or at the point at which they appear to originate.** Because the rays in Figure 26.1 appear to originate at *I*, which is a distance of *q* behind the mirror, this is the location of the image. The distance *q* is called the **image distance.**

Images are classified as real or virtual. **A real image is one in which light actually intersects, or passes through, the image point; a virtual image is one in which the light does not really pass through the image point but appears to diverge from that point.** The image formed by the mirror in Figure 26.1 is a virtual image. The images seen in flat mirrors are *always virtual.* Real images can usually be displayed on a screen (as at a movie), but virtual images cannot be displayed on a screen.

We examine the properties of the images formed by flat mirrors by using the simple geometric techniques shown in Figure 26.2. To find out where an image is formed, it is necessary to follow at least two rays of light as they reflect from the mirror. One of those rays starts at *P*, follows the horizontal path *PQ* to the mirror, and reflects back on itself. The second ray follows the oblique path *PR* and reflects as shown. An observer to the left of the mirror would trace the two reflected rays back to the point at which they appear to have originated, point *P′*. A continuation of this process for points other than *P* on the object would result in a virtual image (drawn as a yellow arrow) to the right of the mirror. Because triangles *PQR* and *P′QR* are identical, *PQ = P′Q.* Hence, we conclude that **the image formed by an object placed in front of a flat mirror is as far behind the mirror as the object is in front of the mirror.**

Geometry also shows that the object height, *h*, equals the image height, *h′*. Let us define **magnification, *M*,** as follows:

$$M \equiv \frac{\text{image height}}{\text{object height}} = \frac{h'}{h} \qquad [26.1]$$

This is a general definition of the magnification of any type of mirror. $M = 1$ for a flat mirror because $h' = h$ in this case.

Thus, we conclude that the image formed by a flat mirror has the following properties:

• The image is as far behind the mirror as the object is in front.
• The image is unmagnified, virtual, and upright. (By *upright* we mean that if the object arrow points upward, as in Figure 26.2, so does the image arrow. Later we shall discuss situations in which inverted images are formed.)

Finally, it is interesting to note that a flat mirror produces an image having an apparent right–left reversal. The apparent right–left reversal is actually a front–back reversal. This reversal can be seen by standing in front of a mirror and raising your right hand. The image you see raises its left hand. Likewise, your hair appears to be parted on the opposite side and a mole on your right cheek appears to be on your left cheek.

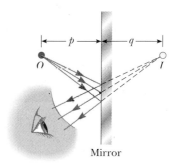

Figure 26.1 An image formed by reflection from a flat mirror. The image point, *I*, is located behind the mirror at a distance *q*, which is equal in magnitude to the object distance, *p*.

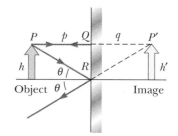

Figure 26.2 Geometric construction used to locate the image of an object placed in front of a flat mirror. Because the triangles *PQR* and *P′QR* are congruent, *p = |q|* and *h = h′*.

Figure 26.3 (Thinking Physics 1)
(Photograph courtesy of Henry Leap and Jim Lehman)

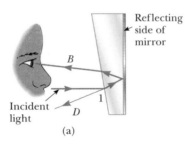

(a)

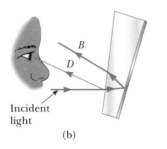

(b)

Figure 26.4 (Thinking Physics 2)

Thinking Physics 1

The professor in the box shown in Figure 26.3 appears to be balancing himself on a few fingers with both of his feet elevated from the floor. The professor can maintain this position for a long time, and he appears to defy gravity. How do you suppose this illusion was created?

Reasoning This is one example of an optical illusion, used by magicians, that makes use of a mirror. The box that the professor is standing in is a cubical frame that contains a flat vertical mirror through a diagonal plane. The professor straddles the mirror so that the foot that you see is in front of the mirror, and the other foot is behind the mirror where you cannot see it. When he raises the foot that you see in front of the mirror, the reflection of this foot also rises, so he appears to float in air.

Thinking Physics 2

Most rearview mirrors on cars have a day setting and a night setting. The night setting greatly diminishes the intensity of the image so that lights from trailing vehicles do not blind the driver. How does such a mirror work?

Reasoning Consider Figure 26.4, which represents a cross-sectional view of the mirror for the two settings. The mirror is a wedge of glass with a reflecting mirror on the back side. When the mirror is in the day setting, as in Figure 26.4a, the light from an object behind the car strikes the mirror at point 1. Most of the light enters the wedge, is refracted, and reflects from the back of the mirror to return to the front surface, where it is refracted again as it reenters the air as ray *B* (for *bright*). In addition, a small portion of the light is reflected at the front surface, as indicated by ray *D* (for *dim*). This dim reflected light is responsible for the image observed when the mirror is in the night setting, as in Figure 26.4b. In this case, the wedge is rotated so that the path followed by the bright light (ray *B*) does not lead to the eye. Instead, the dim light reflected from the front surface travels to the eye, and the brightness of trailing headlights does not become a hazard.

CONCEPTUAL PROBLEM 1

Tape a picture of yourself on a bathroom mirror. Stand several feet away from the mirror. Can you focus your eyes on *both* the picture taped to the mirror *and* your image in the mirror *at the same time*?

CONCEPTUAL PROBLEM 2

In the movies, you sometimes see an actor looking in a mirror. You are able to see the face of the actor. During the filming of this scene, what does the actor see in the mirror?

Example 26.1   Multiple Images Formed by Two Mirrors

Two flat mirrors are at right angles to each other, as in Figure 26.5, and an object is placed at point *O*. In this situation, multiple images are formed. Locate the positions of these images.

Reasoning The image of the object is at I_1 in mirror 1 and at I_2 in mirror 2. In addition, a third image is formed at I_3, which is the image of I_1 in mirror 2 or, equivalently, the image of I_2 in mirror 1. That is, the image at I_1 (or I_2) serves as the

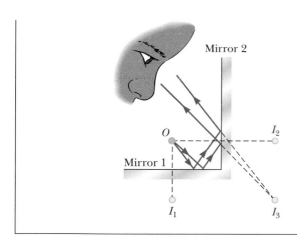

object for I_3. Note that in order to form this image at I_3, the rays reflect twice after leaving the object at O.

EXERCISE 1 Sketch the rays corresponding to viewing the images at I_1 and I_2 and show that the light is reflected only once in these cases.

Figure 26.5 (Example 26.1) When an object is placed in front of two mutually perpendicular mirrors as shown, three images are formed.

26.2 • IMAGES FORMED BY SPHERICAL MIRRORS

Concave Mirrors

A **spherical mirror,** as its name implies, has the shape of a segment of a sphere. Figure 26.6 shows the cross-section of a spherical mirror with light reflecting from its surface, represented by the solid curved line. Such a mirror, in which light is reflected from the inner, concave surface, is called a **concave mirror.** The mirror's radius of curvature is R, and its center of curvature is at point C. Point V is the center of the spherical segment, and a line drawn from C to V is called the **principal axis** of the mirror.

Now consider a point source of light placed at point O in Figure 26.6, on the principal axis and outside point C. Several diverging rays that originate at O are shown. After reflecting from the mirror, these rays converge and meet at I, called the **image point.** They then continue to diverge from I as if there were an object there. As a result, a real image is formed. **Whenever reflected light actually passes through a point, a real image is formed there.**

In what follows, we shall assume that all rays that diverge from an object make a small angle with the principal axis. Such rays, called **paraxial rays,** always reflect through the image point, as in Figure 26.6. Rays that are far from the principal axis, as in Figure 26.7, converge at other points on the principal axis, producing a blurred image. This effect, called **spherical aberration,** is produced to some extent by any spherical mirror and will be discussed in Section 26.5.

We can use the geometry shown in Figure 26.8 to calculate the image distance, q, from the object distance, p, and radius of curvature, R. By convention, these distances are measured from point V. Figure 26.8 shows two rays of light leaving the tip of the object. One passes through the center of curvature, C, of the mirror, hitting the mirror (perpendicular to the mirror surface), and reflecting back on itself. The second ray strikes the mirror at the center, point V, and reflects as shown, obeying the law of reflection. The image of the tip of the arrow is at the point at which these two rays intersect. From the largest right triangle in Figure 26.8 (colored gold), we see that $\tan \theta = h/p$, whereas the light blue triangle gives $\tan \theta = -h'/q$. The negative sign signifies that the image is inverted, so h' is negative. Thus,

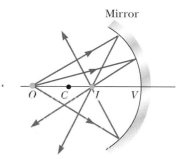

Figure 26.6 A point object at O, outside the center of curvature of a concave spherical mirror, forms a real image at I. If the rays diverge from O at small angles, they all reflect through the same image point.

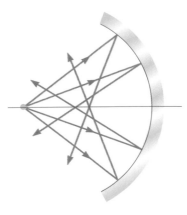

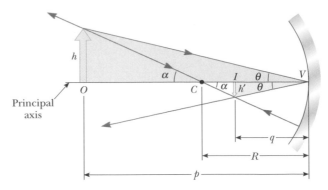

Figure 26.8   The image formed by a spherical concave mirror where the object *O* lies outside the center of curvature *C*.

Figure 26.7   Rays at large angles from the principal axis reflect from a spherical concave mirror to intersect the principal axis at different points, resulting in a blurred image. This is called *spherical aberration.*

from Equation 26.1 and these results, we find that the magnification of the mirror is

$$M = \frac{h'}{h} = -\frac{q}{p} \qquad [26.2]$$

We also note, from two other right triangles in the figure, that

$$\tan \alpha = \frac{h}{p - R} \quad \text{and} \quad \tan \alpha = -\frac{h'}{R - q}$$

from which we find that

$$\frac{h'}{h} = -\frac{R - q}{p - R} \qquad [26.3]$$

If we compare Equations 26.2 and 26.3, we see that

$$\frac{R - q}{p - R} = \frac{q}{p}$$

Simple algebra reduces this to

$$\frac{1}{p} + \frac{1}{q} = \frac{2}{R} \qquad [26.4]$$

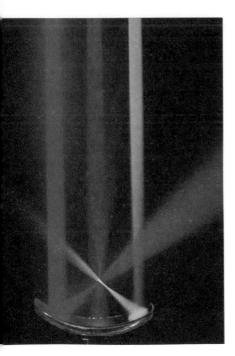

Red, blue, and green light rays are reflected by a parabolic mirror. Note that the focal point at which the three colors meet is white light.   *(Ken Kay/Fundamental Photographs)*

This expression is called the **mirror equation.** It is applicable only to paraxial rays. If the object is very far from the mirror—that is, if the object distance, *p*, is great enough compared with *R* that *p* can be said to approach infinity, then $1/p \approx 0$, and we see from Equation 26.4 that $q \approx R/2$. In other words, **when the object is very far from the mirror, the image point is halfway between the center of curvature and the center of the mirror**, as in Figure 26.9a. The rays are essentially parallel in this figure because their source, the object, is assumed to be very far from the mirror. In this special case, the rays focus at a point called the **focal point** of the mirror, which is a distance *f* from the mirror, which we call the **focal length. The image forms at the focal point only if the incoming rays**

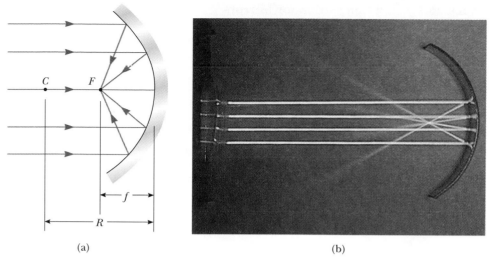

(a)   (b)

Figure 26.9   (a) Light rays from a distant object ($p \approx \infty$) reflect from a concave mirror through the focal point, F. In this case, the image distance $q = R/2 = f$, where f is the focal length of the mirror. (b) Photograph of the reflection of parallel rays from a concave mirror.   *(Henry Leap and Jim Lehman)*

are parallel. The focal length is a parameter associated with the mirror and is given by

$$f = \frac{R}{2}$$

[26.5]   • *Focal length*

The mirror equation can therefore be expressed in terms of the focal length:

$$\frac{1}{p} + \frac{1}{q} = \frac{1}{f}$$

[26.6]   • *Mirror equation*

Convex Mirrors

Figure 26.10 shows the formation of an image by a **convex mirror**—a mirror that is silvered so that light is reflected from the outer, convex surface. Convex mirrors are sometimes called **diverging mirrors** because the rays from any point on a real

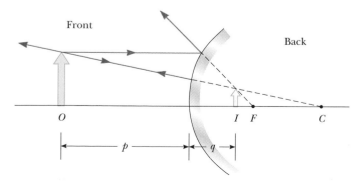

Figure 26.10   Formation of an image by a spherical convex mirror. The image formed by the real object is virtual and upright.

TABLE 26.1   Sign Convention for Mirrors

p is $+$ if the object is in front of the mirror (real object).
p is $-$ if the object is in back of the mirror (virtual object).
q is $+$ if the image is in front of the mirror (real image).
q is $-$ if the image is in back of the mirror (virtual image).
Both f and R are $+$ if the center of curvature is in front of the mirror (concave mirror).
Both f and R are $-$ if the center of curvature is in back of the mirror (convex mirror).
If M is positive, the image is upright.
If M is negative, the image is inverted.

Note: p = object distance; q = image distance; R = radius of curvature, f = focal length; M = magnification.

object diverge after reflection as though they were coming from some point behind the mirror. The image in Figure 26.10 is virtual rather than real because it lies behind the mirror at the point at which the reflected rays appear to originate. In general, as shown in the figure, the image formed by a convex mirror is always upright, virtual, and smaller than the object.

We shall not derive any equations for convex spherical mirrors. If we did, we would find that the equations developed for concave mirrors can be used with convex mirrors if we adhere to a particular sign convention. Let us refer to the region in which light rays move as the *front side* of the mirror, and the other side, where virtual images are formed, as the *back side*. For example, in Figures 26.8 and 26.10, the side to the left of the mirror is the front side, and that to the right of the mirror is the back side. Table 26.1 summarizes the sign conventions for all the necessary quantities.

Ray Diagrams for Mirrors

The positions and sizes of images formed by mirrors can be conveniently determined by constructing **ray diagrams.** These graphical constructions tell us the overall nature of the image and can be used to check parameters calculated from the mirror and magnification equations. To construct a ray diagram, we need to know the position of the object and the location of the center of curvature. Three rays are constructed as shown by the examples in Figure 26.11. All three rays start from the same object point; in these examples, the top of the arrow was chosen. For the concave mirrors in Figure 26.11a and b, the rays are drawn as follows:

- Ray 1 is drawn parallel to the principal axis and is reflected back through the focal point, F.
- Ray 2 is drawn through the focal point. Thus, it is reflected parallel to the principal axis.
- Ray 3 is drawn through the center of curvature, C, and is reflected back on itself.

Note that only two rays are needed in order to locate the image. The third ray serves as a check on your construction. The image point obtained in this fashion must always agree with the value of q calculated from the mirror equation.

Convex cylindrical mirror: reflection of parallel lines. The image of any object in front of the mirror is virtual, upright, and diminished in size. *(© Richard Megna 1990, Fundamental Photographs)*

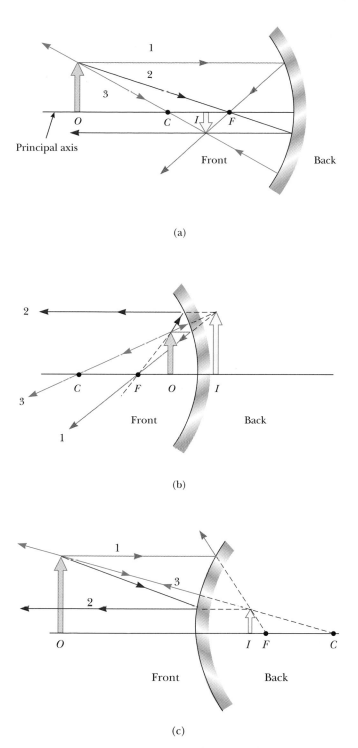

(a)

(b)

(c)

Figure 26.11 Ray diagrams for spherical mirrors. Different colors are used for the three rays so they can easily be followed. (a) An object outside the center of curvature of a spherical concave mirror. (b) An object between the spherical concave mirror and the focal point, *F*, gives a virtual, upright image. (c) An object located anywhere in front of a spherical convex mirror gives a virtual, upright image.

With concave mirrors, note what happens as the object is moved closer to the mirror. The real, inverted image in Figure 26.11a moves to the left as the object approaches the focal point. When the object is at the focal point, the image is infinitely far to the left. However, when the object lies between the focal point and the mirror surface, as in Figure 26.11b, the image is virtual and upright.

For the convex mirror shown in Figure 26.11c, the image of a real object is always virtual and upright. As the object distance increases, the virtual image gets smaller and approaches the focal point as p approaches infinity. You should construct other diagrams to verify how the image position varies with object position.

Thinking Physics 3

For a concave mirror, a virtual image can be anywhere behind the mirror. For a convex mirror, however, there is a maximum distance at which the image can exist behind the mirror. Why?

Reasoning Let us consider the concave mirror first, and imagine two different light rays leaving a tiny object and striking the mirror. If the object is inside the focal point, but infinitesimally close to it, the light rays reflecting from the mirror will be parallel to the mirror axis. They can be interpreted as forming a virtual image infinitely far away behind the mirror. As the object is brought closer to the mirror, the reflected rays will diverge through larger and larger angles, resulting in their extensions converging closer and closer to the back of the mirror. When the object is brought right up to the mirror, the image is just behind the mirror. Conceptually, we can argue that when the object is much closer to the mirror than the focal length, it looks like a flat mirror, so the image is just as far behind the mirror as the object is in front of it. Thus, the image can be anywhere from infinitely far away to right at the surface of the mirror.

For the convex mirror, an object at infinity produces a virtual image at the focal point. As the object is brought closer, the reflected rays diverge more sharply and the image moves closer to the mirror. Thus, the virtual image is restricted to the region between the mirror and the focal point.

Example 26.2 The Image for a Concave Mirror

Assume that a certain concave spherical mirror has a focal length of 10.0 cm. Find the location of the image for object distances of (a) 25.0 cm, (b) 10.0 cm, and (c) 5.00 cm. Describe the image in each case.

Solution (a) For an object distance of 25.0 cm, we find the image distance using the mirror equation:

$$\frac{1}{p} + \frac{1}{q} = \frac{1}{f}$$

$$\frac{1}{25.0 \text{ cm}} + \frac{1}{q} = \frac{1}{10.0 \text{ cm}}$$

$$q = 16.7 \text{ cm}$$

The magnification is given by Equation 26.2:

$$M = -\frac{q}{p} = -\frac{16.7 \text{cm}}{25.0 \text{ cm}} = -0.668$$

The value for M less than unity tells us that the image is smaller than the object. The negative sign for M tells us that the image is inverted. Finally, because q is positive, the image is located on the front side of the mirror and is real. This situation is pictured in Figure 26.11a.

(b) When the object distance is 10.0 cm, the object is located at the focal point. Substituting the values $p = 10.0$ cm and $f = 10.0$ cm into the mirror equation, we find

$$\frac{1}{10.0 \text{ cm}} + \frac{1}{q} = \frac{1}{10.0 \text{ cm}}$$

$$q = \infty$$

Thus, we see that rays of light originating from an object located at the focal point of a mirror are reflected so that the image is formed at an infinite distance from the mirror—that is, the rays travel parallel to one another after reflection.

(c) When the object is at the position $p = 5.00$ cm, it lies between the focal point and the mirror surface. In this case, the mirror equation gives

$$\frac{1}{5.00 \text{ cm}} + \frac{1}{q} = \frac{1}{10.0 \text{ cm}}$$

$$q = -10.0 \text{ cm}$$

That is, the image is virtual because it is located behind the mirror. The magnification is

$$M = -\frac{q}{p} = -\left(\frac{-10.0 \text{ cm}}{5.00 \text{ cm}}\right) = 2.00$$

From this, we see that the image is magnified by a factor of 2, and the positive sign for M indicates that the image is upright (Fig. 26.11b).

Note the characteristics of the images formed by a concave spherical mirror. When the focal point lies between the object and mirror surface, the image is inverted and real; with the object at the focal point, the image is formed at infinity; with the object between the focal point and mirror surface, the image is upright and virtual.

EXERCISE 2 If the object distance is 20.0 cm, find the image distance and the magnification of the mirror.
Answer $q = 20.0$ cm, $M = -1.00$

Example 26.3 **The Image for a Convex Mirror**

An object 3.00 cm high is placed 20.0 cm from a convex mirror having a focal length of 8.00 cm. Find (a) the position of the final image and (b) the magnification.

Solution (a) Because the mirror is convex, its focal length is negative. To find the image position, we use the mirror equation:

$$\frac{1}{p} + \frac{1}{q} = \frac{1}{f} = -\frac{1}{8.00 \text{ cm}}$$

$$\frac{1}{q} = -\frac{1}{8.00 \text{ cm}} - \frac{1}{20.0 \text{ cm}}$$

$$q = -5.71 \text{ cm}$$

The negative value of q indicates that the image is virtual, or behind the mirror, as in Figure 26.11c.

(b) The magnification is

$$M = -\frac{q}{p} = -\left(\frac{-5.71 \text{ cm}}{20.0 \text{ cm}}\right) = 0.286$$

The image is about 30% of the size of the object and upright because M is positive.

EXERCISE 3 Find the height of the image.
Answer 0.857 cm

26.3 • IMAGES FORMED BY REFRACTION

In this section we shall describe how images are formed by the refraction of rays at a spherical surface of a transparent material. Consider two transparent media with indices of refraction n_1 and n_2, where the boundary between the two media is a spherical surface with radius of curvature R (Fig. 26.12). We shall assume that the object at point O is in the medium with index of refraction n_1. Furthermore, let us consider only paraxial rays leaving O. As we shall see, all such rays are refracted at the spherical surface and focus at a single point, I, the image point.

Let us proceed by considering the geometric construction in Figure 26.13, which shows a single ray leaving point O and passing through point I. Snell's law applied to this refracted ray gives

$$n_1 \sin \theta_1 = n_2 \sin \theta_2$$

Because the angles θ_1 and θ_2 are assumed to be small, we can use the approximation $\sin \theta \approx \theta$ (angles in radians). Therefore, Snell's law becomes

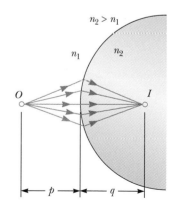

Figure 26.12 An image formed by refraction at a spherical surface. Rays making small angles with the optic axis diverge from a point object at O and pass through the image point, I.

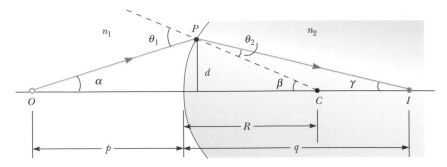

Figure 26.13   Geometry used to derive Equation 26.8.

$$n_1\theta_1 = n_2\theta_2$$

Now we make use of the fact that an exterior angle of any triangle equals the sum of the two opposite interior angles. Applying this to the triangles *OPC* and *PIC* in Figure 26.13 gives

$$\theta_1 = \alpha + \beta$$

$$\beta = \theta_2 + \gamma$$

If we combine the last three equations and eliminate θ_1 and θ_2, we find

$$n_1\alpha + n_2\gamma = (n_2 - n_1)\beta \qquad [26.7]$$

In the small angle approximation, $\tan\theta \approx \theta$, and so we can write the approximate relations

$$\alpha = \frac{d}{p} \qquad \beta = \frac{d}{R} \qquad \gamma = \frac{d}{q}$$

where d is the distance shown in Figure 26.13. We substitute these into Equation 26.7 and divide through by d to give

$$\frac{n_1}{p} + \frac{n_2}{q} = \frac{n_2 - n_1}{R} \qquad [26.8]$$

For a fixed object distance of p, the image distance, q, is independent of the angle that the ray makes with the axis. This tells us that all paraxial rays focus at the same point, *I*. The magnification of a refracting surface is

$$M = \frac{h'}{h} = -\frac{n_1 q}{n_2 p} \qquad [26.9]$$

As we did with mirrors, we must use a sign convention if we are to apply Equation 26.8 to a variety of circumstances. First note that real images are formed on the side of the surface that is *opposite* the side from which the light comes. This is in contrast to mirrors, where real images are formed on the side where the light is. Therefore, **the sign conventions for spherical refracting surfaces are similar to the conventions for mirrors, recognizing the change in sides of the surface for real and virtual images.** For example, in Figure 26.13, p, q, and R are all positive.

TABLE 26.2   Sign Convention for Refracting Surfaces

p is + if the object is in front of the surface (real object).
p is − if the object is in back of the surface (virtual object).

q is + if the image is in back of the surface (real image).
q is − if the image is in front of the surface (virtual image).

R is + if the center of curvature is in back of the surface.
R is − if the center of curvature is in front of the surface.

Note: p = object distance; q = image distance; R = radius of curvature of the surface.

The sign conventions for spherical refracting surfaces are summarized in Table 26.2. The same conventions will be used for thin lenses, discussed in the next section. As with mirrors, we assume that the front of the refracting surface is the side from which the light approaches the surface.

Flat Refracting Surfaces

If the refracting surface is flat, then R approaches infinity and Equation 26.8 reduces to

$$\frac{n_1}{p} = -\frac{n_2}{q}$$

or

$$q = -\frac{n_2}{n_1}p \qquad [26.10]$$

From Equation 26.10 we see that the sign of q is opposite that of p. Thus, **the image formed by a flat refracting surface is on the same side of the surface as the object.** This is illustrated in Figure 26.14 for the situation in which n_1 is greater than n_2, where a virtual image is formed between the object and the surface. Note that the refracted ray bends *away* from the normal in this case, because $n_1 > n_2$.

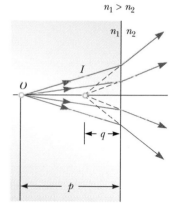

Figure 26.14 The image formed by a flat refracting surface is virtual; that is, it forms to the left of the refracting surface. All rays are assumed to be paraxial.

Example 26.4   Gaze into the Crystal Ball

A coin 2.00 cm in diameter is embedded in a solid glass ball of radius 30.0 cm (Fig. 26.15). The index of refraction of the ball is $n_1 = 1.5$, and the coin is 20.0 cm from the surface. Find the position of the image.

Solution Because $n_1 > n_2$, where $n_2 = 1.00$ is the index of refraction for air, the rays originating from the object are refracted away from the normal at the surface and diverge outward. Hence, the image is formed in the glass and is virtual. Applying Equation 26.8, we get

$$\frac{n_1}{p} + \frac{n_2}{q} = \frac{n_2 - n_1}{R}$$

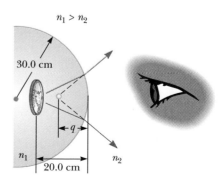

Figure 26.15 (Example 26.4) A coin embedded in a glass ball forms a virtual image between the coin and the glass surface. All rays are assumed to be paraxial.

$$\frac{1.50}{20.0 \text{ cm}} + \frac{1}{q} = \frac{1.00 - 1.50}{-30.0 \text{ cm}}$$

$$q = -17.1 \text{ cm}$$

The negative sign indicates that the image is in the same medium as the object (the side of incident light), in agreement with our ray diagram. Being in the same medium as the object, the image must be virtual.

Example 26.5   The One That Got Away

A small fish is swimming at a depth d below the surface of a pond (Fig. 26.16). What is the apparent depth of the fish as viewed from directly overhead?

Solution   In this example, the refracting surface is flat, and so R is infinite. Hence, we can use Equation 26.10 to determine the location of the image. Using the facts that $n_1 = 1.33$ for water and $p = d$ gives

$$q = -\frac{n_2}{n_1} p = -\frac{1}{1.33} d$$

$$= -0.750d$$

Again, because q is negative, the image is virtual, as indicated in Figure 26.16. The apparent depth is three fourths the actual depth.

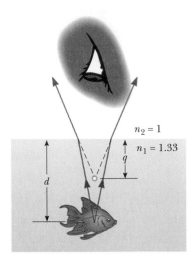

Figure 26.16   (Example 26.5) The apparent depth, q, of the fish is less than the true depth, d. All rays are assumed to be paraxial.

26.4 • THIN LENSES

A typical **thin lens** consists of a piece of glass or plastic, ground so that its two surfaces are either segments of spheres or planes. Lenses are commonly used in optical instruments, such as cameras, telescopes, and microscopes, to form images by refraction. The equation that relates object distances and image distances for a lens is virtually identical to the mirror equation derived earlier, and the method used to derive it is also similar.

Figure 26.17 shows some representative shapes of lenses. Note that we have placed these lenses in two groups. Those in Figure 26.17a are thicker at the center than at the rim, and those in Figure 26.17b are thinner at the center than at the rim. The lenses in the first group are examples of **converging lenses,** and those in the second group are called **diverging lenses.** The reason for these names will become apparent shortly.

As we did with mirrors, it is convenient to define a point called the **focal point** for a lens. For example, in Figure 26.18a a group of rays parallel to the principal axis passes through the focal point, F, after being converged by the lens. The distance from the focal point to the lens is again called the **focal length, f. The focal length is the image distance that corresponds to an infinite object distance.** Recall that we are considering the lens to be very thin. As a result, it makes no difference whether we take the focal length to be the distance from the focal point to the surface of the lens or the distance from the focal point to the center of the lens, because the difference in these two lengths is negligible. A thin lens has one focal length and *two* focal points, as illustrated in Figure 26.18b, corresponding to parallel light rays traveling from the left or right.

Rays parallel to the axis diverge after passing through a lens of the shape shown in Figure 26.18b. In this case, the focal point is defined as the point at which the diverged rays appear to originate, labeled *F* in the figure. Figures 26.18a and 26.18b indicate why the names *converging* and *diverging* are applied to these lenses.

Consider a ray of light passing through the center of a lens, shown as ray 1 in Figure 26.19. If we apply Snell's law at both surfaces, we find that this ray is deflected from its original direction of travel by a distance of δ, shown in Figure 26.20. Here, in order to avoid the complications arising from this deflection, we shall make what is called the **thin-lens approximation: The thickness of the lens is assumed to be negligible.** As a result, δ becomes vanishingly small and we see that the ray passes through the lens undeflected. Ray 2 in Figure 26.17 is parallel to the principal axis of the lens (the horizontal axis passing through *O*), and as a result it passes through the focal point, *F*, after refraction. The point at which these two rays intersect is the image point.

We first note that the tangent of the angle α can be found by using the shaded triangles in Figure 26.19:

$$\tan \alpha = \frac{h}{p} \qquad \text{or} \qquad \tan \alpha = -\frac{h'}{q}$$

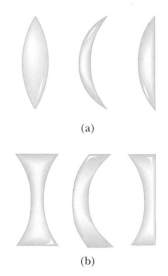

Figure 26.17 Various lens shapes: (a) Converging lenses have a positive focal length and are thickest at the middle. From left to right are biconvex, convex-concave, and plano-convex lenses. (b) Diverging lenses have a negative focal length and are thickest at the edges. From left to right are biconcave, convex-concave, and plano-concave lenses.

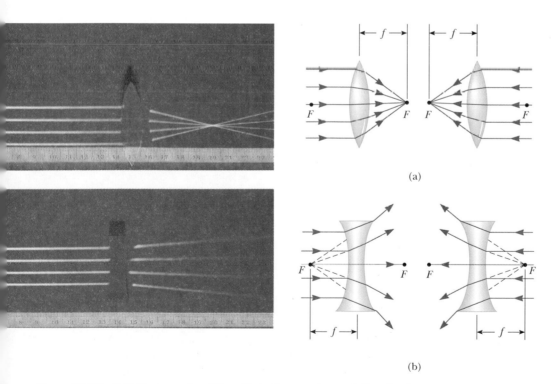

Figure 26.18 *(left)* Photographs of the effect of converging and diverging lenses on parallel rays. *(Henry Leap and Jim Lehman)* *(right)* The object and image focal points of (a) the biconvex lens and (b) the biconcave lens.

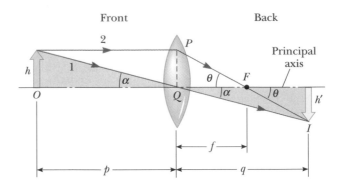

Figure 26.19   A geometric construction for developing the thin-lens equation.

from which

$$M = \frac{h'}{h} = -\frac{q}{p} \qquad \text{[26.11]}$$

Thus, the equation for magnification by a lens is the same as the equation for magnification by a mirror (Eq. 26.2). We also note from Figure 26.19 that the tangent of θ is

$$\tan \theta = \frac{PQ}{f} \qquad \text{or} \qquad \tan \theta = -\frac{h'}{q - f}$$

However, the height PQ is the same as h. Therefore,

$$\frac{h}{f} = -\frac{h'}{q - f}$$

$$\frac{h'}{h} = -\frac{q - f}{f}$$

Using this expression in combination with Equation 26.11 gives

$$\frac{q}{p} = \frac{q - f}{f}$$

which reduces to

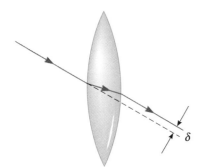

Figure 26.20   A ray passing through the center of the lens is deviated from its path by a distance of δ. In the thin-lens approximation, this deviation is ignored.

Thin-lens equation •

$$\frac{1}{p} + \frac{1}{q} = \frac{1}{f} \qquad \text{[26.12]}$$

This equation, called the **thin-lens equation,** can be used with either converging or diverging lenses if we adhere to a set of sign conventions. Figure 26.21 is useful for obtaining the signs of p and q, and the complete sign conventions for lenses are provided in Table 26.3. Note that **a converging lens has a positive focal length** under this convention and **a diverging lens has a negative focal length.** Hence, the names *positive* and *negative* are often given to these lenses.

The focal length for a lens in air is related to the curvatures of its surfaces and to the index of refraction, n, of the lens material by

$$\frac{1}{f} = (n - 1)\left(\frac{1}{R_1} - \frac{1}{R_2}\right)$$ [26.13] • *Lens makers' equation*

where R_1 is the radius of curvature of the front surface and R_2 is the radius of curvature of the back surface. (As with mirrors, we arbitrarily call the side from which the light approaches the *front* of the lens.) Equation 26.13 enables us to calculate the focal length from the known properties of the lens. It is called the **lens makers' equation.**

Ray Diagrams for Thin Lenses

Ray diagrams are very convenient for locating the image of a thin lens or system of lenses. They should also help clarify the sign conventions we have already discussed. Figure 26.22 illustrates this method for three single-lens situations. To locate the image of a converging lens (Figs. 26.22a and b), the following three rays are drawn from the top of the object:

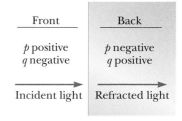

Figure 26.21 A diagram for obtaining the signs of p and q for a thin lens or a refracting surface. Although the interface between the two media is shown flat here, the diagram is also valid for convex and concave surfaces.

- The first ray is drawn parallel to the principal axis. After being refracted by the lens, this ray passes through (or appears to come from) one of the focal points.
- The second ray is drawn through the center of the lens. This ray continues in a straight line.
- The third ray is drawn through the focal point, F, and emerges from the lens parallel to the principal axis.

A similar construction is used to locate the image of a diverging lens, as shown in Figure 26.22c. The point of intersection of *any two* of the rays in these diagrams can be used to locate the image. The third ray serves as a check of construction.

For the converging lens in Figure 26.22a, where the object is *outside* the front focal point ($p > f$), the image is real and inverted. When the real object is *inside* the front focal point ($p < f$), as in Figure 26.22b, the image is virtual and upright. For the diverging lens of Figure 26.22c, the image is virtual and upright.

TABLE 26.3   Sign Convention for Thin Lenses

p is + if the object is in front of the lens.
p is − if the object is in back of the lens.

q is + if the image is in back of the lens.
q is − if the image is in front of the lens.

R_1 and R_2 are + if the center of curvature is in back of the lens.
R_1 and R_2 are − if the center of curvature is in front of the lens.

f is + for a converging lens.
f is − for a diverging lens.

Note: p = object distance; q = image distance; R_1 = radius of curvature of front surface; R_2 = radius of curvature of back surface; f = focal length.

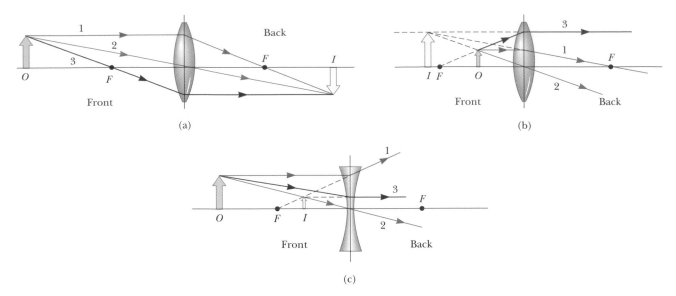

Figure 26.22   Ray diagrams for locating the image formed by a thin lens. (a) The object is located to the left of the focal point of a converging lens. (b) The object is located between the focal point and a converging lens. (c) The object is located to the left of the focal point of a diverging lens.

PROBLEM-SOLVING STRATEGY • Lenses and Mirrors

Success or failure in working lens and mirror problems is largely determined by whether or not you make sign errors when substituting into the equations, and the only way to ensure that you don't make such errors is to become adept at using the sign conventions. The best way to do this is to work a multitude of problems on your own. Watching your instructor or reading the example problems is no substitute for practice.

Thinking Physics 4

Diving masks often have a lens built into the glass for divers who do not have perfect vision. This allows the individual to dive without the necessity of glasses, because the lenses in the faceplate perform the necessary refraction to provide clear vision. Normal glasses have lenses that are curved on both the front and rear surfaces. The lenses in a diving mask faceplate often only have curved surfaces on the *inside* of the glass. Why is this design desirable?

Reasoning The main reason for curving only the inner surface of the lenses in the diving mask faceplate is so that the diver can see clearly while underwater *and* in the air. If there were curved surfaces on both the front and the back of the diving lens, there would be two refractions. The lens could be designed so that these two refractions would give clear vision while the diver is in air. When the diver goes underwater, however, the refraction between the water and the glass at the first interface is now different, because the index of refraction of water is different from that of air. Thus, the vision will not be clear underwater.

By making the outer surface of the lens flat, there is no refraction at the outer surface *in either air or water*—all of the refraction occurs at the inner glass–air surface. Thus, there is the same refractive correction in water and in air, and the diver can see clearly in both environments.

Thinking Physics 5

Consider a glass plano-convex lens—flat on one side and convex on the other. You project three parallel laser beams through it, as shown in Figure 26.23a, and measure the focal length, f—that is, the distance from the lens to the point at which the three beams cross. Now you hold the flat side of the lens against the glass of an aquarium filled with water, as in Figure 26.23b. When you shine the laser beams through the lens from the outside of the aquarium, will the beams cross at a point closer to the lens than before, farther away, or at the same distance? What if you direct the laser beams through the lens from the other side of the aquarium, as in Figure 26.23c? What happens in this case?

Reasoning The laser beam going through the center of the lens is unaffected by the lens in all three cases. We need to look at what happens to the outer beams. In Figure 26.23a, these two laser beams will refract toward the normal as they pass from the air into the glass of the lens. Thus, they will be deviated *toward* the central beam. As they reach the other side, and pass from the glass back into the air, they will deviate away from the normal, which causes additional deviation toward the central beam. As a result, all three beams cross at the focal point. When the flat side of the lens is held against the glass side of the aquarium, as in Figure 26.23b, the outgoing laser beams pass through the glass of the aquarium without additional refraction, but then enter *water*. The change in index of refraction in going from glass to water is much smaller than that for going from glass to air. Thus, the outgoing refraction is smaller than in the previous case, and the deviation toward the central beam is less. As a result, the crossing point for the three laser beams is farther from the lens. When the laser beams are directed through the water and then through the lens, as in Figure 26.23c, we must consider two factors. First, the focal length of a lens is independent of which way the light passes through it. Second, in the situation shown, the laser beams experience no refraction on entering the flat side of the lens, because they strike the glass at normal incidence. Thus, all of the refraction occurs at the interface of the air with the curved side of the lens. This is exactly the same situation as if the laser beams had struck the flat side of the lens while it was in air. Thus, there is no effect of the water in the aquarium. The laser beams cross at the same distance from the lens as they did in the first diagram. This third situation is very similar to the curvature of a lens on the inside of a diving mask, discussed in Thinking Physics 4.

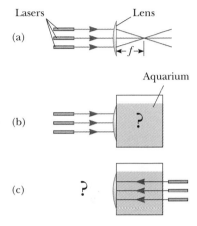

Figure 26.23   (Thinking Physics 5)

CONCEPTUAL PROBLEM 3

You are taking a picture of yourself with a camera that uses an ultrasonic range finder to measure the distance to the subject. When you take a picture of yourself in a mirror with this camera, your image is out of focus. Why?

CONCEPTUAL PROBLEM 4

A virtual image is often described as one through which light rays do not actually travel, as they do for a real image. Can a virtual image be photographed?

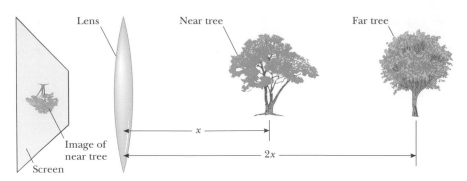

Figure 26.24   (Conceptual Problem 5)

CONCEPTUAL PROBLEM 5

Suppose you want to use a converging lens to project the image of two trees onto a screen. One tree is at a distance x from the lens, the other is at $2x$, as in Figure 26.24. You adjust the screen so that the near tree is in focus. If you now want to move the screen so that the far tree is in focus, do you move the screen toward, or away from, the lens?

CONCEPTUAL PROBLEM 6

Can a converging lens be made to diverge light if placed in a liquid? How about a converging mirror?

CONCEPTUAL PROBLEM 7

A plastic sandwich bag filled with water can act as a crude converging lens in air. If the bag is filled with air and placed under water, is it a lens? If so, is it converging or diverging?

CONCEPTUAL PROBLEM 8

Why does the focal length of a mirror not depend on the mirror material when the focal length of a lens *does* depend on the lens material?

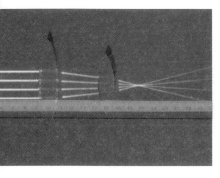

Light from a distant object brought into focus by two converging lenses.   *(Henry Leap and Jim Lehman)*

Example 26.6   The Lens Makers' Equation

The biconvex lens of Figure 26.25 has an index of refraction of 1.50. The curvature of the front surface is 10 cm, and that of the back surface is 15 cm. Find the focal length of the lens.

Solution   From the sign conventions in Table 26.3 we find that $R_1 = +10$ cm and $R_2 = -15$ cm. Thus, using the lens makers' equation, we have

$$\frac{1}{f} = (n-1)\left(\frac{1}{R_1} - \frac{1}{R_2}\right) = (1.5 - 1)\left(\frac{1}{10\text{ cm}} - \frac{1}{-15\text{ cm}}\right)$$

$$f = 12\text{ cm}$$

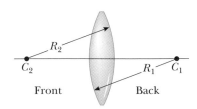

Figure 26.25   (Example 26.6) This lens has two curved surfaces with radii of curvature R_1 and R_2.

Example 26.7   An Image Formed by a Converging Lens

A converging lens of focal length 10.0 cm forms an image of an object placed (a) 30.0 cm, (b) 10.0 cm, and (c) 5.00 cm from the lens. Find the image distance and describe the image in each case.

Solution   (a) The thin lens equation, Equation 26.12, can be used to find the image distance:

$$\frac{1}{p} + \frac{1}{q} = \frac{1}{f}$$

$$\frac{1}{30.0 \text{ cm}} + \frac{1}{q} = \frac{1}{10.0 \text{ cm}}$$

$$q = 15.0 \text{ cm}$$

The positive sign tells us that the image is real. The magnification is

$$M = -\frac{q}{p} = -\frac{15.0 \text{ cm}}{30.0 \text{ cm}} = -0.500$$

Thus, the image is reduced in size by one half, and the negative sign for M tells us that the image is inverted. The situation is like that pictured in Figure 26.22a.

(b) No calculation is necessary for this case because we know that when the object is placed at the focal point, the image is formed at infinity. This is readily verified by substituting $p = 10.0$ cm into the lens equation.

(c) We now move inside the focal point, to an object distance of 5.00 cm. In this case, the lens equation gives

$$\frac{1}{5.00 \text{ cm}} + \frac{1}{q} = \frac{1}{10.0 \text{ cm}}$$

$$q = -10.0 \text{ cm}$$

$$M = -\frac{q}{p} = -\left(\frac{-10.0 \text{ cm}}{5.00 \text{ cm}}\right) = 2.00$$

The negative image distance tells us that the image is virtual. The image is enlarged, and the positive sign for M tells us that the image is upright, as in Figure 26.22b.

There are two general cases for a converging lens. When the object distance is greater than the focal length ($p > f$), the image is real and inverted. When the object is between the focal point and lens ($p < f$), the image is virtual, upright, and enlarged.

Combination of Thin Lenses

If two thin lenses are used to form an image, the system can be treated in the following manner. First, the image of the first lens is calculated as though the second lens were not present. The light then approaches the second lens *as if* it had come from the image formed by the first lens. Hence, the image of the first lens is treated as the object of the second lens. The image of the second lens is the final image of the system. If the image of the first lens lies on the back side of the second lens, then the image is treated as a virtual object for the second lens (that is, p is negative). The same procedure can be extended to a system of three or more lenses. The overall magnification of a system of thin lenses equals the *product* of the magnifications of the separate lenses.

Example 26.8   Where Is the Final Image?

Two thin converging lenses of focal lengths 10.0 cm and 20.0 cm are separated by 20.0 cm, as in Figure 26.26. An object is placed 15.0 cm to the left of the first lens. Find the position of the final image and the magnification of the system.

Solution   First, we find the image position for the first lens and neglect the second lens:

$$\frac{1}{p_1} + \frac{1}{q_1} = \frac{1}{15.0 \text{ cm}} + \frac{1}{q_1} = \frac{1}{10.0 \text{ cm}}$$

$$q_1 = 30.0 \text{ cm}$$

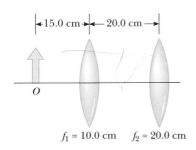

Figure 26.26   (Example 26.8) A combination of two converging lenses.

where q_1 is measured from the first lens.

Because q_1 is greater than the separation between the two lenses, the image of the first lens lies 10.0 cm to the right of the second lens. We take this as the object distance for the second lens. That is, we apply the thin lens equation to the second lens with $p_2 = -10.0$ cm, where distances are now measured from the second lens, whose focal length is 20.0 cm:

$$\frac{1}{p_2} + \frac{1}{q_2} = \frac{1}{f_2}$$

$$\frac{1}{-10.0 \text{ cm}} + \frac{1}{q_2} = \frac{1}{20.0 \text{ cm}}$$

$$q_2 = 6.67 \text{ cm}$$

That is, the final image lies 6.67 cm to the right of the second lens.

The magnification of each lens separately is given by

$$M_1 = \frac{-q_1}{p_1} = -\frac{30.0 \text{ cm}}{15.0 \text{ cm}} = -2.00$$

$$M_2 = \frac{-q_2}{p_2} = -\frac{6.67 \text{ cm}}{-10.0 \text{ cm}} = 0.667$$

The total magnification M of the two lenses is the product $M_1 M_2 = (-2.00)(0.667) = -1.33$. Hence, the final image is real, inverted, and enlarged.

26.5 • LENS ABERRATIONS

One of the basic problems of lenses and lens systems is the imperfect quality of the images. The simple theory of mirrors and lenses assumes that rays make small angles with the principal axis and that all rays reaching the lens or mirror from a point source are focused at a single point, producing a sharp image. This is not always true in the real world. Where the approximations used in this theory do not hold, imperfect images are formed.

If one wishes to precisely analyze image formation, it is necessary to trace each ray, using Snell's law, at each refracting surface. This procedure shows that there is no single point image; instead, the image is *blurred*. The departures of real (imperfect) images from the ideal predicted by the simple theory are called **aberrations.** Two types of aberrations will now be described.

Spherical Aberrations

Spherical aberrations result from the fact that the focal points of light rays far from the principal axis of a spherical lens (or mirror) are different from the focal points of rays of the same wavelength passing near the axis. Figure 26.27 illustrates spherical aberration for parallel rays passing through a converging lens. Rays near the middle of the lens are imaged farther from the lens than rays at the edges. Hence, there is no single focal length for a lens.

Most cameras are equipped with an adjustable aperture to control the light intensity and reduce spherical aberration when possible. (An aperture is an opening that controls the amount of light transmitted through the lens.) Sharper images are produced as the aperture size is reduced, because only the central portion of the lens is exposed to the incident light when the aperture is very small. At the same time, however, less light is imaged. To compensate for this loss in light intensity, a longer exposure time is used. An example of the good results obtained with a small aperture is the sharp image produced by a "pinhole" camera, with an aperture size of approximately 1 mm.

In the case of mirrors used for very distant objects, one can eliminate, or at least minimize, spherical aberration by employing a parabolic surface rather than

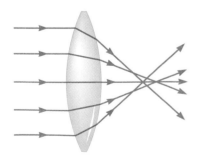

Figure 26.27 Spherical aberration caused by a converging lens. Does a diverging lens cause spherical aberration?

a spherical surface. Parabolic surfaces are not used in many applications, however, because they are very expensive to make with high-quality optics. Parallel light rays incident on such a surface focus at a common point. Parabolic reflecting surfaces are used in many astronomical telescopes in order to enhance the image quality. They are also used in searchlights, where a nearly parallel light beam is produced from a small lamp placed at the focus of the reflecting surface.

Chromatic Aberrations

The fact that different wavelengths of light refracted by a lens focus at different points gives rise to *chromatic aberrations*. In Chapter 25 we described how the index of refraction of a material varies with wavelength. When white light passes through a lens, one finds, for example, that violet light rays are refracted more than red light rays (Fig. 26.28). From this we see that the focal length is larger for red light than for violet light. Other wavelengths (not shown in Fig. 26.28) would have intermediate focal points. The chromatic aberration for a diverging lens is opposite that for a converging lens. Chromatic aberration can be greatly reduced by using a combination of a converging and diverging lens made from two different types of glass.

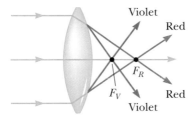

Figure 26.28 Chromatic aberration caused by a converging lens. Rays of different wavelengths focus at different points.

SUMMARY

The **magnification,** M, of a mirror or lens is defined as the ratio of the image height, h', to the object height, h:

$$M = \frac{h'}{h} = -\frac{q}{p}$$ [26.2, 26.11]

In the paraxial ray approximation, the object distance, p, and image distance, q, for a spherical mirror of radius R are related by the **mirror equation,**

$$\frac{1}{p} + \frac{1}{q} = \frac{2}{R} = \frac{1}{f}$$ [26.4, 26.6]

where $f = R/2$ is the **focal length** of the mirror.

An image can be formed by refraction from a spherical surface of radius R. The object and image distances for refraction from such a surface are related by

$$\frac{n_1}{p} + \frac{n_2}{q} = \frac{n_2 - n_1}{R}$$ [26.8]

where the light is incident in the medium of index of refraction n_1 and is refracted in the medium the index of refraction of which is n_2.

For a thin lens, and in the paraxial ray approximation, the object and image distances are related by the **thin lens equation:**

$$\frac{1}{p} + \frac{1}{q} = \frac{1}{f}$$ [26.12]

The **focal length,** *f,* of a thin lens in air is related to the curvature of its surfaces and to the index of refraction, *n,* of the lens material by

$$\frac{1}{f} = (n - 1)\left(\frac{1}{R_1} - \frac{1}{R_2}\right)$$ [26.13]

Converging lenses have positive focal lengths, and **diverging lenses** have negative focal lengths.

CONCEPTUAL QUESTIONS

1. Why do some emergency vehicles have the symbol ECNAJUBMA written on the front?

2. The side-view mirror on late-model cars warns the user that objects may be closer than they appear. What kind of mirror is being used and why was this type selected?

3. Consider a concave spherical mirror with a real object. Is the image always inverted? Is the image always real? Give conditions for your answers.

4. Why does a clear stream always appear to be shallower than it actually is?

5. A person spearfishing from a boat sees a fish that is 3 m from the boat at a depth of 1 m. In order to spear the fish, should the person aim at, above, or below the image of the fish?

6. Explain why a fish in a spherical goldfish bowl appears larger than it really is.

7. If a cylinder of solid glass or clear plastic is placed above the words LEAD OXIDE and viewed from the side, as shown in Figure Q26.7, the LEAD appears inverted but the OXIDE does not. Explain.

Figure Q26.7   *(Courtesy of Henry Leap and Jim Lehman)*

8. A mirage is formed when the air gets gradually cooler as the height above the ground increases. What might happen if the air grows gradually warmer as the height is increased? This often happens over bodies of water or snow-covered ground: the effect is called *looming*.

9. Lenses used in eyeglasses, whether converging or diverging, are always designed such that the middle of the lens curves away from the eye, like the center lenses of Figure 26.17a and 26.17b. Why?

10. Describe lenses that can be used to start a fire.

11. Explain this statement: "The focal point of a lens is the location of the image of a point object at infinity." Discuss the notion of infinity in real terms as it applies to object distances. Based on this statement, can you think of a quick and dirty method for determining the focal length of a positive lens?

12. Consider the image formed by a thin converging lens. Under what conditions is the image (a) inverted, (b) upright, (c) real, (d) virtual, (e) larger than the object, and (f) smaller than the object?

13. Discuss the proper position of a slide relative to the lens in a slide projector. What type of lens must the slide projector have?

14. Discuss why each of the following statements is true or false. (a) A virtual image can be a virtual object. (b) A virtual image can be a real object. (c) A real image can be a virtual object. (d) A real image can be a real object.

15. Discuss the type of aberration involved in each of the following situations. (a) The edges of the image appear reddish. (b) A clear focus of the image's central portion cannot be obtained. (c) A clear focus of the image's outer portion cannot be obtained. (d) The central portion of the image is enlarged relative to the outer portions.

16. In a Jules Verne novel, a piece of ice is shaped to form a magnifying lens to focus sunlight to start a fire. Is this possible?

17. A solar furnace can be constructed by using a concave mirror to reflect and focus sunlight into a furnace enclosure. What factors in the design of the reflecting mirror would guarantee very high temperatures?

18. A pinhole camera can be constructed by punching a small hole in one side of a cardboard box. If the opposite side is cut out and replaced with tissue paper, the image of a distant object is formed by light passing through the pinhole onto the tissue paper (the screen). No lens is involved here. In effect, the pinhole replaces the lens of the camera. Explain why an image is formed, and describe the nature of the image.

19. If you want to examine the fine detail of an object with a magnifying lens of focal length 15 cm, where should the

object be placed in order to observe a magnified image of the object?

20. Large telescopes are usually reflecting rather than refracting. List some reasons for this choice.

21. A classic science fiction novel, *The Invisible Man*, tells of a person who becomes invisible by changing the index of refraction of his body to that of air. This story has been criticized by students who know how the eye works; they claim the invisible man himself would be unable to see. On the basis of your knowledge of the eye, could he see or not?

PROBLEMS

Section 26.1   Images Formed by Flat Mirrors

1. Does your bathroom mirror show you older or younger than your actual age? Compute an order-of-magnitude estimate for the age difference, based on data that you specify.

2. In a physics laboratory experiment, a torque is applied to a small-diameter wire suspended vertically under tensile stress. The small angle through which the wire turns as a consequence of the net torque is measured by attaching a small mirror to the wire and reflecting a beam of light off the mirror and onto a circular scale. Such an arrangement is known as an optical lever and is shown from a top view in Figure P26.2. Show that when the mirror turns through an angle θ, the reflected beam turns through angle 2θ.

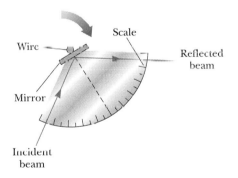

Figure P26.2

3. Determine the minimum height of a vertical flat mirror in which a person 5′10″ in height can see his or her full image. (A ray diagram would be helpful.)

4. Two flat mirrors have their reflecting surfaces facing one another, with the edge of one mirror in contact with an edge of the other, so that the angle between the mirrors is α. When an object is placed between the mirrors, a number of images are formed. In general, if the angle α is such that $n\alpha = 360°$, where n is an integer, the number of images formed is $n - 1$. Graphically, find all the image positions for the case $n = 6$ when a point object is between the mirrors (but not on the angle bisector).

5. A person walks into a room with two flat mirrors on opposite walls, which produce multiple images. When the person is 5.00 ft from the mirror on the left wall and 10.0 ft from the mirror on the right wall, find the distance from the person to the first three images seen in the mirror on the left.

Section 26.2   Images Formed by Spherical Mirrors

6. A concave spherical mirror has a radius of curvature of 20.0 cm. Find the location of the image for object distances of (a) 40.0 cm, (b) 20.0 cm, and (c) 10.0 cm. For each case, state whether the image is real or virtual and upright or inverted. Find the magnification in each case.

7. A convex mirror monitors the aisles in a store, as in Figure P26.7. The mirror has a radius of curvature of 0.550 m. Locate and describe the image of a customer 10.0 m from the mirror. Determine the magnification.

Figure P26.7   Convex mirrors, often used for security in department stores, provide wide-angle viewing.   (© *Paul Silverman 1990, Fundamental Photographs*)

8. A large church has a niche in one wall. On the floor plan it appears as a semicircular indentation of radius 2.50 m. A worshiper stands on the center line of the niche, 2.00 m out from its deepest point, and whispers a prayer. Where is the sound concentrated after reflection from the back wall of the niche?

9. A spherical convex mirror has a radius of curvature of 40.0 cm. Determine the position of the virtual image and magnification for object distances of (a) 30.0 cm and (b) 60.0 cm. (c) Are the images upright or inverted?

10. An object 2.00 cm in height is placed 3.00 cm in front of a concave mirror. If the image is 5.00 cm in height and virtual, what is the focal length of the mirror?

11. A spherical mirror is to be used to form, on a screen located 5.00 m from the object, an image five times the size of the object. (a) Describe the type of mirror required. (b) Where should the mirror be positioned relative to the object?

12. (a) A concave mirror forms an inverted image four times larger than the object. Find the focal length of the mirror if the distance between object and image is 0.600 m. (b) A convex mirror forms a virtual image half the size of the object. If the distance between image and object is 20.0 cm, determine the radius of curvature of the mirror.

13. A concave mirror has a focal length of 40.0 cm. Determine the object position for which the resulting image is upright and four times the size of the object.

Section 26.3   Images Formed by Refraction

14. A simple model of the human eye ignores its lens entirely. Most of what the eye does to light happens at the transparent cornea. Assume that this outer surface has a radius of curvature 6.00 mm, and assume that the eyeball contains just one fluid of index of refraction 1.40. Prove that a very distant object will be imaged on the retina, 21.0 mm behind the cornea. Describe the image.

15. A cubical block of ice 50.0 cm on a side is placed on a level floor over a speck of dust. Find the location of the image of the speck if the index of refraction of ice is 1.309.

16. A colored marble is dropped into a large tank filled with benzene ($n = 1.50$). (a) What is the depth of the tank if the apparent depth of the marble when viewed from directly above the tank is 35.0 cm? (b) If the marble has a diameter of 1.50 cm, what is its apparent diameter when viewed from directly above the tank?

17. A glass sphere ($n = 1.50$) of radius 15.0 cm has a tiny air bubble located 5.00 cm from the center. The sphere is viewed along a direction parallel to the radius containing the bubble. What is the apparent depth of the bubble below the surface of the sphere?

18. A flint glass plate ($n = 1.66$) rests on the bottom of an aquarium tank. The plate is 8.00 cm thick (vertical dimension) and covered with water ($n = 1.33$) to a depth of 12.0 cm. Calculate the apparent thickness of the plate as viewed from above the water. (Assume nearly normal incidence.)

19. A glass hemisphere is used as a paperweight with its flat face resting on a stack of papers. The radius of the circular cross-section is 4.00 cm, and the index of refraction of the glass is 1.55. The center of the hemisphere is directly over a letter "O" that is 2.50 mm in diameter. What is the diameter of the image of the letter as seen looking along a vertical radius?

20. A transparent sphere of unknown composition is observed to form an image of the Sun on the surface of the sphere opposite the Sun. What is the refractive index of the sphere material?

Section 26.4   Thin Lenses

21. The left face of a biconvex lens has a radius of curvature of 12.0 cm, and the right face has a radius of curvature of 18.0 cm. The index of refraction of the glass is 1.44. (a) Calculate the focal length of the lens. (b) Calculate the focal length if the radii of curvature of the two faces are interchanged.

22. A contact lens is made of plastic with an index of refraction of 1.50. The lens has an outer radius of curvature of $+2.00$ cm and an inner radius of curvature of $+2.50$ cm. What is the focal length of the lens?

23. A thin lens has a focal length of 25.0 cm. Locate the image when the object is placed (a) 26.0 cm and (b) 24.0 cm in front of the lens. Describe the image in each case.

24. A converging lens has a focal length of 20.0 cm. Locate the image for object distances of (a) 40.0 cm, (b) 20.0 cm, and (c) 10.0 cm. For each case, state whether the image is real or virtual and upright or inverted. Find the magnification in each case.

25. The nickel's image in Figure P26.25 has twice the diameter of the nickel and is 2.84 cm from the lens. Determine the focal length of the lens.

26. An object located 32.0 cm in front of a lens forms an image on a screen 8.00 cm behind the lens. (a) Find the focal length of the lens. (b) Determine the magnification. (c) Is the lens converging or diverging?

Figure P26.25

27. A magnifying glass is a converging lens of focal length 15.0 cm. At what distance from a postage stamp should you hold this lens to get a magnification of +2.00?

28. A microscope slide is placed in front of a converging lens with a focal length of 2.44 cm. The lens forms an image of the slide 12.9 cm from the slide. How far is the lens from the slide if the image is (a) real? (b) virtual?

29. A person looks at a gem with a jeweler's loupe—a converging lens that has a focal length of 12.5 cm. The loupe forms a virtual image 30.0 cm from the lens. (a) Determine the magnification. Is the image upright or inverted? (b) Construct a ray diagram for this arrangement.

30. An object is placed 30.0 cm from a lens, and a virtual image is formed 15.0 cm from the lens. (a) What is the focal length of the lens? (b) Is the lens converging or diverging?

31. Suppose an object has thickness dp so that it extends from object distance p to $p + dp$. Prove that the thickness dq of its image is given by $(-q^2/p^2)\,dp$, so that the longitudinal magnification $dq/dp = -M^2$, where M is the lateral magnification.

32. An object is placed 50.0 cm from a screen. Where should a converging lens with a 10.0-cm focal length be placed in order to form an image on the screen? Find the magnification(s).

33. The projection lens in a certain slide projector is a single thin lens. A slide 24.0 mm high is to be projected so that its image fills a screen 1.80 m high. The slide-to-screen distance is 3.00 m. (a) Determine the focal length of the projection lens. (b) How far from the slide should the lens of the projector be placed in order to form the image on the screen?

34. A runner passes a spectator at 10.0 m/s. The spectator photographs the runner using a camera having a focal length of 5.00 cm and an exposure time of 20.0 ms. For a good photograph, the maximum image blurring allowed on the film is 0.500 mm. Find the minimum distance between runner and camera in order to obtain a good photograph.

Additional Problems

35. The lens and mirror in Figure P26.35 have focal lengths of +80.0 cm and −50.0 cm, respectively. An object is placed

1.00 m to the left of the lens as drawn. Locate the final image, formed by light that has gone through the lens twice. State whether the image is upright or inverted, and determine the overall magnification.

36. The object in Figure P26.36 is midway between the lens and the mirror. The mirror's radius of curvature is 20.0 cm, and the lens has a focal length of −16.7 cm. Considering only the light that leaves the object and travels first toward the mirror, locate the final image formed by this system. Is this image real or virtual? Is it upright or inverted? What is the overall magnification?

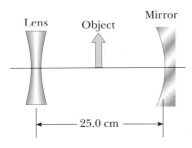

Figure P26.36

37. The distance between an object and its upright image is 20.0 cm. If the magnification is 0.500, what is the focal length of the lens being used to form the image?

38. The distance between an object and its upright image is d. If the magnification is M, what is the focal length of the lens being used to form the image?

39. An object placed 10.0 cm from a concave spherical mirror produces a real image 8.00 cm from the mirror. If the object is moved to a new position 20.0 cm from the mirror, what is the position of the image? Is the latter image real or virtual?

40. Two rays traveling parallel to the principal axis strike a plano-convex lens having a refractive index of 1.60 (Fig. P26.40). If the convex face is spherical, a ray near the edge

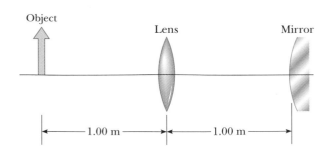

Figure P26.35

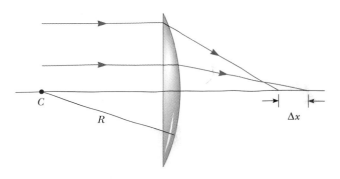

Figure P26.40

does not strike the focal point (spherical aberration). If this face has a radius of curvature of 20.0 cm and the two rays are $h_1 = 0.500$ cm and $h_2 = 12.0$ cm from the principal axis, find the difference in the positions where each strikes the principal axis.

41. A real object is located at the zero end of a meter stick. A large concave mirror at the 100-cm end of the meter stick forms an image of the object at the 70.0-cm position. A small convex mirror placed at the 20.0-cm position forms a final image at the 10.0-cm point. What is the radius of curvature of the convex mirror?

42. Derive the lens-makers' equation as follows: Consider an object in vacuum at $p_1 = \infty$ from a first refracting surface of radius of curvature R_1. Locate its image. Use this image as the object for the second refracting surface, which has nearly the same location as the first, because the lens is thin. Locate the final image, proving it is at the image distance q_2 given by

$$\frac{1}{q_2} = (n - 1)\left(\frac{1}{R_1} - \frac{1}{R_2}\right)$$

43. A parallel beam of light enters a glass hemisphere perpendicular to the flat face, as shown in Figure P26.43. The radius is $R = 6.00$ cm, and the index of refraction is $n = 1.560$. Determine the point at which the beam is focused. (Assume paraxial rays.)

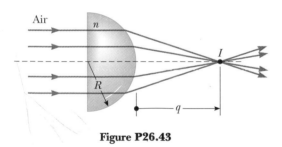

Figure P26.43

44. An object 2.00 cm high is placed 40.0 cm to the left of a converging lens having a focal length of 30.0 cm. A diverging lens having a focal length of -20.0 cm is placed 110 cm to the right of the converging lens. (a) Determine the final position and magnification of the final image. (b) Is the image upright or inverted? (c) Repeat parts (a) and (b) for the case where the second lens is a converging lens having a focal length of $+20.0$ cm.

45. An object is placed 12.0 cm to the left of a diverging lens of focal length -6.00 cm. A converging lens of focal length 12.0 cm is placed a distance d to the right of the diverging lens. Find the distance d so that the final image is at infinity. Draw a ray diagram for this case.

46. The cornea of an eye has a radius of curvature of 0.800 cm. (a) What is the focal length of the reflecting surface of the

eye? (b) If a \$20 gold piece 3.40 cm in diameter is held 25.0 cm from the cornea, what are the size and location of the reflected image?

47. The disk of the Sun subtends an angle of 0.533° at the Earth. What are the position and diameter of the solar image formed by a concave spherical mirror of radius 3.00 m?

48. A floating coin illusion consists of two parabolic mirrors, each having a focal length 7.50 cm, facing each other so that their centers are 7.50 cm apart (Fig. P26.48). If a few coins are placed on the lower mirror, an image of the coins is formed at the small opening at the center of the top mirror. Show that the final image is formed at that location and describe its characteristics. (*Note:* A very startling effect is to shine a flashlight beam on these images. Even at a glancing angle, the incoming light beam is seemingly reflected off the images! Do you understand why?)

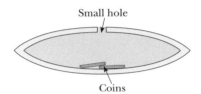

Figure P26.48

49. In a darkened room, a burning candle is placed 1.50 m from a white wall. A lens is placed between candle and wall at a location that causes a larger, inverted image to form on the wall. When the lens is moved 90.0 cm toward the wall, another image of the candle is formed. Find (a) the two object distances that produce the images stated above and (b) the focal length of the lens. (c) Characterize the second image.

50. The fixed lens L_1 in Figure P26.50 has a focal length of 15.0 cm, whereas lens L_2 has a focal length of 13.3 cm. The distance d of lens L_2 from the film plane can be varied from 5.00 cm to 10.0 cm. Determine the range of distances for which objects can be focused on the film.

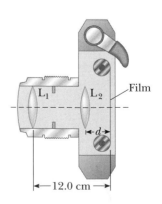

Figure P26.50

Spreadsheet Problem

S1. The equation for refraction by a spherical surface separating two media (Fig. 26.13) having indices of refraction n_1 and n_2 is

$$\frac{n_1}{p} + \frac{n_2}{q} = \frac{n_2 - n_1}{R}$$

This equation was derived using small-angle approximations. Spreadsheet 26.1 tests the range of validity of these approximations. The spreadsheet calculates the image distance for a series of incident angles α without making small-angle approximations. Input parameters are the indices of refraction, the radius of curvature of the surface, and the object distance. (a) Take $n_1 = 1.00$, $n_2 = 1.52$, $R = 10.0$ cm, and $p = 30.0$ cm. If the maximum acceptable error in the image distance is 3.00%, for what maximum angle α can the refraction equation be used? (b) Repeat part (a) using $p = 50.0$ cm. (c) Choose other values for the input parameters and repeat (a). (d) Is the error linearly proportional to α?

ANSWERS TO CONCEPTUAL PROBLEMS

1. You will not be able to focus your eyes on both the picture and your image at the same time. To focus on the picture, you must adjust your eyes so that an object several feet away (the picture) is in focus. Thus, you are focusing *on the mirror surface.* But your image in the mirror is behind the mirror, as far behind the mirror as you are in front of it. Thus, you must focus your eyes beyond the mirror, twice as far away as the picture, to bring the image into focus.

2. The actor will see *the movie camera.* Light from the actor's face must reflect from the mirror in the direction of the camera lens, so that his face is recorded on film. If we reverse that light path, light from the camera reflects from the mirror into the eyes of the actor.

3. The ultrasonic range finder sends out a sound wave and measures the time for the echo to return. Using this information, the camera calculates the distance to the subject and sets the camera lens. When facing a mirror, the ultrasonic signal reflects from the mirror surface and the camera adjusts its focus so that the mirror surface is at the correct focusing distance from the camera. But your image in the mirror is twice this distance from the camera, so it is blurry.

4. Light rays diverge from the position of a virtual image just as they do from an actual object. Thus, a virtual image can be as easily photographed as any object can. Of course, the camera would have to be placed near the axis of the lens or mirror in order to intercept the light rays.

5. We consider the two trees to be two separate objects. The far tree is an object that is farther from the lens than the near tree. Thus, the image of the far tree will be closer to the lens than the image of the near tree. The screen must be moved closer to the lens to put the far tree in focus.

6. If a converging lens is placed in a liquid the index of refraction of which is larger than that of the lens material, the relative size of the angles at the interfaces will be reversed, and the lens will diverge light. A mirror depends only on reflection, which is independent of the surrounding material, so a converging mirror will be converging in any liquid.

7. The bag of air will act as a crude lens underwater. The direction of refraction at the bag surface as light goes from water to air will be the opposite of the directions for the light entering from air into the water filled bag, so the air lens will diverge light underwater.

8. The focal length for a mirror is determined by the law of reflection from the mirror surface. The law of reflection is independent of the material of which the mirror is made and of the surrounding medium. Thus, the focal length depends only on the radius of curvature and not the material. The focal length of a lens depends on the law of refraction, and refraction is highly dependent on the indices of refraction of the lens material and surrounding medium. Thus, the focal length of a lens depends on the lens material.

27

Wave Optics

In the preceding chapter, we used the concept of light rays to examine what happens when light passes through a lens or reflects from a mirror. This chapter is concerned with the subject of **wave optics,** which addresses the related optical phenomena of interference and diffraction. These phenomena cannot be adequately explained with ray (geometric) optics, but we shall discuss how the wave nature of light leads to satisfying descriptions of such events.

27.1 • CONDITIONS FOR INTERFERENCE

In our discussion of wave interference of mechanical waves in Chapter 14, we found that two waves could add together either constructively or destructively. In constructive interference, the amplitude of the resultant wave is greater than that of either individual wave, whereas in destructive interference, the resultant amplitude is less than that of either individual wave. Electromagnetic waves also undergo interference. All interference associated with electromagnetic waves arises fundamentally as a result of combining the electromagnetic fields that constitute the individual waves.

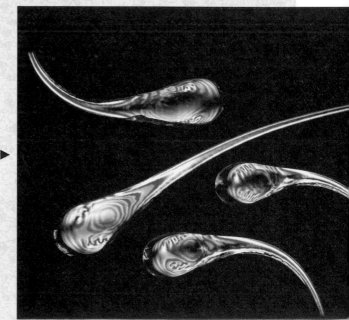

These glass objects, called ▶ Prince Rupert drops, are made by dropping molten glass into water. The photograph was made by placing the objects between two crossed polarizers. The patterns observed represent the strain distribution in the glass. Studies of such patterns led to the development of tempered glass. *(James L. Amos, Peter Arnold, Inc.)*

Interference effects in visible electromagnetic waves are not easy to observe because of their short wavelengths (from about 4×10^{-7} m to 7×10^{-7} m). In order to observe sustained interference in light waves, three conditions must be met:

- The sources must be **coherent**—that is, they must maintain a constant phase with respect to each other.
- The sources must be of identical wavelength.
- The superposition principle must apply.

- *Conditions for interference*

Let us examine the characteristics of coherent sources. Two sources (producing two traveling waves) are needed to create interference. However, in order to produce a stable interference pattern, **the individual waves must maintain a constant phase with one another.** When this situation prevails, the sources are said to be **coherent.** As an example, the sound waves emitted by two side-by-side loudspeakers driven by a single amplifier can produce interference because the two speakers respond to the amplifier in the same way at the same time.

Now, if two light sources are placed side by side, no interference effects are observed because the light waves from one source are emitted independently of the other source; hence, the emissions from the two sources do not maintain a constant phase relationship with each other over the time of observation. An ordinary light source undergoes random changes about once every 10^{-8} s. Therefore, the conditions for constructive interference, destructive interference, or some intermediate state last for times on the order of 10^{-8} s. The result is that no interference effects are observed, because the eye cannot follow such short-term changes. Such light sources are said to be **noncoherent.**

A common method for producing two coherent light sources is to use one single wavelength source to illuminate a screen containing two small slits. The light emerging from the two slits is coherent because a single source produces the original light beam and the two slits serve only to separate the original beam into two parts (which is exactly what was done to the sound signal just discussed). A random change in the light emitted by the source will occur in the two separate beams at the same time, and interference effects can still be observed.

27.2 • YOUNG'S DOUBLE-SLIT EXPERIMENT

Interference in light waves from two slits was first demonstrated by Thomas Young in 1801. A schematic diagram of the apparatus used in this experiment is shown in Figure 27.1a. (Young used pinholes in his original experiments, rather than slits.) Light is incident on a screen in which there is a narrow slit, S_0. The light waves emerging from this slit arrive at a second screen that contains two narrow, parallel slits, S_1 and S_2. These two slits serve as a pair of coherent light sources because waves emerging from them originate from the same source, S_0, and therefore maintain a constant phase relationship. The light from the two slits produces a visible pattern on screen C; the pattern consists of a series of bright and dark parallel bands called **fringes** (Fig. 27.1b). When the light from S_1 and S_2 arrives at a point on the screen so that constructive interference occurs at that location, a bright fringe appears. When the light from the two slits combines destructively at any location, a dark fringe results.

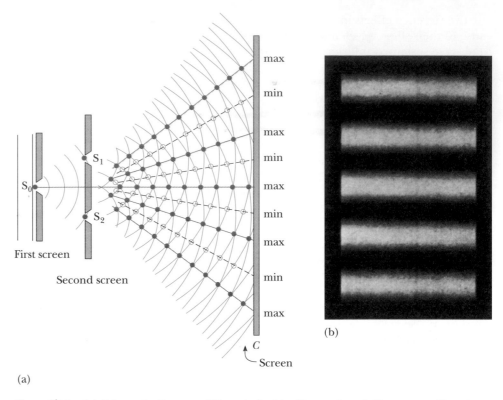

(a)

(b)

Figure 27.1   (a) Schematic diagram of Young's double-slit experiment. The narrow slits act as wave sources. Slits S_1 and S_2 behave as coherent sources that produce an interference pattern on screen C. (Note that this drawing is not to scale.) (b) The fringe pattern formed on screen C could look like this.

Figure 27.2 is a schematic diagram of some of the ways the two waves in Young's experiment can combine at screen C. In Figure 27.2a, the two waves, which leave the two slits in phase, strike the screen at the central point, P. Because these waves travel equal distances, they arrive in phase at P, and as a result constructive interference occurs at this location and a bright fringe is observed. In Figure 27.2b, the

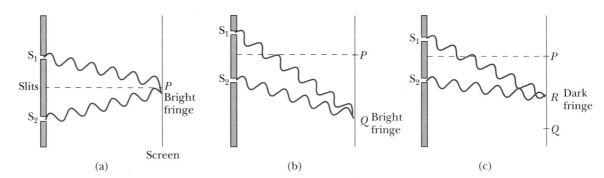

(a)   (b)   (c)

Figure 27.2   (a) Constructive interference occurs at P when the waves combine. (b) Constructive interference also occurs at Q. (c) Destructive interference occurs at R because the wave from the upper slit falls half a wavelength behind the wave from the lower slit. (Note that these figures are not drawn to scale.)

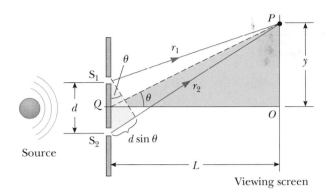

Figure 27.3 Geometric construction for describing Young's double-slit experiment. The path difference between the two rays is $r_2 - r_1 = d \sin \theta$. (Note that this figure is not drawn to scale.)

two light waves again start in phase, but the upper wave has to travel one wavelength farther to reach point Q on the screen. Because the upper wave falls behind the lower one by exactly one wavelength, they still arrive in phase at Q, and so a second bright fringe appears at this location. Now consider point R, midway between P and Q in Figure 27.2c. At this location, the upper wave has fallen half a wavelength behind the lower wave. This means that the trough from the lower wave overlaps the crest from the upper wave, giving rise to destructive interference at R. For this reason, one observes a dark fringe at this location.

We can obtain a quantitative description of Young's experiment with the help of Figure 27.3. Consider point P on the viewing screen; the screen is located a perpendicular distance of L from the screen containing slits S_1 and S_2, which are separated by a distance of d, and r_1 and r_2 are the distances the waves travel from slit to screen. Let us assume that the source is monochromatic. Under these conditions, the waves emerging from S_1 and S_2 have the same frequency and amplitude and are in phase. The light intensity on the screen at P is the resultant of the light coming from both slits. Note that a wave from the lower slit travels farther than a wave from the upper slit by an amount equal to $d \sin \theta$. This distance is called the **path difference, δ** (lowercase Greek delta), where

$$\delta = r_2 - r_1 = d \sin \theta \qquad \text{[27.1]}$$

• *Path difference*

This equation assumes that the two waves are parallel to each other, which is approximately true because L is much greater than d. As noted earlier, the value of this path difference determines whether or not the two waves are in phase or out of phase when they arrive at P. If the path difference is either zero or some integral multiple of the wavelength, the two waves are in phase at P and constructive interference results. Therefore, the condition for bright fringes, or **constructive interference,** at P is

$$\delta = d \sin \theta = m\lambda \qquad (m = 0, \pm 1, \pm 2, \ldots) \qquad \text{[27.2]}$$

• *Conditions for constructive interference*

The number m is called the **order number.** The central bright fringe at $\theta = 0$ ($m = 0$) is called the **zeroth-order maximum.** The first maximum on either side, when $m = \pm 1$, is called the **first-order maximum,** and so forth.

In a similar way, when the path difference is an odd multiple of $\lambda/2$, the two waves arriving at P are 180° out of phase and give rise to destructive interference. Therefore, the condition for dark fringes, or **destructive interference,** at P is

Conditions for destructive •
interference

$$\delta = d \sin \theta = \left(m + \frac{1}{2} \right) \lambda \qquad (m = 0, \pm 1, \pm 2, \ldots) \qquad \textbf{[27.3]}$$

It is useful to obtain expressions for the positions of the bright and dark fringes measured vertically from O to P. In addition to our assumption that $L \gg d$, we shall assume that $d \gg \lambda$—that is, the distance between the two slits is much larger than the wavelength. This situation prevails in practice because L is often on the order of 1 m whereas d is a fraction of a millimeter and λ is a fraction of a micrometer for visible light. Under these conditions, θ is small, and so we can use the approximation $\sin \theta \approx \tan \theta$. From the triangle OPQ in Figure 27.3, we see that

$$\sin \theta \approx \tan \theta = \frac{y}{L} \qquad \textbf{[27.4]}$$

Using this result and making the substitution $\sin \theta = m\lambda/d$ from Equation 27.2, we see that the positions of the bright fringes measured from O are given by

$$y_{\text{bright}} = \frac{\lambda L}{d} m \qquad \textbf{[27.5]}$$

Likewise, using Equations 27.3 and 27.4, we find that the dark fringes are located at

$$y_{\text{dark}} = \frac{\lambda L}{d} \left(m + \tfrac{1}{2} \right) \qquad \textbf{[27.6]}$$

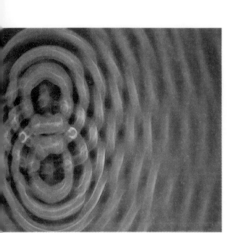

An interference pattern involving water waves is produced by two vibrating sources at the water's surface. The pattern is analogous to that observed in Young's double-slit experiment. Note the regions of constructive and destructive interference. *(Richard Megna, Fundamental Photographs)*

As we shall demonstrate in Example 27.1, Young's double-slit experiment provides a method for measuring the wavelength of light. In fact, Young used this technique to make the first measurement of the wavelength of light. In addition, the experiment gave the wave model of light a great deal of credibility. Today we still use the phenomenon of interference to describe many observations of wavelike behavior.

Thinking Physics 1

Consider a double-slit experiment in which a laser beam is passed through a pair of very closely spaced slits, and a clear interference pattern is displayed on a distant screen. Now, suppose you place smoke particles between the double slit and the screen. With the presence of the smoke particles, will you see the effects of the interference in the space between the slits and the screen, or will you only see the effects on the screen?

Reasoning You will see the effects in the area filled with smoke. There will be bright lines directed toward the bright areas on the screen and dark lines directed toward the dark areas on the screen. In Figure 27.3, the geometrical construction is important for developing the mathematical description of interference. It is subject to misinterpretation, however, because it might suggest that the interference does not occur until the light rays from the two slits strike the screen at the same position. The more fruitful diagram for this situation is Figure 27.1, in which it is clear that there are *paths* of destructive and constructive interference all the way from the slits to the screen. These paths will be made visible by the smoke.

Thinking Physics 2

If your stereo speakers are connected "out of phase"—that is, with one speaker connected correctly and the other with its wires reversed, the bass in the music tends to be weak. Why does this happen, and why is it a problem for the bass and not the treble notes?

Reasoning This is an acoustic analog to double-slit interference. The two speakers act as sources of waves, just like the two slits in a Young's double-slit experiment. If the speakers are connected correctly, and the same sound signal is fed to each speaker, both speakers move inward and outward at the same time in response to the signal. Thus, the sound waves are in phase as they leave the speakers. If you are sitting at a point in front of the speakers and midway between them, you will be located at the zero-order maximum—the interference is constructive and the sound will be loud. If one speaker is wired backward, then one speaker will be moving outward while the other is moving inward. The sound leaves the two speakers half a wavelength out of phase. Thus, if you are sitting at the same place, you will be at an interference minimum. This is a particular problem for the bass due to the long wavelength of low frequency notes. This results in a very large region of destructive interference in front of the speakers, on the order of the size of the room. The much shorter wavelengths of the high-frequency notes result in closely spaced maxima and minima. The spacing can be on the order of the size of the head and smaller. Thus, if one ear is at a minimum, the other might be at a maximum. What's more, small movements of the head will result in a shift from the position of a minimum to that of a maximum.

Thinking Physics 3

Suppose you are watching television by means of an antenna rather than a cable system. If an airplane flies near your location, you may notice wavering ghost images in the television picture. What might cause this?

Reasoning Your television antenna will receive two signals—the direct signal from the transmitting antenna and a reflected signal from the surface of the airplane. As the airplane changes position, there are some times when these two signals are in phase and other times when they are out of phase. As a result, there is a variation in the intensity of the combined signal received at your antenna. This variation is evidenced by the wavering ghost images of the picture.

Example 27.1 Measuring the Wavelength of a Light Source

A viewing screen is separated from a double-slit source by 1.2 m. The distance between the two slits is 0.030 mm. The second-order bright fringe ($m = 2$) is 4.5 cm from the center line. (a) Determine the wavelength of the light.

Solution We can use Equation 27.5, with $y_2 = 4.5 \times 10^{-2}$ m, $m = 2$, $L = 1.2$ m, and $d = 3.0 \times 10^{-5}$ m:

$$\lambda = \frac{dy_2}{mL} = \frac{(3.0 \times 10^{-5} \text{ m})(4.5 \times 10^{-2} \text{ m})}{2 \times 1.2 \text{ m}}$$

$$= 5.6 \times 10^{-7} \text{ m} = \boxed{560 \text{ nm}}$$

(b) Calculate the distance between adjacent bright fringes.

Solution From Equation 27.5 and the results to (a), we get

$$y_{m+1} - y_m = \frac{\lambda L(m + 1)}{d} - \frac{\lambda L m}{d}$$

$$= \frac{\lambda L}{d} = \frac{(5.6 \times 10^{-7} \text{ m})(1.2 \text{ m})}{3.0 \times 10^{-5} \text{ m}}$$

$$= 2.2 \times 10^{-2} \text{ m} = \boxed{2.2 \text{ cm}}$$

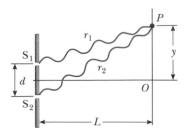

Figure 27.4 Construction for analyzing the double-slit interference pattern. A bright region, or intensity maximum, is observed at *O*.

Intensity Distribution of the Double-Slit Interference Pattern

We shall now calculate the distribution of light intensity (the energy delivered by the wave per unit area per unit time) associated with the double-slit interference pattern. Again, suppose that the two slits represent coherent sources of sinusoidal waves. Hence, the two waves have the same angular frequency, ω, and a constant phase difference, ϕ. The total electric field at the point P on the screen in Figure 27.4 is the *vector superposition* of the two waves. Assuming the two waves have the same amplitude, E_0, we can write the electric field at P due to each wave separately as

$$E_1 = E_0 \sin \omega t \qquad \text{and} \qquad E_2 = E_0 \sin(\omega t + \phi) \qquad [27.7]$$

Although the waves have equal phase at the slits, **their phase difference, ϕ, at P depends on the path difference, $\delta = r_2 - r_1 = d \sin \theta$.** Because a path difference of λ corresponds to a phase difference of 2π rad (constructive interference), whereas a path difference of $\lambda/2$ corresponds to a phase difference of π rad (destructive interference), we obtain the ratio

$$\frac{\delta}{\phi} = \frac{\lambda}{2\pi}$$

Phase difference •

$$\phi = \frac{2\pi}{\lambda} \delta = \frac{2\pi}{\lambda} d \sin \theta \qquad [27.8]$$

This equation tells us precisely how the phase difference ϕ depends on the angle θ.

Using the superposition principle and Equation 27.7, we can obtain the resultant electric field at the point P:

$$E_P = E_1 + E_2 = E_0[\sin \omega t + \sin(\omega t + \phi)] \qquad [27.9]$$

To simplify this expression, we use the trigonometric identity

$$\sin A + \sin B = 2 \sin\left(\frac{A + B}{2}\right) \cos\left(\frac{A - B}{2}\right)$$

Taking $A = \omega t + \phi$ and $B = \omega t$, we can write Equation 27.9 in the form

$$E_P = 2E_0 \cos\left(\frac{\phi}{2}\right) \sin\left(\omega t + \frac{\phi}{2}\right) \qquad [27.10]$$

Hence, the electric field at P has the same frequency ω as the original two waves, but its amplitude is multiplied by the factor $2 \cos(\phi/2)$. To check the consistency of this result, note that if $\phi = 0, 2\pi, 4\pi, \ldots$, the amplitude at P is $2E_0$, corresponding to the condition for constructive interference. Referring to Equation 27.8, we find that our result is consistent with Equation 27.2. Likewise, if $\phi = \pi, 3\pi, 5\pi, \ldots$, the amplitude at P is zero, which is consistent with Equation 27.3 for destructive interference.

Finally, to obtain an expression for the light intensity at P, recall that **the intensity of a wave is proportional to the square of the resultant electric field at that point** (Chapter 24, Section 24.6). Using Equation 27.10, we can therefore express the intensity at P as

$$I \propto E_P^2 = 4E_0^2 \cos^2(\phi/2) \sin^2\left(\omega t + \frac{\phi}{2}\right)$$

Because most light-detecting instruments measure the time-averaged light intensity, and the time-averaged value of $\sin^2(\omega t + \phi/2)$ over one cycle is $1/2$, we can write the average intensity at P as

$$I_{av} = I_0 \cos^2(\phi/2) \qquad \textbf{[27.11]}$$

where I_0 is the *maximum* possible time-averaged light intensity. [Note that $I_0 \propto (E_0 + E_0)^2 = (2E_0)^2 = 4E_0^2$.] Substituting Equation 27.8 into Equation 27.11, we find that

$$I_{av} = I_0 \cos^2\left(\frac{\pi d \sin\theta}{\lambda}\right) \qquad \textbf{[27.12]}$$

• *Average light intensity for the double-slit interference pattern*

Alternatively, because $\sin\theta \approx y/L$ for small values of θ, we can write Equation 27.12 in the form

$$I_{av} = I_0 \cos^2\left(\frac{\pi d}{\lambda L}y\right) \qquad \textbf{[27.13]}$$

Constructive interference, which produces intensity maxima, occurs when the quantity $\pi y d/\lambda L$ is an integral multiple of π, corresponding to $y = (\lambda L/d)m$. This is consistent with Equation 27.5. Intensity distribution versus θ is plotted in Figure 27.5. Note that the interference pattern consists of equally spaced fringes of equal intensity. However, the result is valid only if the slit to screen distance, L, is large relative to the slit separation, and only for small values of θ.

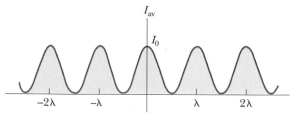

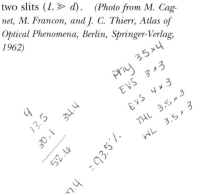

Figure 27.5 Intensity distribution versus $d \sin\theta$ for the double-slit pattern when the screen is far from the two slits $(L \gg d)$. *(Photo from M. Cagnet, M. Francon, and J. C. Thierr, Atlas of Optical Phenomena, Berlin, Springer-Verlag, 1962)*

We have seen that the interference phenomena arising from two coherent sources depend on the relative phase of the waves at a given point. Furthermore, the phase difference at a given point depends on the *path difference* between the two waves. **The resultant intensity at a point is proportional to the square of the resultant amplitude.** That is, the intensity is proportional to $(E_1 + E_2)^2$. It would be *incorrect* to calculate the resultant intensity by adding the intensities of the individual waves. This procedure would give a different quantity, namely $E_1{}^2 + E_2{}^2$. Finally, $(E_1 + E_2)^2$ has the same *average* value as $E_1{}^2 + E_2{}^2$, when the time average is taken over all values of the phase difference between E_1 and E_2. Hence, the principle of energy conservation is not violated.

CONCEPTUAL PROBLEM 1

Consider a dark fringe in a two-slit interference pattern, at which almost no light energy is arriving. Waves from both slits travel to this point, but the waves cancel. Where does the energy go?

27.3 • CHANGE OF PHASE DUE TO REFLECTION

Young's method of producing two coherent light sources involves illuminating a pair of slits with a single source. Another simple arrangement for producing an interference pattern with a single light source is known as *Lloyd's mirror*. A light source is placed at point *S* close to a mirror, as illustrated in Figure 27.6. Waves can reach the viewing point, *P*, either by the direct path *SP* or by the indirect path involving reflection from the mirror. The reflected ray can be treated as a ray originating from a source at *S'*, located behind the mirror. This source *S'*, which is the image of *S*, can be considered a virtual source.

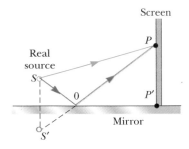

Figure 27.6 Lloyd's mirror. An interference pattern is produced on a screen at *P* as a result of the combination of the direct ray (blue) and the reflected ray (brown). The reflected ray undergoes a phase change of 180°.

At points far from the source, one would expect an interference pattern due to waves from *S* and *S'*, just as is observed for two real coherent sources. An interference pattern is indeed observed. However, the positions of the dark and bright fringes are *reversed* relative to the pattern of two real coherent sources (Young's experiment). This is because the coherent sources at *S* and *S'* differ in phase by 180°. This 180° phase change is produced on reflection.

To illustrate this further, consider the point *P'* at which the mirror meets the screen. This point is equidistant from *S* and *S'*. If path difference alone were responsible for the phase difference, one would expect to see a bright fringe at *P'* (because the path difference is zero for this point), corresponding to the central fringe of the two-slit interference pattern. Instead, one observes a *dark* fringe at *P'* because of the 180° phase change produced by reflection. In general, **an electromagnetic wave undergoes a phase change of 180° on reflection from a medium of higher index of refraction than the one in which it was traveling.**

It is useful to draw an analogy between reflected light waves and the reflections of a transverse wave on a stretched string when the wave meets a boundary (Chapter 13, Section 13.7), as in Figure 27.7. The reflected pulse on a string undergoes a phase change of 180° when it is reflected from the boundary of a denser medium, such as a heavier string, and no phase change when it is reflected from the boundary of a less dense medium. In a similar way, an electromagnetic wave undergoes a 180° phase change when reflected from the boundary of a medium of higher index of refraction than the one in which it was traveling. There is no phase change when

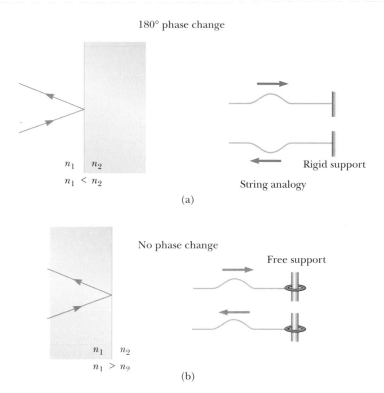

(a)

(b)

Figure 27.7   (a) A ray traveling in medium 1 reflecting from the surface of medium 2 undergoes a 180° phase change. The right side shows the analogy with a reflected pulse on a string fixed at one end. (b) A ray traveling in medium 1 reflecting from the surface of medium 2 with $n_1 > n_2$ undergoes no phase change. The right side shows the analogy with a reflected pulse on a string the end of which is free.

the wave is reflected from a boundary leading to a medium of lower index of refraction. The part of the wave that crosses the boundary undergoes no phase change.

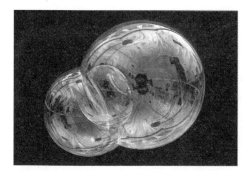
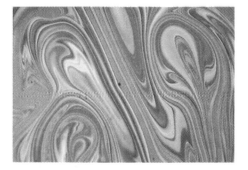

(left) A layer of bubbles on water, produced by soap film. The colors, produced just before the bubbles burst, are due to interference between light rays reflected from the front and back of the thin film of water making the bubble. The colors depend on the thickness of the film, ranging from black where the film is at its thinnest to red as the film gets thicker.   *(Dr. Jeremy Burgess/ Science Photo Library)*   *(right)* Thin film interference. A thin film of oil on water displays interference, as shown by the pattern of colors when white light is incident on the film. The film thickness varies, thereby producing the interesting color pattern.   *(© Tom Branch 1984, Photo Researchers)*

27.4 • INTERFERENCE IN THIN FILMS

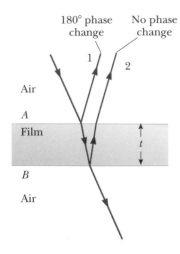

Figure 27.8 Interference in light reflected from a thin film is due to a combination of rays reflected from the upper and lower surfaces of the film.

Interference effects are commonly observed in thin films, such as soap bubbles and thin layers of oil on water. The varied colors observed with ordinary white light result from the interference of waves reflected from the opposite surfaces of the film.

Consider a film of uniform thickness t and index of refraction n, as in Figure 27.8. Let us assume that the light rays traveling in air are nearly normal to the two surfaces of the film. To determine whether the reflected rays interfere constructively or destructively, we must first note the following facts:

- An electromagnetic wave traveling from a medium of index of refraction n_1 toward a medium of index of refraction n_2 undergoes a 180° phase change on reflection when $n_2 > n_1$. There is no phase change in the reflected wave if $n_2 < n_1$.
- The wavelength of light, λ_n, in a medium of index of refraction n is

$$\lambda_n = \frac{\lambda}{n} \qquad \textbf{[27.14]}$$

where λ is the wavelength of light in free space.

Let us apply these rules to the film of Figure 27.8. According to the first rule, ray 1, which is reflected from the upper surface (A), undergoes a phase change of 180° with respect to the incident wave. Ray 2, which is reflected from the lower surface (B), undergoes no phase change with respect to the incident wave. Therefore, ray 1 is 180° out of phase with respect to ray 2, a situation that is equivalent to a path difference of $\lambda_n/2$. However, we must also consider that ray 2 travels an extra distance equal to $2t$ before the waves recombine. For example, if $2t = \lambda_n/2$, rays 1 and 2 will recombine in phase and constructive interference will result. In general, the condition for constructive interference is

$$2t = (m + \tfrac{1}{2})\lambda_n \qquad (m = 0, 1, 2, \ldots) \qquad \textbf{[27.15]}$$

This condition takes into account two factors: (1) the difference in optical path length for the two rays (the term $m\lambda_n$) and (2) the 180° phase change on reflection (the term $\lambda_n/2$). Because $\lambda_n = \lambda/n$, we can write Equation 27.15 in the form

Constructive interference in •
thin films

$$2nt = (m + \tfrac{1}{2})\lambda \qquad (m = 0, 1, 2, \ldots) \qquad \textbf{[27.16]}$$

If the extra distance $2t$ traveled by ray 2 corresponds to a multiple of λ_n, the two waves will combine out of phase and destructive interference will result. The general equation for destructive interference is

Destructive interference in •
thin films

$$2nt = m\lambda \qquad (m = 0, 1, 2, \ldots) \qquad \textbf{[27.17]}$$

It is important that you realize that two factors influence interference: (1) phase changes and (2) differences in travel distance. The preceding conditions for constructive and destructive interference are valid when the medium above the top surface of the film is the same as the medium below the bottom surface. The surrounding medium may have a refractive index less than or greater than that of the film. In either case, the rays reflected from the two surfaces will be out of phase by

180°. If the film is placed between two different media, one with $n < n_{film}$ and the other with $n > n_{film}$, the conditions for constructive and destructive interference are reversed. In this case, either there is a phase change of 180° for both ray 1 reflecting from surface A and ray 2 reflecting from surface B, or there is no phase change for either ray; hence, the net change in relative phase due to the reflections is zero.

CONCEPTUAL PROBLEM 2

In a laboratory accident, you spill two liquids onto water, neither of which mixes with the water. They both form thin films on the water surface. You notice, as the films become very thin as they spread, that one film becomes bright and the other black in reflected light. Why might this be?

CONCEPTUAL PROBLEM 3

In Chapter 9, the Michelson interferometer was discussed. In this device, a light beam is split into two perpendicular paths and then reflected from mirrors in order to recombine, forming an interference pattern. Suppose you are observing the interference pattern in your laboratory and a joking colleague holds a lit match in the light path of one arm of the interferometer. Will this have an effect on the interference pattern?

CONCEPTUAL PROBLEM 4

In our discussion of thin film interference, we looked at light *reflecting* from a thin film. Consider now the light *transmitting* through a thin film, with air on both sides of the film. Consider one light ray, the direct ray, that transmits through the film without reflecting. Consider a second ray, the reflected ray, that transmits through the first surface, reflects back from the second, reflects again from the first, and then transmits out into the air, parallel to the direct ray. For normal incidence, how thick must the film be, in terms of the wavelength of the light, for the outgoing rays to interfere destructively? Is it the same thickness as for reflected destructive interference?

PROBLEM-SOLVING STRATEGY • **Thin Film Interference**

The following features should be kept in mind while working thin-film interference problems:

1. Identify the thin film causing the interference.
2. The type of interference that occurs is determined by the phase relationship between the portion of the wave reflected at the upper surface of the film and the portion reflected at the lower surface.
3. Phase differences between the two portions of the wave have two causes: (a) differences in the distances traveled by the two portions and (b) phase changes occurring on reflection. *Both* causes must be considered when you are determining which type of interference occurs.
4. When distance and phase changes on reflection are both taken into account, the interference will be constructive if the phase difference between the two waves is an integral multiple of λ, and destructive if the phase difference is $\lambda/2$, $3\lambda/2$, $5\lambda/2$, and so forth.

Example 27.2   Interference in a Soap Film

Calculate the minimum thickness of a soap bubble film ($n = 1.33$) that results in constructive interference in the reflected light if the film is illuminated with light the wavelength in free space of which is 600 nm.

$$t = \frac{\lambda}{4n} = \frac{600 \text{ nm}}{4(1.33)} = 113 \text{ nm}$$

Solution   The minimum film thickness for constructive interference in the reflected light corresponds to $m = 0$ in Equation 27.16. This gives $2nt = \lambda/2$, or

EXERCISE 1   What other film thicknesses produce constructive interference?   Answer   338 nm, 564 nm, 789 nm, and so on

Example 27.3   Nonreflecting Coatings for Solar Cells

Semiconductors such as silicon are used to fabricate solar cells—devices that generate electricity when exposed to sunlight. Solar cells are often coated with a transparent thin film, such as silicon monoxide (SiO, $n = 1.45$), in order to minimize reflective losses from the surface. A silicon solar cell ($n = 3.5$) is coated with a thin film of silicon monoxide for this purpose (Fig. 27.9). Determine the minimum film thickness that produces the least reflection at a wavelength of 550 nm, which is the center of the visible spectrum.

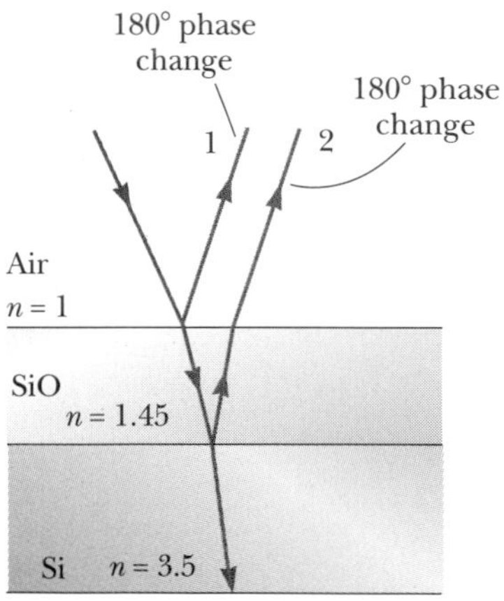

Reasoning   The reflected light is a minimum when rays 1 and 2 in Figure 27.9 meet the condition of destructive interference. Note that both rays undergo a 180° phase change on reflection in this case, one from the upper and one from the lower surface. Hence, the net change in phase is zero due to reflection, and the condition for reflection minimum requires a path difference of $\lambda_n/2$; hence, $2t = \lambda/2n$.

Solution   Because $2t = \lambda/2n$, the required thickness is

$$t = \frac{\lambda}{4n} = \frac{550 \text{ nm}}{4(1.45)} = 94.8 \text{ nm}$$

Figure 27.9   (Example 27.3) Reflective losses from a silicon solar cell are minimized by coating it with a thin film of silicon monoxide.

Typically, such antireflecting coatings reduce the reflective loss from 30% (with no coating) to 10% (with coating), thereby increasing the cell's efficiency, because more light is available to create charge carriers in the cell. In reality, the coating is never perfectly nonreflecting because the required thickness is wavelength-dependent and the incident light covers a wide range of wavelengths.

Glass lenses used in cameras and other optical instruments are usually coated with a transparent thin film, such as magnesium fluoride (MgF_2), to reduce or eliminate unwanted reflection, and thus such coatings enhance the transmission of light through the lenses.

Example 27.4   Interference in a Wedge-Shaped Film

A thin, wedge-shaped film of refractive index n is illuminated with monochromatic light of wavelength λ, as illustrated in Figure 27.10. Describe the interference pattern observed for this case.

Reasoning and Solution   The interference pattern is that of a thin film of variable thickness surrounded by air. Hence, the pattern is a series of alternating bright and dark parallel bands. A dark band corresponding to destructive interference appears at point O, the apex, because the upper reflected ray undergoes a 180° phase change and the lower one does not. According to Equation 27.17, other dark bands appear when $2nt = m\lambda$, so that $t_1 = \lambda/2n$, $t_2 = \lambda/n$, $t_3 = 3\lambda/2n$, and so on. In a similar way, bright bands are observed when the thickness satisfies the condition $2nt = (m + \frac{1}{2})\lambda$, corresponding to thicknesses of $\lambda/4n$, $3\lambda/4n$, $5\lambda/4n$, and so on. If white light is used, bands of different colors are observed at different points, corresponding to the different wavelengths of light.

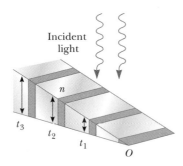

Figure 27.10 (Example 27.4) Interference bands in reflected light can be observed by illuminating a wedge-shaped film with monochromatic light. The dark blue areas correspond to positions of destructive interference.

27.5 • DIFFRACTION

Suppose a light beam is incident on two slits, as in Young's double-slit experiment. If the light truly traveled in straight-line paths after passing through the slits, as in Figure 27.11a, the waves would not overlap and no interference pattern would be seen. Instead, Huygens' principle requires that the waves spread out from the slits, as shown in Figure 27.11b. In other words, the light deviates from a straight-line path and enters the region that would otherwise be shadowed. This divergence of light from its initial line of travel is called **diffraction.**

In general, diffraction occurs when waves pass through small openings, around obstacles, or by sharp edges. For example, when a narrow slit is placed between a distant light source (or a laser beam) and a screen, the light produces a diffraction pattern like that in Figure 27.12. The pattern consists of a broad, intense central band, the **central maximum,** flanked by a series of narrower, less intense secondary

Figure 27.12 The diffraction pattern that appears on a screen when light passes through a narrow vertical slit. The pattern consists of a broad central band and a series of less intense and narrower side bands.

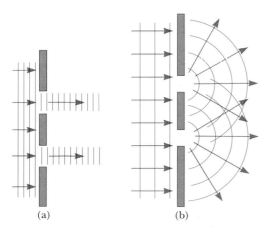

(a)   (b)

Figure 27.11 (a) If light waves did not spread out after passing through the slits, no interference would occur. (b) The light waves from the two slits overlap as they spread out, filling the expected shadowed regions with light and producing interference fringes.

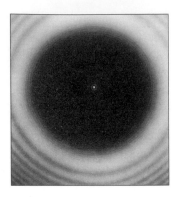

Figure 27.13   Diffraction pattern of a penny, taken with the penny midway between screen and source.   *(Courtesy of P. M. Rinard, from Am. J. Phys. 44:70, 1976)*

Figure 27.14   (a) Fraunhofer diffraction pattern of a single slit. The pattern consists of a central bright region flanked by much weaker maxima alternating with dark bands. (Note that this is not to scale.) (b) Photograph of a single-slit Fraunhofer diffraction pattern.   *(From M. Cagnet, M. Francon, and J. C. Thierr, Atlas of Optical Phenomena, Berlin, Springer-Verlag, 1962, plate 18)*

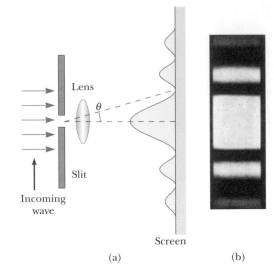

bands (called **secondary maxima**) and a series of dark bands, or **minima.** This cannot be explained within the framework of geometric optics, which says that light rays traveling in straight lines should cast a sharp rendition of the slit on the screen.

Figure 27.13 shows the diffraction pattern and shadow of a penny. The pattern consists of the shadow, a bright spot at its center, and a series of bright and dark bands of light near the edge of the shadow. The bright spot at the center (called the *Arago bright spot* after its discoverer, Dominique Arago) can be explained through the wave theory of light, which predicts constructive interference at this point. In contrast, from the viewpoint of geometric optics, the center of the pattern would be completely screened by the penny, and so one would never observe a central bright spot.

Fraunhofer diffraction occurs when the rays reaching the observing screen are approximately parallel. This can be achieved experimentally either by placing the observing screen far from the slit or by using a converging lens to focus the parallel rays on the screen, as in Figure 27.14a. A bright fringe is observed along the axis at $\theta = 0$, with alternating dark and bright fringes on each side of the central bright fringe. Figure 27.14b is a photograph of a single-slit Fraunhofer diffraction pattern.

Single-Slit Diffraction

Until now we have assumed that slits are point sources of light. In this section we shall determine how their finite widths are the basis for understanding the nature of the Fraunhofer diffraction pattern produced by a single slit.

We can deduce some important features of this problem by examining waves coming from various portions of the slit, as shown in Figure 27.15. According to Huygens' principle, **each portion of the slit acts as a source of waves. Hence, light from one portion of the slit can interfere with light from another portion,** and the resultant intensity on the screen depends on the direction θ.

To analyze the diffraction pattern, it is convenient to divide the slit into two halves, as in Figure 27.15. All the waves that originate at the slit are in phase. Con-

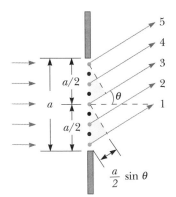

Figure 27.15   Diffraction of light by a narrow slit of width *a*. Each portion of the slit acts as a point source of waves. The path difference between rays 1 and 3 or between rays 2 and 4 is $(a/2)\sin\theta$. (Note that this is not to scale.)

sider waves 1 and 3, which originate at the bottom and center of the slit, respectively. To reach the same point on the viewing screen, wave 1 travels farther than wave 3 by an amount equal to the path difference $(a/2) \sin \theta$, where a is the width of the slit. In a similar way, the path difference between waves 3 and 5 is also $(a/2) \sin \theta$. If the path difference is exactly one half of a wavelength (corresponding to a phase difference of 180°), the two waves cancel each other and destructive interference results. This is true, in fact, for any two waves that originate at points separated by half the slit width, because the phase difference between two such points is 180°. Therefore, waves from the upper half of the slit interfere *destructively* with waves from the lower half of the slit when

$$\frac{a}{2} \sin \theta = \frac{\lambda}{2}$$

or when

$$\sin \theta = \frac{\lambda}{a}$$

If we divide the slit into four parts rather than two and use similar reasoning, we find that the screen is also dark when

$$\sin \theta = \frac{2\lambda}{a}$$

Likewise, we can divide the slit into six parts and show that darkness occurs on the screen when

$$\sin \theta = \frac{3\lambda}{a}$$

Therefore, the general condition for **destructive interference** is

$$\sin \theta = m \frac{\lambda}{a} \qquad (m = \pm 1, \pm 2, \pm 3, \ldots) \qquad \text{[27.18]}$$

• *Condition for destructive interference*

Equation 27.18 gives the values of θ for which the diffraction pattern has zero intensity—that is, a dark fringe is formed. However, Equation 27.18 tells us nothing about the variation in intensity along the screen. The general features of the intensity distribution are shown in Figure 27.16: a broad central bright fringe flanked by

Figure 27.16 Positions of the minima for the Fraunhofer diffraction pattern of a single slit of width a. The pattern is obtained only if $L \gg a$. (Note that this is not to scale.)

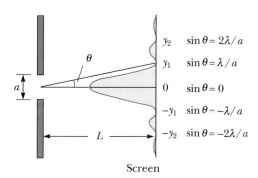

$y_2 \quad \sin \theta = 2\lambda/a$

$y_1 \quad \sin \theta = \lambda/a$

$0 \quad \sin \theta = 0$

$-y_1 \quad \sin \theta = -\lambda/a$

$-y_2 \quad \sin \theta = -2\lambda/a$

Screen

much weaker, alternating bright fringes. The various dark fringes (points of zero intensity) occur at the values of θ that satisfy Equation 27.18. The position of the points of constructive interference lie approximately halfway between the dark fringes. Note that the central bright fringe is twice as wide as the weaker maxima.

Thinking Physics 4

If a classroom door is open even just a small amount, you can hear sounds coming from the hallway. Yet you cannot see what is going on in the hallway. Why is there this difference?

Reasoning   The space between the slightly open door and the wall is acting as a single slit for waves. Sound waves have wavelengths larger than the slit width, so sound is effectively diffracted by the opening and spread throughout the room. Light wavelengths are much smaller than the slit width, so there is virtually no diffraction for the light. You must have a direct line of sight view to detect the light waves.

Example 27.5   Where Are the Dark Fringes?

Light of wavelength 580 nm is incident on a slit of width 0.300 mm. The observing screen is 2.00 m from the slit. Find the positions of the first dark fringes and the width of the central bright fringe.

Solution   The first dark fringes that flank the central bright fringe correspond to $m = \pm 1$ in Equation 27.18. Hence, we find that

$$\sin \theta = \pm \frac{\lambda}{a} = \pm \frac{5.80 \times 10^{-7} \text{ m}}{0.300 \times 10^{-3} \text{ m}} = \pm 1.933 \times 10^{-3}$$

From the triangle in Figure 27.16, note that $\tan \theta = y_1/L$. Because θ is very small, we can use the approximation $\sin \theta \approx \tan \theta$, so that $\sin \theta \approx y_1/L$. Therefore, the positions of the first minima measured from the central axis are given by

$$y_1 \approx L \sin \theta = \pm L \frac{\lambda}{a} = \pm 3.87 \times 10^{-3} \text{ m}$$

The positive and negative signs correspond to the dark fringes on either side of the central bright fringe. Hence, the width of the central bright fringe is equal to $2|y_1| = 7.73 \times 10^{-3}$ m = 7.73 mm. Note that this value is much larger than the width of the slit. However, as the width of the slit is increased, the diffraction pattern narrows, corresponding to smaller values of θ. In fact, for large values of a, the various maxima and minima are so closely spaced that the only thing observed is a large central bright area, which resembles the geometric image of the slit. This matter is of great importance in the design of lenses used in telescopes, microscopes, and other optical instruments.

EXERCISE 2   Determine the width of the first-order bright fringe.   Answer   3.87 mm

27.6 • RESOLUTION OF SINGLE-SLIT AND CIRCULAR APERTURES

The ability of optical systems to distinguish between closely spaced objects is limited because of the wave nature of light. To understand this difficulty, consider Figure 27.17, which shows two light sources far from a narrow slit of width a. The sources can be considered as two point sources, S_1 and S_2, that are noncoherent. For example, they could be two distant stars. If no diffraction occurred, one would observe two distinct bright spots (or images) on the screen at the right in the figure. However, because of diffraction, each source is imaged as a bright central region flanked

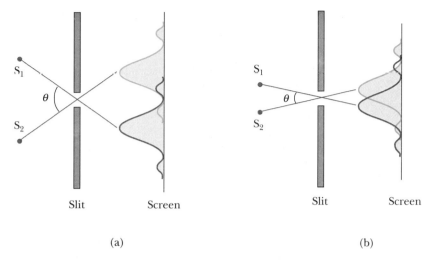

Figure 27.17 Two point sources far from a small aperture each produce a diffraction pattern. (a) The angle subtended by the sources at the aperture is large enough so that the diffraction patterns are distinguishable. (b) The angle subtended by the sources is so small that their diffraction patterns overlap and the images are not well resolved. (Note that the angles are greatly exaggerated.)

by weaker bright and dark bands. What is observed on the screen is the sum of two diffraction patterns, one from S_1 and the other from S_2.

If the two sources are far enough apart to ensure that their central maxima do not overlap, as in Figure 27.17a, their images can be distinguished and are said to be *resolved*. If the sources are close together, however, as in Figure 27.17b, the two central maxima may overlap and the images are *not resolved*. To decide when two images are resolved, the following condition is often used:

> When the central maximum of one image falls on the first minimum of another image, the images are said to be *just resolved*. This limiting condition of resolution is known as **Rayleigh's criterion.**

• *Rayleigh's criterion*

Figure 27.18 shows the diffraction patterns for three situations. When the objects are far apart, their images are well resolved (Fig. 27.18a). The images are just resolved when their angular separation satisfies Rayleigh's criterion (Fig. 27.18b). Finally, the images are not resolved in Figure 27.18c.

From Rayleigh's criterion, we can determine the minimum angular separation, θ_{min}, subtended by the sources at the slit such that their images are just resolved. In Section 27.4, we found that the first minimum in a single-slit diffraction pattern occurs at the angle that satisfies the relationship

$$\sin \theta = \frac{\lambda}{a}$$

where a is the width of the slit. According to Rayleigh's criterion, this expression gives the smallest angular separation for which the two images are resolved. Be-

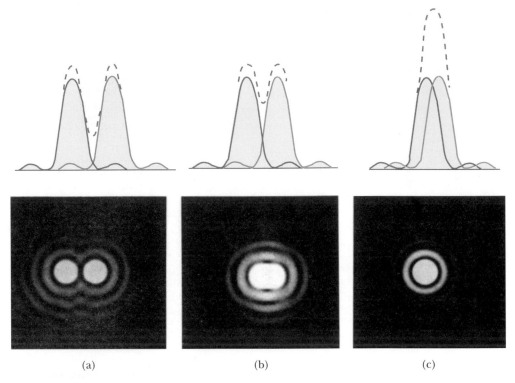

Figure 27.18   The diffraction patterns of two point sources (solid curves) and the resultant pattern (dashed curves), for various angular separations of the sources. In each case, the dashed curve is the sum of the two solid curves. (a) The sources are far apart, and the patterns are well resolved. (b) The sources are closer together, and the patterns are just resolved. (c) The sources are so close together that the patterns are not resolved.   *(From M. Cagnet, M. Francon, and J. C. Thierr, Atlas of Optical Phenomena, Berlin, Springer-Verlag, 1962, plate 16)*

cause $\lambda \ll a$ in most situations, $\sin \theta$ is small and we can use the approximation $\sin \theta \approx \theta$. Therefore, the limiting angle of resolution for a slit of width a is

Limiting angle of resolution •
for a slit

$$\theta_{\min} = \frac{\lambda}{a} \qquad \textbf{[27.19]}$$

where $\theta_{\min}$ is expressed in radians. Hence, the angle subtended by the two sources at the slit must be *greater* than λ/a if the images are to be resolved.

   Many optical systems use circular apertures rather than slits. The diffraction pattern of a circular aperture, illustrated in Figure 27.19, consists of a central circular bright disk surrounded by progressively fainter rings. Analysis shows that the limiting angle of resolution of the circular aperture is

$$\theta_{\min} = 1.22 \frac{\lambda}{D} \qquad \textbf{[27.20]}$$

where D is the diameter of the aperture. Note that Equation 27.20 is similar to Equation 27.19 except for the factor of 1.22, which arises from a complex mathematical analysis of diffraction from the circular aperture.

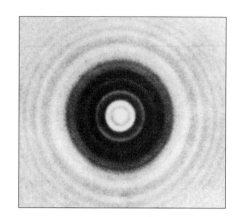

Figure 27.19 The diffraction pattern of a circular aperture consists of a central bright disk surrounded by concentric bright and dark rings. *(From M. Cagnet, M. Francon, and J. C. Thierr, Atlas of Optical Phenomena, Berlin, Springer-Verlag, 1962, plate 34)*

Thinking Physics 5

Cats' eyes have vertical pupils when in dim light. Which would cats be more successful at resolving at night—headlights on a distant car or vertically separated running lights on a distant boat's mast?

Reasoning The effective slit width in the vertical direction of the cat's eye is larger than that in the horizontal direction. Thus, it has more resolving power for lights separated in the vertical direction and would be more effective at resolving the mast lights on the boat. ∎

CONCEPTUAL PROBLEM 5

Suppose you are observing a binary star with a telescope and are having difficulty resolving the two stars. You decide to use a colored filter to help you. Should you choose a blue filter or a red filter?

Example 27.6 Resolution of a Telescope

The Hale telescope at Mount Palomar in California has a diameter of 200 in. What is its limiting angle of resolution for 600-nm light?

Solution Because $\lambda = 6.00 \times 10^{-7}$ m and $D = 200$ in. = 5.08 m, Equation 27.20 gives

$$\theta_{min} = 1.22 \frac{\lambda}{D} = 1.22 \left(\frac{6.00 \times 10^{-7} \text{ m}}{5.08 \text{ m}} \right)$$

$$= 1.44 \times 10^{-7} \text{ rad} \cong 0.03 \text{ s of arc}$$

Therefore, any two stars that subtend an angle greater than or equal to this value are resolved (assuming ideal atmospheric conditions).

The Hale telescope can never reach its diffraction limit. Instead, its limiting angle of resolution is always set by atmospheric blurring. This seeing limit is usually on the order of 1 s of arc and is never smaller than 0.1 s of arc. (The Hubble Space Telescope can take higher quality photographs because it does not suffer from atmospheric limitations.)

EXERCISE 3 The large radio telescope at Arecibo, Puerto Rico, has a diameter of 305 m and is designed to detect 0.75-m radio waves. Calculate the minimum angle of resolution for this telescope and compare your answer with that of the Hale telescope.   Answer 3.0×10^{-3} rad (10 min of arc), more than 10 000 times larger than the Hale minimum

27.7 • THE DIFFRACTION GRATING

The diffraction grating, a useful device for analyzing light sources, consists of a large number of equally spaced parallel slits. A grating can be made by cutting parallel, equally spaced grooves on a glass or metal plate with a precision ruling machine. In a transmission grating, the spaces between lines are transparent to the light and hence act as separate slits. Gratings with many lines very close to each other can have very small slit spacings. For example, a diffraction grating ruled with 5000 lines/cm has a slit spacing of $d = (1/5000)$ cm $= 2 \times 10^{-4}$ cm.

Figure 27.20 is a schematic diagram of a section of a flat diffraction grating. A plane wave is incident from the left, normal to the plane of the grating. A converging lens can be used to bring the rays together at the point P. The pattern observed on the screen is the result of the combined effects of interference and diffraction. Each slit produces diffraction, and the diffracted beams interfere with each other to produce the final pattern. Moreover, each slit acts as a source of waves, with all waves starting at the slits in phase. However, for some arbitrary direction θ measured from the horizontal, the waves must travel different path lengths before reaching a particular point P on the screen. From Figure 27.20, note that the path difference between waves from any two adjacent slits is equal to $d \sin \theta$. If this path difference equals one wavelength or some integral multiple of a wavelength, waves from all slits will be in phase at P and a bright line will be observed. Therefore, when the light is incident normal to the plane of the grating, the condition for *maxima* in the interference pattern at the angle θ is

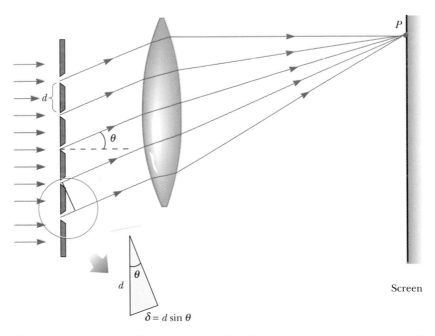

Figure 27.20   Side view of a diffraction grating. The slit separation is d, and the path difference between adjacent slits is $d \sin \theta$.

$$d \sin \theta = m\lambda \qquad (m = 0, 1, 2, 3, \ldots) \qquad \textbf{[27.21]}$$

This expression can be used to calculate the wavelength from a knowledge of the grating spacing and the angle of deviation, θ. If the incident radiation contains several wavelengths, the mth-order maximum for each wavelength occurs at a specific angle. All wavelengths are seen at $\theta = 0$, corresponding to $m = 0$.

The intensity distribution for a diffraction grating is shown in Figure 27.21. If the source contains various wavelengths, a spectrum of lines at different positions for different order numbers will be observed. Note the sharpness of the principal maxima and the broad range of dark areas. This is in contrast to the broad, bright fringes characteristic of the two-slit interference pattern (Fig. 27.5).

A simple arrangement for measuring angles in a diffraction pattern is shown in Figure 27.22. This is a form of a diffraction-grating spectrometer. The light to be analyzed passes through a slit, and a parallel beam of light exits from the collimator perpendicular to the grating. The diffracted light leaves the grating at angles that satisfy Equation 27.21. A telescope is used to view the image of the slit. The wavelength can be determined by measuring the precise angles at which the images of the slit appear for the various orders.

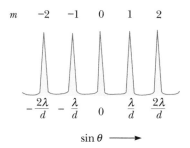

Figure 27.21 Intensity versus $\sin \theta$ for a diffraction grating. The zeroth-, first-, and second-order maxima are shown.

Thinking Physics 6

Light reflected from the surface of a compact disc has a multicolored appearance, as shown in Figure 27.23. Furthermore, the observation depends on the orientation of the disc relative to the eye and the position of the light source. Explain how this works.

Reasoning The surface of a compact disc has a spiral grooved track (with a spacing of approximately 1 μm) that acts as a reflection grating. The light scattered by these closely spaced grooves interferes constructively only in certain directions that depend on the wavelength and on the direction of the incident light. Any one section of the disc serves as a diffraction grating for white light, sending different colors in different directions. The different colors you see when viewing one section of the disc changes as the light source, the disc, or you move to change the angles of incidence or diffraction.

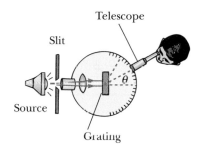

Figure 27.22 Diagram of a diffraction grating spectrometer. The collimated beam incident on the grating is diffracted into the various orders at the angles θ that satisfy the equation $d \sin \theta = m\lambda$, where $m = 1, 2, \ldots$.

Figure 27.23 (Thinking Physics 6) A compact disc acts as a diffraction grating when observed under white light. (© *Kristen Brochmann 1991, Fundamental Photographs*)

Thinking Physics 7

White light enters through an opening in an opaque box, exits through an opening on the other side of the box, and a spectrum of colors appears on the wall. How could you determine whether the box contains a prism or a diffraction grating?

Reasoning   The determination could be made by noticing the order of the colors in the spectrum relative to the direction of the original beam of white light. For a prism, in which the separation of light is a result of dispersion, the violet light will be refracted more than the red light, so the order of the spectrum will be from red, closest to the original direction, to violet. For a diffraction grating, the angle of diffraction increases with wavelength. Thus, the spectrum from the diffraction grating will have colors in the order of violet, closest to the original direction, to red.

CONCEPTUAL PROBLEM 6

If laser light is reflected from a phonograph record or a compact disc, a diffraction pattern appears. This is due to the fact that both devices contain parallel tracks of information that act as a reflection diffraction grating. Which device, record or compact disc, will result in diffraction maxima that are farther apart?

CONCEPTUAL PROBLEM 7

Astronomers often observe *occultations*, in which a star passes behind another object such as the Moon. When such an event occurs, it is found that the intensity of light from the star does not suddenly diminish as it passes behind the edge of the Moon. Instead, the intensity *fluctuates* for a short time before dropping to zero. Why would this happen?

CONCEPTUAL PROBLEM 8

When you receive a chest x-ray at a hospital, the x-rays pass through a series of parallel ribs in your chest. Do the ribs act as a diffraction grating for x-rays?

Example 27.7   The Orders of a Diffraction Grating

Monochromatic light from a helium–neon laser ($\lambda = 632.8$ nm) is incident normally on a diffraction grating containing 6000 lines/cm. Find the angles at which the first-order, second-order, and third-order maxima can be observed.

Solution   First, we must calculate the slit separation, which is equal to the inverse of the number of lines per cm:

$$d = (1/6000) \text{ cm} = 1.667 \times 10^{-4} \text{ cm} = 1667 \text{ nm}$$

For the first-order maximum ($m = 1$), we get

$$\sin \theta_1 = \frac{\lambda}{d} = \frac{632.8 \text{ nm}}{1667 \text{ nm}} = 0.3797$$

$$\theta_1 = 22.31°$$

For $m = 2$, we find

$$\sin \theta_2 = \frac{2\lambda}{d} = \frac{2(632.8 \text{ nm})}{1667 \text{ nm}} = 0.7592$$

$$\theta_2 = 49.41°$$

For $m = 3$, we find $\sin \theta_3 = 1.139$. Because $\sin \theta$ cannot exceed unity, this does not represent a realistic solution. Hence, only zeroth-, first-, and second-order maxima are observed for this situation.

Resolving Power of the Diffraction Grating

The diffraction grating is most useful for taking accurate wavelength measurements. Like the prism, the diffraction grating can be used to disperse a spectrum into its components. Of the two devices, the grating is more precise if one wants to distinguish between two closely spaced wavelengths.

If λ_1 and λ_2 are the two nearly equal wavelengths between which the spectrometer can just barely distinguish, the **resolving power,** R, is

$$R \equiv \frac{\lambda}{\lambda_2 - \lambda_1} = \frac{\lambda}{\Delta\lambda} \qquad \text{[27.22]}$$

• *Resolving power*

where $\lambda = (\lambda_1 + \lambda_2)/2$ and $\Delta\lambda = \lambda_2 - \lambda_1$. It is left to you to show (Problem 27.52) that if N lines of the grating are illuminated, the resolving power in the mth-order diffraction equals the product Nm:

$$R = Nm \qquad \text{[27.23]}$$

• *Resolving power of a grating*

Thus, resolving power increases with increasing order number.

Furthermore, R is large for a grating with a large number of illuminated slits. Note that for $m = 0$, $R = 0$, which signifies that **all wavelengths are indistinguishable for the zeroth-order maximum.** However, consider the second-order diffraction pattern ($m = 2$) of a grating that has 5000 rulings illuminated by the light source. The resolving power of such a grating in second order is $R = 5\,000 \times 2 = 10\,000$. Therefore, the *minimum* wavelength separation between two spectral lines that can be just resolved, assuming a mean wavelength of 600 nm, is $\Delta\lambda = \lambda/R = 6 \times 10^{-2}$ nm. For the third-order principal maximum, we find that $R = 15\,000$ and $\Delta\lambda = 4 \times 10^{-2}$ nm, and so on.

Example 27.8   Resolving the Sodium Spectral Lines

Two strong lines in the spectrum of sodium have wavelengths of 589.00 nm and 589.59 nm. (a) What must the resolving power of a grating be in order to distinguish these wavelengths?

Solution

$$R = \frac{\lambda}{\Delta\lambda} = \frac{589.30 \text{ nm}}{589.59 \text{ nm} - 589.00 \text{ nm}} = \frac{589.30}{0.59} = 999$$

(b) In order to resolve these lines in the second-order spectrum ($m = 2$), how many lines of grating must be illuminated?

Solution   From Equation 27.23 and the results to part (a), we find

$$N = \frac{R}{m} = \frac{999}{2} = 500 \text{ lines}$$

27.8 • DIFFRACTION OF X-RAYS BY CRYSTALS

O P T I O N A L

In principle, the wavelength of any electromagnetic wave can be determined if a grating of the proper spacing (on the order of λ) is available. **X-rays,** discovered in 1895 by W. Roentgen (1845–1923), are electromagnetic waves with very short wavelengths (on the order of 10^{-10} m = 0.1 nm). Obviously, it would be impossible to construct a grating with such a small spacing. However, the atomic spacing in a solid is known to be about 10^{-10} m. In 1913, Max von Laue (1879–1960) suggested

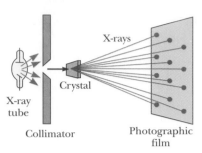

Figure 27.24   Schematic diagram of the technique used to observe the diffraction of x-rays by a crystal. The array of spots formed on the film is called a Laue pattern.

that the regular array of atoms in a crystal could act as a three-dimensional diffraction grating for x-rays. Subsequent experiments confirmed his prediction. The diffraction patterns that are observed are complicated because of the three-dimensional nature of the crystal. Nevertheless, x-ray diffraction is an invaluable technique for elucidating crystalline structures and for understanding the structure of matter.

Figure 27.24 is one experimental arrangement for observing x-ray diffraction from a crystal. A collimated beam of x-rays with a continuous range of wavelengths is incident on a crystal. The diffracted beams are very intense in certain directions, corresponding to constructive interference from waves reflected from layers of atoms in the crystal. The diffracted beams can be detected by a photographic film, and they form an array of spots known as a Laue pattern. The crystalline structure is deduced by analyzing the positions and intensities of the various spots in the pattern.

The arrangement of atoms in a crystal of NaCl is shown in Figure 27.25. The red spheres represent Na$^+$ ions, and the blue spheres represent Cl$^-$ ions. Each unit cell (the set of atoms that repeats through the crystal) contains four Na$^+$ and four Cl$^-$ ions. The unit cell is a cube the edge length of which is a.

The ions in a crystal lie in various planes, as shown by the shaded areas in Figure 27.25. Now suppose an incident x-ray beam makes an angle of θ with one of the planes, as in Figure 27.26. The beam can be reflected from both the upper plane and the lower one. However, the geometric construction in Figure 27.26 shows that the beam reflected from the lower surface travels farther than the beam reflected from the upper surface. The path difference between the two beams is $2d \sin \theta$. The two beams will reinforce each other (constructive interference) when this path difference equals some integral multiple of the wavelength λ. The same

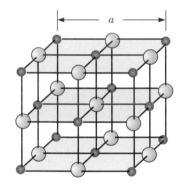

Figure 27.25   A model of the crystalline structure of sodium chloride. The blue spheres represent the Cl$^-$ ions, and the red spheres represent the Na$^+$ ions. The length of the cube edge is $a = 0.562737$ nm.

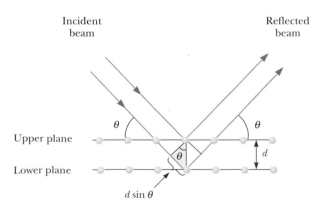

Figure 27.26   A two-dimensional description of the reflection of an x-ray beam from two parallel crystalline planes separated by a distance d. The beam reflected from the lower plane travels farther than the one reflected from the upper plane by a distance equal to $2d \sin \theta$.

is true of reflection from the entire family of parallel planes. Hence, the condition for constructive interference (maxima in the reflected wave) is

$$2d \sin \theta = m\lambda \qquad (m = 1, 2, 3, \ldots) \qquad \textbf{[27.24]}$$

• *Bragg's law*

This condition is known as **Bragg's law** after W. L. Bragg (1890–1971), who first derived the relationship. If the wavelength and diffraction angle are measured, Equation 27.24 can be used to calculate the spacing between atomic planes.

SUMMARY

Interference of light waves is the result of the linear superposition of two or more waves at a given point. A sustained interference pattern is observed if (1) the sources are coherent, (2) the sources have identical wavelengths, and (3) the superposition principle is applicable.

In Young's double-slit experiment, two slits separated by a distance of d are illuminated by a monochromatic light source. An interference pattern consisting of bright and dark fringes is observed on a screen that is a distance of L from the slits. The condition for **constructive interference** is

$$d \sin \theta = m\lambda \qquad (m = 0, \pm 1, \pm 2, \ldots) \qquad \textbf{[27.2]}$$

The condition for **destructive interference** is

$$d \sin \theta = (m + \tfrac{1}{2})\lambda \qquad (m = 0, \pm 1, \pm 2, \ldots) \qquad \textbf{[27.3]}$$

The number m is called the **order number** of the fringe.

The **average intensity** of the double-slit interference pattern is

$$I_{av} = I_0 \cos^2 \left(\frac{\pi d \sin \theta}{\lambda} \right) \qquad \textbf{[27.12]}$$

where I_0 is the maximum intensity on the screen.

An electromagnetic wave traveling from a medium of index of refraction n_1 toward a medium of index of refraction n_2 undergoes a 180° phase change on reflection when $n_2 > n_1$. There is no phase change in the reflected wave if $n_2 < n_1$.

In a medium of refractive index n, the wavelength of light is

$$\lambda_n = \frac{\lambda}{n} \qquad \textbf{[27.14]}$$

where λ is the wavelength of light in free space.

The condition for constructive interference in a film of thickness t and refractive index n with the same medium on both sides of the film is given by

$$2nt = (m + \tfrac{1}{2})\lambda \qquad (m = 0, 1, 2, \ldots) \qquad \textbf{[27.16]}$$

In a similar way, the condition for destructive interference is

$$2nt = m\lambda \qquad (m = 0, 1, 2, \ldots) \qquad \textbf{[27.17]}$$

Diffraction, defined as the deviation of light from a straight-line path when the light passes through an aperture or around obstacles, arises from the interference of a large number or continuous distribution of coherent sources.

The **Fraunhofer diffraction pattern** produced by a *single slit* of width a on a distant screen consists of a central, bright maximum and alternating bright and dark regions of much lower intensities. The angles θ at which the diffraction pattern has *zero* intensity are given by

$$\sin \theta = m \frac{\lambda}{a} \qquad (m = \pm 1, \pm 2, \pm 3, \ldots) \qquad \text{[27.18]}$$

Rayleigh's criterion, which is a limiting condition of resolution, says that two images formed by an aperture are just distinguishable if the central maximum of the diffraction pattern for one image falls on the first minimum of the other image. The limiting angle of resolution for a slit of width a is given by $\theta_{min} = \lambda/a$, and the limiting angle of resolution for a circular aperture of diameter D is given by $\theta_{min} = 1.22\lambda/D$.

A **diffraction grating** consists of a large number of equally spaced, identical slits. The condition for intensity maxima in the interference pattern of a diffraction grating for normal incidence is

$$d \sin \theta = m\lambda \qquad (m = 0, 1, 2, 3, \ldots) \qquad \text{[27.21]}$$

where d is the spacing between adjacent slits and m is the order number of the diffraction pattern. The resolving power of a diffraction grating in the mth order of the diffraction pattern is $R = Nm$, where N is the number of rulings in the grating.

CONCEPTUAL QUESTIONS

1. What is the necessary condition on the path-length difference between two waves that interfere (a) constructively and (b) destructively?

2. Explain why two flashlights held close together do not produce an interference pattern on a distant screen.

3. If Young's double-slit experiment were performed under water, how would the observed interference pattern be affected?

4. In Young's double-slit experiment, why do we use monochromatic light? If white light were used, how would the pattern change?

5. As a soap bubble evaporates, it appears black just before it breaks. Explain this phenomenon in terms of the phase changes that occur on reflection from the two surfaces of the soap film.

6. An oil film on water appears brightest at the outer regions, where it is thinnest. From this information, what can you say about the index of refraction of oil relative to that of water?

7. A soap film on a wire loop and held in air appears dark in the thinnest regions when observed by reflected light and shows a variety of colors in thicker regions, as in Figure Q27.7. Explain.

8. A simple way of observing an interference pattern is to look at a distant light source through a stretched handkerchief or an opened umbrella. Explain how this works.

9. In order to observe interference in a thin film, why must the film not be very thick (on the order of a few wavelengths)?

Figure Q27.7

10. A lens with outer radius of curvature R and index of refraction n rests on a flat glass plate and the combination is illuminated with white light from above. Is there a dark spot or a light spot at the center of the lens? What does it mean if the observed rings are noncircular?

11. Why is the lens on a good-quality camera coated with a thin film?

12. Would it be possible to place a nonreflective coating on an airplane to cancel radar waves of wavelength 3 cm?

13. Why is it so much easier to perform interference experiments with a laser than with an ordinary light source?

14. Observe the shadow of your book when it is held a few inches above a table with a lamp several feet above the book. Why is the shadow somewhat fuzzy at the edges?

15. Although we can hear around corners, we cannot see around corners. How can you explain this in view of the fact that sound and light are both waves?

16. Assuming that the headlights of a car are point sources, estimate the maximum distance from an observer to the car at which the headlights are distinguishable from each other.

17. A laser beam is incident at a shallow angle on a machinist's ruler that has a finely calibrated scale. The rulings on the scale give rise to a diffraction pattern on a screen. Discuss how you can use this technique to obtain a measure of the wavelength of the laser light.

18. If a coin is glued to a glass sheet and this arrangement is held in front of a laser beam, the projected shadow has diffraction rings around its edge and a bright spot in the center. How is this possible?

19. If a fine wire is stretched across the path of a laser beam, is it possible to produce a diffraction pattern?

20. Describe the change in width of the central maximum of the single-slit diffraction pattern as the width of the slit is made smaller.

21. Suppose that reflected white light is used to observe a thin, transparent coating on glass as the coating material is gradually deposited by evaporation in a vacuum. Describe possible color changes that might occur during the process of building up the thickness of the coating.

22. The brilliant colors of peacock feathers (Fig. Q27.22) are due to a phenomenon known as *iridescence*. The melanin fibers in the feathers act as a natural diffraction grating. How do you explain the different colors? Why do the colors often change as the bird moves?

Figure Q27.22 Iridescence in peacock feathers. (© *Diane Schiumo 1988, Fundamental Photographs*)

PROBLEMS

Section 27.2 Young's Double-Slit Experiment

1. A laser beam ($\lambda = 632.8$ nm) is incident on two slits 0.200 mm apart. How far apart are the bright interference lines on a screen 5.00 m away from the double slits?

2. A Young's interference experiment is performed with monochromatic light. The separation between the slits is 0.500 mm and the interference pattern on a screen 3.30 m away shows the first maximum 3.40 mm from the center of the pattern. What is the wavelength?

3. Two radio antennas separated by 300 m, as in Figure P27.3, simultaneously broadcast identical signals at the same wavelength. A radio in a car traveling due north receives the signals. (a) If the car is at the position of the second maximum, what is the wavelength of the signals? (b) How much farther must the car travel to encounter the next minimum in reception? (*Caution*: Do not use the small-angle approximation in this problem.)

4. In a location where the speed of sound is 354 m/s, a 2000 Hz sound wave impinges on two slits 30.0 cm apart. (a) At what angle is the first maximum located? (b) If the sound wave is replaced by 3.00-cm microwaves, what slit separation gives the same angle for the first maximum? (c) If the slit separation is 1.00 μm, what frequency light gives the same first maximum angle?

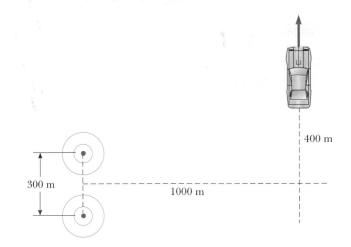

Figure P27.3

5. Young's double-slit experiment is performed with 589-nm light and a slits-to-screen distance of 2.00 m. The tenth interference minimum is observed 7.26 mm from the central maximum. Determine the spacing of the slits.

6. An oscillator drives two loudspeakers 35.0 cm apart, which vibrate in phase at a frequency of 2.00 kHz. At what angles,

measured from the perpendicular bisector of the line joining the speakers, would a distant observer hear maximum
sound intensity? Minimum? (Take the speed of sound as
340 m/s.)

7. In Figure 27.3, let $L = 120$ cm and $d = 0.250$ cm. The slits
are illuminated with coherent 600-nm light. Calculate the
distance y above the central maximum for which the average
intensity on the screen is 75.0% of the maximum.

8. The intensity on the screen at a certain point in a double-
slit interference pattern is 64.0% of the maximum value.
(a) What minimum phase difference (in radians) between
sources produces this result? (b) Express this phase difference as a path difference for 486.1-nm light.

Section 27.4   Interference in Thin Films

9. A soap bubble ($n = 1.33$) is floating in air. If the thickness
of the bubble wall is 115 nm, what is the wavelength of the
light that is most strongly reflected?

10. A thin film of oil ($n = 1.25$) is located on a smooth wet
pavement. When viewed perpendicular to the pavement,
the film appears to be predominantly red (640 nm) and has
no blue color (512 nm). How thick is the oil film?

11. An oil film ($n = 1.45$) floating on water is illuminated by
white light at normal incidence. The film is 280 nm thick.
Find (a) the dominant observed color in the reflected light
and (b) the dominant color in the transmitted light. Explain your reasoning.

12. A possible means for making an airplane invisible to radar
is to coat the plane with an antireflective polymer. If radar
waves have a wavelength of 3.00 cm and the index of refraction of the polymer is $n = 1.50$, how thick would you
make the coating?

13. An air wedge is formed between two glass plates separated
at one edge by a very fine wire as in Figure P27.13. When
the wedge is illuminated from above by 600-nm light, 30
dark fringes are observed. Calculate the radius of the wire.

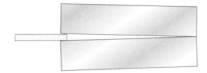

Figure P27.13

Section 27.5   Diffraction

14. Helium–neon laser light ($\lambda = 632.8$ nm) is sent through a
0.300-mm-wide single slit. What is the width of the central
maximum on a screen 1.00 m from the slit?

15. A screen is placed 50.0 cm from a single slit, which is illuminated with 690-nm light. If the distance between the first

and third minima in the diffraction pattern is 3.00 mm, what
is the width of the slit?

16. A beam of green light is diffracted by a slit of width
0.550 mm. The diffraction pattern forms on a wall 2.06 m
beyond the slit. The distance between the positions of zero
intensity on both sides of the central bright fringe is
4.10 mm. Calculate the wavelength of the laser light.

17. Coherent microwaves of wavelength 5.00 cm enter a long,
narrow window in a building otherwise essentially opaque
to the microwaves. If the window is 36.0 cm wide, what is
the distance from the central maximum to the first-order
minimum along a wall 6.50 m from the window?

18. Sound of frequency 650 Hz from a distant source passes
through a doorway 1.10 m wide in a sound-absorbing wall.
Find the number and approximate directions of the diffrac-
tion-maximum beams radiated into the space beyond.

Section 27.6   Resolution of Single-Slit and
Circular Apertures

19. The pupil of a cat's eye narrows to a slit of width 0.500 mm
in daylight. What is the angular resolution? (Use 500-nm
light in your calculation.)

20. Find the radius of a star image formed on the retina of the
eye if the aperture diameter (the pupil) at night is 0.700 cm
and the length of the eye is 3.00 cm. Assume the wavelength
of starlight in the eye is 500 nm.

21. A helium–neon laser emits light that has a wavelength of
632.8 nm. The circular aperture through which the beam
emerges has a diameter of 0.500 cm. Estimate the diameter
of the beam 10.0 km from the laser.

22. On the night of April 18, 1775, a signal was to be sent from
the Old North Church steeple to Paul Revere, who was 1.80
miles away: "One if by land, two if by sea." At what minimum
separation did the sexton have to set the lanterns so that
Revere could receive the correct message? Assume that Revere's pupils had a diameter of 4.00 mm at night, and that
the lantern light had a predominant wavelength of 580 nm.

23. A binary star system in the constellation Orion has an angular separation between the two stars of 1.00×10^{-5} rad.
If $\lambda = 500$ nm, what is the smallest diameter the telescope
can have and just resolve the two stars?

24. The Impressionist painter Georges Seurat created paintings
with an enormous number of dots of pure pigment approximately 2.00 mm in diameter. The idea was to have colors
such as red and green next to each other to form a scintil-
lating canvas. Outside what distance would one be unable
to discern individual dots on the canvas? (Assume $\lambda =$
500 nm and a pupil diameter of 4.00 mm.)

25. A circular radar antenna on a navy ship has a diameter of
2.10 m and radiates at a frequency of 15.0 GHz. Two small
boats are located 9.00 km away from the ship. How close
together could the boats be and still be detected as two objects?

Section 27.7 The Diffraction Grating

In the following problems, assume that the light is incident normally on the gratings.

26. Light from an argon laser strikes a diffraction grating that has 5310 lines per centimeter. The central and first-order principal maxima are separated by 0.488 m on a wall 1.72 m from the grating. Determine the wavelength of the laser light.

27. The hydrogen spectrum has a red line at 656 nm and a violet line at 434 nm. What is the angular separation between these two spectral lines obtained with a diffraction grating that has 4500 lines/cm?

28. A helium–neon laser ($\lambda = 632.8$ nm) is used to calibrate a diffraction grating. If the first-order maximum occurs at 20.5°, what is the groove spacing, d?

29. Three discrete spectral lines occur at angles of 10.09°, 13.71°, and 14.77° in the first-order spectrum of a grating spectroscope. (a) If the grating has 3660 slits/cm, what are the wavelengths of the light? (b) At what angles are these lines found in the second-order spectra?

30. A diffraction grating has 800 rulings per millimeter. A beam of light containing wavelengths from 500 to 700 nm hits the grating. Do the spectra of different orders overlap? Explain.

31. A diffraction grating of length 4.00 cm contains 6000 rulings over a width of 2.00 cm. (a) What is the resolving power of this grating in the first three orders? (b) If two monochromatic waves incident on this grating have a mean wavelength of 400 nm, what is their wavelength separation if they are just resolved in the third order?

32. A source emits 531.62-nm and 531.81-nm light. (a) What minimum number of lines is required for a grating that resolves the two wavelengths in the first-order spectrum? (b) Determine the slit spacing for a grating 1.32 cm wide that has the required minimum number of lines.

Section 27.8 Diffraction of X-Rays by Crystals (Optional)

33. Potassium iodide (KI) has a crystalline structure with an interplanar distance of 0.353 nm. A monochromatic x ray beam shows a diffraction maximum when the grazing angle is 7.60°. Calculate the x-ray wavelength. (Assume first-order.)

34. A wavelength of 0.129 nm characterizes K_β x-rays from zinc. When a beam of these x-rays is incident on the surface of a crystal the structure of which is similar to that of NaCl, a first-order maximum is observed at 8.15°. Calculate the interplanar spacing based on this information.

35. If the interplanar spacing of NaCl is 0.281 nm, what is the predicted angle at which 0.140-nm x-rays are diffracted in a first-order maximum?

36. The first-order diffraction is observed at 12.6° for a crystal in which the interplanar spacing is 0.240 nm. How many other orders can be observed?

Additional Problems

37. Raise your hand and hold it flat. Think of the space between your index finger and your middle finger as one slit, and think of the space between middle finger and ring finger as a second slit. (a) Consider the interference resulting from sending coherent visible light perpendicularly through this pair of openings. Compute an order-of-magnitude estimate for the angle between adjacent zones of constructive interference. (b) To make the angles in the interference pattern easy to measure with a plastic protractor, you should use an electromagnetic wave with frequency of what order of magnitude? How is this wave classified on the electromagnetic spectrum?

38. A certain crude oil has an index of refraction $n = 1.25$. A ship dumps 1.00 m^3 of this oil into the ocean, and the oil spreads into a thin slick. If the film produces a first-order maximum of light of wavelength 500 nm normally incident on it, how much surface area of the ocean does the oil slick cover? Assume the index of refraction of the ocean water is 1.34.

39. Interference effects are produced at point P on a screen as a result of direct rays from a 500-nm source and reflected rays off the mirror, as in Figure P27.39. If the source is 100 m to the left of the screen, and 1.00 cm above the mirror, find the distance y (in millimeters) to the first dark band above the mirror.

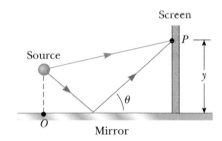

Figure P27.39

40. The waves from a radio station can reach a home receiver by two paths. One is a straight-line path from transmitter to home, a distance of 30.0 km. The second path is by reflection from the ionosphere (a layer of ionized air molecules near the top of the atmosphere). Assume this reflection takes place at a point midway between receiver and transmitter. If the wavelength broadcast by the radio station is 350 m, find the minimum height of the ionospheric layer that produces destructive interference between the direct

and reflected beams. (Assume no phase changes on reflection.)

41. Astronomers observed a 60.0-MHz radio source both directly and by reflection from the sea. If the receiving dish is 20.0 m above sea level, what is the angle of the radio source above the horizon at first maximum?

42. Measurements are made of the intensity distribution in a Young's interference pattern (Fig. 27.5). At a particular value of y, it is found that $I/I_0 = 0.810$ when 600-nm light is used. What wavelength of light should be used to reduce the relative intensity at the same location to 64.0%?

43. In a Young's interference experiment, the two slits are separated by 0.150 mm and the incident light includes light of wavelengths $\lambda_1 = 540$ nm and $\lambda_2 = 450$ nm. The overlapping interference patterns are formed on a screen 1.40 m from the slits. Calculate the minimum distance from the center of the screen to the point at which a bright line of the λ_1 light coincides with a bright line of the λ_2 light.

44. An air wedge is formed between two glass plates in contact along one edge and slightly separated at the opposite edge. When the plates are illuminated with monochromatic light from above, the reflected light has 85 dark fringes. Calculate the number of dark fringes that would appear if water ($n = 1.33$) were to replace the air between the plates.

45. Our discussion of the techniques for determining constructive and destructive interference by reflection from a thin film in air has been confined to rays striking the film at nearly normal incidence. Assume that a ray is incident at an angle of 30.0° (relative to the normal) on a film with index of refraction 1.38. Calculate the minimum thickness for constructive interference if the light is sodium light with a wavelength of 590 nm.

46. Light from a helium–neon laser ($\lambda = 632.8$ nm) is incident on a single slit. What is the maximum width for which no diffraction minima are observed?

47. (a) Both sides of a uniform film that has index of refraction n and thickness d are in contact with air. For normal incidence of light, an intensity minimum is observed in the reflected light at λ_2 and an intensity maximum is observed at λ_1, where $\lambda_1 > \lambda_2$. If there are no intensity minima observed between λ_1 and λ_2, show that the integer m in Equations 27.16 and 27.17 is given by $m = \lambda_1/2(\lambda_1 - \lambda_2)$. (b) Determine the thickness of the film if $n = 1.40$, $\lambda_1 = 500$ nm, and $\lambda_2 = 370$ nm.

48. What are the approximate dimensions of the smallest object on Earth that astronauts can resolve by eye when they are orbiting 250 km above the Earth? Assume $\lambda = 500$ nm light and a pupil diameter of 5.00 mm.

49. A beam of 541-nm light is incident on a diffraction grating that has 400 lines/mm. (a) Determine the angle of the second-order ray. (b) If the entire apparatus is immersed in water, determine the new second-order angle of diffraction. (c) Show that the two diffracted rays of parts (a) and (b) are related through the law of refraction.

50. Consider the double-slit arrangement shown in Figure P27.50, where the separation d is 0.300 mm and the distance L is 1.00 m. A sheet of transparent plastic ($n = 1.50$) 0.0500 mm thick (about the thickness of this page) is placed over the upper slit. As a result, the central maximum of the interference pattern moves upward a distance y'. Find this distance.

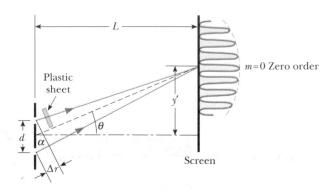

Figure P27.50

51. Consider the double-slit arrangement shown in Figure P27.50, where the slit separation is d and the slit-to-screen distance is L. A sheet of transparent plastic having an index of refraction n and thickness t is placed over the upper slit. As a result, the central maximum of the interference pattern moves upward a distance y'. Find y'.

52. Derive Equation 27.23 for the resolving power of a grating, $R = Nm$, where N is the number of lines illuminated and m is the order in the diffraction pattern. Remember that Rayleigh's criterion (Section 27.6) states that two wavelengths will be resolved when the principal maximum for one falls on the first minimum for the other.

Spreadsheet Problems

S1. To calculate the intensity distribution of the interference pattern for equally spaced sources, it is necessary to add a series of terms such as: $E_1 = A_0 \sin \alpha$, $E_2 = A_0 \sin(\alpha + \phi)$, $E_3 = A_0 \sin(\alpha + 2\phi)$, where ϕ is the phase difference caused by different path lengths. For N sources the time-averaged relative intensity as a function of the phase angle ϕ is

$$I_{rel} = f_N^2(\phi) + g_N^2(\phi)$$

where

$$f_N(\phi) = \sum_{n=0}^{N-1} \cos(n\phi) \qquad g_N(\phi) = \sum_{n=0}^{N-1} \sin(n\phi)$$

and I_{rel} is the ratio of the average intensity for N sources to that of one source. Spreadsheet 27.1 calculates and plots

I_{rel} versus phase angle ϕ for up to a maximum of six sources. The coefficients a, b, c, d, e, and f used in the spreadsheet determine the number of sources. For two sources, for example, set $a = 1$, $b = 1$, and $c = d = e = f = 0$. For three sources set $a = b = c = 1$, $d = e = f = 0$, and so on. (a) For three sources, what is the ratio of the intensity of the principal maxima to that of a single source? (b) What is the ratio of the intensity of the secondary maxima to that of the principal maximum?

S2. Use Spreadsheet 27.1 to calculate the intensity pattern for four equally spaced sources. (a) What is the ratio of the intensity of the principal maxima to that of a single source? (b) What is the ratio of the intensity of the secondary maxima to that of the principal maxima? (c) Repeat parts (a) and (b) for five and six sources. (d) Are all the secondary maxima of equal intensity?

S3. Figure S27.3 shows the relative intensity of a single-slit Fraunhofer diffraction pattern as a function of the parameter $\beta/2 = \pi a \sin \theta/\lambda$. Spreadsheet 27.2 plots the relative intensity I/I_0 as a function of θ, where θ is defined in Figure 27.16. Examine the cases where $\lambda = a$, $\lambda = 0.5a$, $\lambda = 0.1a$, and $\lambda = 0.05a$, and explain your results.

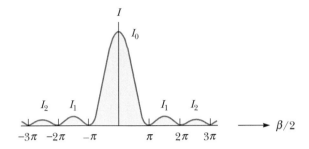

Figure S27.3   A plot of the intensity I versus $\beta/2$ for the single-slit Fraunhofer diffraction pattern.

ANSWERS TO CONCEPTUAL PROBLEMS

1. The result of the double slit is to *redistribute* the energy arriving at the screen. Although there is no energy at the location of a dark fringe, there is more energy at the location of a bright fringe than there would be without the double slit. The total amount of energy arriving at the screen is the same as without the slit, as it must be according to the conservation of energy principle.

2. One of the materials has a higher index of refraction than water, the other lower. The material with a higher index of refraction than water will appear black as it approaches zero thickness. There will be a 180° phase shift for the light reflected from the upper surface, but no such phase change from the lower surface, because the index of refraction for water on the other side is lower than that of the film. Thus, the two reflections will be out of phase and will interfere destructively. The material with index of refraction lower than water will have a phase change for the light reflected from both upper and lower surfaces, so that the reflections from the zero-thickness film will be back in phase, and the film will appear bright.

3. There will be an effect on the interference pattern—it will be distorted. The high temperature of the flame will change the index of refraction of the air for the arm of the interferometer in which the match is held. As the index of refraction varies turbulently, the wavelength of the light in that region will also vary turbulently. As a result, the effective optical path length difference between the two arms will vary, resulting in a wildly varying interference pattern.

4. For normal incidence, the extra path length followed by the reflected ray is twice the thickness of the film. For destructive interference, this must be a distance of half a wavelength of the light in the material of the film. Since no 180° phase change will occur in these reflections, the thickness of the film must be one quarter wavelength, which is the same as that for reflected destructive interference.

5. We would like to reduce the minimum angular separation for two objects below the angle subtended by the two stars in the binary system. We can do that by reducing the wavelength of the light—this in essence makes the aperture larger, relative to the light wavelength, increasing the resolving power. Thus, we would choose a blue filter.

6. The tracks of information on a compact disc are much closer together than on a phonograph record. As a result, the diffraction maxima from the compact disc will be farther apart than those from the record.

7. As the edge of the Moon cuts across the light from the star, edge diffraction effects occur. Thus, as the edge of the Moon moves relative to the star, the observed light from the star proceeds through a series of maxima and minima.

8. Strictly speaking, the ribs do act as a diffraction grating, but the separation distance of the ribs is so much larger than the wavelength of the x-rays that there are no observable effects.

28

Quantum Physics

Although many problems were resolved by Einstein's special theory of relativity in the early part of the 20th century, many other problems remained unanswered. Attempts to apply the laws of classical physics to the behavior of matter on the atomic scale were consistently unsuccessful. Various phenomena, such as black-body radiation, the photoelectric effect, and the emission of sharp spectral lines by atoms in a gas discharge, could not be understood within the framework of classical physics. Between 1900 and 1930, however, a modern version of mechanics called *quantum mechanics* was highly successful in explaining the behavior of atoms, molecules, and nuclei. Moreover, the quantum theory reduces to classical physics when applied to macroscopic systems. Like relativity, the quantum theory requires a modification of our ideas concerning the physical world.

An extensive study of quantum theory is certainly beyond the scope of this book, and therefore this chapter is simply an introduction to its underlying ideas. We shall also discuss two simple applications of quantum theory: the photoelectric effect and the Compton effect.

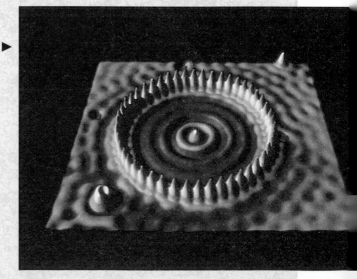

This photograph is a demonstration of a "quantum-corral" consisting of a ring of 48 iron atoms located on a copper surface. The diameter of the ring is 143 nm, and the photograph was obtained using a low-temperature scanning tunneling microscope (STM). Corrals and other structures are able to confine surface-electron waves. The study of such structures will play an important role in determining the future of small electronic devices. *(IBM Corporation Research Division)*

28.1 • BLACK-BODY RADIATION AND PLANCK'S THEORY

An object at any temperature emits radiation that is sometimes referred to as **thermal radiation.** The characteristics of this radiation depend on the temperature and properties of the object. At low temperatures, the wavelengths of the thermal radiation are mainly in the infrared region and hence not observed by the eye. As the temperature of the object increases, the object eventually begins to glow red. At sufficiently high temperatures, it appears to be white, as in the glow of the hot tungsten filament of a lightbulb. A careful study of thermal radiation shows that it consists of a continuous distribution of wavelengths from the infrared, visible, and ultraviolet portions of the spectrum.

From a classical viewpoint, thermal radiation originates from accelerated charged particles near the surface of the object; those charges emit radiation much as small antennas do. The thermally agitated charges can have a distribution of accelerations, which accounts for the continuous spectrum of radiation emitted by the object. By the end of the 19th century, it had become apparent that this classical explanation of thermal radiation was inadequate. The basic problem was in understanding the observed distribution of wavelengths in the radiation emitted by a black body. As we learned in Chapter 17, Section 17.7, a black body is an ideal system that absorbs all radiation incident on it. A good approximation of a black body is a small hole leading to the inside of a hollow object, as shown in Figure 28.1. The nature of the radiation emitted through the small hole leading to the cavity depends only on the temperature of the cavity walls. Note that the hole itself is the black body.

Experimental data for the distribution of energy in black-body radiation at three temperatures are shown in Figure 28.2. The radiated energy varies with wavelength and temperature. As the temperature of the black body increases, the total

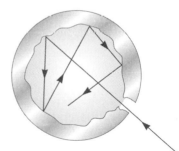

Figure 28.1   A small hole that leads to a cavity with rough walls is a good approximation of a black body. Light entering the hole strikes the far wall, where some of it is absorbed but some is reflected at some random angle. The light continues to be reflected, and each time it is, part of it is absorbed by the cavity walls. After many reflections, essentially all of the incident energy is absorbed.

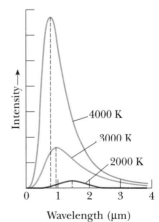

Figure 28.2   Intensity of black-body radiation versus wavelength at three temperatures. Note that the amount of radiation emitted (the area under a curve) increases with increasing temperature.

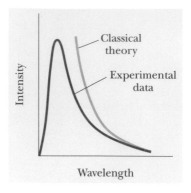

Figure 28.3 Comparison of the experimental results with the curve predicted by classical theory for the distribution of black-body radiation.

Max Planck (1858–1947)

Planck introduced the concept of a "quantum of action" (Planck's constant, *h*) in an attempt to explain the spectral distribution of black-body radiation, which laid the foundations for quantum theory. In 1918 he was awarded the Nobel Prize for this discovery of the quantized nature of energy. The work leading to the "lucky" black-body radiation formula was described by Planck in his Nobel Prize acceptance speech (1920): "But even if the radiation formula proved to be perfectly correct, it would after all have been only an interpolation formula found by lucky guesswork and, thus, would have left us rather unsatisfied." *(Photo courtesy of AIP Niels Bohr Library, W. F. Meggers Collection)*

amount of energy it emits increases. Also, with increasing temperatures, the peak of the distribution shifts to shorter wavelengths. This shift was found to obey the following relationship, called **Wien's displacement law:**

$$\lambda_{max} T = 0.2898 \times 10^{-2} \text{ m} \cdot \text{K} \qquad \text{[28.1]}$$

where λ_{max} is the wavelength at which the curve peaks and T is the absolute temperature of the surface of the object emitting the radiation.

Early attempts to use classical ideas to explain the shapes of the curves in Figure 28.2 failed. Figure 28.3 shows an experimental plot of the black-body radiation spectrum (red) together with the curve predicted by classical theory (blue). At long wavelengths, classical theory is in good agreement with the experimental data. At short wavelengths, however, major disagreement exists between classical theory and experiment. This contradiction is often called the **ultraviolet catastrophe.**

In 1900 Planck developed a formula for black-body radiation that was in complete agreement with experiment at all wavelengths. Planck's analysis led to the red curve shown in Figure 28.3. In his theory, Planck made two bold and controversial assumptions concerning the nature of the oscillating charges, which he called resonators, at the surface of the black body:

- The resonators can have only certain *discrete* amounts of energy E_n,

$$E_n = nhf \qquad \text{[28.2]}$$

where, as we learned in Chapter 10, Section 10.10, n is a positive integer called a quantum number, f is the frequency of oscillation of the resonators, and h is Planck's constant. Because the energy of each resonator can have only discrete values given by Equation 28.2, we say the energy is *quantized.* Each discrete energy value represents a different *quantum state,* with each value of n representing a specific quantum state. When the resonator is in the $n = 1$ quantum state, its energy is hf; when it is in the $n = 2$ quantum state, its energy is $2hf$; and so on.

- The resonators emit energy in discrete units which were later called **photons.** A suggestive image of several photons, not to be taken too literally, is shown in Figure 28.4. The resonators emit or absorb these quanta by "jumping" from one quantum state to another. If the jump is from one state to an adjacent state—say, from the $n = 3$ state to the $n = 2$ state—Equation 28.2 shows that the amount of energy radiated by the resonator equals hf. Hence, the energy of one photon corresponds to the energy difference between two adjacent levels and is given by

$$E = hf \qquad \text{[28.3]}$$

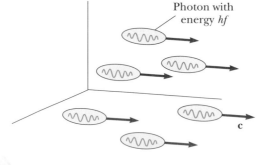

Figure 28.4 A representation of light quanta. Each photon has a discrete energy, *hf*.

A resonator radiates or absorbs energy only when it changes quantum states. If it remains in one quantum state, no energy is absorbed or emitted. Figure 28.5 shows the quantized energy levels and allowed transitions proposed by Planck.

The key point in Planck's theory is the radical assumption of quantized energy states. This development marked the birth of the quantum theory. When Planck presented his theory, most scientists (including Planck!) did not consider the quantum concept to be realistic. Hence, Planck and others continued to search for a more rational explanation of black-body radiation. However, subsequent developments showed that a theory based on the quantum concept (rather than on classical concepts) had to be used to explain a number of other phenomena at the atomic level.

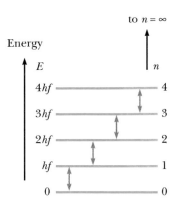

Figure 28.5 Allowed energy levels for a resonator that oscillates at a natural frequency, f. Allowed transitions are indicated by vertical arrows.

Thinking Physics 1

You are observing a yellow candle flame, and your laboratory partner claims that the light from the flame is atomic in origin. You disagree, claiming that the candle flame is hot, so the radiation must be thermal in origin. Before this disagreement leads to fisticuffs, how could you determine who is correct?

Reasoning A simple determination could be made by observing the light from the candle flame through a spectrometer, which is a slit and diffraction grating combination, and was discussed in Section 27.7. If the spectrum of the light is continuous, then it is thermal in origin. If the spectrum shows discrete lines, it is atomic in origin. The results of the experiment show that the light is indeed thermal in origin and originates in the hot particles of soot in the candle flame.

CONCEPTUAL PROBLEM 1

If you observe objects inside a very hot kiln, it is difficult to discern the shape of the objects. Why?

CONCEPTUAL PROBLEM 2

Are black bodies black?

CONCEPTUAL PROBLEM 3

All objects radiate energy. Why, then, are we not able to see all objects in a dark room?

Example 28.1   Thermal Radiation from the Human Body

The temperature of your skin is approximately 35°C. What is the peak wavelength of the radiation it emits?

Solution From Wien's displacement law (Eq. 28.1), we have

$$\lambda_{max} T = 0.2898 \times 10^{-2}\ \text{m} \cdot \text{K}$$

Solving for λ_{max}, noting that 35°C corresponds to an absolute temperature of 308 K, we have

$$\lambda_{max} = \frac{0.2898 \times 10^{-2}\ \text{m} \cdot \text{K}}{308\ \text{K}} = 9.40\ \mu\text{m}$$

This radiation is in the infrared region of the spectrum.

Example 28.2   **The Quantized Oscillator**

A 2.0-kg mass is attached to a massless spring of force constant $k = 25$ N/m. The spring is stretched 0.40 m from its equilibrium position and released. (a) Find the total energy and frequency of oscillation according to classical calculations.

Solution   From Chapter 12, Equation 12.22, the total energy of a simple harmonic oscillator having an amplitude of A is $\frac{1}{2}kA^2$. Therefore,

$$E = \tfrac{1}{2}kA^2 = \tfrac{1}{2}(25 \text{ N/m})(0.40 \text{ m})^2 = 2.0 \text{ J}$$

The frequency of oscillation is, from Equation 12.19,

$$f = \frac{1}{2\pi}\sqrt{\frac{k}{m}} = \frac{1}{2\pi}\sqrt{\frac{25 \text{ N/m}}{2.0 \text{ kg}}} = 0.56 \text{ Hz}$$

(b) Assume that the energy is quantized and find the quantum number, n, for the system.

Solution   If the energy is quantized, we have $E_n = nhf$, and from the result of (a) we have

$$E_n = nhf = n(6.626 \times 10^{-34} \text{ J·s})(0.56 \text{ Hz}) = 2.0 \text{ J}$$

Therefore,

$$n = 5.4 \times 10^{33}$$

(c) How much energy is carried away in a one-quantum change?

Solution   The energy carried away in a one-quantum change of energy is

$$E = hf = (6.63 \times 10^{-34} \text{ J·s})(0.56 \text{ Hz})$$
$$= 3.7 \times 10^{-34} \text{ J}$$

The energy carried away by a one-quantum change is such a small fraction of the total energy of the oscillator that we could never expect to see such a small change in the system. Thus, even though the decrease in energy of a spring–mass system is quantized and proceeds by small quantum jumps, our senses perceive the decrease as continuous. Quantum effects become important and measurable only on the submicroscopic level of atoms and molecules.

EXERCISE 1   A tungsten filament is heated to 800°C. What is the wavelength of the most intense radiation?   Answer   2.70 μm

28.2 • THE PHOTOELECTRIC EFFECT

In the latter part of the 19th century, experiments showed that light incident on certain metallic surfaces caused electrons to be emitted from the surfaces. As mentioned in Chapter 25, Section 25.1, this phenomenon is known as the **photoelectric effect,** and the emitted electrons are called **photoelectrons.** The first discovery of the photoelectric effect was made by Hertz.

Figure 28.6 is a schematic diagram of a photoelectric effect apparatus. An evacuated glass tube contains a metal plate, E, connected to the negative terminal of a battery. Another metal plate, C, is maintained at a positive potential by the battery. When the tube is kept in the dark, the ammeter reads zero, indicating that there is no current in the circuit. However, when monochromatic light of the appropriate wavelength shines on plate E, a current is detected by the ammeter, indicating a flow of charges across the gap between E and C. This current arises from electrons emitted from the negative plate (the emitter) and collected at the positive plate (the collector).

Figure 28.7 is a plot of the photoelectric current versus the potential difference, ΔV, between E and C for two light intensities. For large values of ΔV, the current reaches a maximum value. In addition, the current increases as the incident light intensity increases, as you might expect. Finally, when ΔV is negative—that is, when the battery polarity is reversed to make E positive and C negative—the current drops to a very low value because most of the emitted photoelectrons are repelled by the negative collecting plate, C. Only those electrons having a kinetic energy

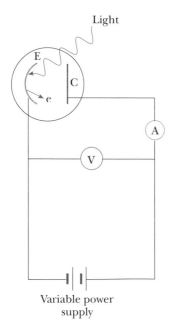

Figure 28.6   A circuit diagram for observing the photoelectric effect. When light strikes the plate, E (the emitter), photoelectrons are ejected from the plate. Electrons collected at C (the collector) constitute a current in the circuit.

greater than $e \Delta V$ will reach C, where e is the charge on the electron. When ΔV is greater than or equal to ΔV_s, the **stopping potential,** no electrons reach C and the current is zero. The stopping potential is *independent* of the radiation intensity. The maximum kinetic energy of the photoelectrons is related to the stopping potential through the relation

$$K_{\max} = e \Delta V_s \qquad [28.4]$$

Several features of the photoelectric effect cannot be explained with classical physics or with the wave theory of light:

- No electrons are emitted if the incident light frequency falls below some **cutoff frequency,** f_c, which is characteristic of the material being illuminated. This is inconsistent with the wave theory, which predicts that the photoelectric effect should occur at any frequency, provided the light intensity is high enough.
- The maximum kinetic energy of the photoelectrons is independent of light intensity.
- The maximum kinetic energy of the photoelectrons increases with increasing light frequency.
- Electrons are emitted from the surface almost instantaneously (less than 10^{-9} s after the surface is illuminated), even at low light intensities. Classically, one would expect that the electrons would require some time to absorb the incident radiation before they acquired enough kinetic energy to escape from the metal.

A successful explanation of the photoelectric effect was given by Einstein in 1905, the same year he published his special theory of relativity. As part of a general paper on electromagnetic radiation, for which he received the Nobel Prize in 1921, Einstein extended Planck's concept of quantization to electromagnetic waves. He assumed that light (or any other electromagnetic wave) of frequency f can be considered a stream of photons. Each photon has an energy E given by Equation 28.3.

Einstein's view was that a photon of the incident light was so localized that it gave *all* its energy, hf, directly to a single electron in the metal. Electrons emitted from the surface of the metal possess the maximum kinetic energy, $K_{\max}$. According to Einstein, the maximum kinetic energy for these emitted electrons is

$$K_{\max} = hf - \phi \qquad [28.5]$$

where ϕ is called the **work function** of the metal. **The work function represents the minimum energy with which an electron is bound in the metal** and is on the order of a few electron volts. Table 28.1 lists selected values.

With the photon theory of light, one can explain the features of the photoelectric effect that cannot be understood using classical concepts:

- That the effect is not observed below a certain cutoff frequency follows from the fact that the photon energy must be $\geq \phi$. If the energy of the incoming photon is not $\geq \phi$, the electrons will never be ejected from the surface, regardless of light intensity.
- That $K_{\max}$ is independent of the light intensity can be understood with the following argument. If the light intensity is doubled, the number of photons is doubled, which doubles the number of photoelectrons emitted. However,

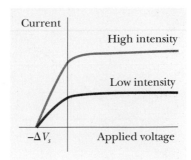

Figure 28.7 Photoelectric current versus applied voltage for two light intensities. The current increases with intensity but reaches a saturation level for large values of ΔV. At voltages equal to or less than $-\Delta V_s$, the current is zero.

TABLE 28.1
Work Functions of Selected Metals

Metal	ϕ (eV)
Na	2.46
Al	4.08
Cu	4.70
Zn	4.31
Ag	4.73
Pt	6.35
Pb	4.14
Fe	4.50

• *Photoelectric effect equation*

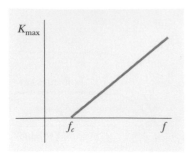

Figure 28.8 A sketch of K_{max} versus frequency of incident light for photoelectrons in a typical photoelectric effect experiment. Photons with frequency less than f_c do not have sufficient energy to eject an electron from the metal.

their kinetic energy, which equals $hf - \phi$, depends only on the light frequency and the work function, not on the light intensity.
- That K_{max} increases with increasing frequency is easily understood with Equation 28.5.
- That the electrons are emitted almost instantaneously is consistent with the particle theory of light, in which the incident energy appears in small packets and there is a one-to-one interaction between photons and electrons. This is in contrast to the energy of the photons being distributed uniformly over a large area.

Experimental observation of a linear relationship between f and K_{max} would be a final confirmation of Einstein's theory. Indeed, such a linear relationship is observed, as sketched in Figure 28.8. The slope of this curve is h. The intercept on the horizontal axis is the cutoff frequency, which is related to the work function through the relation $f_c = \phi/h$. This corresponds to a **cutoff wavelength** of

$$\lambda_c = \frac{c}{f_c} = \frac{c}{\phi/h} = \frac{hc}{\phi} \qquad [28.6]$$

where c is the speed of light. Wavelengths *greater* than λ_c incident on a material with a work function of ϕ do not result in the emission of photoelectrons.

CONCEPTUAL PROBLEM 4

A student claims that he is going to eject electrons from a piece of metal by placing a radio transmitter antenna adjacent to the metal and sending a strong AM radio signal into the antenna. The work function of a metal is typically a few electron volts. Will this work?

CONCEPTUAL PROBLEM 5

In the photoelectric effect, explain why the stopping potential depends on the frequency of the light but not on the intensity.

Example 28.3 The Photoelectric Effect for Sodium

A sodium surface is illuminated with light of wavelength 300 nm. The work function for sodium metal is 2.46 eV. Find (a) the maximum kinetic energy of the ejected photoelectrons and (b) the cutoff wavelength for sodium.

Solution (a) The energy of a photon in the illuminating light beam is

$$E = hf = \frac{hc}{\lambda} = \frac{(6.626 \times 10^{-34}\,\text{J}\cdot\text{s})(3.00 \times 10^8\,\text{m/s})}{300 \times 10^{-9}\,\text{m}}$$
$$= 6.63 \times 10^{-19}\,\text{J} = 4.14\,\text{eV}$$

Use of Equation 28.5 gives

$$K_{max} = hf - \phi = 4.14\,\text{eV} - 2.46\,\text{eV} = 1.68\,\text{eV}$$

(b) The cutoff wavelength can be calculated from Equation 28.6 after we convert ϕ from electron volts to joules:

$$\phi = 2.46\,\text{eV} = 3.94 \times 10^{-19}\,\text{J}$$
$$\lambda_c = \frac{hc}{\phi} = \frac{(6.626 \times 10^{-34}\,\text{J}\cdot\text{s})(3.00 \times 10^8\,\text{m/s})}{3.94 \times 10^{-19}\,\text{J}}$$
$$= 5.05 \times 10^{-7}\,\text{m} = 505\,\text{nm}$$

This wavelength is in the green region of the visible spectrum.

EXERCISE 2 Calculate the maximum speed of the photoelectrons under the conditions described in this example.
Answer 7.68×10^5 m/s

EXERCISE 3 The work function for potassium is 2.24 eV. If potassium metal is illuminated with 480-nm light, find (a) the maximum kinetic energy of the photoelectrons and (b) the cutoff wavelength.   Answer   (a) 0.350 eV   (b) 554 nm

28.3 • THE COMPTON EFFECT

In 1919 Einstein concluded that a photon of energy E travels in a single direction (unlike a spherical wave) and carries a momentum equal to $E/c = hf/c$. In his own words, "If a bundle of radiation causes a molecule to emit or absorb an energy packet hf, then momentum of quantity hf/c is transferred to the molecule, directed along the line of the bundle for absorption and opposite the bundle for emission." In 1923 Arthur Holly Compton (1892–1962) and Peter Debye (1884–1966) independently carried Einstein's idea of photon momentum farther. They realized that the scattering of x-ray photons from electrons could be explained by treating photons as point-like particles with energy hf and momentum hf/c and by assuming that the energy and momentum of the photon–electron pair are conserved in a collision.

Arthur Holly Compton (1892–1962), American physicist. *(Courtesy of AIP Niels Bohr Library)*

Prior to 1922, Compton and his coworkers had accumulated evidence that showed that the classical wave theory of light failed to explain the scattering of x-rays from electrons. According to classical theory, incident electromagnetic waves of frequency f_0 should have two effects: (1) the electrons will accelerate in the direction of propagation of the x-ray by radiation pressure (see Section 24.7), and (2) the oscillating electric field will set the electrons into oscillation at the apparent frequency of the radiation as seen by the moving electron. Both of these effects are included in Figure 28.9a. This apparent frequency will change due to the Doppler effect (see Section 13.10) since the electron absorbs as a moving particle. The electron will then re-radiate as a moving particle, exhibiting another Doppler shift in the frequency of radiation. Since different electrons will move at different speeds, depending on the amount of energy absorbed from the electromagnetic waves, the scattered wave frequency at a given angle should show a distribution of Doppler-shifted values. Contrary to this prediction, Compton's experiment showed that, at a given angle, only *one* frequency of radiation was observed. Figure 28.9b shows the quantum picture of the exchange of momentum and energy between an individual x-ray photon and an electron. Another major distinction between the classical and quantum models is shown in Figure 28.9b. In the classical model, the electron is pushed along the direction of propagation of the incident x-ray by radiation pressure. In the quantum model, the electron is scattered through an angle ϕ with respect to this direction, as if this were a billiard-ball type collision.

Figure 28.10a is a schematic diagram of the apparatus used by Compton. In his original experiment Compton measured how scattered x-ray intensity depended on wavelength at three scattering angles. The incident beam consisted of monochromatic x-rays of wavelength $\lambda_0 = 0.071$ nm. The experimental plots of intensity versus wavelength obtained by Compton for four scattering angles are shown in Figure 28.10b. They show two peaks, one at λ_0 and one at a longer wavelength λ'. The peak at λ_0 is caused by x-rays scattered from electrons that are tightly bound to the target atoms, and the shifted peak at λ' is caused by x-rays scattered from free electrons in the target. In his analysis, Compton predicted that the shifted peak should depend on scattering angle θ as

Compton shift equation •

$$\lambda' - \lambda_0 = \frac{h}{m_e c}(1 - \cos\theta) \qquad [28.7]$$

In this expression, known as the **Compton shift equation,** m_e is the mass of the electron; $h/m_e c$ is called the **Compton wavelength,** λ_C, of the electron and has a currently accepted value of

Compton wavelength •

$$\lambda_C = \frac{h}{m_e c} = 0.00243 \text{ nm}$$

Compton's measurements were in excellent agreement with the predictions of Equation 28.7. It is fair to say that these were the first experimental results to convince most physicists of the fundamental validity of the quantum theory!

Figure 28.9 X-ray scattering from an electron: (a) the classical model and (b) the quantum model.

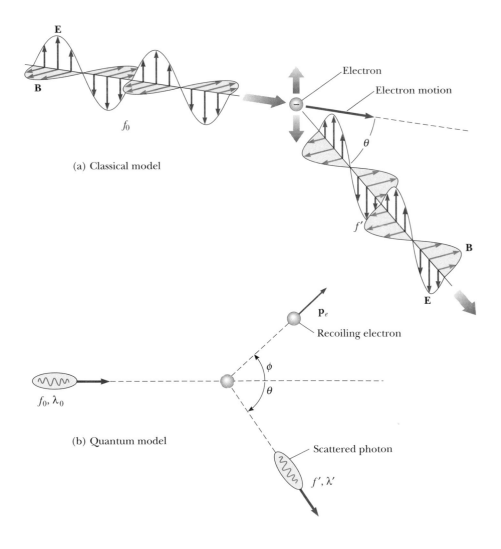

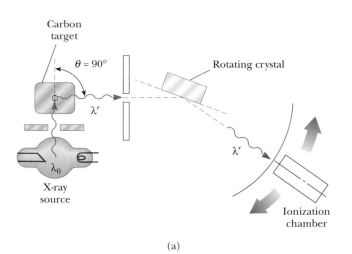

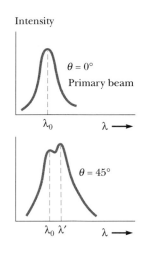

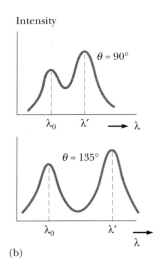

(a)

(b)

Figure 28.10 (a) Schematic diagram of Compton's apparatus. The wavelength was measured with a rotating crystal spectrometer using carbon as the target. The intensity was determined by a movable ionization chamber that generated a current proportional to the x-ray intensity. (b) Scattered intensity versus wavelength of Compton scattering at $\theta = 0°$, $45°$, $90°$, and $135°$.

Thinking Physics 2

The Compton effect involves a change in wavelength as photons are scattered through different angles. Suppose we illuminate a piece of material with a beam of light and then view the material from different angles relative to the beam of light. Will we see a *color change* corresponding to the change in wavelength of the scattered light?

Reasoning There will be a wavelength change for visible light scattered by the material, but the change will be far too small to detect as a color change. The largest possible wavelength change, at 180° scattering, will be twice the Compton wavelength, about 0.005 nm. This represents a change of less than 0.001% of the wavelength of red light. The Compton effect is only detectable for wavelengths that are very short to begin with, so that the Compton wavelength is an appreciable fraction of the incident wavelength. As a result, the usual radiation for observing the Compton effect is in the x-ray range of the electromagnetic spectrum.

CONCEPTUAL PROBLEM 6

Suppose a photoelectric effect experiment is attempted, but the frequency of the radiation is less than the cutoff frequency for the target material. In such a situation, the photon energy is not sufficient to eject electrons. Where does the photon energy go in this case?

Example 28.4   **Compton Scattering at 45°**

X-rays of wavelength $\lambda_0 = 0.200\,000$ nm are scattered from a block of material. The scattered x-rays are observed at an angle of 45.0° to the incident beam. Calculate the wavelength of the x-rays scattered at this angle.

Solution   The shift in wavelength of the scattered x-rays is given by Equation 28.7:

$$\Delta\lambda = \frac{h}{m_e c}(1 - \cos\theta)$$

$$= \frac{6.626 \times 10^{-34}\,\text{J·s}}{(9.11 \times 10^{-31}\,\text{kg})(3.00 \times 10^8\,\text{m/s})}(1 - \cos 45°)$$

$$= 7.10 \times 10^{-13}\,\text{m} = 0.000\,710\,\text{nm}$$

Hence, the wavelength of the scattered x-rays at this angle is

$$\lambda' = \Delta\lambda + \lambda_0 = 0.200\,710\,\text{nm}$$

EXERCISE 4   Find the fraction of energy lost by the photon in this collision.   Answer   Fraction = $\Delta E/E = 0.003\,54$

EXERCISE 5   X-rays of wavelength 0.200 nm are scattered from a carbon target. If the scattered radiation is detected at 60° to the incident beam, find (a) the Compton shift $\Delta\lambda$ and (b) the kinetic energy imparted to the recoiling electron.
Answer   (a) 0.001 21 nm   (b) 37.4 eV

28.4 • PHOTONS AND ELECTROMAGNETIC WAVES

Explanations of phenomena such as the photoelectric effect and the Compton effect offer ironclad evidence that when light and matter interact, the light behaves as if it were composed of particles with energy hf and momentum h/λ. An obvious question at this point is, "How can light be considered a photon when it exhibits wave-like properties?" On the one hand, we describe light in terms of photons having energy and momentum. On the other hand, we recognize that light and other electromagnetic waves exhibit interference and diffraction effects, which are consistent only with a wave interpretation.

Which model is correct? Is light a wave or a particle? The answer depends on the phenomenon being observed. Some experiments can be explained better, or solely, with the photon concept, whereas others are best described, or can be described only, with a wave model. The end result is that **we must accept both models and admit that the true nature of light is not describable in terms of any single classical picture.** However, you should recognize that the same light beam that can eject photoelectrons from a metal can also be diffracted by a grating. In other words, **the photon theory and the wave theory of light complement each other.**

All forms of electromagnetic radiation can be described from two points of view. At one extreme, electromagnetic waves describe the overall interference pattern or probability distribution, formed by photons. At the other extreme, the photon description is natural in the context of a highly energetic photon of very short wavelength. Both views are necessary and complementary. Hence, **light has a dual nature: It exhibits both wave and photon characteristics.**

The dual nature of light •

The success of the particle model of light in explaining the photoelectric effect and the Compton effect raises many other questions. Because the photon is a particle, what is the meaning of its "frequency" and "wavelength," and which determines its energy and momentum? Is light in some sense simultaneously a wave and a particle? Although photons have no rest energy (a nonobservable quantity), is there a simple expression for the effective mass-energy of a "moving" photon? If a "moving" photon has mass-energy, do photons experience gravitational attraction? What is the spatial extent of a photon, and how does an electron absorb or scatter one photon? Some of these questions can be answered, but others demand a view of atomic processes that is too pictorial and literal. Furthermore, many of these questions stem from classical analogies such as colliding billiard balls and water waves breaking on a shore. Quantum mechanics gives light a more fluid and flexible nature by treating the particle model and wave model of light as both necessary and complementary. Neither model can be used exclusively to describe all properties of light. A complete understanding of the observed behavior of light can be attained only if the two models are combined in a complementary manner.

We can perhaps understand why photons are compatible with electromagnetic waves in the following manner. We may suspect that long-wavelength waves do not exhibit particle characteristics. Consider, for instance, radio waves at a frequency of 2.5 MHz. The energy of a photon having this frequency is only about 10^{-8} eV. From a practical viewpoint, this energy is too small to be detected as a single photon. A sensitive radio receiver might require as many as 10^{10} of these photons to produce a detectable signal. Such a large number of photons would appear, on the average, as a continuous wave. With so many photons reaching the detector every second, it is unlikely that any graininess would appear in the detected signal. That is, with 2.5-MHz waves one would not be able to detect the individual photons striking the antenna.

Now consider what happens with shorter wavelengths. In the visible region, it is possible to observe both the photon and the wave characteristics of light. As we mentioned earlier, a beam of visible light shows interference phenomena (thus, it is a wave) and at the same time can produce photoelectrons (thus, it is a particle). At even shorter wavelengths, the momentum and energy of the photon increase. As a consequence, the photon nature of light becomes more evident than its wave nature. For example, absorption of an x-ray photon is easily detected as a single event. However, as the wavelength decreases, wave effects become more difficult to observe.

Thinking Physics 3

Suppose you are observing a photon while moving toward its source at a high speed. According to the Doppler effect, you should detect a higher frequency. According to the equation for the energy of the photon, $E = hf$, this should result in an increased energy. How can the energy of a photon be increased simply by moving toward it? Does this violate the conservation of energy principle?

Reasoning There is no violation of conservation of energy, because energy had to be added to the *system of the photon and the observer* to represent the motion of the observer. According to the observer, the increased energy appears in the form of a higher frequency of the photon. According to a second observer, external to the system, the increased energy appears as kinetic energy of the first observer.

28.5 · THE WAVE PROPERTIES OF PARTICLES

Louis de Broglie (1892–1987), a French physicist, was awarded the Nobel Prize in 1929 for his discovery of the wave nature of electrons. "It would seem that the basic idea of quantum theory is the impossibility of imaging an isolated quantity of energy without associating with it a certain frequency."
(AIP Niels Bohr Library)

Students who are being introduced to the dual nature of light often find the concept very difficult to accept. Even more disconcerting, however, is that *matter* has a dual nature as well!

In 1923, in his doctoral dissertation, Louis Victor de Broglie postulated that, **because photons have wave and particle characteristics and electrons have particle characteristics and properties typical of waves, perhaps all forms of matter have wave as well as particle properties.** This was a highly revolutionary idea with no experimental confirmation at that time. According to de Broglie, electrons have a dual particle–wave nature. An electron in motion exhibits wave character or wave properties. De Broglie explained the source of this assertion in his 1929 Nobel Prize acceptance speech:

> On the one hand the quantum theory of light cannot be considered satisfactory since it defines the energy of a light corpuscle by the equation $E = hf$ containing the frequency f. Now a purely corpuscular theory contains nothing that enables us to define a frequency; for this reason alone, therefore, we are compelled, in the case of light, to introduce the idea of a corpuscle and that of periodicity simultaneously. On the other hand, determination of the stable motion of electrons in the atom introduces integers, and up to this point the only phenomena involving integers in physics were those of interference and of normal modes of vibration. This fact suggested to me the idea that electrons too could not be considered simply as corpuscles, but that periodicity must be assigned to them also.

In Chapter 9 we found that the relationship between energy and momentum for a photon is $p = E/c$. We also know from Equation 28.3 that the energy of a photon is $E = hf = hc/\lambda$. Thus, the momentum of a photon can be expressed as

Momentum of a photon •

$$p = \frac{E}{c} = \frac{hc}{c\lambda} = \frac{h}{\lambda} \qquad [28.8]$$

From this equation we see that the photon wavelength can be specified by its momentum, or $\lambda = h/p$. De Broglie suggested that material particles of momentum p should also have wave properties and a corresponding wavelength. Because the momentum of a particle of mass m and speed[1] v is $p = mv$, the **de Broglie wavelength** of a particle is

De Broglie wavelength •

$$\lambda = \frac{h}{p} = \frac{h}{mv} \qquad [28.9]$$

Furthermore, in analogy with photons, de Broglie postulated that the frequencies of matter waves obey the Einstein relation $E = hf$, so that

Frequency of matter waves •

$$f = \frac{E}{h} \qquad [28.10]$$

---

[1]This speed, v, the speed of the particle, is the group speed of the wave (the speed at which a signal may travel). It is not the same as the phase speed of the wave, related to the index of refraction and wavelength. See L. de Broglie, *Matter and Light*, Dover, New York, pp. 168–171.

The dual nature of matter is apparent in these two equations because each contains both particle concepts (mv and E) and wave concepts (λ and f). The fact that these relationships are established experimentally for photons makes the de Broglie hypothesis that much easier to accept.

Quantization of Angular Momentum in the Bohr Model

Bohr's model of the atom, discussed in Chapter 11, had many shortcomings and problems. For example, as the electrons revolve around the nucleus, how can one understand the fact that only certain electron energies are allowed? Why do all atoms of a given element have precisely the same physical properties regardless of the infinite variety of starting velocities and positions of the electrons in each atom?

De Broglie's great insight was to recognize that wave theories of matter use interference phenomena to explain these observations. Recall from Chapter 14 that a plucked guitar string, although initially subjected to a wide range of wavelengths, will support only those standing wave patterns that have nodes at its ends. Any free vibration of the string consists of a superposition of varied amounts of many standing waves. This same reasoning can be applied to electron matter waves bent into a circle around the nucleus. All of the possible states of the electron are standing wave states, each with its own wavelength, speed, and energy. Figure 28.11a illustrates this point when three complete wavelengths are contained in one circumference of the orbit. Similar patterns can be drawn for orbits containing two wavelengths, four wavelengths, and so forth. This situation is analogous to that of standing waves on a string that has preferred (resonant) frequencies of vibration. Figure 28.11b shows a standing wave pattern containing three wavelengths for a string that is fixed at each end. Now imagine that the vibrating string is removed from its supports at A and B and bent into a circular shape such that points A and B are connected. The end result is a pattern like that shown in Figure 28.11a.

Another aspect of the Bohr theory that is easier to visualize by using de Broglie's hypothesis is the quantization of angular momentum. We simply assume that the allowed Bohr orbits arise because the electron matter waves form standing waves when an integral number of wavelengths exactly fits into the circumference of a circular orbit. Thus,

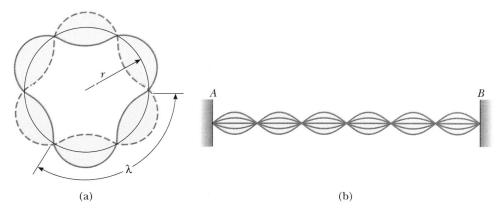

(a)   (b)

Figure 28.11   (a) The standing wave pattern for an electron wave in a stable orbit of hydrogen. There are three full wavelengths in this orbit. (b) The standing wave pattern for a vibrating, stretched string fixed at its ends. This pattern has three full wavelengths.

$$n\lambda = 2\pi r$$

where r is the radius of the orbit and $n = 1, 2, 3, \ldots$ Because the de Broglie wavelength of an electron is $\lambda = h/m_e v$, and because $\hbar = h/2\pi$, we can write the preceding condition as $n(h/m_e v) = 2\pi r$, or

$$m_e v r = n\hbar$$

Note that this is precisely the Bohr condition of quantization of angular momentum (see Chapter 11, Section 11.5).[2] Thus, electron waves that fit the orbits are standing waves because of the boundary conditions imposed. These standing waves have discrete frequencies, corresponding to the allowed wavelengths. If $n\lambda \neq 2\pi r$, a standing-wave pattern can never form a closed circular orbit, as in Figure 28.11a.

The Davisson–Germer Experiment

De Broglie's proposal that any kind of particle exhibits both wave and particle properties was first regarded as pure speculation. If particles such as electrons had wave-like properties, then under the correct conditions they should exhibit diffraction effects. In 1927, 3 years after de Broglie published his work, Clinton J. Davisson (1881–1958) and Lester H. Germer (1896–1971) of the United States succeeded in measuring the wavelength of electrons. Their important discovery provided the first experimental confirmation of the matter waves proposed by de Broglie.

It is interesting to note that the intent of the initial Davisson–Germer experiment was not to confirm the de Broglie hypothesis. In fact, their discovery was made by accident (as is often the case). The experiment involved the scattering of low-energy electrons (about 54 eV) shot at a nickel target in a vacuum. During one experiment, the nickel surface was badly oxidized because of an accidental break in the vacuum system. After the nickel target was heated in a flowing stream of hydrogen to remove the oxide coating, electrons scattered by it exhibited intensity maxima and minima at specific angles. The experimenters finally realized that the nickel had formed large crystal regions on heating and that the regularly spaced planes of atoms in the crystalline regions served as a diffraction grating for electron matter waves.

Shortly thereafter, Davisson and Germer performed more extensive diffraction measurements on electrons scattered from single-crystal targets. Their results showed conclusively the wave nature of electrons and confirmed the de Broglie relation $p = h/\lambda$. A year later in 1928, George P. Thomson (1892–1975) of Scotland observed electron diffraction patterns by passing electrons through very thin gold foils. Diffraction patterns have since been observed for helium atoms, hydrogen atoms, and neutrons. Hence, the universal nature of matter waves has been established in a variety of ways.

Thinking Physics 4

Electron microscopes take advantage of the wave nature of particles. Electrons are accelerated to high speeds, giving them a short de Broglie wavelength. Imagine an

---

[2]Note that de Broglie's analysis still failed to explain the fact that states having an orbital angular momentum quantum number of zero do exist. Furthermore, the actual states of the hydrogen atom are not planar but have spherical symmetry.

electron microscope using electrons with a de Broglie wavelength of 0.2 nm. Why don't we design a microscope using 0.2 nm *photons* to do the same thing?

Reasoning Because electrons are charged particles, they will interact with the sample in the microscope and scatter according to the shape and density of various portions of the sample, providing a means of viewing the sample. Photons of wavelength 0.2 nm are uncharged and in the x-ray region of the spectrum. They will tend to simply pass through the sample without interacting. It is also much easier to image electrons because, unlike x-rays, electrons can be guided by electrostatic focusing fields.

CONCEPTUAL PROBLEM 7

Why is an electron microscope more suitable than an optical microscope for "seeing" objects of an atomic size?

Example 28.5   The Wavelength of an Electron

Calculate the de Broglie wavelength for an electron ($m_e = 9.11 \times 10^{-31}$ kg) moving at 1.00×10^7 m/s.

Solution Equation 28.9 gives

$$\lambda = \frac{h}{m_e v} = \frac{6.63 \times 10^{-34} \text{ J} \cdot \text{s}}{(9.11 \times 10^{-31} \text{ kg})(1.00 \times 10^7 \text{ m/s})}$$

$$= 7.28 \times 10^{-11} \text{ m}$$

This wavelength corresponds to that of typical x-rays.

EXERCISE 6   Find the de Broglie wavelength of a proton moving with a speed of 1.00×10^7 m/s.
Answer 3.97×10^{-14} m

Example 28.6   The Wavelength of a Rock

A rock of mass 50 g is thrown with a speed of 40 m/s. What is its de Broglie wavelength?

Solution From Equation 28.9 we have

$$\lambda = \frac{h}{mv} = \frac{6.63 \times 10^{-34} \text{ J} \cdot \text{s}}{(50 \times 10^{-3} \text{ kg})(40 \text{ m/s})} = 3.32 \times 10^{-34} \text{ m}$$

This wavelength is much smaller than any aperture through which the rock could possibly pass. This means that we could not observe diffraction effects, and as a result the wave properties of large-scale objects cannot be observed.

Example 28.7   An Accelerated Charge

A particle of charge q and mass m is accelerated from rest through a potential difference ΔV. Find its de Broglie wavelength.

Solution When a charge is accelerated from rest through a potential difference of ΔV, its gain in kinetic energy, $\frac{1}{2}mv^2$, must equal the loss in potential energy, $q \, \Delta V$:

$$\tfrac{1}{2}mv^2 = q \, \Delta V$$

Because $p = mv$, we can express this in the form

$$\frac{p^2}{2m} = q \, \Delta V \qquad \text{or} \qquad p = \sqrt{2mq \, \Delta V}$$

Substituting this expression for p in the de Broglie relation $\lambda = h/p$ gives

$$\lambda = \frac{h}{\sqrt{2mq \, \Delta V}}$$

EXERCISE 7   Calculate the de Broglie wavelength of an electron accelerated through a potential difference of 50 V.
Answer   0.174 nm

28.6 • THE DOUBLE-SLIT EXPERIMENT REVISITED

One way to crystallize our ideas about the electron's wave–particle duality is to consider a double-slit electron diffraction experiment. This experiment shows the impossibility of *simultaneously* measuring both wave and particle properties and illustrates the use of the wave function in the determination of interference effects.

Consider a parallel beam of monoenergetic electrons that is incident on a double slit, as in Figure 28.12. Because the slit openings are much smaller than the slit separation D, no structure is expected from single-slit diffraction. An electron detector is positioned far from the slits at a distance much greater than D. If the detector collects electrons for a long enough time, one finds a typical wave interference pattern for the counts per minute or probability of arrival of electrons. Such an interference pattern would not be expected if the electrons behaved as classical particles.

If the experiment is carried out at lower beam intensities, the interference pattern is still observed if the time of the measurement is sufficiently long. This is illustrated by the computer-simulated patterns in Figure 28.13. Note that the interference pattern becomes clearer as the number of electrons reaching the screen increases.

If one imagines a single electron producing in-phase "wavelets" as it reaches one of the slits, standard wave theory can be used to find the angular separation,

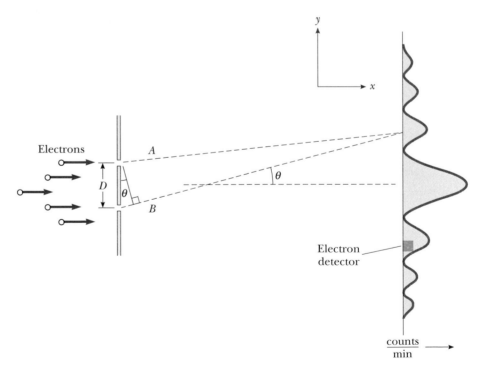

Figure 28.12 Electron diffraction. The slit separation D is much greater than the individual slit widths and much less than the distance between the slit and the detector. The electron detector is movable along the y-direction in the drawing and so can detect electrons diffracted at different values of θ. The detector acts like the "viewing screen" of Young's double-slit experiment with light, discussed in Chapter 27.

θ, between the central probability maximum and its neighboring minimum. The minimum occurs when the path length difference between A and B is half a wavelength, or

$$D \sin \theta = \frac{\lambda}{2}$$

As an electron's wavelength is given by $\lambda = h/p_x$, we see that, for small θ,

$$\sin \theta \approx \theta = \frac{h}{2p_x D}$$

Thus, the dual nature of the electron is clearly shown in this experiment: **The electrons are detected as particles, each at a localized spot at some instant of time, but the probability of arrival at that spot is determined by finding the intensity of two interfering matter waves.**

And there is more. If one slit is covered during the experiment, and the slit width is small compared to the wavelength of the electrons, a symmetric curve peaked around the center of the open slit is observed. Plots of the counts per minute (probability of arrival of electrons) with the lower or upper slit closed are shown as blue curves in the central part of Figure 28.14. These are expressed as the square of the absolute value of some wave function, $|\psi_1|^2 = \psi_1^* \psi_1$ or $|\psi_2|^2 = \psi_2^* \psi_2$, where ψ_1 and ψ_2 represent the electron passing through slit 1 and slit 2, respectively.

If an experiment is now performed with slit 2 blocked half of the time and then slit 1 blocked during the remaining time, the accumulated pattern of counts/min shown by the blue curve on the right side of Figure 28.14 is completely different

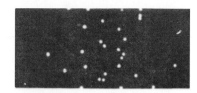

(a) After 28 electrons

(b) After 1000 electrons

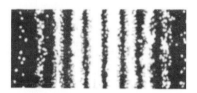

(c) After 10,000 electrons

(d) Two-slit electron pattern

Figure 28.13 (a, b, c) Computer-simulated interference patterns for a beam of electrons incident on a double slit. *(From E. R. Huggins, Physics I, New York, W. A. Benjamin, 1968)* (d) Photograph of a double-slit interference pattern produced by electrons. *(From C. Jönsson, Zeitschrift für Physik 161:454, 1961; used with permission)*

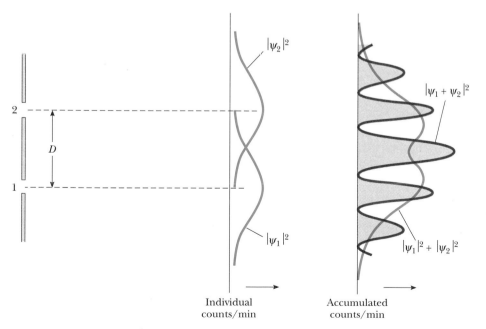

Figure 28.14 Accumulated results of the two-slit electron diffraction experiment with each slit closed half the time. The result with both slits open is shown in red.

from the case with both slits open (red curve). There is no longer a maximum probability of arrival of an electron at $\theta = 0$. In fact, **the interference pattern has been lost and the accumulated result is simply the sum of the individual results.** When only one slit is open at a time, we know the electron has the same localizability and indivisibility at the slits as we measure at the detector, because the electron clearly goes through slit 1 or slit 2. Thus, the total must be analyzed as the sum of those electrons that come through slit 1, $|\psi_1|^2$, and those that come through slit 2, $|\psi_2|^2$.

When both slits are open, it is tempting to assume that the electron goes through either slit 1 or slit 2, and that the counts/min are again given by $|\psi_1|^2 + |\psi_2|^2$. We know, however, that the experimental results indicated by the red interference pattern in Figure 28.14 contradict this. Hence, our assumption that the electron is localized and goes through only one slit when both slits are open must be wrong (a painful conclusion!). Because it exhibits interference, we must conclude that—somehow—*the electron must be simultaneously present at both slits.* In order to find the probability of detecting the electron at a particular point on the screen with both slits open, we may say that the electron is in a *superposition state* given by

$$\psi = \psi_1 + \psi_2$$

Thus, the probability of detecting the electron at the screen is equal to the quantity $|\psi_1 + \psi_2|^2$ and not $|\psi_1|^2 + |\psi_2|^2$. Because matter waves that start out in phase at the slits in general travel different distances to the screen (see Fig. 28.12), ψ_1 and ψ_2 possess a relative phase difference of ϕ at the screen. Using a vector diagram (Fig. 28.15) to find $|\psi_1 + \psi_2|^2$ immediately yields

$$|\psi|^2 = |\psi_1 + \psi_2|^2 = |\psi_1|^2 + |\psi_2|^2 + 2|\psi_1||\psi_2|\cos\phi$$

where $|\psi_1|^2$ is the probability of detection if slit 1 is open and slit 2 is closed, and $|\psi_2|^2$ is the probability of detection if slit 2 is open and slit 1 is closed. The term $2|\psi_1||\psi_2|\cos\phi$ is the interference term, which arises from the relative phase, ϕ, of the waves.

In order to interpret these results, we are forced to conclude that **an electron interacts with both slits simultaneously.** If we attempt to determine experimentally which slit the electron goes through, the act of measuring will destroy the interference pattern. It is impossible to determine which slit the electron will go through. In effect, **we can say only that the electron passes through both slits!** The same arguments apply to photons.

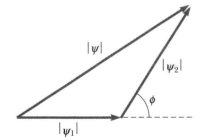

Figure 28.15 A vector diagram representing the addition of two complex quantities, ψ_1 and ψ_2.

28.7 · THE UNCERTAINTY PRINCIPLE

Whenever one measures the position and velocity of a particle at any instant, experimental uncertainties are built into the measurements. According to classical mechanics, there is no fundamental barrier to an ultimate refinement of the apparatus or experimental procedures. In other words, it is possible, in principle, to make such measurements with arbitrarily small uncertainty. Quantum theory predicts, however, that **it is fundamentally impossible to make simultaneous measurements of a particle's position and momentum with infinite accuracy.**

In 1927, Werner Heisenberg (1901–1976) introduced this notion, which is now known as the **Heisenberg uncertainty principle:**

If a measurement of position is made with precision Δx and a simultaneous measurement of momentum in the x direction is made with precision Δp_x, then the product of the two uncertainties can never be smaller than $\hbar/2$, where $\hbar = h/2\pi$.

$$\Delta x \, \Delta p_x \geqslant \frac{\hbar}{2} \qquad [28.11]$$

• *Uncertainty principle*

That is, **it is physically impossible to simultaneously measure the exact position and exact momentum of a particle.** Heisenberg was careful to point out that the inescapable uncertainties Δx and Δp_x do not arise from imperfections in practical measuring instruments. Rather, they arise from the quantum structure of matter, from effects such as the unpredictable recoil of an electron when struck by an indivisible photon, or the diffraction of light or electrons passing through a small opening.

Another important uncertainty relation involves the uncertainty in the energy of a particle ΔE and the time Δt taken to measure that energy. This energy–time uncertainty principle is

$$\Delta E \, \Delta t \geqslant \frac{\hbar}{2} \qquad [28.12]$$

Equation 28.12 states that the precision with which we can know the energy of some system is limited by the time available for measuring the energy. A common application of the energy–time uncertainty is in calculating the lifetimes of very short-lived subatomic particles whose lifetimes cannot be measured directly, but whose uncertainty in energy or mass can be measured.

To understand the uncertainty principle, consider the following thought experiment introduced by Heisenberg. Suppose you wish to measure the position and momentum of an electron as accurately as possible. You might be able to do this by viewing the electron with a powerful light microscope. In order to see the electron and thus determine its position, at least one photon of light must bounce off the electron, as in Figure 28.16a, and then it must pass through the microscope into your eye, as in Figure 28.16b. When it strikes the electron, however, the photon transfers some unknown amount of its own momentum to the electron. Thus, in the process of locating the electron's position very accurately (that is, making Δx very small), the very light that allows you to succeed changes the electron's momentum to some undeterminable extent (making Δp_x very large).

Let us analyze the collision by first noting that the incoming photon has momentum h/λ. As a result of the collision, the photon transfers all of its momentum along the x axis to the electron. Thus, the uncertainty in the electron's momentum after the collision is as large as the momentum of the incoming photon. That is, $\Delta p_x = h/\lambda$. Furthermore, because light also has wave properties, we would expect to be able to determine the position of the electron to within one wavelength of

Werner Heisenberg (1901–1976)

A German theoretical physicist, Heisenberg obtained his PhD in 1923 at the University of Munich, where he studied under Arnold Sommerfeld. Although physicists such as de Broglie and Schrödinger tried to develop physical models of the atom, Heisenberg developed an abstract mathematical model called *matrix mechanics* to explain the wavelengths of spectral lines. The more successful wave mechanics by Schrödinger, announced a few months later, was shown to be equivalent to Heisenberg's approach. Heisenberg made many other significant contributions to physics, including his famous uncertainty principle, for which he received the Nobel Prize in 1932: the prediction of two forms of molecular hydrogen; and theoretical models of the nucleus. *(Courtesy of University of Hamburg)*

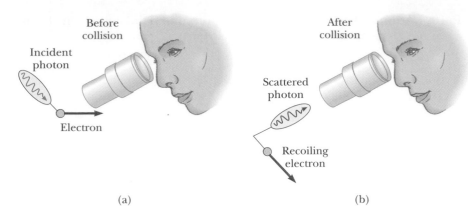

Figure 28.16   A thought experiment for viewing an electron with a powerful microscope. (a) The electron is viewed before it collides with the photon. (b) The electron recoils (is disturbed) as a result of the collision with the photon.

the light being used to view it, so that $\Delta x = \lambda$. Multiplying these two uncertainties gives

$$\Delta x \, \Delta p_x = \lambda \left(\frac{h}{\lambda}\right) = h$$

This represents the minimum in the products of the uncertainties. Because the uncertainty can always be greater than this minimum, we have

$$\Delta x \, \Delta p_x \gtrsim h$$

This agrees with Equation 28.11 (apart from a small numerical factor introduced by Heisenberg's more precise analysis).

Heisenberg's uncertainty principle enables us to better understand the dual wave–particle nature of light and matter. We have seen that the wave description is quite different from the particle description. Therefore, if an experiment (such as the photoelectric effect) is designed to reveal the particle character of an electron, its wave character will become less apparent. If the experiment (such as diffraction from a crystal) is designed to accurately measure the electron's wave properties, its particle character will become less apparent.

Thinking Physics 5

A common, but erroneous, description of the absolute zero of temperature is *that temperature at which all molecular motion ceases.* How can the uncertainty principle be used to argue against this description?

Reasoning   Let us imagine molecules in a finite piece of material. The molecules are constrained to be within the piece of material, so there is a fixed uncertainty in their position along one axis, Δx, which is the size of the piece of material along this axis. If the molecular motion were to cease at absolute zero, we would be claiming that the molecular velocity is *zero*, with no uncertainty, $\Delta v = 0$. The product of zero uncertainty in velocity and a nonzero uncertainty in position will be zero, violating the uncertainty principle. Thus, even at absolute zero, there must be some molecular motion.

Thinking Physics 6

According to the Bohr model of the hydrogen atom, the electron in the ground state moves in a circular orbit of radius 0.529×10^{-10} m, and the speed of the electron in this state is 2.2×10^6 m/s. What is the approximate uncertainty in the electron's radial position when the electron is in the ground state?

Reasoning Because the Bohr model assumes that the electron is located exactly at $r = 0.0529$ nm, the uncertainty Δr in its radial position in this model is zero. But according to the uncertainty principle, the product $\Delta p_r \Delta r \geq \hbar/2$, where Δp_r is the uncertainty in the momentum of the electron in the radial direction. Because the momentum of the electron is $m_e v$, we can assume that the uncertainty in its momentum is less than this value. That is,

$$\Delta p_r < m_e v = (9.11 \times 10^{-31} \text{ kg})(2.2 \times 10^6 \text{ m/s}) = 2.0 \times 10^{-24} \text{ kg} \cdot \text{m/s}$$

From the uncertainty principle, we can now estimate the minimum uncertainty in the radial position:

$$\Delta r_{\min} = \frac{\hbar}{2 \, \Delta p_r} = \frac{1.05 \times 10^{-34} \text{ J} \cdot \text{s}}{2 \times 2 \times 10^{-24} \text{ kg} \cdot \text{m/s}} = 0.26 \times 10^{-10} \text{ m}$$

This uncertainty in the radial position of the electron is half the size of the Bohr radius, and is a measure of the fuzziness of the orbit.

Example 28.8 Locating an Electron

The speed of an electron is measured to be 5.00×10^3 m/s $\pm$ 0.003%. Within what limits could one determine the position of this electron?

Solution The momentum of the electron is

$$p = m_e v = (9.11 \times 10^{-31} \text{ kg})(5.00 \times 10^3 \text{ m/s})$$
$$= 4.56 \times 10^{-27} \text{ kg} \cdot \text{m/s}$$

Because the uncertainty is 0.003% of this value, we get

$$\Delta p = 0.00003 p = 1.37 \times 10^{-31} \text{ kg} \cdot \text{m/s}$$

The uncertainty in position can now be calculated by using this value of Δp and Equation 28.11:

$$\Delta x \geq \frac{\hbar}{2 \, \Delta p_x} = \frac{1.05 \times 10^{-34} \text{ J} \cdot \text{s}}{2.74 \times 10^{-31} \text{ kg} \cdot \text{m/s}}$$
$$= 0.384 \times 10^{-3} \text{ m} = 0.384 \text{ mm}$$

This is the *lower limit* on the uncertainty with which we could have determined the electron's position.

EXERCISE 8 A monoenergetic beam of electrons is incident on a single slit of width 0.500 nm. A diffraction pattern is formed on a screen 20.0 cm from the slit. If the distance between successive minima of the diffraction pattern is 2.10 cm, what is the energy of the incident electrons? Answer 546 eV

28.8 • AN INTERPRETATION OF QUANTUM MECHANICS

Matter waves are described by the complex-valued wave function ψ, introduced in Section 28.6. The absolute square $|\psi|^2 = \psi^* \psi$ gives the probability density of finding a particle at a given point at some instant, where ψ^* is the complex conjugate of ψ. The wave function contains within it all the information that can be known about the particle.

This interpretation of de Broglie's matter waves was first suggested by Max Born (1882–1970) in 1928. In 1926, Erwin Schrödinger (1887–1961) proposed a wave equation that described the manner in which matter waves change in space and time. The *Schrödinger wave equation*, which we will examine in Section 28.10, represents a key element in the theory of quantum mechanics.

The concepts of quantum mechanics, strange as they sometimes may seem, developed from classical ideas. In fact, when the techniques of quantum mechanics are applied to macroscopic systems, the results are essentially identical to those of classical physics. This blending of the two theories occurs when the de Broglie wavelength is small compared with the dimensions of the system. The situation is similar to the agreement between relativistic mechanics and classical mechanics when $v \ll c$.

A number of experiments show that matter has both a wave nature and a particle nature. A question that arises quite naturally in this regard is the following: If we are describing a particle, how do we view what is waving? In the cases of waves on strings, water waves, and sound waves, the wave is represented by some quantity that varies with time and position. In a similar manner, matter waves (de Broglie waves) are represented by the **wave function,** Ψ. In general, Ψ depends on the positions of all the particles in a system and on time, and therefore is often written Ψ (*x, y, z, t*). If Ψ is known for a particle, then the particular properties of that particle can be described. In fact, a fundamental problem of quantum mechanics is this: Given the wave function at some instant, find the wave function at some later time, *t*.

In Section 28.5, we found that the de Broglie equation relates the momentum of a particle to its wavelength through the relation $p = h/\lambda$. If a free particle has a precisely known momentum, p_x, its wave function is an uninterrupted sinusoidal wave of wavelength $\lambda = h/p_x$, and the particle has equal probability of being at any point along the *x* axis. The wave function for such a free particle moving along the *x* axis can be written as

$$\psi(x) = A \sin\left(\frac{2\pi x}{\lambda}\right) = A \sin(kx) \qquad \text{[28.13]}$$

where $k = 2\pi/\lambda$ is the angular wave number and A is a constant amplitude.[3] As we mentioned earlier, the wave function is generally a function of both position and time. Equation 28.13 represents the part of the wave function that depends on position only. For this reason, one can view $\psi(x)$ as a snapshot of the wave function at a given instant, as shown in Figure 28.17a. The wave function for a particle the wavelength of which is not precisely defined is shown in Figure 28.17b. Because the wavelength is not precisely defined, it follows that the momentum is only approximately known. That is, if one were to measure the momentum of the particle, the result would have any value over some range, determined by the spread in wavelength.

Although we cannot measure ψ, we can measure the quantity $|\psi|^2$, which can be interpreted as follows. If ψ represents a single particle, then $|\psi|^2$—called the **probability density**—is the relative probability per unit volume that the particle will be found at any point in the volume. This interpretation, first suggested by

$|\psi|^2$ *equals the probability* •
density

---

[3]In general, ψ has a real part and an imaginary part, and we would write the wave function for the particle in the form Ae^{ikx}, where the imaginary part of the function describes the phase of the wave.

$\psi(x)$

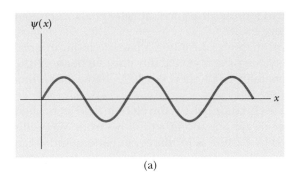

(a)

$\psi(x)$

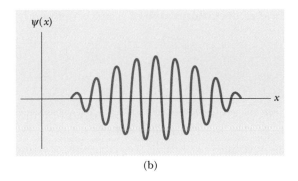

(b)

Figure 28.17 (a) A portion of the wave function for a particle the wavelength of which is precisely known. (b) The wave function for a particle the wavelength of which is not precisely known and hence the momentum of which is known only over some range of values. This pattern is formed by a sum of functions as in (a) having a continuous distribution of wavelengths.

Born in 1928, can also be stated in the following manner. If dV is a small volume element surrounding some point, then the probability of finding the particle in that volume element is $|\psi|^2\, dV$. In this section we deal only with one-dimensional systems, in which the particle must be located along the x axis, and thus we replace dV with dx. In this case, the probability, $P(x)\, dx$, that the particle will be found in the infinitesimal interval dx around the point x is

$$P(x)\, dx = |\psi|^2\, dx$$

Because the particle must be somewhere along the x axis, the sum of the probabilities over all values of x must be 1:

$$\int_{-\infty}^{\infty} |\psi|^2\, dx = 1$$

[28.14] • *Normalization condition on ψ*

Any wave function satisfying Equation 28.14 is said to be **normalized.** Normalization is simply a statement that the particle exists at some point at all times. If the probability were zero, the particle would not exist. Therefore, although it is not possible to specify the position of a particle with complete certainty, it is possible, through $|\psi|^2$, to specify the probability of observing it. Furthermore, **the probability of finding the particle in the interval $a \leq x \leq b$ is**

$$P_{ab} = \int_{a}^{b} |\psi|^2\, dx$$

[28.15]

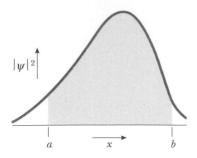

Figure 28.18 The probability that a particle is in the interval $a \leq x \leq b$ is represented by the area under the curve from a to b of the probability density function $|\psi(x, t)|^2$.

That is, the probability P_{ab} is just the area under the curve of $|\psi|^2$ versus x between the points $x = a$ and $x = b$, as in Figure 28.18.

Experimentally, there is a finite probability of finding a particle in an interval near some point at some instant. The value of that probability must lie between the limits 0 and 1. For example, if the probability is 0.3, there is a 30% chance of finding the particle.

The wave function ψ satisfies a wave equation, just as the electric field associated with an electromagnetic wave satisfies a wave equation that follows from Maxwell's equations. The wave equation satisfied by ψ is the Schrödinger equation, and ψ can be computed from it. Although ψ is not a measurable quantity, all the measurable quantities of a particle, such as its energy and momentum, can be derived from a knowledge of ψ. For example, once the wave function for a particle is known, it is possible to calculate the average position, x, of the particle after many experimental trials. This average position is called the **expectation value** of x and is defined by the equation

Expectation value of x •

$$\langle x \rangle \equiv \int_{-\infty}^{\infty} x |\psi|^2 \, dx \qquad \text{[28.16]}$$

(Brackets $\langle \ldots \rangle$ are used to denote expectation values.) This expression implies that the particle is in a definite state, so that the probability density is time independent. Note that the expectation value is equivalent to the average value of x that one would obtain when dealing with a large number of particles in the same state. Furthermore, one can find the expectation value of any function $f(x)$ by using Equation 28.16 with the integrand replaced by $\psi^* f(x) \psi$.

28.9 • A PARTICLE IN A BOX

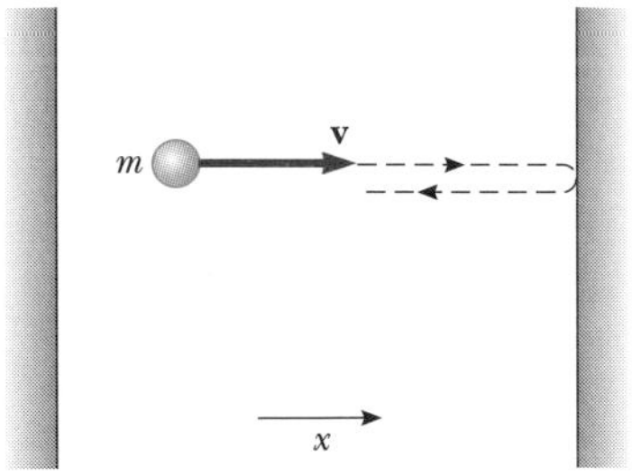

From a classical viewpoint, if a particle is confined to moving along the x axis and to bouncing back and forth between two impenetrable walls (Fig. 28.19), its motion is easy to describe. If the speed of the particle is v, then the magnitude of its momentum (mv) remains constant, as does its kinetic energy. Furthermore, classical physics places no restrictions on the values of a particle's momentum and energy. The quantum mechanics approach to this problem is quite different and requires that we find the appropriate wave function consistent with the conditions of the situation.[4] As we shall see, this description leads to the interesting phenomenon of energy quantization.

Figure 28.19 A particle of mass m and velocity $\mathbf{v}$, confined to moving parallel to the x axis and bouncing between two impenetrable walls.

Because the walls are impenetrable, the particle can *never* be found outside the box, which requires $\psi(x)$ to be zero at the walls and outside the walls. That is, $\psi(x) = 0$ for $x \leq 0$ and for $x \geq L$, where L is the distance between the two walls. Only those wave functions that satisfy this condition are allowed. In analogy with standing waves on a string (Eq. 14.3), the allowed wave functions for the particle in the box are sinusoidal and are given by

Allowed wave functions for a •
particle in a box

$$\psi(x) = A \sin\left(\frac{n\pi x}{L}\right) \qquad n = 1, 2, 3, \ldots \qquad \text{[28.17]}$$

[4]Before continuing, you should review Chapter 14, Sections 14.2 and 14.3 on standing mechanical waves.

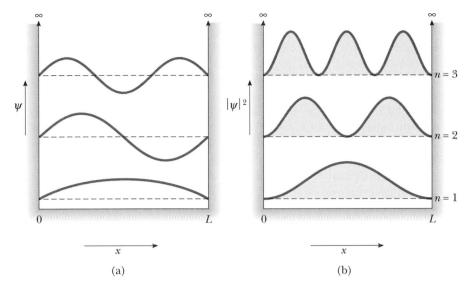

Figure 28.20   The first three allowed stationary states for a particle confined to a one-dimensional box. (a) The wave functions for $n = 1$, 2, and 3. (b) The probability distributions for $n = 1$, 2, and 3.

where A is the maximum value of the wave function. This expression shows that, for a particle confined to a box and having a well-defined de Broglie wavelength, ψ is represented by a sinusoidal wave. The allowed wavelengths are those for which the length L is equal to an integral number of half-wavelengths—that is, $L = n\lambda/2$. These allowed states of the systems are called **stationary states** because they are constant with time.

Figures 28.20a and b are graphs of ψ versus x and $|\psi|^2$ versus x for $n = 1$, 2, and 3. As we shall soon see, these correspond to the three lowest allowed energies for the particle. Note that although ψ can be positive or negative, $|\psi|^2$ is always positive. From any viewpoint, a negative value for $|\psi|^2$ is meaningless.

Further inspection of Figure 28.20b shows that $|\psi|^2$ is always zero at the boundaries, indicating that it is impossible to find the particle at these points. In addition, $|\psi|^2$ is zero at other points, depending on the value of n. For $n = 2$, $|\psi|^2 = 0$ at $x = L/2$; for $n = 3$, $|\psi|^2 = 0$ at $x = L/3$ and at $x = 2L/3$. For $n = 2$, $|\psi|^2$ has maxima at $x = L/4$ and at $x = 3L/4$, and so on.

Because the wavelengths of the particle are restricted by the condition $\lambda = 2L/n$, the magnitude of the momentum is also restricted to specific values:

$$p = \frac{h}{\lambda} = \frac{h}{2L/n} = \frac{nh}{2L}$$

Using $p = mv$, we find that the allowed values of the kinetic energy are

$$E_n = \tfrac{1}{2}mv^2 = \frac{p^2}{2m} = \frac{(nh/2L)^2}{2m}$$

$$E_n = \left(\frac{h^2}{8mL^2}\right)n^2 \qquad n = 1, 2, 3 \ldots \qquad \text{[28.18]}$$

• *Allowed energies for a particle in a box*

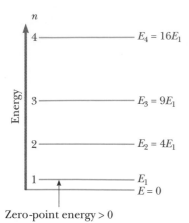

Zero-point energy > 0

Figure 28.21 An energy level diagram for a particle confined to a one-dimensional box of width L. The lowest allowed energy is E_1 and has the value $E_1 = h^2/8mL^2$.

As we see from this expression, **the energy of the particle is quantized,** as we would expect. The lowest allowed energy corresponds to $n = 1$, for which $E_1 = h^2/8mL^2$. Because $E_n = n^2E_1$, the excited states corresponding to $n = 2$, 3, 4, . . . have energies given by $4E_1$, $9E_1$, $16E_1$, . . . Figure 28.21 is an energy level diagram describing the positions of the allowed states. Note that the state $n = 0$ is not allowed. This means that according to quantum mechanics, the particle can never be at rest. The least energy it can have, corresponding to $n = 1$, is called the **zero-point energy.** This result is clearly contradictory to the classical viewpoint, in which $E = 0$ is an acceptable state, as are all positive values of E. In this situation, only positive values of E are considered, because the total energy E equals the kinetic energy, and the potential energy is zero.

The energy levels are of special importance for the following reason. If the particle is electrically charged, it can emit a photon when it drops from an excited state, such as E_3, to one of the lower-lying states, such as E_2. It can also absorb a photon the energy of which matches the difference in energy between two allowed states. For example, if the photon frequency is f, the particle will jump from the state E_1 to the state E_2 if $hf = E_2 - E_1$. The processes of photon emission and absorption can be observed by spectroscopy, in which spectral wavelengths are a direct measure of such energy differences.

Example 28.9   A Bound Electron

An electron is confined between two impenetrable walls 0.200 nm apart. Determine the energy levels for $n = 1, 2$, and 3.

Solution We can apply Equation 28.18, taking $n = 1$ and using $m_e = 9.11 \times 10^{-31}$ kg. This gives

$$E_1 = \frac{h^2}{8m_eL^2} = \frac{(6.63 \times 10^{-34}\,\text{J}\cdot\text{s})^2}{8(9.11 \times 10^{-31}\,\text{kg})(2 \times 10^{-10}\,\text{m})^2}$$

$$= 1.51 \times 10^{-18}\,\text{J} = \boxed{9.42\ \text{eV}}$$

For $n = 2$ and $n = 3$, we find that $E_2 = 4E_1 = \boxed{37.7\ \text{eV}}$ and $E_3 = 9E_1 = \boxed{84.8\ \text{eV}}$. Although this is a rather primitive model, it can be used to describe an electron trapped in a vacant crystal site.

Example 28.10   Energy Quantization for a Macroscopic Object

A 1.00-mg object is confined to moving between two rigid walls separated by 1.00 cm. (a) Calculate its minimum speed.

Solution The minimum speed corresponds to the state for which $n = 1$. Using Equation 28.18 with $n = 1$ gives the zero-point energy:

$$E_1 = \frac{h^2}{8mL^2} = \frac{(6.63 \times 10^{-34}\,\text{J}\cdot\text{s})^2}{8(1.00 \times 10^{-6}\,\text{kg})(1.00 \times 10^{-2}\,\text{m})^2}$$

$$= 5.49 \times 10^{-58}\,\text{J}$$

Because $E = \frac{1}{2}mv^2$, we can find v as follows:

$$\tfrac{1}{2}mv^2 = 5.49 \times 10^{-58}\,\text{J}$$

$$v = \left[\frac{2(5.49 \times 10^{-58}\,\text{J})}{1.00 \times 10^{-6}\,\text{kg}}\right]^{1/2} = 3.31 \times 10^{-26}\,\text{m/s}$$

This result is so small that the object can be considered to be at rest, which is what one would expect for a macroscopic object.

(b) If the speed of the object is 3.00×10^{-2} m/s, find the corresponding value of n.

Solution The kinetic energy of the object is

$$E = \tfrac{1}{2}mv^2 = \tfrac{1}{2}(1.00 \times 10^{-6}\,\text{kg})(3.00 \times 10^{-2}\,\text{m/s})^2$$

$$= 4.50 \times 10^{-10}\,\text{J}$$

Because $E_n = n^2E_1$ and $E_1 = 5.49 \times 10^{-58}$ J, we find that

$$n^2E_1 = 4.50 \times 10^{-10}\,\text{J}$$

$$n = \left(\frac{4.50 \times 10^{-10}\,\text{J}}{E_1}\right)^{1/2} = \left(\frac{4.50 \times 10^{-10}\,\text{J}}{5.49 \times 10^{-58}\,\text{J}}\right)^{1/2}$$

$$\approx \boxed{9.05 \times 10^{23}}$$

This value of n is so large that we would never be able to distinguish the quantized nature of the energy levels. That is, the difference in energy between the two states $n_1 = 9.05 \times 10^{23}$ and $n_2 = (9.05 \times 10^{23}) + 1$ is too small to be detected experimentally. This is another example that illustrates the working of the correspondence principle; that is, as m and/or L become large, the quantum description must agree with the classical result.

EXERCISE 9   An alpha particle in a nucleus can be modeled as a particle moving in a box of width 1.0×10^{-14} m (the approximate diameter of a nucleus). Using this model, estimate (a) the energy and (b) the momentum of an alpha particle in its lowest energy state. ($m_\alpha = 4 \times 1.66 \times 10^{-27}$ kg)

Answer   (a) 0.52 MeV   (b) 3.3×10^{-20} kg·m/s

28.10 · THE SCHRÖDINGER EQUATION

As we mentioned earlier, the wave function for de Broglie waves must satisfy an equation developed by Schrödinger in 1926. One of the methods of quantum mechanics is to determine a solution to this equation, which in turn yields the allowed wave functions and energy levels of the system under consideration. Proper manipulation of the wave functions enables one to calculate all measurable features of the system.

The Schrödinger equation as it applies to a particle confined to moving along the x axis is

$$\frac{d^2\psi}{dx^2} = -\frac{2m}{\hbar^2}(E - U)\psi \qquad \text{[28.19]}$$

Because this equation is independent of time, it is commonly referred to as the **time-independent Schrödinger equation.** (We shall not discuss the time-dependent Schrödinger equation in this text.)

In principle, if the potential energy $U(x)$ for the system is known, one can solve Equation 28.19 and obtain the wave functions and energies for the allowed states. Because U may vary discontinuously with position, it may be necessary to solve the equation in pieces. In the process, the wave functions for the different regions must join smoothly at the boundaries. In the language of mathematics, we require that $\psi(x)$ be *continuous*. Furthermore, in order that $\psi(x)$ obey the normalization condition, we require that $\psi(x)$ approach zero as x approaches $\pm \infty$. Finally, $\psi(x)$ must be *single-valued* and $d\psi/dx$ must also be continuous for finite values of $U(x)$.

The task of solving the Schrödinger equation may be very difficult, depending on the form of the potential energy function. As it turns out, the Schrödinger equation has been extremely successful in explaining the behavior of atomic and nuclear systems, whereas classical physics has failed to do so. Furthermore, when quantum mechanics is applied to macroscopic objects, the results agree with classical physics, as required by the correspondence principle.

Erwin Schrödinger (1887–1961)

An Austrian theoretical physicist, Schrödinger is best known as the creator of wave mechanics. He also produced important papers in the fields of statistical mechanics, color vision, and general relativity. Schrödinger did much to hasten the universal acceptance of quantum theory by demonstrating the mathematical equivalence between his wave mechanics and the more abstract matrix mechanics developed by Heisenberg. In 1927, Schrödinger accepted the chair of theoretical physics at the University of Berlin, where he formed a close friendship with Max Planck. In 1933 he left Germany and eventually settled at the Dublin Institute of Advanced Study, where he spent 17 creative years working on problems in general relativity, cosmology, and the application of quantum physics to biology. In 1956 he returned home to Austria, where he died in 1961.

The Particle in a Box

Let us solve the Schrödinger equation for our particle in a one-dimensional box of width L (Fig. 28.22). The walls are infinitely high, corresponding to $U(x) = \infty$ for $x = 0$ and $x = L$. The potential energy is constant within the box, and it is conve-

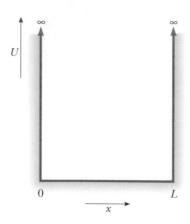

Figure 28.22 Diagram of a one-dimensional box of width L and infinitely high walls.

nient to choose $U = 0$ as its value. Hence, in the region $0 < x < L$, we can express the Schrödinger equation in the form

$$\frac{d^2\psi}{dx^2} = -\frac{2mE}{\hbar^2}\,\psi = -k^2\psi \qquad [28.20]$$

$$k = \frac{\sqrt{2mE}}{\hbar}$$

Because the walls are infinitely high, the particle cannot exist outside the box. Consequently, $\psi(x)$ must be zero outside the box and at the walls. The solution of Equation 28.20 that meets the boundary conditions $\psi(x) = 0$ at $x = 0$ and $x = L$ is

$$\psi(x) = A\sin(kx) \qquad [28.21]$$

This can easily be verified by substitution into Equation 28.20. Note that the first boundary condition, $\psi(0) = 0$, is satisfied by Equation 28.21 because $\sin 0° = 0$. The second boundary condition, $\psi(L) = 0$, is satisfied only if kL is an integral multiple of π, that is, if $kL = n\pi$, where n is an integer. Because $k = \sqrt{2mE}/\hbar$, we get

$$kL = \frac{\sqrt{2mE}}{\hbar}\,L = n\pi$$

Solving for the allowed energies E gives

$$E_n = \left(\frac{h^2}{8mL^2}\right)n^2 \qquad [28.22]$$

Likewise, the allowed wave functions are given by

$$\psi_n(x) = A\sin\left(\frac{n\pi x}{L}\right) \qquad [28.23]$$

These results agree with those obtained in the preceding section. Normalizing this relationship shows that $A = (2/L)^{1/2}$ (see Problem 28.36).

OPTIONAL

28.11 • TUNNELING THROUGH A BARRIER

A very interesting and peculiar phenomenon occurs when a particle strikes a barrier of finite height and width. Consider a particle of energy E that is incident on a rectangular barrier of height U and width L, where $E < U$ (Fig. 28.23). Classically, the particle is reflected by the barrier, because it does not have sufficient energy to cross or even penetrate it. Thus, regions II and III are classically *forbidden* to the particle. According to quantum mechanics, however, **all regions are accessible to the particle, regardless of its energy,** because the amplitude of the matter wave associated with the particle is nonzero everywhere (except at certain points). A typical wavefunction for this case, illustrated in Figure 28.23, shows the penetration of the wave into the barrier and beyond. The wave functions are sinusoidal to the left (region I) and right (region III) of the barrier and join smoothly with an exponentially decaying function within the barrier (region II). Because the probability of locating the particle is proportional to $|\psi|^2$, we conclude that the chance of finding the particle beyond the barrier in region III is nonzero. This barrier penetration is in complete disagreement with classical physics. The possibility of finding the particle on the far side of the barrier is called **tunneling or barrier penetration.** Any attempt to observe the particle inside the barrier and confirm the value of its energy is frustrated by the uncertainty principle. If tunneling is to take place, the barrier must be sufficiently narrow and low, to allow a measurement of the particle's position and momentum consistent with expectations.

The probability of tunneling can be described with a **transmission coefficient, T,** and a **reflection coefficient, R. The transmission coefficient measures the probability that the particle penetrates to the other side of the barrier, and the reflection coefficient is the probability that the particle is reflected by the barrier.** Because the incident particle is either reflected or transmitted, we must require that $T + R = 1$. An approximate expression for the transmission coefficient that is obtained when $T \ll 1$ (a high or wide barrier) is

$$T \cong e^{-2KL} \qquad \text{[28.24]}$$

where

$$K = \frac{\sqrt{2m(U - E)}}{\hbar} \qquad \text{[28.25]}$$

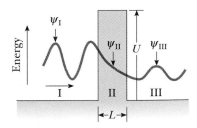

Figure 28.23   Wave function for a particle incident from the left on a barrier of height U. Note that the wave function is sinusoidal in regions I and III but is exponentially decaying in region II. Both amplitude of ψ and energy are plotted along the vertical axis.

Example 28.11   Transmission Coefficient for an Electron

A 30-eV electron is incident on a square barrier of height 40 eV. What is the probability that the electron will tunnel through the barrier if its width is (a) 1.0 nm? (b) 0.10 nm?

Solution   (a) In this situation, the quantity $U - E$ has the value

$$U - E = (40 \text{ eV} - 30 \text{ eV}) = 10 \text{ eV} = 1.6 \times 10^{-18} \text{ J}$$

Using Equation 28.25, and given that $L = 1.0$ nm, the quantity $2KL$ is

$$2KL = 2 \frac{\sqrt{2(9.11 \times 10^{-31} \text{ kg})(1.6 \times 10^{-18} \text{ J})}}{1.054 \times 10^{-34} \text{ J} \cdot \text{s}} (1 \times 10^{-9} \text{ m})$$

$$= 32.4$$

Thus, the probability of tunneling through the barrier is

$$T \cong e^{-2KL} = e^{-32.4} = \boxed{8.5 \times 10^{-15}}$$

That is, the electron has only about 1 chance in 10^{14} to tunnel through the 1.0-nm-wide barrier.

(b) For $L = 0.10$ nm, we find $2KL = 3.24$, and

$$T \cong e^{-2KL} = e^{-3.24} = \boxed{0.039}$$

This result shows that the electron has a high probability (4% chance) of penetrating the 0.10-nm barrier. Thus, reducing the width of the barrier by only one order of magnitude has increased the probability of tunneling by about 12 orders of magnitude!

Applications of Tunneling

As we have seen, tunneling is a quantum phenomenon, a manifestation of the wave nature of matter. There are many examples in nature on the atomic and nuclear scales that may be understood only on the basis of tunneling.

- *Tunnel diode.* The tunnel diode is a semiconductor device consisting of two oppositely charged regions separated by a very narrow neutral region. The current in this device is largely due to tunneling of electrons through the neutral region. The current, or rate of tunneling, can be controlled over a wide range by varying the bias voltage, which changes the height of the barrier.

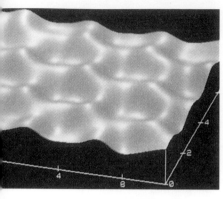

The surface of graphite as "viewed" with a scanning tunneling microscope. This technique enables scientists to see small details on surfaces with a lateral resolution of about 0.2 nm and a vertical resolution of 0.001 nm. The contours seen here represent the arrangement of individual carbon atoms on the crystal surface.

- *Josephson junction.* A Josephson junction consists of two superconductors separated by a thin insulating oxide layer, 1 to 2 nm thick. Under appropriate conditions, electrons in the superconductors travel as pairs and tunnel from one superconductor to the other through the oxide layer. Several effects have been observed in this type of junction. For example, **a dc current is observed across the junction in the absence of electric and magnetic fields.** The current is proportional to sin ϕ, where ϕ is the phase difference between the wave functions in the two superconductors. When a bias voltage, ΔV, is applied across the junction, one observes the current oscillating with a frequency $f = 2e\,\Delta V/h$, where e is the charge on the electron.

- *Alpha decay.* One form of radioactive decay is the emission of alpha particles (the nuclei of helium atoms) by unstable, heavy nuclei. In order for the alpha particle to escape from the nucleus, it must penetrate a barrier several times larger than its energy. The barrier is due to a combination of the attractive nuclear force and the Coulomb repulsion between the alpha particle and the rest of the nucleus. Occasionally an alpha particle tunnels through the barrier, which explains the basic mechanism for this type of decay and the large variations in the mean lifetimes of various radioactive nuclei.

- *Scanning tunneling microscope.* The scanning tunneling microscope, or STM, is a remarkable device that uses tunneling to create images of surfaces with resolution comparable to the size of a single atom. A small probe with a very fine tip is made to scan close to the surface of a specimen. A tunneling current is maintained between the probe and specimen; the current is sensitive to the separation between the tip and specimen. With the maintenance of a constant tunneling current, a feedback signal is obtained that is used to raise and lower the probe as the surface is scanned. Because the vertical motion of the probe follows the contour of the specimen's surface, an image of the surface is obtained.

- *Decay of black holes.* Once inside the event horizon, nothing—not even light—can escape the gravitational pull of a black hole. That was the view held until 1974, when the brilliant British astrophysicist Stephen Hawking proposed that black holes are indeed radiant objects, emitting a variety of particles by a mechanism involving tunneling through the (gravitational) potential barrier surrounding the black hole. The thickness of this barrier is proportional to the size of the black hole, so that the likelihood of a tunneling event initially may be extremely small. As the black hole emits particles, however, its mass and size steadily decrease, making it easier for particles to tunnel out. In this way emission continues at an ever increasing rate, until eventually the black hole radiates itself out of existence in an explosive climax! Thus, Hawking's scenario leads to the decay and eventual demise of any black hole. Calculations indicate that a black hole with the mass of our Sun would survive against decay by tunneling for about 10^{66} years. On the other hand, a black hole with the mass of only a billion tons and roughly the size of a proton should have almost completely evaporated in the 10 billion years that have elapsed since the time of creation. (Such mini black holes are believed to have been formed just after the Big Bang origin of the Universe.)

SUMMARY

The characteristics of *black-body radiation* cannot be explained by classical concepts. Planck introduced the *quantum concept* when he assumed that the atomic oscillators responsible for this radiation existed only in discrete states.

In the **photoelectric effect,** electrons are ejected from a metallic surface when light is incident on that surface. Einstein provided a successful explanation of this effect by extending Planck's quantum theory to the electromagnetic field. In this model, light is viewed as a stream of particles called *photons*, each with energy $E = hf$, where f is the frequency and h is Planck's constant. The maximum kinetic energy of the ejected photoelectron is given by

$$K_{\max} = hf - \phi \qquad \text{[28.5]}$$

where ϕ is the work function of the metal.

X-rays striking a target are scattered at various angles by electrons in the target. A shift in wavelength is observed for the scattered x-rays, and the phenomenon is known as the **Compton effect.** Classical physics does not explain this effect. If the x-ray is treated as a photon, conservation of energy and momentum applied to the photon–electron collisions yield for the Compton shift the expression

$$\lambda' - \lambda_0 = \frac{h}{m_e c} (1 - \cos \theta) \qquad \text{[28.7]}$$

where m_e is the mass of the electron, c is the speed of light, and θ is the scattering angle.

Every object of mass m and momentum p has wave-like properties, with a wavelength given by the de Broglie relation

$$\lambda = \frac{h}{p} \qquad \text{[28.9]}$$

By applying this wave theory of matter to electrons in atoms, de Broglie was able to explain the appearance of quantization in the Bohr model of hydrogen as a standing wave phenomenon.

The **uncertainty principle** states that if a measurement of position is made with precision Δx and a *simultaneous* measurement of momentum is made with precision Δp_x, then the product of the two uncertainties can never be less than a number on the order of $\hbar$.

$$\Delta x \, \Delta p_x \geqslant \frac{\hbar}{2} \qquad \text{[28.11]}$$

Matter waves are represented by the wave function $\Psi(x, y, z, t)$. The probability per unit volume that a particle will be found at a point is $|\psi|^2$. If the particle is confined to moving along the x axis, then the probability that it will be located in an interval dx is given by $|\psi|^2 \, dx$. Furthermore,

$$\int_{-\infty}^{\infty} |\psi|^2 \, dx = 1 \qquad \text{[28.14]}$$

The measured position x of the particle, averaged over many trials, is called the **expectation value** of x and is defined by

$$\langle x \rangle \equiv \int_{-\infty}^{\infty} x |\psi|^2 \, dx \qquad \text{[28.16]}$$

If a particle of mass m is confined to moving in a one-dimensional box of width L the walls of which are perfectly rigid, the **allowed wave functions** for the particle are

$$\psi(x) = A \sin\left(\frac{n\pi x}{L}\right) \qquad n = 1, 2, 3, \ldots \qquad [28.17]$$

where A is the maximum value of ψ. The particle has a well-defined wavelength λ the values of which are such that the width of the box L is equal to an integral number of half wave-lengths—that is, $L = n\lambda/2$. These allowed states are called **stationary states** of the system. The energies of a particle in a box are quantized and are given by

$$E_n = \left(\frac{h^2}{8mL^2}\right) n^2 \qquad n = 1, 2, 3, \ldots \qquad [28.18]$$

The wave function must satisfy the **Schrödinger equation.** The time-independent Schrödinger equation for a particle confined to moving along the x axis is

$$\frac{d^2\psi}{dx^2} = -\frac{2m}{\hbar^2}(E - U)\psi \qquad [28.19]$$

where E is the total energy of the system and U is the potential energy.

When a particle of energy E meets a barrier of height U, where $E < U$, the particle has a finite probability of penetrating the barrier. Part of the incident wave is transmitted through the barrier and part is reflected. This process, called **tunneling,** is the basic mechanism that explains the operation of the Josephson junction and the phenomenon of alpha decay in some radioactive nuclei.

CONCEPTUAL QUESTIONS

1. Why is it impossible to simultaneously measure with infinite accuracy the position and speed of a particle?

2. If matter has a wave nature, why is this wave-like characteristic not observable in our daily experiences?

3. An electron and a proton are accelerated from rest through the same potential difference. Which particle has the longer wavelength?

4. In describing the passage of electrons through a slit and arriving at a screen, the physicist Richard Feynman said that "electrons arrive in lumps, like particles, but the probability of arrival of these lumps is determined as the intensity of the waves would be. It is in this sense that the electron behaves sometimes like a particle and sometimes like a wave." Elaborate on this point in your own words. (For a further discussion of this point, see R. Feynman, *The Character of Physical Law*, Cambridge, Mass., MIT Press, 1980, chap. 6.)

5. In the photoelectric effect, explain why the photocurrent depends on the intensity of the light source but not on the frequency.

6. Why does the existence of a cutoff frequency in the photoelectric effect favor a particle theory of light rather than a wave theory?

7. Is light a wave or a particle? Support your answer by citing specific experimental evidence.

8. Suppose a photograph were made of a person's face using only a few photons. Would the result be simply a very faint image of the face? Discuss.

9. Which has more energy, a photon of ultraviolet radiation or a photon of yellow light?

10. The brightest star in the constellation Lyra is the bluish star Vega, whereas the brightest star in Bootes is the reddish star Arcturus. How do you account for the difference in color of the two stars?

11. An x-ray photon is scattered by an electron. What happens to the frequency of the scattered photon relative to that of the incident photon?

12. Using Wien's displacement law, estimate the wavelength of highest intensity of radiation given off by a human body. Using this information, explain why an infrared detector would be a useful alarm for security work.

13. Why was the Davisson–Germer diffraction of electrons an important experiment?

14. What is the significance of the wave function ψ?

15. What is the Schrödinger equation? How is it useful in describing atomic phenomena?

16. If the photoelectric effect is observed for one metal, can you conclude that the effect will also be observed for another metal under the same conditions? Explain.

PROBLEMS

Section 28.1 Black-Body Radiation and Planck's Theory

1. The human eye is most sensitive to 560-nm light. What temperature black body would radiate most intensely at this wavelength?

2. An FM radio transmitter has a power output of 150 kW and operates at a frequency of 99.7 MHz. How many photons per second does the transmitter emit?

3. Calculate the energy, in electron volts, of a photon the frequency of which is (a) 6.20×10^{14} Hz, (b) 3.10 GHz, (c) 46.0 MHz. (d) Determine the corresponding wavelengths for these photons.

4. The Sun's light reaches the Earth at an average intensity of 1340 W/m². Estimate the number of photons that reach the Earth per second if the temperature of the Sun is 6000 K.

5. The average threshold of dark-adapted (scotopic) vision is 4.00×10^{-11} W/m² at a central wavelength of 500 nm. If light having this intensity and wavelength enters the eye and the pupil is open to its maximum diameter of 8.50 mm, how many photons per second enter the eye?

6. A simple pendulum has a length of 1.00 m and a mass of 1.00 kg. If the amplitude of oscillations of the pendulum is 3.00 cm, estimate the quantum number for the pendulum.

Section 28.2 The Photoelectric Effect

7. Molybdenum has a work function of 4.20 eV. (a) Find the cutoff wavelength and threshold frequency for the photoelectric effect. (b) Calculate the stopping potential if the incident light has a wavelength of 180 nm.

8. Electrons are ejected from a metallic surface with speeds ranging up to 4.60×10^5 m/s when light with a wavelength of $\lambda = 625$ nm is used. (a) What is the work function of the surface? (b) What is the cutoff frequency for this surface?

9. A student studying the photoelectric effect from two different metals records the following information: (a) the stopping potential for photoelectrons released from metal 1 is 1.48 V larger than that for metal 2, and (b) the threshold frequency for metal 1 is 40% smaller than that for metal 2. Determine the work function for each metal.

10. From the scattering of sunlight, Thomson calculated the classical radius of the electron as having a value of 2.82×10^{-15} m. If sunlight with an intensity of 500 W/m² falls on a disk with this radius, estimate the time required to accumulate 1.00 eV of energy. Assume that light is a classical wave and that the light striking the disk is completely absorbed. How does your estimate compare with the observation that photoelectrons are emitted promptly (within 10^{-9} s)?

Section 28.3 The Compton Effect

11. Calculate the energy and momentum of a photon of wavelength 700 nm.

12. X-rays having an energy of 300 keV undergo Compton scattering from a target. If the scattered rays are detected at 37.0° relative to the incident rays, find (a) the Compton shift at this angle, (b) the energy of the scattered x-ray, and (c) the energy of the recoiling electron.

13. A 0.00160-nm photon scatters from a free electron. For what (photon) scattering angle will the kinetic energy of the recoiling electron and the energy of the scattered photon be the same?

14. A 0.110-nm photon collides with a stationary electron. After the collision, the electron moves forward and the photon recoils backward. Find the momentum and kinetic energy of the electron.

15. A 0.880-MeV photon is scattered by a free electron initially at rest such that the scattering angle of the scattered electron equals that of the scattered photon ($\phi = \theta$ in Fig. 28.9b). (a) Determine the common value of these angles. (b) Determine the energy and momentum of the scattered photon. (c) Determine the kinetic energy and momentum of the scattered electron.

Section 28.4 Photons and Electromagnetic Waves
Section 28.5 The Wave Properties of Particles

16. Calculate the de Broglie wavelength for a proton moving with a speed of 1.00×10^6 m/s.

17. Calculate the de Broglie wavelength for an electron that has kinetic energy (a) 50.0 eV and (b) 50.0 keV.

18. (a) An electron has kinetic energy 3.00 eV. Find its wavelength. (b) A photon has energy 3.00 eV. Find its wavelength.

19. For an electron to be confined to a nucleus, its de Broglie wavelength would have to be less than 1.00×10^{-14} m. (a) What would be the kinetic energy of an electron confined to this region? (b) On the basis of this result, would you expect to find an electron in a nucleus? Explain.

20. In the Davisson–Germer experiment, 54.0-eV electrons were diffracted from a nickel lattice. If the first maximum in the diffraction pattern was observed at $\phi = 50.0°$ (Fig. P28.20), what was the lattice spacing a?

21. Robert Hofstadter won the 1961 Nobel Prize in physics for his pioneering work in scattering 20-GeV electrons from nuclei. (a) What is the γ-factor for a 20.0-GeV electron, where $\gamma = (1 - v^2/c^2)^{-1/2}$? What is the momentum of the electron in kg·m/s? (b) What is the wavelength of a 20.0-GeV electron and how does it compare with the size of a nucleus?

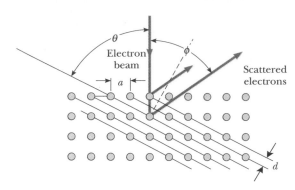

Figure P28.20

22. (a) Show that the frequency f, and wavelength λ, of a particle are related by the expression

$$\left(\frac{f}{c}\right)^2 = \frac{1}{\lambda^2} + \frac{1}{\lambda_C^2}$$

where $\lambda_C = h/mc$ is the Compton wavelength of the particle. (b) Is it ever possible for a photon and a particle (having nonzero mass) to have the same wavelength *and* frequency? Explain.

23. The resolving power of a microscope depends on the wavelength used. If one wished to use a microscope to "see" an atom, a resolution of approximately 1.00×10^{-11} m would have to be obtained. (a) If electrons are used (in an electron microscope), what minimum kinetic energy is required for the electrons? (b) If photons are used, what minimum photon energy is needed to obtain the required resolution?

Section 28.6   The Double-Slit Experiment Revisited

24. A modified oscilloscope is used to perform an electron interference experiment. Electrons are incident on a pair of narrow slits 0.0600 μm apart. The bright bands in the interference pattern are separated by 0.400 mm on a screen 20.0 cm from the slits. Determine the potential difference through which the electrons were accelerated to give this pattern.

25. Neutrons traveling at 0.400 m/s are directed through a double slit having a 1.00-mm separation. An array of detectors is placed 10.0 m from the slit. (a) What is the de Broglie wavelength of the neutrons? (b) How far off axis is the first zero-intensity point on the detector array? (c) When a neutron reaches a detector, can we say which slit the neutron passed through? Explain.

Section 28.7   The Uncertainty Principle

26. Suppose Fuzzy, a quantum-mechanical duck, lives in a world in which $h = 2\pi$ J·s. Fuzzy has a mass of 2.00 kg and is initially known to be within a pond 1.00 m wide. (a) What is

the minimum uncertainty in his speed? (b) Assuming this uncertainty in speed to prevail for 5.00 s, determine the uncertainty in position after this time.

27. An electron ($m_e = 9.11 \times 10^{-31}$ kg) and a bullet ($m = 0.0200$ kg) each have a speed of 500 m/s, accurate to within 0.0100%. Within what limits could we determine the position of the objects?

28. A woman on a ladder drops small pellets toward a spot on the floor. (a) Show that, according to the uncertainty principle, the average miss distance must be at least

$$\Delta x = \left(\frac{\hbar}{2m}\right)^{1/2} \left(\frac{H}{2g}\right)^{1/4}$$

where H is the initial height of each pellet above the floor and m is the mass of each pellet. (b) If $H = 2.00$ m and $m = 0.500$ g, what is Δx?

29. Use the uncertainty principle to show that an electron confined inside a nucleus of diameter 2×10^{-15} m is moving relativistically, whereas a proton confined to the same nucleus can be moving non-relativistically.

Section 28.8   An Interpretation of Quantum Mechanics

30. An electron in an infinite square well has a wave function given by

$$\psi_2(x) = \sqrt{\frac{2}{L}} \sin\left(\frac{2\pi x}{L}\right)$$

for $0 \le x \le L$ and zero elsewhere. What are the most probable positions of the electron?

31. A free electron has a wave function

$$\psi(x) = A \sin(5.00 \times 10^{10}\, x)$$

where x is in meters. Find (a) the de Broglie wavelength, (b) the momentum, and (c) the energy in electron volts.

32. The wave function for a particle in an infinite square well is

$$\psi_2(x) = \sqrt{\frac{2}{L}} \sin\left(\frac{2\pi x}{L}\right)$$

for $0 \le x \le L$ and zero elsewhere. (a) Determine the expectation value of x. (b) Determine the probability of finding the particle near $L/2$, by calculating the probability the particle lies in the range $0.49L \le x \le 0.51L$. (c) Determine the probability of finding the particle near $L/4$ by calculating the probability the particle lies in the range $0.24L \le x \le 0.26L$. (d) Reconcile these probabilities with the result for the average value of x found in part (a).

33. The wave function for a particle is

$$\psi(x) = \sqrt{\frac{a}{\pi(x^2 + a^2)}}$$

for $a > 0$ and $-\infty < x < +\infty$. Determine the probability that the particle is located somewhere between $x = -a$ and $x = +a$.

Section 28.9 A Particle in a Box

34. An electron that has an energy of approximately 6 eV moves between rigid walls 1.00 nm apart. Find (a) the quantum number n for the energy state that the electron occupies and (b) the precise energy of the electron.

35. An electron is contained in a one-dimensional box of width 0.100 nm. (a) Draw an energy-level diagram for the electron for levels up to $n = 4$. (b) Find the wavelengths of all photons that can be emitted by the electron in making transitions that will eventually get it from the $n = 4$ state to the $n = 1$ state.

36. The wave function for a particle confined to moving in a one-dimensional box is

$$\psi(x) = A \sin\left(\frac{n\pi x}{L}\right)$$

Use the normalization condition on ψ to show that

$$A = \sqrt{\frac{2}{L}}$$

Hint: Because the box width is L, the normalization condition (Eq. 28.14) is

$$\int_0^L |\psi|^2\, dx = 1$$

37. A ruby laser emits 694.3-nm light. If this light were due to a transition of an electron in a box from the $n = 2$ state to the $n = 1$ state, find the width of the box.

38. A laser emits light of wavelength λ. If this light were due to a transition of an electron in a box from the $n = 2$ state to the $n = 1$ state, find the width of the box.

39. The nuclear potential energy that binds protons and neutrons in a nucleus is often approximated by a square well. Imagine a proton confined in an infinitely high square well of width 1.00×10^{-5} nm, a typical nuclear diameter. Calculate the wavelength and energy associated with the photon emitted when the proton moves from the $n = 2$ state to the ground state. In what region of the electromagnetic spectrum does this wavelength belong?

Section 28.10 The Schrödinger Equation

40. The wave function of a particle is

$$\psi(x) = A\cos(kx) + B\sin(kx)$$

where A, B, and k are constants. Show that ψ is a solution of the Schrödinger equation (Eq. 28.19), assuming the particle

is free ($U = 0$), and find the corresponding energy E of the particle.

41. Show that the wave function $\psi = Ae^{i(kx - \omega t)}$ is a solution to the Schrödinger equation (Eq. 28.19), where $k = 2\pi/\lambda$ and $U = 0$.

42. A particle of mass m moves in a potential well of width $2L$ (from $x = -L$ to $x = +L$), and in this well is a potential given by

$$U(x) = \frac{-\hbar^2 x^2}{mL^2(L^2 - x^2)}$$

In addition, the particle is in a stationary state described by the wave function, $\psi(x) = A(1 - x^2/L^2)$ for $-L < x < +L$, and $\psi(x) = 0$ elsewhere. (a) Determine the energy of the particle in terms of $\hbar$, m, and L. (*Hint:* Use the Schrödinger equation, Eq. 28.19.) (b) Show that $A = (15/16L)^{1/2}$. (c) Determine the probability that the particle is located between $x = -L/3$ and $x = +L/3$.

28.11 Tunneling Through a Barrier (Optional)

43. An electron with kinetic energy $E = 5.00$ eV is incident on a barrier with thickness $L = 0.200$ nm and height $U = 10.0$ eV (Fig. P28.43). What is the approximate probability that the electron (a) will tunnel through the barrier and (b) will be reflected?

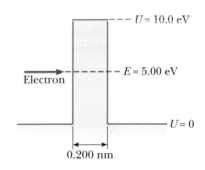

$$--- U = 10.0 \text{ eV}$$

$$--- E = 5.00 \text{ eV}$$

Electron

$$U = 0$$

0.200 nm

Figure P28.43

44. An electron having total energy $E = 4.50$ eV approaches a rectangular energy barrier with $U = 5.00$ eV and $L = 950$ pm, as in Figure P28.44. Classically, the electron could not pass through the barrier because $E < U$. However, quan-

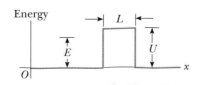

Energy

L

E

U

O

x

Figure P28.44

tum-mechanically there is a finite probability of tunneling. Calculate this probability, which is the transmission coefficient.

45. In Problem 44, by how much would the width L of the potential barrier have to be increased for the chance of an incident 4.50-eV electron tunneling through the barrier to be one in one million?

Additional Problems

46. Figure P28.46 shows the stopping potential versus incident photon frequency for the photoelectric effect for sodium. Use these data points to find (a) the work function, (b) the ratio h/e, and (c) the cutoff wavelength. (Data taken from R. A. Millikan, *Phys. Rev.* 7:362, 1916.)

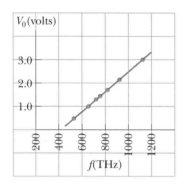

Figure P28.46

47. Photons of wavelength 450 nm are incident on a metal. The most energetic electrons ejected from the metal are bent into a circular arc of radius 20.0 cm by a magnetic field the strength of which is 2.00×10^{-5} T. What is the work function of the metal?

48. A two-slit electron diffraction experiment is done with slits of *unequal* widths. When only slit 1 is open, the number of electrons reaching the screen per second is 25.0 times the number of electrons reaching the screen per second when only slit 2 is open. When both slits are open, an interference pattern results in which the destructive interference is not complete. Find the ratio of the probability of an electron arriving at an interference maximum to the probability of an electron arriving at an adjacent interference minimum. (*Hint:* Use the superposition principle.)

49. The neutron has a mass of 1.67×10^{-27} kg. Neutrons emitted in nuclear reactions can be slowed down via collisions with matter. They are referred to as thermal neutrons once they come into thermal equilibrium with their surroundings. The average kinetic energy $(3k_B T/2)$ of a thermal neutron is approximately 0.04 eV. Calculate the de Broglie wavelength of a neutron with a kinetic energy of 0.0400 eV. How does it compare with the characteristic atomic spacing in a

crystal? Would you expect thermal neutrons to exhibit diffraction effects when scattered by a crystal?

50. A particle of mass 2.00×10^{-28} kg is confined to a one-dimensional box of width 1.00×10^{-10} m (1 Å). For $n = 1$, (a) What is the particle wavelength? (b) its momentum? (c) its ground-state energy?

51. The table below shows data obtained in a photoelectric experiment. (a) Using these data, make a graph similar to Figure 28.8 that plots as a straight line. From the graph, determine (b) an experimental value for Planck's constant (in joule-seconds) and (c) the work function (in electron volts) for the surface. (Two significant figures for each answer are sufficient.)

Wavelength (nm)	Maximum Kinetic Energy of Photoelectrons (eV)
588	0.67
505	0.98
445	1.35
399	1.63

52. A particle that has kinetic energy $E = 7.00$ eV moves from a region where the potential energy is zero into one in which $U_0 = 5.00$ eV (Fig. P28.52). Classically, one would expect the particle to continue on, although with less kinetic energy. According to quantum mechanics, the particle has a probability of being transmitted and a probability of being reflected. What are these probabilities?

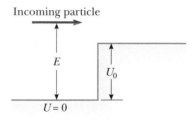

Figure P28.52

53. Particles incident from the left are confronted with a step potential energy shown in Figure P28.52. The step has a height U_0, and the particles have energy $E > U_0$. Classically, all the particles would pass into the region of higher potential energy at the right. However, according to quantum mechanics, a fraction of the particles are reflected at the barrier. The reflection coefficient R for this case is

$$R = \frac{(k_1 - k_2)^2}{(k_1 + k_2)^2}$$

where $k_1 = 2\pi/\lambda_1$ and $k_2 = 2\pi/\lambda_2$ are the wave numbers for the incident and transmitted particles, respectively. If $E = 2U_0$, what fraction of the incident particles are reflected? (This situation is analogous to the partial reflection and transmission of light striking an interface between two different media.)

54. An electron is confined to one-dimensional motion between two rigid walls separated by a distance L. (a) What is the probability of finding the electron within the interval $x = 0$ to $x = L/3$ from one wall if the electron is in its $n = 1$ state? (b) Compare this value with the classical probability.

55. An electron is represented by the time-independent wave function

$$\psi(x) = \begin{cases} Ae^{-\alpha x} & \text{for } x > 0 \\ Ae^{+\alpha x} & \text{for } x < 0 \end{cases}$$

(a) Sketch the wave function as a function of x. (b) Sketch the probability that the electron is found between x and $x + dx$. (c) Why could this be a physically reasonable wave function? (d) Normalize the wave function. (e) Determine the probability of finding the electron somewhere in the range

$$x_1 = -\frac{1}{2\alpha} \quad \text{to} \quad x_2 = \frac{1}{2\alpha}$$

56. An electron is trapped at a defect in a crystal. The defect may be modeled as a one-dimensional, rigid-walled box of width 1.00 nm. (a) Sketch the wave functions and probability densities for the $n = 1$ and $n = 2$ states. (b) For the $n = 1$ state, calculate the probability of finding the electron between $x_1 = 0.150$ nm and $x_2 = 0.350$ nm, where $x = 0$ is the left side of the box. (c) Repeat part (b) for the $n = 2$ state. (d) Calculate the energies in electron volts of the $n = 1$ and $n = 2$ states. *Hint:* For parts (b) and (c), use Equation 28.15 and note that

$$\int \sin^2 ax \, dx = \tfrac{1}{2}x - \frac{1}{4a}\sin 2ax$$

Spreadsheet Problem

S1. Compton scattering involves the scattering of an incident photon of wavelength λ_0 from a free electron. The scattered photon has a decreased wavelength λ' that depends on its scattering angle θ. The electron recoils with kinetic energy K_e at a scattering angle ϕ relative to the incident direction. Spreadsheet 28.1 calculates and plots both K_e and ϕ as functions of θ for a given λ. (a) For $\lambda = 1 \times 10^{-10}$ m, examine the values of K_e and ϕ at $\theta = 30°$, $90°$, $120°$, and $180°$. Do your results make sense? (Consider conservation of energy and momentum.) (b) Repeat part (a) for $\lambda = 1 \times 10^{-9}$ m, 1×10^{-11} m, and 1×10^{-12} m. Explain the overall variation with incident wavelength.

ANSWERS TO CONCEPTUAL PROBLEMS

1. The shape of an object is determined by observing the light reflecting from its surface. In a kiln, the objects will be very hot and will be glowing red. The emitted radiation is far stronger than the reflected radiation, and the thermal radiation emitted is only slightly dependent on the material from which the objects are made. Thus, we have a collection of objects glowing equally with emitted radiation, and only weak reflected light compared to the emitted light. This will result in indistinct outlines of the objects when viewed.

2. The "blackness" of a black body refers to its ideal property of absorbing all radiation incident on it. If an observed room-temperature object in everyday life absorbs all radiation, we describe it as (visibly) black. The black appearance, however, is due to the fact that our eyes are sensitive only to visible light. If we could detect infrared light with our eyes, we would see the object emitting radiation. If the temperature of the black body is raised, Wien's law tells us that the emitted radiation will move into the visible. Thus, the black body could appear possibly as red, white, or blue, depending on its temperature.

3. All objects do radiate energy, but at room temperature, this energy is primarily in the infrared region of the electromagnetic spectrum, which our eyes cannot detect. (Pit vipers have sensory organs that are sensitive to infrared radiation; thus they can seek out their warm-blooded prey in what we would consider complete darkness.)

4. Most metals have cutoff frequencies corresponding to photons in or near the visible range of the electromagnetic spectrum. AM radio wave photons will have far too little energy to eject electrons from the metal.

5. We can picture higher frequency light as a stream of photons of higher energy. In a collision, one photon can give all of its energy to a single electron. The kinetic energy of such an electron is measured by the stopping potential. The reverse voltage (stopping potential) required to stop the current is proportional to the frequency of the incoming light. More intense light consists of more photons striking a unit area each second, but atoms are so small that one emitted electron never gets a "kick" from more than one photon. Increasing the light intensity will increase the size of the current but will not change the energy of the individual ejected electrons. Thus, the stopping potential remains constant.

6. There are several possibilities for the fate of the photon energy. The photon could simply pass through the material and carry the energy away. It could also reflect from the surface of the material and carry the energy away. The photon could be absorbed into an atom, raising it to a higher energy state. The photon could Compton scatter off many electrons, dissipating its energy as it moves. The photon could also scatter off an atom, exciting it to a higher level and carry a reduced amount of energy off toward another interaction.

7. A microscope can see details no smaller than the wavelength of the waves it uses to produce images. Electrons with kinetic energies of several electron volts have wavelengths of less than a nanometer, which is much smaller than the wavelength of visible light (having wavelengths ranging from about 400 nm to 700 nm.) Therefore, an electron microscope can resolve details of much smaller size as compared to an optical microscope.

29

Atomic Physics

In Chapter 28 we introduced some of the basic concepts and techniques used in quantum mechanics, along with their applications to various simple systems. This chapter describes the application of quantum mechanics to the real world of atomic structure.

A large portion of this chapter is concerned with the study of the hydrogen atom from the viewpoint of quantum mechanics. Although the hydrogen atom is the simplest atomic system, it is an especially important system to understand, for several reasons:

- Much of what we learn about the hydrogen atom, with its single electron, can be extended to such single-electron ions as He⁺ and Li²⁺.
- The hydrogen atom is an ideal system for performing precise tests of theory against experiment and for improving our overall understanding of atomic structure.

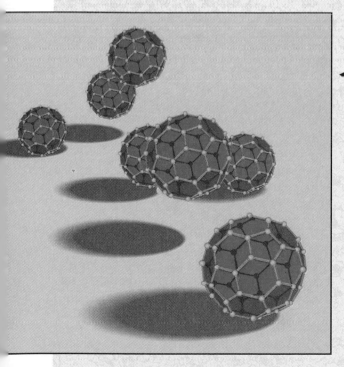

◀ An artist's rendition of "bucky-balls," short for the molecule buckminsterfullerene. These nearly spherical molecular structures that look like soccer balls were named for R. Buckminster Fuller, inventor of the geodesic dome. This new form of carbon, C_{60}, was discovered by astrophysicists while investigating the carbon gas that exists between stars. Scientists are actively studying the properties and potential uses of buckminsterfullerene and related molecules. For example, samples containing C_{60} doped with rubidium exhibit superconductivity at approximately 30 K.

- The quantum numbers used to characterize the allowed states of hydrogen can be used to describe approximately the allowed states of more complex atoms. This characterization enables us to understand the periodic table of the elements, which is one of the greatest triumphs of quantum mechanics.
- The basic ideas about atomic structure must be well understood before we attempt to deal with the complexities of molecular structures and the electronic structures of solids.

The full mathematical solution of the Schrödinger equation as applied to the hydrogen atom gives a complete and beautiful description of the atom's properties. However, the mathematical procedures that make up the solution are beyond the scope of this book, and so the details are omitted. The solutions for some states of hydrogen are discussed, together with the quantum numbers used to characterize allowed stationary states. We also discuss the physical significance of the quantum numbers and the effect of a magnetic field on certain quantum states.

The *exclusion principle,* also presented in this chapter, is extremely important to an understanding of the properties of multielectron atoms and the arrangement of elements in the periodic table. The implications of the exclusion principle are almost as far reaching as those of the Schrödinger equation. Finally, we apply our knowledge of atomic structure to describe the mechanisms involved in the production of x-rays and in the operation of a laser.

29.1 • EARLY MODELS OF THE ATOM

The model of the atom in Newton's day was a tiny, hard, indestructible sphere. Although this model was a good basis for the kinetic theory of gases, new models had to be devised when later experiments revealed the electrical nature of atoms. J. J. Thomson suggested a model that describes the atom as a volume of positive charge with electrons embedded throughout it, much like the seeds in a watermelon (Fig. 29.1).

In 1911 Ernest Rutherford and his students Hans Geiger and Ernest Marsden performed a critical experiment that showed that Thomson's model could not be correct. In this experiment, a beam of positively charged alpha particles was projected into a thin metal foil, as in Figure 29.2a. The results of the experiment were astounding. Most of the particles passed through the foil as if it were empty space. But many particles deflected from their original direction of travel were scattered through large angles. Some particles were even deflected backward, reversing their direction of travel. When Geiger informed Rutherford of these results, Rutherford wrote, "It was quite the most incredible event that has ever happened to me in my life. It was almost as incredible as if you fired a 15-inch shell at a piece of tissue paper and it came back and hit you."

Such large deflections were not expected on the basis of Thomson's model. According to this model, a positively charged alpha particle would never come close enough to a large positive charge to cause any large-angle deflections. Rutherford explained his astounding results by assuming that the positive charge was concentrated in a region that was small relative to the size of the atom. He called this concentration of positive charge the **nucleus** of the atom. Any electrons belonging to the atom were assumed to be in the relatively large volume outside the nucleus. In order to explain why these electrons were not pulled into the nucleus, the elec-

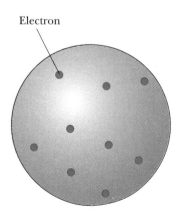

Electron

Figure 29.1   Thomson's model of the atom with the electrons embedded inside the positive charge like seeds in a watermelon.

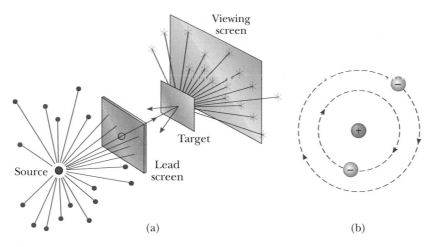

Figure 29.2   (a) Rutherford's technique for observing the scattering of alpha particles from a thin foil target. The source is a naturally occurring radioactive substance, such as radium. (b) Rutherford's planetary model of the atom.

trons were viewed as moving in orbits about the positively charged nucleus in the same manner as the planets orbit the Sun, as in Figure 29.2b.

There are two basic difficulties with Rutherford's planetary model. As we saw in Chapter 11, an atom emits certain characteristic frequencies of electromagnetic radiation and no others; the Rutherford model is unable to explain this phenomenon. A second difficulty is that Rutherford's electrons are undergoing a centripetal acceleration. According to Maxwell's theory of electromagnetism, centripetally accelerated charges revolving with frequency *f* should radiate electromagnetic waves having the same frequency. Unfortunately, this classical model leads to disaster when applied to the atom. As the electron radiates energy, the radius of its orbit steadily decreases and its frequency of revolution increases. This leads to an ever increasing frequency of emitted radiation and a rapid collapse of the atom as the electron plunges into the nucleus (Fig. 29.3).

Now the stage was set for Bohr. In order to circumvent the erroneous deductions of electrons falling into the nucleus and a continuous emission spectrum from elements, Bohr postulated that classical radiation theory does not hold for atomic-sized systems. He overcame the problem of a classical electron that continuously loses energy by applying Planck's ideas of quantized energy levels to orbiting atomic electrons. Thus, as described in Section 11.5, Bohr postulated that electrons in atoms are generally confined to stable, nonradiating energy levels and orbits called stationary states. Furthermore, he applied Einstein's concept of the photon to arrive at an expression for the frequency of light emitted when the electron jumps from one stationary state to another.

One of the first indications that there was a need to modify the Bohr theory arose when improved spectroscopic techniques were used to examine the spectral lines of hydrogen. It was found that many of the lines in the Balmer and other series were not single lines at all. Instead, each was a group of lines spaced very close together. An additional difficulty arose when it was observed that, in some situations, certain single spectral lines were split into three closely spaced lines when the atoms were placed in a strong magnetic field.

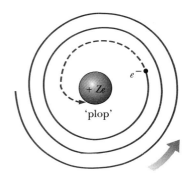

Figure 29.3   The classical model of the nuclear atom.

Efforts to explain these deviations from the Bohr model led to improvements in the theory. One of the changes introduced was the classical concept that the electron could spin on its axis. Also, Arnold Sommerfeld improved the Bohr theory by introducing the theory of relativity into the analysis of the electron's motion. An electron in an elliptical orbit has a continuously changing speed, with an average value that depends on the eccentricity of the orbit. Different orbits have different average speeds and slightly different relativistic energies.

29.2 • THE HYDROGEN ATOM REVISITED

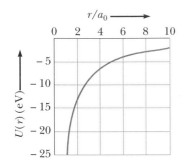

Figure 29.4   The potential energy $U(r)$ versus the ratio r/a_0 for the hydrogen atom. The constant a_0 is the Bohr radius, and r is the electron–proton separation.

In Chapter 11, we described how the Bohr model of the hydrogen atom views the electron as a particle orbiting around the nucleus in nonradiating, quantized energy levels. The de Broglie model gave electrons a wave nature by viewing them as standing waves in allowed orbits. This standing wave description removed the objections to Bohr's postulates, which were quite arbitrary. However, the de Broglie model created other problems. The exact nature of the de Broglie wave was unspecified, and the model implied unlikely electron densities at great distances from the nucleus. Fortunately, these difficulties are removed when the methods of quantum mechanics are used to describe atoms.

The potential energy function for the hydrogen atom is

$$U(r) = -k_e \frac{e^2}{r} \qquad \text{[29.1]}$$

where k_e is the Coulomb constant and r is the radial distance from the proton (situated at $r = 0$) to the electron. Figure 29.4 is a plot of this function versus r/a_0 where a_0 is the Bohr radius, 0.0529 nm.

The formal procedure for solving the problem of the hydrogen atom would be to substitute $U(r)$ into the Schrödinger equation and find appropriate solutions to the equation, as we did for the particle in the box in Chapter 28. The current problem is more complicated, however, because it is three-dimensional and because U depends on the radial coordinate r. We shall not attempt to carry out these solutions. Rather, we shall simply describe their properties and some of their implications with regard to atomic structure.

According to quantum mechanics, the energies of the allowed states for the hydrogen atom are

Allowed energies for the •
hydrogen atom

$$E_n = -\left(\frac{k_e e^2}{2a_0}\right)\frac{1}{n^2} = -\frac{13.6}{n^2} \text{ eV} \qquad n = 1, 2, 3, \ldots \qquad \text{[29.2]}$$

This result is in exact agreement with the Bohr theory. Note that the allowed energies depend only on the quantum number n.

In one-dimensional problems, only one quantum number is needed to characterize a stationary state. In the three-dimensional hydrogen atom, three quantum numbers are needed for each stationary state, corresponding to three independent degrees of freedom for the electron. The three quantum numbers that emerge from the theory are represented by the symbols n, ℓ, and m_ℓ. The quantum number n is called the **principal quantum number,** ℓ is called the **orbital quantum number,** and m_ℓ is called the **orbital magnetic quantum number,** where all three quantum numbers are integers and

TABLE 29.1   Three Quantum Numbers for the Hydrogen Atom

Quantum Number	Name	Allowed Value	Number of Allowed States
n	Principal quantum number	1, 2, 3, . . .	Any number
ℓ	Orbital quantum number	0, 1, 2, . . . , $n-1$	n
m_ℓ	Orbital magnetic quantum number	$-\ell, -\ell+1, \ldots, 0,$ $\ldots, \ell-1, \ell$	$2\ell+1$

- n can range from 1 to ∞
- ℓ can range from 0 to $n-1$
- m_ℓ can range from $-\ell$ to ℓ

Table 29.1 summarizes the rules for determining the allowed values of ℓ and m_ℓ for a given value of n.

For historical reasons, all states with the same principal quantum number are said to form a shell. Shells are identified by the letters $K, L, M, \ldots$, which designate the states for which $n = 1, 2, 3, \ldots$. Likewise, **all states with given values of n and ℓ are said to form a subshell.** The letters $s, p, d, f, g, h, \ldots$ are used to designate the subshells for which $\ell = 0, 1, 2, 3, \ldots$. For example, the state designated by $3p$ has the quantum numbers $n = 3$ and $\ell = 1$; the $2s$ state has the quantum numbers $n = 2$ and $\ell = 0$. These notations are summarized in Table 29.2.

States that violate the rules given in Table 29.1 cannot exist. For instance, the $2d$ state, which would have $n = 2$ and $\ell = 2$, cannot exist because the highest allowed value of ℓ is $n - 1$, or 1 in this case. Thus, for $n = 2$, $2s$ and $2p$ are allowed states but $2d$, $2f$, . . . are not. For $n = 3$, the allowed states are $3s$, $3p$, and $3d$.

TABLE 29.2   Atomic Shell and Subshell Notations

n	Shell Symbol	ℓ	Subshell Symbol
1	K	0	s
2	L	1	p
3	M	2	d
4	N	3	f
5	O	4	g
6	P	5	h
. . .		. . .	

CONCEPTUAL PROBLEM 1

In the hydrogen atom, the quantum number n can increase without limit. Because of this, does the frequency of possible spectral lines from hydrogen also increase without limit?

Example 29.1   The $n = 2$ Level of Hydrogen

For a hydrogen atom, determine the number of orbital states corresponding to the principal quantum number $n = 2$, and calculate the energies of these states.

Solution   When $n = 2$, ℓ can be 0 or 1. For $\ell = 0$, m_ℓ can only be 0; for $\ell = 1$, m_ℓ can be -1, 0, or 1. Hence, we have a state designated as the $2s$ state associated with the quantum numbers $n = 2$, $\ell = 0$, and $m_\ell = 0$, and three orbital states designated as $2p$ states for which the quantum numbers are $n = 2$, $\ell = 1$, $m_\ell = -1$; $n = 2$, $\ell = 1$, $m_\ell = 0$; and $n = 2$, $\ell = 1$, $m_\ell = 1$.

Because all these states have the same principal quantum number, they also have the same energy, which can be calculated with Equation 29.2, taking $n = 2$:

$$E_2 = -\frac{13.6}{2^2} \text{ eV} = -3.40 \text{ eV}$$

EXERCISE 1   How many possible orbital states are there for the $n = 3$ level of hydrogen? For the $n = 4$ level?
Answer   9 and 16

EXERCISE 2   When the principal quantum number is $n = 4$, how many different values of (a) ℓ and (b) m_ℓ are possible?   Answer   (a) 4   (b) 7

Spin up

Spin down

(a)   (b)

Figure 29.5 The spin of an electron can be either (a) up or (b) down.

29.3 • THE SPIN MAGNETIC QUANTUM NUMBER

Example 29.1 was presented to give you practice in manipulating quantum numbers, but, as we shall see in this section, there are *eight* electron states for $n = 2$ rather than the four we found. These extra states can be explained by requiring a fourth quantum number for each state; the **spin magnetic quantum number, m_s.**

The need for this new quantum number came about because of an unusual feature in the spectra of certain gases such as sodium vapor. Close examination of one of the prominent lines of sodium shows that it is, in fact, two very closely spaced lines called a *doublet*. The wavelengths of these lines occur in the yellow region at 589.0 nm and 589.6 nm. In 1925, when this doublet was first noticed, atomic theory could not explain it. To resolve this dilemma, Samuel Goudsmidt and George Uhlenbeck, following a suggestion by the Austrian physicist Wolfgang Pauli, proposed a new quantum number, called the *spin quantum number*.

In order to describe the spin quantum number, it is convenient (but incorrect) to think of the electron as spinning on its axis as it orbits the nucleus, just as the Earth spins on its axis as it orbits the Sun. There are only two directions that the electron can spin as it orbits the nucleus, as shown in Figure 29.5. If the direction of spin is as shown in Figure 29.5a, the electron is said to have "spin up." If the direction of spin is reversed, as in Figure 29.5b, the electron is said to have "spin down." In the presence of a magnetic field, the energy of the electron is slightly different for the two spin directions, and this energy difference accounts for the sodium doublet. The quantum numbers associated with electron spin are $m_s = \frac{1}{2}$ for the spin-up state and $m_s = -\frac{1}{2}$ for the spin-down state. As we shall see in Example 29.2, this added quantum number doubles the number of allowed states specified by the quantum numbers n, ℓ, and m_ℓ.

The classical description of electron spin as due to a spinning electron is incorrect, because quantum mechanics tells us that a rotational degree of freedom would require too many quantum numbers, and more recent theory indicates that the electron is a point particle, without spatial extent. The electron cannot be considered to be spinning, as pictured in Figure 29.5. In spite of this conceptual difficulty, all experimental evidence supports the fact that the electron does have some intrinsic property that can be described by the spin quantum number. The origin of this fourth quantum number was shown by Sommerfeld and Dirac to lie

Wolfgang Pauli and Niels Bohr watch a spinning top.
(Courtesy of AIP Niels Bohr Library, Margarethe Bohr Collection)

in the relativistic properties of the electron, which requires four quantum numbers to describe its location in four-dimensional space–time.

Example 29.2   Adding Some Spin on Hydrogen

For a hydrogen atom, determine the quantum numbers associated with the possible states that correspond to the principal quantum number $n = 2$.

Solution   With the addition of the spin quantum number, we have the possibilities given in the table below.

EXERCISE 3   Show that for $n = 3$, there are 18 possible states.

n	ℓ	m_ℓ	m_s	Subshell	Shell	Maximum Number of Electrons in Subshell
2	0	0	$\frac{1}{2}$	$2s$	L	2
2	0	0	$-\frac{1}{2}$			
2	1	1	$\frac{1}{2}$			
2	1	1	$-\frac{1}{2}$			
2	1	0	$\frac{1}{2}$	$2p$	L	6
2	1	0	$\frac{1}{2}$			
2	1	-1	$\frac{1}{2}$			
2	1	-1	$-\frac{1}{2}$			

29.4 • THE WAVE FUNCTIONS FOR HYDROGEN

Neglecting electron spin for the present, the potential energy of the hydrogen atom depends only on the radial distance r between nucleus and electron. We therefore expect that some of the allowed states for this atom can be represented by wave functions that depend only on r. This indeed is the case. The simplest wave function for hydrogen is the one that describes the $1s$ ground state and is designated $\psi_{1s}(r)$:

$$\psi_{1s}(r) = \frac{1}{\sqrt{\pi a_0{}^3}}\, e^{-r/a_0} \qquad [29.3]$$

• *Wave function for hydrogen in its ground state*

where a_0 is the Bohr radius. This wave function satisfies the condition that it approach zero as r approaches ∞ and is normalized as presented. Furthermore, because ψ_{1s} depends only on r, it is spherically symmetric. In fact, **all s states have spherical symmetry.**

Recall that the probability of finding the electron in any region is equal to an integral of $|\psi|^2$ over the region, if ψ is normalized. The probability density for the $1s$ state is

$$|\psi_{1s}|^2 = \left(\frac{1}{\pi a_0{}^3}\right) e^{-2r/a_0} \qquad [29.4]$$

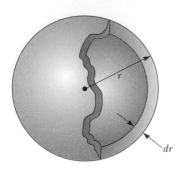

Figure 29.6   A spherical shell of radius r and thickness dr has a volume equal to $4\pi r^2\ dr$.

The probability of finding the electron in a volume element, dV, is $|\psi|^2\ dV$. It is convenient to define the **radial probability density function,** $P(r)$, as the probability per unit radial distance of finding the electron in a spherical shell of radius r and thickness dr. The volume of such a shell equals its surface area, $4\pi r^2$, multiplied by the shell thickness, dr (Fig. 29.6), so that we get

$$P(r)\ dr = |\psi|^2\ dV = |\psi|^2 4\pi r^2\ dr \qquad \text{[29.5]}$$

$$P(r) = 4\pi r^2 |\psi|^2 \qquad \text{[29.6]}$$

Substituting Equation 29.4 into Equation 29.6 gives the radial probability density function for the hydrogen atom in its ground state:

$$P_{1s}(r) = \left(\frac{4r^2}{a_0{}^3}\right) e^{-2r/a_0} \qquad \text{[29.7]}$$

A plot of the function $P_{1s}(r)$ versus r is presented in Figure 29.7a. The peak of the curve corresponds to the most probable value of r for this particular state. The spherical symmetry of the distribution function is shown in Figure 29.7b.

Example 29.3   The Ground State of Hydrogen

Calculate the most probable value of r for an electron in the ground state of the hydrogen atom.

Solution   The most probable value of r corresponds to the peak of the plot of $P(r)$ versus r. The slope of the curve at this point is zero, and so we can evaluate the most probable value of r by setting $dP/dr = 0$ and solving for r. Using Equation 29.7, we get

$$\frac{dP}{dr} = \frac{d}{dr}\left[\left(\frac{4r^2}{a_0{}^3}\right) e^{-2r/a_0}\right] = 0$$

Carrying out the derivative operation and simplifying the expression, we get

$$e^{-2r/a_0}\frac{d}{dr}(r^2) + r^2\frac{d}{dr}(e^{-2r/a_0}) = 0$$

$$2re^{-2r/a_0} + r^2(-2/a_0)e^{-2r/a_0} = 0$$

$$2r[1 - (r/a_0)]e^{-2r/a_0} = 0$$

This expression is satisfied if

$$1 - \frac{r}{a_0} = 0$$

$$r = a_0$$

Figure 29.7   (a) The probability of finding the electron as a function of distance from the nucleus for the hydrogen atom in the $1s$ (ground) state. Note that the probability has its maximum value when r equals the Bohr radius, a_0.
(b) The spherical electron cloud for the hydrogen atom in its $1s$ state.

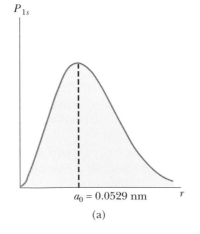

$a_0 = 0.0529$ nm

(a)

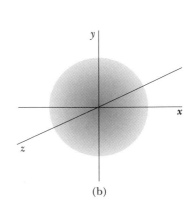

(b)

Example 29.4   **Probabilities for the Electron in Hydrogen**

Calculate the probability that the electron in the ground state of hydrogen will be found outside the Bohr radius.

Solution   The probability is found by integrating the radial probability density for this state, $P_{1s}(r)$, from the Bohr radius, a_0, to ∞. Using Equation 29.7, we get

$$P = \int_{a_0}^{\infty} P_{1s}(r)\ dr = \frac{4}{a_0{}^3} \int_{a_0}^{\infty} r^2 e^{-2r/a_0}\ dr$$

We can put the integral in dimensionless form by changing variables from r to $z = 2r/a_0$. Noting that $z = 2$ when $r = a_0$, and that $dr = (a_0/2)\ dz$, we get

$$P = \frac{1}{2} \int_{2}^{\infty} z^2 e^{-z}\ dz = -\frac{1}{2}(z^2 + 2z + 2)e^{-z}\Big|_{2}^{\infty}$$

$$P = 5e^{-2} = 0.677 \qquad \text{or} \qquad 67.7\%$$

Example 29.3 shows that, for the ground state of hydrogen, the most probable value of r equals the Bohr radius, a_0. It turns out that the average value of r for the ground state of hydrogen is $\frac{3}{2}a_0$, which is 50% larger than the most probable value of r (see Problem 33). The reason for this is the large asymmetry in the radial distribution function shown in Figure 29.7a. According to quantum mechanics, there is no sharply defined boundary to the atom. The probability distribution for the electron can be viewed as being a diffuse region of space, commonly referred to as an "electron cloud."

The next-simplest wave function for the hydrogen atom is the one corresponding to the 2s state ($n = 2$, $\ell = 0$). The normalized wave function for this state is

$$\psi_{2s}(r) = \frac{1}{4\sqrt{2\pi}} \left(\frac{1}{a_0}\right)^{3/2} \left[2 - \frac{r}{a_0}\right] e^{-r/2a_0} \qquad [29.8]$$

• *Wave function for hydrogen in the 2s state*

Like the ψ_{1s} function, ψ_{2s} depends only on r and is spherically symmetric. The energy corresponding to this state is $E_2 = -(13.6/4)$ eV $= -3.4$ eV. This energy level represents the first excited state of hydrogen. Plots of the radial distribution function for this state and several other states of hydrogen are shown in Figure 29.8. The plot for the 2s state, which applies to a single electron, has two peaks. In this case, the most probable value corresponds to that value of r with the higher value of $P(\approx 5a_0)$. An electron in the 2s state would be much farther from the nucleus (on the average) than an electron in the 1s state. The average value of r is even greater for the 3d, 3p, and 4d states.

As we have mentioned, all s states have spherically symmetric wave functions. The other states are not spherically symmetric. For example, the three wave functions corresponding to the states for which $n = 2$, $\ell = 1$ ($m_\ell = 1$, 0, or -1) can be expressed as appropriate linear combinations of the three p states. Although quantum mechanics limits our knowledge of angular momentum to the projection along any one axis at a time, these p states may be described mathematically as linear combinations of mutually perpendicular functions p_x, p_y, and p_z, as represented in Figure 29.9, where *only* the angular dependence of these functions is shown. Note that the three clouds have identical structure but differ in orientation with respect to the x, y, and z axes. The nonspherical wave functions for these states are

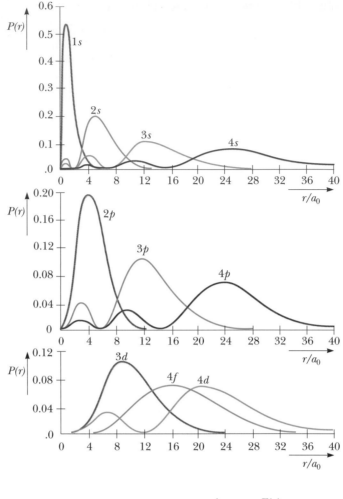

Figure 29.8 The radial probability density function versus r/a_0 for several states of the hydrogen atom. *(From E. U. Condon and G. H. Shortley, The Theory of Atomic Spectra, Cambridge, Cambridge University Press, 1953; used with permission)*

$$\psi_{2p_x} = xF(r)$$

$$\psi_{2p_y} = yF(r)$$

Wave functions for the 2p •
state

$$\psi_{2p_z} = zF(r) \qquad\qquad [29.9]$$

where $F(r)$ is some exponential function of r. Wave functions with a highly directional character, such as these, are convenient for bookkeeping descriptions of chemical bonding, the formation of molecules, and chemical properties.

Figure 29.9 Angular dependence of the electron charge distribution for an electron in a p state. The three charge distributions p_x, p_y, and p_z have the same structure and differ only in their orientation in space.

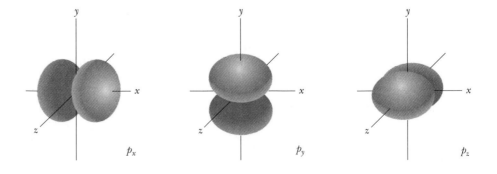

29.5 • THE "OTHER" QUANTUM NUMBERS

The energy of a particular state depends primarily on the principal quantum number. Now let us see what the other three quantum numbers contribute to our atomic model.

The Orbital Quantum Number

If a particle moves in a circle of radius r, the magnitude of its angular momentum relative to the center of the circle is $L = mvr$. The direction of **L** is perpendicular to the plane of the circle, and the sense of **L** is given by a right-hand rule.[1] According to classical physics, L can have any value. However, the Bohr model of hydrogen postulates that the angular momentum is restricted to multiples of $\hbar$; that is, $mvr = n\hbar$. This model must be modified because it predicts (incorrectly) that the ground state of hydrogen ($n = 1$) has one unit of angular momentum. Furthermore, if L is taken to be zero in the Bohr model, one is forced to accept the electron as a particle oscillating along a straight line through the nucleus. This is a physically unacceptable situation.

These difficulties are resolved with the quantum mechanical model of the atom. According to quantum mechanics, an atom in a state the principal quantum number of which is n can take on the following *discrete* values of orbital angular momentum:

$$L = \sqrt{\ell(\ell + 1)}\,\hbar \qquad \ell = 0, 1, 2, \ldots, n - 1 \qquad \text{[29.10]}$$

• *Allowed values of L*

Because ℓ is restricted to these values, $L = 0$ (corresponding to $\ell = 0$) is an acceptable value of the angular momentum. The fact that L can be zero in this model points out the difficulties inherent in any attempt to describe results based on quantum mechanics in terms of a purely particle-like model. In the quantum mechanical interpretation, the electron cloud for the $L = 0$ state is spherically symmetric and has no fundamental axis of revolution.

Example 29.5   Calculating L for a p State

Calculate the orbital angular momentum of an electron in a p state of hydrogen.

Solution Because we know that $\hbar = 1.054 \times 10^{-34}$ J·s, we can use Equation 29.10 to calculate L. With $\ell = 1$ for a p state, we have

$$L = \sqrt{1(1 + 1)}\,\hbar = \sqrt{2}\,\hbar = 1.49 \times 10^{-34} \text{ J·s}$$

This number is extremely small relative to the orbital angular momentum of the Earth orbiting the Sun, which is about 2.7×10^{40} J·s. The quantum number that describes L for macroscopic objects, such as the Earth, is so large that the separation between adjacent states cannot be measured. Once again, the correspondence principle is upheld.

The Magnetic Orbital Quantum Number

Because angular momentum is a vector, its direction must also be specified. Recall from Chapter 22 that an orbiting electron can be considered an effective current loop with a corresponding magnetic moment. Such a moment placed in a magnetic field, **B**, will interact with the field. Suppose a weak magnetic field is applied along

[1]See Sections 10.8 and 10.9 to review details on angular momentum.

the z axis so that it defines a direction in space. According to quantum mechanics, L^2 and L_z, the projection of **L** along the z axis, can have discrete values. The magnetic orbital quantum number m_ℓ specifies the allowed values of L_z according to the expression

Allowed values of L_z •

$$L_z = m_\ell \hbar \qquad \qquad \text{[29.11]}$$

Space quantization •

The fact that the direction of **L** is quantized with respect to an external magnetic field is often referred to as **space quantization.**

Let us look at the possible orientations of **L** for a given value of ℓ. Recall that m_ℓ can have values ranging from $-\ell$ to ℓ. If $\ell = 0$, then $m_\ell = 0$ and $L_z = 0$. In order for L_z to be zero, **L** must be perpendicular to **B**. If $\ell = 1$, then the possible values of m_ℓ are -1, 0, and 1, so that L_z may be $-\hbar$, 0, or $\hbar$. If $\ell = 2$, m_ℓ can be -2, -1, 0, 1, or 2, corresponding to L_z values of $-2\hbar$, $-\hbar$, 0, $\hbar$, or $2\hbar$, and so on.

A vector model describing space quantization for $\ell = 2$ is shown in Figure 29.10a. Note that **L** can never be aligned parallel or antiparallel to **B** because L_z must be smaller than the total angular momentum, L. From a three-dimensional viewpoint, **L** must lie on the surface of a cone that makes an angle of θ with the z axis, as shown in Figure 29.10b. From the figure, we see that θ is also quantized and that its values are specified through the relation

θ is quantized •

$$\cos \theta = \frac{L_z}{|\mathbf{L}|} = \frac{m_\ell}{\sqrt{\ell(\ell + 1)}} \qquad \qquad \text{[29.12]}$$

Note that m_ℓ is never greater than ℓ, and therefore θ can never be zero. (Classically, θ can have any value.)

Because of the uncertainty principle, **L** does not point in a specific direction but rather traces out a cone in space. If **L** had a definite value, then all three components L_x, L_y, and L_z would be exactly specified. For the moment, let us assume that this is the case, and let us suppose that the electron moves in the xy plane, so that **L** is in the z direction and $p_z = 0$. This means that p_z is precisely known, which is

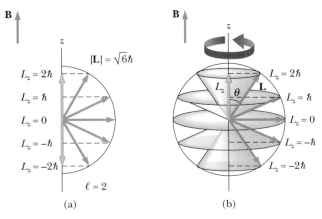

Figure 29.10   (a) The allowed projections of the orbital angular momentum for the case $\ell = 2$. (b) The orbital angular momentum vector lies on the surface of a cone and precesses about the z axis when a magnetic field **B** is applied in this direction.

in violation of the uncertainty principle, $\Delta p_z \, \Delta z \geqslant \hbar/2$. In reality, only the magnitude of **L** and one component (say, L_z) can have definite values. In other words, quantum mechanics allows us to specify L and L_z but not L_x and L_y. Because the direction of **L** is constantly changing as it precesses about the z axis, the average values of L_x and L_y are zero and L_z maintains a fixed value of $m_\ell \hbar$.

Example 29.6 Space Quantization for Hydrogen

For the hydrogen atom in the $\ell = 3$ state, calculate the magnitude of **L** and the allowed values of L_z and θ.

Solution We use Equation 29.10 with $\ell = 3$:

$$L = \sqrt{\ell(\ell + 1)}\,\hbar = \sqrt{3(3 + 1)}\,\hbar = 2\sqrt{3}\,\hbar$$

The allowed values of L_z are $L_z = m_\ell \hbar$ with $m_\ell = -3$, $-2, -1, 0, 1, 2$, and 3:

$$L_z = -3\hbar, -2\hbar, -\hbar, 0, \hbar, 2\hbar, 3\hbar$$

Finally, we use Equation 29.12 to calculate the allowed values of θ. Because $\ell = 3$, $\sqrt{\ell(\ell + 1)} = 2\sqrt{3}$, and we have

$$\cos \theta = \frac{m_\ell}{2\sqrt{3}}$$

Substitution of the allowed values of m_ℓ gives

$$\cos \theta = \pm 0.866, \pm 0.577, \pm 0.289, 0$$

$$\theta = 30.0°, 54.8°, 73.2°, 90.0°, 107°, 125°, 150°$$

Electron Spin

In 1921 Stern and Gerlach performed an experiment that demonstrated space quantization. However, their results were not in quantitative agreement with the theory that existed at that time. In their experiment, a beam of neutral silver atoms sent through a nonuniform magnetic field was split into two components (Fig. 29.11). The experiment was repeated using other atoms, and in each case the beam split into two or more components. The classical argument is as follows: If the z direction is chosen to be the direction of the maximum inhomogeneity of **B**, the net magnetic force on the atom is along the z axis and is proportional to the mag-

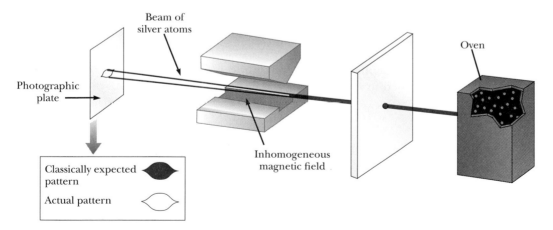

Figure 29.11 The apparatus used by Stern and Gerlach to verify space quantization. A beam of neutral silver atoms is split into two components by a nonuniform magnetic field, as shown by the actual pattern in the box.

netic moment μ_z in the z direction. Classically, $\boldsymbol{\mu}$ can have any orientation, and so the deflected beam should be spread out continuously. According to quantum mechanics, however, the deflected beam has several components, and the number of components determines the possible values of μ_z. Hence, because the Stern–Gerlach experiment showed split beams, space quantization was at least qualitatively verified.

For the moment, let us assume that μ_z is due to the orbital angular momentum. Because μ_z is proportional to m_ℓ, the number of possible values of μ_z is $2\ell + 1$. Furthermore, because ℓ is an integer, the number of values of μ_z is always odd. This prediction is clearly not consistent with the observations of Stern and Gerlach, who observed only two components in the deflected beam of silver atoms. Hence, one is forced to conclude that either quantum mechanics is incorrect or the model is in need of refinement.

In 1927 Phipps and Taylor repeated the Stern–Gerlach experiment using a beam of hydrogen atoms. This experiment is important because it deals with an atom with a single electron in its ground state, for which the theory makes reliable predictions. Recall that $\ell = 0$ for hydrogen in its ground state, and so $m_\ell = 0$. Hence, one would not expect the beam to be deflected by the field, because μ_z would be zero. However, the beam in the Phipps–Taylor experiment is again split into two components. On the basis of this result, one can conclude only one thing: There is some contribution to the magnetic moment other than the orbital motion.

In 1925 Goudsmit and Uhlenbeck proposed that the electron has an intrinsic angular momentum apart from its orbital angular momentum. From a classical viewpoint, this intrinsic angular momentum is attributed to the charged electron spinning about its own axis, and hence is called **electron spin.**[2] In other words, the total angular momentum of the electron in a particular electronic state contains both an orbital contribution, **L**, and a spin contribution, **S**. There is a quantum number s for spin, which is analogous to ℓ for orbital angular momentum. The value for s, however, is always $s = \frac{1}{2}$.

The magnitude of the **spin angular momentum, S,** for the electron is

Spin angular momentum of •
an electron

$$S = \sqrt{s(s + 1)}\,\hbar = \frac{\sqrt{3}}{2}\,\hbar \qquad\qquad [\mathbf{29.13}]$$

Like orbital angular momentum, spin angular momentum is quantized in space, as described in Figure 29.12. It can have two orientations, specified by the

[2]Physicists often use the word *spin* when referring to *spin angular momentum.* For example, it is common to use the statement "The electron has a spin of one half." The spin angular momentum of the electron *never changes.* This notion contradicts classical laws, which would hold that a rotating charge slows down in the presence of an applied magnetic field because of the Faraday emf that accompanies the changing field. Furthermore, if the electron is viewed as a spinning ball of charge subject to classical laws, parts of it near its surface would be rotating with velocities exceeding the speed of light. Thus, the classical picture must not be pressed too far; ultimately, the spinning electron is a quantum entity defying any simple classical description. It is more accurate to describe spin as a relativistic effect (which was treated classically by Sommerfeld and quantum mechanically by Dirac).

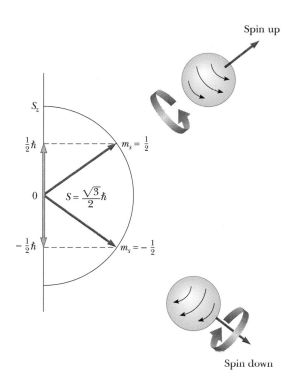

Spin up

Spin down

Figure 29.12   The spin angular momentum also exhibits space quantization. This figure shows the two allowed orientations of the spin vector S for a spin $\frac{1}{2}$ particle, such as the electron.

spin magnetic quantum number m_s, where m_s has two possible values, $\pm\frac{1}{2}$. The z component of spin angular momentum is

$$S_z = m_s\hbar = \pm\tfrac{1}{2}\hbar \qquad\qquad \text{[29.14]}$$

The two values $\pm\hbar/2$ for S_z correspond to the two possible orientations for **S** shown in Figure 29.12.

The spin magnetic moment of the electron, $\boldsymbol{\mu}_s$, is related to its spin angular momentum, **S**, by the expression

$$\boldsymbol{\mu}_s = -\frac{e}{m_e}\,\mathbf{S} \qquad\qquad \text{[29.15]}$$

Because $S_z = \pm\frac{1}{2}\hbar$, the z component of the spin magnetic moment can have the values

$$\mu_{sz} = \pm\frac{e\hbar}{2m_e} \qquad\qquad \text{[29.16]}$$

The quantity $e\hbar/2m_e$ is the **Bohr magneton,** μ_B, and has the numerical value 9.27×10^{-24} J/T (see Eq. 22.28). Note that the spin contribution to the angular momentum is *twice* the contribution of the orbital motion.

Today physicists explain the outcome of the Stern–Gerlach experiment as follows. The observed moments for both silver and hydrogen are due to spin angular momentum and not to orbital angular momentum. A single-electron atom such as hydrogen has its electron quantized in the magnetic field in such a way that its z component of spin angular momentum is either $\frac{1}{2}\hbar$ or $-\frac{1}{2}\hbar$, corresponding to $m_s = \pm\frac{1}{2}$.

Electrons with spin $+\frac{1}{2}$ are deflected in one direction, and those with spin $-\frac{1}{2}$ are deflected in the opposite direction. The Stern–Gerlach experiment provided two important results. First, it verified the concept of space quantization. Second, it showed that spin angular momentum existed even though this property was not recognized until long after the experiments were performed.

Thinking Physics 1

Angular momentum is a vector in three dimensions. Thus, it requires three pieces of information to be completely specified, such as the x-, y-, and z-components. In an atom, however, the orbital angular momentum vector **L** is described by only *two* pieces of information—the quantum numbers ℓ and m_ℓ. Why is there a discrepancy?

Reasoning The clue in the question is the phrase, "to be completely specified." The notion of complete specification is not appropriate in the quantum world. Imagine a completely specified vector for an electron following a circular Bohr orbit (but remember that this is inconsistent with the quantum mechanical description of the electron). The plane of the orbit would be fixed in space. Thus, in the z-direction *perpendicular* to the plane of the orbit (and parallel to **L**), the position of the electron is completely specified—it is in the plane, so there is no uncertainty in position and $\Delta z = 0$. The velocity of the electron perpendicular to the plane is also completely specified—it is zero, again with no uncertainty. Thus, the uncertainty principle is violated. The uncertainty principle is satisfied by stating only two pieces of information. The length of the angular momentum vector and its z-component are known, but there is no information about the x- or y-components.

Thinking Physics 2

Does the Stern–Gerlach experiment differentiate between orbital angular momentum and spin angular momentum?

Reasoning There is a magnetic force on the magnetic moment arising from both orbital angular momentum and spin angular momentum. In this sense, the experiment does not differentiate between the two. The number of lines on the screen will tell us something, however, due to the fact that orbital angular momenta are described by an integral quantum number, whereas spin angular momentum depends on a half-integral quantum number. If there are an odd number of lines on the screen, then there are three possibilities: the atom has orbital angular momentum only, an even number of electrons with spin angular momentum, or a combination of orbital angular momentum and an even number of electrons with spin angular momentum. If there is an even number of lines on the screen, then there is at least one unpaired spin angular momentum, possibly in combination with orbital angular momentum. The only numbers of lines for which we can specify the type of angular momentum is one line (no orbital, no spin) and two lines (spin one half). Once we see more than two lines, there are multiple possibilities. For example, three lines could result from two spin one half electrons or one electron with orbital angular momentum quantum number $\ell = 1$.

___

CONCEPTUAL PROBLEM 2

Imagine a single free electron floating in the atmosphere. Does the spin angular momentum vector of the electron align with the magnetic field lines due to the Earth's magnetic field?

29.6 • THE EXCLUSION PRINCIPLE AND THE PERIODIC TABLE

Because any electron in any atom is specified by four quantum numbers, n, ℓ, m_ℓ, and m_s, an obvious and important question is "How many electrons can have a particular set of quantum numbers?" Pauli provided an answer in 1925 in a powerful statement known as the **exclusion principle:**

> No two electrons in an atom can ever be in the same quantum state; that is, no two electrons in the same atom can have the same set of quantum numbers.

It is interesting to note that if this principle were not valid, every electron would end up in the lowest energy state of the atom and the chemical behavior of the elements would be grossly modified. Nature as we know it would not exist! In reality, we can view the electronic structure of complex atoms as a succession of filled levels increasing in energy, where the outermost electrons are primarily responsible for the chemical properties of the element.

As a general rule, the order of filling of an atom's subshells with electrons is as follows. Once one subshell is filled, the next electron goes into the vacant subshell that is lowest in energy. One can understand this principle by recognizing that if the atom were not in the lowest energy state available to it, it would radiate energy until it reached this state.

Before we discuss the electronic configurations of some elements, it is convenient to define an **orbital** as the state of an electron characterized by the quantum numbers n, ℓ, and m_ℓ. From the exclusion principle, we see that **there can be only two electrons in any orbital.** One of these electrons has $m_s = +\frac{1}{2}$, and the other has $m_s = -\frac{1}{2}$. Because each orbital is limited to two electrons, the numbers of electrons that can occupy the levels are also limited.

Table 29.3 shows the numbers of allowed quantum states for an atom up to $n = 3$. Each square in the bottom row of the table represents one orbital, with the $\uparrow$ arrows representing $m_s = \frac{1}{2}$ and the $\downarrow$ arrows representing $m_s = -\frac{1}{2}$. The $n = 1$ shell can accommodate only two electrons, because only one orbital is allowed with $m_\ell = 0$. The $n = 2$ shell has two subshells, with $\ell = 0$ and $\ell = 1$. The $\ell = 0$ subshell is limited to only two electrons, because $m_\ell = 0$. The $\ell = 1$ subshell

Wolfgang Pauli (1900–1958)

An extremely talented Austrian theoretical physicist who made important contributions in many areas of modern physics, Pauli gained public recognition at the age of 21 with a masterful review article on relativity, which is still considered one of the finest and most comprehensive introductions to the subject. Other major contributions were the discovery of the exclusion principle, the explanation of the connection between particle spin and statistics, and theories of relativistic quantum electrodynamics, the neutrino hypothesis, and the hypothesis of nuclear spin. *(Photo taken by S. A. Goudsmit, AIP Niels Bohr Library)*

Chapter **29** *Atomic Physics*

TABLE 29.3 Allowed Quantum Numbers for an Atom up to $n = 3$

n	1	2				3								
ℓ	0	0	1			0	1			2				
m_ℓ	0	0	1	0	-1	0	1	0	-1	2	1	0	-1	-2
m_s	↑↓	↑↓	↑↓	↑↓	↑↓	↑↓	↑↓	↑↓	↑↓	↑↓	↑↓	↑↓	↑↓	↑↓

has three allowed orbitals, corresponding to $m_\ell = 1, 0$, and -1. Because each orbital can accommodate two electrons, the $\ell = 1$ subshell can hold six electrons (and the $n = 2$ shell can hold eight). The $n = 3$ shell has three subshells and nine orbitals and can accommodate up to 18 electrons. Each shell can accommodate up to $2n^2$ electrons.

The exclusion principle can be illustrated by an examination of the electronic arrangement in a few of the lighter atoms. **Hydrogen** has only one electron, which, in its ground state, can be described by either of two sets of quantum numbers: $1, 0, 0, \frac{1}{2}$, or $1, 0, 0, -\frac{1}{2}$. The electronic configuration of this atom is often designated as $1s^1$. The notation $1s$ refers to a state for which $n = 1$ and $\ell = 0$, and the superscript indicates that one electron is present in the s subshell.

Neutral **helium** has two electrons. In the ground state, the quantum numbers of these two electrons are $1, 0, 0, \frac{1}{2}$, and $1, 0, 0, -\frac{1}{2}$. There are no other possible combinations of quantum numbers for this level, and we say that the K shell is filled. Helium is designated as $1s^2$.

The electronic configurations of some successive elements are given in Figure 29.13. Neutral **lithium** has three electrons. In the ground state, two of these are in the $1s$ subshell and the third is in the $2s$ subshell, because this subshell is lower in energy than the $2p$ subshell. Hence, the electronic configuration for lithium is $1s^2 2s^1$.

Note that the electronic configuration of **beryllium,** with its four electrons, is $1s^2 2s^2$, and **boron** has a configuration of $1s^2 2s^2 2p^1$. The $2p$ electron in boron may be described by one of six sets of quantum numbers, corresponding to six states of equal energy.

Carbon has six electrons, and a question arises concerning how to assign the two $2p$ electrons. Do they go into the same orbital with paired spins (↑ ↓), or do they occupy different orbitals with unpaired spins (↑ ↑)? Experimental data show that the most stable configuration (that is, the one that is energetically preferred) is the latter, in which the spins are unpaired. Hence, the two $2p$ electrons in carbon and the three $2p$ electrons in nitrogen have unpaired spins (Fig. 29.13). The general

Hund's rule •

rule that governs such situations, called **Hund's rule,** states that **when an atom has orbitals of equal energy, the order in which they are filled by electrons is such that a maximum number of electrons will have unpaired spins.** Some exceptions to this rule occur in elements having sublevels close to being filled or half filled.

A complete list of electronic configurations is provided in Table 29.4. An early attempt to find some order among the elements was made by a Russian chemist, Dmitri Mendeleev, in 1871. He arranged the atoms in a table (similar to that in Appendix C) according to their atomic masses and chemical similarities. The

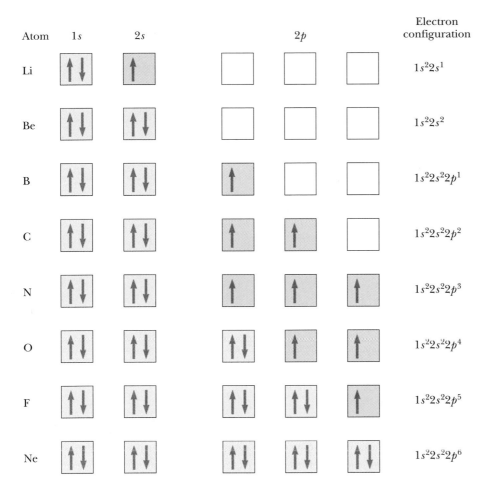

Atom	$1s$	$2s$	$2p$			Electron configuration

Li $\uparrow\downarrow$ $\uparrow$ $1s^2 2s^1$

Be $\uparrow\downarrow$ $\uparrow\downarrow$ $1s^2 2s^2$

B $\uparrow\downarrow$ $\uparrow\downarrow$ $\uparrow$ $1s^2 2s^2 2p^1$

C $\uparrow\downarrow$ $\uparrow\downarrow$ $\uparrow$ $\uparrow$ $1s^2 2s^2 2p^2$

N $\uparrow\downarrow$ $\uparrow\downarrow$ $\uparrow$ $\uparrow$ $\uparrow$ $1s^2 2s^2 2p^3$

O $\uparrow\downarrow$ $\uparrow\downarrow$ $\uparrow\downarrow$ $\uparrow$ $\uparrow$ $1s^2 2s^2 2p^4$

F $\uparrow\downarrow$ $\uparrow\downarrow$ $\uparrow\downarrow$ $\uparrow\downarrow$ $\uparrow$ $1s^2 2s^2 2p^5$

Ne $\uparrow\downarrow$ $\uparrow\downarrow$ $\uparrow\downarrow$ $\uparrow\downarrow$ $\uparrow\downarrow$ $1s^2 2s^2 2p^6$

Figure 29.13   The filling of electronic states must obey the Pauli exclusion principle and Hund's rule.

first table Mendeleev proposed contained many blank spaces, and he boldly stated that the gaps were there only because the elements had not yet been discovered. By noting the columns in which these missing elements should be located, he was able to make rough predictions about their chemical properties. Within 20 years of Mendeleev's announcement, the missing elements were indeed discovered.

The elements in the periodic table are arranged so that all those in a vertical column have similar chemical properties. For example, consider the elements in the last column: He (helium), Ne (neon), Ar (argon), Kr (krypton), Xe (xenon), and Rn (radon). The outstanding characteristic of these elements is that they do not normally take part in chemical reactions—that is, they do not join with other atoms to form molecules—and they are therefore classified as being *inert*. Because of this aloofness, they are referred to as the **noble gases.**

We can partially understand this behavior by looking at the electronic configurations in Table 29.4. This table also lists the ionization energies for certain ele-

TABLE 29.4   Electronic Configuration of the Elements

Z	Symbol	Ground Configuration	Ionization Energy (eV)	Z	Symbol	Ground Configuration	Ionization Energy (eV)
1	H	$1s^1$	13.595	25	Mn	$3d^5 4s^2$	7.432
2	He	$1s^2$	24.581	26	Fe	$3d^6 4s^2$	7.87
				27	Co	$3d^7 4s^2$	7.86
3	Li	[He] $2s^1$	5.39	28	Ni	$3d^8 4s^2$	7.633
4	Be	$2s^2$	9.320	29	Cu	$3d^{10} 4s^1$	7.724
5	B	$2s^2 2p^1$	8.296	30	Zn	$3d^{10} 4s^2$	9.391
6	C	$2s^2 2p^2$	11.256	31	Ga	$3d^{10} 4s^2 4p^1$	6.00
7	N	$2s^2 2p^3$	14.545	32	Ge	$3d^{10} 4s^2 4p^2$	7.88
8	O	$2s^2 2p^4$	13.614	33	As	$3d^{10} 4s^2 4p^3$	9.81
9	F	$2s^2 2p^5$	17.418	34	Se	$3d^{10} 4s^2 4p^4$	9.75
10	Ne	$2s^2 2p^6$	21.559	35	Br	$3d^{10} 4s^2 4p^5$	11.84
				36	Kr	$3d^{10} 4s^2 4p^6$	13.996
11	Na	[Ne] $3s^1$	5.138				
12	Mg	$3s^2$	7.644	37	Rb	[Kr] $5s^1$	4.176
13	Al	$3s^2 3p^1$	5.984	38	Sr	$5s^2$	5.692
14	Si	$3s^2 3p^2$	8.149	39	Y	$4d^1 5s^2$	6.377
15	P	$3s^2 3p^3$	10.484	40	Zr	$4d^2 5s^2$	
16	S	$3s^2 3p^4$	10.357	41	Nb	$4d^4 5s^1$	6.881
17	Cl	$3s^2 3p^5$	13.01	42	Mo	$4d^5 5s^1$	7.10
18	Ar	$3s^2 3p^6$	15.755	43	Tc	$4d^5 5s^2$	7.228
				44	Ru	$4d^7 5s^1$	7.365
19	K	[Ar] $4s^1$	4.339	45	Rh	$4d^8 5s^1$	7.461
20	Ca	$4s^2$	6.111	46	Pd	$4d^{10}$	8.33
21	Sc	$3d^1 4s^2$	6.54	47	Ag	$4d^{10} 5s^1$	7.574
22	Ti	$3d^2 4s^2$	6.83	48	Cd	$4d^{10} 5s^2$	8.991
23	V	$3d^3 4s^2$	6.74	49	In	$4d^{10} 5s^2 5p^1$	5.78
24	Cr	$3d^5 4s^1$	6.76	50	Sn	$4d^{10} 5s^2 5p^2$	7.342

Note: The bracket notation is used as a shorthand method to avoid repetition in indicating inner-shell electrons. Thus, [He] represents $1s^2$, [Ne] represents $1s^2 2s^2 2p^6$, [Ar] represents $1s^2 2s^2 2p^6 3s^2 3p^6$, and so on.

ments. The element helium is one in which the electronic configuration is $1s^2$— in other words, one shell is filled. In addition, it is found that the energy of the electrons in this filled shell is considerably lower than the energy of the next available level, the $2s$ level. Next, look at the electronic configuration for neon, $1s^2 2s^2 2p^6$. Again, the outermost shell is filled and there is a gap in energy between the $2p$ level and the $3s$ level. Argon has the configuration $1s^2 2s^2 2p^6 3s^2 3p^6$. Here, the $3p$ subshell is filled and there is a gap in energy between the $3p$ subshell and the $3d$ subshell. We could continue this procedure through all the noble gases; the pattern remains the same. A noble gas is formed when a shell is filled and there is a gap in energy before the next possible level is encountered.

It is interesting to plot ionization energy versus the atomic number Z, as in Figure 29.14. Note the pattern 2, 8, 8, 18, 18, 32 for the ionization energies. This pattern follows from the Pauli exclusion principle and helps explain why the elements repeat their chemical properties in groups. For example, the peaks at $Z = 2$, 10, 18, and 36 correspond to the elements He, Ne, Ar, and Kr,

TABLE 29.4   Electronic Configuration of the Elements (Continued)

Z	Symbol	Ground Configuration	Ionization Energy (eV)	Z	Symbol	Ground Configuration	Ionization Energy (eV)
51	Sb	$4d^{10}5s^2 5p^3$	8.639	79	Au	[Xe] $4f^{14}5d^{10}6s^1$	9.22
52	Te	$4d^{10}5s^2 5p^4$	9.01	80	Hg	$6s^2$	10.434
53	I	$4d^{10}5s^2 5p^5$	10.454	81	Tl	$6s^2 6p^1$	6.106
54	Xe	$4d^{10}5s^2 5p^6$	12.127	82	Pb	$6s^2 6p^2$	7.415
				83	Bi	$6s^2 6p^3$	7.287
55	Cs	[Xe] $6s^1$	3.893	84	Po	$6s^2 6p^4$	8.43
56	Ba	$6s^2$	5.210	85	At	$6s^2 6p^5$	
57	La	$5d^1 6s^2$	5.61	86	Rn	$6s^2 6p^6$	10.745
58	Ce	$4f^1 5d^1 6s^2$	6.54				
59	Pr	$4f^3 6s^2$	5.48	87	Fr	[Rn] $7s^1$	
60	Nd	$4f^4 6s^2$	5.51	88	Ra	$7s^2$	5.277
61	Pm	$4f^5 6s^2$		89	Ac	$6d^1 7s^2$	6.9
62	Sm	$4f^6 6s^2$	5.6	90	Th	$6d^2 7s^2$	
63	Eu	$4f^7 6s^2$	5.67	91	Pa	$5f^2 6d^1 7s^2$	
64	Gd	$4f^7 5d^1 6s^2$	6.16	92	U	$5f^3 6d^1 7s^2$	4.0
65	Tb	$4f^9 6s^2$	6.74	93	Np	$5f^4 6d^1 7s^2$	
66	Dy	$4f^{10} 6s^2$		94	Pu	$5f^6 7s^2$	
67	Ho	$4f^{11} 6s^2$		95	Am	$5f^7 7s^2$	
68	Er	$4f^{12} 6s^2$		96	Cm	$5f^7 6d^1 7s^2$	
69	Tm	$4f^{13} 6s^2$		97	Bk	$5f^8 6d^1 7s^2$	
70	Yb	$4f^{14} 6s^2$	6.22	98	Cf	$5f^{10} 7s^2$	
71	Lu	$4f^{14} 5d^1 6s^2$	6.15	99	Es	$5f^{11} 7s^2$	
72	Hf	$4f^{14} 5d^2 6s^2$	7.0	100	Fm	$5f^{12} 7s^2$	
73	Ta	$4f^{14} 5d^3 6s^2$	7.88	101	Mv	$5f^{13} 7s^2$	
74	W	$4f^{14} 5d^4 6s^2$	7.98	102	No	$5f^{14} 7s^2$	
75	Re	$4f^{14} 5d^5 6s^2$	7.87	103	Lw	$5f^{14} 6d^1 7s^2$	
76	Os	$4f^{14} 5d^6 6s^2$	8.7	104	Ku	$5f^{14} 6d^2 7s^2$	
77	Ir	$4f^{14} 5d^7 6s^2$	9.2				
78	Pt	$4f^{14} 5d^8 6s^2$	8.88				

which have filled shells. These elements have similar energies and chemical behavior.

Thinking Physics 3

As one moves from left to right across one row of the periodic table, the effective size of the atoms first decreases and then increases. What would cause this behavior?

Reasoning As one begins at the left side of the periodic table and moves toward the middle, the nuclear charge is increasing. As a result, there is an increasing Coulomb attraction between the nucleus and the electrons, and the electrons are pulled into an average position that is closer to the nucleus. Although the additional electrons repel one another, that repulsion is diffused throughout their orbital volume. From the middle of a row to the right side, the increasing number of electrons being placed in proximity to each other results in an accumulated stronger repulsion which increases their average distance from the nucleus and causes the atomic size to grow.

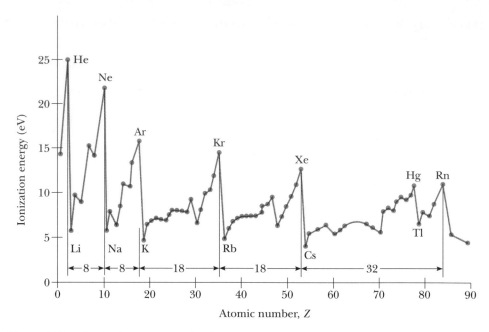

Figure 29.14   Ionization energy of the elements versus atomic number *Z*.   *(Adapted from J. Orear, Physics, New York, Macmillan, 1979)*

29.7 • ATOMIC SPECTRA: VISIBLE AND X-RAY

In Chapter 11 we briefly discussed the origin of the spectral lines for hydrogen and hydrogen-like ions. Recall that an atom will emit electromagnetic radiation if an electron in an excited state makes a transition to a lower energy state.

   The energy level diagram for hydrogen is shown in Figure 29.15. The diagonal lines represent allowed transitions between stationary states. Whenever an electron makes a transition from a higher energy state to a lower one, a photon of light is emitted. The frequency of this photon is $f = \Delta E/h$, where ΔE is the energy differ-

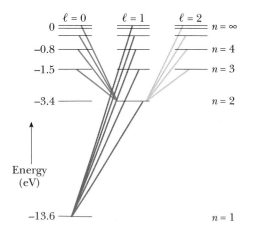

Figure 29.15   Some allowed electronic transitions for hydrogen, represented by the colored lines. These transitions must obey the selection rule $\Delta\ell = \pm 1$.

ence between the two levels and h is Planck's constant. The **selection rules for the allowed transitions are**

$$\Delta \ell = \pm 1 \quad \text{and} \quad \Delta m_\ell = 0 \text{ or } \pm 1 \qquad \text{[29.17]}$$

Transitions that do not obey these selection rules are said to be **forbidden.** (Such transitions can occur, but their probability is negligible relative to the probability of the allowed transitions.)

Because the orbital angular momentum of an atom changes when a photon is emitted or absorbed (that is, as a result of a transition) and because angular momentum must be conserved, we conclude that **the photon involved in the process must carry angular momentum.** In fact, the photon has an angular momentum equivalent to that of a particle with a spin of 1. Also, the angular momentum of the photon is consistent with the classical description of electromagnetic radiation. Hence, **a photon has energy, linear momentum, and angular momentum.**

As a generalization from Chapter 11, Equation 11.23, the allowed energies for one-electron atoms, such as hydrogen and He$^+$, are

$$E_n = -\frac{13.6Z^2}{n^2} \text{ eV} \qquad \text{[29.18]}$$

For multi-electron atoms, the nuclear charge Ze is largely canceled or shielded by the negative charge of the inner-core electrons. Hence, the outer electrons interact with a net charge of the order of the electronic charge. The expression for the allowed energies for multi-electron atoms has the same form as Equation 29.18, with Z replaced by an effective atomic number, Z_{eff}. That is,

$$E_n = -\frac{13.6Z_{\text{eff}}^2}{n^2} \text{ eV} \qquad \text{[29.19]}$$

where Z_{eff} depends on n and ℓ. For the higher energy states, this reduction in charge increases and $Z_{\text{eff}} \to 1$.

X-Ray Spectra

X-rays are emitted from a metal target when it is bombarded by high-energy electrons. The x-ray spectrum typically consists of a broad continuous band and a series of sharp lines that depend on the type of material used for the target, as shown in Figure 29.16. These lines, called **characteristic x-rays,** were discovered in 1908, but their origin remained unexplained until the details of atomic structure were brought to light.

The continuous x-ray spectrum arises due to the change in velocity of the electrons as they strike the target. In Section 24.5, we stated that electromagnetic radiation is emitted whenever a charged particle accelerates. As the electrons in an x-ray tube are slowed down by interacting with the target atoms, they emit x-rays. The accelerations of the electrons vary over a wide range, since they interact with the atoms over a variety of separation distances. As a result, the frequencies of the emitted x-rays vary continuously.

The first step in the production of characteristic x-rays occurs when a bombarding electron collides with an electron in an inner shell of a target atom with sufficient energy to remove the electron from the atom. The vacancy created in the

• *Selection rule for allowed atomic transitions*

• *The photon carries angular momentum.*

• *Allowed energies for one-electron atoms*

• *Allowed energies for multi-electron atoms*

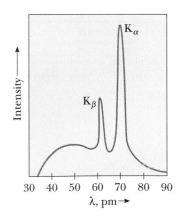

Figure 29.16 The x-ray spectrum of a metal target consists of a broad continuous spectrum plus a number of sharp lines, which are due to *characteristic x-rays.* The data shown were obtained when 35-keV electrons bombarded a molybdenum target. Note that 1 pm = 10^{-12} m = 10^{-3} nm.

shell is filled when an electron in a higher level drops down into the level containing the vacancy. The time it takes for this to happen is very short, less than 10^{-9} s. This transition is accompanied by the emission of a photon the energy of which equals the difference in energy between the two levels. Typically, the energy of such transitions is greater than 1000 eV, and the emitted x-ray photons have wavelengths in the range of 0.01 nm to 1 nm.

Let us assume that the incoming electron has dislodged an atomic electron from the innermost shell, the K shell. If the vacancy is filled by an electron dropping from the next higher shell, the L shell, the photon emitted in the process has an energy corresponding to the K_α line on the curve of Figure 29.16. If the vacancy is filled by an electron dropping from the M shell, the line produced is called the K_β line.

Other characteristic x-ray lines are formed when electrons drop from upper levels to vacancies other than those in the K shell. For example, L lines are produced when vacancies in the L shell are filled by electrons dropping from higher shells. An L_α line is produced as an electron drops from the M shell to the L shell, and an L_β line is produced by a transition from the N shell to the L shell.

We can estimate the energy of the emitted x-rays as follows. Consider two electrons in the K shell of an atom the atomic number of which is Z. Each electron partially shields the other from the charge of the nucleus, Ze, and so each electron is subject to an effective nuclear charge of $Z_{\text{eff}} = (Z - 1)e$. We can now use Equation 29.18 to estimate the energy of either electron:

$$E_K \approx -(Z - 1)^2(13.6 \text{ eV}) \qquad [\textbf{29.20}]$$

As we shall show in the following example, one can estimate the energy of an electron in an L or M shell in a similar fashion. Taking the energy difference between these two levels, one can then calculate the energy and wavelength of the emitted photon.

In 1914, Henry G. J. Moseley plotted the Z values for a number of elements versus $\sqrt{1/\lambda}$, where λ is the wavelength of the K_α line of each element. He found that the plot is a straight line, as in Figure 29.17. This is consistent with rough calculations of the energy levels given by Equation 29.20. From this plot, Moseley was able to determine the Z values of some missing elements, which provided a periodic chart in excellent agreement with the known chemical properties of the elements.

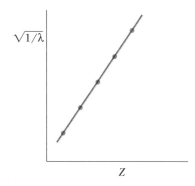

Figure 29.17 A Moseley plot for the K_α x-ray lines of a number of elements.

Thinking Physics 4

The Bohr theory gives good theoretical agreement with experimentally measured frequencies of spectral lines for the hydrogen atom, but it fails completely to predict the correct optical frequencies for heavier atoms. Moseley, however, used the Bohr theory to predict the frequencies of characteristic x-rays from heavy atoms. Why would the Bohr theory give good predictions for x-rays?

Reasoning Characteristic x-rays have their origin in transitions between the lowest energy levels in the atom. An electron in a lower energy shell is ejected from the atom by the accelerated electron in the x-ray tube, leaving a vacancy in the shell. Let us consider an electron ejected from the K shell and the resulting electrical interactions of an L shell electron with the other electrons. If we look first at the electrons in higher shells than L, these electrons spend most of their time farther from the nucleus than the L electrons. What's more, when averaged over time, the charge of these outer electrons is isotropic—there is no preferred direction. Thus, the L electrons find themselves *inside* a set of almost perfectly symmetric shells of charge. According to Gauss's law, then, **the L electron is largely unaffected by the outer electrons**—the time averaged electric field from these electrons is zero. What remains for the L electrons is a nuclear charge shielded by the one remaining K electron and the other electrons in the L shell. If we ignore the other L electrons, we can model the L electron as belonging to a hydrogen atom, with a nuclear charge of $(Z - 1)$, where the nuclear charge Z is reduced by 1 due to the shielding effect of the remaining K electron. As a result, the Bohr theory works relatively well in describing the energy of the L electron, although it is not perfect, because we have ignored the other L electrons.

CONCEPTUAL PROBLEM 3

In an x-ray tube, if the energy with which the electrons strike the metal target is increased, the wavelengths of the characteristic x-rays do not change. Why not?

CONCEPTUAL PROBLEM 4

Is it possible for a spectrum from an x-ray tube to show the continuous spectrum of x-rays without the presence of the characteristic x-rays?

CONCEPTUAL PROBLEM 5

The transition from the $3d$ state in hydrogen to the $1s$ state is not seen. Why?

Example 29.7 Estimating the Energy of an X-Ray

Estimate the energy of the characteristic x-ray emitted from a tungsten target when an electron drops from an M shell ($n = 3$ state) to a vacancy in the K shell ($n = 1$ state).

Solution The atomic number for tungsten is $Z = 74$. Using Equation 29.20, we get

$$E_K \approx -(74 - 1)^2(13.6 \text{ eV}) = -72\,500 \text{ eV}$$

The electron in the M shell is subject to an effective nuclear charge that depends on the number of electrons in the $n = 1$ and $n = 2$ states, which shield the nucleus. Because there are eight electrons in the $n = 2$ state and one electron in the $n = 1$ state, roughly nine electrons shield the nucleus, and so $Z_{eff} = Z - 9$. Hence, the energy of an electron in the M shell, following Equation 29.19, is

$$E_M \approx -Z_{eff}^2 E_3 - -(Z - 9)^2 \frac{E_0}{3^2}$$

$$= -(74 - 9)^2 \frac{(13.6 \text{ eV})}{9} = -6380 \text{ eV}$$

where E_3 is the energy of an electron in the M shell of the hydrogen atom and E_0 is the ground-state energy. Therefore, the emitted x-ray has an energy equal to $E_M - E_K \approx -6380 \text{ eV} - (-72\,500 \text{ eV}) = 66\,100 \text{ eV}$. Note that this energy difference is also equal to $hf = hc/\lambda$, where λ is the wavelength of the emitted x-ray.

EXERCISE 4 Calculate the wavelength of the emitted x-ray for this transition.   Answer   0.0188 nm

---

29.8 • ATOMIC TRANSITIONS

We have seen that an atom emits radiation only at certain frequencies that correspond to the energy separations between the allowed states. Consider an atom with many allowed energy states, labeled E_1, E_2, E_3, . . . in Figure 29.18. When light is incident on the atom, only those photons the energy hf of which matches the energy

E_4
E_3

E_2

E_1

Figure 29.18   Energy level diagram of an atom with various allowed states. The lowest energy state, E_1, is the ground state. All others are excited states.

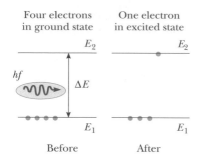

Four electrons in ground state | One electron in excited state

E_2 hf ΔE E_2

E_1 E_1

Before   After

Figure 29.19   Diagram representing the *stimulated absorption* of a photon by an atom. The dots represent electrons. One electron is transferred from the ground state to the excited state when the atom absorbs a photon of energy $hf = E_2 - E_1$.

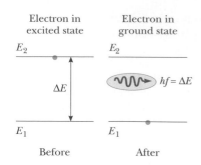

Electron in excited state | Electron in ground state

E_2 ΔE E_2 $hf = \Delta E$

E_1 E_1

Before   After

Figure 29.20   Diagram representing the *spontaneous emission of* a photon by an atom that is initially in the excited state E_2. When the electron falls to the ground state, the atom emits a photon of energy $hf = E_2 - E_1$.

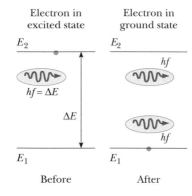

Electron in excited state | Electron in ground state

E_2 E_2 hf

$hf = \Delta E$ ΔE hf

E_1 E_1

Before   After

Figure 29.21   Diagram representing the *stimulated emission* of a photon by an incoming photon of energy hf. Initially, the atom is in the excited state. The incoming photon stimulates the atom to emit a second photon of energy $hf = E_2 - E_1$.

separation, ΔE, between two levels can be absorbed by the atom. Figure 29.19 is a schematic diagram representing this **stimulated absorption.** At ordinary temperatures, most of the atoms are in the ground state. If a vessel containing many atoms of a gaseous element is illuminated with a light beam containing all possible photon frequencies (that is, a continuous spectrum), only those photons of energies $E_2 - E_1$, $E_3 - E_1$, $E_4 - E_1$, and so on, can be absorbed. As a result of this absorption, some atoms are raised to allowed higher energy levels, called **excited states.**

Once an atom is in an excited state, there is a certain probability that the excited electron will jump back to a lower level by emitting a photon, as shown in Figure 29.20. This process is known as **spontaneous emission.** Typically, an atom remains in an excited state for only about 10^{-8} s.

Finally, there is a third process, **stimulated emission,** that is of importance in lasers. Suppose an atom is in the excited state E_2, as in Figure 29.21, and a photon with energy $hf = E_2 - E_1$ is incident on it. The excited state must be a metastable state having a lifetime much longer than 10^{-8} s. The incoming photon increases the probability that the excited electron will return to the ground state and thereby emit a second photon having the same energy hf. Note that the incident photon is not absorbed, so after the stimulated emission, there are two nearly identical photons—the incident photon and the emitted photon. The emitted photon is in phase with the incident photon. These photons can stimulate other atoms to emit photons in a chain of similar processes. The many photons produced in this fashion are the source of the intense, coherent light in a laser.

Thinking Physics 5

A physics student is watching a meteor shower in the early morning hours. She notices that the streaks of light from the meteoroids entering the very high regions of the atmosphere last for as long as 2 or 3 seconds before fading. She also notices a lightning storm off in the distance. The streaks of light from the lightning fade away almost immediately after the flash, certainly in much less than 1 second. Both lightning and meteors cause the air to turn into a plasma because of the very high temperatures generated. The light is given off when the stripped electrons in the plasma recombine with the ionized atoms. Why would this light last longer for meteors than for lightning?

Reasoning The answer lies in the subtle phrase in the description of the meteoroids— " . . . entering the very high regions of the atmosphere." In the very high regions of the atmosphere, the pressure is very low. In addition, the *density* is very low, so that atoms of the gas are relatively far apart. Thus, after the air is ionized by the passing meteoroid, the probability of freed electrons finding an ionized atom with which to recombine is relatively low. As a result, the recombination process occurs over a relatively long time, measured in seconds.

However, lightning occurs in the lower regions of the atmosphere (the troposphere) where the pressure and density are relatively high. After the ionization by the lightning flash, the electrons and ionized atoms are much closer together than in the upper atmosphere. The probability of a recombination is much higher, and the time for the recombination to occur is much shorter.

EXERCISE 5 The wavelength of coherent ruby laser light is 694.3 nm. What is the energy difference (in electron volts) between the upper, excited state and the lower, unexcited state?

Answer 1.789 eV

29.9 · LASERS AND HOLOGRAPHY

We have described how an incident photon can cause atomic transitions either upward (stimulated absorption) or downward (stimulated emission). The two processes are equally probable. When light is incident on a system of atoms, there is usually a net absorption of energy, because when the system is in thermal equilibrium, there are many more atoms in the ground state than in excited states. However, if one can invert the situation so that there are more atoms in an excited state than in the ground state, a net emission of photons can result. Such a condition is called **population inversion.**

This is the fundamental principle involved in the operation of a **laser,** an acronym for **l**ight **a**mplification by **s**timulated **e**mission of **r**adiation. The amplification corresponds to a buildup of photons in the system as the result of a chain reaction of events. The following three conditions must be satisfied in order to achieve laser action:

1. The system must be in a state of population inversion (more atoms in an excited state than in the ground state).
2. The excited state of the system must be a *metastable state,* which means its lifetime must be long compared with the usually short lifetimes of excited states. When such is the case, stimulated emission will occur before spontaneous emission.
3. The emitted photons must be confined in the system long enough to enable them to stimulate further emission from other excited atoms. This is

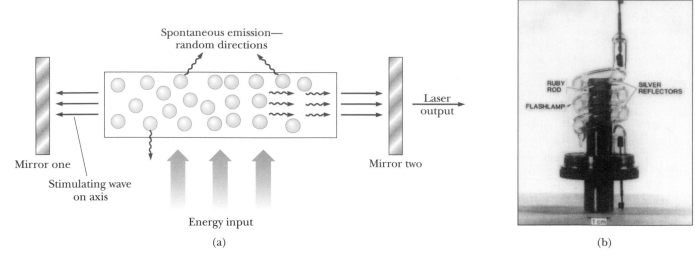

Figure 29.22   (a) A schematic diagram of a laser design. The tube contains atoms, which represent the active medium. An external source of energy (optical, electrical, etc.) is needed to "pump" the atoms to excited energy states. The parallel end mirrors provide the feedback of the stimulating wave. (b) Photograph of the first ruby laser showing the flash lamp surrounding the ruby rod.   *(Courtesy of Hughes Aircraft Company)*

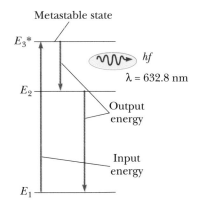

Figure 29.23   Energy level diagram for the neon atom, which emits photons at a wavelength of 632.8 nm through stimulated emission. The photon at this wavelength arises from the transition $E_3^* \rightarrow E_2$. This is the source of coherent light in the helium–neon gas laser.

achieved by the use of reflecting mirrors at the ends of the system. One end is made totally reflecting and the other is slightly transparent to allow the laser beam to escape (Fig. 29.22a).

One device that exhibits stimulated emission of radiation is the helium–neon gas laser. Figure 29.23 is an energy level diagram for the neon atom in this system. The mixture of helium and neon is confined to a glass tube that is sealed at the ends by mirrors. An oscillator connected to the tube causes electrons to sweep through the tube, colliding with the atoms of the gases and raising them into excited states. Neon atoms are excited to state E_3 through this process and also as a result of collisions with excited helium atoms. Stimulated emission occurs as the neon atoms make a transition to state E_2 and neighboring excited atoms are stimulated. This results in the production of coherent light at a wavelength of 632.8 nm.

Since the development of the first laser in 1960, laser technology has experienced tremendous growth. Lasers that cover wavelengths in the infrared, visible, and ultraviolet regions are now available. Applications include surgical "welding" of detached retinas, precision surveying and length measurement, a potential source for inducing nuclear fusion reactions, precision cutting of metals and other materials, and telephone communication along optical fibers. These and other applications are possible because of the unique characteristics of laser light. In addition to its being highly monochromatic, laser light is also highly directional and can therefore be sharply focused to produce regions of extremely intense light energy (with energy densities approaching those inside the laser tube).

Holography

One interesting application of the laser is **holography,** the production of three-dimensional images of objects. Figure 29.24a shows how a hologram is made. Light from the laser is split into two parts by a half-silvered mirror at *B*. One part of the beam reflects off the object to be photographed and strikes an ordinary photographic film. The other half of the beam is diverged by lens L_2, reflects from mirrors M_1 and M_2, and finally strikes the film. The two beams overlap to form an extremely complicated interference pattern on the film. Such an interference pattern can be produced only if the phase relationship of the two waves is constant throughout the exposure of the film. This condition is met by illuminating the scene with light coming through a pinhole or with coherent laser radiation. The hologram records not only the intensity of the light scattered from the object (as in a conventional photograph) but also the phase difference between the reference beam and the beam scattered from the object. Because of this phase difference, an interference pattern is formed that produces an image with full three-dimensional perspective.

A hologram is best viewed by allowing coherent light to pass through the developed film as one looks back along the direction from which the beam comes. Figure 29.24b is a photograph of a hologram made using a cylindrical film.

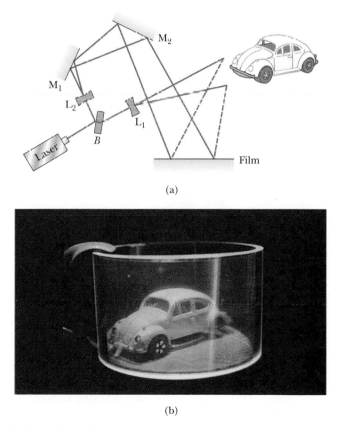

(a)

(b)

Figure 29.24 (a) Experimental arrangement for producing a hologram. (b) Photograph of a hologram that uses a cylindrical film. Note the detail of the Volkswagen image. *(Courtesy of Central Scientific Company)*

SUMMARY

The methods of quantum mechanics can be applied to the hydrogen atom using the appropriate potential energy function, $U(r) = -k_e e^2/r$, in the Schrödinger equation. The solution to this equation yields the wave functions for the allowed states and the allowed energies, given by

$$E_n = -\left(\frac{k_e e^2}{2a_0}\right)\frac{1}{n^2} = -\frac{13.6}{n^2}\ \text{eV} \qquad n = 1, 2, 3, \ldots \qquad \textbf{[29.2]}$$

This is precisely the result obtained in the Bohr theory. The allowed energy depends only on the **principal quantum number, n.** The allowed wave functions depend on three quantum numbers, n, ℓ, and m_ℓ, where ℓ is the **orbital quantum number** and m_ℓ is the **orbital magnetic quantum number.** The restrictions on the quantum numbers are as follows:

$$n = 1, 2, 3, \ldots$$
$$\ell = 0, 1, 2, \ldots, (n-1)$$
$$m_\ell = -\ell, -\ell + 1, \ldots, \ell - 1, \ell$$

All states with the same principal quantum number n form a **shell,** identified by the letters K, L, M, ... (corresponding to $n = 1, 2, 3, \ldots$). All states with given values of both n and ℓ form a **subshell,** designated by the letters, s, p, d, f, ... (corresponding to $\ell = 0, 1, 2, 3, \ldots$).

In order to completely describe a quantum state of the hydrogen atom, it is necessary to include a fourth quantum number, m_s, called the **spin magnetic quantum number.** This quantum number can have only two values, $\pm\frac{1}{2}$. In effect, this doubles the number of allowed states specified by the quantum numbers n, ℓ, and m_ℓ.

An atom in a state characterized by a specific n can have the following values of **orbital angular momentum, L:**

$$L = \sqrt{\ell(\ell + 1)}\,\hbar \qquad \ell = 0, 1, 2, \ldots, n - 1 \qquad \textbf{[29.10]}$$

The allowed values of the projection of **L** along the z axis are given by

$$L_z = m_\ell \hbar \qquad \textbf{[29.11]}$$

where m_ℓ is restricted to integer values lying between $-\ell$ and ℓ. Only discrete values of L_z are allowed, and these are determined by the restrictions on m_ℓ. This quantization of L_z is referred to as **space quantization.**

The electron has an intrinsic angular momentum called **spin angular momentum.** That is, the total angular momentum of an electron in an atom can have two contributions, one arising from the spin of the electron (**S**) and one arising from the orbital motion of the electron (**L**).

Electronic spin can be described by a single quantum number, $s = \frac{1}{2}$. The **magnitude of the spin angular momentum** is

$$S = \frac{\sqrt{3}}{2}\,\hbar \qquad \textbf{[29.13]}$$

and the z component of **S** is

$$S_z = m_s \hbar = \pm\tfrac{1}{2}\hbar \qquad \textbf{[29.14]}$$

That is, the spin angular momentum is also quantized in space, as specified by the **spin magnetic quantum number, $m_s = \pm\frac{1}{2}$.**

The magnetic moment $\boldsymbol{\mu}_s$ associated with the spin angular momentum of an electron is

$$\boldsymbol{\mu}_s = -\frac{e}{m_e}\mathbf{S} \qquad\qquad \textbf{[29.15]}$$

which is *twice* as large as the orbital magnetic moment. The z component of μ_s can have the values

$$\mu_{sz} = \pm\frac{e\hbar}{2m_e} \qquad\qquad \textbf{[29.16]}$$

The **exclusion principle** states that **no two electrons in an atom can ever have the same set of quantum numbers** n, ℓ, m_ℓ, and m_s. Using this principle and the principle of minimum energy, one can determine the electronic configuration of the elements. This serves as a basis for understanding atomic structure and the chemical properties of the elements.

The allowed electronic transitions between any two levels in an atom are governed by the selection rules

$$\Delta\ell = \pm 1 \qquad \text{and} \qquad \Delta m_\ell = 0 \text{ or } \pm 1 \qquad\qquad \textbf{[29.17]}$$

X-rays are emitted by atoms when an electron undergoes a transition from an outer shell into an electron vacancy in one of the inner shells. Transitions into a vacant state in the K shell give rise to the K series of spectral lines; transitions into a vacant state in the L shell create the L series of lines; and so on. The x-ray spectrum of a metal target consists of a set of sharp characteristic lines superimposed on a broad, continuous spectrum.

CONCEPTUAL QUESTIONS

1. Does the light emitted by a neon sign constitute a continuous spectrum or only a few colors? Defend your answer.
2. Must an atom first be ionized before it can emit light?
3. Discuss why the term *electron clouds* is used to describe the electronic arrangement in the quantum mechanical view of the atom.
4. It was stated in the chapter that, if the exclusion principle were not valid, every electron would end up in the lowest energy state of the atom and the chemical behavior of the elements would be grossly modified. Explain this statement.
5. When a hologram is produced, the system (including light source, object, beam splitter, and so on) must be held motionless within a quarter of a wavelength. Why?
6. Why is an inhomogeneous magnetic field used in the Stern–Gerlach experiment?
7. Could the Stern–Gerlach experiment be performed with ions rather than neutral atoms? Explain.
8. Describe some experiments that support the conclusion that the spin quantum number for electrons can only have the values $\pm\frac{1}{2}$.
9. Discuss some of the consequences of the exclusion principle.
10. Why do lithium, potassium, and sodium exhibit similar chemical properties?

11. From Table 29.4, we find that the ionization energies for Li, Na, K, Rb, and Cs are 5.390, 5.138, 4.339, 4.176, and 3.893 eV, respectively. Explain why these values are to be expected in terms of the atomic structures.
12. Does the intensity of light from a laser fall off as $1/r^2$?
13. How is it possible that electrons, which have a probability distribution around a nucleus, can exist in states of definite energy (e.g., $1s$, $2p$, $3d$, . . .)?
14. It is easy to understand how two electrons (one spin up, one spin down) can fill the $1s$ shell for a helium atom. How is it possible that eight more electrons can fit into the $2s$, $2p$ level to complete the $1s^2 2s^2 2p^6$ shell for a neon atom?
15. In 1914 Henry Moseley was able to define the atomic number of an element from its characteristic x-ray spectrum. How was this possible? (*Hint:* See Figs. 29.16 and 29.17.)
16. Why is stimulated emission so important in the operation of a laser?
17. The efficiencies of most solid-state lasers are on the order of 1% to 2%. Although the laser output is monochromatic and highly directional, can you use Figures 29.22 and 29.23 to determine why the energy input must exceed laser energy output by a factor of 50 to 100? Explain your answer.

PROBLEMS

Section 29.2   The Hydrogen Atom Revisited

1. A photon with energy 2.28 eV is barely capable of causing a photoelectric effect when it strikes a sodium plate. Suppose that the photon is instead absorbed by hydrogen. Find (a) the minimum n for a hydrogen atom that can be ionized by such a photon and (b) the speed of the released electron far from the nucleus.

2. The Balmer series for the hydrogen atom corresponds to electronic transitions that terminate in the state with quantum number $n = 2$, as shown in Figure P29.2. (a) Consider the photon of longest wavelength; determine its energy and wavelength. (b) Consider the spectral line of shortest wavelength; find its photon energy and wavelength.

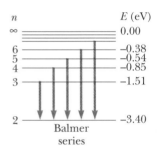

Figure P29.2   An energy-level diagram for hydrogen showing the Balmer series (not to scale).

3. A general expression for the energy levels of one-electron atoms and ions is

$$E_n = -\mu k_e^2 q_1^2 q_2^2 / 2\hbar^2 n^2$$

where k_e is the Coulomb constant, q_1 and q_2 are the charges of the two particles, and μ is the reduced mass, given by $\mu = m_1 m_2 / (m_1 + m_2)$. In Problem 2 we found that the wavelength for the $n = 3$ to $n = 2$ transition of the hydrogen atom is 656.3 nm (visible red light). What are the wavelengths for this same transition in (a) positronium, which consists of an electron and a positron, and (b) singly ionized helium? (*Note:* A positron is a positively charged electron.)

4. (a) Determine the possible values of the quantum numbers ℓ and m_ℓ for the He$^+$ ion in the state corresponding to $n = 3$. (b) What is the energy of this state?

Section 29.3   The Spin Magnetic Quantum Number

Section 29.4   The Wave Functions for Hydrogen

5. List the possible sets of quantum numbers for electrons in (a) the $3d$ subshell and (b) the $3p$ subshell.

6. Plot the wave function $\psi_{1s}(r)$ (Eq. 29.3) and the radial probability density function $P_{1s}(r)$ (Eq. 29.7) for hydrogen. Let r range from 0 to $1.5a_0$, where a_0 is the Bohr radius.

7. The wave function for an electron in the $2p$ state of hydrogen is

$$\psi_{2p} = \frac{1}{\sqrt{3}(2a_0)^{3/2}} \frac{r}{a_0} e^{-r/2a_0}$$

What is the most likely distance from the nucleus to find an electron in the $2p$ state? (See Fig. 29.8.)

8. During a particular period of time, an electron in the ground state of a hydrogen atom is "observed" 1000 times at a distance $a_0/2$ from the nucleus. How many times is this electron observed at a distance $2a_0$ from the nucleus during this period of "observation"?

9. Show that the $1s$ wave function for an electron in hydrogen,

$$\psi(r) = \frac{1}{\sqrt{\pi a_0^3}} e^{-r/a_0}$$

satisfies the radially symmetric Schrödinger equation,

$$-\frac{\hbar^2}{2m}\left(\frac{d^2\psi}{dr^2} + \frac{2}{r}\frac{d\psi}{dr}\right) - \frac{k_e e^2}{r}\psi = E\psi$$

Section 29.5   The "Other" Quantum Numbers

10. Calculate the angular momentum for an electron in (a) the $4d$ state and (b) the $6f$ state.

11. If an electron has orbital angular momentum 4.714×10^{-34} J·s, what is the orbital quantum number for the state of the electron?

12. A hydrogen atom is in its fifth excited state. The atom emits a 1090-nm wavelength photon. Determine the maximum possible orbital angular momentum of the electron after emission.

13. Find all possible values of L, L_z, and θ for an electron in a $3d$ state of hydrogen.

14. Two electrons in the same atom both have $n = 3$ and $\ell = 1$. List the possible states. (b) How many states would be possible if the exclusion principle were inoperative?

15. How many sets of quantum numbers are possible for an electron for which   (a) $n = 1$,   (b) $n = 2$,   (c) $n = 3$, (d) $n = 4$, and (e) $n = 5$? Check your results to show that they agree with the general rule that the number of sets of quantum numbers is equal to $2n^2$.

16. The z component of the electron's spin magnetic moment is given by the Bohr magneton, $\mu_B = e\hbar/2m_e$. Show that μ_B has the numerical value 9.27×10^{-24} J/T $= 5.79 \times 10^{-5}$ eV/T.

17. (a) Find the mass density of a proton, picturing it as a solid sphere of radius 1.00×10^{-15} m. (b) Think of a classical

model of an electron as a solid sphere with the same density as the proton. Find its radius. (c) If this electron possesses spin angular momentum $I\omega = \hbar/2$ because of classical rotation about the z axis, determine the speed of a point on the equator of the electron, and (d) compare this speed to the speed of light.

18. All objects, large and small, behave quantum mechanically. (a) Estimate the quantum number ℓ for the Earth in its orbit about the Sun. (b) What energy change (in joules) would occur if the Earth made a transition to an adjacent allowed state?

19. Like the electron, the nucleus of an atom has spin angular momentum and a corresponding magnetic moment. The z-component of the spin magnetic moment for a nucleus is characterized by the *nuclear magneton* $\mu_n = e\hbar/2m_p$, where m_p is the proton mass. (a) Calculate the value of μ_n in J/T and in eV/T. (b) Determine the ratio μ_n/μ_B and comment on your result.

Section 29.6 The Exclusion Principle and the Periodic Table

20. (a) Write out the electronic configuration for the ground state of oxygen ($Z = 8$). (b) Write out the values for the set of quantum numbers n, ℓ, m_ℓ, and m_s for each electron in oxygen.

21. What would you predict for the electronic configuration of element 110?

22. Devise a table similar to that shown in Figure 29.13 for atoms containing 11 through 19 electrons. Use Hund's rule and educated guesswork.

23. (a) Scanning through Table 29.4 in order of increasing atomic number, note that the electrons fill the subshells in such a way that those subshells with the lowest values of $n + \ell$ are filled first. If two subshells have the same value of $n + \ell$, the one with the lower value of n is filled first. Using these two rules, write the order in which the subshells are filled through $n + \ell = 7$. (b) Predict the chemical valence for the elements that have atomic numbers 15, 47, and 86 and compare your predictions with the actual valences.

24. Which electronic configuration has the lesser energy and the greater number of unpaired spins, [Kr]$4d^95s^1$ or [Kr]$4d^{10}$? Identify the element and discuss Hund's rule in this case. (*Note:* The notation [Kr] represents the filled configuration for krypton.)

Section 29.7 Atomic Spectra: Visible and X-Ray

25. If you wish to produce 10.0-nm x-rays in the laboratory, what is the minimum voltage you must use in accelerating the electrons?

26. A tungsten target is struck by electrons that have been accelerated from rest through a 40.0-kV potential difference. Find the shortest wavelength of the radiation emitted.

27. Use the method illustrated in Example 29.7 to calculate the wavelength of the x-ray emitted from a molybdenum target ($Z = 42$) when an electron moves from the L shell ($n = 2$) to the K shell ($n = 1$).

28. The wavelength of characteristic x-rays corresponding to the K_β line is 0.152 nm. Determine the material in the target.

29. The K series of the discrete spectrum of tungsten contains wavelengths of 0.0185 nm, 0.0209 nm, and 0.0215 nm. The K-shell ionization energy is 69.5 keV. Determine the ionization energies of the L, M, and N shells. Sketch the transitions.

Section 29.8 Atomic Transitions

30. The familiar yellow light from a sodium-vapor street lamp results from the $3p \rightarrow 3s$ transition in ^{11}Na. Evaluate the wavelength of this light given that the energy difference $E_{3p} - E_{3s} = 2.10$ eV.

31. A ruby laser delivers a 10.0-ns pulse of 1.00 MW average power. If the photons have a wavelength of 694.3 nm, how many are contained in the pulse?

32. A Nd:YAG laser used in eye surgery emits a 3.00-mJ pulse in 1.00 ns, focused to a spot 30.0 μm in diameter on the retina. (a) Find (in SI units) the power per unit area at the retina. (This quantity is called the irradiance.) (b) What energy is delivered to an area of molecular size—say, a circular area 0.600 nm in diameter?

Additional Problems

33. Show that the average value of r for the 1s state of hydrogen has the value $3a_0/2$. (*Hint:* Use Eq. 29.7.)

34. (a) How much energy is required to cause an electron in hydrogen to move from the $n = 1$ state to the $n = 2$ state? (b) Suppose the electrons gain this energy by collisions among hydrogen atoms at a high temperature. At what temperature would the thermal energy $3k_BT/2$, where k_B is the Boltzmann constant, be large enough to excite the electrons?

35. If a muon (a negatively charged particle having a mass 206 times the electron's mass) is captured by a lead nucleus, $Z = 82$, the resulting system behaves like a one-electron atom. (a) What is the "Bohr radius" for a muon captured by a lead nucleus? (*Hint:* Follow the treatment in Section 11.5.) (b) Using Equation 29.2 with one factor of e replaced by Ze, calculate the ground state energy of a muon captured by a lead nucleus. (c) What is the transition energy for a muon descending from the $n = 2$ to the $n = 1$ level in a muonic lead atom?

36. Show that the wave function for an electron in the 2s state in hydrogen

$$\psi(r) = \frac{1}{4\sqrt{2\pi}}\left(\frac{1}{a_0}\right)^{3/2}\left(2 - \frac{r}{a_0}\right)e^{-r/2a_0}$$

satisfies the radially symmetric Schrödinger equation given in Problem 9.

37. A pulsed ruby laser emits light at 694.3 nm. For a 14.0-ps pulse containing 3.00 J of energy, find (a) the physical length of the pulse as it travels through space and (b) the number of photons in it. (c) If the beam has a circular cross section of 0.600 cm diameter, find the number of photons per cubic millimeter.

38. A pulsed laser emits light having wavelength λ. For a pulse of duration Δt having energy E, find (a) the physical length of the pulse as it travels through space and (b) the number of photons in it. (c) If the beam has a circular cross section having diameter d, find the number of photons per unit volume.

39. The number N of atoms in a particular state is called the population of that state. This number depends on the energy of that state and the temperature. In thermal equilibrium, the population of atoms in a state of energy E_n is given by a Boltzmann distribution expression:

$$N = N_g e^{-(E_n - E_g)/k_B T}$$

where T is the absolute temperature and N_g is the population of the ground state of energy E_g. (a) Find the equilibrium ratio of populations of the states E_3^* to E_2 for the laser in Figure 29.23, assuming $T = 27.0°C$. (b) Find the equilibrium ratio of the populations of the two states in a ruby laser that produces a light beam of wavelength 694.3 nm at 4.00 K.

40. The force on a magnetic moment μ_z in a nonuniform magnetic field B_z is given by $F_z = \mu_z(dB_z/dz)$. If a beam of silver atoms travels a horizontal distance of 1.00 m through such a field and each atom has a speed of 100 m/s, how strong must be the field gradient dB_z/dz in order to deflect the beam 1.00 mm?

41. (a) Find the most probable position for an electron in the 2s state of hydrogen. (*Hint:* Let $x = r/a_0$, find an equation for x, and show that $x = 5.236$ is a solution to this equation.) (b) Show that the wave function given by Equation 29.8 is normalized.

42. An electron in chromium moves from the $n = 2$ state to the $n = 1$ state without emitting a photon. Instead, the excess energy is transferred to an outer electron (one in the $n = 4$ state), which is then ejected by the atom. (This is called an Auger [pronounced "ohjay"] process, and the ejected electron is referred to as an Auger electron.) Use the Bohr theory to find the kinetic energy of the Auger electron.

43. Suppose the ionization energy of an atom is 4.10 eV. In the spectrum of this same atom, we observe emission lines with wavelengths 310 nm, 400 nm, and 1377.8 nm. Use this information to construct the energy-level diagram with the least number of levels. Assume the higher levels are closer together.

44. All atoms have the same size, to an order of magnitude. (a) To show this, estimate the diameters for aluminum,

molar atomic mass = 27.0 g/mol and density 2.70 g/cm^3, and uranium, molar atomic mass = 238 g/mol and density 18.9 g/cm^3. (b) What do the results imply about the wave functions for inner-shell electrons as we progress to higher and higher atomic mass atoms? (*Hint:* The molar volume is approximately $D^3 N_A$, where D is the atomic diameter and N_A is Avogadro's number.)

45. For hydrogen in the 1s state, what is the probability of finding the electron farther than $2.50a_0$ from the nucleus?

46. In interstellar space, atomic hydrogen produces the sharp spectral line called the 21-cm radiation, which astronomers find most helpful in detecting clouds of hydrogen between stars. This radiation is useful because interstellar dust that obscures visible wavelengths is transparent to these radio wavelengths. The radiation is not generated by an electron transition between energy states characterized by n. Instead, in the ground state ($n = 1$), the electron and proton spins may be parallel or antiparallel, with a resultant slight difference in these energy states. (a) Which condition has the higher energy? (b) The line is actually at 21.11 cm. What is the energy difference between the states? (c) The average lifetime in the excited state is about 10^7y. Calculate the associated uncertainty in energy of this excited energy level.

47. According to classical physics, an accelerated charge e radiates at a rate

$$\frac{dE}{dt} = -\frac{1}{6\pi\epsilon_0}\frac{e^2 a^2}{c^3}$$

(a) Show that an electron in a classical hydrogen atom (see Fig. 29.3) spirals into the nucleus at a rate

$$\frac{dr}{dt} = -\frac{e^4}{12\pi^2 \epsilon_0^2 r^2 m_e^2 c^3}$$

(b) Find the time it takes the electron to reach $r = 0$, starting from $r_0 = 2.00 \times 10^{-10}$ m.

48. When electron clouds overlap, a detailed calculation of the effective charge exerting electrical force on another electron may be made using quantum mechanics. For the case of the lithium atom, the effective charge on each inner electron is $-0.85e$. Use this to find (a) the effective charge on the nucleus as seen by the outer valence electron and (b) the ionization energy (compare this with 5.4 eV).

49. Light from a certain He–Ne laser has a power output of 1.00 mW and a cross-sectional area of 10.0 mm^2. The entire beam is incident on a metal target that requires 1.50 eV to remove an electron from its surface. (a) Perform a classical calculation to determine how long it takes one atom in the metal to absorb 1.50 eV from the incident beam. (*Hint:* Assume the face area of an atom is 1.00×10^{-20} m^2, and first calculate the energy incident on each atom per second.) (b) Compare the (wrong) answer obtained in (a) to the actual response time for photoelectric emission ($\sim 10^{-9}$ s), and discuss the reasons for the large discrepancy.

50. Use the energy levels for hydrogen to show that when the electron falls from state n to state $n - 1$, the frequency of the emitted light is

$$f = \left(\frac{2\pi^2 m_e k_e^2 e^4}{h^3 n^2} \right) \frac{(2n - 1)}{(n - 1)^2}$$

51. Calculate the frequency of the light that a hydrogen atom should emit according to classical theory. To do so, note that the frequency of revolution of the electron is $v/2\pi r$, where r is the radius of its Bohr orbit. Show that as n approaches infinity in the equation in Problem 50, the expression varies as $1/n^3$ and reduces to the classical frequency the atom is expected to emit. This is one example of the correspondence principle, which requires that the classical and quantum models agree for large values of n.

ANSWERS TO CONCEPTUAL PROBLEMS

1. If the energy of the hydrogen atom were proportional to n (or any power of n), then the energy would become infinite as n grew to infinity. But the energy of the atom is *inversely* proportional to n^2. Thus, as n grows to infinity, the energy of the atom approaches zero from the negative side. As a result, the maximum frequency of emitted radiation approaches a value determined by the difference in energy between zero and the energy of the ground state.

2. A magnetic field exerts a torque on a macroscopic magnetic moment, in the direction such that the magnetic moment tends to align with the field. This torque will also exist on the microscopic magnetic moment associated with the spin of an electron. According to quantum physics, however, the spin angular momentum cannot line up with the magnetic field. It will point in a direction such that the component of the angular momentum along the magnetic field is $\pm \hbar/2$.

3. The characteristic x-rays originate from transitions within the atoms of the target, such as an L shell electron making a transition to a vacancy in the K shell. This vacancy in the K shell is caused when an accelerated electron in the x-ray tube supplies energy to the K electron to eject it from the atom. If the energy of the bombarding electrons were to be increased, the K electron will be ejected from the atom with more remaining kinetic energy. But the energy difference between the K and L shell has not changed, so the emitted x-ray has exactly the same wavelength.

4. A continuous spectrum without characteristic x-rays is possible. At a low accelerating potential difference for the electron, the electron may not have enough energy to eject an electron from a target atom. As a result, there will be no characteristic x-rays. The change in speed of the electron as it enters the target will result in the continuous spectrum.

5. This transition is forbidden by the selection rules, because the orbital angular momentum quantum number changes by two units.

30

Nuclear Physics

In 1896, the year that marked the birth of nuclear physics, Antoine Henri Becquerel (1852–1908) discovered radioactivity in uranium compounds. A great deal of research followed as scientists attempted to understand the radiation emitted by radioactive nuclei. Pioneering work by Rutherford showed that the radiation was of three types, which he called alpha, beta, and gamma rays. These types are classified according to the natures of their electric charges and according to their ability to penetrate matter and ionize air. Later experiments showed that alpha rays are helium nuclei, beta rays are electrons, and gamma rays are high-energy photons.

As we saw in Chapter 29, Section 29.1, the 1911 experiments of Rutherford and his students Geiger and Marsden established that the nucleus of an atom could be regarded as essentially a point mass and point charge and that most of the atomic mass is contained in the nucleus. Furthermore,

Bubble chamber photograph of electron (green) and positron (red) tracks produced by energetic gamma rays. The highly curved tracks at the top are due to an electron–positron pair that bend in opposite direction in the magnetic field. The lower tracks are produced by more energetic electrons and a positron. *(Lawrence Berkeley Laboratory/Science Photo Library, Photo Researchers, Inc.)*

such studies demonstrated a new type of force, the **nuclear force,** which is predominant at distances of less than about 10^{-14} m and zero at great distances. That is, the nuclear force is a short-range force.

Other milestones in the development of nuclear physics include

- The observation of nuclear reactions by Cockroft and Walton in 1930,
- The discovery of the neutron by Chadwick in 1932,
- The discovery of artificial radioactivity by Joliot and Irene Curie in 1933,
- The discovery of nuclear fission by Hahn and Strassman in 1938,
- The development of the first controlled fission reactor by Fermi and his collaborators in 1942.

In this chapter we discuss the properties and structure of the atomic nucleus. We start by describing the basic properties of nuclei, then discuss nuclear forces and binding energy, nuclear models, and the phenomenon of radioactivity. We also discuss nuclear reactions and the processes by which nuclei decay.

30.1 • SOME PROPERTIES OF NUCLEI

All nuclei are composed of two types of particles: protons and neutrons. The only exception is the ordinary hydrogen nucleus, which is a single proton. In describing the atomic nucleus, we must talk about the following quantities:

- The **atomic number, Z,** which equals the number of protons in the nucleus (the atomic number is sometimes called the charge number),
- The **neutron number, N,** which equals the number of neutrons in the nucleus,
- The **mass number, A,** which equals the number of nucleons (neutrons plus protons) in the nucleus.

In representing nuclei, it is convenient to have a symbolic way that shows how many protons and neutrons are present. The symbol used is $^A_Z X$, where X represents the chemical symbol for the element. For example, $^{56}_{26} Fe$ (iron) has a mass number of 56 and an atomic number of 26; therefore, it contains 26 protons and 30 neutrons. When no confusion is likely to arise, we omit the subscript Z because the chemical symbol can always be used to determine Z.

The nuclei of all atoms of a particular element contain the same number of protons but often contain different numbers of neutrons. Nuclei that are related in this way are called **isotopes. The isotopes of an element have the same Z value but different N and A values.** The natural abundances of isotopes can differ substantially. For example, $^{11}_6 C$, $^{12}_6 C$, $^{13}_6 C$, and $^{14}_6 C$ are four isotopes of carbon. The natural abundance of the $^{12}_6 C$ isotope is about 98.9%, whereas that of the $^{13}_6 C$ isotope is only about 1.1%. Even the simplest element, hydrogen, has isotopes: $^1_1 H$, the ordinary hydrogen nucleus; $^2_1 H$, deuterium; and $^3_1 H$, tritium. Some isotopes do not occur naturally but can be produced in the laboratory through nuclear reactions.

Ernest Rutherford (1871–1937)

Rutherford, a New Zealander, was awarded the Nobel prize in 1908 for his studies of radioactivity. When he discovered that atoms can be broken apart by alpha rays, he remarked, "On consideration, I realized that this scattering backward must be the result of a single collision, and when I made calculations I saw that it was impossible to get anything of that order of magnitude unless you took a system in which the greater part of the mass of the atom was concentrated in a minute nucleus. It was then that I had the idea of an atom with a minute massive center carrying a charge." *(Photo courtesy of AIP Niels Bohr Library)*

TABLE 30.1   Masses of the Proton, Neutron, Hydrogen Atom, and Electron in Various Units

	Mass		
Particle	kg	u	MeV/c^2
Proton	1.6726×10^{-27}	1.007 276	938.28
Neutron	1.6750×10^{-27}	1.008 665	939.57
Hydrogen atom	1.6735×10^{-27}	1.007 825	938.78
Electron	9.109×10^{-31}	5.486×10^{-4}	0.511

Charge and Mass

The proton carries a single positive charge, $+ e$, the electron carries a single negative charge, $- e$, where $e = 1.6 \times 10^{-19}$ C, and the neutron is electrically neutral. Because the neutron has no charge, it is difficult to detect.

In Chapter 1, we defined the atomic mass unit, u, in such a way that the mass of the isotope ^{12}C is exactly 12 u, where 1 u = 1.660 540 $\times 10^{-27}$ kg. The proton and neutron each have a mass of approximately 1 u, and the electron has a mass that is only a small fraction of an atomic mass unit:

$$\text{Mass of proton} = 1.007\ 276 \text{ u}$$

$$\text{Mass of neutron} = 1.008\ 665 \text{ u}$$

$$\text{Mass of electron} = 0.000\ 5486 \text{ u}$$

Because the rest energy of a particle is given by $E_R = mc^2$ (Chapter 9), it is often convenient to express the atomic mass unit in terms of its rest energy equivalent. For one atomic mass unit, we have

$$E_R = mc^2 = (1.660\ 540 \times 10^{-27} \text{ kg})(2.997\ 92 \times 10^8 \text{ m/s})^2$$
$$= 931.494 \text{ MeV}$$

Nuclear physicists often express mass in terms of the unit MeV/c^2, where

$$1 \text{ u} \equiv 931.494 \text{ MeV}/c^2$$

The masses of several simple particles are given in Table 30.1. The masses and some other properties of selected isotopes are provided in Table 30.4 and Appendix A.3.

The Size of Nuclei

The size and structure of nuclei were first investigated in the scattering experiments of Rutherford, discussed in Section 29.1. Using the principle of conservation of energy, Rutherford found an expression for how close an alpha particle moving directly toward the nucleus can come to the nucleus before being turned around by Coulomb repulsion.

In such a head-on collision, the kinetic energy of the incoming alpha particle must be converted completely to electrical potential energy when the particle stops at the point of closest approach and turns around (Fig. 30.1). If we equate the initial kinetic energy of the alpha particle to the maximum electrical potential energy of the system (alpha particle plus target nucleus), we have

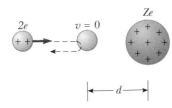

Figure 30.1 An alpha particle on a head-on collision course with a nucleus of charge *Ze*. Because of the Coulomb repulsion between the like charges, the alpha particle stops instantaneously at a distance *d* from the target nucleus, called the distance of closest approach.

$$\tfrac{1}{2}mv^2 = k_e \frac{q_1 q_2}{r} = k_e \frac{(2e)(Ze)}{d}$$

where d is the distance of closest approach. Solving for d, we get

$$d = \frac{4k_e Z e^2}{mv^2}$$

From this expression, Rutherford found that alpha particles approached to within 3.2×10^{-14} m of a nucleus when the foil was made of gold. Thus, the radius of the gold nucleus must be less than this value. For silver atoms, the distance of closest approach was found to be 2×10^{-14} m. From these results, Rutherford concluded that the positive charge in an atom is concentrated in a small sphere, which he called the nucleus, with a radius of no greater than about 10^{-14} m. Because such small lengths are common in nuclear physics, a convenient unit of length is the *femtometer* (fm), sometimes called the **fermi,** defined as

$$1 \text{ fm} \equiv 10^{-15} \text{ m}$$

Since the time of Rutherford's scattering experiments, a multitude of other experiments have shown that most nuclei are approximately spherical and have an average radius of

$$r = r_0 A^{1/3} \qquad\qquad \text{[30.1]} \qquad \bullet \; \textit{Radius of a nucleus}$$

where A is the mass number and r_0 is a constant equal to 1.2×10^{-15} m. Because the volume of a sphere is proportional to the cube of the radius, it follows from Equation 30.1 that the volume of a nucleus (assumed to be spherical) is directly proportional to A, the total number of nucleons. This suggests that **all nuclei have nearly the same density.** Nucleons combine to form a nucleus as though they were tightly packed spheres (Fig. 30.2).

Figure 30.2 A nucleus can be visualized as a cluster of tightly packed spheres, where each sphere is a nucleon.

Example 30.1 Nuclear Volume and Density

Find (a) an approximate expression for the mass of a nucleus of mass number A, (b) an expression for the volume of this nucleus in terms of the mass number, and (c) a numerical value for its density.

Solution (a) The mass of the proton is approximately equal to that of the neutron. Thus, if the mass of one of these particles is m, the mass of the nucleus is approximately Am.

(b) Assuming the nucleus is spherical and using Equation 30.1, we find that the volume is

$$V = \tfrac{4}{3}\pi r^3 = \tfrac{4}{3}\pi r_0^3 A$$

(c) The nuclear density is

$$\rho_n = \frac{\text{mass}}{\text{volume}} = \frac{Am}{\tfrac{4}{3}\pi r_0^3 A} = \frac{3m}{4\pi r_0^3}$$

$$= \frac{3(1.67 \times 10^{-27} \text{ kg})}{4\pi(1.2 \times 10^{-15} \text{ m})^3} = 2.3 \times 10^{17} \text{ kg/m}^3$$

Note that the nuclear density is about 2.3×10^{14} times greater than the density of water (10^3 kg/m^3)!

Nuclear Stability

Because the nucleus consists of a closely packed collection of protons and neutrons, you might be surprised that it can exist. The very large repulsive electrostatic forces between protons in close proximity should cause the nucleus to fly apart. However, nuclei are stable because of the presence of another force, the **nuclear force.** This force, which is short-range (about 2 fm), is an attractive force that acts between all nuclear particles. The protons attract each other via the nuclear force, and at the same time they repel each other through the Coulomb force. The nuclear force also acts between pairs of neutrons and between neutrons and protons.

The nuclear force dominates the Coulomb repulsive force within the nucleus (at short ranges). If this were not the case, stable nuclei would not exist. Moreover, the strong nuclear force is independent of charge. In other words, the forces associated with the proton–proton, proton–neutron, and neutron–neutron interactions are the same, apart from the additional repulsive Coulomb force for the proton–proton interaction.

There are about 260 stable nuclei; hundreds of others have been observed but are unstable. A plot of N versus Z for a number of stable nuclei is given in Figure 30.3. Note that light nuclei are most stable if they contain equal numbers of protons and neutrons—that is, if $N = Z$—but heavy nuclei are more stable if $N > Z$. This can be partially understood by recognizing that as the number of protons increases, the strength of the Coulomb force increases, which tends to break the nucleus apart. As a result, more neutrons are needed to keep the nucleus stable, because neutrons experience only the attractive nuclear forces. Eventually, when $Z = 83$, the repulsive forces between protons cannot be compensated by the addition of more neutrons. In effect, the additional neutrons "dilute" the nuclear charge density. Elements that contain more than 83 protons do not have stable nuclei.

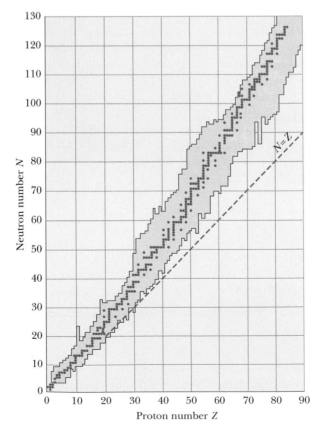

Figure 30.3 A plot of neutron number, N, versus atomic number, Z, for the stable nuclei (solid points). The dashed line corresponds to the condition $N = Z$. The shaded area shows radioactive nuclei.

It is interesting to note that most stable nuclei have even values of A. In fact, certain values of Z and N correspond to nuclei with unusually high stability. These values of N and Z, called **magic numbers,** are

$$Z \text{ or } N = 2, 8, 20, 28, 50, 82, 126 \qquad \textbf{[30.2]}$$

For example, the helium nucleus (two protons and two neutrons), which has $Z = 2$ and $N = 2$, is very stable.

Nuclear Spin and Magnetic Moment

In Chapter 29 we discussed the fact that an electron has an intrinsic angular momentum, which is called its spin. Nuclei, like electrons, also have an intrinsic angular momentum that arises from relativistic properties. The magnitude of the **nuclear angular momentum** is $\sqrt{I(I + 1)}\hbar$, where I is a quantum number called the **nuclear spin** and may be an integer or a half-integer. The maximum component of the nuclear angular momentum projected along any direction is $I\hbar$. Figure 30.4 illustrates the possible orientations of the nuclear spin and its projections along the z axis for the case where $I = \frac{3}{2}$.

The nuclear angular momentum has a nuclear magnetic moment associated with it. The magnetic moment of a nucleus is measured in terms of the **nuclear magneton,** μ_n, a unit of magnetic moment defined as

$$\mu_n \equiv \frac{e\hbar}{2m_p} = 5.05 \times 10^{-27} \text{ J/T} \qquad \textbf{[30.3]}$$

This definition is analogous to Equation 22.28 for the Bohr magneton, μ_B. Note that μ_n is smaller than μ_B by a factor of about 2000, due to the large difference in masses of the proton and electron.

The magnetic moment of a free proton is not μ_n but 2.7928 μ_n. Unfortunately, there is no general theory of nuclear magnetism that explains this value. Another surprising point is the fact that a neutron also has a magnetic moment, which has a value of 1.9135 μ_n. The minus sign indicates that the neutron's magnetic moment is opposite its spin angular momentum.

Nuclear Magnetic Resonance and MRI

It is interesting that nuclear magnetic moments (as well as electronic magnetic moments) precess in an external magnetic field. The frequency at which they precess, called the **Larmor precessional frequency,** ω_p, is directly proportional to the magnetic field. This is described schematically in Figure 30.5a, where the magnetic field is along the z axis. For example, the Larmor frequency of a proton in a magnetic field of 1 T is equal to 42.577 MHz. The potential energy of a magnetic dipole moment in an external magnetic field is $-\boldsymbol{\mu} \cdot \mathbf{B}$. When the projection of $\boldsymbol{\mu}$ is along the field, the potential energy of the dipole moment is $-\mu B$; that is, it has its minimum value. When the projection of $\boldsymbol{\mu}$ is against the field, the potential energy is μB and it has its maximum value. These two energy states for a nucleus with a spin of $\frac{1}{2}$ are shown in Figure 30.5b.

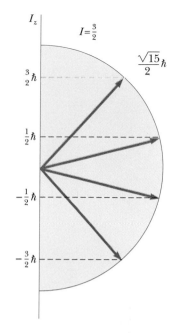

Figure 30.4   The possible orientations of the nuclear spin and its projections along the z axis for the case $I = \frac{3}{2}$.

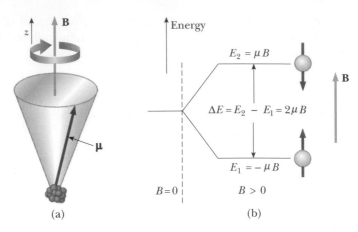

$E_2 = \mu B$

$\Delta E = E_2 - E_1 = 2\mu B$

$E_1 = -\mu B$

$B = 0$ $B > 0$

(a)   (b)

Figure 30.5   (a) When a nucleus is placed in an external magnetic field, the nuclear magnetic moment precesses about the magnetic field with a frequency that is proportional to the field. (b) A proton, the spin of which is $\frac{1}{2}$, can occupy one of two energy states when placed in an external magnetic field. The lower energy state, E_1, corresponds to the case in which the spin is aligned with the field, and the higher energy state, E_2, corresponds to the case in which the spin is opposite the field. The reverse is true for electrons.

Nuclear magnetic resonance •

It is possible to observe transitions between these two spin states by using a technique known as **nuclear magnetic resonance.** A dc magnetic field is introduced to align the magnetic moments (Fig. 30.5a), along with a second, weak, oscillating magnetic field oriented perpendicular to **B**. When the frequency of the oscillating field is adjusted to match the Larmor precessional frequency, a torque acting on the precessing moments causes them to "flip" between the two spin states. These transitions result in a net absorption of energy by the spin system. A diagram of the apparatus used to detect a nuclear magnetic resonance signal is illustrated in Figure 30.6. The absorbed energy is supplied by the generator producing the oscillating magnetic field. Nuclear magnetic resonance and a related technique called electron spin resonance are extremely important methods for studying nuclear and atomic systems and how these systems interact with their surroundings. A typical NMR spectrum is shown in Figure 30.7.

A widely used diagnostic technique called **MRI,** for **magnetic resonance imaging,** is based on nuclear magnetic resonance. In MRI, the patient is placed inside a large solenoid that supplies a spatially varying magnetic field. Because of the

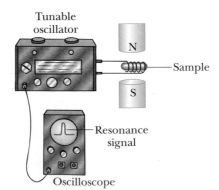

Tunable oscillator

N

Sample

S

Resonance signal

Oscilloscope

Figure 30.6   An experimental arrangement for nuclear magnetic resonance. The radio-frequency magnetic field of the coil, provided by the variable-frequency oscillator, must be perpendicular to the dc magnetic field. When the nuclei in the sample meet the resonance condition, the spins absorb energy from the field of the coil, which changes the response of the circuit in which the coil is included. Most modern NMR spectrometers use superconducting magnets at fixed field strengths and operate at frequencies of approximately 200 MHz.

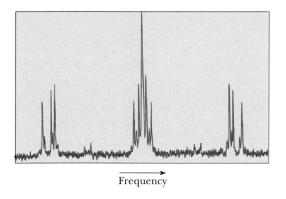

Figure 30.7 An NMR spectrum of ^{31}P in a bridged metallic complex containing platinum. The lines that flank the central strong peak are due to the interaction between ^{31}P and other distant ^{31}P nuclei. The outermost set of lines is due to the interaction between ^{31}P and neighboring platinum nuclei. The spectrum was recorded at a fixed field of about 4 T, and the mean frequency was 200 MHz.

Frequency

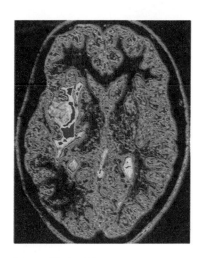

Figure 30.8 A computer-digitized, color-enhanced MRI of a brain with a glioma tumor. (© *Scott Camazine/Science Source*)

gradient in the magnetic field, protons in different parts of the body precess at different frequencies, so that the resonance signal can be used to provide information on the positions of the protons. A computer is used to analyze the position information to provide data for constructing a final image. An MRI scan taken on a human head is shown in Figure 30.8. The main advantage of MRI over other imaging techniques in medical diagnostics is that it causes minimal damage to cellular structures. Photons associated with the radio-frequency signals used in MRI have energies of only about 10^{-7} eV. Because molecular bond strengths are much larger (approximately 1 eV), the radio-frequency radiation causes little cellular damage. In comparison, x-rays or γ-rays have energies ranging from 10^4 to 10^6 eV and can cause considerable cellular damage.

30.2 · BINDING ENERGY

The total mass of a nucleus is always less than the sum of the masses of its nucleons. Because mass is another manifestation of energy, **the total energy of the bound system (the nucleus) is less than the combined energy of the separated nucleons.** This difference in energy is called the **binding energy** of the nucleus and can be thought of as the energy that must be added to a nucleus to break it apart into its components. Therefore, in order to separate a nucleus into protons and neutrons, energy must be put into the system.

Conservation of energy and the Einstein mass–energy equivalence relationship show that the binding energy of any nucleus ^A_ZX is

$$E_b(\text{MeV}) = [ZM(\text{H}) + Nm_n - M(^A_ZX)] \times 931.494 \text{ MeV/u} \qquad [30.4]$$

• *Binding energy of a nucleus*

where $M(\text{H})$ is the atomic mass of the neutral hydrogen atom, $M(^A_ZX)$ represents the atomic mass of the element (^A_ZX), m_n is the mass of the neutron, and the masses are all in atomic mass units. Note that the mass of the Z electrons included in the first term in Equation 30.4 cancels with the mass of the Z electrons included in the term $M(^A_ZX)$ within a small difference associated with the atomic binding energy of the electrons. Because atomic binding energies are typically several eV and nuclear binding energies are several MeV, this difference is negligible.

Example 30.2   The Binding Energy of the Deuteron

Calculate the binding energy of the deuteron (the nucleus of a deuterium atom), which consists of a proton and a neutron, given that the atomic mass of deuterium is 2.014 102 u.

Solution   From Table 30.1, we see that masses of a hydrogen atom and a neutron are $M(H) = 1.007\ 825$ u, and $m_n = 1.008\ 665$ u. Therefore,

$$M(H) + m_n = 2.016\ 490\ \text{u}$$

To calculate the mass difference, we subtract the deuterium mass from this value:

$$\Delta m = [M(H) + m_n] - M(D)$$
$$= 2.016\ 490\ \text{u} - 2.014\ 102\ \text{u}$$
$$= 0.002\ 388\ \text{u}$$

Using Equation 30.4, we find that the binding energy is

$$E_b = \Delta mc^2 = (0.002\ 388\ \text{u})(931.494\ \text{MeV/u}) = \boxed{2.224\ \text{MeV}}$$

This result tells us that, to separate a deuteron into its constituent proton and neutron, it is necessary to add 2.224 MeV of energy to the deuteron. One way of supplying the deuteron with this energy is by bombarding it with energetic particles.

If the binding energy of a nucleus were zero, the nucleus would separate into its constituent protons and neutrons without the addition of any energy; that is, it would spontaneously break apart.

A plot of binding energy per nucleon, E_b/A, as a function of mass number for various stable nuclei is shown in Figure 30.9. Except for the lighter nuclei, the average binding energy per nucleon is about 8 MeV. For the deuteron, the average binding energy per nucleon is $E_b/A = (2.224\ \text{MeV})/2 = 1.112$ MeV. Note that the curve in Figure 30.9 peaks in the vicinity of $A = 60$. That is, nuclei with mass numbers greater or less than 60 are not as strongly bound as those near the middle of the periodic table. The higher values of binding energy near $A = 60$ imply that energy is released when a heavy nucleus with $A \approx 200$ splits or fissions into several lighter nuclei which lie near $A = 60$. Energy is released in fission because the final state consisting of two lighter fragments is more tightly bound or lower in energy

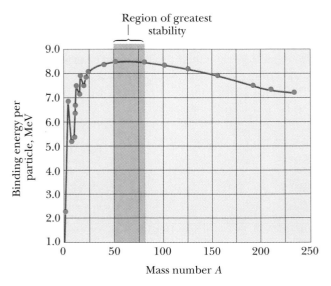

Figure 30.9   A plot of binding energy per nucleon versus mass number for the stable nuclei in Figure 30.3.

than the original nucleus. In a similar way, energy can be released when two light nuclei with $A \leq 20$ combine or fuse to form one heavier nucleus.

Another important feature of Figure 30.9 is that the binding energy per nucleon is approximately constant for $A > 20$. In this case the nuclear forces between a particular nucleon and all the other nucleons in the nucleus are said to be saturated; that is, a particular nucleon forms attractive bonds with only a limited number of other nucleons. Because of the short-range character of the nuclear force, these other nucleons can be viewed as being the nearest neighbors in the close packed structure shown in Figure 30.2.

The general features of the nuclear force responsible for the binding energy of nuclei have been revealed in a wide variety of experiments and are as follows.

- The attractive nuclear force is the strongest force in nature.
- The nuclear force is a short-range force that rapidly falls to zero when the separation between nucleons exceeds several fermis. Evidence for the limited range of nuclear forces comes from scattering experiments and from the saturation of nuclear forces previously mentioned. The short range of the nuclear force is shown in the neutron–proton (n–p) potential energy plot of Figure 30.10a obtained by scattering neutrons from a target containing hydrogen. The depth of the n–p potential energy curve is 40 MeV to 50 MeV and contains a strong repulsive component that prevents the nucleons from approaching much closer than 0.4 fm. Another interesting feature of the nuclear force is that its size depends on the relative spin orientations of the nucleons, as shown by scattering experiments using spin polarized beams and targets.

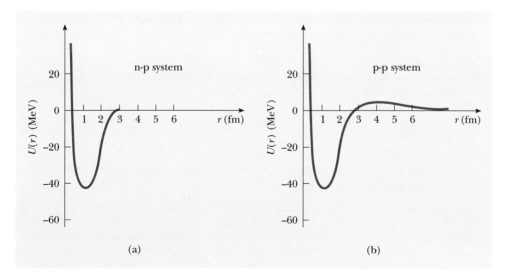

(a)

(b)

Figure 30.10 (a) Potential energy versus separation for the neutron–proton system. (b) Potential energy versus separation for the proton–proton system. The difference in the two curves is due mainly to the large Coulomb repulsion in the case of the proton–proton interaction.

- From n–n, n–p, and p–p scattering experiments and other indirect evidence, it is found that the nuclear force is independent of the electric charge of the interacting nucleons. As might be expected from this "charge-blind" character of the nuclear force, the nuclear force does not affect electrons, enabling energetic electrons to serve as point-like probes of the charge density of nuclei. The charge independence of the nuclear force also means that the main difference between the n–p and p–p interactions is that the p–p potential energy consists of a *superposition* of nuclear and Coulomb interactions as shown in Figure 30.10b. At distances less than 2 fm, both p–p and n–p potential energies are nearly identical, but for distances greater than this, the p–p potential has a positive energy barrier with a maximum of about 1 MeV at 4 fm.

Thinking Physics 1

Figure 30.9 shows a graph of the amount of energy necessary to remove a nucleon from the nucleus. Figure 29.15 in the previous chapter shows the energy necessary to remove an electron from an atom. Why does Figure 30.9 show an *approximately constant* amount of energy necessary to remove a nucleon (above about $A = 56$), whereas Figure 29.15 shows *widely varying* amounts of energy necessary to remove an electron from the atom?

Reasoning In the case of Figure 30.9, the approximately constant value of the nuclear binding energy is a result of the short-range nature of the nuclear strong force. A given nucleon interacts only with its few nearest neighbors, rather than with all of the nucleons in the nucleus. Thus, no matter how many nucleons are present in the nucleus, pulling one nucleon out involves separating it only from its nearest neighbors. The energy to do this, therefore, is approximately independent of how many nucleons are present.

However, the electrical force holding the electrons to the nucleus in an atom is a long-range force. An electron in the atom interacts with all of the protons in the nucleus. When the nuclear charge increases, there is a stronger attraction between the nucleus and the electrons. As a result, as the nuclear charge increases, more energy is necessary to remove an electron. This is demonstrated by the upward tendency of the ionization energy in Figure 29.15 for each period.

Thinking Physics 2

When a heavy nucleus undergoes fission into just two lighter nuclei, the product nuclei are invariably very unstable. Why is this?

Reasoning According to Figure 30.3, the ratio of the number of neutrons to the number of protons increases with Z. As a result, when a heavy nucleus splits in a fission reaction to two lighter nuclei, the lighter nuclei tend to have too many neutrons. This leads to instability, as the nucleus returns to the curve in Figure 30.3 by decay processes that reduce the number of neutrons.

CONCEPTUAL PROBLEM 1

Isotopes of a given element can have different masses, different magnetic moments, but the same chemical behavior. Why is this?

30.3 · RADIOACTIVITY

In 1896 Henri Becquerel accidentally discovered that uranyl potassium sulfate crystals emitted an invisible radiation that could darken a photographic plate when the plate was covered to exclude light. After a series of experiments, he concluded that the radiation emitted by the crystals was of a new type, one that required no external stimulation and was so penetrating that it could darken protected photographic plates and ionize gases. This spontaneous emission of radiation was soon to be called **radioactivity.** Subsequent experiments by other scientists showed that other substances were also radioactive.

The most significant investigations of this type were conducted by Marie Curie (1867–1934) and Pierre Curie (1859–1906). After several years of careful and laborious chemical separation processes on tons of pitchblende, a radioactive ore, the Curies reported the discovery of two previously unknown radioactive elements. These were named polonium and radium. Subsequent experiments, including Rutherford's famous work on alpha-particle scattering, suggested that radioactivity was the result of the decay, or disintegration, of unstable nuclei.

Three types of radiation can be emitted by a radioactive substance: alpha (α) rays, where the emitted particles are ^{4}He nuclei; beta (β) rays, in which the emitted particles are either electrons or positrons; and gamma (γ) rays, in which the emitted rays are high-energy photons. A **positron** is a particle similar to the electron in all respects except that it has a charge of $+e$. (The positron is said to be the **antiparticle** of the electron.) The symbol e$^-$ is used to designate an electron and e$^+$ designates a positron.

It is possible to distinguish these three forms of radiation using the scheme illustrated in Figure 30.11. The radiation from a radioactive sample is directed into a region in which there is a magnetic field. The beam splits into three components, two bending in opposite directions and the third experiencing no change in direction. From this simple observation, one can conclude that the radiation of the undeflected beam carries no charge (the gamma ray), the component deflected upward corresponds to positively charged particles (alpha particles), and the component deflected downward corresponds to negatively charged particles (e$^-$). If the beam includes a positron (e$^+$), it is deflected upward.

Marie Curie (1867–1934)

A Polish scientist, Marie Curie shared the Nobel prize in 1903 with her husband, Pierre, and with Becquerel for their work on spontaneous radioactivity and the radiation emitted by radioactive substances. "I persist in believing that the ideas that then guided us are the only ones which can lead to the true social progress. We cannot hope to build a better world without improving the individual. Toward this end, each of us must work toward his own highest development, accepting at the same time his share of responsibility in the general life of humanity."

Figure 30.11   The radiation from a radioactive source can be separated into three components by a magnetic field. The photographic plate at the right records the events. The gamma ray is not deflected by the magnetic field.

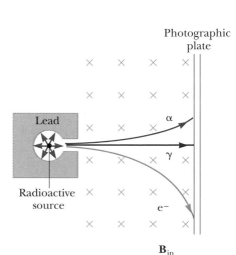

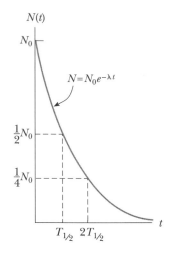

$N(t)$

N_0

$N = N_0 e^{-\lambda t}$

$\frac{1}{2}N_0$

$\frac{1}{4}N_0$

$T_{1/2}$ $2T_{1/2}$ t

Figure 30.12   A plot of the exponential decay law for radioactive nuclei. The vertical axis represents the number of radioactive nuclei present at any time t, and the horizontal axis is time. The time $T_{1/2}$ is the half-life of the sample.

The three types of radiation have quite different penetrating powers. Alpha particles barely penetrate a sheet of paper, beta particles can penetrate a few millimeters of aluminum, and gamma rays can penetrate several centimeters of lead.

The rate at which a particular decay process occurs in a radioactive sample is proportional to the number of radioactive nuclei present (that is, those nuclei that have not yet decayed). If N is the number of radioactive nuclei present at some instant, the rate of change of N is

$$\frac{dN}{dt} = -\lambda N \qquad \text{[30.5]}$$

where λ is called either the **decay constant** or the **disintegration constant.** The minus sign indicates that dN/dt is negative; that is, N is decreasing in time.

If we write Equation 30.5 in the form

$$\frac{dN}{N} = -\lambda \, dt$$

we can integrate

$$\int_{N_0}^{N} \frac{dN}{N} = -\lambda \int_0^t dt$$

$$\ln\left(\frac{N}{N_0}\right) = -\lambda t$$

Exponential decay •

$$N = N_0 e^{-\lambda t} \qquad \text{[30.6]}$$

The constant N_0 represents the number of radioactive nuclei at $t = 0$ and e is the base of the natural logarithm.

The **decay rate,** $R(= |dN/dt|)$, can be obtained by differentiating Equation 30.6 with respect to time:

Decay rate •

$$R = \left|\frac{dN}{dt}\right| = N_0 \lambda e^{-\lambda t} = R_0 e^{-\lambda t} \qquad \text{[30.7]}$$

where $R = N\lambda$ and $R_0 = N_0\lambda$ is the decay rate at $t = 0$. The decay rate of a sample is often referred to as its **activity.** Note that both N and R decrease exponentially with time. The plot of N versus t in Figure 30.12 illustrates the exponential decay law.

Another parameter that is useful for characterizing radioactive decay is the half-life, $T_{1/2}$. **The half-life of a radioactive substance is the time it takes half of a given number of radioactive nuclei to decay.** Setting $N = N_0/2$ and $t = T_{1/2}$ in Equation 30.6 gives

$$\frac{N_0}{2} = N_0 e^{-\lambda T_{1/2}}$$

Writing this in the form $e^{\lambda T_{1/2}} = 2$ and taking the natural logarithm of both sides, we get

$$T_{1/2} = \frac{\ln 2}{\lambda} = \frac{0.693}{\lambda}$$ [30.8] • *Half-life equation*

This is a convenient expression relating the half-life to the decay constant. Note that after an elapsed time of one half-life, $N_0/2$ radioactive nuclei remain (by definition); after two half-lives, half of these will have decayed and $N_0/4$ radioactive nuclei will be left; after three half-lives, $N_0/8$ will be left; and so on. In general, after n half-lives, the number of radioactive nuclei remaining is $N_0/2^n$. Thus, we see that nuclear decay is independent of the past history of the sample.

Activity can be measured with a unit called the **curie (Ci),** defined as

$$1 \text{ Ci} \equiv 3.7 \times 10^{10} \text{ decays/s}$$ • *The curie*

This unit was selected as the original activity unit because it is the approximate activity of 1 g of radium. The SI unit of activity is called the **becquerel (Bq):**

$$1 \text{ Bq} \equiv 1 \text{ decay/s}$$ • *The becquerel*

Therefore, 1 Ci = 3.7×10^{10} Bq. The most commonly used units of activity are millicuries and microcuries.

Thinking Physics 3

The isotope $^{14}_{6}\text{C}$ is radioactive and has a half-life of 5730 years. If you start with a sample of 1000 carbon-14 nuclei, how many will still be around in 17 190 years?

Reasoning In 5730 years, half the sample will have decayed, leaving 500 radioactive $^{14}_{6}\text{C}$ nuclei. In another 5730 years (for a total elapsed time of 11 460 years), the number will be reduced to 250 nuclei. After another 5730 years (total time 17 190 years), 125 remain.

These numbers represent ideal circumstances. Radioactive decay is an averaging process over a very large number of atoms, and the actual outcome depends on statistics. Our original sample in this example contained only 1000 nuclei, certainly not a very large number. Thus, if we were actually to count the number remaining after one half-life for this small sample, it probably would not be exactly 500.

Example 30.3 The Activity of Radium

The half-life of the radioactive nucleus $^{226}_{86}\text{Ra}$ is 1.6×10^3 years. If a sample contains 3×10^{16} such nuclei, determine the activity at this initial time $t = 0$.

Solution First, let us convert the half-life to seconds:

$$T_{1/2} = (1.6 \times 10^3 \text{ years})(3.16 \times 10^7 \text{ s/year}) = 5.0 \times 10^{10} \text{ s}$$

Now we can use this value in Equation 30.8 to get the decay constant:

$$\lambda = \frac{0.693}{T_{1/2}} = \frac{0.693}{5.0 \times 10^{10} \text{ s}} = 1.4 \times 10^{-11} \text{ s}^{-1}$$

We can calculate the activity of the sample at $t = 0$ using $R_0 = \lambda N_0$, where R_0 is the decay rate at $t = 0$ and N_0 is the number of radioactive nuclei present at $t = 0$:

$$R_0 = \lambda N_0 = (1.4 \times 10^{-11} \text{ s}^{-1})(3 \times 10^{16})$$
$$= 4.1 \times 10^5 \text{ decays/s}$$

Because 1 Ci = 3.7×10^{10} decays/s, the activity, or decay rate, at $t = 0$ is

$$R_0 = 11.1 \text{ } \mu\text{Ci}$$

The hands and numbers of this luminous watch contain minute amounts of radium salt. The radioactive decay of radium causes the watch to glow in the dark.   *(© Richard Megna 1990, Fundamental Photographs)*

Example 30.4   The Activity of Carbon

A radioactive sample contains 3.50 μg of pure $^{11}_{6}$C, which has a half-life of 20.4 min. (a) Determine the number of nuclei present initially.

Solution   The atomic mass of $^{11}_{6}$C is approximately 11.0, and therefore 11.0 g contains Avogadro's number of nuclei. Hence, 3.50 μg contains N nuclei, where

$$\frac{N}{6.02 \times 10^{23} \text{ nuclei/mol}} = \frac{3.50 \times 10^{-6} \text{ g}}{11.0 \text{ g/mol}}$$

$$N = 1.92 \times 10^{17} \text{ nuclei}$$

(b) What is the activity of the sample initially and after 8.00 h?

Solution   Because $T_{1/2} = 20.4$ min $= 1224$ s, the decay constant is

$$\lambda = \frac{0.693}{T_{1/2}} = \frac{0.693}{1224 \text{ s}} = 5.66 \times 10^{-4} \text{ s}^{-1}$$

Therefore, the initial activity of the sample is

$$R_0 = \lambda N_0 = (5.66 \times 10^{-4} \text{ s}^{-1})(1.92 \times 10^{17})$$

$$= 1.08 \times 10^{14} \text{ decays/s}$$

We can use Equation 30.7 to find the activity at any time t. For $t = 8.00$ h $= 2.88 \times 10^4$ s, we see that $\lambda t = 16.3$, and so

$$R = R_0 e^{-\lambda t} = (1.09 \times 10^{14} \text{ decays/s}) e^{-16.3}$$

$$= 8.96 \times 10^6 \text{ decays/s}$$

EXERCISE 1   Calculate the number of radioactive nuclei remaining after 8.00 h.   Answer $N = 1.58 \times 10^{10}$ nuclei

30.4 • THE DECAY PROCESSES

As we stated in the preceding section, a radioactive nucleus spontaneously decays via alpha, beta, or gamma decay. Let us discuss these three processes in more detail.

Alpha Decay

If a nucleus emits an alpha particle ($^4_2$He), it loses two protons and two neutrons. Therefore, N decreases by 2, Z decreases by 2, and A decreases by 4. The decay can be written symbolically as

Alpha decay •

$$^A_Z X \longrightarrow ^{A-4}_{Z-2} Y + ^4_2 He \qquad [30.9]$$

where X is called the **parent nucleus** and Y the **daughter nucleus.** As examples, ^{238}U and ^{226}Ra are both alpha emitters and decay according to the schemes

$$^{238}_{92}\text{U} \longrightarrow {}^{234}_{90}\text{Th} + {}^{4}_{2}\text{He} \qquad\qquad \text{[30.10]}$$

$$^{226}_{88}\text{Ra} \longrightarrow {}^{222}_{86}\text{Rn} + {}^{4}_{2}\text{He} \qquad\qquad \text{[30.11]}$$

The half-life for ^{238}U decay is 4.47×10^9 years, and the half-life for ^{226}Ra decay is 1.60×10^3 years. In both cases, note that the mass number A of the daughter nucleus is 4 less than that of the parent nucleus. Likewise, Z is reduced by 2. The differences are accounted for in the emitted alpha particle (the ^{4}He nucleus).

The decay of ^{226}Ra is shown in Figure 30.13. When one element changes into another, as happens in alpha decay, the process is called **spontaneous decay.** As a general rule, (1) the sum of the mass numbers A must be the same on both sides of the equation, and (2) the sum of the atomic numbers Z must be the same on both sides of the equation. In addition, the total energy must be conserved. If we call M_X the mass of the parent nucleus, M_Y the mass of the daughter nucleus, and M_α the mass of the alpha particle, we can define the **disintegration energy, Q:**

$$Q \equiv (M_X - M_Y - M_\alpha)\,c^2 \qquad\qquad \text{[30.12]}$$

The quantity given by Equation 30.12 is sometimes referred to as the Q value of the nuclear reaction. Note that Q will be in joules if the masses are in kilograms, and c is the usual 3.00×10^8 m/s. We make this expression more convenient for calculations by including in all terms the masses of enough electrons to make neutral atoms. Also, we convert atomic mass units for mass to MeV units for energy:

$$Q = [M(\text{X}) - M(\text{Y}) - M(^4\text{He})] \times 931.494 \text{ MeV/u} \qquad \text{[30.13]}$$

The disintegration energy Q appears in the form of kinetic energy of the daughter nucleus and the alpha particle. In the case of the ^{226}Ra decay described in Figure 30.13, if the parent nucleus decays at rest, the residual kinetic energy of the products is 4.87 MeV. Most of the kinetic energy is associated with the alpha particle because this particle is much less massive than the recoiling daughter nucleus. That is, because momentum must be conserved, the lighter alpha particle recoils with a much higher speed than the daughter nucleus. Generally, light particles carry off most of the energy in nuclear decays.

Finally, it is interesting to note that if one assumed that ^{238}U (or other alpha emitters) decayed by emitting protons and neutrons, the mass of the decay products would exceed that of the parent nucleus, corresponding to negative Q values. Therefore, such spontaneous decays do not occur.

Figure 30.13 Alpha decay of radium. The radium nucleus is initially at rest. After the decay, the radon nucleus has kinetic energy K_{Rn} and momentum $\mathbf{p}_{\text{Rn}}$, and the alpha particle has kinetic energy K_α and momentum $\mathbf{p}_\alpha$.

Example 30.5   The Energy Liberated When Radium Decays

The ^{226}Ra nucleus undergoes alpha decay according to Equation 30.11. Calculate the Q value for this process. Take the mass of ^{226}Ra to be 226.025 406 u, the mass of ^{222}Rn to be 222.017 574 u, and the mass of $^4_2$He to be 4.002 603 u, as found in Table A.3.

Solution   Using Equation 30.13, we see that

$$Q = [M(^{226}\text{Ra}) - M(^{222}\text{Rn}) - M(^4\text{He})] \times 931.494 \text{ MeV/u}$$
$$= (226.025\ 406\ \text{u} - 222.017\ 574\ \text{u} - 4.002\ 603\ \text{u})$$
$$\times 931.494 \text{ MeV/u}$$

$$= (0.005\ 229\ \text{u}) \times \left(931.494 \frac{\text{MeV}}{\text{u}}\right) = 4.87 \text{ MeV}$$

It is left to end-of-chapter Problem 49 to show that the kinetic energy of the alpha particle is about 4.8 MeV, whereas that of the recoiling daughter nucleus is only about 0.1 MeV.

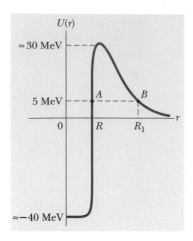

Figure 30.14 Potential energy versus separation for the alpha particle–nucleus system. Classically, the energy of the alpha particle is not sufficient to overcome the barrier, and so the particle should not be able to escape the nucleus.

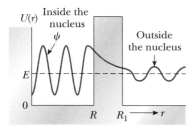

Figure 30.15 The nuclear potential energy is modeled as a square barrier. The energy of the alpha particle is E, which is less than the height of the barrier. According to quantum mechanics, the alpha particle has some chance of tunneling through the barrier, as indicated by the finite size of the wave function for $r > R_1$.

We now turn to the mechanism of alpha decay. Figure 30.14 is a plot of the potential energy versus distance r from the nucleus for the alpha particle–nucleus system, where R is the range of the nuclear force. The curve represents the combined effects of (1) the Coulomb repulsive energy, which gives the positive peak for $r > R$, and (2) the nuclear attractive force, which causes the curve to be negative for $r < R$. As we saw in Example 30.5, the disintegration energy is about 5 MeV, which is the approximate kinetic energy of the alpha particle, represented by the lower dotted line in Figure 30.14. According to classical physics, the alpha particle is trapped in the potential well. How, then, does it ever escape from the nucleus?

The answer to this question was provided by Gamow and, independently, Gurney and Condon in 1928, using quantum mechanics. In brief, the view of quantum mechanics is that there is always some probability that the particle can penetrate (or tunnel) through the barrier (Section 28.11). Recall that the probability of locating the particle depends on its wave function, ψ, and that the probability of tunneling is measured by $|\psi|^2$. Figure 30.15 is a sketch of the wave function for a particle of energy E, meeting a square barrier of finite height, which approximates the nuclear barrier. Note that the wave function is oscillating both inside and outside the barrier but is greatly reduced in amplitude because of the barrier. As the energy E increases, the probability of escaping also increases. Furthermore, the probability increases as the width of the barrier is decreased.

Beta Decay

When a radioactive nucleus undergoes beta decay, the daughter nucleus has the same number of nucleons as the parent nucleus, but the atomic number is changed by 1:

$$^{A}_{Z}\text{X} \longrightarrow \,^{A}_{Z+1}\text{Y} + \text{e}^- \qquad [30.14]$$

$$^{A}_{Z}\text{X} \longrightarrow \,^{A}_{Z-1}\text{Y} + \text{e}^+ \qquad [30.15]$$

Again, note that the nucleon number and total charge are both conserved in these decays. However, as we shall see later, these processes are not described completely by these expressions. We shall give reasons for this shortly.

It is important to note that the electron or positron involved in these decays is created within the nucleus as an initial step in the decay process. This is equivalent to saying that, during beta decay, a neutron in the nucleus is transformed into a proton. Indeed, the process $\text{n} \rightarrow \text{p} + \text{e}^-$ does occur, although it is extremely improbable outside the nucleus.

Now consider the energy of the system before and after the decay. As with alpha decay, energy must be conserved. One finds experimentally that the beta particles are emitted over a continuous range of energies (Fig. 30.16). The kinetic energy of the electrons must be balanced by the decrease in mass of the system—that is, the Q value. However, because all decaying nuclei have the same initial mass, **the Q value must be the same for each decay.** In view of this, why do the emitted electrons have different kinetic energies? The principle of conservation of energy seems to be violated! Further analysis shows that, according to the decay processes given by Equations 30.14 and 30.15, the principles of conservation of both angular momentum (spin) and linear momentum are also violated!

After a great deal of experimental and theoretical study, Pauli in 1930 proposed that a third particle must be present to carry away the "missing" energy and momentum. Fermi later named this particle the **neutrino** (little neutral one), because it had to be electrically neutral and have little or no rest mass. Although it eluded detection for many years, the neutrino (symbol ν) was finally detected experimentally in 1956. It has the following properties:

- Zero electric charge;
- A mass smaller than that of the electron, and it is possibly zero (recent experiments suggest that mass of the neutrino is less than 7 eV/c^2);
- A spin of $\frac{1}{2}$, which allows the law of conservation of angular momentum to be satisfied in beta decay;
- Very weak interaction with matter, which, therefore, makes it quite difficult to detect.

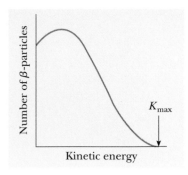

Figure 30.16 A typical beta-decay curve. The maximum kinetic energy observed for the beta particles corresponds to the value of Q for the reaction.

We can now write the beta decay processes in their correct form for carbon-14, nitrogen-12, and an isolated neutron:

$$_{6}^{14}\text{C} \longrightarrow _{7}^{14}\text{N} + e^- + \bar{\nu} \qquad [30.16]$$

$$_{7}^{12}\text{N} \longrightarrow _{6}^{12}\text{C} + e^+ + \nu \qquad [30.17]$$

$$\text{n} \longrightarrow \text{p} + \text{c}^- + \bar{\nu} \qquad [30.18]$$

where $\bar{\nu}$ represents the **antineutrino,** the antiparticle to the neutrino. We shall discuss antiparticles further in Chapter 31. For now, it suffices to say that **a neutrino is emitted in positron decay, and an antineutrino is emitted in electron decay.** Tables list isotopic masses of neutral atoms. To refer to neutral atoms in the first two processes above, we must add six electrons to each side of Equation 30.16 and seven electrons to each side of Equation 30.17.

A process that competes with e^+ decay is called **electron capture.** This occurs when a parent nucleus captures one of its own orbital electrons and emits a neutrino. The final product after decay is a nucleus the charge of which is $Z - 1$:

$$_{Z}^{A}\text{X} + _{-1}^{0}e \longrightarrow _{Z-1}^{A}\text{Y} + \nu \qquad [30.19]$$

• *Electron capture process*

In most cases, it is an inner K-shell electron that is captured, and this is referred to as **K capture.**

Carbon Dating

The beta decay of ^{14}C given by Equation 30.16 is used to date organic samples. Cosmic rays (high-energy particles from outer space) in the upper atmosphere cause nuclear reactions that create ^{14}C. The ratio of ^{14}C to ^{12}C in the carbon dioxide molecules of our atmosphere has a constant value of about 1.3×10^{-12}. All living organisms have the same ratio of ^{14}C to ^{12}C because they continuously exchange carbon dioxide with their surroundings. When an organism dies, however, it no longer absorbs ^{14}C from the atmosphere, and so the ratio of ^{14}C to ^{12}C decreases as the result of the beta decay of ^{14}C, which has a half-life of 5730 years. It is therefore possible to determine the age of a material by measuring its activity per unit mass due to the decay of ^{14}C. Using carbon dating, samples of wood, charcoal, bone, and shell have been identified as having lived 1000 to 25 000 years ago.

A particularly interesting example is the dating of the Dead Sea Scrolls, a group of manuscripts discovered by a shepherd in 1947. Translation showed them to be religious documents, including most of the books of the Old Testament. Carbon dating applied to the material in which they were wrapped established their age at approximately 1950 years at the time of discovery.

Thinking Physics 4

In 1991, a German tourist discovered the well-preserved remains of the Ice Man trapped in a glacier in the Italian Alps. Radioactive dating of a sample of Ice Man revealed an age of 5300 years. Why did scientists date the sample using the isotope ^{14}C rather than ^{11}C, a beta emitter with a half-life of 20.4 min?

Reasoning ^{14}C has a long half-life of 5730 years, so the fraction of ^{14}C nuclei remaining after one half-life is high enough to measure accurate changes in the sample's activity. The ^{11}C isotope, which has a very short half-life, is not useful because its activity decreases to a vanishingly small value over the age of the sample, making it impossible to detect.

The Ice Man was discovered in 1991 when an Italian glacier melted enough to expose his remains. His possessions, particularly his tools, have shed light on the way people lived in the Bronze Age. Archaeologists used ^{14}C dating to determine the time frame of this discovery. *(Paul Hanny/Gamma Liaison)*

Thinking Physics 5

In comparing alpha decay energies from a number of radioactive nuclides, it is found that the half life of the decay goes down as the energy of the decay goes up. Why is this?

Reasoning As discussed in the text, the alpha particle in the nucleus is trapped in a potential well and escapes by tunneling through the barrier. In Figure 30.14, it is clear that the higher the energy of the alpha particle, the thinner is the potential barrier. The thinner barrier translates to a higher probability of escape. The higher probability of escape translates to a faster rate of decay, which appears as a shorter half-life.

Thinking Physics 6

A wooden coffin is found that contains a skeleton holding a gold statue. Which of the three objects (coffin, skeleton, and gold statue) can be carbon dated to find out how old it is?

Reasoning Only the coffin and the skeleton can be carbon dated. These two items exchanged air with the environment during their lifetimes, resulting in a fixed ratio of carbon-14 to carbon-12. Once the human and the tree died, the amount of carbon-14 began to decrease due to radioactive decay. The gold in the statue was never alive and did not breathe air; therefore, there is no carbon-14 to detect.

CONCEPTUAL PROBLEM 2

Can carbon-14 dating be used to measure the age of a stone?

Example 30.6   Radioactive Dating

A piece of charcoal of mass 25 g is found in the ruins of an ancient city. The sample shows a ^{14}C ($T_{1/2} = 5730$ years) activity of 250 decays/min. How long has the tree from which this charcoal came been dead?

Solution   First, let us calculate the decay constant for ^{14}C:

$$\lambda = \frac{0.693}{T_{1/2}} = \frac{0.693}{(5730\ \text{y})(3.16 \times 10^7\ \text{s/y})} = 3.83 \times 10^{-12}\ \text{s}^{-1}$$

The number of ^{14}C nuclei can be calculated in two steps. First, the number of ^{12}C nuclei in 25 g of carbon is

$$N(^{12}\text{C}) = \frac{6.02 \times 10^{23}\ \text{nuclei/mol}}{12\ \text{g/mol}}\ (25\ \text{g})$$
$$= 1.25 \times 10^{24}\ \text{nuclei}$$

Assuming that the ratio of ^{14}C to ^{12}C is 1.3×10^{-12}, we see that the number of ^{14}C nuclei in 25 g *before* decay is

$$N_0(^{14}\text{C}) = (1.3 \times 10^{-12})(1.26 \times 10^{24}) = 1.63 \times 10^{12}\ \text{nuclei}$$

Hence, the initial activity of the sample is

$$R_0 = N_0\lambda = (1.63 \times 10^{12}\ \text{nuclei})(3.83 \times 10^{-12}\ \text{s}^{-1})$$
$$- 6.25\ \text{decays/s} = 375\ \text{decays/min}$$

We can now calculate the age of the charcoal using Equation 30.7, which relates the activity R at any time t to the initial activity, R_0:

$$R = R_0 e^{-\lambda t} \qquad \text{or} \qquad e^{-\lambda t} = \frac{R}{R_0}$$

Because it is given that $R = 250$ decays/min, and because we found that $R_0 = 375$ decays/min, we can calculate t by taking the natural logarithm of both sides of the last equation.

$$-\lambda t = \ln\left(\frac{R}{R_0}\right) = \ln\left(\frac{250}{375}\right) = -0.405$$

$$t = \frac{0.405}{\lambda} = \frac{0.405}{3.84 \times 10^{-12}\ \text{s}^{-1}}$$
$$= 1.06 \times 10^{11}\ \text{s} = \boxed{3350\ \text{years}}$$

As discussed previously, this technique of radioactive-carbon dating has been successfully used to measure the ages of many organic relics up to 25 000 years old.

Gamma Decay

Very often, a nucleus that undergoes radioactive decay is left in an excited energy state. It can then undergo a second decay to a lower energy state, perhaps to the ground state, by emitting a photon. Photons emitted in such a de-excitation process are called *gamma rays*. Such photons have very high energy (in the range of 1 MeV to 1 GeV) relative to the energy of visible light (about 1 eV). Recall from Chapter 29 that the energy of photons emitted (or absorbed) by an atom equals the difference in energy between the two electronic states involved in the transition. In a similar way, a gamma ray photon has an energy, *hf*, that equals the energy difference, ΔE, between two nuclear energy levels. When a nucleus decays by emitting a gamma ray, the nucleus does not change, apart from the fact that it ends up in a lower energy state. We can represent gamma decay as

Gamma decay •

$$\underset{Z}{\overset{A}{}}X^* \longrightarrow \underset{Z}{\overset{A}{}}X + \gamma \qquad \text{[30.20]}$$

where X* indicates a nucleus in an excited state.

A nucleus may reach an excited state as the result of a violent collision with another particle. However, it is more common for a nucleus to be in an excited state after it has undergone an alpha or beta decay. The following sequence of events represents a typical situation in which gamma decay occurs:

$$\underset{5}{\overset{12}{}}B \longrightarrow \underset{6}{\overset{12}{}}C^* + \underset{-1}{\overset{0}{}}e + \bar{\nu} \qquad \text{[30.21]}$$

$$\underset{6}{\overset{12}{}}C^* \longrightarrow \underset{6}{\overset{12}{}}C + \gamma \qquad \text{[30.22]}$$

Figure 30.17 shows the decay scheme for ^{12}B, which undergoes beta decay to either of two levels of ^{12}C. It can either (1) decay directly to the ground state of ^{12}C by emitting an electron with energy up to 13.4 MeV or (2) undergo beta decay to an excited state of ^{12}C*, followed by gamma decay to the ground state. The latter process results in the emission of an electron with energy up to 9.0 MeV and a 4.4-MeV photon.

The pathways by which a radioactive nucleus can undergo decay are summarized in Table 30.2.

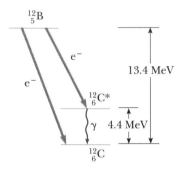

Figure 30.17 The ^{12}B nucleus undergoes beta decay to two levels of ^{12}C. The decay to the excited level, ^{12}C*, is followed by gamma decay to the ground state.

TABLE 30.2   Decay Pathways

Alpha decay	$\underset{Z}{\overset{A}{}}X \longrightarrow \underset{Z-2}{\overset{A-4}{}}X + \underset{2}{\overset{4}{}}He$
Beta decay (e^-)	$\underset{Z}{\overset{A}{}}X \longrightarrow \underset{Z+1}{\overset{A}{}}X + e^- + \bar{\nu}$
Beta decay (e^+)	$\underset{Z}{\overset{A}{}}X \longrightarrow \underset{Z-1}{\overset{A}{}}X + e^+ + \nu$
Electron capture	$\underset{Z}{\overset{A}{}}X + \underset{-1}{\overset{0}{}}e \longrightarrow \underset{Z-1}{\overset{A}{}}X + \nu$
Gamma decay	$\underset{Z}{\overset{A}{}}X^* \longrightarrow \underset{Z}{\overset{A}{}}X + \gamma$

CONCEPTUAL PROBLEM 3

In beta decay, the energy of the electron or positron emitted from the nucleus lies somewhere in a relatively large range of possibilities. In alpha decay, however, the alpha particle energy can only have discrete values. Why is there this difference?

CONCEPTUAL PROBLEM 4

In positron decay, a proton in the nucleus becomes a neutron, and the positive charge is carried away by the positron. But a neutron has a larger rest energy than a proton. How is this possible?

CONCEPTUAL PROBLEM 5

A student claims that a heavy form of hydrogen decays by alpha emission. How do you respond?

TABLE 30.3   The Four Radioactive Series

Series	Starting Isotope	Half-Life (years)	Stable End Product
Uranium (natural)	$^{238}_{92}\text{U}$	4.47×10^9	$^{206}_{82}\text{Pb}$
Actinium (natural)	$^{235}_{92}\text{U}$	7.04×10^8	$^{207}_{82}\text{Pb}$
Thorium (natural)	$^{232}_{90}\text{Th}$	1.41×10^{10}	$^{208}_{82}\text{Pb}$
Neptunium	$^{237}_{93}\text{Np}$	2.14×10^6	$^{209}_{83}\text{Bi}$

30.5 • NATURAL RADIOACTIVITY

Radioactive nuclei are generally classified into two groups: (1) unstable nuclei that are found in nature, which give rise to what is called **natural radioactivity,** and (2) nuclei produced in the laboratory through nuclear reactions, which exhibit **artificial radioactivity.**

Three series of radioactive nuclei occur naturally (Table 30.3). Each series starts with a specific long-lived radioactive isotope the half-life of which exceeds that of any of its descendents. The fourth series in Table 30.3 begins with ^{237}Np, a transuranic element (one having an atomic number greater than that of uranium) not found in nature. This element has a half-life of "only" 2.14×10^6 years.

The two uranium series are somewhat more complex than the ^{232}Th series (Fig. 30.18). Also, there are several naturally occurring radioactive isotopes, such as ^{14}C and ^{40}K, that are not part of either decay series.

Natural radioactivity constantly resupplies our environment with radioactive elements that would otherwise have disappeared long ago. For example, because the Solar System is about 5×10^9 years old, the supply of ^{226}Ra (the half-life of which is only 1600 years) would have been depleted by radioactive decay long ago if it were not for the decay series that starts with ^{238}U, with a half-life of 4.47×10^9 years.

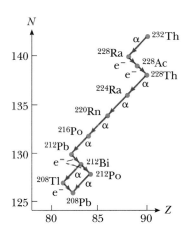

Figure 30.18 Successive decays for the ^{232}Th series.

30.6 • NUCLEAR REACTIONS

It is possible to change the structures and properties of nuclei by bombarding them with energetic particles. Such changes are called **nuclear reactions.** In 1919 Rutherford was the first to observe nuclear reactions, using naturally occurring radioactive sources for the bombarding particles. Since then, thousands of nuclear reactions have been observed following the development of charged-particle accelerators in the 1930s. With today's advanced technology in particle accelerators and particle detectors, it is possible to achieve particle energies of at least 1000 GeV = 1 TeV. These high-energy particles are used to create new particles the properties of which are helping to solve the mysteries of the nucleus.

Consider a reaction in which a target nucleus, X, is bombarded by a particle, a, resulting in a nucleus, Y, and a particle, b:

$$a + X \longrightarrow Y + b \qquad \qquad \textbf{[30.23]}$$

• *Nuclear reaction*

Sometimes this reaction is written in the more compact form

$$X(a, b)Y$$

Enrico Fermi (1901–1954)

An Italian physicist, Fermi was awarded the Nobel prize in 1938 for producing the transuranic elements by neutron irradiation and for his discovery of nuclear reactions bought about by slow neutrons. He made many other outstanding contributions to physics including his theory of beta decay, the free electron theory of metals, and the development of the world's first fission reactor in 1942. Fermi was truly a gifted theoretical and experimental physicist. He was also well known for his ability to present physics in a clear and exciting manner. "Whatever Nature has in store for mankind, unpleasant as it may be, men must accept, for ignorance is never better than knowledge."

In Section 30.4, the Q value, or disintegration energy, associated with radioactive decay was defined as the energy released during the decay process. Likewise, we define the **reaction energy, Q_r,** associated with a nuclear reaction as **the total energy released as the result of the reaction:**

$$Q_r = (M_a + M_X - M_Y - M_b)\,c^2 \qquad \text{[30.24]}$$

A reaction such as this, for which Q_r is positive, is called **exothermic.** After the reaction, this energy appears as an increase in kinetic energy of (Y, b). The loss of energy (and mass) of the system is balanced by an increase in the kinetic energy (and mass) of the escaping particles. A reaction for which Q_r is negative is called **endothermic.** An endothermic reaction will not occur unless the bombarding particle has a kinetic energy greater than Q_r. The minimum energy necessary for such a reaction to occur is called the **threshold energy.**

Nuclear reactions must obey the principle of conservation of linear momentum. That is, the total linear momentum of the system of interacting particles must be the same before and after the reaction. This assumes that the only force acting on the interacting particles is their mutual force of interaction; that is, no external accelerating electric fields are present near the colliding particles.

If a nuclear reaction occurs in which particles a and b are identical, so that X and Y are also necessarily identical, the reaction is called a **scattering event.** If kinetic energy is constant as a result of the reaction (that is, if $Q_r = 0$), the event is classified as **elastic scattering.** However, if $Q_r \neq 0$, kinetic energy is not constant and the reaction is called **inelastic scattering.** This terminology is identical to that used in dealing with collisions between macroscopic objects (Chapter 8, Section 8.4).

In addition to energy and momentum, the total charge and total number of nucleons must be conserved in any nuclear reaction. For example, consider the reaction $^{19}\text{F}(\text{p}, \alpha)\,^{16}\text{O}$, which has a Q_r value of 8.124 MeV. We can show this reaction more completely as

$$^{1}_{1}\text{H} + {}^{19}_{9}\text{F} \longrightarrow {}^{16}_{8}\text{O} + {}^{4}_{2}\text{He}$$

We see that the total number of nucleons before the reaction $(1 + 19 = 20)$ is equal to the total number after the reaction $(16 + 4 = 20)$. Furthermore, the total charge $(Z = 10)$ is the same before and after the reaction.

SUMMARY

A nuclear species can be represented by $^{A}_{Z}\text{X}$ where A is the **mass number,** the total number of nucleons, and Z is the **atomic number,** the total number of protons. The total number of neutrons in a nucleus is the **neutron number, N,** where $A = N + Z$. Elements with the same Z but different A and N values are called **isotopes.**

Assuming that a nucleus is spherical, its radius is

$$r = r_0 A^{1/3} \qquad \text{[30.1]}$$

where $r_0 = 1.2$ fm.

Nuclei are stable because of the **nuclear force** between nucleons. This short-range force dominates the Coulomb repulsive force at distances of less than about 2 fm and is independent of charge.

Light nuclei are most stable when the number of protons equals the number of neutrons. Heavy nuclei are most stable when the number of neutrons exceeds the number of protons. In addition, many stable nuclei have Z and N values that are both even. Nuclei with unusually high stability have Z or N values of 2, 8, 20, 28, 50, 82, and 126, called **magic numbers.**

Nuclei have an intrinsic spin angular momentum of magnitude $\sqrt{I(I+1)}\hbar$, where I is the **nuclear spin.** The magnetic moment of a nucleus is measured in terms of the **nuclear magneton,** μ_n, where

$$\mu_n = 5.05 \times 10^{-27}\,\text{J/T} \qquad \textbf{[30.3]}$$

The difference in mass between the separate nucleons and the nucleus containing these nucleons, when multiplied by c^2, gives the **binding energy, E_b,** of the nucleus: $E_b = \Delta m c^2$. We can calculate the binding energy of any nucleus $^A_Z X$ using the expression

$$E_b(\text{MeV}) = [ZM(\text{H}) + Nm_n - M(^A_Z X)] \times 931.494\,\text{MeV/u} \qquad \textbf{[30.4]}$$

A radioactive substance can undergo alpha decay, beta decay, or gamma decay. An alpha particle is the ^{4}He nucleus; a beta particle is either an electron (e^-) or a positron (e^+); a gamma particle is a high-energy photon.

If a radioactive material contains N_0 radioactive nuclei at $t = 0$, the number, N, of nuclei remaining after a time t has elapsed is

$$N = N_0 e^{-\lambda t} \qquad \textbf{[30.6]}$$

where λ is the **decay constant,** or **disintegration constant.** The *decay rate,* or *activity,* of a radioactive substance is given by

$$R = \left|\frac{dN}{dt}\right| = R_0 e^{-\lambda t} \qquad \textbf{[30.7]}$$

where $R_0 = N_0\lambda$ is the activity at $t = 0$. The **half-life,** $T_{1/2}$, is defined as the time it takes half of a given number of radioactive nuclei to decay, where

$$T_{1/2} = \frac{0.693}{\lambda} \qquad \textbf{[30.8]}$$

Alpha decay can occur because, according to quantum mechanics, some nuclei have barriers that can be penetrated by the alpha particles (the tunneling process). This process is energetically more favorable for those nuclei having large excesses of neutrons. A nucleus can undergo beta decay in two ways. It can emit either an electron (e^-) and an antineutrino ($\bar{\nu}$) or a positron (e^+) and a neutrino (ν). In the electron capture process, the nucleus of an atom absorbs one of its own electrons (usually from the K shell) and emits a neutrino. In gamma decay, a nucleus in an excited state decays to its ground state and emits a gamma ray.

Nuclear reactions can occur when a target nucleus, X, is bombarded by a particle, a, resulting in a nucleus, Y, and a particle, b:

$$a + X \longrightarrow Y + b \qquad \text{or} \qquad X(a, b)Y \qquad \textbf{[30.23]}$$

The energy released in such a reaction, called the **reaction energy, Q_r,** is

$$Q_r = (M_a + M_X - M_Y - M_b)c^2 \qquad \textbf{[30.24]}$$

CONCEPTUAL QUESTIONS

1. In Rutherford's experiment, assume that an alpha particle is headed directly toward the nucleus of an atom. Why doesn't the alpha particle make physical contact with the nucleus?

2. Explain the main differences between alpha, beta, and gamma rays.

3. Why do heavier elements require more neutrons in order to maintain stability?

4. A proton precesses with a frequency ω_p in the presence of a magnetic field. If the magnetic field intensity is doubled, what happens to the precessional frequency?

5. Why do nearly all the naturally occurring isotopes lie above the $N = Z$ line in Figure 30.3?

6. If a nucleus has a half-life of 1 year, does this mean it will be completely decayed after 2 years? Explain.

7. What fraction of a radioactive sample has decayed after two half-lives have elapsed?

8. Two samples of the same radioactive nuclide are prepared. Sample A has twice the initial activity of sample B. How does the half-life of A compare with the half-life of B? After each has passed through five half-lives, what is the ratio of their activities?

9. If no more people were to be born, the world's population would strongly resemble radioactive decay behavior. Discuss this statement.

10. If a nucleus captures a slow-moving neutron, the product is left in a highly excited state with an energy approximately 8.0 MeV above the ground state. Explain the source of the excitation energy.

11. Use the analogy of a bullet and a rifle to explain why the recoiling nucleus carries off only a very small fraction of the disintegration energy in the α decay of a nucleus.

12. If a nucleus such as ^{226}Ra initially at rest undergoes alpha decay, which has more kinetic energy after the decay, the alpha particle or the daughter nucleus?

13. Explain why many heavy nuclei undergo alpha decay but do not spontaneously emit neutrons or protons.

14. If an alpha particle and an electron have the same kinetic energy, which undergoes the greater deflection when passed through a magnetic field?

15. If film is kept in a wooden box, alpha particles from a radioactive source outside the box cannot expose the film, but beta particles can. Explain.

16. Suppose it could be shown that the cosmic ray intensity at the Earth's surface was much greater 10 000 years ago. How would this difference affect what we accept as valid carbon-dated values of the age of ancient samples of once-living matter?

17. The radioactive nucleus $^{226}_{88}$Ra has a half-life of approximately 1.6×10^3 years. Being that the solar system is about 5 billion years old, why do we still find this nucleus in nature?

18. From Table A.3, identify the four stable nuclei that have magic numbers in both Z and N.

PROBLEMS

(Table 30.4 will be useful for many of these problems. A more complete list of atomic masses is given in Appendix A.3.)

Section 30.1   Some Properties of Nuclei

1. What is the order of magnitude of the number of protons in your body? Of the number of neutrons? Of the number of electrons?

2. Find the radius of (a) a nucleus of $^4_2$He and (b) a nucleus of $^{238}_{92}$U.

3. Determine the atomic mass number of isotopes that have a radius approximately equal to one-half the radius of uranium, $^{238}_{92}$U.

4. What would be the gravitational force between two golf balls (each with a 4.30-cm diameter), 1.00 meter apart, if they were made of nuclear matter?

5. (a) Find the speed an alpha particle must have in order to come within 3.20×10^{-14} m of a gold nucleus. (b) Find the energy of the alpha particle in MeV units.

TABLE 30.4   Some Atomic Masses

Element	Atomic Mass (u)	Element	Atomic Mass (u)
$^1_1$H	1.007 825	$^{27}_{13}$Al	26.981 541
$^4_2$He	4.002 603	$^{30}_{15}$P	29.978 310
$^7_3$Li	7.016 004	$^{40}_{20}$Ca	39.962 591
$^9_4$Be	9.012 182	$^{42}_{20}$Ca	41.958 63
$^{10}_5$B	10.012 938	$^{43}_{20}$Ca	42.958 770
$^{12}_6$C	12.000 000	$^{56}_{26}$Fe	55.934 939
$^{13}_6$C	13.003 355	$^{64}_{30}$Zn	63.929 145
$^{14}_7$N	14.003 074	$^{64}_{29}$Cu	63.929 599
$^{15}_7$N	15.000 109	$^{93}_{41}$Nb	92.906 378
$^{15}_8$O	15.003 065	$^{197}_{79}$Au	196.966 560
$^{17}_8$O	16.999 131	$^{202}_{80}$Hg	201.970 632
$^{18}_8$O	17.999 159	$^{216}_{84}$Po	216.001 790
$^{18}_9$F	18.000 937	$^{220}_{86}$Rn	220.011 401
$^{20}_{10}$Ne	19.992 439	$^{234}_{90}$Th	234.043 583
$^{23}_{11}$Na	22.989 770	$^{238}_{92}$U	238.050 786
$^{23}_{12}$Mg	22.994 127		

6. An α particle ($Z = 2$, mass 6.64×10^{-27} kg) approaches to within 1.00×10^{-14} m of a carbon nucleus ($Z = 6$). What are (a) the maximum Coulomb force on the α particle, (b) the acceleration of the α particle at this point, and (c) the potential energy of the α particle at this point?

7. Certain stars at the end of their lives are thought to collapse, combining their protons and electrons to form a neutron star. Such a star could be thought of as a gigantic atomic nucleus. If a star of mass equal to that of the Sun ($M = 1.99 \times 10^{30}$ kg) collapsed into neutrons ($m_n = 1.67 \times 10^{-27}$ kg), what would the radius be? (*Hint:* Use $r = r_0 A^{1/3}$.)

8. The Larmor precessional frequency is

$$f_p = \frac{\Delta E}{h} = \frac{2\mu B}{h}$$

Calculate the radio-wave frequency at which resonance absorption will occur for (a) free neutrons in a magnetic field of 1.00 T, (b) free protons in a magnetic field of 1.00 T, and (c) free protons in the Earth's magnetic field at a location where the field strength is 50.0 μT.

9. How much energy (in MeV units) must an α particle have to reach the surface of a gold nucleus ($Z = 79$, $A = 197$)?

10. In a Rutherford scattering experiment, alpha particles having kinetic energy of 7.70 MeV are fired toward a gold foil. (a) Use energy conservation to determine the distance of closest approach between the alpha particle and a gold nucleus. Imagine a head-on collision with a nucleus that remains at rest. (b) Calculate the de Broglie wavelength for the 7.70-MeV alpha particle and compare it to the distance obtained in part (a). (c) Based on this comparison, why is it proper to treat the alpha particle as a particle and not as a wave in the Rutherford scattering experiment?

Section 30.2 Binding Energy

11. Calculate the binding energy per nucleon for (a) ^{2}H, (b) ^{4}He, (c) ^{56}Fe, and (d) ^{238}U.

12. The peak of the stability curve occurs at ^{56}Fe. This is why iron is prominent in the spectrum of the Sun and stars. Show that ^{56}Fe has a higher binding energy per nucleon than its neighbors ^{55}Mn and ^{59}Co. Compare your results with Figure 30.9.

13. Calculate the minimum energy required to remove a neutron from the $^{43}_{20}$Ca nucleus.

14. Two nuclei having the same mass number are known as *isobars*. Calculate the difference in binding energy per nucleon for the isobars $^{23}_{11}$Na and $^{23}_{12}$Mg. How do you account for the difference?

15. The isotope $^{139}_{57}$La is stable. A radioactive isobar (see Problem 14) of this lanthanum isotope, $^{139}_{59}$Pr, is located below the line of stable nuclei in Figure 30.3 and decays by e^+ emission. Another radioactive isobar of $^{139}_{57}$La, $^{139}_{55}$Cs, decays by e^- emission and is located above the line of stable nuclei in Figure 30.3. (a) Which of these three isobars has the highest neutron-to-proton ratio? (b) Which has the greatest binding energy per nucleon? (c) Which do you expect to be heavier, ^{139}Pr or ^{139}Cs?

Section 30.3 Radioactivity

16. Ordinary potassium from milk contains 0.0117% of the radioactive isotope ^{40}K having a half-life of 1.227×10^9 y. If you drink 1 glass of milk (0.250 liter) a day, what is its activity just after you drink the milk? Assume that 1 liter of milk contains 2.00 g of potassium.

17. A sample of radioactive material contains 1.00×10^{15} atoms and has an activity of 6.00×10^{11} Bq. What is its half-life?

18. How much time elapses before 90.0% of the radioactivity of a sample of $^{72}_{33}$As disappears as measured by its activity? The half-life of $^{72}_{33}$As is 26 h.

19. The half-life of ^{131}I is 8.0 days. On a certain day, the activity of an iodine-131 sample is 6.40 mCi. What is its activity 40.0 days later?

20. Determine the activity of 1.00 g of ^{60}Co. The half-life of ^{60}Co is 5.24 y.

21. A freshly prepared pure sample of a certain radioactive isotope has an activity of 10.0 mCi. After 4.00 h, its activity is 8.00 mCi. (a) Find the decay constant and half-life. (b) How many atoms of the isotope were contained in the freshly prepared sample? (c) What is the sample's activity 30.0 h after it is prepared?

22. A building has become accidentally contaminated with radioactivity. The longest-lived material in the building is strontium-90 ($^{90}_{38}$Sr has an atomic mass 89.9077, and its half-life is 28.8 y). If the building initially contained 5.00 kg of this substance uniformly distributed throughout the building (a very unlikely situation) and the safe level is less than 10.0 decays/min, how long will the building be unsafe?

23. The most common isotope of radon is ^{222}Rn, which has a half-life of 3.8 days. (a) What fraction of the nuclei that were on Earth 1 week ago are now undecayed? (b) What fraction of those that existed 1 year ago? (c) In view of these results, explain why radon remains a problem in modern society. (*Hint:* Radon-222 is the decay product of radium-226.)

24. Consider a radioactive sample. Determine the ratio of the number of atoms decayed during the first half of its half-life to the number of atoms decayed during the second half of its half-life.

25. The radioactive isotope ^{198}Au has a half-life of 64.8 h. A sample containing this isotope has an initial activity (at $t = 0$) of 40.0 μCi. Calculate the number of nuclei that decay in the time interval between $t_1 = 10.0$ h and $t_2 = 12.0$ h.

26. A radioactive nucleus has a half-life of $T_{1/2}$. A sample containing these nuclei has an initial activity R_0. Calculate the number of nuclei that decay in the time interval between the times t_1 and t_2.

Section 30.4   The Decay Processes

27. Enter the correct isotope symbol in each open square in Figure P30.27, which shows the sequences of decays starting with uranium-235 and ending with the stable isotope lead-207.

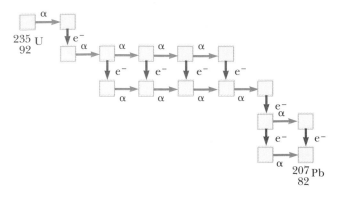

Figure P30.27

28. Identify the missing nuclide or particle (X):
    (a) $X \rightarrow {}^{65}_{28}\text{Ni} + \gamma$
    (b) ${}^{215}_{84}\text{Po} \rightarrow X + \alpha$
    (c) $X \rightarrow {}^{55}_{26}\text{Fe} + e^+ + \nu$
    (d) ${}^{109}_{48}\text{Cd} + X \rightarrow {}^{109}_{47}\text{Ag} + \nu$
    (e) ${}^{14}\text{N}(\alpha, X){}^{17}\text{O}$

29. Find the energy released in the alpha decay

$$ {}^{238}_{92}\text{U} \longrightarrow {}^{234}_{90}\text{Th} + {}^{4}_{2}\text{He} $$

You will find the following mass values useful:

$$ M({}^{238}_{92}\text{U}) = 238.050\ 786\ \text{u} $$

$$ M({}^{234}_{90}\text{Th}) = 234.043\ 583\ \text{u} $$

$$ M({}^{4}_{2}\text{He}) = 4.002\ 603\ \text{u} $$

30. A living specimen in equilibrium with the atmosphere contains 1 atom of ${}^{14}\text{C}$ (half-life = 5730 y) for every 7.7×10^{11} stable carbon atoms. An archeological sample of wood (cellulose, $C_{12}H_{22}O_{11}$) contains 21.0 mg of carbon. When the sample is placed inside a beta counter the counting efficiency of which is 88.0%, 837 counts are accumulated in 1 week. Assuming that the cosmic-ray flux and the Earth's atmosphere have not changed appreciably since the sample was formed, find the age of the sample.

31. The nucleus ${}^{15}_{8}\text{O}$ decays by electron capture. Write (a) the basic nuclear process and (b) the decay process referring to neutral atoms. (c) Determine the energy of the neutrino. Disregard the daughter's recoil.

32. Is it energetically possible for a ${}^{8}_{4}\text{Be}$ nucleus to spontaneously decay into two alpha particles? Explain. (${}^{8}_{4}\text{Be}$ has an atomic mass of 8.005 305 u.)

Section 30.6   Nuclear Reactions

33. The reaction ${}^{27}_{13}\text{Al}(\alpha, \text{n}){}^{30}_{15}\text{P}$, achieved in 1934, was the first known in which the product nucleus is radioactive. Calculate the Q_r value of this reaction.

34. Find the energy released in the fission reaction

$$ \text{n} + {}^{235}_{92}\text{U} \longrightarrow {}^{98}_{40}\text{Zr} + {}^{135}_{52}\text{Te} + 3\text{n} $$

The atomic masses of the fission products are: ${}^{98}_{40}\text{Zr}$, 97.912 0 u; ${}^{135}_{52}\text{Te}$, 134.908 7 u.

35. Natural gold has only one isotope, ${}^{197}_{79}\text{Au}$. If natural gold is irradiated by a flux of slow neutrons, electrons are emitted. (a) Write the appropriate reaction equation. (b) Calculate the maximum energy of the emitted electrons. The mass of ${}^{198}_{80}\text{Hg}$ is 197.966 75 u.

36. A beam of 6.61 MeV protons is incident on a target of ${}^{27}_{13}\text{Al}$. Those that collide produce the reaction

$$ \text{p} + {}^{27}_{13}\text{Al} \longrightarrow {}^{27}_{14}\text{Si} + \text{n} $$

(${}^{27}_{14}\text{Si}$ has a mass of 26.986 721 u.) Neglecting any recoil of the product nucleus, determine the kinetic energy of the emerging neutrons.

37. (a) Suppose ${}^{10}_{5}\text{B}$ is struck by an alpha particle, releasing a proton and a product nucleus in the reaction. What is the product nucleus? (b) An alpha particle and a product nucleus are produced when ${}^{13}_{6}\text{C}$ is struck by a proton. What is the product nucleus?

Additional Problems

38. One method of producing neutrons for experimental use is to bombard ${}^{7}_{3}\text{Li}$ with protons. The neutrons are emitted according to the following reaction:

$$ {}^{1}_{1}\text{H} + {}^{7}_{3}\text{Li} \longrightarrow {}^{7}_{4}\text{Be} + {}^{1}_{0}\text{n} $$

What is the minimum kinetic energy the incident proton must have if this reaction is to occur?

39. A by-product of some fission reactors is the isotope ${}^{239}_{94}\text{Pu}$, an alpha emitter having a half-life of 24 000 years:

$$ {}^{239}_{94}\text{Pu} \longrightarrow {}^{235}_{92}\text{U} + \alpha $$

Consider a sample of 1.00 kg of pure ${}^{239}_{94}\text{Pu}$ at $t = 0$. Calculate (a) the number of ${}^{239}_{94}\text{Pu}$ nuclei present at $t = 0$ and (b) the initial activity in the sample. (c) How long does the sample have to be stored if a "safe" activity level is 0.100 Bq?

40. (a) Can ${}^{57}\text{Co}$ decay by positron (e^+) emission? Explain. (b) Can ${}^{14}\text{C}$ decay by electron (e^-) emission? Explain. (c) If either answer is yes, what is the range of kinetic energies available for the positron or electron?

41. During the manufacture of a steel engine component, radioactive iron (${}^{59}\text{Fe}$) is included in the total mass of 0.200 kg. The component is placed in a test engine when the activity due to this isotope is 20.0 μCi. After a 1000-h test period, oil is removed from the engine and found to

contain enough ^{59}Fe to produce 800 disintegrations/min per liter of oil. The total volume of oil in the engine is 6.50 liters. Calculate the total mass worn from the engine component per hour of operation. (The half-life of ^{59}Fe is 45.1 days.)

42. The activity of a radioactive sample was measured over 12 h, with the following background-corrected count rates:

Time (h)	Counting Rate (counts/min)
1.00	3100
2.00	2450
4.00	1480
6.00	910
8.00	545
10.0	330
12.0	200

(a) Plot the activity curve on semilog paper. (b) Determine the disintegration constant and half-life of the radioactive nuclei in the sample. (c) What counting rate would you expect for the sample at $t = 0$? (d) Assuming the efficiency of the counting instrument to be 10.0%, calculate the number of radioactive atoms in the sample at $t = 0$.

43. A large nuclear power reactor produces about 3000 MW of thermal power in its core. Three months after a reactor is shut down, the core power from radioactive by-products is 10.0 MW. Assuming that each emission delivers 1.00 MeV of energy to the thermal power, estimate the activity in becquerels 3 months after the reactor is shut down.

44. In a piece of rock from the Moon, the ^{87}Rb content is assayed to be 1.82×10^{10} atoms per gram of material, and the ^{87}Sr content is found to be 1.07×10^9 atoms per gram. (The relevant decay is ^{87}Rb $\rightarrow$ ^{87}Sr $+$ e$^-$. The half-life of the decay is 4.8×10^{10} years.) (a) Determine the age of the rock. (b) Could the material in the rock actually be much older? What assumption is implicit in using the radioactive dating method?

45. (a) Find the radius of the $^{12}_{6}$C nucleus. (b) Find the force of repulsion between a proton at the surface of a $^{12}_{6}$C nucleus and the remaining five protons. (c) How much work (in MeV) has to be done to overcome this electrostatic repulsion in order to put the last proton into the nucleus? (d) Repeat parts (a), (b), and (c) for $^{238}_{92}$U.

46. (a) Why is the inverse beta decay p $\rightarrow$ n $+$ e$^+$ $+$ ν forbidden for a free proton? (b) Why is the same reaction possible if the proton is bound in a nucleus? For example, the following reaction occurs:

$$^{13}_{7}\text{N} \longrightarrow {}^{13}_{6}\text{C} + e^+ + \nu$$

(c) How much energy is released in the reaction given in part (b)? [$m(e^+) = m(e^-) = 0.000\ 549$ u, $M(^{13}\text{C atom}) = 13.003\ 355$ u, $M(^{13}\text{N atom}) = 13.005\ 739$ u]

47. Carbon detonations are powerful nuclear reactions that temporarily tear apart the cores of massive stars late in their lives. These blasts are produced by carbon fusion, which requires a temperature of about 6×10^8 K to overcome the strong Coulomb repulsion between carbon nuclei. (a) Estimate the repulsive energy barrier to fusion, using the required ignition temperature for carbon fusion. (In other words, what is the average kinetic energy for a carbon nucleus at 6×10^8 K?) (b) Calculate the energy (in MeV) released in each of these "carbon-burning" reactions:

$$^{12}\text{C} + {}^{12}\text{C} \longrightarrow {}^{20}\text{Ne} + {}^{4}\text{He}$$

$$^{12}\text{C} + {}^{12}\text{C} \longrightarrow {}^{24}\text{Mg} + \gamma$$

(c) Calculate the energy (in kWh) given off when 2.00 kg of carbon completely fuses according to the first reaction.

48. When a material of interest is irradiated by neutrons, radioactive atoms are produced continually and some decay according to their characteristic half-lives. (a) If radioactive atoms are produced at a constant rate R and their decay is governed by the conventional radioactive decay law, show that the number of radioactive atoms accumulated after an irradiation time t is

$$N = \frac{R}{\lambda}(1 - e^{-\lambda t})$$

(b) What is the maximum number of radioactive atoms that can be produced?

49. The decay of an unstable nucleus by alpha emission is represented by Equation 30.9. The disintegration energy Q given by Equation 30.12 must be shared by the alpha particle and the daughter nucleus in order to conserve both energy and momentum in the decay process. (a) Show that Q and K_α, the kinetic energy of the alpha particle, are related by the expression

$$Q = K_\alpha \left(1 + \frac{M_\alpha}{M}\right)$$

where M is the mass of the daughter nucleus. (b) Use the result of part (a) to find the energy of the alpha particle emitted in the decay of ^{226}Ra. (See Example 30.5 for the calculation of Q.)

50. The ^{145}Pm nucleus decays by alpha emission. (a) Determine the daughter nucleus. (b) Using the values given in Table A.3, determine the energy released in this decay. (c) What fraction of this energy is carried away by the alpha particle when the recoil of the daughter is taken into account?

51. A very slow neutron (speed approximately equal to zero) can initiate the reaction

$$\text{n} + {}^{10}_{5}\text{B} \longrightarrow {}^{7}_{3}\text{Li} + {}^{4}_{2}\text{He}$$

If the alpha particle moves away with a speed of 9.30×10^6 m/s, calculate the kinetic energy of the lithium nucleus. Use nonrelativistic formulas.

52. When, after a reaction or disturbance of any kind, a nucleus is left in an excited state, it can return to its normal (ground) state by emission of a gamma-ray photon (or several photons). This process is illustrated by Equation 30.20. The emitting nucleus must recoil in order to conserve both energy and momentum. (a) Show that the recoil energy of the nucleus is

$$E_r = \frac{(\Delta E)^2}{2Mc^2}$$

where ΔE is the difference in energy between the excited and ground states of a nucleus of mass M. (b) Calculate the recoil energy of the ^{57}Fe nucleus when it decays by gamma emission from the 14.4-keV excited state. For this calculation, take the mass to be 57 u. (*Hint:* When writing the equation for conservation of energy, use $(Mv)^2/2M$ for the kinetic energy of the recoiling nucleus. Also, assume that $hf \ll Mc^2$ and use the binomial expansion.)

53. The theory of nuclear astrophysics proposes that all the heavy elements, such as uranium, are formed in the cores of massive stars or in supernova explosions ending the lives of these stars. If we assume that at the time of the explosion there were equal amounts of ^{235}U and ^{238}U, how long ago did the star(s) explode that released the elements that formed our Earth? The present ^{235}U/^{238}U ratio is 0.007. (The half-lives of ^{235}U and ^{238}U are 0.7×10^9 years and 4.47×10^9 years.)

Spreadsheet Problems

S1. Soon after the discovery of radioactivity, investigators realized that some radioactive nuclides decayed into nuclides that were themselves radioactive. Entire chains of radioactive nuclides would form. For example, with a half-life of 3.83 days, ^{222}Rn decays to ^{218}Po, which has a half-life of 3.05 min and decays to ^{214}Pb. For the case in which the parent and daughter both decay, the rate of decay of the parent is

$$\frac{dN_1}{dt} = -\lambda_1 N_1$$

where N_1 is the number of parent nuclei present and λ_1 is the decay constant for the parent. The rate of formation of

the daughter is just

$$\frac{dN_2}{dt} = \lambda_1 N_1 \qquad (\text{formation})$$

where N_2 is the number of daughter nuclei present. The rate of decay of the daughter is

$$\frac{dN_2}{dt} = -\lambda_2 N_2 \qquad (\text{decay})$$

where λ_2 is the decay constant for the daughter. The net production of the daughter is then

$$\frac{dN_2}{dt} = \lambda_1 N_1 - \lambda_2 N_2$$

Spreadsheet 30.1 calculates these rates of decay and formation of any mother–daughter pair of radioactive nuclides by integrating these coupled differential equations. (a) Let's assume that we start with 5000 atoms of the parent nuclei, which have a half-life of 5 min. The daughter nuclei further decay with a half-life of 1 min. Enter these data in the spreadsheet. How many nuclei of the parent and daughter are present after 1 min? 2 min? 30 min? (b) If the half-life of the parent is changed to 2 min and the daughter's half-life is changed to 5 min, how many nuclei of each are present at the times listed in part (a)? (c) Change the parent's half-life to 1000 min and the daughter's half-life to 2 min. Discuss the number of daughter nuclei present as a function of time.

S2. Modify Spreadsheet 30.1 to take into account the decay of the granddaughter nuclei, which have a half-life of 10 min. Set the parent half-life to 5 min and the daughter half-life to 2 min. The net rate of production of the granddaughter is

$$\frac{dN_3}{dt} = \lambda_2 N_2 - \lambda_3 N_3$$

where N_3 is the number of granddaughter nuclei present and λ_3 is the decay constant for the granddaughter. How many of each species of nuclei are present at 1 min, 2 min, 5 min, 10 min, 30 min, and 60 min?

ANSWERS TO CONCEPTUAL PROBLEMS

1. Isotopes of a given element correspond to nuclei with different numbers of neutrons. This will result in different masses of the atom, and different magnetic moments, because the neutron, despite being uncharged, does have a magnetic moment. The chemical behavior, however, is governed by the electrons. All isotopes of a given element have

the same number of electrons and, therefore, the same chemical behavior.

2. Carbon dating cannot generally be used to estimate the age of a stone. The stone was not alive and breathing air containing carbon-14. Only the ages of artifacts that were once alive can be estimated with carbon dating.

3. In alpha decay, there are only two final particles—the alpha particle and the daughter nucleus. There are also two conservation principles—energy and momentum. As a result, the alpha particle must be ejected with a discrete energy to satisfy both conservation principles. There are a small number of discrete energies of the alpha particle, as the daughter nucleus can be left in various excited states, but the allowed energies of the alpha particle are not continuous.

However, beta decay is a three-particle decay—the beta particle, the neutrino (or antineutrino), and the daughter nucleus. As a result, the energy and momentum can be shared in a variety of ways among the three particles while still satisfying the two conservation principles. This allows a continuous range of energies for the beta particle.

4. The larger rest energy of the neutron means that a free proton in space will not spontaneously decay into a neutron and a positron. When the proton is in the nucleus, however, the important question is that of the total rest energy *of the nucleus*. If it is energetically lower for the nucleus to have one less proton and one more neutron, then the beta-positive decay process will occur to achieve this lower energy.

5. An alpha particle contains two protons and two neutrons. Because a hydrogen nucleus only contains one proton, it cannot emit an alpha particle.

31

Particle Physics and Cosmology

In this concluding chapter, we examine the properties and classifications of the various known subatomic particles and the fundamental interactions that govern their behavior. We also discuss the current theory of elementary particles, in which all matter is believed to be constructed from only two families of particles, quarks and leptons. Finally, we discuss how clarifications of such models might help scientists understand cosmology, which deals with the evolution of the Universe.

The word "atom" is from the Greek *atomos*, which means "indivisible." At one time, atoms were thought to be the indivisible constituents of matter; that is, they were regarded to be elementary particles. After 1932, physicists viewed all matter as consisting of only three constituent particles: electrons, protons, and neutrons. With the exception of the free neutron, these particles are very stable. Beginning in

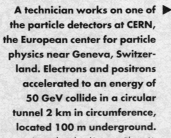

A technician works on one of ▶ the particle detectors at CERN, the European center for particle physics near Geneva, Switzerland. Electrons and positrons accelerated to an energy of 50 GeV collide in a circular tunnel 2 km in circumference, located 100 m underground. *(David Parker/Science Photo Library, Photo Researchers, Inc.)*

1945, many new particles were discovered in experiments involving high-energy collisions between known particles. These new particles are characteristically very unstable and have very short half-lives, ranging between 10^{-6} and 10^{-23} s. So far more than 300 of these unstable, temporary particles have been catalogued.

During the past 35 years, many powerful particle accelerators have been constructed throughout the world, making it possible to observe collisions of highly energetic particles under controlled laboratory conditions in an effort to reveal the subatomic world in finer detail. Until the 1960s, physicists were bewildered by the large number and variety of subatomic particles being discovered. They wondered if the particles were like animals in a zoo, having no systematic relationship connecting them or whether a pattern was emerging that would provide a better understanding of the elaborate structure in the subnuclear world. In the past 35 years, physicists have advanced our knowledge of the structure of matter tremendously by recognizing that all particles except electrons, photons, and a few related particles are made of smaller particles called *quarks*. Thus, protons and neutrons, for example, are not truly elementary but are systems of tightly bound quarks.

31.1 · THE FUNDAMENTAL FORCES IN NATURE

As we learned in Section 5.6, all natural phenomena can be described by four fundamental forces between elemental particles. In order of decreasing strength, these are the strong force, the electromagnetic force, the weak force, and the gravitational force.

The strong force, as we mentioned in Chapter 30, represents the glue that holds nucleons together. It is very short-range and is negligible for separations greater than about 10^{-15} m (about the size of the nucleus). The electromagnetic force, which binds atoms and molecules together to form ordinary matter, has about 10^{-2} times the strength of the strong force. It is a long-range force that decreases in strength as the inverse square of the separation between interacting particles. The weak force is a short-range force that tends to produce instability in certain nuclei. It is responsible for most radioactive decay processes such as beta decay, and its strength is only about 10^{-5} times that of the strong force. Finally, the gravitational force is a long-range force that has a strength of only about 10^{-39} times that of the strong force (in SI units). Although this familiar interaction is the force that holds the planets, stars, and galaxies together, its effect on elementary particles is negligible. In modern physics, interactions between particles are often described in terms of the exchange of field particles, or quanta. In the case of the familiar electromagnetic interaction, for instance, the field particles are photons. In the language of modern physics, it is said that the electromagnetic force is *mediated* by photons and that photons are the quanta of the electromagnetic field. Likewise, the strong force is mediated by field particles called **gluons,** the weak force is mediated by particles called the W and Z **bosons,** and the gravitational force is mediated by quanta of the gravitational field called **gravitons.** These interactions, their ranges, and their relative strengths are summarized in Table 31.1.

TABLE 31.1   Particle Interactions

Interaction (Force)	Relative Strength[a]	Typical Lifetimes for Decays via a Given Interaction	Range of Force	Mediating Field Particle
Strong	1	$\leqslant 10^{-20}$ s	Short ($\sim$1 fm)	Gluon
Electromagnetic	$\sim 10^{-2}$	$\sim 10^{-16}$ s	Long ($\propto 1/r^2$)	Photon
Weak	$\sim 10^{-6}$	$\geqslant 10^{-10}$ s	Short ($\sim 10^{-3}$ fm)	$W^{\pm}$, Z^0 bosons
Gravitational	$\sim 10^{-43}$	?	Long ($\propto 1/r^2$)	Graviton

[a] For two u quarks at 3×10^{-17} m.

31.2 • POSITRONS AND OTHER ANTIPARTICLES

In the 1920s, the theoretical physicist Paul Adrien Maurice Dirac (1902–1984) developed a version of quantum mechanics that incorporated special relativity. Dirac's theory automatically explained the origin of the electron's spin and its magnetic moment. However, it also presented a major difficulty. Dirac's relativistic wave equation required solutions corresponding to negative energy states even for free particles. But if negative energy states existed, one would expect an electron in a state of positive energy to make a rapid transition to one of these states, emitting a photon in the process. Dirac avoided this difficulty by postulating that all negative energy states are filled. The multitude of electrons that occupy these negative energy states is called the *Dirac sea*. Electrons in the Dirac sea are not directly observable because the Pauli exclusion principle (see Section 29.6) does not allow them to react to external forces. However, if one of these negative energy states is vacant, leaving a hole in the sea of filled states, the hole can react to external forces and would be observable.

The second, equivalent interpretation of this theory is that **for every particle, there is an antiparticle.** The antiparticle would have the same mass as the particle, but their charges would be opposite. For example, the electron's antiparticle (called a *positron*) would have a mass of 0.511 MeV/c^2 and a positive charge of 1.60×10^{-19} C. Usually we denote an antiparticle with a bar over the symbol for the particle. For example, $\bar{e}$ denotes the antielectron or positron (although in this book the notation e^+ is preferred), and $\bar{p}$ denotes the antiproton.

The positron was discovered by Carl Anderson in 1932, and in 1936 he was awarded the Nobel prize for that achievement. Anderson discovered the positron while examining tracks created by electron-like particles of positive charge in a cloud chamber. (These early experiments used cosmic rays—mostly energetic protons passing through interstellar space—to initiate high-energy reactions in the upper atmosphere, which resulted in the production of positrons at ground level.) To discriminate between positive and negative charges, Anderson placed the cloud chamber in a magnetic field, causing moving charges to follow curved paths. He noted that some of the electron-like tracks deflected in a direction corresponding to a positively charged particle.

Paul Adrien Maurice Dirac (1902–1984), winner of the Nobel prize for physics in 1933. *(Courtesy AIP Emilio Segrè Visual Archives)*

Since Anderson's initial discovery, the positron has been observed in a number of experiments. Perhaps the most common process for producing positrons is **pair production.** In this process, a gamma ray photon with sufficiently high energy collides with a nucleus and an electron-positron pair is created. Because the total rest energy of the electron–positron pair is $2m_ec^2 = 1.02$ MeV (where m_e is the mass of the electron), the photon must have at least this much energy to create an electron–positron pair. Thus, electromagnetic energy in the form of a gamma ray is converted to mass in accordance with Einstein's famous relationship $E = mc^2$. Figure 31.1 shows tracks of electron–positron pairs created by 300-MeV gamma rays striking a lead sheet.

The reverse process can also occur. Under the proper conditions, an electron and positron can annihilate each other to produce two gamma-ray photons that have a combined energy of at least 1.02 MeV:

$$e^- + e^+ \longrightarrow 2\gamma$$

Prior to 1955, on the basis of the Dirac theory, it was expected that every particle had a corresponding antiparticle, but antiparticles such as the antiproton and antineutron had not been detected experimentally. In 1955 a team led by Emilio Segrè and Owen Chamberlain used the Bevatron particle accelerator at the University of California, Berkeley, to produce antiprotons and antineutrons, and thus established with certainty the existence of antiparticles. For this work Segrè and Chamberlain received the Nobel prize in 1959.

It is now established that every particle has a corresponding antiparticle of equal mass and spin and of equal and opposite charge, magnetic moment, and strangeness. (The property of *strangeness* is explained in Section 31.6.) The only exceptions to these rules for particles and antiparticles are the neutral photon, pion, and eta, each of which is its own antiparticle.

An intriguing aspect of antiparticle theory is that if we replace every proton, neutron, and electron in an atom with its antiparticle, we should create a stable

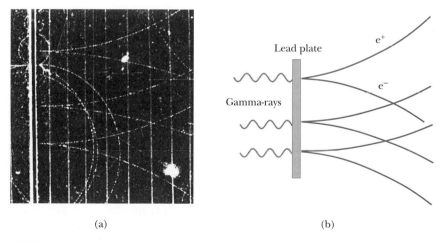

(a)   (b)

Figure 31.1 (a) Bubble-chamber tracks of electron–positron pairs produced by 300-MeV gamma rays striking a lead sheet. *(Courtesy Lawrence Berkeley Laboratory, University of California)*
(b) Sketch of the pertinent pair-production events. Note that the positrons deflect upward and the electrons deflect downward in an applied magnetic field directed into the page.

antiatom; combinations of antiatoms should form antimolecules and eventually antiworlds. As far as we know, everything would behave in the same way in an antiworld as in our world. In principle, it is possible that some distant antimatter galaxies exist, separated from normal-matter galaxies by millions of lightyears. Unfortunately, because the photon is its own antiparticle, there is no difference in the light emitted from matter and antimatter galaxies, and so astronomical observations cannot tell the two apart. Although no evidence of antimatter galaxies exists at present, it is awe-inspiring to imagine the cosmic spectacle that would result if matter and antimatter worlds were to collide: a gigantic eruption of jets of annihilation radiation, transforming the entire galactic mass into neutrinos, antineutrinos, and gamma rays, all fleeing the collision point at the speed of light.

Thinking Physics 1

When an electron and a positron meet at low speeds in free space, why are two 0.511-MeV gamma rays produced, rather than one gamma ray with an energy of 1.022 MeV?

Reasoning Gamma rays are photons, and photons carry momentum. If only one photon were produced, momentum would not be conserved, because the total momentum of the electron–positron system is approximately zero, whereas a single photon of energy 1.0022 MeV would have a very large momentum. However, the two gamma-ray photons that are produced travel off in opposite directions, so their total momentum is zero.

31.3 • MESONS AND THE BEGINNING OF PARTICLE PHYSICS

In the mid-1930s, physicists had a fairly simple view of the structure of matter. The building blocks were the proton, the electron, and the neutron. Three other particles were known or had been postulated at the time: the photon, the neutrino, and the positron. These six particles were considered the fundamental constituents of matter. With this marvelously simple picture of the world, however, no one was able to provide an answer to the following important question: Because many protons in proximity in any nucleus should strongly repel each other because of their like charges, what is the nature of the force that holds the nucleus together? Scientists recognized that this mysterious force must be much stronger than anything encountered in nature up to that time.

In 1935 the Japanese physicist Hideki Yukawa (1907–1981) proposed the first theory to successfully explain the nature of the nuclear force, an effort that later earned him the Nobel prize. To understand Yukawa's theory, it is useful to first recall that in the modern view of electromagnetic interactions, **charged particles interact by exchanging photons.** Yukawa used this idea to explain the nuclear force by proposing a new particle whose exchange between nucleons in the nucleus produces the nuclear force. Furthermore, he established that the range of the force is inversely proportional to the mass of this particle and predicted that the mass would be about 200 times the mass of the electron. Because the new particle would have a mass between that of the electron and that of the proton, it was called a **meson** (from the Greek *meso*, "middle").

Hideki Yukawa (1907–1981), a Japanese physicist, was awarded the Nobel prize in 1949 for predicting the existence of mesons. This photograph of Yukawa at work was taken in 1950 in his office at Columbia University.   *(UPI/Bettmann)*

In an effort to substantiate Yukawa's predictions, physicists began an experimental search for the meson by studying cosmic rays entering the Earth's atmosphere. In 1937, Carl Anderson and his collaborators discovered a particle of mass 106 MeV/c^2, about 207 times the mass of the electron. However, subsequent experiments showed that the particle interacted very weakly with matter and hence could not be the carrier of the nuclear force. The puzzling situation inspired several theoreticians to propose that there are two mesons with slightly different masses. This idea was confirmed by the discovery in 1947 of the pi (π) meson, or simply **pion,** by Cecil Frank Powell (1903–1969) and Guiseppe P. S. Occhialini (b. 1907). The particle discovered by Anderson in 1937, the one thought to be a meson, is not really a meson. Instead, it takes part in the weak and electromagnetic interactions only, and is now called the **muon.**

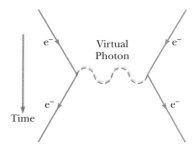

Figure 31.2 Feynman diagram showing how a photon mediates the electromagnetic force between two interacting electrons. The *t*-arrow shows the direction of increasing time.

The pion, Yukawa's carrier of nuclear force, comes in three varieties, corresponding to three charge states: π^+, π^-, and π^0. The π^+ and π^- particles have masses of 139.6 MeV/c^2, whereas the π^0 has a mass of 135.0 MeV/c^2. Pions and muons are very unstable particles. For example, the π^-, which has a mean lifetime of 2.6 × 10^{-8} s, first decays to a muon and an antineutrino. The muon, which has a mean lifetime of 2.2 μs, then decays into an electron, a neutrino, and an antineutrino:

$$\pi^- \longrightarrow \mu^- + \bar{\nu} \qquad\qquad \text{[31.1]}$$
$$\mu^- \longrightarrow e^- + \nu + \bar{\nu}$$

The interaction between two particles can be represented in a simple diagram called a *Feynman diagram*, developed by the American physicist Richard P. Feynman (1918–1988). Figure 31.2 is such a diagram for the electromagnetic interaction between two electrons. In this simple case, a photon is the field particle that mediates the electromagnetic force between the electrons. The photon transfers energy and momentum from one electron to the other in this interaction. The photon is called a *virtual photon* because it is reabsorbed by the emitting particle, without having been detected. Virtual photons can violate the law of conservation of energy by ΔE for a very short time, Δt, provided that $\Delta E \, \Delta t > \hbar/2$, and E is greater than the photon energy.

Now consider the pion exchange between a proton and a neutron via the nuclear force (Fig. 31.3). We can reason that the energy ΔE needed to create a pion of mass m_π is given by Einstein's equation $\Delta E = m_\pi c^2$. Again, the very existence of the pion would violate conservation of energy if it lasted for a time greater than $\Delta t \approx \hbar/2\,\Delta E$, (from the minimum form of the uncertainty principle), where ΔE is the energy of the pion and Δt is the time it takes the pion to transfer from one nucleon to the other. Therefore,

$$\Delta t \approx \frac{\hbar}{2\,\Delta E} = \frac{\hbar}{2m_\pi c^2} \qquad\qquad \text{[31.2]}$$

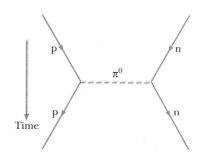

Figure 31.3 Feynman diagram representing a proton interacting with a neutron via the nuclear force. In this case, the pion mediates the nuclear force. The *t*-arrow shows the direction of increasing time.

Because the pion cannot travel faster than the speed of light, the maximum distance d it can travel in a time Δt is $c\,\Delta t$. Using Equation 31.2 and $d = c\,\Delta t$, we find this maximum distance to be

$$d \approx \frac{\hbar}{2m_\pi c} \qquad\qquad \text{[31.3]}$$

Richard Feynman (1918–1988)

with his son, Carl, in 1965. Feynman, together with Julian S. Schwinger and Shinichiro Tomonaga, won the 1965 Nobel prize for physics for fundamental work in the principles of quantum electrodynamics. His many important contributions to physics include the invention of simple diagrams to represent particle interactions graphically, the theory of the weak interaction of subatomic particles, a reformulation of quantum mechanics, the theory of superfluid helium, and his contribution to physics education through the magnificent three-volume text *The Feynman Lectures on Physics.*   *(UPI Telephotos)*

From Chapter 30, we know that the range of the strong force is approximately 1.5×10^{-15} m. Using this value for d in Equation 31.3, we calculate the rest energy of the pion to be

$$m_\pi c^2 \approx \frac{\hbar c}{2d} = \frac{(1.05 \times 10^{-34}\,\text{J}\cdot\text{s})(3.00 \times 10^8\,\text{m/s})}{2(1.5 \times 10^{-15}\,\text{m})}$$

$$\sim 100\ \text{MeV}$$

This corresponds to a mass of 100 MeV/c^2 (about 250 times the mass of the electron), a value in reasonable agreement with the observed pion mass.

The concept we have just described is quite revolutionary. In effect, it says that a proton can change into a proton plus a pion, as long as it returns to its original state in a very short time. Physicists often say that a nucleon undergoes "fluctuations" as it emits and absorbs pions. As we have seen, these fluctuations are a consequence of a combination of quantum mechanics (through the uncertainty principle) and special relativity (through Einstein's energy–mass relationship $E = mc^2$).

This section has dealt with the particles that mediate the strong force, namely the pions, and the mediators of the electromagnetic force, photons. Current ideas indicate that the nuclear force is more accurately described as an average or residual effect of the strong color force between quarks, as will be explained in Section 31.10. The graviton, which is the mediator of the gravitational force, has yet to be observed. The $W^\pm$ and Z^0 particles that mediate the weak force were discovered in 1983 by the Italian physicist Carlo Rubbia (b. 1934) and his associates using a proton–antiproton collider. Rubbia and Simon van der Meer, both at CERN, shared the 1984 Nobel prize for the discovery of the $W^\pm$ and Z^0 particles and the development of the proton–antiproton collider. In this accelerator, protons and antiprotons that have a momentum of 270 GeV/c undergo head-on collisions with each other. In some of the collisions $W^\pm$ and Z^0 particles are produced, which in turn are identified by their decay products.

31.4 • CLASSIFICATION OF PARTICLES

All particles other than photons can be classified into two broad categories, hadrons and leptons, according to their interactions.

Hadrons

Particles that interact through the strong force are called **hadrons.** There are two classes of hadrons, *mesons* and *baryons,* distinguished by their masses and spins.

Mesons all have zero or integer spin (0 or 1), with masses that lie between the mass of the electron and the mass of the proton. All mesons are known to decay finally into electrons, positrons, neutrinos, and photons. The pion is the lightest of known mesons; it has a mass of about 140 MeV/c^2 and a spin of 0. Another is the K meson, having a mass of about 500 MeV/c^2 and a spin of 0.

Baryons, the second class of hadrons, have masses equal to or greater than the proton mass (the name *baryon* means "heavy" in Greek), and their spin is always a noninteger value (1/2 or 3/2). Protons and neutrons are baryons, as are many other particles. With the exception of the proton, all baryons decay in such a way that the end products include a proton. For example, the baryon called the Ξ^-

TABLE 31.2   Some Particles and Their Properties

Category	Particle Name	Symbol	Anti-particle	Mass (MeV/c^2)	B	I_e	I_μ	L_τ	S	Lifetime(s)	Principal Decay Modes[a]
Leptons	Electron	e^-	e^+	0.511	0	+1	0	0	0	Stable	
	Electron–Neutrino	ν_e	$\bar{\nu}_e$	<7 eV/c^2	0	+1	0	0	0	Stable	
	Muon	μ^-	μ^+	105.7	0	0	+1	0	0	2.20×10^{-6}	$e^- \bar{\nu}_e \nu_\mu$
	Muon–Neutrino	ν_μ	$\bar{\nu}_\mu$	<0.3	0	0	+1	0	0	Stable	
	Tau	τ^-	τ^+	1784	0	0	0	+1	0	$<4 \times 10^{-13}$	$\mu^- \bar{\nu}_\mu \nu_\tau,\ e^- \bar{\nu}_e \nu_\tau$
	Tau–Neutrino	ν_τ	$\bar{\nu}_\tau$	<30	0	0	0	+1	0	Stable	
Hadrons											
Mesons	Pion	π^+	π^-	139.6	0	0	0	0	0	2.60×10^{-8}	$\mu^+ \nu_\mu$
		π^0	Self	135.0	0	0	0	0	0	0.83×10^{-16}	2γ
	Kaon	K^+	K^-	493.7	0	0	0	0	+1	1.24×10^{-8}	$\mu^+ \nu_\mu,\ \pi^+ \pi^0$
		K_S^0	$\bar{K}_S^0$	497.7	0	0	0	0	+1	0.89×10^{-10}	$\pi^+ \pi^-,\ 2\pi^0$
		K_L^0	$\bar{K}_L^0$	497.7	0	0	0	0	+1	5.2×10^{-8}	$\pi^\pm e^\mp \bar{\nu}_e,\ 3\pi^0$ $\pi^\pm \mu^\mp \bar{\nu}_\mu$
	Eta	η	Self	548.8	0	0	0	0	0	$<10^{-18}$	$2\gamma,\ 3\pi$
		η'	Self	958	0	0	0	0	0	2.2×10^{-21}	$\eta\pi^+ \pi^-$
Baryons	Proton	p	$\bar{p}$	938.3	+1	0	0	0	0	Stable	
	Neutron	n	$\bar{n}$	939.6	+1	0	0	0	0	920	$pe^- \bar{\nu}_e$
	Lambda	Λ^0	$\bar{\Lambda}^0$	1115.6	+1	0	0	0	-1	2.6×10^{-10}	$p\pi^-,\ n\pi^0$
	Sigma	Σ^+	$\bar{\Sigma}^-$	1189.4	+1	0	0	0	-1	0.80×10^{-10}	$p\pi^0,\ n\pi^+$
		Σ^0	$\bar{\Sigma}^0$	1192.5	+1	0	0	0	-1	6×10^{-20}	$\Lambda^0 \gamma$
		Σ^-	$\bar{\Sigma}^+$	1197.3	+1	0	0	0	-1	1.5×10^{-10}	$n\pi^-$
	Xi	Ξ^0	$\bar{\Xi}^0$	1315	+1	0	0	0	-2	2.9×10^{-10}	$\Lambda^0 \pi^0$
		Ξ^-	Ξ^+	1321	+1	0	0	0	-2	1.64×10^{-10}	$\Lambda^0 \pi^-$
	Omega	Ω^-	Ω^+	1672	+1	0	0	0	-3	0.82×10^{-10}	$\Xi^0 \pi^0,\ \Lambda^0 K^-$

[a] Notations in this column such as $p\pi-$, $n\pi^0$ mean two possible decay modes. In this case, the two possible decays are $\Lambda^0 \rightarrow p + \pi^-$ and $\Lambda^0 \rightarrow n + \pi^0$.

hyperon first decays to the Λ^0 baryon in about 10^{-10} s. The Λ^0 then decays to a proton and a π^- in approximately 3×10^{-10} s.

Today it is believed that hadrons are composed of more elemental units called quarks. Some of the important properties of hadrons are listed in Table 31.2. The symbols B, L_e, L_μ, L_τ, and S stand for baryon, electron, muon, and tau numbers, and strangeness, respectfully, and are explained in Sections 31.5 and 31.6.

Leptons

Leptons (from the Greek *leptos* meaning "small" or "light") are a group of particles that participate in the electromagnetic (if charged) and weak interactions. All leptons have spins of 1/2. Included in this group are electrons, muons, and neutrinos, which are less massive than the lightest hadron. Although hadrons have size and structure, **leptons appear to be truly elementary, point-like particles with no structure.**

Quite unlike hadrons, the number of known leptons is small. Currently, scientists believe there are only six leptons: the electron, the muon, and the tau and a neutrino associated with each. We now classify the six known leptons into three pairs called families:

$$\begin{pmatrix} e^- \\ \nu_e \end{pmatrix} \qquad \begin{pmatrix} \mu^- \\ \nu_\mu \end{pmatrix} \qquad \begin{pmatrix} \tau^- \\ \nu_\tau \end{pmatrix}$$

Current evidence suggests that neutrinos travel with the speed of light and, therefore have zero mass. As we shall see later, a firm knowledge of the neutrino's mass (it may be small, not exactly zero as in Table 31.2) could have great significance in cosmological models and predictions of the future of the Universe. Finally, note that each of the six leptons has a corresponding antiparticle.

31.5 • CONSERVATION LAWS

Conservation laws are important in understanding why certain decays and reactions occur and others do not. In general, the laws of conservation of energy, linear momentum, angular momentum, and electric charge provide us with a set of rules that all processes must follow.

Certain new conservation laws are important in the study of elementary particles. Two of these laws, concerning baryon number and lepton number, are described in this section, and others will be discussed later in this chapter. Although the two described here have no theoretical foundation, they are supported by abundant empirical evidence.

Baryon Number

Conservation of baryon number tells us that whenever a baryon is created in a reaction or decay, an antibaryon is also created. This scheme can be quantified by assigning a baryon number $B = +1$ for all baryons, $B = -1$ for all antibaryons, and $B = 0$ for all other particles. Thus, the **law of conservation of baryon number** states that **whenever a nuclear reaction or decay occurs, the sum of the baryon numbers before the process must equal the sum of the baryon numbers after the process.** An equivalent statement is that the net number of baryons remains constant in any process.

Conservation of baryon • number

If baryon number is absolutely conserved, the proton must be absolutely stable. If it were not for the law of conservation of baryon number, the proton could decay to a positron and a neutral pion. However, such a decay has never been observed. At the present, we can say only that the proton has a half-life of at least 10^{31} years (the estimated age of the Universe is only 10^{10} years). In one recent version of a grand unified theory, physicists predicted that the proton is unstable. According to this theory, the baryon number is not absolutely conserved.

Example 31.1   Checking Baryon Numbers

Determine whether or not the following reactions can occur based on the law of conservation of baryon number.

$$(1) \qquad p + n \longrightarrow p + p + n + \overline{p}$$

$$(2) \qquad p + n \longrightarrow p + p + \overline{p}$$

Solution   For reaction 1, recall that $B = +1$ for baryons and $B = -1$ for antibaryons. Hence the left side of reaction 1

gives a total baryon number of $1 + 1 = 2$. The right side of reaction 1 gives a total baryon number of $1 + 1 + 1 + (-1) = 2$. Thus the reaction can occur provided the incoming proton has sufficient energy.

The left side of reaction 2 gives a total baryon number of $1 + 1 = 2$. However, the right side gives $1 + 1 + (-1) = 1$. Because the baryon number is not conserved, the reaction cannot occur.

Lepton Number

There are three conservation laws involving lepton numbers, one for each variety of lepton. The **law of conservation of electron–lepton number** states that **the sum of the electron–lepton numbers before a reaction or decay must equal the sum of the electron–lepton numbers after the reaction or decay.**

- *Conservation of lepton number*

The electron and the electron neutrino are assigned a positive lepton number $L_e = +1$, the antileptons e^+ and $\bar{\nu}_e$ are assigned a negative lepton number $L_e = -1$, and all others have $L_e = 0$. For example, consider the decay of the neutron

$$n \longrightarrow p + e^- + \bar{\nu}_e$$

Before the decay, the electron–lepton number is $L_e = 0$; after the decay it is $0 + 1 + (-1) = 0$. Thus, the electron–lepton number is conserved. It is important to recognize that the baryon number must also be conserved. This can easily be checked by noting that before the decay $B = +1$, and after the decay B is $+1 + 0 + 0 = +1$.

In a similar way, when a decay involves muons, the muon–lepton number, L_μ, is conserved. The μ^- and the ν_μ are assigned positive numbers, $L_\mu = +1$; the antimuons μ^+ and $\bar{\nu}_\mu$ are assigned negative numbers, $L_\mu = -1$; and all others have $L_\mu = 0$. Finally, the tau–lepton number, L_τ, is conserved, and similar assignments can be made for the tau–lepton and its neutrino.

Thinking Physics 2

A student claims to have observed a decay of an electron into two neutrinos, traveling in opposite directions. What conservation laws would be violated by this decay?

Reasoning Several conservation laws are violated. Conservation of electric charge is violated, because the negative charge of the electron has disappeared. Conservation of angular momentum is violated, because the original angular momentum is that of the electron, with spin $1/2$, and there are two spin-$1/2$ particles after the decay. Conservation of electron lepton number is also violated, because there is one lepton before the decay and two afterward. If both neutrinos were electron–neutrinos, electron–lepton number conservation would be violated in the final state. However, if one of the product neutrinos were other than an electron–neutrino, then another lepton conservation law would be violated, since there were no other leptons in the initial state.

Other conservation laws are obeyed by this decay. Energy can be conserved—the rest energy of the electron appears as the kinetic energy (and possibly some small rest energy) of the neutrinos. The opposite directions of the velocities of the two neutrinos allows for conservation of momentum. Conservation of baryon number and conservation of other lepton numbers are also upheld in the decay.

Example 31.2 Checking Lepton Numbers

Determine which of the following decay schemes can occur on the basis of conservation of electron–lepton number.

$$(1) \qquad \mu^- \longrightarrow e^- + \bar{\nu}_e + \nu_\mu$$

$$(2) \qquad \pi^+ \longrightarrow \mu^+ + \nu_\mu + \nu_e$$

Solution Because decay 1 involves both a muon and an electron, L_μ and L_e must both be conserved. Before the decay, $L_\mu = +1$ and $L_e = 0$. After the decay, $L_\mu = 0 + 0 + 1 = +1$, and $L_e = +1 - 1 + 0 = 0$. Thus, both numbers are conserved, and on this basis the decay mode is possible.

Before decay 2 occurs, $L_\mu = 0$ and $L_e = 0$. After the decay, $L_\mu = -1 + 1 + 0 = 0$, but $L_e = +1$. Thus, the decay is not possible because the electron–lepton number is not conserved.

EXERCISE 1   Determine whether the decay $\mu^- \rightarrow e^- + \bar{\nu}_e$ can occur.   Answer   No. The muon–lepton number is $+1$ before the decay and 0 after.

31.6 • STRANGE PARTICLES AND STRANGENESS

Many particles discovered in the 1950s were produced by the nuclear interaction of pions with protons and neutrons in the atmosphere. A group of these particles—namely the kaon (K), lambda (Λ), and sigma (Σ) particles—exhibited unusual properties in production and decay and hence were called *strange particles*.

One unusual property is that these particles were always produced in pairs. For example, when a pion collides with a proton, two neutral strange particles were produced with high probability (Fig. 31.4):

$$\pi^- + \mathrm{p} \longrightarrow \mathrm{K}^0 + \Lambda^0$$

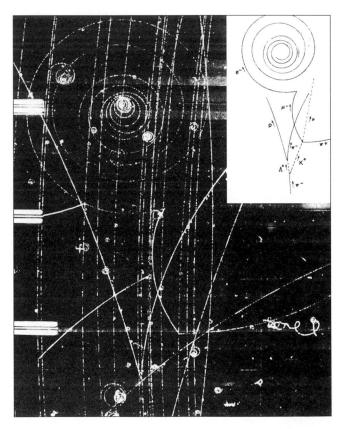

Figure 31.4   This bubble-chamber photograph shows many events, and the inset is a drawing of identified tracks. The strange particles Λ^0 and K^0 are formed at the bottom as the π^- interacts with a proton according to $\pi^- + \mathrm{p} \rightarrow \Lambda^0 + \mathrm{K}^0$. (Note that the neutral particles leave no tracks, as indicated by the dashed lines.) The Λ^0 and K^0 may then decay according to $\Lambda^0 \rightarrow \pi^- + \mathrm{p}$ and $\mathrm{K}^0 \rightarrow \pi^+ + \mu^- + \bar{\nu}_\mu$.   *(Courtesy Lawrence Berkeley Laboratory, University of California, Photographic Services.)*

However, the reaction $\pi^- + p \rightarrow K^0 + n$ never occurred, even though no known conservation laws would have been violated and the energy of the pion was sufficient to initiate the reaction.

The second peculiar feature of strange particles is that, although they are produced by the strong interaction at a high rate, they do not decay into particles that interact via the strong force at a very high rate. Instead, they decay very slowly, which is characteristic of the weak interaction, as shown in Table 31.1. Their half-lives are in the range 10^{-10} s to 10^{-8} s; most other particles that interact via the strong force have lifetimes on the order of 10^{-20} s or less.

To explain these unusual properties of strange particles, a law called conservation of strangeness was introduced, together with a new quantum number S, called **strangeness.** The strangeness numbers for some particles are given in Table 31.2. The production of strange particles in pairs is explained by assigning $S = +1$ to one of the particles and $S = -1$ to the other. All nonstrange particles are assigned strangeness $S = 0$. The **law of conservation of strangeness** states that **whenever a nuclear reaction or decay occurs, the sum of the strangeness numbers before the process must equal the sum of the strangeness numbers after the process.**

• *Conservation of strangeness number*

The slow decay of strange particles can be explained by assuming that the strong and electromagnetic interactions obey the law of conservation of strangeness, but the weak interaction does not. Because the decay reaction involves the loss of one strange particle, it violates strangeness conservation and, hence, proceeds slowly via the weak interaction.

Example 31.3   Is Strangeness Conserved?

(a) Determine whether the following reaction occurs on the basis of conservation of strangeness.

$$\pi^0 + n \longrightarrow K^+ + \Sigma^-$$

Solution   The initial state has strangeness $S = 0 + 0 = 0$. Because the strangeness of the K^+ is $S = +1$ and the strangeness of the Σ^- is $S = -1$, the strangeness of the final state is $+1 - 1 = 0$. Thus, strangeness is conserved and the reaction is allowed.

(b) Show that the following reaction does not conserve strangeness.

$$\pi^- + p \longrightarrow \pi^- + \Sigma^+$$

Solution   The initial state has strangeness $S = 0 + 0 = 0$, and the final state has strangeness $S = 0 + (-1) = -1$. Thus, strangeness is not conserved.

EXERCISE 2   Show that the reaction $p + \pi^- \rightarrow K^0 + \Lambda^0$ obeys the law of conservation of strangeness.

31.7 • HOW ARE ELEMENTARY PARTICLES PRODUCED AND PARTICLE PROPERTIES MEASURED?

Examination of the bewildering array of entries in Table 31.2 leaves one yearning for firm ground. In fact it is natural to wonder about an entry that shows a particle that exists for 10^{-20} s and has a mass of 1192.5 MeV/c^2. How is it possible to detect a particle that exists for only 10^{-20} s, and furthermore how can its mass be measured? If a standard attribute of a particle is some type of permanence or stability, in what sense is a fleeting entity that exists for 10^{-20} s a particle? In this section we attempt to answer such questions and explain how elementary particles are produced and how their properties are measured.

Elementary particles, most of which are unstable and created naturally only rarely in cosmic ray showers, are abundantly created in human-made collisions of high energy particles with a suitable target. Because very high energy beams of incident particles are desirable, stable charged particles such as electrons or protons generally make up the incident beam. It takes a considerable time to accelerate particles to high energies by the use of electromagnetic fields so we want to accelerate particles with long lifetimes. In a similar way, targets must be simple and stable and the simplest target, hydrogen, serves nicely as both a target and a detector. In a liquid hydrogen bubble chamber, which is basically a large container filled with hydrogen near its boiling point, a charged particle traversing the chamber ionizes the atoms along its path, and the ionization causes a visible track of tiny bubbles. The liquid hydrogen also serves as an efficient source of target protons, with a proton density sufficient to ensure many incident particle–target collisions within a reasonable time. Figure 31.4 shows a typical event in which a bubble chamber has served as both target source and detector. In this figure many parallel tracks of negative pions are visible entering the photograph from the bottom. One of the pions has hit a stationary proton in the hydrogen and produced two strange particles, the Λ^0 and K^0 according to the reaction

$$\pi^- + p \longrightarrow \Lambda^0 + K^0$$

Neither neutral strange particle leaves a track, but their subsequent decays into charged particles can be clearly seen as indicated in Figure 31.4. A magnetic field directed into the plane of the photograph causes the track of each charged particle to curve, and from the measured curvature one can determine the particle's charge and linear momentum. If the mass and momentum of the incident particle are known, we can then usually calculate the product particle mass, kinetic energy, and speed from conservation of momentum and energy. Finally, combining a product particle's speed with a measurable decay track length, we can calculate the product particle's lifetime. Figure 31.4 shows that sometimes one can use this lifetime technique even for a neutral particle that leaves no track. As long as the start and finish of the missing track are known as well as the particle speed, one can infer the missing track length and find the lifetime of the neutral particle.

Resonance Particles

With clever experimental technique and much effort, decay track lengths as short as 1 micron (10^{-6} m) can be measured. This means that lifetimes as short as 10^{-16} s can be measured with this technique for the case of high energy particles traveling at about the speed of light. We arrive at this result by assuming that a decaying particle travels 1 micron in the lab at a speed of $0.99c$ yielding a lab lifetime of $\Delta t_{\text{lab}} = 10^{-6}$ m$/0.99c \approx 0.33 \times 10^{-14}$ s. Now, however, relativity helps us. Because the proper lifetime as <u>measured in</u> the decaying particle's restframe is shorter than Δt_{lab} by a factor of $\sqrt{1 - (v^2/c^2)}$, we can actually measure lifetimes of duration

$$\Delta t_{\text{p}} = \Delta t_{\text{lab}} \sqrt{1 - \frac{v^2}{c^2}} = (0.33 \times 10^{-14} \text{ s}) \sqrt{1 - \frac{(0.99 \ c)^2}{c^2}} = 4 \times 10^{-16} \text{ s}$$

Unfortunately, even with Einstein's help, we are several orders of magnitude away from 10^{-20} to 10^{-23} s lifetimes with the best efforts of the track-length method.

How, then, can we detect the presence of particles that exist for as short a time as 10^{-23} s? As we shall see, these very short-lived particles, known as **resonance particles,** can only have their mass, lifetime, and very existence inferred indirectly from peaks (resonances) in the cross section versus energy plots describing their decay products.

Let's consider this in more detail by looking at the case of the resonance particle called the delta plus (Δ^+) which has a mass of 1232 MeV/c^2 and a lifetime of about 10^{-23} s. The Δ^+ is produced in the reaction

$$e^- + p \longrightarrow e^- + \Delta^+ \qquad \text{[31.4]}$$

followed in 10^{-23} s by the decay of the delta plus according to

$$\Delta^+ \longrightarrow \pi^+ + n \qquad \text{[31.5]}$$

Because the Δ^+ lifetime is so short it leaves no measurable track and it might seem impossible to distinguish the reactions given in Equations 31.4 and 31.5 from the net direct reaction in which no Δ^+ is produced:

$$e^- + p \longrightarrow e^- + \pi^+ + n \qquad \text{[31.6]}$$

In fact we can tell whether a Δ^+ was formed by measuring the momentum and energy of the suspected decay products (pion and neutron) and using the conservation of momentum and energy to see if these values combine to give a Δ^+ mass of 1232 MeV/c^2.

The typical method for showing the existence of a resonance particle involves analyzing a large number of events in which a π^+ and a neutron are produced and then plotting a histogram of these events. (A *histogram* in this case is a plot of the number of events in a given energy range versus energy.) Following this procedure, we obtain a slowly varying curve with sharp peaks superimposed showing the existence of resonance particles. Figure 31.5 is an experimental histogram for the Δ^+

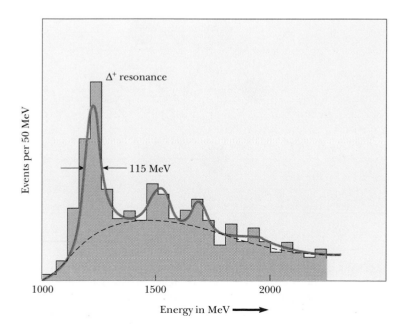

Figure 31.5   Experimental evidence for the existence of the Δ^+ particle. The sharp peak at 1232 MeV was produced by the events in which a Δ^+ formed and promptly decayed into a π^+ and neutron.

particle. The broad background (dashed curve) is produced by direct events in which no Δ^+ was created; the sharp peak at 1232 MeV containing many events was produced by all those events in which a Δ^+ was formed and decayed into a pion and neutron with just the right energies and momenta to make up a delta particle. Peaks corresponding to two other resonance particles of larger mass can also be seen in Figure 31.5.

Histograms can tell us not only the mass of a short-lived particle but also the lifetime of the particle. The width of the resonance peak and the uncertainty relation $\Delta E\,\Delta t \cong \hbar/2$ are used to infer the lifetime Δt of the particle. The measured width of 115 MeV from Figure 31.5 leads to a lifetime of 5.7×10^{-24} s for the delta particle. In this incredibly short lifetime, a delta particle moving at the highest possible speed of c only travels 10^{-15} m, or about one nuclear diameter.

31.8 • THE EIGHTFOLD WAY

As we have seen, we associate conserved quantities such as spin, baryon number, lepton number, and strangeness with particles. Many classification schemes have been proposed that group particles into families. Consider the first eight baryons listed in Table 31.2, all having a spin of 1/2. If we plot their strangeness versus their charge using a sloping coordinate system, as in Figure 31.6a, a fascinating pattern is observed. Six of the baryons form a hexagon with the remaining two at its center.

Now consider the family of mesons listed in Table 31.2, which have spins of zero. If we count both particles and antiparticles, there are nine such mesons. Figure 31.6b is a plot of strangeness versus charge for this family. Again, a fascinating hexagonal pattern emerges. In this case, the particles on the perimeter of the

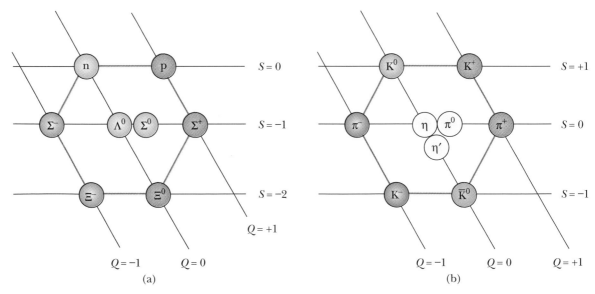

Figure 31.6   (a) The hexagonal eightfold-way pattern for the eight spin-1/2 baryons. This strangeness-versus-charge plot uses a horizontal axis for the strangeness values, *S*, but a sloping axis for the charge number, *Q*. (b) The eightfold-way pattern for the nine spin-zero mesons.

hexagon lie opposite their antiparticles, and the remaining three (which form their own antiparticles) are at its center. These and related symmetric patterns, called the **eightfold way,** were proposed independently in 1961 by Murray Gell-Mann and Yuval Ne'eman.

The groups of baryons and mesons can be displayed in many other symmetrical patterns within the framework of the eightfold way. For example, the family of spin-3/2 baryons contains ten particles arranged in a pattern like the tenpins in a bowl-ing alley. When the pattern was proposed in 1961, one of the particles was missing—it had yet to be discovered. Gell-Mann predicted that the missing particle, which he called the omega minus (Ω^-), should have a spin of 3/2, a charge of -1, a strangeness of -3, and a rest energy of about 1680 MeV. Shortly thereafter, in 1964, scientists at the Brookhaven National Laboratory found the missing particle through careful analyses of bubble chamber photographs and confirmed all its predicted properties.

The patterns of the eightfold way in particle physics have much in common with the periodic table. Whenever a vacancy (a missing particle or element) occurs in the organized patterns, experimentalists have a guide for their investigations.

31.9 · QUARKS

As we have noted, leptons appear to be truly elementary particles because they have no measurable size or internal structure, are limited in number, and do not seem to break down into smaller units. Hadrons, however, are complex particles having size and structure. The existence of the eightfold-way patterns suggests that baryons and mesons—in other words, hadrons—have a more elemental substructure. Fur-thermore, we know that hadrons decay into other hadrons and are many in number. Table 31.2 lists only those hadrons that are stable against hadronic decay; hundreds of others have been discovered. These facts strongly suggest that hadrons cannot be truly elementary. In this section we show that the complexity of hadrons can be explained by a simpler substructure.

The Original Quark Model

In 1963 Gell-Mann and George Zweig independently proposed that hadrons have a more elemental substructure. According to their model, all hadrons are compos-ite systems of two or three fundamental constituents called **quarks** (pronounced to rhyme with *forks*). Gell-Mann borrowed the word *quark* from the passage "Three quarks for Muster Mark" in James Joyce's *Finnegan's Wake*. In the original quark model, there were three types of quarks designated by the symbols u, d, and s. These were given the arbitrary names *up, down,* and *sideways* (or now more com-monly, *strange*).

An unusual property of quarks is that they have fractional electronic charges. The u, d, and s quarks have charges of $+2e/3$, $-e/3$, and $-e/3$, respectively. Each quark has a baryon number of 1/3 and a spin of 1/2. (The spin 1/2 means that all quarks are **fermions,** defined as any particle having half-integral spin.) The u and d quarks have strangeness 0, and the s quark has strangeness -1. Other properties of quarks and antiquarks are given in Table 31.3. Associated with each quark is an antiquark of opposite charge, baryon number, and strangeness.

American physicist Murray Gell-Mann was awarded the Nobel prize in 1969 for his theoretical studies dealing with subatomic particles. *(Photo courtesy of Michael R. Dressler)*

TABLE 31.3   Properties of Quarks and Antiquarks

				Quarks				
Name	Symbol	Spin	Charge	Baryon Number	Strangeness	Charm	Bottomness	Topness
Up	u	$\frac{1}{2}$	$+\frac{2}{3}e$	$\frac{1}{3}$	0	0	0	0
Down	d	$\frac{1}{2}$	$-\frac{1}{3}e$	$\frac{1}{3}$	0	0	0	0
Strange	s	$\frac{1}{2}$	$-\frac{1}{3}e$	$\frac{1}{3}$	-1	0	0	0
Charmed	c	$\frac{1}{2}$	$+\frac{2}{3}e$	$\frac{1}{3}$	0	$+1$	0	0
Bottom	b	$\frac{1}{2}$	$-\frac{1}{3}e$	$\frac{1}{3}$	0	0	$+1$	0
Top	t	$\frac{1}{2}$	$+\frac{2}{3}e$	$\frac{1}{3}$	0	0	0	$+1$

				Antiquarks				
Name	Symbol	Spin	Charge	Baryon Number	Strangeness	Charm	Bottomness	Topness
Anti-up	$\bar{u}$	$\frac{1}{2}$	$-\frac{2}{3}e$	$-\frac{1}{3}$	0	0	0	0
Anti-down	$\bar{d}$	$\frac{1}{2}$	$+\frac{1}{3}e$	$-\frac{1}{3}$	0	0	0	0
Anti-strange	$\bar{s}$	$\frac{1}{2}$	$+\frac{1}{3}e$	$-\frac{1}{3}$	$+1$	0	0	0
Anti-charmed	$\bar{c}$	$\frac{1}{2}$	$-\frac{2}{3}e$	$-\frac{1}{3}$	0	-1	0	0
Anti-bottom	$\bar{b}$	$\frac{1}{2}$	$+\frac{1}{3}e$	$-\frac{1}{3}$	0	0	-1	0
Anti-top	$\bar{t}$	$\frac{1}{2}$	$-\frac{2}{3}e$	$-\frac{1}{3}$	0	0	0	-1

TABLE 31.4
Quark Composition of Several Hadrons

Particle	Quark Composition
Mesons	
π^+	$u\bar{d}$
π^-	$\bar{u}d$
K^+	$u\bar{s}$
K^-	$\bar{u}s$
K^0	$d\bar{s}$
Baryons	
p	uud
n	udd
Λ^0	uds
Σ^+	uus
Σ^0	uds
Σ^-	dds
Ξ^0	uss
Ξ^-	dss
Ω^-	sss

The composition of all hadrons known when Gell-Mann and Zweig presented their models could be completely specified by three simple rules:

1. Mesons consist of one quark and one antiquark, giving them a baryon number of 0 as required.
2. Baryons consist of three quarks.
3. Antibaryons consist of three antiquarks.

Table 31.4 lists the quark compositions of several mesons and baryons. Note that just two of the quarks, u and d, are contained in all hadrons encountered in ordinary matter (protons and neutrons). The third quark, s, is needed only to construct strange particles with a strangeness number of either $+1$ or -1. Figure 31.7 is a pictorial representation of the quark composition of several particles.

Charm and Other Recent Developments

Although the original quark model was highly successful in classifying particles into families, there were some discrepancies between predictions of the model and certain experimental decay rates. Consequently, a fourth quark was proposed by several physicists in 1967. They argued that if there are four leptons (as was thought at the time), then there should also be four quarks because of an underlying symmetry in nature. The fourth quark, designated by c, was given a property called **charm.** A *charmed* quark would have charge $+2e/3$, but its charm would distinguish it from the other three quarks. The new quark would have a charm of $C = +1$, its antiquark

Mesons   Baryons
π^+ p

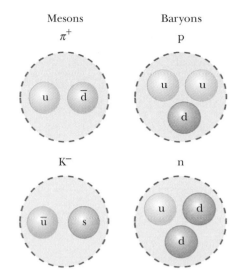

K⁻   n

Figure 31.7 Quark compositions of two mesons and two baryons. Note that the mesons on the left each contain two quarks, whereas the baryons on the right each contain three quarks.

would have a charm of $C = -1$, and all other quarks would have $C = 0$, as indicated in Table 31.3. Charm, like strangeness, would be conserved in strong and electromagnetic interactions but not in weak interactions.

In 1974, a new heavy meson called the J/Ψ particle (or simply Ψ) was discovered independently by two groups: one led by Burton Richter at the Stanford Linear Accelerator (SLAC) and the other led by Samuel Ting at the Brookhaven National Laboratory. Richter and Ting were awarded the Nobel prize in 1976 for this work. The J/Ψ particle did not fit into the three-quark model but had the properties of a combination of a charmed quark and its antiquark (cc). It was much more massive than the other known mesons (~ 3100 MeV/c^2), and its lifetime was much longer than those that decay via the strong force. Soon, related charmed mesons were discovered corresponding to such quark combinations as $\overline{c}d$ and $c\overline{d}$, all of which have large masses and long lifetimes. In 1975 researchers at Stanford University reported strong evidence for the tau (τ) lepton, mass 1784 MeV/c^2. Such discoveries led to more elaborate quark models, and the proposal of two new quarks, *top* (t) and *bottom* (b). (Some physicists prefer the designators *truth* and *beauty*.) To distinguish these quarks from the old ones, quantum numbers called *topness* and *bottomness* were assigned to these new particles and are included in Table 31.3. In 1977, researchers at the Fermi National Laboratory, under the direction of Leon Lederman, reported the discovery of a very massive new meson, Υ (the Greek letter upsilon), the composition of which is considered to be b$\overline{b}$. In March of 1995, researchers at Fermilab announced the discovery of the top quark (supposedly the last of the quarks to be found) having mass 173 GeV/c^2.

You are probably wondering whether or not such discoveries will ever end. How many "building blocks" of matter really exist? At the present, physicists believe that the fundamental particles in nature are six quarks and six leptons (together with their antiparticles). Table 31.5 lists some of the properties of these particles as well as the properties of the fundamental field particles.

Despite many extensive experimental efforts, no isolated quark has ever been observed. Physicists now believe that quarks are permanently confined inside ordinary particles because of an exceptionally strong force that prevents them from

TABLE 31.5   **Properties of the Fundamental Point Particles**

Particle		Mass in Terms of Proton Mass	Interactions Experienced[a]	Interaction Mediated	Electric Charge in Units of |e|	Color Charge
Matter Particles						
Leptons	e^-, μ, τ	$\frac{1}{1836}, \frac{1}{9}, 1.9$	EM, W	None	-1	No
	ν_e, ν_μ, ν_τ	$<10^{-8}, <\frac{1}{3500}, <\frac{1}{30}$	W	None	0	No
Quarks	u, c, t	$\frac{1}{235}, 1.6, 165$	EM, W, S	None	$+\frac{2}{3}$	Yes
	d, s, b	$\frac{1}{135}, \frac{1}{6}, 5.2$	EM, W, S	None	$-\frac{1}{3}$	Yes

All are fermions with spin $\frac{1}{2}$.

Particle		Mass in Terms of Proton Mass	Interactions Experienced[a]	Interaction Mediated	Electric Charge in Units of |e|	Color Charge
Field Particles						
Photon	γ	0	EM	EM	0	No
Intermediate	$W^\pm$	85	W, EM	W	± 1	No
bosons	Z^0	97	W	W	0	No
Gluons	g	0	S	S	0	Yes

All are bosons with spin 1.

[a] EM = electromagnetic; W = weak; S = strong.

escaping. This force, called the "color" force (discussed in Section 31.10), increases with separation distance (similar to the force of a spring). The great strength of the force between quarks has been described by one author as follows:[1]

> Quarks are slaves of their own color charge . . . bound like prisoners of a chain gang . . . Any locksmith can break the chain between two prisoners, but no locksmith is expert enough to break the gluon chains between quarks. Quarks remain slaves forever.

Thinking Physics 3

We have seen a law of conservation of *lepton number* and a law of conservation of *baryon number*. Why isn't there a law of conservation of *meson number*?

Reasoning We can argue this from the point of view of creating particle–antiparticle pairs from available energy. If energy is converted to rest energy of a lepton–antilepton pair, then there is no net change in lepton number, because the lepton has a lepton number of +1 and the antilepton −1. Energy could also be transformed into rest energy of a baryon–antibaryon pair. The baryon has baryon number +1, the antibaryon −1, and there is no net change in baryon number.

But now, suppose energy is transformed into rest energy of a quark–antiquark pair. By definition in quark theory, a quark–antiquark pair *is a meson*. Thus, we have created a meson from energy—there was no meson before; now there is. Thus, there is no conservation of meson number. With more energy, we can create more mesons, with no restriction from a conservation law other than that of energy.

[1]H. Fritzsch, *Quarks, The Stuff of Matter*, London, Allen Lane, 1983.

CONCEPTUAL PROBLEM 1

If high energy electrons, with de Broglie wavelengths smaller than the size of the nucleus, are scattered from nuclei, the behavior of the electrons is consistent with scattering from very massive structures much smaller in size than the nucleus—quarks. How is this similar to another classic experiment that detected small structures in an atom?

CONCEPTUAL PROBLEM 2

Doubly charged baryons, such as the Δ^{++}, are known to exist. Why are there no doubly charged mesons?

31.10 · COLORED QUARKS

Shortly after the concept of quarks was proposed, scientists recognized that certain particles had quark compositions that were in violation of the Pauli exclusion principle. Because all quarks are fermions with spins of 1/2, they are expected to follow the exclusion principle. One example of a particle that violates the exclusion principle is the Ω^- (sss) baryon that contains three s quarks having parallel spins, giving it a total spin of 3/2. Other examples of baryons that have identical quarks with parallel spins are the Δ^{++} (uuu) and the Δ^- (ddd). To resolve this problem, Moo-Young Han and Yoichiro Nambu suggested in 1965 that quarks possess a new property called **color.** This property is similar in many respects to electric charge except that it occurs in three varieties called red, green, and blue. (The antiquarks have the colors antired, antigreen, and antiblue.) To satisfy the exclusion principle, all three quarks in a baryon must have different colors. Just as a combination of actual colors of light can produce the neutral color white, a combination of three quarks with different colors is also white, or colorless. A meson consists of a quark of one color and an antiquark of the corresponding anticolor. The result is that baryons and mesons are always colorless (or white).

Although the concept of color in the quark model was originally conceived to satisfy the exclusion principle, it also provided a better theory for explaining certain experimental results. For example, the modified theory correctly predicts the lifetime of the π^0 meson. The theory of how quarks interact with each other is called **quantum chromodynamics,** or QCD, to parallel quantum electrodynamics (the theory of interaction between electric charges). In QCD, the quark is said to carry a **color charge,** in analogy to electric charge. The strong force between quarks is often called the **color force.**

As stated earlier, the strong interaction between hadrons is mediated by massless particles called **gluons** (g), which are analogous to photons for the electromagnetic force. According to QCD, there are eight gluons, all with color charge. When a quark emits or absorbs a gluon, its color changes. For example, a blue quark that emits a gluon may become a red quark, and the red quark that absorbs this gluon becomes a blue quark. The color force between quarks is analogous to the electric force between charges; like colors repel and opposite colors attract. Therefore, two red quarks repel each other, but a red quark will be attracted to an antired quark. The attraction between quarks of opposite color to form a meson ($q\overline{q}$) is indicated in Figure 31.8a. Differently colored quarks also attract each other, but with less intensity than opposite colors of quark and antiquark. For example, a cluster of red, blue, and green quarks all attract each other to form baryons, as

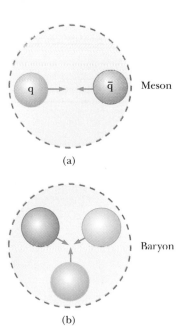

Meson

(a)

Baryon

(b)

Figure 31.8 (a) A red quark is attracted to an antired quark. This forms a meson the quark structure of which is ($q\overline{q}$). (b) Three different colored quarks attract each other to form a baryon.

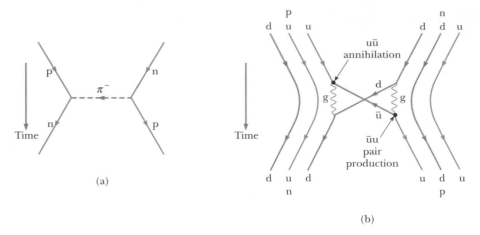

Figure 31.9   (a) A nuclear interaction between a proton and a neutron explained in terms of Yukawa's pion exchange model. (b) The same interaction as in (a), explained in terms of quarks and gluons. Note that the exchanged $\bar{u}d$ quark pair makes up a π^- meson.

indicated in Figure 31.8b. Thus, every baryon contains three quarks of three different colors.

Although the color force between two color-neutral hadrons is negligible at large separations, the strong color force between their constituent quarks does not exactly cancel at small separations. This residual strong force is in fact the nuclear force that binds protons and neutrons to form nuclei. In Section 31.3 we explained the nuclear interaction of a proton with a neutron using Yukawa's theory of pion exchange. According to QCD, a more basic explanation of nuclear force can be given in terms of quarks and gluons as shown by contrasting Feynman diagrams of the same process in Figure 31.9. Each quark within the neutron and proton is continually emitting and absorbing virtual gluons and creating and annihilating virtual $(q\bar{q})$ pairs. When the neutron and proton approach within 1 fm of each other, these virtual gluons and quarks can be exchanged between the two nucleons, and such exchanges produce the nuclear force. Figure 31.9b depicts one likely possibility or contribution to the process shown in Figure 31.9a. A down quark emits a virtual gluon (represented by a wavy line, g), which creates a $u\bar{u}$ pair. Both the recoiling d quark and the $\bar{u}$ are transmitted to the proton where the $\bar{u}$ annihilates a proton u quark (with the creation of a gluon) and the d is captured.

31.11 · ELECTROWEAK THEORY AND THE STANDARD MODEL

Recall that the weak interaction is an extremely short range force with an interaction distance of approximately 10^{-18} m. Such a short range interaction implies that the quantized particles that carry the weak field (the spin one W^+, W^-, and Z^0 bosons) are quite massive as is indeed the case (see Table 31.5). These bosons are especially amazing when we think of them as structureless, pointlike particles as massive as krypton atoms. As mentioned earlier, the weak interaction is responsible for neutron decay and the beta decay of other heavier baryons. More important, the weak interaction is responsible for the decay of the c, s, b, t quarks into lighter, more stable u and d quarks as well as the decay of the massive μ and τ leptons into (lighter) electrons. Thus, **the weak interaction is very important because it governs the stability of the basic matter particles.**

A mysterious feature of the weak interaction is its lack of symmetry, especially compared to the high degree of symmetry shown by the strong, electromagnetic, and gravitational interactions. For example, the weak interaction, unlike the strong interaction, is not symmetric under mirror reflection or charge exchange. (Mirror reflection means that all the quantities in a given particle reaction are exchanged as in a mirror reflection—left for right, an inward motion toward the mirror for an outward motion. Charge exchange means that all the electric charges in a particle reaction are converted to their opposites—all positives to negatives and vice versa.) When we say that the weak interaction is not symmetric, we mean that the reaction with all quantities changed occurs less frequently than the direct reaction. For example, the decay of the K^0, which is governed by the weak interaction, is not symmetric under charge exchange, because $K^0 \rightarrow \pi^- + e^+ + \nu_e$ occurs much more frequently than $K^0 \rightarrow \pi^+ + e^- + \bar{\nu}_e$.

In 1979, Sheldon Glashow, Abdus Salam, and Steven Weinberg won a Nobel prize for developing a theory that unified the electromagnetic and weak interactions. This **electroweak theory** postulates that the weak and electromagnetic interactions have the same strength at very high particle energies. Thus, the two interactions are viewed as two different manifestations of a single unifying electroweak interaction. The photon and the three massive bosons ($W^\pm$ and Z^0) play a key role in the electroweak theory. The theory makes many concrete predictions, but perhaps the most spectacular is the prediction of the masses of the W and Z particles at about 82 GeV/c^2 and 93 GeV/c^2, respectively. The 1984 Nobel prize was awarded to Carlo Rubbia and Simon van der Meer for their work leading to the discovery of these particles with just those masses at the CERN Laboratory in Geneva, Switzerland.

The combination of the electroweak theory and QCD for the strong interaction form what is called the **Standard Model** in high-energy physics. Although the details of the Standard Model are complex, its essential ingredients can be summarized with the help of Figure 31.10. The strong force, mediated by gluons, holds

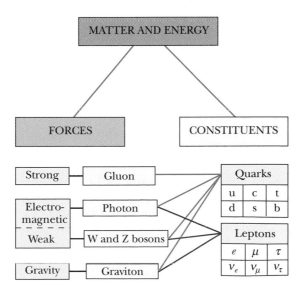

Figure 31.10 The Standard Model of particle physics.

quarks together to form composite particles such as protons, neutrons, and mesons. Leptons participate in the electromagnetic, weak, and gravitational interactions. The electromagnetic force is mediated by photons, and the weak force is mediated by W and Z bosons. Note that all fundamental forces are mediated by spin 1 bosons with properties given, to a large extent, by symmetries involved in the theories.

The Standard Model does not answer all questions, however. A major question is why the photon has no mass, whereas the W and Z bosons do have mass. Because of this mass difference, the electromagnetic and weak forces are quite distinct at low energies but become similar at very high energies, where the rest energies of the W and Z bosons are insignificant fractions of their total energies. This behavior in the transition from high to low energies, called **symmetry breaking,** leaves open the question of the origin of particle masses. To resolve this problem, a hypothetical particle called the **Higgs boson** has been proposed. The Higgs boson provides a mechanism for breaking the electroweak symmetry. The Standard Model, including the Higgs mechanism, provides a logically consistent explanation of the massive nature of the W and Z bosons. Unfortunately, the Higgs boson has not yet been found, but physicists know that its mass should be less than $1 \text{ TeV}/c^2$ $(10^{12} \text{ eV}/c^2)$.

In order to determine whether the Higgs boson exists, two quarks of at least 1 TeV of energy must collide, but calculations show that this requires injecting 40 TeV of energy within the volume of a proton. Scientists are convinced that, because of the limited energy available in conventional accelerators using fixed targets, it is necessary to build colliding-beam accelerators called **colliders.** The concept of colliders is straightforward. Particles with equal masses and kinetic energies, traveling in opposite directions in an accelerator ring, collide head-on to produce the required reaction and the formation of new particles. Because the total momentum of the interacting particles is zero, all of their kinetic energy is available for the reaction. The Large Electron–Positron collider (LEP) at CERN near Geneva, Switzerland, and the Stanford Linear Collider in California collide both electrons and protons. The Super Proton Synchrotron at CERN accelerates protons and antiprotons to energies of 270 GeV, and the world's highest-energy proton accelerator, the Tevatron, at the Fermi National Laboratory in Illinois, produces protons at almost 1000 GeV (or 1 TeV). The Superconducting Super Collider (SSC), which was being built in Texas, was an accelerator designed to produce 20-TeV protons in a ring 52 mi in circumference. After much debate in Congress, and an investment of almost $2 billion, the SSC project was canceled by the U.S. Department of Energy in October 1993. CERN recently approved the development of the Large Hadron Collider (LHC), a proton–proton collider that will provide a center of mass energy of 14 TeV and allow an exploration of Higgs-boson physics. The accelerator will be constructed in the same 27-km circumference tunnel as CERN's Large Electron–Positron collider, and many countries are expected to participate in the project.

Following the success of the electroweak theory, scientists attempted to combine it with QCD in a **grand unification theory** known as GUT. In this model, the electroweak force was merged with the strong color force to form a grand unified force. One version of the theory considers leptons and quarks as members of the same family that are able to change into each other by exchanging an appropriate particle. Many GUT theories predict that protons are unstable and will decay with a lifetime of about 10^{31} years. Attempts to detect such proton decays have so far been unsuccessful.

Thinking Physics 4

Consider a car making a head-on collision with an identical car moving in the opposite direction at the same speed. Compare that collision to one of the cars making a collision with the second car at rest. In which collision is there the larger transformation of kinetic energy to other forms during the collision? How does this relate to particle accelerators?

Reasoning In the head-on collision with both cars moving, conservation of momentum causes most if not all of the kinetic energy to be transformed to other forms. In the collision between a moving car and a stationary car, the cars are still moving after the collision, in the direction of the moving car, but with reduced speed. Thus, *only part* of the kinetic energy is transformed to other forms.

This suggests the advantage of using colliding beams in a particle accelerator, as opposed to firing a beam into a stationary target. When particles moving in opposite directions collide, all of the kinetic energy is available for transformation into other forms—in this case, the creation of new particles. When a beam is fired into a stationary target, only part of the energy is available for transformation, so higher mass particles cannot be created.

31.12 • THE COSMIC CONNECTION

In this section we shall describe one of the most fascinating theories in all of science—the Big Bang theory of the creation of the Universe—and the experimental evidence that supports it. This theory of cosmology states that the Universe had a beginning and, further, that the beginning was so cataclysmic that it is impossible to look back beyond it. According to this theory, the Universe erupted from a point-like singularity about 10 billion to 20 billion years ago. Such extremes of energy occurred in the first few minutes after the Big Bang that it is believed that all four interactions of physics were unified and that all matter melted down into an undifferentiated "quark soup."

The evolution of the four fundamental forces from the Big Bang to the present is shown in Figure 31.11. During the first 10^{-43} s (the ultra-hot epoch during which $T \approx 10^{32}$ K), it is presumed that the strong, electroweak, and gravitational forces were joined to form a completely unified force. In the first 10^{-32} s following the Big Bang (the hot epoch, $T \approx 10^{29}$ K), gravity broke free of this unification and the strong and electroweak forces remained joined, as described by a grand unification theory. This was a period when particle energies were so great ($> 10^{16}$ GeV) that very massive particles as well as quarks, leptons, and their antiparticles existed. Then the Universe rapidly expanded and cooled during the warm epoch when the temperatures ranged from 10^{29} to 10^{15} K; the strong and electroweak forces parted company, and the grand unification scheme was broken. As the Universe continued to cool, the electroweak force split into the weak force and the electromagnetic force about 10^{-10} s after the Big Bang.

After the first half-hour of creation, the temperature was probably about 10^{10} K and the Universe could be viewed as a cosmic thermonuclear bomb fusing protons and neutrons into deuterium and then helium. Until about 700 000 years after the Big Bang, the Universe was dominated by radiation: Ions absorbed and re-emitted photons, thereby ensuring thermal equilibrium of radiation and matter. Energetic radiation also prevented matter from forming clumps or even single hy-

George Gamow (1904–1968)

Gamow and two of his students, Ralph Alpher and Robert Herman, were the first to take the first half hour of the Universe seriously. In a mostly overlooked paper published in 1948, they made truly remarkable cosmological predictions. They correctly calculated the abundances of hydrogen and helium after the first half hour (75% H and 25% He) and predicted that radiation from the Big Bang should still be present and have an apparent temperature of about 5 K.

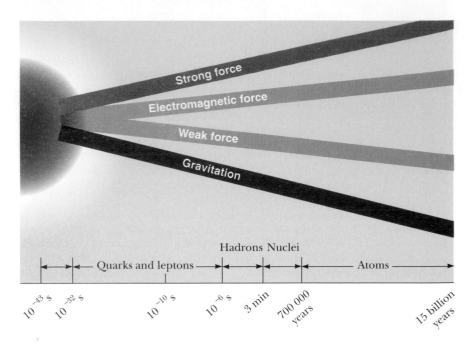

Hadrons Nuclei

|← Quarks and leptons →| |← Atoms →|

10^{-43} s 10^{-32} s 10^{-10} s 10^{-6} s   3 min   700 000 years   15 billion years

Figure 31.11   A brief history of the Universe from the Big Bang to the present. The four forces became distinguishable during the first microsecond. Following this, all the quarks combined to form particles that interact via the strong force. However, the leptons remained separate and exist as individually observable particles to this day.

drogen atoms. When the Universe was about 700 000 years old, it had expanded and cooled to about 3000 K, and protons could bind to electrons to form neutral hydrogen atoms. Because neutral atoms do not appreciably scatter photons, the Universe suddenly became transparent to photons. Radiation no longer dominated the Universe and clumps of neutral matter steadily grew—first atoms, followed by molecules, gas clouds, stars, and finally galaxies.

Observation of Radiation from the Primordial Fireball

In 1965, Arno A. Penzias and Robert W. Wilson of Bell Labs were testing a sensitive microwave receiver and made an amazing discovery. A pesky signal producing a faint background hiss was interfering with their satellite communications experiments. Despite their valiant efforts, the signal remained. Ultimately, it became clear that they were perceiving microwave background radiation (at a wavelength of 7.35 cm) representing the leftover glow from the Big Bang.

The microwave horn that served as their receiving antenna is shown in Figure 31.12. The intensity of the detected signal remained unchanged as the antenna was pointed in different directions. The fact that the radiation had equal strengths in all directions suggested that the entire Universe was the source of this radiation. Eviction of a flock of pigeons from the 20-foot horn and cooling the microwave detector both failed to remove the "spurious" signal. Through a casual conversation, Penzias and Wilson discovered that a group at Princeton had predicted the residual radiation from the Big Bang and were planning an experiment seeking to confirm the theory. The excitement in the scientific community was high when Penzias and

Figure 31.12   Robert W. Wilson (left) and Arno A. Penzias (right) with Bell Telephone Laboratories horn-reflector antenna.   *(AT&T Bell Laboratories)*

Wilson announced that they had already observed an excess microwave background compatible with a 3-K black-body source.

Because the measurements of Penzias and Wilson were taken at a single wavelength, they did not completely confirm the radiation as 3-K blackbody radiation. Subsequent experiments by other groups added intensity data at different wavelengths, as shown in Figure 31.13. The results confirm that the radiation is that of a black body at 2.9 K. This figure is, perhaps, the most clearcut evidence for the Big Bang theory. It also presents the earliest view of the Universe, a much earlier view than that available with the largest optical and radio telescopes. The 1978 Nobel prize was awarded to Penzias and Wilson for their most important discovery.

Other Evidence for the Expanding Universe

Most of the key discoveries supporting the theory of an expanding Universe, and indirectly the Big Bang theory of cosmology, were made in the 20th century. In 1913 Vesto Melvin Slipher, an American astronomer, reported that most nebulae are receding from the Earth at speeds up to several million miles per hour. Slipher was one of the first to use the methods of Doppler shifts in spectral lines to measure velocities.

In the late 1920s, Edwin P. Hubble made the bold assertion that the whole Universe is expanding. From 1928 to 1936, he and Milton Humason toiled at Mount Wilson to prove this assertion until they reached the limits of the 100-inch telescope. The results of this work and its continuation on a 200-inch telescope in the 1940s showed that the speeds of galaxies increase in direct proportion to their distance R from us (Fig. 31.14). This linear relationship, known as **Hubble's law,** may be written

$$v = HR \qquad [31.7]$$

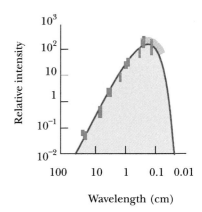

Figure 31.13   Radiation spectrum of the Big Bang. The blue areas are experimental results. The red line is the spectrum calculated for a black body at 2.9 K.

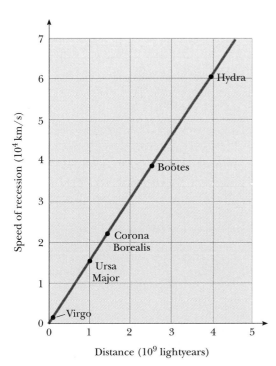

Figure 31.14   Hubble's law: The speed of recession is directly proportional to distance. The speeds of galaxies in five clusters are plotted versus distance.

where H, called the **Hubble parameter,** has the approximate value

$$H = 17 \times 10^{-3} \text{ m/(s} \cdot \text{lightyear)}$$

with an uncertainty of about 50%. Note that a lightyear is the distance light travels in one Earth year (9.46×10^{15} m).

Example 31.4   Recession of a Quasar

A quasar is a star-like object that is very distant from the Earth. Its speed can be measured from Doppler shift measurements in the light it emits. A certain quasar recedes from the Earth at a speed of $0.55c$. How far away is it?

Solution   We can find the distance from Hubble's law:

$$R = \frac{v}{H} = \frac{(0.55)(3.00 \times 10^8 \text{ m/s})}{17 \times 10^{-3} \text{ m/(s} \cdot \text{lightyear)}}$$

$$= 9.7 \times 10^9 \text{ lightyears}$$

$$= 9.2 \times 10^{25} \text{ m}$$

EXERCISE 3   Assuming that the quasar has moved with the speed $0.55c$ ever since the Big Bang, estimate the age of the Universe.   Answer $t = R/v = 1/H \approx 18$ billion years, which is in reasonable agreement with other estimates of the age of the Universe.

Will the Universe Expand Forever?

During the 1950s and 1960s Allan R. Sandage used the 200-inch telescope at Mount Palomar in California to measure the speeds of galaxies at distances of up to 6 billion lightyears. His measurements showed that these very distant galaxies were moving about 10 000 km/s faster than the Hubble law predicted. According to this result, the Universe must have been expanding more rapidly 1 billion years ago and, consequently, the expansion is slowing[2] (Fig. 31.15). Today astronomers and

Figure 31.15   Red shift, or speed of recession, versus apparent magnitude of 18 faint clusters. Curve A is the trend suggested by the six faintest clusters of galaxies. Curve C corresponds to a Universe having a constant rate of expansion. If the data fall between B and C, the expansion slows but never stops. If the data fall to the left of B, expansion stops and contraction occurs.

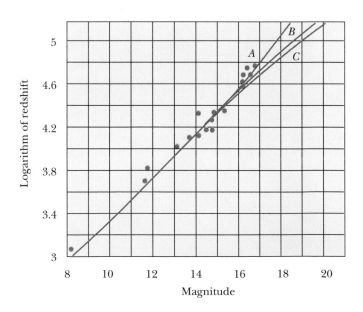

[2]The data at large distances have large observational uncertainties and may be systematically in error from selection effects such as abnormal brightness in the most distant visible clusters.

physicists are trying to determine the rate of slowing. If the average mass density of the Universe is less than some critical density ($\rho_c \approx 4$ hydrogen atoms/m^3), the galaxies will slow in their outward rush but still escape to infinity. If the average density exceeds the critical value, the expansion will eventually stop and contraction will begin, possibly leading to a superdense state and another expansion or an oscillating Universe.

Example 31.5 The Critical Density of the Universe

Estimate the critical mass density of the Universe, ρ_c, using energy considerations.

Reasoning and Solution Figure 31.16 shows a large section of the Universe with radius R, containing galaxies with a total

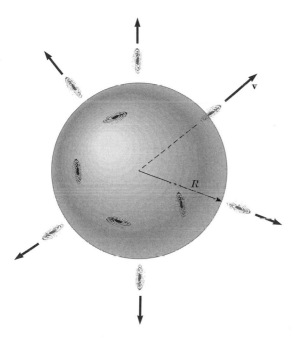

Figure 31.16 A galaxy escaping from a large cluster contained within radius R. Only the mass within R slows the mass m.

mass M. A galaxy of mass m and speed v at R will just escape to infinity with zero speed if the sum of its kinetic energy and gravitational potential energy is zero. Thus,

$$E_{\text{total}} = 0 = K + U = \tfrac{1}{2}mv^2 - \frac{GmM}{R}$$

$$\tfrac{1}{2}mv^2 = \frac{Gm\frac{4}{3}\pi R^3 \rho_c}{R}$$

$$(1) \qquad v^2 = \frac{8\pi G}{3}R^2 \rho_c$$

Because the galaxy of mass m obeys the Hubble law, $v = HR$, (1) becomes

$$H^2 = \frac{8\pi G}{3}\rho_c$$

$$(2) \qquad \rho_c = \frac{3H^2}{8\pi G}$$

Using $H = 17 \times 10^{-3}$ m/(s·lightyear), where 1 lightyear = 9.46×10^{15} m, and $G = 6.67 \times 10^{-11}$ N·m^2/kg^2 yields $\rho_c = 5.8 \times 10^{-27}$ kg/m^3. As the mass of a hydrogen atom is 1.67×10^{-27} kg, ρ_c corresponds to about 4 hydrogen atoms per cubic meter.

Missing Mass in the Universe(?)

The visible matter in galaxies averages out to 5×10^{-30} kg/m^3. The radiation in the Universe has a mass equivalent of approximately 2% of the visible matter. Non-luminous matter (such as interstellar gas or black holes) can be estimated from the speeds of galaxies orbiting each other in the cluster. The higher the galaxy speeds, the more mass in a galaxy cluster. Results from measurements on the Coma cluster of galaxies indicate, surprisingly, that the amount of invisible matter is 20 to 30 times the amount present in stars and luminous gas clouds. Yet even this large invisible component, if applied to the Universe as a whole, leaves the observed mass density a factor of 10 less than ρ_c. This so-called *missing mass* (or *dark matter*) has been the subject of intense theoretical and experimental work. Exotic massive particles such as axions, photinos, and superstring particles have been suggested as candidates for the missing mass. More mundane proposals have been that the missing mass is present in certain galaxies as neutrinos. In fact, neutrinos are so abundant that a tiny neutrino mass on the order of 20 eV/c^2 could furnish the missing mass and "close" the Universe.

Although we are a bit more sure about the beginning of the Universe, we are uncertain about its end. Will the Universe expand forever? Will it collapse and repeat its expansion in an endless series of oscillations? Results and answers to these questions remain inconclusive, and the exciting controversy continues.

CONCEPTUAL PROBLEM 3

Atoms did not exist until hundreds of thousands of years after the Big Bang. Why?

31.13 • PROBLEMS AND PERSPECTIVES

While particle physicists have been exploring the realm of the very small, cosmologists have been exploring cosmic history back to the first instants of the Big Bang. Observation of events that occur when two particles collide in an accelerator is essential to a reconstruction of the early moments in cosmic history. Perhaps the key to understanding the early Universe is first to understand the world of elementary particles. Cosmologists and physicists now find they have many common goals and are joining hands to attempt to understand the physical world at its most fundamental level.

Our understanding of physics at short distances is far from complete. Particle physics is faced with many questions. Why is there so little antimatter in the Universe? Do neutrinos have a small mass, and if so, how do they contribute to the "dark matter" of the Universe? Is it possible to unify the strong and electroweak theories in a logical and consistent manner? Why do quarks and leptons form three similar but distinct families? Are muons and taus the same as electrons (apart from their different masses), or do they have other subtle differences that have not been detected? Why are some particles charged and others neutral? Why do quarks carry a fractional charge? What determines the masses of the fundamental leptons and quarks? Can isolated quarks exist? The questions go on and on. Given the rapid advances and new discoveries in the field of particle physics, some of these questions will likely be resolved by the time you read this book.

An important and obvious question that remains is whether leptons and quarks have a substructure. If they have a substructure, we can envision an infinite number of deeper structure levels. However, if leptons and quarks are indeed the ultimate constituents of matter, as physicists today tend to believe, we should be able to construct a final theory of the structure of matter as Einstein dreamed. In the view of many physicists, the end of the road is in sight, but how long it will take to reach that goal is anyone's guess.

SUMMARY

There are four fundamental forces in nature: strong, electromagnetic, weak, and gravitational. The strong, or color force, is the force between quarks that binds them into neutrons, protons, and other hadrons. The weak force is responsible for beta decay. The electromagnetic and weak forces are now considered to be manifestations of a single force called the electroweak force. Every fundamental interaction is said to be mediated by the exchange of field particles. The electromagnetic interaction is mediated by the photon; the weak interaction is mediated by the $W^{\pm}$ and Z^0 bosons; the gravitational interaction is mediated by gravitons; the strong interaction is mediated by gluons.

An antiparticle and a particle have the same mass but opposite charge; other properties may also have opposite values such as lepton number and baryon number. It is possible to produce particle–antiparticle pairs in nuclear reactions if the available energy is greater than $2mc^2$, where m is the mass of the particle (or antiparticle).

Many particles are classified as hadrons or leptons. **Hadrons** interact through the strong force. They have size and structure and are not elementary particles. There are two types of hadrons, *baryons* and *mesons*. Mesons have baryon number zero and have either zero or integral spin. Baryons, which generally are the most massive particles, have non-zero baryon number and a spin of 1/2 or 3/2. The neutron and proton are examples of baryons.

Leptons have no structure or size and are considered truly elementary. They interact only through the weak and electromagnetic forces. There are six leptons, the electron e^-, the muon μ^-, the tau τ^-; and their neutrinos ν_e, ν_μ, and ν_τ.

In all reactions and decays, quantities such as energy, linear momentum, angular momentum, electric charge, baryon number, and lepton number are strictly conserved. Certain particles have properties called **strangeness** and **charm.** These unusual properties are conserved only in those reactions and decays that occur via the strong force.

Theories in elementary particle physics have postulated that all hadrons are composed of smaller units known as **quarks.** Quarks have fractional electric charge, baryon numbers of 1/3, and come in six "flavors": up (u), down (d), strange (s), charmed (c), top (t), and bottom (b). Each baryon contains three quarks, and each meson contains one quark and one antiquark.

According to the theory of **quantum chromodynamics,** quarks have a property called **color,** and the strong force between quarks is referred to as the **color force.**

The background microwave radiation discovered by Penzias and Wilson strongly suggests that the Universe started with a Big Bang about 15 billion years ago. The background radiation is equivalent to that of a black body at approximately 3 K.

Astronomical measurements show that the Universe is expanding. According to **Hubble's law,** distant galaxies are receding from the Earth at a speed $v = HR$, where R is the distance from Earth to the galaxy and H is **Hubble's parameter,** $H \approx 17 \times 10^{-3}$ m/ (s·lightyear).

CONCEPTUAL QUESTIONS

1. Name the four fundamental interactions and the field particle that mediates each.
2. Describe the quark model of hadrons, including the properties of quarks.
3. What are the differences between hadrons and leptons?
4. Describe the properties of baryons and mesons and the important differences between them.
5. Particles known as resonances have very short lifetimes, of the order of 10^{-23} s. From this information, would you guess they are hadrons or leptons? Explain.

6. Kaons all decay into final states that contain no protons or neutrons. What is the baryon number of kaons?

7. The Ξ^0 particle decays by the weak interaction according to the decay mode $\Xi^0 \rightarrow \Lambda^0 + \pi^0$. Would you expect this decay to be fast or slow? Explain.

8. Identify the particle decays listed in Table 31.2 that occur by the weak interaction. Justify your answers.

9. Identify the particle decays listed in Table 31.2 that occur by the electromagnetic interaction. Justify your answers.

10. Two protons in a nucleus interact via the strong interaction. Are they also subject to the weak interaction?

11. Discuss the following conservation laws: energy, linear momentum, angular momentum, electric charge, baryon number, lepton number, and strangeness. Are all of these laws based on fundamental properties of nature? Explain.

12. An antibaryon interacts with a meson. Can a baryon be produced in such an interaction? Explain.

13. Describe the essential features of the Standard Model of particle physics.

14. How many quarks are there in (a) a baryon, (b) an antibaryon, (c) a meson, (d) an antimeson? How do you account for the fact that baryons have half-integral spins and mesons have spins of 0 or 1? (*Hint:* Quarks have spin 1/2.)

15. In the theory of quantum chromodynamics, quarks come in three colors. How would you justify the statement that "all baryons and mesons are colorless"?

16. Which baryon did Murray Gell-Mann predict in 1961? What is the quark composition of this particle?

17. What is the quark composition of the Ξ^- particle? (See Tables 31.2 and 31.3.)

18. The W and Z bosons were first produced at CERN in 1983 (by having a beam of protons and a beam of antiprotons meet at high energy). Why was this an important discovery?

19. How did Edwin Hubble in 1928 determine that the Universe is expanding?

PROBLEMS

Section 31.2   Positrons and Other Antiparticles

1. A photon produces a proton–antiproton pair according to the reaction $\gamma \rightarrow p + \bar{p}$. What is the minimum possible frequency of the photon? What is its wavelength?

2. Two photons are produced when a proton and antiproton annihilate each other. What is the minimum frequency and corresponding wavelength of each photon?

3. A photon with an energy of $E_\gamma = 2.09$ GeV creates a proton–antiproton pair in which the proton has a kinetic energy of 95.0 MeV. What is the kinetic energy of the antiproton? ($m_p c^2 = 938.3$ MeV.)

Section 31.3   Mesons and the Beginning of Particle Physics

4. Occasionally, high-energy muons collide with electrons and produce two neutrinos according to the reaction $\mu^+ + e^- \rightarrow 2\nu$. What kind of neutrinos are these?

5. One of the mediators of the weak interaction is the Z^0 boson, with a mass of 96 GeV/c^2. Use this information to find an approximate value for the range of the weak interaction.

6. A free neutron beta decays by creating a proton, electron, and antineutrino according to the reaction $n \rightarrow p + e^- + \bar{\nu}$. Pretend that a free neutron beta decays by creating a proton and electron according to the reaction $n \rightarrow p +$

e^- and assume that the neutron is initially at rest in the laboratory. (a) Determine the energy released in this reaction. (b) Determine the speed of the proton and electron after the reaction. (Energy and momentum are conserved in the reaction.) (c) Are any of these particles moving at relativistic speeds? Explain.

7. When a high-energy proton or pion traveling near the speed of light collides with a nucleus, it travels an average distance of 3×10^{-15} m before interacting. From this information, estimate the duration of the strong interaction.

8. Calculate the range of the force that might be produced by the virtual exchange of a proton.

9. A neutral pion at rest decays into two photons according to

$$\pi^0 \longrightarrow \gamma + \gamma$$

Find the energy, momentum, and frequency of each photon.

Section 31.4   Classification of Particles

10. Identify the unknown particle on the left side of the following reaction:

$$? + p \longrightarrow n + \mu^+$$

11. Name one possible decay mode (see Table 31.2) for Ω^+, $\overline{K}_S^0$, $\overline{\Lambda}^0$, and $\overline{n}$.

Section 31.5   Conservation Laws

12. Each of the following reactions is forbidden. Determine a conservation law that is violated for each reaction.
    (a) $p + \overline{p} \rightarrow \mu^+ + e^-$
    (b) $\pi^- + p \rightarrow p + \pi^+$
    (c) $p + p \rightarrow p + \pi^+$
    (d) $p + p \rightarrow p + p + n$
    (e) $\gamma + p \rightarrow n + \pi^0$

13. (a) Show that baryon number and charge are conserved in the following reactions of a pion with a proton.

$$\pi^- + p \longrightarrow K^- + \Sigma^+ \qquad \textbf{[1]}$$

$$\pi^- + p \longrightarrow \pi^- + \Sigma^+ \qquad \textbf{[2]}$$

   (b) The first reaction is observed, but the second never occurs. Explain.

14. For the following two reactions, the first may occur but the second cannot. Explain.

$$K_S^0 \longrightarrow \pi^+ + \pi^- \quad \text{(can occur)}$$

$$\Lambda^0 \longrightarrow \pi^+ + \pi^- \quad \text{(cannot occur)}$$

15. The following reactions or decays involve one or more neutrinos. Supply the missing neutrinos (ν_e, ν_μ, or ν_τ).
    (a) $\pi^- \rightarrow \mu^- + ?$
    (b) $K^+ \rightarrow \mu^+ + ?$
    (c) $? + p \rightarrow n + e^+$
    (d) $? + n \rightarrow p + e^-$
    (e) $? + n \rightarrow p + \mu^-$
    (f) $\mu^- \rightarrow e^- + ? + ?$

16. A K_S^0 particle at rest decays into a π^+ and a π^-. What will be the speed of each of the pions? The mass of the K_S^0 is 497.7 MeV/c^2, and the mass of each π is 139.6 MeV/c^2.

17. Determine which of the reactions below can occur. For those that cannot occur, determine the conservation law (or laws) violated:
    (a) $p \rightarrow \pi^+ + \pi^0$
    (b) $p + p \rightarrow p + p + \pi^0$
    (c) $p + p \rightarrow p + \pi^+$
    (d) $\pi^+ \rightarrow \mu^+ + \nu_\mu$
    (e) $n \rightarrow p + e^- + \overline{\nu}_e$
    (f) $\pi^+ \rightarrow \mu^+ + n$

Section 31.6   Strange Particles and Strangeness

18. The neutral ρ meson decays by the strong interaction into two pions: $\rho^0 \rightarrow \pi^+ + \pi^-$, half-life 10^{-23} s. The neutral kaon also decays into two pions: $K_S^0 \rightarrow \pi^+ + \pi^-$, half-life 10^{-10} s. How do you explain the large difference in half-lives?

19. Determine whether or not strangeness is conserved in the following decays and reactions.
    (a) $\Lambda^0 \rightarrow p + \pi^-$
    (b) $\pi^- + p \rightarrow \Lambda^0 + K^0$
    (c) $\overline{p} + p \rightarrow \overline{\Lambda}^0 + \Lambda^0$
    (d) $\pi^- + p \rightarrow \pi^- + \Sigma^+$
    (e) $\Xi^- \rightarrow \Lambda^0 + \pi^-$
    (f) $\Xi^0 \rightarrow p + \pi^-$

20. The following decays are forbidden. Determine a conservation law that each violates.
    (a) $\mu^- \rightarrow e^- + \gamma$
    (b) $n \rightarrow p + e^- + \nu_e$
    (c) $\Lambda^0 \rightarrow p + \pi^0$
    (d) $p \rightarrow e^+ + \pi^0$
    (e) $\Xi^0 \rightarrow n + \pi^0$

Section 31.9   Quarks

21. The quark composition of the proton is uud and that of the neutron is udd. Show that the charge, baryon number, and strangeness of the particle equal the sum of these numbers for their quark constituents.

22. (a) Find the number of electrons and the number of each species of quarks, in one liter of water. (b) Make an order-of-magnitude estimate of the number of each kind of fundamental matter particle in your body. State your assumptions and the quantities you take as data.

23. The quark compositions of the K^0 and Λ^0 particles are d$\overline{s}$ and uds, respectively. Show that the charge, baryon number, and strangeness of these particles equal the sums of these numbers for the quark constituents.

24. Neglect binding energies and estimate the masses of the u and d quarks from the masses of the proton and neutron.

25. Analyze each reaction in terms of constituent quarks:
    (a) $\pi^- + p \rightarrow K^0 + \Lambda^0$
    (b) $\pi^+ + p \rightarrow K^+ + \Sigma^+$
    (c) $K^- + p \rightarrow K^+ + K^0 + \Omega^-$
    (d) $p + p \rightarrow K^0 + p + \pi^+ + ?$
    In the last reaction, identify the mystery particle.

26. A Σ^0 particle traveling through matter strikes a proton and a Σ^+ and a gamma ray emerges, as well as a third particle. Use the quark model of the Σ^+ and the gamma ray to determine the identity of the third particle.

Section 31.12   The Cosmic Connection

27. Using Hubble's law (Eq. 31.7), estimate the wavelength of the 590-nm sodium line emitted from galaxies (a) 2×10^6

lightyears away from Earth, (b) 2×10^8 lightyears away, and (c) 2×10^9 lightyears away. *Hint:* Use the relativistic Doppler formula for wavelength λ' of light emitted from a moving source:

$$\lambda' = \lambda \sqrt{\frac{1 + v/c}{1 - v/c}} \; .$$

28. A distant quasar is moving away from Earth at such high speed that the violet 434-nm hydrogen line is observed at 650 nm in the red portion of the spectrum. (a) How fast is the quasar receding? (See the hint in the preceding problem.) (b) Using Hubble's law, determine the distance from Earth to this quasar.

29. The various spectral lines observed from a distant receding galaxy or quasar have longer wavelengths λ'_n than the wavelengths λ_n measured in a stationary laboratory. It is found that the fractional change of wavelengths toward the red are the same for all spectral lines. That is,

$$\frac{\lambda'_n - \lambda_n}{\lambda_n} = Z$$

where Z is the redshift parameter common to all spectral lines for a given object. In terms of Z, determine (a) the speed of recession of the quasar and (b) the distance from Earth to this quasar. Use the hint in Problem 27 and Hubble's law (Eq. 31.7).

Additional Problems

30. Take the range of the strong interaction as approximately 1.4×10^{-15} m. It is thought that an elementary particle is exchanged between the protons and neutrons in the nucleus, leading to an attractive force. (a) Utilize the uncertainty principle $\Delta E \, \Delta t \geqslant \hbar/2$ to estimate the mass of the elementary particle if it moves at nearly the speed of light. (b) Using Table 31.2, identify the particle.

31. Name at least one conservation law that prevents each of the following reactions:
    (a) $\pi^- + p \rightarrow \Sigma^+ + \pi^0$
    (b) $\mu^- \rightarrow \pi^- + \nu_e$
    (c) $p \rightarrow \pi^+ + \pi^+ + \pi^-$

32. Two protons approach each other with 70.4 MeV of kinetic energy and engage in a reaction in which a proton and positive pion emerge at rest. What third particle, obviously uncharged and therefore difficult to detect, must have been created?

33. The energy flux carried by neutrinos from the Sun is estimated to be on the order of 0.4 W/m² at Earth's surface. Estimate the fractional mass loss of the Sun over 10^9 years due to the radiation of neutrinos. (The mass of the Sun is 2×10^{30} kg. The Earth–Sun distance is 1.5×10^{11} m.)

34. Supernova 1987A, located about 170 000 lightyears from the Earth, is estimated to have emitted a burst of $\sim 10^{46}$ J of neutrinos. Suppose the average neutrino energy was 6 MeV and your body presented a target area of 5000 cm². To an order of magnitude, how many of these neutrinos passed through you?

35. A gamma-ray photon strikes a stationary electron. Determine the minimum gamma-ray energy needed to make this reaction occur:

$$\gamma + e^- \longrightarrow e^- + e^- + e^+$$

36. An unstable particle, initially at rest, decays into a proton (rest energy 938.3 MeV) and a negative pion (rest energy 139.5 MeV). A uniform magnetic field of 0.250 T exists with the field perpendicular to the velocities of the created particles. The radius of curvature of each track is found to be 1.33 m. What is the mass of the original unstable particle?

37. Calculate the kinetic energies of the proton and pion resulting from the decay of a Λ^0 at rest:

$$\Lambda^0 \longrightarrow p + \pi^-$$

38. A Σ^0 particle at rest decays according to

$$\Sigma^0 \longrightarrow \Lambda^0 + \gamma$$

Find the gamma-ray energy.

39. If a K_S^0 meson at rest decays in 0.900×10^{-10} s, how far will a K_S^0 meson travel if it is moving at $0.960c$ through a bubble chamber?

40. Two protons approach each other with equal and opposite velocities. What is the minimum kinetic energy of each of the protons if they are to produce a π^+ meson at rest in the following reaction?

$$p + p \longrightarrow p + n + \pi^+$$

41. A π-meson at rest decays according to $\pi^- \rightarrow \mu^- + \bar{\nu}_\mu$. What is the energy carried off by the neutrino? (Assume that the neutrino moves off with the speed of light.) $m_\pi c^2 = 139.5$ MeV, $m_\mu c^2 = 105.7$ MeV, $m_\nu = 0$.

42. What processes are described by the Feynman diagrams in Figure P31.42? What is the exchanged particle in each process?

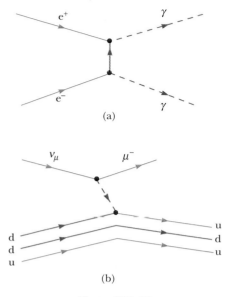

(a)

(b)

Figure P31.42

"Particles, particles, particles."

43. Identify the mediators for the two interactions described in the Feynman diagrams shown in Figure P31.43.

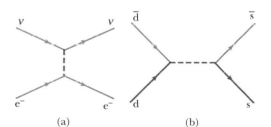

(a)   (b)

Figure P31.43

ANSWERS TO CONCEPTUAL PROBLEMS

1. The experiment described is a nice analogy to the Rutherford scattering experiment described in Section 29.1. In the Rutherford experiment, alpha particles were scattered from atoms, and the scattering was consistent with a small structure in the atom containing the positive charge.

2. The largest quark charge is $2e/3$, so a combination of only a quark and an antiquark, forming a meson, could not possibly have electric charge adding up to $+2e$. The Δ^{++} particle has three up quarks, each with charge $2e/3$, for a total charge of $+2e$.

3. Until about 700 000 years after the Big Bang, the temperature of the Universe was high enough for any atoms that formed to be ionized by ambient radiation. Once the average radiation energy dropped below the hydrogen ionization energy of 13.6 eV, hydrogen atoms could form and remain as neutral atoms for a relatively long period of time.

The Meaning of Success

To earn the respect of intelligent people and to win the affection of children;

To appreciate the beauty in nature and all that surrounds us;

To seek out and nurture the best in others;

To give the gift of yourself to others without the slightest thought of return, for it is in giving that we receive;

To have accomplished a task, whether it be saving a lost soul, healing a sick child, writing a book, or risking your life for a friend;

To have celebrated and laughed with great joy and enthusiasm and sung with exaltation;

To have hope even in times of despair, for as long as you have hope, you have life;

To love and be loved;

To be understood and to understand;

To know that even one life has breathed easier because you have lived;

This is the meaning of success.

RALPH WALDO EMERSON
and modified by Ray Serway
December 1989

APPENDIX A

TABLE A.1 **Conversion Factors**

			Length			
	m	**cm**	**km**	**in.**	**ft**	**mi**
1 meter	1	10^2	10^{-3}	39.37	3.281	6.214×10^{-4}
1 centimeter	10^{-2}	1	10^{-5}	0.3937	3.281×10^{-2}	6.214×10^{-6}
1 kilometer	10^3	10^5	1	3.937×10^4	3.281×10^3	0.6214
1 inch	2.540×10^{-2}	2.540	2.540×10^{-5}	1	8.333×10^{-2}	1.578×10^{-5}
1 foot	0.3048	30.48	3.048×10^{-4}	12	1	1.894×10^{-4}
1 mile	1609	1.609×10^5	1.609	6.336×10^4	5280	1

		Mass		
	kg	**g**	**slug**	**u**
1 kilogram	1	10^3	6.852×10^{-2}	6.024×10^{26}
1 gram	10^{-3}	1	6.852×10^{-5}	6.024×10^{23}
1 slug	14.59	1.459×10^4	1	8.789×10^{27}
1 atomic mass unit	1.660×10^{-27}	1.660×10^{-24}	1.137×10^{-28}	1

			Time		
	s	**min**	**h**	**day**	**year**
1 second	1	1.667×10^{-2}	2.778×10^{-4}	1.157×10^{-5}	3.169×10^{-8}
1 minute	60	1	1.667×10^{-2}	6.994×10^{-4}	1.901×10^{-6}
1 hour	3600	60	1	4.167×10^{-2}	1.141×10^{-4}
1 day	8.640×10^4	1440	24	1	2.738×10^{-3}
1 year	3.156×10^7	5.259×10^5	8.766×10^3	365.2	1

		Speed		
	m/s	**cm/s**	**ft/s**	**mi/h**
1 meter/second	1	10^2	3.281	2.237
1 centimeter/second	10^{-2}	1	3.281×10^{-2}	2.237×10^{-2}
1 foot/second	0.3048	30.48	1	0.6818
1 mile/hour	0.4470	44.70	1.467	1

Note: 1 mi/min = 60 mi/h = 88 ft/s

Force			
	N	**dyn**	**lb**
1 newton	1	10^5	0.2248
1 dyne	10^{-5}	1	2.248×10^{-6}
1 pound	4.448	4.448×10^5	1

Work, Energy, Heat			
	J	**erg**	**ft·lb**
1 joule	1	10^7	0.7376
1 erg	10^{-7}	1	7.376×10^{-8}
1 ft·lb	1.356	1.356×10^7	1
1 eV	1.602×10^{-19}	1.602×10^{-12}	1.182×10^{-19}
1 cal	4.186	4.186×10^7	3.087
1 Btu	1.055×10^{-3}	1.055×10^{10}	7.779×10^2
1 kWh	3.600×10^6	3.600×10^{13}	2.655×10^6

	eV	**cal**	**Btu**	**kWh**
1 joule	6.242×10^{18}	0.2389	9.481×10^{-4}	2.778×10^{-7}
1 erg	6.242×10^{11}	2.389×10^{-8}	9.481×10^{-11}	2.778×10^{-14}
1 ft·lb	8.464×10^{18}	0.3239	1.285×10^{-3}	3.766×10^{-7}
1 eV	1	3.827×10^{-20}	1.519×10^{-22}	4.450×10^{-26}
1 cal	2.613×10^{19}	1	3.968×10^{-3}	1.163×10^{-6}
1 Btu	6.585×10^{21}	2.520×10^2	1	2.930×10^{-4}
1 kWh	2.247×10^{25}	8.601×10^5	3.413×10^2	1

Pressure			
	Pa	**dyn/cm^2**	**atm**
1 pascal = 1 newton/meter^2	1	10	9.869×10^{-6}
1 dyne/centimeter^2	10^{-1}	1	9.869×10^{-7}
1 atmosphere	1.013×10^5	1.013×10^6	1
1 centimeter mercury*	1.333×10^3	1.333×10^4	1.316×10^{-2}
1 pound/inch^2	6.895×10^3	6.895×10^4	6.805×10^{-2}
1 pound/foot^2	47.88	4.788×10^2	4.725×10^{-4}

	cm Hg	**lb/in.^{2}**	**lb/ft^2**
1 pascal = newton/meter^2	7.501×10^{-4}	1.450×10^{-4}	2.089×10^{-2}
1 dyne/centimeter^2	7.501×10^{-5}	1.450×10^{-5}	2.089×10^{-3}
1 atmosphere	76	14.70	2.116×10^3
1 centimeter mercury*	1	0.1943	27.85
1 pound/inch^2	5.171	1	144
1 pound/foot^2	3.591×10^{-2}	6.944×10^{-3}	1

* At 0°C and at a location at which the free-fall acceleration has its "standard" value, 9.80665 m/s^2.

TABLE A.2   Symbols, Dimensions, and Units of Physical Quantities

Quantity	Common Symbol	Unit*	Dimensions[†]	Unit in Terms of Base SI Units
Acceleration	**a**	m/s^2	L/T^2	m/s^2
Amount of substance	n	mole		mol
Angle	θ, ϕ	radian (rad)	1	
Angular acceleration	$\boldsymbol{\alpha}$	rad/s^2	T^{-2}	s^{-2}
Angular frequency	ω	rad/s	T^{-1}	s^{-1}
Angular momentum	**L**	kg·m^2/s	ML^2/T	kg·m^2/s
Angular velocity	$\boldsymbol{\omega}$	rad/s	T^{-1}	s^{-1}
Area	A	m^2	L^2	m^2
Atomic number	Z			
Capacitance	C	farad (F) (=C/V)	Q^{2}T^2/ML^2	A^2·s^4/kg·m^2
Charge	q, Q, e	coulomb (C)	Q	A·s
Charge density				
Line	λ	C/m	Q/L	A·s/m
Surface	σ	C/m^2	Q/L^2	A·s/m^2
Volume	ρ	C/m^3	Q/L^3	A·s/m^3
Conductivity	σ	1/Ω·m	Q^{2}T/ML^3	A^2·s^3/kg·m^3
Current	I	AMPERE	Q/T	A
Current density	**J**	A/m^2	Q/T^2	A/m^2
Density	ρ	kg/m^3	M/L^3	kg/m^3
Dielectric constant	κ			
Displacement	**s**	METER	L	m
Distance	d, h			
Length	ℓ, L			
Electric dipole moment	**p**	C·m	QL	A·s·m
Electric field	**E**	V/m (=N/C)	ML/QT^2	kg·m/A·s^3
Electric flux	Φ	V·m	ML^3/QT^2	kg·m^3/A·s^3
Electromotive force	$\mathcal{E}$	volt (V)	ML^2/QT^2	kg·m^2/A·s^3
Energy	E, U, K	joule (J)	ML^2/T^2	kg·m^2/s^2
Entropy	S	J/K	ML^2/T^2·K	kg·m^2/s^2·K
Force	**F**	newton (N)	ML/T^2	kg·m/s^2
Frequency	f	hertz (Hz)	T^{-1}	s^{-1}
Heat	Q	joule (J)	ML^2/T^2	kg·m^2/s^2
Inductance	L	henry (H)	ML^2/Q^2	kg·m^2/A^2·s^2
Magnetic dipole moment	$\boldsymbol{\mu}$	N·m/T	QL^2/T	A·m^2
Magnetic field	**B**	tesla (T) (=Wb/m^2)	M/QT	kg/A·s^2
Magnetic flux	Φ_B	weber (Wb)	ML^2/QT	kg·m^2/A·s^2
Mass	m, M	KILOGRAM	M	kg
Molar specific heat	C	J/Mol·K		kg·m^2/s^2·mol·K
Moment of inertia	I	kg·m^2	ML^2	kg·m^2
Momentum	**p**	kg·m/s	ML/T	kg·m/s
Period	T	s	T	s
Permeability of space	μ_0	N/A^2 (=H/m)	ML/Q^{2}T	kg·m/A^2·s^2
Permittivity of space	ε_0	C^2/N·m^2 (=F/m)	Q^{2}T^2/ML^3	A^2·s^4/kg·m^3
Potential (voltage)	V	volt (V) (=J/C)	ML^2/QT^2	kg·m^2/A·s^3
Power	P	watt (W) (=J/s)	ML^2/T^3	kg·m^2/s^3
Pressure	P, p	pascal (Pa) = (N/m^2)	M/LT^2	kg/m·s^2

continued

* The base SI units are given in uppercase letters.

[†] The symbols M, L, T, and Q denote mass, length, time, and charge, respectively.

TABLE A.2   *(Continued)*

Quantity	Common Symbol	Unit*	Dimensions[†]	Unit in Terms of Base SI Units
Resistance	R	ohm $(\Omega)(=V/A)$	ML^2/Q^2T	$kg \cdot m^2/A^2 \cdot s^3$
Specific heat	c	$J/kg \cdot K$	$L^2/T^2 \cdot K$	$m^2/s^2 \cdot K$
Temperature	T	KELVIN	K	K
Time	t	SECOND	T	s
Torque	τ	$N \cdot m$	ML^2/T^2	$kg \cdot m^2/s^2$
Speed	v	m/s	L/T	m/s
Volume	V	m^3	L^3	m^3
Wavelength	λ	m	L	m
Work	W	joule $(J)(=N \cdot m)$	ML^2/T^2	$kg \cdot m^2/s^2$

* The base SI units are given in uppercase letters.

[†] The symbols M, L, T, and Q denote mass, length, time, and charge, respectively.

TABLE A.3   Table of Selected Atomic Masses[a]

Z	Element	Symbol	Chemical Atomic Mass (u)	Mass Number (* Indicates Radioactive) A	Atomic Mass (u)	Percentage Abundance	Half-Life (if Radioactive) $T_{1/2}$
0	(Neutron)	n		1*	1.008 665		10.4 m
1	Hydrogen	H	1.0079	1	1.007 825	99.985	
	Deuterium	D		2	2.014 102	0.015	
	Tritium	T		3*	3.016 049		12.33 y
2	Helium	He	4.00260	3	3.016 029	0.00014	
				4	4.002 602	99.99986	
3	Lithium	Li	6.941	6	6.015 121	7.5	
				7	7.016 003	92.5	
4	Beryllium	Be	9.0122	7*	7.016 928		53.3 d
				9	9.012 174	100	
5	Boron	B	10.81	10	10.012 936	19.9	
				11	11.009 305	80.1	
6	Carbon	C	12.011	11*	11.011 433		20.4 m
				12	12.000 000	98.90	
				13	13.003 355	1.10	
				14*	14.003 242		5730 y
7	Nitrogen	N	14.0067	13*	13.005 738		9.96 m
				14	14.003 074	99.63	
				15	15.000 108	0.37	
8	Oxygen	O	15.9994	15*	15.003 065		122 s
				16	15.994 915	99.761	
				18	17.999 160	0.20	
9	Fluorine	F	18.99840	19	18.998 404	100	
10	Neon	Ne	20.180	20	19.992 435	90.48	
				22	21.991 383	9.25	
11	Sodium	Na	22.98987	22*	21.994 434		2.61 y
				23	22.989 770	100	
				24*	23.990 961		14.96 h

TABLE A.3 *(Continued)*

Z	Element	Symbol	Chemical Atomic Mass (u)	Mass Number (* Indicates Radioactive) A	Atomic Mass (u)	Percentage Abundance	Half-Life (if Radioactive) $T_{1/2}$
12	Magnesium	Mg	24.305	24	23.985 042	78.99	
				25	24.985 838	10.00	
				26	25.982 594	11.01	
13	Aluminum	Al	26.98154	27	26.981 538	100	
14	Silicon	Si	28.086	28	27.976 927	92.23	
				29	28.976 495	4.67	
				30	29.973 770	3.10	
15	Phosphorus	P	30.97376	31	30.973 762	100	
				32*	31.973 908		14.26 d
16	Sulfur	S	32.066	32	31.972 071	95.02	
				33	32.971 459	0.75	
				34	33.967 867	4.21	
				35*	34.969 033		87.5 d
17	Chlorine	Cl	35.453	35	34.968 853	75.77	
				37	36.965 903	24.23	
18	Argon	Ar	39.948	36	35.967 547	0.337	
				40	39.962 384	99.600	
19	Potassium	K	39.0983	39	38.963 708	93.2581	
				40*	39.964 000	0.0117	1.28×10^9 y
				41	40.961 827	6.7302	
20	Calcium	Ca	40.08	40	39.962 591	96.941	
				44	43.955 481	2.086	
21	Scandium	Sc	44.9559	45	44.955 911	100	
22	Titanium	Ti	47.88	46	45.952 630	8.0	
				47	46.951 765	7.3	
				48	47.947 947	73.8	
				49	48.947 871	5.5	
				50	49.944 792	5.4	
23	Vanadium	V	50.9415	50*	49.947 161	0.25	1.5×10^{17} y
				51	50.943 962	99.75	
24	Chromium	Cr	51.996	50	49.946 047	4.345	
				52	51.940 511	83.79	
				53	52.940 652	9.50	
				54	53.938 883	2.365	
25	Manganese	Mn	54.93805	55	54.938 048	100	
26	Iron	Fe	55.847	54	53.939 613	5.9	
				56	55.934 940	91.72	
				57	56.935 396	2.1	
				58	57.933 278	0.28	

continued

[a] The masses in the sixth column are atomic masses, which include the mass of Z electrons. Data are from the National Nuclear Data Center, Brookhaven National Laboratory, prepared by Jagdish K. Tuli, July 1990. The data are based on experimental results reported in *Nuclear Data Sheets* and *Nuclear Physics* and also from *Chart of the Nuclides,* 14th cd. Atomic masses are based on those by A. H. Wapstra, G. Audi, and R. Hoekstra. Isotopic abundances are based on those by N. E. Holden.

TABLE A.3   *(Continued)*

Z	Element	Symbol	Chemical Atomic Mass (u)	Mass Number (* Indicates Radioactive) A	Atomic Mass (u)	Percentage Abundance	Half-Life (if Radioactive) $T_{1/2}$
27	Cobalt	Co	58.93320	59	58.933 198	100	
				60*	59.933 820		5.27 y
28	Nickel	Ni	58.693	58	57.935 346	68.077	
				60	59.930 789	26.223	
				62	61.928 346	3.634	
29	Copper	Cu	63.54	63	62.929 599	69.17	
				65	64.927 791	30.83	
30	Zinc	Zn	65.39	64	63.929 144	48.6	
				66	65.926 035	27.9	
				67	66.927 129	4.1	
				68	67.924 845	18.8	
31	Gallium	Ga	69.723	69	68.925 580	60.108	
				71	70.924 703	39.892	
32	Germanium	Ge	72.61	70	69.924 250	21.23	
				72	71.922 079	27.66	
				73	72.923 462	7.73	
				74	73.921 177	35.94	
				76	75.921 402	7.44	
33	Arsenic	As	74.9216	75	74.921 594	100	
34	Selenium	Se	78.96	74	73.922 474	0.89	
				76	75.919 212	9.36	
				77	76.919 913	7.63	
				78	77.917 307	23.78	
				80	79.916 519	49.61	
				82*	81.916 697	8.73	1.4×10^{20} y
35	Bromine	Br	79.904	79	78.918 336	50.69	
				81	80.916 287	49.31	
36	Krypton	Kr	83.80	78	77.920 400	0.35	
				80	79.916 377	2.25	
				82	81.913 481	11.6	
				83	82.914 136	11.5	
				84	83.911 508	57.0	
				86	85.910 615	17.3	
37	Rubidium	Rb	85.468	85	84.911 793	72.17	
				87*	86.909 186	27.83	4.75×10^{10} y
38	Strontium	Sr	87.62	86	85.909 266	9.86	
				87	86.908 883	7.00	
				88	87.905 618	82.58	
				90*	89.907 737		29.1 y
39	Yttrium	Y	88.9058	89	88.905 847	100	
40	Zirconium	Zr	91.224	90	89.904 702	51.45	
				91	90.905 643	11.22	
				92	91.905 038	17.15	
				94	93.906 314	17.38	
				96	95.908 274	2.80	
41	Niobium	Nb	92.9064	93	92.906 376	100	

TABLE A.3 *(Continued)*

Z	Element	Symbol	Chemical Atomic Mass (u)	Mass Number (* Indicates Radioactive) A	Atomic Mass (u)	Percentage Abundance	Half-Life (if Radioactive) $T_{1/2}$
42	Molybdenum	Mo	95.94	92	91.906 807	14.84	
				94	93.905 085	9.25	
				95	94.905 841	15.92	
				96	95.904 678	16.68	
				97	96.906 020	9.55	
				98	97.905 407	24.13	
				100	99.907 476	9.63	
43	Technetium	Tc		98*	97.907 215		4.2×10^6 y
				99*	98.906 254		2.1×10^5 y
44	Ruthenium	Ru	101.07	96	95.907 597	5.54	
				99	98.905 939	12.7	
				100	99.904 219	12.6	
				101	100.905 558	17.1	
				102	101.904 348	31.6	
				104	103.905 428	18.6	
45	Rhodium	Rh	102.9055	103	102.905 502	100	
46	Palladium	Pd	106.42	104	103.904 033	11.14	
				105	104.905 082	22.33	
				106	105.903 481	27.33	
				108	107.903 893	26.46	
				110	109.905 158	11.72	
47	Silver	Ag	107.868	107	106.905 091	51.84	
				109	108.904 754	48.16	
48	Cadmium	Cd	112.41	110	109.903 004	12.49	
				111	110.904 182	12.80	
				112	111.902 760	24.13	
				113*	112.904 401	12.22	9.3×10^{15} y
				114	113.903 359	28.73	
				116	115.904 755	7.49	
49	Indium	In	114.82	113	112.904 060	4.3	
				115*	114.903 876	95.7	4.4×10^{14} y
50	Tin	Sn	118.71	116	115.901 743	14.53	
				117	116.902 953	7.68	
				118	117.901 605	24.22	
				119	118.903 308	8.58	
				120	119.902 197	32.59	
				122	121.903 439	4.63	
				124	123.905 274	5.79	
51	Antimony	Sb	121.76	121	120.903 820	57.36	
				123	122.904 215	42.64	
52	Tellurium	Te	127.60	122	121.903 052	2.59	
				124	123.902 817	4.79	
				125	124.904 429	7.12	
				126	125.903 309	18.93	
				128*	127.904 463	31.70	$>8 \times 10^{24}$ y
				130*	129.906 228	33.87	$\leq 1.25 \times 10^{21}$ y

continued

TABLE A.3   *(Continued)*

Z	Element	Symbol	Chemical Atomic Mass (u)	Mass Number (* Indicates Radioactive) A	Atomic Mass (u)	Percentage Abundance	Half-Life (if Radioactive) $T_{1/2}$
53	Iodine	I	126.9045	127	126.904 474	100	
				129*	128.904 984		1.6×10^7 y
54	Xenon	Xe	131.29	129	128.904 779	26.4	
				130	129.903 509	4.1	
				131	130.905 069	21.2	
				132	131.904 141	26.9	
				134	133.905 394	10.4	
				136*	135.907 215	8.9	$\geq 2.36 \times 10^{21}$ y
55	Cesium	Cs	132.9054	133	132.905 436	100	
56	Barium	Ba	137.33	134	133.904 492	2.42	
				135	134.905 671	6.593	
				136	135.904 559	7.85	
				137	136.905 816	11.23	
				138	137.905 236	71.70	
57	Lanthanum	La	138.905	139	138.906 346	99.9098	
58	Cerium	Ce	140.12	140	139.905 434	88.43	
				142*	141.909 241	11.13	$>5 \times 10^{16}$ y
59	Praseodymium	Pr	140.9076	141	140.907 647	100	
60	Neodymium	Nd	144.24	142	141.907 718	27.13	
				143	142.909 809	12.18	
				144*	143.910 082	23.80	2.3×10^{15} y
				145	144.912 568	8.30	
				146	145.913 113	17.19	
				148	147.916 888	5.76	
				150*	149.920 887	5.64	$>1 \times 10^{18}$ y
61	Promethium	Pm		145*	144.912 745		17.7 y
62	Samarium	Sm	150.36	144	143.911 996	3.1	
				147*	146.914 894	15.0	1.06×10^{11} y
				148*	147.914 819	11.3	7×10^{15} y
				149*	148.917 180	13.8	$>2 \times 10^{15}$ y
				150	149.917 273	7.4	
				152	151.919 728	26.7	
				154	153.922 206	22.7	
63	Europium	Eu	151.96	151	150.919 846	47.8	
				153	152.921 226	52.2	
64	Gadolinium	Gd	157.25	155	154.922 618	14.80	
				156	155.922 119	20.47	
				157	156.923 957	15.65	
				158	157.924 099	24.84	
				160	159.927 050	21.86	
65	Terbium	Tb	158.9253	159	158.925 345	100	
66	Dysprosium	Dy	162.50	161	160.926 930	18.9	
				162	161.926 796	25.5	
				163	162.928 729	24.9	
				164	163.929 172	28.2	
67	Holmium	Ho	164.9303	165	164.930 316	100	

TABLE A.3 *(Continued)*

Z	Element	Symbol	Chemical Atomic Mass (u)	Mass Number (* Indicates Radioactive) A	Atomic Mass (u)	Percentage Abundance	Half-Life (if Radioactive) $T_{1/2}$
68	Erbium	Er	167.26	166	165.930 292	33.6	
				167	166.932 047	22.95	
				168	167.932 369	27.8	
				170	169.935 462	14.9	
69	Thulium	Tm	168.9342	169	168.934 213	100	
70	Ytterbium	Yb	173.04	170	169.934 761	3.05	
				171	170.936 324	14.3	
				172	171.936 380	21.9	
				173	172.938 209	16.12	
				174	173.938 861	31.8	
				176	175.942 564	12.7	
71	Lutecium	Lu	174.967	175	174.940 772	97.41	
				176*	175.942 679	2.59	3.78×10^{10} y
72	Hafnium	Hf	178.49	176	175.941 404	5.206	
				177	176.943 218	18.606	
				178	177.943 697	27.297	
				179	178.945 813	13.629	
				180	179.946 547	35.100	
73	Tantalum	Ta	180.9479	181	180.947 993	99.988	
74	Tungsten (Wolfram)	W	183.85	182	181.948 202	26.3	
				183	182.950 221	14.28	
				184	183.950 929	30.7	
				186	185.954 358	28.6	
75	Rhenium	Re	186.207	185	184.952 951	37.40	
				187*	186.955 746	62.60	4.4×10^{10} y
76	Osmium	Os	190.2	188	187.955 832	13.3	
				189	188.958 139	16.1	
				190	189.958 439	26.4	
				192	191.961 468	41.0	
77	Iridium	Ir	192.2	191	190.960 585	37.3	
				193	192.962 916	62.7	
78	Platinum	Pt	195.08	194	193.962 655	32.9	
				195	194.964 765	33.8	
				196	195.964 926	25.3	
				198	197.967 867	7.2	
79	Gold	Au	196.9665	197	196.966 543	100	
80	Mercury	Hg	200.59	198	197.966 743	9.97	
				199	198.968 253	16.87	
				200	199.968 299	23.10	
				201	200.970 276	13.10	
				202	201.970 617	29.86	
				204	203.973 466	6.87	
81	Thallium	Tl	204.383	203	202.972 320	29.524	
				205	204.974 400	70.476	
		(Th C″)		208*	207.981 992		3.053 m

continued

TABLE A.3   *(Continued)*

Z	Element	Symbol	Chemical Atomic Mass (u)	Mass Number (* Indicates Radioactive) A	Atomic Mass (u)	Percentage Abundance	Half-Life (if Radioactive) $T_{1/2}$
82	Lead	Pb	207.2	204*	203.973 020	1.4	$\geq 1.4 \times 10^{17}$ y
				206	205.974 440	24.1	
				207	206.975 871	22.1	
				208	207.976 627	52.4	
		(Ra D)		210*	209.984 163		22.3 y
		(Ac B)		211*	210.988 734		36.1 m
		(Th B)		212*	211.991 872		10.64 h
		(Ra B)		214*	213.999 798		26.8 m
83	Bismuth	Bi	208.9803	209	208.980 374	100	
		(Th C)		211*	210.987 254		2.14 m
84	Polonium	Po					
		(Ra F)		210*	209.982 848		138.38 d
		(Ra C′)		214*	213.995 177		164 μs
85	Astatine	At		218*	218.008 685		1.6 s
86	Radon	Rn		222*	222.017 571		3.823 d
87	Francium	Fr					
		(Ac K)		223*	223.019 733		22 m
88	Radium	Ra		226*	226.025 402		1600 y
		(Ms Th$_1$)		228*	228.031 064		5.75 y
89	Actinium	Ac		227*	227.027 749		21.77 y
90	Thorium	Th	232.0381				
		(Rd Th)		228*	228.028 716		1.913 y
				232*	232.038 051	100	1.40×10^{10} y
91	Protactinium	Pa		231*	231.035 880		32.760 y
92	Uranium	U	238.0289	232*	232.037 131		69 y
				233*	233.039 630		1.59×10^5 y
		(Ac U)		235*	235.043 924	0.720	7.04×10^8 y
				236*	236.045 562		2.34×10^7 y
		(UI)		238*	238.050 784	99.2745	4.47×10^9 y
93	Neptunium	Np		237*	237.048 168		2.14×10^6 y
94	Plutonium	Pu		239*	239.052 157		24,120 y
				242*	242.058 737		3.73×10^5 y
				244*	244.064 200		8.1×10^7 y

APPENDIX B

Mathematics Review

These mathematics appendices are intended as a brief review of operations and methods. Early in your course, you should be facile with basic algebraic techniques, analytic geometry, and trigonometry. The appendices on differential and integral calculus are more detailed and are intended for those students who have difficulty applying calculus concepts to physical situations. Mathematical symbols used in the text are given in the following table.

Mathematical Symbols Used in the Text and Their Meanings

Symbol	Meaning
$=$	is equal to
$\equiv$	is defined as
$\neq$	is not equal to
$\propto$	is proportional to
$>$	is greater than
$<$	is less than
$\gg (\ll)$	is much greater (less) than
$\approx$	is approximately equal to
$\sim$	is on the order of
Δx	the change in x
$\displaystyle\sum_{i=1}^{N} x_i$	the sum of all quantities x_i from $i = 1$ to $i = N$
$\lvert x \rvert$	the magnitude of x (always a nonnegative quantity)
$\Delta x \rightarrow 0$	Δx approaches zero
$\dfrac{dx}{dt}$	the derivative of x with respect to t
$\dfrac{\partial x}{\partial t}$	the partial derivative of x with respect to t
$\displaystyle\int$	integral

B.1 · SCIENTIFIC NOTATION

Many quantities that scientists deal with often have very large or very small values. For example, the speed of light is about 300 000 000 m/s, and the ink required to make the dot over an i in this textbook has a mass of about 0.000 000 001 kg.

Obviously, it is very cumbersome to read, write, and keep track of numbers such as these. We avoid this problem by using the powers of the number 10:

$$10^0 = 1$$
$$10^1 = 10$$
$$10^2 = 10 \times 10 = 100$$
$$10^3 = 10 \times 10 \times 10 = 1000$$
$$10^4 = 10 \times 10 \times 10 \times 10 = 10\ 000$$
$$10^5 = 10 \times 10 \times 10 \times 10 \times 10 = 100\ 000$$

and so on. The number of zeros corresponds to the power to which 10 is raised, called the exponent of 10. For example, the speed of light, 300 000 000 m/s, can be expressed as 3×10^8 m/s.

For numbers less than one, we note the following:

$$10^{-1} = \frac{1}{10} = 0.1$$

$$10^{-2} = \frac{1}{10 \times 10} = 0.01$$

$$10^{-3} = \frac{1}{10 \times 10 \times 10} = 0.001$$

$$10^{-4} = \frac{1}{10 \times 10 \times 10 \times 10} = 0.0001$$

$$10^{-5} = \frac{1}{10 \times 10 \times 10 \times 10 \times 10} = 0.00001$$

In these cases, the number of places that the decimal point lies to the left of the digit 1 equals the value of the (negative) exponent. Numbers that are expressed as some power of 10 multiplied by another number between 1 and 10 are said to be in **scientific notation.** For example, the scientific notation for 5 943 000 000 is 5.943×10^9, and that for 0.0000832 is 8.32×10^{-5}.

When numbers expressed in scientific notation are being multiplied, the following general rule is very useful:

$$10^n \times 10^m = 10^{n+m} \qquad\qquad \text{[B.1]}$$

where n and m can be *any* numbers (not necessarily integers). For example, $10^2 \times 10^5 = 10^7$. The rule also applies if one of the exponents is negative. For example, $10^3 \times 10^{-8} = 10^{-5}$.

When dividing numbers expressed in scientific notation, note that

$$\frac{10^n}{10^m} = 10^n \times 10^{-m} = 10^{n-m} \qquad\qquad \text{[B.2]}$$

EXERCISES   With help from these rules, verify the answers to the following:

1. $86\ 400 = 8.64 \times 10^4$
2. $9\ 816\ 762.5 = 9.8167625 \times 10^6$

3. $0.0000000398 = 3.98 \times 10^{-8}$

4. $(4 \times 10^8)(9 \times 10^9) = 3.6 \times 10^{18}$

5. $(3 \times 10^7)(6 \times 10^{-12}) = 1.8 \times 10^{-4}$

6. $\dfrac{75 \times 10^{-11}}{5 \times 10^{-9}} = 1.5 \times 10^{-7}$

7. $\dfrac{(3 \times 10^6)(8 \times 10^{-2})}{(2 \times 10^{17})(6 \times 10^5)} = 2 \times 10^{-18}$

B.2 · ALGEBRA

Some Basic Rules

When algebraic operations are performed, the laws of arithmetic apply. Symbols such as x, y, and z are usually used to represent quantities that are not specified—the **unknowns.**

First, consider the equation

$$8x = 32$$

If we wish to solve for x, we can divide (or multiply) each side of the equation by the same factor without destroying the equality. In this case, if we divide both sides by 8, we have

$$\frac{8x}{8} = \frac{32}{8}$$

$$x = 4$$

Next consider the equation

$$x + 2 = 8$$

In this type of expression, we can add or subtract the same quantity from each side. If we subtract 2 from each side, we get

$$x + 2 - 2 = 8 - 2$$

$$x = 6$$

In general, if $x + a = b$, then $x = b - a$.

Now consider the equation

$$\frac{x}{5} = 9$$

If we multiply each side by 5, we are left with x on the left by itself and 45 on the right:

$$\left(\frac{x}{5}\right)(5) = 9 \times 5$$

$$x = 45$$

In all cases, **whatever operation is performed on the left side of the equality must also be performed on the right side.**

The following rules for multiplying, dividing, adding, and subtracting fractions should be recalled, where a, b, and c are three numbers:

	Rule	**Example**
Multiplying	$\left(\dfrac{a}{b}\right)\left(\dfrac{c}{d}\right) = \dfrac{ac}{bd}$	$\left(\dfrac{2}{3}\right)\left(\dfrac{4}{5}\right) = \dfrac{8}{15}$
Dividing	$\dfrac{(a/b)}{(c/d)} = \dfrac{ad}{cb}$	$\dfrac{2/3}{4/5} = \dfrac{(2)(5)}{(4)(3)} = \dfrac{10}{12}$
Adding	$\dfrac{a}{b} \pm \dfrac{c}{d} = \dfrac{ad \pm bc}{bd}$	$\dfrac{2}{3} - \dfrac{4}{5} = \dfrac{(2)(5) - (4)(3)}{(3)(5)} = -\dfrac{2}{15}$

EXERCISES   In the following exercises, solve for x:

Answers

1. $a = \dfrac{1}{1 + x}$ $x = \dfrac{1 - a}{a}$

2. $3x - 5 = 13$ $x = 6$

3. $ax - 5 = bx + 2$ $x = \dfrac{7}{a - b}$

4. $\dfrac{5}{2x + 6} = \dfrac{3}{4x + 8}$ $x = -\dfrac{11}{7}$

Powers

When powers of a given quantity x are multiplied, the following rule applies:

$$x^n x^m = x^{n+m} \qquad \text{[B.3]}$$

For example, $x^2 x^4 = x^{2+4} = x^6$.

When dividing the powers of a given quantity, the rule is

$$\frac{x^n}{x^m} = x^{n-m} \qquad \text{[B.4]}$$

For example, $x^8/x^2 = x^{8-2} = x^6$.

A power that is a fraction, such as $\frac{1}{3}$, corresponds to a root as follows:

$$x^{1/n} = \sqrt[n]{x} \qquad \text{[B.5]}$$

For example, $4^{1/3} = \sqrt[3]{4} = 1.5874$. (A scientific calculator is useful for such calculations.)

Finally, any quantity x^n that is raised to the mth power is

$$(x^n)^m = x^{nm} \qquad \text{[B.6]}$$

Table B.1 summarizes the rules of exponents.

TABLE B.1   Rules of Exponents

$x^0 = 1$
$x^1 = x$
$x^n x^m = x^{n+m}$
$x^n/x^m = x^{n-m}$
$x^{1/n} = \sqrt[n]{x}$
$(x^n)^m = x^{nm}$

EXERCISES   Verify the following:

1. $3^2 \times 3^3 = 243$
2. $x^5 x^{-8} = x^{-3}$
3. $x^{10}/x^{-5} = x^{15}$
4. $5^{1/3} = 1.709975$ (Use your calculator.)
5. $60^{1/4} = 2.783158$ (Use your calculator.)
6. $(x^4)^3 = x^{12}$

Factoring

Some useful formulas for factoring an equation are

$$ax + ay + az = a(x + y + z) \qquad \text{Common factor}$$

$$a^2 + 2ab + b^2 = (a + b)^2 \qquad \text{Perfect square}$$

$$a^2 - b^2 = (a + b)(a - b) \qquad \text{Difference of squares}$$

Quadratic Equations

The general form of a quadratic equation is

$$ax^2 + bx + c = 0 \qquad \qquad \textbf{[B.7]}$$

where x is the unknown quantity and a, b, and c are numerical factors referred to as **coefficients** of the equation. This equation has two roots, given by

$$x = \frac{-b \pm \sqrt{b^2 - 4ac}}{2a} \qquad \qquad \textbf{[B.8]}$$

If $b^2 > 4ac$, the roots are real.

Example 1

The equation $x^2 + 5x + 4 = 0$ has the following roots, corresponding to the two signs of the square-root term:

$$x = \frac{-5 \pm \sqrt{5^2 - (4)(1)(4)}}{2(1)} = \frac{-5 \pm \sqrt{9}}{2} = \frac{-5 \pm 3}{2}$$

That is,

$$x_+ = \frac{-5 + 3}{2} = -1 \qquad x_- = \frac{-5 - 3}{2} = -4$$

where x_+ denotes the root corresponding to the positive sign and x_- denotes the root corresponding to the negative sign.

EXERCISES   Solve the following quadratic equations:

		Answers	
1. $x^2 + 2x - 3 = 0$	$x_+ = 1$	$x_- = -3$
2. $2x^2 - 5x + 2 = 0$	$x_+ = 2$	$x_- = \frac{1}{2}$
3. $2x^2 - 4x - 9 = 0$	$x_+ = 1 + \sqrt{22}/2$	$x_- = 1 - \sqrt{22}/2$

Linear Equations

A linear equation has the general form

$$y = ax + b \qquad \qquad \textbf{[B.9]}$$

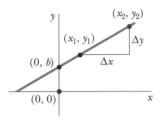

Figure B.1

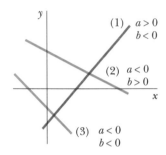

Figure B.2

where a and b are constants. This equation is referred to as linear because the graph of y versus x is a straight line, as shown in Figure B.1. The constant b, called the **intercept,** represents the value of y at which the straight line intersects the y axis. The constant a is equal to the slope of the straight line. If any two points on the straight line are specified by the coordinates (x_1, y_1) and (x_2, y_2), as in Figure B.1, then the **slope** of the straight line can be expressed as

$$\text{Slope} = \frac{y_2 - y_1}{x_2 - x_1} = \frac{\Delta y}{\Delta x} \qquad \text{[B.10]}$$

Note that a and b can have either positive or negative values. If $a > 0$, the straight line has a *positive* slope, as in Figure B.1. If $a < 0$, the straight line has a *negative* slope. In Figure B.1, both a and b are positive. Three other possible situations are shown in Figure B.2: $a > 0$, $b < 0$; $a < 0$, $b > 0$; and $a < 0$, $b < 0$.

EXERCISES

1. Draw graphs of the following straight lines:
   (a) $y = 5x + 3$ (b) $y = -2x + 4$ (c) $y = -3x - 6$.
2. Find the slopes of the straight lines described in Exercise 1.
   Answers (a) 5 (b) -2 (c) -3
3. Find the slopes of the straight lines that pass through the following sets of point:
   (a) $(0, -4)$ and $(4, 2)$; (b) $(0, 0)$ and $(2, -5)$; (c) $(-5, 2)$ and $(4, -2)$.
   Answers (a) $3/2$ (b) $-5/2$ (c) $-4/9$

Solving Simultaneous Linear Equations

Consider an equation such as $3x + 5y = 15$, which has two unknowns, x and y. Such an equation does not have a unique solution. That is, $(x = 0, y = 3)$, $(x = 5, y = 0)$, and $(x = 2, y = 9/5)$ are all solutions to this equation.

If a problem has two unknowns, a unique solution is possible only if there are *two* independent equations. In general, if a problem has n unknowns, its solution requires n equations. In order to solve two simultaneous equations involving two unknowns, x and y, we solve one of the equations for x in terms of y and substitute this expression into the other equation.

Example 2

Solve the following two simultaneous equations:

$$(1)\ 5x + y = -8$$
$$(2)\ 2x - 2y = 4$$

Solution From (2), $x = y + 2$. Substitution of this into (1) gives

$$5(y + 2) + y = -8$$
$$6y = -18$$
$$y = -3$$
$$x = y + 2 = -1$$

Alternative Solution Multiply each term in (1) by the factor 2 and add the result to (2):

$$10x + 2y = -16$$
$$\underline{2x - 2y = 4}$$
$$12x = -12$$
$$x = -1$$
$$y = x - 2 = -3$$

Two linear equations with two unknowns can also be solved by a graphical method. If the straight lines corresponding to the two equations are plotted in a conventional coordinate system, the intersection of the two lines represents the solution. For example, consider the two equations

$$x - y = 2$$

$$x - 2y = -1$$

These are plotted in Figure B.3. The intersection of the two lines has the coordinates $x = 5$, $y = 3$. This represents the solution to the equations. You should check this solution by the analytical technique discussed earlier.

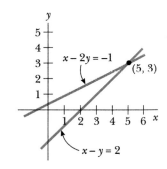

Figure B.3

EXERCISES   Solve the following pairs of simultaneous equations involving two unknowns:

Answers

1. $x + y = 8$ $x = 5, y = 3$
      $x - y = 2$
2. $98 - T = 10a$ $T = 65, a = 3.27$
      $T - 49 = 5a$
3. $6x + 2y = 6$ $x = 2, y = -3$
      $8x - 4y = 28$

Logarithms

Suppose that a quantity, x, is expressed as a power of some quantity a:

$$x = a^y \qquad \text{[B.11]}$$

The number a is called the **base** number. The **logarithm** of x with respect to the base a is equal to the exponent to which the base must be raised in order to satisfy the expression $x = a^y$:

$$y = \log_a x \qquad \text{[B.12]}$$

Conversely, the **antilogarithm** of y is the number x:

$$x = \text{antilog}_a y \qquad \text{[B.13]}$$

In practice, the two bases most often used are base 10, called the *common* logarithm base, and base $e = 2.718 \ldots$, called the *natural* logarithm base. When common logarithms are used,

$$y = \log_{10} x \qquad (\text{or } x = 10^y) \qquad \text{[B.14]}$$

When natural logarithms are used,

$$y = \log_e x = \ln x \qquad (\text{or } x = e^y) \qquad \text{[B.15]}$$

For example, $\log_{10} 52 = 1.716$, and so $\text{antilog}_{10} 1.716 = 10^{1.716} = 52$. Likewise, $\ln 52 = 3.951$, so $\text{antiln } 3.951 = e^{3.951} = 52$.

In general, note that you can convert between base 10 and base e with the equality

$$\ln x = (2.302585) \log_{10} x \qquad \text{[B.16]}$$

Finally, some useful properties of logarithms are as follows:

$$\log(ab) = \log a + \log b$$

$$\log(a/b) = \log a - \log b$$

$$\log(a^n) = n \log a$$

$$\ln e = 1$$

$$\ln e^a = a$$

$$\ln\left(\frac{1}{a}\right) = -\ln a$$

B.3 · GEOMETRY

The **distance,** d, between two points the coordinates of which are (x_1, y_1) and (x_2, y_2) is

$$d = \sqrt{(x_2 - x_1)^2 + (y_2 - y_1)^2} \qquad \text{[B.17]}$$

The arc length, s, of a circular arc (Fig. B.4) is proportional to the radius, r, for a fixed value of θ (in radians) — the **radian measure:**

$$s = r\theta$$

$$\qquad \text{[B.18]}$$

$$\theta = \frac{s}{r}$$

Table B.2 gives the areas and volumes for several geometric shapes used throughout this book:

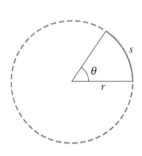

Figure B.4

TABLE B.2   Useful Information for Geometry

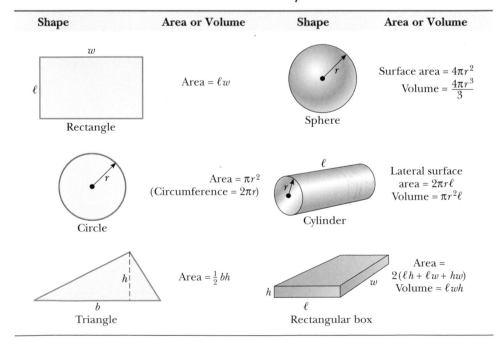

Shape	Area or Volume	Shape	Area or Volume
Rectangle	Area $= \ell w$	Sphere	Surface area $= 4\pi r^2$ Volume $= \frac{4\pi r^3}{3}$
Circle	Area $= \pi r^2$ (Circumference $= 2\pi r$)	Cylinder	Lateral surface area $= 2\pi r \ell$ Volume $= \pi r^2 \ell$
Triangle	Area $= \frac{1}{2} bh$	Rectangular box	Area $= 2(\ell h + \ell w + hw)$ Volume $= \ell w h$

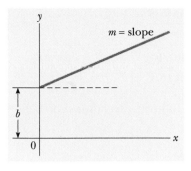

Figure B.5

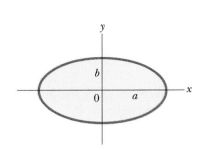

Figure B.6

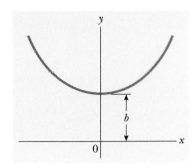

Figure B.7

The equation of a **straight line** (Fig. B.5) is

$$y = mx + b \qquad\qquad \text{[B.19]}$$

where b is the y intercept and m is the slope of the line.

The equation of a **circle** of radius R centered at the origin is

$$x^2 + y^2 = R^2 \qquad\qquad \text{[B.20]}$$

The equation of an **ellipse** with the origin at its center (Fig. B.6) is

$$\frac{x^2}{a^2} + \frac{y^2}{b^2} = 1 \qquad\qquad \text{[B.21]}$$

where a is the length of the semimajor axis and b is the length of the semiminor axis.

The equation of a **parabola** the vertex of which is at $y = b$ (Fig. B.7) is

$$y = ax^2 + b \qquad\qquad \text{[B.22]}$$

The equation of a **rectangular hyperbola** (Fig. B.8) is

$$xy = \text{constant} \qquad\qquad \text{[B.23]}$$

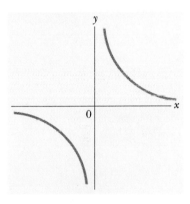

Figure B.8

B.4 · TRIGONOMETRY

That portion of mathematics based on the special properties of the right triangle is called trigonometry. By definition, a right triangle is one containing a 90° angle. Consider the right triangle in Figure B.9, where side a is opposite the angle θ, side b is adjacent to the angle θ, and side c is the hypotenuse of the triangle, The three basic trigonometric functions defined by such a triangle are the sine (sin), cosine (cos), tangent (tan) functions. In terms of the angle θ, these functions are defined by

$$\sin \theta \equiv \frac{\text{side opposite } \theta}{\text{hypotenuse}} = \frac{a}{c} \qquad\qquad \text{[B.24]}$$

a = opposite side
b = adjacent side
c = hypotenuse

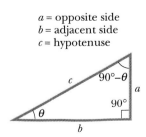

Figure B.9

$$\cos \theta \equiv \frac{\text{side adjacent to } \theta}{\text{hypotenuse}} = \frac{b}{c} \qquad \text{[B.25]}$$

$$\tan \theta \equiv \frac{\text{side opposite } \theta}{\text{side adjacent to } \theta} = \frac{a}{b} \qquad \text{[B.26]}$$

The Pythagorean theorem provides the following relationship between the sides of a right triangle:

$$c^2 = a^2 + b^2 \qquad \text{[B.27]}$$

From the preceding definitions and the Pythagorean theorem, it follows that

$$\sin^2 \theta + \cos^2 \theta = 1$$

$$\tan \theta = \frac{\sin \theta}{\cos \theta}$$

The cosecant, secant, and cotangent functions are defined by

$$\csc \theta \equiv \frac{1}{\sin \theta} \qquad \sec \theta \equiv \frac{1}{\cos \theta} \qquad \cot \theta \equiv \frac{1}{\tan \theta}$$

The following relations come directly from the right triangle shown in Figure B.9:

$$\begin{cases} \sin \theta = \cos(90° - \theta) \\ \cos \theta = \sin(90° - \theta) \\ \cot \theta = \tan(90° - \theta) \end{cases}$$

Some properties of trigonometric functions are as follows:

$$\begin{cases} \sin(-\theta) = -\sin \theta \\ \cos(-\theta) = \cos \theta \\ \tan(-\theta) = -\tan \theta \end{cases}$$

The following relations apply to *any* triangle, as shown in Figure B.10:

$$\alpha + \beta + \gamma = 180°$$

$$\text{Law of cosines} \quad \begin{cases} a^2 = b^2 + c^2 - 2bc \cos \alpha \\ b^2 = a^2 + c^2 - 2ac \cos \beta \\ c^2 = a^2 + b^2 - 2ab \cos \gamma \end{cases}$$

$$\text{Law of sines} \quad \begin{cases} \dfrac{a}{\sin \alpha} = \dfrac{b}{\sin \beta} = \dfrac{c}{\sin \gamma} \end{cases}$$

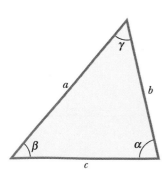

Figure B.10

Table B.3 lists a number of useful trigonometric identities.

TABLE B.3 Some Trigonometric Identities

$$\sin^2\theta + \cos^2\theta = 1 \qquad \csc^2\theta = 1 + \cot^2\theta$$

$$\sec^2\theta = 1 + \tan^2\theta \qquad \sin^2\frac{\theta}{2} = \tfrac{1}{2}(1 - \cos\theta)$$

$$\sin 2\theta = 2\sin\theta\cos\theta \qquad \cos^2\frac{\theta}{2} = \tfrac{1}{2}(1 + \cos\theta)$$

$$\cos 2\theta = \cos^2\theta - \sin^2\theta \qquad 1 - \cos\theta = 2\sin^2\frac{\theta}{2}$$

$$\tan 2\theta = \frac{2\tan\theta}{1 - \tan^2\theta} \qquad \tan\frac{\theta}{2} = \sqrt{\frac{1 - \cos\theta}{1 + \cos\theta}}$$

$$\sin(A \pm B) = \sin A\cos B \pm \cos A\sin B$$

$$\cos(A \pm B) = \cos A\cos B \mp \sin A\sin B$$

$$\sin A \pm \sin B = 2\sin[\tfrac{1}{2}(A \pm B)]\cos[\tfrac{1}{2}(A \mp B)]$$

$$\cos A + \cos B = 2\cos[\tfrac{1}{2}(A + B)]\cos[\tfrac{1}{2}(A - B)]$$

$$\cos A - \cos B = 2\sin[\tfrac{1}{2}(A + B)]\sin[\tfrac{1}{2}(B - A)]$$

Example 3

Consider the right triangle in Figure B.11, in which $a = 2$, $b = 5$, and c is unknown. From the Pythagorean theorem, we have

$$c^2 = a^2 + b^2 = 2^2 + 5^2 = 4 + 25 = 29$$

$$c = \sqrt{29} = \boxed{5.39}$$

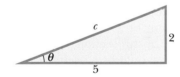

Figure B.11

To find the angle θ, note that

$$\tan\theta = \frac{a}{b} = \frac{2}{5} = 0.400$$

From a table of functions or from a calculator, we have

$$\theta = \tan^{-1}(0.400) = \boxed{21.8°}$$

where $\tan^{-1}(0.400)$ is the notation for "angle the tangent of which is 0.400," sometimes written as arctan (0.400).

EXERCISES

1. In Figure B.12, find (a) the side opposite θ, (b) the side adjacent to ϕ, (c) $\cos\theta$, (d) $\sin\phi$, and (e) $\tan\phi$.
   Answers   (a) 3,   (b) 3,   (c) $\frac{4}{5}$,   (d) $\frac{4}{5}$,   (e) $\frac{4}{3}$
2. In a certain right triangle, the two sides that are perpendicular to each other are 5m and 7m long. What is the length of the third side of the triangle?
   Answer   8.60 m
3. A right triangle has a hypotenuse of length 3m, and one of its angles is 30°. What are the lengths of (a) the side opposite the 30° angle and (b) the side adjacent to the 30° angle?
   Answers   (a) 1.5 m,   (b) 2.60 m

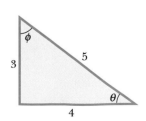

Figure B.12

B.5 · SERIES EXPANSIONS

$$(a + b)^n = a^n + \frac{n}{1!} a^{n-1} b + \frac{n(n-1)}{2!} a^{n-2}b^2 + \cdots$$

$$(1 + x)^n = 1 + nx + \frac{n(n-1)}{2!} x^2 + \cdots$$

$$e^x = 1 + x + \frac{x^2}{2!} + \frac{x^3}{3!} + \cdots$$

$$\ln(1 \pm x) = \pm x - \tfrac{1}{2}x^2 \pm \tfrac{1}{3}x^3 - \cdots$$

$$\left.
\begin{aligned}
\sin x &= x - \frac{x^3}{3!} + \frac{x^5}{5!} - \cdots \\[2mm]
\cos x &= 1 - \frac{x^2}{2!} + \frac{x^4}{4!} - \cdots \\[2mm]
\tan x &= x + \frac{x^3}{3} + \frac{2x^5}{15} + \cdots \qquad |x| < \pi/2
\end{aligned}
\right\} \quad x \text{ in radians}$$

For $x \ll 1$, the following approximations can be used:

$$(1 + x)^n \approx 1 + nx \qquad \sin x \approx x$$

$$e^x \approx 1 + x \qquad \cos x \approx 1$$

$$\ln(1 \pm x) \approx \pm x \qquad \tan x \approx x$$

B.6 · DIFFERENTIAL CALCULUS

In various branches of science, it is sometimes necessary to use the basic tools of calculus, first invented by Newton, to describe physical phenomena. The use of calculus is fundamental in the treatment of a variety of problems in newtonian mechanics, electricity, and magnetism. In this section, we simply state some basic properties and "rules of thumb," which should be a useful review.

First, a **function relation** must be specified that relates one variable to another (such as a coordinate as a function of time). Suppose one of the variables is called y (the dependent variable), and the other, x (the independent variable). We might have a function relation such as

$$y(x) = ax^3 + bx^2 + cx + d$$

If a, b, c, and d are specified constants, then y can be calculated for any value of x. We usually deal with continuous functions, that is, those for which y varies "smoothly" with x.

The **derivative** of y with respect to x is defined as the limit of the slopes of chords drawn between two points on the y versus x curve as Δx approaches zero. Mathematically, we write this definition as

$$\frac{dy}{dx} = \lim_{\Delta x \to 0} \frac{\Delta y}{\Delta x} = \lim_{\Delta x \to 0} \frac{y(x + \Delta x) - y(x)}{\Delta x} \qquad \text{[B.28]}$$

where Δy and Δx are defined as $\Delta x = x_2 - x_1$ and $\Delta y = y_2 - y_1$ (see Fig. B.13).

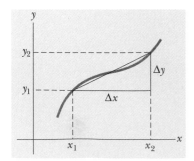

Figure B.13

A useful expression to remember when $y(x) = ax^n$, where a is a *constant* and n is *any* positive or negative number (integer or fraction), is

$$\frac{dy}{dx} = nax^{n-1} \qquad\qquad \textbf{[B.29]}$$

If $y(x)$ is a polynomial or algebraic function of x, we apply Equation B.29 to *each* term in the polynomial and take $da/dx = 0$. It is important to note that dy/dx *does not* mean dy divided by dx but is simply a notation of the limiting process of the derivative, as defined by Equation B.28. In Examples 4 through 7, we evaluate the derivatives of several well-behaved functions.

Example 4

Suppose $y(x)$ (that is, y as a function of x) is given by

$$y(x) = ax^3 + bx + c$$

where a and b are constants. Then it follows that

$$y(x + \Delta x) = a(x + \Delta x)^3 + b(x + \Delta x) + c$$
$$= a(x^3 + 3x^2\,\Delta x + 3x\,\Delta x^2 + \Delta x^3)$$
$$+ b(x + \Delta x) + c$$

so

$$\Delta y = y(x + \Delta x) - y(x) = a(3x^2\,\Delta x + 3x\,\Delta x^2 + \Delta x^3) + b\,\Delta x$$

Substituting this into Equation B.28 gives

$$\frac{dy}{dx} = \lim_{\Delta x \to 0} \frac{\Delta y}{\Delta x} = \lim_{\Delta x \to 0} [3ax^2 + 3x\,\Delta x + \Delta x^2] + b$$
$$\frac{dy}{dx} = 3ax^2 + b$$

Example 5

$$y(x) = 8x^5 + 4x^3 + 2x + 7$$

Solution Applying Equation B.29 to each term independently and remembering that d/dx (constant) $= 0$, we have

$$\frac{dy}{dx} = 8(5)x^4 + 4(3)x^2 + 2(1)x^0 + 0$$
$$\frac{dy}{dx} = 40x^4 + 12x^2 + 2$$

Special Properties of the Derivative

A. Derivative of the Product of Two Functions. If a function, y, is given by the product of two functions—say, $g(x)$ and $h(x)$—then the derivative of y is defined as

$$\frac{d}{dx} f(x) = \frac{d}{dx} [g(x)h(x)] = g\frac{dh}{dx} + h\frac{dg}{dx} \qquad\qquad \textbf{[B.30]}$$

B. Derivative of the Sum of Two Functions. If a function, y, is equal to the sum of two functions, then the derivative of the sum is equal to the sum of the derivatives:

$$\frac{d}{dx} f(x) = \frac{d}{dx} [g(x) + h(x)] = \frac{dg}{dx} + \frac{dh}{dx} \qquad\qquad \textbf{[B.31]}$$

C. Chain Rule of Differential Calculus. If $y = f(x)$ and x is a function of some other variable, z, then dy/dx can be written as the product of two derivatives:

$$\frac{dy}{dx} = \frac{dy}{dz}\frac{dz}{dx}$$
[B.32]

D. The Second Derivative. The second derivative of y with respect to x is defined as the derivative of the function dy/dx (that is, the derivative of the derivative). It is usually written

$$\frac{d^2y}{dx^2} = \frac{d}{dx}\left(\frac{dy}{dx}\right)$$
[B.33]

$y(x) = \dfrac{x^3}{(x+1)^2}$ $\qquad \dfrac{3x^2}{2x+1}$

Example 6

Find the first derivative of $y(x) = x^3/(x+1)^2$ with respect to x.

Solution   We can rewrite this function in the alternate form $y(x) = x^3(x+1)^{-2}$ and apply Equation B.30 directly:

$$\frac{dy}{dx} = (x+1)^{-2}\frac{d}{dx}(x^3) + x^3\frac{d}{dx}(x+1)^{-2}$$

$$= (x+1)^{-2}\,3x^2 + x^3(-2)(x+1)^{-3}$$

$$\frac{dy}{dx} = \frac{3x^2}{(x+1)^2} - \frac{2x^3}{(x+1)^3}$$

Example 7

A useful formula that follows from Equation B.30 is the derivative of the quotient of two functions. Show that the expression is given by

$$\frac{d}{dx}\left[\frac{g(x)}{h(x)}\right] = \frac{h\dfrac{dg}{dx} - g\dfrac{dh}{dx}}{h^2}$$

Solution   We can write the quotient as gh^{-1} and then apply Equations B.29 and B.30:

$$\frac{d}{dx}\left(\frac{g}{h}\right) = \frac{d}{dx}(gh^{-1}) = g\frac{d}{dx}(h^{-1}) + h^{-1}\frac{d}{dx}(g)$$

$$= -gh^{-2}\frac{dh}{dx} + h^{-1}\frac{dg}{dx}$$

$$= \frac{h\dfrac{dg}{dx} - g\dfrac{dh}{dx}}{h^2}$$

TABLE B.4   Derivatives for Several Functions

$\dfrac{d}{dx}(a) = 0$	$\dfrac{d}{dx}(\cos ax) = -a\sin ax$	$\dfrac{d}{dx}(\sec x) = \tan x\sec x$
$\dfrac{d}{dx}(ax^n) = nax^{n-1}$	$\dfrac{d}{dx}(\tan ax) = a\sec^2 ax$	$\dfrac{d}{dx}(\csc x) = -\cot x\csc x$
$\dfrac{d}{dx}(e^{ax}) = ae^{ax}$	$\dfrac{d}{dx}(\cot ax) = -a\csc^2 ax$	$\dfrac{d}{dx}(\ln ax) = \dfrac{1}{x}$
$\dfrac{d}{dx}(\sin ax) = a\cos ax$		

Note: The letters a and n are constants.

Some of the most commonly used derivatives of functions are listed in Table B.4.

B.7 · INTEGRAL CALCULUS

We think of integration as the inverse of differentiation. As an example, consider the expression

$$f(x) = \frac{dy}{dx} = 3ax^2 + b$$

which was the result of differentiating the function

$$y(x) = ax^3 + bx + c$$

in Example 4. We can write the first expression, $dy = f(x)\, dx = (3ax^2 + b)\, dx$, and obtain $y(x)$ by "summing" over all values of x. Mathematically, we write this inverse operation as

$$y(x) = \int f(x)\, dx$$

For the function $f(x)$ already given,

$$y(x) = \int (3ax^2 + b)\, dx = ax^3 + bx + c$$

where c is a constant of the integration. This type of integral is called an *indefinite integral* because its value depends on the choice of the constant c.

A general **indefinite integral,** $I(x)$, is defined as

$$I(x) = \int f(x)\, dx \qquad \qquad \textbf{[B.34]}$$

where $f(x)$ is called the *integrand* and $f(x) = \dfrac{dI(x)}{dx}$.

For a *general continuous* function $f(x)$, the integral can be described as the area under the curve bounded by $f(x)$ and the x axis, between two specified values of x—say, x_1 and x_2—as in Figure B.14.

The area of the shaded element is approximately $f_i\, \Delta x_i$. If we sum all these area elements from x_1 to x_2 and take the limit of this sum as $\Delta x_i \to 0$, we obtain the *true* area under the curve bounded by $f(x)$ and x, between the limits x_1 and x_2:

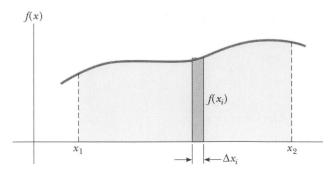

Figure B.14

$$\text{Area} = \lim_{\Delta x_i \to 0} \sum_i f(x_i)\, \Delta x_i = \int_{x_1}^{x_2} f(x)\, dx \qquad \text{[B.35]}$$

Integrals of the type defined by Equation B.35 are called **definite integrals.**

    One of the common types of integrals that arise in practical situations has the form

$$\int x^n\, dx = \frac{x^{n+1}}{n+1} + c \qquad (n \neq -1) \qquad \text{[B.36]}$$

This result is obvious because differentiation of the right-hand side with respect to x gives $f(x) = x^n$ directly. If the limits of the integration are known, this integral becomes a *definite integral* and is written

$$\int_{x_1}^{x_2} x^n\, dx = \frac{x_2^{\,n+1} - x_1^{\,n+1}}{n+1} \qquad (n \neq -1) \qquad \text{[B.37]}$$

Examples

1. $\displaystyle \int_0^a x^2\, dx = \frac{x^3}{3}\bigg]_0^a = \frac{a^3}{3}$

2. $\displaystyle \int_0^b x^{3/2}\, dx = \frac{x^{5/2}}{5/2}\bigg]_0^b = \frac{2}{5}\, b^{5/2}$

3. $\displaystyle \int_3^5 x\, dx = \frac{x^2}{2}\bigg]_3^5 = \frac{5^2 - 3^2}{2} = 8$

Partial Integration

Sometimes it is useful to apply the method of *partial integration* to evaluate certain integrals. The method uses the property that

$$\int u \, dv = uv - \int v \, du \qquad\qquad \text{[B.38]}$$

where u and v are *carefully* chosen in order to reduce a complex integral to a simpler one. In many cases, several reductions have to be made. Consider the example

$$I(x) = \int x^2 e^x \, dx$$

This can be evaluated by integrating by parts twice. First, if we choose $u = x^2$, $v = e^x$, we get

$$\int x^2 e^x \, dx = \int x^2 \, d(e^x) = x^2 e^x - 2 \int e^x x \, dx + c_1$$

Now, in the second term, choose $u = x$, $v = e^x$, which gives

$$\int x^2 e^x \, dx = x^2 e^x - 2xe^x + 2 \int e^x \, dx + c_1$$

or

$$\int x^2 e^x \, dx = x^2 e^x - 2xe^x + 2e^x + c_2$$

The Perfect Differential

Another useful method to remember is the use of the *perfect differential*. That is, we should sometimes look for a change of variable such that the differential of the function is the differential of the independent variable appearing in the integrand. For example, consider the integral

$$I(x) = \int \cos^2 x \sin x \, dx$$

This becomes easy to evaluate if we rewrite the differential as $d(\cos x) = -\sin x \, dx$. The integral then becomes

$$\int \cos^2 x \sin x \, dx = -\int \cos^2 x \, d(\cos x)$$

If we now change variables, letting $y = \cos x$, we get

$$\int \cos^2 x \sin x \, dx = -\int y^2 \, dy = -\frac{y^3}{3} + c = -\frac{\cos^3 x}{3} + c$$

TABLE B.5   **Some Indefinite Integrals (an arbitrary constant should be added to each of these integrals)**

$$\int x^n \, dx = \frac{x^{n+1}}{n+1} \qquad \text{(provided } n \neq -1)$$

$$\int \frac{dx}{x} = \int x^{-1} \, dx = \ln x$$

$$\int \frac{dx}{a+bx} = \frac{1}{b} \ln(a+bx)$$

$$\int \frac{dx}{(a+bx)^2} = -\frac{1}{b(a+bx)}$$

$$\int \frac{dx}{a^2+x^2} = \frac{1}{a} \tan^{-1} \frac{x}{a}$$

$$\int \frac{dx}{a^2-x^2} = \frac{1}{2a} \ln \frac{a+x}{a-x} \qquad (a^2-x^2 > 0)$$

$$\int \frac{dx}{x^2-a^2} = \frac{1}{2a} \ln \frac{x-a}{x+a} \qquad (x^2-a^2 > 0)$$

$$\int \frac{x \, dx}{a^2 \pm x^2} = \pm \tfrac{1}{2} \ln(a^2 \pm x^2)$$

$$\int \frac{dx}{x(x+a)} = -\frac{1}{a} \ln \frac{x+a}{x}$$

$$\int \frac{dx}{\sqrt{a^2-x^2}} = \sin^{-1} \frac{x}{a} = -\cos^{-1} \frac{x}{a} \qquad (a^2-x^2 > 0)$$

$$\int \frac{dx}{\sqrt{x^2 \pm a^2}} = \ln(x + \sqrt{x^2 \pm a^2})$$

$$\int \frac{x \, dx}{\sqrt{a^2-x^2}} = \sqrt{a^2-x^2}$$

$$\int \frac{x \, dx}{\sqrt{x^2 \pm a^2}} = \sqrt{x^2 \pm a^2}$$

$$\int \sqrt{a^2-x^2} \, dx = \tfrac{1}{2}\left(x\sqrt{a^2-x^2} + a^2 \sin^{-1} \frac{x}{a} \right)$$

$$\int x\sqrt{a^2-x^2} \, dx = -\tfrac{1}{3}(a^2-x^2)^{3/2}$$

$$\int \sqrt{x^2 \pm a^2} \, dx = \tfrac{1}{2}[x\sqrt{x^2 \pm a^2} \pm a^2 \ln(x + \sqrt{x^2 \pm a^2})]$$

$$\int x(\sqrt{x^2 \pm a^2}) \, dx = \tfrac{1}{3}(x^2 \pm a^2)^{3/2}$$

$$\int e^{ax} \, dx = \frac{1}{a} e^{ax}$$

$$\int \ln ax \, dx = (x \ln ax) - x$$

$$\int xe^{ax} \, dx = \frac{e^{ax}}{a^2}(ax - 1)$$

$$\int \frac{dx}{a+be^{cx}} = \frac{x}{a} - \frac{1}{ac} \ln(a+be^{cx})$$

$$\int \sin ax \, dx = -\frac{1}{a} \cos ax$$

$$\int \cos ax \, dx = \frac{1}{a} \sin ax$$

$$\int \tan ax \, dx = -\frac{1}{a} \ln(\cos ax) = \frac{1}{a} \ln(\sec ax)$$

$$\int \cot ax \, dx = \frac{1}{a} \ln(\sin ax)$$

$$\int \sec ax \, dx = \frac{1}{a} \ln(\sec ax + \tan ax) = \frac{1}{a} \ln \left[\tan \left(\frac{ax}{2} + \frac{\pi}{4} \right) \right]$$

$$\int \csc ax \, dx = \frac{1}{a} \ln(\csc ax - \cot ax) = \frac{1}{a} \ln \left(\tan \frac{ax}{2} \right)$$

$$\int \sin^2 ax \, dx = \frac{x}{2} - \frac{\sin 2ax}{4a}$$

$$\int \cos^2 ax \, dx = \frac{x}{2} + \frac{\sin 2ax}{4a}$$

$$\int \frac{dx}{\sin^2 ax} = -\frac{1}{a} \cot ax$$

$$\int \frac{dx}{\cos^2 ax} = \frac{1}{a} \tan ax$$

$$\int \tan^2 ax \, dx = \frac{1}{a}(\tan ax) - x$$

$$\int \cot^2 ax \, dx = -\frac{1}{a}(\cot ax) - x$$

$$\int \sin^{-1} ax \, dx = x(\sin^{-1} ax) + \frac{\sqrt{1-a^2x^2}}{a}$$

$$\int \cos^{-1} ax \, dx = x(\cos^{-1} ax) - \frac{\sqrt{1-a^2x^2}}{a}$$

$$\int \frac{dx}{(x^2+a^2)^{3/2}} = \frac{x}{a^2\sqrt{x^2+a^2}}$$

$$\int \frac{x \, dx}{(x^2+a^2)^{3/2}} = -\frac{1}{\sqrt{x^2+a^2}}$$

Table B.5 lists some useful indefinite integrals. Table B.6 gives Gauss's probability integral and other definite integrals. A more complete list can be found in various handbooks, such as *The Handbook of Chemistry and Physics,* CRC Press.

TABLE B.6   **Gauss's Probability Integral and Related Integrals**

$$\int_0^\infty x^n e^{-ax} \, dx = \frac{n!}{a^{n+1}}$$

$$I_0 = \int_0^\infty e^{-ax^2} \, dx = \frac{1}{2}\sqrt{\frac{\pi}{a}} \qquad \text{(Gauss's probability integral)}$$

$$I_1 = \int_0^\infty xe^{-ax^2} \, dx = \frac{1}{2a}$$

$$I_2 = \int_0^\infty x^2 e^{-ax^2} \, dx = -\frac{dI_0}{da} = \frac{1}{4}\sqrt{\frac{\pi}{a^3}}$$

$$I_3 = \int_0^\infty x^3 e^{-ax^2} \, dx = -\frac{dI_1}{da} = \frac{1}{2a^2}$$

$$I_4 = \int_0^\infty x^4 e^{-ax^2} \, dx = \frac{d^2 I_0}{da^2} = \frac{3}{8}\sqrt{\frac{\pi}{a^5}}$$

$$I_5 = \int_0^\infty x^5 e^{-ax^2} \, dx = \frac{d^2 I_1}{da^2} = \frac{1}{a^3}$$

$$\cdot$$
$$\cdot$$
$$\cdot$$

$$I_{2n} = (-1)^n \frac{d^n}{da^n} I_0$$

$$I_{2n+1} = (-1)^n \frac{d^n}{da^n} I_1$$

Periodic Table of the Elements

Group I	Group II	Transition elements						
H 1 1.0080 $1s^1$								
Li 3 6.94 $2s^1$	**Be** 4 9.012 $2s^2$							
Na 11 22.99 $3s^1$	**Mg** 12 24.31 $3s^2$							
K 19 39.102 $4s^1$	**Ca** 20 40.08 $4s^2$	**Sc** 21 44.96 $3d^14s^2$	**Ti** 22 47.90 $3d^24s^2$	**V** 23 50.94 $3d^34s^2$	**Cr** 24 51.996 $3d^54s^1$	**Mn** 25 54.94 $3d^54s^2$	**Fe** 26 55.85 $3d^64s^2$	**Co** 27 58.93 $3d^74s^2$
Rb 37 85.47 $5s^1$	**Sr** 38 87.62 $5s^2$	**Y** 39 88.906 $4d^15s^2$	**Zr** 40 91.22 $4d^25s^2$	**Nb** 41 92.91 $4d^45s^1$	**Mo** 42 95.94 $4d^55s^1$	**Tc** 43 (99) $4d^55s^2$	**Ru** 44 101.1 $4d^75s^1$	**Rh** 45 102.91 $4d^85s^1$
Cs 55 132.91 $6s^1$	**Ba** 56 137.34 $6s^2$	57-71*	**Hf** 72 178.49 $5d^26s^2$	**Ta** 73 180.95 $5d^36s^2$	**W** 74 183.85 $5d^46s^2$	**Re** 75 186.2 $5d^56s^2$	**Os** 76 190.2 $5d^66s^2$	**Ir** 77 192.2 $5d^76s^2$
Fr 87 (223) $7s^1$	**Ra** 88 (226) $7s^2$	89-103**	**Rf‡** 104 (261) $6d^27s^2$	**Ha** 105 (262) $6d^37s^2$	**Sg** 106 (263)	**Ns** 107 (262)	**Hs** 108 (265)	**Mt** 109 (266)

Key:

Symbol — **Ca** 20 — Atomic number
Atomic mass † — 40.08
$4s^2$ — Electron configuration

*Lanthanide series

La 57 138.91 $5d^16s^2$	**Ce** 58 140.12 $5d^14f^16s^2$	**Pr** 59 140.91 $4f^36s^2$	**Nd** 60 144.24 $4f^46s^2$	**Pm** 61 (147) $4f^56s^2$	**Sm** 62 150.4 $4f^66s^2$

**Actinide series

Ac 89 (227) $6d^17s^2$	**Th** 90 (232) $6d^27s^2$	**Pa** 91 (231) $5f^26d^17s^2$	**U** 92 (238) $5f^36d^17s^2$	**Np** 93 (239) $5f^46d^17s^2$	**Pu** 94 (239) $5f^66d^07s^2$

† Atomic mass values given are averaged over isotopes in the percentages in which they exist in nature. For an unstable element, mass number of the most stable known isotope is given in parentheses.
‡ The names for elements 104–109 are controversial. The names given here are those recommended by the discoverers.

	Group III	Group IV	Group V	Group VI	Group VII	Group 0
					H 1 1.0080 $1s^1$	**He** 2 4.0026 $1s^2$
	B 5 10.81 $2p^1$	**C** 6 12.011 $2p^2$	**N** 7 14.007 $2p^3$	**O** 8 15.999 $2p^4$	**F** 9 18.998 $2p^5$	**Ne** 10 20.18 $2p^6$
	Al 13 26.98 $3p^1$	**Si** 14 28.09 $3p^2$	**P** 15 30.97 $3p^3$	**S** 16 32.06 $3p^4$	**Cl** 17 35.453 $3p^5$	**Ar** 18 39.948 $3p^6$

Ni 28 58.71 $3d^8 4s^2$	**Cu** 29 63.54 $3d^{10} 4s^2$	**Zn** 30 65.37 $3d^{10} 4s^2$	**Ga** 31 69.72 $4p^1$	**Ge** 32 72.59 $4p^2$	**As** 33 74.92 $4p^3$	**Se** 34 78.96 $4p^4$	**Br** 35 79.91 $4p^5$	**Kr** 36 83.80 $4p^6$
Pd 46 106.4 $4d^{10}$	**Ag** 47 107.87 $4d^{10} 5s^1$	**Cd** 48 112.40 $4d^{10} 5s^2$	**In** 49 114.82 $5p^1$	**Sn** 50 118.69 $5p^2$	**Sb** 51 121.75 $5p^3$	**Te** 52 127.60 $5p^4$	**I** 53 126.90 $5p^5$	**Xe** 54 131.30 $5p^6$
Pt 78 195.09 $5d^9 6s^1$	**Au** 79 196.97 $5d^{10} 6s^1$	**Hg** 80 200.59 $5d^{10} 6s^2$	**Tl** 81 204.37 $6p^1$	**Pb** 82 207.2 $6p^2$	**Bi** 83 208.98 $6p^3$	**Po** 84 (210) $6p^4$	**At** 85 (218) $6p^5$	**Rn** 86 (222) $6p^6$
110 Discovered Nov. 1994	111 Discovered Dec. 1994							

Eu 63 152.0 $4f^7 6s^2$	**Gd** 64 157.25 $5d^1 4f^7 6s^2$	**Tb** 65 158.92 $5d^1 4f^8 6s^2$	**Dy** 66 162.50 $4f^{10} 6s^2$	**Ho** 67 164.93 $4f^{11} 6s^2$	**Er** 68 167.26 $4f^{12} 6s^2$	**Tm** 69 168.93 $4f^{13} 6s^2$	**Yb** 70 173.04 $4f^{14} 6s^2$	**Lu** 71 174.97 $5d^1 4f^{14} 6s^2$
Am 95 (243) $5f^7 6d^0 7s^2$	**Cm** 96 (245) $5f^7 6d^1 7s^2$	**Bk** 97 (247) $5f^8 6d^1 7s^2$	**Cf** 98 (249) $5f^{10} 6d^0 7s^2$	**Es** 99 (254) $5f^{11} 6d^0 7s^2$	**Fm** 100 (253) $5f^{12} 6d^0 7s^2$	**Md** 101 (255) $5f^{13} 6d^0 7s^2$	**No** 102 (255) $6d^0 7s^2$	**Lr** 103 (257) $6d^1 7s^2$

APPENDIX D

SI Units

TABLE D.1 SI Base Units

Base Quantity	SI Base Unit	
	Name	Symbol
Length	meter	m
Mass	kilogram	kg
Time	second	s
Electric current	ampere	A
Temperature	kelvin	K
Amount of substance	mole	mol
Luminous intensity	candela	cd

TABLE D.2 Some Derived SI Units

Quantity	Name	Symbol	Expression in Terms of Base Units	Expression in Terms of Other SI Units
Plane angle	radian	rad	m/m	
Frequency	hertz	Hz	s^{-1}	
Force	newton	N	$kg \cdot m/s^2$	J/m
Pressure	pascal	Pa	$kg/m \cdot s^2$	N/m^2
Energy:work	joule	J	$kg \cdot m^2/s^2$	$N \cdot m$
Power	watt	W	$kg \cdot m^2/s^3$	J/s
Electric charge	coulomb	C	$A \cdot s$	
Electric potential (emf)	volt	V	$kg \cdot m^2/A \cdot s^3$	W/A
Capacitance	farad	F	$A^2 \cdot s^4/kg \cdot m^2$	C/V
Electric resistance	ohm	Ω	$kg \cdot m^2/A^2 \cdot s^2$	V/A
Magnetic flux	weber	Wb	$kg \cdot m^2/A \cdot s^2$	$V \cdot s$
Magnetic field intensity	tesla	T	$kg/A \cdot s^2$	Wb/m^2
Inductance	henry	H	$kg \cdot m^2/A^2 \cdot s^2$	Wb/A

APPENDIX E

Spreadsheet Problems

Overview

Students come to introductory physics courses with a wide variety of computing experience. Many are already accomplished programmers in one or more programming languages (BASIC, Pascal, FORTRAN, and so forth). Others have never even turned on a computer. To further complicate matters, a wide variety of hardware environments exists, although the vast majority can be classified as IBM/compatible (MS-DOS) or Macintosh environments. We have designed the spreadsheet problems and *Spreadsheet Investigations in Physics* to be usable by and useful to students in all these diverse situations. Our goal is to enable students to investigate a range of physical phenomena and obtain a feel for the physics. Merely "getting the right answer" by plugging numbers into a formula and comparing the result to the answer in the back of the book is discouraged.

Spreadsheets are particularly valuable in exploratory investigations. Once you have constructed a spreadsheet, you can simply vary the parameters and see instantly how things change. Even more important is the ease with which you can construct accurate graphs of relations between physical variables. When you change a parameter, you can view the effects of the change upon the graphs simply by pressing a key or clicking a mouse button. "What if" questions can be easily addressed and depicted graphically.

How to Use the Templates

The computer spreadsheet problems are arranged by difficulty level. The least difficult problems are coded in black. For most of these problems, spreadsheets are provided on disk. Only the input parameters need to be changed. Problems of moderate difficulty are coded in blue. These require additional analysis and the provided spreadsheets will need to be modified to solve them. The most challenging problems are coded in magenta. For most of these, you must develop your own spreadsheets. The emphasis should be on understanding what the results mean rather than just getting an answer. For example, one spreadsheet problem explores the strategies for fuel burn rates in rocket propulsion. How long and at what rate should the rocket fuel be burned to obtain the highest velocity or the greatest distance?

Software Requirements

The spreadsheet template files are provided on a high-density (1.44 MB) 3.5″ diskette, which can be obtained in either of two forms. The disk for the Macintosh environment will contain files in the file format used by the program Microsoft Excel 5.0 for the Macintosh.

Since there are many spreadsheet programs used in the IBM/compatible environment, the disk for Windows and DOS users will contain three different versions of the spreadsheet template files. These versions are chosen to allow the diskette to be usable by as many students as possible. The file ***123WIN.EXE*** contains templates in the format suitable for use in the program Lotus 1-2-3 for Windows, Release 4 or 5. Files in this format have filenames in the form *.WK4, where * stands for a filename of eight letters or less. The file ***EXCELWIN.EXE*** contains templates in the format used by Microsoft Excel for Windows, Version 5.0; files in this format have the form *.XLS. Finally, for students using an earlier version of Lotus 1-2-3 or Excel or using other popular spreadsheet programs (including non-Windows programs), a third file, ***123DOS.EXE,*** contains templates in the format introduced with the versions 2.x of DOS-based Lotus 1-2-3; files in this format have the form *.WK1. The Lotus WK1 format can be read directly by almost all spreadsheet programs, including Lotus 1-2-3 (versions 2.0 and later), Excel, Microsoft Works, and Quattro Pro. The f(g) Scholar program can also import WK1 spreadsheets; however, some minor format changes of the templates are needed.

To fit these versions of the template files onto one diskette, the diskette is supplied with the files in compressed form. These files are self-extracting; they can be decompressed by a simple procedure without using any external programs. Choose the version you want, copy the compressed file to a subdirectory on a hard-disk, and execute the file. For example, to install the Lotus for Windows templates in the subdirectory C:\INVEST, do the following steps:

A. Make sure that the subdirectory C:\INVEST exists. If it does not exist, make the subdirectory.
B. Copy the file 123WIN.EXE from the distribution disk to the subdirectory C:\INVEST.
C. Run or launch the file 123WIN.EXE. The templates will be extracted and are then ready to be used.

The templates on the Macintosh diskette are for Excel 5.0 for the Macintosh. The templates are not compressed. Copy these templates to a folder; they are ready to be used just as they are.

Hardware Requirements

You will need a microcomputer that can run one of the spreadsheet programs in one of their many versions. Your computer should be connected to a printer that can print text and graphics. Older versions of the software will run on a 8086/8088 MS-DOS system with just a single floppy disk drive or on a 512K Macintosh with two floppy drives. Newer versions require a more powerful computer to run effectively. For example, to run Excel For Windows, Version 5.0, you must have a hard disk with about 15 megabytes of disk memory available and 4 to 8 megabytes of RAM memory. Your software manuals will tell you exactly what you need to run the particular version of your spreadsheet program. However, all the problems require only a minimal computer system.

There are many different software versions and many different computer configurations. You might have one floppy drive, two floppy drives, a hard disk, a local area network, and so on. The combinations are almost endless. Our best suggestion is to read your software manual and ask your instructor or computer laboratory personnel how to start your spreadsheet program.

Spreadsheet Tutorial

Some students will have the needed computer and spreadsheet skills to start working with the templates immediately. Other students, who have not used a spreadsheet program before, will need some instruction. We have written *Spreadsheet In vestigations in Physics* for both these groups of students. The first part contains a spreadsheet tutorial that the novice student can use *independently* to gain the required spreadsheet skills. A two- or three-hour initial session with the tutorial and the computer is all that most students need to get started. Once the students have mastered the basic operations of the spreadsheet program, they should try one or two of the easier problems.

Because very few introductory physics students have studied numerical methods, we have also included a brief introduction to numerical methods in *Spreadsheet Investigations in Physics*. This section covers numerical interpolation, differentiation, integration, and the solution of simple differential equations. The student should not try to master all of this material at one time; only study the sections that are needed to solve the currently assigned problems.

The templates supplied on the distribution diskettes with *Spreadsheet Investigations in Physics* constitute an outline. You must enter the appropriate data and parameters. The parameters must be adjusted to fit the needs of your problem. Feel free to change any parameters, to expand or decrease the number of rows of output, and to change the size of increments (for example, in time or distance). Most templates have graphs associated with them. You may have to adjust the ranges of variables plotted and the scales for the axes. See the tutorials for how to adjust the appearance of your graphs.

Answers to Odd-Numbered Problems

Chapter 1

1. (a) $4.00 \text{ u} = 6.64 \times 10^{-24}$ g
   (b) $55.9 \text{ u} = 9.28 \times 10^{-23}$ g
   (c) $207 \text{ u} = 3.44 \times 10^{-22}$ g
3. 2.86 cm
5. (a) 623 kg/m^3
   (b) Yes, because this is less than 1000 kg/m^3
7. $0.579 \, t \, \text{ft}^3/\text{s} + 1.19 \times 10^{-9} t^2 \, \text{ft}^3/\text{s}^2$
11. (a) 6.31×10^4 AU   (b) 1.33×10^{11} AU
13. $151 \, \mu\text{m}$
15. (a) $1.609 \text{ km/h} = 1 \text{ mi/h}$ (b) 88 km/h   (c) 16 km/h
17. 1.19×10^{57}
19. $\sim 10^2$ tuners
21. (a) 797   (b) 11   (c) 18
23. (a) $346 \text{ m}^2 \pm 13 \text{ m}^2$ (b) (66.0 ± 1.3) m
25. $(-2.75, -4.76)$ m
27. (a) $(2.17 \text{ m}, 1.25 \text{ m})$ and $(-1.90 \text{ m}, 3.29 \text{ m})$
    (b) 4.55 m
29. (a) 10.0 m   (b) 15.7 m   (c) 0
31. (a) 5.2 m at 60°   (b) 3.0 m at 330°
    (c) 3.0 m at 150°   (d) 5.2 m at 300°
33. 47.2 units at 122°
35. 227 paces at 165°
37. (a) $R_x = 49.5$ units; $R_y = 27.1$ units
    (b) 56.4 units at 28.7°
39. 240 m at 237°
41. 196 cm at $-14.7°$
43. 0.449%
45. 5×10^9 gal
47. 6×10^{20} kg
49. 2.29 km
51. $\sim 10^{11}$
S1. (a) A plot of log T versus log L for the given data shows
    an approximate linear relationship. Applying the least-
    squares method to the data yields 0.503 for the slope
    and 0.302 for the intercept of the straight line that
    best fits the data. The data points and the best-fit
    line, log T = 0.503 log L + 0.302 are shown in the
    figure at the top of the second column.

Chapter 2

1. (a) 2.30 m/s   (b) 16.1 m/s   (c) 11.5 m/s
3. (a) 5 m/s   (b) 1.2 m/s   (c) -2.5 m/s
   (d) -3.3 m/s   (e) 0

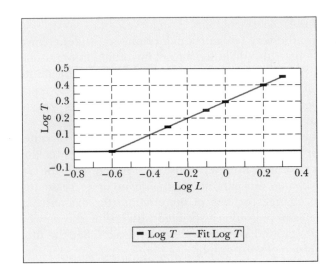

5. (a) $2v_1 v_2/(v_1 + v_2)$ (b) 0
7. (a) -2.4 m/s   (b) -3.2 m/s   (c) 4 s
9. (a) 5 m/s   (b) -2.5 m/s   (c) 0
   (d) 5 m/s
11. $1.34 \times 10^4 \text{ m/s}^2$
13. (a) 2.0 m   (b) -3.0 m/s   (c) -2.0 m/s^2
15. (a) 1.3 m/s^2 (b) 2 m/s^2 at 3 s
    (c) at $t = 6$ s and for $t > 10$ s   (d) -1.5 m/s^2 at 8 s
17. -16.0 cm/s^2
19. 160 ft
21. (a) 20.0 s   (b) no
23. 3.10 m/s
25. (a) -202 m/s^2 (b) 198 m
27. (a) 10.0 m/s up   (b) 4.68 m/s down
29. (a) 29.4 m/s   (b) 44.1 m
31. (a) 7.82 m   (b) 0.782 s
33. (a) 3.00 m/s   (b) 6.00 s   (c) -0.300 m/s^2
    (d) 2.05 m/s
35. Yes, at 11.4 s and 212 m
37. 0.509 s
39. (a) 5.43 m/s^2 and 3.83 m/s^2
    (b) 10.9 m/s and 11.5 m/s   (c) Maggie by 2.62 m
41. $\sim 10^3 \text{ m/s}^2$
43. (a) 2.99 s   (b) -15.4 m/s
    (c) 31.3 m/s down and 34.9 m/s down
45. (a) 26.4 m   (b) 6.88%

47. (a) 5.46 s   (b) 73.0 m   (c) 26.7 m/s and 22.6 m/s
49. (c) $v = 0$ and $a = v_0^2/h$ (d) $v = v_0$ and $a = 0$
51. $27\sqrt{3}$
S3. (a) At approximately $t = 37.0$ s after the police car starts, or 42.0 s after the first police officer sees the speeder.
    (b) 74.0 m/s   (c) 1370 m

Chapter 3

1. (a) 4.87 km at 209° from east   (b) 23.3 m/s
   (c) 13.5 m/s at 209°
3. (a) $18.0t\,\mathbf{i} + (4.00t - 4.90t^2)\mathbf{j}$
   (b) $18.0\mathbf{i} + (4.00 - 9.80t)\mathbf{j}$ (c) $-9.80\mathbf{j}$ m/s²
   (d) $(54.0\mathbf{i} - 32.1\mathbf{j})$ m   (e) $(18.0\mathbf{i} - 25.4\mathbf{j})$ m/s
   (f) $-9.80\mathbf{j}$ m/s²
5. (a) $(2.00\mathbf{i} + 3.00\mathbf{j})$ m/s²
   (b) $(3.00t + t^2)\mathbf{i}$ m $+ (1.50t^2 - 2.00t)\mathbf{j}$ m
7. (a) $a_x = 0.800$ m/s²; $a_y = -0.300$ m/s²   (b) 339°
   (c) $(360\mathbf{i} - 72.7\mathbf{j})$ m, $-15.2°$
9. (a) $3.34\mathbf{i}$ m/s   (b) $-50.9°$
11. 74.5 m/s
13. 48.6 m/s
15. 0.600 m/s²
17. (a) 2.67 s   (b) $29.9\mathbf{i}$ m/s   (c) $(29.9\mathbf{i} - 26.2\mathbf{j})$ m/s
19. (a) The ball clears by 0.889 m   (b) while descending
21. 67.8°
23. (a) 1.03 s   (b) $(8.67\mathbf{i} + 4.20\mathbf{j})$ m/s   (c) 25.8°
25. 377 m/s²
27. (a) 6.00 rev/s   (b) 1.52 km/s²   (c) 1.28 km/s²
29. (a) 7.90 km/s   (b) 5.07 ks = 1.41 h
31. 1.48 m/s²
33. (a) 13.0 m/s²   (b) 5.70 m/s   (c) 7.50 m/s²
35. (a) $4.00\mathbf{i}$ m/s   (b) $(4.00\mathbf{i} + 6.00\mathbf{j})$ m
37. (a) 41.7 m/s   (b) 3.81 s
   (c) $v_x = 34.1$ m/s; $v_y = -13.4$ m/s; 36.6 m/s
39. 8.94 m/s at $-63.4°$
41. 20.0 m
43. (a) 6.80 km
   (b) 3.00 km vertically above the impact point
   (c) 66.2°
45. (a) 46.5 m/s   (b) $-77.6°$ (c) 6.34 s
47. (a) 20.0 m/s, 5.00 s   (b) $(16.0\mathbf{i} - 27.1\mathbf{j})$ m/s
   (c) 6.54 s   (d) $24.6\mathbf{i}$ m
49. (a) 43.2 m   (b) $v_x = 9.66$ m/s; $v_y = -25.6$ m/s
51. Less than 265 m and more than 3.48 km
S1. There are an infinite number of solutions to the problem; however, there is one practical limitation. We can estimate that the maximum speed that the punter can give the ball is about 20 to 30 m/s. Use a speed in this range.
S3. There are an infinite number of solutions. For example if $v_0 = 22.32$ m/s at 45°, then the ball clears the crossbar in 3.01 s. Or if $v_0 = 24.60$ m/s at 30°, then the ball clears the crossbar in 2.23 s. The time it takes the ball to clear

the crossbar is immaterial, since in all likelihood time will run out before the clock is stopped.

Chapter 4

3/4

1. (a) 1/3   (b) 0.750 m/s²
3. $(6.00\mathbf{i} + 15.0\mathbf{j})$ N; 16.2 N
5. (a) 1.44 m   (b) $(50.8\mathbf{i} + 1.40\mathbf{j})$ N
7. (a) $(2.50\mathbf{i} + 5.00\mathbf{j})$ N   (b) 5.59 N
9. (a) 3.64×10^{-18} N
   (b) 8.93×10^{-30} N, which is 408 billion times smaller
11. (a) 534 N down   (b) 54.4 kg
13. (a) 5.00 m/s² at 36.9°   (b) 6.08 m/s² at 25.3°
15. (a) $\sim 10^{-22}$ m/s²   (b) $\sim 10^{-23}$ m
17. (a) 0.200 m/s² forward   (b) 10.0 m   (c) 2.00 m/s
19. (a) 15.0 lb up   (b) 5.00 lb up   (c) 0
21. (a) 31.5 N, 37.5 N, 49.0 N   (b) 113 N, 56.6 N, 98.0 N
23. (b) 514 N, 557 N, 325 N
25. (a) 5.10 kN   (b) 3620 kg
27. 8.66 N east
29. 3.73 m
31. $a = F/(m_1 + m_2)$; $T = Fm_1/(m_1 + m_2)$
33. (a) $a_1 = 2a_2$
   (b) $T_1 = m_1 m_2 g/(2m_1 + \tfrac{1}{2}m_2)$ and
      $T_2 = m_1 m_2 g/(m_1 + \tfrac{1}{4}m_2)$
   (c) $a_1 = m_2 g/(2m_1 + \tfrac{1}{2}m_2)$ and
      $a_2 = m_2 g/(4m_1 + m_2)$
35. (a) 706 N   (b) 814 N   (c) 706 N   (d) 648 N
37. (a) 3.00 s   (b) $(18.0\mathbf{i} - 9.00\mathbf{j})$ m   (c) 20.1 m
39. 1.66 MN
41. (a)

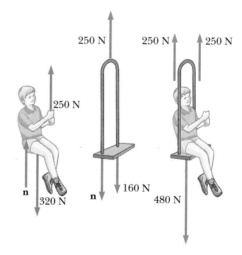

250 N   250 N   250 N

250 N

160 N

320 N   n   480 N

   (b) 0.408 m/s²   (c) 83.3 N
43. 1.18 kN
45. (a) $Mg/2$, $Mg/2$, $Mg/2$, $3Mg/2$, Mg (b) $Mg/2$
47. $(M + m_1 + m_2)(m_2 g/m_1)$
49. (a) 30.7°   (b) 0.843 N

Chapter 5

1. $\mu_s = 0.306$; $\mu_k = 0.245$
3. (a) 256 m   (b) 42.7 m
5. (a) 1.78 m/s^2 (b) 0.368   (c) 9.37 N
   (d) 2.67 m/s
7. (a) 0.161   (b) 1.01 m/s^2
9. 37.8 N
11. (a) 0.931 m/s^2 (b) 6.10 cm
13. Any speed up to 8.08 m/s
15. (a) 9.13×10^{22} m/s^2 (b) 83.2 nN
17. (a) static friction   (b) 0.0850
19. (a) 68.6 N toward the center of the circle and 784 N up
   (b) 0.857 m/s^2
21. No. The jungle-lord needs a vine of minimal tensile
   strength 1.38 kN.
23. (a) 6670 N up   (b) 20.3 m/s
25. 3.13 m/s
27. (a) 32.7 s^{-1} (b) 9.80 m/s^2 down
   (c) 4.90 m/s^2 down
29. (a) 0.0347 s^{-1} (b) 2.50 m/s   (c) $a = -cv$
31. (a) 13.7 m/s down

   (b)

t(s)	x(m)	v(m/s)
0	0	0
0.2	0	-1.96
0.4	-0.392	-3.88
. . .	. . .	. . .
1.0	-3.77	-8.71
2.0	-14.4	-12.56
4.0	-41.0	-13.67

33. (a) 49.5 m/s down and 4.95 m/s down

   (b)

t(s)	y(m)	v(m/s)
0	1000	0
1.00	995	-9.70
2.00	980	-18.6
10.0	674	-47.7
10.1	671	-16.7
12.0	659	-4.95
145	0	-4.95

35. 2.97 nN
37. 0.612 m/s^2 toward the Earth
39. (a) 19.3°   (b) 4.21 N
41. 2.14 rev/min
43. (b) 732 N down at the Equator and 735 N down at the
   poles
45. 20.1°
47. (b) 2.54 s; 23.6 rev/min
49. 12.8 N
51. 1.26×10^{32} kg
53. $\sim 10^{-7}$ N toward you

Chapter 6

1. 15.0 MJ
3. (a) 32.8 mJ   (b) -32.8 mJ
5. 4.70 kJ

7. 14.0
9. (a) 16.0 J   (b) 36.9°
11. (a) 11.3°   (b) 156°   (c) 82.3°
13. (a) 7.50 J   (b) 15.0 J   (c) 7.50 J   (d) 30.0 J
15. 50.0 J
17. (a) 575 N/m   (b) 46.0 J
19. 12.0 J
21. (a) 1.20 J   (b) 5.00 m/s   (c) 6.30 J
23. (a) 60.0 J   (b) 60.0 J
25. (a) 79.4 N   (b) 1.49 kJ   (c) $W_f = -1.49$ kJ
27. 2.04 m
29. (a) -168 J   (b) 184 J   (c) 500 J
   (d) 148 J   (e) 5.64 m/s
31. 234 J
33. 875 W
35. 685 bundles
37. (a) 20.6 kJ   (b) 686 W
39. 90.0 J
43. (a) $(2.0 + 24t^2 + 72t^4)$ J   (b) $12t$ m/s^2; $48t$ N
   (c) $(48t + 288t^3)$ W   (d) 1250 J
45. 878 kN
47. (a) 4.12 m   (b) 3.35 m
49. 1.68 m/s
S1. (a) A plot of F as the independent variable and L as
   the dependent variable shows that the last four points
   tend to vary the most from a straight line. However,
   because the first point has the largest percentage deviation from a straight line, we probably are not justified in
   throwing any of the data points out. The slope of the
   best straight line obtained from the least-squares fit is
   8.654 545 5 mm/N.   (b) $k = 0.116$ N/mm = 116 N/m.
   (c) The least-squares fit we used was of the form
   $L = aF + b$; solving for F gives $F = L/a - b/a =$
   $0.116L - 0.561$. Hence, for $L = 105$ mm, $F = 12.7$ N.

Chapter 7

1. (a) 259 kJ, 0, 259 kJ   (b) 0, -259 kJ, 259 kJ
3. (a) 80.0 J   (b) 10.7 J   (c) 0
5. (a) 40.0 J   (b) -40.0 J   (c) 62.5 J
7. (a) 22.0 J, 40.0 J   (b) Yes.
9. $v = (3gR)^{1/2}$, 0.0980 N down
11. (a) 4.43 m/s   (b) 5.00 m
13. 1.84 m
15. (a) 18.5 km, 51.0 km   (b) 10.0 MJ
17. At $h = 2H/3$ or at $h = R$, whichever is smaller
19. 2.00 m/s, 2.79 m/s, 3.19 m/s
21. 3.74 m/s
23. (a) -160 J   (b) 73.5 J   (c) 28.8 N   (d) 0.679
25. (a) 24.5 m/s   (b) Yes   (c) 206 m
   (d) unrealistic
29. 289 m
31. (a) 1.84×10^9 kg/m^3 (b) 3.27 Mm/s^2
   (c) -2.08×10^{13} J

33. (a) -1.67×10^{-14} J   (b) At the center
35. (a) $+$ at B, $-$ at D, 0 at A, C, and E
   (b) C stable; A and E unstable
   (c)

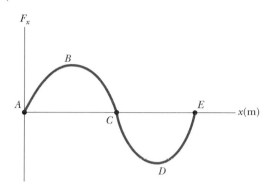

37. A/r^2
39. (a) 0.225 J   (b) $W_{nc} = -0.363$ J   (c) No; the normal force changes in a complicated way because of the curved slope.
41. 0.328
43. 0.115
45. 1.24 m/s
47. (a) 0.400 m   (b) 4.10 m/s   (c) The block stays on the track
49. 914 N/m
51. 2.06 m/s
S1. If the particle has an initial energy less than 2471 J, it will be trapped in the potential well; for example, if $E_T = 1000$ J, it will be confined approximately to -3.15 m $\leq x \leq 1.58$ m.
S3. $x = 0$ is a point of stable equilibrium; $x = 8.33$ m is a point of unstable equilibrium.

Chapter 8
1. $(9.00\mathbf{i} - 12.0\mathbf{j})$ kg·m/s; 15.0 kg·m/s
3. 6.25 cm/s west
5. (a) 6.00 m/s   (b) 8.40 J
7. (a) 13.5 N·s   (b) 9.00 kN   (c) 18.0 kN
9. 260 N normal to the wall
11. 87.5 N
13. 301 m/s
15. $(4M/m)(g\ell)^{1/2}$
17. (a) 20.9 m/s east   (b) 8.68 kJ into thermal energy
19. (a) 0.284   (b) 115 fJ and 45.4 fJ
21. 91.2 m/s
23. 0.556 m
25. (a) 2.88 m/s at 32.3° north of east   (b) 783 J into thermal energy
27. 2.50 m/s at $-60.0°$
29. (a) 1.07 m/s at $-29.7°$ (b) 0.318
31. (a) $(-9.33\mathbf{i} - 8.33\mathbf{j})$Mm/s   (b) 439 fJ
33. $\mathbf{r}_{CM} = (0\mathbf{i} + 1.00\mathbf{j})$m

35. $\mathbf{r}_{CM} = (11.7\mathbf{i} + 13.3\mathbf{j})$cm
37. 0.00673 nm from the oxygen nucleus along the bisector of the angle
39. 0.700 m
41. (a) 39.0 MN   (b) 3.20 m/s² up
43. (a) 442 metric tons   (b) 19.2 metric tons
45. 1.39 km/s
47. (a) -0.667 m/s   (b) 0.952 m
49. 240 s
51. (a) 100 m/s   (b) 374 J
53. $2v_0$ and 0
55. $\sim 10^{-23}$ m/s to 10^{-22} m/s
S1. (a) The maximum acceleration is 100 m/s². It occurs at the end of the burn time of 80 s, when the rocket has its smallest mass. The maximum speed that the rocket reaches is 3.22 km/s.   (b) The speed reaches half its maximum after 55.5 s; if the acceleration were constant, it would reach half its maximum speed at 40 s (half the burn time), but the acceleration is always increasing during the burn time.
S3. The disadvantages are that the ship has to withstand twice the acceleration and that it has only traveled half as far when the fuel burns out. The advantage is that it takes half as long to reach its final speed; hence, it travels farther in 100 s.

Chapter 9
3. 0.866c
7. 64.9/min; 10.6/min
9. 1.54 ns
11. 0.800c
13. (a) 2.18 μs   (b) 649 m
15. 0.696c
17. 0.960c
19. 0.0880
21. (a) 2.73×10^{-24} kg·m/s   (b) 1.58×10^{-22} kg·m/s   (c) 5.64×10^{-22} kg·m/s
23. (a) 0.850c (b) No change
25. 0.285c
27. (a) 939 MeV   (b) 3.01 GeV   (c) 2.07 GeV
29. (a) 0.582 MeV   (b) 2.45 MeV
31. $\sim 10^{-15}$
33. 42.1 g/cm³
37. 0.687%
39. (a) a few hundred seconds   (b) $\sim 10^8$ km
41. (a) 83.3 ns   (b) 23.7 m
45. Yes, with 18.8 m to spare
47. (b) For v small compared to c, the relativistic agrees with the classical expression. As v approaches c, the acceleration approaches zero, so the object can never reach or surpass the speed of light.
   (c) Perform $\int (1 - v^2/c^2)^{-3/2}\, dv = (qE/m) \int dt$ to obtain $v = qEct(m^2c^2 + q^2E^2t^2)^{-1/2}$ and then

$\int dx = \int qEct(m^2c^2 + q^2E^2t^2)^{-1/2}\, dt$ to obtain
$x = (c/qE)[(m^2c^2 + q^2E^2t^2)^{1/2} - mc]$

S1. (a)

Z	0.2	0.5	1.0	2.0
v/c	0.180	0.385	0.600	0.800

  (b) For $Z = 3.8$, $v/c = 0.870$

S3. For $Z = 0.2$, $r = 2710$ Mly; $Z = 0.5$, $r = 5770$ Mly;
  $Z = 1.0$, $r = 9000$ Mly; and $Z = 2.0$, $r = 12\,000$ Mly and
  $Z = 3.8$, $r = 13\,800$ Mly

Chapter 10
1. (a) 1.99×10^{-7} rad/s   (b) 2.65×10^{-6} rad/s
3. (a) 5.24s   (b) 27.4 rad
5. (a) 822 rad/s^2 (b) 4.21×10^3 rad
7. 50.0 rev
9. (a) 25.0 rad/s   (b) 39.8 rad/s^2 (c) 0.628 s
11. (a) 126 rad/s   (b) 3.77 m/s   (c) 1.26 km/s^2
  (d) 20.1 m
13. 0.545
15. (a) 143 kg·m^2 (b) 2.57 kJ
17. -3.55 N·m
19. (a) $-17.0\mathbf{k}$ (b) 70.5°
21. No. The cross product must be perpendicular to each of
  its factors, having zero dot product with either of them.
23. 2.94 kN on each rear wheel and 4.41 kN on each front
  wheel
25. 1.46 kN; 1.33 kN (to the right) and 2.58 kN (upward)
27. (a) 24.0 N·m   (b) 0.0356 rad/s^2 (c) 1.07 m/s^2
29. $17.5\mathbf{k}$ kg·m^2/s
31. $60.0\mathbf{k}$ kg·m^2/s
33. $-m\ell gt\,\cos\theta\mathbf{k}$
35. (a) 0.360 rad/s counterclockwise   (b) 99.9 J
37. (a) 7.20×10^{-3} kg·m^2/s   (b) 9.47 rad/s
39. $\sim 10^0$ kg·m^2
41. (a) 500 J   (b) 250 J   (c) 750 J
43. $3g/2L$ and $3g/2$
45. (a) $2(Rg/3)^{1/2}$ (b) $4(Rg/3)^{1/2}$ (c) $(Rg)^{1/2}$
47. 1.21×10^{-4} kg·m^2
49. (a) 118 N and 156 N   (b) 1.17 kg·m^2
51. (a) $(2mgd\sin\theta + kd^2)^{1/2}\,(I + mR^2)^{-1/2}$
  (b) 1.74 rad/s
53. (a) 3750 kg·m^2/s   (b) 1.88 kJ   (c) 3750 kg·m^2/s
  (d) 10.0 m/s   (e) 7.50 kJ   (f) 5.62 kJ
55. $3w/8$
S1. The answer is not unique because the torque $\tau = FR$.

Chapter 11
1. The sun: 5.90 mN versus 33.3 μN
3. 2/3
5. 346 Mm
7. 1.27

9. 1.90×10^{27} kg
11. 8.92×10^7 m
13. 16.6 km/s
15. 469 MJ
19. 15.6 km/s
21. 11.8 km/s
23. (a) 2.19 Mm/s   (b) 2.18 aJ   (c) -4.36 aJ
25. (a) 0.212 nm   (b) 9.95×10^{-25} kg·m/s
  (c) 2.11×10^{-34} kg·m^2/s   (d) 3.40 eV
  (e) -6.80 eV   (f) -3.40 eV
29. 492 m/s
31. 0.0572 rad/s
33. 7.41×10^{-10} N
35. (a) $m_2(2G/d)^{1/2}(m_1 + m_2)^{-1/2}$ and
  $m_1(2G/d)^{1/2}(m_1 + m_2)^{-1/2}$ and $(2G/d)^{1/2}(m_1 + m_2)^{1/2}$
  (b) 1.07×10^{32} J and 2.67×10^{31} J
37. (a) 850 MJ   (b) 2.71 GJ
39. (a) 7.34×10^{22} kg   (b) 1.63 km/s
  (c) 1.32×10^{10} J
41. 119 km
43. (a) B   (b) A   (c) B and C
47. 44.2 km/s
S5. The kinetic energy is

$$K = \tfrac{1}{2}mv^2 = \tfrac{1}{2}m(v_x^2 + v_y^2)$$

The potential energy is

$$U = \frac{GM_Em}{R_E} - \frac{GM_Em}{r}$$

where G is the universal gravitation constant, M_E is the
mass of the Earth, R_E is the radius of the Earth, m is the
mass of the satellite, and r is the distance from the center of the Earth to the satellite. The total energy, $K + U$,
is constant. There may be a small change in the numerical value of the total energy because of numerical errors
during the integration of the differential equations.

Chapter 12
1. (a) 1.50 Hz, 0.667 s   (b) 4.00 m   (c) π rad
  (d) 2.83 m
3. 18.8 m/s, 7.11 km/s^2
5. (b) 18.8 cm/s, 0.333 s   (c) 178 cm/s^2, 0.500 s
  (d) 12.0 cm
7. (a) 575 N/m   (b) 46.0 J
9. (a) 40.0 cm/s, 160 cm/s^2
  (b) 32.0 cm/s, -96.0 cm/s^2
  (c) 0.232 s
11. 0.628 m/s
13. 2.23 m/s
15. (a) 126 N/m   (b) 0.178 m
17. (a) 28.0 mJ   (b) 1.02 m/s   (c) 12.2 mJ
  (d) 15.8 mJ
19. (a) 1.55 m   (b) 6.06 s

21. (a) 0.820 m/s   (b) 2.57 rad/s^2 (c) 0.641 N
25. (a) 2.09 s   (b) 4.08% larger
27. 1.00×10^{-3} s^{-1}
31. (a) 1.00 s   (b) 5.09 cm
33. (a) 38.6 N/m   (b) 1.32 kg
35. $f = (2\pi L)^{-1} (gL + kh^2/M)^{1/2}$
37. 6.62 cm
39. (a) 3.00 s   (b) 14.3 J   (c) 25.5°
41. (a) $\frac{1}{2}(M + m/3)v^2$ (b) $2\pi(M + m/3)^{1/2} k^{-1/2}$
47. After 42.1 minutes if the hole is frictionless
S3. The periods increase as θ_0 increases. These periods are always greater than those calculated using $T_0 = 2\pi\sqrt{L/g}$. For example, if $\theta_0 = 45° = \pi/4$ rad, and $L = 1$ m, then $T_0 = 2.00\ 709$ s, but the actual period is 2.08 s. (*Note:* Due to numerical errors in the integration of the differential equations, the amplitudes may tend to increase. If this occurs, try smaller time steps.)

Chapter 13
1. $y = 6 [(x - 4.5t)^2 + 3]^{-1}$
3. (a) left   (b) 5.00 m/s
5. 0.319 m
7. (a) $y = (8.00$ cm$)\sin(7.85x + 6.00\pi t)$
   (b) $y = (8.00$ cm$)\sin(7.85x + 6.00\pi t - 0.785)$
9. 2.00 cm, 2.98 m, 0.576 Hz, 1.72 m/s
11. (a) 2.15 cm   (b) 1.95 rad   (c) 541 cm/s
    (d) $y = (2.15$ cm$)\sin(8.38x + 80\pi t + 1.95)$
13. (a) 79.5°   (b) 0.050 cm
15. 80.0 N
17. 631 N
19. 1.64 m/s^2
21. 55.1 Hz
23. $(\sqrt{2})P_0$
25. 7.82 m
29. 5.81 m
31. (a) A drop by 75.7 Hz   (b) 0.948 m
33. 26.4 m/s
35. 439 Hz to 441 Hz
37. (a) 39.2 N   (b) 0.892 m   (c) 83.6 m/s
39. (a) 3.33**i** m/s   (b) −5.48 cm   (c) 0.667 m, 5.00 Hz
    (d) 11.0 m/s
41. 6.01 km
45. (a) 55.8 m/s   (b) 2500 Hz
47. Bat is gaining at 1.69 m/s

Chapter 14
1. (a) 9.24 m   (b) 600 Hz
3. 5.66 cm
5. At 0.0891 m, 0.303 m, 0.518 m, 0.732 m, 0.947 m, 1.16 m from one speaker
9. (a) $x = \pi, 3\pi, 5\pi, \ldots$ (b) 0.0294 m
11. 150 Hz

13. (a) 163 N   (b) 660 Hz
15. 19.976 kHz
17. 5.00 Hz, 10.0 Hz, 15.0 Hz; the fifth state, 25.0 Hz
19. 0.786 Hz, 1.57 Hz, 2.36 Hz, 3.14 Hz
21. (a) 0.357 m   (b) 0.715 m
23. (a) 531 Hz   (b) 42.5 mm
25. (a) $L = nv/2f$ where $n = 1, 2, 3, \ldots$
    (b) $L = (2n - 1)v/4f$
27. $n(206$ Hz$)$ and $(n + 1)(84.5$ Hz$)$ for $n = 1, 2, 3, \ldots$
29. 239 s
31. (a) 162 Hz   (b) 1.06 m
33. 5.64 beats/s
35. (a) 0.655 m   (b) 11.7°C
37. (a) 34.8 m/s   (b) 0.977 m
39. 3.85 m/s away from the station or 3.77 m/s toward the station
41. 4.85 m
43. 15.7 Hz
45. (a) 59.9 Hz   (b) 20.0 cm
47. (a) 1/2   (b) $[n/(n + 1)]^2 F$ (c) 9/16
49. 50.0 Hz, 1.70 m
51. (a) $2A \sin(2\pi x/\lambda)\cos(2\pi vt/\lambda)$
    (b) $2A \sin(\pi x/L)\cos(\pi vt/L)$
    (c) $2A \sin(2\pi x/L)\cos(2\pi vt/L)$
    (d) $2A \sin(n\pi x/L)\cos(n\pi vt/L)$
S4. The following steps can be used to modify the spreadsheet.
    1. MOVE the entire Y_T column—one column to the right.
    2. COPY the Y_2 column—one column to the right. EDIT the heading label to Y_3.
    3. COPY the second wave input data block to the right and EDIT the labels to reflect the third wave.
    4. EDIT the Y_3 column to reflect the third wave data block addresses.
    5. EDIT the Y_T column to include Y_1, Y_2, and Y_3 in the sum.
S6. Plot $Y(t)$ versus ωt. You may want to start with three terms in the series and then add additional terms and watch how $Y(t)$ changes.

Chapter 15
1. 6.24 MPa
3. 5.27×10^{18} kg
5. 1.62 m
7. 7.74×10^{-3} m^2
9. 98.6 kPa
11. 10.5 m; no
13. $m/[(\rho_w - \rho_s)h]$
15. (a) 7.00 cm   (b) 2.80 kg
17. 1250 kg/m^3 and 500 kg/m^3
19. 1430 m^3

21. (a) 4.24 m/s　(b) 17.0 m/s
23. (a) 17.7 m/s　(b) 1.73 mm
25. (a) 1 atm + 15.0 MPa　(b) 2.98 m/s
　　(c) 4.45 kPa
27. 68.0 kPa
29. 103 m/s
31. (b) $2\sqrt{h(h_0 - h)}$
33. 0.258 N
35. 1.91 m
37. 8.01 km; yes
39. 709 kg/m^3
41. 2.00 MN
45. 90.04%
47. 4.43 m/s
51. (b) 1.40 s

Chapter 16
1. (a) 37.0°C = 310 K　(b) −20.6°C = 253 K
3. (a) −274°C　(b) 1.27 atm　(c) 1.74 atm
5. (a) −320°F　(b) 77.3 K
9. From 40.6°C to −31.6°C
11. 3.27 cm
13. 55.0°C
15. 0.109 cm^2
17. (a) 0.176 mm　(b) 8.78 μm　(c) 0.0930 cm^3
19. (a) 3.00 mol　(b) 1.80 × 10^{24} molecules
21. 1.50 × 10^{29} molecules
23. 471 K
25. (a) 400 kPa　(b) 449 kPa
27. Between 10 kg and 100 kg
29. $(P_0 VM/R)(1/T_1 - 1/T_2)$
31. 6.64 × 10^{-27} kg
33. (a) 8.76 × 10^{-21} J for both　(b) 1.62 km/s for helium
　　and 514 m/s for argon
35. 17.6 kPa
37. 1.51 × 10^{-20} J
39. 575°F
41. 0.522 kg
43. 3.55 cm
45. $\alpha\Delta T$ is much less than 1.
47. (a) Expansion makes density drop.　(b) 5 × 10^{-5}/°C
49. (a) $h = nRT/(mg + P_{atm}A)$　(b) 0.661 m

Chapter 17
1. 0.105°C
3. 0.234 kJ/kg·°C
5. 29.6°C
7. (a) 0.435 cal/g·°C　(b) beryllium
9. (a) 25.8°C　(b) No
11. 59.4°C
13. 0.414 kg
15. (a) 0°C　(b) 114 g
17. Liquid lead at 805°C, if the specific heat is constant

19. 1.18 MJ
21. (a) $4P_0V_0$　(b) $T = (P_0/nRV_0)V^2$
23. (a) −567 J　(b) 167 J
25. (a) 12.0 kJ　(b) −12.0 kJ
27. (a) 7.50 kJ　(b) 900 K
29. 3.10 kJ; 37.6 kJ
31. (a) 0.0410 m^3　(b) −5.48 kJ　(c) −5.48 kJ
33. 2.47 L
37. 51.2°C
39. 1.34 kW
41. 47.5 g
43. (a) 16.8 L　(b) 0.351 L/s
45. (a) $-P_0V_0/2$　(b) $-P_0V_0 \ln 4$　(c) 0

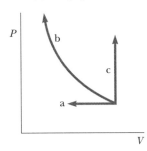

47. (a) $4P_0V_0$　(b) $4P_0V_0$　(c) 9.08 kJ
49. 5.31 h
51. 466 J
S1. By varying h until $T = 45°C$ at $t = 180$ s, one finds $h =$ 0.010 855 cal/s·cm^2·°C. Examine the associated graph for each choice of h.
S2. (b) $\Delta U = 4669$ J

Chapter 18
1. (a) 6.94%　(b) 335 J
3. (a) 10.7 kJ　(b) 0.533 s
5. (a) 67.2%　(b) 58.8 kW
7. (a) 26.8%　(b) 42.3%
9. (a) 741 J　(b) 459 J
11. (a) 5.12%　(b) 5.27 TJ/h
13. 453 K
17. (a) 24.0 J　(b) 144 J
19. −610 J/K
21. 195 J/K
23. 1.02 kJ/K
25. 716 J/K
27. 5.76 J/K; temperature is constant if the gas is ideal
29. (a) 1　(b) 6
31. (a)

Result	Number of draws
all red	1
two red, one green	3
one red, two green	3
all green	1

(b)

Result	Number of draws
5R	1
4R 1G	5
3R 2G	10
2R 3G	10
1R 4G	5
5G	1

33. 8.36 MJ/K
35. 32.9 kJ
37. (a) 5.00 kW   (b) 763 W
39. 77.8 W
41. 5.97×10^4 kg/s
43. 29.3 J
45. (a) $2nRT_0 \ln 2$ (b) 0.273

Chapter 19

1. (a) 21.6 μN toward the other   (b) 0.899 μN away from the other
3. The force is $\sim 10^{26}$ N
5. 2.51×10^{-9}
7. K/ed in the direction of motion
9. 1.82 m to the left of the negative charge
11. $-k_e\lambda_0\mathbf{i}/x_0$
13. (a) at the center   (b) 1.73 $k_e q\,\mathbf{j}/a^2$
15. 1.59 MN/C toward the rod
17. $-21.6\mathbf{i}$ MN/C
19. 4.14 MN/C
21. 226 kN·m^2/C
23. (a) $+Q/2\epsilon_0$ (b) $-Q/2\epsilon_0$
25. 2.33×10^{21} N/C radially outward
27. $\mathbf{E} = 0$ inside and $\mathbf{E} = k_e Q/r^2$ radially away, outside
29. $\mathbf{E} = \rho r/2\epsilon_0$ away from the axis
31. (a) 0   (b) 5400 N/C outward
    (c) 540 N/C outward
33. (a) $+708$ nC/m^2 and -708 nC/m^2
    (b) $+177$ nC and -177 nC
35. 2.00 N
37. (a) 496 nC/m^2 (b) 248 nC/m^2
39. 5.25 μC
41. (a) 10.9 nC   (b) 5.43 mN
43. (a) $\rho r/3\epsilon_0$; $Q/4\pi\epsilon_0 r^2$; 0; $Q/4\pi\epsilon_0 r^2$, all radially outward
    (b) $-Q/4\pi b^2$ and $+Q/4\pi c^2$
45. For $r < a$, $\mathbf{E} = \lambda/2\pi\epsilon_0 r$ radially outward. For $a < r < b$,
    $\mathbf{E} = [\lambda + \rho\pi(r^2 - a^2)]/2\pi\epsilon_0 r$ radially outward. For
    $r > b$, $\mathbf{E} = [\lambda + \rho\pi(b^2 - a^2)]/2\pi\epsilon_0 r$ radially outward
47. (a) σ/ϵ_0 away from both plates   (b) 0
    (c) σ/ϵ_0 away from both plates
49. 205 nC
51. $-707\mathbf{j}$ mN

S5. The dipole approximation of the electric field
$E = 2k_e p/x^3$ is within 20% of the actual value when
$x > 6.2$ cm. It is within 5% when $x > 12.6$ cm.

Chapter 20

1. (a) -600 μJ   (b) -50.0 V
3. (a) 152 km/s   (b) 6.49 Mm/s
5. (a) 81.6 Mm/s   (b) 83.9 Mm/s, too large by 2.91%
7. (a) 59.0 V   (b) 4.56 Mm/s
9. 260 V
11. (a) 144 nV   (b) -71.9 nV   (c) -144 nV and 71.9 nV
13. -11.0 MV
15. (a) -27.3 eV   (b) -6.81 eV   (c) 0
19. (a) 10.8 m/s and 1.55 m/s   (b) Larger
21. $\mathbf{E} = (-5 + 6xy)\mathbf{i} + (3x^2 - 2z^2)\mathbf{j} - 4yz\mathbf{k}$; 7.08 N/C
23. $-0.553\, k_e Q/R$
25. (a) C/m^2 (b) $k_e\alpha[L - d\ln(1 + L/d)]$
27. (a) 45.0 MV/m away and 30.0 MV/m away
    (b) 1.80 MV
29. (a) 1.00 μF   (b) 100 V
33. (a) 11.1 kV/m   (b) 98.3 nC/m^2 (c) 3.74 pF
    (d) 74.8 pC
35. (a) 17.0 μF   (b) 9.00 V   (c) 15.0 μC and 108 μC
37. 6.00 pF and 3.00 pF
39. (a) 5.96 μF   (b) 89.5 μC on 20.0 μF, 63.2 μC on
    6.00 μF, 26.3 μC on 15.0 μF and on 3.00 μF
41. 120 μC; 80.0 μC and 40.0 μC
43. (a) 216 μJ   (b) 54.0 μJ
45. (a) circuit diagram:

energy stored: 150 mJ
(b) potential difference: 268 V

circuit diagram:

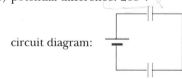

49. (a) 4.00 μF   (b) 8.40 μF   (c) 5.71 V and 48.0 μC
51. $\sim 10^{-6}$ F and $\sim 10^2$ V for two 40-cm by 100-cm sheets of
    aluminum foil sandwiching a thin sheet of plastic
53. 253 MeV
55. (a) 488 V   (b) 7.81×10^{-17} J   (c) 306 km/s
    (d) 3.90×10^{11} m/s^2 (e) 6.51×10^{-16} N
    (f) 4.07 kN/C
57. (a) 13.3 μC   (b) 0.200 m
59. (a) 40.0 μJ   (b) 500 V
61. 0.188 m^2
63. $\epsilon_0 A/(s - d)$

Chapter 21

1. 7.50×10^{15} electrons
3. (a) $0.632 I_0 \tau$　　(b) $0.99995 I_0 \tau$　　(c) $I_0 \tau$
5. 0.130 mm/s
7. 6.43 A
9. $1.33 \, \Omega$
11. 1.98 A
13. 1440°C
15. 0.180 V/m
17. 5.00 A; $24.0 \, \Omega$
19. $28.9 \, \Omega$
21. 36.1%
23. (a) 0.660 kWh　　(b) 3.96¢
25. (a) $6.73 \, \Omega$　　(b) $1.97 \, \Omega$
27. Put the three resistors in parallel
29. (a) 227 mA　　(b) 5.68 V
31. (a) 75.0 V　　(b) 25.0 W, 6.25 W, and 6.25 W; 37.5 W
33. 846 mA down in the $8.00\text{-}\Omega$ resistor; 462 mA down in the middle branch; 1.31 A up in the right-hand branch
35. (a) -222 J and 1.88 kJ　　(b) 687 J, 128 J, 25.6 J, 616 J, 205 J　　(c) 1.66 kJ
37. (a) 0.385 mA, 3.08 mA, 2.69 mA　　(b) c is higher by 69.2 V
39. (a) 5.00 s　　(b) $150 \, \mu$C　　(c) $4.06 \, \mu$A
41. (a) 6.00 V　　(b) $8.29 \, \mu$s
43. 25.5 y
45. 1.56 cm
47. (a) either $3.84 \, \Omega$ or $0.375 \, \Omega$　　(b) impossible
49. (a) 12.5 A, 6.25 A, 8.33 A　　(b) No; together they would require 27.1 A
51. Experimental resistivity $= 1.47 \, \mu\Omega \cdot$m $\pm 4\%$, in agreement with $1.50 \, \mu\Omega \cdot$m
53. (a) $\mathbf{E} = V_0 \mathbf{i}/\ell$　　(b) $R = 4\rho\ell/\pi d^2$
　　(c) $I = V_0 \pi d^2/4\rho\ell$　　(d) $\mathbf{J} = V_0 \mathbf{i}/\rho\ell$
S1. (a) The savings are $254.92.
S3. Kirchhoff's equations for this circuit can be written as

$$\mathcal{E}_1 - I_1 R_1 - I_4 R_4 = 0$$

$$\mathcal{E}_2 - I_2 R_2 - I_4 R_4 = 0$$

$$\mathcal{E}_3 - I_3 R_3 - I_4 R_4 = 0$$

and

$$I_1 + I_2 + I_3 = I_4$$

In matrix form, they are

$$\begin{bmatrix} R_1 & 0 & 0 & R_4 \\ 0 & R_2 & 0 & R_4 \\ 0 & 0 & R_3 & R_4 \\ 1 & 1 & 1 & -1 \end{bmatrix} \begin{bmatrix} I_1 \\ I_2 \\ I_3 \\ I_4 \end{bmatrix} = \begin{bmatrix} \mathcal{E}_1 \\ \mathcal{E}_2 \\ \mathcal{E}_3 \\ 0 \end{bmatrix}$$

In Lotus 1-2-3 use the /DataMatrixInvert and /DataMatrixMultiply commands and in Excel use the array formulas = MINVERSE(array) and = MMULTI(array1, array2) to carry out the calculations. Using the data given in the problem, we find

$$I_1 = -1.26 \text{ A}, I_2 = 0.87 \text{ A}, I_3 = 1.08 \text{ A, and } I_4 = 0.69 \text{ A}.$$

Chapter 22

1. (a) up　　(b) toward you, out of the plane of the paper
　　(c) no deflection　　(d) into the plane of the paper
3. $-20.9\mathbf{j}$ mT
5. 8.93×10^{-30} N down, 1.60×10^{-17} N up, 4.80×10^{-17} N down
7. 3.00 T
9. 0.245 T east
11. $2\pi r I B \sin \theta$ up
13. 9.98 N·m clockwise as seen looking down from above
15. (a) 3.97°　　(b) 3.39 mN·m
17. (a) $28.3 \, \mu$T into the paper
　　(b) $24.7 \, \mu$T into the paper
19. $\mu_0 I/4\pi x$ into the paper
21. $-27.0\mathbf{i} \, \mu$N
23. $20.0 \, \mu$T toward the bottom of the page
25. (a) 3.60 T　　(b) 1.94 T
27. (a) 6.34 mN/m inward　　(b) greater
29. 31.8 mA
31. 464 mT
33. (a) 8.63×10^{45} electrons　　(b) 4.01×10^{20} kg
35. $20.0 \, \mu$T
37. 2.00 GA west
39. (a) 30.0 A　　(b) $88.8 \, \mu$T out of the paper
41. $\mu_0 I \mathbf{k} \ln (1 + w/b)/2\pi w$
43. 143 pT away along the axis
45. (a) The electric current feels a magnetic force.
47. 12 layers, 120 m
49. (a) $-8.00 \times 10^{-21}\mathbf{j}$ kg·m/s　　(b) 8.90°
S3. This problem is somewhat similar to Problem S11.4, which dealt with gravitation and orbits, except that here the acceleration is velocity dependent. The equation of motion is

$$\mathbf{F} = q\mathbf{E} + q(\mathbf{v} \times \mathbf{B}) = m\mathbf{a} = q(E_x - v_z B_y) \, \mathbf{i} + q v_x B_y \, \mathbf{k}$$

The acceleration in the y direction is zero, so the y component of the velocity of the electron is constant.

Chapter 23

1. 160 A
3. (a) 1.60 A　　(b) $20.1 \, \mu$T　　(c) up
5. $(-87.1 \text{ mV}) \cos(200\pi t + \phi)$
7. 1.00 m/s
9. 0.763 V
11. 2.83 mV
13. 9.82 mV

15. (b) Larger R makes current smaller, so the loop must travel faster to maintain equality of magnetic force and weight.   (c) The magnetic force is proportional to the product of field and current, whereas the current is itself proportional to field. If B gets two times smaller, the speed must get four times larger to compensate.

17. 1.80 mN/C upward and to the left, perpendicular to r_1

19. 22.3 μN/C

21. 19.5 mV

23. 18.8 V cos 377t

29. (a) 0.139 s   (b) 0.461 s

31. (a) 0.800   (b) 0

33. 30.0 mH

35. (a) 1.00 A   (b) 12.0 V, 1.20 kV, 1.21 kV
    (c) 7.62 ms

37. (a) 8.06 MJ/m^3 (b) 6.32 kJ

39. 44.2 nJ/m^3 for the E field and 995 μJ/m^3 for the B field

41. $2\pi B_0^2 R^3/\mu_0 = 2.70 \times 10^{18}$ J

43. (a) 7.54 kV   (b) The plane of the coil is parallel to **B**

47. 6.00 A

51. (a) 62.5 GJ   (b) 2000 N

53. (a) $L \approx \pi\mu_0 N^2 R/2$
    (b) $\sim$100 nH   (c) $\sim$1 ns

S5. (a) $\tau = 0.35$ ms,
    (b) $t_{50\%} = 0.24$ ms $= 0.686\tau$, and
    (c) $t_{90\%} = 0.80$ ms $= 2.29\tau$; $t_{99\%} = 1.6$ ms $= 4.57\tau$.

Chapter 24

1. 1.85 aT up

3. 733 nT

5. (a) 6.00 MHz   (b) $(-73.3\mathbf{k})$ nT
    (c) $\mathbf{B} = (-73.3\mathbf{k}$ nT$)$ cos$(0.126x - 3.77 \times 10^7 \, t)$

7. They are, so long as $k = \omega(\mu_0\epsilon_0)^{1/2}$; that is, so long as $\lambda = c/f$

9. 1.09 pF to 1.64 pF

11. (a) $2\pi^2 r^2 f B_{max}$ cos θ, where θ is the angle between the magnetic field and the normal to the loop
    (b) The loop should be in the vertical plane containing the line of sight to the transmitter.

13. (a) away along the perpendicular bisector of the line segment joining the antennas
    (b) along the extensions of the line segment joining the antennas

15. 307 μW/m^2

17. 3.33 $\times$ 10^3 m^2

19. (a) 332 kW/m^2 radially inward
    (b) 1.88 kV/m and 222 μT

21. (a) $\mathbf{E} \cdot \mathbf{B} = 0$ (b) $(11.5\mathbf{i} - 28.6\mathbf{j})$ W/m^2

23. (a) 2.33 mT   (b) 650 MW/m^2 (c) 510 W

25. 83.3 nPa

27. (a) 1.90 kN/C   (b) 50.0 pJ
    (c) 1.67 $\times$ 10^{-19} kg$\cdot$m/s

29. 545 THz

31. (a) 0.690λ (b) 58.9λ

33. 3/8

35. (a) 54.7°   (b) 63.4°   (c) 71.6°

37. (a) 3.74 $\times$ 10^{26} W   (b) 1.01 kV/m and 3.35 μT

39. (a) 6.67 $\times$ 10^{-16} T   (b) 5.31 $\times$ 10^{-17} W/m^2
    (c) 1.67 $\times$ 10^{-14} W   (d) 5.56 $\times$ 10^{-23} N

41. 95.1 mV/m

43. (a) $B_{max} = (583\mathbf{k})$ nT, $k = 419$ rad/m, $\omega = 0.126$ Trad/s
    (b) $S_{av} = \frac{1}{2}S_{max} = 40.6$ W/m^2 (c) 271 nPa
    (d) 406 nm/s^2

45. (a) 6   (b) 7.50°

47. (a) $(4\rho gc/3)(3m/4\pi\rho)^{1/3}$
    (b) $(4\pi\rho gcr^2/3)(3m/4\pi\rho)^{1/3}$

49. (a) 6.16 μPa
    (b) 1.64 $\times$ 10^{10} times smaller than P_0

Chapter 25

1. 25.5°, 442 nm

3. 19.5° above the horizon

5. 30.0° and 19.5° at entry; 19.5° and 30.0° at exit

7. 3.39 m

9. 6 times from mirror 1 and 5 times from mirror 2

11. 3.88 mm

13. 30.4° and 23.3°

15. 24.7°

17. 0.171°

19. 4.61°

21. (a) 27.9°   (b) 48.6°

23. 39.8° for both; 0.0422°

25. 1.000 08

27. 67.2°

29. 4.74 cm

31. 22.0°

33. 2.27 m

35. 77.5°

39. greater than 66.5°

41. 62.2%

43. 82 reflections

45. 27.5°

47. slope = 1.328 $\pm$ 0.8%

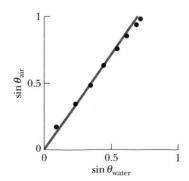

49. (a) It always happens
    (b) 30.3°
    (c) It cannot happen
51. 2.37 cm
S1. Using Snell's Law and plane geometry, the following relationships can be derived:

$$\sin \theta_1 = n \sin \rho_1$$

$$\rho_2 = \Phi - \rho_1$$

$$n \sin \rho_2 = \sin \theta_2$$

$$\delta = \theta_1 + \theta_2 - \Phi$$

These equations are used to find ρ_1, ρ_2, θ_2, and δ in the spreadsheet. IF statements are used in the calculation of θ_2 and δ so that error messages are not printed when ρ_2 exceeds the critical angle for total internal reflection.

Chapter 26

1. $\sim 10^{-9}$ s younger
3. 2'11"
5. 10.0 ft, 30.0 ft, 40.0 ft
7. 0.267 m behind the mirror; virtual, upright, diminished; $M = 0.0267$
9. (a) -12.0 cm; 0.400
   (b) -15.0 cm; 0.250
   (c) upright
11. (a) a concave mirror with radius of curvature 2.08 m
    (b) 1.25 m from the object

45. 8.00 cm

13. 30.0 cm
15. 38.2 cm below the top surface of the ice
17. 8.57 cm
19. 3.88 mm
21. (a) 16.4 cm   (b) 16.4 cm
23. (a) 650 cm from the lens on the opposite side from the object; real, inverted, enlarged
    (b) 600 cm from the lens on the same side as the object; virtual, upright, enlarged
25. 2.84 cm
27. 7.50 cm
29. (a) 3.40, upright
    (b)

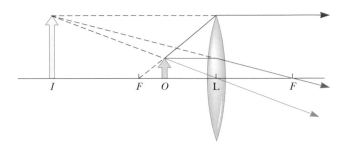

33. (a) 39.0 mm   (b) 39.5 mm
35. 160 cm to the left of the lens; inverted; $M = -0.800$
37. -40.0 cm
39. 5.71 cm in front of the mirror; real
41. -25.0 cm
43. 0.107 m to the right of the vertex of the hemispherical face

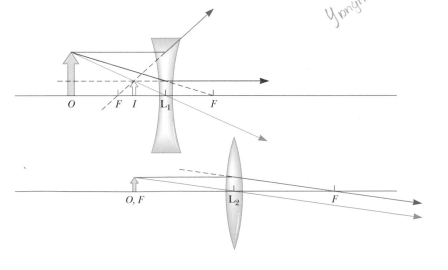

47. 1.50 m in front of the mirror; 1.40 cm (inverted)
49. (a) 30.0 cm and 120 cm   (b) 24.0 cm
    (c) real, inverted, diminished
S1. See Figure 26.13. Applying the law of sines to triangle *OPC*, one finds $R \sin \theta_1 = (R + p) \sin \alpha$. Snell's law gives $n_1 \sin \theta_1 = n_2 \sin \theta_2$. Using plane geometry gives $\gamma = \theta_1 - \alpha - \theta_2$. Applying the law of sines to triangle *PIC* gives $q - R = R \sin \theta_2 / \sin \gamma$. These equations are used to calculate θ_1, θ_2, γ, and q in the spreadsheet. The maximum angle α for which the error in the image distance is 3% or less is 2.35°.   (b) 2.25°   (d) No. (For a plot of the percent error versus α, just call up the associated graph.)

Chapter 27
1. 1.58 cm
3. (a) 55.7 m   (b) 124 m
5. 1.54 mm
7. 0.0480 mm
9. 612 nm
11. (a) green   (b) purple
13. 4.35 μm
15. 0.230 mm
17. 91.2 cm
19. 1.00 mrad
21. 3.09 m
23. 61.0 mm
25. 105 m
27. 5.91° in first order, 13.2° in second order, 26.5° in third order
29. (a) 478.7 nm, 647.6 nm, and 696.6 nm
    (b) 20.51°, 28.30°, and 30.66°
31. (a) 12 000, 24 000, 36 000
    (b) 0.0111 nm
33. 0.0934 nm
35. 14.4°
37. (a) $\sim 10^{-3}$ degree
    (b) $\sim 10^{11}$ Hz, microwave
39. 2.50 mm
41. 3.58°
43. 2.52 cm
45. 115 nm
47. (b) 266 nm
49. (a) 25.6°   (b) 19.0°
51. $y' = (n - 1)\, tL/d$
S1. (a) 9:1   (b) 1:9

Chapter 28
1. 5.18 kK
3. (a) 2.57 eV   (b) 12.8 μeV   (c) 191 neV
    (d) 484 nm, 9.68 cm, 6.52 m
5. 5.71×10^3 photons

7. (a) 296 nm, 1.01 PHz
    (b) 2.71 V
9. 2.22 eV, 3.70 eV
11. 1.78 eV, 9.47×10^{-28} kg·m/s
13. 70.1°
15. (a) 43.0°   (b) 602 keV, 3.21×10^{-22} kg·m/s
    (c) 278 keV, 3.21×10^{-22} kg·m/s
17. (a) 0.174 nm   (b) 5.49 pm
19. (a) 124 MeV or more   (b) With kinetic energy much larger than the magnitude of its negative electric potential energy, the electron would immediately escape.
21. (a) 3.91×10^4; 1.07×10^{-17} kg·m/s   (b) 62.2 am, much smaller than 10^{-14} m
23. (a) 15.1 keV   (b) 124 keV
25. (a) 993 nm   (b) 4.97 mm   (c) If we detect the interference pattern, we cannot say the neutron has passed through a single slit because both slits are open. If we test which slit the neutrons pass through, we will destroy the interference pattern—and are thereby asking a different question.
27. Within 1.16 mm for the electron, 5.28×10^{-32} m for the bullet
29. The electron energy must be $\sim 100 mc^2$ or larger. The proton energy can be as small as $1.001 mc^2$, which is quite classical.
31. (a) 126 pm
    (b) 5.28×10^{-24} kg·m/s
    (c) $K = 95.5$ eV
33. 0.500
35. (a)

n

4 ——————— 603 eV

3 ——————— 339 eV

2 ——————— 151 eV

1 ——————— 37.7 eV

    (b) 2.20 nm, 2.75 nm, 4.71 nm, 4.12 nm, 6.60 nm, 11.0 nm
37. 0.793 nm
39. 202 fm, 6.14 MeV, a gamma ray
43. (a) 0.0103   (b) 0.990
45. By 0.959 nm, to 1.91 nm
47. 1.36 eV
49. 0.143 nm. This is of the same order of magnitude as the spacing between atoms in a crystal, so Bragg diffraction will occur.

51. (a)

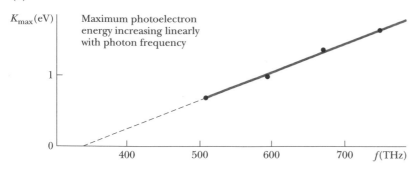

K_{max}(eV)   Maximum photoelectron energy increasing linearly with photon frequency

(b) 6.4×10^{-34} J·s ± 8%   (c) 1.4 eV
53. 0.0294
55. (a)

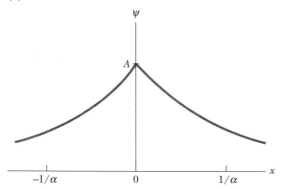

(b)

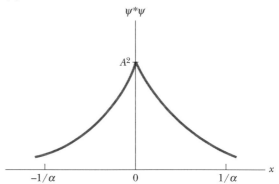

(c) It is continuous, finite, and can be normalized.
(d) $A = \sqrt{\alpha}$ (e) 0.632

Chapter 29

1. (a) 3   (b) 520 km/s
3. (a) 1312 nm   (b) 164 nm
5. (a)

n	ℓ	m_ℓ	m_s
3	2	2	$\frac{1}{2}$
3	2	2	$-\frac{1}{2}$
3	2	1	$\frac{1}{2}$
3	2	1	$-\frac{1}{2}$
3	2	0	$\frac{1}{2}$
3	2	0	$-\frac{1}{2}$
3	2	-1	$\frac{1}{2}$
3	2	-1	$-\frac{1}{2}$
3	2	-2	$\frac{1}{2}$
3	2	-2	$-\frac{1}{2}$

(b)

n	ℓ	m_ℓ	m_s
3	1	1	$\frac{1}{2}$
3	1	1	$-\frac{1}{2}$
3	1	0	$\frac{1}{2}$
3	1	0	$-\frac{1}{2}$
3	1	-1	$\frac{1}{2}$
3	1	-1	$-\frac{1}{2}$

7. $4a_0$
9. It does, with $E = -k_e e^2/2a_0$

11. 4
13. $L = \sqrt{6}\hbar = 2.58 \times 10^{-34}$ J·s;
    $L_z = -2\hbar, -1\hbar, 0, 1\hbar,$ or $2\hbar$;
    $\theta = 145°, 114°, 90.0°, 65.9°,$ or $35.3°$
15. (a) 2   (b) 8   (c) 18   (d) 32   (e) 50
17. (a) 3.99×10^{17} kg/m³   (b) 8.17×10^{-17} m
    (c) 1.77 Tm/s   (d) 5.91×10^3 c
19. (a) 5.05×10^{-27} J/T = 31.6 neV/T
    (b) Here μ_n is 1836 times smaller than μ_B, as a proton is 1836 times more massive than an electron. The electron has a bigger charge-to-mass ratio than anything else, which gives it a greater relative ability for a magnetic field to twist it.
21.
    $1s^2 2s^2 2p^6 3s^2 3p^6 3d^{10} 4s^2 4p^6 4d^{10} 4f^{14} 5s^2 5p^6 5d^{10} 5f^{14} 6s^2 6p^6 6d^8 7s^2$
23. (a) 1s, 2s, 2p, 3s, 3p, 4s, 3d, 4p, 5s, 4d, 5p, 6s, 4f, 5d, 6p, 7s
    (b) Element 15 should have valence +3 or −5, and it does. Element 47 should have valence −1, but it has valence +1. Element 86 should be inert, and it is.

25. 124 V
27. 0.0725 nm
29. 11.8 keV, 10.1 keV, and 2.39 keV

———————————————— 0

N ————————— −2.39 keV

M ————— −10.1 keV

L ———— −11.8 keV

K_α

K_β

K_γ

K ————————— −69.5 keV

31. 3.49×10^{16} photons
35. (a) 3.13 fm (b) −18.8 MeV (c) 14.1 MeV
37. (a) 4.20 mm
(b) 1.05×10^{19} photons
(c) $8.82 \times 10^{16}/mm^3$
39. (a) 1.22×10^{-33}
(b) 10^{-2255}
41. (a) $5.24a_0$
43.

———————————— 0
———————————— −0.100 eV

———————————— −1.00 eV

———————————— −4.10 eV

45. 0.125
47. (b) 0.846 ns
49. (a) 0.240 s
(b) The atoms need not all wait together for energy to accumulate, because the light comes in photon chunks. In the first nanosecond of exposure some photon will hit some atom to knock out a photoelectron.
51. $4\pi^2 m_e k_e^2 e^4 / h^3 n^3$

Chapter 30

1. $\sim 10^{28}$, $\sim 10^{28}$, $\sim 10^{28}$
3. $A = 30$
5. (a) 18.5 Mm/s (b) 7.11 MeV
7. 12.7 km
9. 25.6 MeV
11. (a) 1.11 MeV/nucleon
(b) 7.07 MeV/nucleon
(c) 8.79 MeV/nucleon
(d) 7.57 MeV/nucleon
13. 7.94 MeV
15. (a) Cs (b) La (c) Cs
17. 1.16 ks
19. 0.200 mCi
21. (a) $1.55 \times 10^{-5}/s$, 12.4 h
(b) 2.39×10^{13} atoms
(c) 1.87 mCi
23. (a) 0.279
(b) 1.22×10^{-29}
(c) Radium-226 continuously creates radon
25. 9.47×10^9 nuclei
27.

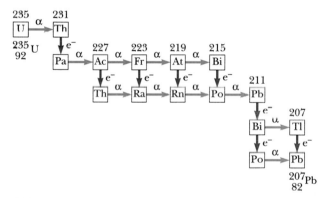

29. 4.28 MeV
31. (a) $e^- + p \rightarrow n + \nu$
(b) $^{15}_{8}O$ atom $\rightarrow$ $^{15}_{7}N$ atom $+ \nu$
(c) 2.75 MeV
33. −2.64 MeV
35. (a) For nuclei, $^{197}_{79}Au + ^{1}_{0}n \rightarrow ^{198}_{80}Hg + _{-1}^{0}e^- + \bar{\nu}$
(b) 7.89 MeV
37. (a) $^{13}_{6}C$ (b) $^{10}_{5}B$
39. (a) 2.52×10^{24} (b) 2.31×10^{12} Bq (c) 1.07 My
41. 4.45×10^{-8} kg/h
43. 6.25×10^{19} Bq
45. (a) 2.75 fm (b) 152 N (c) 2.62 MeV
(d) 7.44 fm, 379 N, 17.6 MeV
47. (a) 0.08 MeV
(b) 4.62 MeV and 13.9 MeV
(c) 1.03×10^7 kWh
49. (b) 4.79 MeV
51. 1.02 MeV
53. 5.94 Gy

S1.   The following tables give the number of parent nuclei (N1) and the number of daughter nuclei (N2) present:

(a)

	N1	N2
$t = 1$ min	4349	478
$t = 2$ min	3782	648
$t = 5$ min	2488	587
$t = 30$ min	76	19

(b)

	N1	N2
$t = 1$ min	3514	1391
$t = 2$ min	2470	2187
$t = 5$ min	857	2718
$t = 30$ min	zero	126

(c)

	N1	N2
$t = 1$ min	4997	3
$t = 2$ min	4993	5
$t = 5$ min	4983	8
$t = 30$ min	4897	9

Chapter 31

1. 4.53×10^{23} Hz, 662 am
3. 118 MeV
5. $\sim 10^{-18}$ m
7. $\sim 10^{-23}$ s
9. 67.5 MeV, 67.5 MeV/c, 1.63×10^{22} Hz
11. $\Omega^+ \to \overline{\Lambda}^0 + K^+$ $\overline{K}^0_S \to \pi^+ + \pi^-$ $\overline{\Lambda}^0 \to \overline{p} + \pi^+$
    $\overline{n} \to \overline{p} + e^+ + \nu_e$
13. (b) The second violates strangeness conservation.
15. (a) $\overline{\nu}_\mu$ (b) ν_μ (c) $\overline{\nu}_e$ (d) ν_e (e) ν_μ
    (f) $\overline{\nu}_e + \nu_\mu$
17. (a), (c), and (f) violate baryon number conservation.
    (b), (d), and (e) can occur. (f) violates muon–lepton number conservation.
19. (b) and (c) conserve strangeness. (a), (d), (e), and (f) violate strangeness conservation.
25. (a) $d\overline{u} + uud \to d\overline{s} + uds$
    (b) $\overline{d}u + uud \to u\overline{s} + uus$
    (c) $\overline{u}s + uud \to u\overline{s} + d\overline{s} + sss$
    (d) $uud + uud \to d\overline{s} + uud + u\overline{d} + uds$; Λ^0
27. (a) 590.07 nm   (b) 597 nm   (c) 661 nm
29. (a) $v = c(Z^2 + 2Z)/(Z^2 + 2Z + 2)$
    (b) $r = (c/H)(Z^2 + 2Z)/(Z^2 + 2Z + 2)$
31. (a) charge   (b) energy
    (c) baryon number
33. one part in 50 000 000
35. 2.04 MeV
37. 5.35 MeV and 32.3 MeV
39. 9.26 cm
41. 29.7 MeV
43. (a) Z^0 (b) gluon

Index

Page numbers in *italics* indicate illustrations; page numbers followed by "n" indicate footnotes; page numbers followed by "t" indicate tables.

Solar System Data

Body	Mass (kg)	Mean Radius (m)	Period (s)	Distance from Sun (m)
Mercury	3.18×10^{23}	2.43×10^6	7.60×10^6	5.79×10^{10}
Venus	4.88×10^{24}	6.06×10^6	1.94×10^7	1.08×10^{11}
Earth	5.98×10^{24}	6.37×10^6	3.156×10^7	1.496×10^{11}
Mars	6.42×10^{23}	3.37×10^6	5.94×10^7	2.28×10^{11}
Jupiter	1.90×10^{27}	6.99×10^7	3.74×10^8	7.78×10^{11}
Saturn	5.68×10^{26}	5.85×10^7	9.35×10^8	1.43×10^{12}
Uranus	8.68×10^{25}	2.33×10^7	2.64×10^9	2.87×10^{12}
Neptune	1.03×10^{26}	2.21×10^7	5.22×10^9	4.50×10^{12}
Pluto	$\approx 1.4 \times 10^{22}$	$\approx 1.5 \times 10^6$	7.82×10^9	5.91×10^{12}
Moon	7.36×10^{22}	1.74×10^6	—	—
Sun	1.991×10^{30}	6.96×10^8	—	—

Physical Data Often Used[a]

Average earth-moon distance	3.84×10^8 m
Average earth-sun distance	1.496×10^{11} m
Average radius of the earth	6.37×10^6 m
Density of air (20°C and 1 atm)	1.20 kg/m^3
Density of water (20°C and 1 atm)	1.00×10^3 kg/m^3
Free-fall acceleration	9.80 m/s^2
Mass of the earth	5.98×10^{24} kg
Mass of the moon	7.36×10^{22} kg
Mass of the sun	1.99×10^{30} kg
Standard atmospheric pressure	1.013×10^5 Pa

[a] These are the values of the constants as used in the text.

Some Prefixes for Powers of Ten

Power	Prefix	Abbreviation	Power	Prefix	Abbreviation
10^{-18}	atto	a	10^1	deka	da
10^{-15}	femto	f	10^2	hecto	h
10^{-12}	pico	p	10^3	kilo	k
10^{-9}	nano	n	10^6	mega	M
10^{-6}	micro	μ	10^9	giga	G
10^{-3}	milli	m	10^{12}	tera	T
10^{-2}	centi	c	10^{15}	peta	P
10^{-1}	deci	d	10^{18}	exa	E